SYNTHETIC APPLICATIONS OF 1,3-DIPOLAR CYCLOADDITION CHEMISTRY TOWARD HETEROCYCLES AND NATURAL PRODUCTS

This is the fifty-ninth volume in the series

THE CHEMISTRY OF HETEROCYCLIC COMPOUNDS

THE CHEMISTRY OF HETEROCYCLIC COMPOUNDS

A SERIES OF MONOGRAPHS

EDWARD C. TAYLOR AND PETER WIPF, *Editors*

ARNOLD WEISSBERGER, *Founding Editor*

SYNTHETIC APPLICATIONS OF 1,3-DIPOLAR CYCLOADDITION CHEMISTRY TOWARD HETEROCYCLES AND NATURAL PRODUCTS

Edited by

Albert Padwa

Department of Chemistry
Emory University

William H. Pearson

Department of Chemistry
University of Michigan

AN INTERSCIENCE® PUBLICATION

JOHN WILEY & SONS, INC.

Published by John Wiley & Sons, Inc., Hoboken, New Jersey.
Published simultaneously in Canada.

For general information on our other products and services please contact our Customer Care Department within the U.S. at 877-762-2974, outside the U.S. at 317-572-3993 or fax 317-572-4002.

Wiley also publishes its books in a variety of electronic formats. Some content that appears in print, however, may not be available in electronic format.

Library of Congress Cataloging-in-Publication Data is available.

ISBN 0-471-28061-5

Printed in the United States of America.

10 9 8 7 6 5 4 3 2 1

The Chemistry of Heterocyclic Compounds
Introduction to the Series

The chemistry of heterocyclic compounds is one of the most complex and intriguing branches of organic chemistry, of equal interest for its theoretical implications, for the diversity of its synthetic procedures, and for the physiological and industrial significance of heterocycles.

The Chemistry of Heterocyclic Compounds has been published since 1950 under the initial editorship of Arnold Weissberger, and later, until his death in 1984, under the joint editorship of Arnold Weissberger and Edward C. Taylor. In 1997, Peter Wipf joined Prof. Taylor as editor. This series attempts to make the extraordinarily complex and diverse field of heterocyclic chemistry as organized and readily accessible as possible. Each volume has traditionally dealt with syntheses, reactions, properties, structure, physical chemistry, and utility of compounds belonging to a specific ring system or class (e.g., pyridines, thiophenes, pyrimidines, three-membered ring systems). This series has become the basic reference collection for information on heterocyclic compounds.

Many broader aspects of heterocyclic chemistry are recognized as disciplines of general significance that impinge on almost all aspects of modern organic chemistry, medicinal chemistry, and biochemistry, and for this reason we initiated several years ago a parallel series entitled General Heterocyclic Chemistry, which treated such topics as nuclear magnetic resonance, mass spectra, and photochemistry of heterocyclic compounds, the utility of heterocycles in organic synthesis, and the synthesis of heterocycles by means of 1,3-dipolar cycloaddition reactions. These volumes were intended to be of interest to all organic, medicinal, and biochemically oriented chemists, as well as to those whose particular concern is heterocyclic chemistry. It has, however, become increasingly clear that the above distinction between the two series was unnecessary and somewhat confusing, and we have therefore elected to discontinue *General Heterocyclic Chemistry* and to publish all forthcoming volumes in this general area in *The Chemistry of Heterocyclic Compounds* series.

It is a major challenge to keep our coverage of this immense field up to date. One strategy is to publish Supplements or new Parts when merited by the amount of new material, as has been done, *inter alia*, with pyridines, purines, pyrimidines, quinazolines, isoxazoles, pyridazines and pyrazines. The chemistry and applications to synthesis of 1,3-dipolar cycloaddition reactions in the broad context of organic chemistry were first covered in a widely cited two-volume treatise edited by Prof. Albert Padwa that appeared in 1984. Since so much has been published on this fascinating and broadly useful subject in the intervening years, we felt that a Supplement would be welcomed by the international chemistry community, and we are immensely grateful to Prof. Padwa and Prof. Pearson for tackling this arduous task. The result is another outstanding contribution to the organic and heterocyclic chemistry literature that we are delighted to publish within *The Chemistry of Heterocyclic Compounds* series.

EDWARD C. TAYLOR

Department of Chemistry
Princeton University
Princeton, New Jersey

PETER WIPF

Department of Chemistry
University of Pittsburgh
Pittsburgh, Pennsylvania

Preface

Cycloaddition reactions figure prominently in both synthetic and mechanistic organic chemistry. The current understanding of the underlying principles in this area has grown from a fruitful interplay between theory and experiment. The monumental work of Rolf Huisgen and co-workers in the early 1960s led to the general concept of 1,3-dipolar cycloaddition. Few reactions rival this process in the number of bonds that undergo transformation during the reaction, producing products considerably more complex than the reactants. Over the years, this reaction has developed into a generally useful method for five-membered heterocyclic ring synthesis, since many 1,3-dipolar species are readily available and react with a wide variety of dipolarophiles.

The last comprehensive survey of this area dates back to 1984, when the two-volume set edited by Padwa, "*1,3-Dipolar Cycloaddition Chemistry,*" appeared. Since then, substantial gains in the synthetic aspects of this chemistry have dominated the area, including both methodology development and a body of creative and conceptually new applications of these [3 + 2]-cycloadditions in organic synthesis. The focus of this volume centers on the utility of this cycloaddition reaction in synthesis, and deals primarily with information that has appeared in the literature since 1984. Consequently, only a selected number of dipoles are reviewed, with a major emphasis on synthetic applications. Both carbonyl ylides and nitronates, important members of the 1,3-dipole family that were not reviewed previously, are now included. Discussion of the theoretical, mechanistic, and kinetic aspects of the dipolar-cycloaddition reaction have been kept to a minimum, but references to important new work in these areas are given throughout the 12 chapters.

Beyond the ability of the 1,3-dipolar cycloaddition reaction to produce heterocycles, its importance extends to two other areas of organic synthesis, both of which are included in the current volume. First, the heteroatom-containing cycloadducts may be transformed into a variety of other functionalized organic molecules, whether cyclic or acyclic. Second, many 1,3-dipolar cycloadditions have the ability to generate rings (and functionality derived from transformations of such rings) containing several contiguous stereocenters in one synthetic operation. The configurations of these new stereocenters arise from the geometry of the dipole and dipolarophile as well as the topography (endo or exo) of the cycloaddition. An additional stereochemical feature arises when the reactive π faces of either of the cycloaddends are diastereotopic. Relative stereocontrol in 1,3-dipolar cycloadditions is dealt with in some detail, and asymmetric versions of these dipolar cycloadditions represent an entirely new aspect of the current reference work.

In recent years, numerous natural and unnatural products have been prepared by synthetic routes that have a 1,3-dipolar cycloaddition as a crucial step in their synthesis. Consequently, this reaction has become recognized as an extremely important transformation in the repertoire of the synthetic organic chemist.

ALBERT PADWA

Department of Chemistry
Emory University
Atlanta, Georgia

WILLIAM H. PEARSON

Department of Chemistry
University of Michigan
Ann Arbor, Michigan

Contents

1	**NITRONES** Jason N. Martin and Raymond C. F. Jones	**1**
2	**NITRONATES** Scott E. Denmark and Jeromy J. Cottell	**83**
3	**AZOMETHINE YLIDES** L. M. Harwood and R. J. Vickers	**169**
4	**CARBONYL YLIDES** Mark C. McMills and Dennis Wright	**253**
5	**THIOCARBONYL YLIDES** Grzegorz Mloston and Heinz Heimgartner	**315**
6	**NITRILE OXIDES** Volker Jager and Pedro A. Colinas	**361**
7	**NITRILE YLIDES AND NITRILE IMINES** John T. Sharp	**473**
8	**DIAZOALKANES** Gerhard Maas	**539**
9	**AZIDES** Chin-Kang Sha and A. K. Mohanakrishnan	**623**
10	**MESOIONIC RING SYSTEMS** Gordon W. Gribble	**681**
11	**EFFECT OF EXTERNAL REAGENTS** Shuji Kanemasa	**755**
12	**ASYMMETRIC REACTIONS** Kurt Vesterager Gothelf and Karl Anker Jorgensen	**817**
INDEX		**901**

SYNTHETIC APPLICATIONS OF 1,3-DIPOLAR CYCLOADDITION CHEMISTRY TOWARD HETEROCYCLES AND NATURAL PRODUCTS

This is the fifty-ninth volume in the series

THE CHEMISTRY OF HETEROCYCLIC COMPOUNDS

CHAPTER 1

Nitrones

Jason N. Martin and Raymond C. F. Jones

Department of Chemistry, Loughborough University, Loughborough, United Kingdom

1.1. Nitrones and the 1,3-Dipolar Cycloaddition Reaction . . . 2
1.2. Toward Natural Products through Nitrone Cycloadditions . . . 3
1.3. Nucleosides . . . 4
1.4. Lactams . . . 8
1.5. Quinolizidines, Indolizidines, and Pyrrolizidines . . . 12
1.6. Peptides and Amino Acids . . . 18
1.7. Sugars . . . 24
1.8. Sulfur- and Phosphorus-Containing Compounds . . . 29
1.9. Catalytic Cycloadditions . . . 34
1.10. Pyrrolidines, Piperidines, and Other Amines . . . 34
1.11. Isoxazolidines . . . 47
1.11.1. Nitrones by the 1,3-ATP Process . . . 48
1.11.2. Intramolecular Oxime–Alkene Cycloaddition . . . 54
1.11.3. Intramolecular Nitrone–Alkene Cycloadditions . . . 55
1.11.4. Isoxazolidines from Intermolecular Nitrone Cycloaddition Reactions . . . 59
1.12. Conclusion . . . 68

The synthetic utility of the 1,3-dipolar cycloaddition reaction is evident from the number and scope of targets that can be prepared by this chemistry. As one of the most thoroughly investigated 1,3-dipoles, nitrones are arguably the most useful through their ability to generate nitrogen- and oxygen-based functionality from the cycloadducts as well as the potential to introduce multiple chiral centers stereoselectively. A comprehensive review of all nitrone cycloadditions would fill many volumes; instead, this chapter will focus upon the highlights of synthetic endeavor through 1,3-dipolar cycloaddition reactions of nitrones since 1984.

The Chemistry of Heterocyclic Compounds, Volume 59: Synthetic Applications of 1,3-Dipolar Cycloaddition Chemistry Toward Heterocycles and Natural Products, Edited by Albert Padwa and William H. Pearson.
ISBN 0-471-38726-6

1.1. NITRONES AND THE 1,3-DIPOLAR CYCLOADDITION REACTION

Nitrones (or azomethine oxides) (1–7) were first prepared by Beckmann in 1890 (8,9) and named from a shortening of "nitrogen–ketones" by Pfeiffer in 1916 to emphasize their similarity to ketones (10). While aromatic N-oxides also contain the nitrone moiety, they retain the name of the N-oxides whose reactivity they more closely resemble. The general terms aldo- and keto-nitrones are used on occasion to distinguish between those with and without a proton on the α-carbon, respectively, and nitrones exist in (*E*)- and (*Z*)-forms that may interconvert. Their chemistry is hugely varied and frequently reviewed, but it is ultimately dominated by their use as 1,3-dipoles for cycloaddition reactions. In 1960, Huisgen proposed the now widely accepted concept of the 1,3-dipolar cycloaddition reaction (11–20) in which the formation of the two new bonds occurs as a concerted (but not simultaneous) process, rejecting Firestone's proposed reaction via a diradical intermediate on the basis of stereospecificity (21–26). Ironically, Huisgen himself then went on to demonstrate the first example of a two-step cycloaddition, using a thiocarbonyl ylide 1,3-dipole (27).

The most common nitrone 1,3-dipolar cycloaddition (DC) reaction (28–35) is the formation of an isoxazolidine using alkene dipolarophiles (Scheme 1.1), although other multiply bonded systems may also be used (alkynes, allenes, isocyanates, nitriles, thiocarbonyls, etc.). The isoxazolidine cycloadduct contains up to three new chiral centers and, as with other 1,3-dipoles, the highly ordered transition state often allows the regio- and stereochemical preference of a given nitrone to be predicted. This prediction is achieved through a consideration of steric and electronic factors, but most significantly through the frontier molecular orbital (FMO) theory proposed by Fukui (36,37), for which he shared the 1981 Nobel Prize.

A number of cyclic nitrones have been developed that avoid the issue of nitrone (*E*/*Z*) isomerization by permitting only a single geometry about the C=N double bond and so reduce the number of possible cycloaddition products. Cyclic nitrones have also become popular as facially differentiated reagents, allowing predictable

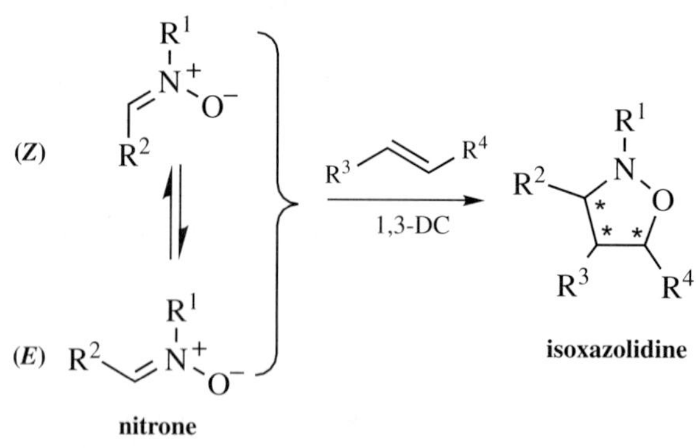

Scheme 1.1

asymmetric induction through their ability to enforce the cycloaddition reaction at one or other face of the 1,3-dipole. In recent years, the effect of catalysis on the rate and selectivity of the nitrone cycloaddition reaction has been examined from which impressive results have begun to emerge. Thus, nitrones represent a powerful tool in modern synthetic chemistry, whose limits are still being explored more than a century after their discovery.

1.2. TOWARD NATURAL PRODUCTS THROUGH NITRONE CYCLOADDITIONS

With a wealth of nitrone-derived cycloadditions reported in the literature, we have sought to arrange this survey according to the synthetic target (e.g., nucleosides or amino acids) or, where more relevant, grouped by the nature of the cycloaddition partners (e.g., those derived from sugars). Where a total synthesis is concerned, this task is straightforward, but with more speculative and developmental papers it would be possible to classify the same work in a number of ways. Apologies, then, to authors who feel *misplaced*. Naturally, there is a degree of overlap between many of our groupings, and in each case we have attempted to direct the reader to relevant work. Section 1.11 on isoxazolidine synthesis covers a particularly broad range of reactions and it is here we have collected some of the most significant reports in which the major aim of the work was to characterize a novel nitrone cycloaddition reaction rather than achieve the total synthesis of a given target molecule.

1.3. NUCLEOSIDES

Nucleosides are potent antibiotic, antitumor, and antiviral agents, vital in chemotherapy for acquired immune deficiency syndrome/human immunodeficiency virus (AIDS/HIV). The polyoxins (e.g., **1a**–**b**), and closely related nikkomycins are pyrimidine nucleoside antibiotics that are potent inhibitors of the biosynthesis of chitin, a major structural component of the cell wall of most fungi (Scheme 1.2). Merino and co-workers (38,39) reported the total synthesis of (+)-polyoxin J **1b** and of the isoxazolidine analogue of thymine polyoxin C, by nucleophilic addition to chiral sugar nitrones. By a 1,3-dipolar cycloaddition route, they have prepared polyoxin analogues **2** in which the furanose ring of the parent nucleoside is substituted by an isoxazolidine (40). Thus, nitrone **3**, prepared in six steps from serine (58% overall yield), afforded four isoxazolidine cycloadducts in excellent yield (93%) in its reaction with vinyl acetate. The product mixture contained predominantly the (3*R*,5*S*)-adduct **4a** and its C(5) epimer the (3*R*,5*R*)-adduct **4b**, separable by chromatography, along with an inseparable mixture of the two C(3) epimers. Isoxazolidines **4a** and **4b** were used as a mixture or separately to glycosylate silylated thymine **5** or uracil **6**. Acidic cleavage of the acetonide of the (3*R*,5*S*)-adducts **7a**–**b** afforded the amino alcohols, which were oxidized to the acids with

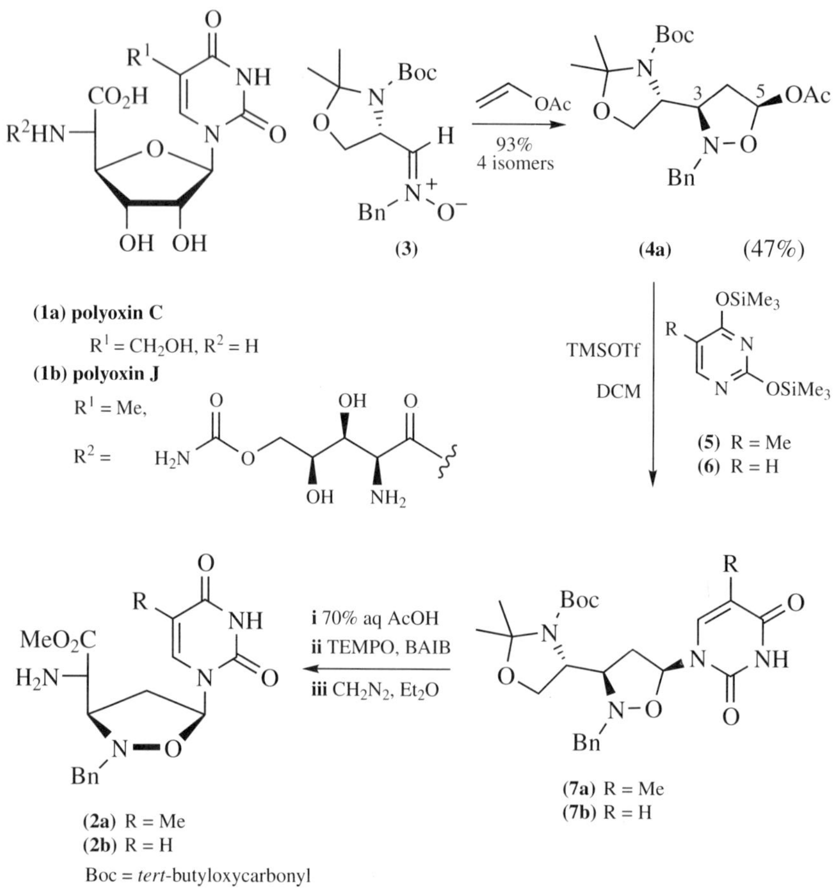

Scheme 1.2

2,2,6,6-tetramethyl-1-piperidinoxyl (TEMPO) and bis(acetoxy)iodobenzene (BAIB) before esterification with diazomethane to give **2a** or **2b**. The C(5) epimer of each N,O-nucleoside was accessed by similar treatment of the diastereomeric isoxazolidine adduct (3*R*,5*R*) **4b**. In earlier work, this group also prepared a related oxazolidinyl thymine nucleoside (side-chain amino acid replaced by $-CH_2OH$) via the addition of the corresponding D-glyceraldehyde-derived nitrone with the sodium enolate of methyl acetate (41). The amino acid side of nikkomycin Bz was synthesized by a nitrone cycloaddition route by Tamura et al. (42).

Chiacchio et al. (43,44) investigated the synthesis of isoxazolidinylthymines by the use of various C-functionalized chiral nitrones in order to enforce enantioselection in their cycloaddition reactions with vinyl acetate (Scheme 1.3). They found, as in the work of Merino et al. (40), that asymmetric induction is at best partial with dipoles whose chiral auxiliary does not maintain a fixed geometry and so cannot completely direct the addition to the nitrone. After poor results with menthol ester- and methyl lactate-based nitrones, they were able to prepare and separate isoxazolidine **8a** and its diastereomer **8b** in near quantitative yield using the *N*-glycosyl

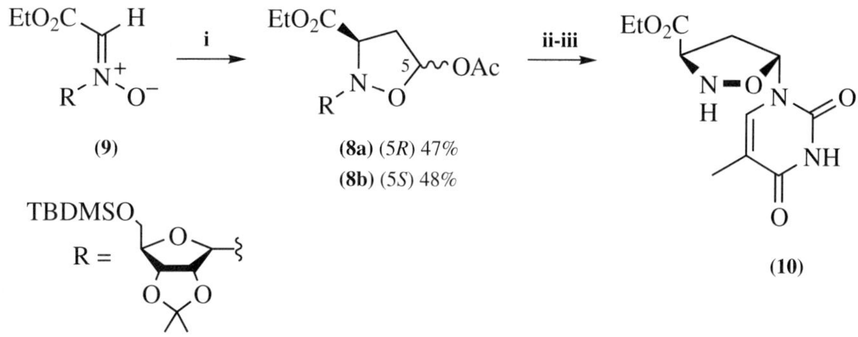

Reagents: [i] vinyl acetate, ethyl glyoxalate, 60 °C, 14h; [ii] bovine serum albumin (BSA), thymine, TMSOTf, MeCN, Δ, 1 h; [iii] 3.7% HCl in EtOH, room temperature (rt), 3 h.

Scheme 1.3

nitrone **9** of Vasella, derived from D-ribofuranose. Adduct **8a** was coupled with thymine before removal of the sugar auxiliary to afford N,O-nucleoside **10**.

Among a number of other homochiral furanosyl- and isoxazolidinylthymine targets, these workers also applied an achiral cycloaddition approach with vinyl acetate to successfully prepare the antiviral agent d4T (**11**) and its 2-methyl analogue (Fig. 1.1) (45). In more recent work, similar nitrones [**9**, R = Me or benzyl (Bn)] were used to prepare hydroxymethyl substituted isoxazolidines [3-(46) and 3,5-substituted (47)] for the preparation of further nucleoside analogues.

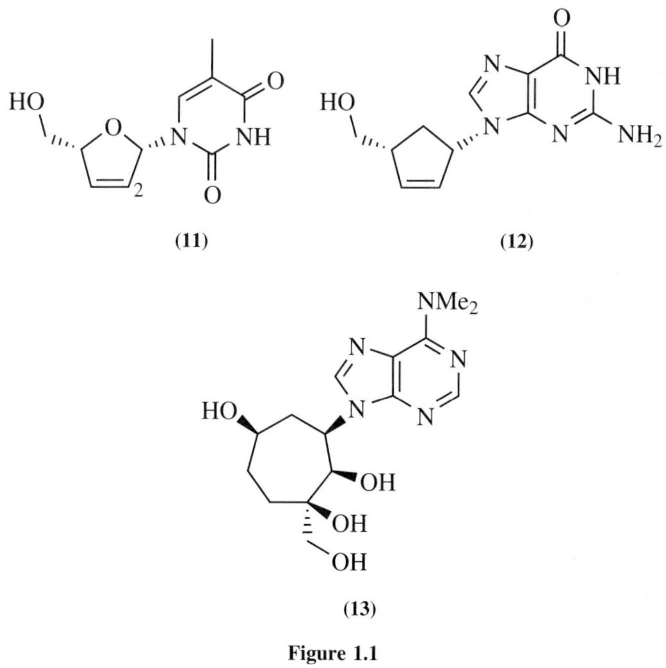

Figure 1.1

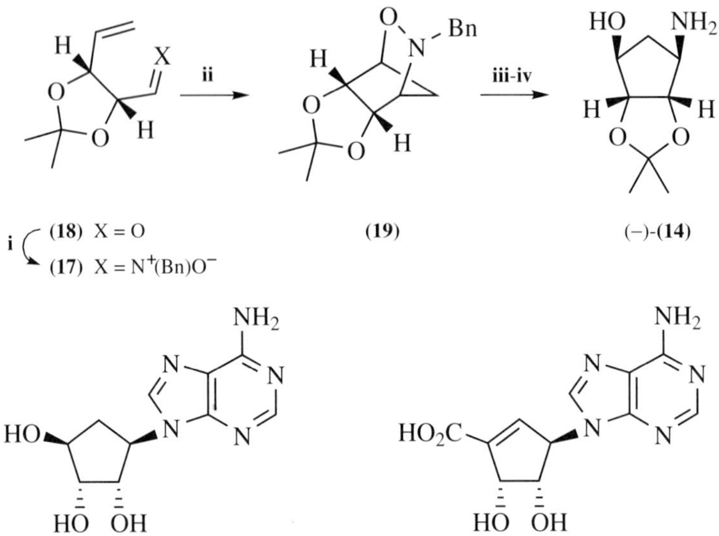

(18) X = O (19) (−)-(14)

(17) X = N⁺(Bn)O⁻

(15) (16)

Reagents: [i] BnNHOH, EtOH, 15 min, 95%; [ii] C_6H_5Cl, Δ, 30 min, 62%; [iii] Zn, AcOH, Et_2O, rt, 48 h, 78%; [iv] NH_4HCO_2, 10% Pd-C, MeOH, Δ, 1 h 75%.

Scheme 1.4

Elsewhere, Langlois and co-worker (48) applied a 3-hydroxyaminoborneol-derived nitrone to the total synthesis of (+)-carbovir (**12**), the enantiomer of a potent reverse transcriptase inhibitor for the treatment of AIDS.

Mandal and co-workers (49,50) (Scheme 1.4) prepared five- and seven-membered carbocyclic nucleosides including the (+)-dimethylaminopurine compound **13** (3.3% from D-glucose) and its enantiomer. Aminocyclopentitol (−)-**14** is an intermediate in the synthesis of the carbocyclic nucleosides (−)-noraristeromycin (**15**) and (−)-nepalocin A (**16**) and has been prepared in enantiopure form by Gallos et al. (50a) from nitrone **17** by condensation of the corresponding D-ribose derived aldehyde **18** with BnNHOH (Scheme 1.4). Thus, intramolecular cycloaddition of nitrone **17** affords tricyclic adduct **19** as a single enantiomer, which is converted to (−)-**14** after N—O bond reduction (Zn/AcOH) and debenzylation (ammonium formate, 10% Pd—C).

In contrast to the installation of the nucleobase via nucleophilic substitution of a suitable leaving group on the isoxazolidinyl cycloadduct, Colacino et al. (51) and Sindona and co-workers (52,53) prepared isoxazolidinyl nucleosides using vinyl nucleobases as the dipolarophile (Scheme 1.5). In Sindona's work, while a three-component reaction of hydroxylamine, formaldehyde, and **20** afforded a complex mixture of cycloadducts and byproducts, the known dipole **21** reacted with *N*-9-vinyladenine (**20**) in benzene at reflux to afford a racemic mixture of adduct **22** and its enantiomer (45%). The ester function was then used to effect a resolution by pig

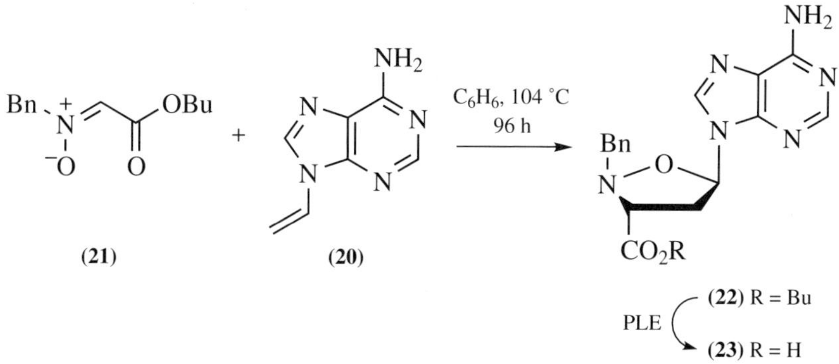

Scheme 1.5

liver esterase (PLE) enzyme-catalyzed hydrolysis to afford the enantiomerically pure acid **23**.

The discovery of spirocyclic nucleosides with anti-HIV-1 activity has prompted Chattopadhyaya and co-workers (54,55) to prepare spiroisoxazolidine nucleosides (Scheme 1.6). Thus, after proving the reactivity of related systems in an

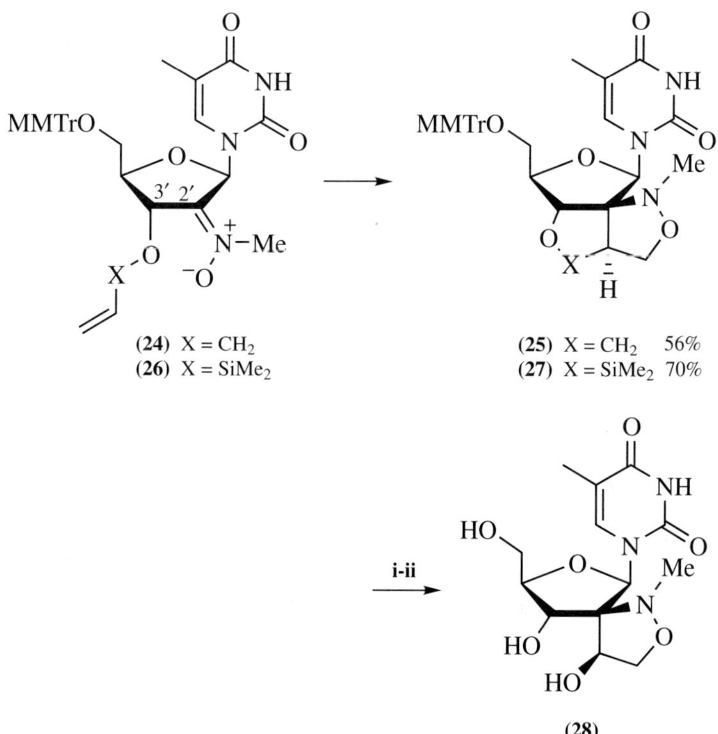

MMTr = para-methoxyphenyldiphenylmethyl

Reagents: [i] H_2O_2, KF, $KHCO_3$, MeOH, tetrahydrofuran (THF); [ii] 80% AcOH, 40 °C, 1 h.

Scheme 1.6

intermolecular sense, nucleoside nitrone **24** (or the isomeric 2′-*O*-allyl-3′-nitrones) were prepared from the corresponding ketones to afford the spirotricyclic cycloadduct **25**. Similarly, a reagent with a vinyl silyl ether tether (**26**) gave the related tricyclic adduct **27**, which was desilylated by hydrogen peroxide mediated Tamao oxidation to afford the spiroisoxazolidine nucleoside **28**. Related nitrones were earlier prepared by Tronchet et al. (56–59) for studies of nucleophilic addition to the nitrone function.

1.4. LACTAMS

The continued importance of β-lactam ring systems in medicine has encouraged a number of research groups to investigate their synthesis via a nitrone cycloaddition protocol. Kametani et al. (60–62) reported the preparation of advanced intermediates of penems and carbapenems including (+)-thienamycin (**29**) and its enantiomer (Scheme 1.7). They prepared the chiral nitrone **30** from (−)-menthyl

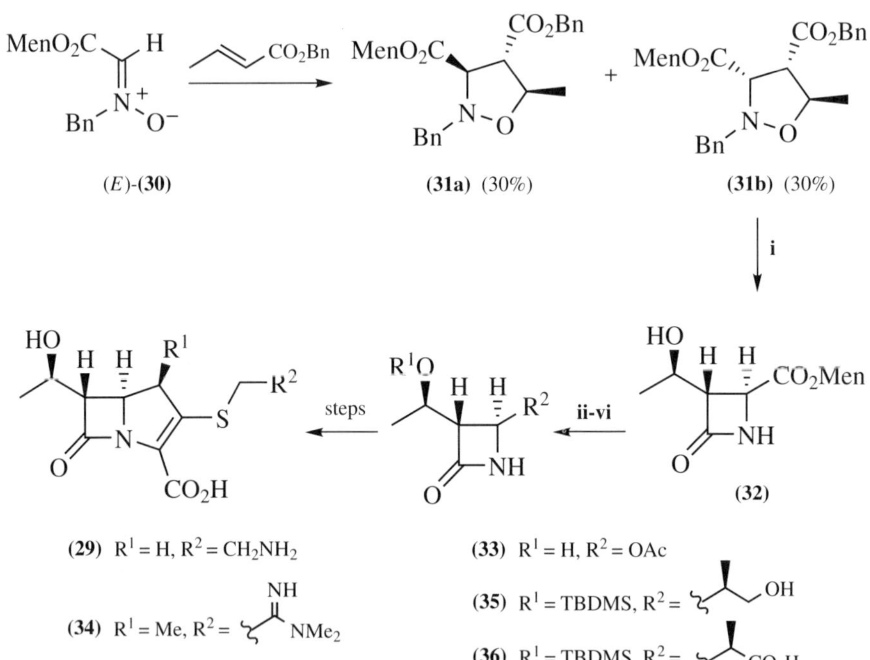

Men = menthyl; TBDMS = *tert*-butyldimethylsilyl

Reagents: [i] H_2, PtO_2, MeOH, 20 h, then DCC, MeCN, 60 °C, 3 h, 39%; [ii] TBDMSCl, Et_3N, dimethyl formamide (DMF), 16 h, 80%; [iii] NaOH aq (1 *M*), THF, MeOH then HCl aq (1 *M*), 81%; [iv] TBDMSCl, Et_3N, DMF, 18 h, 94%; [v] AcOH aq (2.5 *M*), THF, 6 h then $NaHCO_3$ aq (5%); [vi] $Pb(OAc)_4$, KOAc, DMF, 40 °C, 1 h, 70%.

Scheme 1.7

glyoxal hydrate and benzylhydroxylamine but found it exerted incomplete stereocontrol in its cycloaddition to benzyl crotonate. The major isolated products, a 1:1 mixture of isoxazolidines **31a** and **31b**, are rationalized as the consequence of endo or exo addition to the more reactive (*E*) form of nitrone (**30**), respectively. Simultaneous O- and N-debenzylation and N—O bond hydrogenolysis of **31b** gave an amino alcohol intermediate, which was used without purification for dicyclohexylcarbodiimide (DCC)-mediated cyclization to afford the β-lactam **32**. Silylation of the hydroxyl and the amide nitrogen was followed by hydrolysis of the menthyl ester, which also brought about N-desilylation. Reinstallation of the silyl groups through two steps, before insertion of the C(4) acetyl group by oxidative acetoxylation with lead tetraacetate, afforded the lactam **33**, a known intermediate en route to (+)-thienamycin (**29**). An earlier, related total synthesis of **29** by this group using a homochiral *N*-(2-phenylethyl) auxiliary afforded similar low yields of the desired isoxazolidine adducts (60).

The unusually potent and broad spectrum antibacterial action of (+)-thienamycin is tempered by its instability at high concentration and susceptibility to decomposition by renal dehydropeptidase I. In 1984, Shih and co-workers (62a) at Merck reported that the 1-β-methylcarbapenem **34** demonstrated increased chemical and metabolic stability while retaining high antibiotic activity. Work published by Ito et al. (63) described the preparation of a 1-β-methylcarbapenem intermediate (**35**) via a nitrone cycloaddition that gave an equimolar amount of all four possible adducts. Later, intermediate **35** was prepared by Ihara et al. (64,65) by intramolecular cycloaddition of a complex chiral alkenyl nitrone to afford a single stereoisomer (51%). Separately, Kang and Lee (66), then Jung and Vu (67), prepared 1-β-methylcarbapenem intermediate (**36**) and a synthetic precursor respectively, via intramolecular nitrone–alkene 1,3-dipolar cycloaddition reactions with complete diastereocontrol.

Alcaide et al. (68,69) recently published their studies of the intramolecular 1,3-dipolar cycloaddition reactions of alkynyl-β-lactams in which they found that the desired cycloaddition was in competition with a reverse-Cope elimination. The reaction of alkynyl aldehydes **37a–c** with *N*-methylhydroxylamine afforded a mixture of products depending on the reaction conditions and the chain length separating the alkyne and the lactam (Scheme 1.8). Thus, up to three separate

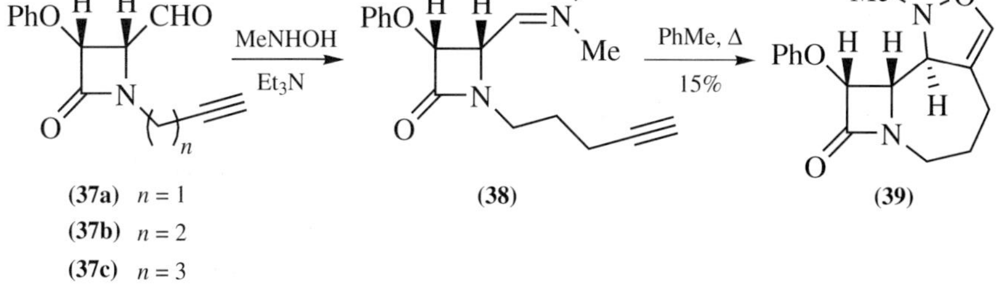

Scheme 1.8

nitrones were identified from the reaction mixture including two from a complex proposed mechanism. Thus, alkynyl nitrone (**38**) was formed from the condensation of **37c** with *N*-methylhydroxylamine in refluxing toluene, and underwent a 1,3-dipolar cycloaddition to afford the homochiral isoxazoline **39** via addition to the less sterically crowded upper face of **38**. Significantly, the isolated yield of the cycloadduct is very low (15%), the product ratio favoring the nitrone (**38:39** = 3:1).

Chmielewski and co-workers (70–73) prepared the β-lactam skeleton via a nitrone cycloaddition to a sugar ene lactone dipolarophile providing latent polyol functionality at C(3) of the lactam (Scheme 1.30, Section 1.7). In other lactam cycloaddition chemistry, Rigolet et al. (74) prepared various spirocyclic adducts, including **40–42** from the corresponding methylene lactams (75) or the unstable methylene isoindolones **43**, the latter showing enhanced yields for the cycloaddition of *N*-benzyl-*C*-phenyl nitrone to the exocyclic double bond under microwave irradiation (Scheme 1.9) (76). Related spiroisoxazolidinyl lactams were reported by Fisera and co-workers (77). Funk and Daggett (78) prepared similar spirocyclic lactams (e.g., **44**) via the cycloaddition reaction of exocyclic nitrone **45** (derived from cyclohexanone) with unsaturated esters (Scheme 1.10). The N—O bond cleavage of isoxazolidine (**46**) makes available the nitrogen for spontaneous lactamization to the spirocyclic product **44**.

Cycloaddition to endocyclic unsaturation has been used by many researchers for the preparation of isoxazolidinyl adducts with γ-lactams derived from pyroglutaminol and is discussed later in this chapter as a synthesis of unusual amino acids (Scheme 1.20, Section 1.6) (79,80). A related α,β-unsaturated lactam has been prepared by a nitrone cycloaddition route in the total synthesis of the fungal metabolite leptosphaerin (81). A report of lactam synthesis from acyclic starting materials is given in the work of Chiacchio et al. (82) who prepared isoxazolidine (**47**) via an intramolecular nitrone cycloaddition reaction (Scheme 1.11).

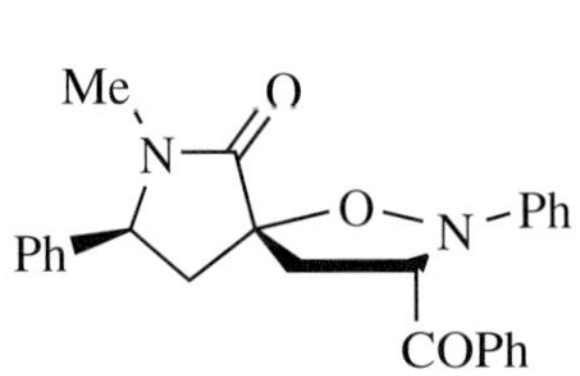

(40)

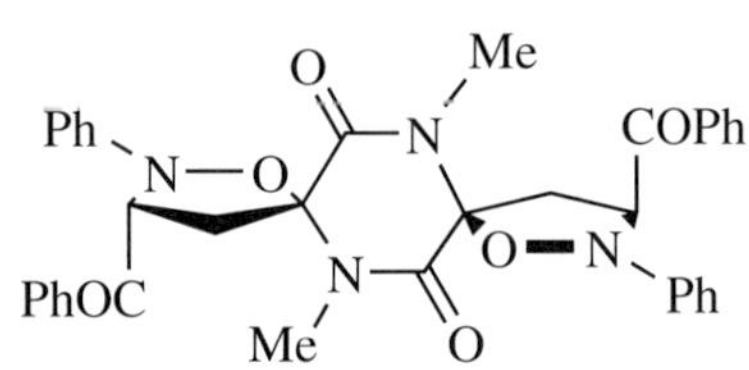

(41)

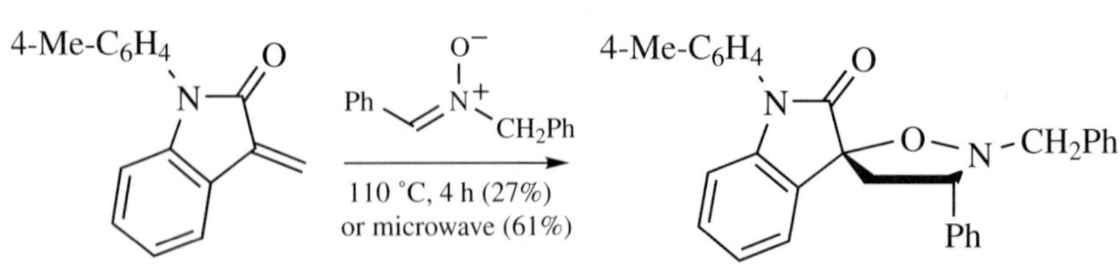

(43) **(42)**

Scheme 1.9

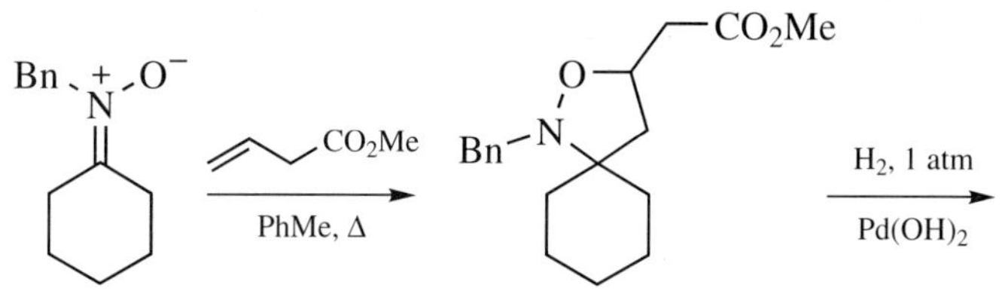

(45) (46) (44)

Scheme 1.10

The acyclic precursor is an α,β-unsaturated amido aldehyde that was condensed with *N*-methylhydroxylamine to generate the nitrone (*E*)-**48**, which then underwent a spontaneous cycloaddition with the alkene to afford the 5,5-ring system of the isoxazolidinyl lactam **47**. The observed product arises via the (*E*)-nitrone transition state **A** [or the (*Z*)-nitrone equivalent] in which the position of the benzyl group α to the nitrone effectively controls the two adjacent stereocenters while a third stereocenter is predicted from the alkene geometry. Both transition states maintain the benzyl auxiliary in an equatorial position and thus avoid the unfavorable 1,3-diaxial interaction with the nitrone methyl or oxygen found in transition state **B**. Semiempirical PM3 calculations confirm the extra stability, predicting exclusive formation of the observed product **47**. Related cycloadducts from the intramolecular reaction of nitrones containing ester- rather than amide-tethered alkene functionality are also known (83-85).

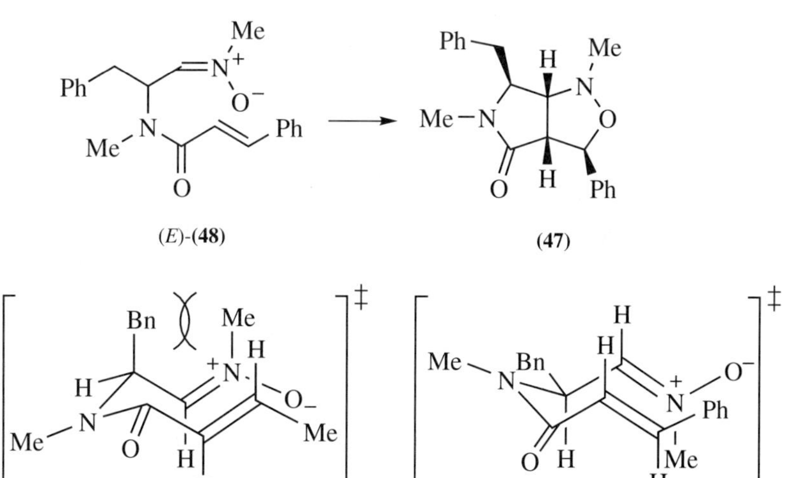

(*E*)-(**48**) (**47**)

(**B**) (**A**)

Scheme 1.11

1.5. QUINOLIZIDINES, INDOLIZIDINES, AND PYRROLIZIDINES

The title compounds, quinolizidines, indolizidines, and pyrrolizidines (86), are characterized by the presence of a bridgehead nitrogen atom in six,six-, six,five-, or five,five-membered bicyclic ring systems, respectively. The polyhydroxylated indolizidines and pyrrolizidines have a range of biological effects through their inhibition of glycosidase enzymes, including some examples of antiviral activity. The nitrogen and nearby oxygen functionality lend themselves to a nitrone cycloaddition strategy, as demonstrated by McCaig et al. (87,88) in their synthesis of each enantiomer of the indolizidine lentiginosine (**49**) and related pyrrolizidines (Scheme 1.12). Chiral cyclic nitrone **50** was prepared from doubly methoxymethyl (MOM) protected diethyl D-tartrate via oxidation of the corresponding pyrrolidine with Davis' reagent. The cycloaddition reaction of nitrone **50** with benzyl but-3-enoate in toluene at reflux gave a single cycloadduct **51** in 44% yield after 4 days. Reductive N—O bond cleavage and concomitant recyclization with the pendant ester function gave a lactam (**52**), which was reduced to the amine with borane–dimethyl sulfide complex. Radical deoxygenation at C-7 of the imidazolylthio-

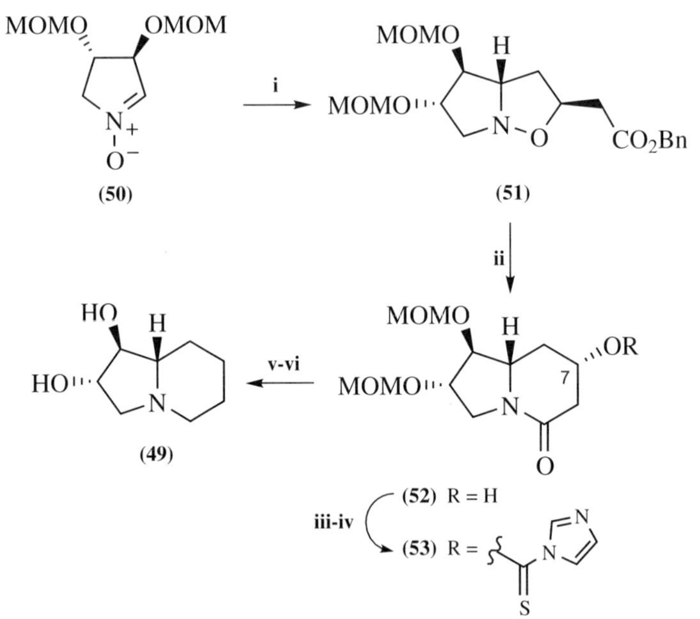

Reagents: [i] $CH_2{=}CHCH_2CO_2Bn$, toluene, Δ, 4 days, 44%; [ii] Zn, AcOH, 60 °C, 2 h, 83%; [iii] $BH_3{\bullet}Me_2S$, THF, rt, 4 h, then EtOH, Δ, 3 h, 95%; [iv] 1,1′-thiocarbonyldiimidazole, $ClCH_2CH_2Cl$, Δ, 2 h, then rt overnight, 83%; [v] Bu_3SnH, AIBN, toluene, Δ, 3 h, 53%; [vi] HCl aq (6M), rt, overnight, 60%.

Scheme 1.12

carbonyl derivative **53** and removal of the MOM protecting groups in acid afforded a single isomer of lentiginosine (+)-**49**. By an identical scheme, these workers prepared (−)-**49** from the enantiomer of isoxazolidine cycloadduct **51**.

A similar approach has been applied by Brandi and co-workers (89–97) using chiral 3- and 3,4-substituted pyrrolidine nitrones. With such dipoles they have prepared a number of hydroxylated indolizidines (89–92,95) including (−)-hastanecine and (−)-croalbinecine (96). As before, N—O bond cleavage was followed by recyclization, this time through nucleophilic substitution of the terminal hydroxyl moiety derived from the dipolarophile, as its tosylate. These workers have recently reported the synthesis of a related monohydroxylated nitrone by oxidation of the *N*-hydroxypyrrolidine to afford an 11:1 mixture of the two separable regioisomers (95). Indolizidine and pyrrolizidine skeletons were then prepared from this material. In another elegant synthesis, Holmes and co-workers (98) prepared the indolizidine core of the allopumiliotoxins (**54**) (Scheme 1.13). Retrosynthetic analysis suggested an isoxazolidinyl intermediate, ultimately derived by an intramolecular cycloaddition reaction of alkenyl nitrone **55**. The desired cycloadduct **56** was the major product isolated from a mixture containing small amounts of three other diastereomers and afforded the target skeleton **54** or its C(3) epimer after extensive synthetic manipulation (98). In other work on the intramolecular nitrone cycloaddition (99), this group has published intermediates in the total synthesis of the indolizidine alkaloid gephyrotoxin (100) as well as the total synthesis of spiropiperidine natural product histrionicotoxin (Scheme 1.49, Section 1.10) (101). Kibayashi and co-worker (102,103) reported two total syntheses of the indolizidine (+)-monomorine I (**57**), both of which rely on the same cycloaddition reaction of an achiral methyl glyoxalate-derived nitrone and a homochiral allyl ether. The resultant mixture of isoxazolidines was a 3:1 mixture in favor of the desired product in 76% combined yield.

The rare reports of quinolizidine formation by a nitrone cycloaddition strategy include the racemic total synthesis of lasubine II (**58**), one of a series of related alkaloid isolated from the leaves of *Lagerstoemia subcostata* Koehne (Scheme 1.14) (104). While these alkaloids were previously accessed by *inter*molecular nitrone cycloaddition reactions, this more recent report uses an *intra*molecular approach to form the desired piperidine ring. Thus, cycloaddition of nitrone **59** affords predominantly the desired bridged adduct **60** along with two related

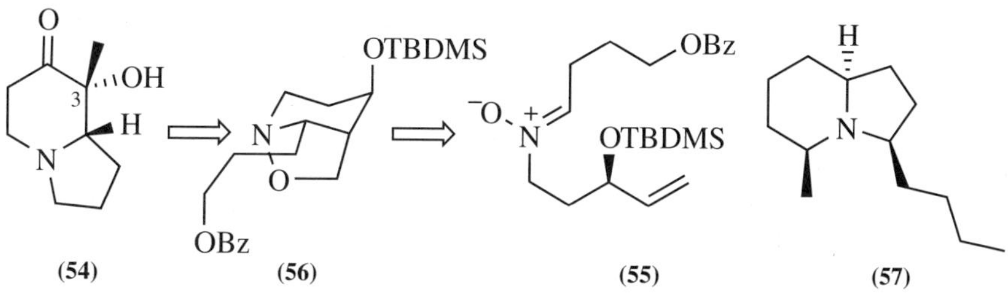

(54) (56) (55) (57)

Scheme 1.13

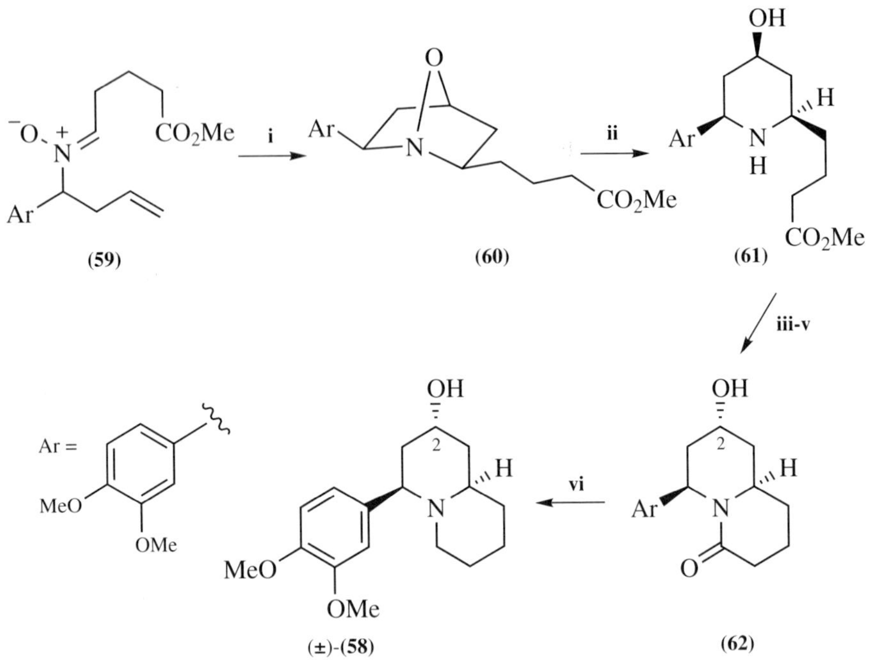

Reagents: [i] PhMe, Δ, 1 h, 60%; [ii] Zn, AcOH, 65 °C, 4 h, 95%; [iii] TMS-imidazole, DCM, 4 h; [iv] 160 °C, 2 h, then TBAF, THF, 2 h, 50% (from **61**); [v] Ph_3P, $PhCO_2H$, diethylazodicarboxylate (DEAD), DCM, rt, 2 days, then KOH, MeOH, rt, 6 h, 74%; [vi] $LiAlH_4$, THF, Δ, 4 h, 76%.

Scheme 1.14

diastereomers. Reductive N—O cleavage of **60** with Zn/AcOH provided a trisubstituted piperidine (**61**) which, after formation of the silyl ether from the hydroxyl group, was cyclized in a melt of 2-hydroxypyridine at 160 °C. The stereochemistry at this position [C(2) in lasubine II numbering] was inverted under Mitsunobu conditions to afford the desilylated lactam **62** and, after reduction of the carbonyl with $LiAlH_4$, afforded the target compound (±)-**58**.

A recent article describes the use of an unusual nitrone–alkene intramolecular cycloaddition–retrocycloaddition–intramolecular cycloaddition strategy (Scheme 1.15). Here, Cordero et al. (83) used a pyrrolidine nitrone to afford the isoxazolidine skeleton before installation of the alkenyl ester side chain of **63** by Mitsunobu methodology employing a polymer supported triphenylphosphine. Thermally induced retrocycloaddition of **63** in *o*-dichlorobenzene at 150 °C afforded an unisolated nitrone (**64**) that underwent an intramolecular cycloaddition to afford a second isoxazolidine (**65**). Removal of the p-methoxybenzyl (PMB) protecting group and mesylation of the revealed hydroxyl was followed by hydrogenolytic N—O bond cleavage, to free the amine nitrogen for nucleophilic attack at the carbon carrying the mesylate to afford indolizidine (**66**).

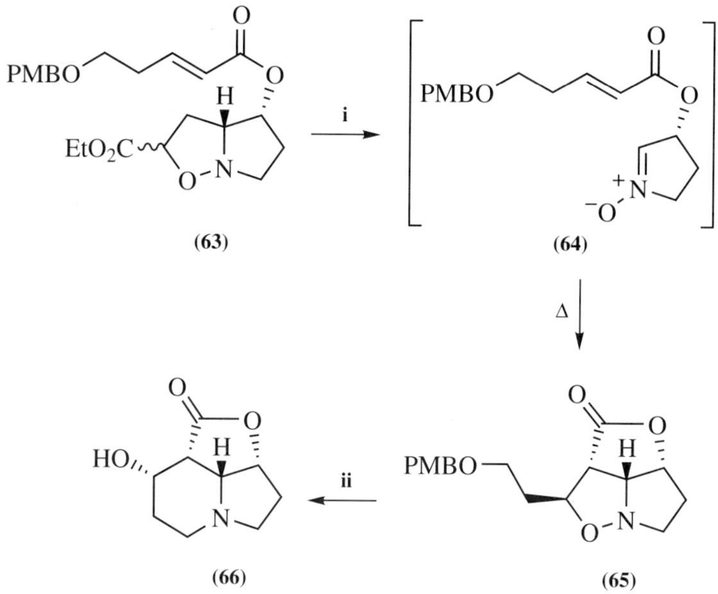

Reagents: [i] o-$Cl_2C_6H_4$, 150 °C, 3 h, 74%; [ii] trifluoroacetic acid (TFA), DCM rt, then MsCl, Et_3N, DCM, 0 °C, then H_2, Pd-C, MeOH.
Ms = methanesulfonyl

Scheme 1.15

These authors also showed that the indolizidine skeleton can be prepared from cyclopropyl dipolarophiles (Scheme 1.16). The cycloaddition of alkylidenecyclopropanes **67** with various nitrones (e.g., **68**) afforded the expected isoxazolidine adducts **69** and **70**, commonly forming the C(5) substituted adducts **70** (97,105–108) predominantly but not exclusively (109–111). Thermally induced rearrangement of the spirocyclopropyl isoxazolidine adduct **70** afforded the piperidinones **71** (107,108). These authors propose reaction via initial N—O bond homolysis of **70** to diradical **72** followed by ring expansion through relief of the cyclopropyl ring strain forming the carbonyl of a second diradical intermediate **73**, which cyclizes to afford the isolated piperidinone **71**.

In this way, spirocyclopropyl adduct **74** (from cycloaddition of **75** and **76**) was used to prepare gephyrotoxins (106) and lentiginosine (**49**) (Scheme 1.17) (105,112). In the latter case, pyrolysis of adduct **74** afforded indolizidinone **77** (45%) along with the amino ketone **78** (55%), the predominance of the latter being accredited to the steric hindrance of the diradical coupling by the bulky TBDPS groups. Reduction of the carbonyl of **77** was achieved with sodium borohydride after conversion to the tosyl hydrazone before final desilylation of **79** with HF afforded **49**, allowing the authors to challenge the published absolute stereochemistry. The presence of a phenyl group (particularly when substituted by electron-donating groups) on the nitrogen atom of the isoxazolidine exerts a powerful activation of the rearrangement, allowing the thermolysis reaction to occur at much

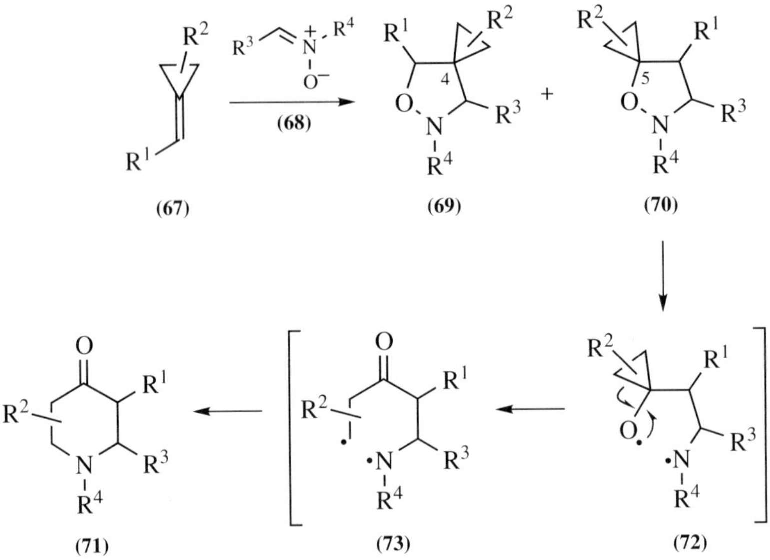

Scheme 1.16

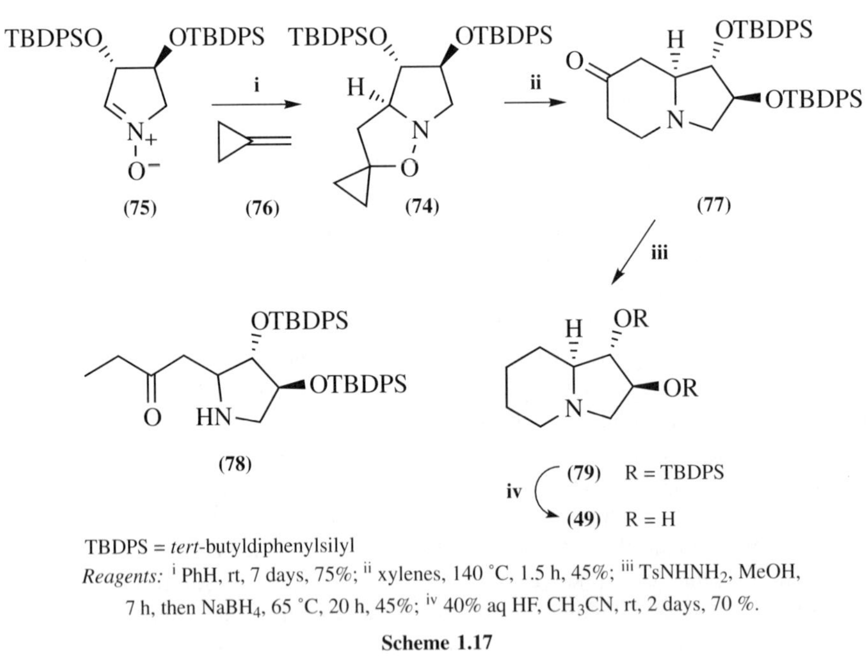

TBDPS = *tert*-butyldiphenylsilyl

Reagents: [i] PhH, rt, 7 days, 75%; [ii] xylenes, 140 °C, 1.5 h, 45%; [iii] $TsNHNH_2$, MeOH, 7 h, then $NaBH_4$, 65 °C, 20 h, 45%; [iv] 40% aq HF, CH_3CN, rt, 2 days, 70 %.

Scheme 1.17

(80)

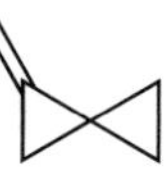
(81)

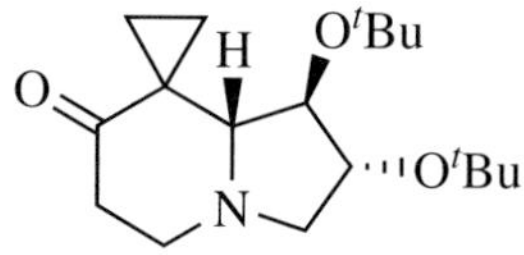

(84)

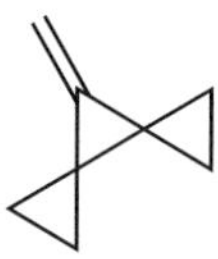
(82)

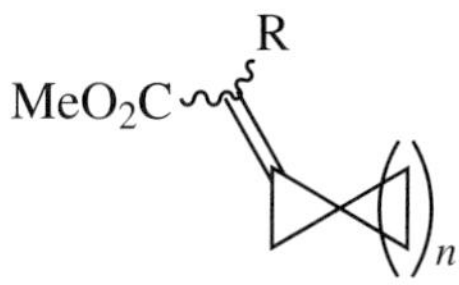

(83) n = 0, 1, R = H, Cl

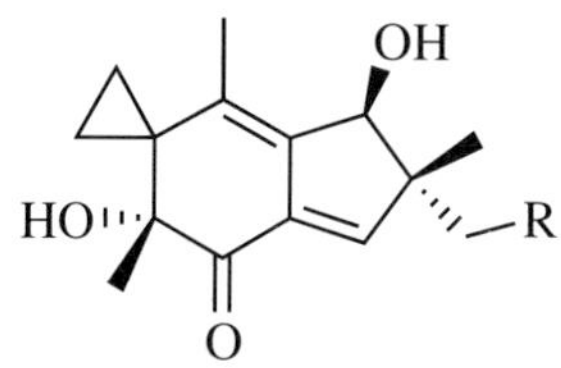

(85a) Illudin S (R = OH)
(85b) Illudin M (R = H)

Figure 1.2

lower temperatures (97). The authors were also able to prepare adducts from bicyclopropylidene (**80**) (113–117) and methylene spirocyclopropyl dipolarophiles (**81–83**) (118–120), which were transformed on heating into spirocyclopropyl piperidones (e.g., **84**) from dipolarophile **81**, as aza-analogues of the illudin family of cytotoxic sesquiterpenes (**85a–b**) (Fig. 1.2). This rearrangement has been applied to the synthesis of racemic indolizidine elaeokanine A and precursors of (±)-lupinine and (±)-epilupinine (121) and related targets (122) as well as (*2S*)-4-oxopipecolic acid, a rare amino acid found in the virginiamycin cyclic peptides (123).

This work has since been extended to cyclobutyl isoxazolidine adducts (e.g., **86**) from the cycloaddition of **87** to methylenecyclopropane (**88**) (Scheme 1.18) (124–127). Thermolysis afforded a mixture of products, of which the bicyclic azepinone (**89**) predominated. Spirocyclic adducts were also prepared from an intramolecular reaction in the synthesis of cyclic amines (Scheme 1.72, Section 1.11.3).

A further rearrangement route to bicyclic aminoketones has been investigated by Padwa et al. (128–134) (Scheme 1.19). Building on the allene–nitrone cycloadditions reported by Tufariello, the alkenylisoxazolidine adducts **90** and **91** were

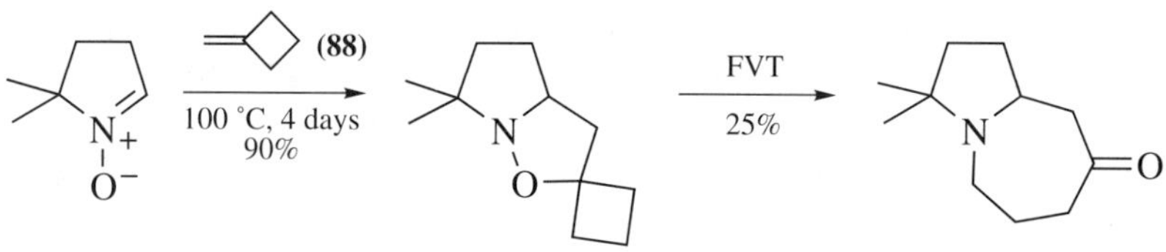

(87) **(86)** **(89)**

FVT = flash vacuum thermolysis

Scheme 1.18

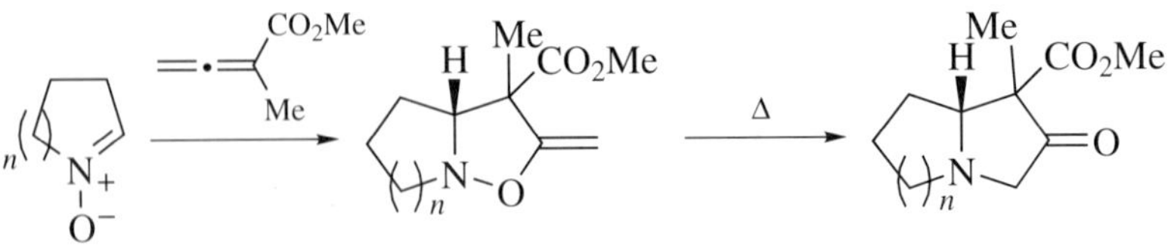

(92) $n = 1$ **(90)** $n = 1$ **(94)** $n = 1$
(93) $n = 2$ **(91)** $n = 2$ **(95)** $n = 2$

Scheme 1.19

prepared from the reaction of the corresponding cyclic nitrones **92** and **93**, respectively, and an electron-deficient allene dipolarophile, with chemoselection for the more hindered, more electron-deficient alkene in almost every case. The pyrrolizidine and indolizidine skeletons were prepared by thermolysis of these adducts at 80–90 °C (sealed tube, 8 h) to afford the bicyclic aminoketones **94** and **95** via a proposed diradical mechanism.

1.6. PEPTIDES AND AMINO ACIDS

3-Hydroxy-4-methylproline (**96**) is a common structural feature of the echinocandins and mulundocandins, which exhibit specific fungicidal activities, and as such this moiety has been retained in a number of synthetic antifungal agents. Langlois and Rakotondradany (80) have prepared the natural (2*S*,3*S*,4*S*) form of **96** by the 1,3-dipolar cycloaddition of (1-ethoxy)ethoxymethyl protected α,β-unsaturated γ-lactam (**97**) [prepared from (*S*)-pyroglutaminol] (79) with excess *N*-methylnitrone, which affords the desired adduct **98** (70%) along with the regioisomer **99** (9%) (Scheme 1.20). Quantitative reduction of the lactam carbonyl with diisobutylaluminum hydride (DIBAL) and sodium cyanoborohydride was followed by a protection exchange, N—O cleavage, Cope elimination, and enantioselective hydrogenation to afford the target amino acid **96**, isolated as the hydrochloride.

Similarly, Kibayashi and co-workers (135) installed both chiral centers in the unusual peptide-like antibiotic (+)-negamycin (**100**) by a cycloaddition strategy (Scheme 1.21). D-Gulose-derived nitrone D-**101** was reacted with carbobenzoxy (Cbz)-protected allylamine to afford an inseparable mixture of two isoxazolidines, **102** and its C(3) epimer. After N-protection exchange, reduction of the ester function allowed separation as the corresponding alcohols **103**. Tosylation, homologation with NaCN and nitrile hydrolysis in methanol afforded the correct chain length at C(3) of the (3*R*,5*R*) isomer **104** before activated ester coupling of the hydrazide moiety and deprotection gave the natural product (+)-**100** (135).

The use by Langlois of an amidoalcohol (79,80) is an unusual strategy for the construction of α-amino acids. More commonly, the required amine and carboxylic acid functionalities are carried into the cycloaddition in the dipolarophile, as a homochiral alkenyl α-amino acid derivative. Importantly, this introduces a second

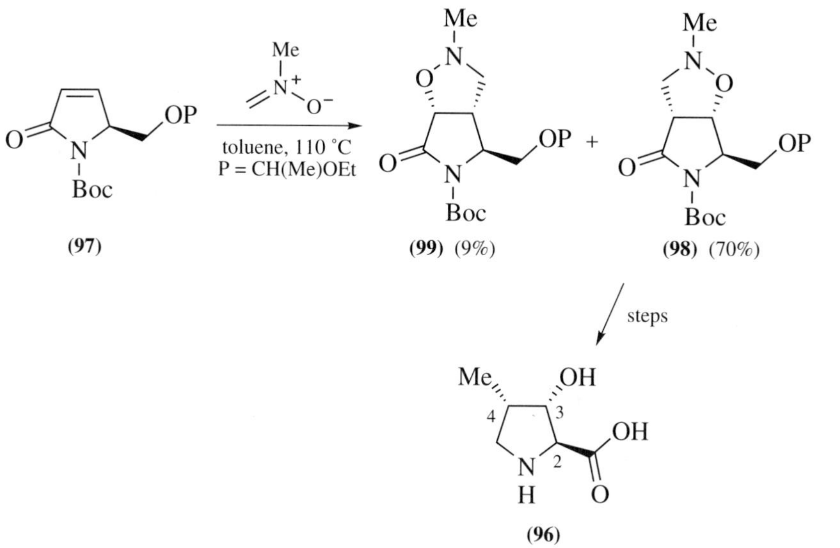

Scheme 1.20

N and O function into the molecule and has been used to prepare hydroxyarginines (136,137), hydroxyornithines (136–138), β-lysine, β-leucine, and β-phenyl-β-alanine (139,140), the low-calorie sweetener aspartame (141) and the antitumor antibiotic acivicin (142–144).

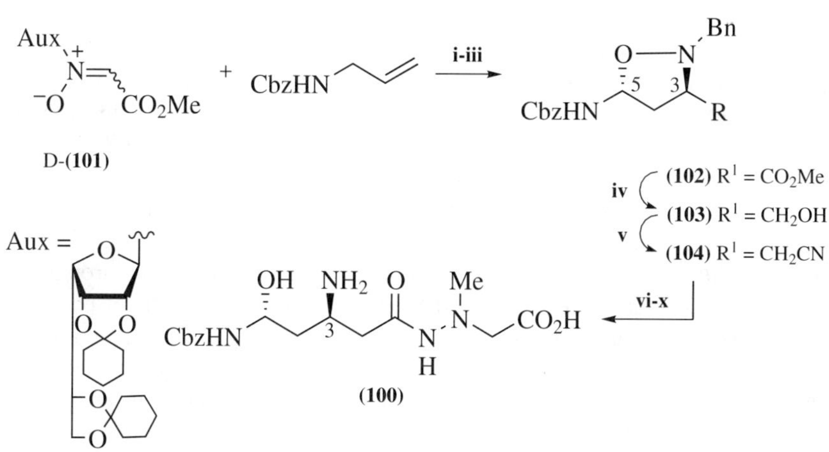

Reagents: [i] PhMe, Δ; [ii] 10% HCl, MeOH, 40 °C; [iii] BnBr, K_2CO_3, DMF, 50 °C; [iv] $LiAlH_4$, Et_2O, 0 °C to rt; [v] TsCl, $EtN(iPr)_2$, DCM, 0 °C to rt then NaCN, dimethyl sulfoxide (DMSO), 80 °C; [vi] HCl, EtOH, rt; [vii] aq NaOH, MeOH, rt; [viii] EtOCOCl, Et_3N, PhMe, 0 °C; [ix] $H_2NN(Me)CH_2CO_2Bn$, PhMe, 0 °C to rt; [x] H_2, 10% Pd-C, MeOH, aq AcOH.

Scheme 1.21

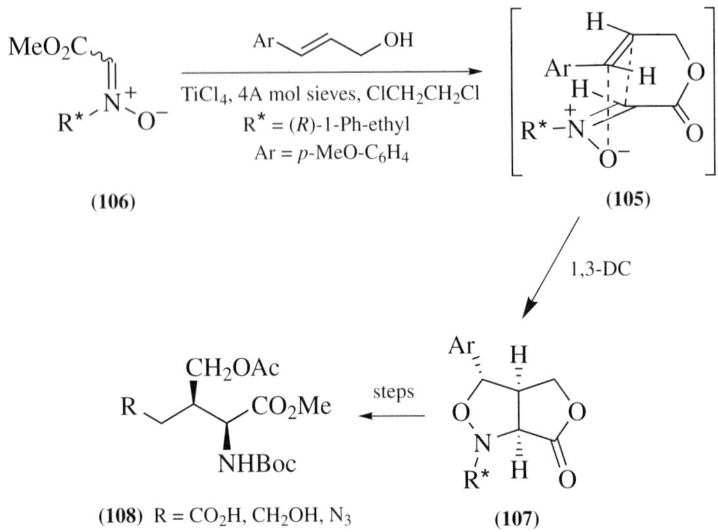

Scheme 1.22

An unusual route was described by Tamura et al. (42,84,85,145–148) in which β-substituted α-amino acid precursors were formed by a tandem transesterification and cycloaddition process (Scheme 1.22). The alkenyl nitrone **105** was formed by the treatment of chiral nitrone **106** (with a carboxylic ester substituent on the nitrone carbon atom) with an unsaturated alcohol in the presence of catalytic $TiCl_4$. Spontaneous intramolecular dipolar cycloaddition of this reagent afforded adduct **107** and a diastereomer with moderate selectivity (3:1) using (*R*)-α-phenylethyl chiral auxiliary on the nitrone nitrogen atom. Thus, after further synthetic manipulation including ruthenium-mediated oxidative cleavage of the aromatic ring, adduct **107** afforded the β-functionalized α-amino esters **108**. Similarly, β-aminoalcohol functionality was introduced with a small measure of stereoselectivity into intermediates of potent oligoamide renin inhibitors through the use of a homochiral alkenylamine dipolarophile (149).

Peptide functionality may be prepared in a homochiral dipolarophile for subsequent cycloaddition reaction, as demonstrated in the synthesis of peptidomimetic isoxazolidine anatagonists of human neurokinin-A by Brandi and co-workers (150) (Scheme 1.23). The 1,3-dipolar cycloaddition of macrocyclic maleic acid diamide (**109**) to a *tert*-butoxypyrrolidine nitrone (**110**) afforded in 86% yield a 25:1 mixture in favor of **111** (via transition state ***exo*-A**) over its regioisomer (from ***exo*-B**) after a 27 h reaction in refluxing toluene. Comparable yields but lower selectivities were observed in DMSO, indicating that, with the existence of an equilibrium excluded, a conformational change in the peptide enforces a change in regioselectivity. The observed products are rationalized in terms of double asymmetric induction. First, the alkene functionality of the macrocyclic dipolarophile **109** can only reasonably be approached via an exo transition state, and so endo

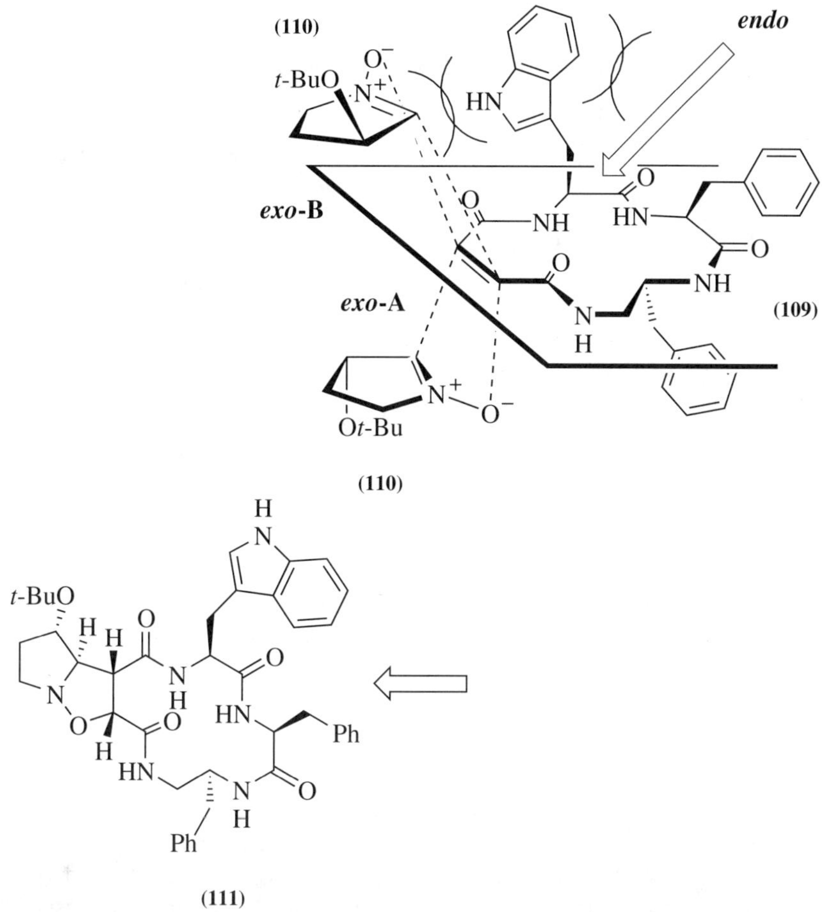

Scheme 1.23

approach (endo approach to the top face of **109** is arrowed) is excluded on steric grounds. Additionally, the dipole **110** is also facially discriminating, only tolerating reaction at the face opposite the bulky C-*tert*-butoxy substituent. Furthermore, the dipolarophile **109** exerts a preference for reaction at its lower face as drawn (via transition state ***exo*-A**), through steric crowding of the upper face by the indole moiety of the tryptophan residue (***exo*-B**).

Alternatively, some of the desired amino acid functionality may be contained within the nitrone fragment, as in the synthesis of homochiral allyl glycines by Katagiri et al. (151), which reveals the carboxylate by hydrolysis of a lactone in the dipole (Scheme 1.24). Here, thermolysis of nitroso Meldrum's acid (**112**) via a nitrosoketene intermediate **113** and reaction with *l*-menthone gave the separable nitrones **114a** (26%) and **114b** (28%) by a [3 + 2] cycloaddition, although a

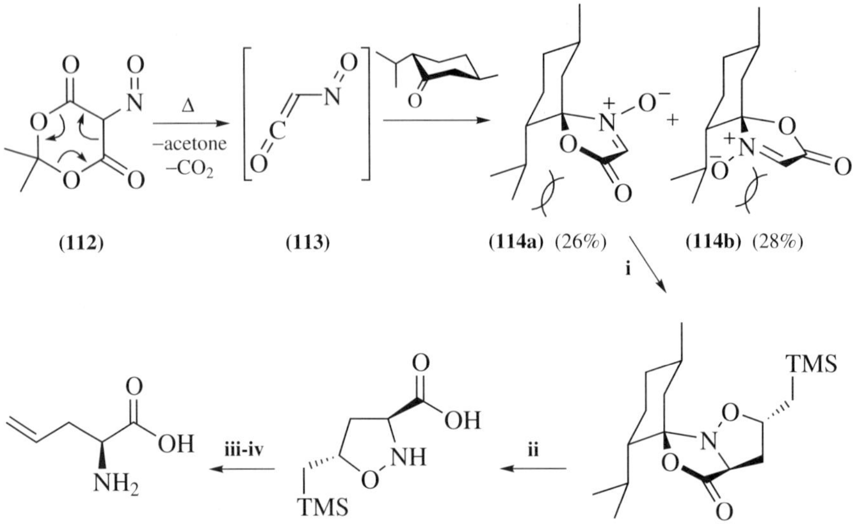

(112) (113) (114a) (26%) (114b) (28%) (S)-(117) (116) (115)

Reagents: [i] Allyltrimethylsilane, 800 MPa, toluene, 40 °C, 90%; [ii] 0.15 *M* aq NaOH, 4 h, 100%; [iii] H_2, Pd-C, MeOH, 88%; [iv] BF_3•Et_2O, MeCN, 3 h, 100%.

Scheme 1.24

possible [4 + 2] addition and 1,2-migration pathway was also acknowledged. Cycloaddition of nitrone **114a** with allyltrimethylsilane under high pressure proceeds to the isoxazolidine **115** as a single isomer in excellent yield (90%), which becomes quantitative with boron trifluoride–diethyl etherate catalysis. The high stereoselection is explained through steric hindrance of the lower face of the nitrone by the pendant isopropyl group of the menthyl auxiliary, allowing addition of the dipolarophile to the upper face only. The carboxylic acid functionality was revealed by hydrolysis of the oxazolidinone to give **116** and afforded the (*S*)-allyl glycine **117** by N—O bond hydrogenolysis followed by a Peterson-type elimination with boron trifluoride. The isomeric nitrone (**114b**) afforded (*R*)-**117** by identical treatment. Another significant use of this strategy is the preparation of pyroglutamic acids (**118**) by Merino et al. (152,153) using the cycloaddition of furfuryl nitrones with acrylate esters or the acrylamide of Oppolzer's bornane-10,2-sultam chiral auxiliary (Scheme 1.25). As a key step in the synthesis, the furfuryl side chain was used as latent carboxylate functionality, conversion being achieved using ruthenium-mediated oxidation (RuO_2—$NaIO_4$). Semiempirical and ab initio calculations supported the experimental findings in which (3*R*,5*R*) isomer **119** was consistently found to be the major product.

As part of their exploration of peptide secondary structure, Hermkens et al. (154) reported an innovative use of the nitrone dipolar cycloaddition

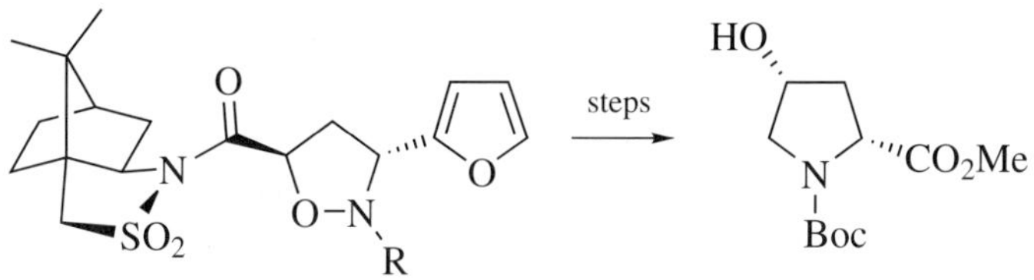

(119) (118)

Scheme 1.25

(Scheme 1.26).The reaction of homochiral amino acid derived nitrone **120** with the complex allyl amide **121** afforded a mixture of three isoxazolidine diastereoismers. Catalytic hydrogenolysis of the benzyl ester and Cbz protecting groups was followed by amide coupling with *O*-(1*H*-benzotriazol-1-yl)-*N,N,N′,N′*-tetramethyluronium tetrafluoroborate (TBTU) and acidolysis of the Boc and *tert*-butyl ester. The resultant macrobicyclic isoxazolidinyllactam β-turn mimics (e.g., **122**) were tested for activity in a human platelet aggregation assay, in which the negative results suggest that a β-turn is not present in the bioactive form of the receptor.

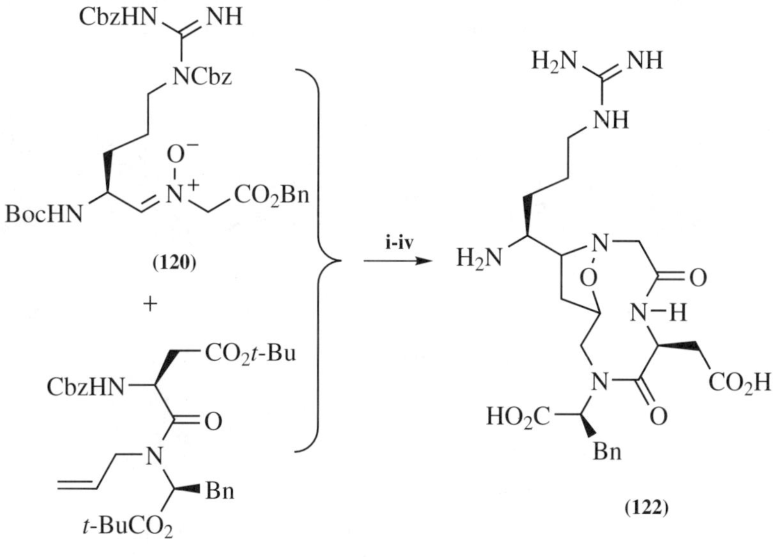

(121)

Reagents: [i] PhMe, 15 kbar, 50 °C, 2 days; [ii] H_2, Pd-C, MeOH, [iii] DMF, TBTU, pH 8.0, rt; [iv] TFA, PhOH, H_2O, $(i\text{-Pr})_3SiH$ (88:5:5:2).

Scheme 1.26

1.7. SUGARS

Carbohydrates are in common use in nitrone cycloaddition chemistry, whether in the synthesis of homochiral dipoles from a readily available chiral pool (155–162) or as *off-the-shelf* homochiral dipolarophiles (163–165), and also constitute important synthetic targets (166–168). The carbohydrate-derived nitrones of Vasella and co-workers (169) are among the earliest examples of homochiral cyclic nitrone 1,3-dipoles and applications include the addition of methyl methacrylate to the spirocyclic nitrone **123** (Scheme 1.27). The isoxazolidine **124** was the major product isolated from a mixture containing the C(2) epimer **125** and two regioisomers (83:2:7.5:7.5, respectively). The assignment of stereochemistry of adduct **124** is supported by X-ray crystallography and, against expectation, the configuration at C(4) of the major product is consistent with the approach of the dipolarophile to the sterically crowded face of the nitrone **123** (*anti* to the C(4) to O bond). The diasteromeric excess (de) at C(2) of 81% (**124:125** = 83:2%) indicates a strong preference for the ester in an endo position in the transition state. Reaction of **123** with methyl acrylate afforded a more complex mixture, which showed that reaction proceeds with a similar regiochemical preference but with little facial or endo–exo selectivity.

Elsewhere, the reaction of styrene with nitrones derived from cyclic acetals of D-erythrose (e.g., **126**) or D-threose has afforded a mixture of diastereomeric isoxazolidines (Scheme 1.28) (170,171). In all cases, nuclear magnetic resonance (NMR) analysis suggests that the major product contains the C(3)/C(4′) *erythro*

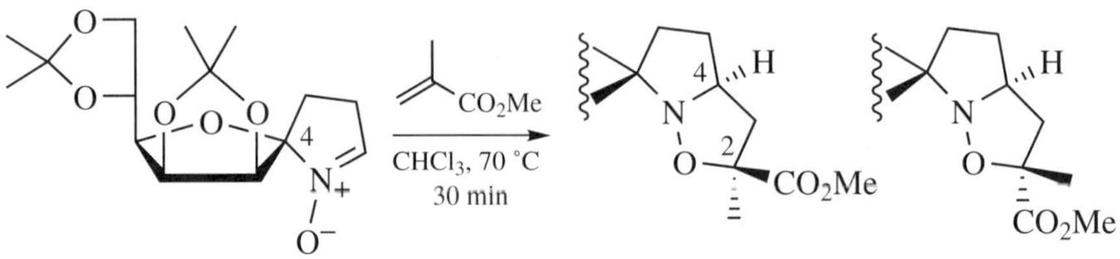

(123) **(124)** **(125)**

Scheme 1.27

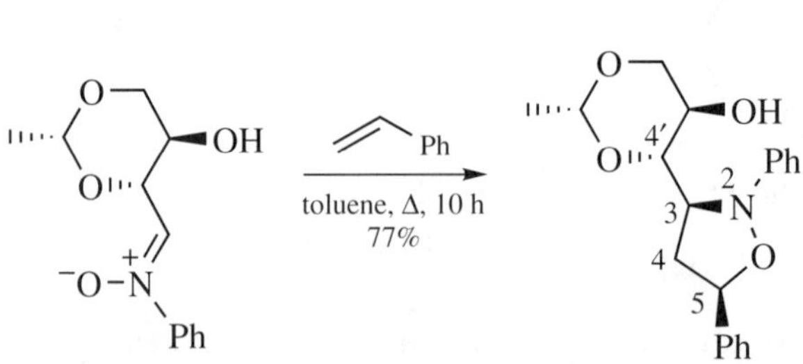

(126) **(127)**

Scheme 1.28

C(3)/C(5) cis product (e.g., **127**) and was confirmed by X-ray crystallography. The stereoselectivity of addition increases with the bulk of the N-substituent of the nitrone, and is rationalized through the less-hindered endo approach of the dipolarophile to the more reactive (Z)-nitrone. While the stereoselection was unpredictable, all of the nitrones exhibited total *regio*selectivity for the 5-phenyl isoxazolidines. This work bears some similarity to the approach of DeShong et al. (172,173) who prepared amino- and deoxy-sugars from the cycloaddition of non-carbohydrate derived nitrone acetals (Scheme 1.29). Reaction of **128** with vinyltrimethylsilane afforded a diastereomeric mixture of isoxazolidine intermediates **129**, which underwent a complex bond cleavage cascade induced by treatment with dilute HF, ultimately to afford a single α,β-unsaturated aldehyde **130**. This intermediate is a latent 5-hydroxyaldehyde and, once revealed by enal reduction and acetonide hydrolysis, underwent a ring closure to afford the trideoxyhexose sugar rhodinose **131** as a mixture of anomers. The ethyl vinyl ether-derived adduct **132** afforded the 3-amino-5-hydroxyaldehyde **133** by a more familiar hydrogenolytic N—O bond cleavage and spontaneously formed the amino sugar daunosamine (**134**), isolated as the protected methyl glycoside (**135**).

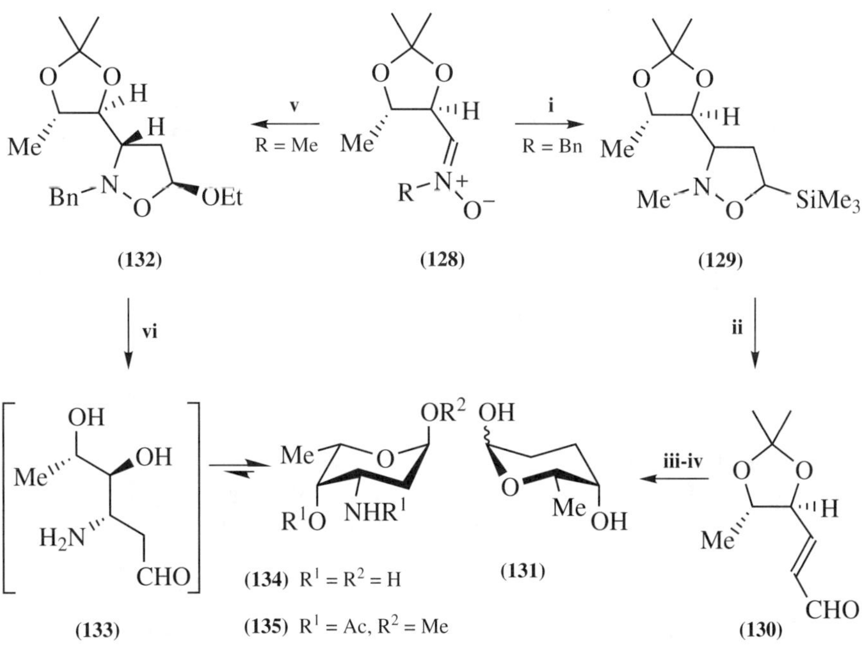

Reagents: i $CH_2{=}CHSiMe_3$, 80 °C, 24 h, 90%; ii 50% aq HF, MeCN, rt, 30 min; iii H_2, 5% Pd-C, EtOH, 2 h; iv 2% aq HCl, acetone, rt, 40 % from **128**; v $CH_2{=}CHOEt$, Δ, 72 h, 93%; vi H_2, 5% $Pd(OH)_2$, 10% HCl / MeOH, 48 h, then Ac_2O, py, 4-(dimethylamino)pyridine (DMAP), 24 h rt

Scheme 1.29

The most commonly reported carbohydrate-derived dipolarophiles are the α,β-unsaturated lactones (70–73,174–177). Chmielewski and co-workers (73) prepared the polyol β-lactam **136** via a nitrone cycloaddition strategy based on the Tufariello approach (178) and took advantage of the regioselectivity of the cycloaddition of nitrones to ene–lactones, in which the nitrone oxygen is added at C(3) of the sugar δ-lactone (Scheme 1.30). Adduct **137** was the sole product of the cycloaddition of *N*-phenyl-*C*-(4-methoxyphenyl)nitrone with enelactone **138**, but unexpectedly, conventional N—O bond hydrogenolysis resulted in deamination. The isoxazolidine ring of **137** was successfully opened after conversion to **139** by ester hydrolysis and protection of the acid (as the benzyl ester) and hydroxy moieties (as silyl ethers). The acyclic aminoester intermediate **140** underwent ring-closure mediated by 2-chloro-1-methylpyridinium iodide to afford the β-lactam skeleton **136**.

In recent work, Chmielewski and co-workers (174) reported the highly stereoselective reaction of ene–lactones with chiral pyrrolidine nitrone (**141**) to afford tricyclic adducts (Scheme 1.31). A 1:1 mixture of ene–lactone **142** and nitrone **141** provided adduct **143** with an uncharacterized isomer (97:3) (91%) while homochiral D-glycero (**138**) gave the adduct **144** as a single diastereomer (88%). A 2:1 mixture of racemic **138** and nitrone **141** afforded a 91:1 mixture of the two possible adducts, representing an effective kinetic resolution of the racemic lactone.

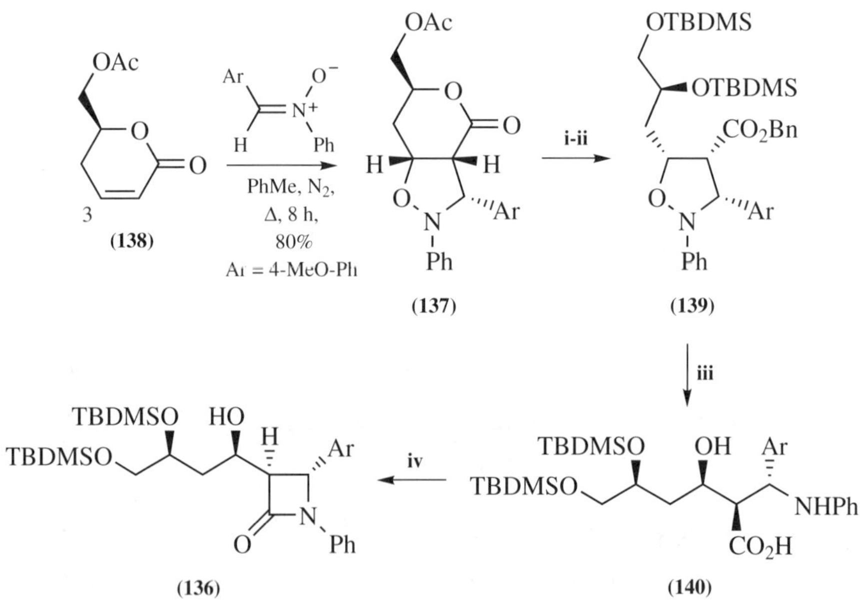

Reagents: [i] KOH, dioxane, H_2O, rt, overnight, then BnBr, 18-crown-6, DMF, 60%; [ii] TBDMSCl, imidazole, DMF, 0 °C, then rt, overnight, 90%; [iii] H_2, 2 atm, 10% Pd-C, EtOH, 2 h, 85%; [iv] 2-chloro-1-methylpyridium iodide, Et_3N, DCM, rt, 2 h, 80%.

Scheme 1.30

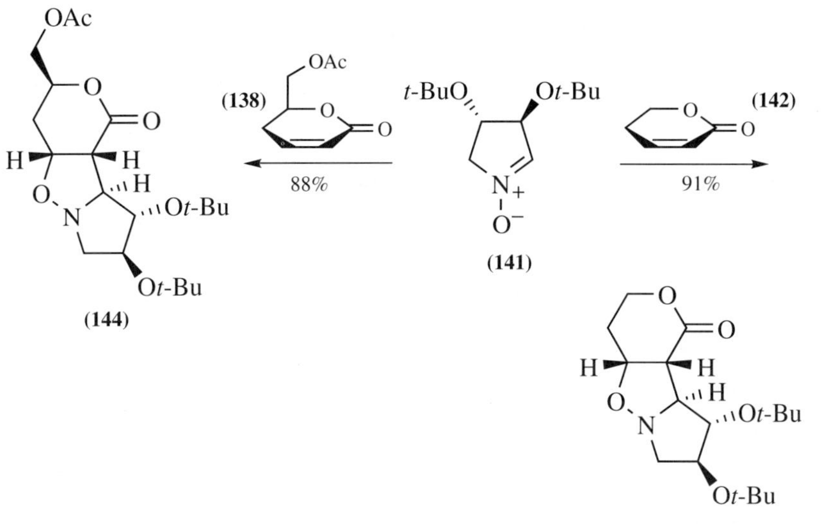

Scheme 1.31

Observed adducts arise through exo addition of the lactone to the *re–re* face of the nitrone **141**, which avoids an unfavorable C=O/O-*tert*-butyl steric clash.

Langlois and co-workers (179) found the same exo stereochemical preference through double asymmetric induction of a related ene–lactone (*R*)-**145** with their well-explored and efficient camphor-derived oxazoline nitrone (1*S*)-**146** (Scheme 1.32). They found the cycloaddition components form a matched pair and allowed kinetic resolution of the racemic lactone in up to 70% enantiomeric excess (ee). They suggest the selectivity for exo adduct **147** arises through destabilization of the endo transition state by a steric clash between dipolarophile ring hydrogens and the bornane moiety.

Borrachero et al. (180) prepared a number of sugar isoxazolidines by the reaction of carbohydrate-functionalized nitrones with nitroalkenes (Scheme 1.33). They found a matched pair of chiral sugar cycloaddition reaction partners to be

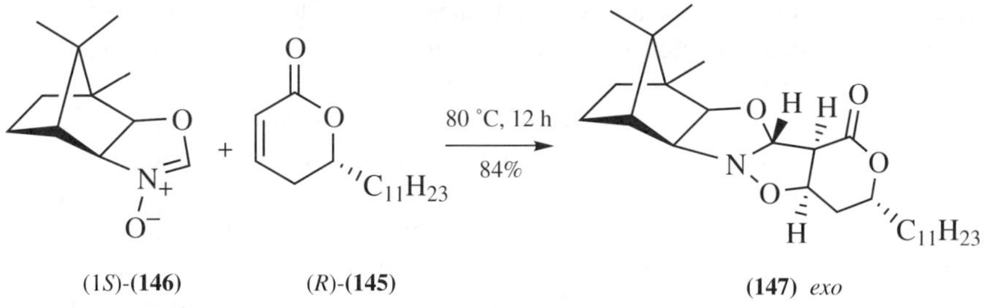

(1*S*)-(**146**) (*R*)-(**145**) (**147**) *exo*

Scheme 1.32

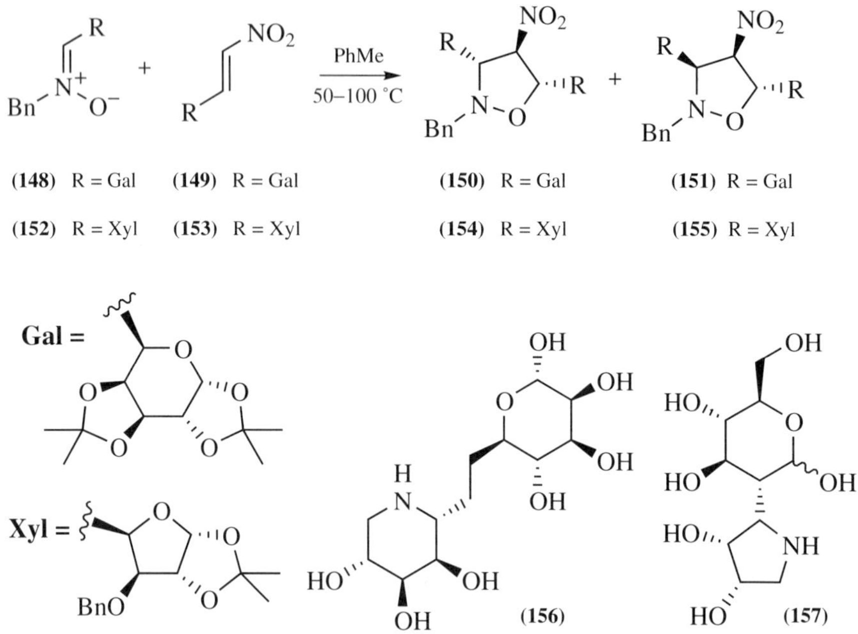

Scheme 1.33

much more effective than a single carbohydrate auxiliary, which also suffered some isomerization on silica gel. In all the isoxazolidine adducts, the nitro group is at C(4) and the trans geometry of the alkene is maintained, as expected. Thus, the reaction of nitrone **148** with alkene **149** (both derived from α-D-galactose) afforded a 7:1 mixture of the two possible exo adducts **150** and **151**. The ratio is improved by using a matched pair of α-D-xylose-derived components (**152** and **153**) to afford a 9:1 mixture of adducts **154** and **155**. Similarly, Wightman and co-workers (181) found excellent yields and stereoselectivies in the addition of a D-lyxose-derived nitrone to D-mannosyl- and D-galactosyl alkenes, although each cycloaddition contributes only one new chiral centre to the target aza-*C*-disaccharides (e.g., **156**). Brandi and co-workers (182,183) used their tartaric and malic acid derived nitrones in the preparation of a number of pseudo-aza-*C*-disaccharides (e.g., **157**) and reported significant rate and yield enhancements under high pressure (184).

The intramolecular cycloaddition of an alkenyl nitrone by Tronchet in 1972 was the first example using a sugar derivative (185). Since then, many researchers investigated the intramolecular reaction of nitrones generated from sugars, in particular those from O-allyl carbohydrates, which can give rise to tetrahydropyanyl- or oxepanyl-isoxazolidine cycloadducts (Scheme 1.34) (186–193). Shing's work demonstrates that the stereochemical outcome depends only on the relative configuration at C(2) and C(3) of the sugar (192). Thus, *threo* D-hexose-derived allenylnitrone **158** (from 3-*O*-allyl-D-glucose **159**) afforded oxepane **160** only (isolated as the tetraacetate) while the regioisomeric tetrahydropyran (**161**) was

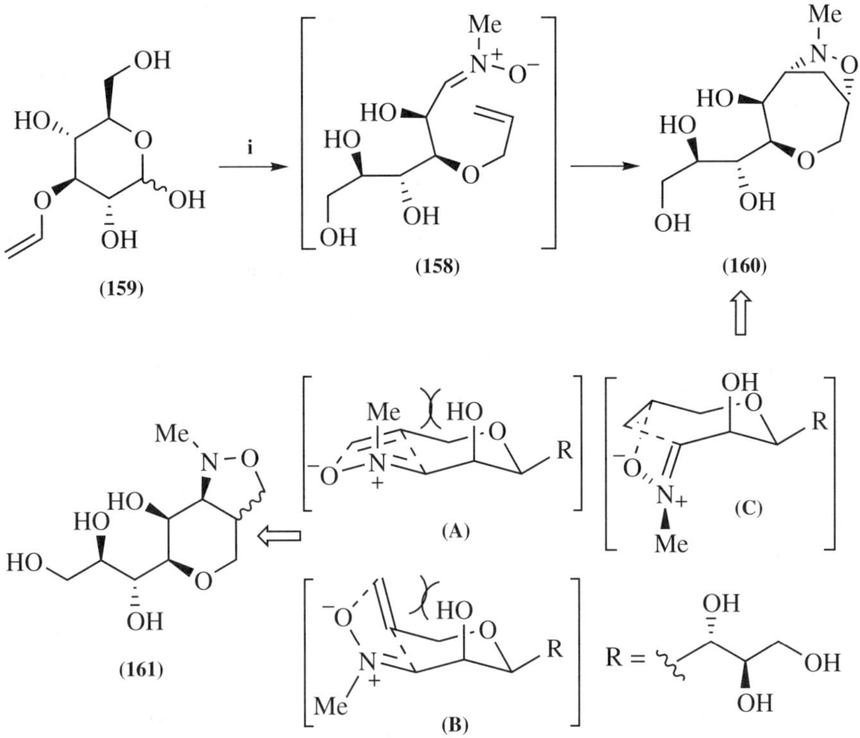

Reagents: [i] MeNHOH•HCl, $NaHCO_3$, 80% aq EtOH, Δ, 48 h.

Scheme 1.34

not observed. Conversely, a tetrahydropyranyl adduct was the sole product of an analogous erythro alkenyl nitrone derived from D-mannose. The authors propose two interconvertible chair-like transition state conformations for tetrahydropyran formation by intramolecular cycloaddition of *threo*-nitrone **158** (transition states **A** and **B**). Both conformations incur unfavorable 1,3-diaxial interactions and the tetrahydropyranyl product is not observed, while no such impediment exists in the erythro series. Exclusive formation of oxepane (**160**) from nitrone (**158**) can thus be rationalized via the relatively favored seven-membered transition state **C**. Yields for the unprotected carbohydrates are moderate and the presence of decomposition products has prompted further work on benzyl–ether protected derivatives.

1.8. SULFUR- AND PHOSPHORUS-CONTAINING COMPOUNDS

While nitrones have demonstrated reactivity toward a number of sulfur-containing dipolarophiles including C=S multiply bonded molecules, by far the most common are the vinyl sulfur compounds (194,195). The known cycloaddition reaction of chiral vinyl sulfoxide dipolarophiles with acyclic nitrones has now been extended to cyclic dipoles by two independent groups. The reaction of tetrahydropyridine N-oxide **93** with (*S*)-*p*-tolyl vinyl sulfoxides (**162**) in ether for 7–10 days afforded exclusively the exo adducts, with **163a** as the major product along with its diastereomer **163b** in 89–98% de and 85–97% yield (Scheme 1.35) (196). The high regio- and stereoselectivity of the addition was first confirmed for selected adducts by reduction with TMSI/NaI to enantiomeric sulfides [e.g., **163b** (R = Me) to **164**]. Reductive cleavage of the N—O bond of adduct **163a** (R = Me) and simultaneous desulfurization with Ni/Al amalgam afforded the known piperidine natural product (+)-sedridine (**165**) along with its C(7) epimer. To overcome this epimerization, selective cleavage of the N—O bond was followed by protection of the amino functionality before desulfurization with Raney nickel. Deprotection with TMSI rapidly and efficiently reveals the enantiomerically pure natural product **165** (97% yield).

The reaction of tetrahydropyridine *N*-oxide (**93**) ($n = 1$) with a chiral sulfinyl maleimide dipolarophile (**166**) has been reported (Scheme 1.36), but afforded the major product **167** with only modest stereoselectivity, despite the use of a

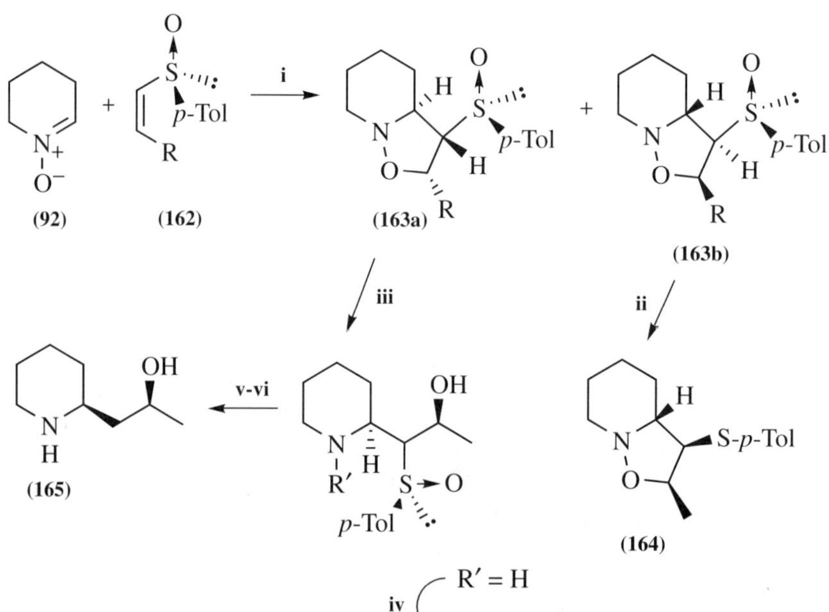

Reagents: [i] Et_2O, rt, 7–10 days, 85–97%; [ii] TMSI, NaI; [iii] Ni/Al, aq KOH, MeOH, rt, 2 h, 93%; [iv] $ClCO_2CH_3$, aq K_2CO_3, rt, 18 h, 98%; [v] W-6 Raney Ni, H_2, MeOH, 18 h, 84%; [vi] TMSI, DCM, Δ, 1 h, then MeOH, rt, 10 min, 97%.

Scheme 1.35

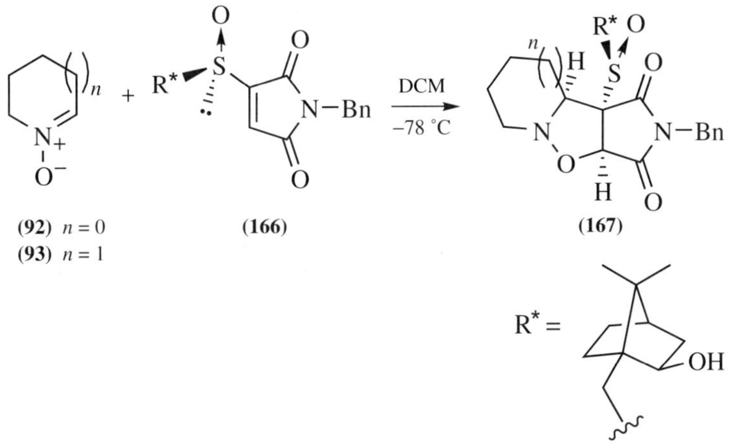

Scheme 1.36

homochiral sulfoxide carrying a bornyl alcohol chiral auxiliary (R*) (197). Conversely, reaction with 1-pyrroline *N*-oxide (**92**) afforded an inseparable mixture of products.

Aggarwal et al. (198) reported nitrone cycloaddition reactions with the unusual disulfoxide dipolarophile *trans*-2-methylene-1,3-dithiolane (**168**) (Scheme 1.37). The question of exo/endo selectivity of the cycloaddition reaction is avoided by the use of this C_2-symmetric dipolarophile for which only two transition states are

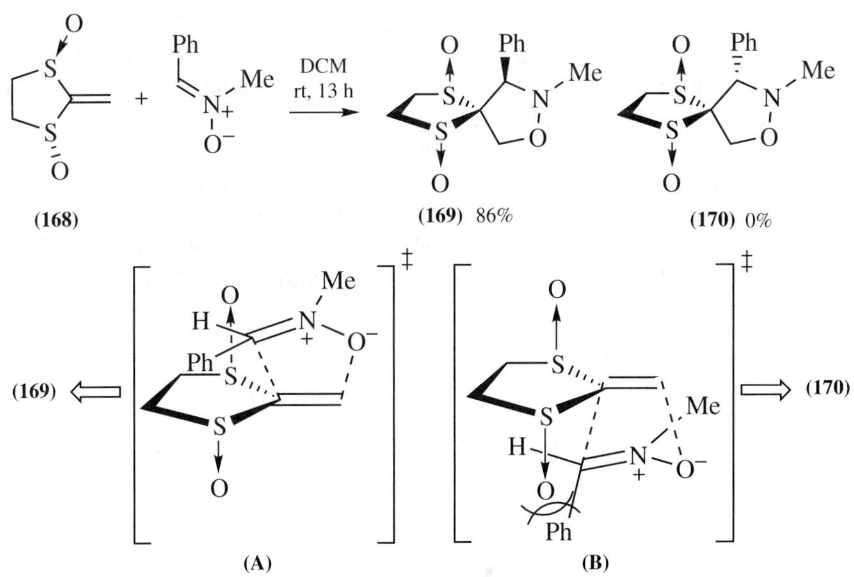

Scheme 1.37

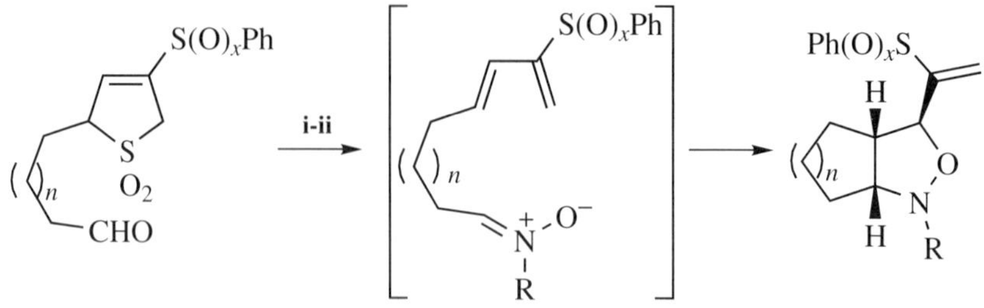

(172a) $n = 1$
(172b) $n = 2$

(171a) $n = 1$
(171b) $n = 2$

(173)

Reagents: [i] RNHOH•HCl, NaOMe, rt; [ii] PhMe, 95 °C

Scheme 1.38

possible. Thus, the reaction of **168** with *N*-methyl-*C*-phenyl nitrone can lead to two diastereomeric products **169** and **170**, but only **169** is observed, via the favored transition state **A**. The disfavored transition state **B** is characterized by a repulsion between the nitrone phenyl moiety and the sulfinyl oxygen. The complete regioselectivity for 4,4-disubstitution in the isoxazolidine cycloadducts is unusual for addition of a nitrone to 1,1-disubstituted alkenes. Similarly, high selectivities were also found in the reaction of **168** with simple cyclic nitrones.

Elsewhere, the intramolecular cycloaddition of a nitrone carrying a vinyl sulfoxide or vinyl sulfone moiety has been examined (Scheme 1.38) (199). The nitrone reagents **171a–b** were generated by condensation of an N-substituted hydroxylamine with the corresponding aldehydes **172a** and **172b** followed by thermal cheletropic removal of SO_2. The cis-fused cycloadducts **173** predominate in the cycloaddition reactions of the 5-dienylnitrones (**171a**, $n = 1$) with either sulfide or sulfone substituents and is explained in terms of a more highly strained transition state for the trans isomer. In contrast, of the 6-dienyl analogues **171b** ($n = 2$), only the sulfones retain stereoselectivity. The phenylsulfide adduct **173** ($x = 0$) has been transformed into a bicyclic azetidine, a rare heterocyclic 4,5-ring system. A four-membered ring system has also been utilized in the more common intermolecular cycloaddition of vinyl sulfones in which electrochemically generated 1-phenylsulfonylcyclobutene (**174**) was reacted with a range of nitrones to afford cyclobutenyl isoxazolidines (**175**) (Scheme 1.39) (200).

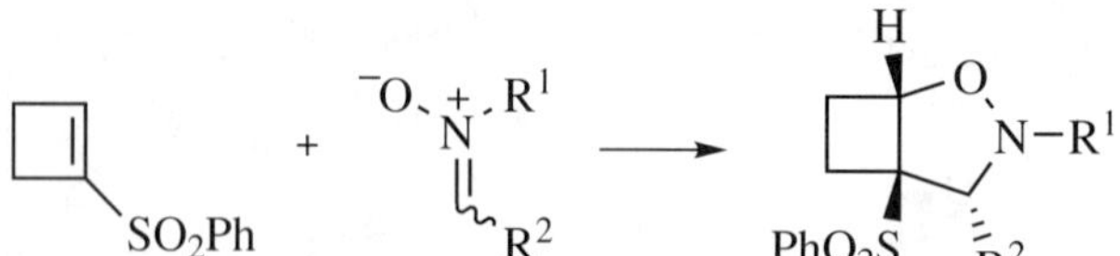

(174) (175)

Scheme 1.39

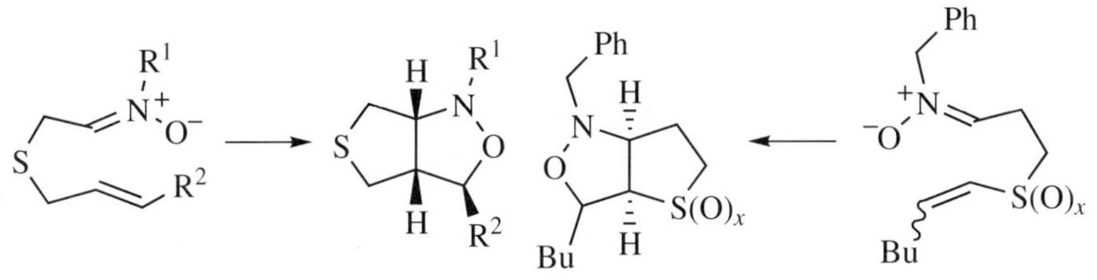

(176) (178) (179) $x = 0, 2$ (177)

Scheme 1.40

The use of *C*-furfuryl nitrones by Merino and co-workers (201) as latent carboxylate functionality (see Section 1.7) has inspired the development of an analogous *C*-thiazolyl nitrone. Thus, after reaction of this dipole with acrylate esters, the authors take advantage of the thiazolyl-to-formyl synthetic equivalence to reveal aldehyde functionality after reduction with sodium cyanoborohydride.

There have been simultaneous reports of the intramolecular cycloaddition of alkenyl nitrones bearing alkenylsulfide (202,203) or alkenylsulfone substitutents (Scheme 1.40) (203). In both cases, the isolated products were the corresponding fused bicyclic isoxazolidine adducts **178** and **179**, respectively.

Brandi and co-workers (204) utilized phosphorus chirality in a dipolarophile for a 1,3-dipolar cycloaddition reaction with 2,2-dimethyl-3,4-pyrroline-*N*-oxide (**87**) (Scheme 1.41). They found diastereoselectivity was possible through judicious choice of the substitution at phosphorus of vinylphosphine oxides and sulfides. The most successful additions were between the dipole **87** and *P*-diallyl-*P*-*tert*-butylphosphine oxide **180** or the corresponding phosphine sulfide **181**. These adducts, isolated in good yields (84 and 80%, respectively) show almost total diasteroselectivities for the isoxazolidines **182** and **183** over the corresponding enantiomeric adducts, with stereochemical assignments based on X-ray crystallographic analysis of sulfide **183**. In later work, an enantiopure nitrone generated from L-tartaric acid was reacted with a racemic dihydro-1-phenylphosphole, to give predominantly the (*S*)-P product over its (*R*)-P enantiomer (3:1, 91% total yield) along with the unreacted (*R*)-phosphole (27%) almost completely resolved (> 96% ee) (205).

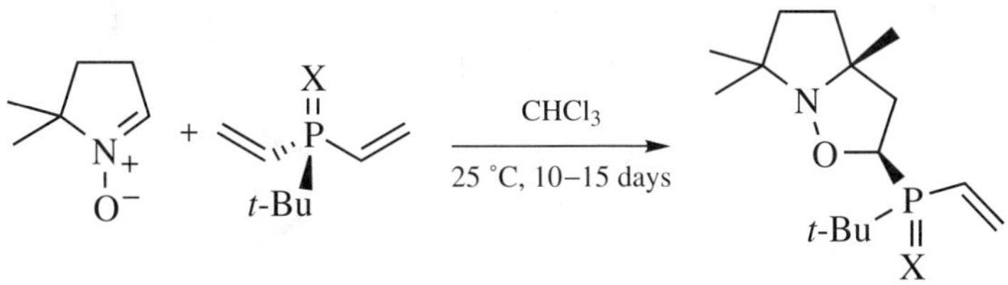

(87) (180) X = O (181) X = S (182) X = O (183) X = S

Scheme 1.41

1.9. CATALYTIC CYCLOADDITIONS

The widespread use of chiral nitrones or dipolarophiles in enantioselective cycloaddition reactions relies on effective asymmetric induction in the synthesis of the homochiral cycloaddition component. Thus, chiral catalysts that could control a cycloaddition reaction at substoichiometric levels would be powerful additions to the synthetic literature. The search for such catalysts, the fastest growing field of cycloaddition chemistry, has made significant progress since the pioneering work of Kanemasa in 1992 and the first enantioselective nitrone cycloaddition reaction in 1994. This field is now regularly reviewed (33,35,206). Much work has concentrated on the use of salts of magnesium, zinc, copper and titanium (84,85,207–216) including the use of complexes of bis(oxazoline) (**184**) (217–220) and tetraaryldioxolanedimethanols (TADDOLs) (**185**) (Fig. 1.3) [221,222]. Elsewhere, nickel catalysis in the presence of an unusual bis(oxazoline) was reported (223) as well as the cycloaddition of a platinum-coordinated nitrile dipolarophile (224). A number of workers have used metal complexes of 2,2′-dihydroxy-1,1′-binaphthyl (BINOL) (**186**) and its derivatives, including those with aluminium (225–227) and lanthanides (228–230). Other use of lanthanides includes the reactions catalyzed by $Eu(fod)_3$ [europium tris(6,6,7,7,8,8,8-heptafluoro-2-2-dimethyl-3,5-octanedionate)] (231) or lanthanide triflates, both in solution (232) and on polymer support (233). Trimethylsilyl triflates have also been employed (234,235) while alternatives to the BINOL ligands include binaphthylphosphine complexes with palladium (236,237). In earlier work, the cycloadditions of chiral chromium tricarbonyl (238–240) or iron tricarbonyl complexed (241) nitrones have been investigated.

1.10. PYRROLIDINES, PIPERIDINES, AND OTHER AMINES

The plethora of piperidine (242) and pyrrolidine alkaloids in Nature have made attractive targets for total synthesis, particularly since many contain a hydroxyl moiety adjacent to the amino group. This functionality is the case with the pseudodistomins–piperidines with antitumor activity isolated from the Okinawan tunicate *Pseudodistoma kanako*. The limited availability of the natural material has prompted

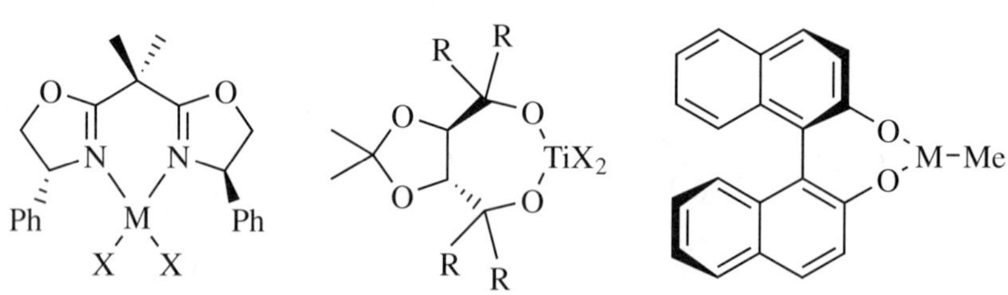

Ts = paratoluenesulfonyl

(**184**) X = OTf, SbF_6, etc. (**185**) X = Cl, OTf, OTs, etc. (**186**) X = Cl, OTf, OTs, etc.

Figure 1.3

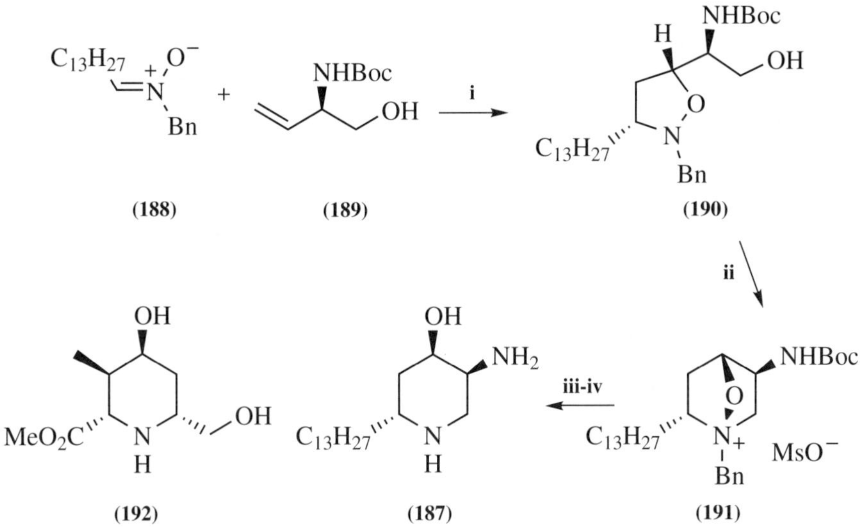

Reagents: [i] PhMe, Δ, 94 h, 14%; [ii] MsCl, py, 0 °C;
[iii] 10% $Pd(OH)_2$-C, H_2, MeOH, rt; [iv] TFA, DCM, rt.
py = Pyridine

Scheme 1.42

a number of total syntheses, including the preparation of tetrahydropseudodistomin (**187**) by Naito et al. (Scheme 1.42) (243). The cycloaddition of the known tetradecanal-derived nitrone **188** with (+)-Boc-vinylglycinol (**189**) afforded a mixture of four diastereoisomers in 79% combined yield. Isomer **190** was converted to the bicyclic structure **191** via ring closure through nitrogen onto the side chain after mesylation of the pendant alcohol. This bicycle was subjected to simultaneous N-benzyl and N—O hydrogenolysis, followed by amine deprotection with TFA, to afford the natural product **187**. These workers have also used the cycloaddition of allyl alcohols to nitrones to prepare the related piperidinol alkaloids (+)-azimic acid and (+)-julifloridine (244). Elsewhere, the formation of a bicyclic system similar to **191** was used to prepare polysubstituted 4-hydroxypiperidines (245,246) and α-amino ester **192**, the A-ring of cylindrospermopsin (247).

A number of other simple cyclic amine targets with oxygen substitution have been prepared using a nitrone cycloaddition strategy to provide both the N- and O-functionalities. These include the pyrrolidines darlinine **193** and analogues (248), (+)-preussin (**194**) (249) and the piperidines (−)-allosedamine (**195**) (250), and the related structure **196** (251) as well as analogous β-aminoketones (Fig. 1.4) (252).

The oxygen atom has also been used to generate other functionalities, such as the aldehyde moiety in Kibayashi's syntheses of (−)-coniine (**197**) and its enantiomer (Scheme 1.43) (253). Here, reaction of tetrahydropyridine N-oxide **93** with a silylated chiral allyl ether dipolarophile **198** delivered the adduct **199** with the desired bridgehead stereochemistry via the "inside alkoxy effect". Desilylation and hydrogenolytic N—O bond rupture with palladium(II) chloride provided the diol **200**

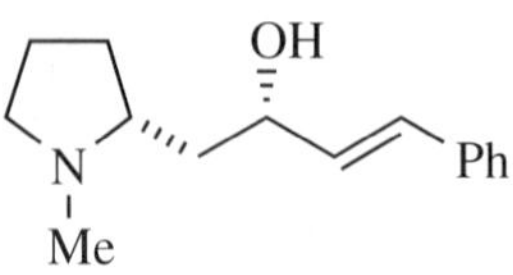

(193)

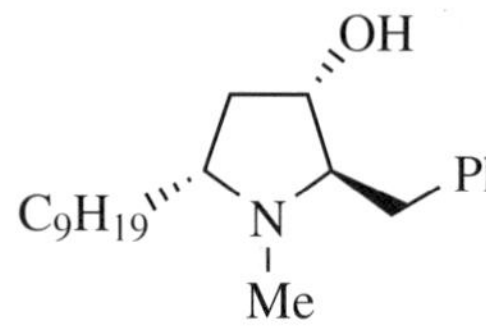

(194)

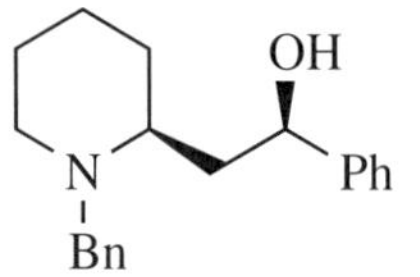

(195)

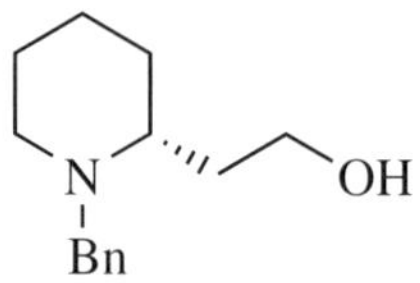

(196)

Figure 1.4

from which the aldehyde function of **201** was formed by periodic acid oxidative cleavage. The side chain was lengthened by one carbon atom using a Wittig reaction followed by hydrogenation to afford the desired product (−)-coniine (**197**). The same aldehyde intermediate was later used by these workers in the synthesis of

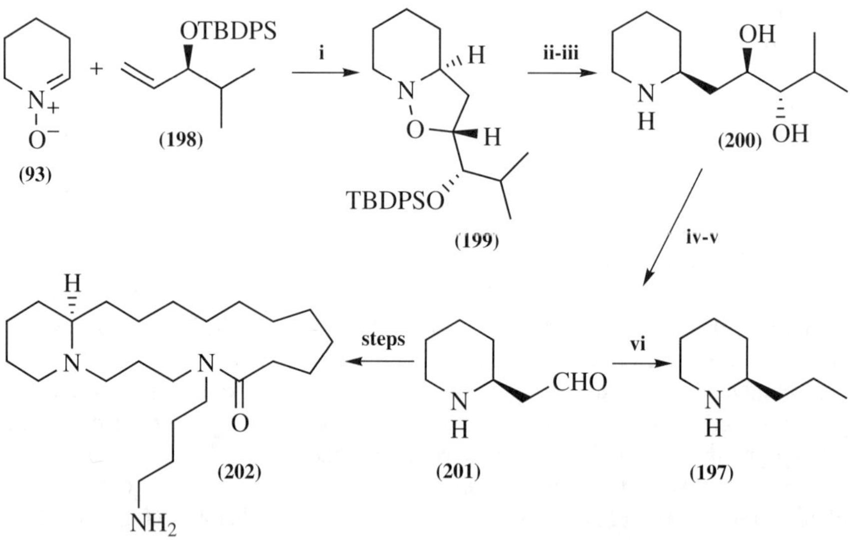

Reagents: [i] PhMe, δ, 14 h, 69%; [ii] 1 *M* aq Bu_4NF, THF, 45 °C, 1.5 h, 86%; [iii] H_2, $PdCl_2$, MeOH, 6.5 atm, 37 h, 91%; [iv] CbzCl, aq Na_2CO_3, DCM, rt, 16 h, 85%; [v] HIO_4, THF, H_2O, 0 °C, 2 h, 84%; [vi] Ph_3PMeBr, BuLi, THF, then H_2, Pd-C, MeOH.

Scheme 1.43

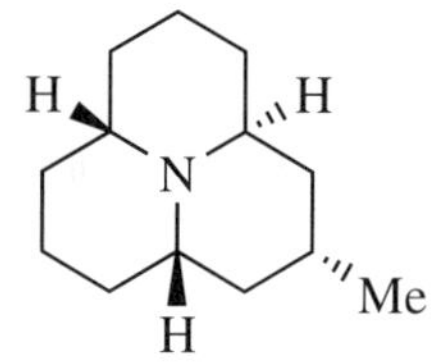

(203)

Figure 1.5

the macrocyclic polyamine lactam alkaloid (−)-oncinotine (**202**) (254). Elsewhere, a substituted form of nitrone (**93**) was prepared by intramolecular 1,3-azaprotio-cyclotransfer (1,3-APT; see Section 1.11.1) and reacted with a diene dipolarophile in the synthesis of the tricyclic coccinelline alkaloid *epi*-hippodamine **203** in racemic form (Fig. 1.5) (255).

The utility of the intramolecular cycloaddition reaction lies in its ability to form a bicyclic system during the addition and so lead to products of greater complexity (19,256). Chiacchio et al. (82) used the intramolecular cycloaddition of a nitrone containing an α,β-unsaturated amide in the same molecule to generate pyrrolidinones or piperidinones after N—O bond hydrogenolysis (Scheme 1.11, Section 1.4). While this strategy has been shown to allow access to substituted piperidines (257), the preparation of the cyclic amine rather than amide in the intramolecular cycloaddition reaction is far more commonplace. LeBel and Whang (258), who published the first example of an intramolecular nitrone cycloaddition in 1959, more recently reported a racemic total synthesis of the *Dendrobatidae* alkaloid pumiliotoxin C **204** (Scheme 1.44) (259). The desired alkenylnitrone **205**, prepared by the condensation of the corresponding *N*-alkylhydroxylamine and phenylsulfonyl aldehyde, underwent a thermally induced intramolecular cycloaddition to afford exclusively the bridged isoxazolidine adduct **206** (74%). After reductive isoxazolidine ring opening of the bridged adduct to give piperidine **207**, the authors took advantage of the newly generated hydroxyl moiety as a potential leaving group. Thus, a carbanion generated α to the sulfone displaced the tosylate to close the carbocyclic ring of the target molecule **204**, which was revealed after desulfonation with sodium amalgam and N-deprotection.

In a related strategy, the dihydropiperidine skeleton of the macrocyclic alkaloid cannabisativine (**208**) (Fig. 1.6), isolated from the leaves and roots of the common cannabis plant, was prepared with regio- and stereoselectivity using an intramolecular allylsilane–nitrone cycloaddition reaction as a key step (260).

It has long been recognized that nitrone cycloadditions may allow access to spirocyclic ring systems. Such systems are inherently difficult to synthesize by conventional methods, yet are a structural component of a number of biologically active natural materials. Two common strategies have emerged for spirocycle generation from exocyclic or endocyclic nitrones (Scheme 1.45). In the exocyclic version, the carbon atom (arrowed) of the nitrone C=N double bond of dipole **209** carries a cyclic substitutent and thus an intermolecular cycloaddition reaction will

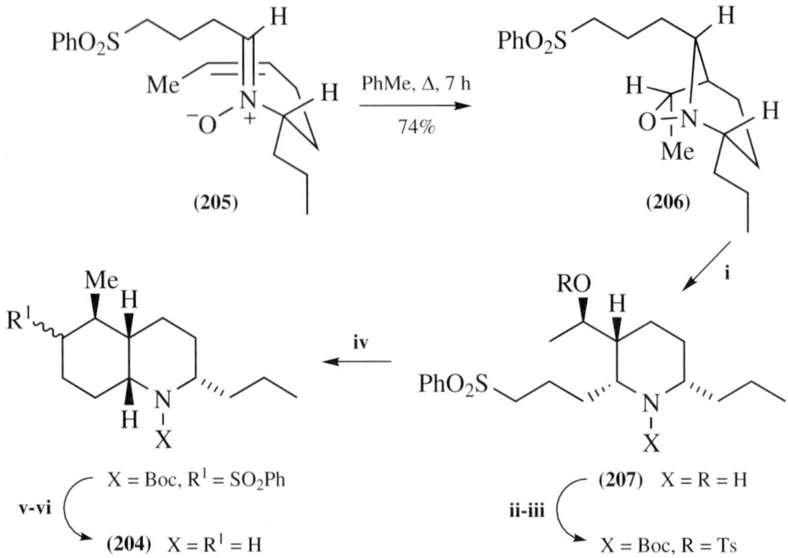

Reagents: [i] Zn, 80% aq AcOH, 87%; [ii] Boc_2O, NaOH, THF, 82%; [iii] TsCl, py, 0 °C, 92%; [iv] *n*-BuLi, HMPA, THF, −78 °C to rt, 2 h, 83%; [v] Na_2HPO_4, Na/Hg, MeOH, 0 °C; [vi] TFA, DCM, 78% (2 steps).

Scheme 1.44

add a new C–C bond here to create the desired spiro stereocenter in the adduct **210** (potentially as a mixture of regioisomers). Alternatively, the spirocycle may be formed from the endocyclic nitrone **211**, which carries the alkene functionality in the substituent on the corresponding carbon atom of the C=N bond in the dipole (arrowed). Thus, the conformational preference of the side-chain alkene with respect to the nitrone function is crucial to the eventual product, which may be a fused spiroisoxazolidine (**212**) or the bridged regioisomer (**213**).

An example of the exocyclic strategy is the recent 22-step synthesis by Snider and Lin (261) of (−)-FR901483 **214**, a novel immunosuppressant isolated from the

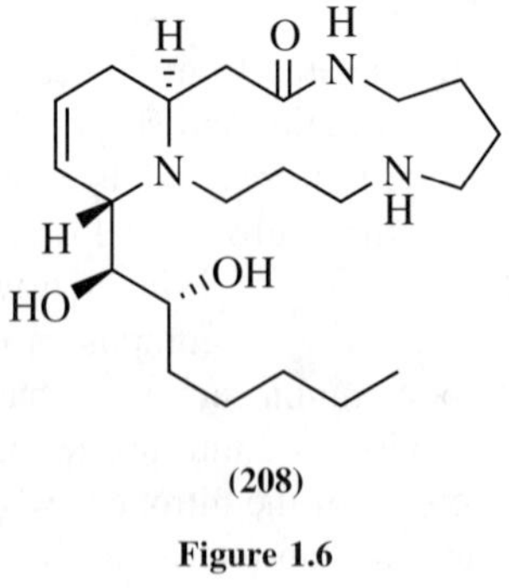

(208)

Figure 1.6

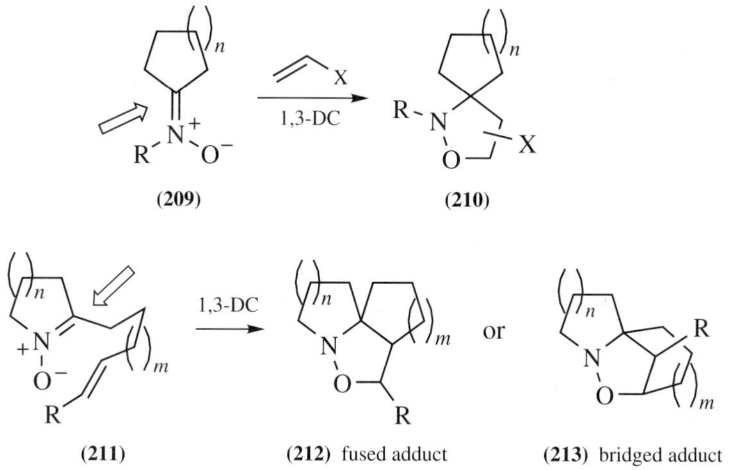

Scheme 1.45

fermentation broth of *Cladobotrym* bacteria (Scheme 1.46). Key steps include spirocyclization through the intermolecular cycloaddition of **215**, and synthesis of the final ring by intramolecular cyclization of ketoaldehyde enolate **216** by an aldol reaction. A chiral hydroxylamine was prepared from *O*-methyltyrosine methyl ester and condensed with a cyclohexanone derivative to afford the desired dipole **215**. The spiroisoxazolidine cycloadduct **217** was formed after a 48 h reaction at 25 °C as an inseparable 6:1 mixture with a second diastereomeric product. Hydrogenolytic N—O bond rupture provided a γ-aminoester that spontaneously condensed to give a mixture of the diastereomeric spirocyclic lactams, of which **218** was the major product. In the aldol condensation step, the enolate **216** afforded the desired tricycle **219** (36%) and the hydroxy epimer (16%) along with a structure (5%) derived from generation of the enolate at the alternative position (arrowed). A further example of the exocyclic nitrone in synthesis is the work of Funk and Daggett (Scheme 1.10, Section 1.4) (78).

An example of spirocyclization with an endocyclic nitrone is the synthesis of the azaspirocyclic core of the piperidine alkaloids halichlorine and pinnaic acid **220** by Lee and Zhao (262) using the tandem 1,3-APT/cycloaddition strategy developed by Grigg et al. (263) (Scheme 1.47). Thus, treatment of the ketone **221** with hydroxylamine afforded the unisolated oxime **222**, which cyclized spontaneously via a 1,3-azaprotio cyclotransfer selectively onto the alkene function of the α,β-unsaturated ester. The racemic cyclic alkenyl nitrone **223** underwent an intramolecular 1,3-dipolar cycloaddition with the pendant alkene to afford the spirocyclic adduct **224**. Reaction occurred with facial selectivity, through enforced alkene addition to the face of the dipole opposite the bulky methyl ester. Similarly, the bridged regioisomer was not observed. Having the undesired stereochemistry at C(5), tricyclic adduct **224** was first ring opened through N—O bond cleavage then converted to the thermodynamically favored C(5) epimer **225** by heating for 24 h in 1,2-dichlorobenzene, via a proposed retro-Michael intermediate.

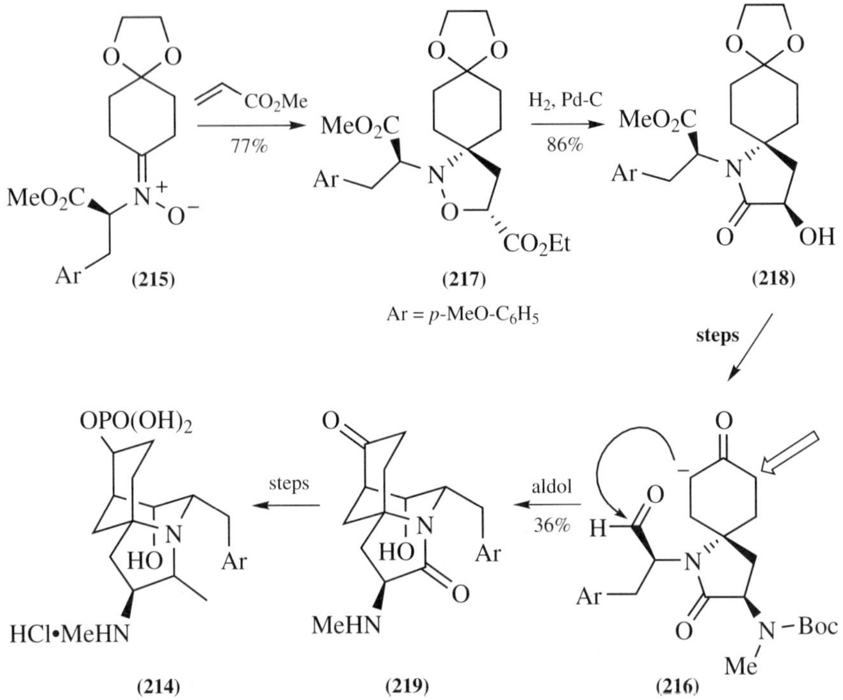

Scheme 1.46

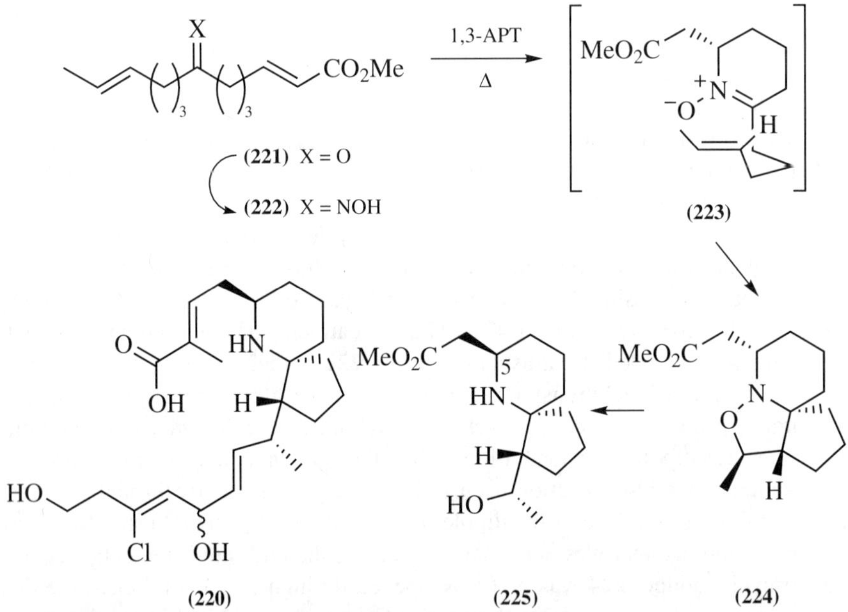

Scheme 1.47

Another synthesis of the pinnaic acid spirocyclic skeleton was recently reported using a transannular strategy by White et al. (264) (Scheme 1.48). Hydroxylamine **226** was generated in situ, by thermolysis of the oxaziridine, and condensed in an intramolecular reaction (arrowed) to afford (*E*)-alkenylnitrone (**227**) [(*E*/*Z*) = 4:1] after careful chromatography. Conformational analysis of dipole **227** suggests that this 12-membered lactone is too small for the oxygen to pass through the ring and so reaction is directed to a single face of the alkene moiety as drawn. Thermal transannular cycloaddition of (*E*)-**227** afforded the polycyclic adduct **228** as a single enantiomer, in which the single stereocenter of the dipole stereoselectively generates three new centers. Methanolysis in the presence of base, then N—O reductive cleavage with samarium iodide afforded the spirocyclic piperidine target **229**.

Further examples of the endocyclic nitrone route to spirocyclic adducts are the total syntheses of (−)-histrionicotoxin (**230**) by Holmes and of cylindricines by Weinreb. Histrionicotoxin is one of many spiropiperidine alkaloids isolated from the poison-arrow frog *Dendrobates histrionicus* and has been the subject of many attempted total syntheses by a nitrone cycloaddition strategy that failed to provide the desired regioisomer, possibly through unfavorable steric interactions (265–268). Unlike these reports, Holmes and co-workers (101) found that the intermolecular reaction of nitrone (**231**), prepared by the 1,3-APT of the corresponding alkynylhydroxylamine carrying Oppolzer's chiral sultam auxiliary, afforded the styrene

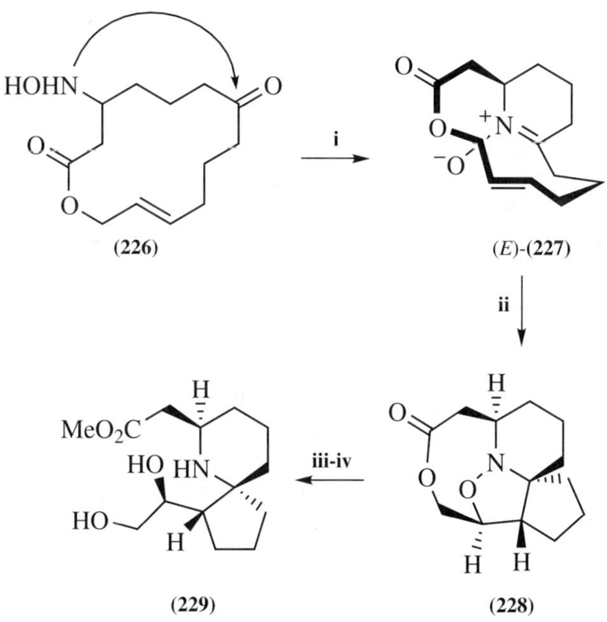

Reagents: [i] *p*-TsOH•H_2O, MeOH, H_2O, Δ, 70%; [ii] PhMe, Δ, 64%; [iii] K_2CO_3, MeOH, 88%; [iv] SmI_2, rt, 64%.

Scheme 1.48

adduct **232** as a single regio- and stereoisomer (85% from the acyclic material) (Scheme 1.49). After elimination of the auxiliary and introduction of the desired (*Z*)-α,β-unsaturated nitrile in the side chain, the nitrone function was regenerated from the isoxazolidine **232** by thermolytic cycloreversion (sealed tube 190 °C, 80%). A second dipolar cycloaddition, this time with the pendant alkenyl nitrile of **233**, afforded the desired spirocyclic ring system in adduct **234**. Conversion to the bis(nitrile) allowed parallel installation of the two (*Z*)-enyne functionalities of the target material by reduction to the aldehydes, before preparation of the bis(iodoalkenes) with Stork's iodophosphorane. Sonagashira coupling with trimethylsilyl acetylene and desilylation afforded **230**. The bis(nitrile) (**235**), an advanced intermediate in Holmes' synthesis, was recently

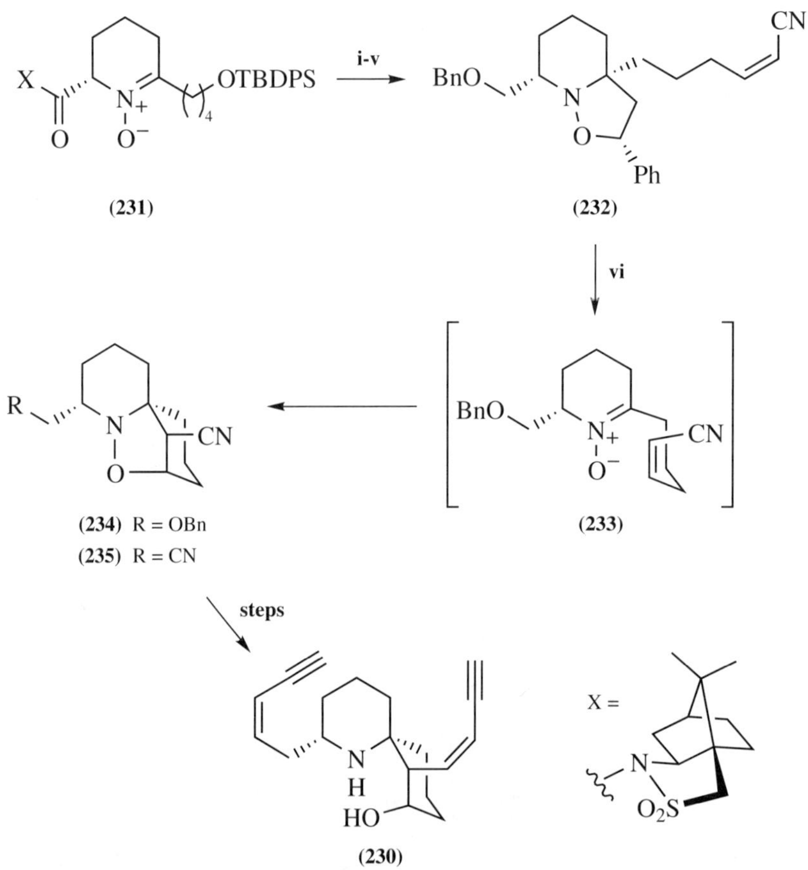

Reagents: [i] styrene, 75 °C; [ii] $LiAlH_4$, THF, 0 °C; [iii] NaH, BnBr, THF, 90% (2 steps); [iv] HF, CH_3CN, 91%; [v] TPAP, NMO, 4 Å mol. sieves, 98%; $TMSCH_2CN$, *n*-BuLi; HF −78 °C, B(*i*-OPr)$_3$, HMPA, 87%; [vi] PhMe, sealed tube, 190 °C, 3.5 h, 80%.

Scheme 1.49

prepared by Stockman from a symmetrical dialdehyde with a related oxime–alkene 1,3-APT and nitrone 1,3-dipolar cycloaddition as key steps (269).

A similar nitrone spirocyclization is at the heart of a recent synthesis by Weinreb of the skeleton of the cylindricine marine alkaloids and related lepadiformine (Scheme 1.50) (270,271). Attempts at electrophile promoted cyclic nitrone formation using Grigg's methodology (265,266) were unsuccessful, with cyclization through oxygen to afford an oxazine. In contrast, conventional nitrone generation afforded the desired dipole **240** by intramolecular condensation of the hydroxylamine function with an adjacent ketone (revealed after acid hydrolysis of the γ-acetal of **241**). Cycloaddition under vigorous conditions (195 °C, 63%) gave the desired fused isoxazolidine adduct **242** as a single stereoisomer. During similar cycloadditions with a related nitrone (**243**) lacking the cylindricine alkyl side chain, the undesired seven-membered bridged adduct **244** was the major product, formed via a chair-like transition state. This result suggests that the fused product **242** was formed via an energetically favored boat conformation, as drawn for **240**, in order to accommodate the bulky alkyl side chain. After reductive N—O bond breaking, the free hydroxyl group was oxidized to the ketone, which underwent a spontaneous cyclization through nitrogen onto the conjugated alkene β-carbon (arrowed) to form the final ring of the natural product. Removal of the phenyl protecting group

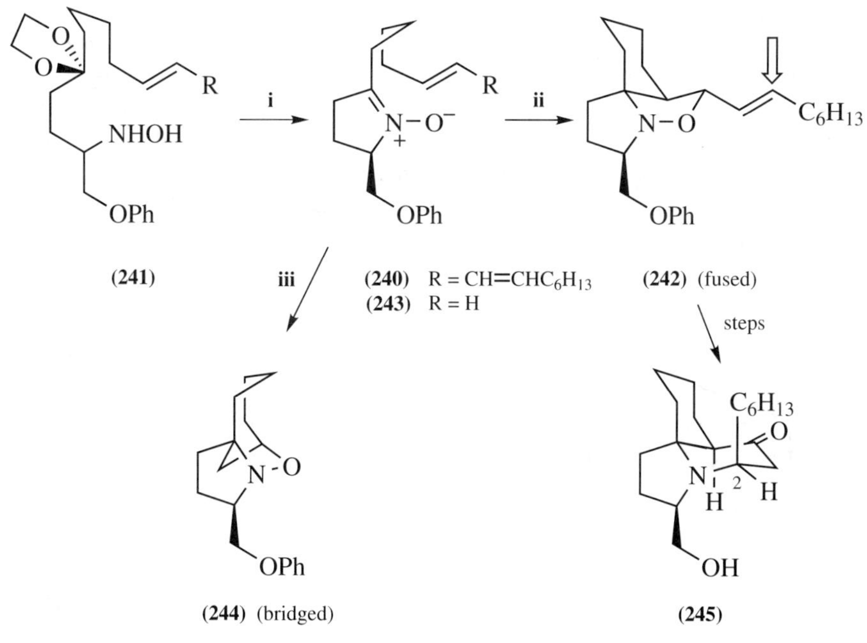

Reagents: [i] 3 *M* aq HCl, THF, rt, 3.5 h, 92%; [ii] DMSO, 195 °C, sealed tube, 16 h, 63%; [iii] PhMe, Δ, 69% (2:1 bridged adduct **244**:fused adduct).

Scheme 1.50

gave 2-*epi*-cylindricine C (**245**), which resisted all attempts at C(2) epimerization to cyclindricine C. However, successful deoxygenation of this material afforded the putative structure of the related alkaloid lepadiformine, although an unfavorable NMR spectroscopic comparison with the natural product implies that the proposed structure must be refined. Elsewhere, Oppolzer and co-workers (271a) used an alkenyl nitrone carrying their bornane-10,2-sultam chiral auxiliary to introduce the spirocyclic stereogenic centre of the cylindricine skeleton, although the desired adduct (**247**) was isolated from a mixture slightly favoring the unwanted regioisomer **248** (Scheme 1.51).

A bis(spirocyclic) ring system is at the heart of the ptilomycalin and related crambescidin family of natural pentacyclic guanidine antiviral cytotoxins [e.g., the isocrambescidin (**249**)]. Nagasawa et al. (272) reported a synthesis of a symmetrical model system of this complex target having the desired stereochemistry, via a pyrrolidine intermediate prepared using two separate dipolar cycloaddition reactions followed by double spiroacetal formation (Scheme 1.52). Reaction of nitrone **92** with a terminal silyloxyhexene (**250**) afforded the bicyclic adduct **251** from which the nitrone function **252** was regenerated with concomitant N—O cleavage by treatment with meta-chloroperbenzoic acid (*m*-CPBA). A second addition of the silyloxyhexene dipolarophile (this time to **252**) and N—O bond hydrogenolysis of the cycloadduct afforded the C_2-symmetric diol **253**, which formed the protected guanidine system by reaction with bis(Boc-thiourea) and $HgCl_2$. Oxidation of the pendant hydroxyl moieties afforded the diketone that underwent double N,O-spiroacetal formation to yield the model system **254**. In related work, a nitrone cycloaddition strategy was the key step in the synthesis of 3-hydroxyspermidine—the polyamine from which the side-chain amide of the crambescidins is formed (273).

The guanidine group is a common structural feature in Nature, primarily due to its ability to stabilize the three-dimensional structure of proteins in enzymes through binding with anionic substrates. Roush and Walts (274) prepared the

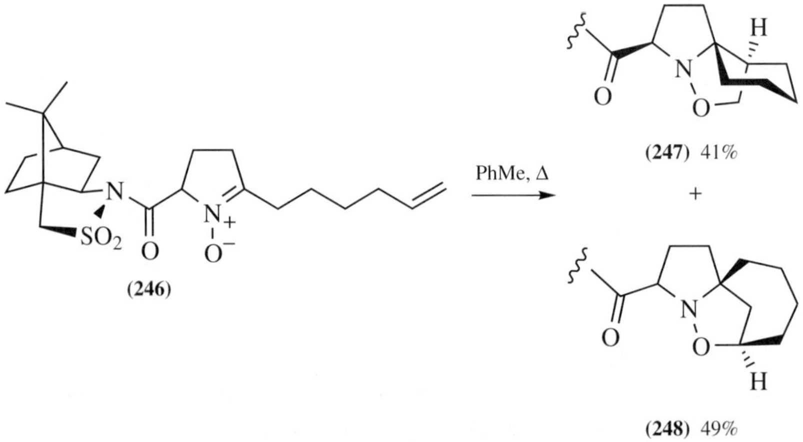

Scheme 1.51

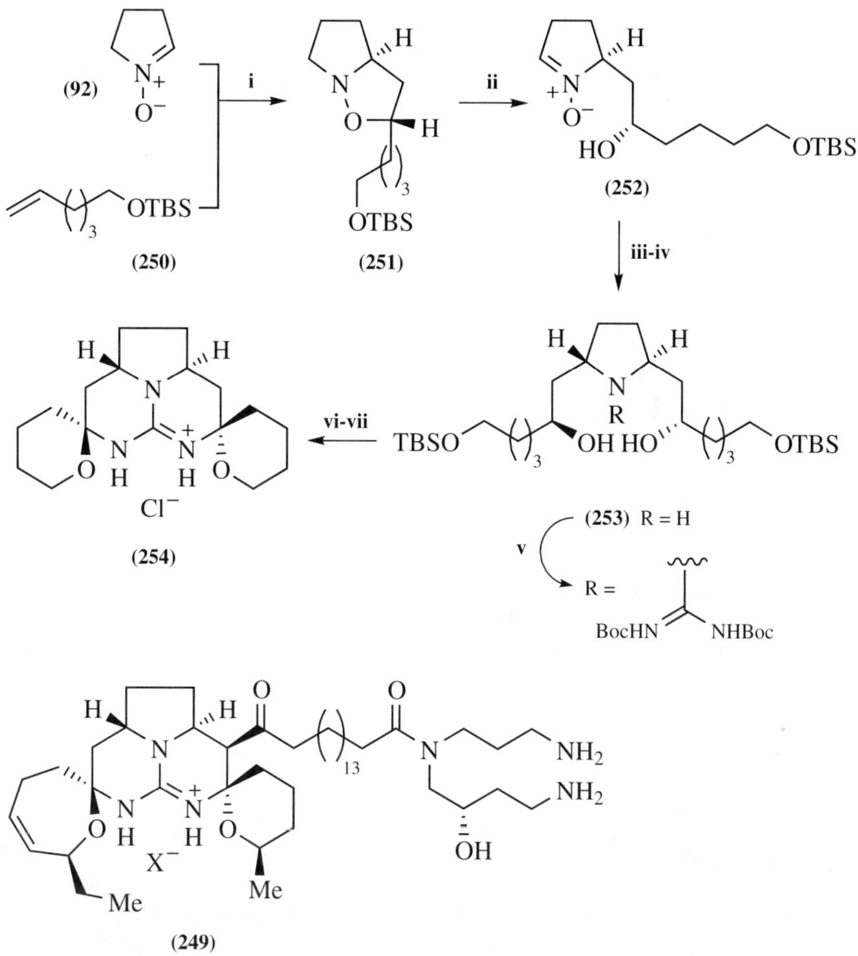

Reagents: [i] PhMe, 95 °C, 72%; [ii] *m*-CPBA, DCM, 0 °C; [iii] PhMe, 100 °C, 60% 2 steps; [iv] H_2, Pd-C, EtOH, 93%; [v] $(BocNH)_2C{=}S$, $HgCl_2$; [vi] TPAP, NMO; [vii] HCl, MeOH, 75% from **253**.

Scheme 1.52

antitumor antibiotic (−)-ptilocaulin (**255**) by the intramolecular cycloaddition of an enantiomerically pure cyclohexenyl nitrone (Scheme 1.53). The aldehyde **256** was condensed with *N*-benzylhydroxylamine to form the dipole **257**, which underwent spontaneous intramolecular reaction with the cyclohexene to form the required tricyclic adduct **258**. After cleavage of the N—O bond with Zn/AcOH, the hydroxyl moiety was oxidized to the ketone **259** with Jones' reagent although with some epimerization at C(8b). After separation of the unwanted isomers, *N*-debenzylation afforded the amino ketone required to form the guanidine functionality. While this afforded a mixture of isomeric products, equilibration to the thermodynamically favored natural product **255** was achieved by extended reflux in benzene.

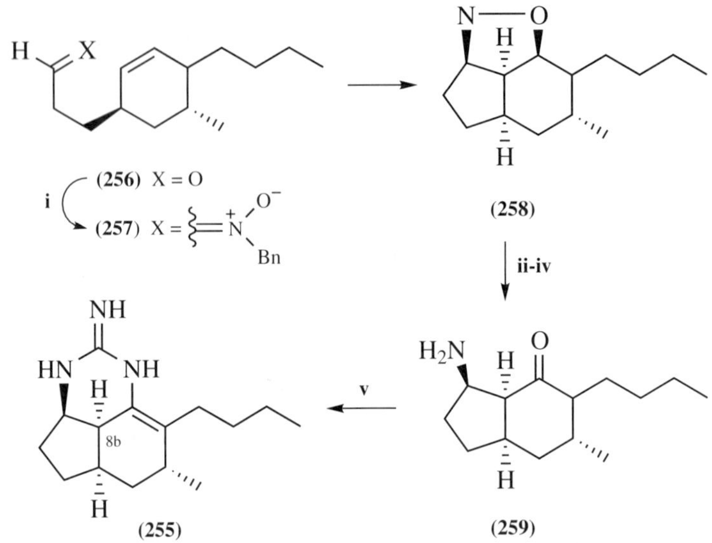

Reagents: [i] BnNHOH, C_6H_6, 80 °C, 8 h, 80%; [ii] Zn, AcOH, 55 °C, 3.5 h, 95%; [iii] Jones' reagent, AcOH, aq HCl, 0 °C, 95%; [iv] Pd black, 10% HCO_2H/MeOH, 23 °C, 1.5 h, 95%; [v] 1-guanyl-3,5-dimethylpyrazole nitrate, 150 °C, 6 h, 58%.

Scheme 1.53

The double cycloaddition strategy used by Nagasawa has previously been applied to the synthesis of other cyclic amines, including racemic andrachamine **260** (Scheme 1.54) (275,276). Carruthers et al. (277) found that the oxidative ring opening of bicyclic isoxazolidinyl cycloadduct **261** with *m*-CPBA afforded a 2:5 mixture of the desired aldonitrone **262** with the isomeric ketonitrone (2:5). However, only the more reactive aldonitrone took part in a subsequent cycloaddition reaction with but-1-ene at 110 °C (sealed tube), to afford the isoxazolidine **263**, which gave the diol natural product **260** after reductive cleavage of the N—O bond with Zn/AcOH. In related work, these researchers produced in racemic form the 17-membered cyclophane alkaloid (**264**) (the C(3), C(11) epimer of lythranidine) by consecutive cycloaddition of alkenes carrying a terminal phenyl group, completing the macrocyclization via nickel-mediated aryl–aryl coupling of the aryl iodides (277). A similar double-cycloaddition–double oxidative cleavage strategy was employed in the synthesis of 2,5-disubstituted pyrrolidines (278). A related strategy was employed in the synthesis of the antimalarial alkaloids (+)-febrifugine and (+)-isofebrifugine (279).

An impressive enantiopure synthesis of *Amaryllidaceae* alkaloids has been achieved through the formation of sugar-derived homochiral alkenyl nitrone **265** (Fig. 1.7).[280] While this reagent required lengthy preparation, it underwent an intramolecular dipolar cycloaddition to establish the required stereochemistry of the polycyclic pyrrolidine skeleton of (−)-haemanthidine (**266**), which was converted to (+)-pretazettine and (+)-tazettine by established procedures (281).

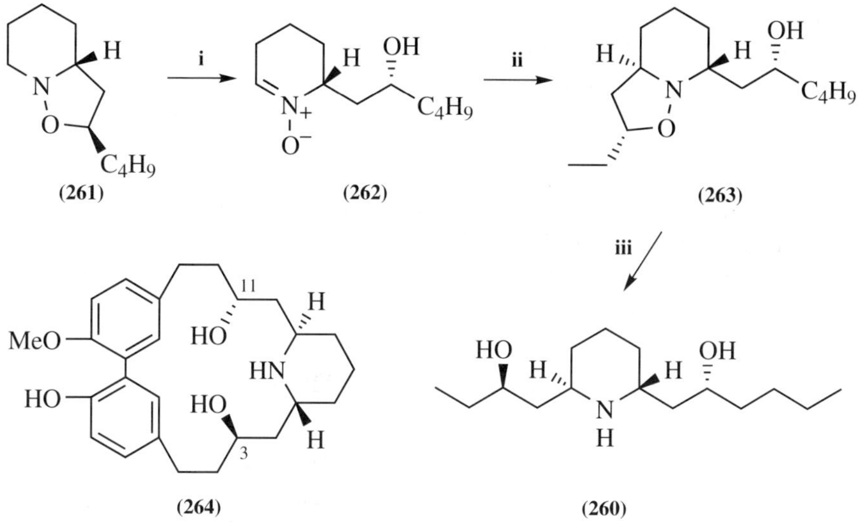

Reagents: [i] *m*-CPBA, 0 °C 10 min; [ii] $CHCl_3$, sealed tube, 110 °C, 18 h; [iii] Zn, ethylenediaminetetraacetic acid (EDTA), EtOH, AcOH, Δ, 45 min, 91%.

Scheme 1.54

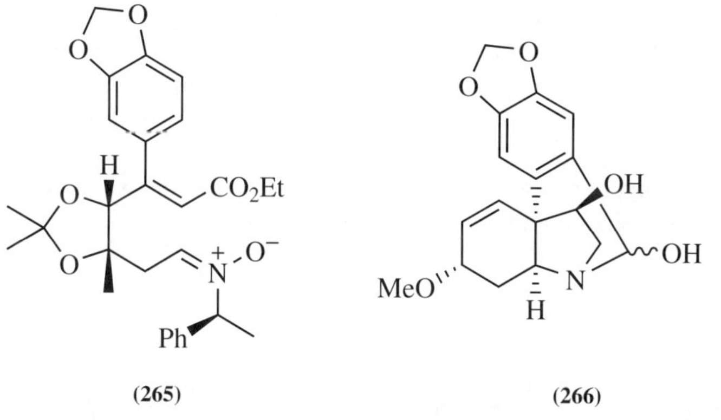

Figure 1.7

1.11. ISOXAZOLIDINES

Many of the recent publications on nitrone cycloadditions have focused on the cycloaddition process itself and on the identification of the character of the reagents involved rather than with a specific synthetic target and thus, this section is potentially vast. To bring order to this wealth of information, we have identified a number of important groups of papers and will present them accordingly. We shall begin with two cascade processes. While these are ostensibly nitrone-formation reactions and therefore outside the scope of this article, they are included for their use of novel and facile in situ nitrone generation/cycloaddition protocols of general synthetic interest. The first deals mainly with the transformation that affords nitrones or amine *N*-oxides from carbon–carbon multiple bonds and oximes or hydroxylamines (Section 1.11.1). Next, we shall consider the use of the oxime–nitrone (Section 1.11.2) equilibrium for intramolecular cycloadditions in the synthetic literature. Finally, we shall outline the highlights of synthetic reports using intra- (Section 1.11.3) and intermolecular (Section 1.11.4) nitrone cycloadditions. Isoxazolidine formation has been reviewed (33,282,283).

1.11.1. Nitrones by the 1,3-APT Process

In 1967, Ochiai et al. (284) described the reaction of oximes with 2 equiv of an alkene to afford isoxazolidines (Scheme 1.55). The first equivalent acts as a Michael acceptor, alkylating the nitrogen of oxime **267** to afford the intermediate nitrone **268**, for subsequent reaction with a second alkene that then behaves as a dipolarophile (to give adduct **269**) (285–287). This process forms part of a family of reactions in which N—O—H adds across a C—C multiple bond, augmented by the discovery by House et al. in 1976 (288,289) of the addition of hydroxylamines to alkenes to generate N-oxides (the retro-Cope elimination or 'EPOC'—from 'Cope' in reverse) (290–294). The family now includes the reaction of oximes with alkenes (263,285,286,295–297), allenes (298–301) and alkynes (302,303) and of hydroxylamines with alkenes (288–292,304) or alkynes (305–309). The initially proposed radical mechanism has been discounted (309) in favor of a concerted ene-like process: Grigg has proposed a common mechanism for the first step, describing it as a 1,3-azaprotiocyclotransfer (1,3-APT) for the reaction of oximes with alkenes, a terminology we shall adopt.

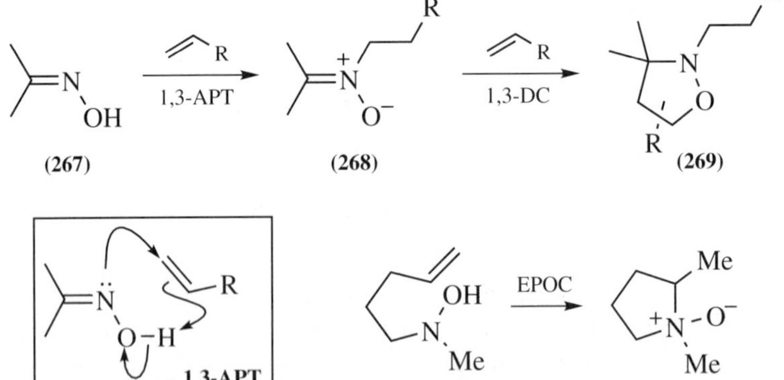

Scheme 1.55

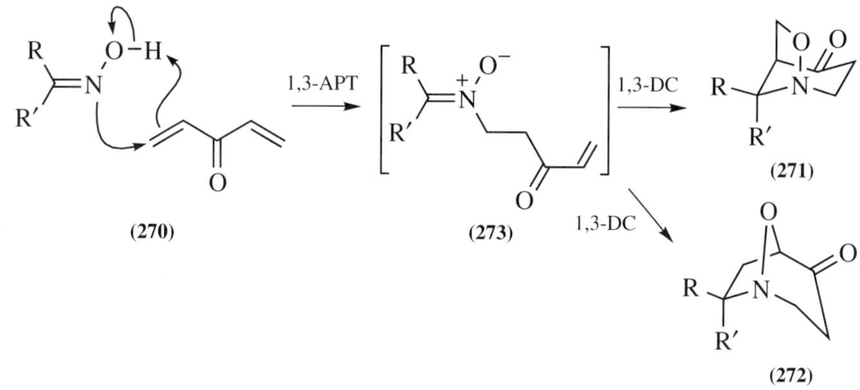

Scheme 1.56

Grigg and co-workers (310) recently examined the 1,3-APT reaction of various aldoximes (**270**) (R or R′ = H) with divinyl ketone (Scheme 1.56). While ketoximes **270** (R′ = R) form a mixture of adducts, **271** and **272** via nitrone **273**, the aldoximes selectively afford **272** (as a mixture of endo and exo diastereoisomers). Under the thermal reaction conditions, the oxime starting materials can undergo (*E*/*Z*) isomerization, while the nitrone intermediate was expected to be unaffected and the isolated cycloadducts showed no interconversion via cycloreversion. Thus, the increasing selectivity for *endo*-**272** [via (*E*)-**273**, R = H] over *exo*-**272** [via (*Z*)-**273**, R′ = H] with the increasing size of the aldoxime substituent was attributed primarily to the inhibition of oxime isomerization by steric clash between R or R′ and the oxime OH. In contrast, Lewis acid catalysis, in particular by hafnium (iv) chloride, of the cycloaddition of various aldoximes with this dipolarophile gave *exo*-**271** exclusively (216).

For the alkynylhydroxylamines **274** studied by Holmes (309,311), initial intramolecular 1,3-azaprotiocyclotransfer affords the N-oxides **275** via a 5- or a 6-*exo-dig* process [Baldwin's terminology (312)], which then tautomerize to the cyclic nitrones (**276**) (Scheme 1.57). The 7-*exo-dig* cyclization required for the formation of a seven-membered nitrone (**277**) by this approach from **278** was found to be disfavored with respect to the alternative reaction of hydroxylamine and alkene (5-*exo-trig* process) to afford a mixture of the alkynylpyrrolidines **279**

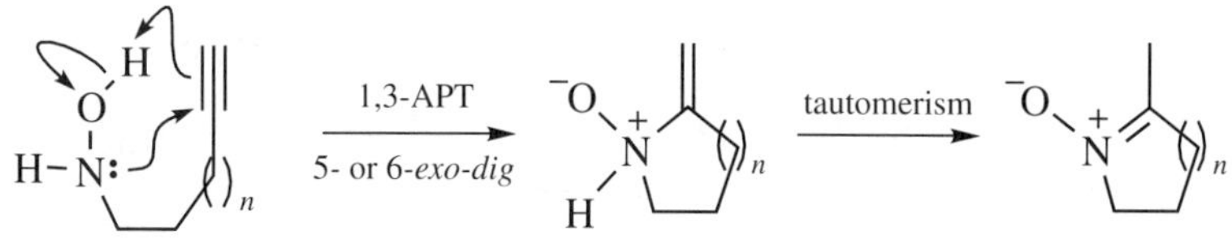

Scheme 1.57

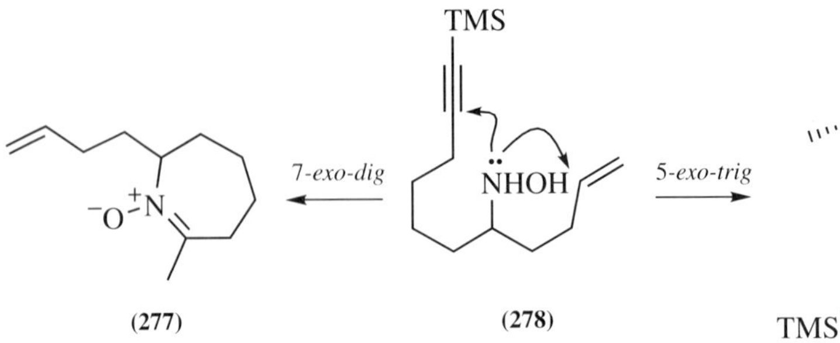

(277) (278) (279) cis (280) trans

Scheme 1.58

and **280** (Scheme 1.58). However, the seven-membered alkenyl nitrone **277** was formed, along with tricyclic adduct **281**, by prolonged heating of the *trans-N*-hydroxypyrrolidine **280** (Scheme 1.59). In the proposed mechanism, an initial 1,3-APT of hydroxylamine and alkyne functions in **280** affords N-oxide **282** and is followed by a Cope elimination (294) to the intermediate hydroxyazepine **283**. Tautomerization and desilylation yields the cyclic alkenyl nitrone **277** and the subsequent intramolecular cycloaddition reaction affords the tricycle **281**. This mechanism is supported by a parallel experiment with the *cis*-pyrrolidine **279**, which cannot undergo a Cope elimination through the anti-disposition of N-oxide and the methyl protons required for reaction. A seven-membered nitrone generated in this way was used to disprove the proposed structure of (±)-acacialactam (**284**)

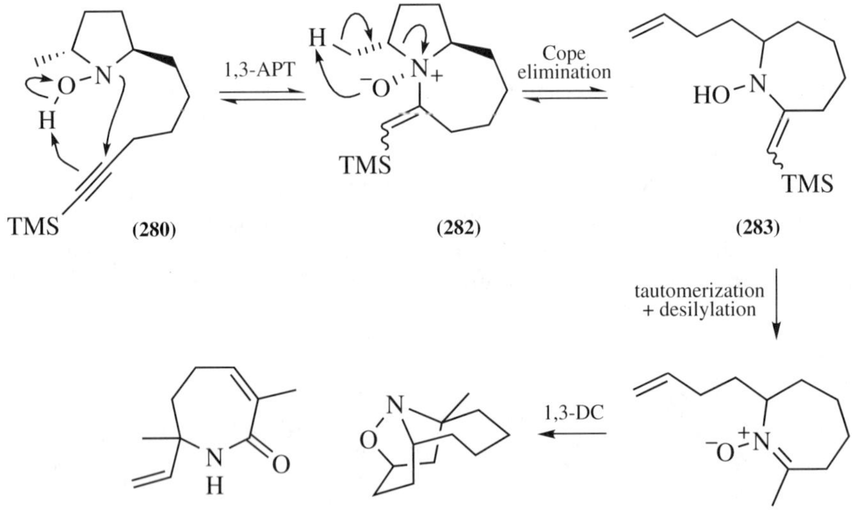

(280) (282) (283) (284) (281) (277)

Scheme 1.59

by a noncycloaddition strategy (308,309). A seven-membered nitrone was prepared in a related reaction by Gallagher and co-workers (301), by the Ag^{I} electrophile-mediated cyclization of oxime functionality onto an allene moiety (Scheme 1.64, Section 1.11.1).

Elsewhere, Heaney et al. (313–315) found that alkenyloximes (e.g., **285**), may react in a number of ways including formation of cyclic nitrones by the 1,3-APT reaction (Scheme 1.60). The benzodiazepinone nitrones (**286**) formed by the intramolecular 1,3-APT will undergo an intermolecular dipolar cycloaddition reaction with an external dipolarophile to afford five,seven,six-membered tricyclic adducts (**287**). Alternatively, the oximes may equilibrate to the corresponding N—H nitrones (**288**) and undergo intramolecular cycloaddition with the alkenyl function to afford five,six,six-membered tricyclic isoxazolidine adducts (**289**, R = H; see also Section 1.11.2). In the presence of an electron-deficient alkene such as methyl vinyl ketone, the nitrogen of oxime **285** may be alkylated via the acyclic version of the 1,3-APT reaction and thus afford the N-alkylated nitrone **290** and the corresponding adduct **291**. In more recent work, they prepared the related pyrimidodiazepine N-oxides by oxime–alkene cyclization for subsequent cycloaddition reactions (316). Related nitrones have been prepared by a number of workers by the more familiar route of condensation with alkylhydroxylamines (Scheme 1.67, Section 1.11.3).

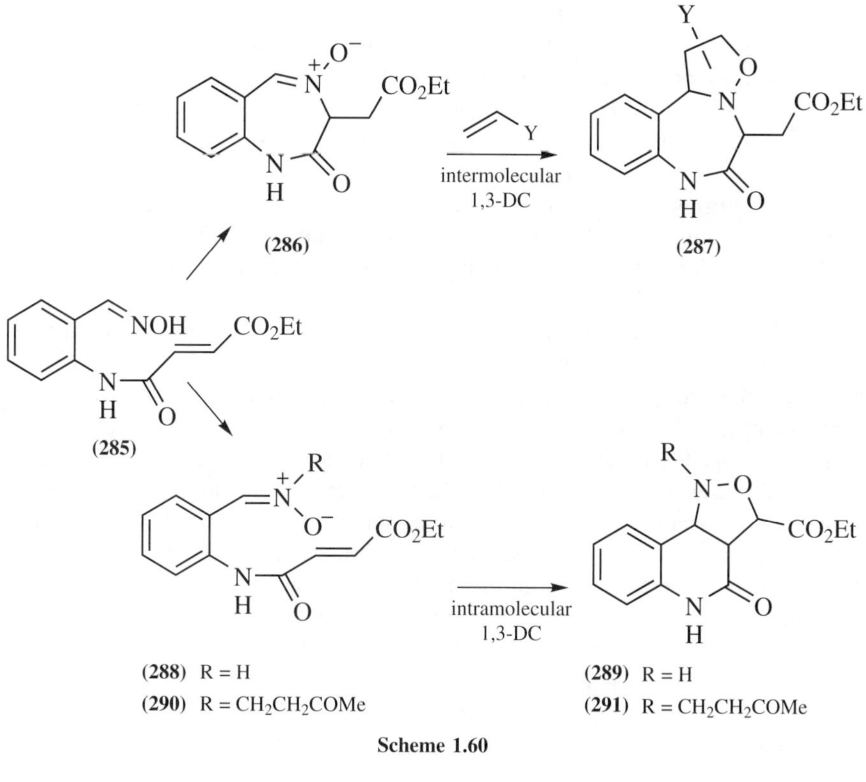

Scheme 1.60

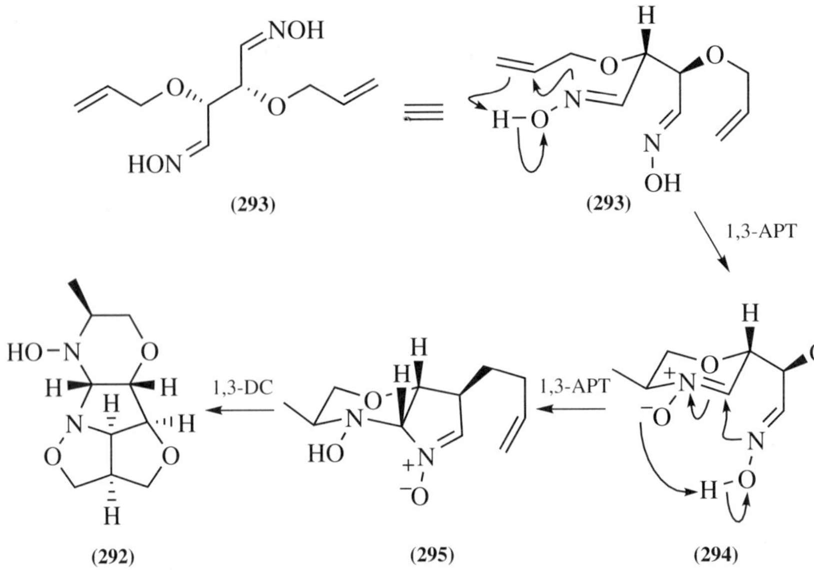

Scheme 1.61

While many researchers have used the 1,3-APT process to generate cyclic nitrones, it is clear that the operating reaction pathway in the oxime to isoxazolidine conversion may not always be predicted. In the work of Aurich and co-workers (317,318), the polycyclic isoxazolidine **292** was isolated as the major product from thermolysis of oxime **293** and may have been formed via two separate reaction paths (Scheme 1.61). In the proven route, initial 1,3-APT of **293** formed a 1,4-oxazine nitrone (**294**), which acted as the acceptor for the second 1,3-APT with the remaining oxime function. The cyclic nitrone **295** so formed underwent a 1,3-dipolar cycloaddition with the allyl ether forming the isolated polycyclic isoxazolidine adduct **292**.

The electrophile-induced cyclization of heteroatom nucleophiles onto an adjacent alkene function is a common strategy in heterocycle synthesis (319,320) and has been extended to electrophile-assisted nitrone generation (Scheme 1.62). The formation of a cyclic cationic species **296** from the reaction of an electrophile (E^+), such as a halogen, with an alkene is well known and can be used to N-alkylate an oxime and so generate a nitrone (**297**). Thus, electrophile-promoted oxime–alkene reactions can occur at room temperature rather than under thermolysis as is common with 1,3-APT reactions. The induction of the addition of oximes to alkenes has been performed in an intramolecular sense with *N*-bromosuccinimide (NBS) (321–323), *N*-iodosuccinimide (NIS) (321), I_2 (321,322), and ICl (321) for subsequent cycloaddition reactions of the cyclic nitrones with alkenes and alkynes.

More recently, cyclizations have been performed via cyclic selenides (265,324–327) and by ring opening of epoxides (328,329). Similar cyclizations have been achieved using alkenes complexed to Hg(II) (265) or Pd(II) salts (267,330) which,

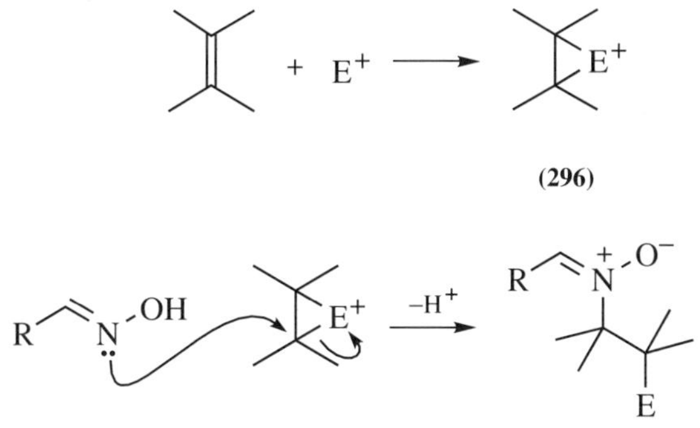

(296)

(296) (297)

Scheme 1.62

in the case of O-allyl oximes, effected the known [2,3]-sigmatropic rearrangement for inter- and intramolecular trapping with dipolarophiles (331).

The electrophile-mediated formation of cyclic nitrones is neither restricted to the reaction of oximes with alkenes, nor to the electrophiles listed above. Gallagher and co-workers (298–301) demonstrated the utility of allenes as the acceptor in Ag(I)-catalyzed cyclization reactions in the preparation of five- and six-membered nitrones from the corresponding oximes (Scheme 1.63). However, the ambident nucleophilicity of oximes hinders many reactions of this type: cyclization through N affords the cyclic nitrones, while nucleophilic cyclization through O affords isomeric oxazines. The product mixture in this case was found to depend heavily on the oxime geometry and the number of atoms separating the oxime and allene functions. Oximes (*E*)-**298** and (*E*)-**299** afforded the desired five- and six-membered nitrones **300** and **301**, respectively on treatment with $AgBF_4$ (300). Similar treatment of (*Z*)-**298** gave the oxazine **302**, while (*Z*)-**299** slowly equilibrated to the (*E*)-isomer and was otherwise inert. Dipolar cycloaddition reactions were observed between the C-unsubstituted nitrones **300** or **301** and alkenes to give adducts **303**, although the analogous C-methyl nitrones were unreactive toward such dipolarophiles.

A related oxime **304** was prepared for the total synthesis of **305**, an unusual pyrrolizidine alkaloid isolated from the venom of a variety of thief ant

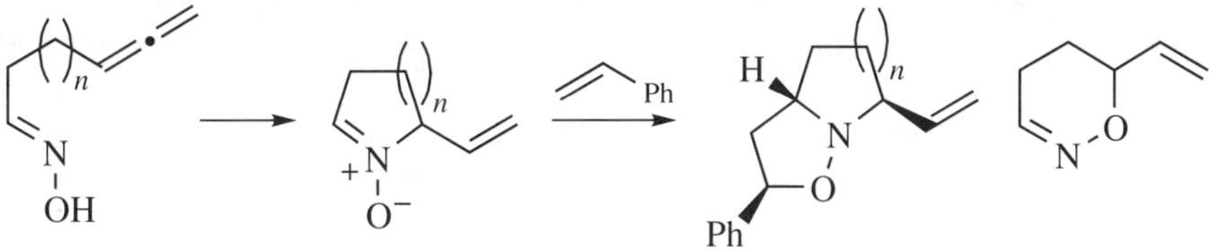

E-(**298**) $n = 1$ (**300**) $n = 1$ (**303**) (**302**)

E-(**299**) $n = 2$ (**301**) $n = 2$

Scheme 1.63

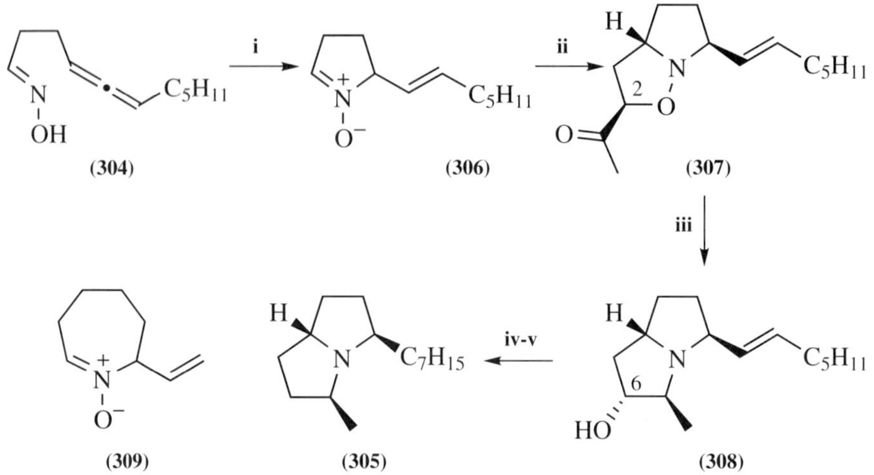

Reagents: i $AgBF_4$, DCM, 3 h; ii $MeCOCH{=}CH_2$, THF, 15 h, 48% [from (*E*)-**304**]; iii H_2, $PdCl_2$, EtOH, 18 h, 76%; iv Jones reagent, $Me_2C{=}O$, 1.5 h then $HSCH_2CH_2SH$, $BF_3{\bullet}OEt_2$, DCM, 2.5 h, 48%; v Raney-Ni, EtOH, 40 min, 61%.

Scheme 1.64

(Scheme 1.64). The Ag(I)-mediated cyclization afforded dipole **306** for 1,3-dipolar cycloaddition with methyl vinyl ketone to yield adducts **307** and the C(2) epimer as a 1:1 mixture (48%). Hydrogenolytic N–O cleavage and simultaneous intramolecular reductive amination of the pendant ketone of the former dipolarophile afforded a mixture of alcohol **308** and the C(6) epimer. Oxidation to a single ketone was followed by carbonyl removal by conversion to the dithiolane and desulfurization with Raney nickel to afford the target compound **305** (299). By this methodology, a seven-membered nitrone (**309**) was prepared for a dipolar cycloaddition reaction with *N*-methyl maleimide or styrene (301).

1.11.2. Intramolecular Oxime–Alkene Cycloaddition

Ochiai's landmark paper of 1967, which outlined the basis of the 1,3-APT (Section 1.11.1), also described the reaction of formaldoxime with alkenes and alkynes (284). The oxime **310**, behaving as a 1,3-dipole, afforded isoxazolidine cycloadducts **311** through a thermally driven equilibrium (heavily favoring the oxime) in which a proton is transferred from O to N (1,2-prototropy) to afford nitrone **312** (Scheme 1.65).

The intramolecular oxime–alkene cycloaddition (IOAC) proceeds via N–H nitrones, in contrast to most other nitrone syntheses, which afford N-alkylated 1,3-dipoles. This process was used by Wildman and co-worker (332) in the synthesis of 6-hydroxybuphanidine and 6-hydroxypowelline, and since then by

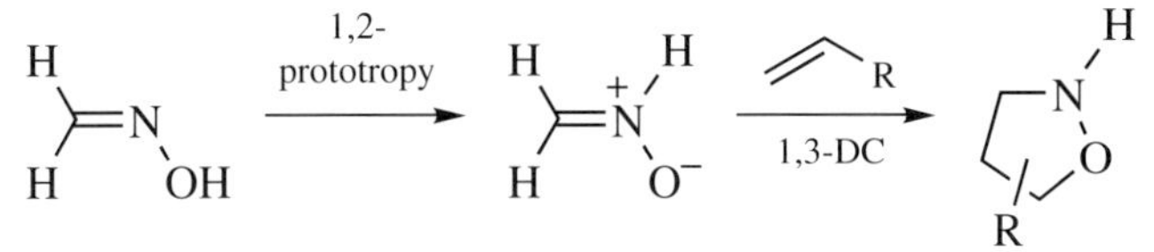

Scheme 1.65

many researchers. The IOAC was recently used in the work of Moutel and Shipman who prepared a single, enantiomerically pure bicyclic adduct (**313**) by thermolysis of the D-glucose-derived oxime **314**. They found less selectivity with other carbohydrate-derived alkenyl oximes (Scheme 1.66) (333,334). This adduct yielded the target aminocyclopentitol **315** after benzoyl ester cleavage and palladium-catalyzed N—O bond hydrogenolysis. As well as such carbocycles (268), the IOAC has been used to prepare related bicyclic isoxazolidines fused to pyrrolidines (335), piperidines (336,337), and THFs (317,318,338). Such heterocyclic ring systems have also been prepared by the intramolecular cycloaddition reaction of conventionally formed alkenyl nitrones (Scheme 1.69, Section 1.11.3). In a recent report, Cheng and co-workers described the microwave-assisted intramolecular oxime–alkene cycloaddition reaction of alkenyl oximes adsorbed onto the surface of silica gel in solvent free conditions (339).

1.11.3. Intramolecular Nitrone–Alkene Cycloadditions

The in situ formation of nitrones from oximes by 1,3-APT or 1,2-prototropy is clearly a powerful synthetic strategy but conventional nitrone generation, in particular hydroxylamine–carbonyl condensation, has been applied to numerous syntheses, in intra- and intermolecular mode (258). Accordingly, the ring systems similar to those synthesized using 1,3-APT/intramolecular nitrone-alkene cycloaddition (INAC) methodology by Heaney (313–315) (see Section 1.11.2) or Padwa and Norman (340) have been made using conventionally prepared nitrones (Scheme 1.67). As with the previous examples, once formed, the nitrones are suitably placed for a spontaneous intramolecular cycloaddition reaction with the

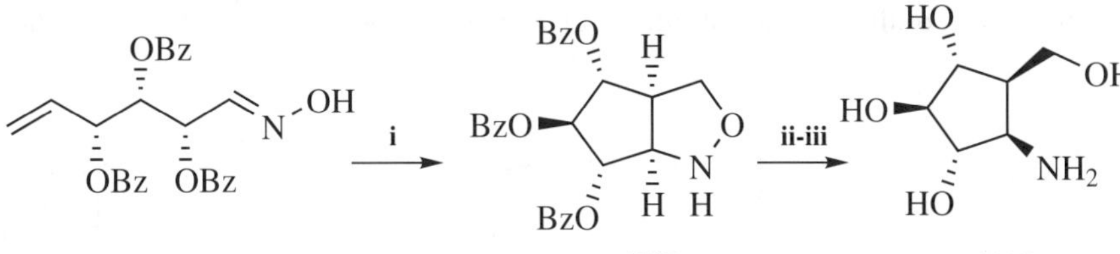

Reagents: [i] PhMe, 110 °C, 15 h, 60%; [ii] NaOMe, MeOH, rt 60%; [iii] 10% Pd-C, H_2, MeOH, 91%.

Scheme 1.66

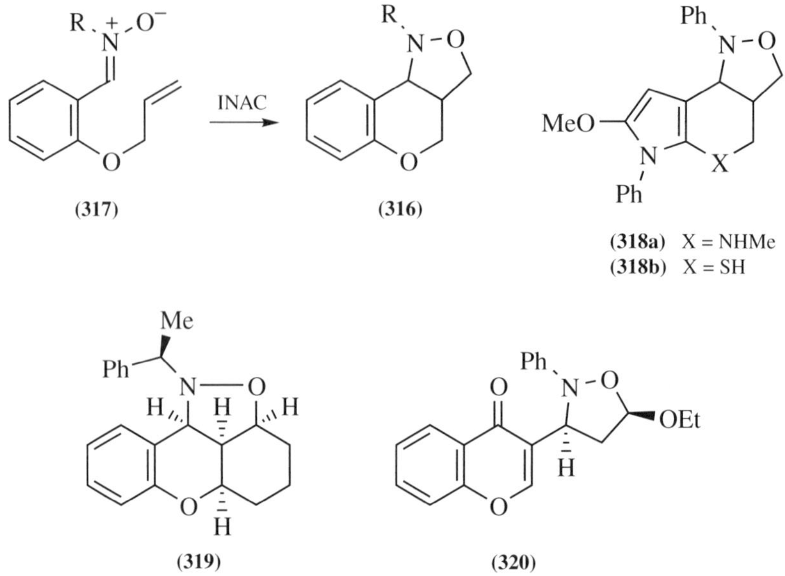

Scheme 1.67

adjacent alkene. In this way, it has been possible to prepare benzopyranoisoxazole adducts **316** from the corresponding alkenyl nitrones **317** (305,341–343) [based on the original work of Oppolzer and co-workers (344,345)] as well as related pyrrolyl–piperidines (**318a**), pyrrolyl-thiopyrans (**318b**) (346), and the tetracyclopyrans (**319**) (347). Benzopyranyl nitrones were prepared for reaction with electron-rich or electron-deficient alkenes to afford the expected isoxazolidinyl adducts (e.g., **320** with an enol ether) (348,349). In their reaction with allenic esters, however, the initially formed cycloadducts undergo a series of spontaneous intramolecular transformations to afford functionalized benzoindolizidines (350). Related polycyclic ring systems based on β-lactams have been reported recently by Alcaide and co-workers (68,69) (Scheme 1.8, Section 1.4). In a similar report, a trifluoromethyl alkenyl nitrone moiety underwent an intramolecular cycloaddition to afford tricyclic adducts (351).

Saito et al. (351a) reported the previously unprecedented formation of a 4,5-fused bicyclic isoxazolidine by an intramolecular cycloaddition reaction (Scheme 1.68).

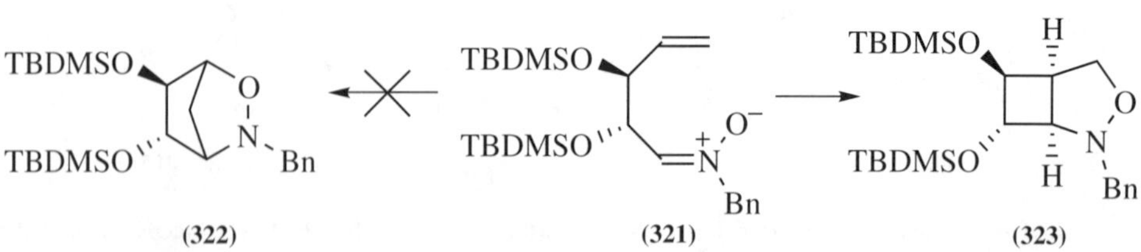

Scheme 1.68

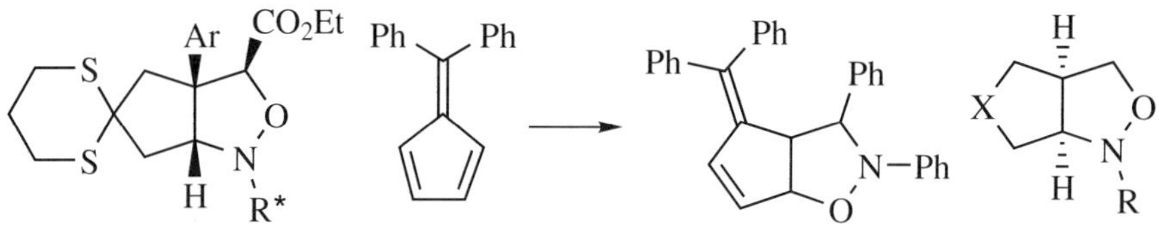

(324) (325) (326) X = O, N, S, OSi

Scheme 1.69

The alkenyl nitrone **321** can conceivably undergo two different cycloaddition modes to afford either the bridged bicyclo[2.2.1] adduct **322** or the fused bicyclo[3.2.0] adduct **323**. Three examples of fused adducts were reported, each formed with total diastereoselectivity, and on reductive cleavage of the isoxazolidine N—O bond (with concomitant N-debenzylation) they afforded the corresponding polysubstituted cyclobutylamines.

Cyclopentyl isoxazolidine cycloadduct **324** was prepared by intramolecular nitrone cycloaddition by Baldwin et al. (280,281,352,353) as part of studies toward a total synthesis of pretazettine (Scheme 1.69). Related adducts have been prepared elsewhere (354–356) including fluorine-substituted carbocycles (357) and the adducts prepared by IOAC by Shipman and co-workers (333,334) who demonstrated their potential as a route to aminocyclopentitols (Scheme 1.66, Section 1.11.2). Such bicyclic structures have been prepared in rather unique intermolecular fashion by Chandrasekhar and co-workers (357a) from the cycloaddition of *C,N*-diphenyl nitrone to fulvene (**325**).

However, this intramolecular cycloaddition methodology has found more widespread use in the preparation of cis-fused heterobicyclic rather than carbobicyclic products, by insertion of a heteroatom into the hydrocarbon *tail* of the alkenyl nitrone. Particular attention has been given to the synthesis of THF adducts (**326**, X = O) (202,317,354,358–365), and to a lesser extent the N (82,361,366–368) S (202,203,369), or silyloxy analogues (54,55,370). Aurich and Biesemeier (359) prepared ether dipoles (**327**), which slowly cyclized at ambient temperature to afford the cis-fused bicyclic adducts **328** (Scheme 1.70). Reductive cleavage of the

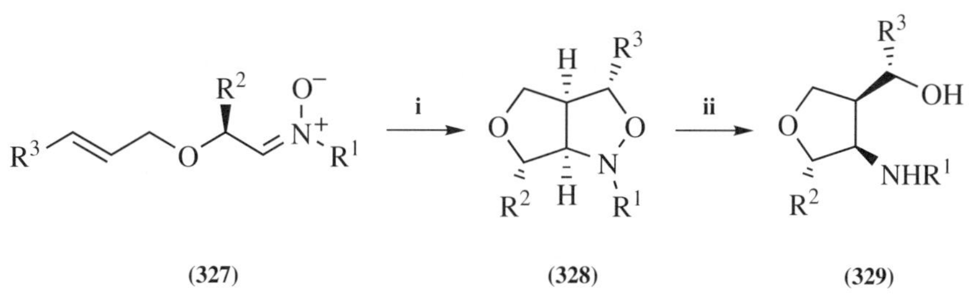

(327) (328) (329)

Reagents: [i] 0-5 °C, 2 h then rt 5 days; [ii] Zn, AcOH, or Ra-Ni, EtOH, or H_2, $Pd(OH)_2$-C, EtOH, 60 °C, 90 bar, 1 days, or H_2, $Pd(OH)_2$-C, EtOH, Me_2CO, 60 °C, 100 bar, 4 days.

Scheme 1.70

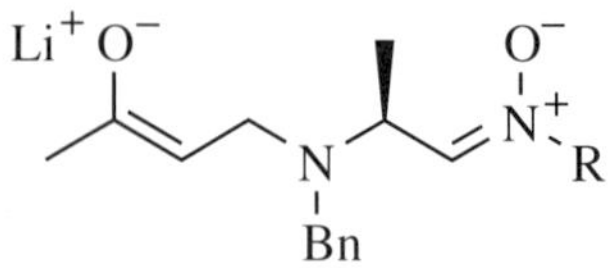

(330)

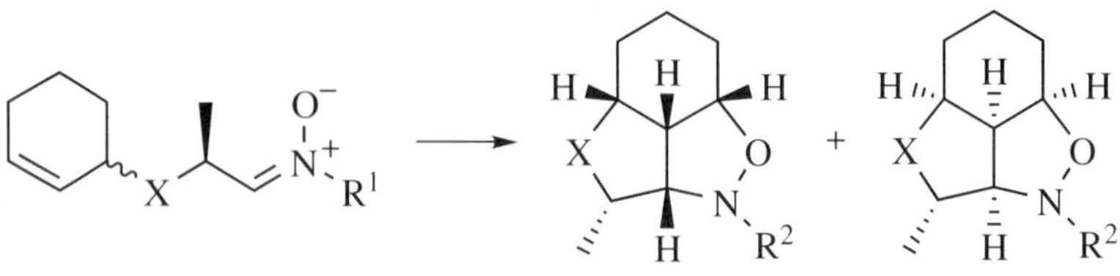

(331) X = O, NR^2 **(332a)** **(332b)**

Scheme 1.71

isoxazolidines afforded polysubstituted chiral THFs **329** for use as ligands in the enantioselective addition of diethylzinc to aldehydes.

In further work, Aurich and Möbus (367) used a lithium enolate reagent (**330**), generated from the corresponding ketone, in related intramolecular cycloadditions (Scheme 1.71). Replacement of the simple acyclic alkene chain terminus with cyclohexene gave reagent **331**, which generated a mixture of diastereomeric cis-fused tricyclic products **332a** and **332b** in its intramolecular cycloaddition reaction (360). By a similar strategy, Peseke and co-workers (371) synthesized novel carbasugar derivatives as potential glycosidase inhibitors using a cyclohexenyl nitrone ultimately derived from D-glucose.

The intermolecular cycloaddition route to spirocyclopropyl isoxazolidines and their subsequent rearrangement, used so widely by Brandi and co-workers (372–375) (Schemes 1.16 and 1.17, Section 1.5), has also been achieved in an intramolecular sense (Scheme 1.72). Cycloaddition of the alkenyl nitrone reagents (**333a–c**) afforded bicyclic isoxazolidinyl adducts **334**, which rearranged under thermolysis in analogous fashion to the earlier work to give piperidinones (**335**) via

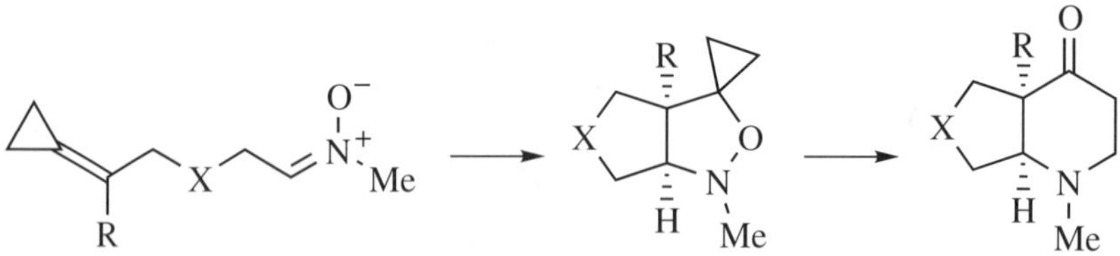

(333a) X = CH_2 **(334)** **(335)**

(333b) X = O

(333c) X = NTs

Scheme 1.72

homolytic cleavage of the isoxazolidine N—O bond and concomitant spirocycle ring opening.

The pyrinodemins are unusual and potent cytotoxic isoxazolidine alkaloids isolated from a marine sponge. Snider and Shi (376) prepared the nitrone **336** by the condensation of aldehyde **337** and hydroxylamine **338,** which underwent an intramolecular cycloaddition reaction to afford the adduct **339a** (Scheme 1.73). By a similar scheme, they also prepared the proposed structure of the natural product **339b** [C(16′)/C(17′) cis alkene] and, from comparison of spectroscopic data, they suggest the revised structure **339a** for pyrinodemin A (376). The position of the double bond was originally tentatively assigned by mass spectrometry (377). Baldwin et al. (378) prepared **339a** and **339b**. They argue that neither represents the true structure of the natural product. Elsewhere, intramolecular nitrone cycloaddition reactions have been used to prepare prostaglandins (379,380), the active forms of vitamins D_2 and D_3 (381–383) and steroid-derived isoxazolidines (302,384).

1.11.4. Isoxazolidines from Intermolecular Nitrone Cycloaddition Reactions

A number of unusual nitrones have been reported from which novel isoxazolidine cycloadducts were prepared. These include the adamantyl dipoles **340,**

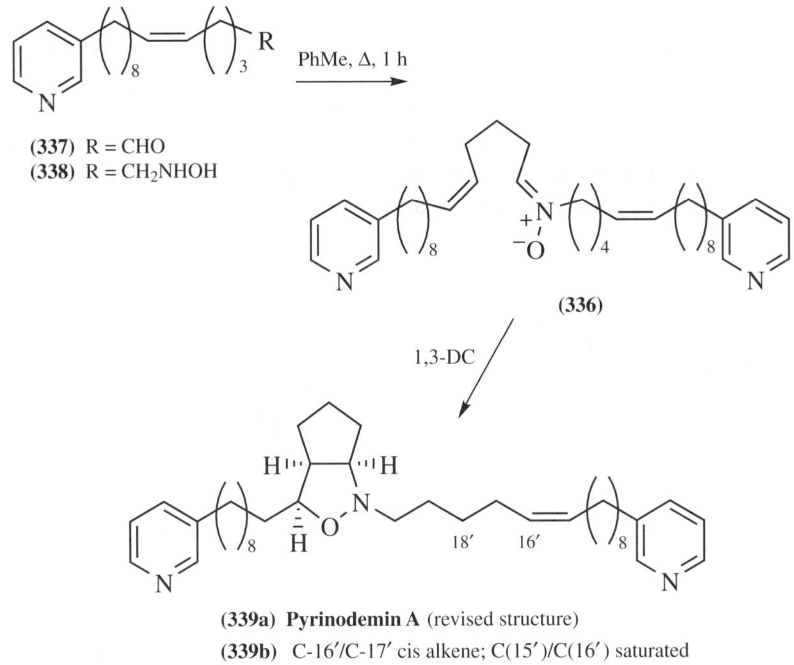

(339a) Pyrinodemin A (revised structure)
(339b) C-16′/C-17′ cis alkene; C(15′)/C(16′) saturated

Scheme 1.73

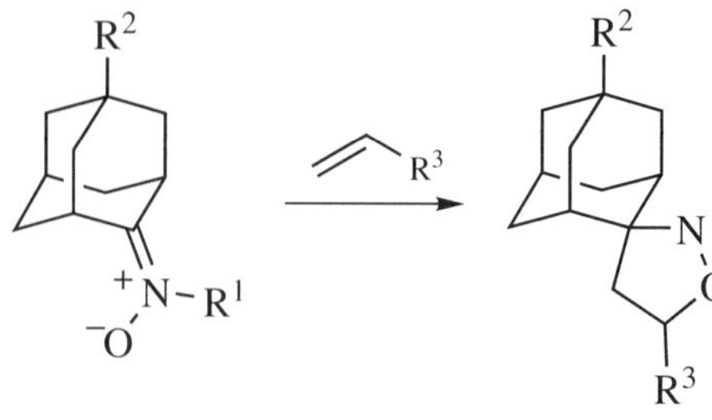

(340) (341)

Scheme 1.74

formed by the reaction of adamantan-2-one with alkylhydroxylamines, which react with alkene dipolarophiles to afford the spirocyclic adducts **341** (Scheme 1.74) (385,386). Reinhoudt and co-workers (387) prepared facially differentiated azetidine nitrones (**342**), from the [4 + 2] cycloadducts (**343**) of nitroalkenes and alkynamines, which spontaneously undergo a [1,3]-sigmatropic rearrangement to afford the isolated racemic mixture of nitrone **342** and the (2*S*,3*S*)-enantiomer (Scheme 1.75). Subsequent dipolar cycloaddition reaction with alkene and alkyne dipolarophiles proceeded with complete regio- and stereoselectivity (388). Related achiral nitrones were prepared by oxidation of the corresponding *N*-hydroxyazetidines with Pb(IV) oxide (389). Tufariello et al. reported the synthesis of highly strained three,five- and four,five-membered bicyclic nitrones (**344**, Scheme 1.76) (390). Reaction with a range of aromatic and aliphatic monosubstituted alkenes afforded tricyclic 5-substituted isoxazolidine cycloadducts (**345**) with high regio- and stereoselectivities. Katritzky et al. (391) have used *N*-bis(benzotriazol-1-ylmethyl)hydroxylamine (**346**) as a nitrone synthon, revealing the dipole **347** by thermolysis in toluene in the presence of a dipolarophile to afford the *N*-(benzotriazolylmethyl)isoxazolidine adducts (**348**, Scheme 1.77).

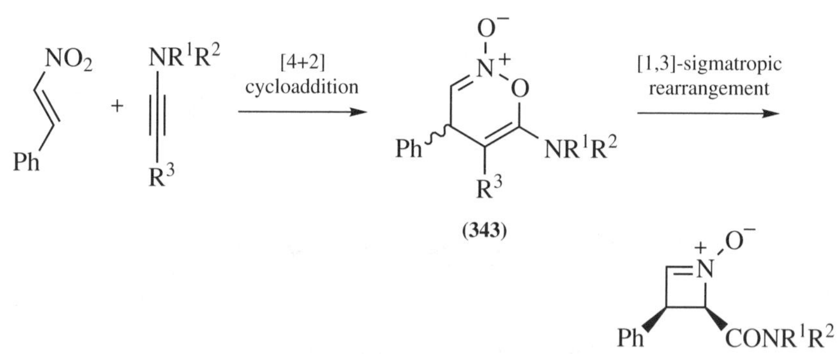

(343)

(342) (2*R*,3*R*)

Scheme 1.75

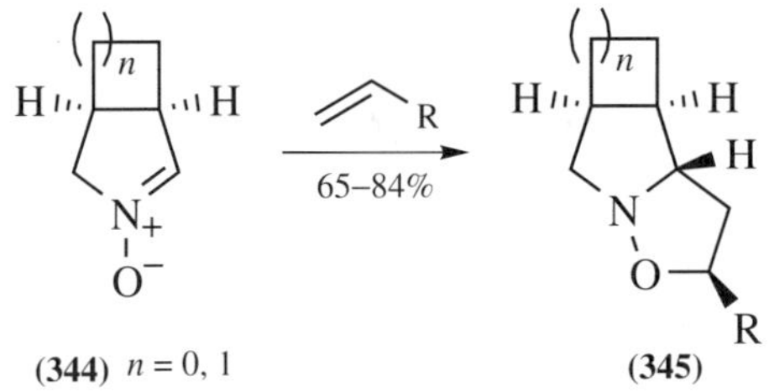

Scheme 1.76

As part of an extensive study of the 1,3-dipolar cycloadditions of cyclic nitrones, Ali et al. (392–397) found that the reaction of the 1,4-oxazine **349** with various dipolarophiles afforded the expected isoxazolidinyloxazine adducts (Scheme 1.78) (398). In line with earlier results (399,400), oxidation of styrene-derived adduct **350** with *m*-CPBA facilitated N—O cleavage and further oxidation as above to afford a mixture of three compounds, an inseparable mixture of ketonitrone **351** and bicyclic hydroxylamine **352**, along with aldonitrone **353** with a solvent-dependent ratio (401). These workers have prepared the analogous nitrones based on the 1,3-oxazine ring by oxidative cleavage of isoxazolidines to afford the hydroxylamine followed by a second oxidation with benzoquinone or Hg(II) oxide (402–404). These dipoles, along with a more recently reported pyrazine nitrone (405), were all used in successful cycloaddition reactions with alkenes. Elsewhere, the synthesis and cycloaddition reactions of related pyrazine-3-one nitrone **354** (406,407) or a benzoxazine-3-one dipolarophile **355** (408) have been reported. These workers have also reported the use of isoxazoles with an exocyclic alkene in the preparation of spiro[isoxazolidine-5,4′-isoxazolines] (409).

The oxidative cleavage of isoxazolidines has been taken to a third generation by Ali and Wazeer (278,400) in which, from their initial nitrone, the cycloaddition then oxidation are each performed twice to generate a third, highly substituted nitrone. Later, de March et al. (410) took this scheme to a third cycloaddition

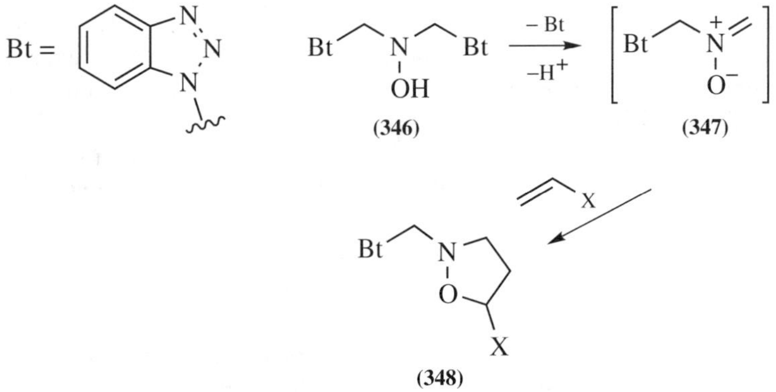

Scheme 1.77

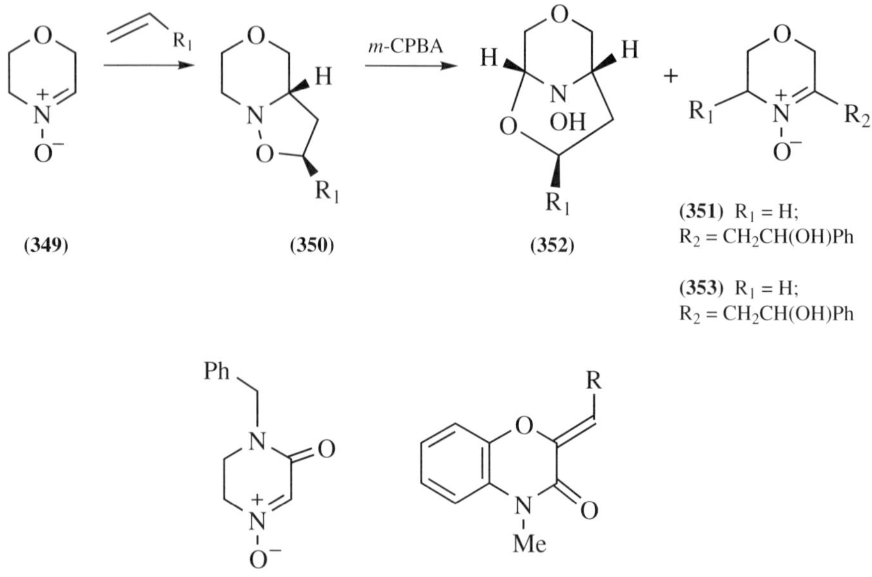

(349) (350) (352)

(351) $R_1 = H$; $R_2 = CH_2CH(OH)Ph$

(353) $R_1 = H$; $R_2 = CH_2CH(OH)Ph$

(354) (355)

Scheme 1.78

(Scheme 1.79). Here, the reaction of nitrone **92** with 2-(5*H*)-furanone (**356**) afforded the adduct **357,** which when treated with *m*-CPBA in dichloromethane, gave the aldonitrone **358** as a single diastereoisomer (70%). Adduct **359** was isolated from a second cycloaddition reaction with the lactone dipolarophile (62%) along with a small amount of the product of dehydration across the lactone C(3′) to C(4′) bond (12%). A further oxidative regeneration of the nitrone for a third cycloaddition, this time with dimethyl acetylenedicarboxylate (DMAD), afforded a single diastereoisomer of **360** (34%) in which the second lactone has been dehydrated [C(3″) to C(4″)]. Earlier, pioneering use of this cycloaddition–oxidative cleavage strategy in synthesis includes the preparation by Tufariello and Puglis (411) of 2,5-dialkylpyrrolidines found in the venom of *Solenopsis* ant species and the potent marine algal toxin anatoxin A (or Very Fast Death Factor) (412).

A number of workers have described the synthesis and cycloaddition reactions of oxazoline nitrone dipoles, (e.g., **361** and **362**, Scheme 1.80) (413–417). A homochiral oxazoline nitrone derived from camphor has been used to great effect by Langlois and co-workers (418,419), from which they have prepared a number of natural targets through enantioselective cycloaddition reactions, including the antiviral carbocyclic nucleoside analogue (+)-carbovir (48) (Fig. 1.1, Section 1.3).

Similarly, variously substituted 3-imidazoline nitrone reagents and the related nitroxides (**363** $R^1 = O^{\cdot}$) are popular targets and have demonstrated their reactivity toward various dipolarophiles, (e.g., styrene, Scheme 1.80) (420–431). However, only one group reports the use of a 2-imidazoline nitrone (432–434), which has demonstrated complete diastereofacial selectivity in its reactions with electron-

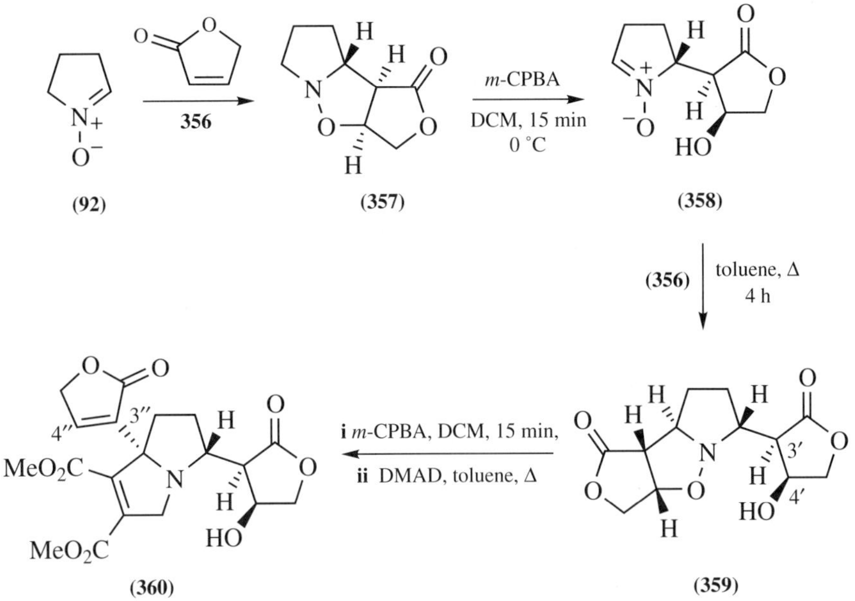

Scheme 1.79

deficient alkene dipolarophiles (Scheme 1.81) (434). The upper face of the dipole **364** is effectively blocked by the bulk of the phenyl moiety, enforcing reaction with the dipolarophiles exclusively at the lower face. In this way, the imidazo[1,2-*b*]isoxazolidine adduct (**365**), was formed as a single diastereoisomer from cycloaddition of **364** and dimethyl maleate. Subsequent N—O bond hydrogenolytic rupture and spontaneous recyclization yielded the related lactam **366**. Reaction of nitrone **364** with alkynes under identical conditions afforded the enediamines (e.g., methyl propiolate adduct **367**), via rearrangement of the initially formed isoxazoline cycloadduct **368** (432).

The utility of solid-phase synthesis for rapid generation of libraries of small molecules is unquestionable and this technique is particularly amenable to automation. It is not surprising then that the 1,3-dipolar cycloaddition reactions of nitrones have been applied to solid-phase synthesis, either through a supported

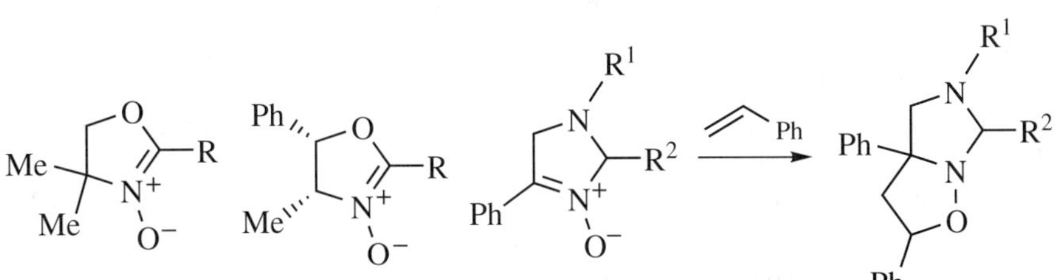

(361) R = Me, Et, Ph **(362)** R = Me, Pr **(363)**

Scheme 1.80

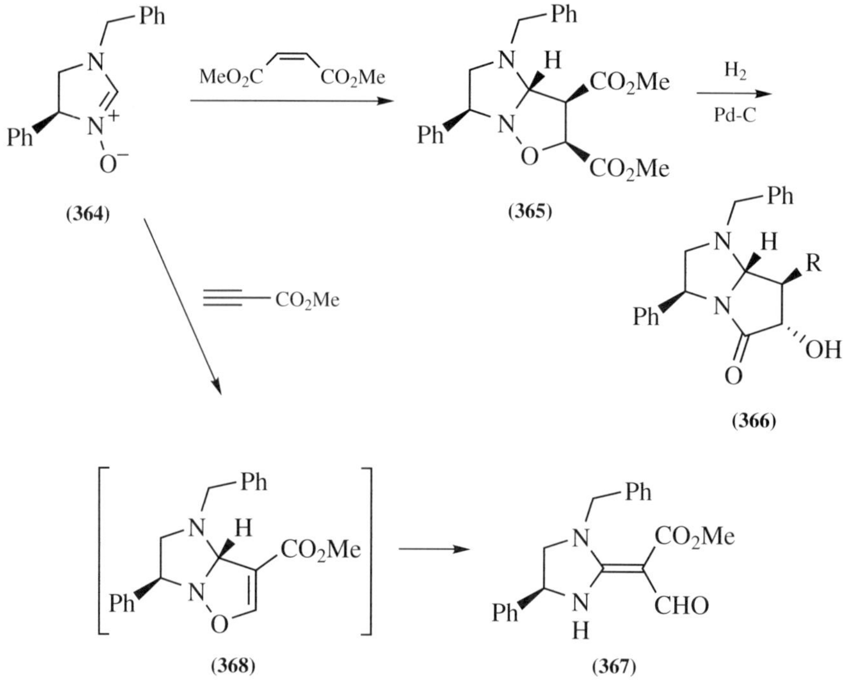

Scheme 1.81

dipole or dipolarophile. In fact, the dipolar cycloaddition may be considered the reaction of three components—initial nitrone generation through the combination of an aldehyde and hydroxylamine, which then reacts with the alkene dipolarophile. All three components have been polymer-bound in an elegant recent report by Jung and co-workers (435), in which they found most success in the combination of hydroxylamines supported on Rink amide resin with aldehydes to generate the dipoles followed by addition of the alkene dipolarophiles (Scheme 1.82). Thus, Rink resin **369** was first coupled to α-fluorenylmethyloxycarboxyl (Fmoc)-L-alanine, the N-protection cleaved with piperidine, then the amino group coupled to α-bromoacetic acid to afford the bromoamide functionalized resin **370**. Rapid nucleophilic substitution of the bromide by the hydroxylamine afforded bound hydroxylamine **371** for condensation with 2,4-dimethylbenzaldehyde in toluene at 80 °C. The polymer-supported nitrone **372** was then trapped by addition of a toluene solution of *N*-methylmaleimide (**373**) and further heating (5–18 h). After intensive washing and drying, the cycloadducts were cleaved from the resin with TFA and through the use of the chiral alanine linker, the diastereomeric adducts separated by high performance liquid chromatography (HPLC) for electrospray MS analysis. It was determined that the cycloaddition of nitrone **372** with dipolarophile **373** forms all four expected isomers of **374**, along with a trace of an azomethine ylide-derived adduct **375** from incomplete coupling of the bromoacetic acid. A number of different aldehydes, bromoacids, and dipolarophiles were used successfully in

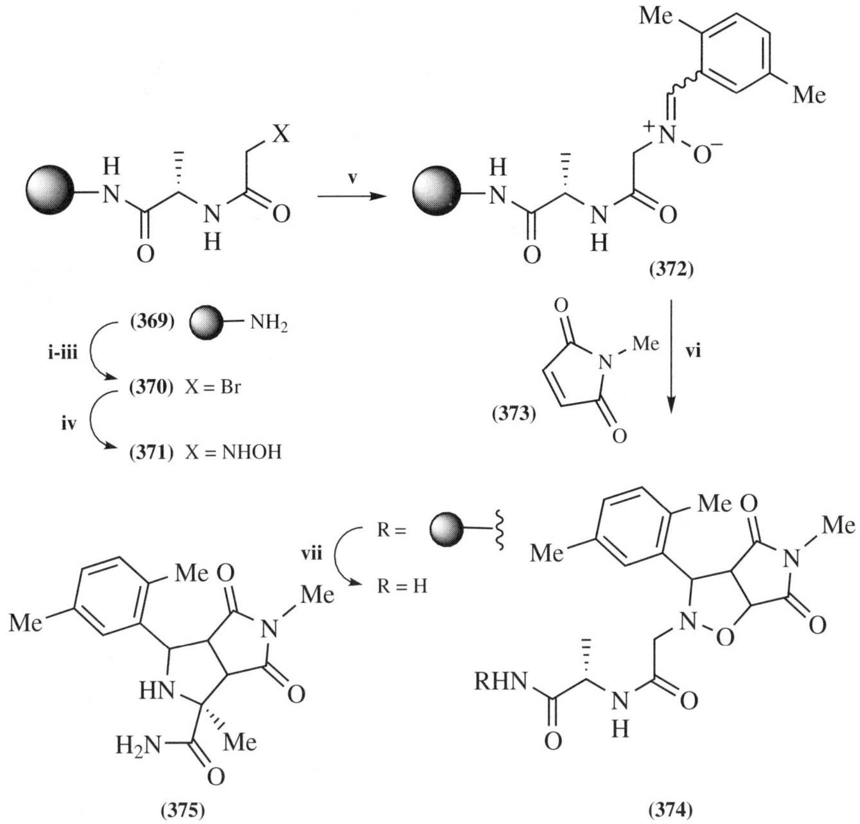

Reagents: i Fmoc-L-alanine; ii piperidine, DMF; iii $BrCH_2CO_2H$, diisopropylcarbodiimide; iv $NH_2OH{\bullet}HCl$; v PhMe, Δ,2,4-Me_2-C_6H_4CHO; vi PhMe, Δ; vii TFA.

Scheme 1.82

this study, demonstrating impressive flexibility. Other bound nitrones were used for an illustrative "split and mix" synthesis of a small model library of nine isoxazolidines by reaction of five bound dipoles with phenylvinylsulfone or *N*-methylmaleimide.

Schreiber and co-workers (436) prepared a library calculated to contain 2.18 million polycyclic compounds through the 1,3-dipolar cycloaddition of a number of nitrones with alkenes supported on TentaGel S NH_2 resin (Scheme 1.83). (−)-Shikimic acid was converted into the polymer bound epoxycyclohexenol carboxylic acid **376** (or its enantiomer), coupled to the resin via a photolabile linker developed by Geysen and co-workers (437) to allow release of the products from the resin in the presence of live cells by ultraviolet (UV)-irradiation. A range of iodoaromatic nitrones (**377**) was then reacted with the α,β-unsaturation of the polymer-bound amide in the presence of an organotin catalyst, using the tandem esterification/dipolar cycloaddition methodology developed by Tamura et al. (84,85) Simultaneous cyclization by PyBrop-mediated condensation of the acid with the alcohol

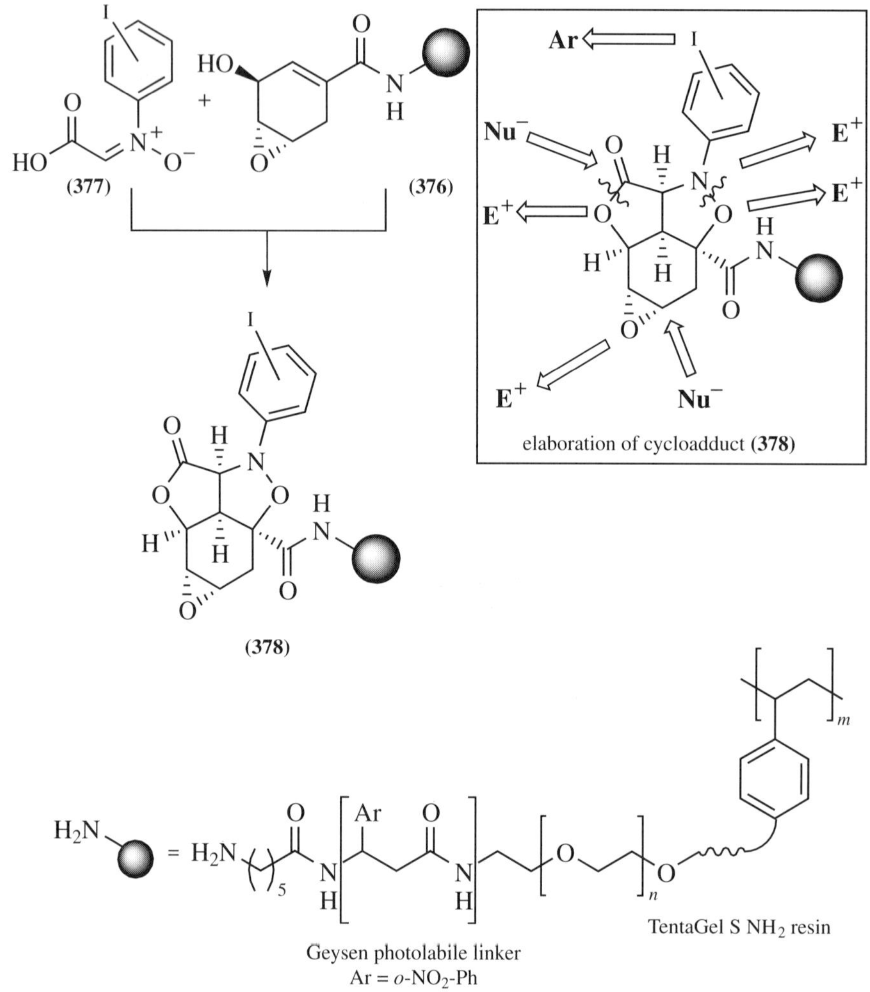

Scheme 1.83

functionality afforded the tetracyclic adducts **378** with complete regio- and stereoselectivity. This template was designed to provide multiple routes to further elaboration including N—O bond cleavage, nucleophilic addition to the lactone or the epoxide and Stille or Suzuki coupling to the iodobenzene moiety as well as multiple electrophilic additions.

Elsewhere, Faita et al. (438) bound the Evans chiral auxiliary to Wang or Merrifield resin for use as a dipolarophile in cycloadditions with *C,N*-diphenylnitrone. Yields on both resins are significantly reduced in comparison to the solution phase reaction (43–20% compared to 95%) but are unaffected by addition of magnesium perchlorate or scandium triflate catalyst. A one-pot process has been reported by Hinzen and Ley (439) that oxidizes secondary hydroxylamines to the

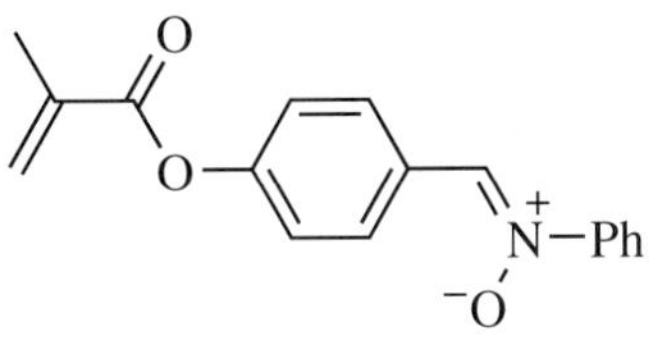

(379)

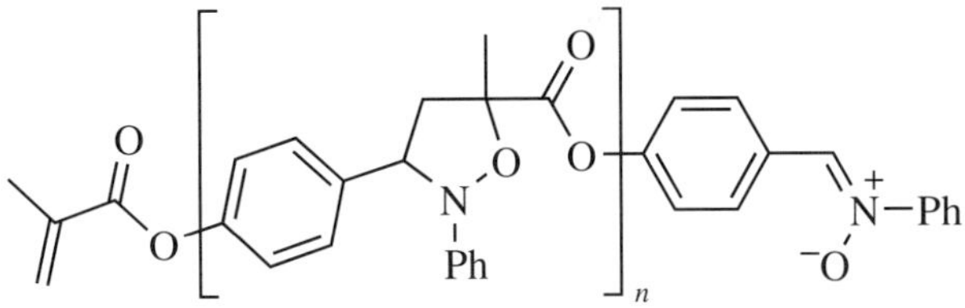

(380)

Figure 1.8

corresponding nitrones with polymer-supported perruthenate for *in situ* condensation with alkene dipolarophiles.

As part of a study of the synthesis of polymeric nitrones, Heineberg and coworkers (440) reported the synthesis of oligomeric isoxazolidines (Fig. 1.8). The nitrone **379** does not undergo the desired homopolymerization under typical free radical polymerization conditions [70 °C, DMF, azobis(isobutyronitrile) (AIBN)], preferring a 1,3-dipolar cycloaddition reaction. The isolated oligomeric products (in near quantitative yields) are the dimer, trimer, and tetramer **380** ($n = 1$–3). Similarly, reaction in the presence of methyl methacrylate produces analogous materials in which only the terminal nitrone functionality has reacted with the acrylate, indicating a strong preference for curiously limited self-condensation. Other unusual nitrone cycloaddition reactions include the high regio- and stereoselectivities found through the use of smectic-type liquid crystals as the reaction media (441) and rate enhancement using ultrasound (442) or ionic dipolarophiles (443).

Ramamoorthy et al. (444) found that α-phenyl-*N*-(4-methylphenyl)nitrone can be the guest molecule in inclusion complexes with a β-cyclodextrin host in 1:1 and 1:2 ratios (guest/host), and that the latter undergoes a 1,3-dipolar cycloaddition reaction with electron-deficient alkenes. In more recent work, they have formed 1:1 inclusion complexes of the bowl-shaped β-cyclodextrin **383** with β-nitrostyrene **381** or 1-nitrocyclohexene **382**, which leave the alkene moiety exposed (Fig. 1.9) (445). Complexes **381** and **382** undergo cycloaddition reaction with α-phenyl-*N*-(4-methylphenyl)nitrone in the solid state after thorough homogenization (60 °C, 3 h) to give the 4-substituted products exclusively in 80 and 85% yield, respectively.

With ever more powerful technology becoming available, computer-based molecular modeling of complex chemical reactions is now commonplace. But while the Diels–Alder reaction has received a great deal of theoretical attention, the dipolar cycloaddition reactions of nitrones were until very recently relatively

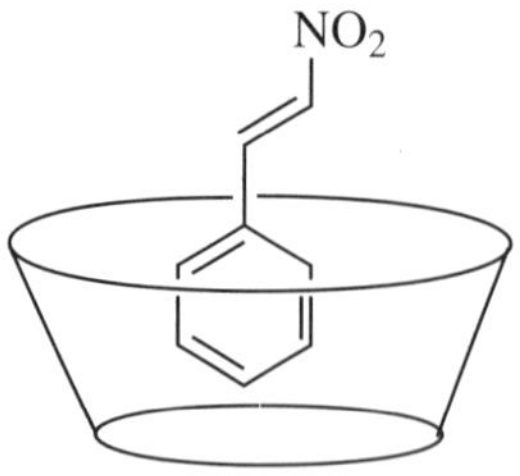

(381)

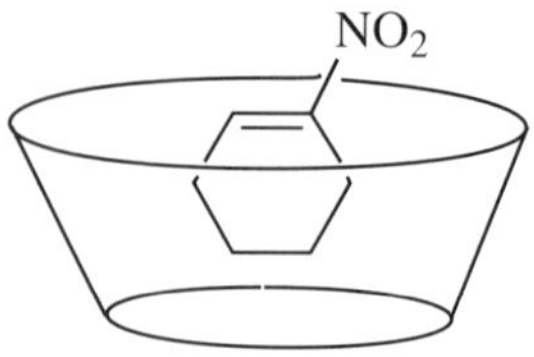

(382)

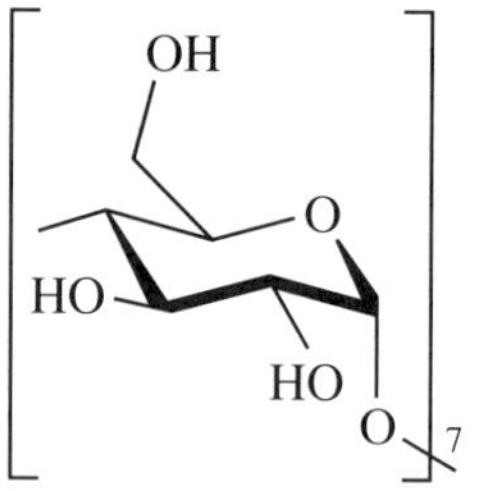

(383)

Figure 1.9

unexplored. Many workers have now compiled data on the theoretical distribution of products from a chosen nitrone cycloaddition reaction, often presented in parallel with the experimental results. Such studies include the reaction of nitrones with alkenes (446), acrylonitrile derivatives (447), cyclobutadienes (448,449), allyl ethers (450), vinyl ethers and vinyl acetates (152), vinyl boranes (451), alkynyl Fischer carbene complexes (452,453) and the metal-catalyzed reaction of nitrones with a crotonyl isoxazolidinone (454). A chiral erythrulose-derived nitrone was the subject of theoretical–experimental parallel studies (455), as was the camphor-based oxazoline nitrone developed by Langlois and co-workers (419). Purely theoretical studies were reported in a joint paper by Sustmann, Sicking, and Huisgen (456) on the high reactivity of the C=S double bond in nitrone cycloadditions.

1.12. CONCLUSION

In this chapter, we have sampled the main areas of synthetic interest in which the 1,3-dipolar cycloaddition of nitrones has been applied since 1984. It has not attempted to provide a comprehensive survey, rather a selective review to illustrate the strategies in which nitrones have found their place. It can be seen that nitrones have found application in the synthesis of a wide range of natural product target types, from sugars and nucleoside analogues, through β-lactams to alkaloids and other nitrogen heterocyclic natural products, both bridgehead bicyclic and monocyclic systems. The stereocontrol possible in the pericyclic transition state of the cycloadditions has been a key feature of many of these applications, and has been used to construct acyclic systems, for example, in the amino acid and peptide arena. Catalysis of the cycloadditions, and the use of solid phase have provide new dimensions.

The wide range of chemistry incorporating nitrone cycloaddition shows no signs of having reached its limits, and we fully expect new and novel applications to continue to appear in the future.

REFERENCES

1. L. I. Smith, *Chem. Rev.*, **1938**, *23*, 193.
2. J. Hamer and A. Macaluso, *Chem. Rev.*, **1964**, *64*, 473.
3. G. R. Delpierre and M. Lamchen, *Q. Rev., Chem. Soc.*, **1965**, *19*, 329.
4. J. J. Tufariello, *Acc. Chem. Res.*, **1979**, *12*, 396.
5. E. Breuer, "Nitrones and Nitronic Acid Derivatives : Their Structure and their Roles in Synthesis", in *The Chemistry of Amino, Nitroso and Nitro Compounds and their Derivatives*, Vol. 1, S. Patai ed., Wiley, Chichester, **1982**, pp. 459–538.
6. J. J. Tufariello, "Nitrones", in *1,3-Dipolar Cycloaddition Chemistry*, Vol. 2, A. Padwa ed., Wiley, New York, **1984**, pp. 83–168.
7. K. B. G. Torsell, *Nitrile Oxides, Nitrones and Nitronates in Organic Synthesis*, VCH, New York, **1988**.
8. E. Beckmann, *Ber. Dtsch. Chem. Ges.*, **1890**, *23*, 3331.
9. E. Beckmann, *Ber. Dtsch. Chem. Ges.*, **1894**, *27*, 1957.
10. P. Pfeiffer, *Annalen*, **1916**, 72.
11. R. Huisgen, *Proc. Chem. Soc.*, **1961**, 357.
12. R. Huisgen, *Angew. Chem., Int. Ed. Engl.*, **1963**, *2*, 565.
13. R. Huisgen, *Angew. Chem., Int. Ed. Engl.*, **1963**, *2*, 633.
14. A. Padwa, *Angew. Chem., Int. Ed. Engl.*, **1976**, *15*, 123.
15. R. Huisgen, "1,3-Dipolar Cycloadditions–Introduction, Survey, Mechanism", in *1,3-Dipolar Cycloaddition Chemistry*, Vol. 1, A. Padwa ed., Wiley, New York, **1984**, pp. 1-176.
16. K. N. Houk and K. Yamaguchi, "Theory of 1,3-Dipolar Cycloadditions", in *1,3-Dipolar Cycloaddition Chemistry*, Vol. 2, A. Padwa ed., Wiley, New York, **1984**, pp. 407–450.
17. K. N. Houk, J. Gonzalez, and Y. Li, *Acc. Chem. Res.*, **1995**, *28*, 81.
18. R. D. Little, "Thermal Cycloadditions", in *Comprehensive Organic Synthesis*, Vol. 5, B. M. Trost and I. Fleming eds., Pergamon Press, Oxford, UK, **1991**, Chapter 4.1, pp. 239–270.
19. P. A. Wade, "Intramolecular 1,3-Dipolar Cycloadditions", in *Comprehensive Organic Synthesis*, Vol. 5, B. M. Trost and I. Fleming eds., Pergamon Press, Oxford, UK, **1991**, Chapter 4.10, pp. 1111–1168.
20. A. Padwa, "Intermolecular 1,3-Dipolar Cycloadditions", in *Comprehensive Organic Synthesis*, Vol. 5, B. M. Trost and I. Fleming eds., Pergamon Press, Oxford, UK, **1991**, Chapter 4.9, pp. 1069–1109.
21. R. A. Firestone, *J. Org. Chem.*, **1968**, *33*, 2285.
22. R. A. Firestone, *J. Chem. Soc.* (A), **1970**, 1570.
23. R. A. Firestone, *J. Org. Chem.*, **1972**, *37*, 2181.
24. R. Huisgen, *J. Org. Chem.*, **1976**, *41*, 403.
25. R. Huisgen, *J. Org. Chem.*, **1968**, *33*, 2291.
26. R. Huisgen and R. Weinberger, *Tetrahedron Lett.*, **1985**, *26*, 5119.
27. R. Huisgen, G. Mloston, and E. Langhals, *J. Org. Chem.*, **1986**, *51*, 4085.
28. D. St. C. Black, R. F. Crozier, and V. C. Davis, *Synthesis*, **1975**, 205.
29. G. Bianchi, C. De Micheli, and R. Gandolfi, "1,3-Dipolar Cycloadditions Involving X = Y Groups", in *The Chemistry of Double-Bonded Functional Groups, Part 1*, S. Patai ed., Wiley, London, **1977**, Chapter 6, pp. 369–532.
30. N. Balasubramanian, *Org. Prep. Proced. Int.*, **1985**, *17*, 25.
31. P. DeShong, S. W. Lander, Jr., J. M. Leginus, and C. M. Dicken, "Dipolar Cycloadditions of Nitrones with Vinyl Ethers and Silane Derivatives", in *Advances in Cycloaddition*, Vol. 1, D. P. Curran ed., JAI Press, **1988**, pp. 87–128.

32. P. N. Confalone and E. M. Huie, *Org. React.*, **1988**, *36*, 1.
33. M. Frederickson, *Tetrahedron*, **1997**, *53*, 403.
34. K. V. Gothelf and K. A. Jorgensen, *Chem. Rev.*, **1998**, *98*, 863.
35. K. V. Gothelf and K. A. Jorgensen, *Chem. Commun.*, **2000**, 1449.
36. K. Fukui, *Acc. Chem. Res.*, **1971**, *4*, 57.
37. I. Fleming, *Frontier Molecular Orbitals and Organic Chemical Reactions*, William Clowes & Sons, London, **1976**.
38. A. Dondoni, S. Franco, F. Junquera, F. L. Merchán, P. Merino, and T. Tejero, *J. Org. Chem.*, **1997**, *62*, 5497.
39. P. Merino, S. Franco, F. L. Merchan, and T. Tejero, *Tetrahedron Lett.*, **1998**, *39*, 6411.
40. P. Merino, S. Franco, F. L. Merchan, and T. Tejero, *J. Org. Chem.*, **2000**, *65*, 5575.
41. P. Merino, S. Franco, N. Garces, F. L. Merchán, and T. Tejero, *Chem. Commun.*, **1998**, 493.
42. O. Tamura, N. Mita, H. Kusaka, H. Suzuki, and M. Sakamoto, *Tetrahedron Lett.*, **1997**, *38*, 429.
43. U. Chiacchio, A. Corsaro, G. Gumina, A. Rescifina, D. Iannazzo, A. Piperno, G. Romeo, and R. Romeo, *J. Org. Chem.*, **1999**, *64*, 9321.
44. U. Chiacchio, A. Corsaro, D. Iannazzo, A. Piperno, A. Procopio, A. Rescifina, R. Romeo and G. Romeo, *Eur. J. Org. Chem.*, **2001**, 1893.
45. U. Chiacchio, A. Rescifina, D. Iannazzo, and G. Romeo, *J. Org. Chem.*, **1999**, *64*, 28.
46. U. Chiacchio, G. Gumina, A. Rescifina, R. Romeo, N. Uccella, F. Casuscelli, A. Piperno, and G. Romeo, *Tetrahedron*, **1996**, *52*, 8889.
47. U. Chiacchio, A. Corsaro, D. Iannazzo, A. Piperno, A. Rescifina, R. Romeo, and G. Romeo, *Tetrahedron Lett.*, **2001**, *42*, 1777.
48. T. Berranger and Y. Langlois, *Tetrahedron Lett.*, **1995**, *36*, 5523.
49. A. Roy, K. Chakrabarty, P. K. Dutta, N. C. Bar, N. Basu, B. Achari, and S. B. Mandal, *J. Org. Chem.*, **1999**, *64*, 2304.
50. N. C. Bar, A. Roy, B. Achari, and S. B. Mandal, *J. Org. Chem.*, **1997**, *62*, 8948.
50a. J. K. Gallos, E. G. Goga, and A. E. Koumbis, *J. Chem. Soc., Perkin Trans 1*, **1994**, 613.
51. E. Colacino, A. Converso, A. De Nino, A. Leggio, A. Liguori, L. Maiuolo, A. Napoli, A. Procopio, C. Siciliano, and G. Sindona, *Nucleosides Nucleotides*, **1999**, *18*, 581.
52. A. Leggio, A. Liguori, L. Maiuolo, A. Napoli, A. Procopio, C. Siciliano, and G. Sindona, *J. Chem. Soc., Perkin Trans. 1*, **1997**, 3097.
53. A. Leggio, A. Liguori, A. Procopio, C. Siciliano, and G. Sindona, *Tetrahedron Lett.*, **1996**, *37*, 1277.
54. N. Hossain, A. Papchikhin, J. Plavec, and J. Chattopadhyaya, *Tetrahedron*, **1993**, *49*, 10133.
55. J. Rong, P. Roselt, J. Plavec, and J. Chattopadhyaya, *Tetrahedron*, **1994**, *50*, 4921.
56. J. M. J. Tronchet, R. Benhamza, N. Dolatshahi, M. Geoffroy, and H. Türler, *Nucleosides Nucleotides*, **1988**, *7*, 249.
57. J. M. J. Tronchet, R. Benhamza, G. Bernardinelli, and M. Geoffroy, *Tetrahedron Lett.*, **1990**, *31*, 531.
58. J. M. J. Tronchet, M. Zsély, K. Capek, and F. de Villedon de Nadie, *Bioorg. Med. Chem. Lett.*, **1992**, *2*, 1723.
59. J. M. J. Tronchet, R. Benhamza, and G. Bernardinelli, *Nucleosides Nucleotides*, **1993**, *12*, 55.
60. T. Kametani, T. Nagahara, and T. Honda, *J. Org. Chem.*, **1985**, *50*, 2327.
61. T. Kametani, S. D. Chu, and T. Honda, *Heterocycles*, **1987**, *25*, 241.
62. T. Kametani, S. D. Chu, and T. Honda, *J. Chem. Soc., Perkin Trans. 1*, **1988**, 1593.
62a. D. H. Shih, F. Baker, L. Cama, and B. G. Christensen, *Heterocycles*, **1984**, *21*, 29.
63. Y. Ito, Y. Kimura, and S. Terashima, *Bull. Chem. Soc. Jpn.*, **1987**, *60*, 3337.
64. M. Ihara, M. Takahashi, K. Fukumoto, and T. Kametani, *Heterocycles*, **1988**, *27*, 327.

65. M. Ihara, M. Takahashi, K. Fukumoto, and T. Kametani, *J. Chem. Soc., Chem. Commun.*, **1988**, 9.

66. S. H. Kang and H. S. Lee, *Tetrahedron Lett.*, **1995**, *36*, 6713.

67. M. E. Jung and B. T. Vu, *J. Org. Chem.*, **1996**, *61*, 4427.

68. B. Alcaide and E. Saez, *Tetrahedron Lett.*, **2000**, *41*, 1647.

69. B. Alcaide, J. M. Alonso, M. F. Aly, E. Saez, M. P. Martinez-Alcazar, and F. Hernandez-Cano, *Tetrahedron Lett.*, **1999**, *40*, 5391.

70. I. Panfil, C. Belzecki, Z. Urbanczyk-Lipkowska, and M. Chmielewski, *Tetrahedron*, **1991**, *47*, 10087.

71. I. Panfil, C. Belzecki, and M. Chmielewski, *J. Carbohydr. Chem.*, **1987**, *6*, 463.

72. I. Panfil, C. Belzecki, M. Chmielewski, and K. Suwinska, *Tetrahedron*, **1989**, *45*, 233.

73. I. Panfil, M. Chmielewski, and C. Belzecki, *Heterocycles*, **1986**, *24*, 1609.

74. S. Rigolet, J. M. Mélot, J. Vébrel, A. Chiaroni, and C. Riche, *J. Chem. Soc., Perkin Trans. 1*, **2000**, 1095.

75. S. Rigolet, P. Goncalo, J. M. Melot, and J. Vebrel, *J. Chem. Res. (S)*, **1998**, 686.

76. For a review of cycloadditions under microwave irradiation see: A. de la Hoz, A. Díaz-Ortis, A. Moreno, and F. Langa, *Eur. J. Org. Chem.*, **2000**, 3659.

77. P. Oravec, L. Fisera, I. Goljer, and P. Ertl, *Monatsh. Chem.*, **1991**, *122*, 977.

78. R. L. Funk and J. U. Daggett, *Heterocycles*, **1987**, *26*, 2175.

79. N. Langlois, N. Van Bac, N. Dahuron, J.-M. Delcroix, A. Deyine, D. Griffart-Brunet, A. Chiaroni, and C. Riche, *Tetrahedron*, **1995**, *51*, 3571.

80. N. Langlois and F. Rakotondradany, *Tetrahedron*, **2000**, *56*, 2437.

81. J. D. White, R. A. Badger, H. S. Kezar III, A. J. Pallenberg, and G. S. Shiehser, *Tetrahedron*, **1989**, *45*, 6631.

82. U. Chiacchio, G. Buemi, F. Casuscelli, A. Procopio, A. Rescifina, and R. Romeo, *Tetrahedron*, **1994**, *50*, 5503.

83. F. M. Cordero, M. Gensini, A. Goti, and A. Brandi, *Org. Lett.*, **2000**, *2*, 2475.

84. O. Tamura, S. Yoshida, H. Sugita, N. Mita, Y. Uyama, N. Morita, M. Ishiguro, T. Kawasaki, H. Ishibashi, and M. Sakamoto, *Synlett*, **2000**, 1553.

85. O. Tamura, N. Mita, T. Okabe, T. Yamaguchi, C. Fukushima, M. Yamashita, Y. Morita, N. Morita, H. Ishibashi, and M. Sakamoto, *J. Org. Chem.*, **2001**, *66*, 2602.

86. G. Broggini and G. Zecchi, *Synthesis*, **1999**, 905.

87. A. E. McCaig, K. P. Meldrum, and R. H. Wightman, *Tetrahedron*, **1998,** *54*, 9429.

88. A. E. McCaig and R. H. Wightman, *Tetrahedron Lett.*, **1993**, *34*, 3939.

89. S. Cicchi, P. Ponzuoli, A. Goti, and A. Brandi, *Tetrahedron Lett.*, **2000**, *41*, 1583.

90. S. Cicchi, I. Hold, and A. Brandi, *J. Org. Chem.*, **1993**, *58*, 5274.

91. S. Cicchi, A. Goti, and A. Brandi, *J. Org. Chem.*, **1995**, *60*, 4743.

92. S. Cicchi, J. Nunes, A. Goti, and A. Brandi, *Eur. J. Org. Chem.*, **1998,** 419.

93. S. Cicchi, F. Cardona, A. Brandi, M. Corsi, and A. Goti, *Tetrahedron Lett.*, **1999**, *40*, 1989.

94. S. Cicchi, M. Corsi, and A. Goti, *J. Org. Chem.*, **1999**, *64*, 7243.

95. A. Goti, M. Cacciarini, F. Cardona, and A. Brandi, *Tetrahedron Lett.*, **1999**, *40*, 2853.

96. A. Goti, S. Cicchi, M. Cacciarini, F. Cardona, V. Fedi, and A. Brandi, *Eur. J. Org. Chem.*, **2000**, 3633.

97. F. M. Cordero, I. Barile, F. De Sarlo, and A. Brandi, *Tetrahedron Lett.*, **1999**, *40*, 6657.

98. C. H. Tan, T. Stork, N. Feeder, and A. B. Holmes, *Tetrahedron Lett.*, **1999**, *40*, 1397.

99. I. Collins, A. Nadin, A. B. Holmes, M. E. Long, J. Man, and R. Baker, *J. Chem. Soc., Perkin Trans. 1*, **1994**, 2205.

100. A. B. Holmes, A. B. Hughes, and A. L. Smith, *J. Chem. Soc., Perkin Trans. 1*, **1993**, 633.

101. G. M. Williams, S. D. Roughley, J. E. Davies, A. B. Holmes, and J. P. Adams, *J. Am. Chem. Soc.*, **1999**, *121*, 4900.
102. M. Ito and C. Kibayashi, *Tetrahedron Lett.*, **1990**, *31*, 5065.
103. M. Ito and C. Kibayashi, *Tetrahedron*, **1991**, *47*, 9329.
104. R. W. Hoffmann and A. Endesfelder, *Liebigs Ann. Chem.*, **1986**, 1823.
105. F. M. Cordero, S. Cicchi, A. Goti, and A. Brandi, *Tetrahedron Lett.*, **1994**, *35*, 949.
106. F. M. Cordero, A. Brandi, C. Querci, A. Goti, F. De Sarlo, and A. Guarna, *J. Org. Chem.*, **1990**, *55*, 1762.
107. A. Brandi, F. M. Cordero, F. De Sarlo, A. Goti, and A. Guarna, *Synlett*, **1993**, 1.
108. A. Brandi, Y. Dürüst, F. M. Cordero, and F. De Sarlo, *J. Org. Chem.*, **1992**, *57*, 5666.
109. A. Brandi, F. M. Cordero, F. De Sarlo, R. Gandolfi, A. Rastelli, and M. Bagatti, *Tetrahedron*, **1992**, *48*, 3323.
110. A. Brandi, A. Guarna, A. Goti, and F. DeSarlo, *Tetrahedron Lett.*, **1986**, *27*, 1727.
111. A. Brandi, S. Carli, and A. Goti, *Heterocycles*, **1988**, *27*, 17.
112. A. Brandi, S. Cicchi, F. M. Cordero, R. Frignoli, A. Goti, S. Picasso, and P. Vogel, *J. Org. Chem.*, **1995**, *60*, 6806.
113. A. Goti, B. Anichini, A. Brandi, S. Kozhushkov, C. Gratkowski, and A. de Meijere, *J. Org. Chem.*, **1996**, *61*, 1665.
114. B. Anichini, A. Goti, A. Brandi, S. I. Kozhushkov, and A. de Meijere, *Chem. Commun.*, **1997**, 261.
115. B. Anichini, A. Goti, A. Brandi, S. I. Kozhushkov, and A. de Meijere, *Synlett*, **1997**, 25.
116. A. Brandi, A. Goti, S. Kozhushkov, and A. Demeijere, *J. Chem. Soc., Chem. Commun.*, **1994**, 2185.
117. For the use of **80** in 1,3-dipolar cycloadditions see: A. de Meijere and S. I. Kozhushkov, *Eur. J. Org. Chem.*, **2000**, 3809.
118. C. Zorn, B. Anichini, A. Goti, A. Brandi, S. I. Kozhushkov, A. de Meijere, and L. Citti, *J. Org. Chem.*, **1999**, *64*, 7846.
119. C. Zorn, A. Goti, A. Brandi, K. Johnsen, M. Noltemeyer, S. I. Kozhushkov, and A. de Meijere, *J. Org. Chem.*, **1999**, *64*, 755.
120. C. Zorn, A. Goti, A. Brandi, K. Johnsen, S. I. Kozhushkov, and A. de Meijere, *Chem. Commun.*, **1998**, 903.
121. F. M. Cordero, B. Anichini, A. Goti, and A. Brandi, *Tetrahedron*, **1993**, *49*, 9867.
122. A. Goti, F. Cardona, A. Brandi, S. Picasso, and P. Vogel, *Tetrahedron: Asymmetry*, **1996**, *7*, 1659.
123. F. Machetti, F. M. Cordero, F. De Sarlo, A. Guarna, and A. Brandi, *Tetrahedron Lett.*, **1996**, *37*, 4205.
124. A. Goti, A. Brandi, F. De Sarlo, and A. Guarna, *Tetrahedron*, **1992**, *48*, 5283.
125. A. Goti, *Tetrahedron*, **1996**, *52*, 9187.
126. F. M. Cordero, A. Goti, F. De Sarlo, A. Guarna, and A. Brandi, *Tetrahedron*, **1989**, *45*, 5917.
127. A. Goti, A. Brandi, F. De Sarlo, and A. Guarna, *Tetrahedron Lett.*, **1986**, *27*, 5271.
128. A. Padwa, S. P. Carter, U. Chiacchio, and D. N. Kline, *Tetrahedron Lett.*, **1986**, *27*, 2683.
129. A. Padwa, D. N. Kline, K. F. Koehler, M. Matzinger, and M. K. Venkatramanan, *J. Org. Chem.*, **1987**, *52*, 3909.
130. A. Padwa, M. Matzinger, Y. Tomioka, and M. K. Venkatramanan, *J. Org. Chem.*, **1988**, *53*, 955.
131. A. Padwa, D. N. Kline, and B. H. Norman, *Tetrahedron Lett.*, **1988**, *29*, 265.
132. A. Padwa, B. H. Norman, and J. Perumattam, *Tetrahedron Lett.*, **1989**, *30*, 663.
133. A. Padwa, D. N. Kline, and B. H. Norman, *J. Org. Chem.*, **1989**, *54*, 810.
134. A. Padwa, W. H. Bullock, D. N. Kline, and J. Perumattam, *J. Org. Chem.*, **1989**, *54*, 2862.
135. K. Kasahara, H. Iida, and C. Kibayashi, *J. Org. Chem.*, **1989**, *54*, 2225.
136. J. Wityak, *J. Org. Chem.*, **1987**, *52*, 5500.

137. J. Wityak, S. J. Gould, S. J. Hein, and D. A. Keszler, *J. Org. Chem.*, **1987**, *52*, 2179.

138. W. J. Krol, S. Mao, D. L. Steele, and C. A. Townsend, *J. Org. Chem.*, **1991**, *56*, 728.

139. D. Keirs, D. Moffat, and K. Overton, *J. Chem. Soc., Chem. Commun.*, **1988**, 654.

140. D. Keirs, D. Moffat, K. Overton, and R. Tomanek, *J. Chem. Soc., Perkin Trans. 1*, **1991**, 1041.

141. D. Keirs and K. Overton, *Heterocycles*, **1989**, *28*, 841.

142. S. Mzengeza and R. A. Whitney, *J. Chem. Soc., Chem. Commun.*, **1984**, 606.

143. S. Mzengeza, C. M. Yang, and R. A. Whitney, *J. Am. Chem. Soc.*, **1987**, *109*, 276.

144. S. Mzengeza and R. A. Whitney, *J. Org. Chem.*, **1988**, *53*, 4074.

145. O. Tamura, T. Yamaguchi, T. Okabe, and M. Sakamoto, *Synlett*, **1994**, 620.

146. O. Tamura, T. Yamaguchi, K. Noe, and M. Sakamoto, *Tetrahedron Lett.*, **1993**, *34*, 4009.

147. O. Tamura, T. Okabe, T. Yamaguchi, K. Gotanda, K. Noe, and M. Sakamoto, *Tetrahedron*, **1995**, *51*, 107.

148. O. Tamura, T. Okabe, T. Yamaguchi, J. Kotani, K. Gotanda, and M. Sakamoto, *Tetrahedron*, **1995**, *51*, 119.

149. G. Benz, R. Henning, and J.-P. Stasch, *Angew. Chem., Int. Ed. Engl.*, **1991**, *30*, 1702.

150. S. Valenza, F. M. Cordero, A. Brandi, A. Guidi, M. Altamura, A. Giolitti, F. Giuntini, F. Pasqui, A. R. Renzetti, and C. A. Maggi, *J. Org. Chem.*, **2000**, *65*, 4003.

151. N. Katagiri, M. Okada, C. Kaneko, and T. Furuya, *Tetrahedron Lett.*, **1996**, *37*, 1801.

152. P. Merino, S. Anoro, F. L. Merchan, and T. Tejero, *Molecules*, **2000**, *5*, 132.

153. P. Merino, S. Anoro, S. Franco, F. L. Merchan, T. Tejero, and V. Tunon, *J. Org. Chem.*, **2000**, *65*, 1590.

154. P. H. H. Hermkens, T. G. van Dinther, C. W. Juokema, G. N. Wagenaars, and H. C. J. Ottenheijm, *Tetrahedron Lett.*, **1994**, *35*, 9271.

155. A. Peer and A. Vasella, *Helv. Chim. Acta*, **1999**, *82*, 1044.

156. S. Majumdar, A. Bhattacharjya, and A. Patra, *Tetrahedron*, **1999**, *55*, 12157.

157. G. V. M. Sharma, I. S. Reddy, V. G. Reddy, and A. V. R. Rao, *Tetrahedron: Asymmetry*, **1999**, *10*, 229.

158. C. W. Holzapfel and R. Crous, *Heterocycles*, **1998**, *48*, 1337.

159. S. D. Jiang, K. J. McCullough, B. Mekki, G. Singh, and R. H. Wightman, *J. Chem. Soc., Perkin Trans. 1*, **1997**, 1805.

160. M. Yokoyama, K. Sujino, M. Irie, N. Yamazaki, T. Hiyama, N. Yamada, and H. Togo, *J. Chem. Soc., Perkin Trans. 1*, **1991**, 2801.

161. P. Gebarowski and W. Sas, *Chem. Commun.*, **2001**, 915.

162. V. Jager, L. Bierer, H.-Q. Dong, A. M. Palmer, D. Shaw, and W. Frey, *J. Heterocycl. Chem.*, **2000**, *37*, 455.

163. S. Baskaran and G. K. Trivedi, *J. Chem. Res. (S)*, **1996**, 126.

164. S. Baskaran and G. K. Trivedi, *J. Chem. Res. (S)*, **1995**, 308.

165. P. Hammann and G. Kretzschmar, *Tetrahedron*, **1990**, *46*, 5603.

166. M. Koszytkowska-Stawinska and W. Sas, *J. Chem. Res. (S)*, **1998**, 298.

167. M. B. Gravestock, D. W. Knight, J. S. Lovell, and S. R. Thornton, *J. Chem. Soc., Perkin Trans. 1*, **1999**, 3143.

168. P. DeShong, W. Li, J. W. Kennington, Jr., and H. L. Ammon, *J. Org. Chem.*, **1991**, *56*, 1364.

169. B. Bernet, E. Krawczyk, and A. Vasella, *Helv. Chim. Acta*, **1985**, *68*, 2299.

170. J. Kuban, I. Blanarikova, L. Fisera, L. Jaroskova, M. Fengler-Veith, V. Jager, J. Kozisek, O. Humpa, N. Pronayova, and V. Langer, *Tetrahedron*, **1999**, *55*, 9501.

171. L. Fisera, V. Ondrus, J. Kuban, P. Micuch, I. Blanarikova, and V. Jager, *J. Heterocycl. Chem.*, **2000**, *37*, 551.

172. P. Deshong and J. M. Leginus, *Tetrahedron Lett.*, **1984**, *25*, 5355.
173. P. DeShong, C. M. Dicken, J. M. Leginus, and R. R. Whittle, *J. Am. Chem. Soc.*, **1984**, *106*, 5598.
174. M. Jurczak, J. Rabiczko, D. Socha, M. Chmielewski, F. Cardona, A. Goti, and A. Brandi, *Tetrahedron: Asymmetry*, **2000**, *11*, 2015.
175. I. Panfil and M. Chmielewski, *Tetrahedron*, **1985**, *41*, 4713.
176. A. Banerji and S. Basu, *Tetrahedron*, **1992**, *48*, 3335.
177. V. Ondrus, M. Orsag, L. Fisera, and N. Pronayova, *Tetrahedron*, **1999**, *55*, 10425.
178. J. J. Tufariello, G. E. Lee, P. A. Senaratne, and M. Al-Nuri, *Tetrahedron Lett.*, **1979**, *45*, 4359.
179. O. Dirat, C. Kouklovsky, Y. Langlois, P. Lesot, and J. Courtieu, *Tetrahedron: Asymmetry*, **1999**, *10*, 3197.
180. P. Borrachero, F. Cabrera, M. J. Dianez, M. D. Estrada, M. Gomez-Guillen, A. Lopez-Castro, J. M. Moreno, J. L. de Paz, and S. Perez-Garrido, *Tetrahedron: Asymmetry*, **1999**, *10*, 77.
181. F. J. Duff, V. Vivien, and R. H. Wightman, *Chem. Commun.*, **2000**, 2127.
182. F. Cardona, S. Valenza, S. Picasso, A. Goti, and A. Brandi, *J. Org. Chem.*, **1998**, *63*, 7311.
183. F. Cardona, S. Valenza, A. Goti, and A. Brandi, *Eur. J. Org. Chem.*, **1999**, 1319.
184. F. Cardona, P. Salanski, M. Chmielewski, S. Valenza, A. Goti, and A. Brandi, *Synlett*, **1998**, 1444.
185. J. M. J. Tronchet and E. Mihaly, *Helv. Chim. Acta*, **1972**, *55*, 1266.
186. S. Datta, P. Chattopadhyay, R. Mukhopadhyay, and A. Bhattacharjya, *Tetrahedron Lett.*, **1993**, *34*, 3585.
187. A. Bhattacharjya, P. Chattopadhyay, A. T. McPhail, and D. R. McPhail, *J. Chem. Soc., Chem. Commun.*, **1990**, 1508.
188. P. M. Collins, M. S. Ashwood, H. Eder, S. H. B. Wright, and D. J. Kennedy, *Tetrahedron Lett.*, **1990**, *31*, 2055.
189. T. K. M. Shing, W.-C. Fung, and C.-H. Wong, *J. Chem. Soc., Chem. Commun.*, **1994**, 449.
190. A. Bhatacharjee, A. Bhattacharjya, and P. Amarendra, *Tetrahedron Lett.*, **1995**, *36*, 4677.
191. T. K. M. Shing, Y.-L. Zhong, T. C. W. Mak, R. Wang, and F. Xue, *J. Org. Chem.*, **1998**, *63*, 414.
192. T. K. M. Shing and Y.-L. Zhong, *Tetrahedron*, **2001**, *57*, 1573.
193. T. K. M. Shing, C.-H. Wong, and T. Yip, *Tetrahedron: Asymmetry*, **1996**, *7*, 1323.
194. P. Bravo, L. Bruche, A. Farina, G. Fronza, S. V. Meille, and A. Merli, *Tetrahedron: Asymmetry*, **1993**, *4*, 2131.
195. H. Tsuge, T. Okana, S. Eguchi, and H. Kimoto, *J. Chem. Soc., Perkin Trans. 1*, **1997**, 1581.
196. C. Louis and C. Hootele, *Tetrahedron: Asymmetry*, **1995**, *6*, 2149.
197. T. Takahashi, A. Fujii, J. Sugita, T. Hagi, K. Kitano, Y. Arai, T. Koizumi, and M. Shiro, *Tetrahedron: Asymmetry*, **1991**, *2*, 1379.
198. V. K. Aggarwal, R. S. Grainger, H. Adams, and P. L. Spargo, *J. Org. Chem.*, **1998**, *63*, 3481.
199. S.-S. P. Chou and Y.-J. Yu, *Tetrahedron Lett.*, **1997**, *38*, 4803.
200. O. Fourets, P. Cauliez, and J. Simonet, *Tetrahedron Lett.*, **1998**, *39*, 565.
201. T. Tejero, A. Dondoni, I. Rojo, F. L. Merchan, and P. Merino, *Tetrahedron*, **1997**, *53*, 3301.
202. H. G. Aurich, M. Boutahar, H. Köster, K.-D. Möbus, and L. Ruiz, *Chem. Ber.*, **1990**, *123*, 1999.
203. R. Annunziata, M. Cinquini, F. Cozzi, P. Giaroni, and L. Raimondi, *Tetrahedron: Asymmetry*, **1990**, *1*, 251.
204. A. Brandi, S. Cicchi, A. Goti, K. M. Pietrusiewicz, M. Zablocka, and W. Wisniewski, *J. Org. Chem.*, **1991**, *56*, 4383.
205. K. M. Pietrusiewicz, W. Holody, M. Koprowski, S. Cicchi, A. Goti, and A. Brandi, *Phosphorus Sulfur Silicon Relat. Elem.*, **1999**, *144–146*, 389.
206. K. V. Gothelf and K. A. Jorgensen, *Chem. Rev.*, **1998**, *98*, 863.

207. O. Tamura, T. Kuroki, Y. Sakai, J. Takizawa, J. Yoshino, Y. Morita, N. Mita, K. Gotanda, and M. Sakamoto, *Tetrahedron Lett.*, **1999**, *40*, 895.

208. S. Kanemasa and T. Tsuruoka, *Chem. Lett.*, **1995**, 49.

209. S. Kanemasa, T. Tsuruoka, and E. Wada, *Tetrahedron Lett.*, **1993**, *34*, 87.

210. P. Bayon, P. de March, M. Figueredo, and J. Font, *Tetrahedron*, **1998**, *54*, 15691.

211. S. Kanemasa, T. Tsuruoka, and H. Yamamoto, *Tetrahedron Lett.*, **1995**, *36*, 5019.

212. S. Kanemasa, T. Uemura, and E. Wada, *Tetrahedron Lett.*, **1992**, *33*, 7889.

213. A. D. Jones, D. W. Knight, and S. R. Thornton, *J. Chem. Soc., Perkin Trans. 1*, **1999**, 3337.

214. G. Desimoni, G. Faita, M. Mella, P. Righetti, and M. Zema, *Tetrahedron*, **1999**, *55*, 8509.

215. K. V. Gothelf and K. A. Jorgensen, *J. Org. Chem.*, **1994**, *59*, 5687.

216. P. J. Dunn, A. B. Graham, R. Grigg, P. Higginson, and I. S. Saba, *Chem. Commun.*, **2000**, 2035.

217. G. Desimoni, G. Faita, A. Mortoni, and P. Righetti, *Tetrahedron Lett.*, **1999**, *40*, 2001.

218. S. Crosignani, G. Desimoni, G. Faita, S. Filippone, A. Mortoni, P. Righetti, and M. Zema, *Tetrahedron Lett.*, **1999**, *40*, 7007.

219. K. V. Gothelf, R. G. Hazell, and K. A. Jorgensen, *J. Org. Chem.*, **1998**, *63*, 5483.

220. K. B. Jensen, R. G. Hazell, and K. A. Jorgensen, *J. Org. Chem.*, **1999**, *64*, 2353.

221. K. B. Jensen, K. V. Gothelf, R. G. Hazell, and K. A. Jorgensen, *J. Org. Chem.*, **1997**, *62*, 2471.

222. K. V. Gothelf, I. Thomsen, and K. A. Jorgensen, *J. Am. Chem. Soc.*, **1996**, *118*, 59.

223. S. Kanemasa, Y. Oderaotoshi, J. Tanaka, and E. Wada, *J. Am. Chem. Soc.*, **1998**, *120*, 12355.

224. G. Wagner, A. J. L. Pombeiro, and V. Y. Kukushkin, *J. Am. Chem. Soc.*, **2000**, *122*, 3106.

225. K. B. Simonsen, P. Bayon, R. G. Hazell, K. V. Gothelf, and K. A. Jorgensen, *J. Am. Chem. Soc.*, **1999**, *121*, 3845.

226. K. B. Jensen, M. Roberson, and K. A. Jorgensen, *J. Org. Chem.*, **2000**, *65*, 9080.

227. K. B. Simonsen, K. A. Jorgensen, Q. S. Hu, and L. Pu, *Chem. Commun.*, **1999**, 811.

228. S. Kobayashi and M. Kawamura, *J. Am. Chem. Soc.*, **1998**, *120*, 5840.

229. H. Kodama, J. Ito, K. Hori, T. Ohta, and I. Furukawa, *J. Organomet. Chem.*, **2000**, *603*, 6.

230. M. Kawamura and S. Kobayashi, *Tetrahedron Lett.*, **1999**, *40*, 3213.

231. O. Tamura, N. Mita, K. Gotanda, K. Yamada, T. Nakano, R. Katagiri, and M. Sakamoto, *Heterocycles*, **1997**, *46*, 95.

232. S. Minakata, T. Ezoe, K. Nakamura, I. Ryu, and M. Komatsu, *Tetrahedron Lett.*, **1998**, *39*, 5205.

233. S. Kobayashi and R. Akiyama, *Tetrahedron Lett.*, **1998**, *39*, 9211.

234. C. Camiletti, D. D. Dhavale, L. Gentilucci, and C. Trombini, *J. Chem. Soc., Perkin Trans. 1*, **1993**, 3157.

235. P. G. M. Wuts and Y.-W. Jung, *J. Org. Chem.*, **1988**, *53*, 1957.

236. K. Hori, J. Ito, T. Ohta, and I. Furukawa, *Tetrahedron*, **1998**, *54*, 12737.

237. K. Hori, H. Kodama, T. Ohta, and I. Furukawa, *J. Org. Chem.*, **1999**, *64*, 5017.

238. C. Baldoli, P. Del Buttero, E. Licandro, S. Maiorana, and A. Papagni, *Tetrahedron: Asymmetry*, **1995**, *6*, 1711.

239. C. Mukai, W. J. Cho, I. J. Kim, and M. Hanaoka, *Tetrahedron Lett.*, **1990**, *31*, 6893.

240. C. Mukai, I. J. Kim, W. J. Cho, M. Kido, and M. Hanaoka, *J. Chem. Soc., Perkin Trans. 1*, **1993**, 2495.

241. M. Freccero, R. Gandolfi, and M. Sarzi-Amade, *Heterocycles*, **1998**, *49*, 415.

242. For a review of piperidine syntheses see: S. Laschat and T. Dickner, *Synthesis*, **2000**, 1781.

243. T. Naito, M. Ikai, M. Shirakawa, K. Fujimoto, I. Ninomiya, and T. Kiguchi, *J. Chem. Soc., Perkin Trans. 1*, **1994**, 773.

244. T. Kiguchi, M. Shirakawa, R. Honda, I. Ninomiya, and T. Naito, *Tetrahedron*, **1998**, *54*, 15589.

245. A. Budzínska, M. Bukowska, and W. Sas, *Tetrahedron Lett.*, **1999**, *40*, 565.
246. A. Budzínska and W. Sas, *Tetrahedron*, **2001**, *57*, 2021.
247. R. E. Looper and R. M. Williams, *Tetrahedron Lett.*, **2001**, *42*, 769.
248. J. J. Tufariello and J. M. Puglis, *Tetrahedron Lett.*, **1986**, *27*, 1265.
249. M. Shimazaki, F. Okazaki, F. Nakajima, T. Ishikawa, and A. Ohta, *Heterocycles*, **1993**, *36*, 1823.
250. W. Oppolzer, J. Deerberg, and O. Tamura, *Helv. Chim. Acta*, **1994**, *77*, 554.
251. W. Carruthers, P. Coggins, and J. B. Weston, *J. Chem. Soc., Chem. Commun.*, **1991**, 117.
252. V. Mancuso and C. Hootelé, *Tetrahedron Lett.*, **1988**, *29*, 5917.
253. M. Ito, M. Maeda, and C. Kibayashi, *Tetrahedron Lett.*, **1992**, *33*, 3765.
254. H. Ina, M. Ito, and C. Kibayashi, *J. Org. Chem.*, **1996**, *61*, 1023.
255. D. R. Adams, W. Carruthers, and P. J. Crowley, *J. Chem. Soc., Chem. Commun.*, **1991**, 1261.
256. A. Padwa, "1,3-Dipolar Cycloaddition Chemistry" in *1,3-Dipolar Cycloaddition Chemistry*, Vol. 2, A. Padwa ed., Wiley-Interscience, New York, **1984**, p. 277.
257. U. Chiacchio, A. Rescifina, F. Casuscelli, A. Piperno, V. Pisani, and R. Romeo, *Tetrahedron*, **1996**, *52*, 14311.
258. N. A. LeBel and J. J. Whang, *J. Am. Chem. Soc.*, **1959**, *81*, 6334.
259. N. A. LeBel and N. Balasubramanian, *J. Am. Chem. Soc.*, **1989**, *111*, 3363.
260. P. G. M. Wuts and Y.-W. Jung, *J. Org. Chem.*, **1988**, *53*, 5989.
261. B. B. Snider and H. Lin, *J. Am. Chem. Soc.*, **1999**, *121*, 7778.
262. S. Lee and Z. S. Zhao, *Tetrahedron Lett.*, **1999**, *40*, 7921.
263. R. Grigg, J. Markandu, S. Surendrakumar, M. Thornton-Pett, and W. J. Warnock, *Tetrahedron*, **1992**, *48*, 10399.
264. J. D. White, P. R. Blakemore, E. A. Korf, and A. F. T. Yokochi, *Org. Lett.*, **2001**, *3*, 413.
265. R. Grigg, M. Hadjisoteriou, P. Kennewell, J. Markandu, and M. Thornton-Pett, *J. Chem. Soc., Chem. Commun.*, **1992**, 1388.
266. R. Grigg, M. Hadjisoteriou, P. Kennewell, and J. Markandu, *J. Chem. Soc., Chem. Commun.*, **1992**, 1537.
267. M. Frederickson, R. Grigg, J. Markandu, and J. Redpath, *J. Chem. Soc., Chem. Commun.*, **1994**, 2225.
268. R. Grigg and J. Markandu, *Tetrahedron Lett.*, **1989**, *30*, 5489.
269. R. A. Stockman, *Tetrahedron Lett.*, **2000**, 9163.
270. K. M. Werner, J. M. de los Santos, S. M. Weinreb, and M. Y. Shang, *J. Org. Chem.*, **1999**, *64*, 4865.
271. K. M. Werner, J. M. de los Santos, S. M. Weinreb, and M. Y. Shang, *J. Org. Chem.*, **1999**, *64*, 686.
271a. W. Oppolzer and C. G. Bochet, *Tetrahedron: Asymmetry*, **2000**, *11*, 4761.
272. K. Nagasawa, A. Georgieva, and T. Nakata, *Tetrahedron*, **2000**, *56*, 187.
273. J. A. Ponasik and B. Ganem, *Tetrahedron Lett.*, **1995**, *36*, 9109.
274. W. R. Roush and A. E. Walts, *J. Org. Chem.*, **1984**, *106*, 721.
275. W. Carruthers, P. Coggins, and J. B. Weston, *J. Chem. Soc., Perkin Trans. 1*, **1990**, 2323.
276. W. Carruthers, P. Coggins, and J. B. Weston, *J. Chem. Soc., Chem. Commun.*, **1990**, 91.
277. W. Carruthers, P. Coggins, and J. B. Weston, *J. Chem. Soc., Perkin Trans. 1*, **1991**, 611.
278. S. A. Ali and M. I. M. Wazeer, *Tetrahedron*, **1993**, *49*, 4339.
279. H. Ooi, A. Urushibara, T. Esumi, Y. Iwabuchi, and S. Hatakeyama, *Org. Lett.*, **2001**, *3*, 953.
280. S. W. Baldwin and J. S. Debenham, *Org. Lett.*, **2000**, *2*, 99.
281. S. W. Baldwin, J. Aubé, and A. T. McPhail, *J. Org. Chem.*, **1991**, *56*, 6546.
282. Y. Takeuchi and F. Furusaki, "The Chemistry of Isoxazolidines", in *Advances in Heterocyclic Chemistry*, Vol. 21, A. R. Katritzky and A. J. Boulton eds., Academic Press, New York, **1977**, pp. 207–251.

283. J. S. A. Lang and Y.-I. Lin, "Isoxazolidines", in *Comprehensive Heterocyclic Chemistry*, Vol. 6, A. R. Katritzky and K. T. Potts eds., Pergamon, Oxford, UK, **1984**, p. 45.

284. M. Ochiai, M. Obayashi, and K. Morita, *Tetrahedron*, **1967**, *23*, 2641.

285. A. Lablache-Combier and M. L. Villaume, *Tetrahedron*, **1968**, *24*, 6951.

286. R. Grigg, M. Jordan, A. Tangthongkum, F. W. B. Einstein, and T. Jones, *J. Chem. Soc., Chem. Commun.*, **1984**, 47.

287. G. Donegan, R. Grigg, F. Heaney, S. Surendrakumar, and W. J. Warnock, *Tetrahedron Lett.*, **1989**, *30*, 609.

288. H. O. House, D. T. Manning, D. G. Melillo, L. F. Lee, O. R. Haynes, and B. E. Wilkes, *J. Org. Chem.*, **1976**, *41*, 855.

289. H. O. House and L. F. Lee, *J. Org. Chem.*, **1976**, *41*, 863.

290. E. Ciganek, *J. Org. Chem.*, **1990**, *55*, 3007.

291. E. Ciganek, *J. Org. Chem.*, **1995**, *60*, 5803.

292. E. Ciganek, J. M. Read, and J. C. Calabrese, *J. Org. Chem.*, **1995**, *60*, 5795.

293. For recent use of the reverse-Cope elimination in synthesis see: M. B. Gravestock, D. W. Knight, K. M. A. Malik, and S. R. Thornton *J. Chem. Soc., Perkin Tran. 1*, **2000**, 3292. J. R. Hanrahan and D. W. Knight, *Chem. Commun.*, **1998**, 2231.

294. For a review of the Cope Elimination see: A. C. Cope and E. R. Trumbull, *Organic Reactions*, **1960**, 11, 317.

295. R. Bishop, P. R. Brooks, and S. C. Hawkins, *Synthesis*, **1988**, 997.

296. R. Grigg, J. Markandu, T. Perrior, S. Surendrakumar, and W. J. Warnock, *Tetrahedron*, **1992**, *48*, 6929.

297. R. Grigg and S. Thianpantangul, *J. Chem. Soc., Perkin Trans. 1*, **1984**, 653.

298. D. Lathbury and T. Gallagher, *Tetrahedron Lett.*, **1985**, *26*, 6249.

299. D. Lathbury and T. Gallagher, *J. Chem. Soc., Chem. Commun.*, **1986**, 1017.

300. D. C. Lathbury, R. W. Shaw, P. A. Bates, M. B. Hursthouse, and T. Gallagher, *J. Chem. Soc., Perkin Trans. 1*, **1989**, 2415.

301. R. Shaw, D. Lathbury, M. Anderson, and T. Gallagher, *J. Chem. Soc., Perkin Trans. 1*, **1991**, 659.

302. R. Grigg, T. R. Perrior, G. J. Sexton, S. Surendrakumar, and T. Suzuki, *J. Chem. Soc., Chem. Commun.*, **1993**, 372.

303. S. K. Pradhan, K. G. Akamanchi, P. P. Divakaran, and P. M. Pradhan, *Heterocycles*, **1989**, *28*, 813.

304. W. Oppolzer, A. C. Spivey, and C. G. Bochet, *J. Am. Chem. Soc.*, **1994**, *116*, 3139.

305. A. Padwa and G. S. K. Wong, *J. Org. Chem.*, **1986**, *51*, 3125.

306. A. B. Holmes, A. L. Smith, S. F. Williams, L. R. Hughes, Z. Lidert, and C. Swithenbank, *J. Org. Chem.*, **1991**, *56*, 1393.

307. M. E. Fox, A. B. Holmes, I. T. Forbes, and M. Thompson, *Tetrahedron Lett.*, **1992**, *33*, 7421.

308. M. E. Fox, A. B. Holmes, I. T. Forbes, M. Thompson, and J. W. Ziller, *Tetrahedron Lett.*, **1992**, *33*, 7425.

309. M. E. Fox, A. B. Holmes, I. T. Forbes, and M. Thompson, *J. Chem. Soc., Perkin Trans. 1*, **1994**, 3379.

310. P. J. Dunn, A. B. Graham, R. Grigg, P. Higginson, and I. Saba, *Chem. Commun.*, **2000**, 2033.

311. E. C. Davison, I. T. Forbes, A. B. Holmes, and J. A. Warner, *Tetrahedron*, **1996**, *52*, 11601.

312. J. E. Baldwin, *J. Chem. Soc., Chem. Commun.*, **1976**, 734.

313. F. Heaney and S. Bourke, *J. Chem. Soc., Perkin Trans. 1*, **1998**, 955.

314. F. Heaney, S. Bourke, D. Cunningham, and P. McArdle, *J. Chem. Soc., Perkin Trans. 2*, **1998**, 547.

315. F. Heaney and C. O′Mahony, *J. Chem. Soc., Perkin Trans. 1*, **1998**, 341.

316. F. Heaney, C. Burke, D. Cunningham, and P. McArdle, *J. Chem. Soc., Perkin Trans. 1*, **2001**, 622.
317. S. Baskaran, H. G. Aurich, F. Biesemeier, and K. Harms, *J. Chem. Soc., Perkin Trans. 1*, **1998**, 3717.
318. S. Baskaran and H. G. Aurich, *Synlett*, **1998**, 277.
319. M. Frederickson and R. Grigg, *Org. Prep. Proced. Int.*, **1997**, *29*, 63.
320. M. Frederickson and R. Grigg, *Org. Prep. Proced. Int.*, **1997**, *29*, 33.
321. H. A. Dondas, R. Grigg, M. Hadjisoteriou, J. Markandu, P. Kennewell, and M. Thornton-Pett, *Tetrahedron*, **2001**, *57*, 1119.
322. R. Grigg, M. Hadjisoteriou, P. Kennewell, J. Markandu, and M. Thornton-Pett, *J. Chem. Soc., Chem. Commun.*, **1993**, 1340.
323. W. R. Bowman, R. V. Davies, A. M. Z. Slawin, G. S. Sohal, R. B. Titman, and D. J. Wilkins, *J. Chem. Soc., Perkin Trans. 1*, **1997**, 155.
324. M. Tiecco, L. Testaferri, M. Tingoli, L. Bagnoli, and F. Marini, *J. Chem. Soc., Perkin Trans. 1*, **1993**, 1989.
325. R. Grigg, J. Markandu, T. Perrior, Z. Qiong, and T. Suzuki, *J. Chem. Soc., Chem. Commun.*, **1994**, 1267.
326. H. A. Dondas, R. Grigg, M. Hadjisoteriou, J. Markandu, W. A. Thomas, and P. Kennewell, *Tetrahedron*, **2000**, *56*, 10087.
327. H. A. Dondas, R. Grigg, and C. S. Frampton, *Tetrahedron Lett.*, **1997**, *38*, 5719.
328. J. Markandu, H. A. Dondas, M. Frederickson, and R. Grigg, *Tetrahedron*, **1997**, *53*, 13165.
329. H. A. Dondas, M. Frederickson, R. Grigg, J. Markandu, and M. Thornton-Pett, *Tetrahedron*, **1997**, *53*, 14339.
330. M. Frederickson, R. Grigg, J. Markandu, M. Thornton-Pett, and J. Redpath, *Tetrahedron*, **1997**, *53*, 15051.
331. R. Grigg and J. Markandu, *Tetrahedron Lett.*, **1991**, *32*, 279.
332. M. R. Slabaugh and W. C. Wildman, *J. Org. Chem.*, **1971**, *36*, 3202.
333. S. Moutel and M. Shipman, *Synlett*, **1998**, 1333.
334. P. J. Dransfield, S. Moutel, M. Shipman, and V. Sik, *J. Chem. Soc., Perkin Trans. 1*, **1999**, 3349.
335. A. Hassner, R. Maurya, and E. Mesko, *Tetrahedron Lett.*, **1988**, *29*, 5315.
336. M. Noguchi, S. Nishimura, H. Fujii, and A. Kakehi, *Heterocycl. Commun.*, **1999**, *5*, 133.
337. M. Noguchi, H. Okada, S. Nishimura, Y. Yamagata, S. Takamura, M. Tanaka, A. Kakehi, and H. Yamamoto, *J. Chem. Soc., Perkin Trans. 1*, **1999**, 185.
338. M. J. Uddin, M. Kikuchi, K. Takedatsu, K. I. Arai, T. Fujimoto, J. Motoyoshiya, A. Kakehi, R. Iriye, H. Shirai, and I. Yamamoto, *Synthesis*, **2000**, 365.
339. Q. Cheng, W. Zhang, Y. Tagami, and T. Oritani, *J. Chem. Soc., Perkin Trans. 1*, **2001**, 452.
340. A. Padwa and B. H. Norman, *Tetrahedron Lett.*, **1988**, *29*, 2417.
341. A. Abiko, *Chem. Lett.*, **1995**, 357.
342. G. Broggini, F. Folcio, N. Sardone, M. Sonzogni, and G. Zecchi, *Tetrahedron: Asymmetry*, **1996**, *7*, 797.
343. A. Abiko, O. Moriya, S. A. Filla, and S. Masamune, *Angew. Chem., Int. Ed. Engl.*, **1995**, *34*, 793.
344. W. Oppolzer and K. Keller, *Tetrahedron Lett.*, **1970**, 1117.
345. W. Oppolzer and H. P. Weber, *Tetrahedron Lett.*, **1970**, 1121.
346. H. A. Abd El-Nabi, *Tetrahedron*, **2000**, *56*, 3013.
347. G. Broggini, F. Folcio, N. Sardone, and G. Zecchi, *Tetrahedron*, **1996**, *52*, 11849.
348. M. P. S. Ishar, G. Singh, K. Kumar, and R. Singh, *Tetrahedron*, **2000**, *56*, 7817.
349. A. K. Baruah, D. Prajapati, and J. S. Sandhu, *J. Chem. Soc., Perkin Trans. 1*, **1987**, 1995.
350. M. P. S. Ishar and K. Kumar, *Tetrahedron Lett.*, **1999**, *40*, 175.
351. A. Arnone, G. Broggini, L. Bruche, G. Molteni, and G. Zecchi, *J. Chem. Res. (S)*, **1998**, 188.

351a. S. Saito, T. Ishikawa, and T. Moriwake, *J. Org. Chem.*, **1994**, *59*, 4375.

352. S. W. Baldwin, R. B. McFadyen, J. Aubé, and J. D. Wilson, *Tetrahedron Lett.*, **1991**, *32*, 4431.

353. S. W. Baldwin and S. C. Gedon, *Synth. Commun.*, **1991**, *21*, 587.

354. G. K. Tranmer, P. Keech, and W. Tam, *Chem. Commun.*, **2000**, 863.

355. J. Marcus, J. Brussee, and A. van der Gen, *Eur. J. Org. Chem.*, **1998**, 2513.

356. H.-Y. Kang, Y. S. Cho, H. Y. Koh, and M. H. Chang, *Tetrahedron Lett.*, **1991**, *32*, 2779.

357. A. Arnone, M. Cavicchioli, A. Donadelli, and G. Resnati, *Tetrahedron*: *Asymmetry*, **1994**, *5*, 1019.

357a. S. Chandrasekhar, M. Ravindranath, B. S. Neela, S. Ramakumar, and M. A. Viswametra, *J. Chem. Res. (S)*, **1989**, 252.

358. H. G. Aurich, F. Biesemeier, and M. Boutahar, *Chem. Ber.*, **1991**, *124*, 2329.

359. H. G. Aurich and F. Biesemeier, *Synthesis*, **1995**, 1171.

360. H. G. Aurich, M. Geiger, C. Gentes, K. Harms, and H. Koster, *Tetrahedron*, **1998**, *54*, 3181.

361. H. G. Aurich, G. Frenzen, and C. Gentes, *Chem. Ber.*, **1993**, *126*, 787.

362. H. G. Aurich, F. Biesemeier, M. Geiger, and K. Harms, *Liebigs Ann./Rec.*, **1997**, 423.

363. R. Annunziata, M. Cinquini, F. Cozzi, and L. Raimondi, *Tetrahedron Lett.*, **1988**, *29*, 2881.

364. R. Annunziata, M. Cinquini, F. Cozzi, and L. Raimondi, *J. Org. Chem.*, **1990**, *55*, 1901.

365. R. Annunziata, M. Cinquini, F. Cozzi, and L. Raimondi, *Tetrahedron*, **1987**, *43*, 4051.

366. H. G. Aurich, M. Soeberdt, and K. Harms, *Tetrahedron*, **1999**, *55*, 1249.

367. H. G. Aurich and H. Köster, *Tetrahedron*, **1995**, *51*, 6285.

368. H. G. Aurich, C. Gentes, and K. Harms, *Tetrahedron*, **1995**, *51*, 10497.

369. H. G. Aurich and K.-D. Möbus, *Tetrahedron Lett.*, **1988**, *29*, 5755.

370. T. Ishikawa, T. Kudo, K. Shigemori, and S. Saito, *J. Am. Chem. Soc.*, **2000**, *122*, 7633.

371. M. Lalk, H. Reinke, and K. Peseke, *Eur. J. Org. Chem.*, **2001**, 759.

372. F. M. Cordero and A. Brandi, *Tetrahedron Lett.*, **1995**, *36*, 1343.

373. F. Pisaneschi, F. M. Cordero, A. Goti, R. Paugam, J. Ollivier, A. Brandi, and J. Salaun, *Tetrahedron*: *Asymmetry*, **2000**, *11*, 897.

374. M. Ferrara, F. M. Cordero, A. Goti, A. Brandi, K. Estieu, R. Paugam, J. Ollivier, and J. Salaün, *Eur. J. Org. Chem.*, **1999**, 2725.

375. K. Estieu, R. Paugam, J. Ollivier, J. Salaün, F. M. Cordero, A. Goti, and A. Brandi, *J. Org. Chem.*, **1997**, *62*, 8276.

376. B. B. Snider and B. Shi, *Tetrahedron Lett.*, **2001**, *42*, 1639.

377. K. Hirano, T. Kubota, M. Tsuda, Y. Mikami, and J. Kobayashi, *Chem. Pharm. Bull.*, **2000**, *48*, 974.

378. J. E. Baldwin, S. P. Romeril, V. Lee, and T. D. W. Claridge, *Org. Lett.*, **2001**, *3*, 1145.

379. J. R. Hwu and J. A. Robl, *J. Chem. Soc.*, *Chem. Commun.*, **1986**, 704.

380. J. R. Hwu, J. A. Robl, and B. A. Gilbert, *J. Am. Chem. Soc.*, **1992**, *114*, 3125.

381. P. M. Wovkulich, F. Barcelos, A. D. Batcho, J. F. Sereno, E. G. Baggiolini, B. M. Hennessy, and M. R. Uskukovic, *Tetrahedron*, **1984**, *40*, 2283.

382. E. G. Baggiolini, J. A. Iacobelli, B. M. Hennessy, A. D. Batcho, J. F. Sereno, and M. R. Uskukovic, *J. Org. Chem.*, **1986**, *51*, 3098.

383. P. M. Wovkulich, E. G. Baggiolini, B. M. Hennessy, and M. R. Uskukovic, *Heterocycles*, **1993**, *35*, 791.

384. M. L. Mihailovic, M. M. Rajkovic, L. B. Lorenc, V. D. Pavlovic, A. Z. Milovanovic, B. Tinant, and J.-P. Declercq, *Tetrahedron*, **1996**, *52*, 11995.

385. G. B. Mullen, P. A. Swift, and V. St. Goergiev, *J. Pharm. Sci.*, **1987**, *76*, 930.

386. B. H. Toder, G. B. Mullen, and V. St. Georgiev, *Helv. Chim. Acta*, **1990**, *73*, 169.

387. P. A. van Elburg, G. W. N. Honig, and D. N. Reinhoudt, *Tetrahedron Lett.*, **1987**, *28*, 6397.

388. P. J. S. S. van Eijk, W. Verboom, F. C. J. M. van Veggel, D. N. Reinhoudt, and S. Harkema, *Recl. Trav. Chim. Pays-Bas*, **1988**, *107*, 142.

389. P. A. van Elburg and D. N. Reinhoudt, *Heterocycles*, **1987**, *26*, 437.

390. J. J. Tufariello, A. S. Milowsky, M. Al-Nuri, and S. Goldstein, *Tetrahedron Lett.*, **1987**, *28*, 267.

391. A. R. Katritzky, G. J. Hitchings, and X. Zhao, *J. Chem. Soc., Perkin Trans. 1*, **1990**, 2371.

392. S. A. Ali and M. I. M. Wazeer, *J. Chem. Soc., Perkin Trans. 2*, **1986**, 1789.

393. S. A. Ali and M. I. M. Wazeer, *Tetrahedron*, **1988**, *44*, 187.

394. S. A. Ali and M. I. M. Wazeer, *J. Chem. Soc., Perkin Trans. 1*, **1988**, 597.

395. S. A. Ali, J. H. Khan, and M. I. M. Wazeer, *Tetrahedron*, **1988**, *44*, 5911.

396. S. A. Ali, J. H. Khan, M. I. M. Wazeer, and H. P. Perzanowski, *Tetrahedron*, **1989**, *45*, 5979.

397. S. A. Ali and M. I. M. Wazeer, *J. Chem. Soc., Perkin Trans. 2*, **1990**, 1035.

398. S. A. Ali and H. A. Almuallem, *Tetrahedron*, **1992**, *48*, 5273.

399. S. A. Ali and M. I. M. Wazeer, *Tetrahedron Lett.*, **1992**, *33*, 3219.

400. S. A. Ali and M. I. M. Wazeer, *Tetrahedron Lett.*, **1993**, *34*, 137.

401. S. A. Ali and H. A. Al-Muallem, *Tetrahedron*, **1993**, *49*, 7373.

402. S. A. Ali, S. M. A. Hashmi, and M. I. M. Wazeer, *Tetrahedron Lett.*, **1998**, *39*, 1255.

403. S. M. A. Hashmi, S. A. Ali, and M. I. M. Wazeer, *Tetrahedron*, **1998**, *54*, 12959.

404. S. A. Ali and S. M. A. Hashmi, *J. Chem. Soc., Perkin Trans. 2*, **1998**, 2699.

405. M. I. M. Wazeer, H. P. Perzanowski, S. I. Qureshi, M. B. Al-Murad, and S. A. Ali, *Tetrahedron*, **2000**, *56*, 7229.

406. R. C. Bernotas and G. Adams, *Tetrahedron Lett.*, **1996**, *37*, 7343.

407. R. C. Bernotas and G. Adams, *Tetrahedron Lett.*, **1996**, *37*, 7339.

408. P. Dalla Croce, R. Ferraccioli, and C. La Rosa, *Heterocycles*, **1995**, *40*, 349.

409. P. Dalla Croce, C. La Rosa, and R. Destro, *J. Chem. Soc., Perkin Trans. 2*, **1990**, 679.

410. P. de March, M. Figueredo, J. Font, S. Milán, A. Alvarez-Larena, J. F. Piniella, and E. Molins, *Tetrahedron*, **1997**, *53*, 2979.

411. J. J. Tufariello and J. M. Puglis, *Tetrahedron Lett.*, **1986**, *27*, 1489.

412. J. J. Tufariello, H. Meckler, K. Pushpananda, and A. Senaratne, *Tetrahedron*, **1985**, *41*, 3447.

413. S. P. Ashburn and R. M. Coates, *J. Org. Chem.*, **1984**, *49*, 3127.

414. S. P. Ashburn and R. M. Coates, *J. Org. Chem.*, **1985**, *50*, 3076.

415. T. Hisano, K. Harano, T. Matsuoka, S. Watanabe, and T. Matsuzaki, *Chem. Pharm. Bull.*, **1989**, *37*, 907.

416. M. Kobayakawa and Y. Langlois, *Tetrahedron Lett.*, **1992**, *33*, 2353.

417. T. Berranger, C. Andre-Barres, M. Kobayakawa, and Y. Langlois, *Tetrahedron Lett.*, **1993**, *34*, 5079.

418. T. Berranger and Y. Langlois, *J. Org. Chem.*, **1995**, *60*, 1720.

419. C. Kouklovsky, O. Dirat, T. Berranger, Y. Langlois, M. E. Tran-Huu-Dau, and C. Riche, *J. Org. Chem.*, **1998**, *63*, 5123.

420. N. Coskun and D. Sumengen, *Synth. Commun.*, **1993**, *23*, 1699.

421. N. Coskun, O. Asutay, and D. Sumengen, *Chim. Acta Turc.*, **1996**, *24*, 165.

422. N. Coskun, *Tetrahedron*, **1997**, *53*, 13873.

423. N. Coskun, *Tetrahedron Lett.*, **1997**, *38*, 2299.

424. N. Coskun and M. Ay, *Heterocycles*, **1998**, *48*, 537.

425. T. A. Berezina, V. V. Martin, L. B. Volodarsky, V. V. Khramtsov, and L. M. Weiner, *Bioorganich. Khim.*, **1990**, *16*, 262.

426. T. A. Berezina, V. A. Reznikov, and L. B. Volodarsky, *Tetrahedron*, **1993**, *49*, 10693.

427. T. A. Berezina, V. A. Reznikov, V. I. Mamatyuk, P. A. Butakov, Y. V. Gatilov, I. Y. Bagryanskaya, and L. B. Volodarsky, *Russ. Chem. Bull.*, **1994**, *43*, 838.

428. T. A. Berezina, V. A. Reznikov, L. B. Volodarsky, I. Y. Bagryanskaya, and Y. V. Gatilov, *Russ. Chem. Bull.*, **2000**, *49*, 116.

429. V. V. Martin, L. B. Volodarsky, M. A. Voinov, T. A. Berezina, and T. F. Leluch, *Izv. Akad. Nauk, Ser. Khim.*, **1988**, 1875.

430. L. B. Volodarsky, V. V. Martin, and T. F. Leluch, *Tetrahedron Lett.*, **1985**, *26*, 4801.

431. M. Y. Balakirev, V. V. Khramtsov, T. A. Berezina, V. V. Martin, and L. B. Volodarsky, *Synthesis*, **1992**, 1223.

432. R. C. F. Jones, J. N. Martin, P. Smith, T. Gelbrich, M. E. Light, and M. B. Hursthouse, *Chem. Commun.*, **2000**, 1949.

433. R. C. F. Jones, J. N. Martin, and P. Smith, *J. Heterocycl. Chem.*, **2000**, *37*, 481.

434. R. C. F. Jones, J. N. Martin, and P. Smith, *Synlett*, **2000**, 967.

435. W. J. Haap, D. Kaiser, T. B. Walk, and G. Jung, *Tetrahedron*, **1998**, *54*, 3705.

436. D. S. Tan, M. A. Foley, B. R. Stockwell, M. D. Shair, and S. L. Schreiber, *J. Am. Chem. Soc.*, **1999**, *121*, 9073.

437. B. B. Brown, D. S. Wagner, and H. M. Geysen, *Mol. Diversity*, **1995**, *1*, 4.

438. G. Faita, A. Paio, P. Quadrelli, F. Rancati, and P. Seneci, *Tetrahedron Lett.*, **2000**, *41*, 1265.

439. B. Hinzen and S. V. Ley, *J. Chem. Soc., Perkin Trans. 1*, **1998**, 1.

440. M. Heinenberg and H. Ritter, *Macromol. Chem. Phys.*, **1999**, *200*, 1792.

441. H. Kansui and T. Kunieda, *Tetrahedron Lett.*, **1995**, *36*, 5899.

442. D. R. Borthakur and J. S. Sandhu, *J. Chem. Soc., Chem. Commun.*, **1988**, 1444.

443. P. J. Smith, D. J. Soose, and C. S. Wilcox, *J. Am. Chem. Soc.*, **1991**, *113*, 7412.

444. V. Ramamoorthy, A. Ramasubbu, S. Muthusubramanian, and S. Sivasubramanian, *J. Inclusion Phenom. Macrocyclic Chem.*, **1999**, *33*, 69.

445. V. Ramamoorthy, C. Meenakshi, S. Muthusubramanian, and S. Sivasubramanian, *J. Inclusion Phenom. Macrocyclic Chem.*, **2000**, *36*, 425.

446. Y. L. Pascal, J. Chanet-Ray, R. Vessiere, and A. Zeroual, *Tetrahedron*, **1992**, *48*, 7197.

447. A. Rastelli, R. Gandolfi, and M. Sarzi-Amadé, "Regioselectivity and Diastereoselectivity in the 1,3-dipolar Cycloadditions of Nitrones with Acrylonitrile and Maleonitrile. The Origin of ENDO/EXO Selectivity", in *Advances in Quantum Chemistry*, Vol. 36, Academic Press, New York, **1999**, pp. 151–167.

448. M. Freccero, R. Gandolfi, M. Sarzi-Amade, and A. Rastelli, *J. Chem. Soc., Perkin Trans. 2*, **1998**, 2413.

449. C. Di Valentin, M. Freccero, R. Gandolfi, and A. Rastelli, *J. Org. Chem.*, **2000**, *65*, 6112.

450. R. Annunziata, M. Benaglia, M. Cinquini, and L. Raimondi, *Tetrahedron*, **1993**, *49*, 8629.

451. A. Rastelli, R. Gandolfi, M. S. Sarzi-Amadé, and B. Carboni, *J. Org. Chem.*, **2001**, *66*, 2449.

452. M. L. Yeung, W.-K. Li, H.-J. Liu, Y. Wang, and K. S. Chan, *J. Org. Chem.*, **1998**, *63*, 7670.

453. For experimental reports of Fischer carbene complexes as dipolarophiles see: K. S. Chan, M. L. Yeung, W. Chan, R.-J. Wang, and T. C. W. Mak, *J. Org. Chem*, **1995**, 60, 1741; K. S. Chan, *J. Chem. Soc., Perkin Trans. 1*, **1991**, 2602.

454. K. V. Gothelf, R. G. Hazell, and K. A. Jorgensen, *J. Org. Chem.*, **1996**, *61*, 346.

455. R. Carda, R. Portolés, J. Murga, S. Uriel, J. A. Marco, L. R. Domingo, R. J. Zaragozá, and H. Röper, *J. Org. Chem.*, **2000**, *65*, 7000.

456. R. Sustmann, W. Sicking, and R. Huisgen, *J. Am. Chem. Soc.*, **1995**, *117*, 9679.

CHAPTER 2

Nitronates

Scott E. Denmark and Jeromy J. Cottell

Department of Chemistry, University of Illinois, Urbana, IL

2.1. Structure . . . 86
2.1.1. Silyl Nitronates . . . 86
2.1.1.1. Stability . . . 86
2.1.1.2. Bonding . . . 87
2.1.1.3. Spectroscopy . . . 88
2.1.1.4. Silyl Exchange . . . 90
2.1.1.5. Spectroscopic Data of Silyl Nitroso Acetals . . . 93
2.1.2. Alkyl Nitronates . . . 95
2.1.2.1. Stability . . . 95
2.1.2.2. Bonding . . . 96
2.1.2.3. Spectroscopy . . . 97
2.1.2.4. Spectroscopic Data of Alkyl Nitroso Acetals . . . 102
2.2. General Reactivity . . . 105
2.2.1. Regeoselectivity . . . 108
2.2.2. Stereoselectivity . . . 111
2.2.2.1. Intermolecular Cycloadditions . . . 111
2.2.2.2. Intramolecular Cycloadditions . . . 112
2.2.3. Mechanism . . . 114
2.3. Silyl Nitronates . . . 114
2.3.1. Preparation . . . 114
2.3.2. Cycloadditions of Silyl Nitronates . . . 117
2.3.2.1. Intermolecular Cycloadditions . . . 117
2.3.2.2. Facial Selectivity . . . 120
2.3.2.3. Intramolecular Cycloadditions . . . 123
2.3.3. Nitroso Acetal Functionalization . . . 125
2.3.4. Applications . . . 127
2.4. Alkyl Nitronates . . . 130
2.4.1. Preparation of Acyclic Nitronates . . . 130
2.4.1.1. Alkylation . . . 130
2.4.1.2. Trinitromethane Derivatives . . . 132
2.4.1.3. Acylation . . . 133
2.4.2. Preparation of Cyclic Nitronates . . . 134

The Chemistry of Heterocyclic Compounds, Volume 59: Synthetic Applications of 1,3-Dipolar Cycloaddition Chemistry Toward Heterocycles and Natural Products, Edited by Albert Padwa and William H. Pearson.
ISBN 0-471-38726-6

2.4.2.1. Alkylation 134
2.4.2.2. [4+2] Cycloaddition of Nitroalkenes 136
2.4.2.3. Radical Cyclization 137
2.4.3. Cycloadditions of Alkyl Nitronates 138
2.4.3.1. Intermolecular Cycloadditions. 138
2.4.3.2. Facial Selectivity. 146
2.4.3.3. Intramolecular Cycloadditions. 148
2.4.4. Nitroso Acetal Functionalization 153
2.4.5. Applications 154
2.5 Conclusion and Outlook 159

The chemistry of the nitro functional group has had a long and rich history in organic synthesis (1–4). The high oxidation state of the nitro group imparts a number of physical properties that have important chemical consequences. Foremost among these is the powerful electron-withdrawing capability through both dipolar and mesomeric effects. This property imparts the high electrophilicity associated with aromatic and olefinic nitro compounds and their susceptibility to nucleophilic addition and reduction.

A second and related consequence in aliphatic nitro compounds is the acidification of the directly bonded CH_n unit through the attendant stabilization of the derived conjugate bases (5,6). As with all delocalized anions, reprotonation gives rise to tautomers, the original C-nitro compound (**I**) and the *aci*-nitro or isonitro form (**II**), Eq. 2.1. The *aci*-nitro tautomers are typically present in very minor concentrations, with equilibrium constants (K_{eq}) between 10^{-5} and 10^{-7} (7). Alkylation of the delocalized anion leads to both α-substituted nitro compounds and the regioisomeric nitronic esters (nitronates). Nitronates were described as early as 1894 (8), however, the first isolated nitronic ester was obtained several years later upon the addition of diazomethane to phenylazonitromethane (**1**), Eq. 2.2 (9).

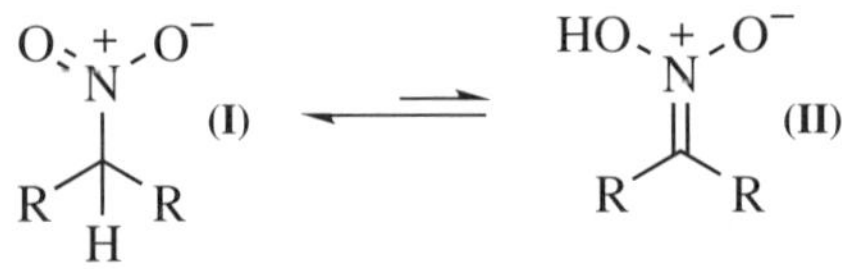

nitro compound — nitronic acid (*aci*-nitro or isonitro compound)

(2.1)

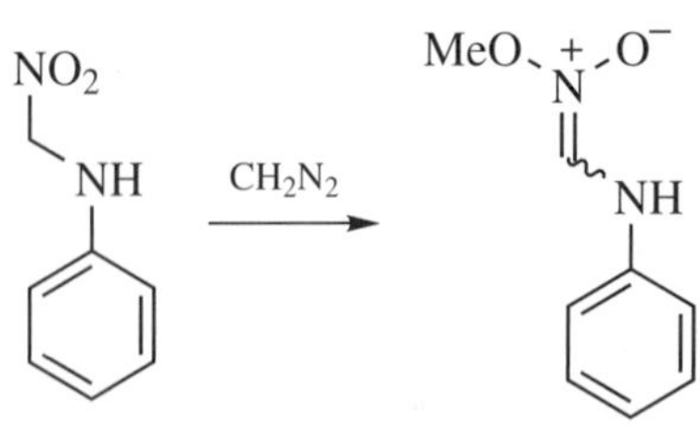

(**1**) (**2**)

(2.2)

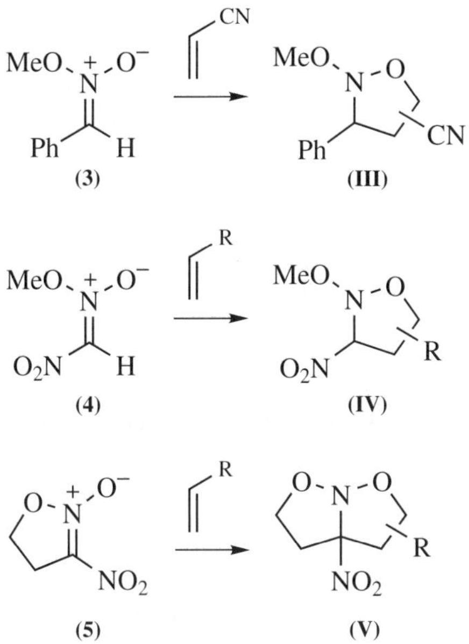

Scheme 2.1

Aside from the relatively trivial conversions of nitronates to the corresponding oxime and carbonyl compounds (10,11), the chemistry of nitronates remained relatively unexplored for much of the early 1900s. However, in 1964, Tartakovskii et al. (12) demonstrated that alkyl nitronate esters were competent partners in the newly discovered class of dipolar cycloadditions with alkenes (Scheme 2.1). Both cyclic and acyclic nitronates participated, thus providing a new functional group were the nitrogen atom existed at the center of an acetal (13). These compounds were subsequently referred to as nitroso acetals (14) or nitrosals (15).

Several years later, Ioffe et al. (16) demonstrated that silyl nitronates also could be engaged in the dipolar cycloaddition with alkenes. These silylated isoxazolidine cycloadducts were then converted to the corresponding isoxazolines by treatment with sodium methoxide (Scheme 2.2).

This chapter summarizes the [3+2]-dipolar cycloaddition chemistry of nitronates that has developed since the first reported discovery in 1964 up to the

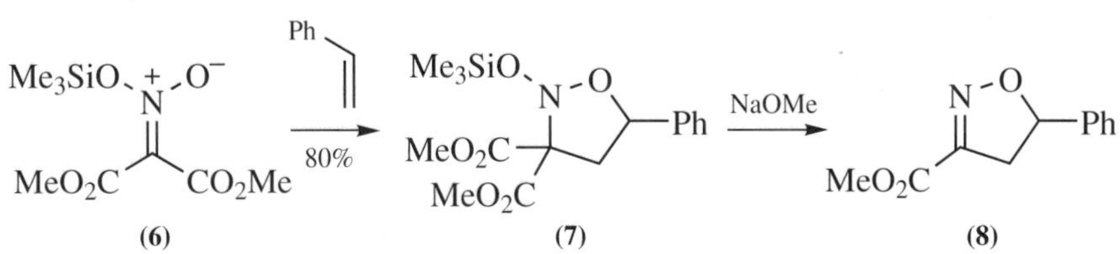

Scheme 2.2

beginning of 2001, almost 40 years of nitronate chemistry. It is worth noting that despite a considerable body of work from the Tartakovskii and Torssell schools, the first volume of this compendium did not contain a section on nitronate cycloadditions. Fortunately, advances in the past two decades have brought these important, early contributions to the fore.

This chapter is divided into four major sections. The first (Section 2.1) will deal with the structure of both alkoxy and silyl nitronates. Specifically, this section will include physical, structural, and spectroscopic properties of nitronates. The next section (Section 2.2) describes the mechanistic aspects of the dipolar cycloaddition including both experimental and theoretical investigations. Also discussed in this section are the regio- and stereochemical features of the process. Finally, the remaining sections will cover the preparation, reaction, and subsequent functionalization of silyl nitronates (Section 2.3) and alkyl nitronates (Section 2.4), respectively. This will include discussion of facial selectivity in the case of chiral nitronates and the application of this process to combinatorial and natural product synthesis.

2.1. STRUCTURE

2.1.1. Silyl Nitronates

2.1.1.1. Stability

The first attempted silylation of nitromethane resulted in the production of a silylated dimeric nitronate due to self-condensation (17). Subsequent investigations, however, have shown that silyl nitronates are stable compounds that can be isolated and distilled at reduced pressures, Table 2.1 (18–21). Bulkier silyl groups help increase the stability of the resulting nitronate (21).

TABLE 2.1. BOILING POINTS OF SELECTED SILYL NITRONATES

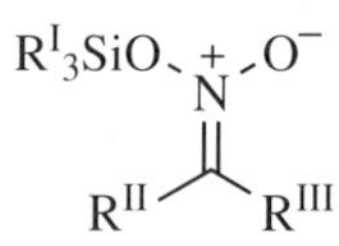

R^I	R^{II}	R^{III}	bp (°C/mmHg)	Reference
Me	Me	H	64 / 25	18
Me	Et	H	58 / 12	18
t-BuMe$_2$	Et	H	50 / 0.15	19
Me	Ph	H	85 / 0.5	20
t-BuMe$_2$	Me	Me	130 / 5	19
t-BuMe$_2$	–(CH$_2$)$_4$–		90 / 0.01	21
t-BuMe$_2$	–(CH$_2$)$_5$–		150 / 0.01	19

Though the silyl nitronates themselves are stable, they are also very labile. Whereas they are unreactive toward mild base, silyl nitronates are readily hydrolyzed back to the corresponding nitro compounds in the presence of water, alcohol, and mild acid (16). In the presence of stronger acids, the elimination of R_3SiOH from monosubstituted nitronates is possible, forming small amounts of the corresponding nitrile oxide in addition to the original nitro compound (18). Although triethylamine is recommended as an effective stabilizer, the handling of the silyl nitronate is often avoided by generating the nitronate *in situ* and using it directly without isolation (22).

2.1.1.2. Bonding

There are only two X-ray crystallographic structures of silyl nitronates, Tables 2.2 and 2.3 (21). In both of these structures the nitronate group is planar, as expected for an sp^2 hybridized nitrogen atom. However, the bond angles are quite different than the 120° normally associated with sp^2 centers. The O(1)—N—C bond angle is opened to 129.7 and 130.0° for **9** and **10**, respectively. The two remaining bond angles for **9** [O(2)—N—C and O(1)—N—O(2)] are 114.1 and 116.2°, while the same angles for nitronate **10** are 114.7 and 115.3°. These values fall close to those observed for α-4-chlorophenyl *N*-methyl nitrone (**11**), Figure 2.1 (23). However, the O(1)—N—C bond angle is 4 to 5° larger in the case of the nitronates. This abnormally large bond angle is ascribed to nonbonding repulsion of the ortho-hydrogen atom of the phenyl ring with O(1). In the case of **10**, the phenyl ring cis in relation to the silyl moiety is also in plane with the nitronate, while the other phenyl ring is in a conformation orthogonal to the nitronate. Therefore the same steric argument holds for the large bond angle as observed in **9**.

TABLE 2.2. SELECTED X-RAY STRUCTURAL DATA FOR SILYL NITRONATE (**9**)

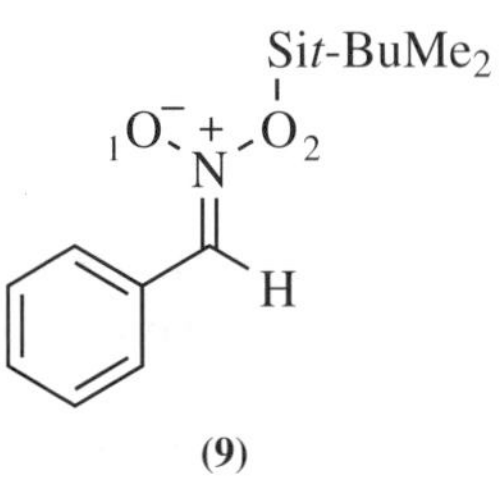

(9)

Bond/Bond Angle	Length (Å)	Angle (°)
N—O(1)	1.259	
N—O(2)	1.411	
N—C	1.302	
O(1)—N—C		129.7
O(2)—N—C		114.1
O(1)—N—O(2)		116.2

TABLE 2.3. SELECTED X-RAY STRUCTURAL DATA FOR SILYL NITRONATE (**10**)

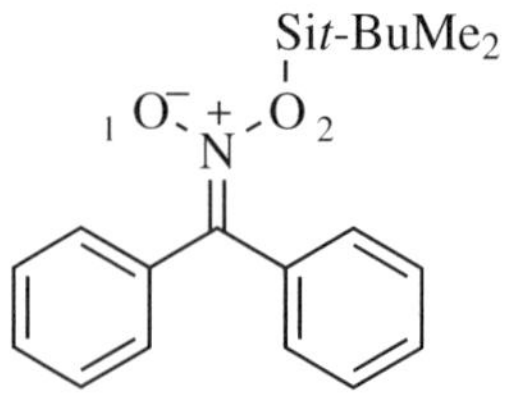

(10)

Bond/Bond Angle	Length (Å)	Angle (°)
N—O(1)	1.271	
N—O(2)	1.400	
N—C	1.309	
O(1)—N—C		130.0
O(2)—N—C		114.7
O(1)—N—O(2)		115.3

Both the C—N bond and the N—O(1) bond show significant double-bond character. For **9**, theses values are 1.302 and 1.259 Å, respectively. The values of 1.309 and 1.271 Å are observed for **10**. The length of the C—N bond agrees well with that of the nitrone **11** in which a value of 1.309 Å is observed. However, the N—O bond of **3** is significantly longer at 1.284 Å. The N—O(2) bond in both **9** and **10** shows single-bond character with lengths of 1.411 and 1.400 Å. The bond length is similar to that observed for the N—O bond of oxime **12**, which is 1.408 Å (23).

Interestingly, the silyl moiety is observed to be only slightly distorted out of the plane of the nitronate. In the case of **9**, there is a 31° cant of the silyl group, where as this group is only 8° out of the plane in **10**.

2.1.1.3. Spectroscopy

Infrared (IR), carbon-13 nuclear magnetic resonance (^{13}C NMR), and proton NMR (^{1}H NMR) spectroscopy provide diagnostic resonances for the identification

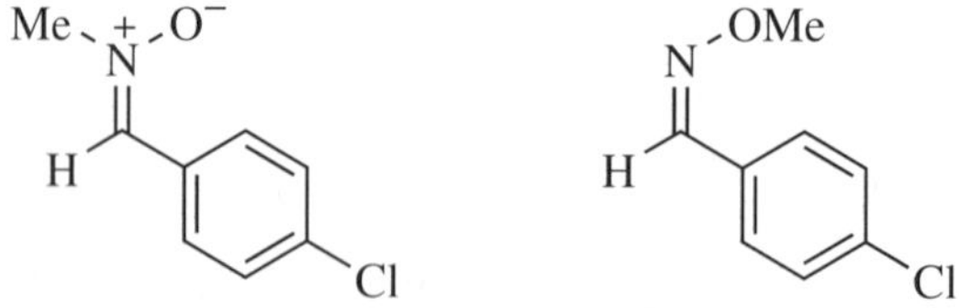

(11) **(12)**

Figure 2.1. Related N—O systems.

TABLE 2.4. CHARACTERISTIC IR ABSORBANCES FOR SILYL NITRONATES

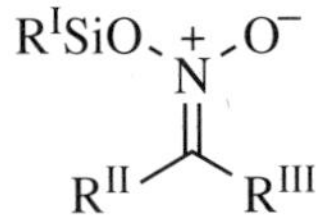

R^I	R^{II}	R^{III}	IR (C=N, cm^{-1})	Reference
Me_3	Me	H	1622	20
Me_3	Et	H	1617	20
t-$BuMe_2$	n-Pent	H	1615	21
t-$BuMe_2$	n-Oct	H	1615	21
Me_3	CF_3	H	1620	24
t-$BuMe_2$	CF_3	H	1620	24
Me_3	Ph	H	1590	20
Me_3	CO_2Me	H	1591	25
Me_3	NO_2	H	1600	25
Me_3	Me	Me	1626	20
t-$BuMe_2$	—$(CH_2)_4$—		1645	21
t-$BuMe_2$	n-Oct	Me	1615	21
t-$BuMe_2$	CF_3	Me	1610	24
t-$BuMe_2$	Ph	Ph	1550	21
Me_3	NO_2	NO_2	1560	26
Me_3	CO_2Me	CO_2Me	1580	16

of nitronates. In the case of disubstituted nitronates, ^1H NMR spectroscopy is less useful due to the lack of a hydrogen atom attached to the nitronate carbon. The IR stretching frequencies for the C=N bond ranges from 1550 to 1645 cm^{-1}, Table 2.4 (16,20,21, 24–26). Typically acyclic, alkyl substituted nitronates fall in the middle of the range, whereas cyclic nitronates are observed at higher frequencies. The incorporation of electron withdrawing or conjugating groups leads to a lowering of the resonance frequency.

The observed C=N stretching frequencies for aromatic nitronates are very similar to those of the corresponding nitrones, which appear at 1550–1600 cm^{-1} (27). The observed absorbances for aliphatic nitrones are between 1600 and 1620 cm^{-1}, which is slightly lower that the corresponding nitronates. Moreover, the resonances observed for the nitronate function are slightly lower than those observed for oximes (1650–1685 cm^{-1}) (28). This data suggests a decrease in the C=N bond character in the order of oximes > nitronates > nitrones.

The ^{13}C NMR resonance of the nitronate group is very diagnostic and typically appears between 105 and 135 ppm, Table 2.5 (21,24,29). The expected downfield shift is observed for disubstituted nitronates in comparison to the monosubstituted silyl nitronates. As is also observed with cyclic ketones, a larger downfield shift is seen for cyclic nitronates. The carbon resonances are similar to those observed for nitrones (~130 ppm) (30), however, they are shifted significantly upfield in comparison to the corresponding oximes (150–160 ppm) (31). The resonances for silylated oximes appear ~127 ppm (29).

TABLE 2.5. CHARACTERISTIC ^{13}C NMR RESONANCES FOR SILYL NITRONATES

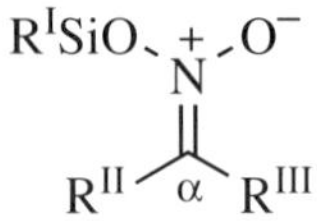

R^{I}	R^{II}	R^{III}	^{13}C [C(α), δ, ppm]	Reference
Me_3	Me	H	109.8	29
t-$BuMe_2$	CF_3	H	105.4	24
Me_3	Me	Me	118.3	29
t-$BuMe_2$	Me	Me	120.1	21
t-$BuMe_2$	CF_3	Me	113.9	24
t-$BuMe_2$	$-(CH_2)_4-$		135.2	21
Me_3	CO_2Me	CO_2Me	114.2	29

In the ^{1}H NMR spectra of monosubstituted silyl nitronates, the HC(α) resonance is typically observed between 5.5 and 6.7 ppm, Table 2.6 (20,24,25,29). As expected, alkyl nitronates are shifted upfield relative to those bearing conjugated and electron-withdrawing groups. In general, the resonances for monosubstituted silyl nitronates are found at slightly higher field compared to the corresponding nitrones which appear between 6.3 and 7.9 ppm (30). This upfield shift is also true in comparison to alkyl and aryl oximes, which are observed between 6.8 and 7.9 ppm and 7.2 and 8.6 ppm, respectively (32). Both the ^{1}H and the ^{13}C NMR data shows that the electron density at the nitronate carbon is greater than either the carbon of the nitrone or oxime, hence the upfield resonances observed with the nitronate compounds.

2.1.1.4. Silyl Exchange

Despite the potential for geometrical isomers, in almost all cases, a single set of signals is observed by both ^{1}H and ^{13}C NMR spectroscopy for silyl nitronates. This

TABLE 2.6. CHARACTERISTIC ^{1}H NMR RESONANCES FOR SILYL NITRONATES

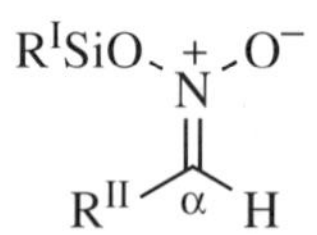

R^{I}	R^{II}	^{1}H [HC(α), δ, ppm]	Reference
Me_3	H	5.55	29
Me_3	Me	6.05	29
Me_3	Et	5.87	20
Me_3	Ph	6.69	29
Me_3	CO_2Me	6.28	25
t-$BuMe_2$	CF_3	6.51 (br)	24
Me_3	NO_2	8.25	25

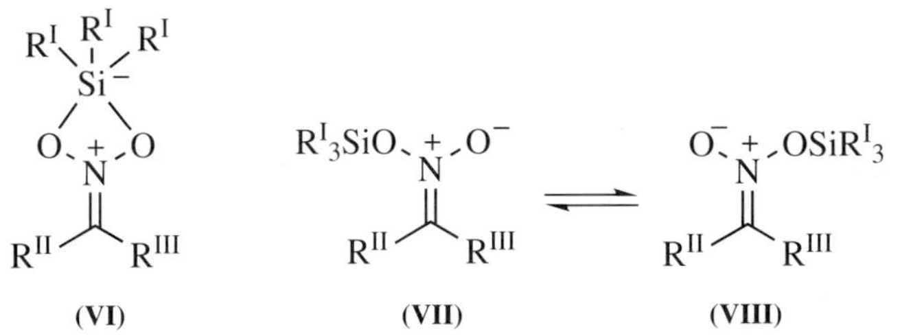

Figure 2.2. Proposed modes of silyl group exchange.

behavior can result from three possibilities: (1) stereoselective silylation, (2) existence of a hypervalent silicon species (**VI**) involved in bonding with both nitronate oxygens, or (3) a rapid exchange of the silyl group (Fig. 2.2). However, for nitronate **13**, there is only a single ^{1}H NMR resonance observed for the two methyl groups attached to the nitronate (Fig. 2.3) (21). Even if the silylation were stereoselective, this alone cannot explain the equivalency of these methyl groups. Interestingly, in the solid state, only the (*E*) isomer is observed for nitronate **9** (Table 2.2). Stereoselective silylation notwithstanding, this is likely due to a more favorable packing of the (*E*) isomer compared to the (*Z*) isomer in the solid state. The X-ray crystallographic structure of **9** also reveals a tetracoordinate silicon atom, suggesting that a hypervalent silicon species is not present in the solid state for this system.

In the solution state, the possibility of a pentacoordinate silicon species can also be discounted by ^{29}Si NMR spectroscopy (29). Comparison of the silyl nitronates **14** and **6** to the corresponding silyl oximes **15** and **16** show that the silicon resonances are almost identical (Fig. 2.4). If there were a significant interaction with both oxygens in the nitronate, the resonance would be shifted from that of the silyl oxime.

The rapid exchange of the silyl group is the most reasonable explanation for these observations. From low-temperature ^{1}H and ^{13}C NMR investigations, the energy barrier for this isomerization is calculated to be 9–12 kcal/mol, independent of the silyl group. The calculated energy barriers and corresponding coalescence temperatures are shown in Figure 2.5 (21,24,33). This exchange is likely an intramolecular process since the ^{1}H spectra are independent of concentration. Studies on

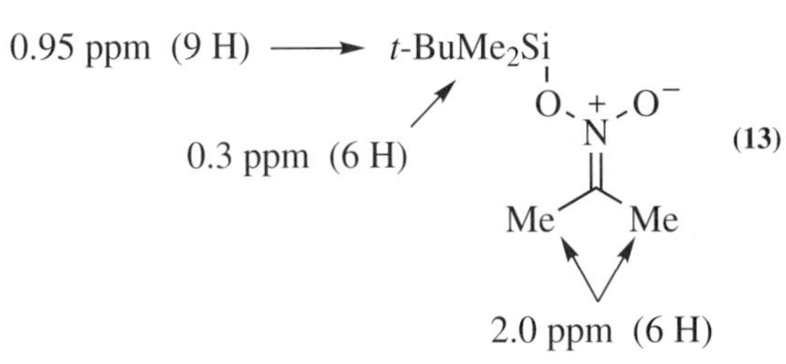

Figure 2.3. The ^{1}H NMR data of the silyl nitronate (**13**).

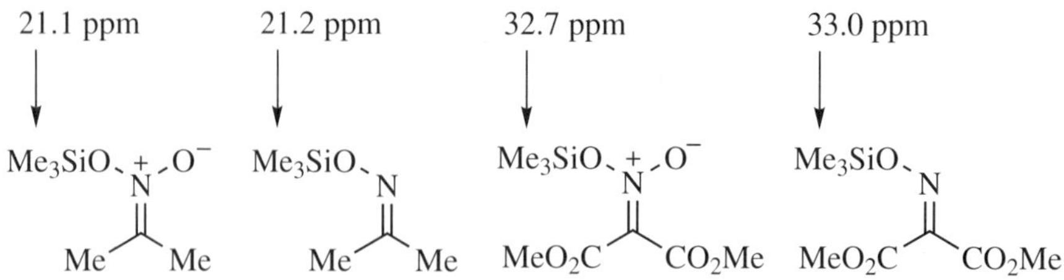

(14) (15) (6) (16)

Figure 2.4. The ^{29}Si NMR data of silyl nitronates and corresponding silyl oximes.

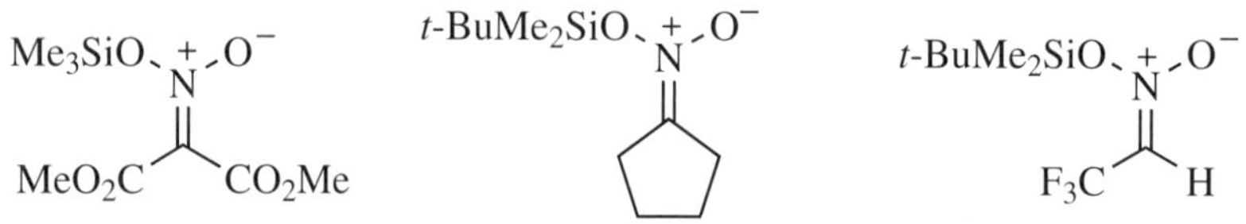

(6)	(17)	(18)
11.8 kcal/mol	10.0 kcal/mol	9.3 kcal/mol
(T_c = −17 °C, toluene)	(T_c = −65 °C, $CHCl_2F$)	[T_c = 35 °C, CS_2/THF]

Figure 2.5. Isomerization barrier and coalescence temperature for silyl nitronates.

the analogous diethylboryl nitronate of **6** suggest that the isomerization proceeds though a four coordinated transition structure (**IX**) and not through a dissociation and recombination of the boryl cation (Fig. 2.6) (33). This conclusion is supported by the lack of a solvent effect on the rate of isomerization. Additionally, in the case of **9** thc nonbonding interaction between the oxygen and silicon results in a distorted tetrahedron at silicon in the solid state (Fig. 2.7) (21). The O(2)—Si—*t*-Bu bond angle is only 100.6°. As a result the threefold axis generated by averaging the three Si—C vectors creates an 8° angle with the Si—O(2) bond, distorted toward O(1).

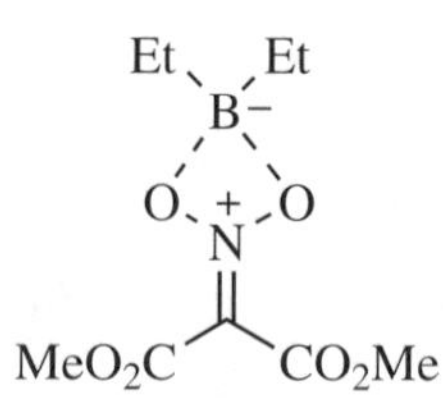

(IX)

Figure 2.6. Proposed transition state for the isomerization of boryl nitronates.

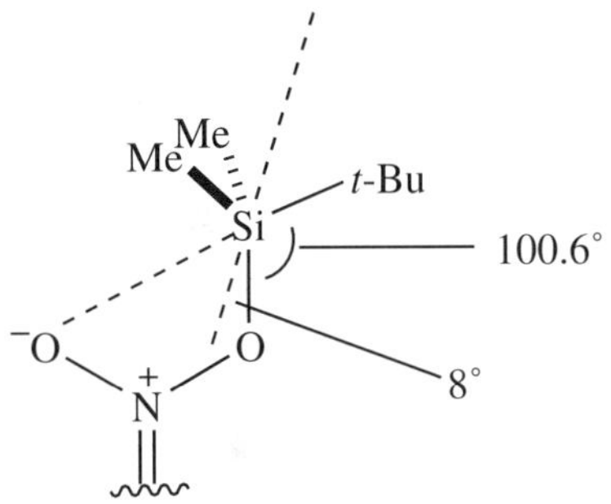

Figure 2.7. Selected angles from the X-ray crystallographic data of **9**.

2.1.1.5. *Spectroscopic Data of Silyl Nitroso Acetals*

The primary cycloadduct from combination of a dipolarophile with a silyl nitronate is an isoxazolidine. The ^{1}H and ^{13}C NMR spectra are highly informative for the structural determination of these products, Tables 2.7 and 2.8 (21,25,34,35). Both the ^{1}H and ^{13}C NMR data show that HC(5) are shifted downfield relative to HC(3). An expected downfield shift is also observed with electron-withdrawing or conjugating groups. In the absence of functionalization at C(3), there is a significant upfield shift of the corresponding ^{13}C resonance. The IR data is less reliable. The O–N–O stretch is reported to be ~1055 cm^{-1} (Fig. 2.8), however, this stretching

TABLE 2.7. CHARACTERISTIC ^{1}H NMR RESONANCES FOR *N*-SILYLOXY NITROSO ACETALS

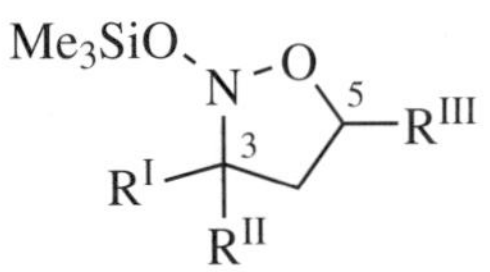

R^{I}	R^{II}	R^{III}	^{1}H (δ, ppm) HC(3)	HC(5)	Reference
CF_3	H	*n*-Bu	3.82	4.62	21
			3.21	4.85	
CF_3	H	Ph	4.13	5.62	21
			not obtained	5.36	
CF_3	H	CO_2Me	3.99	5.01	21
			not obtained	4.82	
NO_2	NO_2	Ph		6.09	34
				5.55	
CO_2Me	H	Ph	4.18	5.48	25
			4.11	5.40	
CO_2Me	H	CO_2Me	4.01	4.88	25
			4.09	4.89	

TABLE 2.8. CHARACTERISTIC ^{13}C NMR RESONANCES FOR *N*-SILYLOXY NITROSO ACETALS

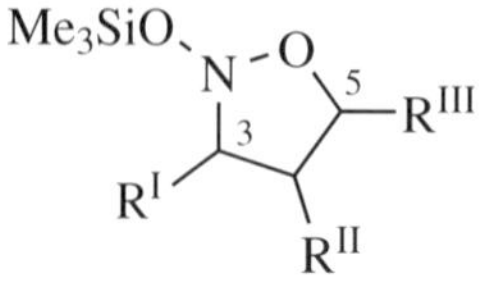

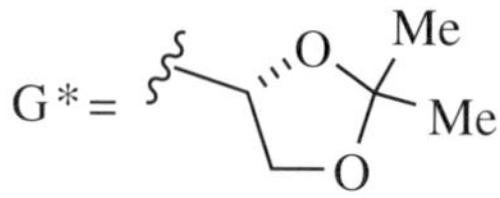

R^{I}	R^{II}	R^{III}	^{13}C (δ, ppm) C(3)	C(5)	Reference
CF_3	H	*n*-Bu	71.6	79.8	21
CF_3	H	Ph	72.2	81.1	21
CF_3	H	CO_2Me	71.4	77.5	21
H	G*	COMe	65.0	84.9	35
Me	G*	COMe	72.8	85.3	35
H	G*	CO_2Et	64.7	78.5	35

band is moderate in strength and frequently obscured for more functionalized nitronates.

Several diastereomers may result from the [3+2] cycloaddition of silyl nitronates. In the case of the trifluoroisoxazolidine, the major isomers are epimers at nitrogen due to the high inversion barrier (24). Diastereomers from different modes of approach of the dipolarophile are observed in less significant amounts. Equilibration of the nitrogen epimers is accomplished upon heating in toluene at 110 °C for several hours.

N-Silyloxy isoxazolidines can be converted to the corresponding isoxazolines by elimination of R_3SiOH. The ^{1}H NMR spectra of the corresponding isoxazolines retain the chemical shift of the HC(5), however, a downfield shift is observed for HC(3), as well as C(3), Tables 2.9 and 2.10 (18,36). The IR spectra return to a range similar to that of the original silyl nitronate (Fig. 2.8).

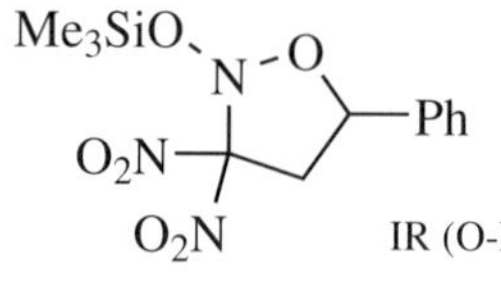

(19)

IR (O-N-O) : 1055 cm^{-1}

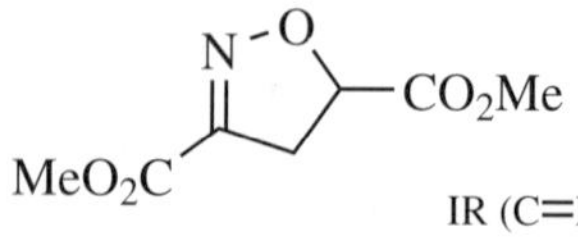

(20)

IR (C=N) : 1580 cm^{-1}

Figure 2.8. Characteristic infrared data for N-silyloxy isoxazolidines and isoxazolines.

TABLE 2.9. CHARACTERISTIC ^{1}H NMR RESONANCES FOR ISOXAZOLINES

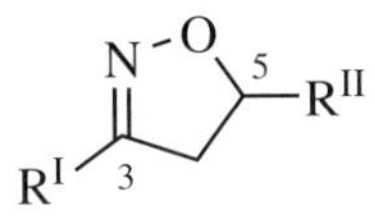

		^{1}H (δ, ppm)		
R^{I}	R^{II}	HC(3)	HC(5)	Reference
H	CO_2Me	7.25	4.99	18
H	Ph	7.17	5.50	18
Et	CN		5.25	18

TABLE 2.10. CHARACTERISTIC ^{13}C NMR RESONANCES FOR ISOXAZOLINES

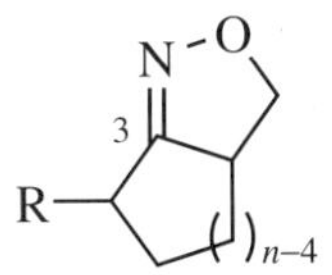

R	*n*	^{13}C [C(3), δ, ppm]	Reference
4-F-C_6H_4	6	161.2	36
$PhCH_2CH_2$	6	162.2	36
4-F-C_6H_4	5	173.4	36
$PhCH_2CH_2$	5	174.3	36

2.1.2. Alkyl Nitronates

2.1.2.1. Stability

Alkyl nitronates derived from simple, acyclic nitroalkanes are less stable than the corresponding silyl nitronates (16). Only a handful of nitronates are amenable to distillation (Table 2.11) (37–39). The majority of monosubstituted nitronates decompose in solution at room temperature over the course of 1–5 days (40,41). Because of this instability, these nitronates are commonly utilized without further purification.

Most cyclic and aryl nitronates are solids at room temperature (41). However, the stability of aryl substituted nitronates is dependent on the configuration of the nitronate (40). For example, only the trans isomer of the ethyl nitronate of 4-nitrophenylnitromethane crystallizes from an analytically pure mixture of cis and trans isomers. The mass recovery reflects the percentage of the trans isomer in the original mixture. As expected, increasing substitution on the nitronate typically

TABLE 2.11. BOILING POINTS OF SIMPLE ALKYL NITRONATES

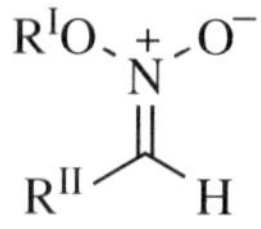

R^I	R^{II}	bp (°C/mmHg)	Reference
Me	CO_2Et	84 / 2.5	37
Et	CO_2Et	81 / 3	38
Me	CN	52 / 0.05	39

leads to greater stability, allowing for purification by recrystallization or chromatography.

2.1.2.2. Bonding

The structure of nitronic acids has been generally accepted to be represented by **II** (Eq. 2.1) since the early 1900s (37,42). Consequently, this formulation was proposed for the analogous nitronic esters. The structure consists of a planar nitrogen atom involved in a double bond with the adjacent carbon atom. INDO calculations of **21** suggest a C=N bond length of 1.27 Å, an N–O(2) bond of 1.38 Å, and a N–O(1) bond of 1.30 Å (Table 2.12). The angles between the substituents of the nitrogen are at an idealized 120° (43).

The structure of alkyl nitronates is confirmed by solid-state investigation. Despite the rather diverse environments in which nitronates are found the structural

TABLE 2.12. CALCULATED BOND DISTANCES AND ANGLES FOR NITRONATES

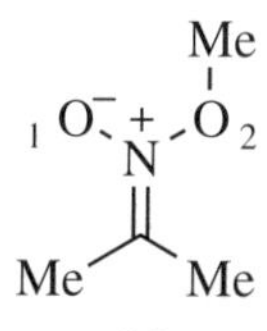

(21)

Bond/Bond Angle	Length (Å)	Angle (°)
N–O(1)	1.30	
N–O(2)	1.38	
N–C	1.27	
O(1)–N–C		120.0
O(2)–N–C		120.0
O(1)–N–O(2)		120.0

TABLE 2.13. SELECTED X-RAY STRUCTURAL DATA FOR NITRONATES **22–24**

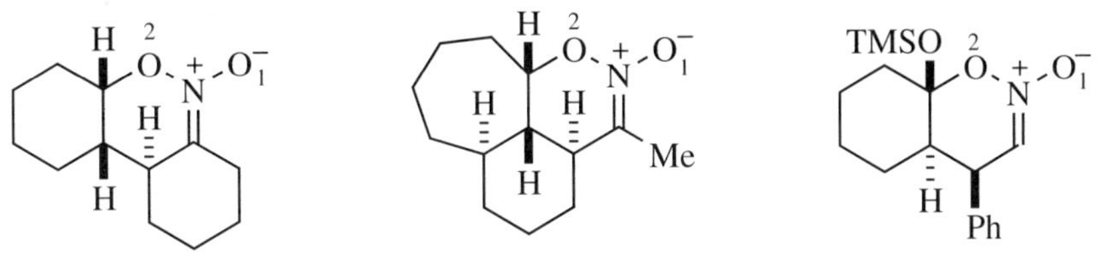

(22) **(23)** **(24)**

Bond/Bond Angle	22 Length (Å)	22 Angle (°)	23 Length (Å)	23 Angle (°)	24 Length (Å)	24 Angle (°)
N—O(1)	1.274		1.271		1.258	
N—O(2)	1.424		1.451		1.434	
N—C	1.293		1.281		1.296	
O(1)—N—C		129.5		133.0		127.8
O(2)—N—C		120.0		114.4		120.6
O(1)—N—O(2)		110.4		112.6		111.5

parameters are fairly consistent (44–55). The N—O(1) bond centers ~1.25 Å with a 0.06-Å range and the N—C bond falls ~1.30 Å with a 0.04-Å range. These parameters are also found to be complimentary, that is, cases with long N—O(1) bonds have shorter N—C bonds and vice versa. The X-ray crystallographic determination of **22** shows that the nitrogen atom is indeed planar, but the angles between the nitronate substituents are distorted from the ideal for an sp^2 center (Table 2.13). The O(1)—N—O(2) angle is narrowed to 110.4°, while the O(1)—N—C bond angle is expanded to 129.5°. The remaining O(2)—N—C bond angle is very close to ideal at 120.0°. The C—N bond distance was found to be 1.293 Å, while the N—O(1) and N—O(2) bonds are 1.274 and 1.424 Å, respectively. The X-ray crystallographic data for **24** corresponds well with those of **22**, however, the bond angles of **23** are drastically different from the other two alkyl nitronates. Interestingly, these bond angles agree better with those observed in the silyl nitronates, however, this may be due to the constraints imposed by the fused seven- and six-membered rings.

The X-ray crystallographic data for an acyl nitronate (**25**) is compiled in Table 2.14 (56). Because the phenyl group attached to the nitronate, the bond angles about the sp^2 hybridized nitrogen are similar to those of nitronates **9** and **10**. However, the O(1)—N bond is slightly shorter than those in either the alkyl or silyl nitronates, while the O(2)—N is slightly longer. This is in line with the observed instability of monosubstituted acyl nitronates, which eliminate readily to the nitrile oxides.

2.1.2.3. Spectroscopy

The spectroscopic data for alkyl nitronates follows very closely to those of silyl nitronates. The C=N stretching frequencies for five- and six-membered, cyclic

TABLE 2.14. SELECTED X-RAY STRUCTURAL DATA FOR ACYL NITRONATE **25**

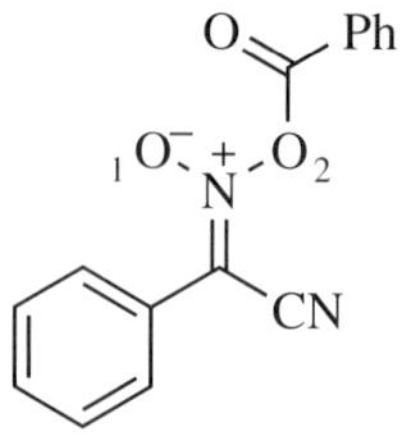

(25)

Bond/Bond Angle	Length (Å)	Angle (°)
N–O(1)	1.232	
N–O(2)	1.455	
N–C	1.318	
O(1)–N–C		130.3
O(2)–N–C		114.0
O(1)–N–O(2)		115.6

nitronates range from 1560 to 1670 cm^{-1} (Tables 2.15 and 2.16) (57–71). As expected, higher frequencies are observed with the five-membered rings compared to six-membered rings. As is seen with silyl nitronates, the substitution of an electron-withdrawing group directly on the nitronate functionality decreases the observed resonance frequency. Interestingly, when the substituent R^{VI} is an ether, the observed absorbance appears at higher frequency, unless there is an electron-withdrawing group at R^{I}. This is not the case when R^{VI} is an amino group.

TABLE 2.15. CHARACTERISTIC IR ABSORBANCES FOR FIVE-MEMBERED CYCLIC NITRONATES

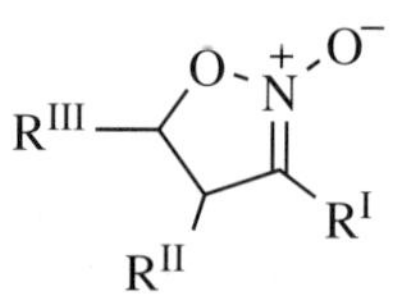

R^{I}	R^{II}	R^{III}	IR (cm^{-1})	Reference
Ph	H	H	1620	57
H	–$(CH_2)_4$–		1628	58
NO_2	OH	H	1640	59
CO_2Et	H	COMe	1630	60
CO_2Et	Me	COEt	1623	60
CO_2Et	–$CO(CH_2)_2$–		1622	60
CO_2Me	CO_2Me	CO_2Me	1632	61
Me	*i*-Pr	CO_2Et	1654	62
Et	*i*-Pr	CO_2Et	1630	62
–$(CH_2)_4$–		CO_2Et	1670	62

TABLE 2.16. CHARACTERISTIC IR ABSORBANCES FOR SIX-MEMBERED CYCLIC NITRONATES

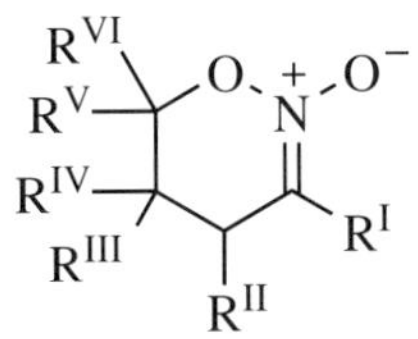

HOG* =

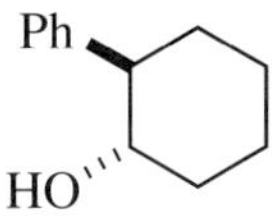

HNG′$_2$ =

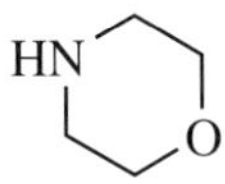

R^I	R^{II}	R^{III}	R^{IV}	R^V	R^{VI}	IR (cm^{-1})	Reference
CO_2Me	H	H	H	H	H	1570	63
Ph	H	H	H	H	H	1590	63
NO_2	H	H	H	H	H	1610	63
CN	H	H	H	H	H	1580	64
Me	Ph	H	$-(CH_2)_3-$		H	1601	65
Me	4-MeC_6H_4	H	$-(CH_2)_3-$		H	1599	65
Me	4-MeC_6H_4	H	$-(CH_2)_5-$		H	1608	65
H	$-(CH_2)_4-$		H	H	Me	1616	58
H	4-MeC_6H_4	H	$-(CH_2)_4-$		$OSiMe_3$	1627	66
H	Me	H	$-(CH_2)_4-$		$OSiMe_3$	1628	66
Me	Ph	OAc	H	H	O*n*-Bu	1618	67
H	Ph	H	H	H	OG*	1626	68
H	OBz	H	H	H	OG*	1622	69
CO_2Me	4-$NO_2C_6H_4$	H	H	H	OEt	1595	70
CO_2Me	4-$NO_2C_6H_4$	H	$-(CH_2)_4-$		OMe	1585	70
Me	Ph	H	$-(CH_2)_3-$		NG'_2	1610	71
Me	Ph	H	$-(CH_2)_4-$		NG'_2	1605	71
Me	Ph	H	$-(CH_2)_5-$		NG'_2	1610	71
Ph	Ph	H	$-(CH_2)_5-$		NG'_2	1560	71

The ^1H NMR spectra of monosubstituted, acyclic nitronates show two resonances in the range of 5.5 and 7.2 ppm (Table 2.17) (39,40). The peaks are nonequilibrating and are assigned to the two possible nitronate diastereomers. This result is in contrast to the silyl nitronates, in which the two diastereomers were observed to be in rapid equilibrium. Based on the measurement of their respective dipole moments (39), and comparison to known nitrones and oxime ethers (72,73), the downfield signal is assigned to the trans isomer.

The chemical shifts of HC(α) are shifted slightly downfield of the corresponding silyl nitronates. As expected, the addition of electron-withdrawing or conjugating groups shifts the resonance downfield.

The ^1H NMR data for HC(5) of five-membered, cyclic nitronates is presented in Table 2.18 (60–62). Though there are no reported examples with a hydrogen at C(3), the influence of the nitronate function can be observed at the second point of connection. The presence of electron-withdrawing groups at R^I provides a

TABLE 2.17. CHARACTERISTIC ^{1}H NMR RESONANCES FOR ACYCLIC ALKYL NITRONATES

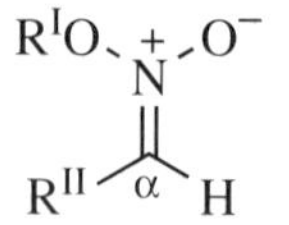

R^I	R^{II}	^{1}H NMR (HC(α), δ, ppm) cis	trans	Reference
Me	Me	5.91	6.25	40
Et	Me	5.57	6.27	40
Et	Et	6.05	6.11	40
Et	*n*-Pr	5.75	6.04	40
Et	4-BrC_6H_4	6.80	7.03	40
Et	4-$NO_2C_6H_4$	6.95	7.22	40
Et	CN	6.09	6.50	39
Et	COMe	6.55	6.83	39
Et	CO_2Me	6.47	6.80	39

slight downfield shift of HC(5), however, the overall range is relatively small, only 4.6–5.1 ppm.

In six-membered, cyclic nitronates, the ^{1}H NMR resonance for HC(3) appears in a similar range to that of HC(α) for alkyl-substituted, acyclic nitronates (Table 2.19) (14,58,65,68,69). Interestingly, there is an downfield shift of HC(6), when there is a ring fused at C(5)/C(6), as opposed to the attachment of an ether or methyl group.

In the case of five-membered, cyclic nitronates, the ^{13}C NMR resonances for C(3) appears in the same region as C(α) in the silyl nitronates (Table 2.20) (60,62,74). In addition, C(5) is observed between 72 and 92 ppm. For six-membered cyclic

TABLE 2.18. CHARACTERISTIC ^{1}H NMR RESONANCES FOR FIVE-MEMBERED, CYCLIC NITRONATES

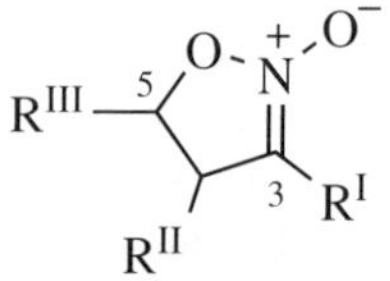

R^I	R^{II}	R^{III}	^{1}H (HC(5), δ, ppm)	Reference
CO_2Et	H	COMe	4.95	60
CO_2Et	Me	COEt	5.05	60
CO_2Et	$-CO(CH_2)_2-$		4.75	60
CO_2Me	CO_2Me	CO_2Me	5.00	61
Me	*i*-Pr	CO_2Et	4.65	62
Et	*i*-Pr	CO_2Et	4.65	62
$-(CH_2)_4-$		CO_2Et	4.57	62

TABLE 2.19. CHARACTERISTIC ^{1}H NMR RESONANCES FOR SIX-MEMBERED, CYCLIC NITRONATES

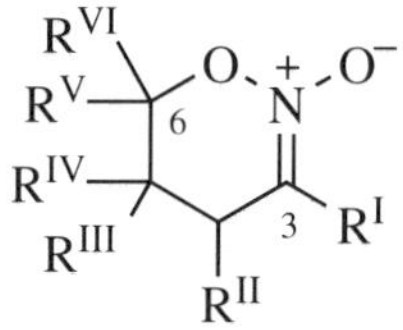

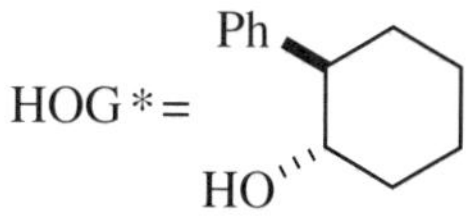

						^{1}H (δ, ppm)		
R^{I}	R^{II}	R^{III}	R^{IV}	R^{V}	R^{VI}	H(C3)	HC(6)	Reference
H	Ph	H	$-(CH_2)_4-$		OSiMe	6.20		14
H	Me	H	$-(CH_2)_4-$		$OSiMe_3$	6.25		14
H	4-$MeOC_6H_4$	H	$-(CH_2)_4-$		$OSiMe_3$	6.27		14
H	$-(CH_2)_4-$		H	H	Me	6.18	4.30	58
H	Ph	H	H	H	OG*	6.31	4.58	68
H	OBz	H	H	H	OG*	6.53	4.64	69
Me	Ph	H	$-(CH_2)_3-$		H		4.91	65
Me	4-$MeOC_6H_4$	H	$-(CH_2)_3-$		H		4.90	65
Me	4-$MeOC_6H_4$	H	$-(CH_2)_4-$		H		4.58	65
Me	4-$MeOC_6H_4$	H	$-(CH_2)_5-$		H		4.59	65

nitronates, the C(3) resonance is observed between 109 and 127 ppm (Table 2.21) (58,65,66,68,69). A downfield shift is observed when the nitronate ring is not substituted with an ether moiety.

The spectroscopic data for alkyl nitronates correlate well to that of silyl nitronates, indicating that the nature of the nitronate substituent has a negligible

TABLE 2.20. CHARACTERISTIC ^{13}C NMR RESONANCES FOR FIVE-MEMBERED, CYCLIC NITRONATES

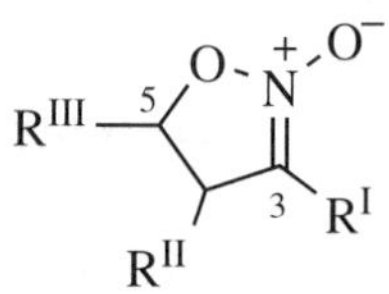

			^{13}C (δ, ppm)		
R^{I}	R^{II}	R^{III}	C(3)	C(5)	Reference
CO_2Et	H	COMe	106.9	77.5	60
CO_2Et	Me	COEt	112.4	82.2	60
CO_2Et	$-CO(CH_2)_2-$		109.5	76.2	60
Et	*i*-Pr	CO_2Et	116.3	71.9	62
$-(CH_2)_4-$		CO_2Et	115.0	77.3	74
$-(CH_2)_4-$		$-(CH_2)_4-$	118.6	92.4	74
$-(CH_2)_4-$		$-(CH_2)_4-$	118.1	82.2	74

TABLE 2.21. CHARACTERISTIC ^{13}C NMR RESONACES FOR SIX-MEMBERED, CYCLIC NITRONATES

R^I	R^{II}	R^{III}	R^{IV}	R^V	R^{VI}	^{13}C (δ, ppm) (C3)	C(6)	Reference
H	4-$MeOC_6H_4$	H	$-(CH_2)_4-$		$OSiMe_3$	113.2	104.9	66
H	Ph	H	H	H	OG*	114.6	102.3	68
H	OBz	H	H	H	OG*	109.8	101.9	69
Me	$-(CH_2)_4-$		H	H	Me	123.8	80.8	58
Me	$-(CH_2)_3-$		H	H	Me	126.2	79.5	58
Me	Ph	H	$-(CH_2)_3-$		H	124.3	83.4	65
Me	4-$MeOC_6H_4$	H	$-(CH_2)_3-$		H	125.0	84.4	65
Me	4-$MeOC_6H_4$	H	$-(CH_2)_4-$		H	119.6	74.9	65
Me	4-$MeOC_6H_4$	H	$-(CH_2)_5-$		H	121.5	79.4	65

effect on the functional group. Therefore the same trends observed with nitrones and oximes, in comparison to silyl nitronates, are applicable to alkyl nitronates.

2.1.2.4. Spectroscopic Data of Alkyl Nitroso Acetals

The nitroso acetals that result from the [3+2] cycloaddition of alkyl nitronates with dipolarophiles typically provide several characteristic spectroscopic signals for identification. As in the silyl counterparts, the O—N—O stretch in the alkyl nitroso acetals is observed in the IR ranges between 1000 and 1030 cm^{-1} (Tables 2.22 and 2.23) (57,75–77). However, this resonance is usually not strong, and can be obscured by other functional group resonances with more substituted nitroso acetals.

TABLE 2.22. CHARACTERISTIC IR ABSORBANCES FOR N-ALKOXY ISOXAZOLIDINES

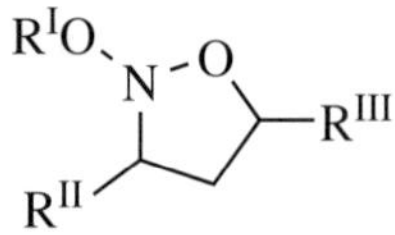

R^I	R^{II}	R^{III}	IR (cm^{-1})	Reference
Me	NO_2	CO_2Me	1020	75
Me	4-$NO_2C_6H_4$	CO_2Me	1030	76
Me	4-$NO_2C_6H_4$	CN	1000	76
Me	4-$MeOC_6H_4$	CO_2Me	1030	76

TABLE 2.23. CHARACTERISTIC IR ABSORBANCES FOR BICYCLIC NITROSO ACETALS

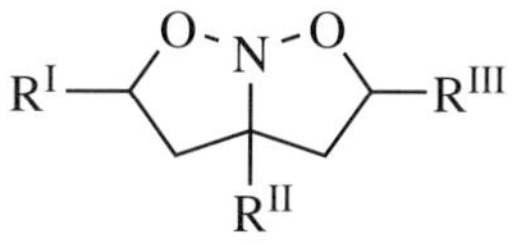

R^I	R^{II}	R^{III}	IR (cm^{-1})	Reference
H	Ph	H	1020	57
H	Ph	Ph	1030	57
H	CO_2Me	Ph	1030	57
H	NO_2	H	1030	77
H	NO_2	Me	1010	77

The ^{1}H NMR spectra of the nitroso acetal cycloadducts display low field resonances for both HC(3) and HC(5) (Table 2.24) (78). As with the corresponding N-silyloxy isoxazolidines, multiple diastereomers are formed. These arise because of the high inversion barrier at nitrogen, as well as different modes of approach of the dipolarophile (79–82). Due to the greater electronegativity of oxygen, with respect to nitrogen, the HC(5) is shifted downfield from that of HC(3). The same relationship is observed in the case of six membered cyclic nitronates where the HC(2) and HC(6) are observed downfield of HC(3a) (Table 2.25) (83,84). The electronegativity effects on diamagnetic shielding are also observed in the ^{13}C NMR spectra, where the resonances for C(2) and C(6) are shifted downfield of C(3a) (Table 2.26) (83,84). The former are typically observed between 81 and 108 ppm, while C(3a) is seen between 72 and 78 ppm.

TABLE 2.24. CHARACTERISTIC ^{1}H NMR RESONANCES FOR N-ALKOXY ISOXAZOLIDINES

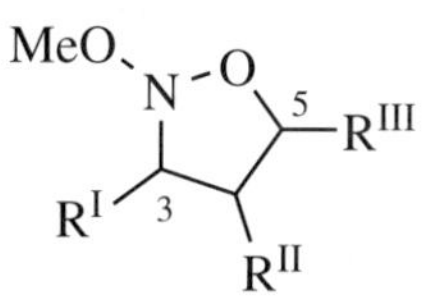

			^{1}H (δ, ppm)		
R^I	R^{II}	R^{III}	HC(3)	HC(5)	Reference
CO_2Me	CO_2Me	CO_2Me	4.34	5.34	78
			4.59	5.24	
			4.74	5.24	
			4.56	4.97	
CO_2Me	CN	CN	4.47	5.32	78
			4.82	5.39	
			4.68	5.16	
			4.86	5.75	

TABLE 2.25. CHARACTERISTIC ¹H NMR RESONANCES FOR BICYCLIC NITROSO ACETALS

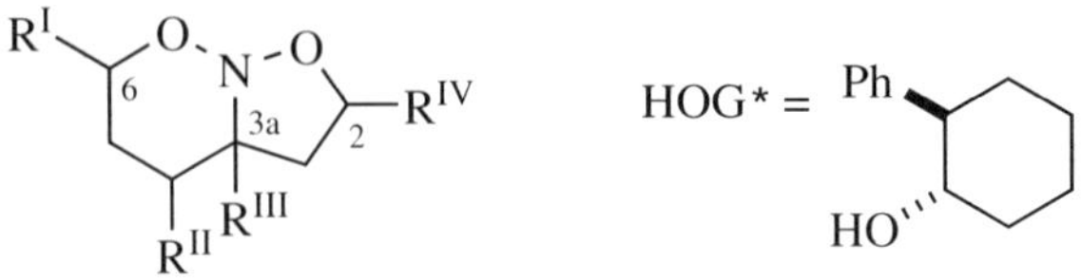

R^I	R^{II}	R^{III}	R^{IV}	¹H (δ, ppm) HC(2)	HC(3a)	HC(6)	Reference
OEt	Ph	H	OEt	5.65	3.84	4.92	83
OEt	Ph	H	CO_2Me	4.94	3.67	4.77	83
OEt	Ph	Me	CO_2Me	5.23		4.98	83
O*n*-Bu	OBz	H	CO_2Me	5.09	3.87	4.97	84
OG*	OBz	H	CH_2OH	4.66	3.65	4.20	84

TABLE 2.26. CHARACTERISTIC ¹³C NMR RESONANCES FOR BICYCLIC NITROSO ACETALS

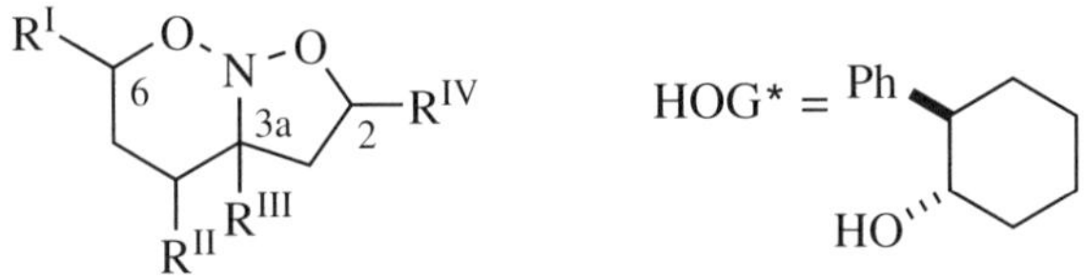

R^I	R^{II}	R^{III}	R^{IV}	¹³C (δ, ppm) C(2)	C(3a)	C(6)	Reference
OEt	Ph	H	OEt	100.0	72.7	107.5	83
OEt	Ph	H	CO_2Me	81.4	75.4	99.9	83
OEt	Ph	Me	CO_2Me	81.4	77.2	100.0	83
O*n*-Bu	OBz	H	CO_2Me	80.5	72.0	98.2	84
OG*	OBz	H	CH_2OH	86.8	73.8	99.1	84

TABLE 2.27. CHARACTERISTIC ¹H NMR RESONANCES FOR TRICYCLIC NITROSO ACETALS

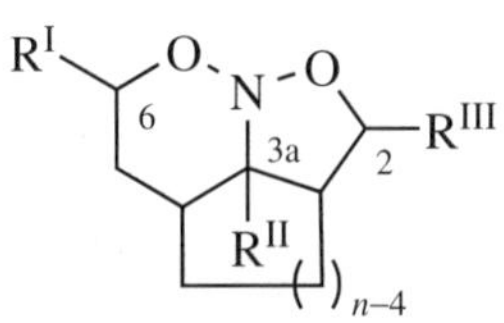

R^I	R^{II}	R^{III}	*n*	¹H (δ, ppm) HC(2)	HC(3a)	HC(6)	Reference
O*n*-Bu	Me	Me	5	4.50		5.06	85
O*n*-Bu	Me	H	6	4.43	3.52	4.76	85
O*n*-Bu	Me	Me	6	4.40		4.94	85

TABLE 2.28. CHARACTERISTIC ^{13}C NMR RESONANCES FOR TRICYCLIC NITROSO ACETALS

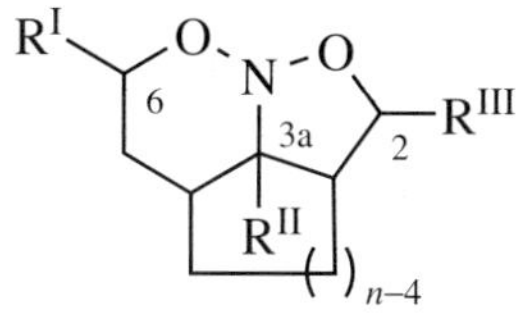

R^I	R^{II}	R^{III}	n	^{13}C (δ, ppm) C(2)	C(3a)	C(6)	Reference
O*n*-Bu	Me	Me	5	86.2	83.1	99.6	85
O*n*-Bu	Me	H	6	85.3	68.3	100.1	85
O*n*-Bu	Me	Me	6	88.9	72.0	99.7	85

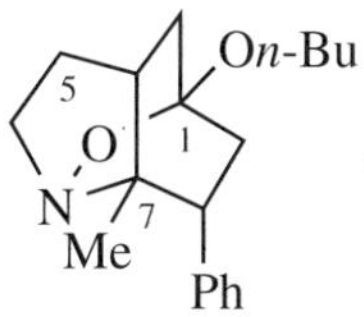

(26)

^1H (δ, ppm)		^{13}C (δ, ppm)	
HC(5)	3.64	C(5)	77.29
		C(7)	68.98
		C(1)	96.66

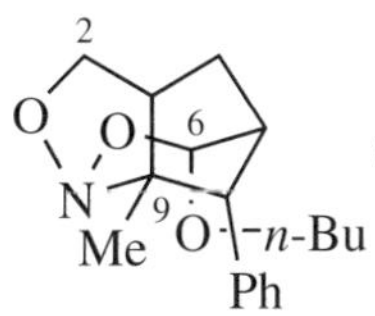

(27)

^1H (δ, ppm)		^{13}C (δ, ppm)	
HC(2)	4.05	C(2)	77.68
		C(9)	81.07
HC(6)	4.69	C(1)	106.32

Figure 2.9. The NMR data for selected tricyclic nitroso acetals.

The spectral data for nitroso acetals arising from intramolecular nitronate cycloadditions mirror those of the previously presented nitroso acetals. Selected examples are collected in Tables 2.27 and 2.28, as well as in Figure 2.9 (85–87).

2.2. GENERAL REACTIVITY

Following the initial report from Tartakovskii et al. (12) describing the successful utilization of a nitronate in a [3+2]-dipolar cycloaddition, there has been much interest in the development and exploration of this process. Since that time a wide variety of nitronates has been prepared and shown to successfully engage in [3+2]-dipolar cycloaddition (30,88,89). In general, these reactions proceed under mild conditions with high regio- and moderate to high relative stereoselectivity. Both electron-rich and electron-deficient alkenes have been used in the cycloaddition of nitronates, however, reactions with the latter dipolarophiles are more facile. Cycloadditions of unactivated alkenes are mainly relegated to intramolecular cases. As expected, additional substituents on either the dipole or the dipolarophile lower the reaction rate because of the increased steric bulk. However, the incorporation electron-withdrawing or electron-donating groups on the nitronate results in an increase in reactivity compared to a similarly substituted dipole.

The relative reactivity of electronically disparate dipolarophiles has been demonstrated in competition experiments between ethyl vinyl ether and a variety of electron deficient alkenes, Eq. 2.3 (90). Capture of the nitronate formed *in situ* from the hetero-[4+2] cycloaddition of ethyl vinyl ether and nitrostyrene is observed only with electron poor alkenes. None of the adduct corresponding to reaction with ethyl vinyl ether was observed. This difference in reactivity is also anticipated by computational analysis (91). The B3LYP/6-31G* calculation of the frontier molecular orbitals (FMO) of methyl vinyl ether (MVE), methyl vinyl ketone (MVK), and the nitronate **31** show a smaller energy gap between the highest occupied molecular orbital of the dipole ($HOMO_{dipole}$) and the lowest occupied molecular orbital of the MVK ($LUMO_{MVK}$) then the $HOMO_{dipole}$ and the $LUMO_{MVE}$ (Fig. 2.10). This suggests a higher reactivity of the electron deficient alkene. Subsequent calculation of the reactive potential energy surface reveals that the transition state energy is lower for the [3+2] cycloaddition of MVK. Interestingly, the nitroso acetal resulting from [3+2] cycloaddition with methyl vinyl ether was found to be 6.4 kcal/mol more stable than the corresponding MVK adduct, presumably due to anomeric stabilization of the product.

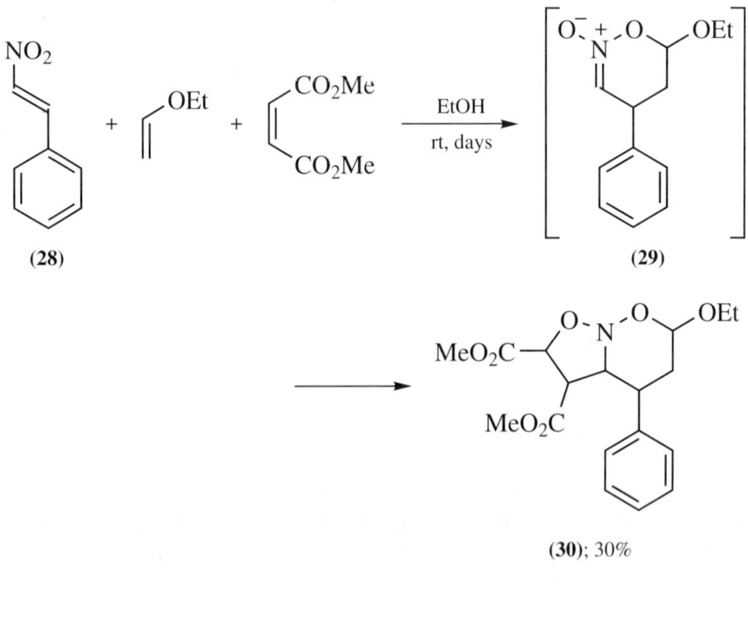

(2.3)

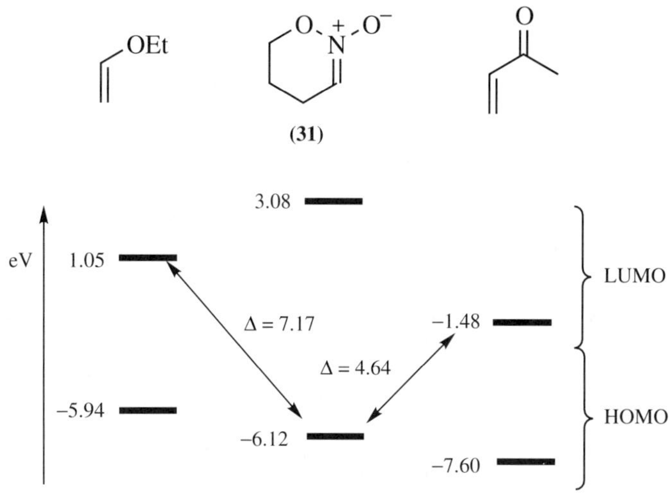

Figure 2.10. Calculated FMO of ethyl vinyl ether, ethyl vinyl ketone, and nitronate (**31**).

As observed in other dipolar cycloadditions, there are three stereochemical issues that must be addressed (Scheme 2.3). These issues include the regioselectivity, the stereoselectivity, and the facial selectivity of the cycloaddition. The first two will be discussed in this section, while the latter will be discussed in Sections 2.3 and 2.4 as it relates to specific examples.

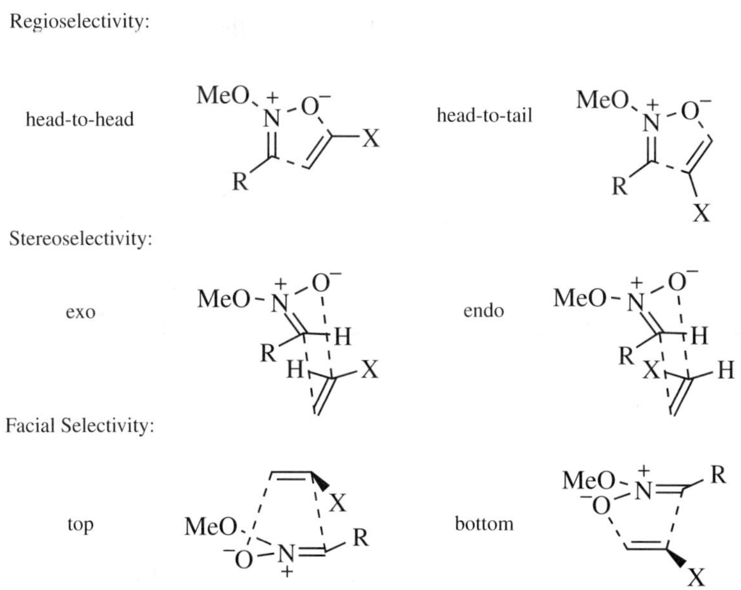

Scheme 2.3

2.2.1. Regioselectivity

Because of the intrinsic structural asymmetry of this dipole, there exist two possible regioisomers resulting from the cycloaddition with unsymmetrical dipolarophiles. The reaction of a monosubstituted dipolarophile with a nitronate, in a head-to-head fashion provides a 5-substituted isoxazolidine (Scheme 2.4). Alternately, the head-to-tail combination of the coupling partners results in a 4-substituted isoxazolidine. With only a few exceptions (92), the 5-substituted isoxazolidine is formed exclusively.

Several groups have attempted to rationalize the observed regioselectivity by use of FMO analysis. INDO calculations of nitronates bearing electron-withdrawing groups provided an energy difference between the two possible combinations that is too small to determine definitively the orbital interactions (78). It is therefore not possible to predict the regioselectivity of the cycloaddition on the basis of these calculations alone. Computational analysis on simpler nitronates have come to similar conclusions, which is the case for both semiempirical calculations (PM3) of electron-rich olefins (93) as well as AM1 calculations of electron-deficient alkenes (94). Interestingly, in the absence of an attached electron-withdrawing group, the larger coefficient for the nitronate resides on the carbon atom in both the HOMO and LUMO. Therefore in the case of electron-deficient dipolarophiles where the larger coefficient resides on the α carbon in both frontier orbitals, both combinations of dipole and dipolarophile result in 5-substituted isoxazolidines (84). However, this is not the case with electron-rich dipolarophiles, where the larger coefficient changes between the HOMO and the LUMO (93).

Two independent calculations of the limiting transition states on the potential energy surface help to confirm the preference for the head-to-head orientation.

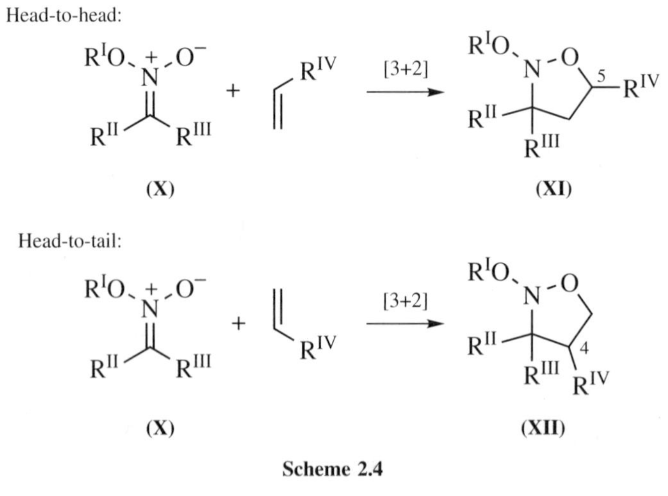

Scheme 2.4

Semiempirical calculations show that the approach of the dipolarophile in a head-to-head manner is 5.91 kcal/mol more favorable than the head-to-tail orientation with electron-rich alkenes (93). This difference is manifest in the ground state of resulting cycloadducts where the head-to-head adduct is 6.6 kcal/mol more stable. Similarly, the head-to-head transition state is more favorable by 3.4 kcal/mol for electron-poor alkenes, and 7.4 kcal/mol for electron-rich alkenes as calculated with density functional theory (91). This preference is also carried over into the ground-state energies of the resulting cycloadducts.

Cycloadditions involving disubstituted dipolarophiles have been mainly limited to either symmetrically substituted alkenes or those delivered intramolecularly by tethers. Neither of the cases embodies a regiochemical issue. However, there has been a systematic study of the regiochemical (and stereochemical) course of intermolecular cycloaddition of unsymmetrical alkenes to cyclic nitronates (84). The substituted dipolarophiles employed in this study are illustrated in Chart 2.1. All dipolarophiles contain an electron-withdrawing substituent (ketone or ester) and the second substituent is a carbon, silicon, or oxygen functional group (in both configurational forms). The results, collected in Table 2.29, reveal a decreasing preference for the head-to-head approach as a function of the β-substituent in the order H > Si > C > O. Since the observed selectivities are inconsistent with the presumed steric effects of the substituents, as well as the FMO analysis, other predictive tools were investigated. There is a good correlation between the regioselectivity of the dipolarophile and its *polarity index* (*P*) (Fig. 2.11) (95). This value is the ratio of the relative charge densities of the α and β carbon of the olefin, based on HF/6-31G* calculations. This model has subsequently been shown to predict the observed regioselectivity for other disubstituted alkenes (84).

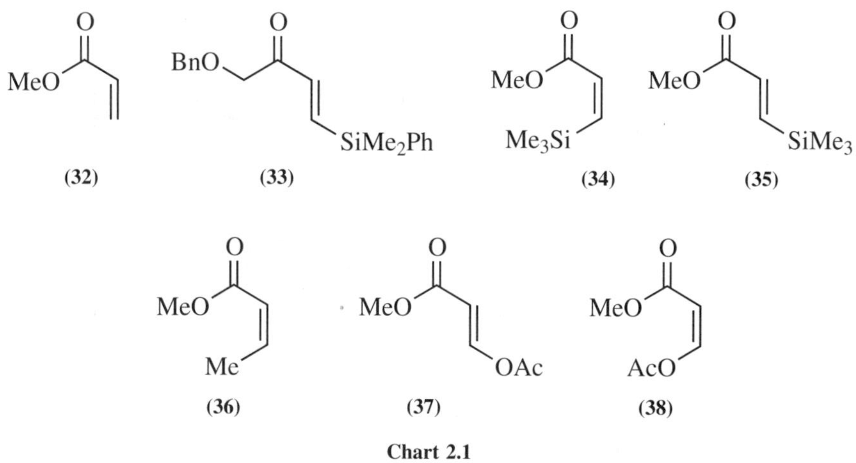

Chart 2.1

TABLE 2.29. REGIOSELECTIVITY AND ELECTROSTATIC CHARGES OF DIPOLAROPHILES **32–38**

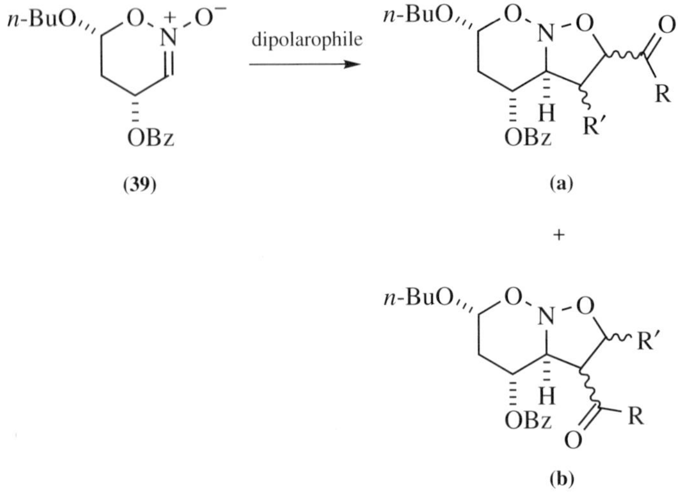

Dipolarophile	Product	a : b	Electrostatic Atomic Charge [C(α)]	Electrostatic Atomic Charge [C(β)]	[C(α) + 1]/ [C(β) + 1] (*P*)	log (**a**/**b**)
32	**40**	1:0				
33	**41**	32:1	−0.22	−0.42	1.34	1.51
34	**42**	2.4:1	−0.42	−0.33	0.87	0.38
35	**43**	2.1:1	−0.51	−0.28	0.67	0.32
36	**44**	1:1.7	−0.64	0.12	0.32	−0.23
37	**45**	1:2.7	−0.68	0.31	0.24	−0.43
38	**46**	0:1	−0.76	0.36	0.18	

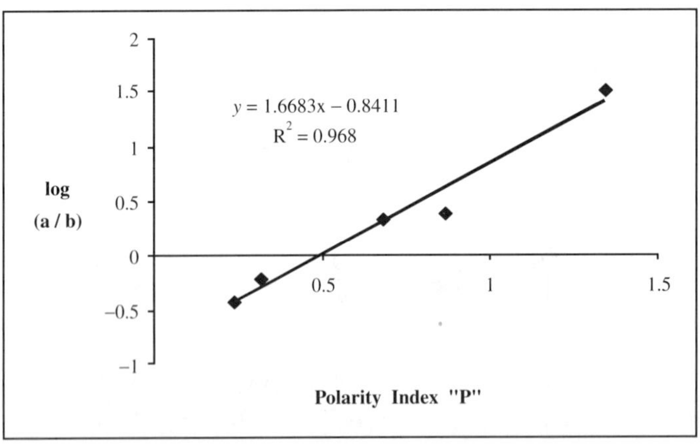

Figure 2.11. Plot of the polarity index vs log (stereoselectivity).

2.2.2. Stereoselectivity

2.2.2.1. *Intermolecular Cycloadditions*

In addition to the regioselectivity of the cycloaddition, there also exists a question of stereoselectivity. This issue involves the location of the substituent X on the dipolarophile in either an exo or endo fashion (Scheme 2.3). In the case of simple alkyl nitronates bearing an electron-withdrawing substituent, the stereoselectivity is highly dependent on both the nature of the dipolarophile and the configuration of the nitronate (Scheme 2.5) (96). In the case of nitronate **47**, either a mixture of diastereomers or only the exo adduct is observed. However, the cycloaddition of maleic anhydride with **47** provides only the endo stereoisomer (97).

The stereoselectivity of monosubstituted dipolarophiles has also been studied with cyclic nitronates (Table 2.30) (84). In most cases, an exo selectivity was observed. The ratio between the endo and exo adducts can be correlated to the size of the substituents on the dipolarophile. Because of the endo preference observed with acrolein, it is believed that there is a slight electronic preference for the endo orientation in the transition state, in the absence of steric hindrance. In line with these results is the observation that, for **49**, maleic anhydride reacts with complete exo selectivity, in contrast the cycloaddition with **47** (69).

INDO and HF/6-31G* calculations suggest that secondary orbital interaction of the dipolarophile and the lone pair of electrons on the dipole nitrogen would favor the endo isomer (43,93). Since this result is not typically observed experimentally,

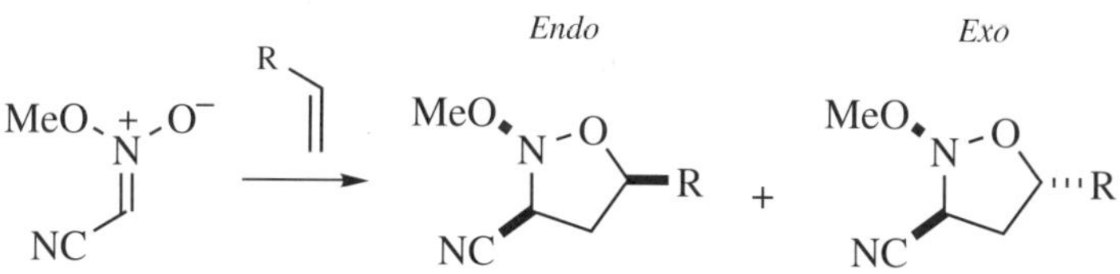

(*Z*)-(**47**) (**48a**) (**b**)

R =	(48a)		(b)
CN	1	/	1
CO_2Me	0	/	1
Ph	0	/	1

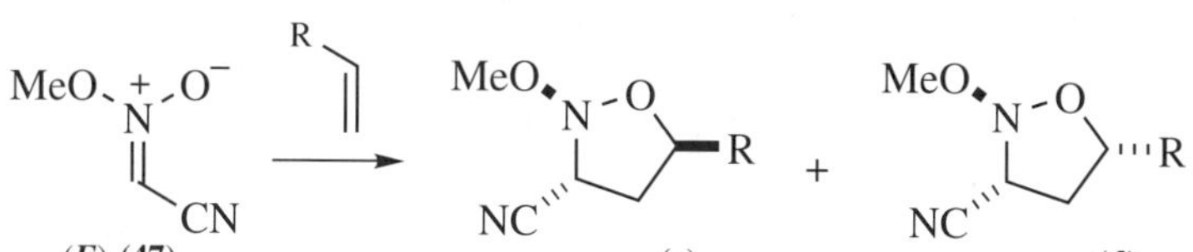

(*E*)-(**47**) (**c**) (**d**)

R =	(c)		(d)
CN	1	/	1
CO_2Me	1	/	1
Ph	0	/	1

Scheme 2.5

TABLE 2.30. STEREOSELECTIVITY OBSERVED FOR THE CYCLOADDITION OF **49** WITH VARIOUS DIPOLAROPHILES

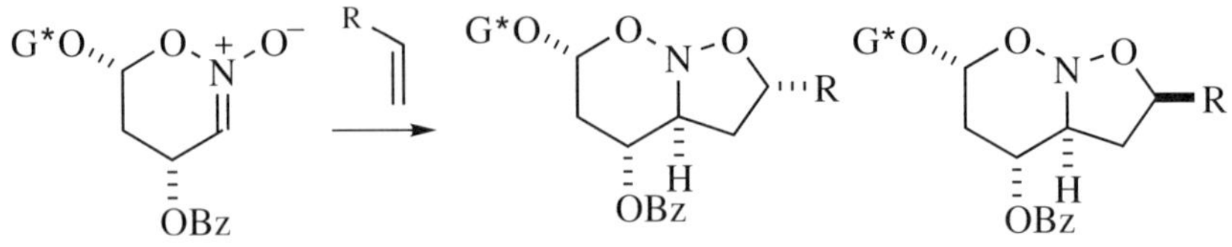

(49) **(a)** **(b)**

G* = (1*S*,2*R*)-2-phenylcyclohexyl

R[a]	Product	dr (**a**:**b**)[b]
CO_2t-Bu	**50**	11.4:1
CO_2Me	**51**	6.5:1
COMe	**52**	6.2:1
$COCH_2OTDS$	**53**	5:1
$COCH_2OBn$	**54**	5.7:1
CH_2OH	**55**	1.3:1
CHO	**56**	1:2

[a]Thexyldimethylsilyl = TDS.
[b]diasteromeric ratio = dr.

it is suggested that stereoselectivity must be sterically driven. Calculation of the enthalpy of formation of the two possible cycloadducts shows less than 1-kcal/mol stabilization of the exo isomers, however, B3LYP/6-31G* calculation shows a preference for the exo transition state by 5.88 kcal/mol (93). This result supports both the belief that the exo approach of the dipolarophile is preferred and that these reactions are not reversible.

2.2.2.2. *Intramolecular Cycloadditions*

The stereoselectivity in intramolecular nitronate cycloadditions is primarily controlled by the nature of the tether to the dipolarophile, in the absence of additional directing groups. In these cases, the exo **(XIII)** and endo **(XIV)** approach of the dipolarophile is described in terms of the fold of the tether (Fig. 2.12). Because of geometrical restrictions and tether–nitronate interactions, the length of the tether has the greatest impact on the approach of the alkene. In the case of acyclic, silyl nitronates, a three-atom tether provides 99:1 stereoselectivity, while the corresponding four-atom tether only provides a 2.6:1 mixture (Table 2.31) (98). Because of the rapid migration of the silyl group, both nitronate configurations must be addressed. For the shorter tether, the major isomer results from an endo approach of the tether to a (*Z*)-configured nitronate, whereas the minor isomer is produced by an exo approach of the tether to an (*E*)-configured nitronate. However, molecular models show no significant differentiation between the two transition structures. By

Stereoselectivity:

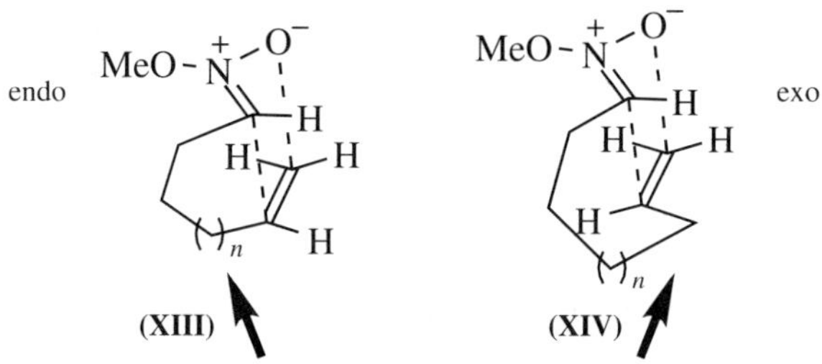

Figure 2.12. Exo and endo modes for intramolecular [3 + 2] cycloadditions.

lengthening the tether to four atoms, the stereoselectivity decreases, but is still proposed to be the result of an endo approach of the dipolarophile.

The combination of the geometrical preference of the tether and the stereochemical preference of the dipolarophile substituent can be seen in the intramolecular cycloadditions of alkyl nitronates, (Scheme 2.6) (99). When the tether is restricted to two atoms, only the endo approach of the tether is observed in up to a 100:1 ratio, independent of the configuration of the disubstituted dipolarophile. However, in the case of a three-atom linker, there exists a matched and mismatched case with respect to the observed stereoselectivities. With a (*Z*)-configured dipolarophile, only the exo isomer was observed since the ester moiety also approaches on the exo to the nitronate. However, with an (*E*)-configured dipolarophile, the ester group is forced to approach in an endo manner to accommodate an exo approach of the tether, thus leading to lower selectivity.

TABLE 2.31. STEREOSELECTIVITY IN THE INTRAMOLECULAR CYCLOADDITION OF SILYL NITRONATES

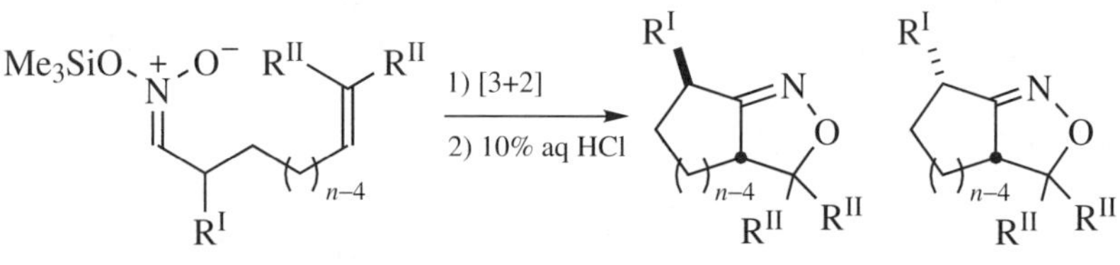

(57)–(62) **(a)** **(b)**

Entry	Dipole	R^I	R^{II}	n	Product	dr (**a**:**b**)
1	**57**	Ph	H	5	**63**	99:1
2	**58**	4-MeOC$_6$H$_4$	H	5	**64**	99:1
3	**59**	*i*-Pr	H	5	**65**	99:1
4	**60**	Ph	Me	5	**66**	95:1
5	**61**	Ph	H	6	**67**	1:2.6
6	**62**	4-MeOC$_6$H$_4$	H	6	**68**	1:2.6

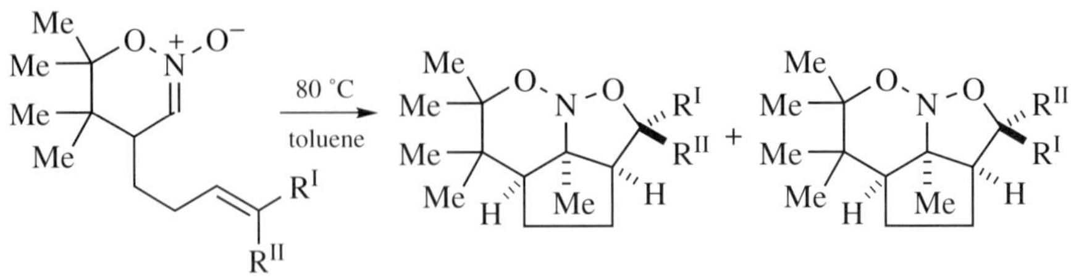

(**69**): $R^{I} = CO_2Me$, $R^{II} = H$ (**70a**) /(**b**) = 100 / 1

(**71**): $R^{I} = H$, $R^{II} = CO_2Me$ (**72a**) /(**b**) = 20 / 1

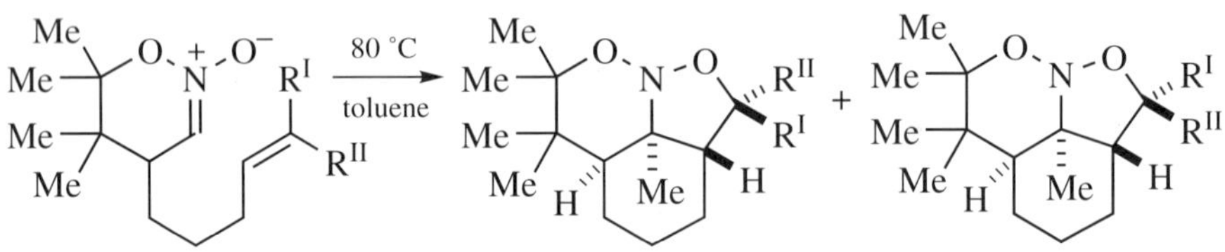

(**73**): $R^{I} = CO_2Me$, $R^{II} = H$ (**74a**) /(**b**) = 2.6 / 1

(**75**): $R^{I} = H$, $R^{II} = CO_2Me$ (**76a**) /(**b**) = 100 / 1

Scheme 2.6

2.2.3. Mechanism

Several computational studies have addressed whether the dipolar cycloaddition of nitronates is a concerted or stepwise process (93,100). Natural population analysis reveals that their is very little zwitterionic character in the transition state. The formation of the C—C bond marginally precedes the C—O bond on the basis of calculated bond lengths and orders in the transition structure. These calculations also show that the reaction is a concerted process that is slightly asynchronous. In addition, the cycloaddition likely proceeds through an early transition state and is overall an exothermic process.

2.3. SILYL NITRONATES

2.3.1. Preparation

The first generation of a silyl nitronate was described in a report on the general silylating ability of trimethylsilyldiphenylurea (**77**, Eq. 2.4) (17). In this case, the silylation of nitromethane resulted in the formation of bis(trimethylsilyl)methazonic acid (**79**). Presumably this dimer arose from the condensation between the intermediate silyl nitronate **78** and excess nitromethane. With substituted nitro

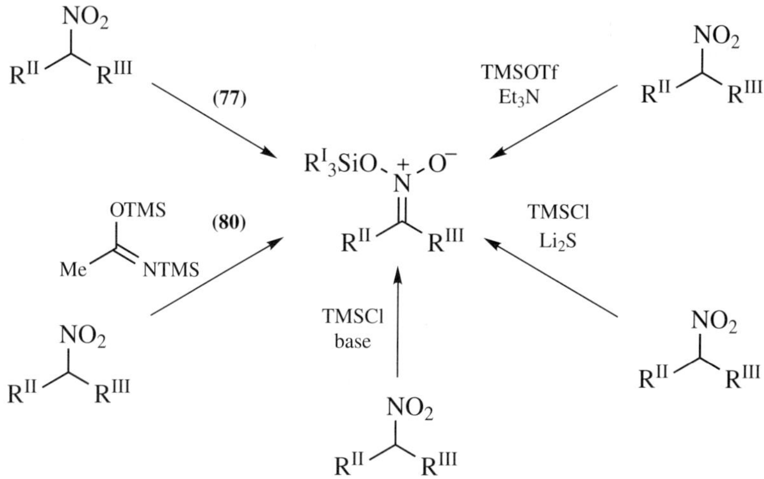

Scheme 2.7

compounds, this self-condensation is not a problem. The silyl urea **77** is effective in synthesizing the silyl nitronates of dimethylnitromalonate, methyl nitroacetate, and dinitromethane (Scheme 2.7) (16,25). However, nitroalkanes lacking these electron-withdrawing groups have been shown to be incompatible. Instead, bis(trimethylsilyl)acetamide (**80**) provides the corresponding silyl nitronates in nearly quantitative yields (20). This method, though, requires elevated temperatures and has not been demonstrated for alkyl groups larger than propyl.

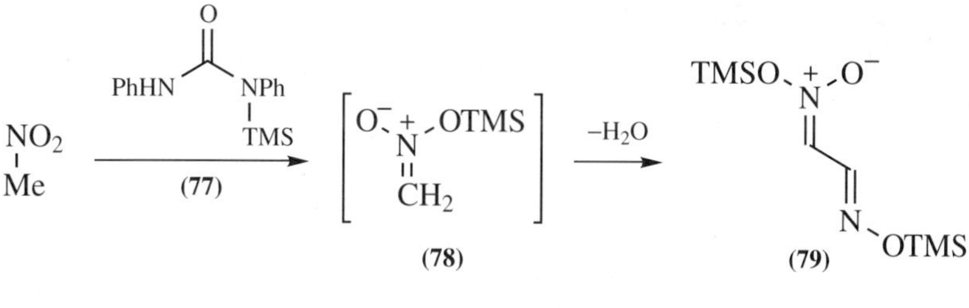

(2.4)

A more common method for the preparation of silyl nitronates is the use of trimethylsilyl chloride (TMSCl) in the presence of a base. With triethylamine, silyl nitronates are prepared from primary nitroalkanes in moderate yields; however, it is necessary to conduct the silylation in acetonitrile for good yields with secondary nitroalkanes (18,101). In several cases, this silylation has been done in the presence of the dipolarophile for both inter- and intramolecular processes, or the nitronate has been used in subsequent reactions without purification (18,22). Employment of 1,8-diazabicyclo(5.4.0)undec-7-ene (DBU) as the base allows this procedure to be general for most nitroalkanes (19).

There are also several variations on this procedure. The use of trimethylsilyl triflate (TMSOTf) provides the silyl nitronate of methyl nitroacetate in good yield. However, for primary nitroalkanes, a second silylation occurs at the α-position of the nitronate (Eq. 2.5) (102). The use of TMSCl in the presence of lithium sulfide provides good yields of silyl nitronates from secondary nitroalkanes (103,104). Unfortunately, the number of examples is limited and this procedure is not applicable to primary nitroalkanes.

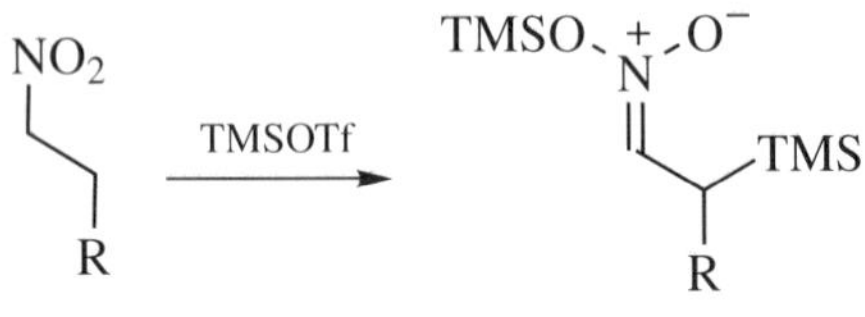

(XV) **(XVI)**

(2.5)

It has also been demonstrated that the nitronate salt can also be trapped with various chlorosilanes. For example, the silver salt of methyl nitromalonate can be effectively silylated with TMSCl (16). Alternatively, deprotonation of a nitroalkane with lithium diisopropylamide (LDA) results in the lithium nitronate, which upon addition of TMSCl, provides the corresponding silyl nitronate (105). This method proceeds in good yields in the cases of primary nitro compounds; however, the yields are significantly lower for secondary nitroalkanes. Nitronate salts have also been prepared by the addition of nucleophiles to nitroalkenes (98,106). Both allylic alkoxides and allylic Grignard reagents have been used for this purpose (Eq. 2.6 and 2.7), which has also been extended to the use of the lithium salts of nitromethane, acetonitrile, acetone, ethyl acetate, and methyl phosphonate (Scheme 2.8) (107). Upon addition of a silylation agent, the corresponding cycloadduct occurs upon warming to room temperature. This approach is hampered by side reactions when bulky nitroalkenes are used.

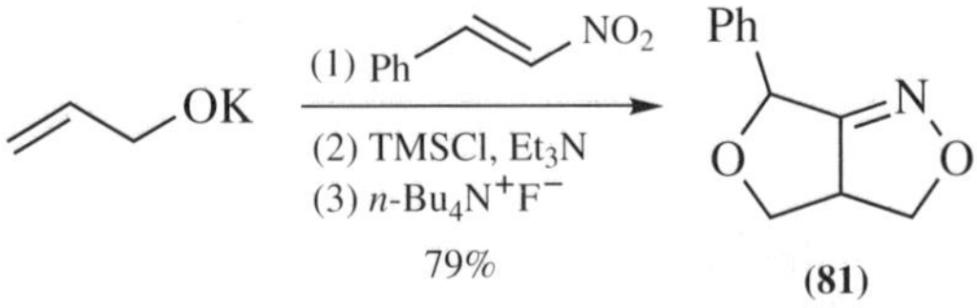

(81)

(2.6)

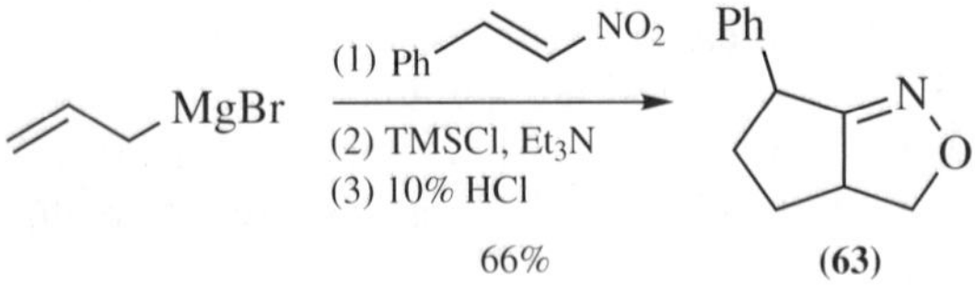

(63)

(2.7)

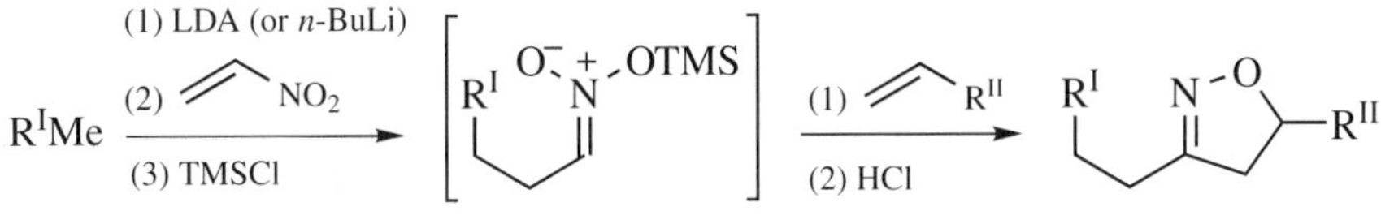

R^I = CN, NO_2, CO_2Et, COMe, P(O)$(OEt)_2$
R^{II} = CO_2Me, COMe

Scheme 2.8

2.3.2. Cycloadditions of Silyl Nitronates

The [3+2] cycloaddition of silyl nitronates has been extensively investigated since the first pioneering studies by Ioffe et al. (16). This transformation is attractive because of the resulting isoxazolidine can be easily converted to the corresponding isoxazoline for which a myriad of transformations are known. Moreover, this procedure provides yields and selectivities different from those of the nitrile oxide [3+2] cycloaddition, which affords the isoxazoline directly. The reaction of silyl nitronates has been briefly reviewed in the context of the chemistry of nitronic acid derivatives (27,30).

This section shall consider the effects of substitution on both the nitronate as well as the dipolarophile, as they relate to both the inter- and intramolecular versions of the dipolar cycloaddition. Also included will be a discussion of facial selectivity in the reaction of a chiral dipolarophile.

2.3.2.1. Intermolecular Cycloadditions

The dipolar cycloaddition of monosubstituted silyl nitronates with methyl acrylate proceeds smoothly to provide the corresponding isoxazolidine (Table 2.32) (18,24,25). In the case of simple alkyl substitution, longer reaction times or elevated temperatures are necessary (entries 2 and 3). However, yields of the resulting isoxazolidines increase as a function of increasing alkyl chain length, which is likely due to increased stability of the nitronate and not greater reactivity. The effect of substitution on the dipole has been investigated only with electron-withdrawing groups (entries 4 and 5). The incorporation of either a methoxycarbonyl or a trifluoromethyl group leads to increased reactivity of the nitronate.

The nitronates derived from secondary nitroalkanes suffer from greatly decreased reactivity (Table 2.33). The reaction of the silyl nitronate of methyl nitromalonate with methyl acrylate proceeds in 96 h at room temperature (entry 1), while the corresponding monosubstituted nitronate (**85**) proceeds in 1 h (entry 5, Table 2.32) (16). Nitronates with dialkyl substitution typically require elevated temperatures for complete reaction (101). For example, the silyl nitronate of 2-nitropropane and nitrocyclopentane both react in under 3 h at 50 °C in acetonitrile

TABLE 2.32. DIPOLAR CYCLOADDITIONS OF MONOSUBSTITUTED SILYL NITRONATES WITH METHYL ACRYLATE

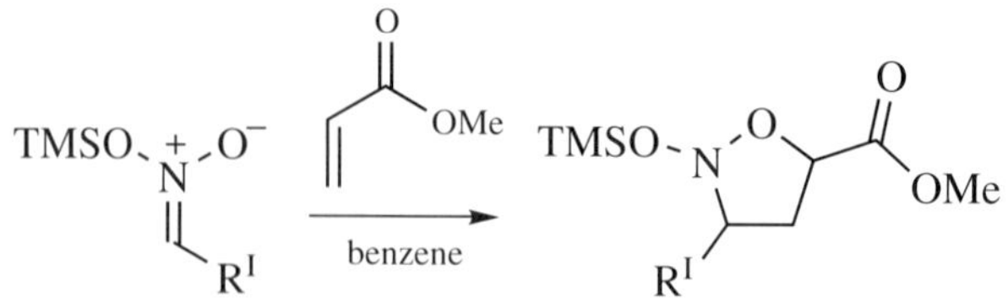

(78),(82)–(85) (86)–(90)

Entry	Dipole	R^I	Product	Temp (°C)	Time (h)	Yield (%)
1	**78**	H	**86**	20	18	57
2	**82**	Me	**87**	80	0.5	64
3	**83**	Et	**88**	20	72	86
4	**84**	CF_3	**89**	20	16	71
5	**85**	CO_2Me	**90**	14	1	96

with methyl acrylate (entries 2 and 3, Table 2.33). Nitronates bearing larger groups such as that of 4-nitropentanoate require 80 °C over several hours, and provide moderate yields of the cycloadduct (entry 4).

The influence of the silyl group on the reactivity of the nitronate was found to be negligible (16). Competition studies between trimethylsilyl and triphenylsilyl nitronates showed the two dipoles to be equally effective in performing [3+2] cycloadditions.

The steric and electronic properties of the dipolarophile have a large impact on the rate and yield of the cycloaddition. In the case of simple monosubstituted silyl nitronates, the [3+2] cycloaddition proceeds smoothly with dipolarophiles bearing electron-withdrawing or conjugating groups (Table 2.34) (20,101,108,109).

TABLE 2.33. DIPOLAR CYCLOADDITION OF DISUBSTITUTED SILYL NITRONATES WITH METHYL ACRYLATE

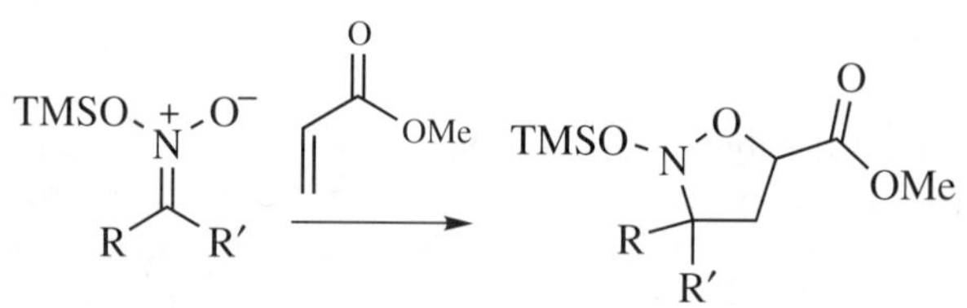

(6), (14), (91)–(92) (93)–(96)

Entry	Dipole	R^I	R^{II}	Product	Temp (°C)	Time (h)	Yield (%)
1	**6**	CO_2Me	CO_2Me	**93**	20	96	80
2	**14**	Me	Me	**94**	50	1	93
3	**91**	$-(CH_2)_4-$		**95**	50	2.5	96
4	**92**	Me	$(CH_2)_2CO_2Me$	**96**	80	3	85

TABLE 2.34. DIPOLAR CYCLOADDITIONS OF ALKYL SUBSTITUTED NITRONATES WITH VARIOUS DIPOLAROPHILES

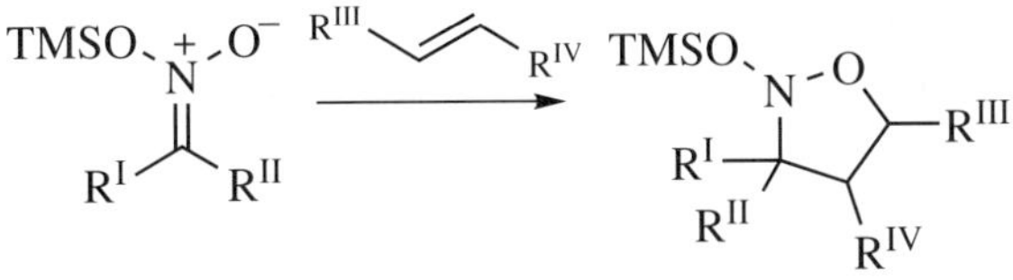

(14), (82)–(83) **(93), (97)–(98), (102)–(104)**

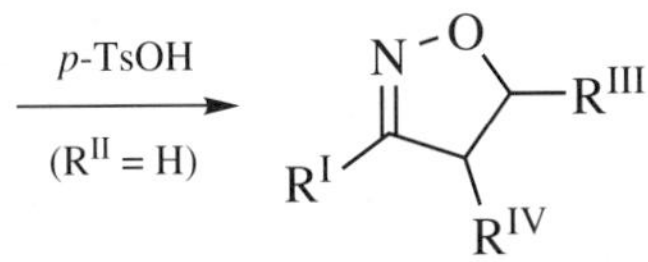

(99)–(101)

Entry	Dipole	R^I	R^{II}	R^{III}	R^{IV}	Product	Temp (°C)	Time (h)	Yield (%)[a]
1	**14**	Me	Me	CO_2Me	H	**93**	50	1	93
2	**14**	Me	Me	CN	H	**97**	50	2	96
3	**14**	Me	Me	CHO	H	**98**	20	1	60
4	**83**	Et	H	COMe	H	**99**	20	12	67[b]
5	**83**	Et	H	COEt	H	**100**	20	12	58[b]
6	**83**	Et	H	$CH{=}CH_2$	H	**101**	20	72	30[b]
7	**82**	Me	H	CO_2Me	H	**102**	80	0.5	64
8	**82**	Me	H	Ph	H	**103**	80	2	69
9	**82**	Me	H	CO_2Me	Me	**104**	50	48	30

[a] Yield of isoxazolidine.
[b] Yield of corresponding isoxazoline.

In addition, phenylsufonylallene (110), α,β-unsaturated phosphonates (111), and alkenes with perfluorinated substituents (112) are all useful dipolarophiles. The yields observed with methyl 2-propenoate are significantly lower than those with the corresponding acrylate (entries 7 and 9), because of the additional substituent. On the other hand, the dipolar cycloadditions with either ethyl vinyl ether, 1-hexene, cyclohexene, or a trisubstituted dipolarophile provide the corresponding isoxazolidines in either low yields or not at all (18).

A similar effect of the dipolarophile substituents is observed in the cycloadditions of the nitronate derived from trifluoronitroethane (**84**, Table 2.35) (24) In the absence of electron-withdrawing groups on the alkene, poor yields of the cycloadducts are obtained (entries 2 and 3). In addition, methyl substitution at either the α or β position of the dipolarophile significantly decreases both the rate and yield (entries 5 and 6). Similar results are observed with nitronates bearing with ester or nitro groups (16,25,26,34).

Heteroatomic dipolarophiles are competent in the dipolar cycloaddition of nitronates. The *in situ* generated thioaldehydes and thioketones react with silyl nitronate **120** to afford the 1,4,2-oxathiazolidine in good yield (Table 2.36) (113–116).

TABLE 2.35. EFFECT OF STERIC BULK ON THE DIPOLAR CYCLOADDITION OF SILYL NITRONATES

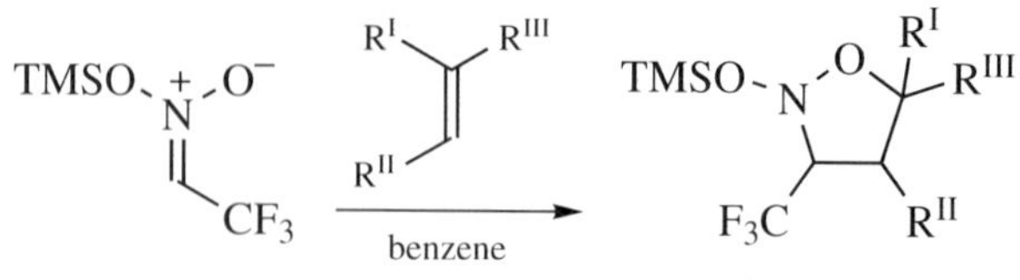

(84) (89), (105)–(109)

Entry	R^I	R^{II}	R^{III}	Product	Time (h)	Yield (%)
1	H	H	CO_2Me	**89**	16	80
2	H	H	*n*-Bu	**105**	120	20
3	H	H	Ph	**106**	40	39
4	H	H	COMe	**107**	16	86
5	Me	H	CO_2Me	**108**	16	48
6	H	Me	CO_2Me	**109**	16	0

2.3.2.2. *Facial Selectivity*

Only a few attempts to control the facial selectivity of this [3+2] process are on record, all dealing with the use of chiral, non-racemic dipolarophiles (117). The reactions of a vinyl substituted cephem (**121**) with the silyl nitronates derived from nitromethane, nitroethane, and nitropropane proceed over 3 days at room temperature to provide a single stereoisomer in moderate yields, Eq. 2.8 (118,119). Approach of the simple nitronate to the dipolarophile is believed to be from the less hindered α-face, however, the configuration of the newly created stereocenter could not be unambiguously assigned.

TABLE 2.36. DIPOLAR CYCLOADDITIONS OF A SILYL NITRONATE WITH HETERODIPOLAROPHILES

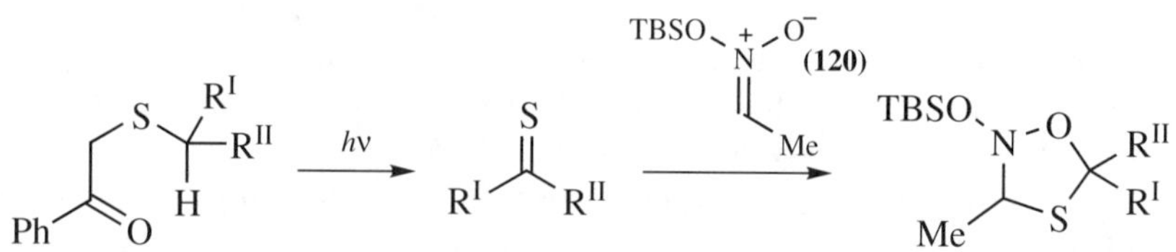

(110)–(114) (115)–(119)

Entry	Dipolarophile	R^I	R^{II}	Product	Yield (%)
1	**110**	H	H	**115**	73
2	**111**	H	*t*-Bu	**116**	93
3	**112**	H	$CH{=}CH_2$	**117**	92
4	**113**	H	$(CH_2)_2Ph$	**118**	91
5	**114**	$-(CH_2)_{11}-$		**119**	90

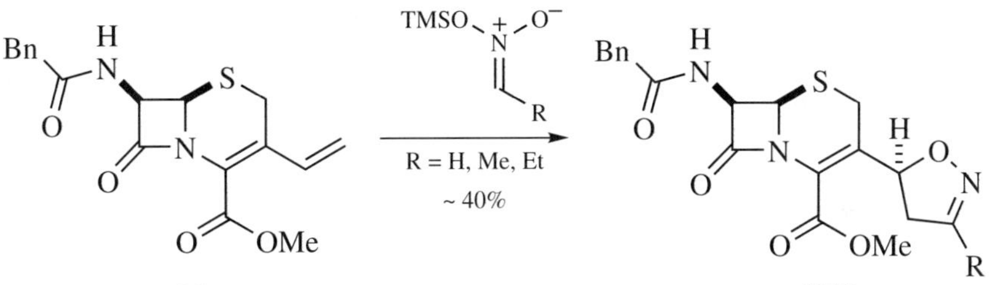

(121) **(122)**

(2.8)

In a second report on the use of chiral dipolarophiles, the cycloadditions of silyl nitronates with **123** and **124** provide modest facial selectivity (Table 2.37) (35). Unfortunately, the yields of the cycloadducts are only moderate because of the steric bulk of the dipolarophile.

TABLE 2.37. DIPOLAR CYCLOADDITIONS WITH A CHIRAL, NON-RACEMIC DIPOLAROPHILE

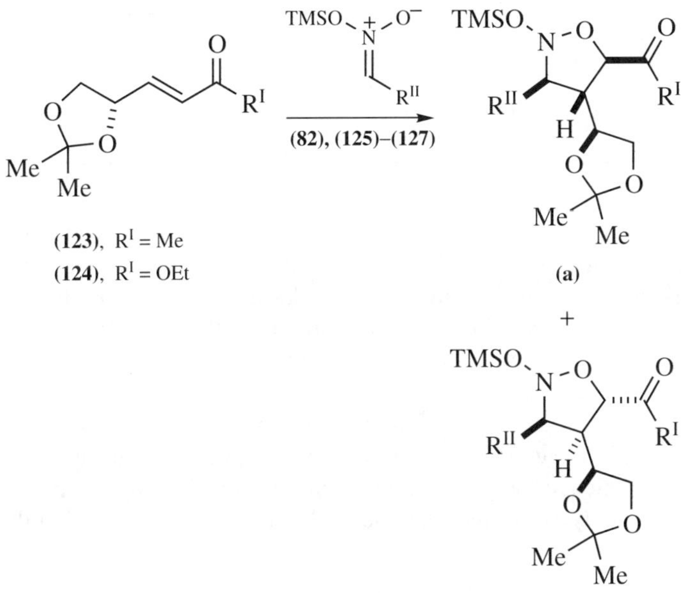

(123), R^I = Me
(124), R^I = OEt

(a) + **(b)**

Entry	R^I	Dipole	R^{II}	Product	Yield (%)	dr (**a**:**b**)
1	Me	**82**	Me	**128**	47	83:17
2	Me	**125**	$HOCH_2$	**129**	71	94:6
3	Me	**126**	$(MeO)_2CH$	**130**	75	88:12
4	OEt	**127**	$HOCH_2$	**131**	42	87:13

TABLE 2.38. SELECTIVITIES OBSERVED IN THE DIPOLAR CYCLOADDITION OF SULTAM DERIVED DIPOLAROPHILES

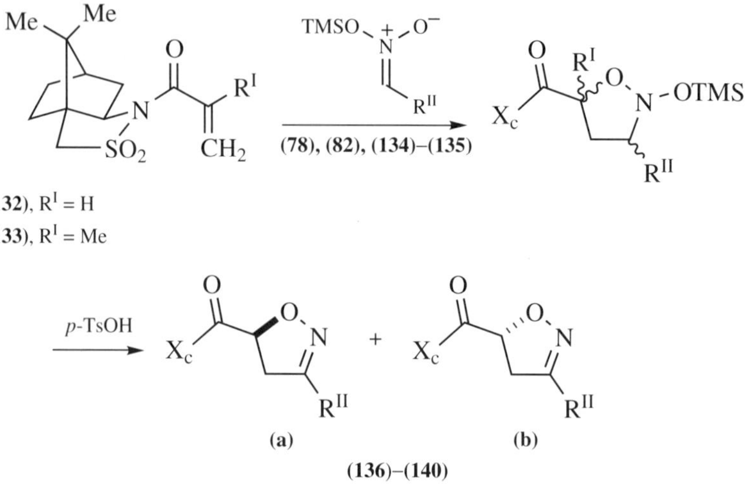

Entry	R^I	Dipole	R^{II}	Product	Yield (%)	dr (**a**:**b**)
1	H	**78**	H	**136**	96	89:11
2	H	**82**	Me	**137**	95	89:11
3	H	**134**	*n*-Pent	**138**	94	90:10
4	H	**135**	Ph	**139**	81	85:15
5	Me	**82**	Me	**140**	[a]	67:33

[a] Yield not reported.

A second strategy to control facial selectivity involves the use of chiral sultams and lactams as auxiliaries for the dipolarophile (120–123). Cycloaddition of **132** with a variety of substituted nitronates provides up to 9:1 selectivity of the major diastereomer (Table 2.38). However, substitution at the α-position of the dipolarophile leads to a reduction in stereoselectivity (entry 5). Assuming an s-cis conformation of the dipolarophile, it is proposed that the major isomer arises from an endo approach of the nitronate to the *Re* face of the dipolarophile (Fig. 2.13). This is supported by X-ray crystallographic analysis of one of the cycloadducts, which resides in a conformation similar to the proposed transition state. However, this analysis assumes that the silyl nitronate is only reacting through the

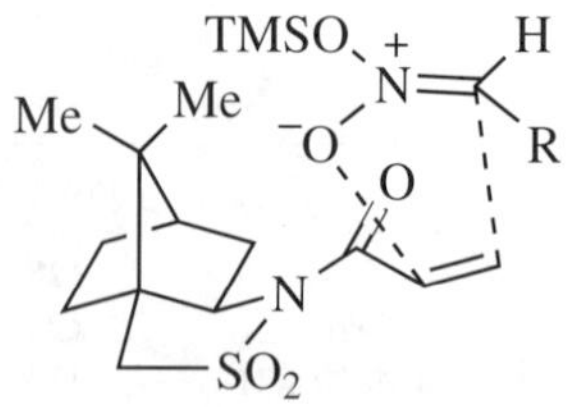

Figure 2.13. Proposed transition state for chral sultam derived dipolarophiles.

(*E*) configuration. The dipolar cycloaddition of **141** with a silyl nitronate shows a slight increase of facial selectivity over **132** (Eq. 2.9). Because the cycloadducts are converted directly to the corresponding isoxazolines, only the facial selectivity can be determined. It is believed that the cycloaddition proceeds on the *Re* face of the dipolarophile due to shielding of the *Si* face by the auxiliary. Both chiral auxiliaries can be liberated from the cycloadduct upon reduction with L-Selectride.

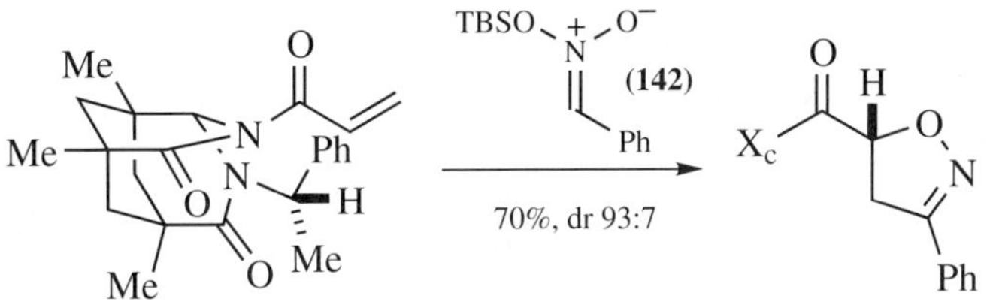

(141) **(143)** (2.9)

2.3.2.3. *Intramolecular Cycloadditions*

The intramolecular cycloaddition of a silyl nitronate bearing a dipolarophilic appendage provides easy access to fused, bicyclic isoxazolidines (22). This process, in general, is very facile, and has allowed the use of unfunctionalized alkenes as dipolarophiles (Table 2.39) (106,124). Thus, a silyl nitronate bearing an allyl group will undergo the [3 + 2] cycloaddition at room temperature over 15 h to provide the corresponding isoxazoline upon acidic workup in moderate yield.

The type of atoms in the tether have little effect on the cycloaddition process. Both alkyl (98,125–127) and ether tethers (128–131) have been extensively investigated and provide similar levels of reactivity. In addition, thioethers (132), amines (133), silanes (134), and silyl ketals (135) are all compatible with the dipolar cycloaddition.

TABLE 2.39. INTRAMOLECULAR DIPOLAR CYCLOADDITION OF SILYL NITRONATES

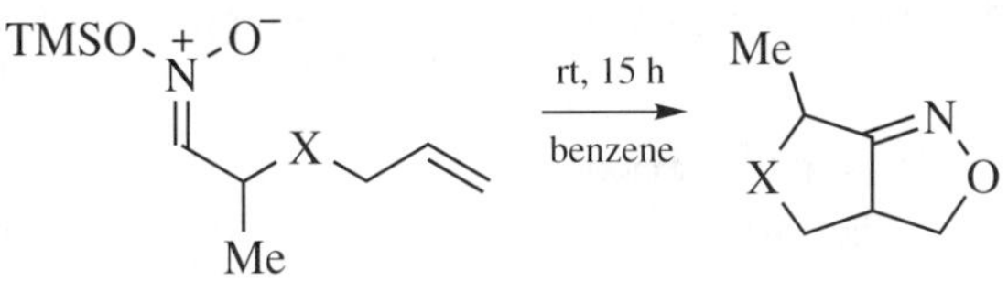

(144)–(146) **(147)–(149)**

Entry	Dipole	X	Product	Yield (%)
1	**144**	O	**147**	74
2	**145**	S	**148**	89
3	**146**	$C(CO_2Me)_2$	**149**	84

TABLE 2.40. SUBSTITUTENT EFFECTS ON THE INTRAMOLECULAR CYCLOADDITION OF SILYL NITRONATES

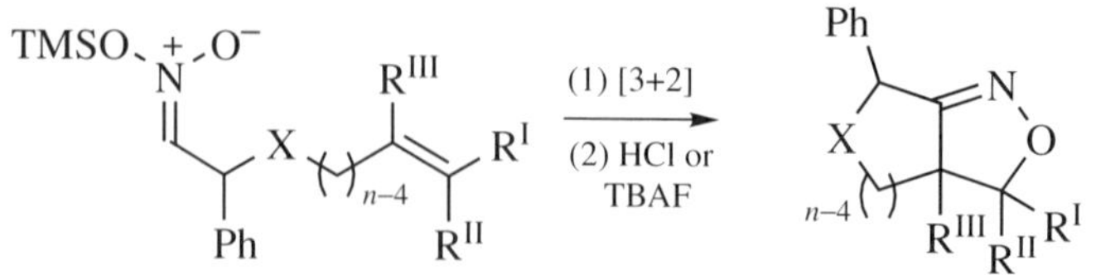

(57), (60)–(61), (150)–(151) (63), (66)–(67), (152)–(153)

Entry	Dipole	R^I	R^{II}	R^{III}	n	X	Temp (°C)	Time (h)	Product	Yield (%)
1	**57**	H	H	H	5	CH_2	20	15	**63**	66
2	**60**	Me	Me	H	5	CH_2	60	48	**66**	53
3	**61**	H	H	H	6	CH_2	60	15	**67**	62
4	**150**	H	H	Me	5	O	20	47	**152**	69
5	**151**	Ph	H	H	5	$OSi(Ph)_2O$	80	312	**153**	23

Both the length of the tether and the substitution of the dipolarophile effect the rate of the dipolar cycloaddition (Table 2.40) (98,106,135). Cyclization of a four atom tether provides the desired 6,5-bicyclic system upon heating to 60 °C for 15 h (entry 3), whereas cyclization to a seven-membered ring with a phenyl-substituted dipolarophile requires 80 °C for 13 days (entry 5). However, methyl substitution on the dipolarophile leads to much lower reactivity in the cycloaddition. The addition of a single methyl group at the terminus of the alkene increases the reaction time up to 48 h at room temperature (entry 4), whereas elevated temperature is necessary for complete reaction when two methyl groups are present (entry 2). Substitution at the internal position of the alkene also leads to longer reaction times. Attempts to incorporate an electron-withdrawing group on the dipolarophile was thwarted by the intervention of a competitive Michael type addition (Eq. 2.10) (106).

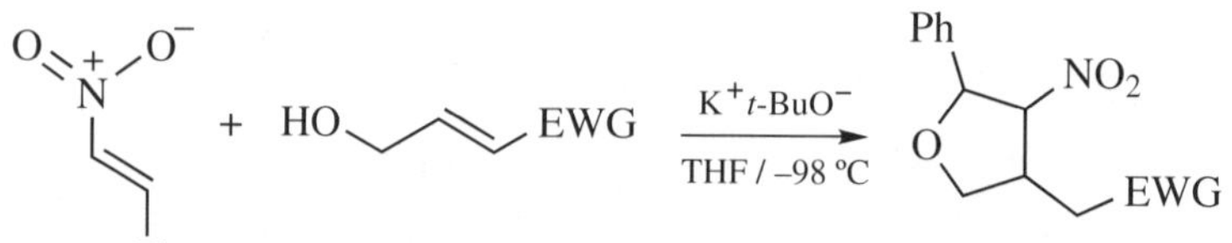

(2.10)

(154), EWG = CO_2Me

(155), EWG = CN

(156), 5 min, 72%

(157), 120 min, 50%

The use of alkynes has been investigated in the context of intramolecular nitronate cycloadditions (130). In this case, the starting material is consumed in 24 h at room temperature, however, the corresponding isoxazoline is not isolated (Table 2.41). Instead, the intermediate cycloadduct undergoes a fragmentation which, following the loss of the nitroso moiety, leads to an α,β-unsaturated aldehyde.

TABLE 2.41. INTRAMOLECULAR SILYL NITRONATE CYCLOADDITIONS WITH ALKYNES

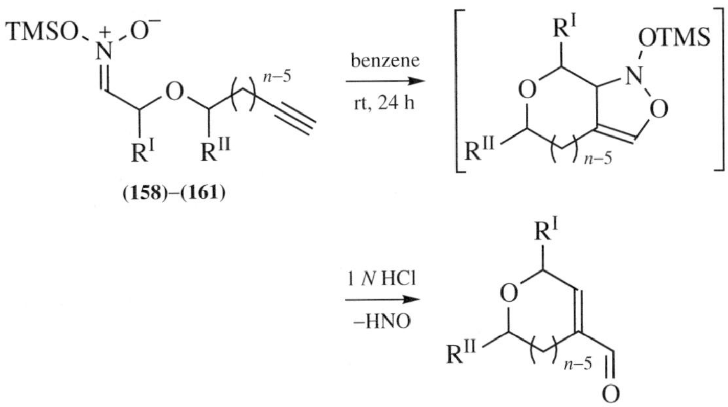

(158)–(161) **(162)–(165)**

Entry	Dipole	R^{I}	R^{II}	n	Product	Yield (%)
1	**158**	Ph	H	5	**162**	98
2	**159**	Ph	Me	5	**163**	92
3	**160**	H	Me	5	**164**	97
4	**161**	Ph	H	6	**165**	21

2.3.3. Nitroso Acetal Functionalization

The *N*-silyloxy-isoxazolidine (**XIX**) provides several opportunities for further elaboration. One of the more common transformations is the elimination of silanol from the isoxazolidine (Scheme 2.9), which results in the formation of isoxazolines (**XX**), which are also prepared from the [3+2] cycloaddition of nitrile oxides

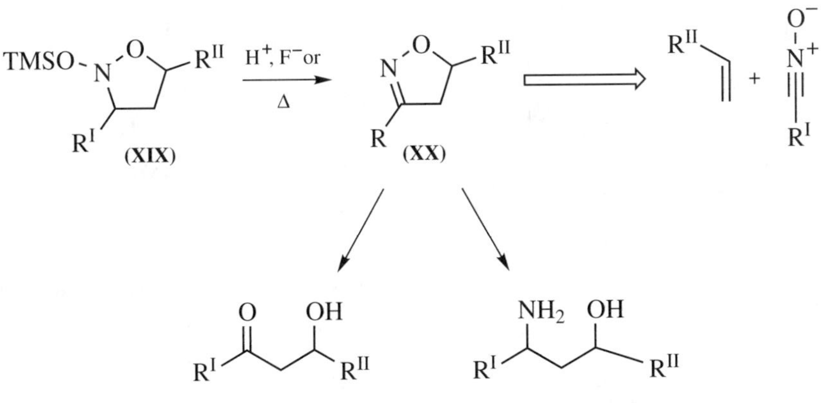

(XIX) (XX) (XXI) (XXII)

Scheme 2.9

(30,136,137). The elimination proceeds rapidly at room temperature with a catalytic amount of acid or fluoride, however, it has also been observed to occur upon distillation of the isoxazolidine (16,18). Because of the facile elimination, several procedures have been developed in which the reaction is quenched with an acidic agents, and the isoxazoline is isolated directly (108). Isoxazolines are useful in the preparation of β-hydroxy ketones and amino alcohols (138).

The convergence of the nitronate and nitrile oxide cycloadditions has allowed for the direct comparisons of yields and stereoselectivities of the two processes. For intramolecular reactions, the nitronate dipole typically required longer reaction times and/or elevated temperatures (22,98,135), however, the nitronate cycloaddition shows considerably higher diastereoselectivity (Table 2.42). Interestingly, the diastereoselectivity is dependent on the placement of a substituent on the tether. In the case of the silyl nitronate derived from **172**, the diastereoselectivity is controlled by the substituent at C(1), while cyclization of the analogous nitrile oxide is governed by the substituent at C(1′) (Scheme 2.10) (124).

Cleavage of the isoxazolidine ring can also be effected directly to give similar products as the isoxazoline. Upon the addition of methoxide to an isoxazolidine bearing a hydrogen at C(3), fragmentation reveals a β-hydroxy oxime, which can be further hydrolyzed to the corresponding ketone (16,20,34,108). The β-hydroxy

TABLE 2.42. DIASTEREOSELECTIVITIES OBSERVED WITH NITRONATES AND NITRILE OXIDE CYCLOADDITIONS

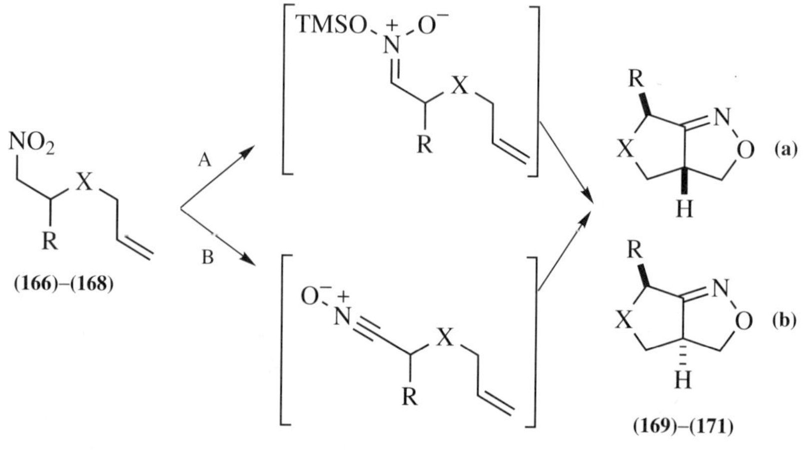

Entry	Dipole	R	X	Product	Method A[a] (dr, **a**:**b**)	Method B[b] (dr, **a**:**b**)
1	**166**	*i*-Pr	S	**169**	99:1	1:1
2	**167**	Ph	S	**170**	99:1	3:2
3	**168**	4-MeOC$_6$H$_4$	C(CO$_2$Me)$_2$	**171**	99:1	1:5

[a] Method A: TMSCl, Et$_3$N, benzene; rt, 48 h.
[b] Method B: phenyl isocyanate, Et$_3$N, benzene; rt, 48 h.

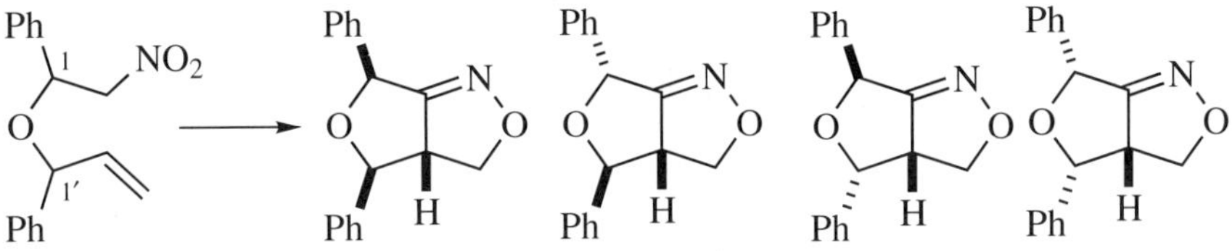

(172)	(173a)	(b)	(c)	(d)
TMSCl, Et_3N, benzene; rt, 48 h	38	-	62	-
phenyl isocyanate, Et_3N, benzene; rt, 48 h	47	53	-	-

Scheme 2.10

ketone can be obtained directly by treatment with aqueous $TiCl_3$ (Scheme 2.11) (18,20,101). Alternatively, reduction with borane or with Raney Ni under an atmosphere of H_2 provides the amino alcohol in moderate yield. In addition, several fragmentations have been observed depending on the substitution of the isoxazolidine. In the case of **XXV**, treatment with fluoride provides the corresponding ketones by fragmentation of the isoxazolidine ring (Scheme 2.12) (114). Alternatively, the treatment of **174** with fluoride provides the oxime **175** by a rearrangement through the adjacent vinyl group (Scheme 2.13) (127).

2.3.4. Applications

As synthetic equivalents of β-hydroxy ketones and amines, isoxazolidines provide access to important synthetic building blocks. However, there have been

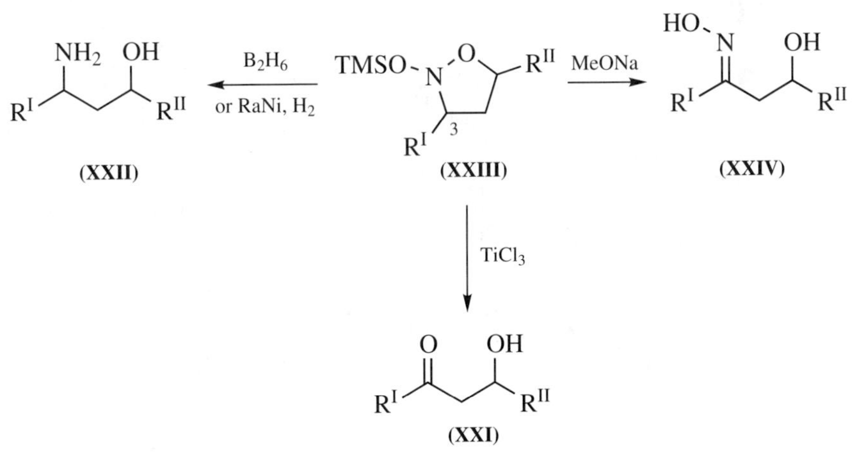

Scheme 2.11

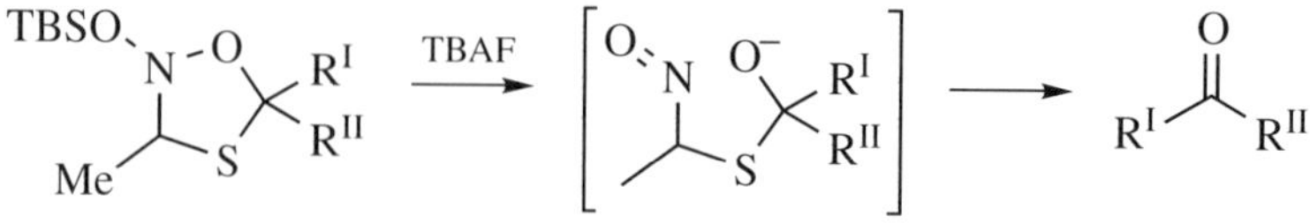

(XXV) (XXVI)

Scheme 2.12

few applications of the silyl nitronate cycloaddition in the context of natural product synthesis. The racemic syntheses and formal syntheses of several naturally occurring cyclopentanones has been accomplished involving the base catalyzed cyclization of the corresponding β-hydroxy ketone (108,139).

The high diastereoselectivity observed in the intramolecular silyl nitronate cycloaddition allows for the synthesis of (+)-iridomyrmecine and (−)-actinidine (Scheme 2.14) (126,140). Cyclization of both silyl nitronates derived from (+)-β-citronellene and (−)-citronellol, respectively, provide a single diastereomer. Fragmentation and rearrangement of the nitroso acetal in the presence of fluoride provides the corresponding oximes. In the case of **180**, the sulfide substituent allows for access to the aldehyde upon fluoride cleavage of the isoxazolidine ring. The oximes **181** and **183** can then be elaborated to the target natural products.

As noted previously, the use of chiral sultam auxiliaries provides access to enantiomerically enriched isoxazolines. These products can be elaborated to provide nonactic acid and 8-*epi*-nonactic acid, the subunits of nonactin (Scheme 2.15) (141–143). Subsequent cleavage of the isoxazoline ring in the presence of Raney

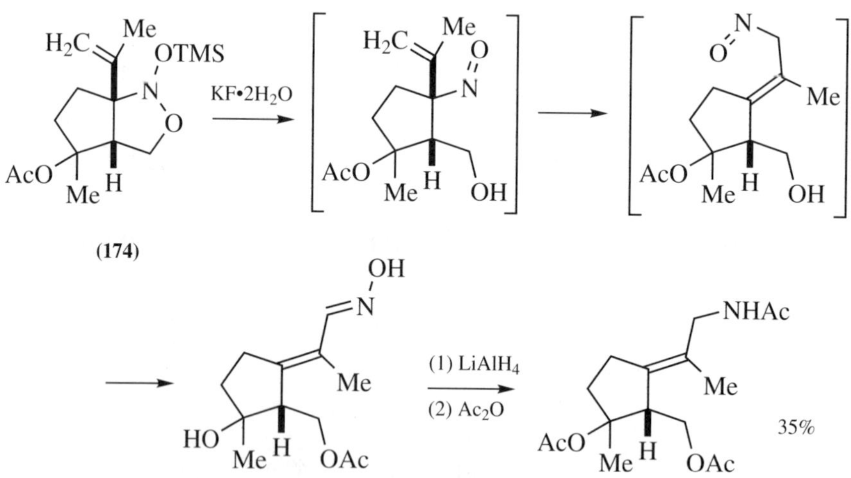

Scheme 2.13

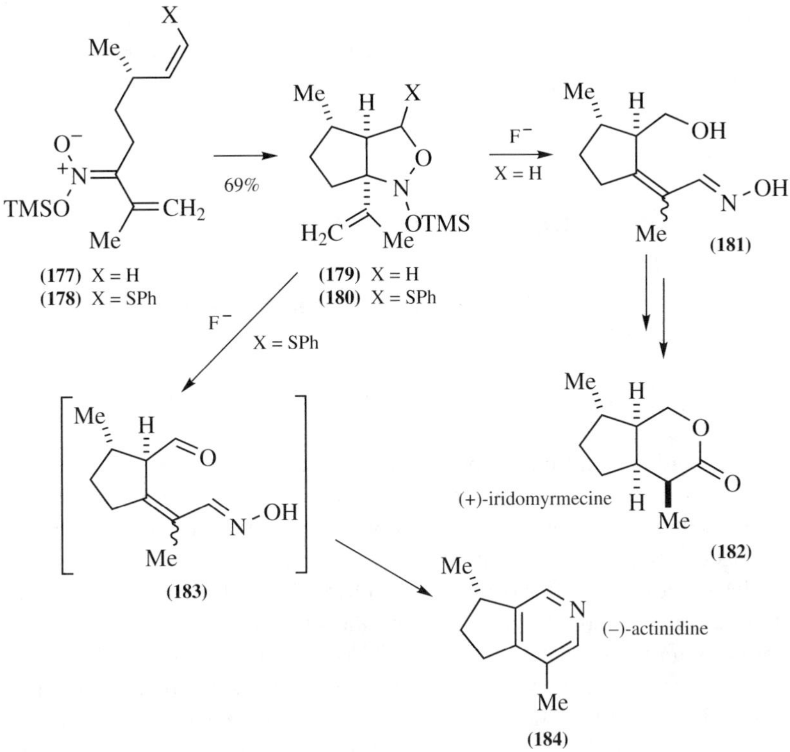

Scheme 2.14

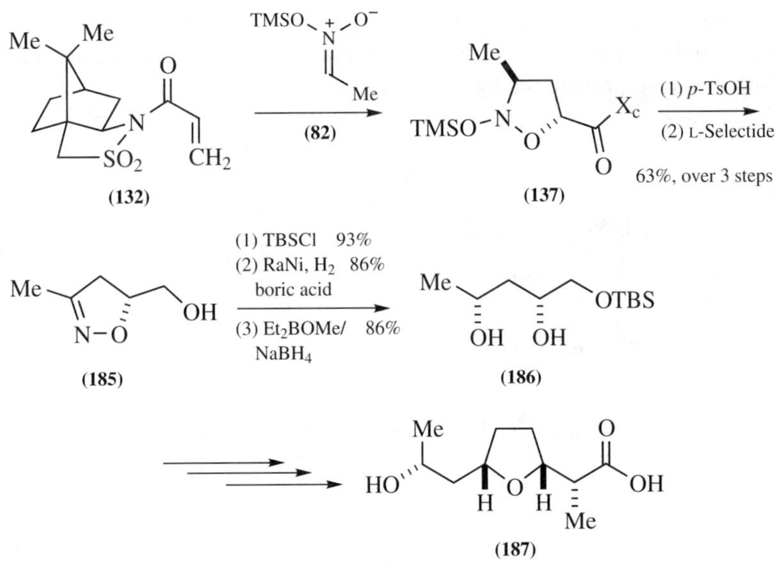

Scheme 2.15

nickel, followed by directed reduction of the ketone moiety, provides the diol **186**. Subsequent functionalization provides 8-*epi*-nonactic acid (**187**). By following the same procedure, (+)-nonactic acid can be prepared from the opposite enantiomer of the chiral sultam. Finally, these units are coupled to provide nonactin.

2.4. ALKYL NITRONATES

2.4.1. Preparation of Acyclic Nitronates

2.4.1.1. Alkylation

The preparation of nitronates by O-alkylation of nitro groups was first reported in 1894, wherein activated nitro compounds were found to react with diazoalkanes (Eq. 2.11) (8,9). However, the use of diazo compounds for the preparation of nitronic esters is limited to highly acidic nitro compounds in combination with either diazomethane or diazoethane (37,39,144,145). The configuration of the resulting nitronate is slightly dependent on the substituent on the nitro compound, but the (Z/E) ratio is typically not > 70:30, usually in favor of the (*Z*) isomer. A more general procedure involves the combination of the sodium nitronate salt (146) with trialkyloxonium tetrafluoroborates (147) to provide the desired alkyl nitronate (Eq. 2.12) (40,144,148). By this method, both strongly acid nitro compounds as well as unfunctionalized nitroalkanes are smoothly converted to nitronates. However, if the process is carried out at elevated temperature, the corresponding oximes are obtained (149). Since the tetrafluoroborate salt must be prepared from the corresponding alkyl ether and its boron trifluoride etherate, this procedure has only been employed to prepare methyl and ethyl nitronic esters.

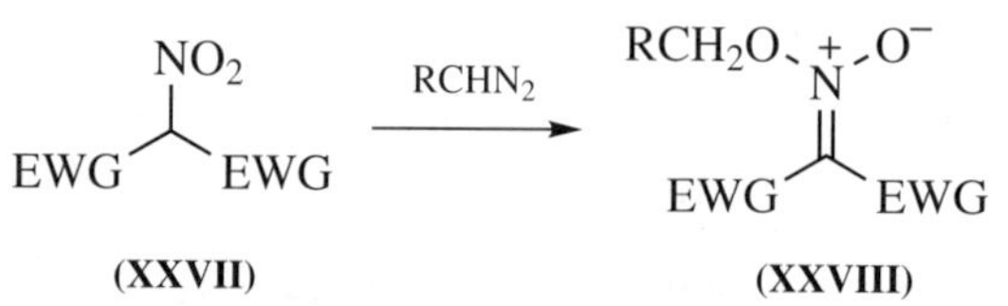

(2.11)

EWG = NO_2, CO_2Me, CN, C(O)NH_2

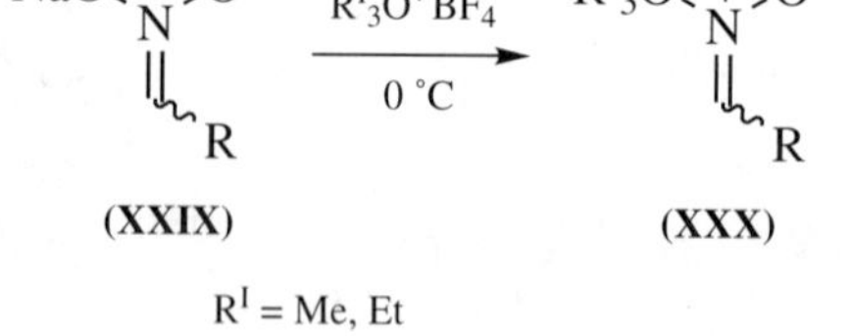

(2.12)

R^1 = Me, Et

The intermolecular alkylation of metallo nitronates with various alkyl halides is limited. The addition of methyl iodide to the silver salt of an aryl nitromethane provides the corresponding methyl nitronate in moderate yield (Eq. 2.13) (150), which has also been extended to the silver salt of trinitromethane (Scheme 2.16) (151–153). However, in the case of primary halides, both O- and C-alkylation are observed. For secondary and tertiary halides, only O-alkylation is observed, but in low yields. Unfortunately, under the reaction conditions, the starting alkyl halide can undergo dehydrohalogenation to provide the corresponding alkene, which then undergoes [3+2] cycloaddition with the alkyl nitronate.

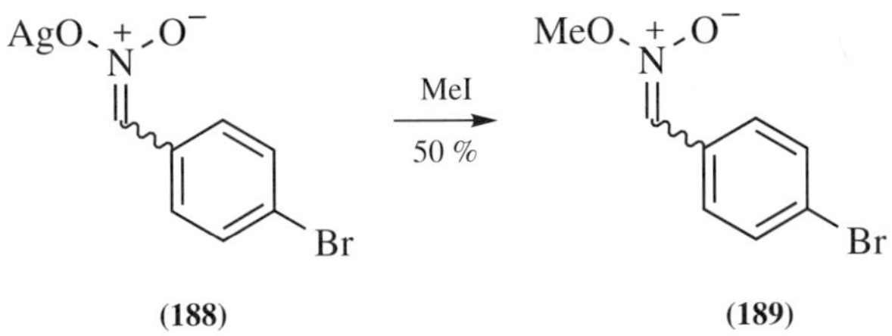

(2.13)

Treatment of sodium and potassium nitronates with alkyl halides typically results in the formation of oximes and carbonyl compounds by cleavage of the N—O bond (11). In one case, however, reaction of *n*-butyl bromide with the potassium salt of nitro ester **191** does afford the *n*-butyl nitronate (**192**, Eq. 2.14) (154).

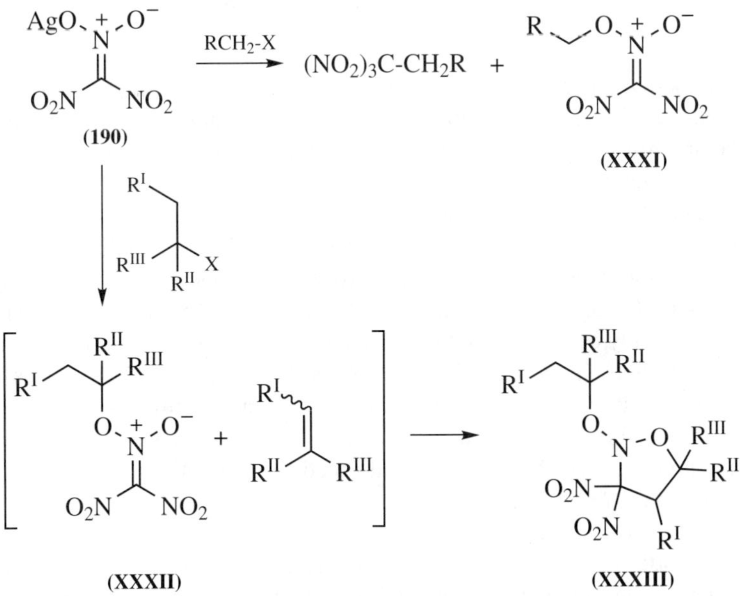

Scheme 2.16

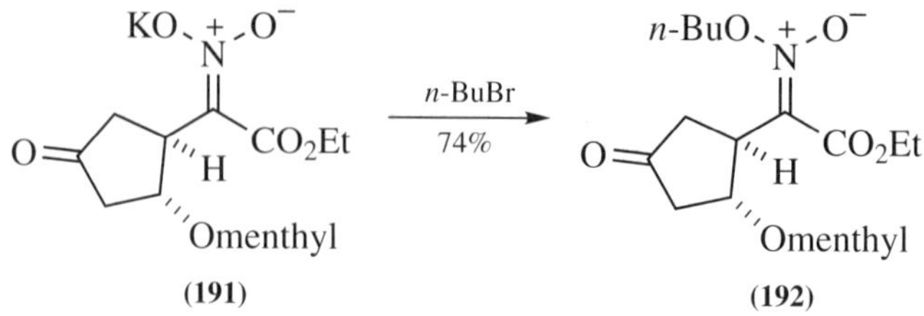

(2.14)

An alternative to activating the nitro moiety by forming the nitronate salt is the activation of an oxygen leaving group under Mitsunobu conditions (Eq. 2.15) (155,156). Treatment of methanol with diethyl azodicarboxylate and triphenylphosphine in the presence of ethyl nitroacetate provides the nitronate **85** in good yield. Unfortunately, only methanol has been demonstrated to be compatible with this procedure.

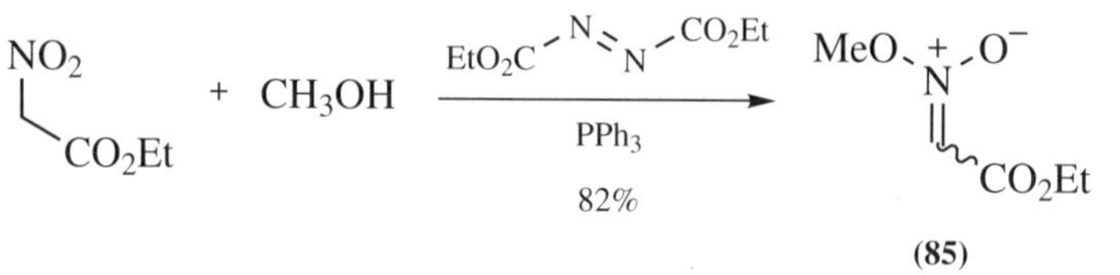

(2.15)

2.4.1.2. Trinitromethane Derivatives

Trinitromethane derivatives have found considerable use for the functionalization of alkenes. For example, tetranitromethane will transfer a nitro group in the presence of ethylene, generating an ion pair, which can collapse by combination of one of the oxygen atoms of the trinitromethide anion with the carbocation, forming **197** (Scheme 2.17). The resulting nitronate then undergoes [3+2] cycloaddition with the ethylene remaining in solution to provide the corresponding isoxazolidine (**198**) in moderate yield (157–159). This process is also successful with vinyl and allyl ethers (160), and other substituted alkenes (161–165). However, this transformation is highly dependent on the structure of the alkene. With trisubstituted and some disubstituted alkenes, C-alkylation of the trinitromethide anion is observed, **199**, along with several other reaction pathways. If the intermediate carbocation is stabilized by one or more substituents, then only C-alkylation is observed (166,167). The same process is operational for trinitromethane derivatives where X=I (168–171), F (172), or CN (173).

The interaction of trinitromethide salts with alkenes has also been investigated (174). In aprotic solvents, trinitromethide anions undergo extrusion of a nitro group to produce a dinitrocarbene (Scheme 2.18). In the presence of monsubstituted alkenes a formal [3+2] cycloaddition produces a nitronate that then further reacts

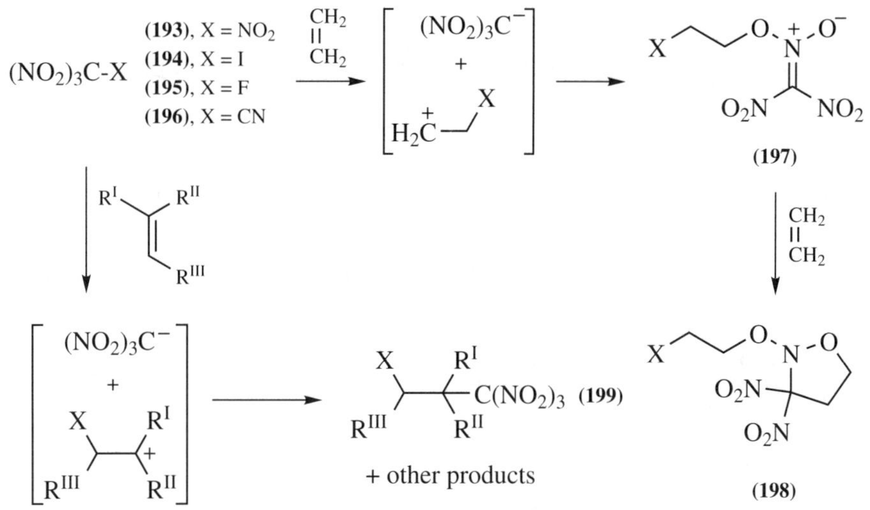

Scheme 2.17

in a second [3+2] cycloaddition to form nitroso acetal **205**. However, with a disubstituted alkene, cyclization leads only to the cyclic nitronate in low yield.

2.4.1.3. Acylation

Nitroalkanes are easily functionalized with a variety of acylating agents such as phenyl isocyanate, acetic anhydride, and ethyl chloroformate in the presence of a

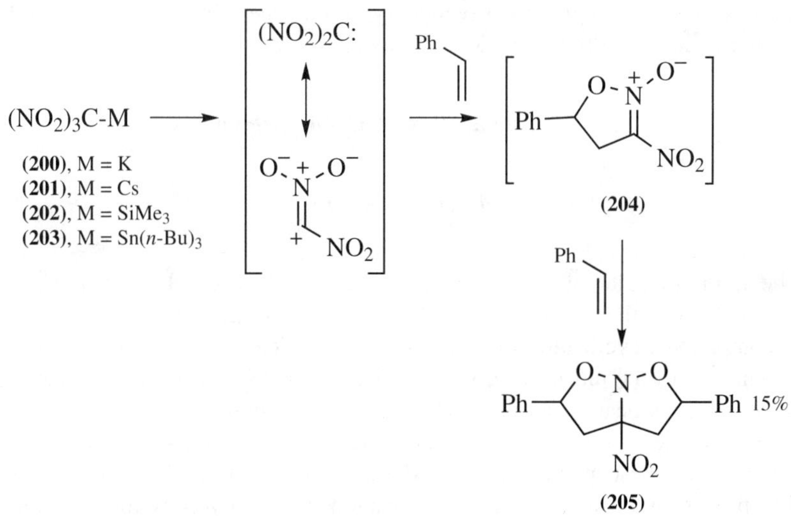

Scheme 2.18

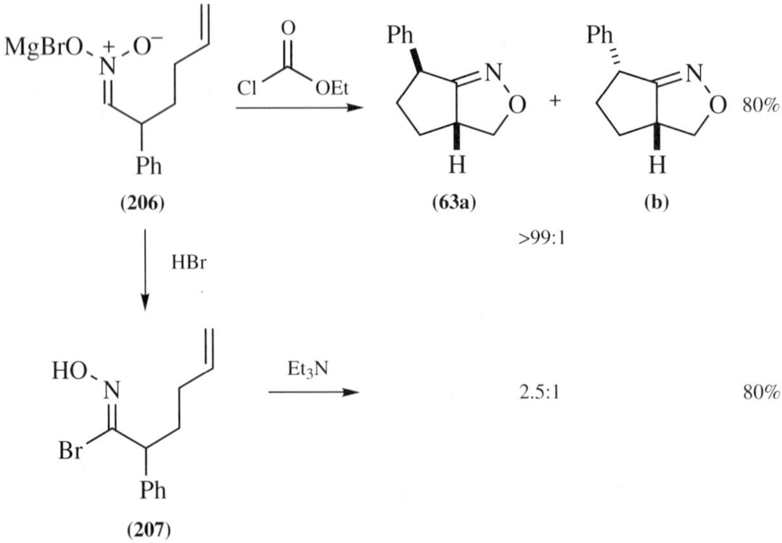

Scheme 2.19

base. With primary nitroalkanes, the intermediate nitronates are converted directly to the corresponding nitrile oxide in most cases (175). One exception is reported where an intramolecular [3+2] cycloaddition of **206** proceeds before elimination of the carbamate, however, only the isoxazoline is isolated (Scheme 2.19). The presence of the nitronate in the cycloaddition is supported by differences in the cis : trans ratios of products in comparison to other methods for generating nitrile oxides (36,176). The acylation of secondary nitroalkanes has been investigated, but not in the context of dipolar cycloadditions (177–186).

2.4.2. Preparation of Cyclic Nitronates

2.4.2.1. Alkylation

The intramolecular alkylation of nitronate salts has found much more widespread application than the intermolecular version. In the case of γ or δ-halogenated nitro compounds, cyclization in basic medium provides the corresponding five- or six-membered cyclic nitronate (Eq. 2.16) (57,63,187–189). This process works best with nitroalkanes bearing an electron-withdrawing group at the α-position. In the absence of an electron-withdrawing group, the resulting nitronates were found to be unstable and are best functionalized *in situ* upon the addition of an alkene (190). A similar process is observed upon the activation of γ- δ-hydroxy nitro compounds under Mitsunobu conditions (155).

(XXXIV) (XXXV) (2.16)

$n = 5,6$ $n = 5,6$

X	Conditions
Br	base
OH	DEAD, PPh_3

Reaction at the C atom of nitronate salts is known with a variety of electrophiles, such as aldehydes (Henry reaction) and epoxides (191–193). Thus the incorporation of the nitro moiety and the cyclization event can be combined into a tandem sequence. Addition of the potassium salt of dinitromethane to an α-haloaldehyde affords a nitro aldol product that can then undergo intramolecular O-alkylation to provide the cyclic nitronate (**208**, Eq. 2.17) (59). This process also has been expanded to α-nitroacetates and unfunctionalized nitroalkanes. Other electrophiles include functionalized α-haloaldehydes (194,195), α-epoxyaldehydes (196), α-haloenones (60), and α-halosulfonium salts (197), (Chart 2.2). In the case of unsubstituted enones, it is reported that the intermediate nitronate salt can undergo formation of a hemiacetal, which can be acetylated in moderate yield (198).

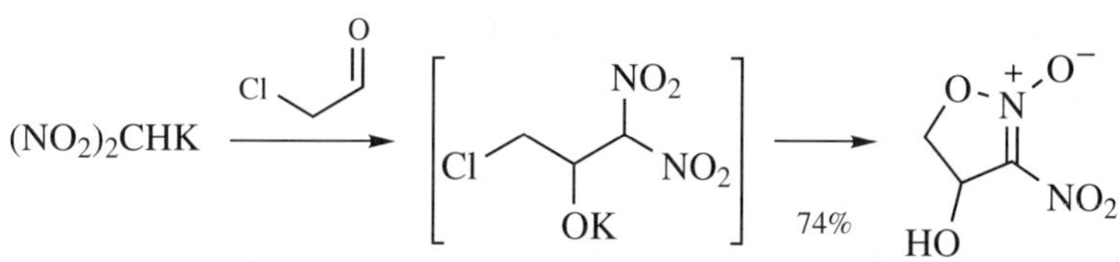

(208) (2.17)

In an analogous manner, nitroalkenes can be utilized as the electrophile in a tandem coupling–cyclization process (Eq. 2.18). Addition of a stabilized sulfonium ylide such as **209**, to a variety of nitroalkenes provides an intermediate nitronic acid, which upon displacement of dimethyl sulfide provides a mixture of the corresponding nitrocyclopropane (**XL**) and cyclic nitronate (**XLI**). The ratio of products is highly dependent on the structure of the nitroalkene. For example, without a substituent on the α-position of the nitroalkene (R^{III}=H), only the

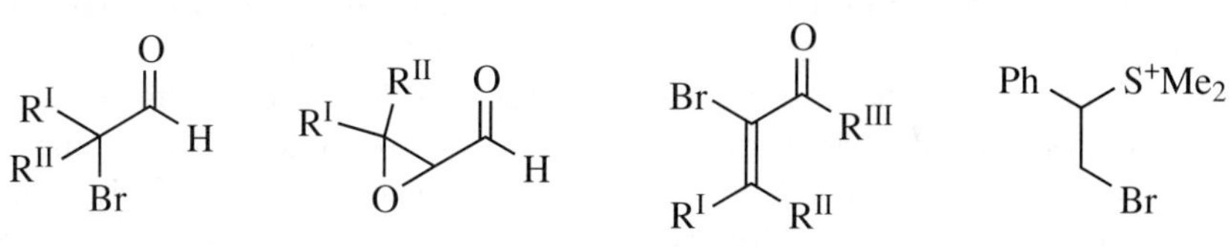

(XXXVI) (XXXVII) (XXXVIII) (XXXIX)

Chart 2.2

cyclopropane is observed. However, with 1-nitrocyclopentene, only the nitronate product is observed (62).

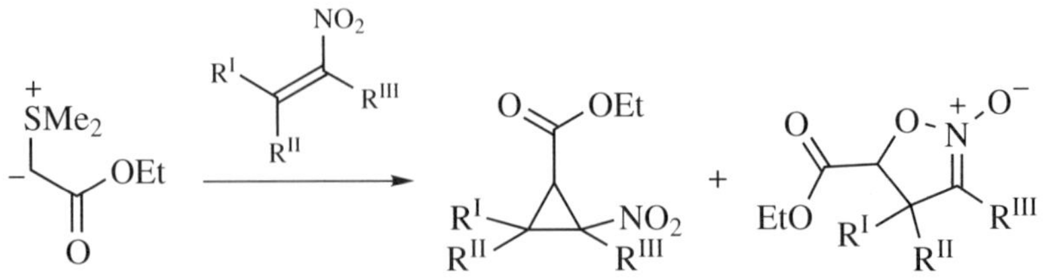

(2.18)

(209) **(XL)** **(XLI)**

2.4.2.2. [4+2] *Cycloaddition of Nitroalkenes*

Nitroalkenes can also be converted to nitronates by direct combination with an alkene. The nitronate is formed as a result of a [4+2] cycloaddition of the electron-deficient nitroalkene, wherein one of the N—O bonds of the nitro group participates as part of the 4π fragment (Eq. 2.19) (89). Because of the electron-deficient nature of the heterodiene, alkenes react in the order: electron rich > electroneutral > electron poor. Therefore, the majority of dienophiles investigated are enamines (52,71,199–207) and vinyl ethers (99,208–213).

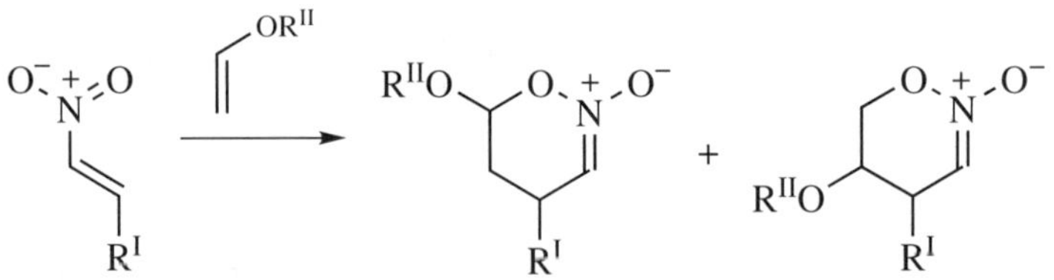

(2.19)

(XLII) head-to-head **(XLIII)** head-to-tail

The thermal cycloaddition generally takes place at room temperature over extended periods of time, however, it was found that this process can be promoted by stoichiometric amounts of a Lewis acid (208). To engage unactivated dienophiles at reasonable temperatures, Lewis acid activation was found to be essential (74).

The reaction of vinyl ethers and enamines with nitroalkenes is highly regioselective, with only the head-to-head adduct observed. The endo approach of the dienophile is preferred in the thermal cycloaddition, however, the mode of approach can be controlled by the choice of the Lewis acid promoter (214). Facial discrimination has been obtained by the use of chiral groups on the both the nitroalkene (215,216) and the enamine (217) or vinyl ether (218), as well as with chiral Lewis acids (46,66,94,219,220).

As a consequence of the complimentary electron demand of the nitroalkene and the product nitronate, there exists the possibility of a one-pot, tandem reaction. In this case, the nitroalkene will react preferentially with the electron-rich alkene to produce an intermediate nitronate. This nitronate can then react with a second alkene bearing an electron-withdrawing substituent. Therefore subjection of the nitroalkene **210** to both ethyl vinyl ether and acrylonitrile provides only the nitroso acetal **211** in moderate yield (Eq. 20) (70). Moreover, this also allows the possibility of intramolecular variants of the process.

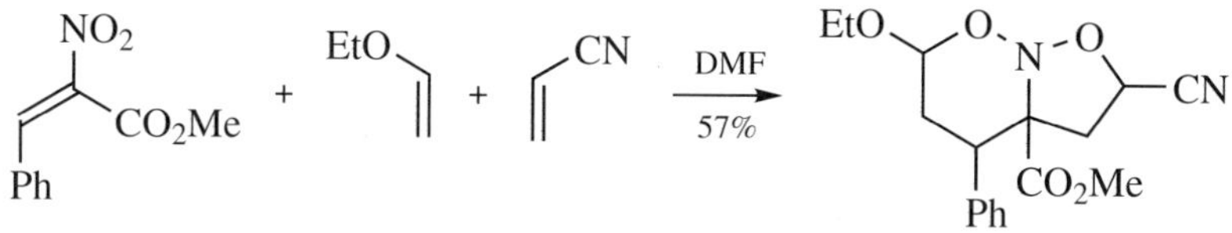

(210) **(211)**

DMF = dimethylformamide

(2.20)

2.4.2.3. *Radical Cyclization*

Cyclization of nitro-stabilized radicals provides another method for the generation of cyclic nitronates (221). Oxidation of the *aci*-form of nitroalkanes with ceric ammonium nitrate generates the α-carbon centered radical, which in the presence of an alkene, leads to the homologation of the α-radical. In the case of a tethered alkene of appropriate length, radical addition leads to a cyclic nitronate (Scheme 2.20).

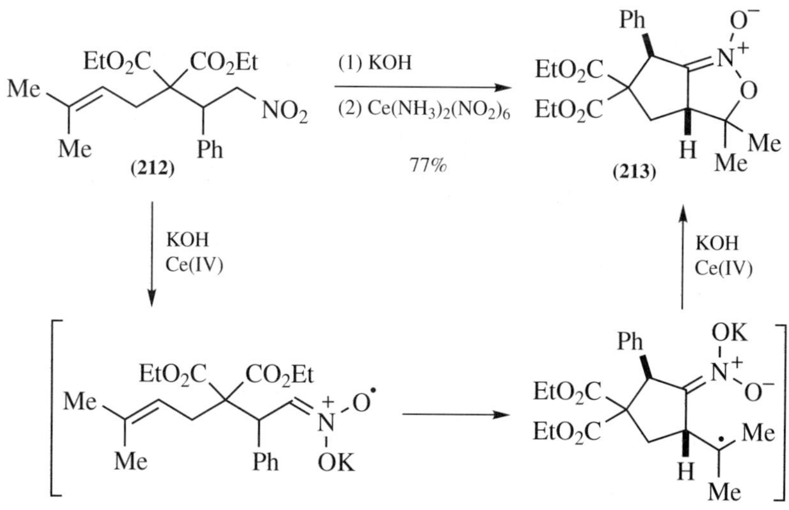

Scheme 2.20

2.4.3. Cycloadditions of Alkyl Nitronates

The wide variety of methods for the preparation of alkyl nitronates, gives rise to a broader diversity of structures compared to silyl nitronates. Alkyl nitronates can be grouped into two subclasses, acyclic and cyclic. Both subclasses participate in dipolar cycloadditions with similar reactivity, however, minor differences are manifest in their stability and stereoselectivity. Additionally, the ability to prepare cyclic nitronates allows access to a wide variety of novel, multicyclic ring structures.

2.4.3.1. Intermolecular Cycloadditions

Despite the availability of methods for the preparation of higher alkyl nitronates from simple nitroalkanes, only the methyl derivative of nitromethane has been investigated. The reaction of methyl acrylate or methyl crotonate with the *in situ* generated nitronate **214** proceeds slowly over extended periods of time to provide the corresponding nitroso acetal in low yield (Eq. 2.21) (148). Similar observations are found with the analogous cyclic nitronate **217**, wherein the reaction with methyl acrylate does not proceed at all, while the addition of styrene provides a small amount of cycloadduct after three days (entry 2, Table 2.43) (190). In both instances, the nitronate likely suffers from a lack of stability. However, the corresponding methyl-substituted nitronate **218** provides increased stability, and therefore higher yields upon the addition of an alkene. The reactivity of the dipole can also be increased by the addition of an alkoxy group at the six-position of the nitronate (entry 5) (222). Upon combination of methyl acrylate with the nitronate

TABLE 2.43. CYCLOADDITIONS OF THE UNFUNCTIONALIZED CYCLIC NITRONATES

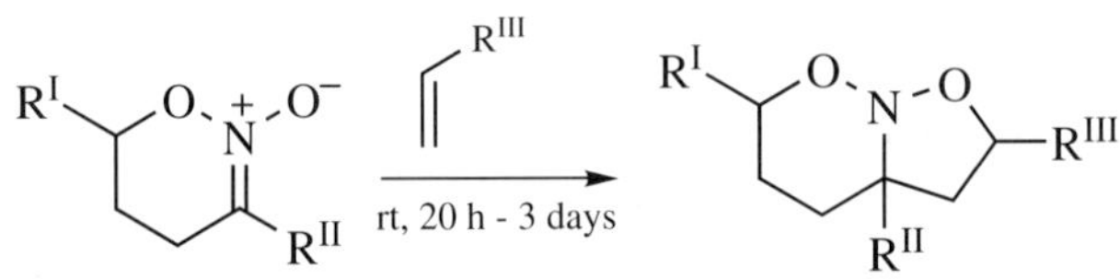

(217)–(219) **(220)–(224)**

G* = (1*S*,2*R*)-2-phenylcyclohexyl

Entry	Dipole	R^{I}	R^{II}	R^{III}	Product	Yield (%)
1	**217**	H	H	CO_2Me	**220**	2
2	**217**	H	H	Ph	**221**	21
3	**218**	H	Me	CO_2Me	**222**	85
4	**218**	H	Me	Ph	**223**	80
5	**219**	OG*	H	CO_2Me	**224**	81

219, the dipolar cycloaddition proceeds in 20 h at room temperature to provide a good yield of the bicyclic cycloadduct as a mixture of diastereomers.

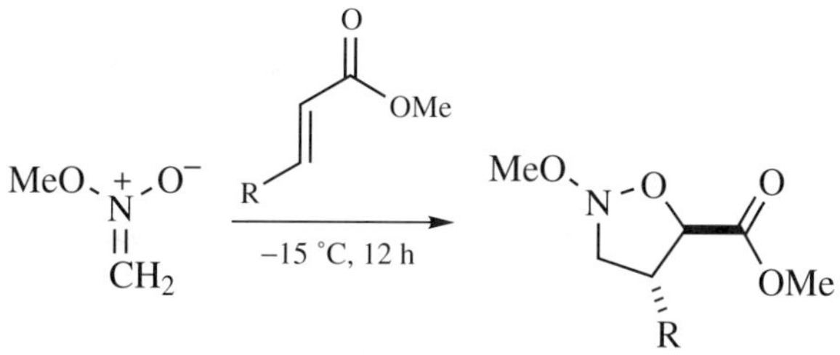

(2.21)

(214) **(215)**, R = H; 8% **(216)**, R = Me; 2%

The influence of electron-withdrawing and electron releasing groups on the reactivity of the corresponding nitronate has been investigated with aryl substituted nitronates (Table 2.44) (76,223). Nitro groups were found to be the most beneficial, providing the cycloadduct in good yield. With other electron-withdrawing groups, an increase in reactivity is also observed. Qualitatively, ester and nitro substituted nitronates react faster than the corresponding aryl nitronates, however, these also suffer slight decreases in yield (Table 2.45) (75,78,224–226). Interestingly, the rate of cycloaddition of a dicarboalkoxy nitronate is greatly enhanced, but only moderate yields are observed (entry 5).

A similar trend of reactivity is observed for the corresponding cyclic nitronates (Table 2.46) (57,227). However, since these are disubstituted nitronates, the reactions are typically slower than the acyclic counterparts.

The structure of the dipolarophile is also a very important component. The most widely utilized dipolarophiles are monosubstituted alkenes bearing an electron-withdrawing group. A survey of dipolarophiles shows good compatibility with activated monosubstituted alkenes (Table 2.47) (224). Less activated alkenes such

TABLE 2.44. SUBSTITUENT EFFECTS OF ARYL NITRONATES

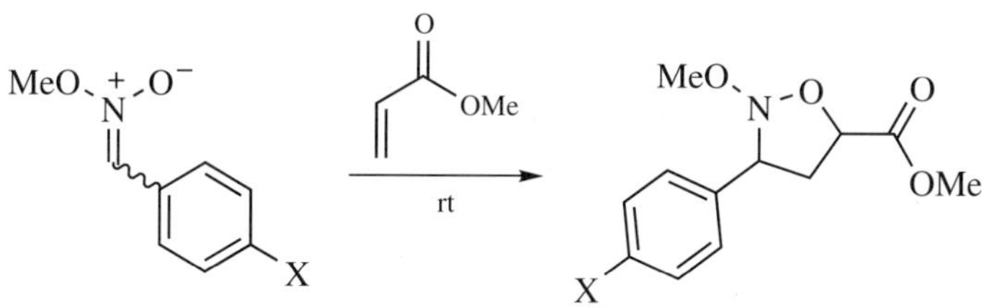

(225)–(227) **(228)–(230)**

Entry	Dipole	X	Time (h)	Product	Yield (%)
1	**225**	OMe	7	**228**	37
2	**226**	H	4	**229**	34
3	**227**	NO_2	7	**230**	77

TABLE 2.45. SUBSTITUENT EFFECTS OF ACYCLIC NITRONATES

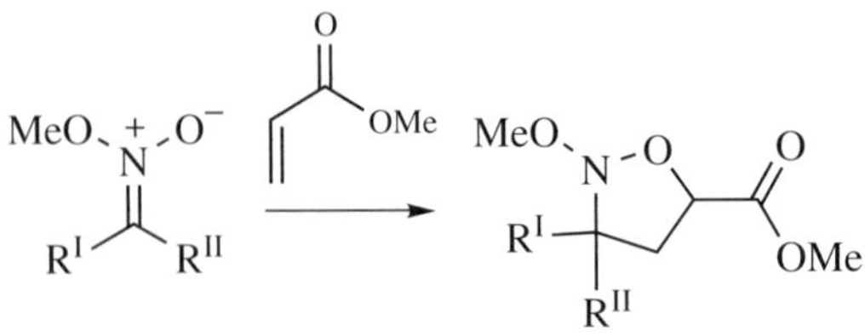

(4), (47), (231)–(233) **(234)–(238)**

Entry	Dipole	R^I	R^{II}	Temp (°C)	Time (h)	Product	Yield (%)
1	**231**	CO_2Me	H	rt	24	**234**	74
2	**4**	NO_2	H	rt	18	**235**	65
3	**232**	SO_2Ph	H	60	4	**236**	70
4	**47**	CN	H	0	96	**237**	48
5	**233**	CO_2Me	CO_2Me	rt	2	**238**	52

TABLE 2.46. SUBSTITUENT EFFECTS OF CYCLIC NITRONATES

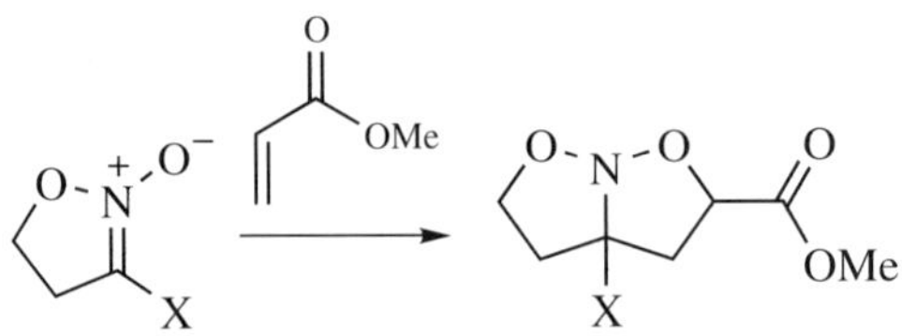

(5), (239)–(240) **(241)–(243)**

Entry	Dipole	X	Temp (°C)	Time (h)	Product	Yield (%)
1	**239**	Ph	80	35	**241**	10
2	**240**	CO_2Me	rt	312	**242**	47
3	**5**	NO_2	rt	12	**243**	73

TABLE 2.47. REACTION OF **231** WITH VARIOUS DIPOLAROPHILES

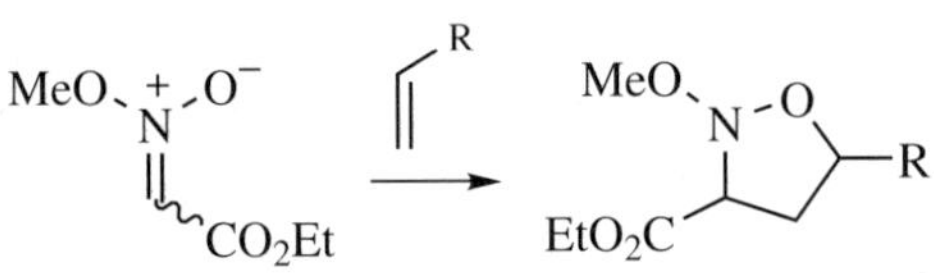

(231) **(234), (244)–(247)**

Entry	R	Time (h)	Product	Yield (%)
1	Ph	70	**244**	65
2	CH_2Cl	70	**245**	76
3	COMe	30	**246**	78
4	CN	30	**247**	76
5	CO_2Me	38	**234**	90

TABLE 2.48. REACTION OF **49** WITH VARIOUS DIPOLAROPHILES

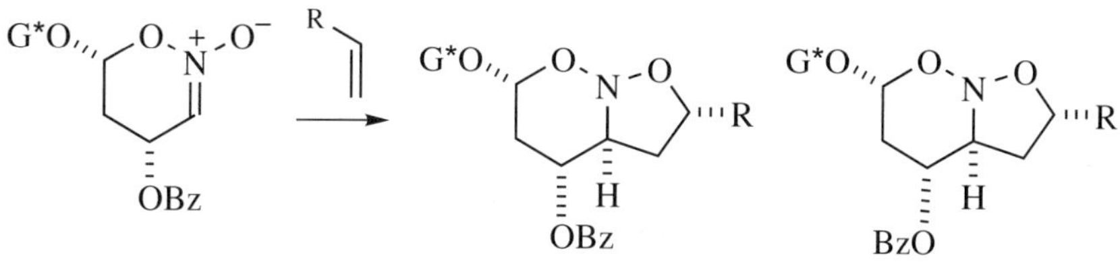

(49) **(a)** **(b)**

G* = (1*S*,2*R*)-2-phenylcyclohexyl

Entry	R	Temp (°C)	Time (h)	Product	Yield (%)
1	CO_2Me	rt	6	**51**	86
2	CO_2t-Bu	rt	2.5	**50**	91
3	COMe	rt	1.5	**52**	88
4	$COCH_2OTDS$	rt	4	**53**	90
5	CHO	rt	3.5	**55**	62[a]
6	CH_2OH	80	1.5	**56**	81

[a] Overall yield after reduction with $NaBH_4$.

as styrene and allyl chloride require longer reaction times and provide decreased yields (entries 1 and 2).

In the case of unactivated cyclic nitronates, elevated temperatures are necessary. Reaction of a variety of enones with the nitronate **49** proceeds smoothly at room temperature over several hours, whereas allyl alcohol requires 1.5 h at 80 °C (Table 2.48) (84). The resulting nitroso acetals are produced in good yields as mixtures of diastereomers.

The cycloaddition of substituted acrylates has been investigated with cyclic nitronate **24** (Table 2.49) (14). The cycloaddition of a 1,1-disubstituted dipolarophile (entry 2), proceeds in good yield, but both 1,2-disubstituted alkenes fail to react. The effect of substitution pattern on the dipolarophile was investigated with a slightly more reactive nitronate (Table 2.50) (228). Less sterically demanding alkenes such as cyclohexene, cyclopentene, and methyl substituted styrenes react, albeit at elevated temperature. The only exception is the 1,1-disubstituted alkene (entry 4), which reacts at room temperature. Both stilbene and dimethyl fumarate fail to provide the desired cycloadduct. In a rare example of the dipolar cycloaddition of tetra-substituted alkenes, tetramethylethylene reacts at 50 °C over 3 days to give a small amount of the cycloadduct (entry 7).

The reactivity of diactivated alkenes depends greatly on the configuration of the dipolarophile. With nitronate **231**, (*Z*)-diactivated alkenes react at 0 °C or room temperature over several days, while an (*E*)-substituted alkene takes 14 days (Table 2.51) (97). By comparison to the cycloaddition of the monosubstituted dipolarophile (entry 5, Table 2.47), there is only a slight increase in reaction time for the (*Z*)-alkenes. This trend can also be seen, though not a drastically, in the dipolar cycloadditions of

TABLE 2.49. INFLUENCE OF DIPOLAROPHILE SUBSTITUTION

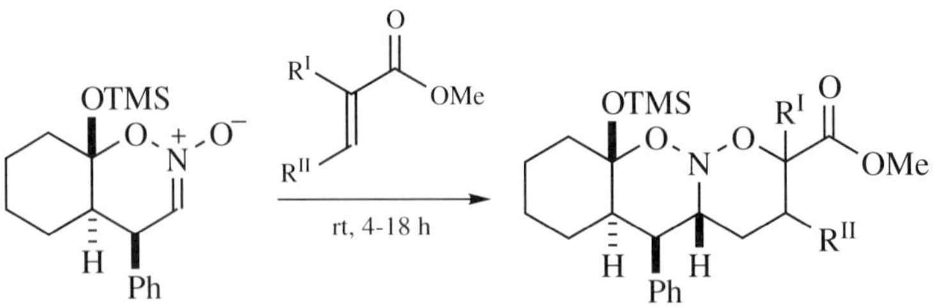

(24) **(248)–(251)**

Entry	R^I	R^{II}	Product	Yield	dr
1	H	H	**248**	59	5:1
2	Me	H	**249**	71	1:1
3	H	Me	**250**	0	
4	H	Ph	**251**	0	

TABLE 2.50. EFFECT OF DIPOLAROPHILE SUBSTITUTION OF THE CYCLOADDITION WITH **5**

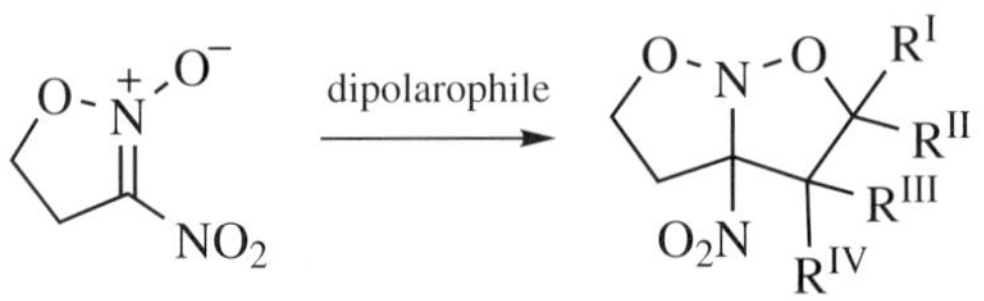

(5) **(252)–(258)**

Entry	Dipolarophile	Temp (°C)	Time (h)	Product	Yield (%)
1		50	48	**252**	37
2		50	4	**253**	53
3	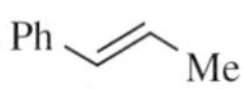	50	15	**254**	38
4	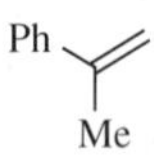	rt	72	**255**	33
5	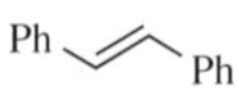			**256**	0
6	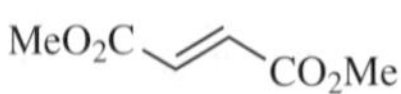			**257**	0
7	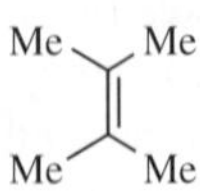	50	72	**258**	12

TABLE 2.51. CYCLOADDITION WITH DIACTIVATED DIPOLAROPHILES

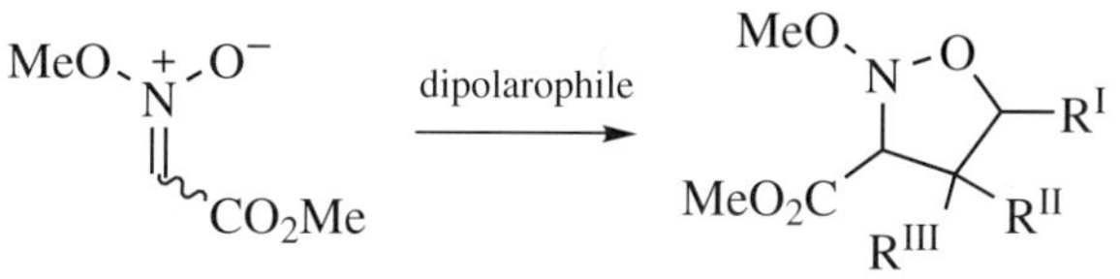

(231) **(259)–(262)**

Entry	Dipolarophile	Temp (°C)	Time (d)	Product	Yield (%)
1	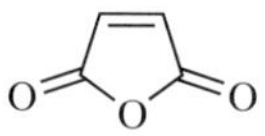	0	3	**259**	67
2	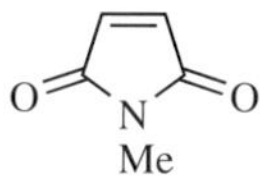	0	3	**260**	59
3	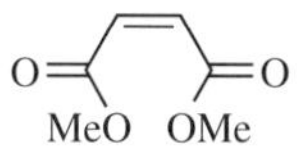	rt	2	**261**	40
4	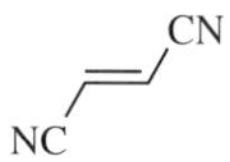	rt	14	**262**	53

49. Reaction with methyl acrylate proceeds in 6 h at room temperature, while dimethyl maleate takes over 10 h to go to completion (Eq. 2.22) (69,84).

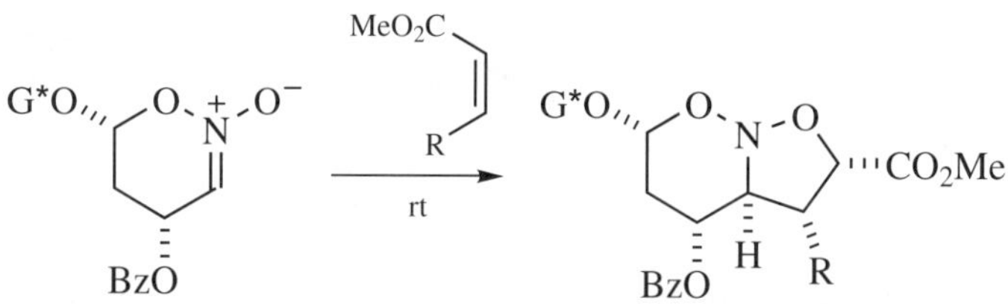

(2.22)

(49)

(51), R = H, 6 h, 86%
(263), R = CO_2Me, 10.5 h, 88%

The [3 + 2] cycloaddition of terminal alkynes has been investigated with several dipoles. These dipolarophiles are competent in the cycloaddition, however, the corresponding isoxazolines cannot be isolated. Instead, the cycloadduct undergoes spontaneous rearrangement to provide acylaziridine products (Table 2.52) (229). Disubstituted alkynes also undergo this process, however, in lower yield. This rearrangement occurs with all nitronates studied (Chart 2.3) (66,230,231).

Because the dipolar cycloaddition is a bimolecular process, there exists the opportunity to accelerate the reaction under high pressure (232,233). The reaction

TABLE 2.52. CYCLOADDITIONS OF ALKYNES

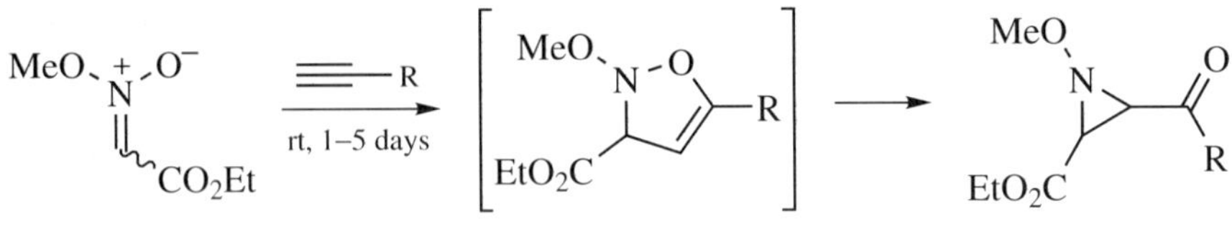

(231) **(264)–(268)**

Entry	R	Product	Yield (%)
1	COMe	**264**	83
2	CO_2Me	**265**	74
3	Ph	**266**	46
4	CH_2Cl	**267**	55
5	CH_2OH	**268**	52

of the nitronate **231** to a steroid derived trisubstituted alkene at 14 kbar provides the desired cycloadduct after 18 h at ambient temperature (Eq. 2.23) (234). The same reaction run at 1 bar provides only 2% of the desired product.

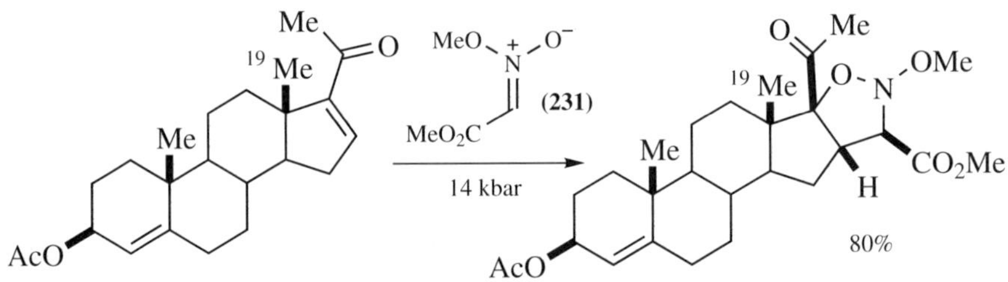

(270) **(271)**

(2.23)

This pressure effect was also observed in the tandem cycloaddition nitroalkene. Combination of methyl acrylate with 2-nitrostyrene and ethyl vinyl ether proceeds at 15 kbar over 1 h to produce a mixture of diastereomeric nitroso acetals (Eq. 2.24) (83,235,236).

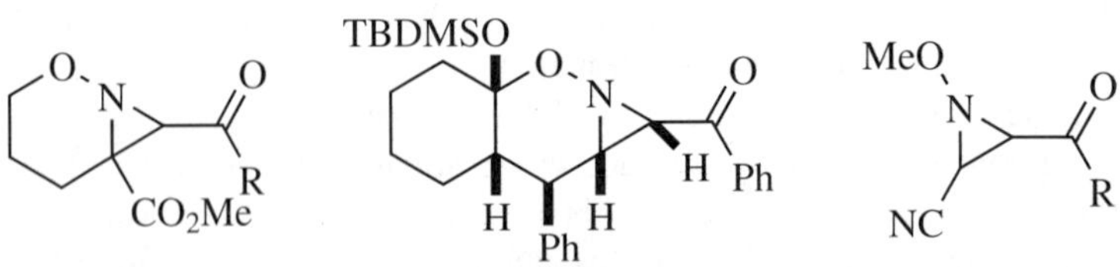

(XLIV) **(269)** **(XLV)**

Chart 2.3

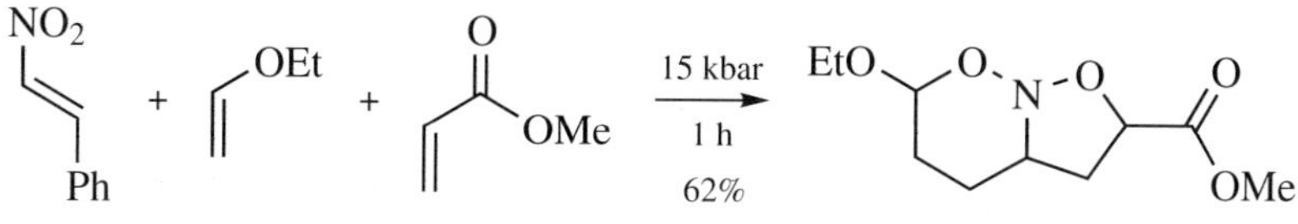

(272)

(2.24)

The only reported [3+2] cycloaddition between a nitronates and a heterodipolarophile results from the decomposition of **3** (Eq. 2.25) (223). Upon elimination of formaldehyde, the resulting oxime can undergo reaction with remaining **3**, which provides the compound **273**, following elimination of water.

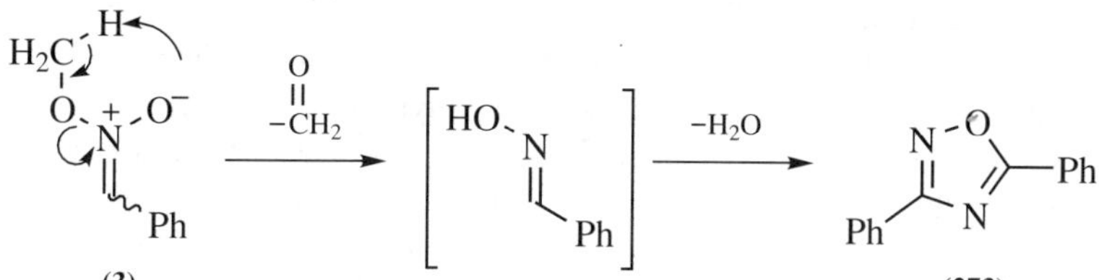

(3) (273) (2.25)

Interestingly, furoxanes (**274**) have also been shown to be competent dipoles for the [3+2] cycloaddition. These compounds result from a dimerization of the corresponding nitrile oxide (Eq. 2.26) (237–241). Under elevated temperatures, the addition of a dipolarophile results in the cycloadduct **275**. This intermediate is unstable under the reaction conditions and undergoes a retro-[3+2] cycloaddition to reveal a nitronate. In the presence of excess dipolarophile, the reaction proceeds to provide bicyclic isoxazolidines in moderate yield.

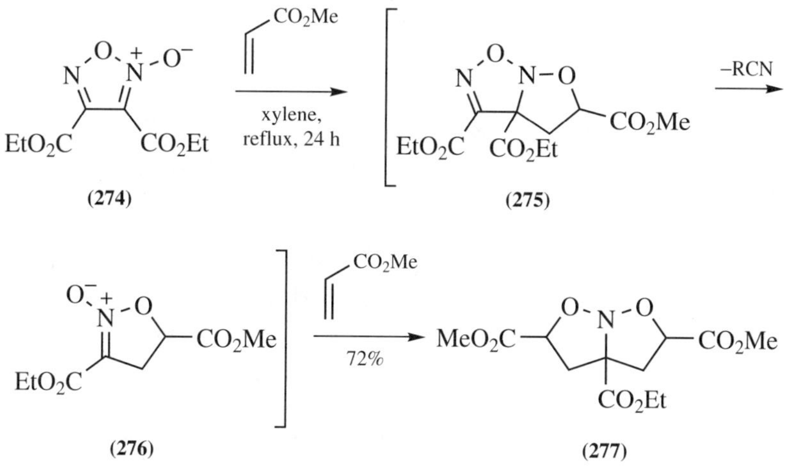

(274) (275) (276) (277)

(2.26)

2.4.3.2. Facial Selectivity

The number of investigations on the enantioselective dipolar cycloaddition of nitronates is still rather limited. In the case of simple alkyl nitronates, the facial selectivity is controlled solely by the steric environment about the two faces of the chiral unit. For example, the reaction of steroid dipolarophile **270** proceeds with the nitronate approaching the *Re* face of the alkene (Eq. 2.23) (234). The facial selectivity is controlled by the C(19) methyl group, which blocks the *Si* face of the dipolarophile. Similarly, exposure of **279** to ethyl acrylate at 40 °C for 24 h, provides a single nitroso acetal (Scheme 2.21) (242). The facial selectivity is presumed to arise from steric shielding by the menthol group, however the full stereostructure has not been established.

The majority of asymmetric dipolar cycloadditions have been investigated in the context of the tandem [4+2]/[3+2]-nitroalkene cycloaddition. The chiral nitronate is prepared by using either a chiral nitroalkene, vinyl ether, or Lewis acid in the first cycloaddition. The acetal center at C(6) of the nitronate provides important steric and electronic effects that control the subsequent dipolar cycloaddition. Subsequently, in the cycloadditions of the chiral nitroalkenes **281** and **284**, the dipolarophile approaches from the side distal to that of the substituent at C(4) and the acetal center at C(6) (Eq. 2.27 and Table 2.53) (90,215).

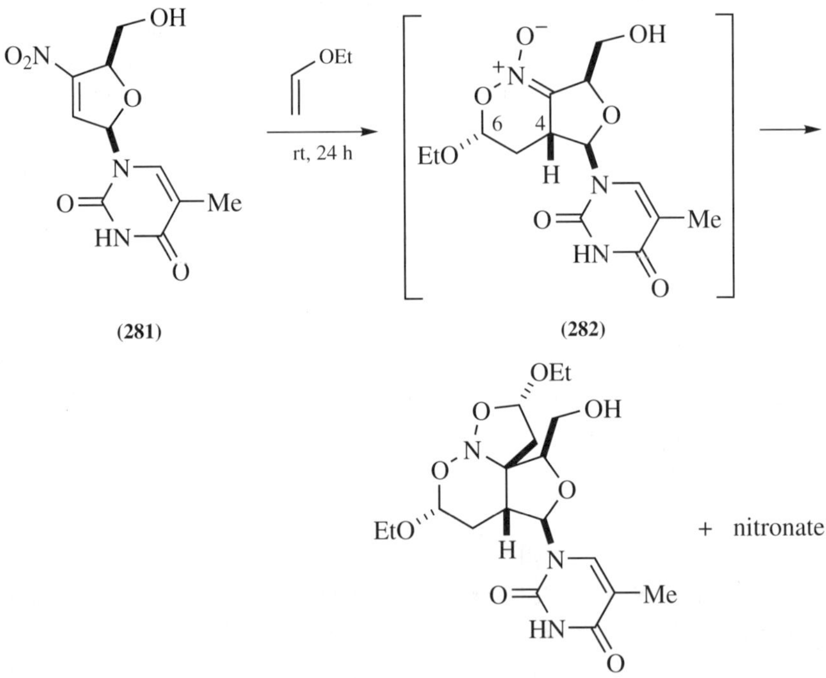

(281) (282) (283)

(2.27)

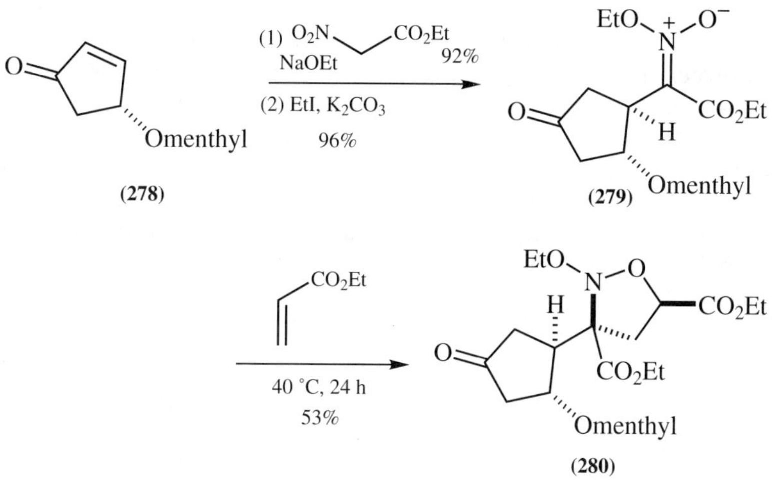

Scheme 2.21

Similar facial selectivity has been observed in nitronates derived from chiral vinyl ethers (69), as well as from nitronates prepared with a chiral Lewis acid, which lack any bias from a chiral auxiliary (66). Even in the absence of a substituent at C(4), as in the nitronate **287**, there remains a high facial selectivity upon the addition of a dipolarophile (Eq. 2.28) (84). Both RHF and B3LYP calculations for the approach of a dipolarophile to the nitronates **289** and **290** show at least a

TABLE 2.53. FACIAL SELECTIVITIES WITH A SUGAR-BASED AUXILIARY

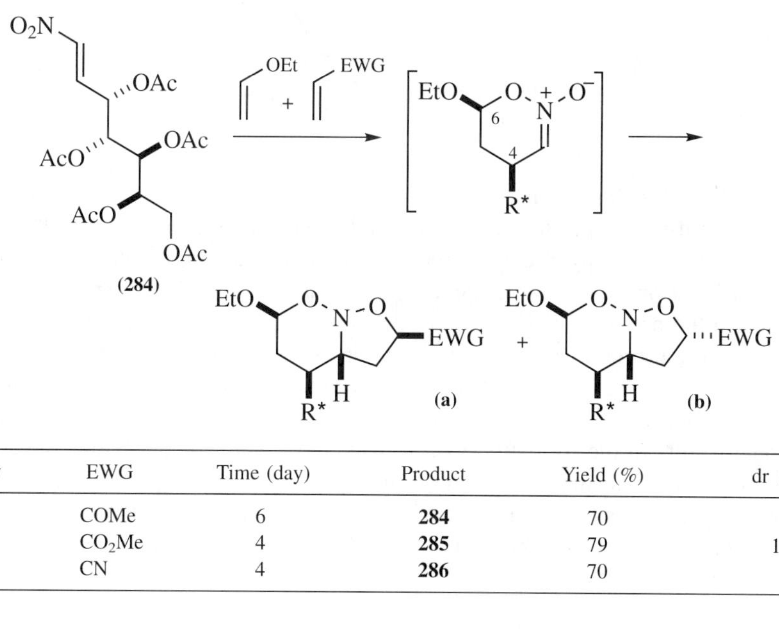

Entry	EWG	Time (day)	Product	Yield (%)	dr (**a**:**b**)
1	COMe	6	**284**	70	1:0
2	CO_2Me	4	**285**	79	19:1
3	CN	4	**286**	70	6:1

0.9 kcal/mol preference for the formation of a chair-like transition structure (Fig. 2.14). Therefore, if the cyclic nitronate is held in one of two possible twist–boat conformations (e.g., by an acetal center as in nitronate **290**), facial selectivity can be observed. In the case of nitronates **281** and **284** (Eq. 2.27, Table 2.53), the steric bias provided by the nitronate enhances the stereoelectronic facial bias. When these two forces oppose each other as with nitronate **291**, the facial selectivity depends on the size of the dipolarophile (Scheme 2.22). With dimethyl maleate, approach is observed distal to the acetal substituent. However, with a bulkier dipolarophile (**292**) the facial selectivity is governed by the substituent at C(4).

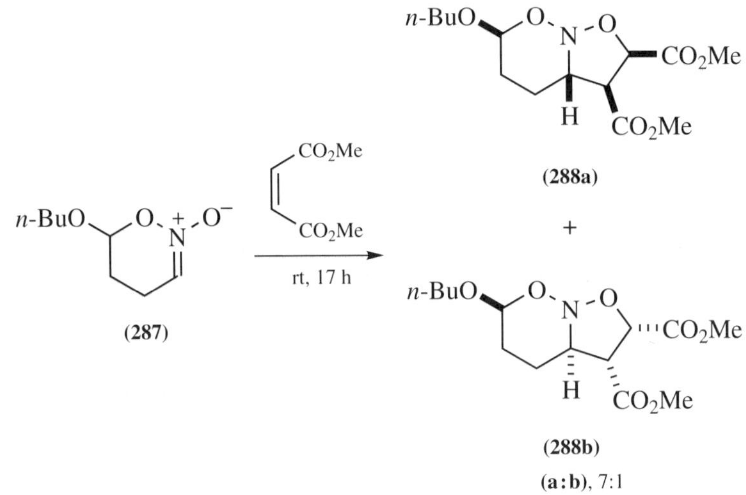

(2.28)

2.4.3.3. *Intramolecular Cycloadditions*

The intramolecular dipolar cycloaddition of nitronates has remained relatively underexplored in comparison to the intermolecular variant. In the case of acyclic nitronates, there are only a few reports of an intramolecular nitronate cycloaddition (36,176,177). However, the intermediate nitroso acetal decomposes to the isoxazoline due to the presence of HCl in the reaction mixture (Scheme 2.19).

The intramolecular cycloadditions of cyclic nitronates have received much more attention. The cyclic nitronate structure provides three basic modes of intramolecular cycloaddition (Fig. 2.15). Attachment of the tether to the C(3) position of the nitronate results in the formation of a spiro system (spiro mode). However, if the tether is appended to the C(4) position of the nitronate, the dipolar cycloaddition yields a fused ring system (fused mode). Finally, if the tether is attached at any other point of the cyclic nitronate, the cycloadducts obtained will consist of bicyclic structures (bridged mode).

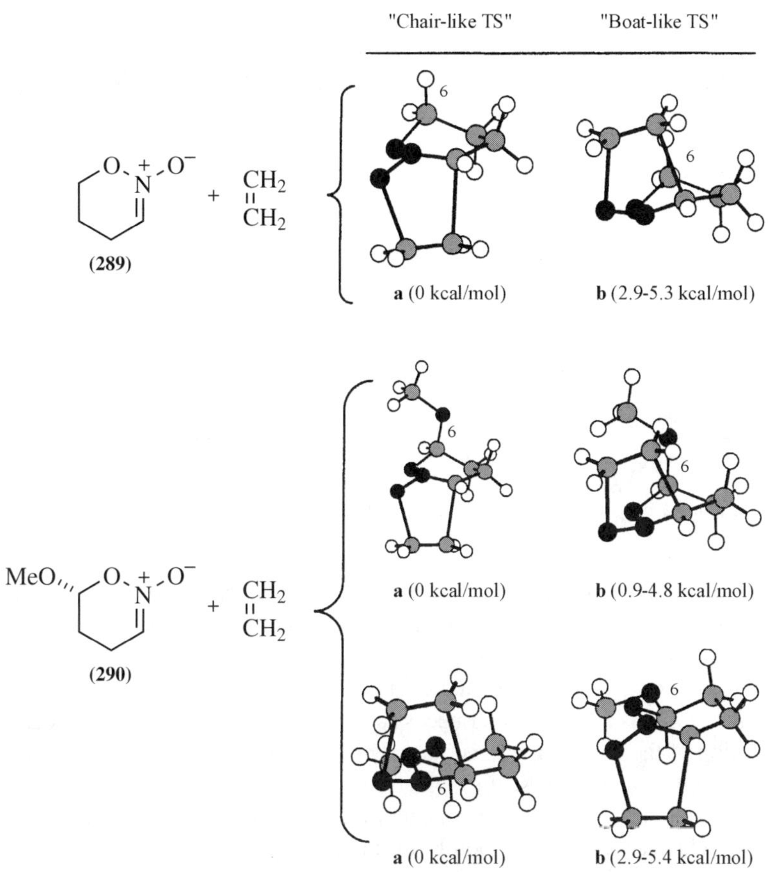

Figure 2.14. Computaional examination of the facial selectivity of nitronates.

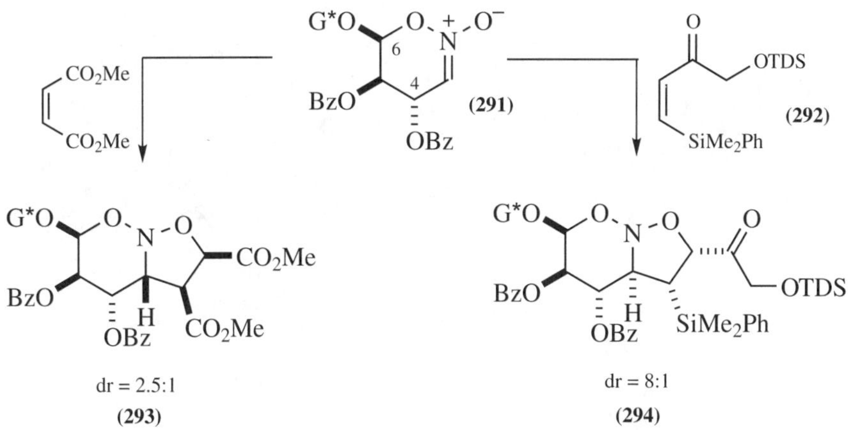

Scheme 2.22

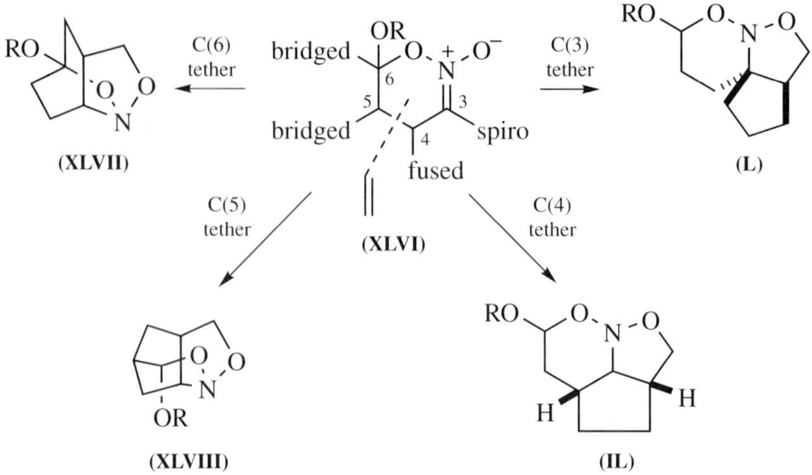

Figure 2.15. Subclasses of the inter [4+2]/intra [3+2]-tandem cycloadditions.

By far the most investigated family of intramolecular cycloadditions is the fused mode. The effect of the tether length and dipolarophile geometry have been examined in the case of nitronate possessing an all-carbon tether (Scheme 2.23) (99). In the case of a two atom tether, the [3+2] cycloaddition proceeds upon quenching the Lewis acid promoter necessary for the [4+2] cycloaddition when the configuration of the dipolarophile is trans. This result is from a matched case for the mode of approach for both the electron-withdrawing group and the tether (see Section 2.2.2.2 for further discussion of stereochemical consequences). The mismatched case reacts smoothly at elevated temperature. The formation of

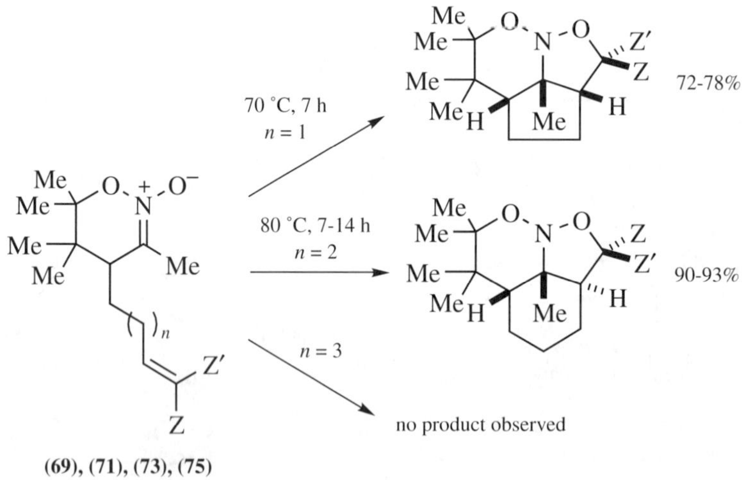

Scheme 2.23

six-membered rings requires elevated temperature, but the same trend is observed for the matched and mismatched cases. Attempts to form seven-membered rings have thus far failed. The use of unactivated dipolarophiles has also been successful, though longer reaction times or elevated temperatures are necessary (85). In addition to carbon tethers, both ester linkages and silyloxy tethers are compatible with the dipolar cycloaddition (194,243–250).

In the case of the bridged mode intramolecular cycloadditions, there are two subsets that have been investigated, differing in the attachment of the tether. Nitronates bearing the dipolarophile attached to C(6) require heating at 110 °C for complete reaction (Scheme 2.24) (251). However, the incorporation of an alkoxy group at C(6) leads to longer reaction times and variable yields (86). The length of time and temperature is dependent on the configuration of the nitronate. For **299**, complete reaction is obtained in 11 h, while 24 h at a higher temperature is necessary for **297**. Since it is likely that the allyl group will only react while in an axial position, the conformational mobility of the nitronate will determine the reaction rate (Fig. 2.16). The nitronate **297** resides mostly in conformer **LI** to position the phenyl group in a pseudo-equatorial position while maintaining the alkoxy group in an axial position to provide anomeric stabilization. These two influences are opposed in **299**, and therefore it is easier to access the reactive conformer **LIII** (i.e., lower reaction temperatures).

The dipolar cycloaddition of nitronates bearing a dipolarophile at C(5) are slightly more facile than their C(6) bridged counterparts (Eq. 2.29) (87,252). Again,

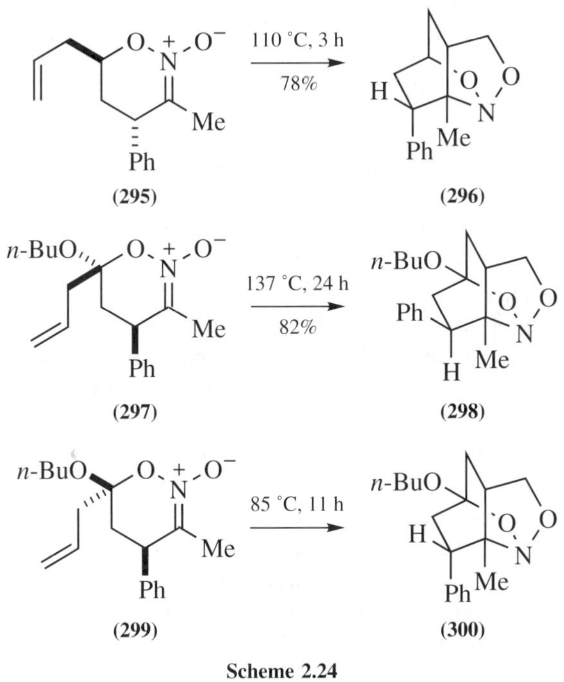

Scheme 2.24

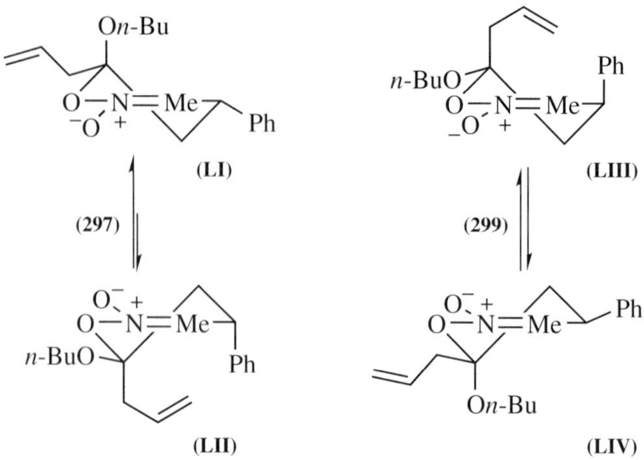

Figure 2.16. Proposed conformations of nitronates bearing and C(6)-bridged tether.

this process is highly dependent on the configuration of the nitronate. However, the observed reactivities cannot be explained by simple conformational analysis, and are dependent on substituents present on the nitronate.

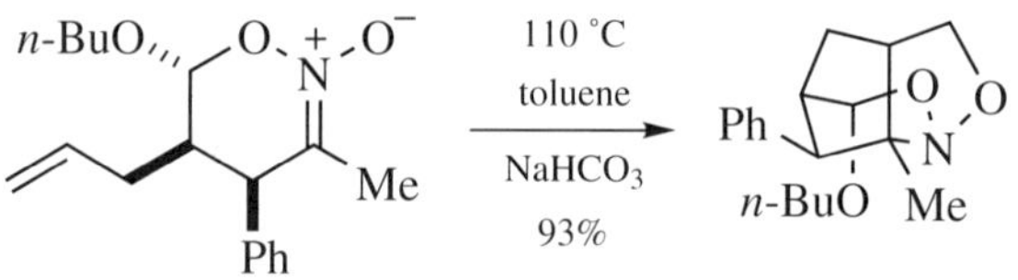

(2.29)

In both bridged cases, chiral vinyl ethers have been utilized to prepare diastereomerically enriched nitronates (86,253). Since only a single atom tether has been investigated for two bridged modes, the facial selectivity of the dipolar cycloaddition is completely controlled by the configuration of the nitronate at the point of attachment.

The placement of the tether at the C(3) position of the nitronate provides access to the spiro mode of dipolar cycloadditions. This class has been evaluated for the preparation of five- and six-membered spiro ring systems (Table 2.54) (254). The dipolar cycloaddition proceeds in good yields with mild heating, however, the three-atom tether proceeds faster than the four-atom tether.

Since the dipolarophile is not predisposed toward one face of the dipole at the point of attachment, the facial selectivity is governed by the configuration of the acetal center on the starting nitronate. As with the intermolecular case (Section 2.4.3.2), the preferred approach of the dipolarophile is distal to the acetal substituent. Lower facial selectivity is observed in the case of a four-atom tether, presumably due to the additional heating involved in driving the reaction to completion.

TABLE 2.54. SPIRO MODE INTRAMOLECULAR CYCLOADDITIONS

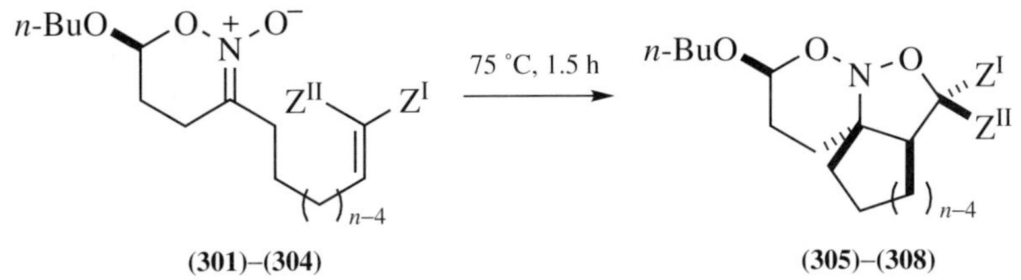

Entry	Dipole	n	Z^I	Z^{II}	Product	Yield (%)	dr (facial)
1	**301**	5	CO_2Me	H	**305**	76	35:1
2	**302**	5	H	CO_2Me	**306**	62	10:1
3	**303**	6	CO_2Me	H	**307**	84	21:1
4	**304**	6	H	CO_2Me	**308**	64	3.5:1

2.4.4. Nitroso Acetal Functionalization

Despite the presence of two, relatively weak, N—O bonds (~53 kcal/mol) (255) nitroso acetals are stable under neutral conditions. Therefore, a number of transformations can be performed on the periphery of the cycloadducts without effecting the nitroso acetal. These include both oxidation of alcohols, dihydroxylation of alkenes (247,248,256), Tamao–Fleming type oxidation (248–250), reductions of ketones and esters (244,257–261), silylation and desilylation (247,257, 260), and activation with a variety of sulfonylating agents (247,248,257–260).

The nitroso acetal is nonetheless a labile function, and several transformations can be carried out selectively on this group (Scheme 2.25). Treatment of a nitroso acetal of the type **LV** with heat, a mild protic source, or boron trifluoride etherate

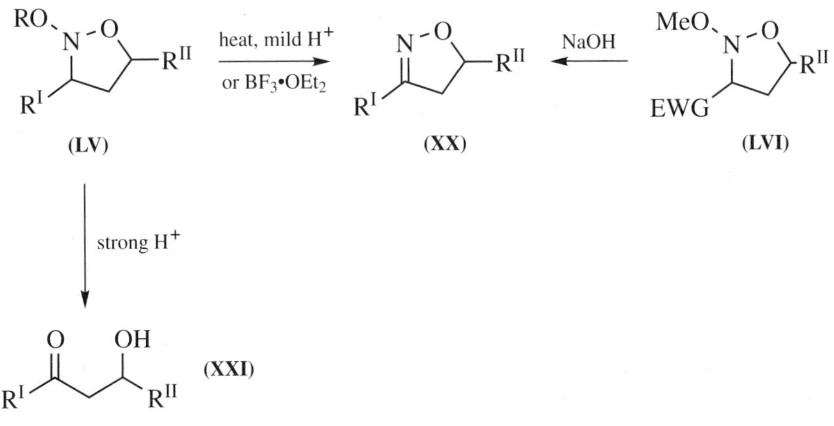

Scheme 2.25

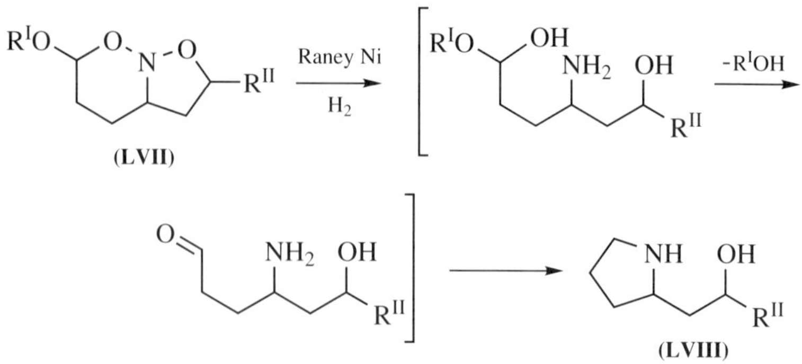

Scheme 2.26

results in the elimination of the alkoxy moiety to form the corresponding isoxazoline **XX** (90,216,224,262). However, the reaction of highly functionalized nitroso acetals with mild acid tends to lead to additional rearrangements (263). If the nitroso acetal bears an electron-withdrawing group at the C(3) position, treatment with aqueous base also leads to elimination of methanol (59,75). Further transformations of **XX** are discussed in Section 2.3.3. The exposure of **LV** to strongly acidic conditions leads directly to the β-hydroxy ketone **XXI** (77,144,223).

The reductive cleavage of the nitroso acetal by catalytic hydrogenolysis is one of the most useful transformations of this functional group. Raney Ni is the most commonly employed catalyst at various pressures of hydrogen (89), however, several other agents have been used successfully (86,250). Cleavage of the N—O bonds provides the corresponding amino alcohol, or amino diol in the case of cyclic nitronates (Scheme 2.26). However, for nitroso acetals of the type **LVII**, collapse of the *in situ* generated hemiacetal, followed by reductive alkylation of the free amine, results in the formation of pyrrolidine type products **LVIII**. If an ester or suitably functionalized leaving group is also present in the molecule, acylation or alkylation of the intermediate amine is also possible, providing bicyclic lactams or amines, respectively (Scheme 2.27). These skeletons correspond to the core structures of pyrrolizidine and indolizidine alkaloids.

2.4.5. Applications

In view of the multicomponent nature of the tandem [4+2] / [3+2] cycloaddition, the potential for a combinatorial approach to the synthesis of nitroso acetals has been investigated on solid-phase supports. The incorporation of either the dipolarophile or the starting nitroalkene on a Wang-type resin is compatible with the tandem cycloaddition promoted at high pressures (Schemes 2.28 and 2.29). The solid-supported nitroso acetals are subsequently liberated (in moderate yields from the staring nitroalkene) upon the addition of a catalytic amount of potassium cyanide in triethylamine and methanol or by reduction with lithium aluminum hydride (LAH) (261,264).

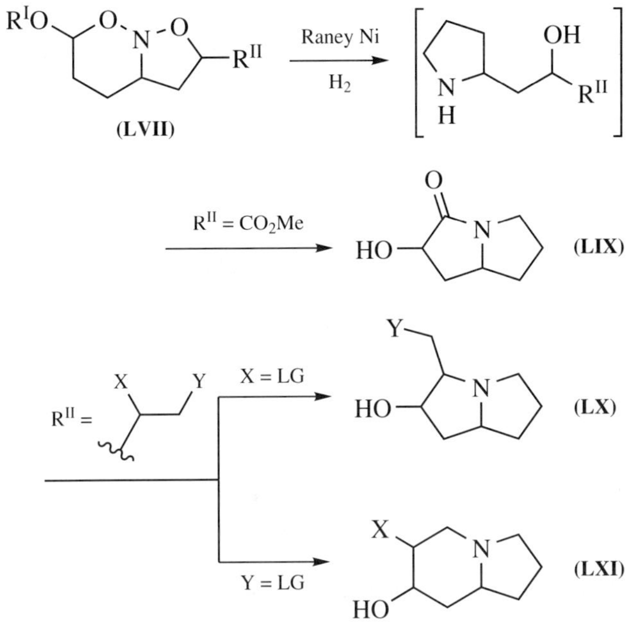

Scheme 2.27

The dipolar cycloaddition of nitronates has been applied to the synthesis of several natural products in the context of the tandem [4+2] / [3+2] nitroalkene cycloaddition process. All of these syntheses have focused on the construction of pyrrolidine, pyrrolizidine, and indolizidine alkaloids. For example, the synthesis of (−)-hastanecine (**316**), a necine alkaloid, involves the elaboration of a β-benzoyloxynitroalkene **311** via [4+2] cycloaddition with a chiral vinyl ether (**312**) in the presence of a titanium based Lewis acid, to provide the nitronate **313** with high diastereo- and facial selectivity (Scheme 2.30) (69). The dipolar cycloaddition of

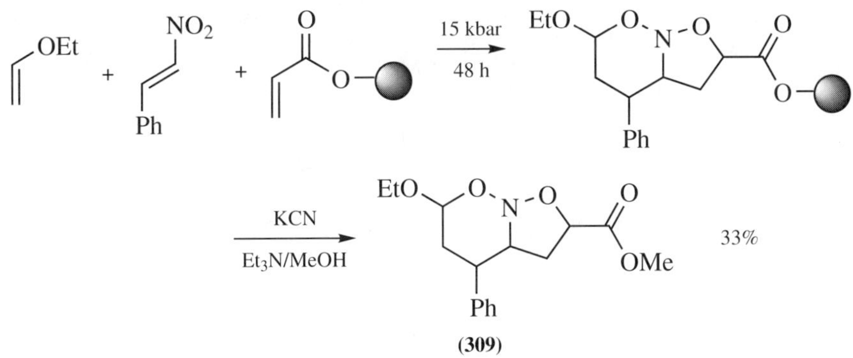

Scheme 2.28

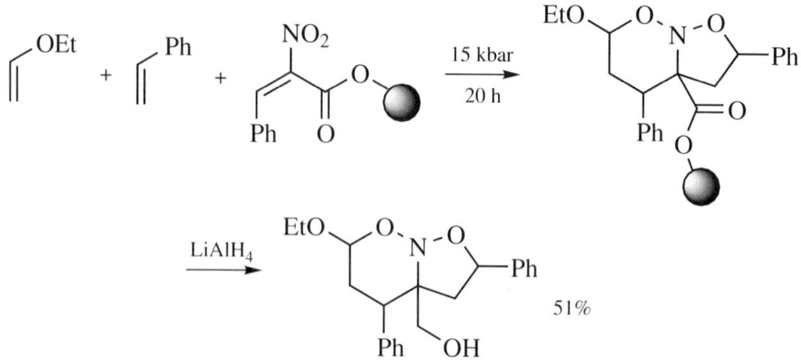

(310)

Scheme 2.29

313 with dimethyl maleate proceeds to give **314** as a single diastereomer. Cleavage of the nitroso acetal with Raney Ni and hydrogen, followed by several functional group manipulations provides the target alkaloid in only six step from available starting materials. Similar intermolecular dipolar cycloadditions have also been

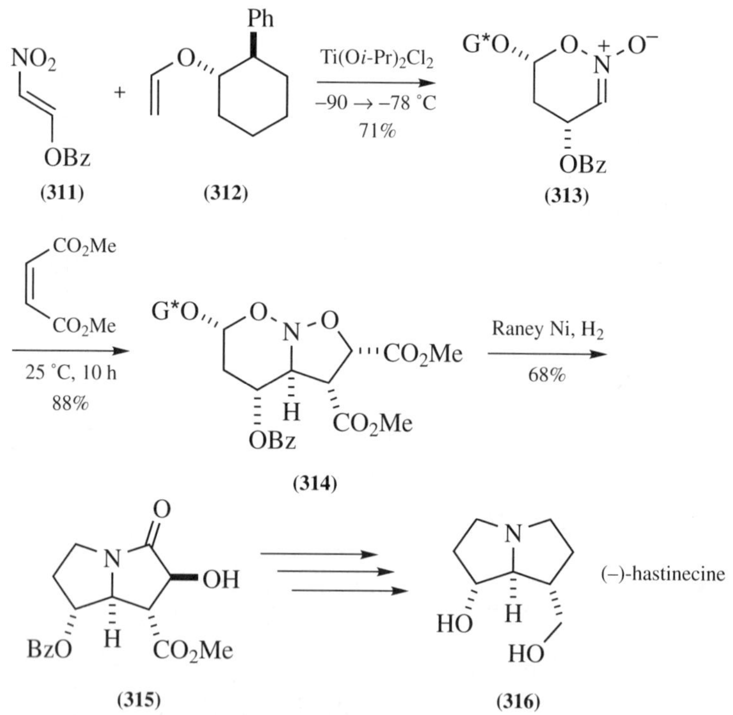

Scheme 2.30

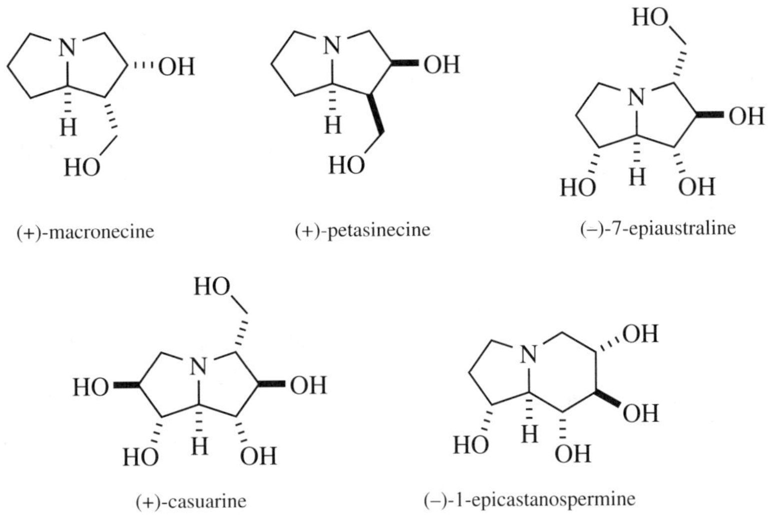

Chart 2.4

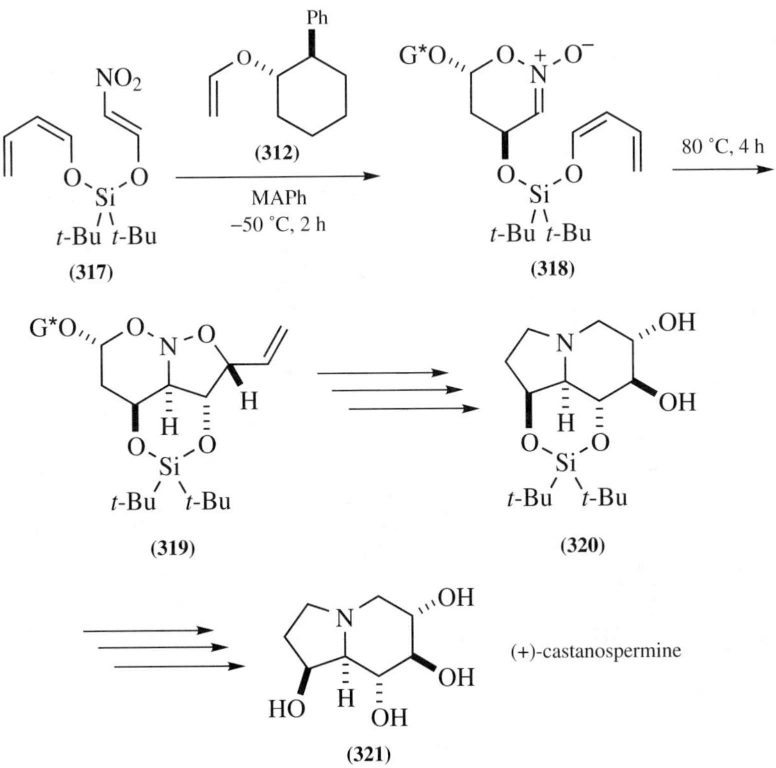

Scheme 2.31

featured as key steps in the syntheses of other natural and nonnatural alkaloids (Chart 2.4) (222,257–260).

The intramolecular dipolar nitronate cycloaddition has also been exploited in several total syntheses of more complex alkaloids. The general strategy is illustrated in the synthesis of (+)-castanospermine (**321**) (Scheme 2.31) (247). In this case, the nitronate is generated by Lewis-acid-promoted cycloaddition of the functionalized nitroalkene **317** with chiral vinyl ether **312**. The regio- and stereochemical course of the [3+2] cycloaddition is controlled by the silylene ketal tether that bears the dipolarophile. The resulting nitroso acetal **318** contains four of the five stereogenic centers in (+)-castanospermine, three of which are created by the complete facial and exo selectivity in the dipolar cycloaddition. Further functionalization of the nitroso acetal involves dihydroxylation of the remaining alkene, followed by activation of the primary alcohol and cyclization initiated by hydrogenolysis with Raney nickel. Simple deprotection then reveals the target alkaloid. Related strategies have been employed using various tethering groups for the successful, enantioselective syntheses of highly functionalized alkaloids (Chart 2.5) (243–248,265,266).

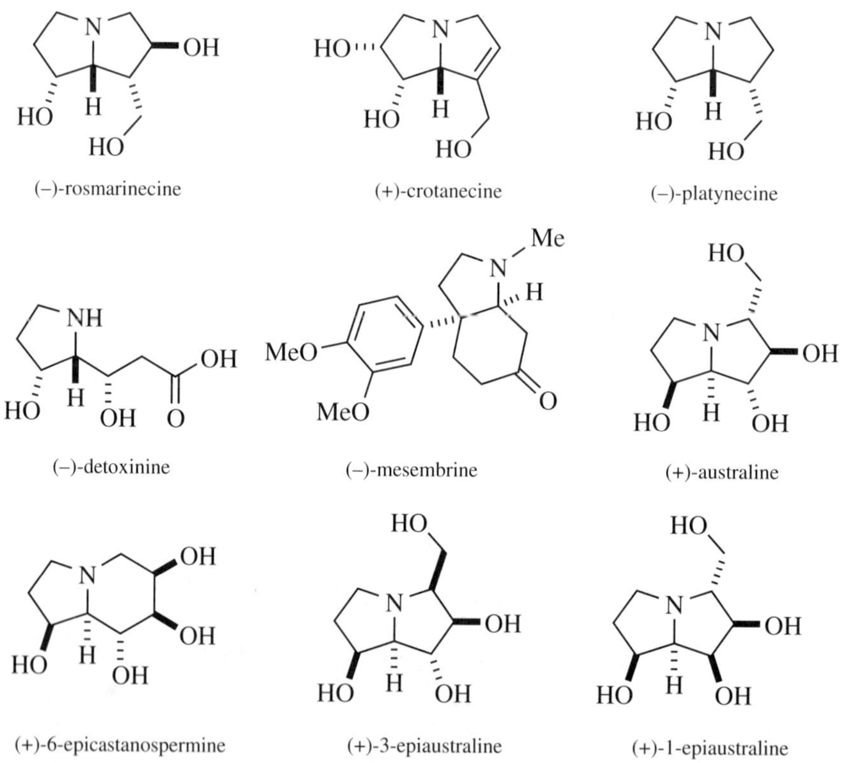

Chart 2.5

2.5. CONCLUSION AND OUTLOOK

Although clearly the most underdeveloped of the C—N—O dipoles compared to nitrones and nitrile oxides, nitronates have emerged in the last 20 years as versatile intermediates in organic synthesis. Perhaps their greatest advantage is the diversity of methods available for the preparation of nitronates from aliphatic and alkeneic nitro compounds. Further, they reside in a unique niche between nitrones and nitrile oxides in stability, but display rather comparable reactivity. Indeed, these important dipoles offer the reactivity of nitrile oxides together with the greater structural versatility associated with nitrones. Finally, the nitroso acetal cycloadducts represent a fascinating class of compounds whose synthetic potential has hardly been explored.

Whereas it is easy to foresee expanded application of nitronate cycloadditions in synthesis, important challenges still remain. For example, the potential of these reactions would benefit from the development of catalysts to accelerate the process. Moreover, asymmetric catalysis has only recently been successful in dipolar cycloaddition chemistry and would have a great impact here. Another important avenue would be the invention of new tandem processes that allow for the creation of nitronates from different precursors in the presence of dipolarophiles.

It is hoped that by chronicling the 30-year history of nitronate cycloadditions, this chapter will stimulate these and still more imaginative applications throughout the synthetic organic enterprise.

REFERENCES

1. N. Ono, *The Nitro Group in Organic Synthesis*, Wiley, New York, **2001**.
2. R. G. Coombes, "Nitro and Nitroso Compounds" in *Comprehensive Organic Chemistry*, Vol. 2, I. O. Sutherland, ed., Pergamon Press, New York, **1979**, pp. 303–382.
3. S. Patai, ed., "The Chemistry of Amino, Nitroso, and Nitro compounds and their derivatives," in *The Chemistry of Functional Groups*, Wiley, New York, **1982**.
4. H. Feuer, ed., "The Chemistry of the Nitro and Nitroso Groups", in *The Chemistry of Functional Groups*, S. Patai, ed., Wiley, New York, **1969**.
5. E. Breuer, "Nitrones and Nitronic Acid Derivatives: Their Strucutre and Their Role in Synthesis" in *The Chemistry of Amino, Nitroso, And Nitro Compounds and Their Derivatives*, S. Patai, ed., Wiley, New York, **1982**, pp. 460–564.
6. A. T. Nielsen, "Nitronic Acids and Esters" in *The Chemistry of the Nitro and Nitroso Groups*, H. Feuer, ed., Wiley, New York, **1969**, pp. 349–486.
7. R. Stewart, "The Proton: Applications to Organic Chemistry", in *Organic Chemistry*, H. H. Wasserman, ed., Vol. 46. Academic Press, New York, **1985**.
8. J. U. Nef, *Justus Liebigs Ann. Chem.*, **1894**, *280*, 263.
9. E. Bamberger and O. Schmidt, *Chem. Ber.*, **1901**, *34*, 574.
10. H. W. Pinnick, *Org. React.*, **1990**, *38*, 655.
11. H. B. Hass and M. L. Bender, *Org. Synth.*, **1950**, *30*, 99; *Coll. Vol IV*, **1963**, 932.
12. V. A. Tartakovskii, I. E. Chlenov, S. S. Smagin, and S. S. Novikov, *Izv. Akad. Nauk SSSR, Ser. Khim.*, **1964**, 583; *Bull. Acad. Sci. USSR, Div. Chem. Sci. (Engl. Transl.)*, **1964**, 549.

13. I. L. Knunyants, E. G. Bykhovskaya, V. N. Frosin, and Y. M. Kisel, *Dokl. Akad. Nauk SSSR*, **1960**, *132*, 123; *Dokl. Chem. (Engl. Transl.)*, **1960**, *132*, 455.

14. M. A. Brook and D. Seebach, *Can. J. Chem.*, **1987**, *65*, 836.

15. V. F. Rudchenko, I. I. Chervin, and R. G. Kostyanovsky, *Mendeleev Commun.*, **1991**, 9.

16. S. L. Ioffe, M. V. Kashutina, V. M. Shitkin, A. A. Levin, and V. A. Tartakovskii, *Izv. Akad. Nauk SSSR, Ser. Khim.*, **1972**, 1341; *Bull. Acad. Sci. USSR, Div. Chem. Sci. (Engl. Transl.)*, **1972**, 1292.

17. J. F. Klebe, *J. Am. Chem. Soc.*, **1964**, *86*, 3399.

18. K. B. G. Torssell and O. Zeuthen, *Acta Chem. Scand. (B)*, **1978**, *32*, 118.

19. J. M. Aizpurua, M. Oiarbide, and C. Palomo, *Tetrahedron Lett.*, **1987**, *28*, 5361.

20. M. V. Kashutina, S. L. Ioffe, and V. A. Tartakovskii, *Dokl. Akad. Nauk SSSR*, **1974**, *218*, 109; *Dokl. Chem. (Engl. Transl.)*, **1974**, *218*, 607.

21. E. W. Colvin, A. K. Beck, B. Bastani, D. Seebach, Y. Kai, and J. D. Dunitz, *Helv. Chem. Acta*, **1980**, *63*, 697.

22. W. Dehaen and A. Hassner, *Tetrahedron Lett.*, **1990**, *31*, 743.

23. K. Folting, B. Jerslev, and W. N. Lipscomb, *Acta Crystallogr.*, **1964**, *17*, 1263.

24. R. E. Marti, J. Heinzer, and D. Seebach, *Liebigs Ann.*, **1995**, 1193.

25. M. V. Kashutina, S. L. Ioffe, V. M. Shitkin, N. O. Cherskaya, V. A. Korenevskii, and V. A. Tartakovskii, *Zh. Obshch. Khim.*, **1973**, *43*, 1715; *J. Gen. Chem. USSR (Engl. Transl.)*, **1973**, *43*, 1699.

26. S. L. Ioffe, L. M. Makarenkova, V. M. Shitkin, M. V. Kashutina, and V. A. Tartakovskii, *Izv. Akad. Nauk SSSR, Ser. Khim.*, **1973**, 203; *Bull. Acad. Sci. USSR, Div. Chem. Sci. (Engl. Transl.)*, **1973**, 212.

27. E. Breuer, "Nitrones and Nitronic Acid Derivatives: Their Structure and their Role in Synthesis" in *Nitrones, Nitronates, and Nitroxides*, S. Patai and Z. Rappoport, ed., Wiley, New York, **1989**, pp. 139–244.

28. K. Nakanishi, *Infrared Absorption Spectroscopy*, Holden-Day, Inc., San Francisco, **1963**, 50.

29. S. L. Ioffe, V. M. Shitkin, B. N. Khasapov, M. V. Kashutina, V. A. Tartakovskii, M. Y. Myagi, and E. T. Lippmaa, *Izv. Akad. Nauk SSSR, Ser. Khim.*, **1973**, 2146; *Bull. Acad. Sci. USSR, Div. Chem. Sci. (Engl. Transl.)*, **1973**, 2100.

30. K. B. G. Torssell, "Nitrile Oxides, Nitrones, and Nitronates in Organic Synthesis", in *Organic Nitro Chemistry*, H. Feuer, ed., VCH Publishers, Inc., New York, **1988**.

31. E. Breitmaier and W. Voelter, *Carbon-13 NMR Spectroscopy*, 3rd ed. VCH, New York, **1990**

32. L. M. Jackman and S. Sternhell, "Applications of Nuclear Magnetic Resonance Spectroscopy in Organic Chemistry", 2nd ed. *Organic Chemistry*, D. H. R. Barton and W. Doering, eds., Vol. 5, Pergamon Press, New York, **1969**.

33. S. L. Ioffe, A. S. Shashkov, A. L. Blyumfel'd, and K. M. Litvak, *Izv, Akad. Nauk SSSR, Ser. Khim.*, **1977**, 2836; *Bull. Acad. Sci. USSR, Div. Chem. Sci. (Engl. Transl.)*, **1977**, 2625.

34. S. L. Ioffe, M. V. Kashutina, V. M. Shitkin, A. A. Levin, and V. A. Tartakovskii, *Zh. Org. Khim.*, **1973**, *9*, 896; *J. Org. Chem. USSR (Engl. Transl.)*, **1973**, *9*, 922.

35. G. Galley, P. G. Jones, and M. Pätzel, *Tetrahedron: Asymmetry*, **1996**, *7*, 2073.

36. J.-T. Liu, W.-W. Lin, J.-J. Jang, J.-Y. Liu, M.-C. Yan, C. Hung, K.-H. Kao, Y. Wang, and C.-F. Yao, *Tetrahedron*, **1999**, *55*, 7115.

37. F. Arndt and J. D. Rose, *J. Chem. Soc.*, **1935**, 1.

38. W. Steinkopf, *Justus Liebigs Ann. Chem.*, **1923**, *434*, 21.

39. R. Grée and R. Carrié, *Tetrahedron Lett.*, **1971**, *12*, 4117.

40. N. Kornblum and R. A. Brown, *J. Am. Chem. Soc.*, **1964**, *86*, 2681.

41. A. T. Nielsen, "Nitronic Acids and Esters" in *Nitrones, Nitronates, and Nitroxides*, S. Patai and Z. Rappoport, eds., Wiley, New York, **1989**, pp. 1–138.

42. J. T. Thurston and R. L. Shriner, *J. Org. Chem.*, **1937**, *2*, 183.

43. R. Grée, F. Tonnard, and R. Carrié, *Bull. Soc. Chim. Fr.*, **1975**, 1325.

44. M. A. Brook, R. Faggiani, C. J. L. Lock, and D. Seebach, *Acta Crystallogr.*, **1988**, *C44*, 1981.

45. S. E. Denmark, C. J. Cramer, and J. A. Sternberg, *Tetrahedron Lett.*, **1986**, *27*, 3693.

46. C. B. W. Senanayake. Ph.D. Thesis, University of Illinois: Urbana-Champaign, **1991**.

47. J. A. Dixon. Ph.D. Thesis, University of Illinois: Urbana-Champaign, **1998**.

48. G. Chiari and D. Viterbo, *Acta Crystallogr.*, **1982**, *B38*, 323.

49. S. C. Kokkou, *Acta Crystallogr.*, **1986**, *C42*, 1750.

50. C. George and R. Gilardi, *Acta Crystallogr.*, **1987**, *C43*, 362.

51. P. G. Jones, O. Kennard, and G. M. Sheldrick, *Acta Crystallogr.*, **1978**, *B34*, 1029.

52. J. W. Huffman, M. M. Cooper, B. B. Miburo, and W. T. Pennington, *Tetrahedron,* **1992**, *48*, 8213.

53. C. P. Butts, L. Eberson, M. P. Hartshorn, W. T. Robinson, D. J. Timmerman-Vaughan, and D. A. W. Young, *Acta Chem. Scand. (B)*, **1996**, *50*, 29.

54. C. P. Butts, L. Eberson, R. Gonzales-Luque, M. C. Hartshorn, M. P. Hartshorn, M. Merchan, W. T. Robinson, B. O. Roos, C. Vallance, and B. R. Wood, *Acta Chem. Scand. (B)*, **1997**, *51*, 984.

55. D. Vichard, J.-C. Halle, B. Huguet, M.-J. Pouet, D. Riou, and F. Terrier, *Chem. Commun.*, **1998**, 791.

56. D. R. Powell, P. E. Hanson, and S. H. Gellman, *Acta Crystallogr.*, **1996**, *C52*, 2945.

57. I. E. Chlenov, V. I. Khudak, V. A. Tartakovskii, and S. S. Novikov, *Isv. Akad. Nauk SSSR, Ser. Khim.*, **1969**, 2266; *Bull. Acad. Sci. USSR, Div. Chem. Sci. (Engl. Transl.)*, **1969**, 2113.

58. S. E. Denmark, Y.-C. Moon, C. J. Cramer, M. S. Dappen, and C. B. W. Senanayake, *Tetrahedron*, **1990**, *46*, 7373.

59. V. A. Tartakovskii, A. A. Onishchenko, I. E. Chlenov, and S. S. Novikov, *Dokl. Akad. Nauk SSSR*, **1966**, *167*, 844; *Dokl. Chem. (Engl. Transl.)*, **1966**, *167*, 406.

60. C. Galli, E. Marotta, P. Righi, and G. Rosini, *J. Org. Chem.*, **1995**, *60*, 6624.

61. Z. A. Krasnaya, T. S. Stytsenko, E. P. Prokof'ev, I. P. Yakovlev, and V. F. Kucherov, *Izv. Akad. Nauk SSSR, Ser. Khim.*, **1974**, 845; *Bull. Acad. Sci. USSR, Div. Chem. Sci. (Engl. Transl.)*, **1974**, 809.

62. G. Kumaran and G. H. Kulkarni, *Synthesis*, **1995**, 1545.

63. I. E. Chlenov, V. I. Khudak, L. N. Kolymagina, N. S. Morozova, and V. A. Tartakovskii, *Izv. Akad. Nauk SSSR; Ser. Khim.*, **1970**, 1867; *Bull. Acad. Sci. USSR, Div. Chem. Sci. (Engl. Transl.)*, **1970**, 1757.

64. I. E. Chlenov, N. S. Morozova, and V. A. Tartakovskii, *Izv. Akad. Nauk SSSR, Ser. Khim.*, **1973**, 216; *Bull. Acad. Sci. USSR, Div. Chem. Sci. (Engl. Transl.)*, **1973**, 228.

65. S. E. Denmark, B. S. Kesler, and Y.-C. Moon, *J. Org. Chem.*, **1992**, *57*, 4912.

66. D. Seebach, I. M. Lyapkalo, and R. Dahinden, *Helv. Chem. Acta*, **1999**, *82*, 1829.

67. S. E. Denmark and M. E. Schnute, *J. Org. Chem.*, **1994**, *59*, 4576.

68. S. E. Denmark and L. R. Marcin, *J. Org. Chem.*, **1995**, *60*, 3221.

69. S. E. Denmark and A. Thorarensen, *J. Org. Chem.*, **1994**, *59*, 5672.

70. Y. Tohda, N. Yamawaki, H. Matsui, T. Kawashima, M. Ariga, and Y. Mori, *Bull. Chem. Soc. Jpn.*, **1988**, *61*, 461.

71. A. T. Nielsen and T. G. Archibald, *Tetrahedron*, **1970**, *26*, 3475.

72. L. E. Sutton and T. W. J. Taylor, *J. Chem. Soc.*, **1931**, 2190.

73. T. W. J. Taylor and L. E. Sutton, *J. Chem. Soc.*, **1933**, 63.

74. S. E. Denmark, C. J. Cramer, and J. A. Sternberg, *Helv. Chem. Acta*, **1986**, *69*, 1971.

75. V. A. Tartakovskii, I. E. Chlenov, N. S. Morozova, and S. S. Novikov, *Izv. Akad. Nauk SSSR, Ser. Khim.*, **1966**, 370; *Bull. Acad. Sci. USSR, Div. Chem. Sci. (Engl. Transl.)*, **1966**, 346.

76. V. A. Tartakovskii, Z. Y. Lapshina, I. A. Savost'yanova, and S. S. Novikov, *Zh. Org. Khim.*, **1968**, *4*, 236–239; *J. Org. Chem. USSR (Engl. Transl.)*, **1968**, *4*, 228.

77. V. A. Tartakovskii, A. A. Onishchenko, I. E. Chlenov, and S. S. Novikov, *Dokl. Akad. Nauk SSSR*, **1965**, *164*, 1081; *Dokl. Chem. (Engl. Transl.)*, **1965**, *164*, 994.
78. R. Grée, F. Tonnard, and R. Carrié, *Tetrahedron*, **1976**, *32*, 675.
79. R. Grée and R. Carrié, *Tetrahedron*, **1976**, *32*, 683.
80. R. Grée and R. Carrié, *Tetrahedron Lett.*, **1972**, *13*, 2987.
81. V. F. Rudchenko, O. A. D'yanchenko, I. I. Chervin, A. B. Zolotoi, L. O. Atovmyan, and R. G. Kostyanovsky, *Izv. Akad. Nauk SSSR, Ser. Khim.*, **1978**, 850; *Bull. Acad. Sci. USSR, Div. Chem. Sci. (Engl. Transl.)*, **1978**, 733.
82. K. Müller and A. Eschenmoser, *Helv. Chem. Acta*, **1969**, *52*, 1823.
83. R. Uittenbogaard, J. P. Seerden, and H. W. Scheeren, *Tetrahedron*, **1997**, *53*, 11929.
84. S. E. Denmark, M. Seierstad, and B. Herbert, *J. Org. Chem.*, **1999**, *64*, 884.
85. S. E. Denmark and C. B. W. Senanayake, *Tetrahedron*, **1996**, *52*, 11579.
86. S. E. Denmark, V. Guagnano, J. A. Dixon, and A. Stolle, *J. Org. Chem.*, **1997**, *62*, 4610.
87. S. E. Denmark and J. A. Dixon, *J. Org. Chem.*, **1998**, *63*, 6167.
88. V. F. Rudchenko, *Chem. Rev.*, **1993**, *93*, 725.
89. S. E. Denmark and A. Thorarensen, *Chem. Rev.*, **1996**, *96*, 137.
90. M. Avalos, R. Babiano, P. Cintas, F. J. Higes, J. L. Jiménez, J. C. Palacios, and M. A. Silva, *J. Org. Chem.*, **1999**, *64*, 1494.
91. L. R. Domingo, *Theor. Chem. Acc.*, **2000**, *104*, 240.
92. S. Kanemasa, S. Kaga, and E. Wada, *Tetrahedron Lett.*, **1998**, *39*, 8865.
93. M. Avalos, R. Babiano, J. L. Bravo, P. Cintas, J. L. Jiménez, J. C. Palacios, and M. A. Silva, *Chem. Eur. J.*, **2000**, *6*, 267.
94. M. E. Schnute. Ph.D. Thesis, Univeristy of Illinois: Urbana-Champaign, **1995**.
95. R. Gompper and H.-U. Wagner, *Angew. Chem., Int. Ed. Engl.*, **1976**, *15*, 321.
96. R. Grée and R. Carrié, *J. Heterocyclic Chem.*, **1997**, *14*, 965.
97. R. Grée and R. Carrié, *Bull. Soc. Chim. Fr.*, **1975**, 1319.
98. I. N. N. Namboothiri, A. Hassner, and H. E. Gottlieb, *J. Org. Chem.*, **1997**, *62*, 485.
99. S. E. Denmark, Y.-C. Moon, and C. B. W. Senanayake, *J. Am. Chem. Soc.*, **1990**, *112*, 311.
100. L. R. Domingo and A. Asensio, *J. Org. Chem.*, **2000**, *65*, 1076.
101. S. K. Mukerji and K. B. G. Torssell, *Acta Chem. Scand. (B)*, **1981**, *35*, 643.
102. H. Feger and G. Simchen, *Synthesis*, **1981**, 378.
103. G. A. Olah, B. G. B. Gupta, S. C. Narang, and R. Malhotra, *J. Org. Chem.*, **1979**, *44*, 4272.
104. G. A. Olah and B. G. B. Gupta, *Synthesis*, **1980**, 44.
105. E. W. Colvin and D. Seebach, *J. Chem. Soc. Chem. Comm.*, **1978**, 689.
106. A. Hassner, O. Friedman, and W. Dehaen, *Liebigs Ann. Rec.*, **1997**, 587.
107. C. Li and C. Yuan, *Synthesis*, **1993**, 471.
108. S. H. Andersen, N. B. Das, R. D. Jørgensen, G. Kjeldsen, J. S. Knudsen, S. C. Sharma, and K. B. G. Torssell, *Acta Chem. Scand. (B)*, **1982**, *36*, 1.
109. N. B. Das and K. B. G. Torssell, *Tetrahedron*, **1983**, *39*, 2247.
110. A. Padwa, S. P. Craig, U. Chiacchio, and D. N. Kline, *J. Org. Chem.*, **1998**, *53*, 2232.
111. C. Yuan and C. Li, *Phosphorus, Sulfur, Silicon*, **1992**, *69*, 75.
112. J. Chen and C.-M. Hu, *J. Fluorine Chem.*, **1995**, *71*, 43.
113. E. Vedejs and D. A. Perry, *J. Am. Chem. Soc.*, **1983**, *105*, 1683.
114. E. Vedejs and D. A. Perry, *J. Org. Chem.*, **1984**, *49*, 573.
115. E. Vedejs, D. A. Perry, and R. G. Wilde, *J. Am. Chem. Soc.*, **1986**, *108*, 2985.
116. E. Vedejs, D. A. Perry, K. N. Houk, and N. G. Rondan, *J. Am. Chem. Soc.*, **1983**, *105*, 6999.

117. K. V. Gothelf and K. A. Jørgensen, *Chem. Rev.*, **1998**, *98*, 863.
118. J. Pitlik, *Synth. Commun.*, **1994**, *24*, 243.
119. J. Pitlik, T. Gunda, and G. Batta, *Magyar Kém. Foly.*, **1995**, *101*, 321.
120. B. H. Kim, J. Y. Lee, K. Kim, and D. Whang, *Tetrahedron: Asymmetry*, **1991** *2*, 27.
121. B. H. Kim and J. Y. Lee, *Tetrahedron: Asymmetry*, **1991**, *2*, 1359.
122. B. H. Kim and D. P. Curran, *Tetrahedron*, **1993**, *49*, 293.
123. J. A. Stack, T. A. Heffner, S. J. Geib, and D. P. Curran, *Tetrahedron*, **1993**, *49*, 995.
124. H. R. Kim, K. M. Kim, J. N. Kim, and E. K. Ryu, *Synth. Commun.*, **1994**, *24*, 1107.
125. S. C. Sharma and K. B. G. Torssell, *Acta Chem. Scand. (B)*, **1979**, *33*, 379.
126. A. V. Stepanov, A. V. Lozanova, and V. V. Veselovsky, *Izv. Akad. Nauk Ser. Khim.*, **1998**, 2357; *Russ. Chem. Bull. (Engl. Transl.)*, **1998**, *47*, 2286.
127. V. V. Veselovsky, A. V. Belyankin, B. I. Maksimov, Y. A. Strelenko, and A. M. Moiseenkov, *Izv. Akad. Nauk SSSR, Ser. Khim.*, **1992**, 377; *Bull. Acad. Sci. USSR, Div. Chem. Sci. (Engl. Transl.)*, **1992**, 292.
128. R. Schneider, P. Gérardin, and B. Loubinoux, *J. Heterocyclic Chem.*, **1994**, *31*, 797.
129. K. S.-L. Huang, E. H. Lee, M. M. Olmstead, and M. J. Kurth, *J. Org. Chem.*, **2000**, *65*, 499.
130. J. L. Duffy and M. J. Kurth, *J. Org. Chem.*, **1994**, *59*, 3783.
131. Q. Cheng, T. Oritani, T. Horiguchi, and Q. Shi, *Eur. J. Org. Chem.*, **1999**, 2689.
132. M. Ahrach, R. Schneider, P. Gérardin, and B. Loubinoux, *Synth. Commun.*, **1997**, *27*, 1865.
133. L. Gottlieb and A. Hassner, *J. Org. Chem.*, **1995**, *60*, 3759.
134. D. G. J. Young, E. Gomez-Bengoa, and A. H. Hoveyda, *J. Org. Chem.*, **1999**, *64*, 692.
135. H. Marrugo, R. Dogbéavou, and L. Breau, *Tetrahedron Lett.*, **1999**, *40*, 8979.
136. S. Kanemasa and O. Tsuge, *Heterocycles*, **1990**, *30*, 719.
137. P. Caramella and P. Grünanger, "Nitrile Oxides and Imines" in *1,3-Dipolar Cycloaddition Chemistry*, A. Padwa, ed., Wiley, New York, **1984**, pp. 291–393.
138. A. P. Kozikowski, *Acc. Chem. Res.*, **1984**, *17*, 410.
139. N. B. Das and K. B. G. Torssell, *Tetrahedron.*, **1983**, *39*, 2227.
140. A. V. Stepanov and V. V. Veselovsky, *Izv. Acad. Nuak, Ser. Khim.*, **1997**, 1683; *Russ. Chem. Bull. (Engl. Transl.)*, **1997**, *46*, 1606.
141. B. H. Kim and J. Y. Lee, *Tetrahedron Lett.*, **1992**, *33*, 2557.
142. J. Y. Lee and B. H. Kim, *Tetrahedron Lett.*, **1995**, *36*, 3361.
143. J. Y. Lee and B. H. Kim, *Tetrahedron*, **1996**, *52*, 571.
144. V. A. Tartakovskii, I. E. Chlenov, G. V. Lagodzinskaya, and S. S. Novikov, *Dokl. Akad. Nauk SSSR*, **1965**, *161*, 136; *Dokl. Chem. (Engl. Transl.)*, **1965**, *161*, 257.
145. M. K. Shakhova, M. I. Budagyants, G. I. Samokhovalov, and N. A. Preobrazhenskii, *Zh. Obshch. Khim.*, **1962**, *32*, 2832; *J. Gen. Chem. USSR. (Engl. Transl.)*, **1962**, *32*, 2790.
146. Caution: The Sodium salt of nitromethane can undergo detonation upon heating and drying (30).
147. H. Meerwein, E. Battenberg, H. Gold, E. Pfeil, and G. Willfang, *J. Prakt. Chem.*, **1939**, *262*, 83.
148. H. Sato, T. Kusumi, K. Imaye, and H. Karisawa, *Chem. Lett.*, **1975**, 965.
149. L. G. Donaruma, *J. Org. Chem.*, **1957**, *22*, 1024.
150. C. H. Dale. Ph.D. Thesis, University of Rochester, **1936**.
151. V. I. Erashko, S. A. Shevelev, and A. A. Fainzil'berg, *Izv. Akad. Nauk SSSR, Ser. Khim.*, **1968**, 2117; *Bull. Acad. Sci. USSR, Div. Chem. Sci. (Engl. Transl.)*, **1968**, 2007.
152. S. A. Shevelev, V. I. Erashko, and A. A. Fainzil'berg, *Izv. Akad. Nauk SSSR, Ser. Khim.*, **1968**, 447; *Bull. Acad. Sci. USSR, Div. Chem. Sci. (Engl. Transl.)*, **1968**, 447.

153. S. A. Shevelev, V. I. Erashko, and A. A. Fainzil'berg, *Izv. Akad. Nauk SSSR, Ser. Khim.*, **1968**, 2113; *Bull. Acad. Sci. USSR, Div. Chem. Sci. (Engl. Transl.)*, **1968**, 2003.

154. F.-A. Kang and C.-L. Yin, *J. Org. Chem.*, **1996**, *61*, 5523.

155. J. R. Falck and J. Yu, *Tetrahedron Lett.*, **1992**, *45*, 6723.

156. D. L. Hughes, *Org. React.*, **1992**, *42*, 335.

157. K. V. Altukhov and V. V. Perekalin, *Zh. Org. Khim.*, **1966**, *2*, 1902; *J. Org. Chem. USSR (Engl. Transl.)*, **1966**, *2*, 1870.

158. K. V. Altukhov, V. A. Tartakovskii, V. V. Perekalin, and S. S. Novikov, Izv. Akad. Nauk SSSR, **1996**, 197; *Bull. Acad. Sci. USSR, Div. Chem. Sci. (Engl. Transl.)*, **1967**, 192.

159. K. V. Altukhov and V. V. Perekalin, *Zh. Org. Khim.*, **1967**, *3*, 2003; *J. Org. Chem. USSR (Engl. Transl.)*, **1967**, *3*, 1953.

160. L. M. Andreeva, K. V. Altukhov, and V. V. Perekalin, *Zh. Org. Khim.*, **1969**, *5*, 1313; *J. Org. Chem. USSR (Engl. Transl.)*, **1969**, *5*, 1281.

161. L. M. Andreeva, K. V. Altukhov, and V. V. Perekalin, *Zh. Org. Khim*, **1969**, *5*, 220; *J. Org. Chem. USSR (Engl. Transl.)*, **1969**, *5*, 212.

162. L. M. Andreeva, K. V. Altukhov, and V. V. Perekalin, *Zh. Org. Khim.*, **1972**, *8* 1419; *J. Org. Chem. USSR (Engl. Transl.)*, **1972**, *8*, 1442.

163. V. A. Buevich, K. V. Altukhov, and V. V. Perekalin, *Zh. Org. Khim.*, **1970**, *6*, 187; *J. Org. Chem. USSR (Engl. Transl.)*, **1970**, *6*, 184.

164. V. A. Buevich, K. V. Altukhov, and V. V. Perekalin, *Zh. Org. Khim.*, **1970**, *6*, 658; *J. Org. Chem. USSR (Engl. Transl.)*, **1970**, *6*, 661.

165. V. A. Buevich, K. V. Altukhov, and V. V. Perekalin, *Zh. Org. Khim.*, **1971**, *7*, 1380; *J. Org. Chem. USSR (Engl. Transl.)*, **1971**, *7*, 1425.

166. E. V. Ratsino, K. V. Altukhov, and V. V. Perekalin, *Zh. Org. Khim.*, **1972**, *8*, 523; *J. Org. Chem. USSR (Engl. Transl.)*, **1972**, *8*, 526.

167. E. V. Ratsino and K. V. Altukhov, *Zh. Org. Khim.*, **1972**, *8*, 2281; *J. Org. Chem. USSR (Engl. Transl.)*, **1972**, *8*, 2327.

168. V. A. Tartakovskii, L. A. Nikonova, and S. S. Novikov, *Izv. Akad. Nauk SSSR, Ser. Khim.*, **1966**, 1290; *Bull. Acad. Sci. USSR, Div. Chem. Sci. (Engl. Transl.)*, **1966**, 1246.

169. V. A. Tartakovskii, G. A. Shvekhgeimer, N. I. Sobtsova, and S. S. Novikov, *Zh. Obshch. Khim.*, **1967**, *37*, 1163; *J. Gen. Chem. USSR (Engl. Transl.)*, **1967**, *37*, 1103.

170. V. A. Tartakovskii, A. A. Fainzil'berg, V. I. Gulevshaya, and S. S. Novikov, *Izv. Akad. Nauk SSSR, Ser Khim.*, **1968**, 621; *Bull. Acad. Sci. USSR, Div. Chem. Sci. (Engl. Transl.)*, **1968**, 598.

171. K. V. Altukhov, E. V. Ratsino, and V. V. Perekalin, *Zh. Org. Khim.*, **1973**, *9*, 269; *J. Org. Chem. USSR (Engl. Transl.)*, **1973**, *9*, 267.

172. E. V. Ratsino, K. V. Altukhov, V. V. Perekalin, and O. N. Fedorishcheva, *Zh. Org. Khim.*, **1977**, *13*, 2495; *J. Org. Chem. USSR (Engl. Transl.)*, **1977**, *13*, 2320.

173. E. V. Ratsino, L. M. Andreeva, K. V. Altukhov, and V. V. Perekalin, *Zh. Org. Khim.*, **1974**, *10*, 728; *J. Org. Chem. USSR (Engl. Transl.)*, **1974**, *10*, 733.

174. I. E. Chlenov, M. V. Kashutina, S. L. Ioffe, S. S. Novikov, and V. A. Tartakovskii, *Izv. Akad. Nauk SSSR, Ser. Khim.*, **1969**, 2085; *Bull. Acad. Sci. USSR, Div. Chem. Sci. (Engl. Transl.)*, **1969**, 1948.

175. T. Mukaiyama and T. Hoshino, *J. Am. Chem. Soc.*, **1960**, *82*, 5339.

176. M.-C. Yan, J.-Y. Liu, W.-W. Lin, K.-H. Kao, J.-T. Liu, J.-J. Jang, and C.-F. Yao, *Tetrahedron*, **1999**, *55*, 12493.

177. P. Camps, D. Muñoz-Torrero, and M.-T. Victoria, *Tetrahedron*, **1995**, *51*, 6587.

178. M. Miyashita, B. Z. E. Awen, and A. Yoshikoshi, *Tetrahedron*, **1990**, *46*, 7569.

179. M. Miyashita, B. Z. E. Awen, and A. Yoshikoshi, *Chem. Lett.*, **1990**, 239.

180. M. Miyashita, B. Z. E. Awen, and A. Yoshikoshi, *J. Chem. Soc., Chem. Commun.*, **1989**, 841.

181. E. G. Janzen and U. M. Oehler, *Tetrahedron Lett.*, **1983**, *24*, 669.

182. V. I. Kornilov, B. B. Paidak, and Y. A, Zhdanov, *Zh. Obshch. Khim.*, **1973**, *34*, 189; *J. Gen. Chem. USSR (Engl. Transl.)*, **1973**, *34*, 185.

183. W. Rank and H. H. Baer, *Can. J. Chem.*, **1971**, *49*, 3236.

184. J. Skramstad, *Tetrahedron Lett.*, **1970**, 955.

185. R. C. Kerber and M. Hodos, *J. Org. Chem.*, **1968**, *33*, 1169.

186. E. P. Stefl and M. F. Dull, *J. Am. Chem. Soc.*, **1947**, *69*, 3037.

187. E. P. Kohler and G. R. Barrett, *J. Am. Chem. Soc.*, **1926**, *48*, 1770.

188. M. S. Gibson, *Tetrahedron*, **1962**, *18*, 1377.

189. V. A. Tartakovskii, B. G. Gribov, and S. S. Novikov, *Izv. Akad. Nauk SSSR, Ser. Khim.*, **1965**, 1074; *Bull. Acad. Sci. USSR, Div. Chem. Sci. (Engl. Transl.)*, **1965**, 1034.

190. I. E. Chlenov, N. S. Morozova, V. I. Khudak, and V. A. Tartakovskii, *Izv. Akad. Nauk SSSR, Ser. Khim.*, **1970**, 2641; *Bull. Acad. Sci. USSR, Div. Chem. Sci. (Engl. Transl.)*, **1970**, 2492.

191. L. Henry, *Bull. Soc. Chim. Fr.*, **1895**, *13*, 911.

192. F. A. Luzzio, *Tetrahedron*, **2001**, *57*, 915.

193. G. Rosini, "The Henry (Nitroaldol) Reaction" in *Comprehensive Organic Synthesis*, Vol. 2, C. H. Heathcock, ed., Pergamon Press, New York, **1991**, pp. 321–340.

194. E. Marotta, P. Righi, and G. Rosini, *Tetrahedron Lett.*, **1998**, *39*, 1041.

195. G. Rosini, E. Marotta, P. Righi, and J. P. Seerden, *J. Org. Chem.*, **1991**, *56*, 6258.

196. G. Rosini, R. Galarini, E. Marotta, and P. Righi, *J. Org. Chem.*, **1990**, *55*, 781.

197. Y. L. Chow, Y. Y. Shu, B. H. Bakker, and K. S. Pillay, *Heterocycles*, **1989**, *29*, 2245.

198. F. Kienzle, J.-Y. Fellmann, and J. Stadlwieser, *Helv. Chem. Acta,* **1984**, *67*, 789.

199. A. Risaliti, M. Forchiassin, and E. Valentin, *Tetrahedron*, **1968**, *24*, 1889.

200. F. P. Colonna, E. Valentin, G. Pitacco, and A. Risaliti, *Tetrahedron*, **1973**, *29*, 3011.

201. R. A. Ferri, G. Pitacco, and E. Valentin, *Tetrahedron*, **1978**, *34*, 2537.

202. F. Benedetti, G. Pitacco, and E. Valentin, *Tetrahedron*, **1979**, *35*, 2293.

203. G. Barbarella, G. Pitacco, C. Russo, and E. Valentin, *Tetrahedron Lett.*, **1983**, *24*, 1621.

204. G. Barbarella, S. Brückner, G. Pitacco, and E. Valentin, *Tetrahedron*, **1984**, *40*, 2441.

205. F. Felluga, P. Nitti, G. Pitacco, and E. Valentin, *Tetrahedron*, **1989**, *45*, 2099.

206. F. Felluga, P. Nitti, G. Pitacco, and E. Valentin, *Tetrahedron*, **1989**, *45*, 5667.

207. F. Felluga, P. Nitti, G. Pitacco, and E. Valentin, *J. Chem. Soc., Perkin Trans. 1*, **1991**, 1645.

208. M. Miyashita, T. Yanami, and A. Yoshikoshi, *J. Am. Chem. Soc.*, **1976**, *98*, 4679.

209. M. Miyashita, T. Yanami, T. Kumazawa, and A. Yoshikoshi, *J. Am. Chem. Soc.*, **1984**, *106*, 2149.

210. A. Yoshikoshi and M. Miyashita, *Acc. Chem. Res.*, **1985**, *18*, 284.

211. D. Seebach and M. A. Brook, *Helv. Chem. Acta*, **1985**, *68*, 319.

212. A. Barco, S. Benetti, G. P. Polllini, G. Spalluto, and V. Zanirato, *Tetrahedron Lett.*, **1991**, *32*, 2517.

213. J.-E. Bäckvall, U. Karlsson, and R. Chinchilla, *Tetrahedron Lett.*, **1991**, *32*, 5607.

214. S. E. Denmark and M. Seierstad, *J. Org. Chem.*, **1999**, *64*, 1610.

215. A. Papchikhin, P. Agback, J. Plavec, and J. Chattopadhyaya, *J. Org. Chem.*, **1993**, *58*, 2874.

216. M. Avalos, R. Babiano, P. Cintas, F. J. Higes, J. L. Jiménez, J. C. Palacios, and M. A. Silva, *J. Org. Chem.*, **1996**, *61*, 1880.

217. R. Chinchilla and J.-E. Bäckvall, *Tetrahedron Lett.*, **1992**, *33*, 5641.

218. S. E. Denmark, C. B. W. Senanayake, and G.-D. Ho, *Tetrahedron*, **1990**, *46*, 4857.

219. B. S. Kesler. Ph.D. Thesis, University of Illinois: Urbana-Champaign, **1993**.

220. J. Vaugeois. Post-Doctoral Report. University of Illinois: Urbana-Champaign, **1999**.

221. N. Arai and K. Narasaka, *Chem. Lett.*, **1995**, 987.

222. S. E. Denmark and A. R. Hurd, *J. Org. Chem.*, **1998**, *63*, 3045.

223. V. A. Tartakovskii, S. S. Smagin, I. E. Chlenov, and S. S. Novikov, *Izv. Akad. Nauk SSSR, Ser. Khim.*, **1965**, 552; *Bull. Acad. Sci. USSR, Div. Chem. Sci. (Engl. Transl.)*, **1965**, 534.

224. V. A. Tartakovskii, I. E. Chlenov, S. L. Ioffe, G. V. Lagodzinskaya, and S. S. Novikov, *Zh. Org. Khim.*, **1966**, *2*, 1593; *J. Org. Chem. USSR (Engl. Transl.)*, **1966**, *2*, 1573.

225. V. A. Tartakovskii, I. A. Savost'yanova, and S. S. Novikov, *Zh. Org. Khim.*, **1968**, *4*, 240; *J. Org. Chem. USSR (Engl. Transl.)*, **1968**, *4*, 232.

226. C. Bellandi, M. de Amici, and C. de Micheli, *Heterocycles*, **1984**, *22*, 2187.

227. V. A. Tartakovskii, A. A. Onishchenko, V. A. Smirnyadin, and S. S. Novikov, *Zh. Org. Khim.*, **1965**, *2*, 2225; *J. Org. Chem. USSR (Engl. Transl.)*, **1966**, *2*, 2183.

228. V. A. Tartakovskii, A. A. Onishchenko, and S. S. Novikov, *Zh. Org. Khim.*, **1967**, *3*, 588; *J. Org. Chem. USSR (Engl. Transl.)*, **1967**, *3*, 564.

229. V. A. Tartakovskii, O. A. Luk'yanov, and S. S. Novikov, *Izv. Akad. Nauk SSSR, Ser. Khim.*, **1966**, 2246; *Bull. Acad. Sci. USSR, Div. Chem. Sci. (Engl. Transl.)*, **1966**, 2186.

230. I. E. Chlenov, I. L. Sokolova, S. S. Novikov, and V. A. Tartakovskii, *Izv. Akad. Nauk SSSR, Ser. Khim.*, **1973**, 473; *Bull. Acad. Sci. USSR, Div. Chem. Sci. (Engl. Transl.)*, **1973**, 460.

231. R. Grée and R. Carrié, *J. Am. Chem. Soc.*, **1977**, *99*, 6667.

232. M. Ciobanu and K. Matsumoto, *Liebigs Ann. Rec.*, **1997**, 623.

233. R. van Eldrik, T. Asano, and W. le Noble, *Chem. Rev.*, **1989**, *89*, 549.

234. A. V. Kamernitzky, I. S. Levina, E. I. Mortikova, V. M. Shitkin, and B. S. El'yanov, *Tetrahedron*, **1977**, *33*, 2135.

235. G. J. Kuster, F. Kalmoua, R. de Gelder, and H. W. Scheeren, *Chem. Commun.*, **1999**, 855.

236. G. J. Kuster, R. H. J. Steeghs, and H. W. Scheeren, *Eur. J. Org. Chem.*, **2001**, 553.

237. P. A. Wade, N. V. Amin, H.-K. Yen, D. T. Price, and G. F. Huhn, *J. Org. Chem.*, **1984**, *49*, 4595.

238. T. Shimizu, Y. Hayashi, and K. Teramura, *J. Org. Chem.*, **1983**, *48*, 3053.

239. T. Shimizu, Y. Hayashi, T. Taniguchi, and K. Teramura, *Tetrahedron*, **1985**, *41*, 727.

240. R. N. Butler, D. Cunningham, E. G. Marren, and P. McArdle, *Tetrahedron Lett.*, **1988**, *29*, 3331.

241. R. N. Butler, D. Cunningham, E. G. Marren, P. McArdle, and D. F. O'Shea, *J. Chem. Research (S)*, **1992**, 256.

242. F.-A. Kang and C.-L. Yin, *Tetrahedron: Asymmetry*, **1997**, *8*, 3585.

243. S. E. Denmark, A. Thorarensen, and D. S. Middleton, *J. Org. Chem.*, **1995**, *60*, 3574.

244. S. E. Denmark, A. Thorarensen, and D. S. Middleton, *J. Am. Chem. Soc.*, **1996**, *118*, 8266.

245. S. E. Denmark, D. L. J. Parker, and J. A. Dixon, *J. Org. Chem.*, **1997**, *62*, 435.

246. S. E. Denmark, A. R. Hurd, and H. J. Sacha, *J. Org. Chem.*, **1997**, *62*, 1668.

247. S. E. Denmark and E. A. Martinborough, *J. Am. Chem. Soc.*, **1999**, *121*, 3046.

248. S. E. Denmark and J. J. Cottell, *J. Org. Chem.*, **2001**, *66*, 4276.

249. P. Righi, E. Marotta, A. Landuzzi, and G. Rosini, *J. Am. Chem. Soc.*, **1996**, *118*, 9446.

250. P. Righi, E. Marotta, and G. Rosini, *Chem. Eur. J.*, **1998**, *4*, 2501.

251. S. E. Denmark, A. Stolle, J. A. Dixon, and V. Guagnano, *J. Am. Chem. Soc.*, **1995**, *117*, 2100.

252. S. E. Denmark and J. A. Dixon, *J. Org. Chem.*, **1997**, *62*, 7086.

253. S. E. Denmark and J. A. Dixon, *J. Org. Chem.*, **1998**, *63*, 6178.

254. S. E. Denmark and D. S. Middleton, *J. Org. Chem.*, **1998**, *63*, 1604.

255. G. C. Pimentel and R. D. Spratley, *Chemical Bonding Clarified Through Quantum Mechanics*, Holden-Day, Inc., San Francisco, **1969**, P. 314.

256. Note 2.

257. S. E. Denmark and B. Herbert, *J. Am. Chem. Soc.*, **1998**, *120*, 7357.

258. S. E. Denmark and A. R. Hurd, *Org. Lett.*, **1999**, *1*, 1311.

259. S. E. Denmark and A. R. Hurd, *J. Org. Chem.*, **2000**, *65*, 2875.

260. S. E. Denmark and B. Herbert, *J. Org. Chem.*, **2000**, *65*, 2887.

261. G. J. Kuster and H. W. Scheeren, *Tetrahedron Lett.*, **2000**, *41*, 515.

262. M. Avalos, R. Babiano, P. Cintas, J. L. Jiménez, J. C. Palacios, and M. A. Silva, *Chem. Commun.*, **1998**, 459.

263. S. E. Denmark, V. Guagnano, and J. Vaugeois, *Can. J. Chem.*, *79*, **2001**, 1606.

264. G. J. Kuster and H. W. Scheeren, *Tetrahedron Lett.*, **1998**, *39*, 3613.

265. S. E. Denmark and L. R. Marcin, *J. Org. Chem.*, **1997**, *62*, 1675.

266. S. E. Denmark and A. Thorarensen, *J. Am. Chem. Soc.*, **1997**, *119*, 125.

CHAPTER 3

Azomethine Ylides

L. M. Harwood and R. J. Vickers

University of Reading, Reading, U.K.

3.1. Azomethine Ylide Generation 170
3.1.1. Silicon-Based Protocols 170
3.1.2. Alternative Generation Procedures 181
3.2. Asymmetric Reactions 193
3.2.1. Chiral Dipolarophiles 193
3.2.2. Chiral Ylides:Stereocontrol by Inherent Molecular Structure 200
3.2.3. Chiral Ylides Derived from Asymmetric Templates and Auxiliaries 208
3.3. Intramolecular Cycloadditions 219
3.3.1. Aziridine Precursors 219
3.3.2. Decarboxylative and Amino Ester Condensation Protocols 228
3.3.3. Alternative Generation Procedures 237
3.4. Metal-Mediated Reactions 240
3.5. Conclusion 248

As the complexity of synthetic targets from both natural and unnatural sources increases, so must the demands for solutions to the synthetic problems their construction poses. As such, development and refinement of established methodologies continues. In particular, cycloaddition chemistry has continued to maintain its place as one of the cornerstones of modern organic chemistry, for the construction of mono- and polycyclic systems, as a consequence of being able to deliver high increases in molecular complexity from relatively simple and accessible precursors. One of the most important classes for 1,3-dipolar cycloaddition involves azomethine ylides. For the construction of nitrogen-containing, five-membered heterocycles, such ylides represent the most conceptually simple and efficient method. Generation of the ylide, usually *in situ*, followed by dipolarophile attack, furnishes pyrrolidines and pyrrolines with operational simplicity. This chapter aims to inform the reader of the major advances made in this field from 1984 to date (1), equipping

The Chemistry of Heterocyclic Compounds, Volume 59: Synthetic Applications of 1,3-Dipolar Cycloaddition Chemistry Toward Heterocycles and Natural Products, Edited by Albert Padwa and William H. Pearson.
ISBN 0-471-38726-6

the synthetically orientated chemist with the requisite knowledge to implement such methodology to the challenges posed by asymmetric and natural product synthesis.

3.1. AZOMETHINE YLIDE GENERATION

Although established methods for azomethine generation have proven to be both general and efficient, new procedures are constantly being divulged. These new methods are not only for ylide generation, but are also for new chemical equivalents and, by careful selection, the synthetic chemist has a wealth of technologies available. In particular, the use of fluorine mediated desilyation strategies has often proven to be the most reliable and probably the most commonly used method, with a plethora of recent publications.

3.1.1. Silicon-Based Protocols

Possibly, the most common protocols used in the generation of azomethine ylides are those based on the *in situ*, fluorine-mediated desilyation of cyanoaminosilanes developed by Padwa et al. (2). Typically, treatment of precursor **1** with AgF, in the presence of dimethyl acetylenedicarboxylate (DMAD), led to the formation of the intermediate cycloadduct **2**, which was subjected to immediate DDQ oxidation to give pyrrole **3**. The mechanistic rationale invokes fluoride-mediated desilyation to form the intermediate anion **4**, which then undergoes loss of cyanide furnishing the corresponding azomethine ylide (Scheme 3.1).

The reaction protocol accommodates a range of singly and doubly activated alkynes and alkenes to furnish the expected cycloadducts in high yield. However,

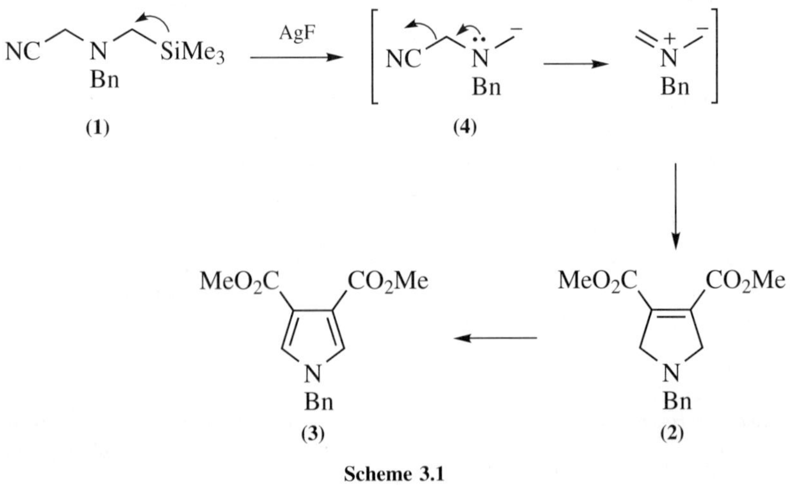

(1) (4) (3) (2)

Scheme 3.1

unactivated dipolarophiles proved unreactive presumably due to too large a highest occupied molecular orbital (HOMO)–lowest unoccupied molecular orbital (LUMO) energy gap between dipole and dipolarophile. Unsymmetrical precursors **5** and **6** also underwent ylide generation and subsequent cycloaddition with methyl propiolate to deliver a mixture of regioisomers **7** and **8** in an 8:1 ratio. Since both precursors furnish the same regioisomeric ratio of products, the implication is that the reaction does indeed proceed via an azomethine ylide intermediate (Scheme 3.2).

The synthetic viability of this protocol was demonstrated by the efficient synthesis of an isoindole alkaloid from the Mexican sponge *Reniera*. α-Cyanosilylamine (**9**) was treated with AgF, in the presence of dipolarophile **10** to furnish the requisite cycloadduct in 68%. An advanced intermediate in the synthesis of nicotine was also prepared by cycloaddition of **11** with phenylvinylsulfone giving the requisite adduct **12** in a 3:1 diastereomic mixture (Schem 3.3).

Further studies into the construction of pyrrolo[1,2-*a*]indoles, however, revealed that the presence of a nitrile functionality was not essential to the reaction (3). Subjecting precursor **13** to ylide formation by AgF (it should be noted that this was the only reagent to induce cycloaddition to any extent) in the presence of *N*-phenylmaleimide surprisingly furnished adduct **14** in which the nitrile functionality was still intact. The reaction pathway was therefore assumed to proceed via initial formation of the silver bonded cation **15**, which after desilyation generated the requisite silver bound azomethine ylide. Cycloaddition followed by sequential loss of silver and a hydrogen delivered the observed products. Replacement of the nitrile moiety with alternative functionalities also generated the expected products in good isolated yields (Scheme 3.4).

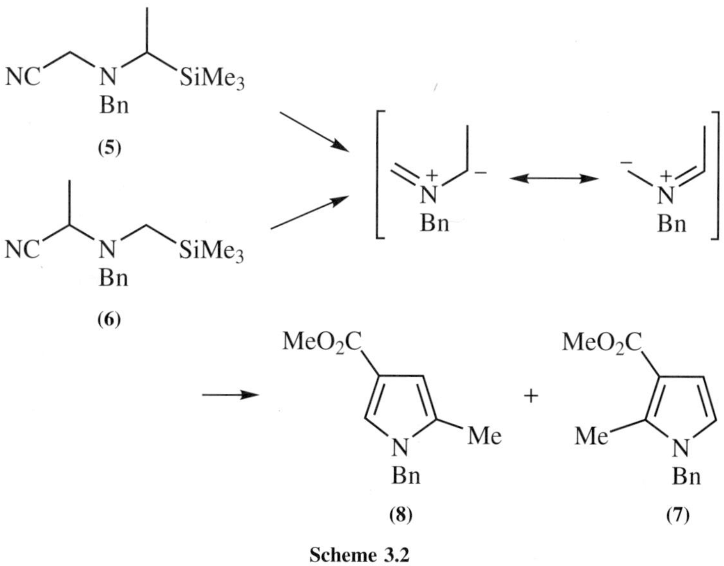

Scheme 3.2

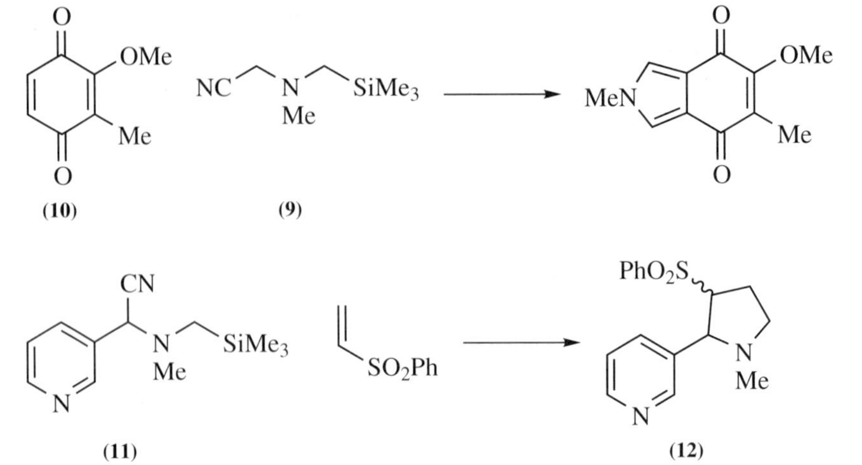

Scheme 3.3

The reaction scope was extended to a range of dipolarophiles (3,4). Ylid generation and cycloaddition with DMAD led to the formation of adduct **16**, which can be reconciled by initial generation of the expected adduct followed by a 1,5-H shift. Reaction with either dimethyl furmarate or dimethyl maleate proceeded with complete stereospecificity, furnishing **17** and **18**, respectively (Fig. 3.1).

Unsymmetrical dipolarophiles led to the formation a single regioisomeric product; acrylonitrile delivered adduct **19** exclusively while maleic anhydride

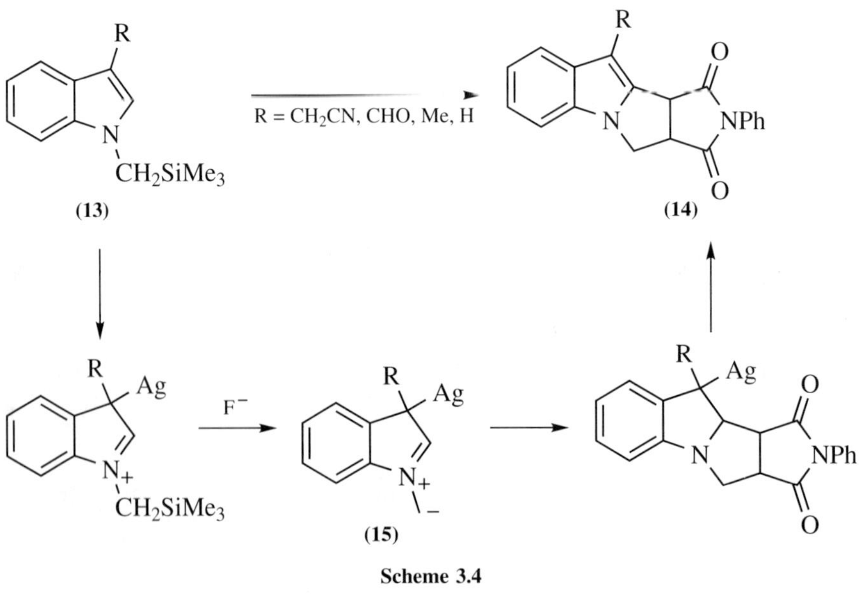

Scheme 3.4

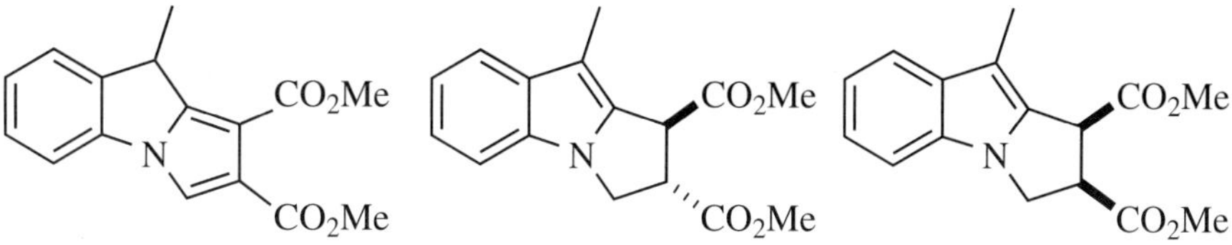

(16) **(17)** **(18)**

Figure 3.1

underwent smooth cycloaddition, with subsequent hydrolysis and decarboxylation *in situ*, to form **20** (4) (Fig. 3.2).

Dodd and co-workers (5) reported the first known synthesis of 11*H*-indolizino[8,7-*b*]indoles by the cycloaddition reaction of a nonstabilized ylide **21** and diethylacetylene dicarboxylate (DEAD). The azomethine ylide, formed by the alkylation of the 3,4-dihydro-β-carboline (**22**) with trimethylsilyl methyl triflate to the triflate salt, followed by *in situ* desilyation with cesium fluoride, underwent cycloaddition with DEAD at low temperature. The expected major cycloadduct **23** was isolated, along with quantities of a minor product **24**, presumed to have been formed by initial reaction of the ylide with 1 equiv of DEAD and the intermediate undergoing reaction with a further equivalent of DEAD before cyclization. Dodd offers no explanation for the unexpected position of the double bond in the newly generated five-membered ring, although it is most likely due to post-reaction isomerization to the thermodynamically more stable β-amino acrylate system (Scheme 3.5).

Komatsu and co-workers (6) developed a more novel use of silicon during their studies on the rearrangements of silicon compounds. *N*-α-(Trimethylsilylbenzyl) amides underwent thermal rearrangement, in toluene, to furnish the azomethine ylides **25**, which underwent subsequent *in situ* cycloaddition. Using DMAD as the dipolarophile resulted in the adducts **26**, after subsequent loss of the silanol in excellent yield. For R = Ph or *p*-$MeOC_6H_4$, yields were in the order of 70–75%, while R = PhCH=CH delivered the adduct in quantitative yield. The presence of the phenyl group α to the nitrogen proved essential, most likely due to its stabilization of the imine. The reaction was also performed with dimethyl maleate, resulting in a single product **27** in high yield. Although the product would seem to indicate that an unfavorable exo transition state to be the predominant reaction pathway, Komatsu and Ohshiro reasoned that the highly acidic nature of the proton

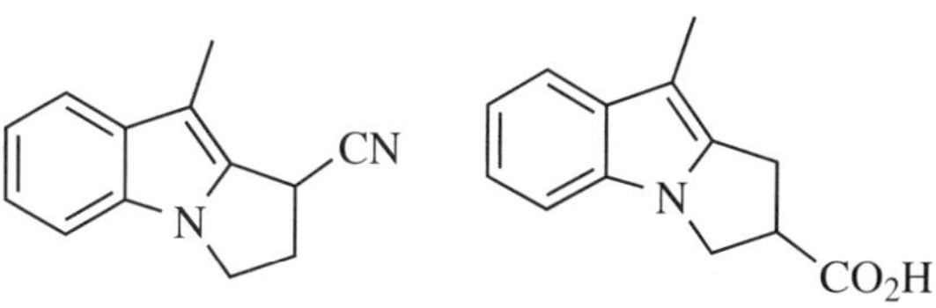

(19) **(20)**

Figure 3.2

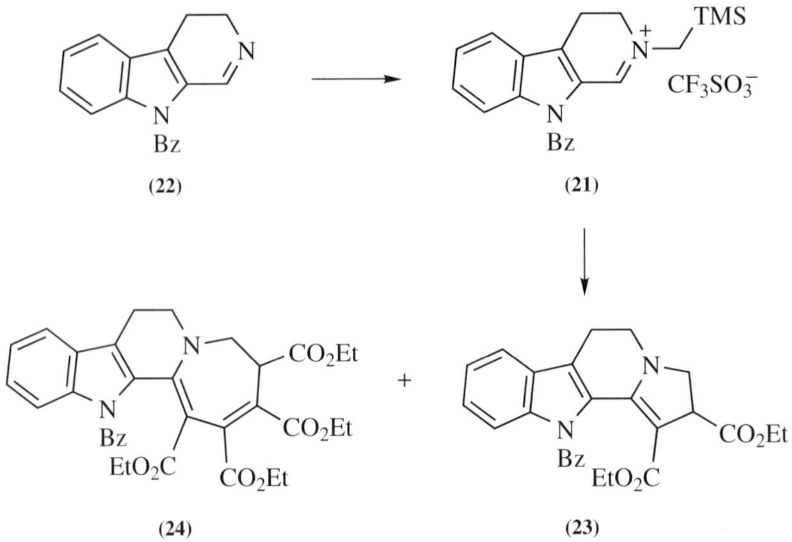

Scheme 3.5

at C(3) could result in facile epimerization. Thus, if the reaction were to proceed by an *endo* transition state, then the anticipated *cis/cis* stereochemistry would ensue. Subsequent epimerization and loss of the silanol would give rise to the observed reaction products **28** (Scheme 3.6).

A more recent report has outlined the use of α-silylimidates for the construction of aromatic pyrroles (7). Treatment of the precursor **29**, with trifluorophenylsilane and DMAD furnished the adduct **30** in 97% yield after purification. The reaction was rationalized *via* quaternization of the imidate and subsequent intramolecular desilylation by fluorine to develop the ylide, which underwent *in situ* cycloaddition and subsequent aromatization delivering **30** (Scheme 3.7).

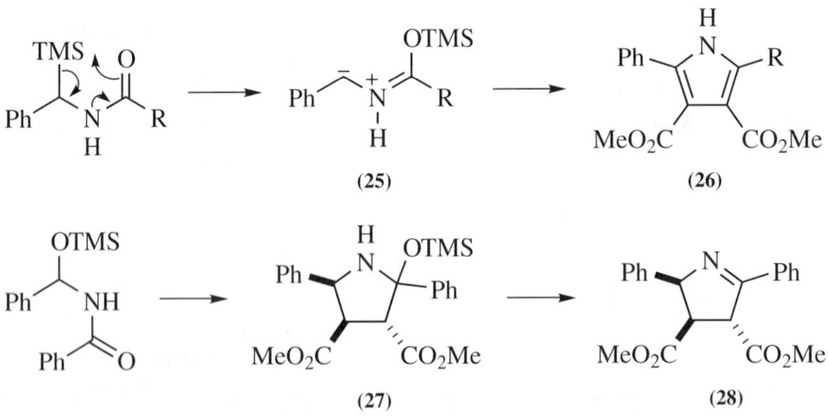

Scheme 3.6

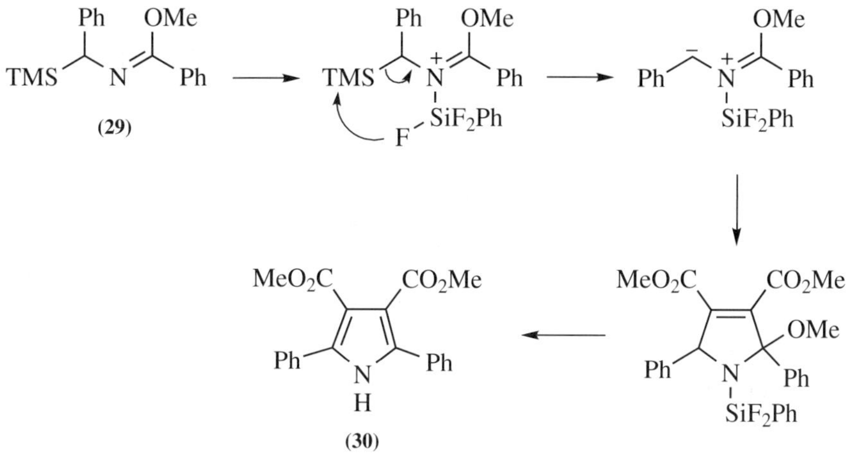

Scheme 3.7

Similarly, replacement of DMAD with either *N*-phenylmaleimide or dimethyl fumarate gave the expected adducts **31** and **32** as 1:1 mixtures of *endo/exo* products (Fig. 3.3).

Similar work by Yoon et al. (8) made elegant use of a carbon–oxygen, silyl rearrangement process in ylide generation. Irradiation of the *N*-silylalkylphthalimide (**33**) in the presence of singly activated alkene dipolarophiles furnished, with total regio- and stereocontrol, cycloadducts of the type **34**, where Z = CN or CO_2Me for R = H or COEt for R = Me. Acetone was also used as a dipolarophile, generating the adduct **35**, again, with complete regiocontrol (Scheme 3.8).

Pandey et al. (9) reported the first known example of a silver fluoride mediated, single-electron oxidation of disilylamines. Precursors **36**, in the presence of AgF at room temperature, generated *in situ*, nonstabilized ylides, which underwent immediate *in situ* reaction with suitable dipolarophiles. The reaction mechanism was rationalized by invoking single electron oxidation of nitrogen, furnishing Ag(0) and a naked fluorine anion, which subsequently removed a TMS group. The process is then repeated, furnishing the azomethine ylide that reacted with dipolarophiles to quantitatively furnish cycloadducts **37**–**40**. The reaction could also be conducted by irradiation of the precursor **36** in the presence of 1,4-dicyanonaphthalene (DCN)

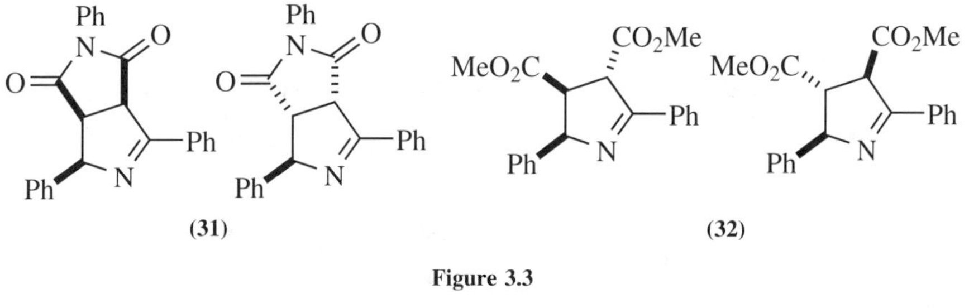

Figure 3.3

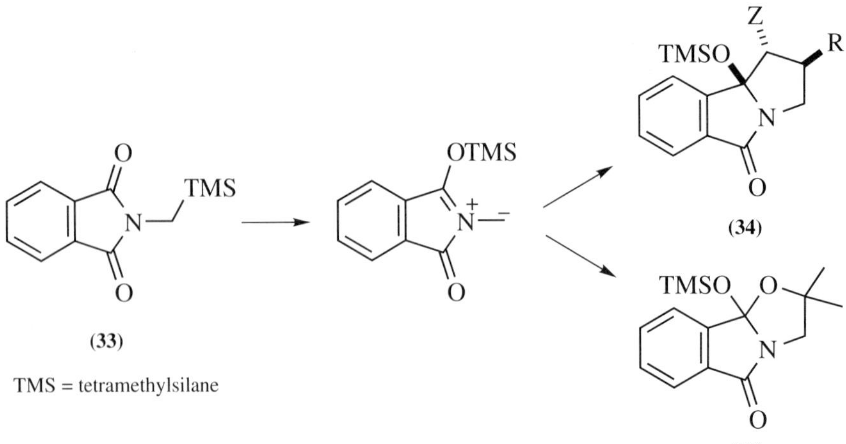

(33) (34) (35)

Scheme 3.8

and an excess of the requisite dipolarophile led to the isolation of the same products **37–40** although in a somewhat reduced yield, typically 55–78% (Scheme 3.9).

Pandey and Lakshmaiah (10) further extended their methodology to the construction of the indolizidine and pyrrolizidine bicyclic skeleta. The basic precursor **41**, which was realized in three straightforward steps, underwent double desilyation and subsequent cycloaddition with ethyl acrylate to furnish the two regioisomers **42** and **43** in essentially quantitative yield and in a 17:3 ratio. The major regioisomer was isolated in a 7:3 *endo/exo* ratio. Further elaboration of the major products where ($n = 1$) and ($n = 2$) delivered the natural products

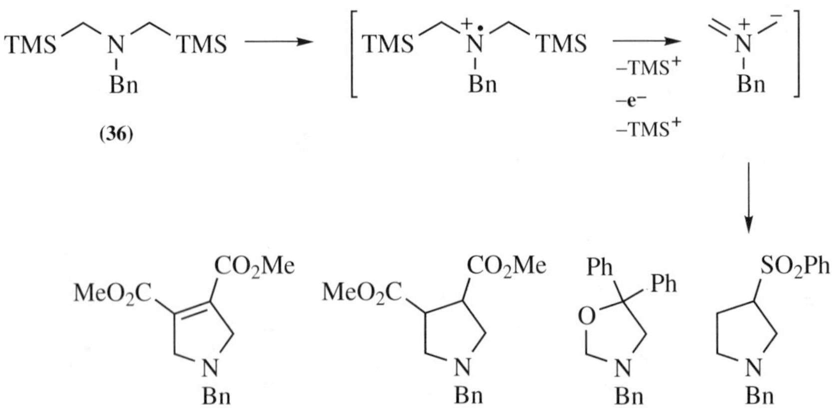

(36) (37) (38) (39) (40)

Scheme 3.9

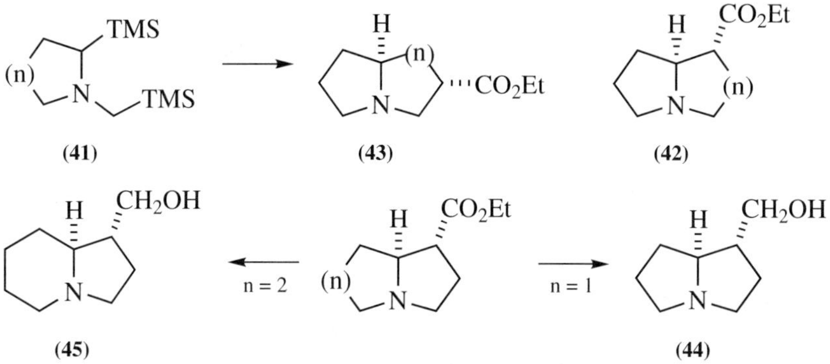

Scheme 3.10

(+/−)-trachelanthamidine (**44**) and 1-aza-7-(hydroxymethyl)bicyclo[4.3.0]nonane (**45**), respectively (Scheme 3.10).

A similar double desilyation method has been reported by Torii et al. (11) and proceeds by single-electron oxidation, which is carried out electrochemically in an undivided cell, utilizing carbon electrodes in dimethylformamide (DMF) solvent with AcONa or KF as the electrolyte. The azomethine ylide was trapped under a variety of conditions, with dimethyl furmarate, smoothly furnishing a single trans-cycloadduct **46** in 47% optimized yield (Scheme 3.11).

Alternative doubly and singly activated olefinic dipolarophiles also underwent cycloaddition, generating the products **47–49** in 27–61% yields, although attempted use of an α,β-unsaturated ketone furnished **50** in only 8% yield, while unactivated dipolarophiles were unreactive (Fig. 3.4).

During extensive studies, the Katritzky group utilized benzotriazole chemistry as an azomethine ylide synthetic equivalent. Initial studies, with the simple *bis* substituted hydroxylamine (**51**) had furnished a wide range of cycloadducts, from

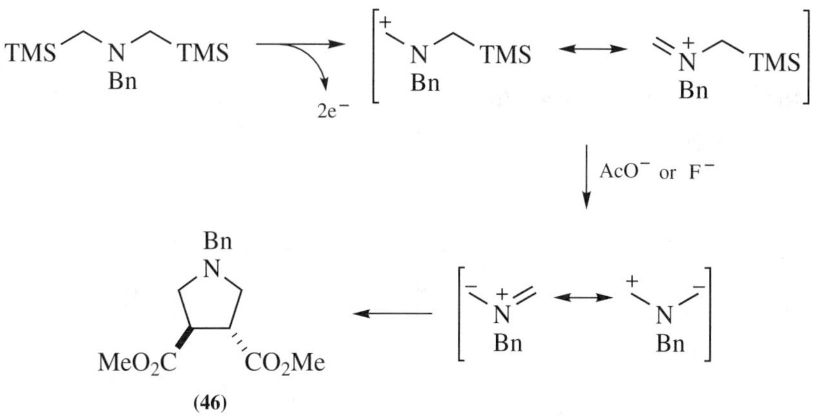

Scheme 3.11

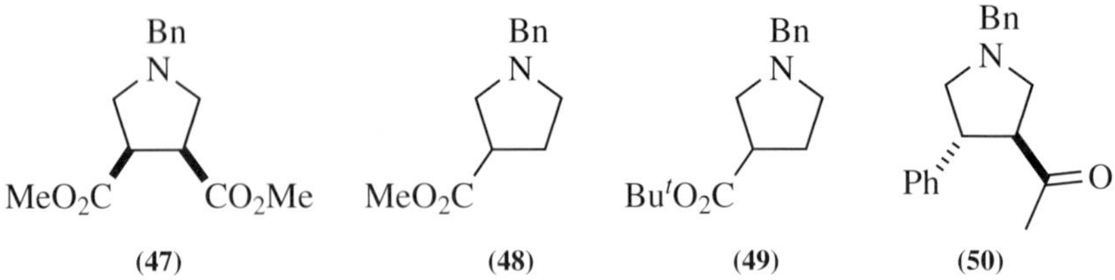

Figure 3.4

the *in situ* generated nitrone 1,3-dipole (12). The *syn* stereochemistry of the product shown confirms that the reaction proceeds in a concerted fashion, indicative of the fact that a genuine cycloaddition process is occurring. However, analogous reactions with **52** as an azomethine ylide precursor generated by oxidation of the equivalent *bis*(benzotriazole) (**53**) furnished only Michael addition products. The lack of cycloaddition products, even under acid catalysis, was explained by the relatively low acidity of the methylene protons. Thus, prototropic tautomerization of the imine to the ylide was not possible, due to the poor subsequent stabilization of negative charge by the benzotriazole group (Scheme 3.12).

Since the dissociation of α-substituted amino benzotriazoles proved unsuccessful, efforts turned toward a more traditional method involving the thermolytic loss of an α-silyl group in refluxing toluene from the easily prepared synthetic precursor **54**, with subsequent loss of the benzotriazole unit to induce neutrality (13). *In situ* trapping of the resultant ylide with electron-deficient alkenes generated a single adduct in high material yield, with retention of the olefinic configuration. The reaction was tolerant of a range of N-substitution in the ylide precursor, with simple alkyl, allyl, and cyclohexyl groups being incorporated into the final pyrrolidines (Scheme 3.13).

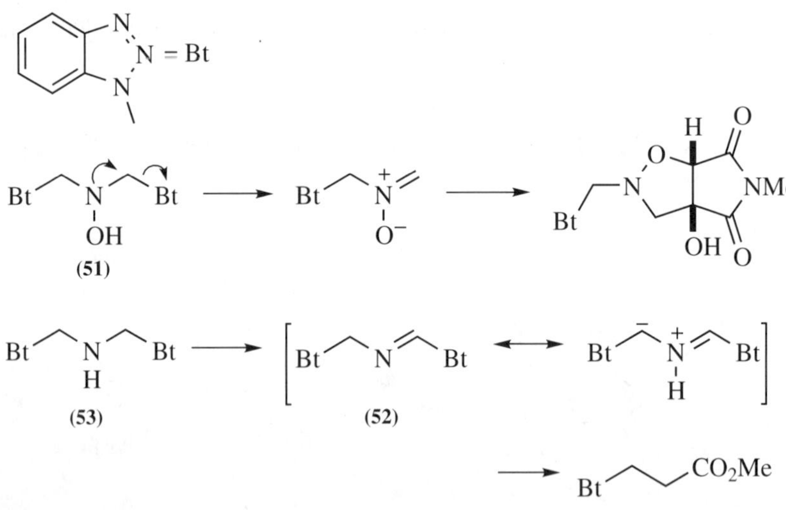

Scheme 3.12

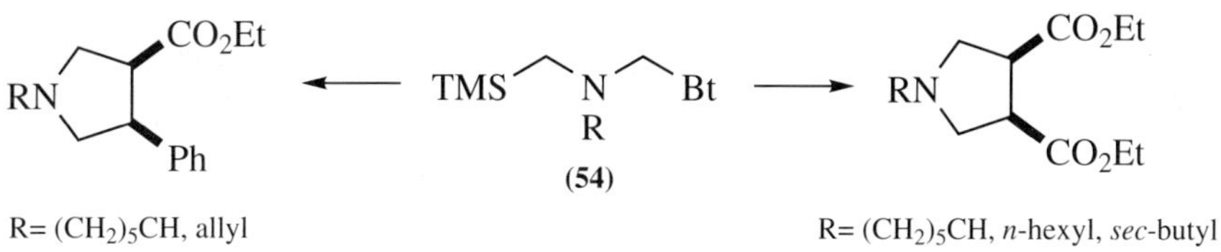

Scheme 3.13

Repetition of the reaction with DEAD as the dipolarophile furnished the desired cycloadduct **55** in 48% yield, but with 25% yield of the enamide **56** also being isolated. This was rationalized by invoking decomposition of the ylide precursor **57** to the (trimethylsilylmethyl)silyl amine **58**, which undergoes subsequent addition to the highly reactive acetylene (Scheme 3.14).

Tsuge et al. (14) outlined the use of *N*-silylmethyl thioureas as azomethine ylide precursors. Treatment of [2-methyl-3-(trimethylsilyl)]thiouronium triflates (**59**, R = H, R′ = Ph) with CsF in 1,2-dimethoxyethane (DME) and *N*-methylmaleimide led to the formation of a cycloadduct in 90% yield. The initial adduct **60**, however, could never be isolated and underwent *in situ* loss of MeSH to furnish the adducts shown, which represent formal aminonitrile ylide cycloadducts (Scheme 3.15).

However, use of a less reactive reagent where [R = R′=$(CH_2)_4$, $(CH_2)_5$, $(CH_2)_2O(CH_2)_2$] led to the isolation of products **61** and **62**, with a reduction in the yields of the desired cycloadducts. The product **62** arises from Michael addition of the liberated methanethiol to *N*-methylmaleimide. The protocol was further extended to olefinic dipolarophiles with dimethyl fumurate, dimethyl maleate, fumaronitrile, and 2-chloroacrylonitrile leading to the corresponding adducts, although these dipolarophiles proved somewhat less reactive with reduced yields being observed. Where applicable, the alkene configuration was reflected in the relative stereochemistry of the cycloadducts (Fig. 3.5).

Similarly, Hosomi and co-workers (15) reported isothioureas as azomethine ylide equivalents. Both *N*-substituted and *N*-unsubstituted *N*-(trimethylsilylmethyl)-isothiourea precursors underwent cycloaddition reactions with carbonyl compounds, when treated with stoichiometric CsF to deliver a range of 2-iminooxazolidine derivatives. Typically, **63** (R^1 = CN, R^2 = H) furnished adduct

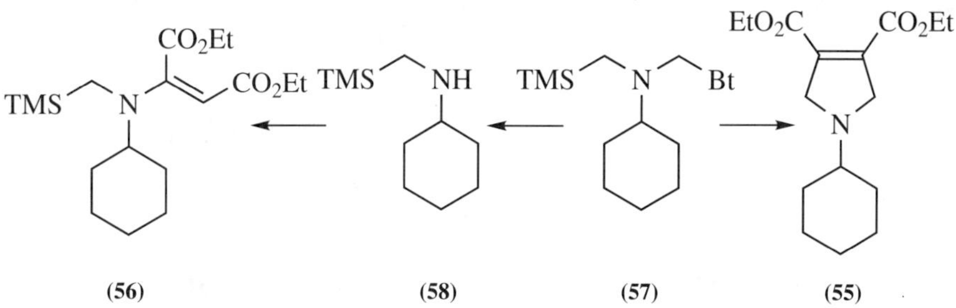

Scheme 3.14

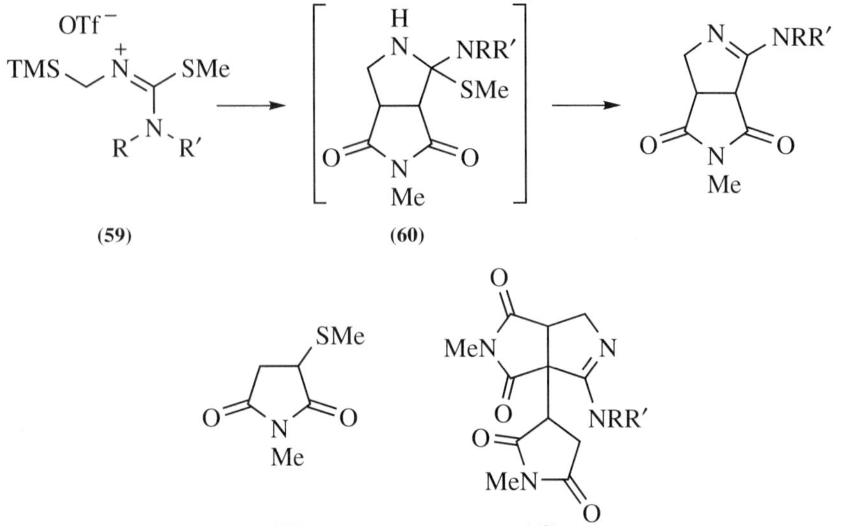

(59) (60) (61) (62)

Scheme 3.15

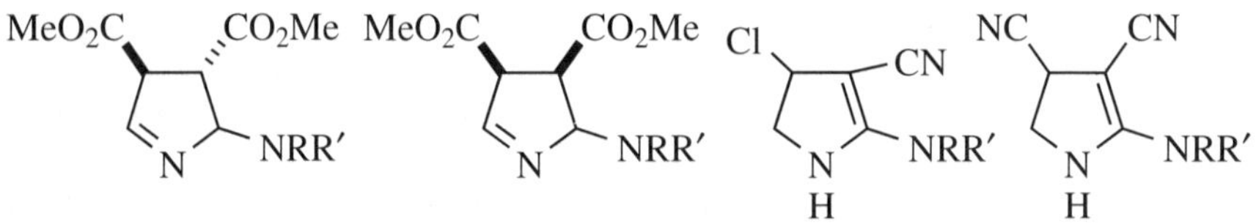

Figure 3.5

64 ($R^1 = CN$, $R^2 = H$, $R^3 = 2,6\text{-}Cl_2C_6H_4$) with 2,6-dichlorobenzaldehyde in 84% yield. The reaction scope was further extended with **63** ($R^1 = CN$ or $4\text{-}MeC_6II_4SO_2$, $R^2 = H$ or Me) and a wide selection of aromatic aldehydes to generate products **64** in 45–86% yield. In each case, the reaction proceeded with complete regioselectivity (Scheme 3.16, Table 3.1).

The protocol also tolerated the use of ketones and thioketones, furnishing products **65** and **66**, respectively, although a reduction in yield was observed (Fig. 3.6).

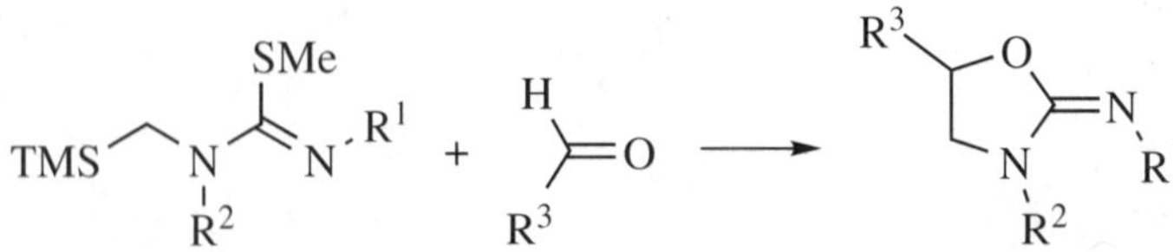

(63) (64)

Scheme 3.16

TABLE 3.1. ISOLATED YIELDS OF **64**

R^1	R^2	R^3	Yield of (64)%
p-$MeOC_6H_4SO_2$	H	p-PhC_6H_4	45
p-$MeOC_6H_4SO_2$	H	2,6-$Cl_2C_6H_3$	61
CN	Me	Ph	77
CN	Me	p-$MeOC_6H_4$	86
CN	Me	1-Napthyl	63

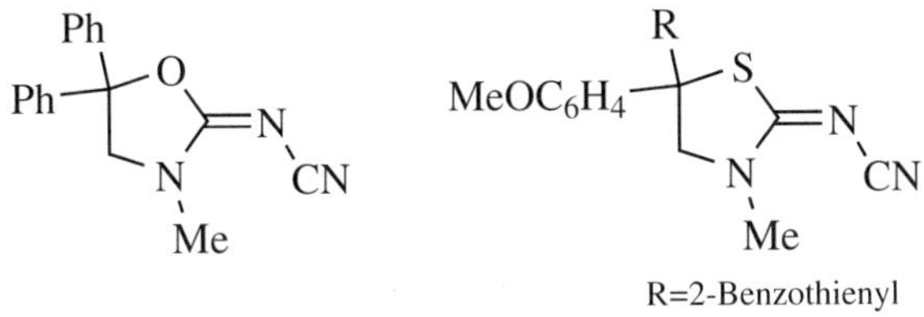

(65) **(66)**

Figure 3.6

Fischwick has detailed a rapid synthetic approach to the imidazole based γ-lactam alkaloid, (+/−) cynometrine (**67**), isolated from the stem bark of *Cynometra hankei*, and which has been shown to be a potential analgesic (16). Cesium fluoride induced formation of the ylide, from precursor **68**, followed by cycloaddition to alkene **69** furnished the required adduct **70** in 71% yield as a 4:1 diastereomeric mixture in favor of the desired isomer. Deprotection of the thioketal followed by $NaBH_4$ reduction delivered the desired racemic product (Scheme 3.17).

3.1.2. Alternative Generation Procedures

A series of publications by Vedejs et al. (17–19) outlined the use of oxazolines in the generation of azomethine ylides. Reduction of the oxazolium salts **71** (R^2 = Ph or Me) with Ph_3SiH/CsF led to formation of the unstable 4-oxazolines (**72**) and their valence bond azomethine ylide tautomers (17), which could be trapped *in situ* with DMAD to afford the bicyclic adducts **73**, where R^1 and R^2 = Ph or Me, in 75% yield *via* [2 + 2] cycloaddition (Scheme 3.18).

However, treatment of the precursor **74**, where there is no substitution at C(4) (*i.e.*, R^2 = Me) led to a single [3 + 2] cycloadduct **75** with methyl acrylate. The unstable oxazolines **75**, are considered to open spontaneously to their valence bond, 1,3-dipole tautomers **76**, which are trapped *in situ* by the dipolarophile. Use of DMAD led to the formation of the expected 2,5-dihydropyrrole (**77**), but difficulties in isolation required DDQ aromatization to pyrrole **78** (Scheme 3.19).

The technology was further extended with formation of the ylide by nucleophiles other than the Ph_3SiH/CsF derived hydride addition (18,19). Cyanide anion, generated from TMSCN/CsF, underwent nucleophilic addition and opening to the

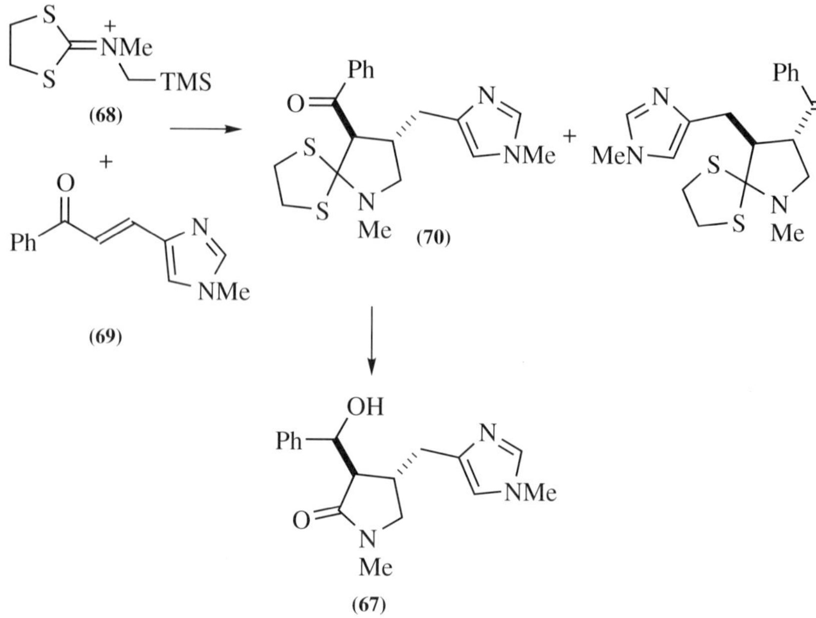

Scheme 3.17

ylide, in accordance with the previously observed results, followed by *in situ* trapping with DMAD to deliver the adducts **79**, which could never be isolated due to immediate loss of HCN and subsequent aromatization to **80**. Material yields were consistently high with a variety of substituents (Scheme 3.20).

Monosubstituted, olefinic dipolarophiles were also successfully employed, although the results were somewhat more complex due to the formation of regioisomers. However, in the case of **81**, a single pyrroline regioisomer was isolated in 73% yield by trapping of the ylide with methyl acrylate. Subsequent DDQ oxidation led to pyrrole **82**. In the case of **81**, the major regioisomer **83** was obtained in 21% yield with 7% of a minor isomer also being isolated. A salient feature of the reaction being the reversal of the regiochemical preference of the reaction although no explanation has been offered (Scheme 3.21).

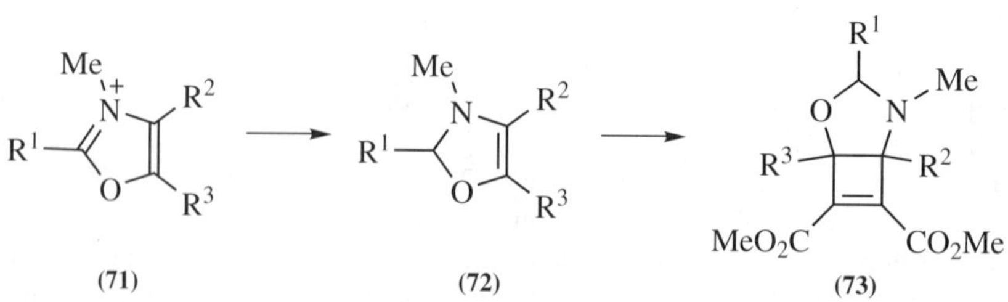

Scheme 3.18

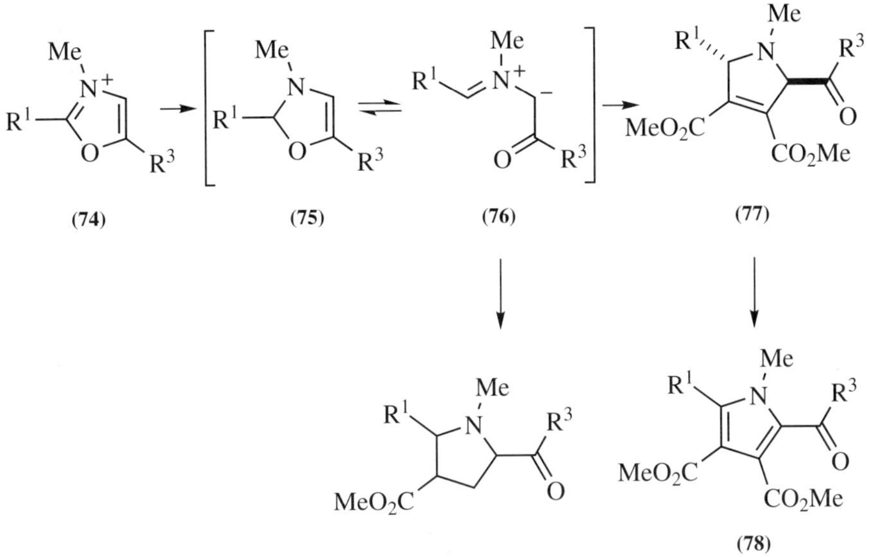

Scheme 3.19

Azomethine ylide generation from oxazolidines has also been achieved by flash vacuum thermolysis (20,21). During synthetic efforts toward alkaloid central skeletal cores, Joucla and co-workers (22) revealed that flash vacuum thermolysis of oxazolidine (**84**) led to an intramolecular [3 + 2] cycloaddition furnishing pyrrolidine **85** in 82% as a single regio- and stereoisomer. Subsequent Dieckmann

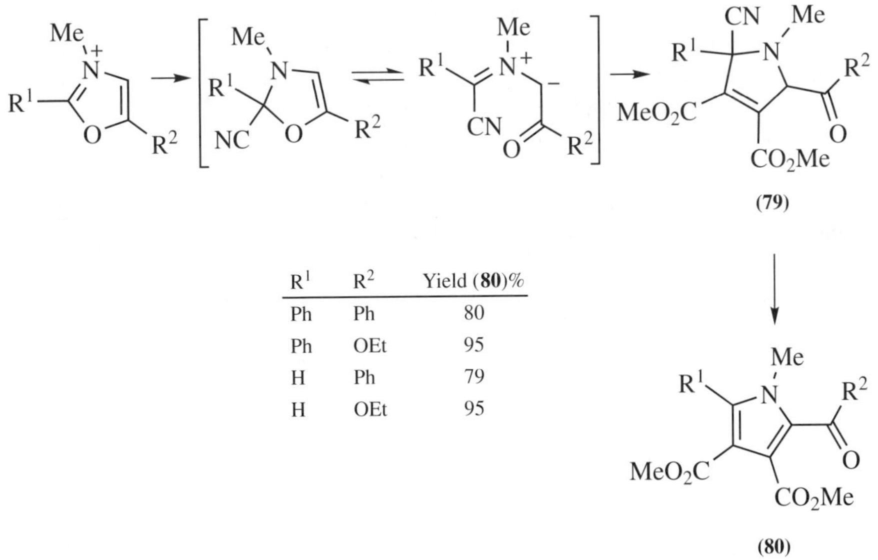

R^1	R^2	Yield (**80**)%
Ph	Ph	80
Ph	OEt	95
H	Ph	79
H	OEt	95

Scheme 3.20

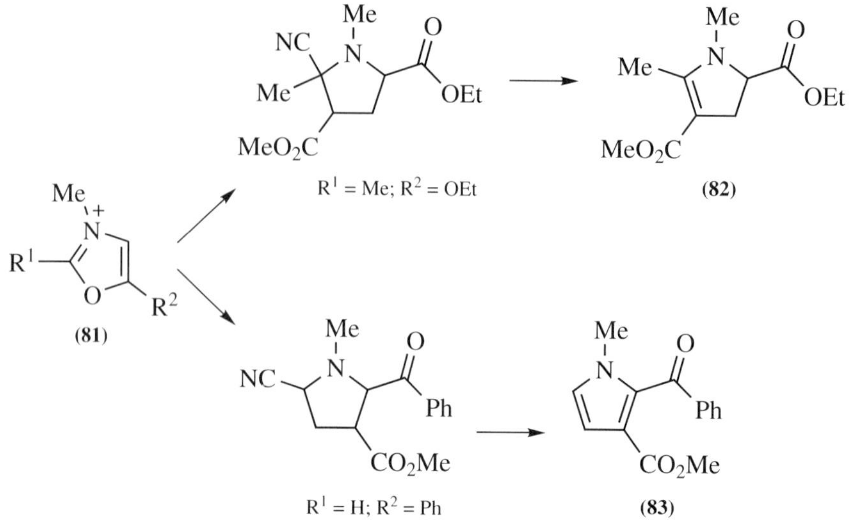

Scheme 3.21

condensation, acid hydrolysis and neutralization furnished the hemiacetal **86** as a single isomer. Although the reaction mechanism was not discussed, the most likely ylide formation and cyclization process is that shown (Scheme 3.22).

Roussi et al. (23) studied the deprotonation of tertiary amine N-oxides to generate azomethine ylides. Typically, treatment of **87** with lithium diisopropylamide (LDA) in the presence of a suitable dipolarophile led to the subsequent formation of the adducts shown (Scheme 3.23). In most cases, material yield is high, although the protocol suffers from formation of isomers for some dipolarophiles.

The formation of isomeric products can be reconciled by consideration of the mechanistic pathway. Deprotonation of the more acidic methyl group leads to formation of iminium salt **88** that can then form two ylides, **89** and **90**, furnishing a mixture of products after subsequent cycloaddition (Scheme 3.24).

Although Padwa et al. (25) extensively studied and developed silicon mediated technologies in azomethine ylide generation, he also developed other entries into azomethine ylides. In particular, the development of rhodium mediated transmuta-

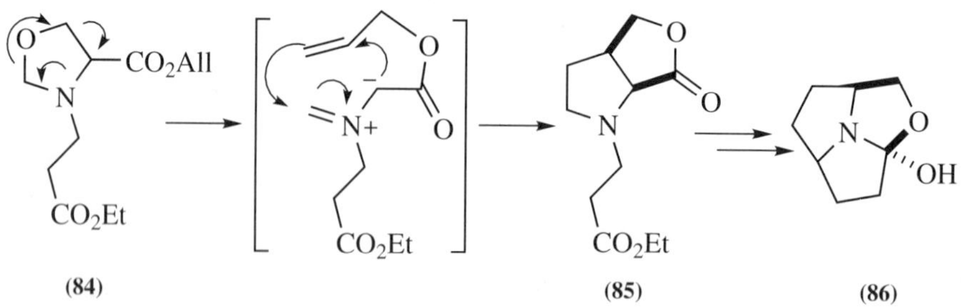

Scheme 3.22

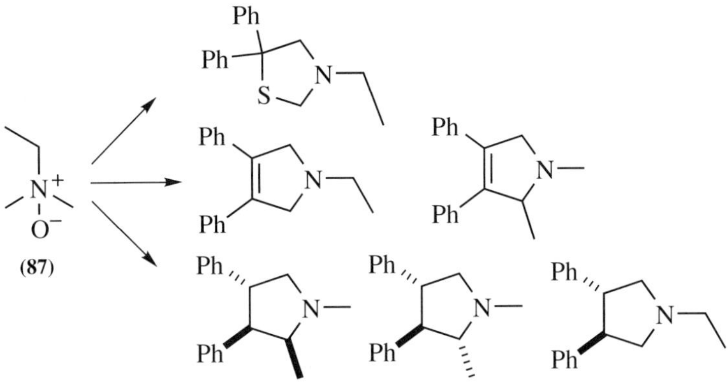

(87)

Scheme 3.23

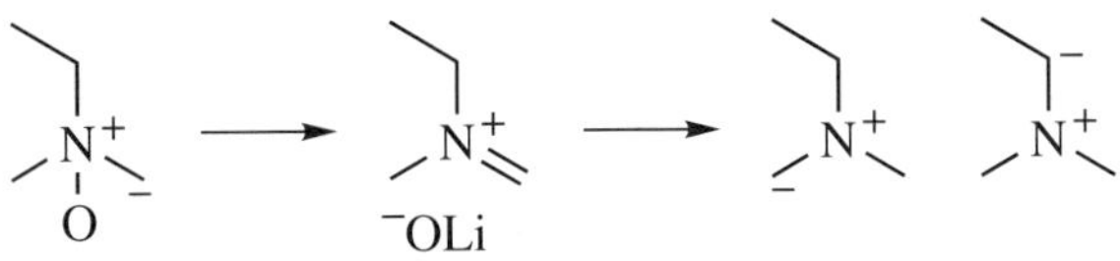

(88) (89) (90)

Scheme 3.24

tion of 1,3-dipoles has provided a method for the construction of a range of polycyclic systems. Treatment of *N*-acyl-2-(1-diazoacetyl)pyrrolidines with catalytic quantities of rhodium(II) carboxylate generated tricyclic dihydropyrrolizines *via* an azomethine ylide cycloaddition. However, the ylide generation proceeded through an initially formed carbonyl ylide that transmuted into the desired azomethine ylide prior to cycloaddition. Such a process has been dubbed a *dipole cascade*. Typically, diazo ketone **91**, in the presence of DMAD and catalytic Rh_2OAc_4, led to the formation of two cycloadducts, **92** and **93**. The product derived from cycloaddition to the initially formed carbonyl ylide was isolated in only 10% yield, while **92**, corresponding to azomethine ylide cycloaddition, was isolated in 87% yield (Scheme 3.25).

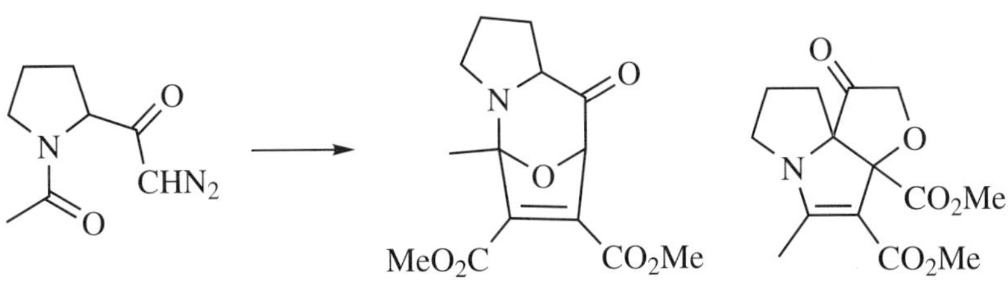

(91) (93) (92)

Scheme 3.25

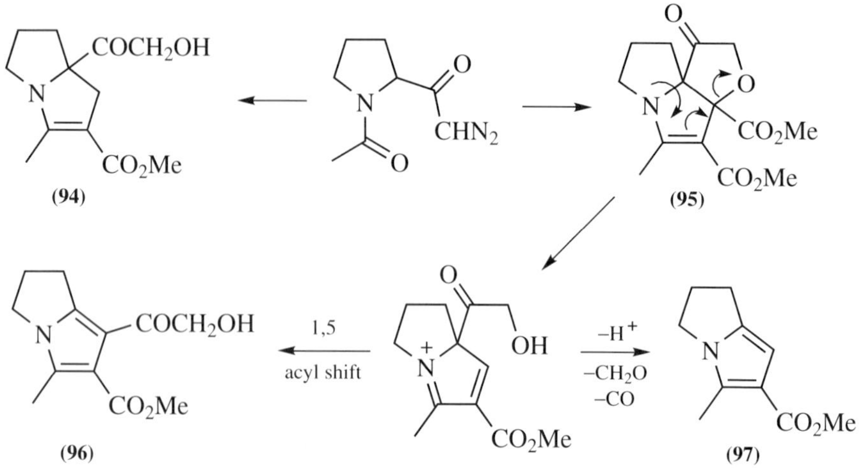

Scheme 3.26

The reaction also proceeded with other dipolarophiles (26). Methyl acrylate and methyl propiolate led to the formation of **94** and **95**, respectively. Acidic treatment of **95** gave rise to a 1:3 mixture of pyrroles **96** and **97**, in which the initial ring-opening step to produce the transient iminium ion occurred, it which was followed by competing loss of CH_2O and CO to deliver **97** with a 1,5 acyl shift furnishing **96**. Precursor **98** underwent cycloaddition with *N*-phenyl maleimide furnishing **99** in 81% yield. Replacement of the *N*-acyl substituent with an *N*-benzoyl group gave rise to the azomethine ylide adduct as the sole product in 95% isolated yield with DMAD as the dipolarophile (Schemes 3.26 and 3.27).

The proposed reaction pathway invokes initial formation of carbonyl ylide **100** by intramolecular cyclization of the intermediate keto carbenoid onto the oxygen atom of the amide. Subsequent isomerization to the azomethine ylide is followed by 1,3-dipolar cycloaddition to DMAD to furnish the intermediate cycloadduct **101**, which undergoes *in situ* alkoxy 1,3-shift to the final dihydropyrrolizine **102** (Scheme 3.28).

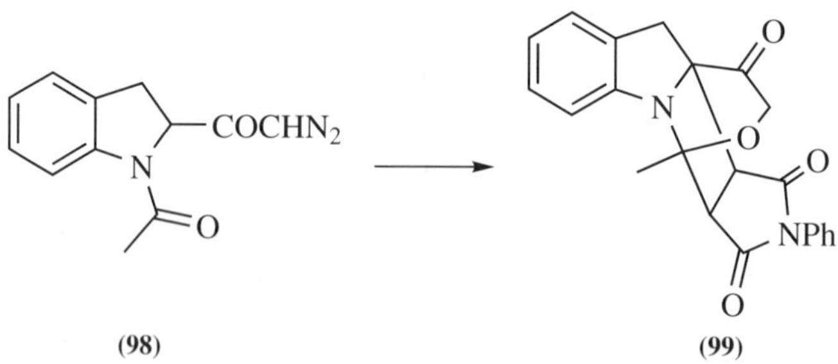

Scheme 3.27

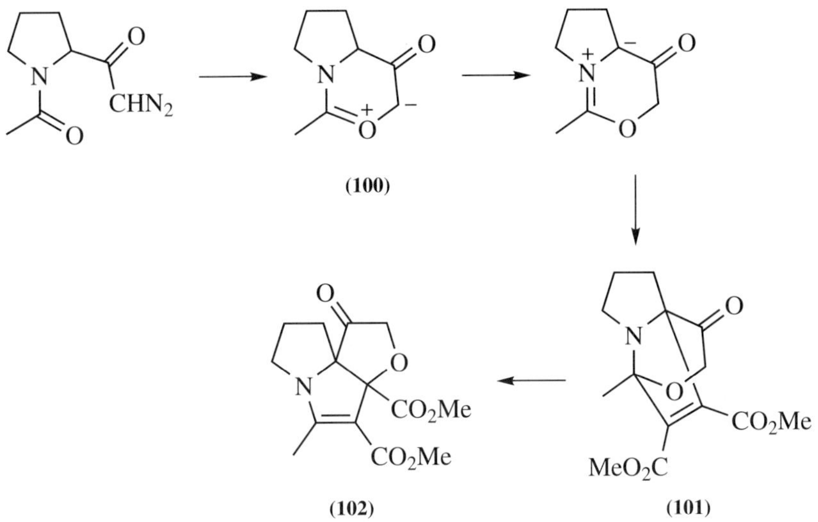

Scheme 3.28

Harwood and co-workers (27), during their extensive studies on morpholinone-based chiral azomethine ylides, detailed ylide generation from carbamate precursors. Initial studies had focused on the derivatives **103**, where elimination of the leaving group, followed by *in situ* trapping of the resultant ylide, led to the expected adducts under Lewis acid catalysis. However, since these precursors were either too unstable to isolate (R = OMe and NMe_2), or furnished only very low yields of cycloadducts (R = phthalamide), an alternative, stable precursor was sought. It was found that treatment of morpholinone (**104**) with the *N*-(dimethylaminomethyl) carbamate **105**, synthesized from Eschenmoser's salt and *tert*-butylcarbamate, delivered the stable ylide precursor in quantitative yield as a crystalline solid. The trifluoroacetyl (TFA) catalyzed decomposition of **105**, in the presence of *N*-methyl or *N*-phenyl maleimide, furnished the desired cycloadducts **106** and **107**, in 57%, *endo/exo* 33:24 (R = Ph) and 34%, *endo/exo* 19:23 (R = Me) (Scheme 3.29).

An exhaustive series of reports by Grigg et al. (28) outlined two basic methods for the generation of azomethine ylides proceeding *via* either a 1,2-prototropic shift, or by a decarboxylative approach (29). The decarboxylative route to azomethine ylides can be exemplified by the condensation of benzaldehyde with the cyclic amino acid tetrahydroisoquinoline (**108**) (30), in DMF at 120 °C, to generate the intermediate *anti*-dipole **109**, which underwent subsequent cycloaddition with *N*-methyl maleimide to furnish a 1:1 *endo/exo* mixture of adducts **110** (R = Ph), in 82% yield (Scheme 3.30).

Similarly, 3-phenyl propionaldehyde gave rise to a 1:1.5 *endo/exo* mixture of adducts **110** (R = CH_2CH_2Ph), derived solely from reaction of the *anti*-dipole. In an analogous fashion, thiazolidine carboxylic acid (**111**) also underwent *anti*-dipole specific cycloaddition, with *N*-phenyl maleimide, after initial condensation with

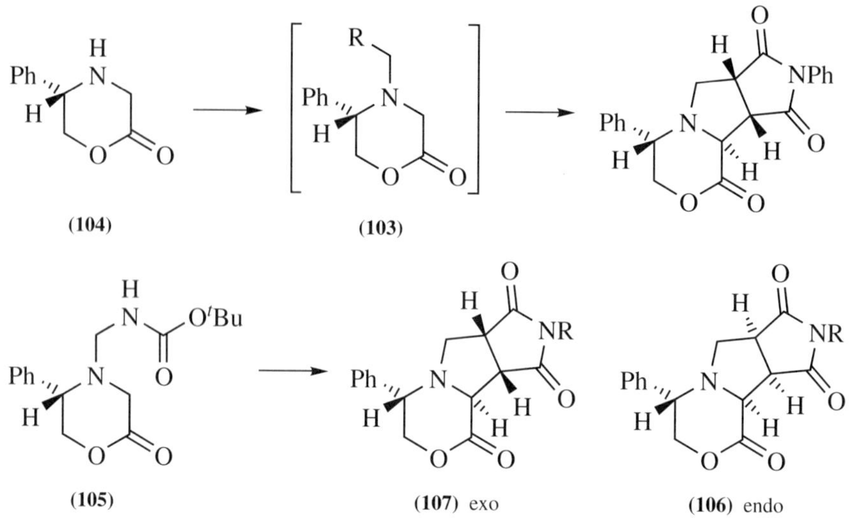

Scheme 3.29

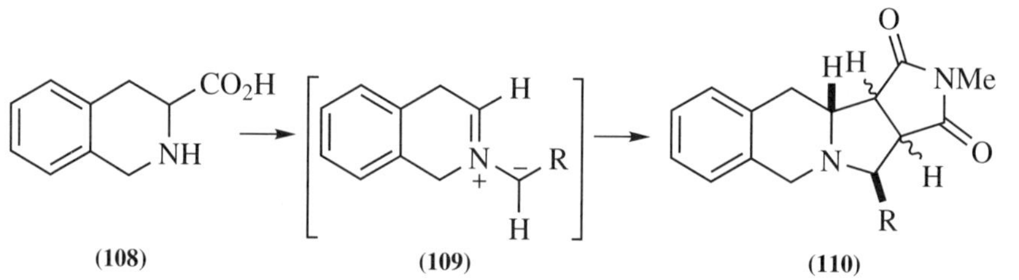

Scheme 3.30

benzaldehyde. The reaction products (**112**) were isolated in 79% overall yield, with an *endo/exo* ratio of 1:1.5 (Scheme 3.31).

However, the highly stereoselective nature of the dipole intermediate, which was observed in the preceding cases, did not translate to azomethine ylide cycloadditions of **113** and **114**, both of which formed mixtures of products derived from *endo*

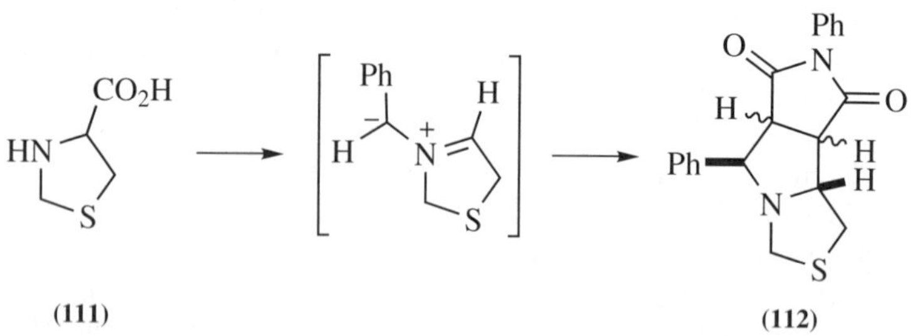

Scheme 3.31

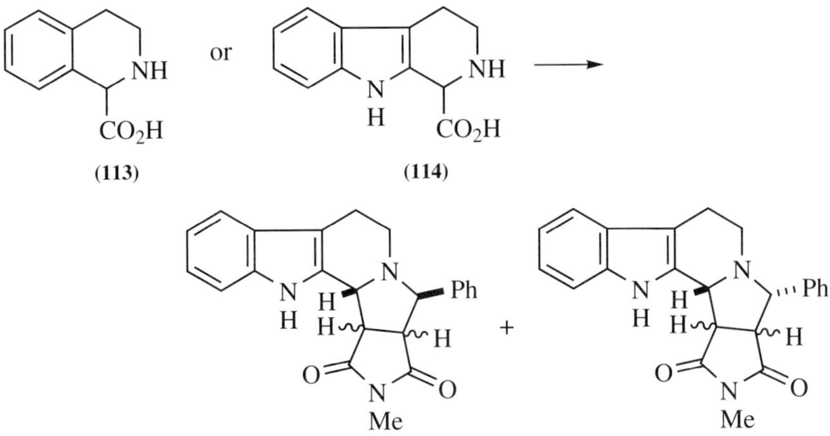

(113) (114) (115) (116)

Scheme 3.32

and *exo* approach of the dipolarophile to a mixture of both *anti* and *syn* dipoles. Typically, tetrahydro-β-carboline-1-carboxylic acid, in the presence of benzaldehyde and *N*-methyl maleimide, led to the isolation of **115**, derived from *anti*-dipole formation, with an *endo/exo* ratio of 2.2:2 and **116** from cycloaddition with the *syn*-dipole in an *endo/exo* ratio of 1:1.2 (Scheme 3.32).

The reaction also showed stereochemical dependence on the aldehyde used in the initial condensation step (31). While *p*-NO_2PhCHO, in the presence of **117** and *N*-methylmaleimide, led to the isolation of **118**, derived from the *anti*-dipole in 66% yield with a 1:1.2 *endo/exo* ratio, *p*-NMe_2PhCHO, **117** and *N*-phenylmaleimide furnished a 10:1 mixture of **119** (*endo/exo* 1.6:1) and **120** arising from *anti*- and *syn*-dipoles, respectively, in 98% overall yield (Scheme 3.33).

An extensive series of reports outlining the use of alternative dipole precursors, aldehydes and dipolarophiles led to the following mechanistic rationale (32). Typically, the initially formed iminium ion from condensation of benzaldehyde with the cyclic, secondary α-amino acid **117** underwent 5-*endo*-trig cyclization to oxazolidin-5-one **118**. Subsequent retro-1,3-dipolar cycloaddition, with concomitant loss of CO_2, formed the requisite dipole **119**. Cycloaddition with a suitable dipolarophile would generate the products outlined in Schemes 3.30 and 3.31. Since 1,3-dipolar cycloreversions have been shown to proceed stereospecifically (33) and that MNDO calculations had shown a *syn* relationship between the phenyl group and proton in **119** to be the more energetically favored, generation of an *anti*-dipole is to be expected. From this, it was assumed that differing electronic and steric effects would affect the stereochemical nature of the intermediate oxazolidin-5-one with subsequent control over the *syn*- and *anti*-dipole ratios. The use of *N*-methylmaleimide as the dipolarophile suggests that no dipole transmutation from the *syn*- to *anti*-dipole occurred, since the highly reactive nature of the dipolarophile would induce immediate cycloaddition (Scheme 3.34).

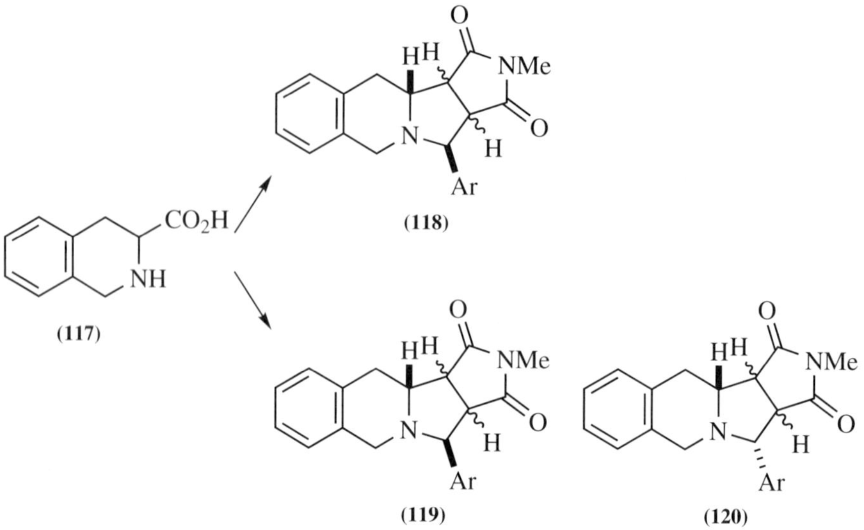

Scheme 3.33

Grigg et al. (34) also conducted extensive studies of the thermal 1,2-prototropic generation of azomethine ylides and this can be exemplified by the diastereofacially selective cycloaddition of 7-aminocephalosprin ylide precursors. Condensation of aryl aldehydes with **120**, in refluxing toluene, furnished imines **121**, which, in the presence of *N*-phenylmaleimide, furnished a mixture of cycloadducts **122** and **123** in essentially quantitative yield in a 2:1 ratio. The only observed products

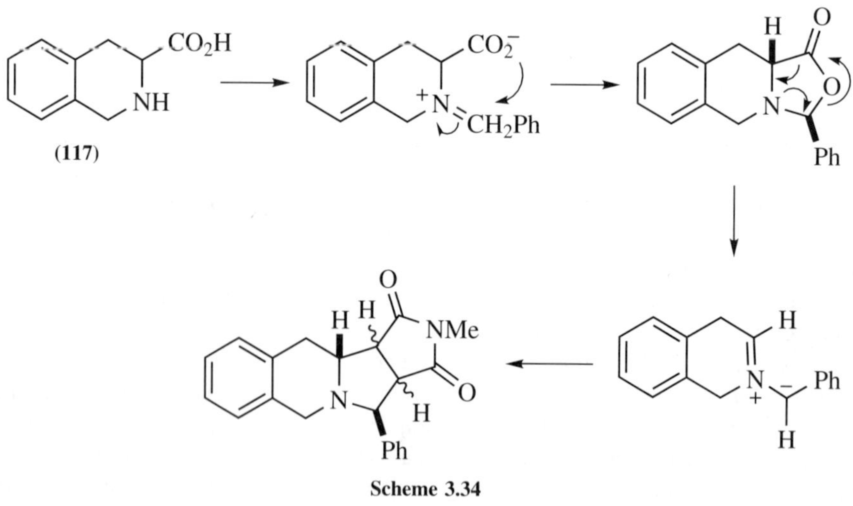

Scheme 3.34

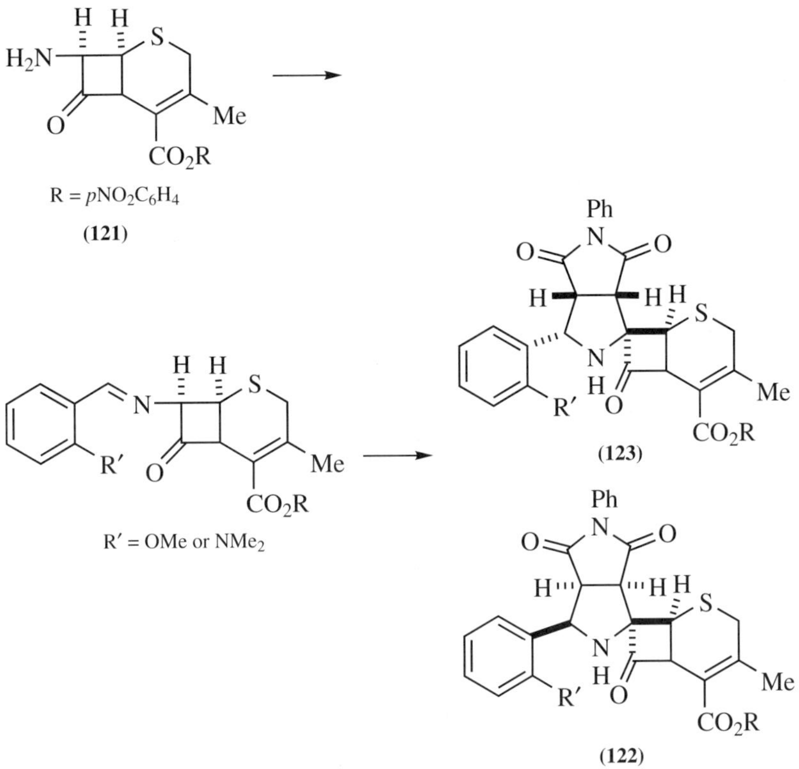

Scheme 3.35

in this instance were those derived from *endo* attack of the dipolarophile (Scheme 3.35).

Ylide generation by this method involves a 1,2-prototropic shift of the proton a to the imine, furnishing, in this instance, a mixture of ylides **124** and **125**, which undergo subsequent *endo*-cycloaddition. Due to the highly reactive nature of the dipolarophile, it is unlikely that a single ylide is initially formed, followed by dipole transmutation (Scheme 3.36).

In addition to the routes previously detailed, Grigg has outlined a dehydrogenation approach to azomethine ylides (35). Typically, treatment of precursor **126** with *N*-methylmaleimide and palladium black in DMF at 110 °C led to the isolation of a 1:1 mixture of **127** and **128**, with no product arising from a *syn*-dipole being observed. The reaction pathway was assumed to proceed *via* initial coordination of the Pd to nitrogen, followed by dehydrogenation and iminium ion formation. The stereospecific formation of an *anti*-dipole suggested subsequent formation of oxazoline (**129**), which opens stereospecifically, as for the decarboxylative approach, to furnish an *anti*-dipole that undergoes subsequent cycloaddition (Scheme 3.37).

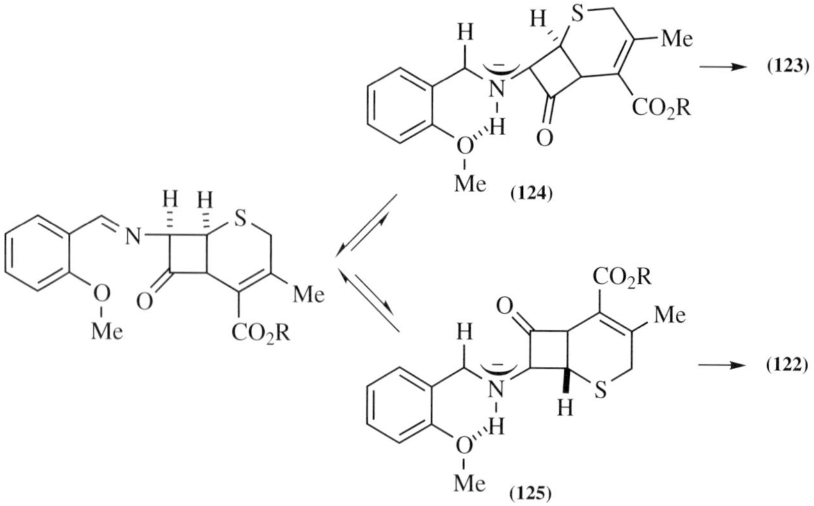

Scheme 3.36

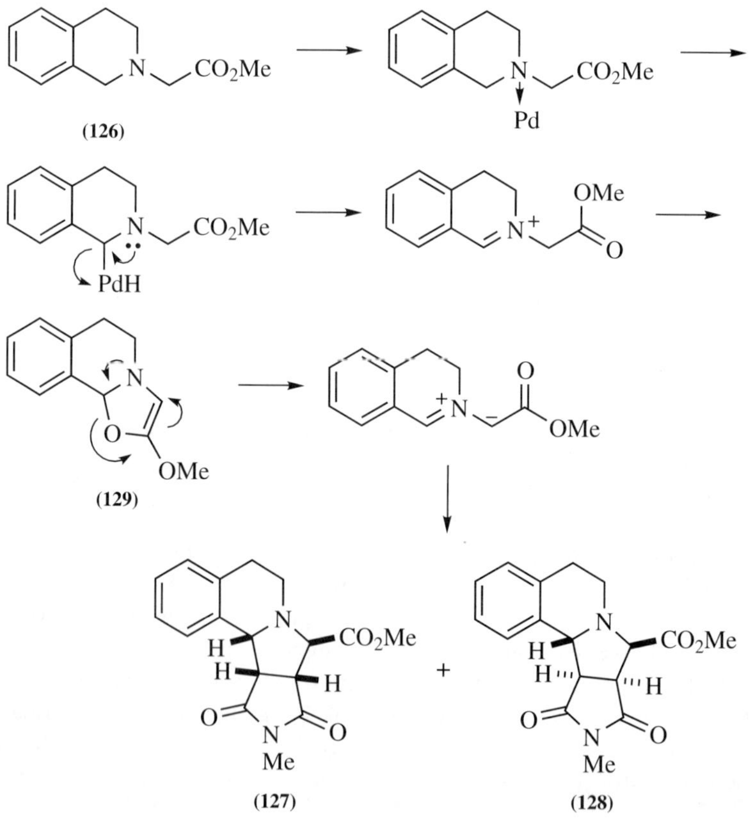

Scheme 3.37

3.2. ASYMMETRIC REACTIONS

Modern synthetic chemistry is increasingly focused on the formation, control, and induction of asymmetry in molecular structure. As the complexity of synthetic targets increases, in particular those derived from natural sources, the demands of the chemist for new and efficient methods for the introduction of new stereogenic centers has generated many elegant solutions. Since azomethine ylides are inherently achiral, the induction of stereocontrol has to be performed by attachment of a suitable stereodirecting unit onto either the dipole, or more recently onto the ylide, and latterly many elegant solutions have been developed.

3.2.1. Chiral Dipolarophiles

Kanemasa reported the use of chiral α, β-unsaturated dipolarophiles in 1,3-dipolar cycloaddition reactions, utilizing a 2-pyrrolidinyl group as the asymmetric controlling element (35,36). An initial communication revealed that treatment of the (benzylideneamino)acetate (**130**) with the readily prepared enantiopure dipolarophile **131** at −78 °C in the presenc of LiBr and 1,8-diazobicyclo(5.4.0)undec-7-ene (DBU) furnished a single cycloadduct in 82% yield. The reaction formed four new stereogenic centers with absolute diastereoselectivity. However, the reaction product was not consistent with the thermodynamically favored *anti*-periplanar ylide conformer **132**, which would involve attack of the dipolarophile to the stericaly less congested *re* face of the ylide, but by unexpected *si* attack to the thermodynamically less stable *syn*-periplanar ylide **133** (Figure 3.7). Removal of the chiral controlling section was easily achieved with no loss of stereochemical integrity, by *N*-tosylation and acetal exchange (Scheme 3.38).

This attractive protocol for the asymmetric addition of α, β-unsaturated esters to *N*-metalated azomethine ylides was further developed using C(2) symmetrical imidazoladine stereodirecting units in an extensive study into the effects of reaction conditions and substituent effects on the facial selectivity of the reaction (36). Both

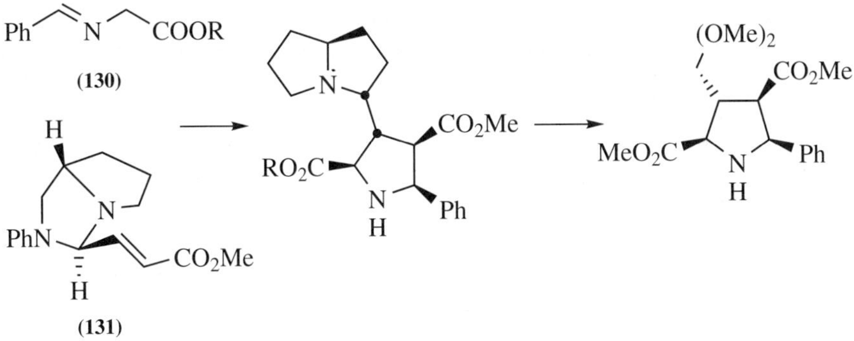

Scheme 3.38

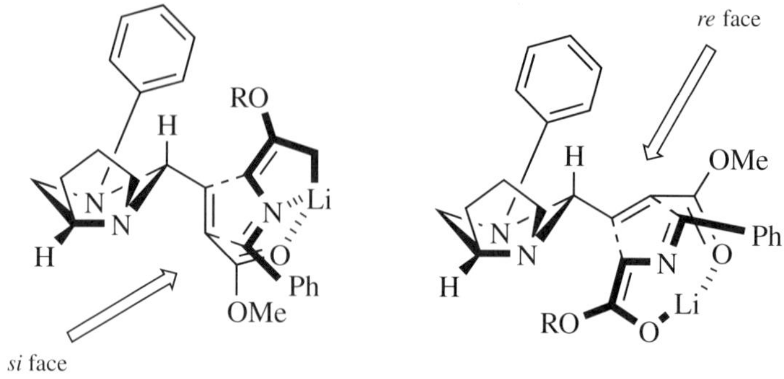

(133) (132)

Figure 3.7

the racemic *N*-phenyl and *N*-methyl dipolarophiles were prepared for a series of studies into the ylide generation from methyl and *tert*-butyl(benzylideneamino) acetates and their subsequent cycloadditions. In the case of the ylide ester **134** (R′ = Me) and dipolarophile **135** (R = Ph) optimum yields were observed with LiBr/DBU at 94% with a diastereomeric ratio of **136**: **137** 85:15 at room temperature, while the use of LDA/*t*-BuOH at −78 °C gave a yield of 89% and an isomeric ratio of 96:4. For ylide **135** (R′ = *t*-Bu) and dipolarophile **134** (R = Ph), *n*-BuLi generation gave a reduced yield of 61% at room temperature but with a reversal of facial selectivity giving a product ratio of **136**: **137** 45:55. At −78 °C, the yield dropped to 20% but the ratio improved to **136**: **137** 20:80. For the olefin **135** (R = Me), ylide **134** (R′ = Me) generated with LDA/MeOH gave a yield of 98% and a single diastereomic product (**136**) at −78 °C, while ylide **134** (R′ = *t*-Bu) gave a yield of 94% and a single product (**137**) at room temperature in the presence of LiBr/DBU (Scheme 3.39).

Having now established the reaction conditions and substituents in both the dipolarophile and N-metalated azomethine ylide, the protocol was extended to an asymmetric process. Treatment of the enantiopure dipolarophile **138** (R = Ph) under optimized conditions, with the ylide **139** (R′ = Me) gave rise to a mixture of enantiomers of **140**. The major isomer was isolated and removal of the chiral controlling unit, by the method previously described, to furnish the N-tosylated pyrrolidine, allowed for unambiguous assignment of the absolute stereochemical outcome of the reaction. The ylide **139** (R′ = Me) gave rise to the major isomer **140** with dipolarophile (−)-**138** (R = Ph), which furnished the pyrrolidine (−)-**141** with the (2*R*, 3*R*, 4*R*, 5*S*) absolute configuration. As with results previously described in the racemic series, ylide **142** (R′ = *t*-Bu) gave the opposite major enantiomer with the same dipolarophile (−)-**138** (R = Ph), delivering the (2*S*,3*S*,4*S*,5*R*) (+)-**144** pyrrolidine (Scheme 3.40). The ylides also showed differing diastereofacial selectivities for the *N*-Me **145** and *N*-Ph **140** dipolarophiles (Scheme 3.41).

This represents an interesting study into the use of a C(2) symmetrical stereocontrolling unit for the azomethine ylide cycloadditions of α, β-unsaturated esters. In particular, consideration of the effect of both N-substituents in the

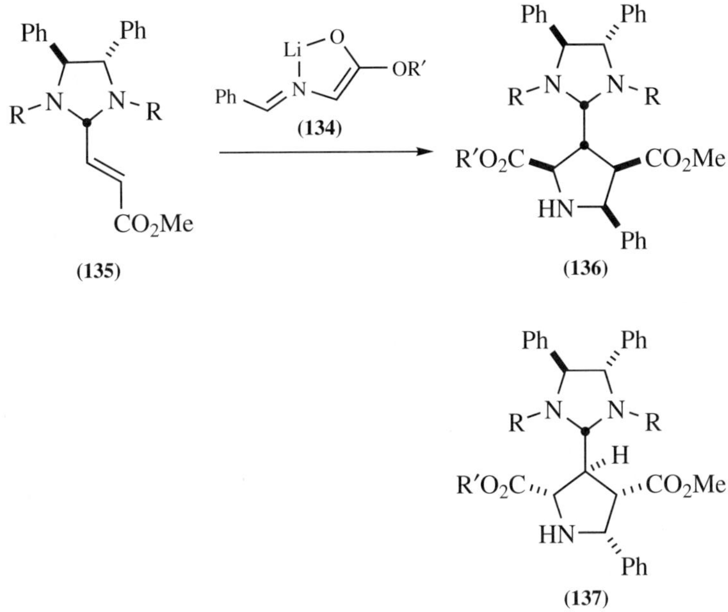

Scheme 3.39

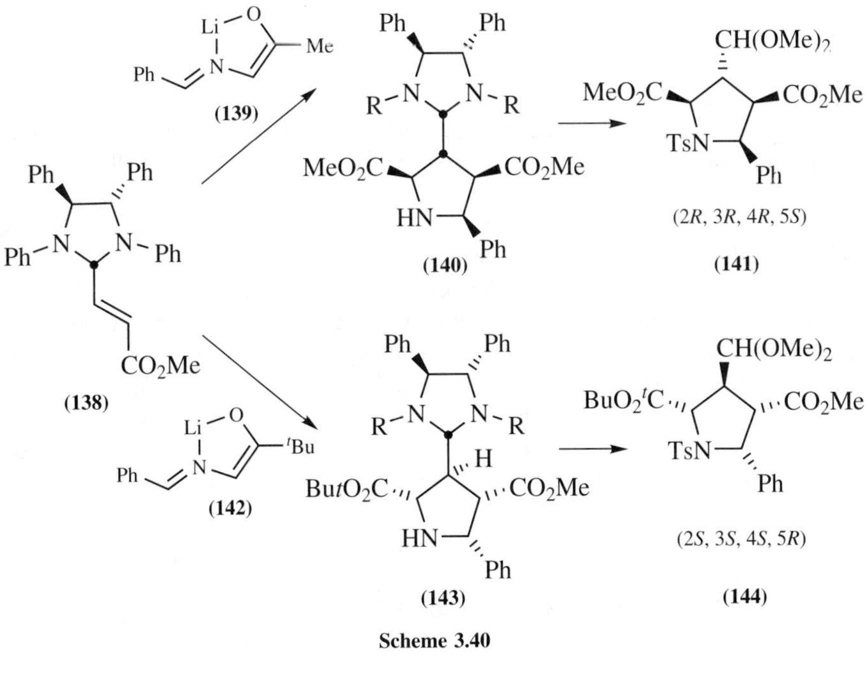

Scheme 3.40

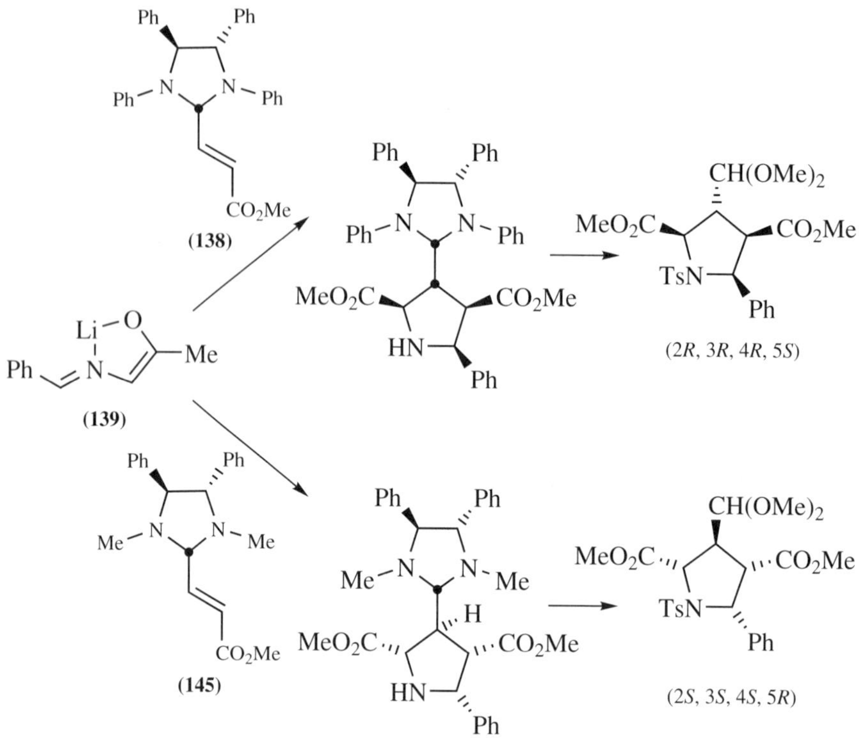

Scheme 3.41

dipolarophile and choice of ylide ester, allows for the imposition of high and exact asymmetric control into the reaction.

Chiral enones have been used in highly diastereoselective ylide reactions (37). In the presence of AgOAc and DBU (a superior combination than LiBr/DBU for this system), a variety of imines underwent cycloaddition to enone **146**, in excellent yields with complete regioselectivity and diastereoselectivities of 95:5 for R = Ph and 3-pyridyl. The reaction also proceeded with aliphatic imines, although selectivities were reduced. Acidic removal of the acetal protection revealed the 5-hydroxy-1, 4-dicarbonyl compounds (**147**) with the *anti* configuration shown between the C(3) substituent and the chiral center of the side chain. Replacement of the ester in the imine with an amide functionality gave a slight reduction in yield and selectivity, although the reaction still proceeded with high efficiency (Scheme 3.42).

Improved selectivities were observed with the enone **148** using the dibenzyl amine as the chiral controlling unit at the γ-position, delivering essentially a single product from the reaction (Scheme 3.43).

Koizumi and co-workers (38) reported the first asymmetric synthesis of (1*S*)-(−)-α-tropanol (**149**) *via* a 1,3-dipolar cycloaddition protocol. Treatment of the chiral dipolarophile **150** with **151** in tetrahydrofuron (THF) delivered cycloadducts *exo*-**152** and *endo*-**153**. Although the reaction proceeded with low facial selectivity,

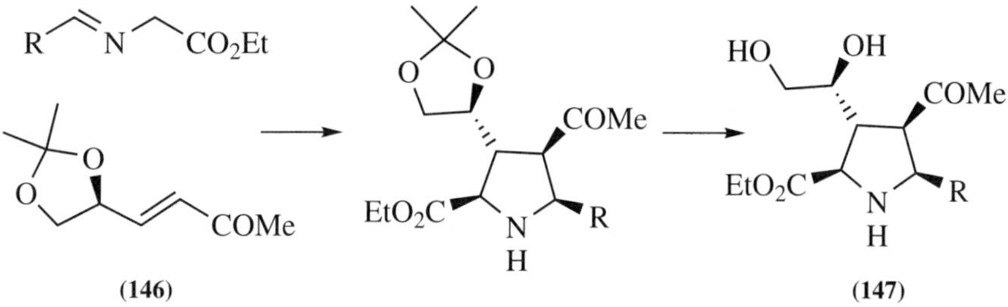

Scheme 3.42

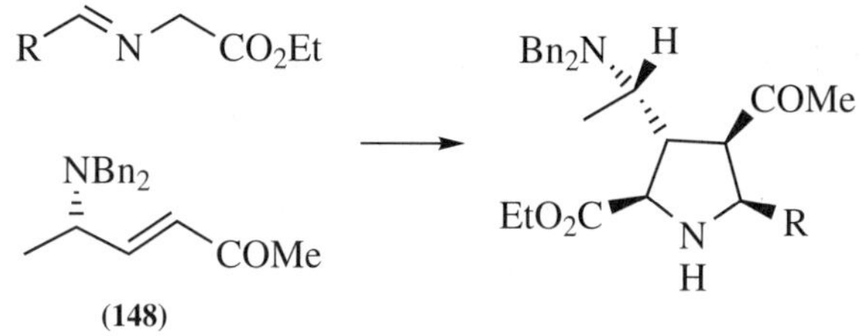

Scheme 3.43

60:40 exo/endo, and in a moderate yield of 68%, the diastereoselectivity of the reaction was high; the exo product (**152**) being isolated with a diastereomeric excess (de) of 68% while the endo product was furnished as a single diastereoisomer. This first case of chiral induction of a 1,3-dipolar cycloaddition by an asymmetric sulfur moiety, after standard chemical elaboration furnished **149** (Scheme 3.44).

Pyridones, as exemplified by ABT-719 (**154**, Figure 3.8), represent a new class of DNA gyrase inhibitors possessing a broad spectrum of antibacterial activity and, in studies toward such compounds, it was revealed that the C(8) functionality was an important part of the DNA binding action. Azomethine ylide cycloadditions were employed to give a range of proline-type derivatives in order to study structure–activity relationships (39).

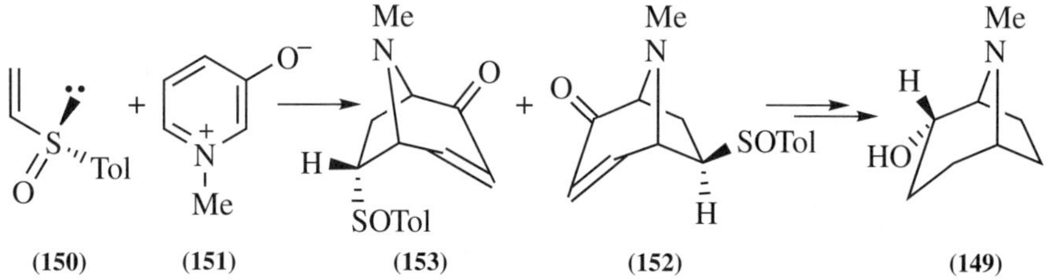

Tol = tolyl

Scheme 3.44

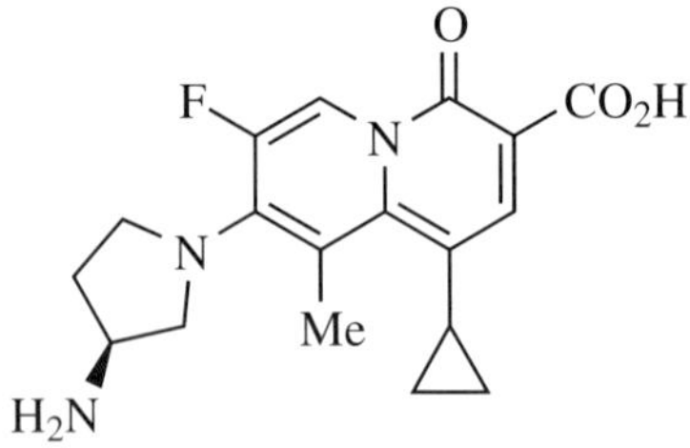

(154)

Figure 3.8

The azomethine ylide was generated by treatment of *N*-benzyl-*N*-(methoxymethyl)-trimethylsilylmethylamine (**155**) with TFA and underwent the required cycloaddition step with chiral dipolarophile **156**, stereocontrol being induced by Evan's auxiliary. The α, β-unsaturated acid dipolarophile was tethered to a chiral oxazoladine in two easy, high-yielding steps. The auxiliary served three purposes; to give asymmetric control to the reaction, to allow for separation of the reaction products by generating column separable diastereoisomers, and finally to activate the olefin in the cycloaddition step (Scheme 3.45).

The diastereomeric products were obtained in a 20:80 **157**: **158** ratio with $R^2 = Ph$ as the chiral control element, whereas the more usual *i*-Pr group in such auxiliaries gave poorer steric control. Template removal was easily achieved with $LiOH/H_2O_2$ in quantitative yield along with recovery of the auxiliary. Curtius rearrangement of the resulting carboxyl group furnished the *N*-butoxycarbonyl (*N*-Boc) protected product **159** in high yield.

Williams and Fegley (40) utilized his chiral template (Section 3.2.3) as a chiral dipolarophile in the concise synthesis of *S*-(−)-cucurbitine, a naturally occurring amino acid isolated from species of pumpkin, that acts as a growth inhibitor to

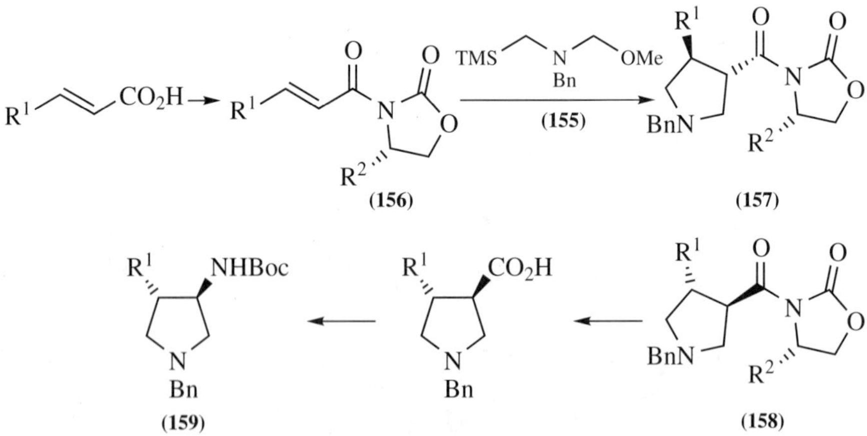

Scheme 3.45

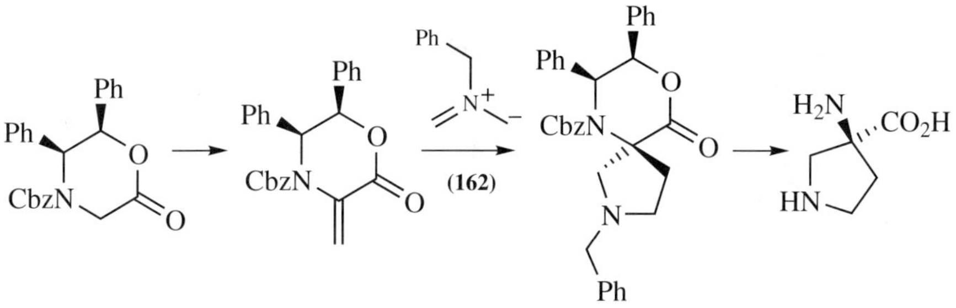

(160) (161) (163)

Scheme 3.46

Schistosoma japonicum. The carbobenzoxy (CBz) protected template **160** was initially converted to the α, β-dehydrolactone **161** *via* the phosphate ester, before undergoing cycloaddition to ylide **162**, generated *in situ* by acidic treatment of *N*-benzyl-*N*-(methoxymethyl)trimethylsilyl amine. The resultant cycloadduct (**163**) was isolated in 94% yield as a single diastereoisomer. Destructive template removal, by catalytic hydrogenation, released (*S*)-(−)-cucurbitine, after ion-exchange chromatography, as the free amino acid in 90% yield (Scheme 3.46).

In synthetic efforts toward the DNA reactive alkaloid naphthyridinomycin (**164**), Garner and Ho (41) reported a series of studies into the construction of the diazobicyclo[3.2.1]octane section. Construction of the five-membered ring, by the photolytic conversion of an aziridine to an azomethine ylide and subsequent alkene 1,3-dipolar cycloaddition, was deemed the best synthetic tactic. Initial studies with menthol- and isonorborneol- tethered chiral dipolarophiles gave no facial selectivity in the adducts formed (42). However, utilizing Oppolzer's sultam as the chiral controlling unit led to a dramatic improvement. Treatment of ylide precursor **165** with the chiral dipolarophile **166** under photochemical conditions led to formation of the desired cycloadducts (Scheme 3.47). The reaction proceeded with an *exo/endo* ratio of only 2.4:1; however, the facial selectivity was good at >25:1 in favor of the desired *re* products. The products derived from *si* attack of the ylide

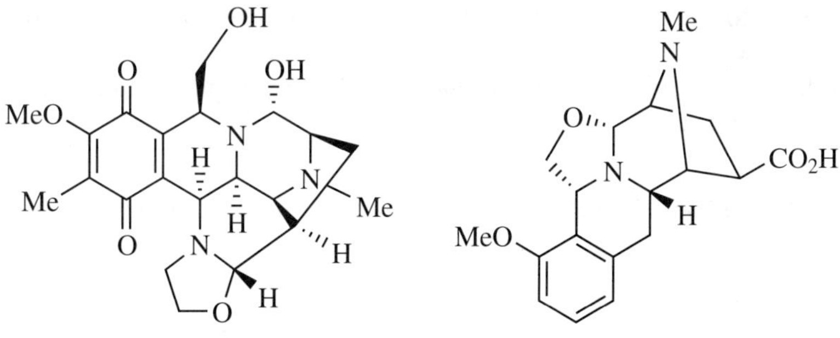

(164) (167)

Figure 3.9

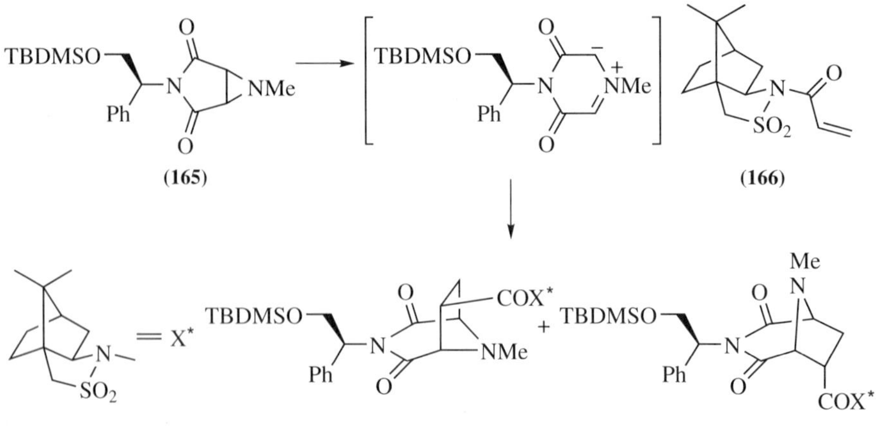

Scheme 3.47

were detected as only minor products in the crude reaction mixture by NMR analysis. Removal and recovery of the auxiliary was easily achieved with $TiCl_4$ in refluxing EtOH. A similar approach was also applied to the first enantiospecific synthesis of (−)-euinocarcin (**167**) (43) (Fig. 3.9).

Waldmann et al. (44) outlined the use of *N*-acroyl-(*S*)-proline benzyl ester (**168**) in the highly stereoselective synthesis of complex, polyfunctionalized prolines. Treatment of the Schiff bases **169** with DBU/LiBr generated *in situ* the N-lithiated azomethine ylides that underwent cycloaddition to dipolarophile **168** to deliver adducts **170** and **171**. In each case, the stereoselectivity of the reaction was exceptionally high. For the cases in which R was sterically influential, such as Ph, *i*-Pr or *t*-Bu, the ratio of both **170**: **171** and *endo/exo* was >99:1. A reduction in the steric bulk of R to either H or Me still gave *endo/exo* selectivities >99:1 although the isolated ratio of **170**: **171** was slightly reduced to 92: 8. Simple hydrolysis removed the stereodirecting unit, releasing the desired, enantiopure proline derivativcs **172** (Scheme 3.48).

The near perfect facial and *endo/exo* selectivities observed were reconciled by assuming that the reaction proceeded through the highly ordered *endo* transition state **173**, in which both the azomethine ylide and the approaching dipolarophile are coordinated to the lithium cation (Fig. 3.10). Alternative *endo* transition state models that would allow for the observed stereochemical outcome of the cycloaddition do not permit an efficient and sterically less imposing environment for the relevant functionalities.

3.2.2. Chiral Ylides:Stereocontrol by Inherent Molecular Structure

In 1985, Padwa et al. (45) reported the first asymmetric 1,3-dipolar cycloaddition reaction of an azomethine ylide. The treatment of α-cyanoaminosilanes with AgF has already been detailed as one of the primary methods for the generation of azomethine ylides (Section 3.1.1). Treatment of the optically active precursor **174**

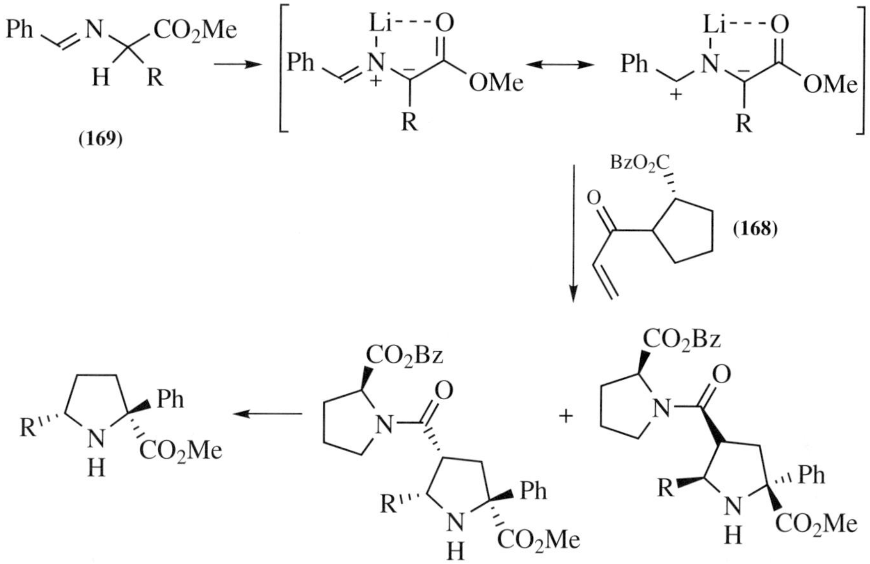

(172) (170) (171)

Scheme 3.48

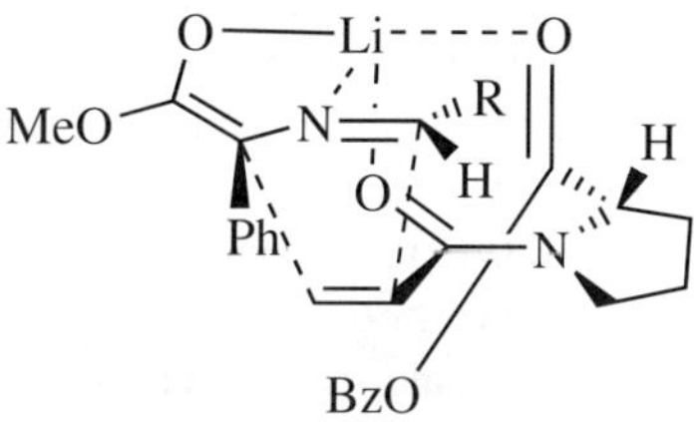

(173)

Figure 3.10

(R = Me) under standard generation conditions, in the presence of activated alkene dipolarophiles led to the expected adducts (Scheme 3.49). In the case of *trans*-1-nitro-2-(3,4-methylenedioxyphenyl) ethene, a 3:2 mixture of diastereoisomers was isolated. Replacement of the methyl group with a methyl ether (R = CH_2OMe),

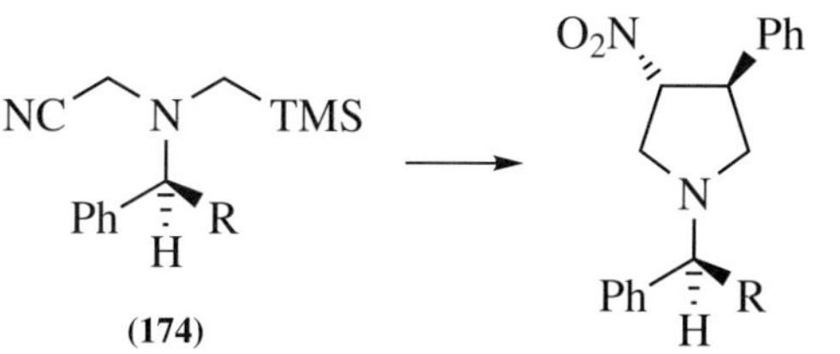

Scheme 3.49

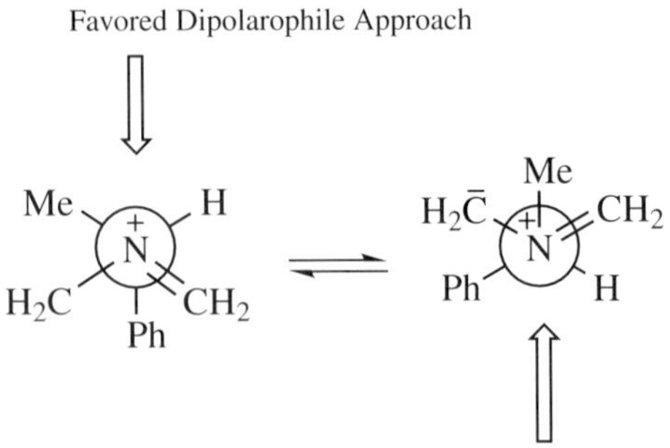

Figure 3.11

gave an increased facial selectivity of 4:1, although the use of benzaldehyde as the dipolarophile gave no diastereofacial selectivity. The reaction outcome was rationalized by invoking the Felkin–Anh model with favored approach of the dipolarophile *anti* to the phenyl group (Fig. 3.11).

Several syntheses of the hepatatoxic alkaloid (+)-retronecine have been reported although the most succinct has utilized a chiral azomethine ylide cycloaddition to construct the bicyclic skeleton. The ylide processor **175**, which was obtained in five efficient steps from commercially available *trans*-(*R*)-4-hydroxy-L-proline, underwent double desilyation in the presence of AgF (described in detail in Section 3.1.1) and *in situ* cycloaddition with methyl propiolate, to deliver a 3:1 mixture of cycloadducts in favor of the desired regioisomer. Diisobutylaluminum (DIBAL) reduction of **176** furnished enantiopure (+)-retronecine (Scheme 3.50).

Husson and co-workers (47–50) have made extensive application of their chiral azomethine ylide precursor. The synthon **177**, prepared in one step in essentially quantitative yield on a multi-gram scale, had previously been used in a wide range of asymmetric processes as a chiral glycine anion equivalent. However, low-temperature treatment with TMS triflate (yet another example of the use of

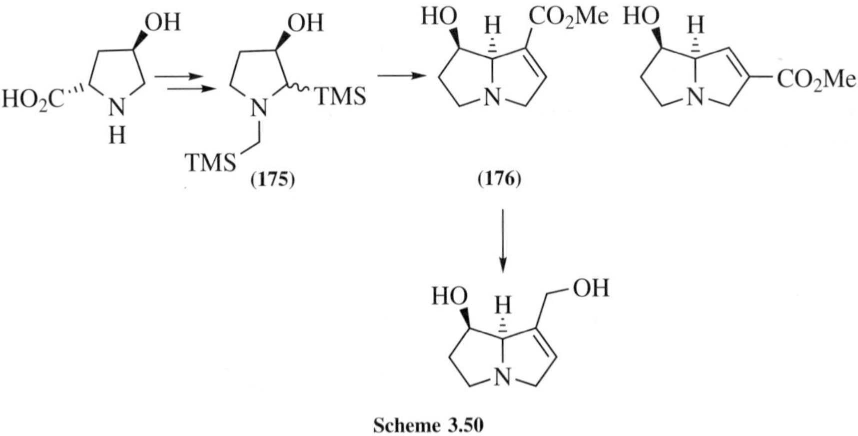

Scheme 3.50

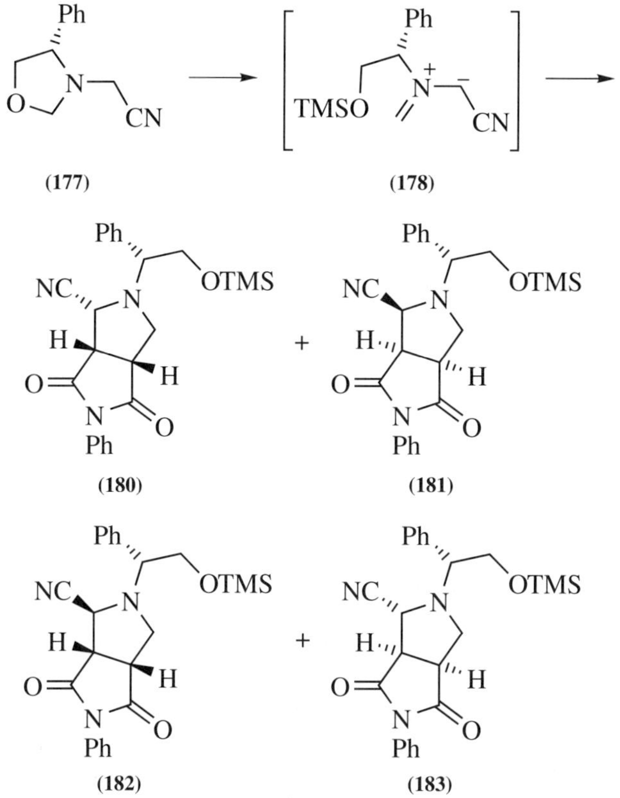

Scheme 3.51

silicon-based reagents in azomethine ylide generation) and Hünig's base, irreversibly opened up the oxazolidine **177** with concomitant alcohol protection to form the azomethine ylide **178** with the α-phenyl substituted amino alcohol section acting as the chiral auxiliary (48). Subsequent trapping of the ylide with *N*-phenyl maleimide led to the formation of four isomers (**180–183**) in a 44:32:16:8 ratio. Although stereoselectivity is poor, the overall material yield for the reaction was almost quantitative (Scheme 3.51).

The reaction protocol was further developed by alterations to the chiral controlling element of the reaction (49). Use of the precursor **183** under the standard ylide generation and cycloaddition conditions gave a greatly improved diastereomeric excess of >95%, an *endo/exo* ratio 1:15 and an isolated yield of 62%, with *N*-phenylmaleimide as the dipolarophile. The improvement in the reaction was rationalized by both *endo* and *exo* attack of the dipolarophile to the *same* diastereomerically favored face of the conformationally restricted U-shaped ylide **184** (Scheme 3.52).

However, the introduction of a second stereocontrolling element into the carboxyl moiety led to a dramatic improvement in the reaction (50). A study into

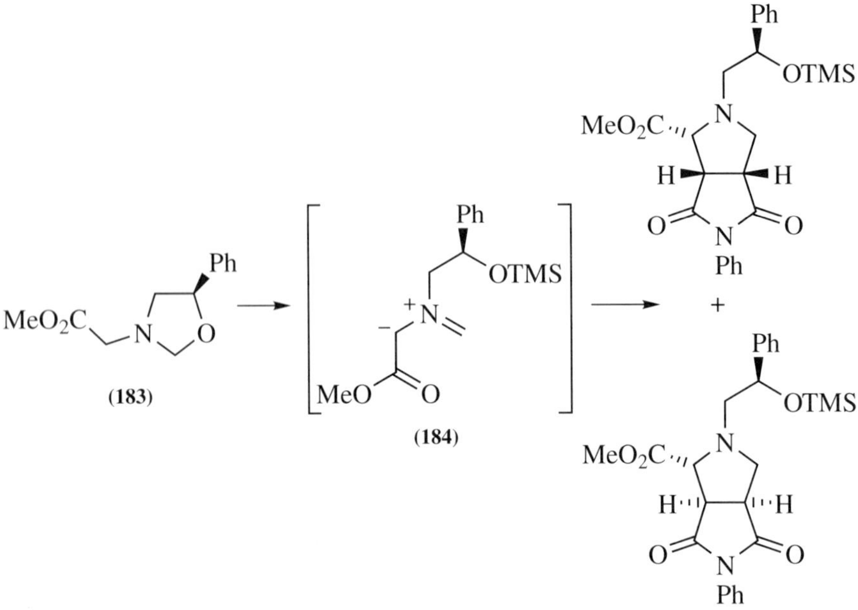

Scheme 3.52

a selection of possible candidates led to the discovery that the (−)-8-phenylmenthyl substituent led to the formation of a reaction product in 86% yield with *N*-phenylmaleimide as the dipolarophile. The adduct was formed exclusively as the *exo* isomer and in a diastereomic ratio >95:5. The mechanistic rationale involved the transition state shown (Fig. 3.12), with the ylide constrained to a U conformation, the chiral center on the nitrogen forcing *exo* attack of the dipolarophile, with the chiral ester group forcing diastereoselectivity.

In a series of reports, Gallagher and co-workers (51,52) outlined the use of azomethine ylide technology in the construction of the bicyclic core of β-lactams. Initially the synthesis of carbapenems was studied. The oxazolidine-based ylide precursor **185** was prepared by a literature precedent and submitted to thermolytic

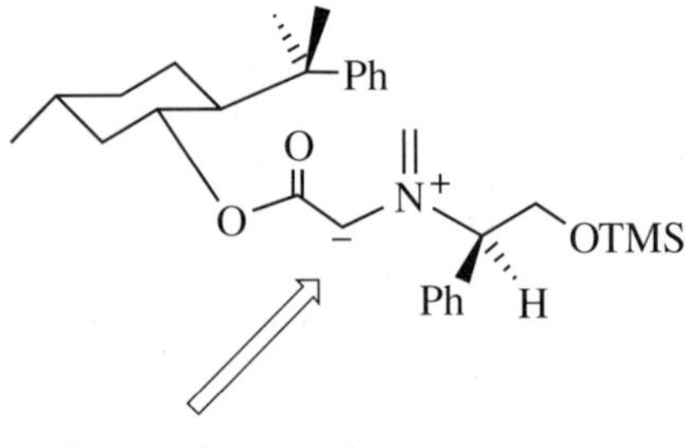

Figure 3.12

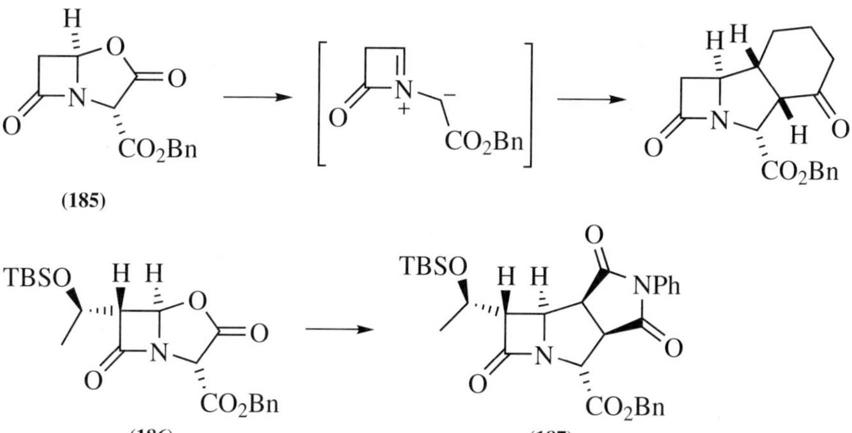

(185) (186) (187)

Scheme 3.53

decarboxylation to form the ylide, which then reacted with a wide range of dipolarophiles in reasonable yields. The reaction proceeded with good regiocontrol and high *endo* selectivity. It could also be conducted asymmetrically with the enantiopure precursor **186**, which underwent ylide formation and subsequent cycloaddition to form the *endo* adduct **187** (Scheme 3.53).

The reaction was further developed to form a wide range of β-lactam-based products (52). Treatment of the racemic ylide processor **186** with suitable sulfur-based thiocarboxylate or thiocarbonate dipolarophiles gave rise to the expected racemic penams **189** and **190** and penems as single regioisomers (Scheme 3.54). Once again, the use of the chiral dipolarophile **186** furnished the cycloaddition product **191** with complete enantiomeric integrity. Similarly, the use of aldehydes

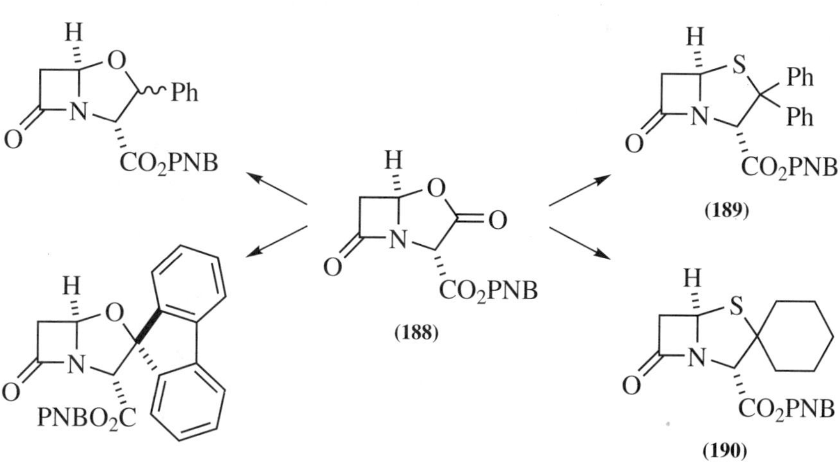

(188) (189) (190)

Scheme 3.54

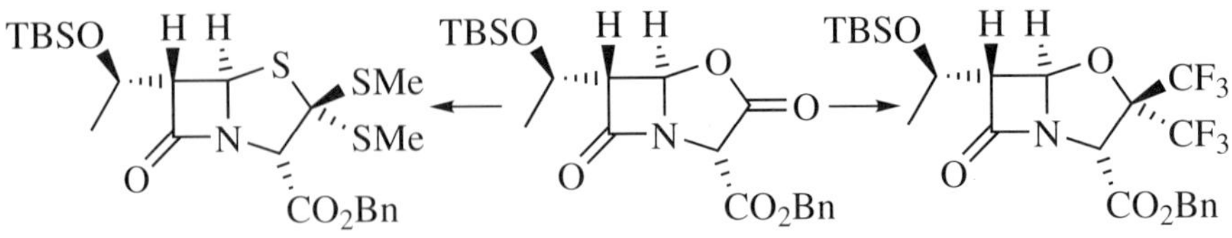

(191) (186) (192)

Scheme 3.55

as dipolarophiles generated 2-substituted ketones, and 2,2-disubstituted oxepams. Again, the use of the asymmetric ylide precursor gave the expected enantiomerically pure adduct **192** with hexafluoroacetone. Note that although the asymmetric reaction proceeds with excellent stereoinduction, the material yields of cycloadducts were reduced, presumably due to increased steric repulsion from the silyl stereocontrolling element (Scheme 3.55).

Enders et al. (53) reported the use of chiral 1,3-dioxan-5-ylamines in condensation reactions with aromatic aldehydes to form ylides *in situ*, which underwent thermal cycloaddition reactions with excellent yields. Treatment of **193** with benzaldehyde or *p*-fluorobenzaldehyde in the presence of excess dimethyl fumarate or fumaronitrile gave rise to the expected adducts in 85% yield with a >96% diastereomeric excess. For nitriles (R = CN), the *endo/exo* selectivity was higher at 70:30 than for the esters (R = CO_2Me) at 55:45 (Scheme 3.56).

Further studies revealed that replacement of the ester functionality of the ylide precursor with a phenyl group, probably acting to destabilize the ylide, gave improved facial selectivity in favour of the *exo* isomer. The use of *N*-phenylmaleimide with a range of both electron-withdrawing and -donating aromatic aldehydes gave rise to the expected products in essentially quantitative yield with an *endo/exo* ratio of 60:40 and a de >96%. The *endo* product **194** was further elaborated by treatment with *p*-toluenesulfonic acid in methanol, quantitatively delivering the tricyclic lactone **195** as a single stereoisomer. (Scheme 3.57).

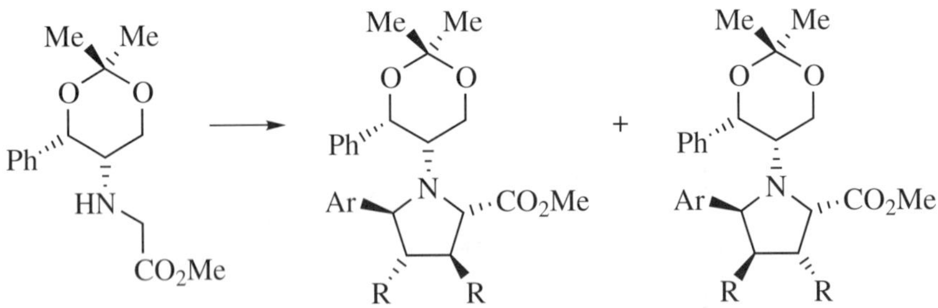

(193) exo endo

R= CN, CO_2Me

Scheme 3.56

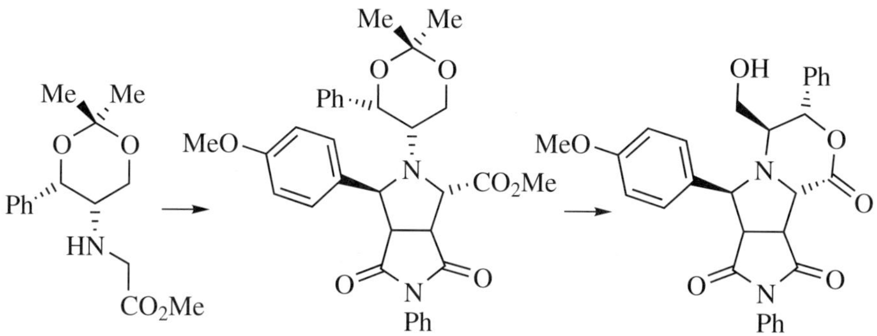

(194) (195)

Scheme 3.57

Seebach reported the use of 2-(*tert*-butyl)-3-imidazolidin-4-one (**196**) as a chiral glycine equivalent, and a report by Peyronel and Carboni outlined its use in chiral azomethine ylide generation (54). Condensation with formaldehyde (R = H), benzaldehyde (R = Ph), or hexenal (R = C_5H_{11}) generated the ylide *in situ*, which underwent subsequent cycloaddition with a range of dipolarophiles to form the cycloadducts **197** in a diastereomeric ratio of 80:20 and **198** in a 60:40 *endo/exo* ratio. The use of either maleimide or DMAD led only to the isolation of Michael adducts. The mechanistic rationale has the dipolarophile approaching *endo* to the ylide and *anti* to the *tert*-butyl group with the R group lying in the sterically least imposing (*E*)-configuration (Scheme 3.58).

As previously described, thermolysis of aziridines is one of the standard methods for the generation of azomethine ylides. A diastereomeric mixture of the aziridines **199** possessing an enantiomerically pure N-substituent underwent ylide formation at 280 °C and subsequent cycloaddition to vinylidine carbonate to form a mixture of four separable compounds (D-**200**, L-**201**, D-**202**, L-**203**) in a 3:3:1:1 ratio (55). Subsequent $LiAlH_4$ reduction and hydrogenolytic N-benzyl cleavage led to all

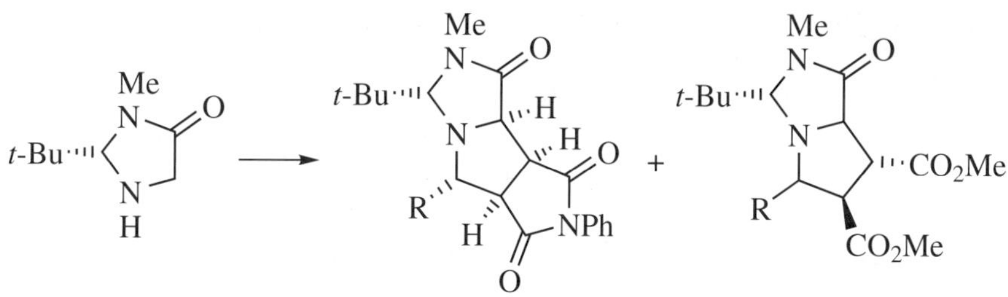

R = Ph, C_5H_{11}, H R = H

(196) (197) (198)

Scheme 3.58

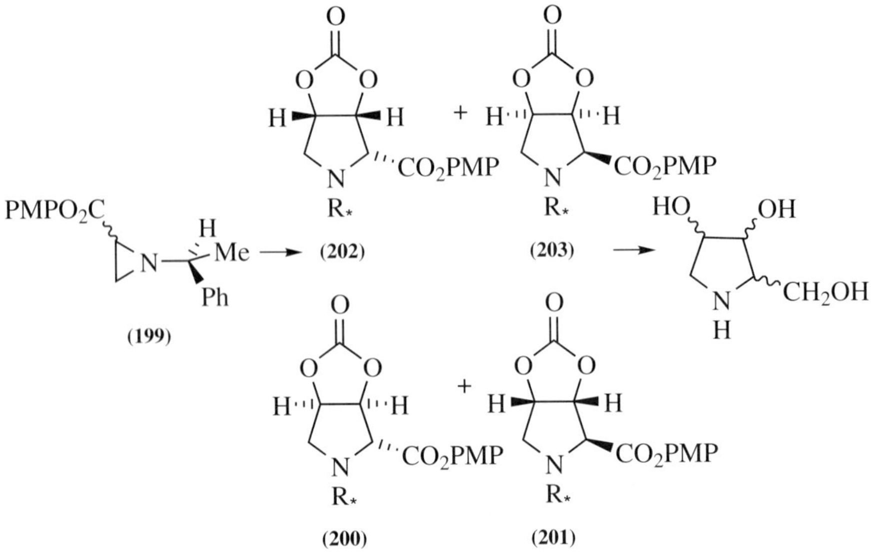

Scheme 3.59

possible isomers of 1,4-dideoxy-1,4-iminolyxitol and 1,4-dideoxy-1,4-iminoribitol, known glucosidase inhibitors (Scheme 3.59).

3.2.3. Chiral Ylides Derived from Asymmetric Templates and Auxiliaries

The chiral dipolarophiles of Garners and Dogan, which were derived from Oppolzer's sultam, have been previously discussed in Section 3.2.1 and, in an extension to these results, the sultam moiety was used as the stereodirecting unit in enantiopure azomethine ylides (56). The ylides were generated either by thermolytic opening of N-substituted aziridines or by the condensation of the amine functionality with benzaldehyde followed by tautomerism. These precursors were derived from the known (+)-*N*-propenoylbornane-2,10-sultam. Subsequent trapping of the ylides with *N*-phenylmaleimide furnished the cycloaddition products shown in Schemes 3.60 and 3.61.

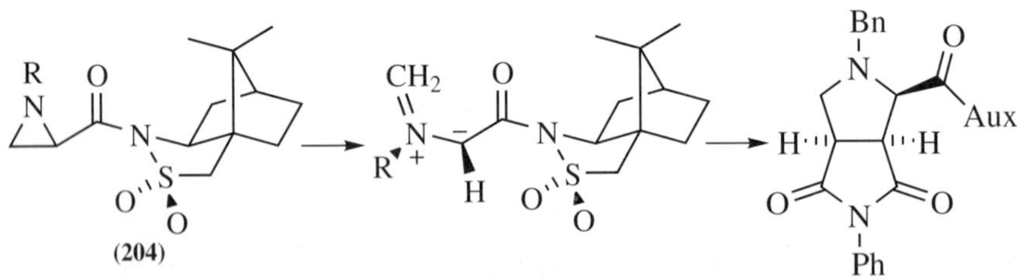

Scheme 3.60

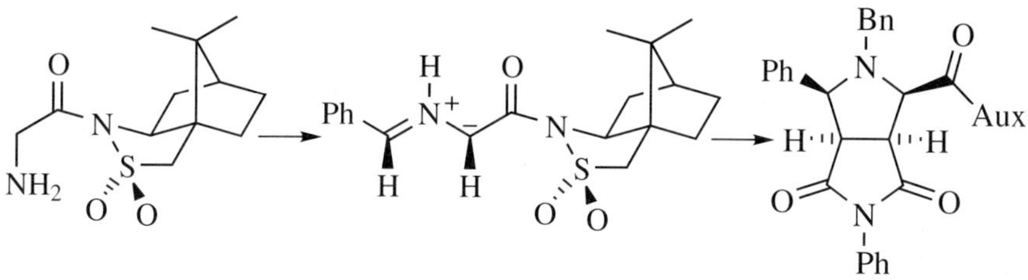

Scheme 3.61

In the case of the template **204**, the *endo/exo* selectivity was only 1.8:1, but the material yield was high with a facial selectivity of 10:1. The ylides were also trapped with methyl acrylate, generating the expected cycloadducts with reasonable regio and diastereocontrol. Although the *endo/exo* selectivity was relatively poor, the facial selectivity and material yield were both reasonable and the removal of the template is nondestructive.

3,4,5,6-Tetrahydro-1*H*-1,4-oxazin-2-one has attracted intense study recently having become the most extensively researched asymmetric template in azomethine ylide cycloadditions (57–65). This template has proved to be one of the most efficient and general chiral substrates for stoichiometric stereocontrol in 1,3-dipolar cycloadditions and has been utilized for the formation of highly functionalized nitrogen-containing heterocycles and amino acids, to name but a few uses. The most in-depth studies and extensive application of chiral templates as chiral dipoles in azomethine chemistry have been those conducted in the laboratories of Harwood and co-workers (27, 57–63, 66–68). By using the monosubstituted morpholinone template **205**, developed by Dellaria, tactics for the highly stereodirective synthesis of prolines, pyrolidines, and amino acids have been developed. As such, they represent one of the more general applications of chirally templated azomethine ylide chemistry.

Amino acids can be used as azomethine ylide precursors, although the stereogenic center is by necessity lost and require reaction with chiral dipolarophiles to circumvent the problem of absence of stereocontrol. Harwood et al. (57) demonstrated that the chirality of the original amino acid could be preserved by derivatization to give back not only the original stereocenter, but further stereoinduction.

Initial studies used the ylide generated *in situ* by condensation with formaldehyde, which was treated with various dipolarophiles in refluxing benzene through a Soxhlet extractor-containing molecular sieves (57). Trapping with *N*-phenylmaleimide gave rise to a 3.5:1 mixture of *endo/exo* products in reasonable material yield. However, most importantly, the absolute configuration at C(3) was consistent in both products, showing that stereodirection by C(5) was apparently total. The use of DMAD gave rise to one product, again, with the C(3) configuration observed earlier (Scheme 3.62).

Destructive removal of the template by catalytic hydrogenation was easily achieved, resulting in the loss of the stereogenic center at C(5), and overall chirality

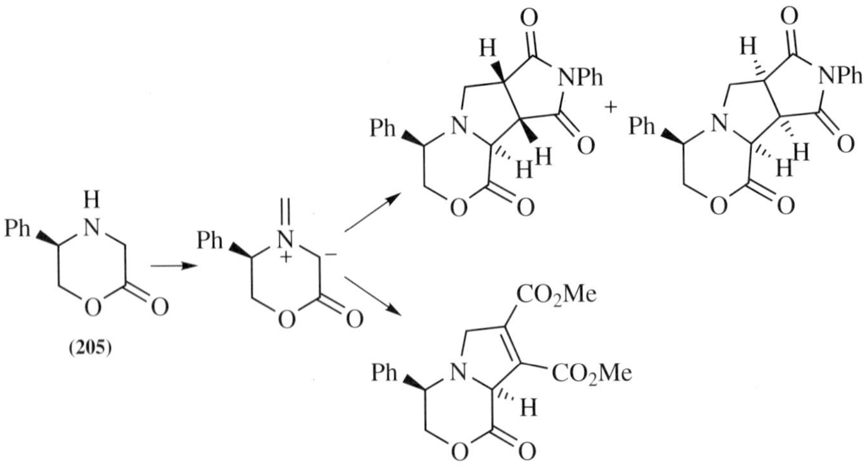

(205)

Scheme 3.62

transfer from one center to the next. Realization of a genuine chiral memory relay was achieved by use of the 3,5-disubstituted template **206** (58). Condensation of the template with paraformaldehyde to generate the ylide with subsequent cycloaddition with the dipolaraphile gave the expected cycloadducts, but with the chirality of C(3) reinstated and an *endo/exo* ratio improved to 5:1. The use of *N*-methylmaleimide gave only the *endo* product. Catalytic hydrogenation of the adducts released the free amino acids in high yield (Scheme 3.63).

The stereochemical outcome of the reaction was rationalized by invoking axial approach of the alkene lowest occupied molecular orbital (LUMO) to the least hindered lower face of the highest occupied molecular orbital (HOMO) of the dipole, in which the morpholinone ring was held in a quasi-chair conformation with the C(5) phenyl substituent in an equatorial environment (Fig. 3.13).

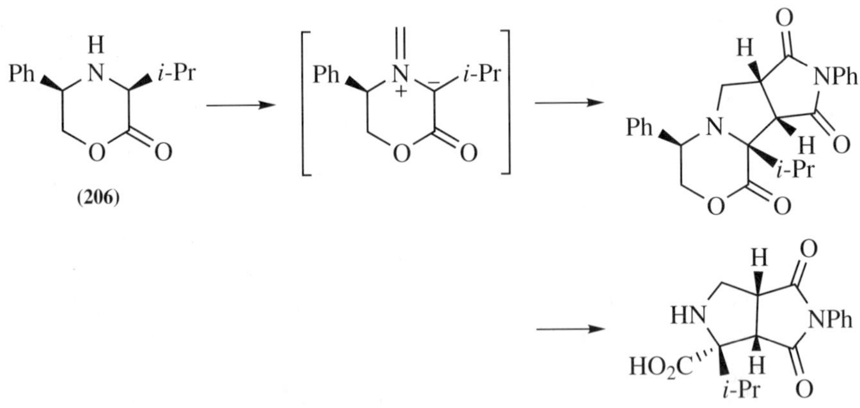

(206)

Scheme 3.63

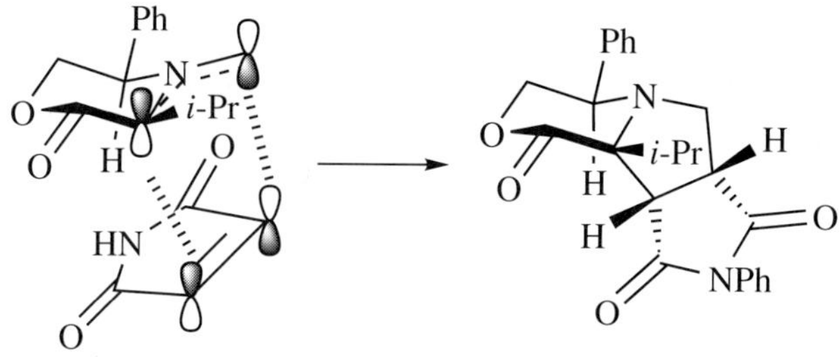

Figure 3.13

Further developments led to the formation of ylides derived from aldehydes other than formaldehyde (58), thus generating products with an additional stereocenter. Condensation of the morpholinone template with benzaldehyde, followed by trapping of the resultant ylide with *N*-methylmaleimide, gave rise to four reaction products with the major *endo* adducts obtained in a 3.5:1 diastereomic ratio and the minor exo products in a 1.5:1 ratio. The ratio of both *endo* products to both *exo* products was 2:1. Invoking the mechanistic rationale described above, the major diastereoisomer is that derived from the *anti*-configured ylide, which would be expected to dominate due to steric repulsion between the phenyl groups. Once again, the configuration at C(3) was consistent in all products (Scheme 3.64).

Addition of freshly prepared $MgBr_2.OEt_2$ to the reaction gave universal improvements in yield of the products (59). Previously poor dipolarophiles now gave good to excellent yields, especially in the case of dimethyl fumarate which,

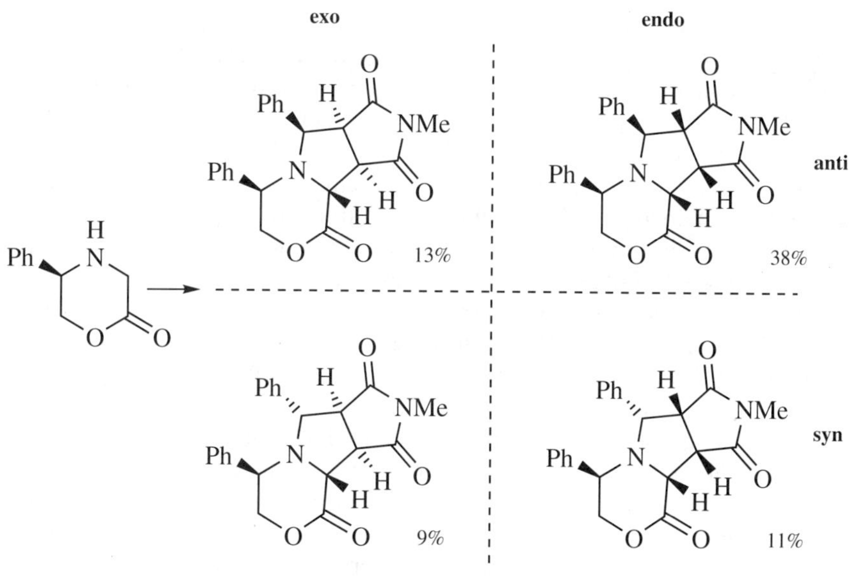

Scheme 3.64

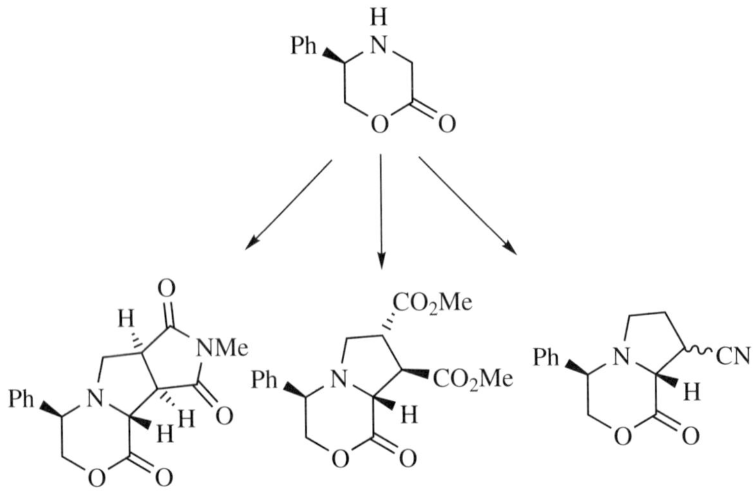

Scheme 3.65

having proved to be unreactive under thermal conditions, now furnished a single product in good yield. What was most noteworthy, however, was that the major isomers, with maleimide dipolarophiles, were now the *exo*-adducts, whereas under the previous thermal conditions the *endo*-products predominated (Scheme 3.65).

Due to the increased reactivity of the reaction in the presence of a Lewis acid, the reaction scope was extended to singly activated alkenes. Previous results had shown either no reaction or extremely poor yields. However, under the Lewis acid catalyzed conditions, acrylonitrile furnished a 1:1, *endo/exo* mixture of products. The addition of the catalyst gave unexpected regiochemistry in the reaction, which is analogous with results described in Grigg's metal catalyzed reactions. These observations in the reversal of regio- and stereocontrol of the reactions were rationalized by a reversal of the dominant, interacting frontier orbitals to a LUMO dipole–HOMO dipolarophile combination due to the ylide-catalyst complex. This complex resulted in a further withdrawal of electrons from the azomethine ylide.

The use of a doubly stabilized ylide derived from condensation of the template with ethyl glyoxylate trimer followed by condensation with suitable dipolarophiles, led to the expected mixture of *exo* and *endo* products under thermal conditions (60). While material yields could be improved by the addition of Lewis acid, the *exo/endo* ratios were reduced due an increase in the formation of *exo* products. Subsequent destructive removal of the template furnished a range of amino acids, exemplified in Scheme 3.66.

The general protocol has been further extended to include cycloadditions using aldehydes both in the condensation step to form the ylide and as the dipolarophile (61,62). Treatment of the template with excess of benzaldehyde in refluxing toluene, with azeotropic removal of water, resulted in a high yield of cycloadduct **207** containing three new stereogenic centers formed from the single stereocontrolling center at C(5) (61) (Scheme 3.67). The mechanistic rationale assumed an

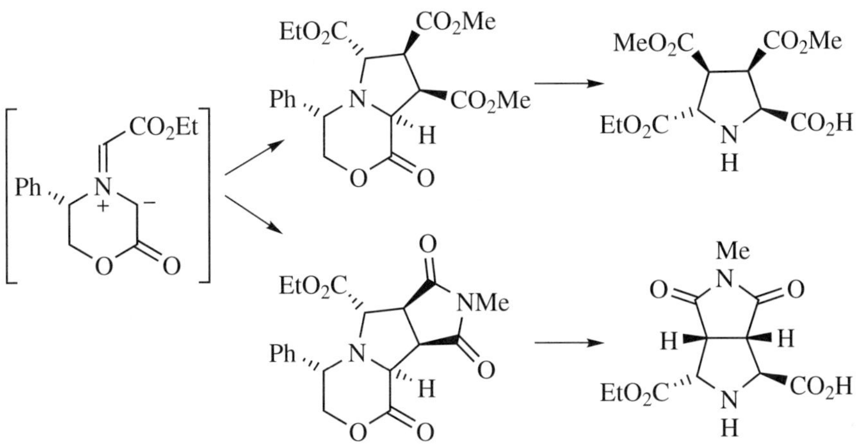

Scheme 3.66

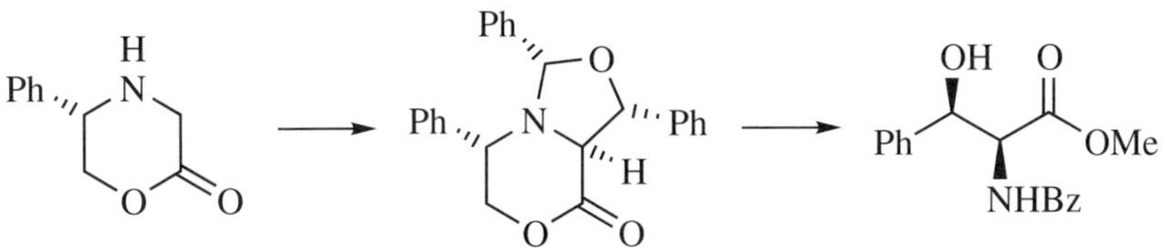

(207)

Scheme 3.67

(*E*)-configured ylide with axial approach of the aldehyde to the least sterically demanding face (Fig. 3.14). Subsequent catalytic hydrogenation of the cycloadduct furnished the *N*-Bz methyl ester of 2(*S*), 3(*R*)-(3-hydroxy) phenylalanine in excellent yield.

The reaction scope was further developed with a range of aromatic and aliphatic aldehydes to furnish a range of β-hydroxy-α-amino acids, which are important constituents of many natural products (62). The resultant bicyclic adducts were obtained in good to excellent yield and represent differentially and orthagonally protected β-hydroxy-α-amino acids. The template removal was easily conducted by

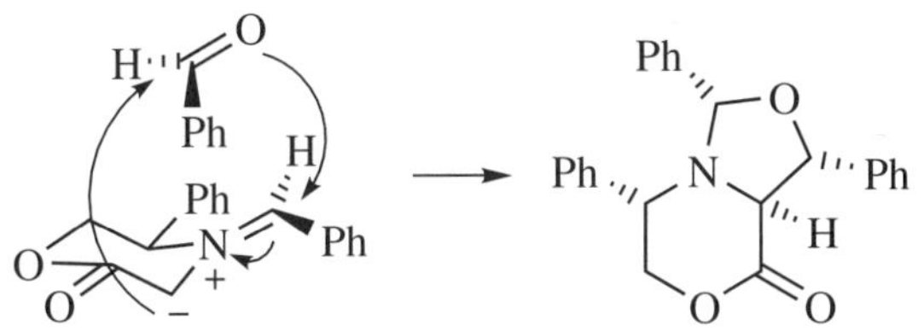

Figure 3.14

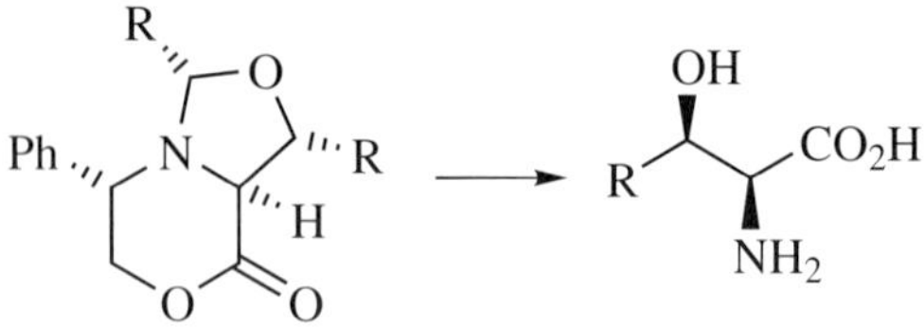

Scheme 3.68

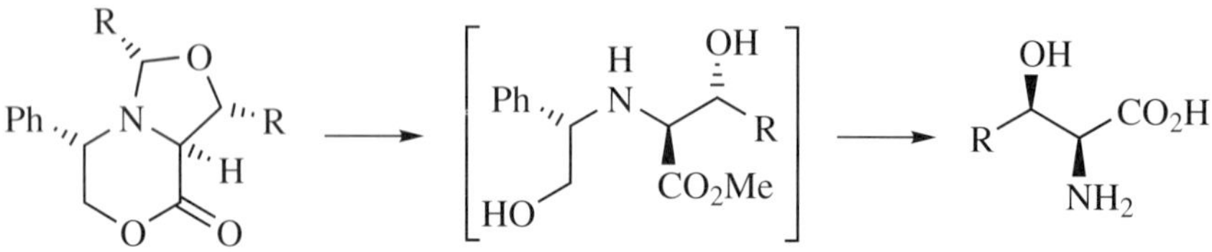

Scheme 3.69

simple hydrogenolysis in the case of adducts derived from aromatic aldehydes, while for those products evolved from aliphatic aldehydes, hydrolysis of the lactone, which was followed by hydrogenation of the resultant amino esters, delivered the desired (2*S*, 3*R*)-threo-products in quantitative overall yield and enantiomerically pure (Scheme 3.68 and 3.69).

The reaction protocol was further extended to the concise synthesis of polyoxamic acid, the unique polyhydroxyamino acid side-chain moiety of the antifungal polyoxin antibiotics (63). Treatment of the template **205** under standard thermal cycloaddition conditions with (*S*)-glyceraldehyde acetonide led to the formation of a single diastereoisomer **208** in 53% yield. Subsequent template removal released polyoxamic acid **209** in essentially quantitative yield. This represents a "*matched*" system, with the "*mismatched*" system leading to more complex reaction mixtures (Scheme 3.70).

Moloney and co-workers (64) also made use of the morpholinone template in the synthesis of analogues of the kainoid group of amino acids. Treatment of the template under the conditions reported by Harwood furnished the expected adducts. In keeping with earlier observations, the reaction proceeded with complete facial selectivity generating the new C(3) stereogenic center with complete control.

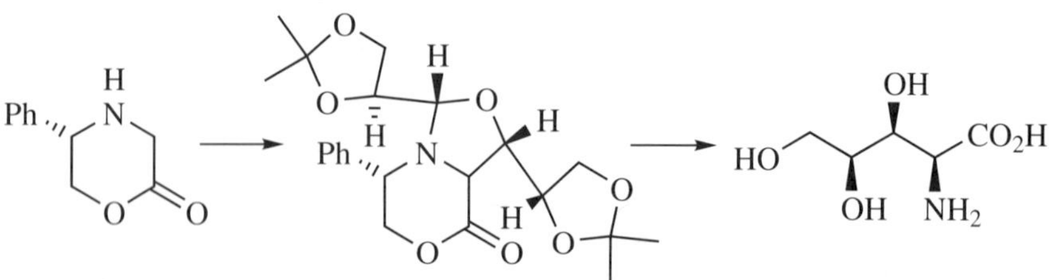

(205) **(208)** **(209)**

Scheme 3.70

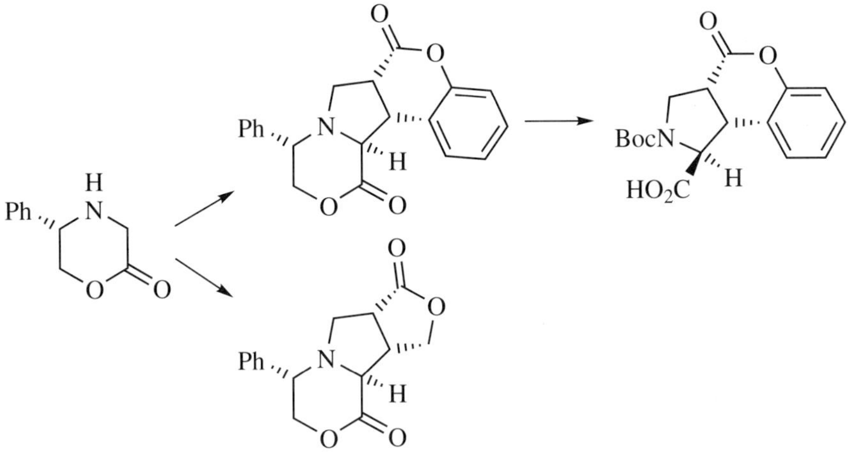

Scheme 3.71

Regiocontrol was also high with unsymmetrical dipolarophiles, as was the *endo/exo* selectivity. Simple destructive template removal followed by Boc protection gave a range of enantiomerically pure polycyclic proline derivatives, as exemplified in Scheme 3.71.

Williams et al. (65) reported studies of the diphenyl-substituted morpholinone **210** in azomethine ylide chemistry for the synthesis of highly substituted pyrrolidinecarboxylic acids. The template (as the crude product after TFA removal of the *N*-Boc group) was treated with the relevant aldehyde, and *p*-TSA in refluxing benzene to form the intermediate Schiff base, and therefore the azomethine ylide. *In situ* trapping with dimethyl maleate furnished the bicyclic adducts shown in good yield. The *endo* control over the reaction was very high, although the diastereoselectivity at C(7) was poor with the best example being that derived from benzaldehyde at 1.7:1. Removal of the template was achieved by catalytic hydrogenation in the case of simple alkyl aldehydes, or, in the case of aromatic aldehydes, by opening of the lactone under acid hydrolysis followed by oxidative cleavage of the resultant amino alcohol with $Pb(OAc)_4$ (Scheme 3.72).

One of the most novel uses of azomethine ylides was reported in the synthesis of fulleroproline compounds that have been shown to exhibit interaction with the

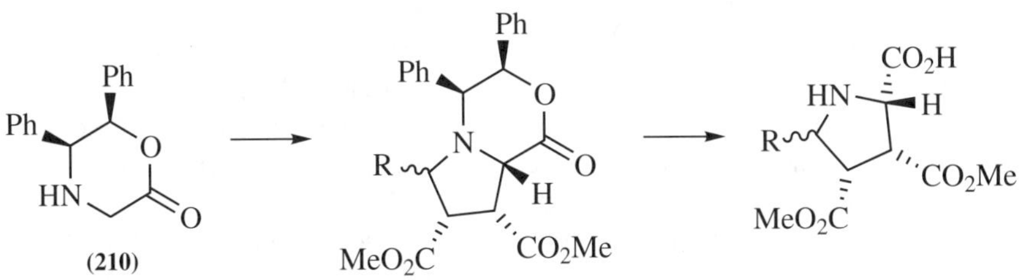

Scheme 3.72

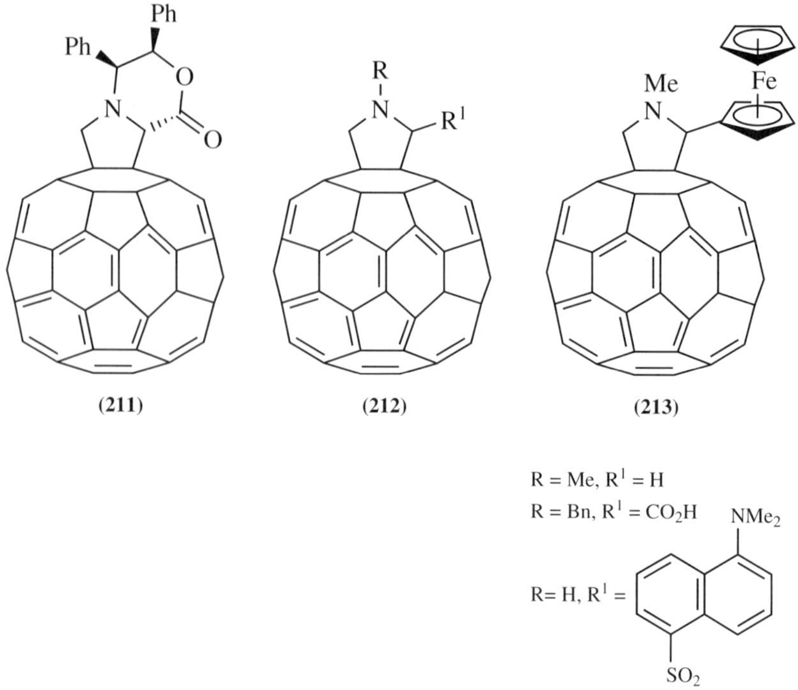

Figure 3.15

active sites of human immunodeficiency virus (HIV)-1 protease. In efforts to explore this chemistry, Prato and co-workers (69) synthesized a series of compounds (**211** and **212**) derived from dipolar cycloaddition of C_{60} with azomethine ylides derived from either aziridines or by the decarboxylation route. Attachment of a photoactive moeity could also be achieved (**213**). Similarly, the reaction could be achieved asymmetrically by reaction with the ylide generated from William's template and chloromethyl octyl ether (Fig. 3.15).

During studies on the use of amidines as azomethine ylide sources, Jones et al. (67–69) reported in a series of papers the application of their general strategy to an asymmetric process. Quaternization of the dihydroimidazole **214** with an α-halo ester followed by DBU-induced ylide formation and subsequent cyclization furnished a range of nitrogen heterocycles in a one pot generation and cyclization protocol (70) (Scheme 3.73).

Application of this one/pot protocol to the preparation of enantiopure amidine (**215**) derived from phenylglycine led to the formation of the desired cycloadducts (71). The reaction proceeded with good stereocontrol with an *endo/exo* ratio in the order of 10:1. In the case of the methyl ester derived product **216** (R = Me), a single isomer was isolated. The minor *exo* products **217** proved to have an unexpected stereochemistry at C(7), which was consistent with the energetically disfavored *anti* attack of the dipolarophile, and it was therefore assumed that the product had

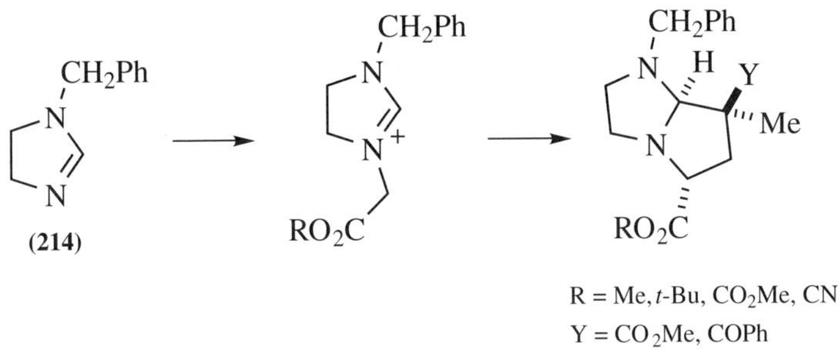

Scheme 3.73

suffered epimerization to the more thermodynamically favored products shown after the cycloaddition process had occurred (Scheme 3.74).

Template removal to release the free amino acids was conducted for the *tert*-butyl ester derivative (72). Aminal reduction with sodium cyanoborohydride in acidified ethanol resulted in cleavage of the cyclic N-benzyl bond. However, the reaction has only been shown to be successful for the *tert*-butyl analogue and gave considerable epimerization at C(7) for monosubstituted pyrroloimidazoles, although the reaction was optimized to give a 3:1 ratio of epimers in favor of the 2,4-trans isomer. For other derivatives, reductive cleavage gave only lactamization. The final section of the template was removed by catalytic hydrogenolysis, followed by deesterification to release the free proline derivatives (Scheme 3.75).

Chiral bicyclic lactams have been successfully utilized by Meyers as chiral dipolarophiles in highly diastereoselective azomethine ylide cycloadditions (73). Treatment of the ylide precursor **218** with the unsaturated, non-racemic dipolarophile **219** in the presence of a catalytic amount of TFA led to the formation of tricyclic adducts **220** and **221** in excellent yields (85–100%). The diastereofacial preference for the reaction was dependent on the nature of R^1 with a methyl group

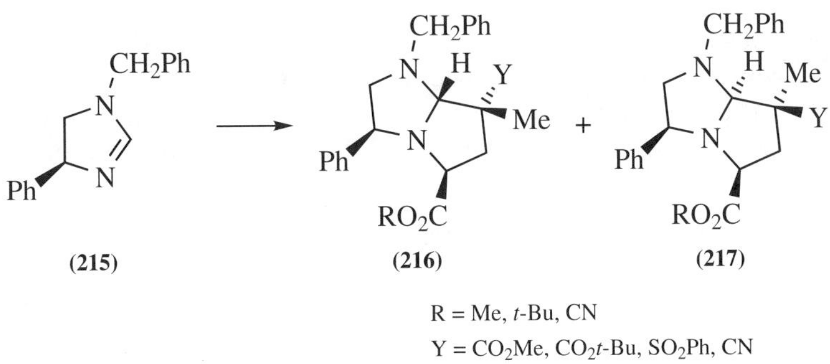

Scheme 3.74

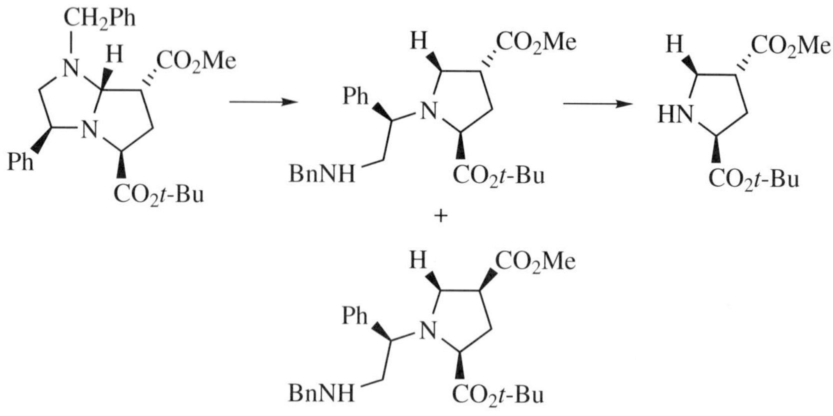

Scheme 3.75

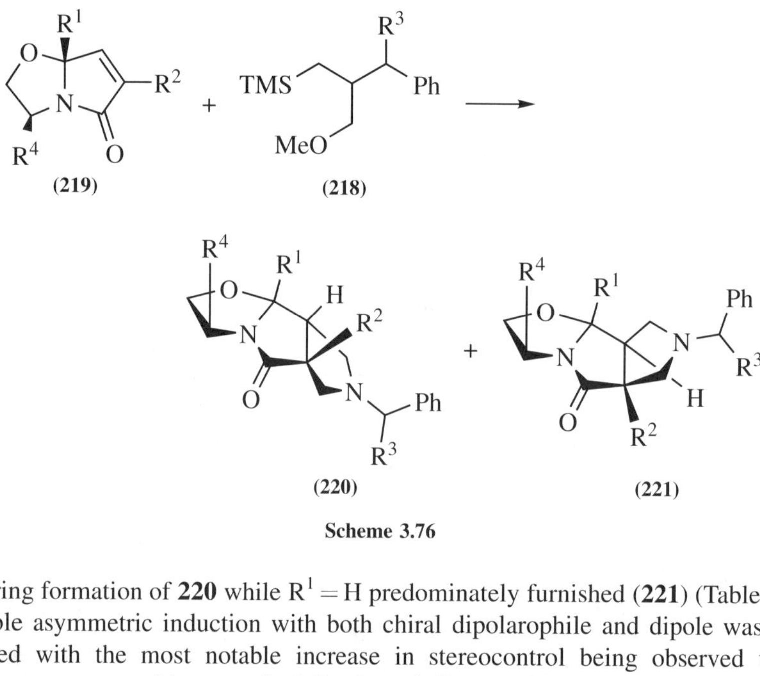

Scheme 3.76

favoring formation of **220** while $R^1 = H$ predominately furnished (**221**) (Table 3.2). Double asymmetric induction with both chiral dipolarophile and dipole was also studied with the most notable increase in stereocontrol being observed when $R^2 =$ ester groups with a *matched* dipole and dipolarophile pair (Scheme 3.76).

TABLE 3.2. DIASTEREOMERIC RATIO OF **220:221**

R^1	R^2	R^4	$R^3 = H$	$R^3 = Me$
Me	CO_2t-Bu	*i*-Pr	72:28	92:8
Me	H	*i*-Pr	91:9	94:6
Me	H	Ph	94:6	
H	H	Ph	17:83	

3.3. INTRAMOLECULAR CYCLOADDITIONS

In this section, those reactions in which the ylide is attached by a tether to the dipolaraphile resulting in an intramolecular cycloaddition will be discussed. To date, such a strategy has proved to be one of the less investigated aspects of azomethine ylide chemistry. However, intramolecular azomethine ylide technology, when combined with the excellent stereocontrol offered by cycloaddition reactions, allows for the rapid construction of complex polycyclic systems from relatively simple precursors. Consequently, it represents a highly attractive synthetic protocol that makes it a candidate for further investigation in the coming years.

3.3.1. Aziridine Precursors

The thermolytic preparation by De Shong et al. (74) of azomethine ylides from aziridines and their intermolecular reactions are the first examples of singly stabilized ylides of this type. However, the protocol has been further extended to include intramolecular processes. Aziridines tethered to both activated and unactivated alkenes were subjected to flash vacuum thermolysis generating cycloadducts in moderate-to-excellent yields. While previously singly activated alkenes had furnished low material yields *via* an intermolecular process, the intramolecular analogue represents a major improvement. Typically, treatment of **222** under standard conditions led to the formation of **223** in 80% yield as a single *cis* isomer. Similarly, the *cis* precursor furnished adduct **224** in 52% yield, although as a 1:1 diastereomeric mixture (Scheme 3.77).

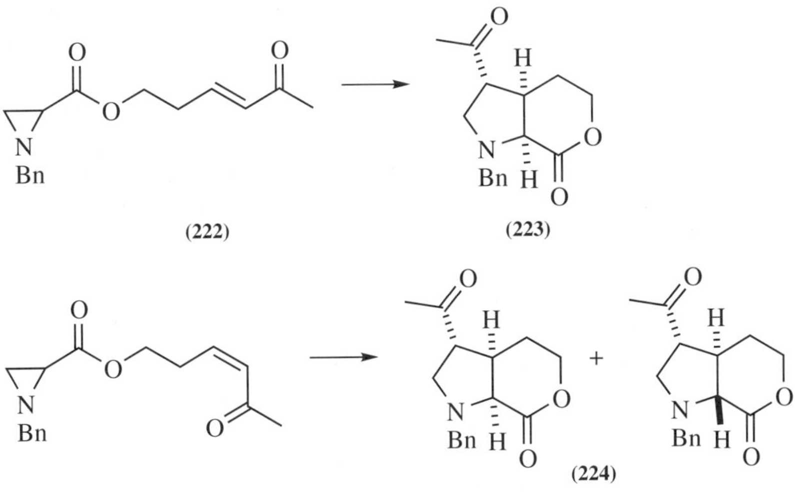

Scheme 3.77

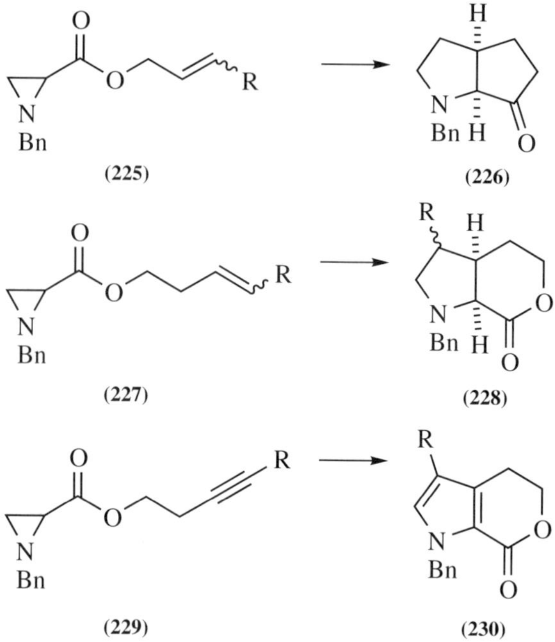

Scheme 3.78

Alkenes without stabilizing functionalities also underwent successful dipolar cycloaddition with allylic ethers **225** and **227** (R = H) furnishing adducts **226** and **228** (R = H) as single stereoisomers, in 67 and 69% yield, respectively. Unfortunately, further substitution of the alkene (R = Et) led to a dramatic reduction in yield to 16%. Alkyne precursor **229** delivered the pyrrole **230** in 63% for R = H in 34% yield for R = Et, with none of the expected dihydropyrrole being isolated due to rapid oxidative aromatization of the initial adduct. Although material recovery is often moderate, this was reported to be the first example of singly stabilized aziridine derived ylides reacting with unactivated dipolarophiles (Scheme 3.78).

At about the same time, Wenkert and c-workers (75) reported a similar study into the intramolecular 1,3-dipolar cycloaddition of 2-alkenoyl-aziridine derived azomethine ylides. Thermolysis of **231** at moderate temperature (85 °C) produced **232** as a single isomer in 58% yield. Similarly, **233** furnished **234** in 67% yield. In each case, the same stereoisomers were produced regardless of the initial stereochemistry of the initial aziridine precursors. However, the reaction proved to be sensitive to both the substituents of the aziridine and tether length, as aziridines **235** and **236** furnished no cycloadducts, even at 200 °C (Scheme 3.79).

During the synthetic efforts of Heathcock and co-workers toward the complex marine alkaloid sarain-A (Scheme 3.80), he outlined an elegant intramolecular, azomethine ylide cycloaddition, as one of the key stages in the construction of the central core (76). Of the generation methods known for azomethine ylides, thermolysis of aziridines was selected in this instance. The azomethine ylide

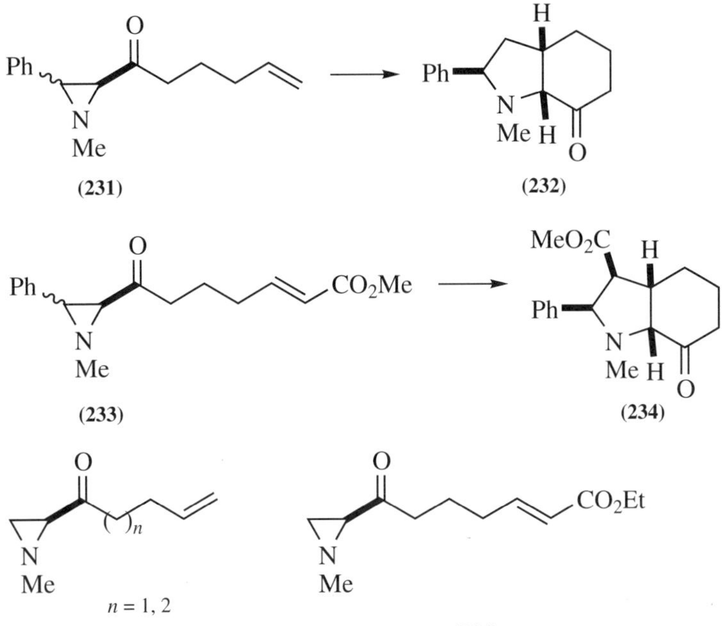

Scheme 3.79

precursor **237** (Scheme 3.81) was synthesized by standard methods and subjected to the flash vacuum thermolysis ylide generation conditions of DeShong at 0.04 Torr and 350 °C, with subsequent intramolecular cycloaddition furnishing the desired product in an exceptional 94% yield, although in a 1:1 diastereomeric ratio. The protocol was developed further to form the bicyclic products **238–240** from their corresponding aziradine precursors in good yields by the standard vacuum thermolysis protocol. In the case of **238** and **240** a single stereoisomer was isolated (Fig 3.16).

However, what is most remarkable about this work is that during attempts to form the diazadicyclo[5.3.0]decane **238**, trace quantities of **241** were isolated, rationalized by cycloaddition with the benzene ring. Although cycloadditions with reactive aromatics such as furans are known, this represents the first example of a phenyl ring undergoing a [2 + 3] cycloaddition. Improvements to this novel cycloaddition were attempted by removal of the competing alkene functionality and treatment of the dibenzylamine **242** under the standard FVT conditions, furnishing the cycloadduct **243** in 67% yield (Scheme 3.82).

Weinreb and co-workers (77) reported a similar protocol in the total synthesis of sarain A. The five-membered ring of the tricyclic central array was constructed *via* thermolysis of aziridine **244** to furnish the intramolecular [3 + 2]-cycloaddition product **245** in 73% as a single regio- and stereoisomer. Further chemical elaboration to **245** followed by $FeCl_3$ induced cyclization delivered the advanced synthetic intermediate **247** (Scheme 3.83).

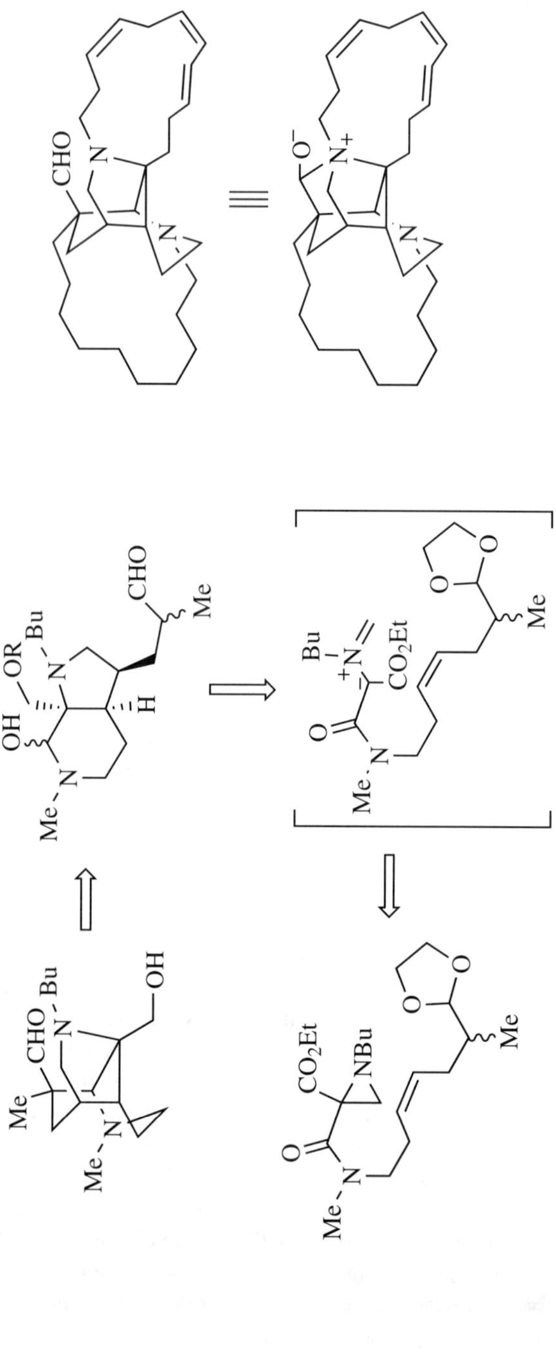

Scheme 3.80

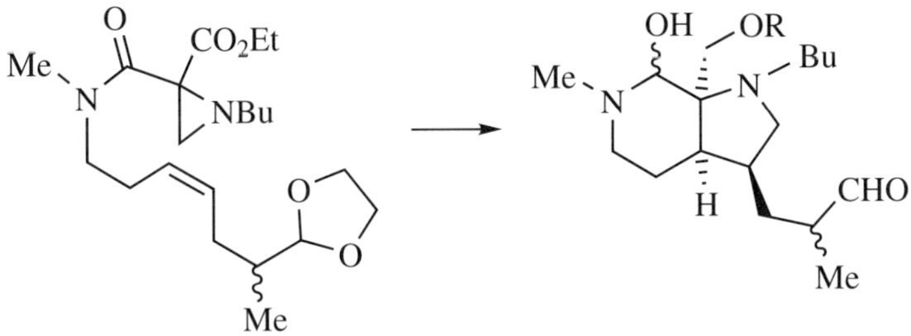

(237)

Scheme 3.81

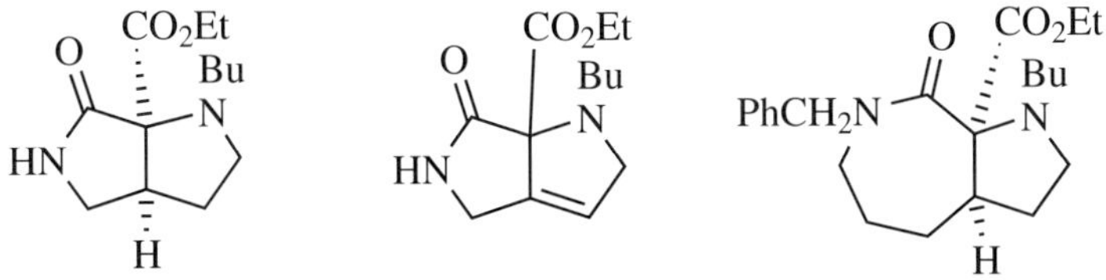

(238) **(239)** **(240)**

Figure 3.16

In a similar approach, Garner et al. (78) made use of silicon-based tethers between ylide and dipolarophile during their program of research into the application of azomethine ylides in the total asymmetric synthesis of complex natural products. In order to form advanced synthetic intermediates of type **248** during the asymmetric synthesis of bioxalomycins (**249**), an intramolecular azomethine ylide reaction from aziridine ylide precursors was deemed the best strategy (Scheme 3.84). Under photochemically induced ylide formation and subsequent cycloaddition, the desired *endo–re* products **250** were formed exclusively. However, due to unacceptably low synthetic yields, this approach was abandoned in favor of a longer tether (Scheme 3.85).

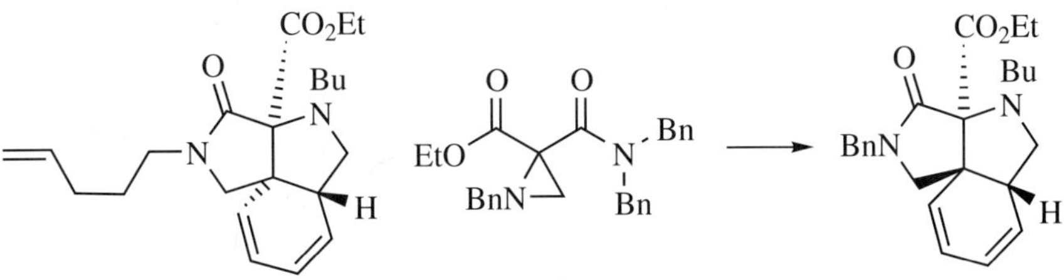

(241) **(242)** **(243)**

Scheme 3.82

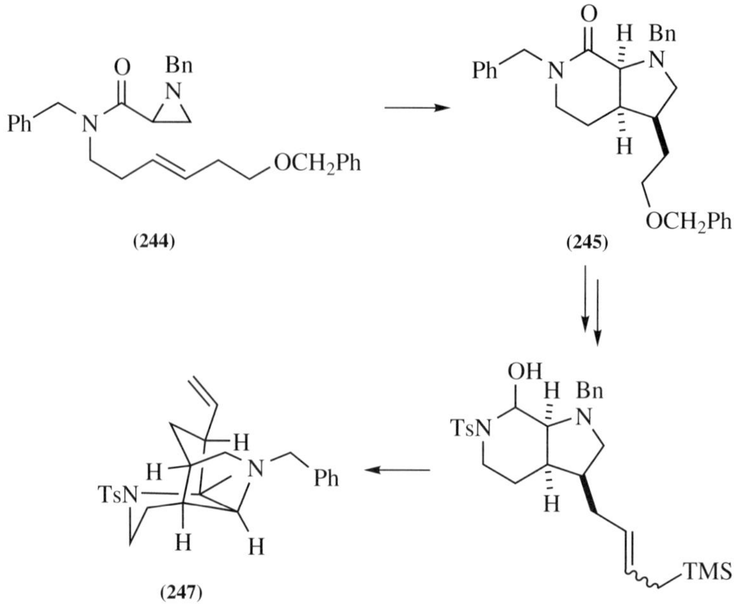

Scheme 3.83

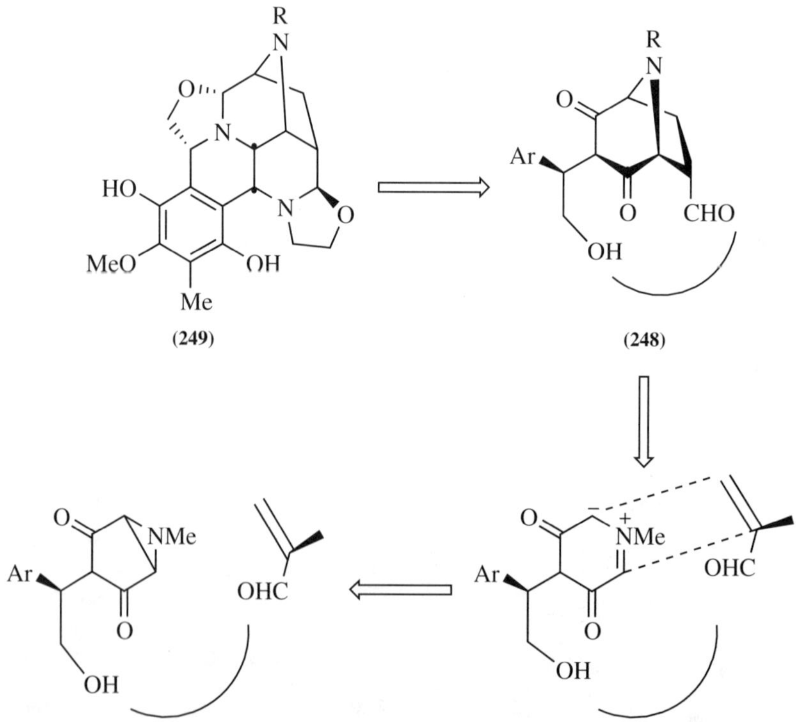

Scheme 3.84

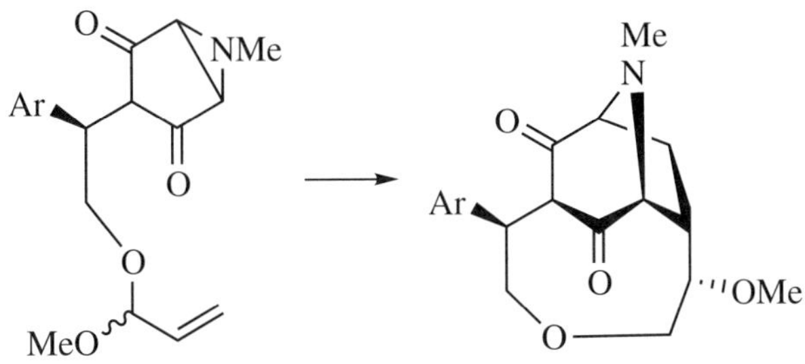

(250)

Scheme 3.85

Silicon-based tethers of the type developed by Shea for Diels–Alder reactions were selected since they appeared to fulfill all the required criteria. Synthesis of the siloxy tethered aziridines was easily achieved and these underwent photochemical formation to the azomethine ylide and *in situ* intramolecular cycloaddition. However, the unexpected and undesired *endo–si* product **251** was generated as the major product in 45% yield. Replacement of the silyl tether to form 12-, 10-, 9- membered rings gave exactly the same unexpected products as obtained with the 13- membered ring. Remarkably, the use of silyl tethered aziridine **252** gave the required cycloadduct **253** in excellent de and yield, and the adduct underwent further elaboration to the desired intermediate **254**. Garner was unable to reconcile these contrasting results, although they do offer extremely interesting potential for stereo- and regiocontrol over such reactions (Scheme 3.86).

Similarly, Takano et al. (79) reported the intramolecular, asymmetric cycloaddition of ylides derived from thermal decomposition of aziridines **255** with the stereoselectivity rationalized by the formation of a postulated nine-membered transition state **256** in which the benzyloxymethyl group was forced into an equatorial disposition. The resultant pyrrolidine **257** contained three new stereogenic centers, each imposed with high control (Scheme 3.87).

In an extension to this work, this group applied the protocol to the synthesis of (−)-mesembrine (**258**). The key ylide formation and cycloaddition sequence was performed in xylenes at 250 °C in a sealed tube, furnishing the desired pyrrolidine lactone **259** as a single isomer in excellent yield, and further chemical elaboration yielded the target compound **258**. Again it can be postulated that the stereochemical outcome is due to the bulky benzyloxymethyl group lying in an equatorial environment in the transition state (Scheme 3.88).

Acromelic acid A (**260**), isolated from *Clitocybe acromelaga*, displays the most potent polarizing effect of all known kainoids. By application of the same intramolecular aziridine thermolysis route described above, Takano et al. (80) outlined a concise entry into such compounds. Thermolysis of the chiral aziridine **261**, derived from (*S*)-*O*-benzylglycidol in 1,2-dichlorobenzene at 200 °C, furnished the single reaction product **262** in 73% yield. The complete control over the

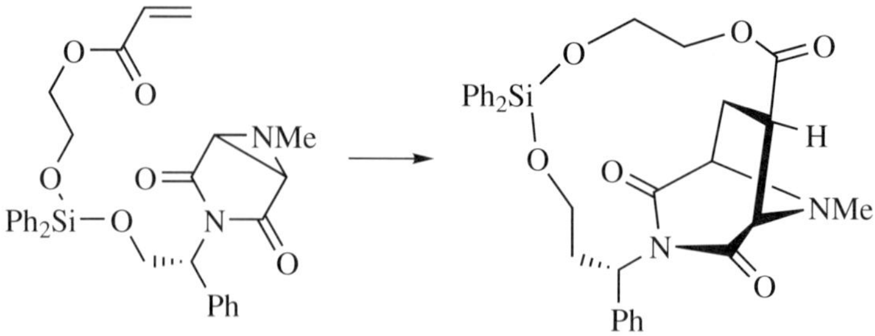

(251)

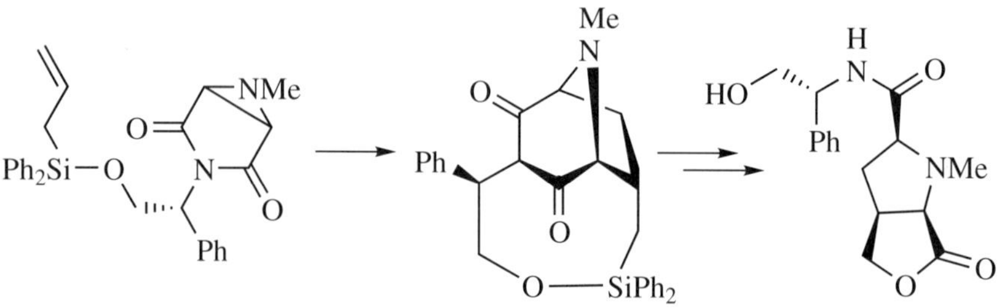

(252) (253) (254)

Scheme 3.86

diastereofacial selectivity, delivering all the substituents in a *syn* orientation, was rationalized by invoking the nine-membered transition state detailed in Scheme 3.89. Standard synthetic manipulation furnished the desired enantiopure product **260**.

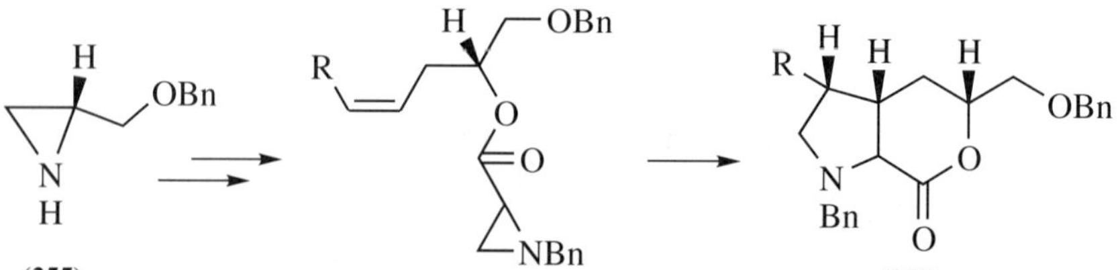

(255) (257)

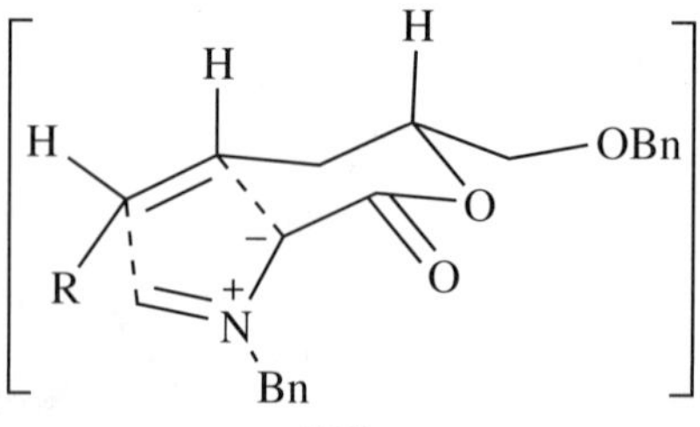

(256)

Scheme 3.87

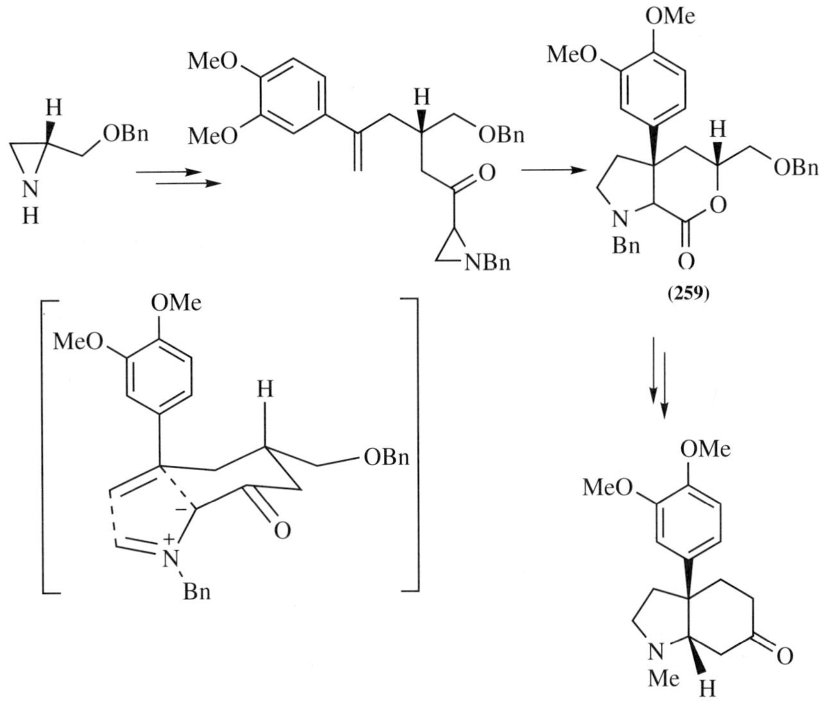

(259) (258)

Scheme 3.88

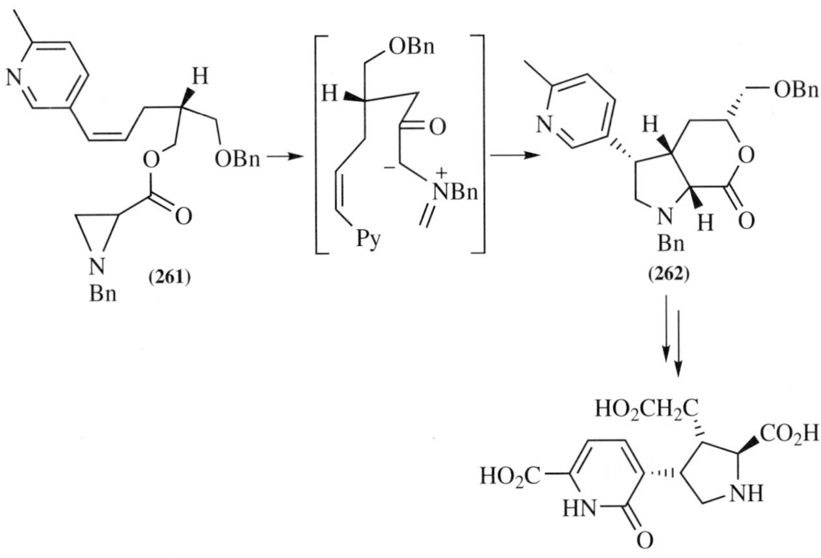

(261) (262) (260)

Scheme 3.89

3.3.2. Decarboxylative and Amino Ester Condensation Protocols

In an extensive study into the application of the decarboxylative approach to azomethine ylides, Grigg reported the construction of numerous, complex polycyclic systems *via* an intramolecular protocol. Thiazolidine-4-carboxylic acid (**263**) was shown to react with **264** in refluxing toluene to furnish a 2:1 mixture of **265** and **266** in 63% yield (81). The reaction is assumed to occur *via* condensation of the aldehyde and amino acid to generate the imine **267**, followed by cyclization to **268**. Subsequent thermal decarboxylation of the ester generates either a *syn* dipole leading to **265** from an *exo* transition state, or an *anti* dipole and *endo* transition state generating adduct **266** (Scheme 3.90).

Likewise, a wide range of complex polycyclic systems was constructed from suitable precursors. Compounds **269** and **270** were synthesized in 79% yield in a 1.6:1 ratio from tetrahydroisoquinoline-1-carboxylic acid, while **271** was obtained as a single stereoisomer in 87% yield from tetrahydro-β-carboline-1-carboxylic acid (Scheme 3.91).

Similar products could be generated *via* azomethine ylides derived by a formal 1,2-H shift from the precusor imine, rather than by the decarboxylative approach outlined above (82,83). For example (83), condensation of aldehyde **272** with the requisite amino ester **273** led to the intermediate ylide, which delivered adducts **274** and **275** in a 1.2:1 ratio in 75% yield. Grigg has once again applied this protocol to the synthesis of a wide range of complex molecular frameworks (**276**

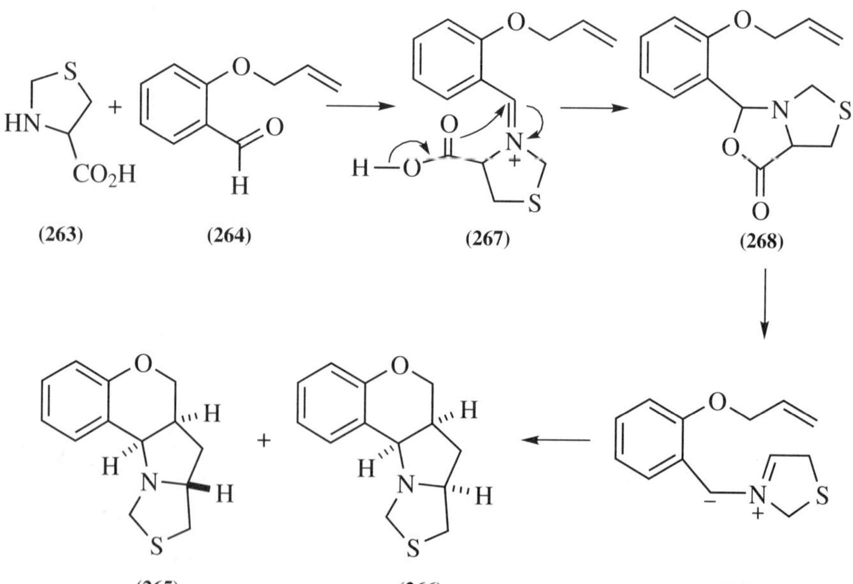

(263) (264) (267) (268) (265) (266) (264)

Scheme 3.90

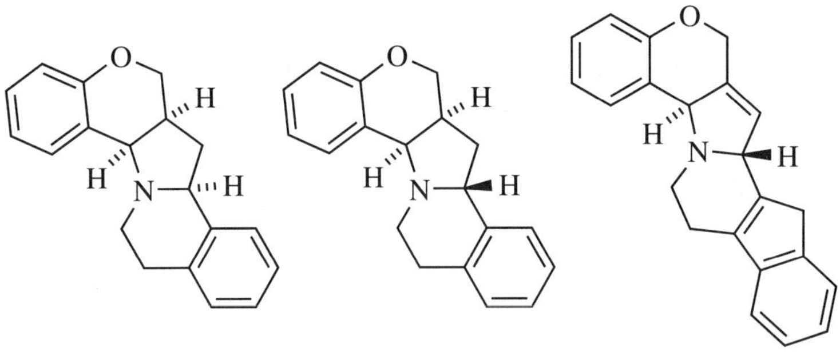

(270) (269) (271)

Scheme 3.91

and **277**) all derived from the condensation of the requisite aldehyde and amino ester (Scheme 3.92).

Grigg et al. (84) also used protoanemonin-based compounds as C(5) synthetic precursors in intramolecular cycloadditions. The 4-substituted protoanemonin **278** was synthesized in seven steps and was primed for intramolecular reactions.

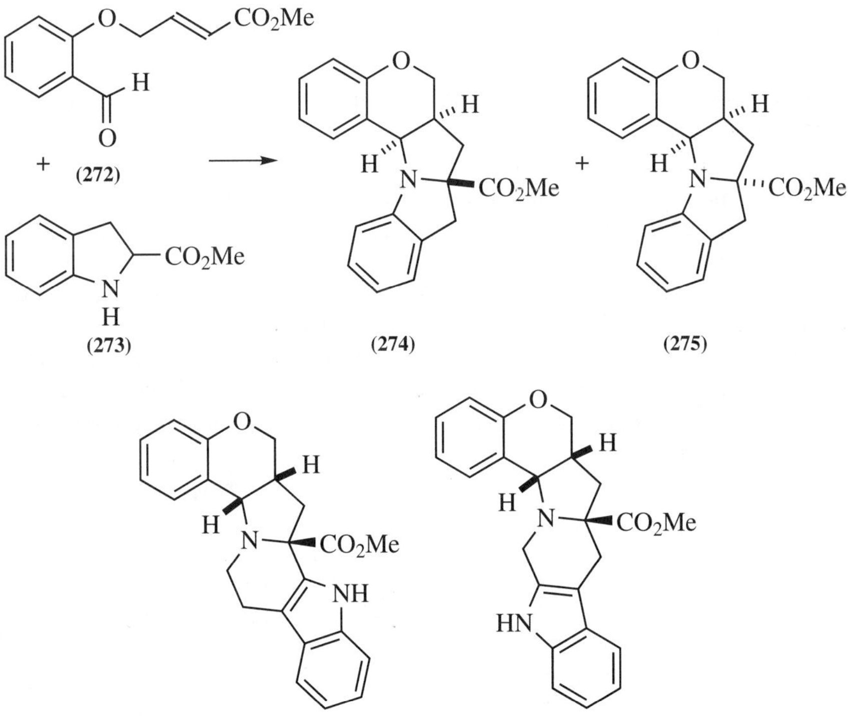

(276) (277)

Scheme 3.92

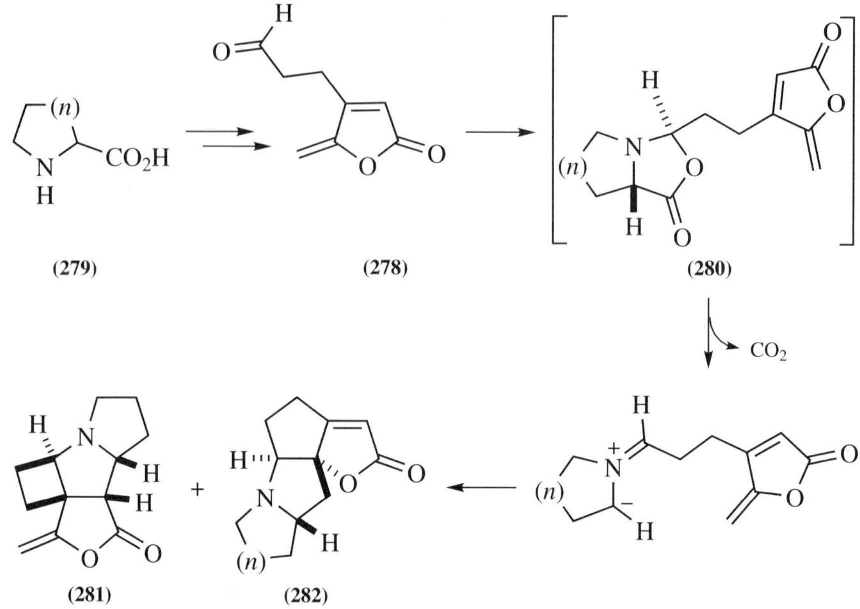

Scheme 3.93

Treatment with amines of the type **279** generated the intermediate oxazolidinone **280**, which underwent thermal decarboxylative formation of the azomethine ylide. Subsequent *in situ* intramolecular cycloaddition formed the products **281** and **282** in 63% yield and in a 1:1.2 ratio for $n = 1$. Replacing toluene for acetonitrile, for $n = 2$, gave comparable yields and an improved ratio of 1:2.1 in favor of **281** (Scheme 3.93).

The stereochemical outcome of the reaction can be reconciled by assuming that the generated ylide lies in such a way as to form an *anti* dipole of the type seen in Figure 3.17, which subsequently undergoes cycloaddition to either the α,β- **283**, or γ,δ- **284** double bond *via endo* transition states. A *syn*-dipole would suffer high steric restriction between the methylene α to the nitrogen and the protons of the iminium bond.

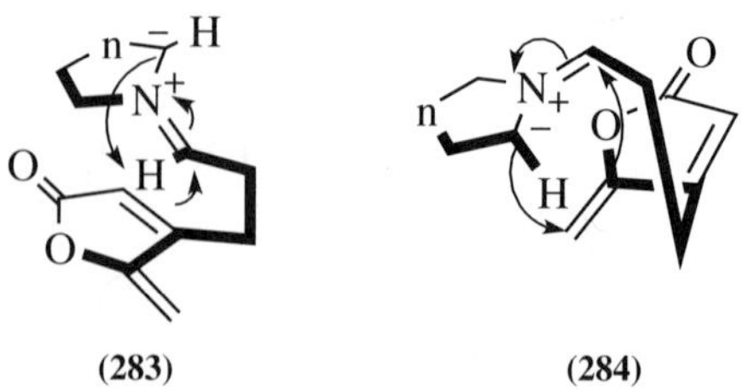

Figure 3.17

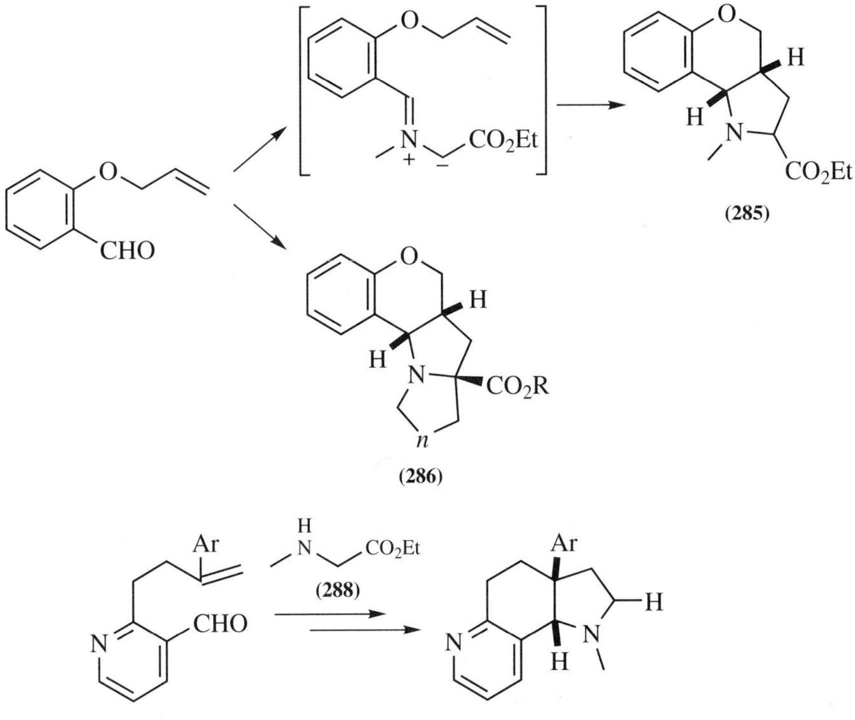

Scheme 3.94

Confalone et al. (85) also made use of an intramolecular cycloaddition step in the construction of a range of tri- and tetracyclic products. Phenyl allyl ethers, of the type shown in Scheme 3.94, underwent dehydrative condensation with the requisite amine to furnish the intermediate ylides, which suffered cycloaddition resulting in **285** and **286** in essentially quantitative yield. The ratio of *cis/trans* fused products was in the range of 10:1. Such a process has been developed to construct the alkaloid (+/−) sceletium A_4 by reaction of the intermediate **287** with amine **288** *via* the cycloaddition protocol already developed, followed by further chemical manipulation, in an efficient five step synthesis (Scheme 3.94).

The same group has also reported the use of dithiane functionalized olefinic aldehydes in an intramolecular process. Treatment of **289** with the secondary amine **290** led to the formation of the intermediate ylide **291**, which delivered a range of tricyclic products (**292**) in high material yield as essentially a single stereoisomer after *in situ* cycloaddition (Scheme 3.95).

A novel application of phosphate-stabilized ylides in intramolecular reactions has been reported by Martin and Cheavens (87). Thus, condensation of aldehydes of the type **293** with the amine **294** furnished the intermediate azomethine ylide, which then underwent a cycloaddition process, leading to pyrrolidines **295**, where

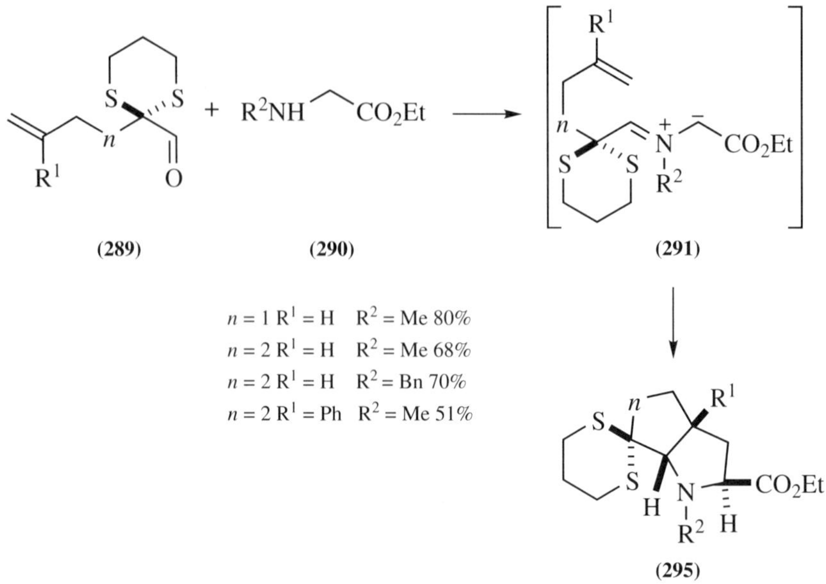

Scheme 3.95

X = O, CH_2, or NSO_2Me. In each case, the only products detected were the *cis*-fused cycloadducts, whereas replacement of the phosphate group with a cyano functionality gave rise to traces of the *trans* product. Subsequent oxidative removal of the epimeric phosphate group provided the lactams shown (Scheme 3.96).

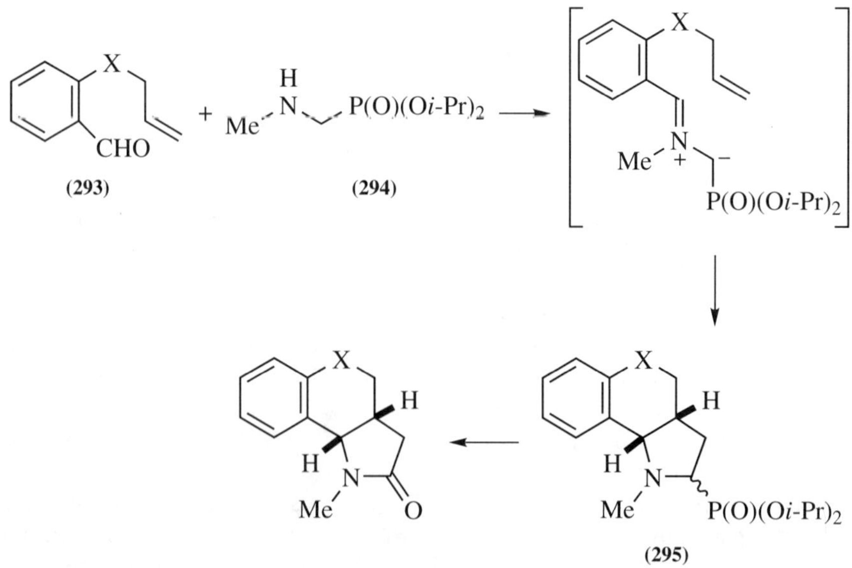

Scheme 3.96

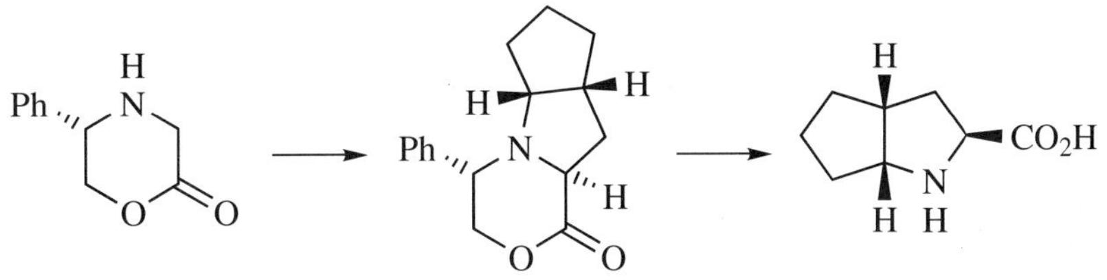

(205) (296) (297)

Scheme 3.97

Harwood and Lilley (87) reported the tandem generation and intramolecular trapping of a stabilized azomethine ylide, derived from the enantiopure template examined in detail in Section 3.2.3. Condensation of 5-hexenal with template **205** under standard conditions led to *in situ* ylide generation and subsequent cycloaddition of the tethered alkene to furnish **296** as a single enantiomer in 95% yield after purification and this despite the fact that the dipolarophile is unactivated. Hydrogenolytic destruction of the template revealed the bicyclic amino acid **297** in 75% yield (Scheme 3.97).

The reaction protocol also accommodated the use of sulfur containing aldehyde 3-thia-5-hexenal, furnishing the expected product **298** in 75% purified yield (66,67). Desulfurization with Raney Ni, followed by hydrogenolysis, resulted in formation of (*1S*, *4R*, *5R*)-4,5-dimethylproline (Scheme 3.98).

Further elaboration of the sulfur cycloadducts could be achieved by the use of a Pummerer rearrangement in the syntheses of 5-(hydroxymethyl)prolines. Oxidation of adduct **298** to sulfoxide **299**, followed by treatment with TFA in DCM and quenching with either methanol or benzyl alcohol, delivered the Pummerer products **300** in 57% yield for R = Me and 38% for R = Bn as single diastereoisomers. Raney Ni desulfurization and Pearlman's catalyst mediated hydrogenolysis, for R = Bn furnished the final enantiopure proline derivative (Scheme 3.99).

In an extension to this work, treatment of the template with 5-hexynal under the standard dehydrating conditions furnished the cycloadduct **301** in good yield (68). However, structural analysis of both the product and the azabicyclo[3.3.0]octane-3-carboxylic acid, derived by hydrogenation of the double bond, followed by

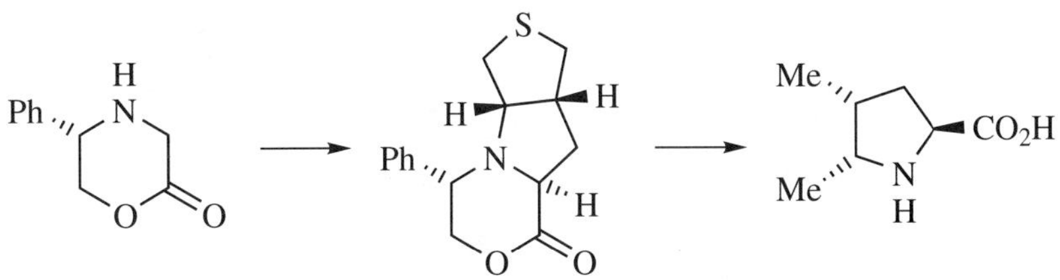

(205) (298)

Scheme 3.98

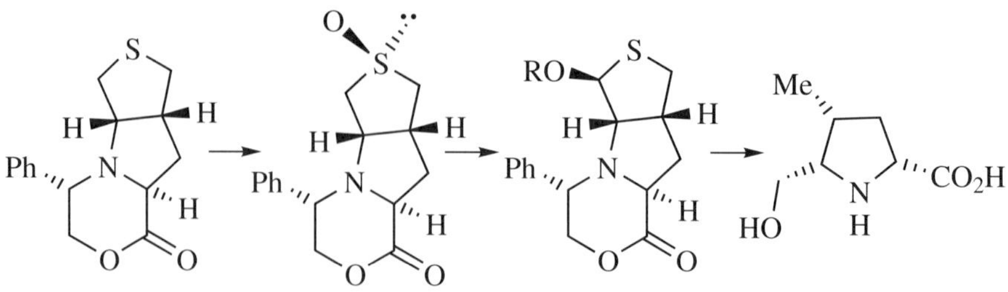

(298) (299) (300)

Scheme 3.99

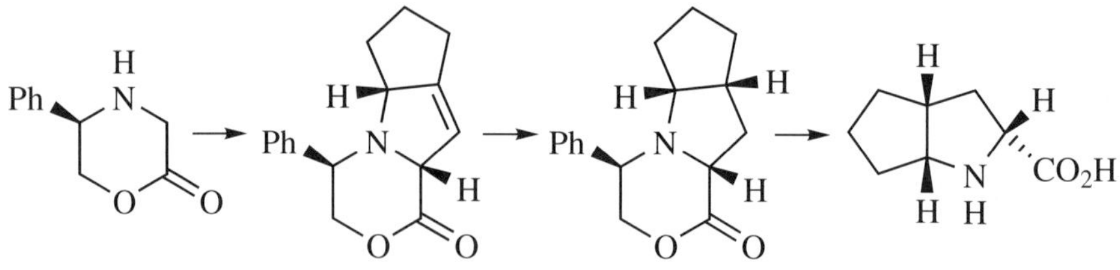

(301)

Scheme 3.100

hydrogenolytic template destruction, revealed that the stereochemistry at C(8) was opposite to that previously observed (Scheme 3.100).

However, by considering models of the anti configured ylide (Fig. 3.18), it was concluded that the inclusion of a three-carbon tether forces the reactive centers to be too sterically constrained to suffer intramolecular cycloaddition with an alkyne dipolarophile. Conversely, the *syn* ylide is able to achieve the correct approach for such a process, despite the steric interaction with the phenyl ring. Extension of the interim chain by one methylene unit using 6-heptynal, introduced a greater degree of flexibility into the system, allowing for the formation of the expected diastereoisomers (Scheme 3.101).

In a similar study to that outlined by Grigg, Kanemasa et al. (68) has demonstrated the intramolecular cycloaddition of azomethine ylides derived from either amino acids or esters. Treatment of the amino methyl ester **302** with

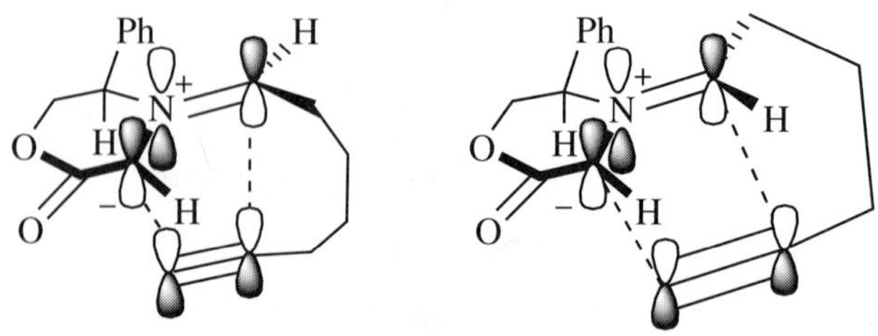

Figure 3.18

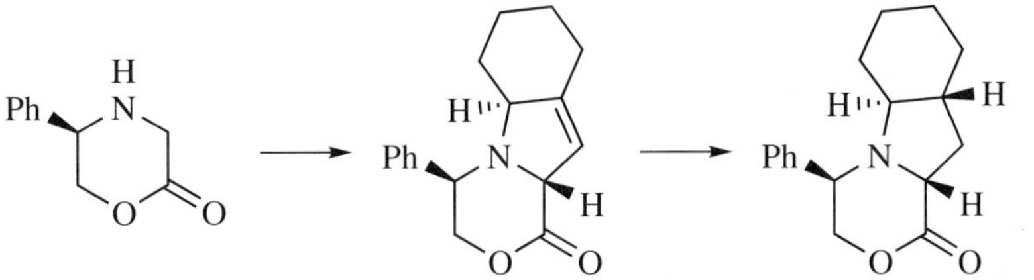

Scheme 3.101

equimolar amounts of (*E*)-7-phenyl-5-oxo-6-heptenal (R = Ph) with azeotropic removal of water led to the formation of cycloadduct **303** as a single stereoisomer in 95% isolated yield. Similarly, **304** furnished **305** (R = Ph) in 90% and **305** (R = Me) in 65%, each as a single *cis* fused stereoisomer (Scheme 3.102).

However, treatment of amino acids **302** and **304** with aldehyde **306** led to the formation of the products **307** and **308**, respectively. Product **307** (R = Ph) was formed in 82% yield, while **308** was isolated in 84% yield (R = Ph) and 70% yield (R = Me), each as a single stereoisomer.

The decarboxylative approach to the ylide formation generated cycloaddition products derived from cycloaddition of the ylide to the carbonyl moiety of the molecule, as opposed to the alkene as seen in previous examples. Kanemasa has reconciled this observation by consideration of the postulated transition state model of the reaction. It was assumed that the steric repulsion of the terminal olefinic substituent and the ylide would favor transition state **309** (Fig. 3.19). Additionally, nonstabilized azomethine ylides have a higher energy HOMO than stabilized ylides, and would therefore prefer the LUMO of the carbonyl than the lower lying alkene LUMO. Formation of fused five-membered rings would also be kinetically favored over construction of six-membered ring (Scheme 3.103).

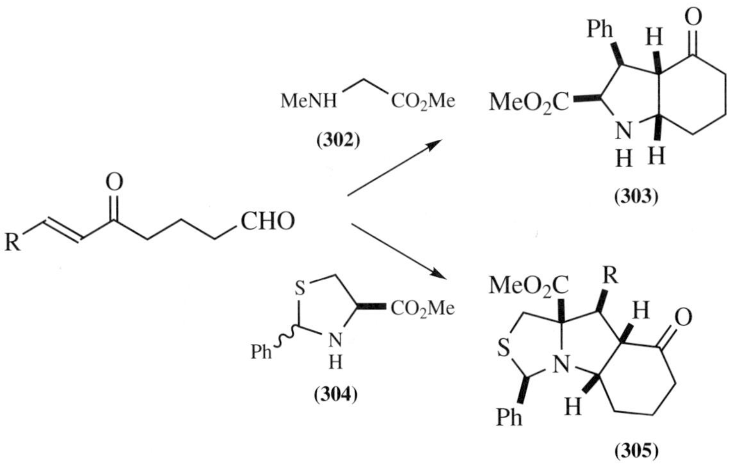

Scheme 3.102

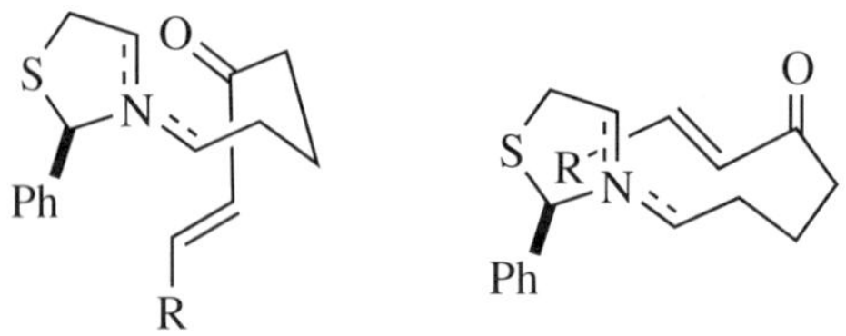

(309)

Figure 3.19

This reaction has also been carried out with 4-oxo-5-hexenal (**310**), where the interim tether between ylide and dipolarophile is one carbon shorter. In this case, **310** delivered two regioisomeric products (**311** and **312**) in a 4:1 ratio in 53% yield. By the same protocol, **310** furnished the two isomers **313** and **314**. For $R^1 = R^2 =$ Ph, the ratio of **313** to **314** was 4:1 in a combined yield of 38%, while for $R^1 =$ Ph and $R^2 =$ Ph-(*E*)-CH=CH, the ratio improved to 7:1 in favor of **314**, although the material yield was still low at 42% (Scheme 3.104).

The central five-membered ring of Manzamine A (**315**), a member of a family of antileukimic and antibacterial marine polycyclic alkaloids, has been constructed using intramolecular azomethine ylide technology (88). Model studies on the construction of the central ABC rings, by condensation of the aldehyde **316** prepared by standard chemistry with sarcosine ethyl ester, furnished the desired ABC ring system as a single diastereoisomer in 45% yield (Scheme 3.105).

The synthesis of oxotetrahydroindoles by an intramolecular reaction of an azomethine ylide and an acetylene functionality has been reported by Martinelli and co-workers (89). Conversion of the precursor ester **317** to the requisite acid by

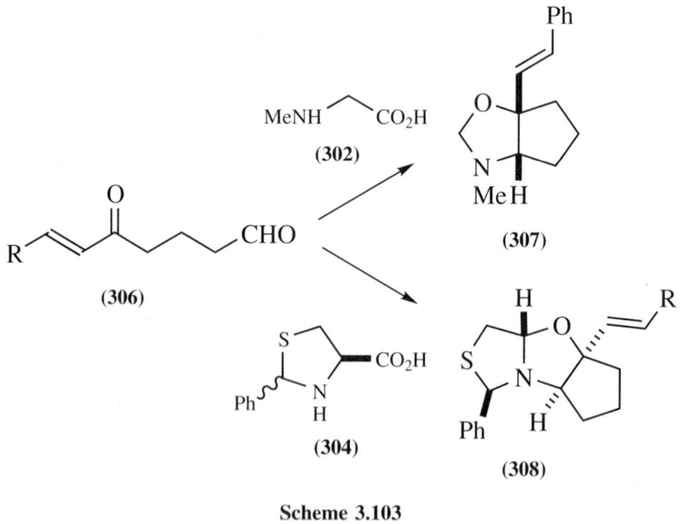

Scheme 3.103

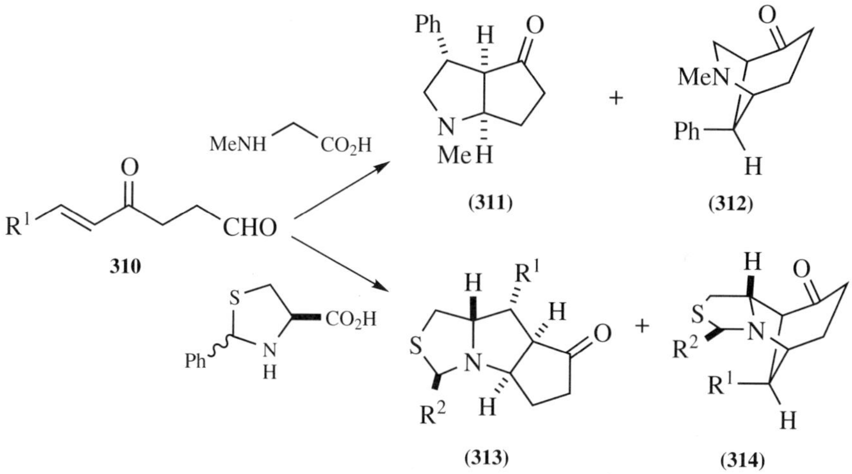

Scheme 3.104

LiI in EtOAc, which underwent subsequent ylide formation and cycloaddition in Ac_2O, furnished the desired cycloadduct **318** in reasonable yield with concomitant loss of the silyl group. The reaction proved to be general, with a range of products (**319–322**) being synthesized from the corresponding acid precursors. The presence of the silyl group appears to be essential to the reaction, since its replacement with other functionalities led to a distinct reduction in reaction efficiency (Scheme 3.106).

3.3.3. Alternative Generation Procedures

4-Oxazolines, as synthetic precursors to azomethine ylides, have been extensively studied by Vedejes and Day (90), as discussed in Section 3.1.2, with the

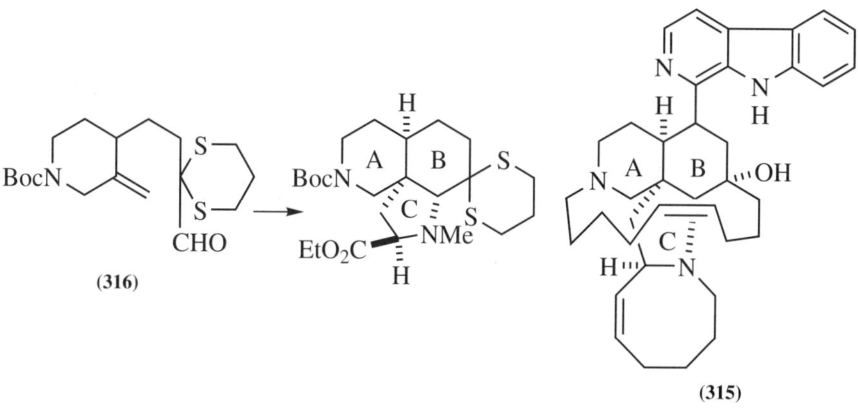

Scheme 3.105

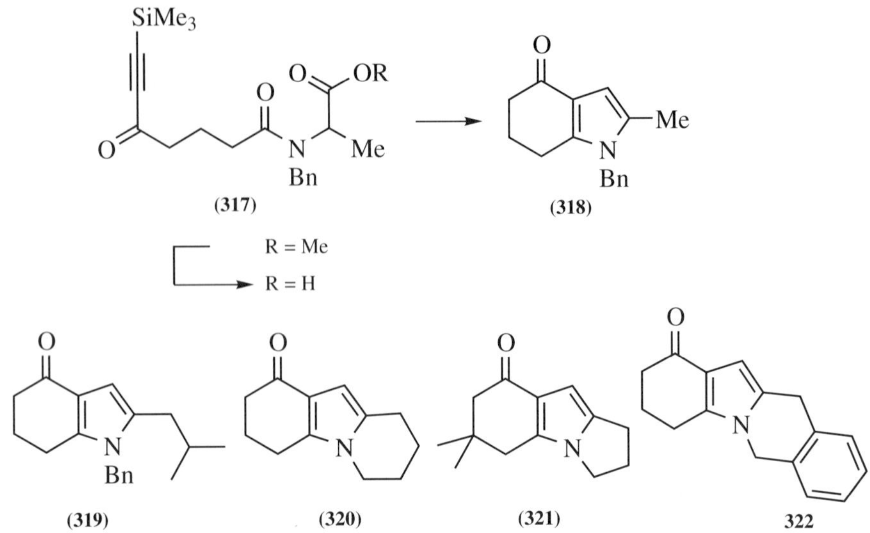

Scheme 3.106

protocol being further extended to the construction of indoquinones. Synthesis of the cycloaddition precursor **323** was achieved in five steps and this was set up for intramolecular cycloaddition by treatment with TMSCN/CsF in acetonitrile to yield the desired product in 68% yield, which subsequently underwent DDQ oxidation to the final indoquinone **324**. The cycloaddition proceeds with total

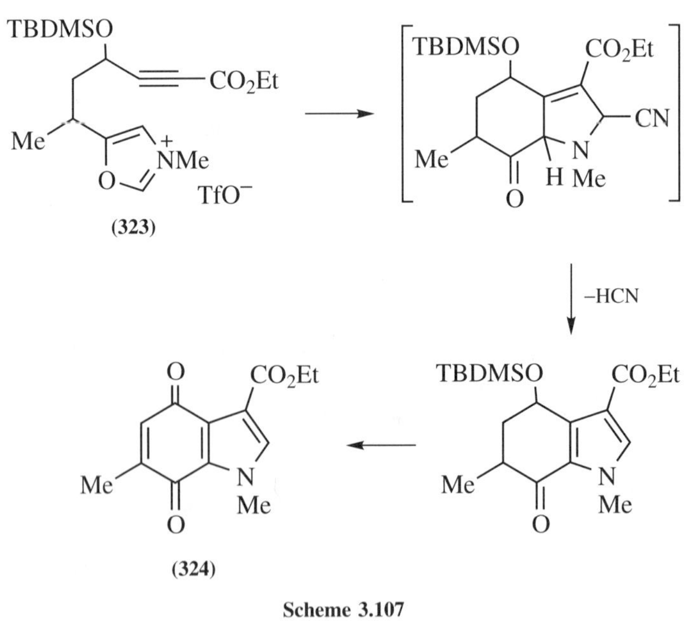

Scheme 3.107

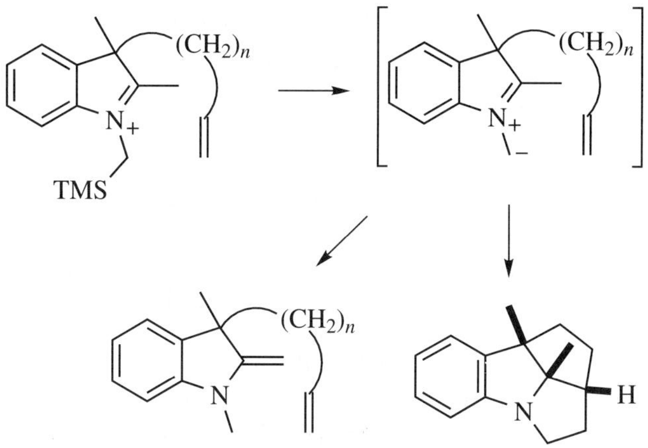

(325)

Scheme 3.108

regiocontrol, a graphic illustration of the benefits of intramolecular methodology (Scheme 3.107).

The desilyation of indolium salts has been shown by Fishwick et al. (91) to provide azomethine ylides, which undergo *in situ*, intramolecular cyclization. The tetracyclic products are formed in moderate yield, but with complete stereocontrol, each ring being *cis*-fused. Unfortunately, the reaction is only successful for two-carbon bridging chains. Elongation of the pendant alkene chain, to either three or four carbons, gave products arising from a 1,4-hydrogen shift, furnishing the exocyclic enamines **325**. Molecular modeling experiments indicated severe steric interactions between the 2-methyl substituents and the methylene group, disfavoring an intramolecular cycloaddition (Scheme 3.108).

The same research group has demonstrated a similar intramolecular process in the construction of bicyclic adducts **326** (92). The CsF desilyation of precursors **327**, after subsequent reaction of the internal dipolarophile, delivered the expected cycloadducts with complete stereocontrol when either thioether and ether tethers or activated and unactivated dipolarophiles were used. In contrast with intermolecular protocols, the reaction was successful with both activated and unactivated alkenes. In addition, unlike the previous example, formation of both six ($n = 2$)- and five ($n = 1$)-membered rings occurs (Scheme 3.109).

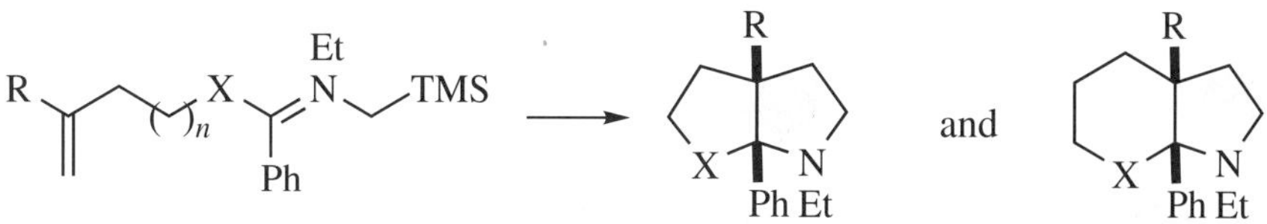

(327) (326)

Scheme 3.109

3.4. METAL-MEDIATED REACTIONS

Metal-mediated processes represent an important advance in azomethine ylide chemistry, allowing reactions to proceed at ambient temperatures. They also allow for the possibility of chiral catalysis, which is a little explored area as yet, but represents an exciting development allowing for the induction of enantiocontrol without the need for stoichiometric stereocontrolling units.

Combinatorial chemistry has attracted intense interest in recent years, particularly from within the pharmaceutical and agrochemical industries, as it allows for the rapid generation of large libraries of small compounds for high throughput screening. Gallop and co-workers (93) at the Affymax Research Institute, utilized resin bound azomethine ylides, generated *via* a metal mediated protocol in the construction of such a library. A series of polystyrene peptide resins, preloaded with Fmoc protected amino acids, underwent condensation with a range of aromatic and heteroaromatic aldehydes to form the requisite imines in almost quantitative yield. The $AgNO_3$/ Et_3N mediated generation and subsequent cycloaddition with singly substituted olefins generated >500 mercaptoacyl prolines.

Split synthesis of the library using four amino acids, four aldehydes, and five olefins in the presence of four mercaptoacyl chlorides (Scheme 3.110) generated the required proline library that was screened, after TFA cleavage of the products from the solid support, for inhibition of angiotensin converting enzyme ACE.

From this, an equimolar mixture of **328** and **329** was shown to display potent activity. Separation of these compounds led to the observation that **329** displayed very little activity, while **328** showed an inhibition of ACE of $K_i \sim 160$ pM, which is among the highest activity of a thiol containing inhibitor known to date (Fig. 3.21).

As part of Tõke's ongoing program toward novel analogues of Cephalotaxus alkaloids, a short and efficient route through to polyfunctionalized pyrrolidines bearing an aromatic substituent at C(3) and a *trans* nitro group at C(4) was required. As such, the 1,3-dipolar cycloaddition route was elected as the most efficient method for the stereoselective formation of such compounds (94). Treatment of the glycine-derived azomethine ylide **331**, formed from imine **330** in the presence of triethylamine and LiBr at ambient temperatures with nitroalkenes, furnished cycloadducts **332** and **333**, each as a single regioisomer. For Ar = **334**, the reaction proceeded in 61% yield for the major isomer with a diastereomic ratio of **332:333** = 2.5:1. Similarly, Ar = **335** yielded **332** and **333** in 47% but with a ratio

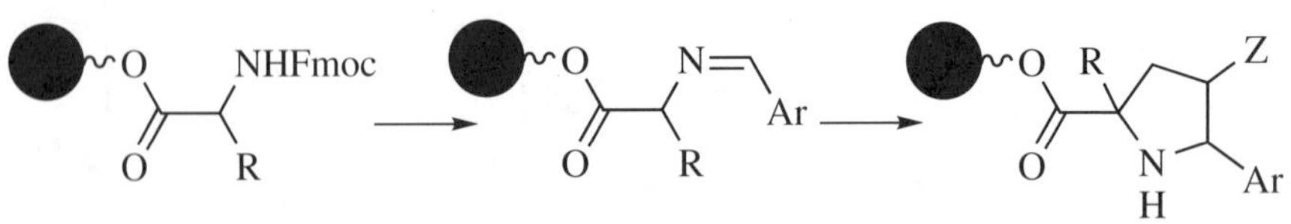

Scheme 3.110

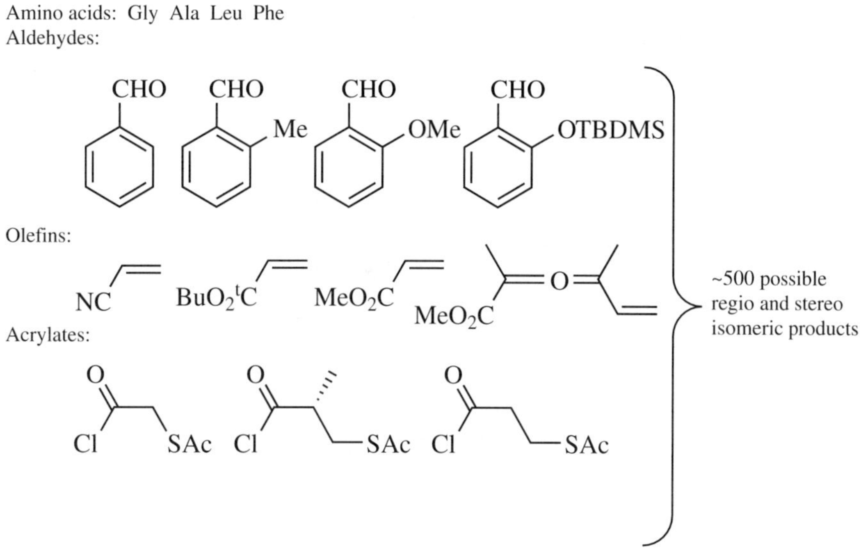

Figure 3.20

of 3.5:1. For Ar = Ph, LiBr proved to be a poor catalyst, furnishing the adducts in only 22% yield (Scheme 3.111).

However, replacement of LiBr with AgOAc inverted the ratio of *exo* to *endo* products. For Ar = **334**, the major adduct was isolated in 42% yield with an *endo/exo* ratio of 1:1.7, while Ar = **335** gave **333** in 36% yield with an *endo/exo* ratio of 1:2.3. Note that attempts at the thermal reaction met with low yields of complex reaction mixtures containing all possible regio- and stereoisomers. This study exemplifies the value of metal mediation in the stereo- and regiocontrol of azomethine ylide cycloadditions.

Pyroglutamates and their analogues are important compounds in the study and treatment of neurological diseases such as epilepsy, Alzheimer's disease, and

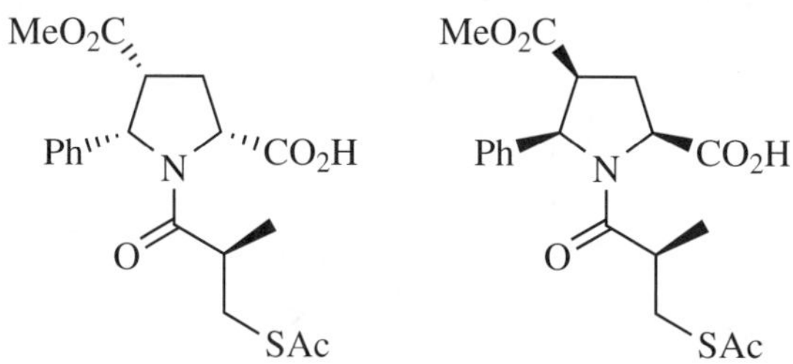

(328) **(329)**

Figure 3.21

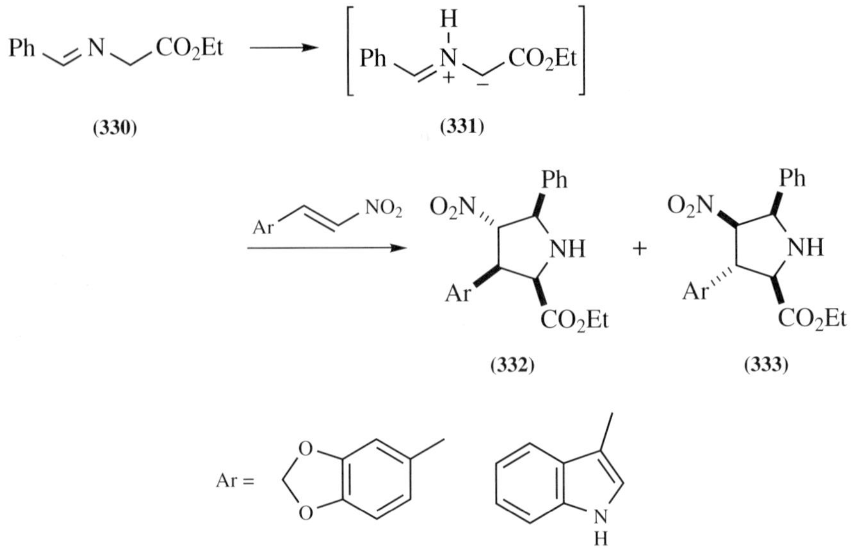

(330) (331) (332) (333) (334) (335)

Scheme 3.111

treatment of stroke. In studies toward the synthesis of α-quaternized analogue, Alvarez–Ibarra reported the use of samarium-mediated dipolar cycloadditions (95). Treatment of the sulfur-containing precursor **336** with SmI_2 in THF, in the presence of an α,β-unsaturated ester followed by *in situ* hydrolysis, furnished the expected pyroglutamates in excellent yields and diastereomeric ratios. For $R^1 = R^2 =$ Me, **337** was isolated in 85% with a de of 80%, while $R^2 = {}^t$Bu gave **337** in 80% yield, de = 80%. Similarly, for $R^1 =$ Bn, $R^2 = {}^t$Bu gave an isolated yield of 80% with a de of 75% (Scheme 3.112).

The mechanistic rationale assumed SmI_2-induced generation of a radical anion by single electron transfer causing ketone **336** to undergo ketyl radical formation. Subsequent unimolecular fragmentation is proposed to be favored over dimerization or C–C bond formation due to the steric bulk of the ketone and also due to resonance stabilization favoring the fragmentation product. Subsequent cycloaddition and hydrolysis of the azathioenolate furnished the α,α-disubstituted γ-carboxy-pyroglutamates (Scheme 3.113).

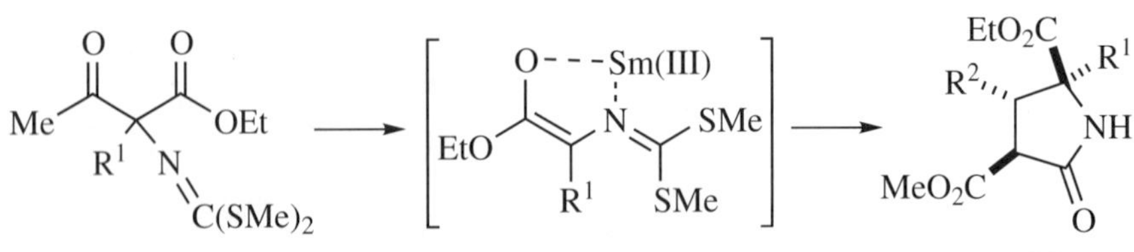

(336) (337)

Scheme 3.112

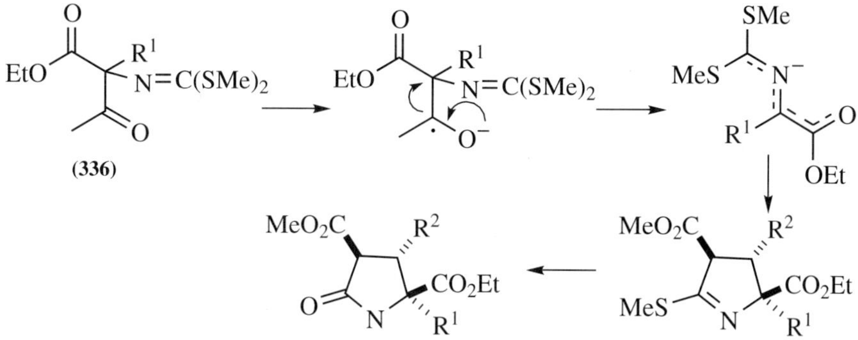

Scheme 3.113

The high diastereofacial selectivity of the reaction is believed to be due to the highly ordered *endo* transition state in which the Sm(II) cation is coordinated to the dipole and dipolarophile, a situation not possible in the *exo* transition state (Fig. 3.22).

Planar chiral arene $Cr(CO)_3$ complexes have been shown to undergo highly diastereoselective cycloadditions and Kündig has extended this protocol to the [3 + 2] cycloadditions of azomethine ylides (96). Enantiopure ortho- substituted η^6-benzaldehyde complex **337** underwent condensation with an α-amino ester to afford imine **338** in the presence of Et_3N. Subsequent treatment with methyl acrylate at ambient temperature in the presence of LiBr and Et_3N delivered cycloadduct **339**, with excellent stereoinduction and high material yield. Photoinduced oxidative decomplexation in air furnished the final arylpyrrolidines (Scheme 3.114).

In each case, only a single diastereoisomer was observed, with enantiomeric excesses (ee) of the final decomplexed products being >98%. For $R^1 = OMe$ and $R^2 = H$ and Me yields were in the order of 75% while $R^1 = OMe$ or Me, $R^2 = Ph$ or H gave yields >80%. However, $R^1 = Cl$ and $R^2 = H$ gave a reduction in yield to 52%, although the reaction still proceeded with complete stereocontrol, with methyl acrylate. The reaction could also be conducted, with complete stereocontrol, in the presence of $TiCl(O\textit{i}\text{-}Pr)_3$ and Et_3N, as outlined by Grigg, although with a

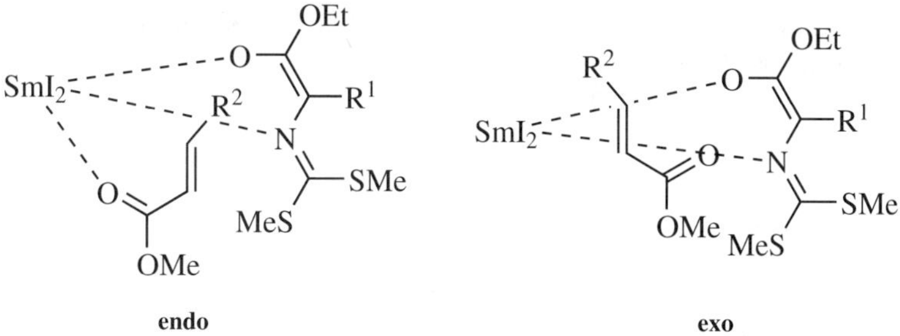

Figure 3.22

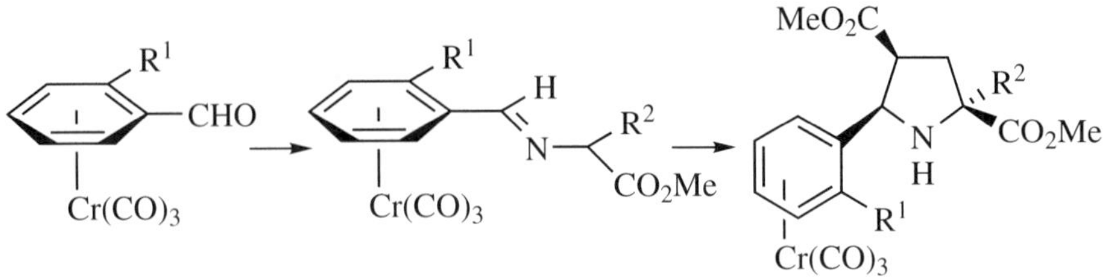

(337) **(338)** **(339)**

Scheme 3.114

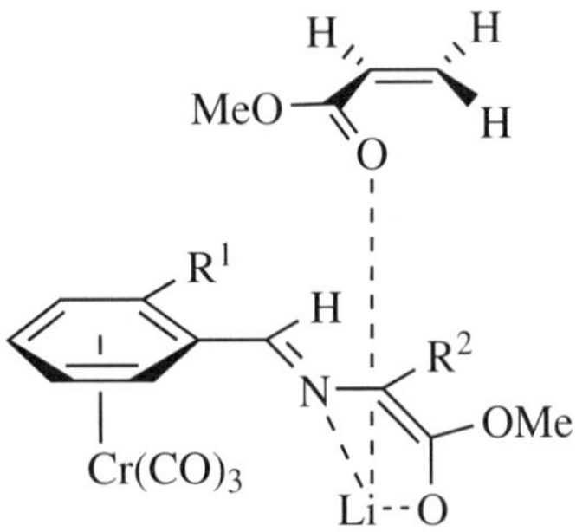

Figure 3.23

reduced yield of 44%. The exclusive *syn-* and *endo*-selectivity of the reaction can be reconciled by assuming the dipolarophile can approach from only one face of the lithiated azomethine ylide (Fig. 3.23).

Although Grigg et al. (97) have been prolific authors in azomethine ylide chemistry, probably their greatest contribution has been in the development of metal-mediated technologies, leading to elegant solutions to the problems of regio-, stereo-, and enantiocontrol over reactions. Typically, Et_3N or DBU deprotonation of amino esters, with subsequent metalation of the anion, led to the formation of stabilized, metallo-1,3-dipolar systems (**340**), often to the extent where isolation is possible (Scheme 3.115).

An initial survey studied the cycloaddition between **341** ($R^1 = Me$) and maleimide in the presence of copper(II) acetate and 1 M equiv of Et_3N with pyridine as solvent, leading to the formation of the expected adduct **342** at room temperature after a few hours. The reaction delivered a single product arising from an *endo* transition state in 60% yield. The reaction protocol was further extended to include

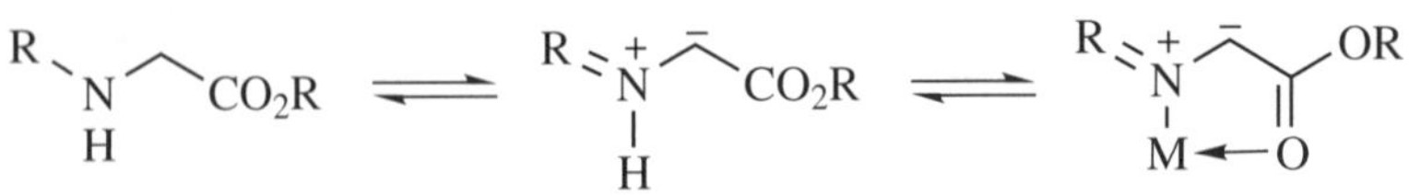

(340)

Scheme 3.115

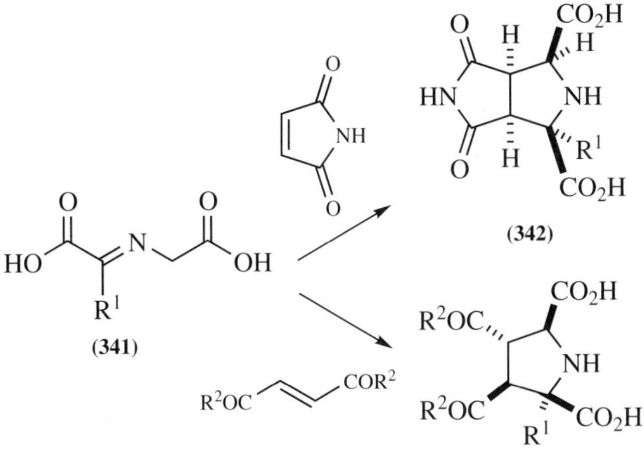

Scheme 3.116

a range of metal ions and olefinic dipolarophiles. Typically, substrate **341** with $R^1 = Me$, $R^2 = OMe$, in the presence of zinc or cadmium as the coordinating metal ion, led to the formation of **343** in 89% and **343** in 90% yield, respectively. While the zinc mediated reaction led to the formation of a single stereoisomer in 80% yield, for $R^1 = Ph$, $R^2 = OMe$, cadmium led to the formation of a 9:1 ratio of diastereoisomers, although it is unclear whether this difference in stereocontrol is due to change in the metal or subsequent product isomerization, due to prolonged reaction times and contact with Dowex resin (Scheme 3.116) (98).

The reaction also proceeds with a high degree of regiospecificity. In particular, treatment of **344** with methyl acrylate ($R = CO_2Me$), acrylonitrile ($R = CN$), and phenylvinyl sulfone ($R = SO_2Ph$) with Zn(II) as the metal salt, led in each case to the formation of a single regio- and stereoisomer (**345**) in 70, 87, and 80% yield, respectively (Scheme 3.117).

Replacement of the metal salts with Ti(II) led to a reversal in the observed regiocontrol of the reaction (99). Treatment of **346** with methyl acrylate, in the presence of $Ti(O\textit{i}\text{-}Pr)_3Cl$ and 1 M equiv of Et_3N, led to the formation of a 1:1.4 mixture of **347** and **348**. The observed transesterification was also regiospecific and was judged

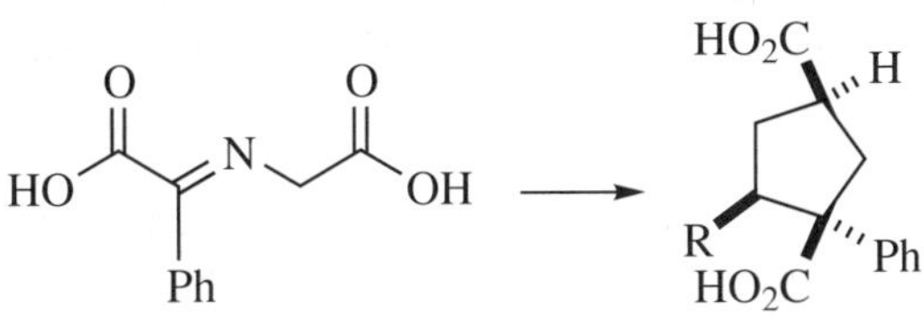

Scheme 3.117

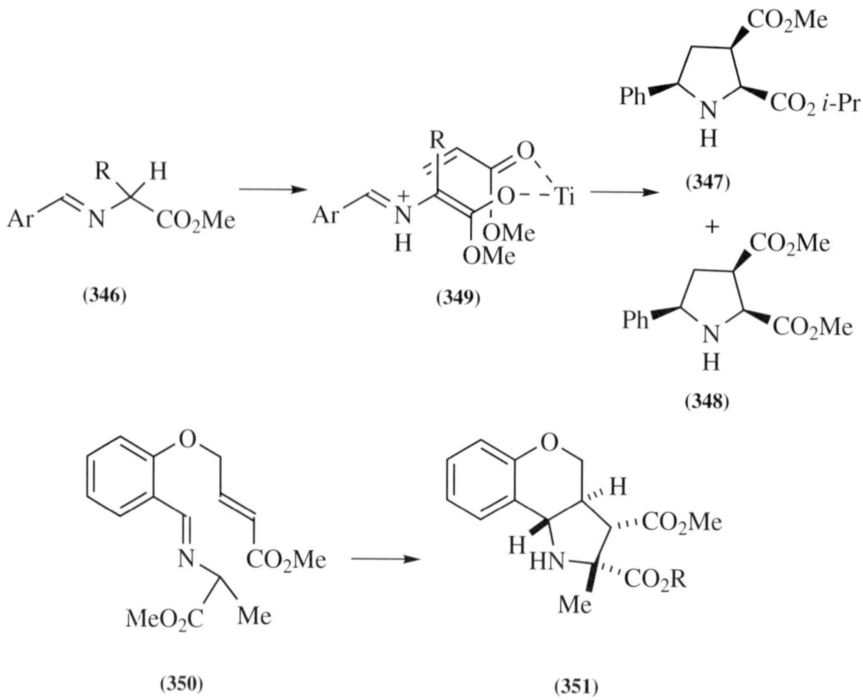

Scheme 3.118

to have occurred on the cycloadducts rather than the initial imines. The observed regiochemistry was reconciled by chelation of both imine and dipolarophile to form an intermediate of the type **349**. The protocol could also be applied intramolecularly, with **350** furnishing **351** as a 6.5:1 mixture of esters R = Me: R = *i*-Pr in a combined yield of 65% (Scheme 3.118).

By attachment of a chiral controlling unit, the reaction could also be carried out asymmetrically (100). Subjecting **352** (R^1 = 2-naphthyl, R^2 = H) to cycloaddition with **353** in the presence of AgOAc (1.5 M equiv) and Et_3N (1.0 M equiv) furnished the enantiopure adduct **354** in 50%, with no other reaction products being observed. The reaction could be improved by alteration of the metal salt. Treatment of **352** ($R^1 = R^2$ = Ph) with dipolarophile **353** in the presence of LiBr and Et_3N delivered the expected, enantiomerically pure adduct **354** in >90% yield, while **352** (R^1 = 2-naphthyl, R^2 = H) gave rise to **354** in quantitative yield with $TlNO_3$ and Et_3N (Scheme 3.119).

The intriguing reversal of regiochemistry observed previously in the Ti(O*i*-Pr)$_3$Cl mediated protocol was also observed in the asymmetric example. Subjection of **355** to cycloaddition with **356** mediated by Ti(O*i*-Pr)$_3$Cl/Et_3N led to the transesterified product **357** as a single enantioenriched diastereoisomer in 75% yield (Scheme 3.120).

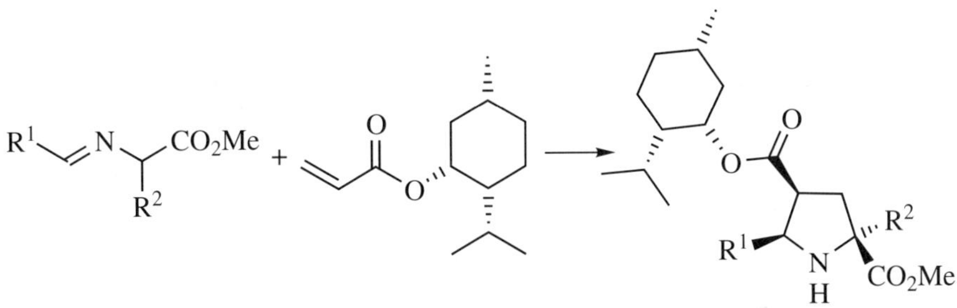

(352) **(353)** **(354)**

Scheme 3.119

Of greater synthetic interest is asymmetric induction by the use of chiral catalysis. Grigg was the first to report chiral catalysis of 1,3-dipolar cycloadditions in 1991 (101). A study of metal salts and chiral ligands revealed that **358** underwent cycloaddition with methyl acrylate to furnish adduct **359** in the presence of $CoCl_2$ and (1*R*, 2*S*)-*N*-methylephedrine as the chiral ligand. The pyrrolidine product was isolated in 55% yield with an ee of 84%. The use of methyl acrylate as solvent led to an improved yield of 84% with an excellent ee of 96% (Scheme 3.121).

The compounds $MnBr_2$, $CoBr_2$, and CoF_2, in the presence of ephedrine, camphor, and other ligands, led to chiral induction, although with less success than the previously detailed example. The reaction outcome is rationalized by the transition state **360** in which the *cis* methyl and phenyl groups of the ligand, gave a pseudo-equatorial 2-naphthyl group, blocking one face of the imine (Fig. 3.24).

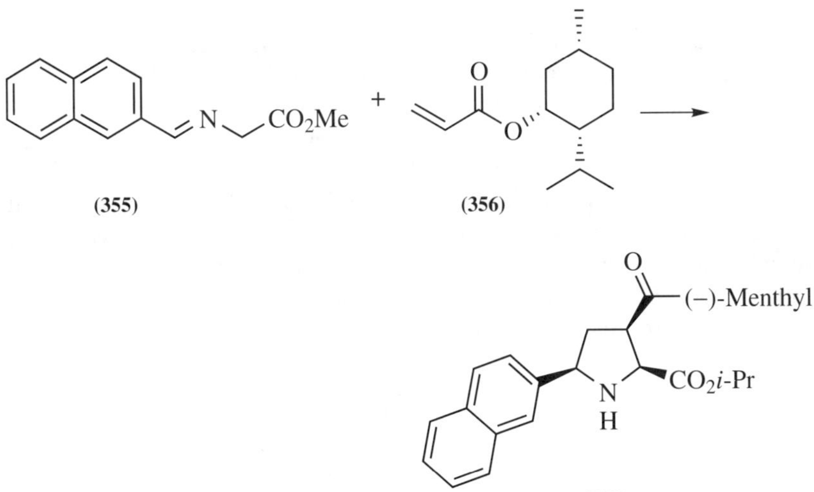

(355) **(356)** **(357)**

Scheme 3.120

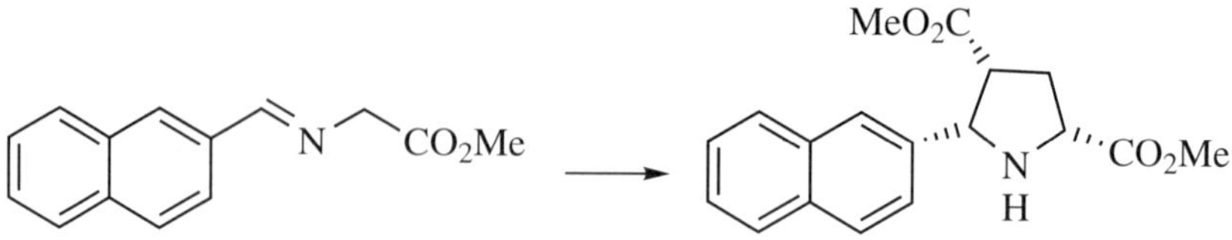

(358) **(359)**

Scheme 3.121

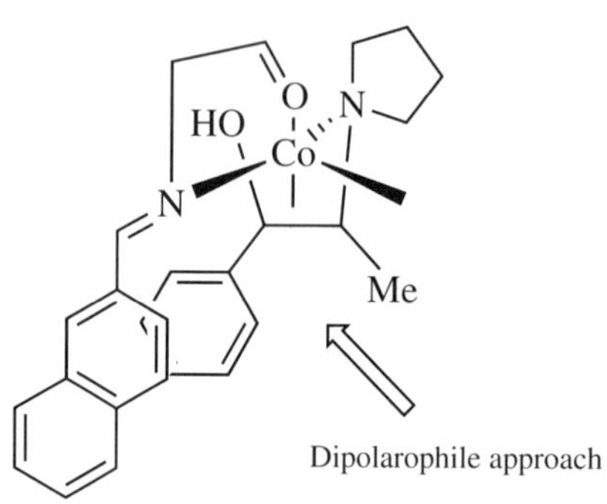

(360)

Figure 3.24

3.5. CONCLUSION

The recent explosion of interest in azomethine ylides is well justified when considering their synthetic utility in the construction of heterocycles and polycyclic systems. Not only does the wealth of ylide generation protocols permit application to a range of heterocyclic structures, but the advent of chiral control over reaction products and intramolecular protocols allows for azomethine ylide cycloadditions to be applied to a wealth of synthetic problems, particularly those developed by natural product synthesis. One of the most exciting areas likely to receive intense scrutiny in the future will be asymmetric catalysis, allowing for the construction of enantioenriched products from achiral precursors without the need for stoichiometric stereocontrolling units. Interest in these reactive intermediates shows no sign of abating.

REFERENCES

1. For earlier work, see A. Padwa, ed., in *1,3-Dipolar Cycloaddition Chemistry*, Vols. 1,2, Wiley-Interscience, New York, **1984**.

2. A. Padwa, Y. Chen, W. Dent, and N. Hildegard, *J. Org. Chem.*, **1985**, *50*, 4006.
3. A. Padwa and J. R. Gasdaska, *J. Am. Chem. Soc.*, **1986**, *108*, 1104.
4. A. Padwa, G. E. Fryxell, J. R. Gasdaska, M. K. Venkataramanan, and G. S. K. Wong, *J. Org. Chem.*, **1989**, *54*, 644.
5. G. Poissonnet, P. Potier, and R. H. Dodd, *Tetrahedron Lett.*, **1989**, *30*, 3423.
6. M. Ohno, M. Komatsu, H. Miyata, and Y. Ohshiro, *Tetrahedron Lett.*, **1991**, *32*, 5813.
7. K. Washizuka, S. Minakata, I. Ryu, and M. Komatsu, *Tetrahedron*, **1999**, *55*, 12969.
8. U. C. Yoon, D. U. Kim, J. C. Kim, J. G. Lee, P. S. Mariano, Y. J. Lee, and H. L. Ammon, *Tetrahedron Lett.*, **1993**, *34*, 5859.
9. G. Pandey, G. Lakshmaiah, and G. Kumaraswamy , *J. Chem. Soc. Chem. Commun.*, **1992**, 1313.
10. G. Pandey and G. Lakshmaiah, *Tetrahedron Lett.*, **1993**, *34*, 4861.
11. S. Torii, H. Okumoto, and A. Genba, *Synlett.*, **1994**, 217.
12. A. R. Katritzky, G. J. Hitchings, and X. Zhao, *J. Chem. Soc., Perkin Trans. 1.*, **1990**, 2371.
13. A. R. Katritzky, J. Köditz, and H, Lang, *Tetrahedron.*, **1994**, *50*, 12571.
14. O. Tsuge, T. Hatta, Y. Kakura, H. Tashiro, H. Maeda, and A. Kakehi, *Chem. Lett.*, **1997**, 945.
15. Y. Tominaga, K. Ogata, S. Kohra, M. Hojo, and A. Hosomi, *Tetrahedron Lett.*, **1991**, *32*, 5987.
16. C. W. G. Fishwick, R. J. Foster, and R. E. Carr, *Tetrahedron Lett.*, **1996**, *37*, 3915.
17. E. Vedejs and J. W. Grissom, *J. Am. Chem. Soc.*, **1986**, *108*, 6433.
18. E. Vedejs and J. W. Grissom, *J. Org. Chem.*, **1988**, *53*, 1876.
19. E. Vedejs and J. W. Grissom, *J. Org. Chem.*, **1988**, *53*, 1882.
20. J. Mortier and M. Joucla, *Tetrahedron Lett.*, **1987**, *28*, 2973.
21. R. Bureau, J. Mortier, and M. Joucla, *Tetrahedron Lett.*, **1987**, *28*, 2975.
22. R. Bureau, J. Mortier, and M. Joucla, *Tetrahedron.*, **1992**, *48*, 8947.
23. G. Roussi and J. Zhang, *Tetrahedron.*, **1991**, *47*, 5161.
24. G. Roussi and J. Chastanet, *J. Org. Chem.*, **1988**, *53*, 3808.
25. A. Padwa, D. C. Dean, and L. Zhi, *J. Am. Chem. Soc.*, **1989**, *111*, 6451.
26. A. Padwa, D. C. Dean, and L. Zhi, *J. Am. Chem. Soc.*, **1992**, *114*, 593.
27. D. Alker, L. M. Harwood, and C. E. Williams, *Tetrahedron.*, **1997**, *53*, 12671.
28. For examples, see; R. Grigg, J. Kemp, and W. J. Warnock, *J. Chem. Soc., Perkin Trans. 1*, **1987**, 2275. R. Grigg, H. Q. N. Gunaratne, and J. Kemp, *Tetrahedron.*, **1990**, *46*, 6467. R. Grigg, P. McMeekin, and V. Sridharan, *Tetrahedron.*, **1995**, *51*, 13331. K. Amornraksa, R. Grigg, H. Q. N. Gunaratne, J. Kemp, and V. Sridharan, *J. Chem. Soc., Perkin Trans. 1*, **1987**, 2285. H. Ardill, M. J. R. Dorrity, R. Grigg, M. Leon-Ling, J. F. Malone, V. Sridharan, and S. Thianpatanagul, *Tetrahedron.*, **1990**, *46*, 6433. R. Grigg, G. Donegan, H. Q. N. Gunaratne, D. A. Kennedy, J. F. Malone, V. Sridharan, and S. Thianpatanagul, *Tetrahedron.*, **1989**, *45*, 1723. R. Grigg, S. Thianatanagul, and J. Kemp, *Tetrahedron.*, **1987**, *44*, 7283. R. Grigg, M. Dowling, M. W. Jordan, and V. Sridharan, *Tetrahedron.*, **1987**, *43*, 5873.
29. R. Grigg, D. Henderson, and A. J. Hudson, *Tetrahedron Lett.*, **1989**, *30*, 2841. T. Coulter, R. Grigg, J. F. Malone, and V. Sridharan, *Tetrahedron Lett.*, **1991**, *32*, 5417. M. F. Aly, G. M. El-Nagger, T. I. El-Emary, R. Grigg, S. A. M. Metwally, and S. Sivagnanam, *Tetrahedron.*, **1994**, *50*, 895.
30. R. Grigg, S. Surendrakumar, S. Thianatanagul, and D. Vipond, *J. Chem. Soc., Chem. Commun.*, **1987**, *47*.
31. R. Grigg, J. Idle, P. McMeekin, and D. Vipond, *J. Chem. Soc., Chem. Commun.*, **1987**, *49*.
32. M. F. Aly, R. Grigg, S. Thianatanagul, and V. Sridharan, *J. Chem. Soc., Perkin Trans. 1*, **1988**, 949. R. Grigg, S. Surendarkumar , S. Thianpatanagul, and D Vipond, *J. Chem. Soc., Perkin Trans. 1*, **1988**, 2693. R. Grigg, J. Idle, P. McMeekin, S. Surendrakumar, and D. Vipond, *J. Chem. Soc.*, **1988**, 2703.
33. G. Bianchi, C. De Micheli, and R. Gandolfi, *Angew. Chem. Int.*, **1979**, *18*, 721.

34. R. Grigg, P. McMeekin, and V. Sridharan, *Tetrahedron*, **1995**, *51*, 13347.
35. R. Grigg and F. Heaney, *J. Chem. Soc., Perkin Trans. 1*, **1989**, 198.
36. S. Kanemasa and H. Yamamoto, *Tetrahedron Lett.*, **1990**, *31*, 3633. S. Kanemasa, T. Hayashi, J. Tanaka, H. Yamamoto, and T. Sakurai, *J. Org. Chem.*, **1991**, *56*, 4473. S. Kanemasa, H. Yamamoto, E. Wada, T. Sakrai, and K. Urushido, *Bull. Chem. Soc. Jpn.*, **1990**, *63*, 2857.
37. A. Chrapkowsky, G. Galley, P. G. Jones, and M. Pätzel, *Tetrahedron Lett.*, **1993**, *34*, 5707. G. Galley, J. Liebcher, and M. Pätzel, *J. Org. Chem.*, **1995**, *60*, 5005. For similar work, see also R. Annunziata, M. Cinquini, F. Cozzi, L. Raimondi, and T. Pilati, *Tetrahedron Asymm.*, **1991**, *2*, 1329.
38. T. Takahashi, K. Kitano, T. Hagi, H. Nihonmatsu, and T. Koizumi, *Chem. Lett.*, **1989**, 597.
39. Z. Ma, S. Wang, C. S. Cooper, A. K. L. Fung, J. K. Lynch, F. Plagge, and D. T. Chu, *Tetrahedron Asymm.*, **1997**, *8*, 883.
40. R. Williams and G. L. Fegley, *Tetrahedron Lett.*, **1992**, *33*, 6755.
41. P. Garner and W. B. Ho, *J. Org. Chem.*, **1990**, *55*, 3973. P. Garner, W. B. Ho, S. K. Grandhee, W. J. Youngs, and V. O. Kennedy, *J. Org. Chem.*, **1991**, *56*, 5893.
42. P. Garner, K. Sunitha, and T. Shanthilal, *Tetrahedron Lett.*, **1988**, *29*, 3525.
43. P. Garner, K. Sunitha, W. B. Ho, W. J. Youngs, V. O. Kennedy, and A. Djebli., *J. Org. Chem.*, **1989**, *54*, 2041. P. Garner, W. B. Ho, and H. Shin, *J. Am. Chem. Soc.*, **1992**, *114*, 2767.
44. H. Waldmann, E. Bläser, M. Jansen, and H. Letschert, *Angew. Chem. Int. Ed. Engl.*, **1994**, *33*, 683.
45. A. Padwa, Y. Y. Chen, U. Chiacchio, and W. Dent, *Tetrahedron.*, **1985**, *41*, 3529.
46. G. Pandey and G. Lakshmaiah, *Synlett.*, **1994**, 277.
47. P. Deprez, J. Royer, and H. P. Husson, *Synthesis*, **1991**, 759. P. Deprez, J. Royer, H. P. Husson, *Tetrahedron*, **1993**, *49*, 3781.
48. J. Rouden., J. Royer, and H. P. Husson, *Tetrahedron Lett.*, **1989**, *30*, 5133.
49. P. Deprez, J. Rouden, A. Chiaroni, C. Riche, J. Royer, and H. P. Husson, *Tetrahedron Lett.*, **1991**, *32*, 7531.
50. P. Deprez, J. Royer, and H. P. Husson, *Tetrahedron Asymm.*, **1991**, *2*, 1189.
51. S. R. Martel, R. Wisedale, T. Gallagher, L. D. Hall, M. F. Mahon, R. H. Bradbury, and N. J. Hales, *J. Am. Chem. Soc.*, **1997**, *119*, 2309.
52. D. Planchenault, R. Wisedale, T. Gallagher, and N. J. Hales, *J. Org. Chem.*, **1997**, *62*, 3438. M. D. Andrews, G. A. Brown, J. P. H. Charmant, T. M. Peakman, A. Rebello, K. E. Walsh, T. Gallagher, and N. J. Hales, *J. Chem. Soc., Chem. Commun.*, **1999**, 249.
53. D. Enders, I. Meyer, J. Runsink, and G. Raabe, *Tetrahedron.*, **1998**, *54*, 10733.
54. J. Peyronel, S. Grisoni, B. Carboni, T. Courgeon, and R. Carrié, *Tetrahedron.*, **1994**, *50*, 189.
55. S. Takano, M. Moriya, and K. Ogasawara, *Tetrahedron Asymm.*, **1992**, *3*, 681.
56. P. Garner and O. Dogan, *J. Org. Chem.*, **1994**, *59*, 4.
57. A. S. Anslow, L. M. Harwood, H. Phillips, and D. Watkin, *Tetrahedron Asymm.*, **1991**, *2*, 169.
58. A. S. Anslow, L. M. Harwood, H. Phillips, and D. Watkin, *Tetrahedron Asymm.*, **1991**, *2*, 997. A. S. Anslow, L. M. Harwood, H. Phillips, D. Watkin, and L. F. Wong, *Tetrahedron Asymm.*, **1991**, *2*, 1343. A. S. Anslow, L. M. Harwood, and I. A. Lilley, *Tetrahedron Asymm.*, **1995**, *6*, 2465.
59. L. M. Harwood, A. C. Manage, S. Robin, S. F. G. Hopes, D. J. Watkin, and C. E. Williams, *Synlett.*, **1993**, 777.
60. L. M. Harwood and I. A. Lilley, *Tetrahedron Asymm.*, **1995**, *6*, 1557.
61. L. M. Harwood, J. Marco, D. Watkin, C. E. Williams, and L. F. Wong, *Tetrahedron Asymm.*, **1992**, *3*, 1127.
62. D. Alker, G. Hamblett, L. M. Harwood, S. M. Robertson, D. Watkin, and C. E. Williams, *Tetrahedron.*, **1998**, *54*, 6089.
63. L. M. Harwood and S. M. Robertson, *J. Chem. Soc., Chem. Commun.*, **1998**, 2641.
64. J. E. Baldwin, S. C. McKenzie Turner, and M. G. Moloney, *Synlett*, **1994**, 925.

65. R. M. Williams, W. Zhai, D. J. Aldous, and S. C. Aldous, *J. Org. Chem.*, **1992**, *57*, 6527.

66. L. M. Harwood and I. A. Lilley, *Tetrahedron Lett.*, **1993**, *34*, 537. L. M. Harwood, L. C. Kitchen, *Tetrahedron Lett.*, **1993**, *34*, 6603.

67. L. M. Harwood and I. A. Lilley, *Synlett.*, **1996**, 1010.

68. S. Kanemasa and K. Doi, E. Wada, *Bull. Chem. Soc. Jpn.*, **1990**, *63*, 2866.

69. M. Maggini, G. Scorrano, and M. Prato, *J. Am. Chem. Soc.*, **1993**, *115*, 9798. M. Maggini, G. Scorrano, A. Bianco, C. Toniolo, R. P. Sijbesma, F. Wudl, and M. Prato, *J. Chem. Soc., Chem. Commun.*, **1994**, 305.

70. R. C. F. Jones, J. R. Nichols, and M. T. Cox, *Tetrahedron Lett.*, **1990**, *31*, 2333. R. C. F. Jones and K. J. Howard, *J. Chem. Soc., Chem. Commun.*, **1993**, 2391.

71. R. C. F. Jones, K. J. Howard, and J. S. Snaith, *Tetrahedron Lett.*, **1996**, *37*, 1707.

72. R. C. F. Jones, K. J. Howard, and J. S. Snaith, *Tetrahedron Lett.*, **1996**, *37*, 1711.

73. A. H. Fray and A. I. Meyers, *Tetrahedron Lett.*, **1992**, *33*, 3575.

74. P. DeShong, D. A. Kell, and D. R. Sidler, *J. Org. Chem.*, **1985**, *50*, 2309.

75. D. Wenkert, S. B. Ferguson, B. Porter, A. Qvarnstrom, and A. McPhail, *J. Org. Chem.*, **1985**, *50*, 4114.

76. B. R. Henke, A. J. Kouklis, and C. H. Heathcock, *J. Org. Chem.*, **1992**, *57*, 7056.

77. J. Sisko and S. M. Weinreb, *J. Org. Chem.*, **1991**, *56*, 3210.

78. P. Garner, P. B. Cox, J. T. Anderson, J. Protasiewicz, and R. Zaniewski, *J. Org. Chem.*, **1997**, *62*, 493.

79. S. Takano, K. Samizu, and K. Ogasawara, *Chem. Lett.*, **1990**, 1239.

80. S. Takano, Y. Iwabuchi, and K. Ogasawara, *J. Am. Chem. Soc.*, **1987**, *109*, 5523.

81. H. Ardill, R. Grigg, V. Sridharan, and S. Surendrakumar, *Tetrahedron.*, **1988**, *44*, 4953.

82. P. Armstrong, R. Grigg, M. W. Jordan, and J. F. Malone, *Tetrahedron.*, **1985**, *41*, 3547. R. Grigg and P. Armstrong, *Tetrahedron.*, **1989**, *45*, 7581.

83. R. Grigg, L. M. Duffy, M. J. Dorrity, J. F. Malone, S. Rajviroongit, and M. Thornton-Pett, *Tetrahedron.*, **1990**, *46*, 2213.

84. R. Grigg, V. Savic, and M. Thornton-Pett, *Tetrahedron.*, **1997** *53*, 10633.

85. P. N. Confalone and E. M. Huie, *J. Am. Chem. Soc.*, **1984**, *106*, 7175.

86. P. N. Confalone and R. A. Earl, *Tetrahedron Lett.*, **1986**, *27*, 2695.

87. S. F. Martin and T. H. Cheavens, *Tetrahedron Lett.*, **1989**, *30*, 7017.

88. I. Coldham, S. J. Coles, K. M. Crapnell, J-C. Fernàndez, T. F. N. Haxell, M. B. Hursthouse, J. D. Moseley, and A. B. Treacy, *J. Chem. Soc., Chem. Commun.*, **1999**, 1757.

89. N. K. Nayyar, D. R. Hutchinson, and M. J. Martinelli, *J. Org. Chem.*, **1997**, *62*, 982.

90. E. Vedejes and S. L. Dax, *Tetrahedron Lett.*, **1989**, *30*, 2627.

91. C. W. G. Fishwick, A. D. Jones, and M. B. Mitchell, *Tetrahedron Lett.*, **1989**, *30*, 4447. See also; C. W. G. Fishwick, A. D. Jones, M. B. Mitchell, and C. Szantay, Jr., *Tetrahedron Lett.*, **1988**, *41*, 5325.

92. A. I. D. Alanine and C. W. G. Fishwick, *Tetrahedron Lett.*, **1989**, *30*, 4443.

93. M. M. Murphy, J. R. Schullek, E. M. Gordon, and M. A. Gallop, *J. Am. Chem. Soc.*, **1995**, *117*, 7029.

94. M. Nyerges, M. Rudas, G. Tóth, B. Herényi, I. Kádas, I. Bitter, and L. Toke, *Tetrahedron.*, **1995**, *51*, 13321. M. Nyerges, L. Balázs , I. Kádas, I. Bitter, I. Kövesdi, and L. Toke, *Tetrahedron.*, **1995**, *51*, 6783.

95. C. Alvarez-Ibarra, A. G. Csákÿ, I. López de Silanes, and M. L. Quioga, *J. Org. Chem.*, **1997**, *62*, 479.

96. B. Schnell, G. Bernardinelli, and E. P. Kündig, *Synlett.*, **1999**, *3*, 348.

97. For examples see; R. Grigg, N. Gunaratne, and V. Sridharan, *Tetrahedron.*, **1987**, *43*, 5887. D. A. Barr, R. Grigg, H. Q. N. Gunaratne, J. Kemp, P. McMeekin, and V. Sridharan, *Tetrahedron.*, **1988**, *44*, 557. K. Amornralsa, D. Barr, G. Donegan, R. Grigg, P. Ratananukul, and V. Sridharan, *Tetrahedron.*,

1989, *45*, 4649. D. A. Barr, G. Donegan, and R. Grigg, *J. Chem. Soc., Chem. Commun.*, **1989**, 1550. R. Grigg, V. Longanathan, S. Sukirthalingam, and V. Sridharan, *Tetrahedron Lett.*, **1990**, *31*, 6573. R. Grigg and T. Coulter, *Tetrahedron Lett.*, **1991**, *32*, 1359. D. M. Cooper, R. Grigg, S. Hargreaves, P. Kennewell, and J. Redpath *Tetrahedron.*, **1995**, *51*, 7791. R. Grigg, M. Thornton-Pett, and G. Yoganathan, *Tetrahedron.*, **1999**, *55*, 8129.

98. R. Grigg, V. Sridharan, and S. Thianpatanagul, *J. Chem. Soc., Perkin Trans. 1*, **1986**, 1669.
99. D. A. Barr, R. Grigg, and V. Sridharan, *Tetrahedron Lett.*, **1989**, *30*, 4727.
100. D. A. Barr, M. J. Dorrity, R. Grigg, J. F. Malone, J. Montgomery, S. Rajviroongit, and P. Stevenson, *Tetrahedron Lett.*, **1990**, *31*, 6569.
101. P. Allway and R. Grigg, *Tetrahedron Lett.*, **1991**, *32*, 5817.

CHAPTER 4

Carbonyl Ylides

Mark C. McMills

Department of Chemistry and Biochemistry, Ohio University, Athens, OH

Dennis Wright

Department of Chemistry, University of Florida, Gainesville, FL

4.1. Synthesis, Structural Studies, and Reactivity of Carbonyl Ylides 254
- 4.1.1. Structural Studies of Carbonyl Ylides 255
- 4.1.2. Generation of Carbonyl Ylides from Nonstabilized Carbenes 259
- 4.1.3. Generation of Carbonyl Ylides from Oxadizolines 262
- 4.1.4. Generation of Carbonyl Ylides from Epoxides 263
- 4.1.5. Carbonyl Ylides from Metallocarbenoides 269
- 4.1.6. Carbonyl Ylides Derived from Ketones 269
- 4.1.7. Carbonyl Ylides Derived from Aldehydes 273
- 4.1.8. Carbonyl Ylides Derived from Esters 274
- 4.1.9. Carbonyl Ylides Derived from Amides and Ureas 276

4.2. Synthetic Aspects of Carbonyl Ylide Cycloaddition Reactions 278
- 4.2.1. Directed Syntheses with Tandem Carbonyl Ylide Formation/[1,3]-Dipolar Cycloaddition 278
  - 4.2.1.1. Preparation of Oxacyclic Compounds 278
- 4.2.2. Preparation of Alkaloid Systems 284
- 4.2.3. The Illudins, Pterosins, and Ptaquilosides 292
- 4.2.4. Approaches to Zaragozic Acids 294
- 4.2.5. Tropolone and Benzotropolone Formation and Reactivity 296
- 4.2.6. Enantioselective Syntheses 298

4.3. Oxidopyrylium Ion Cycloadditions 301

4.4. Synthesis Using Nonmetallocarbenoid Precursors and Novel Metallocarbenoid Processes 306

4.5. Concluding Remarks 309

The Chemistry of Heterocyclic Compounds, Volume 59: Synthetic Applications of 1,3-Dipolar Cycloaddition Chemistry Toward Heterocycles and Natural Products, Edited by Albert Padwa and William H. Pearson.
ISBN 0-471-38726-6

4.1. SYNTHESIS, STRUCTURAL STUDIES, AND REACTIVITY OF CARBONYL YLIDES

Carbonyl ylides (**1**) are highly reactive dipoles that have been proposed as key intermediates in a variety of reactions since the 1960s (Fig. 4.1). Since these early reports, there has been a virtual explosion in the study of these unstable intermediates both at the theoretical level and more recently in their application to organic synthesis. This chapter will focus on the structure, generation, and chemical reactions of carbonyl ylides and will review the literature since 1984.

The structure of the carbonyl ylide reveals that it is a 1,3-dipolar species and is poised to undergo a variety of different reactions. The ability of carbonyl ylides to engage in bond-forming processes has promoted their use in organic synthesis. Although there are several pathways open to these zwitterionic intermediates, there are a few that have been the focus of detailed mechanistic and synthetic investigations (Fig. 4.2).

The carbonyl ylide **1** can undergo an internal cyclization reaction to generate the corresponding epoxide **2**, which is in fact an equilibrium process, and epoxides themselves have frequently served as precursors to carbonyl ylides. Other pathways such as concerted rearrangements and internal proton transfers have also been observed to neutralize the charged ylide intermediate and give substituted ethers as represented by **3**. Perhaps the best known studies and most synthetically useful

Figure 4.1 Structure of carbonyl ylides.

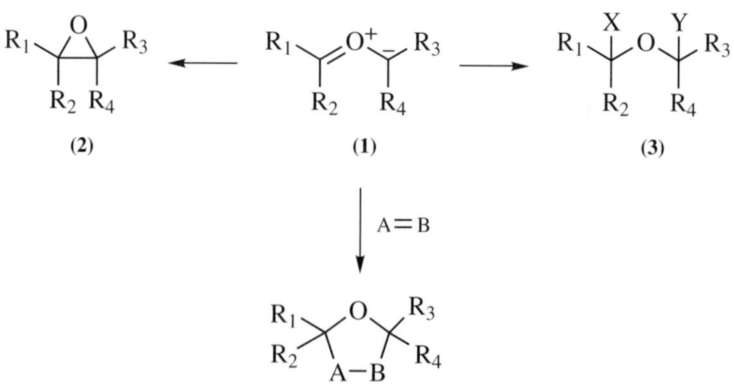

Figure 4.2 Reactions of carbonyl ylides.

reaction of carbonyl ylides arises from their participation in [3 + 2]-dipolar cycloaddition reactions. The carbonyl ylide reacts efficiently as a 1,3-dipolar intermediate with a variety of double- and triple-bond species to lead to the production of five-membered oxacycles such as **4**.

4.1.1. Structural Studies of Carbonyl Ylides

The exact structure of carbonyl ylides has been the subject of a variety of theoretical investigations over the past few decades since their intermediacy was suggested in 1965 during the cycloaddition reaction of substituted epoxides (1). Houk et al. (2) has undertaken a detailed study of the carbonyl ylide structure and reactivity by the application of computational methods (Fig. 4.3).

It was determined that the energy of the carbonyl ylide **6** was 63 kcal/mol higher than the ethylene oxide precursor (**5**). The C—H bond lengths were fixed (1.09 Å), as were the HCO angles (118°), then the optimized geometry was determined (STO-3G level) for the C—O—C bond length and angle. The optimal geometry was determined to be that shown in the planar structure **6** with a C—O—C bond angle of 129° and a C—O bond length of 1.297 Å that would be expected to optimize the allylic-type resonance possible for these charged species. The planar intermediate's C—O bond length of 1.297 Å lies in between the C=O bond length of formaldehyde (1.119 Å) and the C—O bond length of dimethyl ether (1.416 Å). Rotation of the C—O bond, 90° out of plane, gave the twisted ylide structure **7** (with a higher degree of zwitterionic character) that was found to be energetically unfavorable as is the linear structure **8**. Houk et al. (2) proceeded to calculate the geometries of more highly substituted ylides including the highly stabilized push–pull carbonyl ylides.

Although many substituted carbonyl ylides had been generated, the parent carbonyl ylide **6** had not been prepared experimentally until 1986 as reported in an elegant study by Olah and co-workers (3) (Scheme 4.1).

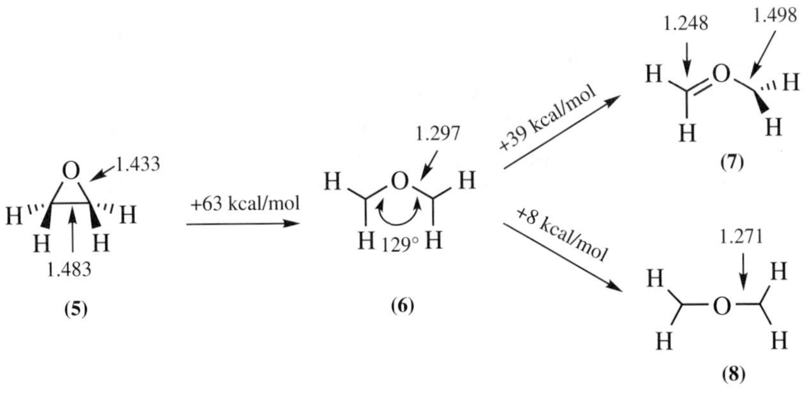

Figure 4.3 Energies of carbonyl ylide geometries, bond length in angstroms (Å), and bond angles in degrees at the 4–31 G level.

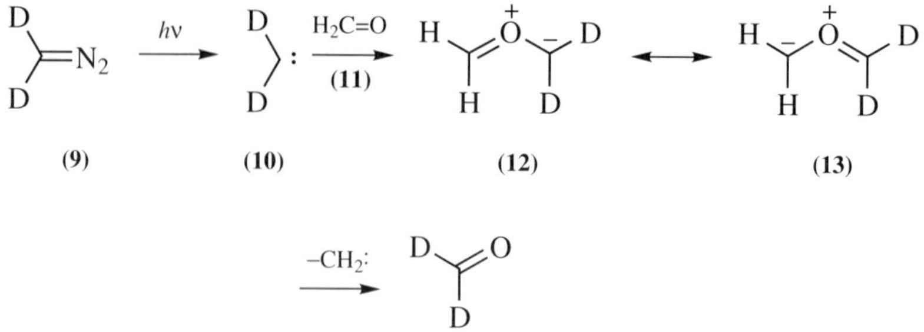

(9) (10) (12) (13)

(14)

Scheme 4.1 Generation of the parent carbonyl ylide.

Photolysis of dideuteriodiazomethane (**9**) generated the dideuteriocarbene (**10**) that was trapped as the carbonyl ylide **12** by addition of a solution of monomeric formaldehyde in dimethyl ether. The position of the deuterium label is fluxional via the resonance form **13** that decomposed through loss of methylene to give dideuterioformaldehyde (**14**) as an exchange product. This product was detected as 10–12% of the total reaction mass by mass spectrometry (MS). Dideuterioethylene oxide, a compound generated by direct insertion of the carbene into the C=O bond of formaldehyde or via collapse of ylide **12**, was not detected by MS. Significantly less exchange was observed without photolysis, which involved a direct dipolar cycloaddition of formaldehyde to dideuteriodiazomethane, followed by the thermal extrusion of nitrogen. These workers take these observations as an indication that they were able to generate the parent ylide, formaldehyde-O-methylidine for the first time.

Turro and Cha (4) conducted ylide-trapping experiments with the simple carbonyl ylide generated from methylene and acetone (Scheme 4.2). A solution of diazomethane and a trapping agent (acrylonitrile) was dissolved in acetone (**15**) and irradiated at low temperatures with an Oriel 1000-W Xe–Hg lamp. Generation of singlet methylene followed by trapping with acetone was proposed as the synthetic route to generate the carbonyl ylide **16**. Subsequent [1,3]-dipolar cycloaddition with acrylonitrile generated the tetrahydrofuran (THF) derivatives **17** and **18** in a 2:1 ratio. The preference for the formation of **17**, as analyzed with frontier molecular orbital (FMO) theory, suggests that in the highest occupied molecular orbital (HOMO) of **16** (typically used in dipolar cycloaddition reactions), the anionic charge is more localized on the unsubstituted position. The terminus of

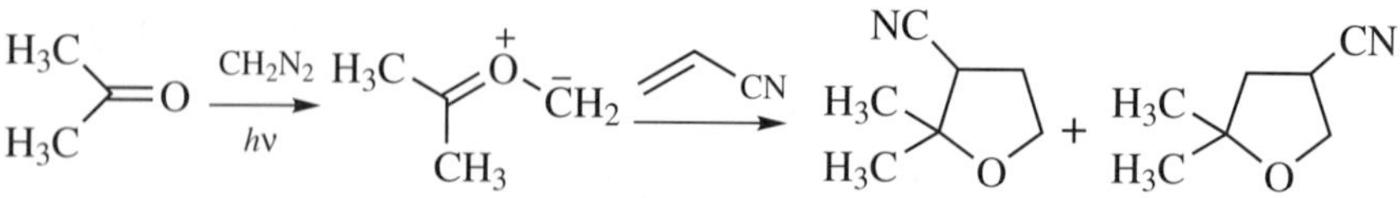

(15) (16) (17) (18)

Scheme 4.2 Ylide trapping experiments.

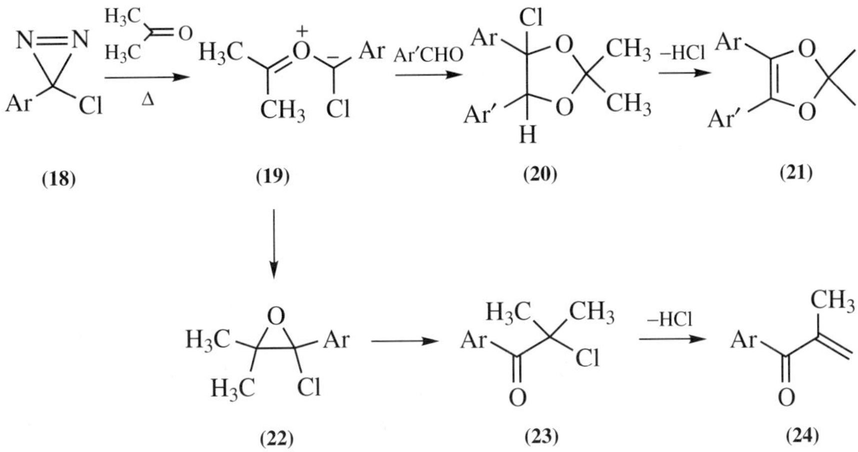

Scheme 4.3 Laser photolysis study of a carbonyl ylide.

the dipole would be expected to interact with the electron-poor β-position of acrylonitrile where the coefficient of the lowest unoccupied molecular orbital (LUMO) would be the largest. Laser flash photolysis of assorted carbonyl ylides was undertaken in 1990 by Bonneau and Liu (5). These authors examined the ylide formed through thermal decomposition of aryl chlorodiazirines in the presence of acetone (Scheme 4.3).

Decomposition of **18** under thermal conditions in the presence of acetone in hydrocarbon solvent led to the formation of **23** and **24**. These were shown to arise through rearrangement of the chloroepoxide (**22**) formed upon collapse of ylide **19**. Thermolysis in the presence of excess acetone and a substituted benzaldehyde led to a 53% yield of **21**, formed by an initial dipolar cycloaddition with the substituted benzaldehyde and ylide **19** to generate dioxolane **20**. Dehydrochlorination of **20** yielded **21**.

Laser flash photolysis of **18** ($Ar = p\text{-}NO_2C_6H_4$) led to the formation of the aryl chlorocarbene detected at 320 nm. In the presence of acetone, a new species was observed at $\lambda = 590$ nm that was assigned the structure of carbonyl ylide **19**. The ylide, formed by attack of acetone on the carbene, was shown to be irreversible, where the lifetime of the ylide (1.35 μs, $k_{\text{cyclization}} = 7.40 \times 10^5\ s^{-1}$) was controlled by cyclization to the aryl epoxide **22**. The rate constant for the cycloaddition of substituted benzaldehydes to produce dehydrodioxolane (**21**) was determined experimentally (e.g., *p*-ClPhCHO=$6.16 \times 10^8\ M^{-1}s^{-1}$).

The formation of carbonyl ylides from esters was also observed by a laser photolysis study conducted by Chateauneuf and Liu (6). The formation of ester ylides is much less common than from aldehydes or ketones (Scheme 4.4).

Photolytic decomposition of diazirine (**25**) in the presence of ethyl acetate led to the formation of the ester derived carbonyl ylide (**26**). Trapping of the ylide with diethyl fumarate led to the formation of the dihydrofuran **28**, generated by dehydrochlorination of the initial cycloadduct **27**. It occurred with a rate constant

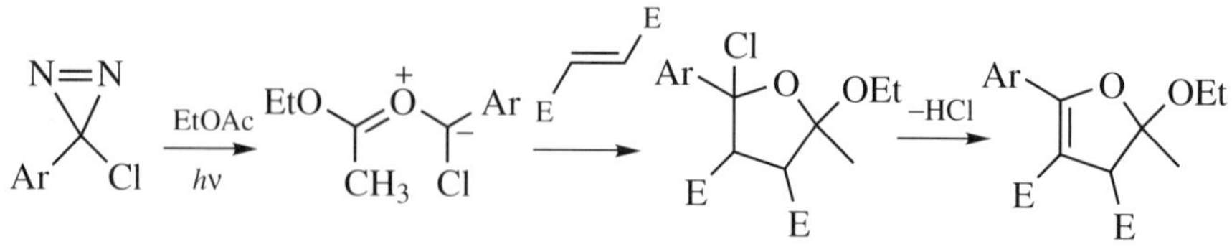

(25) **(26)** **(27)** **(28)**

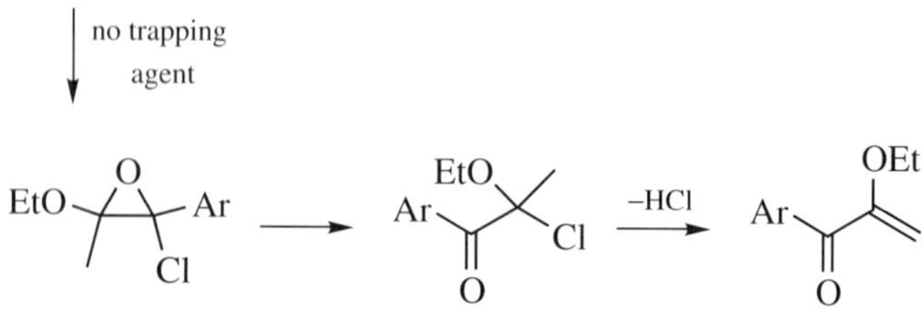

(29) **(30)** **(31)**

Scheme 4.4 Generation of a carbonyl ylides from esters.

of $k = -1.0 \times 10^7\ s^{-1}$. Omission of the trapping agent led to the formation of the chloroepoxide **29** that forms **30** and **31** upon thermolysis of the oxirane via rearrangement. This reaction occurs with a rate constant of $k = 1.3 \times 10^6\ s^{-1}$.

The first report regarding the structure of a stable carbonyl ylide was disclosed in 1983 by Arduengo and co-worker (7). This exciting result was accomplished by taking advantage of the highly stabilized push–pull nature of carbonyl ylides containing substituents that stabilize the zwitterionic intermediate (Scheme 4.5).

Photolytic decomposition of diazotetrakis (trifluoromethyl) cyclopentadiene (**32**) in the presence of tetramethylurea led to trapping of the singlet carbene by the highly nucleophilic oxygen atom of the urea moiety to give ylide **33**. The presence of four trifluoromethyl groups helps to stabilize the aromatic-like cyclopentadienyl anion while the two dimethylamino group help delocalize the cationic terminus of the dipole. The ylide is a remarkably stable example of a push–pull ylide and can be isolated as a crystalline solid (mp 200–202 °C). High-quality crystals, suitable for X-ray analysis, were produced and the solid-state structure of **33** was determined. The central C–O–C angle is 121.1°, similar to the bond angle of an sp^2 hydridized carbon while the two C–O bonds differ in length, 1.348 Å for the cationic terminus and 1.422 Å for the anionic terminus. The high stability of the anion delocalized

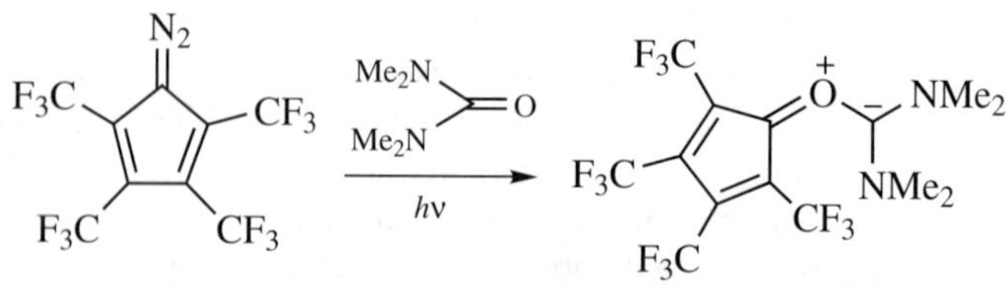

(32) **(33)**

Scheme 4.5 Generation of a stable carbonyl ylide.

through the cyclopentadienyl system allows it to twist out of plane and produce a nonplanar stable intermediate.

4.1.2. Generation of Carbonyl Ylides from Nonstabilized Carbenes

As illustrated in Section 4.1.1, the addition of nonstabilized carbenes to the oxygen atom of a carbonyl derivative can lead to the production of carbonyl ylides. However, these methods are not always practical for preparative scale since many side reactions can accompany the decomposition of alkyl diazo and diazirine derivatives. Landgrebe and co-worker (8) extensively studied the thermal decomposition of organomercurials in the presence of carbonyl compounds for the preparative generation of carbonyl ylides (Scheme 4.6).

Heating organomercurial **34** resulted in the α-elimination of phenyl mercury bromide, generating dichlorocarbene, followed by addition of benzaldehyde to form the corresponding carbonyl ylide **35**. This ylide can be intercepted with dimethylacetylene dicarboxylate (DMAD) to produce dihydrofuran (**36**), which formed furan **37** through a dehydrochlorination process in 46% yield.

Alkene dipolarophiles such as diethyl fumarate were shown to be somewhat less reactive than electron poor acetylenes (9), but were effective for the formation of dihydrofuran derivatives (Scheme 4.7).

Carbonyl ylide **35**, generated from the carbene precursor **34** and benzaldehyde, was trapped with diethylfumarate to give the highly substituted THF **38**. Under the reaction conditions, elimination of hydrochloric acid occurred spontaneously to give the cis and trans substituted dihydrofurans **39** and **40** in a 7:3 ratio and a 30% overall chemical yield. If the trapping agent was omitted from the reaction, complex product mixtures were recovered including products derived from the deoxygenation of the aldehydes (10). Liu and co-workers (11) conducted similar studies, generating the initial carbene via thermolysis of substituted diazirines (Scheme 4.8).

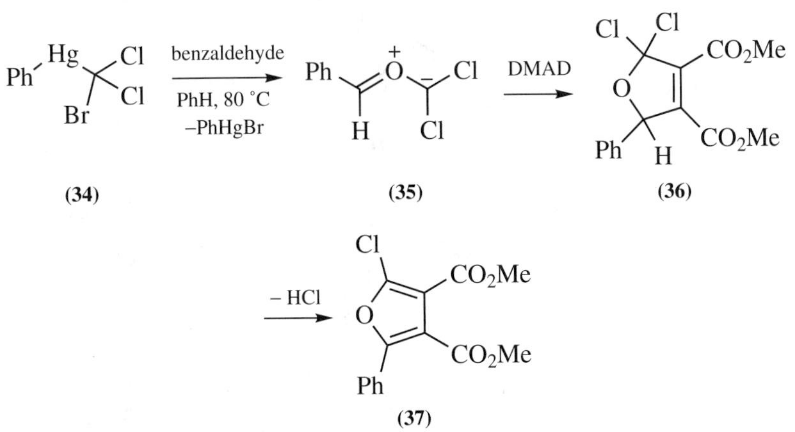

Scheme 4.6 Generation of ylides via organomercurials.

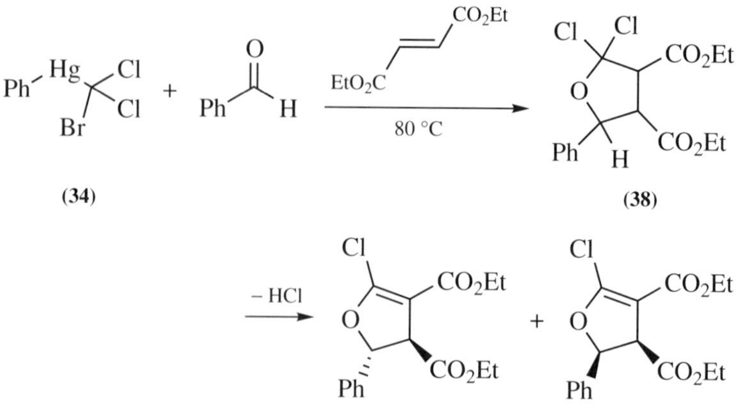

(34) (38) (39) (40)

Scheme 4.7 Generation of dihydrofurans from organomercury compounds.

Thermolysis of aryl chloro diazirine (**18**) in the presence of acetone and a trapping agent such as *N*-phenylmaleimide gave rise to cycloadducts such as **41**. The unstable adduct hydrolyzed during purification resulting in synthesis of bicyclic hemiacetals **42** and **43** as a mixture of endo and exo adducts in 37 and 8% yield, respectively. The exclusive generation of the singlet carbene was confirmed by low-temperature electron spin resonance (ESR) study of the irradiated diazirine.

Hosomi and co-workers (12) recently developed a novel method for generating nonstabilized carbonyl ylides through a samarium-mediated reaction (Scheme 4.9).

This unusual reaction involves the reductive dimerization of protected iodohydrin (**44**) that produced a symmetrically substituted carbonyl ylide (**48**). The mechanism proposed for this interesting process involved initial reduction of the

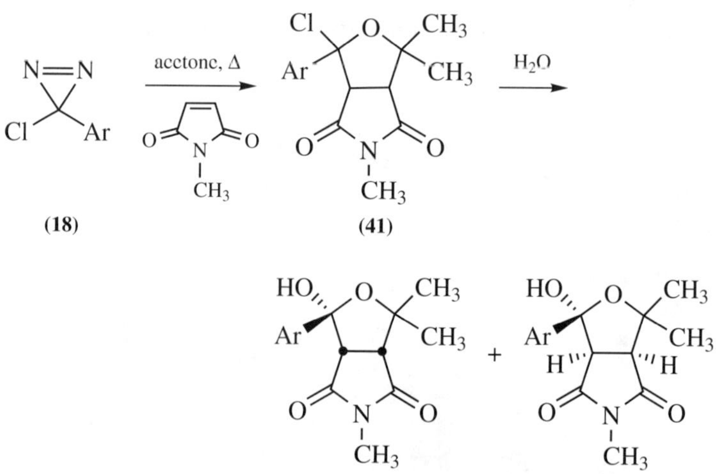

(18) (41) (42) (43)

Scheme 4.8 Trapping of ylides from diazirines.

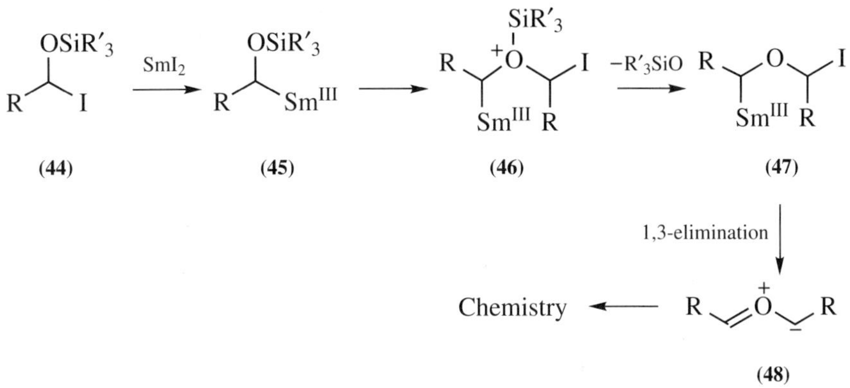

Scheme 4.9 Generation of Nonstabilized ylides.

carbon iodide bond that gave the α-metalated derivative **45** that served as a carbene precursor through α-elimination of the siloxy group. The transient ylide was attacked by a second molecule of **45** producing the metalated oxonium ion **46**, then desilylation formed ylide precursor **47**. α-Iodoether (**47**) was poised to undergo a 1,3-elimination reaction thus generating oxonium ylide **48** that can react in typical carbonyl ylide processes. Hosomi and co-workers showed that these compounds easily participate in dipolar cycloaddition reactions (see Section 4.2).

Earlier, Hosomi and co-workers (13) reported a procedure for generating carbonyl ylides through a 1,3-elimination reaction that converted **47** to ylide **48** (Scheme 4.10).

Silyl substituted chloromethyl ethers such as **49** serve as a convenient precursor to nonstabilized carbonyl ylides. Treatment with a fluoride source promoted

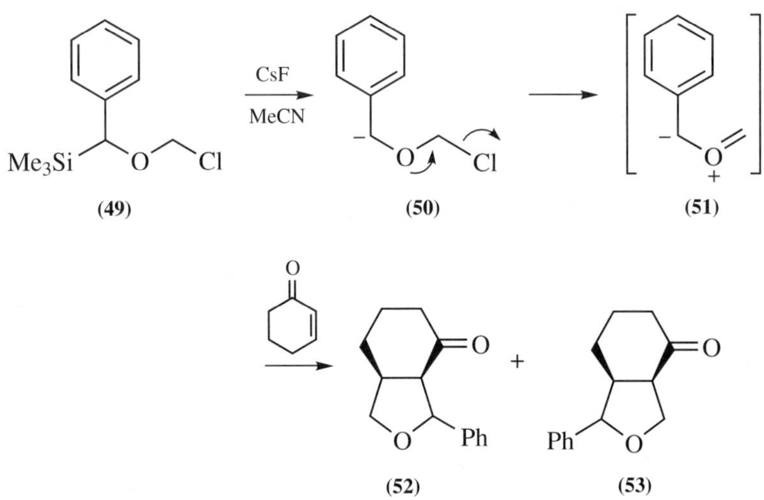

Scheme 4.10 Generation of carbonyl ylides via desilylation.

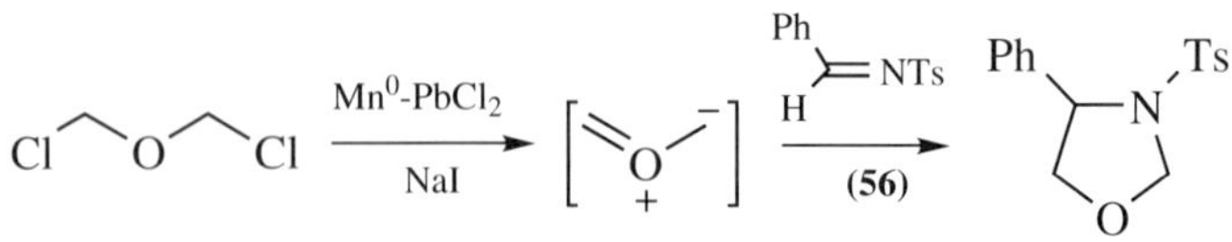

(54) (55) (57)

Scheme 4.11 Generation of carbonyl ylides with a manganese system.

desilylation to give the anion **50** that reacted via a 1,3-elimination of chloride to generate the carbonyl ylide **51**. Addition of the dipolarophile cyclohexenone trapped the ylide in a dipolar cycloaddition process that gave an almost equal mixture of the regioisomeric cycloadducts **52** and **53** in 81% overall yield.

More recently, the same investigators (14) reported a convenient modification of the 1,3-elimination process using a bis(chloromethylether) with a mixed-metal system (Scheme 4.11).

Treatment of bis(chloromethylether) **54** with a mixed-manganese/lead system generated a reactive intermediate equivalent to the carbonyl ylide depicted in **55**. Intermediate **55** behaved as a 1,3-dipole undergoing cycloaddition with a variety of dipolarophiles including *N*-tosylaldimines such as **56**, resulting in the formation of oxazolidine **57** in 73% yield.

4.1.3. Generation of Carbonyl Ylides from Oxadiazolines

Wartenkin and co-workers (15,16) developed a versatile route for the synthesis of carbonyl ylides via the decomposition of 2-methoxy-2,5,5-Δ^3-1,3,4-oxadiazoline (**59**) under thermal conditions (Scheme 4.12).

The carbonyl ylide precursor can be generated by lead tetraacetate oxidation of the hydrazone **58**. Thermolysis of **59** in the presence of perdeuterated acetone led to a variety of products, some of which are shown above. An internal quench of the ylide via a 1,4-proton migration led to enol ether **61,** while cycloaddition with perdeuterated acetone formed the dioxolane **62** and its regioisomer. Interestingly, the presence of products such as acetone and propene-d_6 are proposed to indicate a reversible fragmentation of the ylide to a carbonyl derivative and a carbene.

Other studies with the same ylide precursor showed similar ylide reactivity (Scheme 4.13).

Thermolysis of **59** in chloroform (17) led to formation of carbonyl ylide **64**, subsequent proton abstraction from chloroform ($pK_a = 24.1$), and recombination with the trichloromethyl anion gave acetal **65**. The intermediacy of radicals was discounted since conducting the reaction in neat Bu_3SnH did not change the product distribution.

Trapping of the analogous ylide (18) by dipolarophiles has also been observed (Scheme 4.14). Oxadiazolidine (**66**) generated ylide **67** upon heating, followed by cycloaddition with electron-poor alkenes such as methacrylate and acrylonitrile to generate cyclic acetals such as **68** as a 1:1 mixture of stereoisomers in 25% yield.

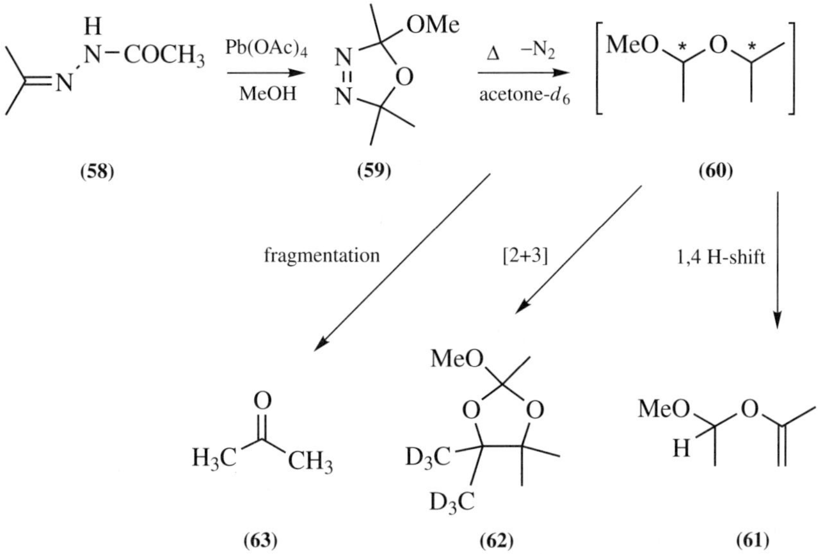

Scheme 4.12 Generation of carbonyl ylides from oxadiazolines.

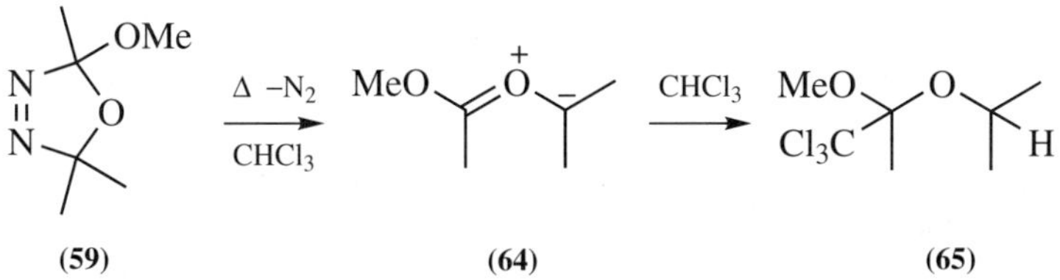

Scheme 4.13 Reaction with chloroform.

4.1.4. Generation of Carbonyl Ylides from Epoxides

It was shown that carbonyl ylides can collapse to the corresponding epoxides through an intramolecular cyclization process. This reaction should be reversible and therefore epoxides could also serve as precursors to carbonyl ylides. Calculations have shown that the ylide is >60 kcal/mol less stable than the corresponding

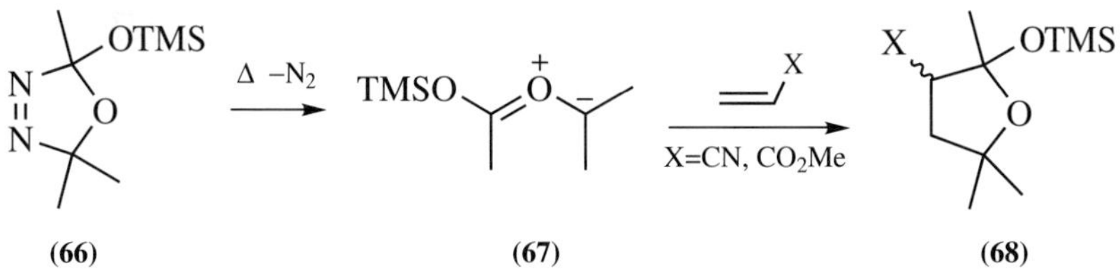

TMS = trimethylsilyl

Scheme 4.14 Reaction with dipolarophiles.

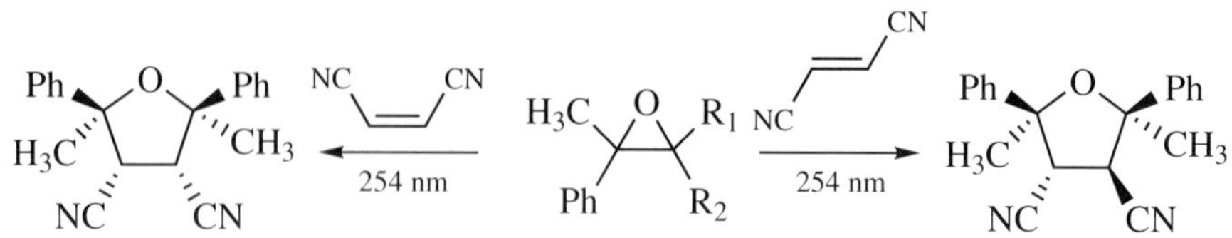

(71) (69a); $R_1 = CH_3$, $R_2 = Ph$ (70)

(69b); $R_1 = Ph$, $R_2 = CH_3$

Scheme 4.15 Epoxide photolysis.

epoxide and would require extreme conditions to cause the isomerization of the oxirane to occur. Griffin and co-workers (19) carried out a systematic study of the photochemical and thermal ring-opening method for substituted oxiranes (Scheme 4.15).

Among the parameters studied, the role of oxirane geometry as it relates to the structure of the ylide was investigated. Photolytic conversion of the isomeric epoxides **69a** and **69b** to the carbonyl ylide followed by trapping with an electron-deficient alkene demonstrated that the geometry of the starting epoxide was independent of the nature of the ylide structure. The stereochemical information contained in the malononitrile and fumaronitrile olefin geometry was conserved in the cycloadduct, suggesting a concerted mechanism for the cycloaddition. Several explanations were offered for the lack of influence of the epoxide, including a rapid isomerization of the ylide after disrotatory ring opening or, recyclization followed by opening, leading to isomerization. Photolysis conducted in a frozen argon matrix at 77 K revealed that the spectra of the ylides derived from **69a** and **69b** were identical. More highly functionalized epoxides have also been studied (Scheme 4.16).

Photolysis of spiro epoxide **72** (20) led to formation of carbonyl ylide **73**. Although the ylide failed to react with either electron-rich or electron-poor

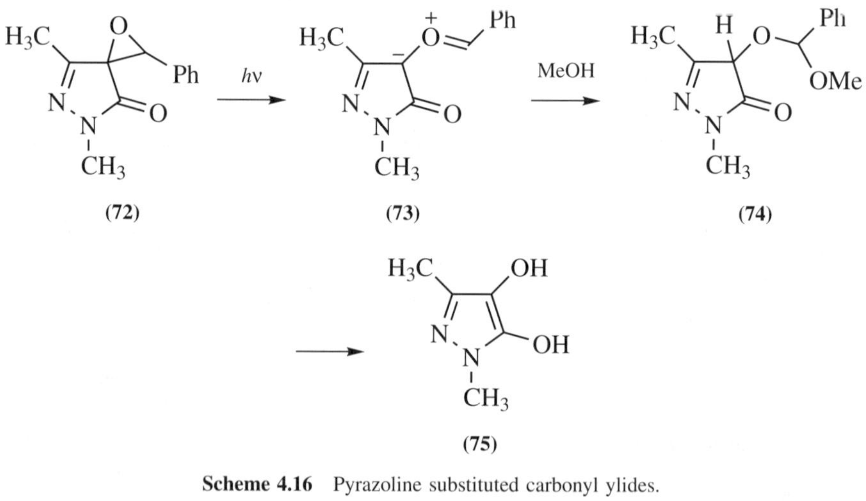

(72) (73) (74) (75)

Scheme 4.16 Pyrazoline substituted carbonyl ylides.

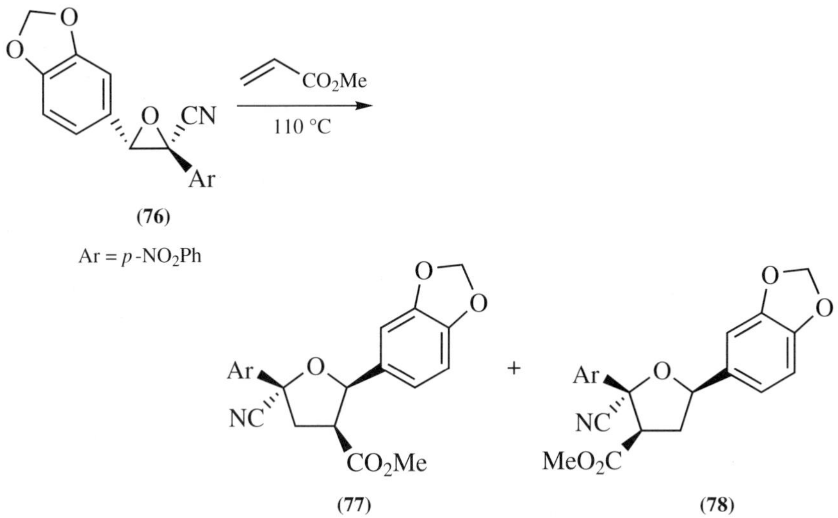

Scheme 4.17 Push–pull ylides for lignan synthesis.

dipolarophiles, the ylide could be trapped by the addition of methanol to give the acetal derivative **74**, which hydrolyzes to the dihydroxypyrazole **75**. The regioselectivity of the methanol addition suggests that **73** best represents the ylide and that the opening of the epoxide is selective to retain anion stabilization by the adjacent carbonyl functionality.

Carbonyl ylides with both electron-donating and electron-withdrawing groups present are highly stablilized ylides known as *push–pull* ylides. The presence of stabilizing groups can control the regioselectivity of the ylide generation that can in turn control the regioselectivity of an ensuing cycloaddition. Clawson and Whiting (21) examined push–pull ylides in a general approach to lignans (Scheme 4.17).

Substituted epoxides such as **76** were designed with an electron-donating aromatic group on one carbon of the pendent ylide to stabilize a cationic charge. Cyano and nitrophenyl groups were substituted at the other end of the epoxide to stabilize the anionic terminus of the dipole. Thermolysis of **76** in the presence of methacrylate led to formation of the corresponding dipolar cycloadducts. Surprisingly, the cycloaddition was non-regioselective, producing both **77** and **78** in a nearly 1:1 ratio. It was proposed that the lack of regioselectivity was related to the extreme polarization of the ylide that could narrow the HOMO and LUMO gap of the FMO that would result in the opposite directing effects.

Carbon-heteroatom multiple bonds can also participate in cycloaddition reactions with carbonyl ylides leading to the synthesis of interesting heterocycles (Scheme 4.18).

Thermolysis of a solution of epoxide **79** and 1,3-thiazole-5(4*H*)-thione derivative **80** in xylene led to the formation of three cycloadducts in 85% overall yield (22). It was expected that the geminal cyano groups of the epoxide would control the regioselectivity of the ring opening such that the anionic terminus of the dipole would be localized on that carbon. Regiochemical control led to a very selective

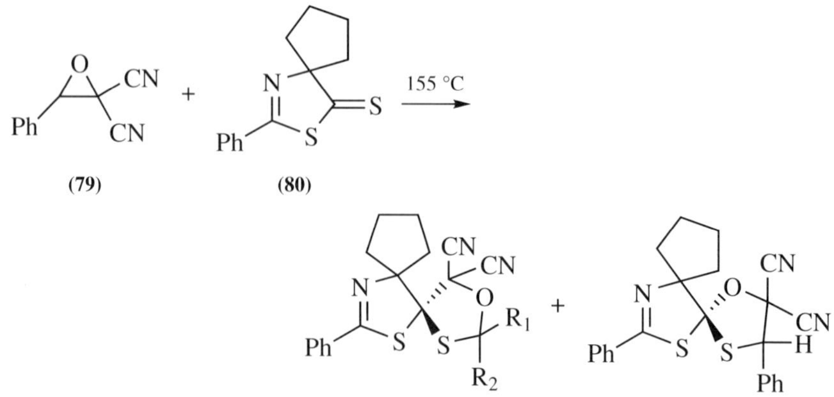

(79) (80)

(81); R_1 = Ph, R_2 = H (83)

(82); R_1 = H, R_1 = Ph

Scheme 4.18 Addition of thiocarbonyl derivatives to carbonyl ylides.

cycloaddition with the thiocarbonyl group generating adducts **81** and **82** in 61 and 21% yields, respectively. The alternative regioisomer **83** would result from the reversed polarity of the dipole and was formed in only 3% yield.

There have been several other investigations relating to the regioselectivity of the cycloaddition step. Eberbach and co-workers (23) undertook a detailed study of a series of trisubstitued epoxides (Scheme 4.19).

Thermal ring opening of epoxide **84** led to the regiospecific generation of carbonyl ylide **85** that subsequently underwent cycloaddition with a variety alkene derivatives to produce two regioisomeric cycloadducts. An electron-poor acrylate dipolarophile showed a surprising lack of control during cycloaddition and led to the formation of the two regioisomers in nearly equivalent amounts. The selectivity improved with a more electron-rich acceptor, 1-hexene, to produce a 4:1 distribution of regioisomers, while the addition of a highly electron-rich olefin, ethyl vinyl ether, gave complete regioselectivity. The highly selective reactions with neutral or

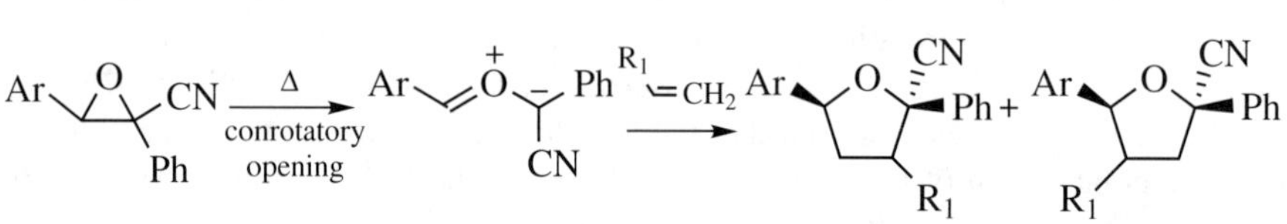

(84) (85) (86) (87)

R_1	**86:87**
CO_2R	50:50
C_4H_9	81:19
OEt	100:0

Scheme 4.19 Regiochemical studies on cycloaddition reactions.

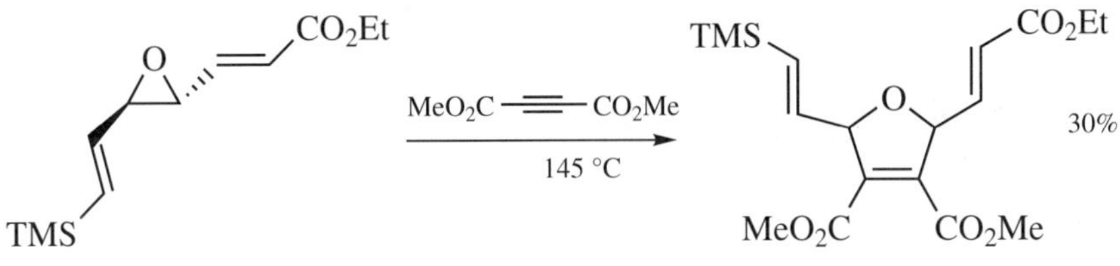

(88) (89)

Scheme 4.20 Carbonyl ylides from divinyl epoxides.

electron-rich olefins were much slower reactions and would seem to be under the control of HOMO–LUMO interactions. White and co-worker (24) found that ylide generation is also possible in more highly conjugated systems prepared from *trans*-divinyl epoxides (Scheme 4.20).

Whereas, *cis*-divinyl epoxides are reactive and well known to undergo thermal Cope rearrangement, the trans isomers are significantly more stable. White and co-worker (24) showed that thermolysis of divinyl epoxides such as **88** could generate the corresponding carbonyl ylide and that it could be intercepted by the addition of an activated acetylene to give the corresponding dihydrofuran **89**, albeit in modest yield.

Ishii and co-workers (25,26) examined the generation of ylides from epoxy dinitrile precursors (Scheme 4.21).

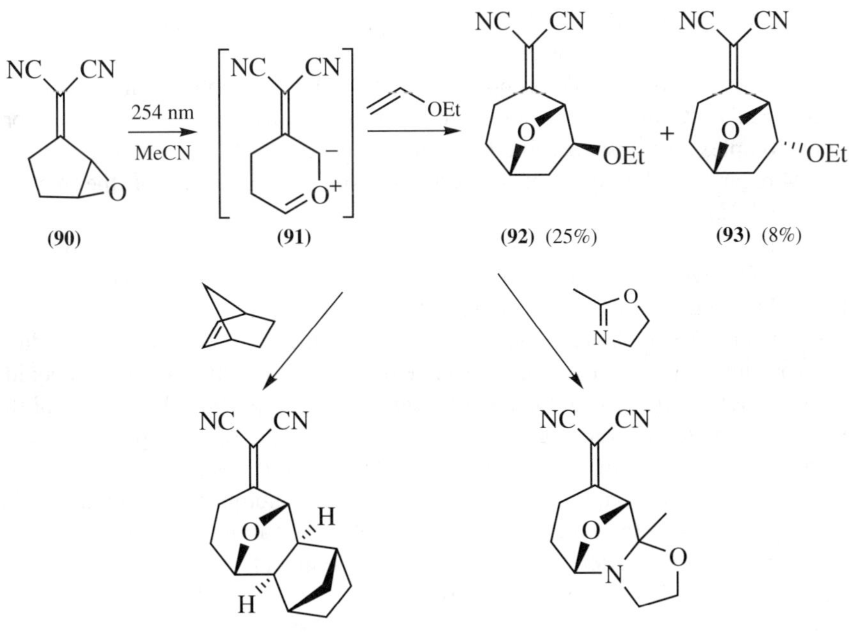

(94) (4%) (95) (10%)

Scheme 4.21 Ylides from epoxy dinitriles.

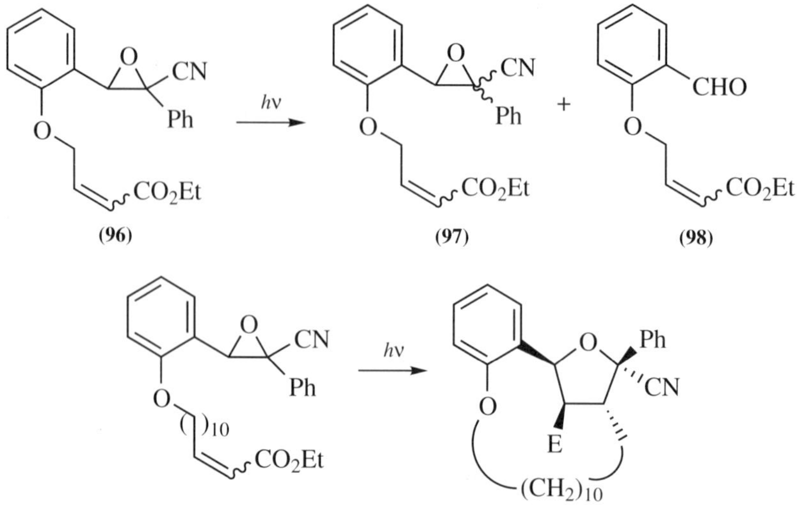

(96) (97) (98) (99) (100)

Scheme 4.22 Intramolecular cycloadditions.

Irradiation of the epoxy dinitrile **90** in a solution of acetonitrile formed the stabilized ylide **91** that was trapped with ethyl vinyl ether, producing both exo and endo adducts **92** and **93** in 25 and 8% yields. Studies with other dipolarophiles (27) such as norbornene and methyloxazoline generated cycloadducts **94** and **95**, respectively, in low overall yield.

The photolysis of epoxides containing a tethered dipolarophile can result in the intramolecular capture of the transient ylide and lead to the synthesis of more highly complex ring systems. Eberbach and co-workers (28–30) conducted several studies regarding this type of process to probe the viability of the method (Scheme 4.22).

Epoxide **96** was prepared such that photolytic conversion to the carbonyl ylide could be followed by an intramolecular cycloaddition with the tethered pendant olefin. However, photolysis of epoxide **96** led only to the formation of the regioisomer **97** and the aldehyde **98** with no evidence of the corresponding cycloadduct. It was presumed that **97** arose from the ylide by thermal recyclization to the epoxide while **98** could form through the loss of a carbene from the ylide. The failure of the tethered alkene to undergo cycloaddition may have resulted from a poor trajectory for the cycloaddition. An extended analogue (**99**) allowed greater flexibility for the dipolarophile to adopt any number of conformations. Photolysis of epoxide **99** did lead to formation of the macrocyclic adduct **100**, albeit in modest yields.

Although cycloaddition reactions are the dominant reaction manifold for carbonyl ylides, other processes such as electrocyclizations (31) are also observed (Scheme 4.23).

Sharp and co-workers (32) observed that carbonyl ylides can participate in electrocyclic ring closures. Flash vacuum pyrolysis of epoxy ester **101** led initially

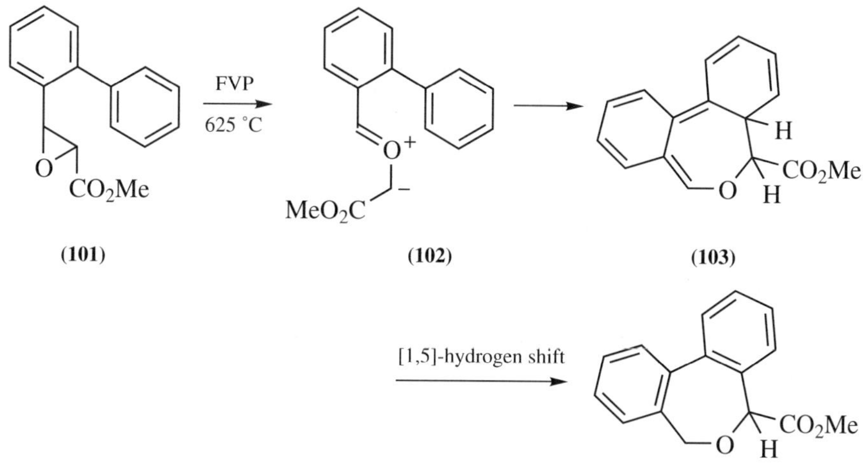

(101) (102) (103) (104)

Scheme 4.23 Electrocyclization of carbonyl ylides.

to the stabilized ylide **102** experienced a 1,7-electrocyclization with the neighboring phenyl ring to give **103**. This compound will further suffer a 1,5-hydrogen shift to regenerate the two aromatic rings found in oxepine **104**. Although epoxides can serve as key precursors to carbonyl ylides, the yields typically associated with these reactions are moderate or low, which makes them unsuitable for organic synthesis.

Over the past decade, a reliable, high-yielding method has been developed for the preparation of carbonyl ylide intermediates, namely, the addition of stabilized metallocarbenoid species to carbonyl compounds.

4.1.5. Carbonyl Ylides from Metallocarbenoids

Carbonyl ylides can be viewed as an adduct between a carbonyl group and a carbene and, in fact, some ylides have been prepared this way (see above). The application of carbonyl ylides to the synthesis of complex natural products has been greatly advanced by the finding that stabilized carbenoids can be generated by the decomposition of α-diazocarbonyl compounds with copper and rhodium complexes. The metallocarbenoids formed by this method are highly electrophilic on carbon and readily add nucleophiles such as the oxygen of many carbonyl derivatives to form carbonyl ylides. This type of reaction is in fact quite old with the first report being the addition of diazomalonate and benzaldehyde (33,34).

4.1.6. Carbonyl Ylides Derived from Ketones

Landgrebe and co-workers (35) studied the reaction of ketones with metallocarbenoids and observed the generation of enol ethers via an intramolecular rearrangement (Scheme 4.24).

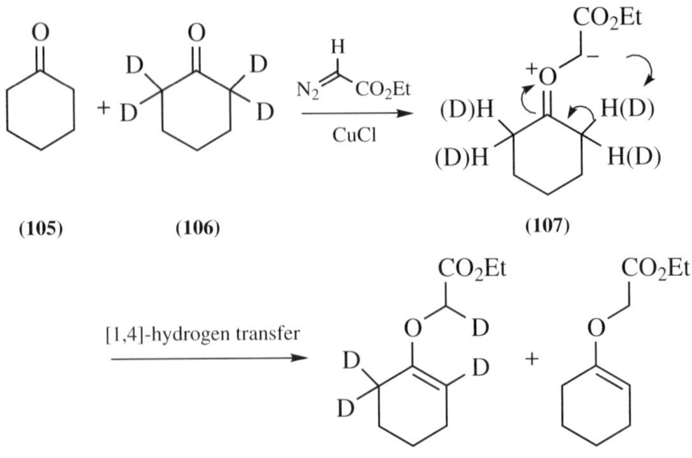

(105) **(106)** **(107)** **(108)** **(109)**

Scheme 4.24 Enol ethers from carbonyl ylides.

Labeling studies were undertaken to observe if quenching of the ylide by proton transfer occurred through an inter- or intramolecular reaction. A cross-over experiment was performed by treatment of an equimolar mixture of cyclohexanone (**105**) and the tetradeuterated derivative (**106**) with ethyl diazoacetate and copper(I)chloride. This reaction led to the predominant formation of **108** and **109**, thus indicating that the proton transfer occurs through an intramolecular [1,4]-hydrogen shift rather than through a bimolecular process. Landgrebe and co-workers (35) extended this study to include a variety of ketones and found that the preference for the [1,4]-migration was $CH_3 > CH_2 \gg CH$ with tertiary hydrogens producing products in extremely low yields.

Padwa and co-workers (36–38) developed a process concerning a tandem carbene cyclization–cycloaddition sequence that led to the formation of products of greater complexity in a single step (Scheme 4.25).

Decomposition of diazoketone **110** with rhodium acetate produced the highly electrophilic rhodium stabilized metallocarbenoid that suffers attack by the Lewis basic oxygen of the pendant ketone, producing cyclic carbonyl ylide **111**. This ylide was trapped by the addition of an activated acetylene such as DMAD to furnish

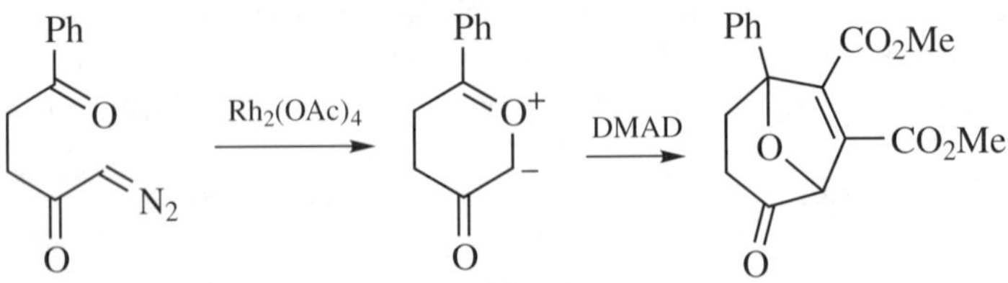

(110) **(111)** **(112)**

Scheme 4.25 Tandem cyclization–cycloaddition reaction.

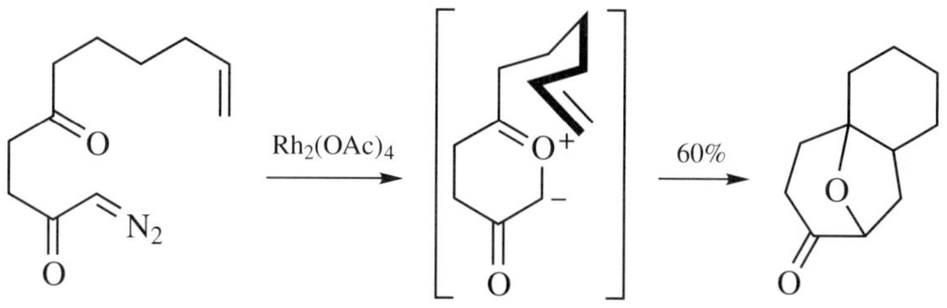

(113) **(114)** **(115)**

Scheme 4.26 Intramolecular cycloadditions.

oxabicyclo[3.2.1]heptane adduct **112** in an excellent overall yield of 93%. More complex ring systems can be envisioned if the dipolarophile were already tethered to the cyclic carbonyl ylide. Padwa et al. (39–41) examined many of these systems (Scheme 4.26).

Decomposition of diazoketone **113** with rhodium acetate led to the formation of a tethered cyclic carbonyl ylide **114** that was poised to undergo an intramolecular cycloaddition, preparing **115** in 60% yield. Interestingly, if DMAD was added to the reaction mixture, the only product arose from intermolecular cycloaddition.

Structurally complex substrates can be realized when an alkyne is inserted between the metallocarbenoid and the carbonyl trap (Scheme 4.27).

Padwa et al. (42,43) investigated the cyclization of diazoalkynyl ketones such as **116** and found that upon exposure to rhodium acetate, the transient metallocarbenoid

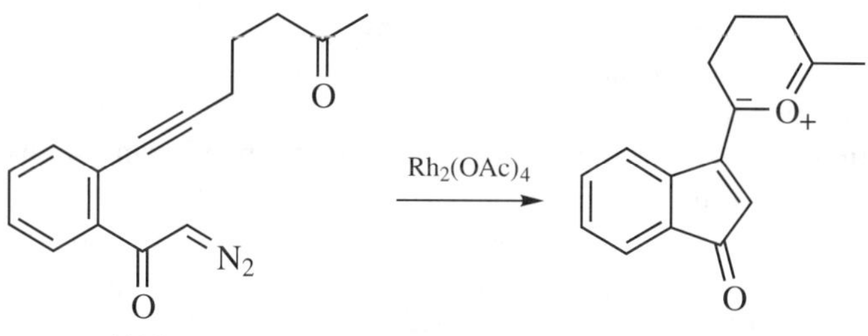

(116) **(117)**

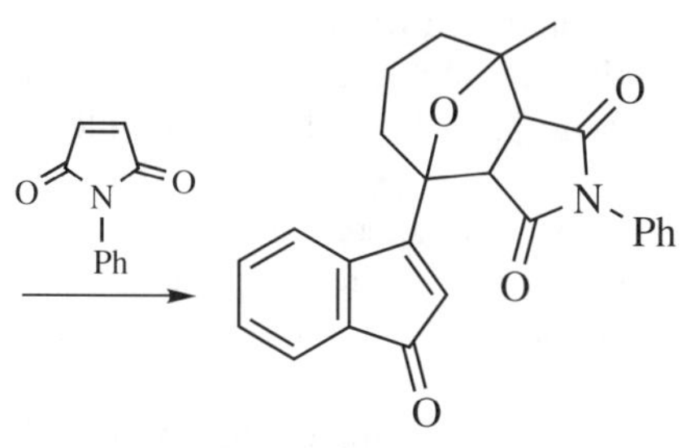

(118)

Scheme 4.27 Cyclization of alkynyl ketone.

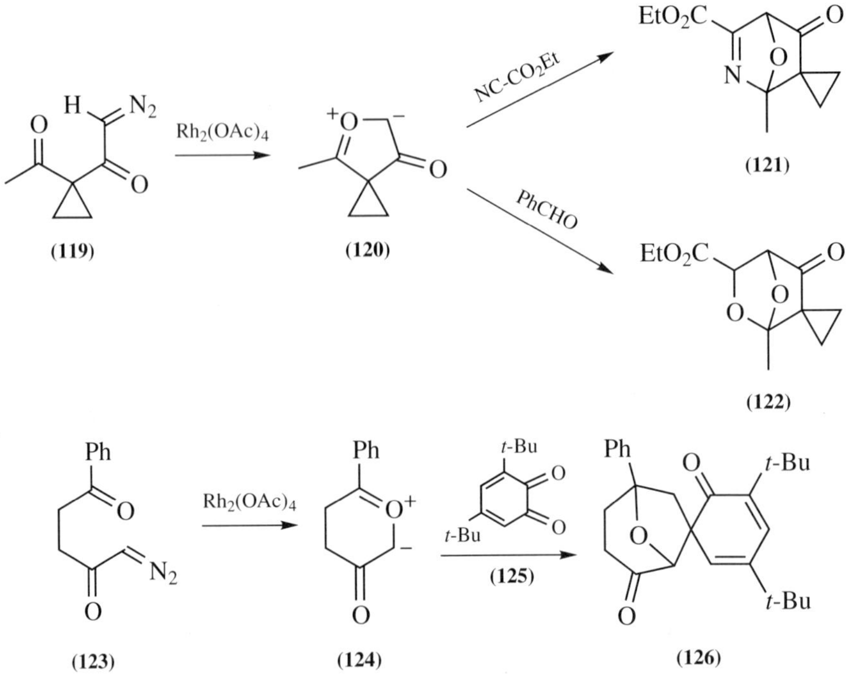

Scheme 4.28 Cycloaddition with hetero dipolarophiles.

experiences an alkynyl insertion reaction to give an intermediate indenyl carbene that was captured by the pendant ketone to give ylide **117**. Once formed, ylide **117** can undergo cycloaddition after the addition of a dipolarophile such as *N*-phenyl maleimide to give tetracyclic adduct **118**.

In addition to olefins, carbon heteroatom multiple bonds can also participate in the cycloaddition with various carbonyl ylides (Scheme 4.28).

Padwa et al. (44) studied the diazo-decomposition of **119** and found that the cyclic ylide **120** could be trapped by a variety of heterodipolarophiles such as ethyl cyanoacetate (Mander's reagent) to provide aminal **121** or with benzaldehyde to generate the bicyclic acetal **122**. In both cases, only a single isomer was formed, with the regiochemistry easily predicted from frontier orbital considerations. Nair et al. (45) were able to employ the highly functionalized *o*-quinone **125** for the trapping of carbonyl ylide **124** to provide the highly complex cycloadduct **126** in 76% yield.

As the reactive substrates become more complex, a number of alternative carbenoid pathways are able to compete effectively with ylide generation. It has been shown that the judicious choice of a catalyst (46,47) can have a remarkable effect on the reaction pathway (Scheme 4.29).

Padwa et al. (48) examined the behavior of diazoketone **127** under rhodium catalysis and found that the ligands associated with rhodium had a dramatic effect on the distribution of products **128** (from the carbonyl ylide) and **129** (from intramolecular C–H insertion). When rhodium acetate was employed there was a

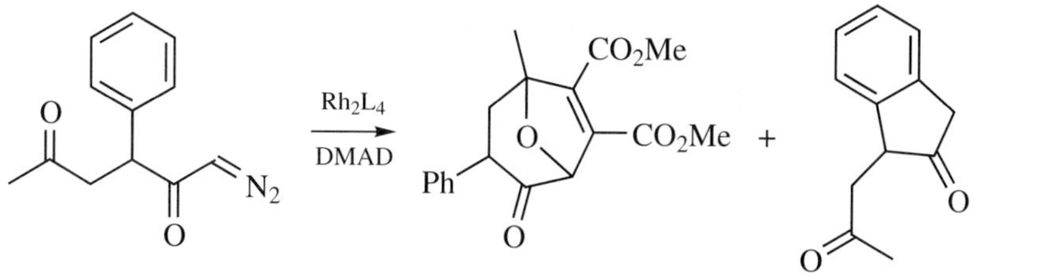

(127) **(128)** **(129)**

L	128:129
OAc	75:25
pfb	0:100
cap	100:0

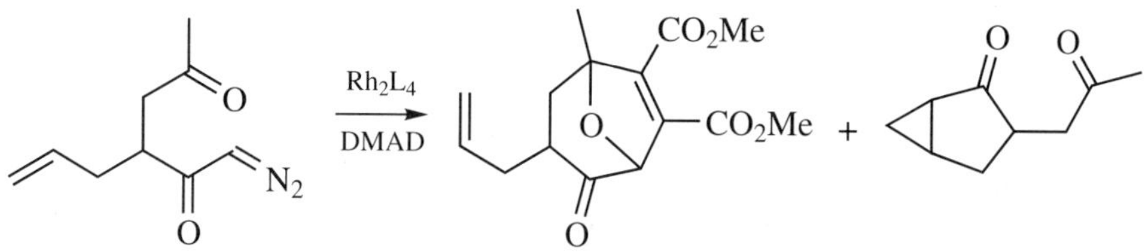

(130) **(131)** **(132)**

L	131:132
OAc	~1:1
pfb	~1:1
cap	~1:1

Scheme 4.29 Ligand effects on ylide generation.

3:1 preference for ylide generation. However, switching to the electron-withdrawing perfluorbutyrate (pfb) ligand reversed the selectivity to the complete exclusion of the ylide derived product. The change was attributed to the generation of a more electrophilic (Lewis acidic) carbene that prefers electrophilic aromatic substitution. The electron-rich caprolactamate (cap) ligand showed a similar product preference as rhodium acetate but changed the selectivity completely to the ylide pathway. The remarkable selectivity is not always operational, as shown by Moody's study (49) of diazoketone **130**, where the competition occurred between ylide generation and cyclopropanation. In this reaction manifold, the same catalyst series produced essentially no effect on selectivity compared to that seen for diazoketone **127**.

4.1.7. Carbonyl Ylides Derived from Aldehydes

Although there are fewer examples of aldehydes serving in the formation of carbonyl ylides, there have been several reports regarding their generation. Maas

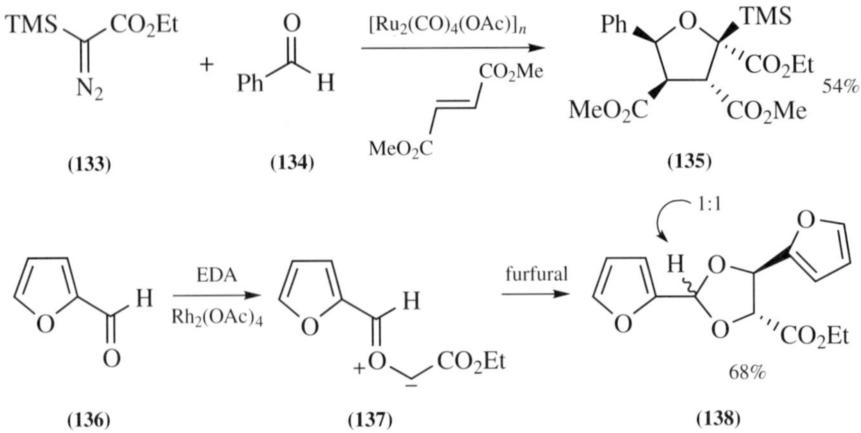

Scheme 4.30 Carbonyl ylides from aldehydes.

and co-worker (50) were able to generate these ylides and trap them with various dipolarophiles (Scheme 4.30).

Decomposition of the trimethylsilyl diazoacetate **133** with a ruthenium cluster in the presence of benzaldehyde and dimethylfumarate led to formation of the THF derivative **135** in 54% isolated yields. The ruthenium catalyst proved superior to all standard rhodium complexes for this transformation.

Wenkert and Khatuya (51) examined the competition between direct insertion of a carbene into furan (via cyclopropanation) and ylide formation with reactive side-chain functionality such as esters, aldehydes, and acetals. They demonstrated the ease of formation of aldehyde derived carbonyl ylides (Scheme 4.30) as opposed to reaction with the electron-rich olefin of the furan. Treatment of 3-furfural (**136**) with ethyl diazoacetate (EDA) and rhodium acetate led to formation of ylide **137**, followed by trapping with a second molecule of furfural to give the acetal **138** as an equal mixture of isomers at the acetal hydrogen position.

Interesting products have been observed from the reaction of aldehydes and vinyl carbenoid species (Scheme 4.31).

Rhodium induced ylide formation (52) between the vinyl diazo derivative **139** and *p*-chlorobenzaldehyde led to the production of **141**. The major products from this ylide were the stereoisomeric epoxides **142** and **143** from collapse of the ylide. The epoxides were accompanied by a significant quantity of dihydrofuran **144** from the cyclization of the ylide through the vinyl substituent.

4.1.8. Carbonyl Ylides Derived from Esters

Ester derivatives are also capable of forming carbonyl ylides and can undergo traditional cycloaddition with activated alkenes and alkynes and can even undergo reactions with heterodipolarophiles. Padwa was able to generate ester derived

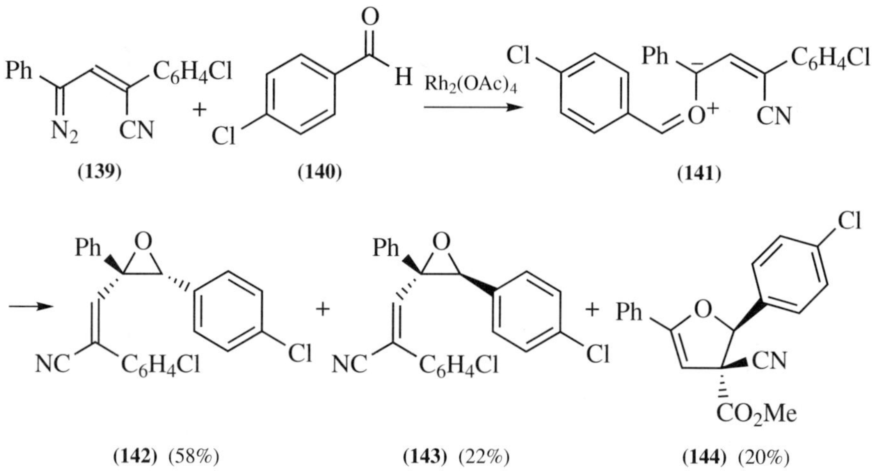

Scheme 4.31 Reaction of aldehydes with vinyl carbenoids.

carbonyl ylides and trap them in both an intramolecular and an intermolecular fashion (Scheme 4.32).

Intramolecular ylide formation with the lactone carbonyl oxygen (53) in **145** provided a carbonyl ylide **146** that was trapped with *N*-phenyl maleimide to give cycloadduct **147**. Likewise (54), carbonyl ylide **149**, derived from ester **148**, suffers intramolecular cycloaddition with the tethered alkene to deliver acetal **150** in 87% yield. An enantioselective version of this process has also been described (Scheme 4.33).

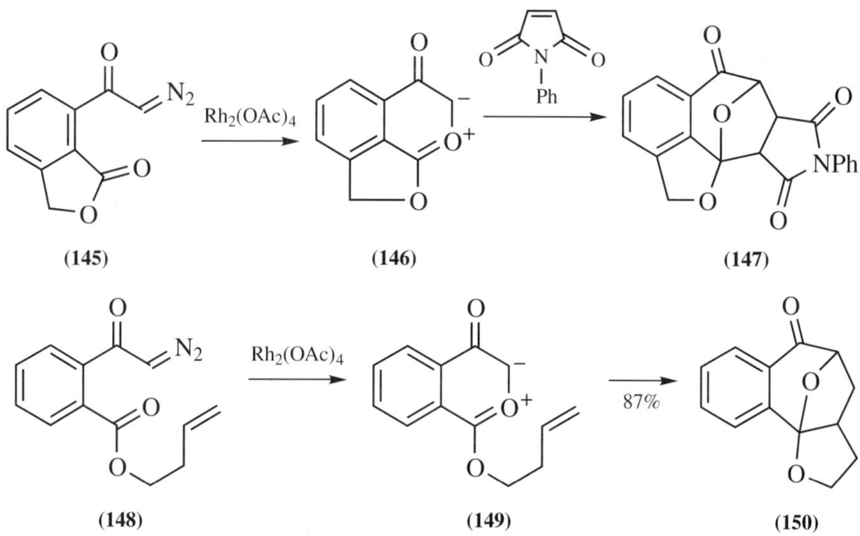

Scheme 4.32 Carbonyl ylides from esters.

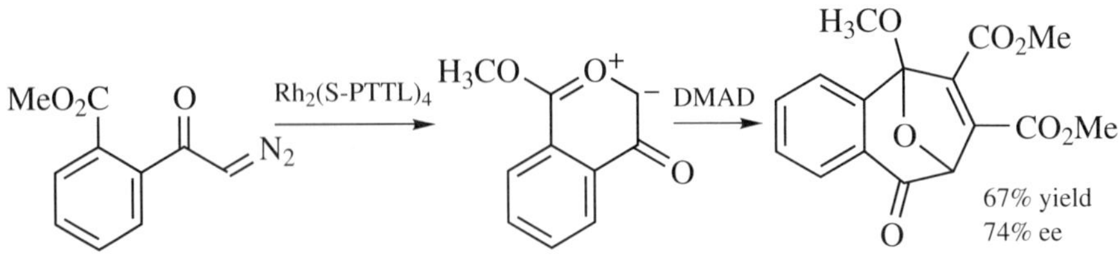

(151) **(152)** **(153)**

Scheme 4.33 Enantioselctive dipolar cycloaddition.

Hashimoto and co-workers (55) reported that generation of ylide **152** from aryl ester **151** in the presence of a chiral rhodium complex $Rh_2(S\text{-}PTTL)_4$, a chiral phthalimide substituted carboxylate, followed by cycloaddition with DMAD, led to the formation of adduct **153** in good yield and in 74% enantiomeric excess (ee).

4.1.9. Carbonyl Ylides Derived from Amides and Ureas

Carbonyl ylides derived from nitrogen-substituted carbonyl moieties provided for the synthesis of very stable *push–pull* dipolar intermediates. Although these compounds are quite stable, they still have sufficient reactivity to engage in cycloaddition and related processes. Carbonyl ylides derived from amides have been trapped in intermolecular cycloadditions to give aminals (Scheme 4.34) (56).

Treatment of proline derivative **154** with rhodium acetate (57) originally led to ylide **155**. However, this ylide (**155**) quickly rearranges to the more stable azomethine ylide **156**, which undergoes cycloaddition with DMAD to give the unusual adduct **157**. Intramolecular trapping experiments (58,59) have also been conducted (Scheme 4.35).

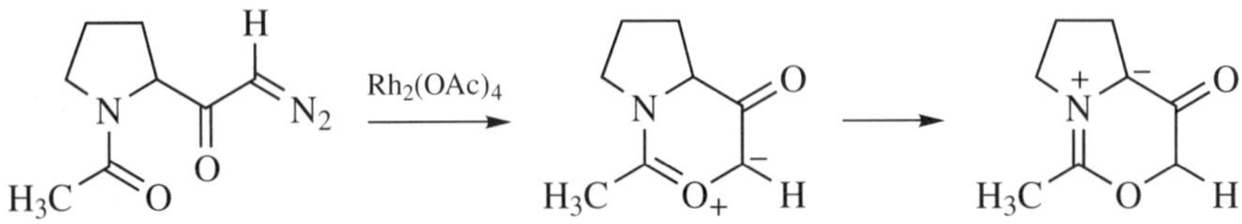

(154) **(155)** **(156)**

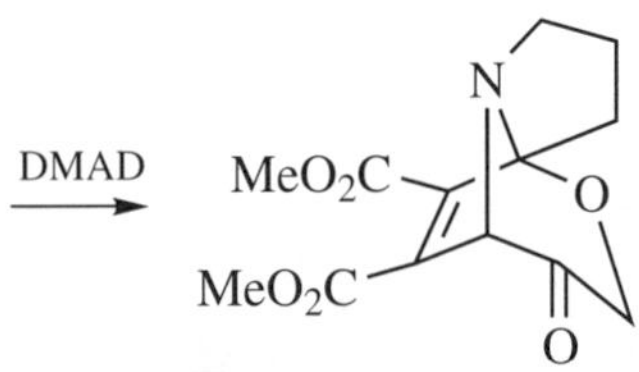

(157)

Scheme 4.34 Intermolecular trapping of an amide derived ylide.

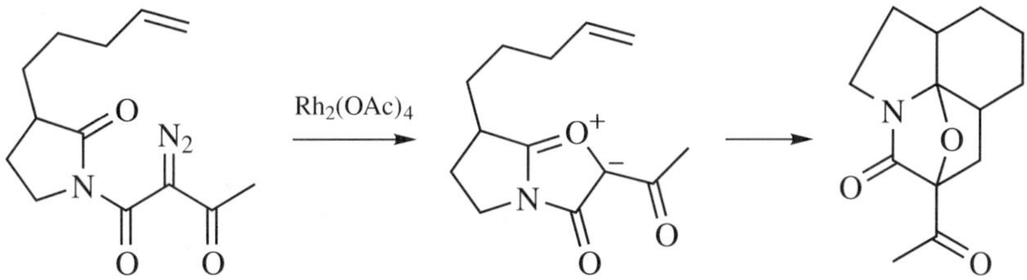

(158) (159) (160)

Scheme 4.35 Intramolecular trapping of an amide derived ylide.

Rhodium-mediated decomposition (60) of diazoamide (**158**) led to formation of the mesoionic oxazolium ylide **159,** which was efficiently trapped by the pendant alkene to produce the oxo-bridged tricyclic amide **160**.

Other carbonyl functionality-containing nitrogen atoms have been examined for the formation and reaction of carbonyl ylides (Schemes 4.36 and 4.37).

Cyclization of the rhodium carbenoid generated from **161** with the carbonyl oxygen of the imide **161** lead to a dipolar intermediate that was trapped with DMAD, giving the bridged adduct **162** in good overall yield (61). Urea derivatives are also capable of undergoing ylide formation (62). Rodgers et al. (63) revealed some interesting chemistry associated with these ylides. Sequential cyclization–cycloaddition of a urea derived ylide led to the initial formation of the cycloadduct **164**. The adduct **164** is unstable and suffered a retro Diels–Alder reaction with concomitant loss of methyl isocyanate to generate the highly substituted furan **165**.

The high levels of reactivity and the wide array of structures that can be accessed from carbonyl ylides has placed them in an excellent position to be used in the

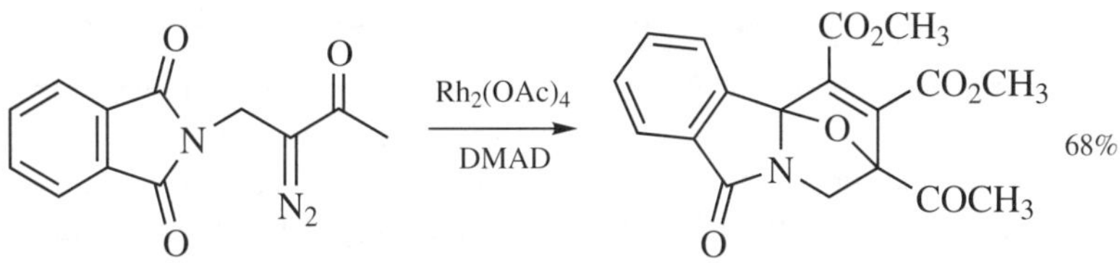

(161) (162)

Scheme 4.36 Carbonyl ylides from imides.

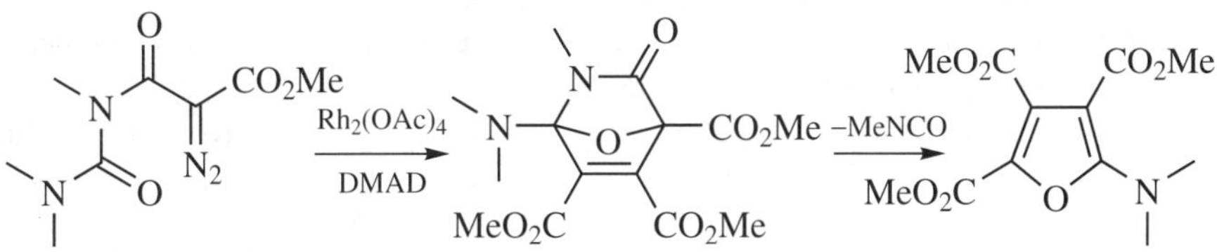

(163) (164) (165)

Scheme 4.37 Carbonyl ylides from ureas.

synthesis of complex molecules. Section 4.2 will focus on specific applications of carbonyl ylide promoted reactions in synthesis.

4.2. SYNTHETIC ASPECTS OF CARBONYL YLIDE CYCLOADDITION REACTIONS

With the ever-increasing need for novel methods to prepare synthetic intermediates, complete total chemical synthesis, and generate analogues of natural products, the application of carbonyl ylide methodology to synthesis has become a highly efficient, stereo- and regiodefined method for preparing carbon–carbon and carbon–heteroatom bonds. This reaction has been an especially powerful method for generating a range of cyclic structures such as cycloheptanoid-containing natural products and polyazacyclic systems found in many bioactive natural products. There has been a literal explosion of synthetic advances using the tandem sequences described here and elsewhere. The chemical systems under scrutiny range from very simple mono- and bicyclic systems to complex *cascade* or *dipole cascade* reactions that involve two, three, or four different reaction manifolds in one sequence. Section 4.1 aimed to describe the novel preparation and reactivity of carbonyl ylides. Section 4.2 aims to describe some of the synthetic uses of carbonyl ylides, particularly as they pertain to cycloaddition reactions.

4.2.1. Directed Syntheses with Tandem Carbonyl Ylide Formation/[1,3]-Dipolar Cycloaddition

4.2.1.1. Preparation of Oxacyclic Compounds

Much of the initial synthetically useful carbonyl ylide work originated from the Ibata group. Exploiting simple disubstituted aromatic diazoketo-esters and structurally diverse dipolarophiles, Ibata and co-workers (64–70) prepared several different cycloadducts **167–169** through an intermolecular ylide cycloaddition (Scheme 4.38).

Note that intramolecular cycloadditions were also plausible with the correct substitution and tether length. Bien and co-workers (71,72) found that addition of rhodium acetate to α-diazoketone **170** produced three products with the major outcome being the oxatricyclic **171** and a minor amount of the diastereomeric intramolecular cyclopropanes **172** (Scheme 4.39).

Over the last 15 years, Padwa et al. (73,74) have been heavily involved with the study and application of carbonyl ylides as cycloaddition precursors in synthesis. Their work has helped make the tandem ylide formation–dipolar cycloaddition process a synthetically accessible transformation. Much of Padwa's early work focused on determining the extent and limitations of this methodology. Many of the early systems were carbocyclic in nature and helped define basic parameters such as

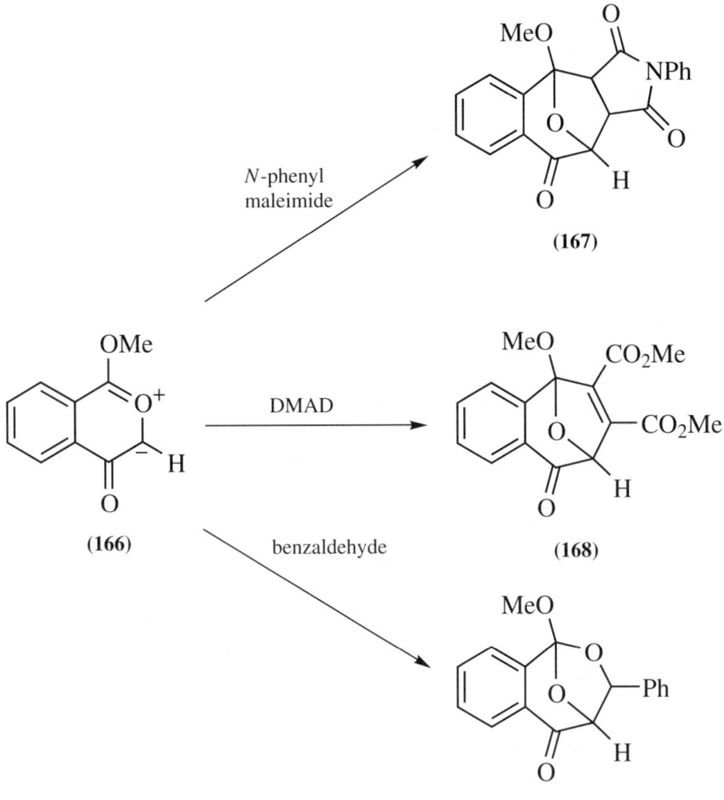

Scheme 4.38 Early Ibata cycloaddition.

the reactivity of five- and six-membered ylides. In 1989, Padwa et al. (75) embarked on a synthesis of the brevicomins, an aggregation pheromone of the Western Pine Beetle (Scheme 4.40).

Padwa synthesized the oxo-brevicomin analogue **174** in 60% yield as a 2:1 mixture of the exo and endo isomers. Dithioacetalization followed by Raney Ni reduction of the dithioketal produced the *exo*- (**175**) and *endo*-brevicomin in good

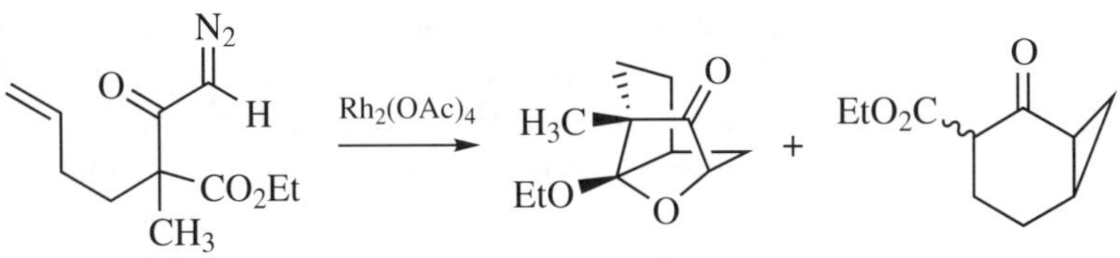

(170) **(171)** **(172)**

Scheme 4.39 Early intramolecular cycloaddition reaction.

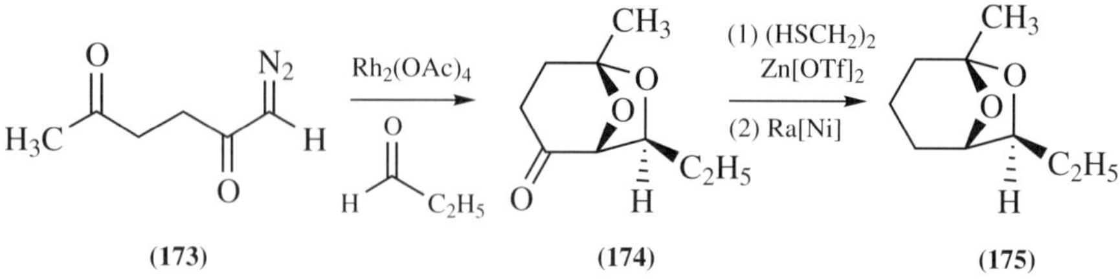

Scheme 4.40 Synthesis of *exo*-brevicomin.

yield. Padwa also utilized this same ylide system with a number of alternative dipolarophiles.

A further study on six-membered ylide formation examined the use of an aliphatic ester in place of a ketone as the Lewis base donor for carbonyl ylide formation. Although the same keto-substituted system underwent an intramolecular cyclization readily, the ester derivative gave no cycloaddition products. Padwa and co-workers (37,76) points to the major electronic differences between the two carbonyl groups to rationalize the disparity in carbonyl ylide formation.

Five-membered carbonyl ylide derivatives form with ease, but tend to suffer from proton-transfer reactions to the carbonyl more readily than their six-membered counterparts. Generally, disubstitution of the position α to the carbonyl led to smooth carbonyl ylide formation and subsequent dipolar cycloaddition (35,77).

Ylide size and tether length have also been issues studied by the Padwa group. They found that it is possible to increase the intervening methylene length to three, thereby generating a seven-membered ylide (44,78). Unfortunately, it was found that the rate of formation of the ylide was somewhat diminished and that secondary reactions such as aromatic insertions or cycloadditions occurred to produce monosubstituted cycloheptatrienes. Generally, Padwa has found that the length of the tether between the olefin and the C=O of the ylide cannot be longer than four methylene units. All attempts at intramolecular cycloaddition with a five methylene unit tether failed to provide any cycloadduct (44,78).

Both Dauben et al. (79) and McMills and co-workers (80,81) used the tandem ylide formation–dipolar cycloaddition methodology to approach the tigliane diterpenes. Phorbol, a tigliane diterpene, and its various esters played a major role in determining the molecular mechanism of carcinogenesis (Fig. 4.4). Although functionally similar to the oxidopyrylium cycloadditions that have been used effectively by Wender for the synthesis of phorbol, the ylide cycloaddition makes it possible to enter the synthesis with an intact A ring and may make for easier analogue preparation and greater structure activity relation determination.

Each group dissected the tigliane core by adding an ether bridge to the C(6) to C(9) positions of the tricyclic B ring. Retrosynthetically, the ether bridge was placed to effect a disconnection to the cyclopentyl substituted cyclohexyl ylide **177** with an olefinic tether at C(9) (phorbol numbering system). The McMills group chose to make a simple phorbol analogue devoid of most oxygen functionality. It

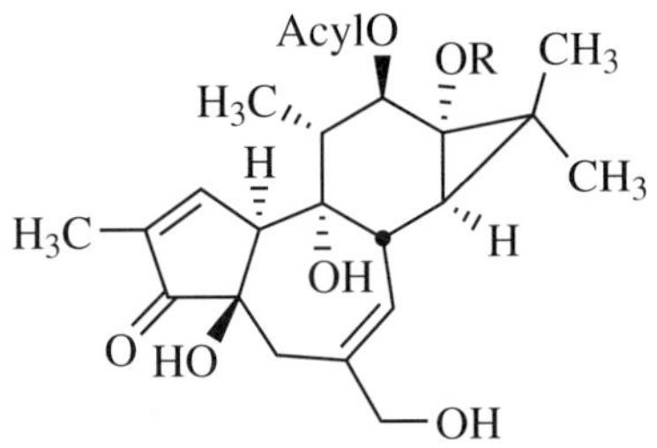

Figure 4.4 Phorbol structure.

was found that both the cis and trans ring junctions of the cyclopentane A ring were proficient in the cyclization. The cycloaddition provided the tricyclic phorbol core **178** in >85–90% yield as a single diastereomer (Scheme 4.41). X-Ray crystallographic analysis of tricyclic **178** gave the correct relative stereochemistry for C(10), C(9), and C(8). Several additional synthetic iterations have been completed.

Dauben's group utilized the same retrosynthetic disconnections, but chose to add more functionality to the cycloaddition precursor. From a simple *trans*-disubstituted cyclopentane, Dauben used an aldol reaction of a cyclopropylvinyl aldehyde to prepare the cycloaddition precursor. The diazo-substituted β-ketoester was completed using a Roskamp–Padwa coupling followed by diazo-transfer. Addition of rhodium acetate to the diazo substituted β-ketoester **179** led to an excellent 86% yield of the correct diastereomer (Scheme 4.42).

Muthusamy et al. (82) prepared a number of oxacyclic ether compounds from the tandem ylide formation–dipolar cycloaddition methodology. Their approach provides a synthetic tactic to compounds such as ambrosic acid, smitopsin, and linearol. Starting with either cyclopentane or cyclohexane templates, they prepared ylide sizes of five or six, which are trapped in an intermolecular cycloaddition reaction by the addition of DMAD. The products are isolated in good overall yield. In a second system, 2,5-disubstituted cyclohexenyl derivatives are utilized to generate the pendent ylide, then, *N*-phenylmaleimide is added in an intermolecular reaction, accessing highly substituted oxatricyclic derivatives such as **182** (Scheme 4.43).

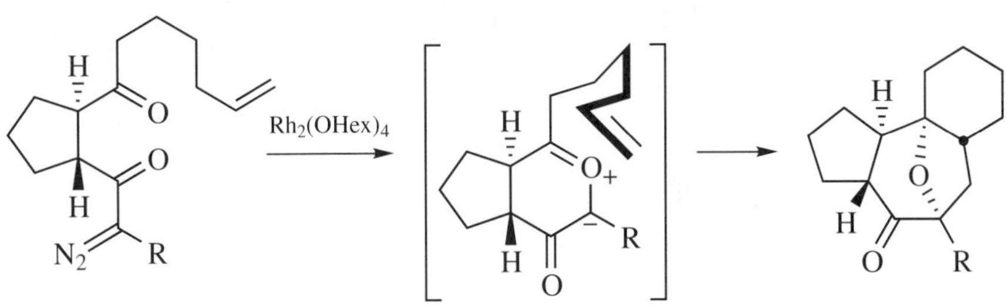

(176) **(177)** **(178)**

Scheme 4.41 McMills approach to phorbol.

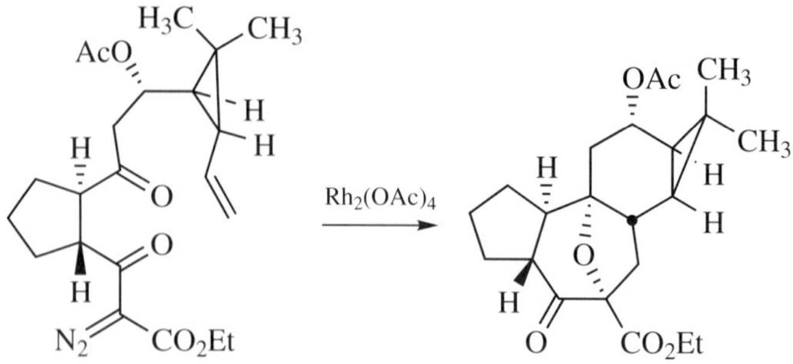

(179) **(180)**

Scheme 4.42 Dauben approach to phorbol.

An interesting change of reactivity was observed (Scheme 4.44) when Muthusamy et al. (83) reacted a similar cyclic ylide system with arylidenetetralones such as **184**. This reaction did not generate the expected olefinic cycloadduct, but instead resulted in addition across the C=O of the tetralone to produce a dioxolane **185**. This behavior is unusual in that normal reactivity would preclude a reaction with the C=O system and addition would occur only with the olefin of the enone.

In a very recent example, Chiu and co-workers (84–86) used the tandem ylide–cycloaddition methodology to prepare advanced intermediates directed toward the synthesis of the pseudolaric acids. Pseudolaric acids are a family of diterpenes isolated from the root bark of *Pseudolarix kaempferi*. These novel compounds have shown antimicrobial activity comparable to that of amphotericin B and have demonstrated cyctotoxicity against several cancer cell lines (Fig. 4.5).

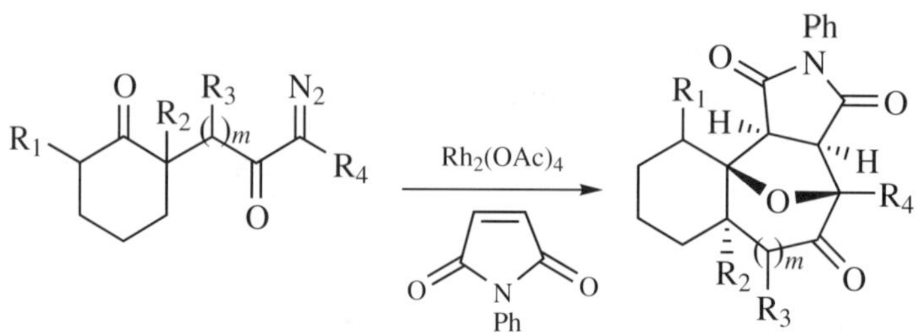

(181) **(182)**

$R_1 = CH_3, C_2H_5, H$
$R_2 = H, CH_3$
$R_3 = H, CH_3$
$R_4 = CO_2Et, CH_3, H$
$m = 0, 1$

Scheme 4.43 Muthusamy oxotricyclic formation.

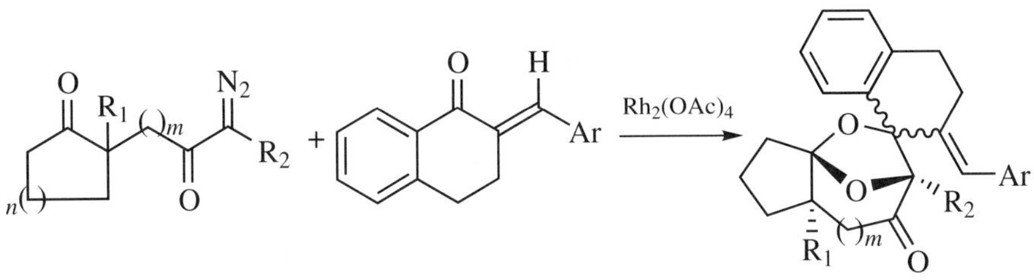

(183) **(184)** **(185)**

$R_1 = CH_3$, H
$R_2 =$ H, CO_2Et
m = 0, 1
n = 1, 2

Scheme 4.44 Muthusamy arylidene cycloaddition.

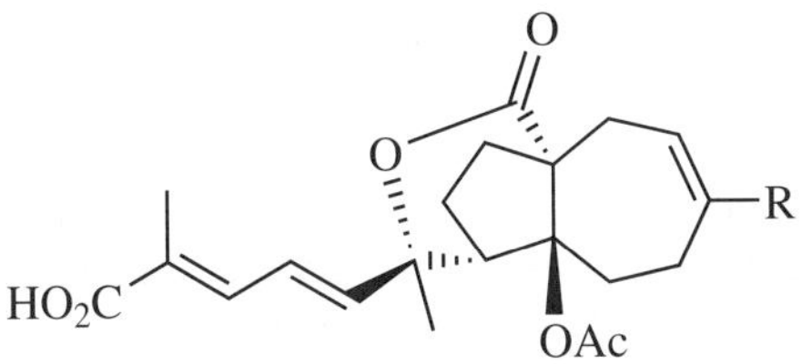

Figure 4.5 Pseudolaric acids.

Chiu's retrosynthetic analysis of the pseudolaric acids proposed a simple intermediate incorporating a bridging ether, similar to that used by Dauben and McMills. Cleavage of the ether bridge further reduced the problem to a tandem ylide formation–cycloaddition through a simple acyclic α-diazoketone with a tethered olefin (Scheme 4.45).

Chiu found that the diastereoselectivity had eroded slightly from an earlier synthetic study that included a methyl group in place of the MEM ether side chain of the A ring. Since it had been demonstrated that many of these catalytic systems are metal associated, Chiu attempted to change rhodium ligands from acetate to the

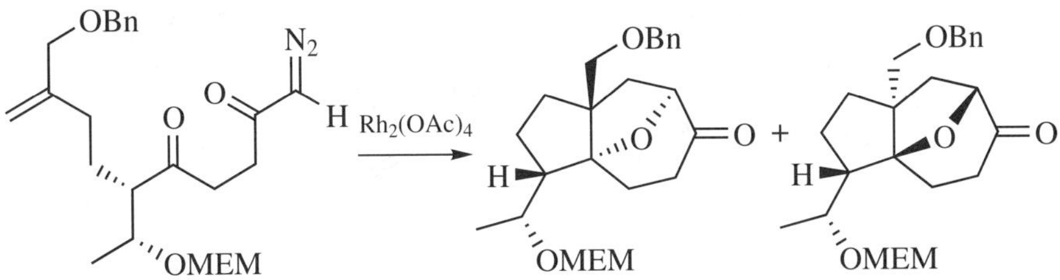

(186) **(187)** **(188)**

Scheme 4.45 Chiu approach to pseudolaric acids.

caprolactam series. There was a noticeable slowing of the reaction, but the diastereoselectivity increased only slightly.

4.2.2. Preparation of Alkaloid Systems

Alkaloids are another family of compounds that are easily accessible from synthetic routes utilizing carbonyl ylides. The complex structure of naturally occurring alkaloids has been the driving force for the generation of new carbonyl ylide methodology. These studies have resulted in the discovery of several new reaction manifolds as well as the total synthesis of several natural products.

Nair et al. (87,88) achieved a synthesis of spirooxindole-containing molecules by adding isatins to various carbonyl ylides (Scheme 4.46). There has been relatively little research regarding the efficiency of C=O of 1,2-dicarbonyl compounds as dipolarophiles relative to their olefinic counterparts. As anticipated, Nair found that the more electrophilic carbonyl of the isatin **187** (non-amide carbonyl) reacted smoothly with the carbonyl ylide formed from diazoketone **186** to give the spirocyclic adduct **188**. Nair's yields were moderate to good (44–83%), but were based on recovered isatin.

Nair et al. (87) was able to extend this methodology with five-, six-, and seven-member carbonyl ylides. The five-membered ylide was the same carbonyl ylide as that used by Padwa for the synthesis of the illudins. The use of the seven-membered ylide was novel due to the fact that ylides greater than six atoms are generally difficult to form and indeed the yield of the cycloaddition with isatin (**187**) suffered and the product was isolated in only 32% yield.

Muthusamy et al. (89) approached the formation of decahydrobenzocarbazoles **191** utilizing an indolic five-membered olefin **190** as the dipolarophile in reaction with a carbonyl ylide derived from **189**. This intermolecular approach is strategically similar to an intramolecular approach to aspidosperma alkaloids developed by Padwa (Scheme 4.47).

Muthusamy found that with the simple cyclohexyl keto-ylide derived from **189** ($R_1 = H$) and an electron-withdrawing group protecting the indole nitrogen, that the indole would add in a regioreversed fashion producing the other possible

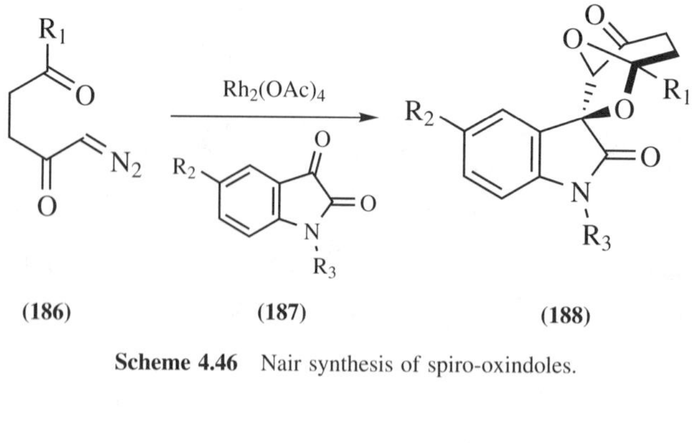

(186) (187) (188)

Scheme 4.46 Nair synthesis of spiro-oxindoles.

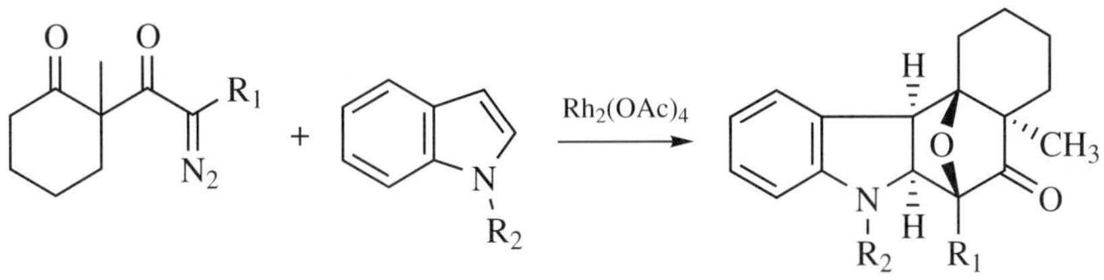

(189) **(190)** **(191)**

Scheme 4.47 Muthusamy approach to decahydrocarbazoles.

regioisomer of **191**. The yields of the cyclohexyl cycloadduct were good, but switching to a cyclopentane-substituted diazoketone moiety reduced the efficiency of the cycloaddition substantially.

Maier and Evertz (90) synthesized an isomünchnone intermediate to generate simple annulated piperidine substrates **193**. The reaction was extended to several ring sizes and substitution patterns about the olefinic tether. The yields were routinely good and produced compounds of known stereochemical integrity (Scheme 4.48).

Padwa and co-workers (91–101) pioneered much of the synthetically useful methodology dealing with applications of isomünchnones. Early on, Padwa generated dipolar intermediates from lactam rings of varied sizes, from four to seven. Cycloaddition of phenyl maleimide produced interesting heterocyclic adducts with multiple rings. In the same manuscript, Padwa used dimethylacetylene dicarboxylate as the dienophile, forming the olefinic cycloadduct, which subsequently achieves a retro-Diels–Alder reaction to prepare a furanisocyanate in excellent yield (102).

Kappe et al. (103,104) approached dihydropyrimidines, a potent group of calcium channel modulators, through the use of an isomünchnone-type cyclization. Kappe prepared the cyclization precursor **195** in the course of a three component Biginelli condensation process (Scheme 4.49).

The pseudoaxial nature of the aryl substituent allows for a favorable placement of the olefinic tether relative to the dipole formed. Semiempirical calculations show that either exo addition of the olefin or the regio-reversed endo transition state

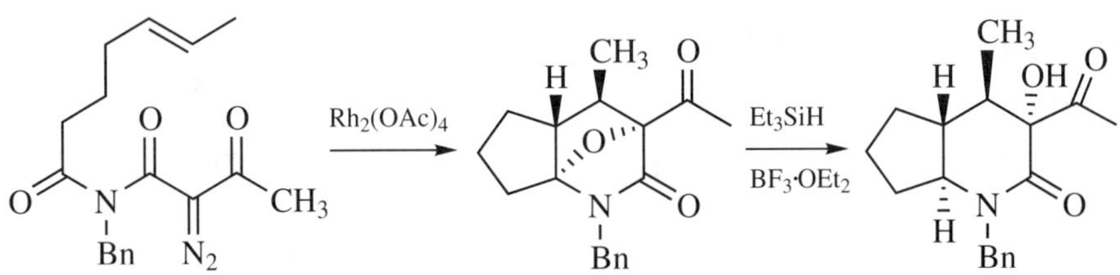

(192) **(193)** **(194)**

Scheme 4.48 Maier approach to complex piperidines.

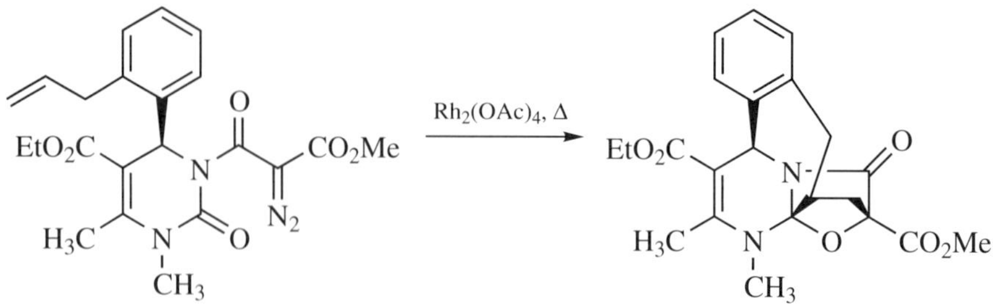

(195) (196)

Scheme 4.49 Kappe approach to dihydropyrimidines.

resulted in significantly higher transition state energy. Compound **195** contains a methyl group bound to nitrogen, in the actual calcium channel modulators this must be an NH moiety. In an attempt to synthesize compounds more closely related to the calcium channel modulators, Kappe attempted to cycloadd a compound devoid of the NMe group. It was found that upon addition of a divalent rhodium catalyst that the original isomünchnone formed readily experiences a 1,5-proton transfer without any attendant cycloaddition.

Harwood and co-workers (105) utilized a phenyloxazine-3-one as a chiral derived template for cycloaddition (Scheme 4.50). An oxazinone template can be formed from phenylglycinol as the template precursor. The diazoamide needed for cycloaddition was generated by addition of diazomalonyl chloride, trimethyl-dioxane-4-one, or succinimidyl diazoacetate, providing the ester, acetyl, or hydrogen R group of the diazoamide **198**. After addition of rhodium acetate, *N*-methylmaleimide was used as the dipolarophile to provide a product that predominantly adds from the less hindered α-face of the template in an endo fashion. The cycloaddition also provided some of the adduct that approaches from the β-face as well. β-Face addition also occurred with complete exo-selectivity. Mono- and disubstituted acetylenic compounds were added as well, providing similar cycloadducts.

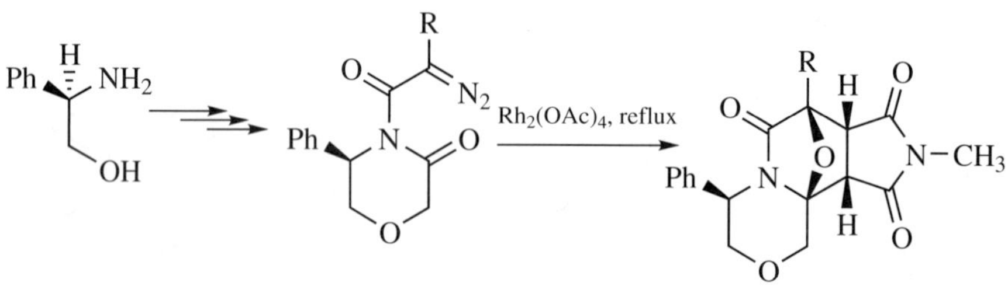

(197) (198) (199)

Scheme 4.50 Chiral templated cycloaddition.

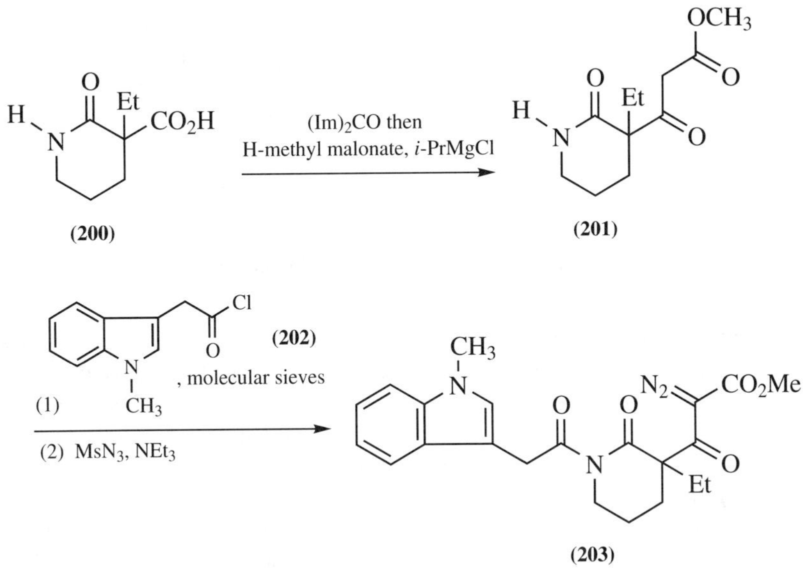

Scheme 4.51 Synthesis of dehydrovindorosin precursor.

Padwa and co-workers (60,106,107) have been highly active in using carbonyl ylides for the synthesis of a number of bioactive alkaloids (Scheme 4.51). In an approach to the aspidosperma alkaloids, a *push–pull* carbonyl ylide was used to generate a bicyclic ylide containing a tethered indole moiety. This strategy ultimatcly allowed for the synthesis of the dehydrovindorosin skeleton (108). Starting from a quaternary substituted piperidone (**200**), elaboration of the 3-carboxylic acid provided β-ketoester amide **201**. Addition of the indole tethered side chain provided a very rapid and efficient method to generate the cycloaddition precursor **203**.

Upon addition of a divalent rhodium catalyst, formation of the putative carbonyl ylide with the piperidine carbonyl occurred, followed by cycloaddition of the tethered olefinic portion of indole (Scheme 4.52). The cycloaddition formed the

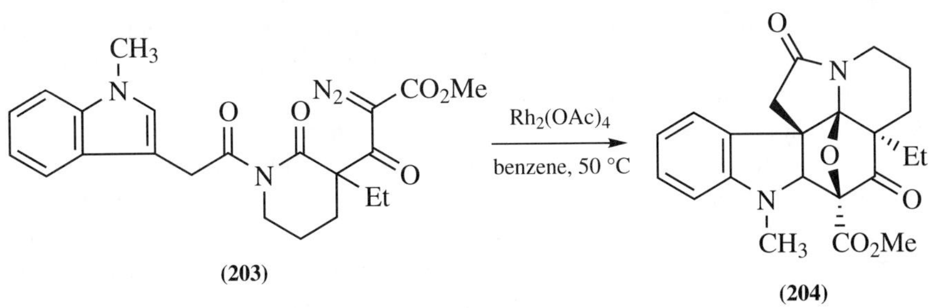

Scheme 4.52 Padwa approach to dehydrovindorosin.

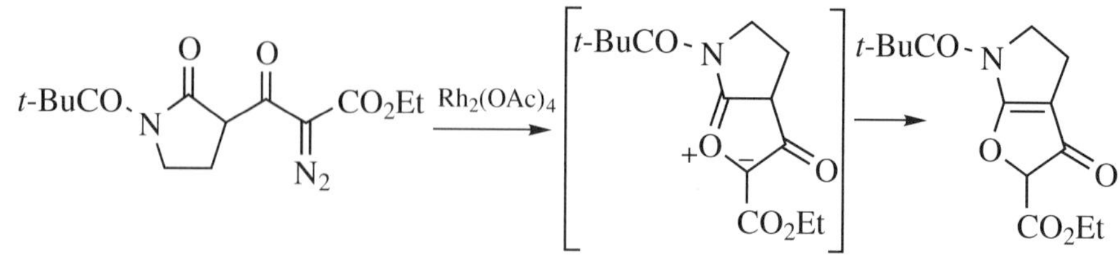

(205) (206) (207)

Scheme 4.53 1,4-Hydrogen transfer of isomunchnones.

endo product **204** with respect to the ylide dipole according to the calculated lowest energy transition state. Padwa found this cycloaddition to be doubly diastereoselective since the indole approaches exclusively from the less encumbered side of the molecule containing the ethyl group. The Padwa group has been able to synthesize desacetoxy-4-oxo-6,7-dihydrovindorosin, but has thus far been unsuccessful in epimerizing the C_4 alcohol.

Lycorine is an alkaloid that has attracted attention from both the synthetic community and pharmacologists. Prior synthetic approaches have included inter- and intramolecular Diels–Alder cycloaddition. Based on a similar retrosynthetic disconnection, Padwa and co-workers (106,109) chose to use a *push–pull* carbonyl ylide cycloaddition with a disubstituted pyrrolidinone core to generate a tricyclic substrate. The major difference for this synthetic study was the availability of a labile proton α to the carbonyl moiety (Scheme 4.53).

During the initial model study, it was found that intramolecular 1,4-transfer of a hydrogen atom was more facile than any inter- or intramolecular cycloaddition process. Although the synthetic scheme shown was not the anticipated path, furanone (**207**) provided the basis for a possible two step path via acylation of **207** to form an amidofuran, followed by Diels–Alder cycloaddition to provide **211** (Scheme 4.54).

Isomünchnones play a large part in the synthetic efforts Padwa and co-workers (91–94,96,98,110,111). The *Lycopodium* alkaloids are a large family of natural products and have inspired numerous synthetic routes to approach these compounds. Interest in this class of alkaloids stems from the myriad of biological properties they exhibit. Padwa utilized the carbonyl ylide methodology in tandem with a cationic π-cyclization to complete a formal synthesis of ($\pm$) lycopodine (Scheme 4.55) (96,112).

Cycloaddition of the isomünchnone occurred anti to the 3-methoxybenzyl moiety, while the formation of a 3:2 ratio of the endo adducts **213** was in accordance with the results of molecular mechanics calculations conducted by Padwa. Analogous to the argument of Stork, Padwa found that the selectivity of the π-cyclization results from the bridgehead hydrogen and tethered aromatic ring conspiring to maintain an anti arrangement in the cyclohexylidene ring of the bicyclic iminium ion during cyclization to produce the bridged adduct **214**. The alternative syn arrangement must adopt an unfavorable boat conformation.

Padwa and co-workers (113–116) provided a synthetic approach to natural products containing the 2(1*H*)-pyridone core. One such target (−)A58365A, is a

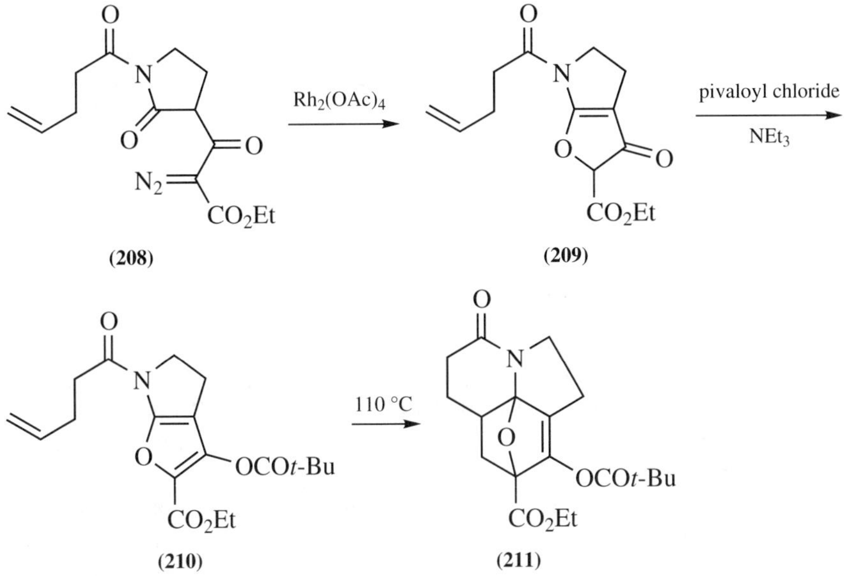

Scheme 4.54 Approach to amaryidaceae alkaloids.

known ACE inhibitor (115). Starting from L-pyroglutamic acid, the synthesis of the cycloaddition precursor followed much of the chemistry that Padwa has used in the preparation of other isomünchnone species. It was found that using a phenylsulfone-substituted diazoamide provided for a tandem cyclization–cycloaddition–ring opening sequence to generate substituted 2(*H*)-pyridones. According to FMO

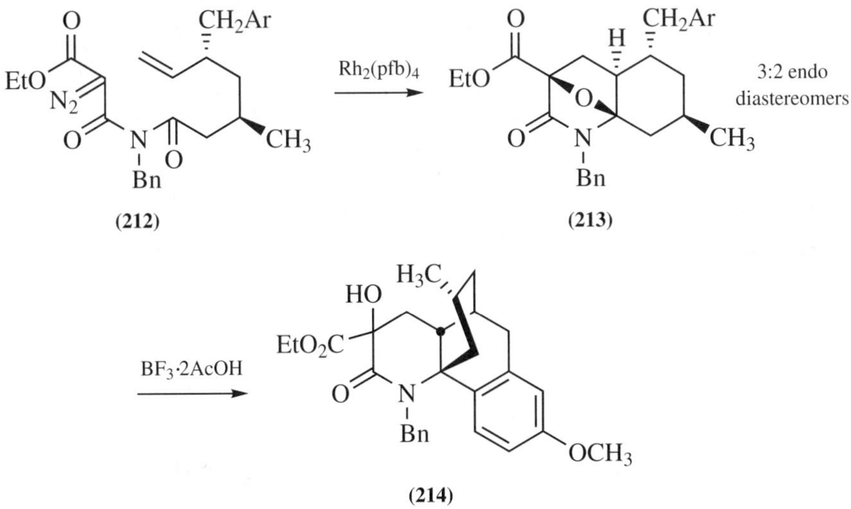

Scheme 4.55 Approach to (±) lycopodine.

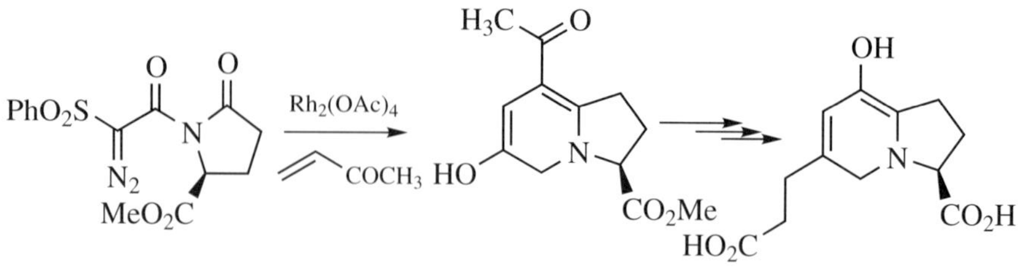

(215) (216) (217)

Scheme 4.56 Approach to A58365A.

theory, the HOMO of the dipole is the dominant molecular orbital set for reaction with electron deficient olefins. The synthesis of A58356A (**217**) is completed via addition of the carboxyethyl side chain using Stille coupling methodology (Scheme 4.56).

Thus far, in the alkaloid series discussed, the nitrogen atom has always been part of the core of the alkaloid structure, rather than acting in a dipolarophilic manner in the cycloaddition of the carbonyl ylide. Recently, Padwa et al. (117) addressed this deficiency by conducting model studies to synthesize the core of ribasine, an alkaloid containing the indanobenzazepine skeleton with a bridging ether moiety (Scheme 4.57). Padwa found that indeed it was possible to use a C=N π-bond as the dipolarophile. In the first generation, a substituted benzylidene imine (**219**) was added after formation of the putative carbonyl ylide from diazoketone **218**. The result was formation of both the endo and exo adduct with the endo adduct favored in an 8:1 ratio. This indicates that the endo transition state was slightly favored as dictated by symmetry controlled HOMO–LUMO interactions.

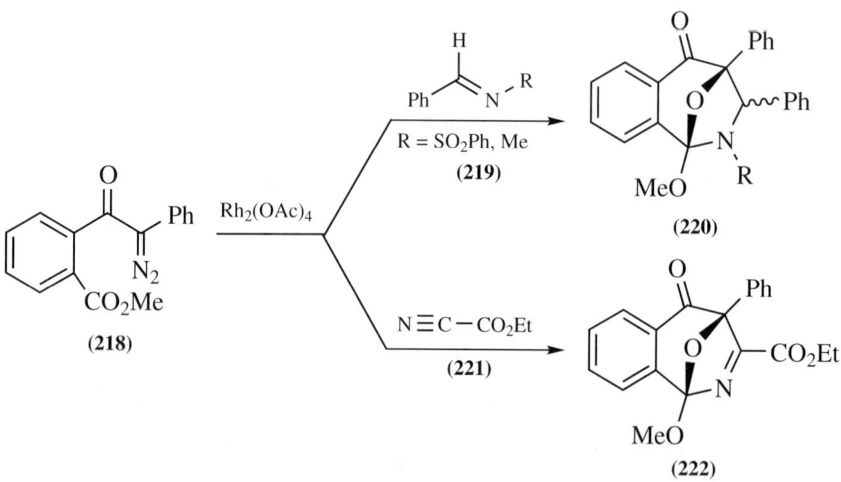

Scheme 4.57 Intermolecular addition of C–N multiple bonds.

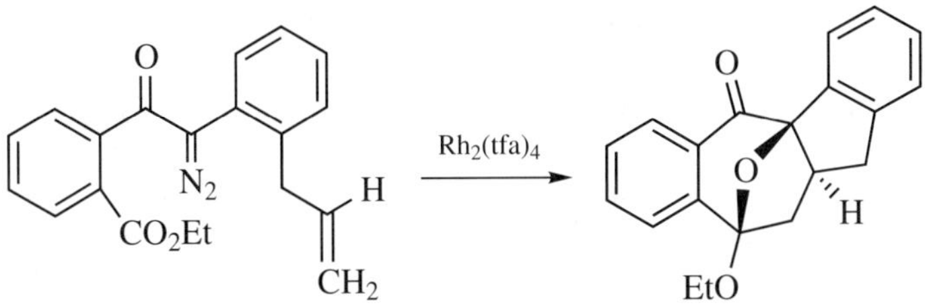

(223) **(224)**

Scheme 4.58 Cycloaddition of deaza–ribasine core.

Mander's reagent (**221**) was also utilized to provide a nitrile CN triple bond as the π component for cycloaddition. Addition of Mander's reagent provided a superb yield of a single regioisomer **222**. The regiochemistry obtained is consistent with addition of the HOMO of the dipole and the LUMO of the dipolarophile as described by Houk (Scheme 4.58).

To test the applicability of the intramolecular cycloaddition to a linear tetracyclic ribasine core, Padwa synthesized bridged ether **224**. Difficulties were encountered after formation of the α-diazocarbonyl moiety (Scheme 4.59). At −10 °C, a noncatalyzed [1,3]-dipolar cycloaddition occurred between the diazo group and the tethered olefin, forming an indenylpyrazole (**225**). Warming the indenylpyrazole to room temperature led to an extrusion of nitrogen followed by formation of two new products, a substituted cyclopropylindane (**226**) and indene (**227**). Good overall yield of the carbonyl ylide cycloaddition product **224** along with some of

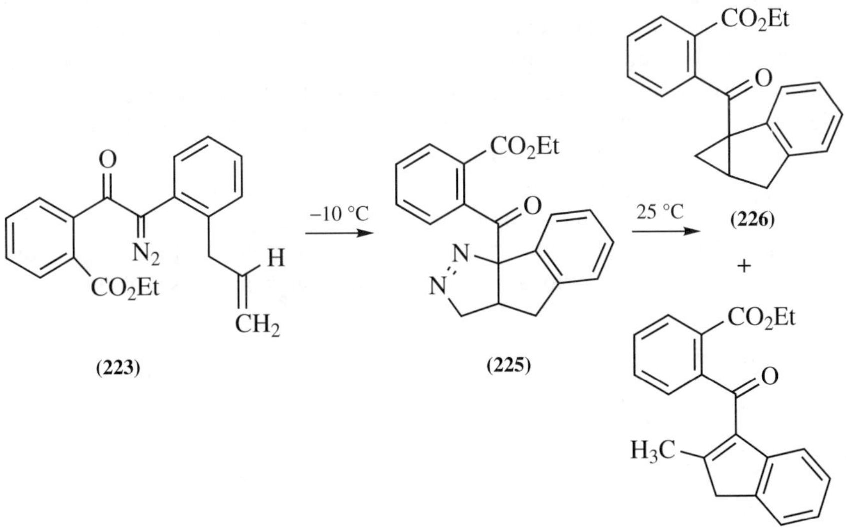

(223) **(225)** **(226)** **(227)**

Scheme 4.59 Ribasine cycloaddition byproducts.

226 and **227** could be obtained by adding $Rh_2(tfa)_4$ to a toluene solution of the diazo and rapidly heating to 110 °C.

In an interesting reaction conducted to support the synthesis of ribasine, the aromatic carboethoxy group was replaced with an aromatic aldehyde to generate the hydrogen-substituted carbonyl ylide. Reaction of aldehydes with diazocarbonyl species can result in formation of a dioxolane along with carbonyl ylide products. Padwa found that formation of the carbonyl ylide was highly dependent on the catalyst used. Addition of rhodium acetate resulted in formation of an intermediate that had inserted into the carbonyl–hydrogen bond. The tautomeric hydroxycyclopentenone was further utilized as the dipolarophile. The use of more electrophilic ligands associated with rhodium promotes ylide formation rather than insertion, and addition of dimethylacetylene dicarboxylate results in cycloaddition occurring in good yield.

4.2.3. The Illudins, Pterosins, and Ptaquilosides

The illudin, ptaquiloside, and the pterosin families of sesquiterpenes are highly toxic compounds that have been known for >40 years. Recently, it was reported that despite the extreme toxicity of these compounds, that they also have a very high level of efficacy against various adenocarcinomas. Indeed, McMorris (118) reported that a third generation of related (hydroxymethyl)-acylfulvalene has entered clinical trials. The illudins (**228**) were isolated from a jack-o'-lantern mushroom and other related fungi. The pterosins (**229**) are isolated from the bracken fern *Pteridium aquilinium*. Ptaquilosin, the aglycon of ptaquiloside, is also isolated from bracken ferns. The bioactivity associated with the illudins was proposed to arise from DNA alkylation of both the enone and the α-hydroxy cyclopropane of the core substrate. In similar fashion, the ptaquilosins (**230**) utilize an addition to the hydroxycyclopropane to alkylate DNA (Fig. 4.6).

In early 1994, Padwa et al. (119) synthesized both mono- and bicyclic core skeletons of the illudins and ptaqualosins. By utilizing a 1,1-disubstituted cyclopropane as the core of the dipolar cycloaddition, Padwa was able to add a number of dipolarophiles including cyclopentenone and cyclohexenone to produce cycloadduct **233** amenable to subsequent transformations to form the illudins and ptaqualosins. The cycloaddition forms the bicyclic constrained ether in

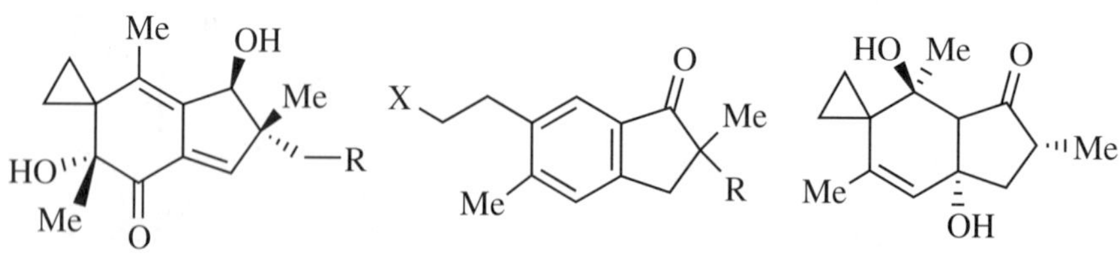

Illudins: (**228**) Pterosins: (**229**) Ptaquilosin: (**230**)

Figure 4.6 Sesquiterpenoid natural products.

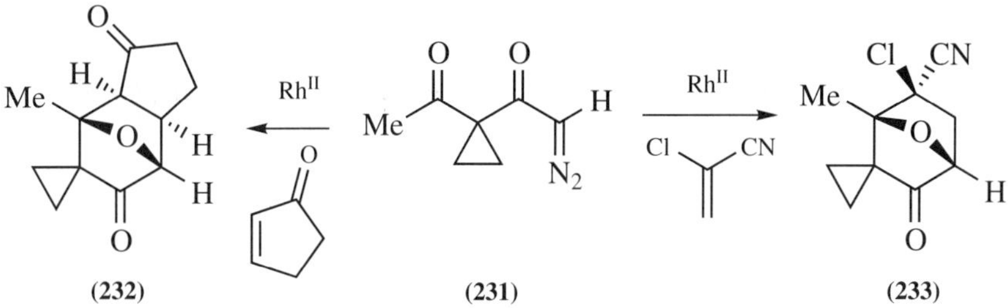

Scheme 4.60 Padwa approach to illudins and ptaquilosins.

excellent yields, good regioselectivity, and good stereoselectivity (4:1 exo/endo) (Scheme 4.60).

Subsequent to their initial synthetic studies, Padwa has also synthesized illudin M and a number of its analogues. To remove the bridging ether moiety, Padwa chose to use a one-electron reductant (Scheme 4.61). Following the work of Molander, Padwa utilized SmI_2 as the reducing agent to cleave the C—O bond α to the ketone. Padwa has used this strategy to regio- and stereospecifically place a hydroxyl group.

Padwa and co-workers (120–122) also utilized this carbonyl ylide cycloaddition strategy to advance to the aromatic pterosin family of compounds. The same intermediates used to approach the nonaromatic illudins and ptaqualosides are also useful for aromatic formation through cleavage and dehydration (Scheme 4.62).

Kinder et al. (123,124) utilized a similar approach to the illudins, more specifically to illudin analogs such as **241** that retain efficacy against various cell lines, but are less toxic. The analog design involves a spirocyclopropyl cyclohexane that contains two electrophilic moieties (Scheme 4.63). Some of the analogs formed are as active as adriamycin against several human tumor cell lines.

McMorris et al. (125–127), one of the original researchers on the biology of the illudins, has utilized the Padwa protocol to synthesize several antitumor acylfulvenes. The acylfulvenes are generally derived from compounds such as illudin S. McMorris's third generation acylfulvenes have been shown to elicit complete tumor regression in animals and increased life span >150%.

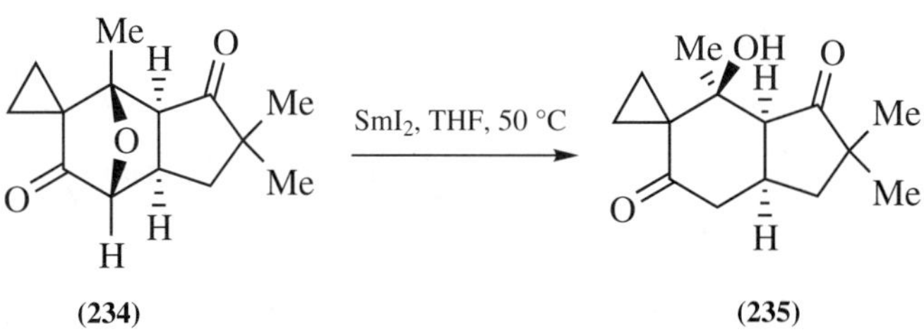

Scheme 4.61 Ether bridge cleavage.

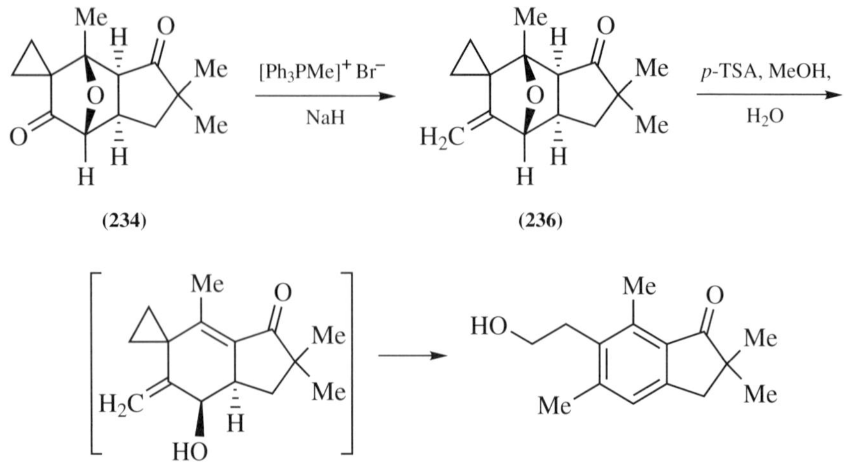

(234) (236) (237) (238)

Scheme 4.62 Synthesis of pterosins.

4.2.4. Approaches to Zaragozic Acids

Zaragozic acid or the squalastatins were discovered as fungal metabolites and have been identified as potent inhibitors of squalene synthase. These compounds have presented themselves as potential cholesterol lowering agents.

A Merck group chose to prepare the bicyclic core of zaragozic acid and some photodegradative products through a carbonyl ylide cycloaddition (128). The Merck group approached the synthesis of the core structure **243** by replacing the C(2) oxygen with a carbon, attempts at the use of a precursor with the C(2) oxygen in place resulted in an extremely low yield of cycloadduct. Attempts at using a 1,2-dioxygenated olefin as the dipolarophile also gave disappointingly low yields of the cycloadduct. These researchers also found that any electron-rich dipolarophiles gave routinely poor results with stabilized diazo-substituted β-ketoesters (Scheme 4.64).

A second generation approach to zaragozic acid has been completed by the Hashimoto group, after encountering difficulties with setting the correct C(5)

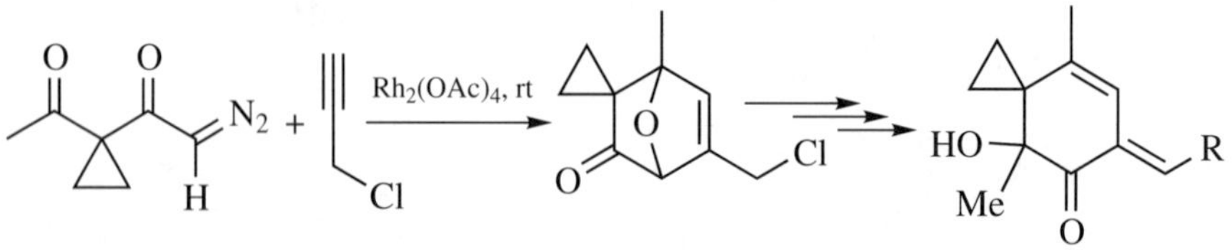

(232) (239) (240) (241)

Scheme 4.63 Kinder illudin analogues.

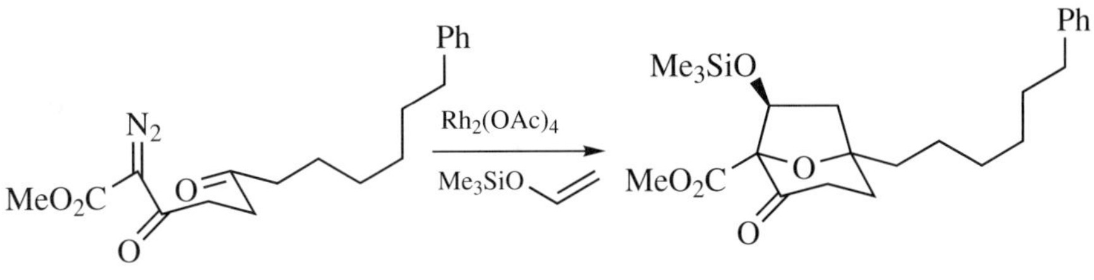

(242) **(243)**

Scheme 4.64 Merck approach to zaragozic acid derivatives.

stereochemistry in a non-ylide approach to the natural product (Scheme 4.65) (129). Despite the difficulties encountered by the Merck group, Hashimoto was able to generate a similar ylide precursor starting from a substituted isopropylidene acetal. After several transformations, the carbonyl ylide precursor **244** was completed. Hashimoto found that addition of dipolarophiles such as (*E*)-vinylene diacetate failed, as Koyama had found in the Merck synthesis. It was established that the dienophile attacked preferentially from the β-face, to avoid unfavorable interaction with the pseudo-axial trimethylsilyloxy group. Interestingly, the epimeric compound, with a pseudo-equatorial trimethylsilyloxy group, produced no cycloadduct after addition of catalyst and the hexenedione.

Hodgson and co-workers (130–132) addressed the synthesis of zaragozic acid derivatives through the use of a cycloaddition precursor that appears to be a hybrid of the Merck and Hashimoto approaches (Scheme 4.66). The Hodgson approach introduces the second ring oxygen by way of the addition of methyl glyoxalate as the dipolarophile. As discovered by Hashimoto, the presence of the silyloxy group prevents cycloaddition from occurring at the face syn to the silyloxy group. It was also found that methyl glyoxylate prefers exo addition compared to the dipole. After cycloaddition and silyl deprotection had occurred, Hodgson found that the use of Evan's conditions at reflux were sufficient to isomerize the core cycloadduct to the zaragozic structure **250**. Initial MM2 calculations predicted that **249** would not be observed at equilibrium, however, it is known that the MM2 force-field ignores

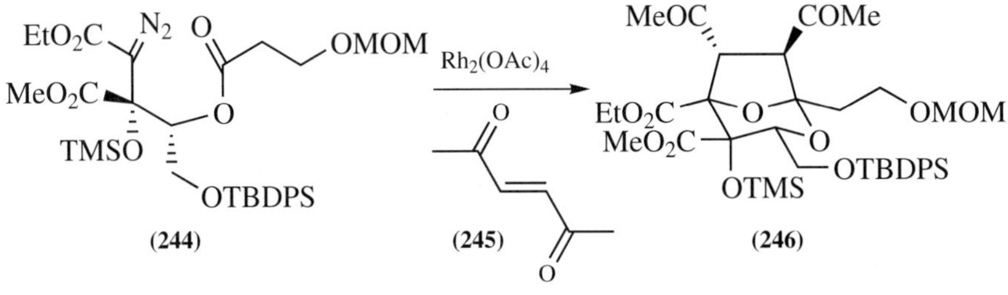

(244) **(245)** **(246)**

MOM = methoxymethyl

Scheme 4.65 Hashimoto approach to zaragozic acid.

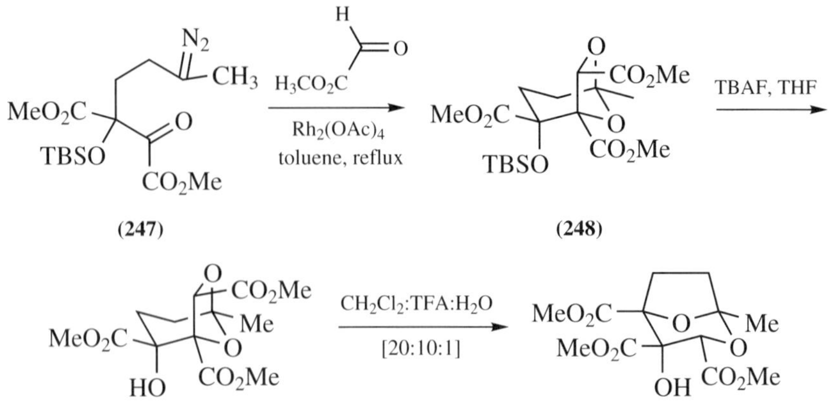

(247) **(248)** **(249)** **(250)**

Scheme 4.66 Hodson approach to zaragozic acid.

O–C destabilizing interactions. Scaling the 1,4-interactions to 70% of full value correctly identifies the equilibrium ratio of this system.

4.2.5. Tropolone and Benzotropolone Formation and Reactivity

Friedrichsen and co-workers (133) approached substituted benzotropolones from an aromatic substituted carbonyl ylide with a tethered alkyne as the intramolecular dipolarophile (Scheme 4.67). Starting from an aromatic anhydride, Friedrichsen was able to make the tethered alkyne via addition of either pentyn-ol or hexyn-ol, then transform the recovered benzoic acid to the α-diazocarbonyl cycloaddition precursor. Addition of rhodium acetate resulted in the tandem formation of cyclic carbonyl ylide followed by cycloaddition of the tethered alkyne producing the tricyclic constrained ether **252**. Addition of $BF_3 \cdot OEt_2$ opened the ether bridge, forming the benzotropylium ion, which subsequently rearranged to form the tricyclic benzotropolone (**253**).

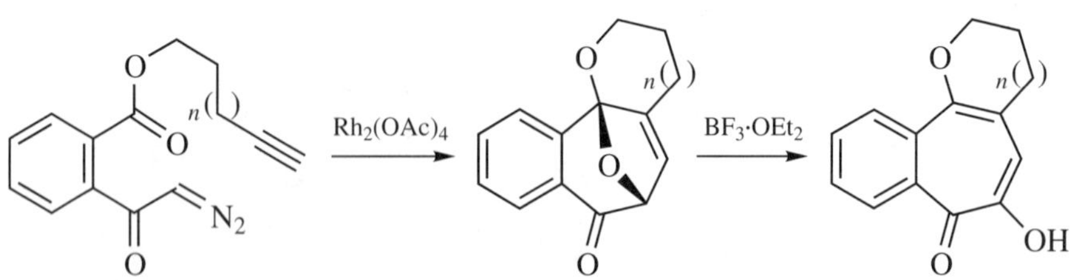

(251) **(252)** **(253)**

Scheme 4.67 Friedrichsen benzotropolone synthesis.

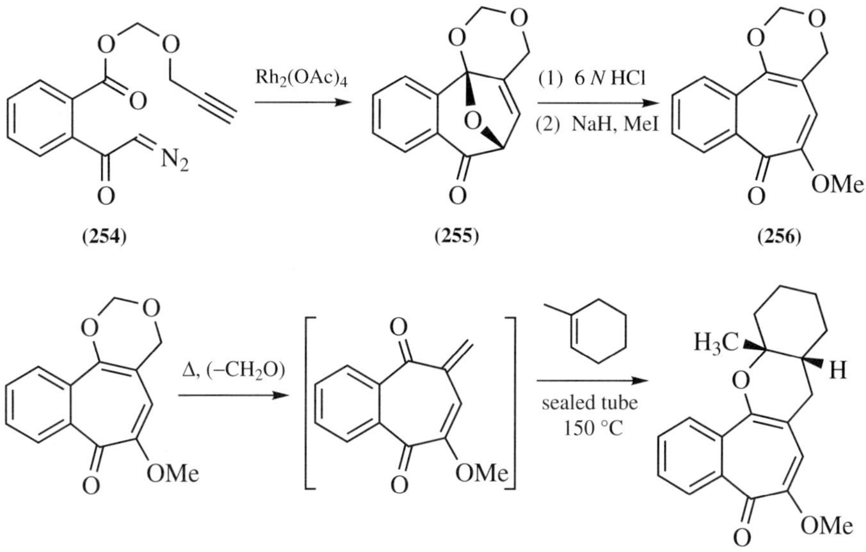

(254) (255) (256) (257) (258) (259)

Scheme 4.68 Baldwin benzotropolone formation.

In a similar vein, Baldwin et al. (134) accessed tropolone fungal metabolites pycnidione, epolone B, and eupenifeldin (Scheme 4.68). Pycnidione has been shown to induce erythropoietin gene expression and inhibit stromelysin. Eupenifeldin has displayed *in vivo* activity against P388 leukemia cells. Starting with commercially available phthalic acid, Baldwin converted the diacid to the monoester using (propargyloxy)methyl chloride, generated the mixed anhydride of the remaining acid moiety, then added diazomethane to generate the cycloaddition precursor **254**. After cleaving the cyclic ether with acid, the cyclic acetonide **257** was used to generate an exocyclic enone for Diels–Alder addition to ultimately deliver tetracycle **259**.

Friedrichsen and co-workers (135), along with Padwa, has utilized the carbonyl ylide cycloaddition to generate reactive furan moieties that can be further used in inter- or intramolecular Diels–Alder reactions to prepare aza- and carbocyclic compounds. Friedrichsen conducted a number of synthetic and theoretical studies on the reactivity, regioselectivity, and stereoselectivity of substituted furan formation and subsequent Diels–Alder reaction (Scheme 4.69).

The alternative regiochemical disposition of the diazo and tether of indole **260** failed to deliver any product upon addition of catalyst. Friedrichsen and co-workers (136) also applied this method to amine substituted tethers to generate polyazacyclic compounds (Scheme 4.70). The presence of the amino substituted furan subsequent to diazo-decomposition made it possible to cleave the ether bridge through the facility of the amino group formed to produce adduct **263**.

Friedrichsen and co-worker (137) used this approach to synthesize polycyclic oxazoles through the efficacy of the *in situ* formation of furo-oxazole, followed by

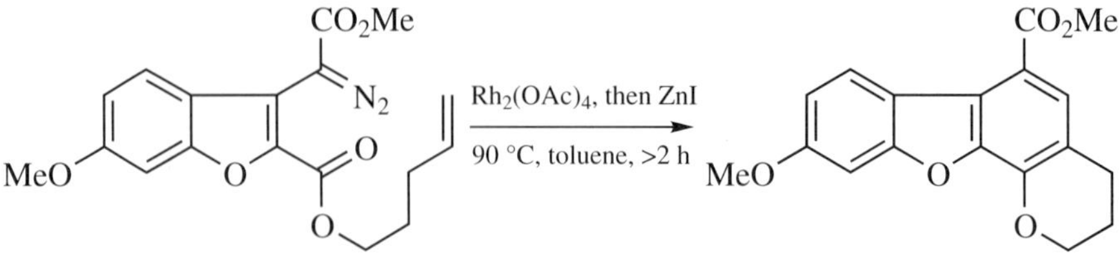

(260) **(261)**

Scheme 4.69 Friedrichsen tandem furan formation/Diels–Alder reaction.

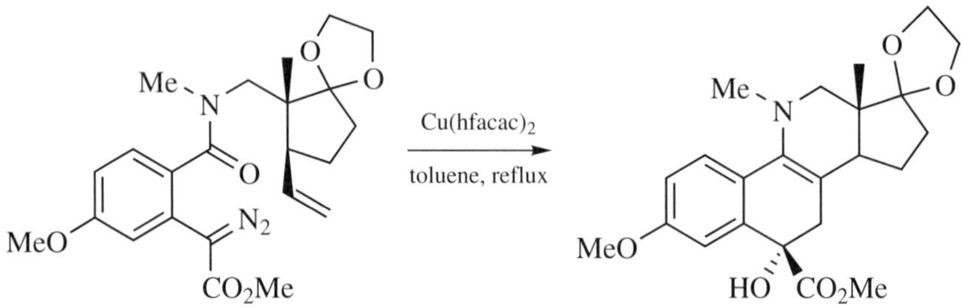

(262) **(263)**

hfacac = hexafluoroacetylacetonate

Scheme 4.70 Friedrichsen approach to azasteroids.

intermolecular Diels–Alder reaction (Scheme 4.71). Addition of dienophiles such as *N*-phenylmaleimide, DMAD, and *p*-benzoquinone all gave products of cycloaddition in rather low yield. Interestingly, all attempts at conducting an intramolecular cycloaddition from the diazocarbonyl ester tether resulted in synthesis of an intramolecular cyclopropane product rather than a product of the tandem ylide formation–cycloaddition chemistry.

4.2.6. Enantioselective Syntheses

One of the major goals set for tandem ylide formation–cycloaddition chemistry has been the application of enantioselective catalysis to form one product in preference to all others. It appears that these transformations must involve a catalyst associated ylide for some degree of enantioselectivity to occur. Generally, if the *free ylide* forms without any catalyst association then enantioselectivity is highly unlikely.

Hodgson et al. (138) chose to investigate a system that had previously been shown to undergo an effective intramolecular addition of a tethered olefin (Scheme 4.72). In his first attempt, using Doyle's $Rh_2[(5R)\text{-MEPY}]_4$, the yield of cycloadduct **270** obtained was comparable to that with rhodium acetate, but no asymmetric induction was observed. Changing to the Davies catalysts in dichloromethane resulted in a

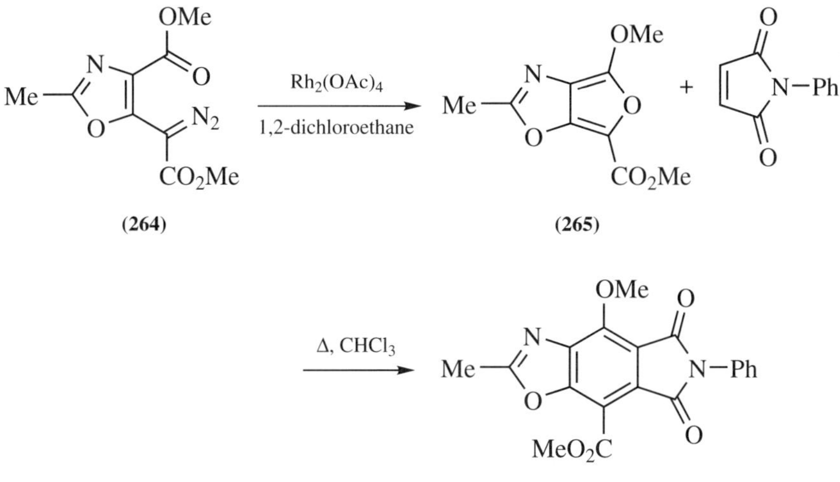

Scheme 4.71 Oxazole cycloaddition.

disappointing 8% ee. Following the work of Davies, the solvent was modified to a hydrocarbon solvent, hexane, resulting in 76% chemical yield with 53% ee. As noted, increasing the reaction temperature lead to an increase in the chemical yield, but an erosion of the enantioselectivity. Decreasing the temperature of the reaction did little to change the ee of the reaction, but caused a decrease in chemical yield.

After completing his initial intramolecular cycloaddition, Hodgson utilized conditions that had been optimized for the intermolecular cycloaddition of DMAD with simple cyclic carbonyl ylides used by Hashimoto and co-workers (139). Hodgson et al. (140) found that the reaction indeed gave excellent overall chemical yield, but the enantioselectivity dropped to ≈1%, giving essentially a racemic mixture. It appeared that ee ratios were sensitive to the electronic nature of the dipole. Hodgson chose to screen several binaphthol derived rhodium catalysts of the type developed by McKervey and Pirrung, due in part to the reports of

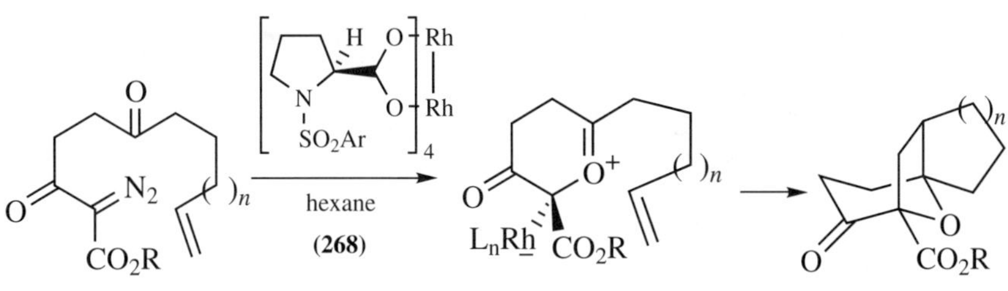

Scheme 4.72 Enantioselective intramolecular cycloaddition.

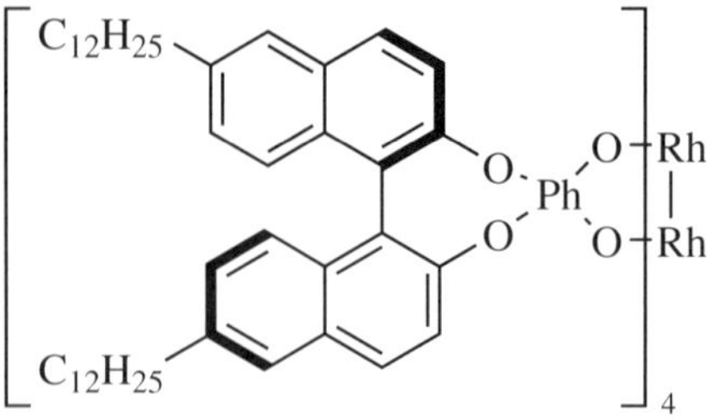

$Rh_2(R\text{-}DDBNP)_4$

Figure 4.7 Hodgson binaphthol catalyst.

enantioselectivities of up to 60% in C—H insertion and cyclopropanation reactions (Fig. 4.7).

By utilizing the same reaction sequence as shown in Scheme 4.72, Hodgson applied his dodecyl-substituted binaphthol phosphate derived catalyst. In this reaction, the chemical yield decreased somewhat from 93% in the original sequence to 81%, but the enantioselectivity increased dramatically from 53 to 88%. Decreasing the temperature $< 0\,°C$ increased the ee ratio slightly, but also dropped the chemical yield.

Hashimoto and co-workers (139) further looked at an intermolecular carbonyl ylide cycloaddition screening several different chiral rhodium catalysts. The Hashimoto group chose to study phthaloyl amino acid derivatives for enantiocontrol of the cycloaddition reactions (Fig. 4.8). Using fluorinated or ethereal solvents with the phthaloyl catalysts gave ee ratios of 20–69%.

To improve these selectivities, Hashimoto studied several catalysts that had been found highly effective for enantioselective C—H insertion reactions. The new catalysts incorporated an additional benzene in the naphthyl system to increase the steric bias of the catalyst. By using the second-generation catalysts in trifluorotoluene as solvent, at 0 °C, and short reaction times gave ee ratios of 68–92%. Lowered reaction temperature generally resulted in reduced chemical yields but did not erode the ee ratio. Tether lengths one smaller or one larger also tended to erode the ee ratio (Scheme 4.73).

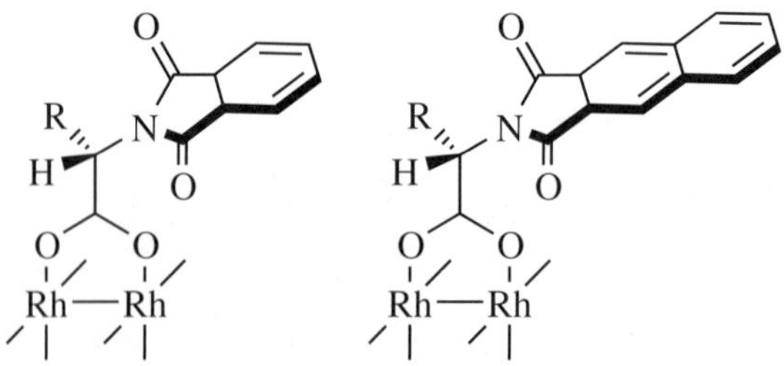

R = Bn, Me, *i*-Pr, *t*-Bu

Figure 4.8 Hashimoto catalysts.

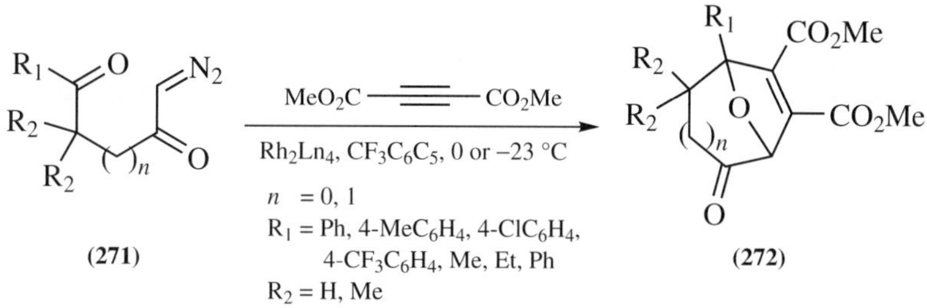

(271) **(272)**

Scheme 4.73 Hashimoto enantioselective reaction.

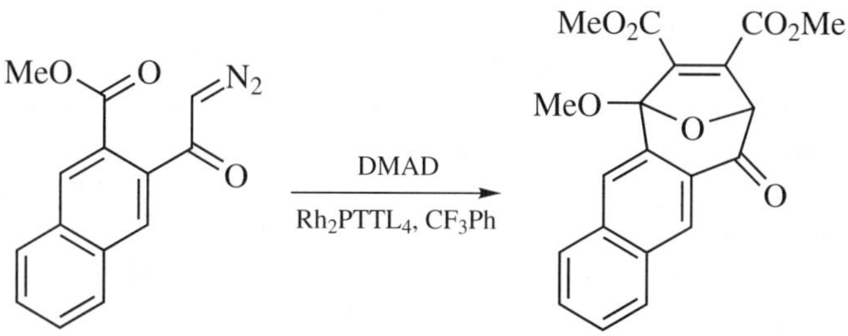

(273) **(274)**

Scheme 4.74 Hashimoto ester cycloaddition.

Hashimoto and co-workers (55) recently extended this study to include tethered ester groups rather than the ketones used in the previous study (Scheme 4.74). Overall the effect of ester replacement was dramatic, with one exception, both yields and ee ratios dropped precipitously. It was found that the σ-bound oxygen within the tether was more deleterious than outside the tether. By using a 3-diazoacetylnaphthoate template as part of the tether was the only substrate found to give both high yield and high enantioselectivity. Use of the 2-diazoacetyl-naphthoate derivative resulted in less than one-half of the chemical yield of the 3-derivative and greatly reduced enantioselectivity.

Prior to the Hashimoto study with aromatic diazoesters, Ibata and co-workers (141) found that in similar cases with aromatic diazoester **275** that addition of

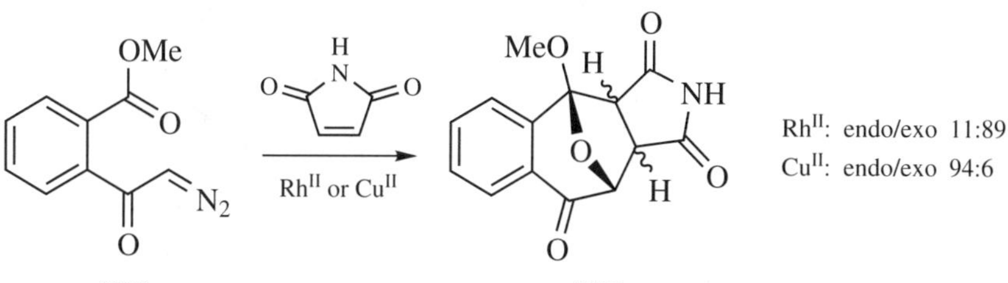

(275) **(276)**

Scheme 4.75 Ibata enantioselective cycloaddition.

phenylmaleimide as the dipolarophile gave a large preference for the exo product **276** with rhodium, but provided little enantioselectivity (Scheme 4.75). Interestingly, the use of copper bis(oxazoline) provided an excess of the endo isomer, but again, the enantioselectivity was extremely low for both isomers.

4.3. OXIDOPYRYLIUM ION CYCLOADDITIONS

Over the past decade, Wender et al. (142–146) utilized the oxidopyrylium ions, a special class of carbonyl ylides with significant aromatic character, for use in concerted [5 + 2]-cycloaddition reactions. This strategy has lead to the synthesis of tigliane natural products such as phorbol and daphnane diterpenes such as resiniferatoxin (146,147). Starting with the cesium salt of kojic acid, O-allylation and Claisen rearrangement was used to produce the cycloaddition precursor **277** (145,148,149). The cycloaddition is initiated from pyrone **277** by thermal group transfer to generate ylide **278**. In the transition state, the side chain adopts a chair-like conformation with the C_{18} methyl assuming an equatorial position to minimize interaction with the C_{10} substituent. This ultimately establishes the correct relative stereochemistry at the three contiguous stereogenic centers (Scheme 4.76).

Wender utilized a similar protocol to achieve a nonracemic phorbol synthesis. Due to the expense and lack of chirality associated with kojic acid, Wender chose to

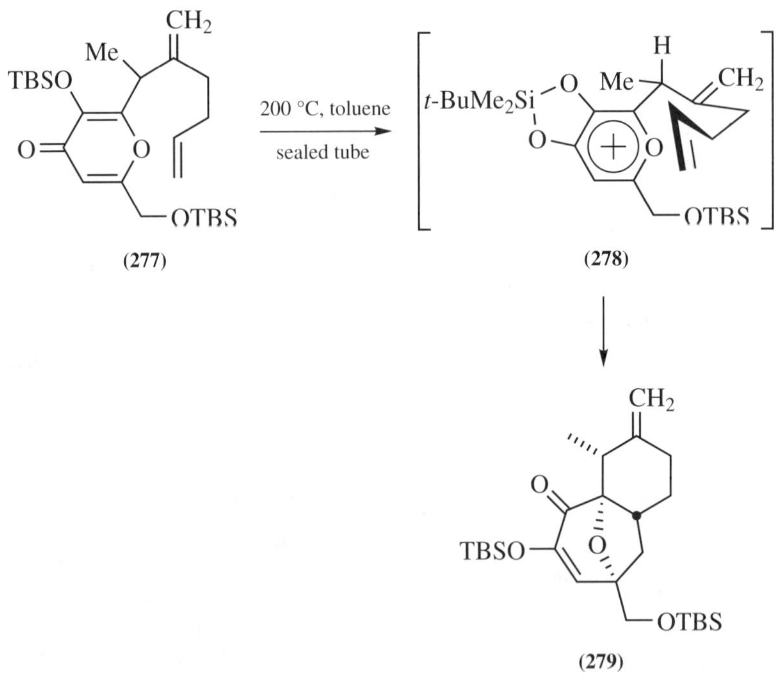

Scheme 4.76 Wender group approach to phorbol.

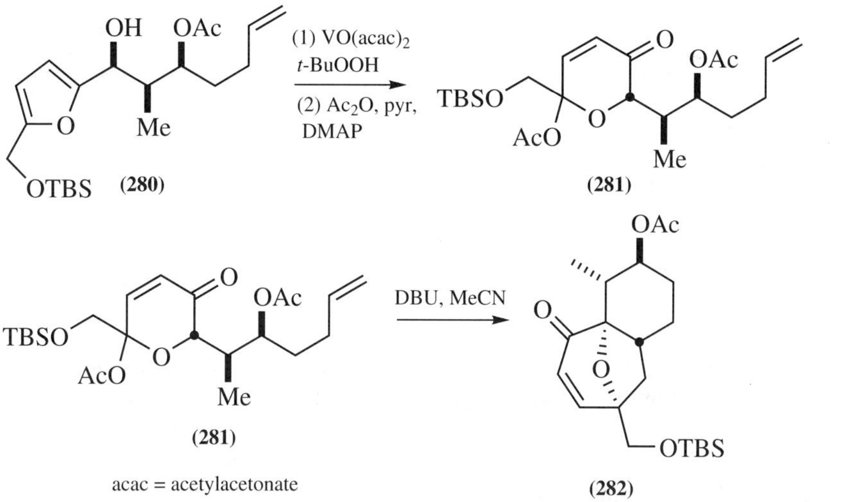

Scheme 4.77 Asymmetric approach to phorbol.

prepare the pyrone cyclization precursor via a furan oxidative ring expansion. Following a similar synthetic pathway, Wender disclosed the first asymmetric synthesis of the daphnane diterpene resiniferatoxin, a potent analgesic agent (147). In this case, the oxidopyrylium ion was derived from dihydropyrone (**281**) (Scheme 4.77).

Over the past several years, Mascarenas and co-workers (150–153) utilized the oxidopyrylium ion with variously hetero-substituted olefin tethers. Mascarenas has used this methodology in tandem with a Diels–Alder reaction to prepare tricyclic cycloheptanoid substrates. Further, Mascarenas and co-workers (154–156) achieved the synthesis of optically active oxabicyclic[3.2.1]octane derivatives through the addition of a homochiral *p*-tolylsulfinyl group substituted at the olefin tether. The Mascarenas group has also used this methodology to prepare the THF portion of (±)-nemorensic acid via oxidative cleavage of the substituted α-hydroxyketone moiety (157) (Scheme 4.78).

Magnus and co-workers (158–162) used the power of the oxidopyrylium cycloaddition to prepare a taxol precursor **288**, generating a synthon for the taxol

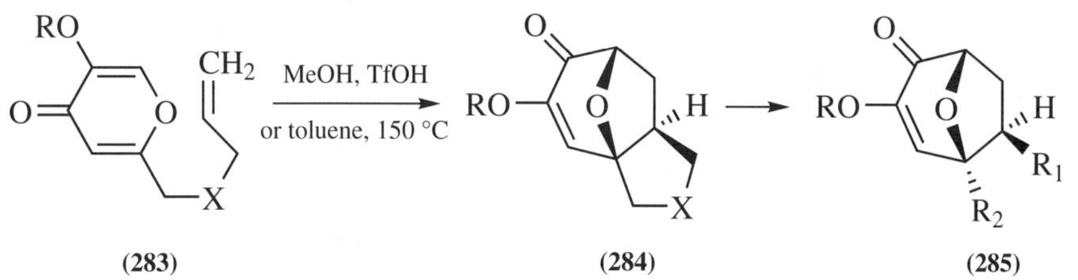

Scheme 4.78 Heteroatom tethered cycloaddition.

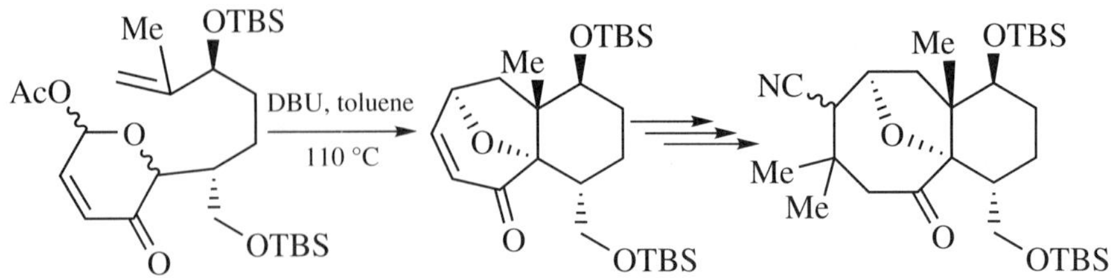

(286) (287) (288)

Scheme 4.79 Magnus approach to taxol.

B and C rings by preparing a bicyclo[5.4.0]undecane (**287**) amenable to cyclopropanation followed by ring opening to generate the cyclooctane B ring (Scheme 4.79).

Magnus and Shen (163) used a similar synthetic strategy to prepare the core structure of the cyathins (Scheme 4.80). This novel group of fused five, seven, six-ring containing species exhibit highly potent nerve growth factor induction, thought to be a possible therapy for neurodegenerative diseases such as Alzheimer's or ALS. Starting from a disubstituted cyclopentane template **289**, Magnus was able to generate the oxidopyrylium ion, then cycloadd to a tethered olefin to give bridged ether **290**.

The Williams approach to bicyclo[5.4.0]undecane containing molecules such as the cyathins, striatins, dolastanes, clavularanes, grayanotoxins, tiglianes, and daphnanes also utilizes an oxidopyrylium cycloaddition to generate the ether constrained bicyclic (Scheme 4.81) (164). The object of the Williams study was to prepare the bicyclo[5.4.0]undecanes and investigate the regio- and stereocontrolled oxidation of the substrate.

Finally, in a very recent disclosure, Lee et al. (165) approached the total synthesis of arteminolide using a [5 + 2] cycloaddition strategy with an oxidopyrylium ion. Despite its long history of use, Lee was the first to utilize an allene moiety both in an intra- and an intermolecular cycloaddition with oxidopyrylium ions. By utilizing a pyrone cycloaddition precursor (**294**) similar to those used in the Wender phorbol synthesis, Lee was able to synthesize various ring sizes and

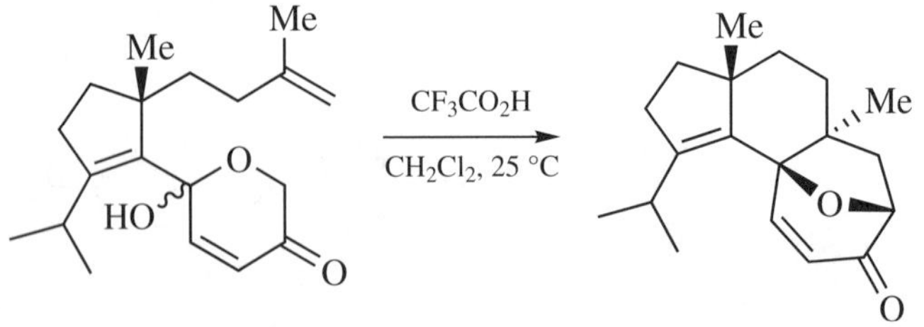

(289) (290)

Scheme 4.80 Magnus approach to the cyathins.

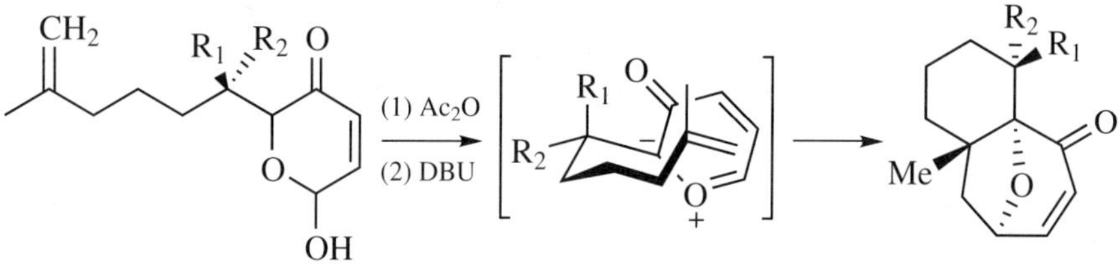

$R_1 = R_2 = OMe$
OCH_2OMe

(291) **(292)** **(293)**

Scheme 4.81 Williams approach to bicyclo[5.4.0]undecanones.

examine the selectivity of the allene. Lee found that only neutral and electron-rich allenes could function effectively as a dipolarophile. In the intermolecular case, only moderate to low yields were obtained and only the terminal olefin was reactive. It was also found in the intramolecular series that tether length and allene reactivity are intimately related (Scheme 4.82).

When the pyrone **297** was exposed to base in dichloromethane, formation of the oxidopyrylium occurs and subsequent cycloaddition transpires. Lee found that when the allene tether was 3 ($n = 0$), cycloaddition occurs to form the bicyclo[5.3.0]decadiene **298** exclusively, rather than the alternative bicyclo[5.2.0] nonadiene product **299**. Increasing the tether length by one resulted in formation of the exo-substituted double bond (**299**) in 45% yield. Increasing the tether length by one carbon, to four, completely retarded the cycloaddition (Scheme 4.83).

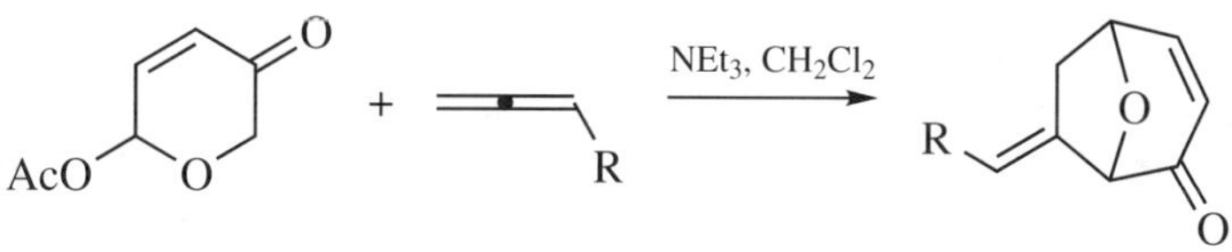

(294) **(295)** **(296)**

Scheme 4.82 Lee initial allene cycloaddition.

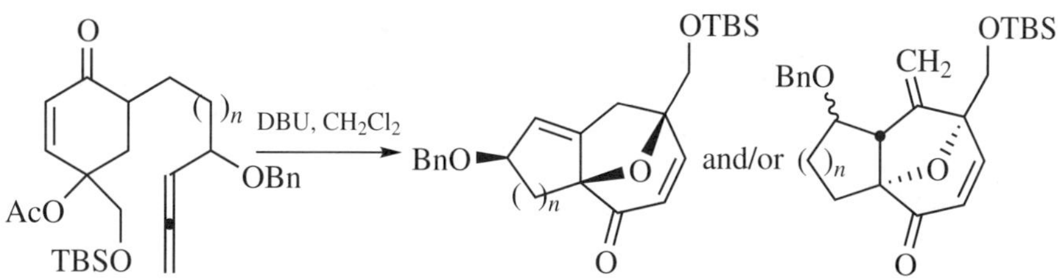

(297) **(298)** **(299)**

Scheme 4.83 Lee approach to arteminolide.

4.4. SYNTHESIS USING NONMETALLOCARBENOID PRECURSORS AND NOVEL METALLOCARBENOID PROCESSES

Although there has been a great deal of progress in designing methodology using stabilized carbonyl ylides over the last two decades, the synthetic uses of nonstabilized ylides are much less complete. Linderman and co-workers (166–169) chose to synthesize furanones from *electrophilic capped carbonyl ylides*, presenting a method of using nonstabilized ylides for synthesis. The Linderman group found that the α-alkoxytin or α-alkoxysilanes could be prepared in good yield via a Mukaiyama aldol reaction of a MOM acetal with a silyloxyenol ether. From the bifunctional reagent formed, transmetalation with butyllithium, followed by internal attack of nucleophile formed on the requisite internal electrophile gave the substituted furanones in good overall yield (Scheme 4.84).

Dittami et al. (170,171) was able to generate a carbonyl ylide via photocyclization of an aryl vinyl ether (Scheme 4.85). The photocyclization proceeded through a six-electron rearrangement providing initially the carbonyl ylide, then subsequent to the cyclization, a dipolar cycloaddition takes place with a pendant olefinic tether.

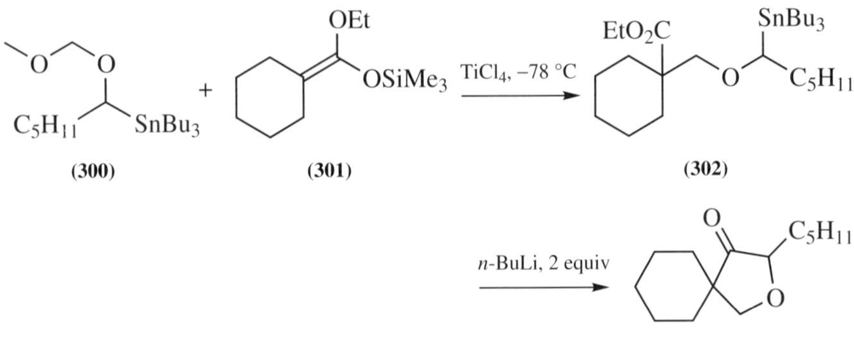

(300) (301) (302) (303)

Scheme 4.84 Linderman nonstabilized carbonyl ylide.

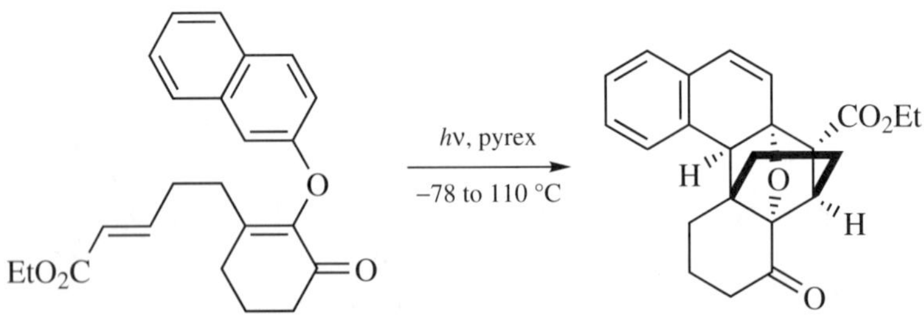

(304) (305)

Scheme 4.85 Dittami photocyclization to carbonyl ylides.

(306) (307)

Scheme 4.86 Oxirane thermolysis to carbonyl ylide.

There are very few examples of photolysis being used for preparation of a carbonyl ylide. The Dittami protocol follows work completed from his lab with aryl vinyl sulfides. Photolysis, followed by cycloaddition, led to the cycloadduct **305** in excellent yield and stereoselectivity. If the aryl vinyl ether **304** was subjected to irradiation in a mixed solution of toluene–methanol at 366 nm rather than a single solvent of toluene, cyclized product was obtained, but no cycloadduct was formed. If a simple phenyl aryl ether was subjected to the same tandem conditions, the cyclized product was generated, but no cycloadduct was detected.

de March and co-workers (172) found that epoxides provided simple precursors to carbonyl ylides. Experimentation showed that to obtain a reasonable yield of cycloaddition product required that the terminal end of the oxirane be substituted with two electron-withdrawing groups to stabilize the incipient anion formed. Thermolysis of **306** in chlorobenzene at 160 °C in a sealed pressure flask provided the dihydrofuran cycloadduct **307** in 76% chemical yield (Scheme 4.86).

Hosami and co-workers (12,173,174) prepared alternative forms of nonstabilized ylides (Scheme 4.87). They generated *tailor-made* carbonyl ylides from substituted 1,3-dichloroethers through a "1,3-elimination" pathway mediated with a samarium reagent. These simple carbonyl ylide intermediates are valuable for preparing tetrahydro- and dihydrofurans. To a mixture of bis(chloromethyl) ethers and

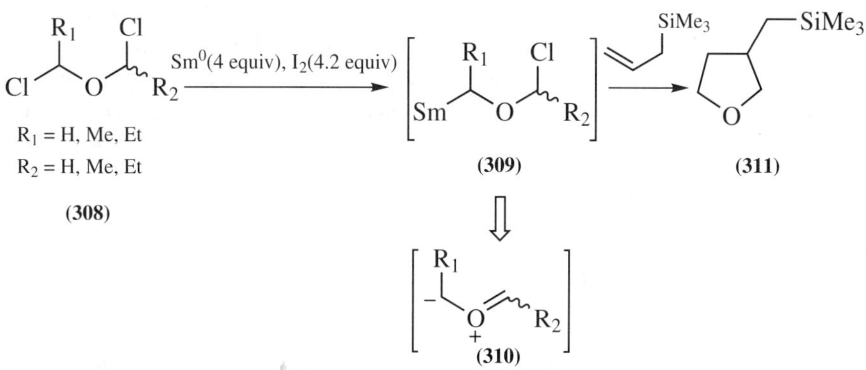

Scheme 4.87 Hosami nonstabilized carbonyl ylide.

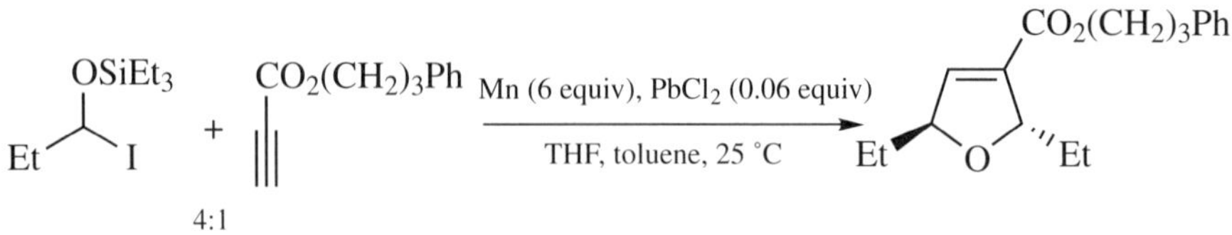

(312) **(313)** **(314)**

Scheme 4.88 Takai cycloaddition of nonstabilized ylides.

samarium/I_2, a number of dipolarophiles were added to complete the carbonyl ylide cycloaddition. Both symmetric and nonsymmetric ylides were possible substrates depending on the substitution about the ether. A number of alkene, alkyne, and allene reagents have been used as the dipolarophile for cycloaddition, providing good-to-excellent yields in all cases. Hosami also added a ketone as the dipolarophile in one case, generating a dioxolane product. The limitation of this methodology is the availability of differentially substituted bis(chloromethyl) ethers.

Takai et al. (175,176) used a similar approach to generate "nonstabilized carbonyl ylides" to that of Hosami. Due to the inherent reactivity and unwanted side reactions of the SmI_2 used by Hosami (reduction of α,β-unsaturated carbonyls or pinacol coupling of aldehydes), Takai chose to use a mixed-metal process that could be tuned depending on the dipolarophile and outcome needed (Scheme 4.88). Takai found that the carbonyl ylide could be generated from either a bis(iodoethyl) ether or from a 1-silyloxy substituted ethyl iodide using Mn metal and $PbCl_2$. Takai assumed that because of similar reactivity profiles of the two ylide precursors, the two could be in equilibrium with one another. In terms of chemoselectivity, both nonstabilized ylides were more reactive to acrylates rather than simple olefins in a competitive study. With the iodoalkyl triethylsilyl ether, it was noted that some diasteroselectivity could be accomplished by the presence or absence of $PbCl_2$.

One novel and interesting method of generating a silacarbonyl ylide occurred through the addition of a carbonyl species with a silylene formed under photolytic conditions. Komatsu and co-workers (177) found that photolysis of trisilane (**315**) in solution with a bulky carbonyl species led initially to the formation of a silacarbonyl ylide followed by a dipolar cycloaddition of an olefinic or carbonyl substrate. Reaction of simple, nonbulky aldehydes led to only moderate yields of cycloadduct, the siladioxolane. One lone ketone example was given, but the cycloadduct from the reaction was prepared in very low yield (Scheme 4.89).

Hudlicky and Barbieri (178) developed a simple two-step strategy to generate vinyl oxiranes, and then rearrange the vinyl oxirane to form functionalized dihydrofurans in a formal [2 + 3]-annulation process. Both pyrolysis and ring opening with TMSI have been used to initiate the rearrangement process (Scheme 4.90).

Hudlicky and Barbieri (178) developed a simple procedure to synthesize the substituted vinyl oxiranes by adding the lithium dienolate of a 1-bromo-4-silyloxyethyl-2-butenoate to substituted aldehydes. From the vinyl oxiranes formed,

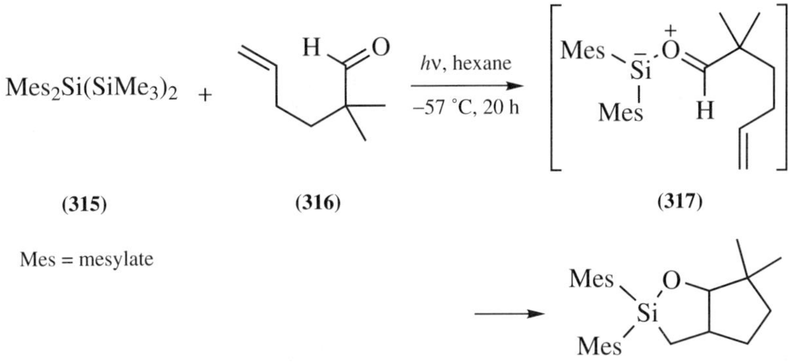

(315) (316) (317) (318)

Mes = mesylate

Scheme 4.89 Reaction of silacarbonyl ylide.

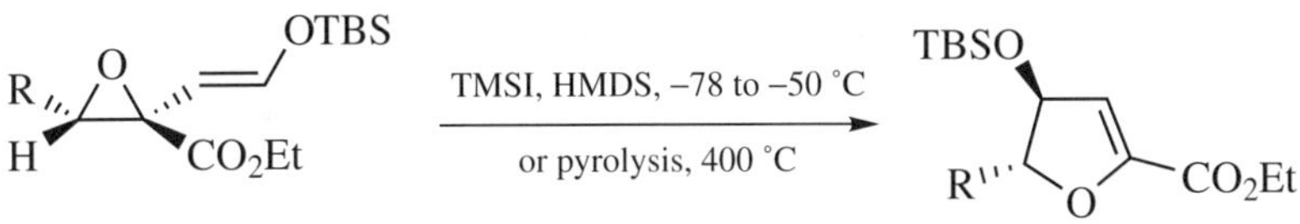

(319) (320)

TBS = tertbutyldimethylsilyl
HMDS = hexamethyldisilayane

Scheme 4.90 [2 + 3] Annulation to dihydrofurans.

Hudlicky found that C—C bond cleavage of the oxirane occurred readily, while the alternative C—O bond cleavage products were not detected.

4.5. CONCLUDING REMARKS

It is tempting to say that the use of carbonyl ylides in tandem with subsequent cycloaddition reactions is the answer to most synthetic problems. This is obviously not the answer for every type of natural product or methodological problem encountered, but it can be argued that these tandem reactions present a highly efficient, regio- and stereospecific way of approaching many synthetic problems. The synthetic systems that have already been addressed, are being addressed, and those problems for which this methodology could provide meaningful access are proof that this methodology adds greatly to the arsenal of the synthetic chemist. Although there are still many problems that have not been addressed and difficulties with the processes that limit it's scope, the speed and efficiency of bond construction, leave this tandem reaction sequence as one who has not reached it's full potential. The strides over the last two decades have been enormous, but there are still a great number of challenges and interesting chemistry left to this sequence.

REFERENCES

1. W. J. Linn and R. E. Benson, *J. Am. Chem. Soc.*, **1965**, *87*, 3657.
2. K. N. Houk, N. G. Rondan, C. Santiago, C. J. Gallo, R. W. Gandour, and G. W. Griffin, *J. Am. Chem. Soc.*, **1980**, *102*, 1504.
3. G. K. S. Prakash, R. W. Ellis, J. D. Felberg, and G. A. Olah, *J. Am. Chem. Soc.*, **1986**, *108*, 1341.
4. N. J. Turro and Y. Cha, *Tetrahedron Lett.*, **1987**, *28*, 1723.
5. R. Bonneau and M. T. H. Liu, *J. Am. Chem. Soc.*, **1990**, *112*, 744.
6. J. E. Chateauneuf and M. T. H. Liu, *J. Am. Chem. Soc.*, **1991**, *113*, 6585.
7. E. P. Janulis and A. Arduengo, *J. Am. Chem. Soc.*, **1983**, *105*, 5929.
8. H. S. Gill and J. A. Landgrebe, *Tetrahedron Lett.*, **1982**, *23*, 5099.
9. H. S. Gill and J. A. Landgrebe, *J. Org. Chem.*, **1983**, *48*, 1051.
10. Z. Huan and J. A. Landgrebe, *J. Org. Chem.*, **1983**, *48*, 4519.
11. T. Ibata, M. T. H. Liu, and J. Toyoda, *Tetrahedron Lett.*, **1986**, *27*, 4383.
12. M. Hojo, H. Aihara, and A. Hosomi, *J. Am. Chem. Soc.*, **1996**, *118*, 3533.
13. M. Hojo, M. Ohkma, N. Ishibashi, and A. Hosomi *Tetrahedron Lett.*, **1993**, *34*, 5943.
14. M. Hojo, H. Aihara, Y. Suginohara, K. Sakata, S.-Y. Nakamura, C. Murakami, and A. Hosomi, *J. Org. Chem.*, **1997**, *62*, 8610.
15. M. Bekhazi and J. Warkentin, *J. Am. Chem. Soc.*, **1983**, *105*, 1289.
16. J. Warkentin, *J. Chem. Soc., Perkin Trans. 1*, **2000**, 2161.
17. M. Bekhazi, W. Lawrynowicz, and J. Warkentin, *Can. J. Chem.-Rev. Can. Chim.*, **1991**, *69*, 1507.
18. P. K. Sharma and J. Warkentin, *Tetrahedron Lett.*, **1995**, *36*, 7591.
19. J. P. K. Wong, A. A. Fahmi, G. W. Griffin, and N. S. Bhacca, *Tetrahedron*, **1981**, *37*, 3345.
20. S. N. Ege, E. J. Gess, A. Thomas, P. Umrigar, G. W. Griffin, P. K. Das, A. M. Trozzolo, and T. M. Leslie, *J. Chem. Soc., Chem. Commun.*, **1980**, 1263.
21. P. Clawson and D. A. Whiting, *Tetrahedron Lett.*, **1987**, *28*, 3155.
22. K. R. Meier, A. Linden, G. Mloston, and H. Heimgartner, *Helv. Chim. Acta*, **1997**, *80*, 1190.
23. J. Brokatzky-Geiger and W. Eberbach, *Heterocycles*, **1981**, *16*, 1907.
24. W. N. Chou and J. B. White, *Tetrahedron Lett.*, **1991**, *32*, 7637.
25. M. Kotera, K. Ishii, O. Tamura, and M. Sakamoto, *J. Chem. Soc., Perkin Trans. 1*, **1994**, 2353.
26. M. Kotera, K. Ishii, O. Tamura, and M. Sakamoto, *J. Chem. Soc., Perkin Trans. 1*, **1998**, 313.
27. M. Kotera, K. Ishii, M. Hiraga, and M. Sakamoto, *Heterocycles*, **1999**, *51*, 2147.
28. J. Brokatzky and W. Eberbach, *Tetrahedron Lett.*, **1980**, *21*, 4909.
29. J. Brokatzky-Geiger and W. Eberbach, *Tetrahedron Lett.*, **1982**, *23*, 4665.
30. J. Brokatzky-Geiger and W. Eberbach, *Heterocycles*, **1983**, *20*, 1519.
31. G. Zecchi, *Synthesis*, **1991**, 181.
32. D. F. O'Shea and J. T. Sharp, *J. Chem. Soc., Perkin Trans. 1*, **1997**, 3025.
33. R. Huisgen and P. De March, *J. Am. Chem. Soc.*, **1982**, *104*, 4953.
34. E. Buchner and T. Curtius, *Ber. Dtsch, Chem. Ges.*, **1885**, *18*, 2371.
35. A. C. Lottes, J. A. Landgrebe, and K. Larsen, *Tetrahedron Lett.*, **1989**, *30*, 4089.
36. A. Padwa, R. L. Chinn, S. F. Hornbuckle, and L. Zhi, *Tetrahedron Lett.*, **1989**, *30*, 301.
37. D. C. Dean, K. E. Krumpe, and A. Padwa, *J. Chem. Soc., Chem. Commun.*, **1989**, 921.
38. A. Padwa, G. E. Fryxell, and L. Zhi, *J. Am. Chem. Soc.*, **1990**, *112*, 3100.
39. A. Padwa, G. E. Fryxell, and L. Zhi, *J. Org. Chem.*, **1988**, *53*, 2875.
40. A. Padwa, S. F. Hornbuckle, G. E. Fryxell, and P. D. Stull, *J. Org. Chem.*, **1989**, *54*, 817.

41. A. Padwa, S. F. Hornbuckle, G. E. Fryxell, and Z. Zhang, *J. Org. Chem.*, **1992**, *57*, 5747.
42. A. Padwa, *J. Organmet. Chem.*, **2000**, *610*, 88.
43. A. Padwa, K. E. Krumpe, Y. Gareau, and Y. Chiacchio, *J. Org. Chem.*, **1991**, *56*, 2523.
44. A. Padwa, R. L. Chinn, S. F. Hornbuckle, and Z. Zhang, *J. Org. Chem.*, **1991**, *56*, 3271.
45. V. Nair, K. C. Sheela, K. V. Radhakrishnan, and N. P. Rath, *Tetrahedron Lett.*, **1998**, *39*, 5627.
46. A. Padwa, D. J. Austin, and S. F. Hornbuckle, *J. Org. Chem.*, **1996**, *61*, 63.
47. A. Padwa, D. J. Austin, S. F. Hornbuckle, and A. T. Price, *Tetrahedron Lett.*, **1992**, *33*, 6427.
48. A. Padwa, D. J. Austin, S. F. Hornbuckle, M. A. Semones, M. P. Doyle, and M. N. Protopopova, *J. Am. Chem. Soc.*, **1992**, *114*, 1874.
49. G. G. Cox, C. J. Moody, D. J. Austin, and A. Padwa, *Tetrahedron*, **1993**, *49*, 5109.
50. M. Alt and G. Maas, *Tetrahedron*, **1994**, *50*, 7435.
51. E. Wenkert and H. Khatuya, *Tetrahedron Lett.*, **1999**, *40*, 5439.
52. M. Hamaguchi, H. Matsubara, and T. Nagai, *Tetrahedron Lett.*, **2000**, *41*, 1457.
53. T. Ibata, H. Nakano, and H. Tamura, *Bull. Chem. Soc. Jpn.*, **1992**, *65*, 1362.
54. A. Padwa and P. D. Stull, *Tetrahedron Lett.*, **1987**, *28*, 5407.
55. S. Kitagaki, M. Yasugahira, M. Anada, M. Nakajima, and S. Hashimoto, *Tetrahedron Lett.*, **2000**, *41*, 5931.
56. A. Padwa and L. Zhi, *J. Am. Chem. Soc.*, **1990**, *112*, 2037.
57. A. Padwa, T. Price, and L. Zhi, *J. Org. Chem.*, **1996**, *61*, 2283.
58. A. Padwa, T. Hasegawa, B. Liu, and Z. J. Zhang, *J. Org. Chem.*, **2000**, *65*, 7124.
59. M. D. Weingarten, M. Prein, A. T. Price, J. P. Snyder, and A. Padwa, *J. Org. Chem.*, **1997**, *62*, 2001.
60. A. Padwa, J. P. Snyder, E. A. Curtis, S. M. Sheehan, K. J. Worsencroft, and C. O. Kappe, *J. Am. Chem. Soc.*, **2000**, *122*, 8155.
61. A. Padwa and Z. J. Zhang, L. Zhi, *J. Org. Chem.*, **2000**, *65*, 5223.
62. C. O. Kappe, *Tetrahedron Lett.*, **1997**, *38*, 3323.
63. J. D. Rodgers, G. W. Caldwell, and A. D. Gauthier, *Tetrahedron Lett.*, **1992**, *33*, 3273.
64. K. Ueda, T. Ibata, and M. Takebayashi, *Bull. Chem. Soc. Jpn.*, **1972**, *45*, 2779.
65. M. Hamaguchi and T. Ibata, *Casreact 132:321687*, **1974**, 4475.
66. T. Ibata, K. Ueda, and M. Takebayashi, *Sci. Rep., Coll. Gen. Educ., Osaka Univ.*, **1976**, *25*, 19.
67. T. Ibata and K. Jitsuhiro, *Bull. Chem. Soc. Jpn.*, **1979**, *52*, 3582.
68. T. Ibata and J. Toyoda, *Chemistry Lett.*, **1983**, 1453.
69. T. Ibata and J. Toyoda, *Bull. Chem. Soc. Jpn.*, **1986**, *59*, 2489.
70. T. Ibata, J. Toyoda, M. Sawada, and T. Tanaka, *J. Chem. Soc.-Chem. Commun.*, **1986**, 1266.
71. A. Gillon, D. Ovadia, M. Kapon, and S. Bien, *Casreact 130:182401*, **1982**, *38*, 1477.
72. S. Bien and A. Gillon, *Casreact 132:321687*, **1974**, 3073.
73. A. Padwa and M. D. Weingarten, *Chem. Rev.*, **1996**, *96*, 223.
74. A. Padwa and S. F. Hornbuckle, *Chem. Rev.*, **1991**, *91*, 263.
75. A. Padwa, R. L. Chinn, and Z. Lin, *Tetrahedron Lett.*, **1989**, *30*, 1491.
76. A. Padwa and K. E. Krumpe, *Tetrahedron*, **1992**, *48*, 5385.
77. M. S. Kharasch, T. Rudy, W. Nudenberg, and G. Buchi, *J. Org. Chem.*, **1953**, *18*, 1030.
78. A. Padwa, R. L. Chinn, S. F. Hornbuckle, and L. Zhi, *Tetrahedron Lett.*, **1989**, *30*, 301.
79. W. G. Dauben, J. Dinges, and T. C. Smith, *J. Org. Chem.*, **1993**, *58*, 7635.
80. M. C. McMills, L. H. Zhuang, D. L. Wright, and W. Watt, *Tetrahedron Lett.*, **1994**, *35*, 8311.
81. D. L. Wright and M. C. McMills, *Abstr. Pap. Am. Chem. Soc.*, **1994**, *208*, 273.

82. S. Muthusamy, S. A. Babu, C. Gunanathan, E. Suresh, P. Dastidar, and R. V. Jasra, *Casreact 130:182401*, **2000**, *56*, 6307.
83. S. Muthusamy, S. A. Babu, and C. Gunanathan, *Tetrahedron Lett.*, **2000**, *41*, 8839.
84. P. Chiu, B. Chen, and K. F. Cheng, *Org. Lett.*, **2001**, *3*, 1721.
85. B. Chen, P. Chiu, and K. F. Cheng, *Abstr. Pap. Am. Chem. Soc.*, **2000**, *219*, 784.
86. P. Chiu, B. Chen, and K. F. Cheng, *Abstr. Pap. Am. Chem. Soc.*, **2000**, *220*, 31.
87. V. Nair, K. C. Sheela, D. Sethumadhavan, S. Bindu, N. P. Rath, and G. K. Eigendorf, *Synlett*, **2001**, 272.
88. V. Nair, K. C. Sheela, and N. P. Rath, *Chem. Lett.*, **2000**, 980.
89. S. Muthusamy, C. Gunanathan, and S. A. Babu, *Tetrahedron Lett.*, **2001**, *42*, 523.
90. M. E. Maier and K. Evertz, *Tetrahedron Lett.*, **1988**, *29*, 1677.
91. D. L. Hertzog, D. J. Austin, W. R. Nadler, and A. Padwa, *Tetrahedron Lett.*, **1992**, *33*, 4731.
92. A. Padwa and D. L. Hertzog, *Tetrahedron*, **1993**, *49*, 2589.
93. A. Padwa, D. L. Hertzog, W. R. Nadler, M. H. Osterhout, and A. T. Price, *J. Org. Chem.*, **1994**, *59*, 1418.
94. A. Padwa, D. L. Hertzog, and W. R. Nadler, *J. Org. Chem.*, **1994**, *59*, 7072.
95. A. Padwa, D. J. Austin, A. T. Price, and M. D. Weingarten, *Tetrahedron*, **1996**, *52*, 3247.
96. A. Padwa, M. A. Brodney, J. P. Marino, and S. M. Sheehan, *J. Org. Chem.*, **1997**, *62*, 78.
97. A. Padwa, M. A. Brodney, J. P. Marino, M. H. Osterhout, and A. T. Price, *J. Org. Chem.*, **1997**, *62*, 67.
98. A. Padwa, in *Topics in Current Chemistry*, Vol. 189, Springer-Verlag, Berlin, **1997**, pp. 121–158.
99. A. Padwa and M. Prein, *J. Org. Chem.*, **1997**, *62*, 6842.
100. A. Padwa and M. Prein, *Tetrahedron*, **1998**, *54*, 6957.
101. A. Padwa, D. J. Austin, and A. T. Price, *Tetrahedron Lett.*, **1994**, *35*, 7159.
102. A. Padwa, D. L. Hertzog, and R. L. Chinn, *Tetrahedron Lett.*, **1989**, *30*, 4077.
103. C. O. Kappe, K. Peters, and E. M. Peters, *J. Org. Chem.*, **1997**, *62*, 3109.
104. C. O. Kappe, K. Peters, and E. M. Peters, *J. Org. Chem.*, **1997**, *62*, 3109.
105. R. Angell, M. FenglerVeith, H. Finch, L. M. Harwood, and T. T. Tucker, *Tetrahedron Lett.*, **1997**, *38*, 4517.
106. W. S. Kissel and A. Padwa, *Tetrahedron Lett.*, **1999**, *40*, 4003.
107. M. D. Weingarten, M. Prein, A. T. Price, J. P. Snyder, and A. Padwa, *J. Org. Chem.*, **1997**, *62*, 2001.
108. A. Padwa and A. T. Price, *J. Org. Chem.*, **1998**, *63*, 556.
109. A. Padwa, M. A. Brodney, and S. M. Lynch, *J. Org. Chem.*, **2001**, *66*, 1716.
110. M. Prein, P. J. Manley, and A. Padwa, *Tetrahedron*, **1997**, *53*, 7777.
111. C. O. Kappe, S. S. Murphree, and A. Padwa, *Tetrahedron*, **1997**, *53*, 14179.
112. J. P. Marino and A. Padwa, *Abstr. Pap. Am. Chem. Soc.*, **1996**, *211*, 46.
113. S. M. Sheehan and A. Padwa, *J. Org. Chem.*, **1997**, *62*, 438.
114. S. M. Sheehan and A. Padwa, *Abstr. Pap. Am. Chem. Soc.*, **1997**, *213*, 597.
115. C. S. Straub and A. Padwa, *Org. Lett.*, *1*, **1999**, 83.
116. A. Padwa, S. M. Sheehan, and C. S. Straub, *J. Org. Chem.*, **1999**, *64*, 8648.
117. A. Padwa, L. Precedo, and M. A. Semones, *J. Org. Chem.*, **1999**, *64*, 4079.
118. T. C. McMorris, in *46 pp.*; Regents of the University of California, USA; Trevor C.: McMorris, PCT Int. Appl.
119. A. Padwa, V. P. Sandanayaka, and E. A. Curtis, *J. Am. Chem. Soc.*, **1994**, *116*, 2667.
120. A. Padwa, E. A. Curtis, and V. P. Sandanayaka, *J. Org. Chem.*, **1996**, *61*, 73.
121. E. A. Curtis, V. P. Sandanayaka, and A. Padwa, *Tetrahedron Lett.*, **1995**, *36*, 1989.
122. A. Padwa, *Farmaco*, **1997**, *52*, 303.

123. F. R. Kinder and K. W. Bair, *J. Org. Chem.*, **1994**, *59*, 6965.

124. F. R. Kinder, Jr., R.-M. Wang, W. E. Bauta, and K. W. Bair, *Synth. Commun.*, **1997**, *27*, 521.

125. T. C. McMorris, J. Yu, Y. Hu, L. A. Estes, and M. J. Kelner, *J. Org. Chem.*, **1997**, *62*, 3015.

126. T. C. McMorris, Y. Hu, J. Yu, and M. J. Kelner, *J. Chem. Soc., Chem. Commun.*, **1997**, 315.

127. T. C. McMorris, J. A. Yu, D. M. Herman, M. J. Kelner, R. Dawe, and A. Minamida, *J. Label. Compd. Radiopharm.*, **1998**, *41*, 279.

128. H. Koyama, R. G. Ball, and G. D. Berger, *Tetrahedron Lett.*, **1994**, *35*, 9185.

129. O. Kataoka, S. Kitagaki, N. Watanabe, J.-I. Kobayashi, S.-I. Nakamura, M. Shiro, and S.-I. Hashimoto, *Casreact 132:321687*, **1998**, *39*, 2371.

130. D. M. Hodgson, J. M. Bailey, C. Villalonga-Barber, M. G. B. Drew, and T. Harrison, *J. Chem. Soc., Perkin Trans. 1*, **2000**, 3432.

131. C. Villalonga-Barber and D. M. Hodgson, *Abstr. Pap. Am. Chem. Soc.*, **2000**, *219*, 778.

132. D. M. Hodgson and C. Villalonga-Barber, *Tetrahedron Lett.*, **2000**, *41*, 5597.

133. C. Plug and W. Friedrichsen, *Tetrahedron Lett.*, **1992**, *33*, 7509.

134. J. E. Baldwin, A. V. W. Mayweg, K. Neumann, and G. J. Pritchard, *Org. Lett.*, **1999**, *1*, 1933.

135. T. Traulsen and W. Friedrichsen, *J. Chem. Soc.-Perkin Trans. 1*, **2000**, *9*, 1387.

136. O. Peters, T. Debaerdemaeker, and W. Friedrichsen, *J. Chem. Soc., Perkin Trans. 1*, **1999**, 59.

137. S. Reck and W. Friedrichsen, *J. Org. Chem.*, **1998**, *63*, 7680.

138. D. M. Hodgson, P. A. Stupple, and C. Johnstone, *Tetrahedron Lett.*, **1997**, *38*, 6471.

139. S. Kitagaki, M. Anada, O. Kataoka, K. Matsuno, C. Umeda, N. Watanabe, and S. Hashimoto, *J. Am. Chem. Soc.*, **1999**, *121*, 1417.

140. D. M. Hodgson, P. A. Stupple, and C. Johnstone, *J. Chem. Soc., Chem. Commun.*, **1999**, 2185.

141. H. Suga, H. Ishida, and T. Ibata, *Tetrahedron Lett.*, **1998**, *39*, 3165.

142. P. A. Wender, H. Kogen, H. Y. Lee, J. D. Munger, R. S. Wilhelm, and P. D. Williams, *J. Am. Chem. Soc.*, **1989**, *111*, 8957.

143. P. A. Wender and F. E. McDonald, *Abstr. Pap. Am. Chem. Soc.*, **1989**, *197*, 61.

144. P. A. Wender and F. E. McDonald, *J. Am. Chem. Soc.*, **1990**, *112*, 4956.

145. P. A. Wender and J. L. Mascarenas, *J. Org. Chem.*, **1991**, *56*, 6267.

146. P. A. Wender, K. D. Rice, and M. E. Schnute, *J. Am. Chem. Soc.*, **1997**, *119*, 7897.

147. P. A. Wender, C. D. Jesudason, H. Nakahira, N. Tamura, A. L. Tebbe, and Y. Ueno, *J. Am. Chem. Soc.*, **1997** *119*, 12976.

148. P. A. Wender and J. L. Mascarenas, *Tetrahedron Lett.*, **1992**, *33*, 2115.

149. P. A. Wender, Y. Martin-Cantalejo, A. J. Carpenter, A. Chiu, J. De Brabander, P. G. Harran, J. M. Jimenez, M. F. T. Koehler, B. Lippa, J. A. Morrison, S. G. Muller, S. N. Muller, C. M. Park, M. Shiozaki, C. Siedenbiedel, D. J. Skalitzky, M. Tanaka, and K. Irie, *Pure Appl. Chem.*, **1998**, *70*, 539.

150. A. Rumbo, L. Castedo, A. Mourino, and J. L. Mascarenas, *J. Org. Chem.*, **1993**, *58*, 5585.

151. A. Rumbo, J. L. Mascarenas, A. Mourino, and L. Castedo, *Abstr. Pap. Am. Chem. Soc.*, **1992**, *203*, 403.

152. A. Rumbo, A. Mourino, L. Castedo, and J. L. Mascarenas, *J. Org. Chem.*, **1996**, *61*, 6114.

153. J. L. Mascarenas, I. Perez, A.Rumbo, and L. Castedo, *Synlett*, **1997**, 81.

154. A. Rumbo, L. Castedo, and J. L. Mascarenas, *Tetrahedron Lett.*, **1997**, *38*, 5885.

155. F. Lopez, L. Castedo, and J. L. Mascarenas, *Org. Lett.*, **2001**, *3*, 623.

156. F. Lopez, L. Castedo, and J. L. Mascarenas, *Org. Lett.*, **2000**, *2*, 1005.

157. J. R. Rodriguez, A. Rumbo, L. Castedo, and J. L. Mascarenas, *J. Org. Chem.*, **1999**, *64*, 4560.

158. W. Bauta, J. Booth, M. E. Bos, M. Deluca, L. Diorazio, T. Donohoe, N. Magnus, P. Magnus, J. Mendoza, P. Pye, J. Tarrant, S. Thom, and F. Ujjainwalla, *Tetrahedron Lett.*, **1995**, *36*, 5327.

159. W. E. Bauta, J.Booth, M. E. Bos, M. DeLuca, L. Diorazio, T. J. Donohoe, C. Frost, N. Magnus, P. Magnus, J. Mendoza, P. Pye, J. G. Tarrant, S. Thom, and F. Ujjainwalla, *Tetrahedron*, **1996**, *52*, 14081.

160. P. Magnus, J. Booth, L. Diorazio, T. Donohoe, V. Lynch, N. Magnus, J. Mendoza, P. Pye, and J. Tarrant, *Tetrahedron*, **1996**, *52*, 14103.

161. P. Magnus, L. Diorazio, T. J. Donohoe, M. Giles, P. Pye, J. Tarrant, and S. Thom, *Tetrahedron*, **1996**, *52*, 14147.

162. P. Magnus, F. Ujjainwalla, N. Westwood, and V. Lynch, *Tetrahedron*, **1998**, *54*, 3069.

163. P. Magnus and L. Shen, *Tetrahedron*, **1999**, *55*, 3553.

164. D. R. Williams, J. W. Benbow, J. G. McNutt, and E. E. Allen, *J. Org. Chem.*, **1995**, *60*, 833.

165. H. Y. Lee, J. H. Sohn, and H. Y. Kim, *Tetrahedron Lett.*, **2001**, *42*, 1695.

166. R. J. Linderman and A. Godfrey, *J. Am. Chem. Soc.*, **1988**, *110*, 6249.

167. W. R. Kwochka and R. J. Linderman, *Abstr. Pap. Am. Chem. Soc.*, **1990**, *200*, 13.

168. R. J. Linderman, D. M. Graves, W. R. Kwochka, A. F. Ghannam, and T. V. Anklekar, *J. Am. Chem. Soc.*, **1990**, *112*, 7438.

169. W. R. Kwochka and R. J. Linderman, *Abstr. Pap. Am. Chem. Soc.*, **1991**, *201*, 261.

170. J. P. Dittami, X. Y. Nie, H. Nie, H. Ramanathan, S. Breining, J. Bordner, D. L. Decosta, J. Kiplinger, P. Reiche, and R. Ware, *J. Org. Chem.*, **1991**, *56*, 5572.

171. J. P. Dittami, Y. Luo, D. Moss, and W. G. McGimpsey, *J. Org. Chem.*, **1996**, *61*, 6256.

172. C. Bernaus, J. Font, and P. Demarch, *Casreact 130:182401*, **1991**, *47*, 7713.

173. M. Hojo, H. Aihara, and A. Hosomi, *Kidorui*, **1997**, *30*, 66.

174. A. Hosomi, *J. Synth. Org. Chem. Jpn.*, **1998**, *56*, 681.

175. K. Takai, *Organomet. News*, **1995**, 134.

176. K. Takai, H. Kaihara, K.-i. Higashiura, and N. Ikeda, *J. Org. Chem.*, **1997**, *62*, 8612.

177. N. Sakai, T. Fukushima, S. Minakata, I. Ryu, and M. Komatsu, *J. Chem. Soc., Chem. Commun.*, **1999**, 1857.

178. T. Hudlicky and G. Barbieri, *J. Org. Chem.*, **1991**, *56*, 4598.

CHAPTER 5

Thiocarbonyl Ylides

Grzegorz Mloston

Department of Organic and Applied Chemistry, University of Lodz, Poland

Heinz Heimgartner

Institute of Organic Chemistry, University of Zurich, Switzerland

5.1. Methods of Generation 319
5.1.1. Cycloreversion Reactions 319
5.1.2. 1,3-Elimination Reactions 320
5.1.3. Deprotonation of Thioxonium Salts 321
5.1.4. Addition of Carbenes and Carbenoids to C=S Groups 322
5.1.5. Photochemical Rearrangements of Thioethers 324
5.2. Chemical Behavior 324
5.2.1. Intramolecular Reactions 324
5.2.1.1. Electrocyclizations 324
5.2.1.2. 1,4-Hydrogen Shift 325
5.2.1.3. Skeleton Rearrangements 326
5.2.2. Intermolecular Reactions 326
5.2.2.1. [3 + 2] Cycloadditions 326
5.2.2.2. 1,3-Addition Reactions 327
5.2.2.3. Dimerization Reactions 328
5.3. Application in the Synthesis of Heterocycles 329
5.3.1. Three-Membered Rings: Thiiranes 329
5.3.2. Five-Membered Rings 331
5.3.2.1. Thiophene Derivatives 331
5.3.2.2. 1,3-Dithiolanes, 1,3-Oxathiolanes, and 1,3-Thiazole Derivatives 339
5.3.2.3. Five-Membered Rings with Three Heteroatoms 344
5.3.2.4. Five-Membered Heterocycles via 1,5-Dipolar Electrocyclization 345
5.3.3. Six-Membered Rings 351
5.3.4. Seven-Membered Rings 351
5.4. Concluding Remarks 354

The Chemistry of Heterocyclic Compounds, Volume 59: Synthetic Applications of 1,3-Dipolar Cycloaddition Chemistry Toward Heterocycles and Natural Products, Edited by Albert Padwa and William H. Pearson.
ISBN 0-471-38726-6

Thiocarbonyl ylides (**1**) belong to the family of sulfur-centered 1,3-dipoles characterized by the presence of two sp^2 C atoms attached to the sulfur atom. Formal replacement of one of the C atoms by heteroatoms such as NR_3, O, and S lead to the other representatives of the family, namely, thiocarbonyl S-imides (**2**), S-oxides (sulfines) (**3**), and S-sulfides (thiosulfines) (**4**), respectively.

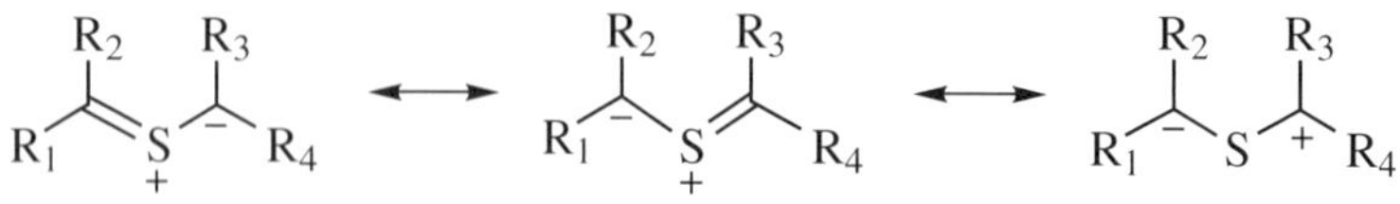

(1A) **(1B)** **(1C)**

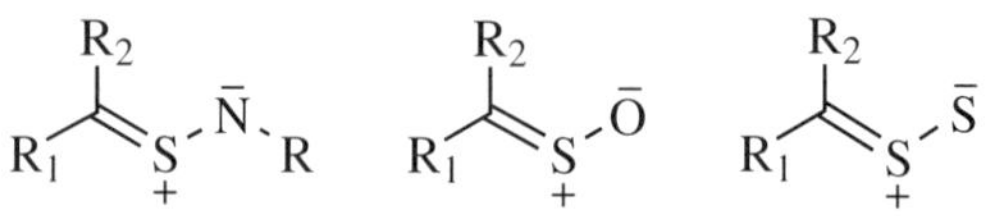

(2) **(3)** **(4)**

The dipolar structure **1** describes the chemical behavior of thiocarbonyl ylides best, although other mesomeric forms have been used for the representation of the electronic structure of these dipoles. The parent compound, thioformaldehyde (*S*)-methylide (**1**), was studied by means of spectroscopic and theoretical methods (2–5), which showed that the molecule possesses a bent allyl-type structure (6). According to theoretical calculations, structures **1A** and **1B** have the largest contribution (31.5% each) in the representation of the electronic structure, whereas **1C**, which reflects the 1,3-dipolar character, has only a 4.2% contribution (5).

In comparison with other 1,3-dipoles that have been extensively explored in organic synthesis (7), sulfur-centered 1,3-dipoles (**1**–**4**) are rather uncommon species. However, within the last two decades, remarkable progress has been made regarding both methods of generation and synthetic applications. In particular, thiocarbonyl ylides (**1**) were established as key intermediates useful for the preparation of sulfur-containing heterocyclic compounds. General methods for the preparation of thiocarbonyl ylides and their chemical reactivity have been reviewed (8–11).

Historically, the first reactions involving thiocarbonyl ylides involve the preparation of thiiranes and 1,3-dithiolanes from diazomethane and thiocarbonyl compounds reported early in the last century by Staudinger and co-workers (12,13). Similar reactions have been studied by Schönberg and co-workers (14–16) during the 1960s, but neither was the reaction mechanism understood nor have thiocarbonyl ylides been recognized as key intermediates. [For some remarks to this subject see (8) and (10) in (17).]

In the middle of the 1950s, Knott reported the synthesis of dyestuffs based on benzothiazole derivatives. Alkylation of *N*-methylbenzo-1,3-thiazole-2-thione with α-bromoacetophenone and deprotonation of the resulting thiocarbonylium salt **5** yielded, after spontaneous desulfurization of the intermediate thiirane (**7**), the alkylidene derivative **8** (18) (Scheme 5.1). In order to rationalize the reaction, thiocarbonyl ylide **6** was proposed as the precursor of thiirane **7**. To the best of our

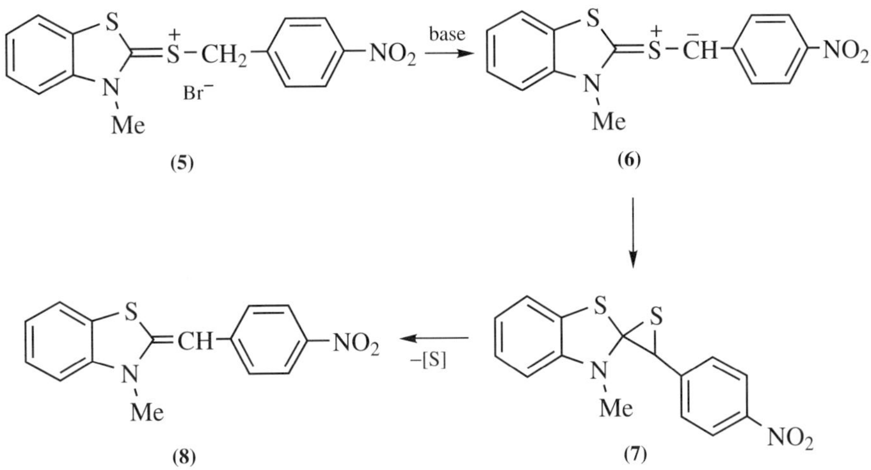

Scheme 5.1

knowledge, this represents the first formulation of a thiocarbonyl-ylide-type intermediate.

Some years later, the first stable thiocarbonyl ylides **9** and **10** were prepared by the reaction of thiourea with cyano-substituted oxiranes (19,20) or by addition of Rh-di(tosyl)carbenoid to benzo-1,2-dithiole-3-thione (21), respectively. Enhanced stability and the low reactivity of **9** and **10**, which enables their isolation in crystalline form, results from the *push–pull* substitution at the two termini [cf. also (22)]. Another class of stable thiocarbonyl ylides that are also able to afford [3 + 2]-cycloaddition products are the mesoionic 1,3-dithiole-4-ones of type **11** (23,24).

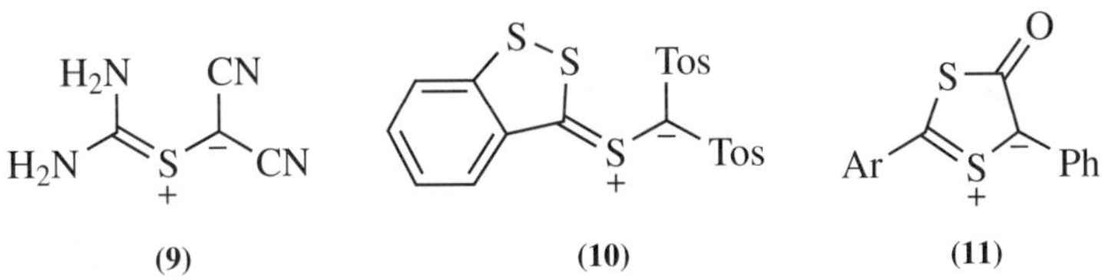

Tos = tosyl

Systematic studies of the nitrogen elimination from 2,5-dihydro-1,3,4-thiadiazoles by Kellogg and co-workers (8,25) established thiocarbonyl ylides as reactive intermediates, which either undergo an electrocyclization reaction to give thiiranes or can be trapped by suitable dipolarophiles to produce sulfur-containing five-membered heterocycles. By using two stereoisomeric precursors (*cis*-**12** and *trans*-**12**), Kellogg showed that the resulting thiiranes (**14**) were trans- and cis-configured, respectively (Scheme 5.2). From this result he concluded that thermal elimination of N_2 leads stereoselectively to thiocarbonyl ylides with retained configuration. A subsequent thermal conrotatory ring closure produces thiiranes, in accordance with the Woodward–Hoffmann rules (26).

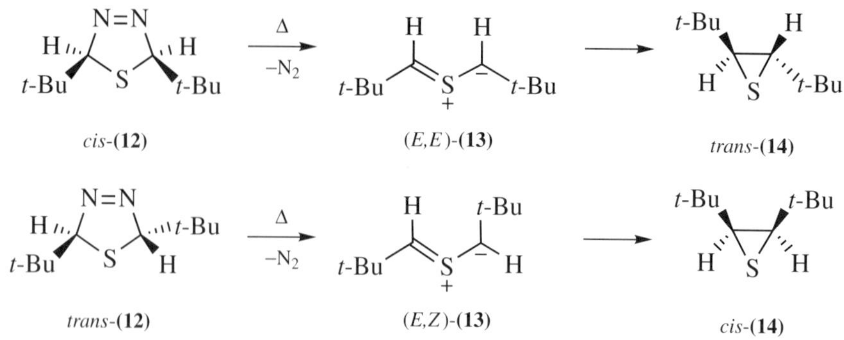

Scheme 5.2

Over the past two decades, important contributions to the chemistry of thiocarbonyl ylides were made by Huisgen et al. (27). By carrying out the reaction of thiobenzophenone with diazomethane at low temperature, formation of 2,5-dihydro-1,3,4-thiadiazole (**15**) with subsequent elimination of N_2 was established as the route to the reactive thiobenzophenone (*S*)-methylide (**16**) (17,28). In the absence of intercepting reagents, **16** undergoes electrocyclization to give **17** or head-to-head dimerization to yield 1,4-dithiane **18** (Scheme 5.3).

The desilylation methodology for the generation of 1,3-dipoles, developed by Vedejs and West (29) with regard to azomethine ylides, was successfully applied by Achiwa and co-workers (30) to the field of thiocarbonyl ylides. This approach allowed the generation of the parent thioformaldehyde (*S*)-methylide (**1a**) and its use for preparative purposes (31,32). Generation of **1a** in the presence of C=C dipolarophiles led to tetrahydrothiophenes (**19**) in high yield (Scheme 5.4).

The goal of this chapter is to summarize the methods used for the generation of thiocarbonyl ylides and their subsequent use in organic synthesis.

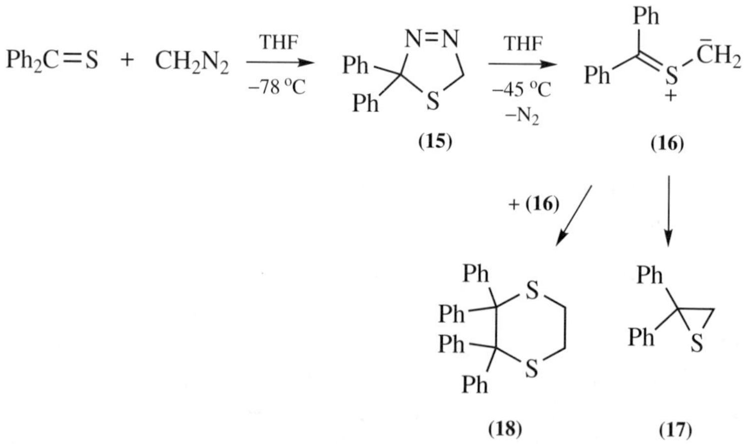

Scheme 5.3

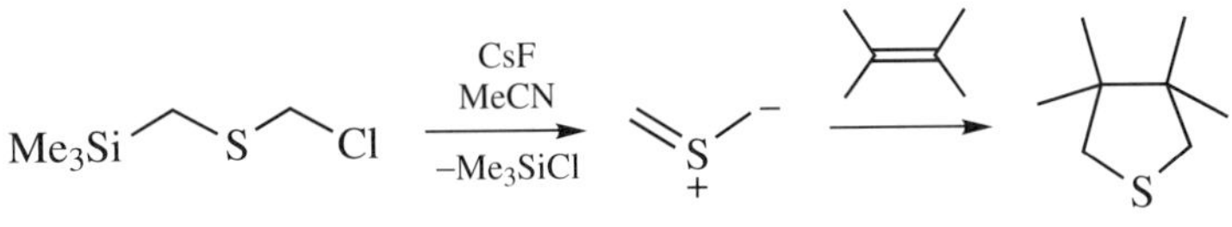

(1a) (19)

Scheme 5.4

5.1. METHODS OF GENERATION

5.1.1. Cycloreversion Reactions

Two types of sulfur-containing five-membered heterocycles are convenient precursors of thiocarbonyl ylides, namely, 2,5-dihydro-1,3,4-thiadiazoles (**20**) and 1,3-oxathiolan-5-ones (**21**) (Scheme 5.5). The precursors **20** are accessible by two different methods.

The first method, leading exclusively to symmetrically substituted products, involves a multistep procedure starting with a carbonyl compound and hydrazine. The resulting symmetrical azine can then be transformed into **20** either *via* addition of H_2S followed by a dehydrogenation with diethyl azodicarboxylate (25,33,34) or by chlorination and subsequent reaction with H_2S in the presence of a base (25). The second approach is based on the reaction of thiocarbonyl derivatives with diazo compounds [cf. (11)], which in the case of diazoalkanes, already occurs at low temperature. This method allows for the synthesis of unsymmetrically substituted precursors **20**. Kinetic studies by Huisgen and co-workers showed that the N_2-elimination step is a first-order reaction, and that the reaction rate is strongly dependent on the type and number of substituents. In general, increasing the number of aliphatic and bulky substituents enhances the thermal stability of **20**. On the other hand, aryl groups tend to accelerate the elimination of nitrogen (17,35–41).

In some instances, sterically encumbered 2,5-dihydro-1,3,4-thiadiazoles do not eliminate nitrogen. Instead, cycloreversion leading to the starting materials or a new pair of diazo- and thiocarbonyl compounds was reported. Thus, a crystalline product of type **20**, obtained from di(*tert*-butyl)diazomethane and 2-benzyl-4,4-dimethyl-1,3-thiazole-5(4*H*)-thione, was found to dissociate in solution to give the starting materials (42). In the case of (*tert*-butyl)(trimethylsilyl)thioketone and

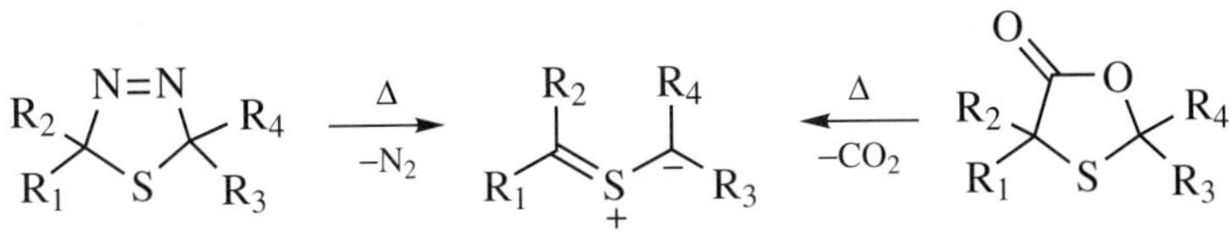

(20) (1) (21)

Scheme 5.5

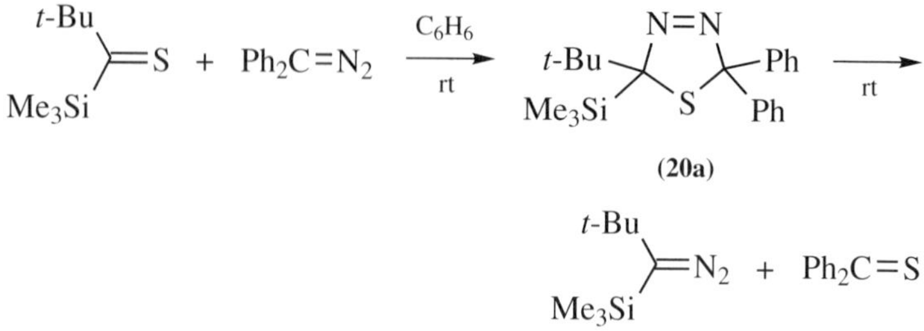

Scheme 5.6

diphenyldiazomethane, the expected dihydrothiadiazole **20a** was formed but rapidly decomposed at room temperature to give thiobenzophenone and (*tert*-butyl)(trimethylsilyl)diazomethane (43) (Scheme 5.6).

Elimination of CO_2 from **21** requires higher temperatures. Reported protocols include the thermolysis without solvent at 130 °C (44), heating in solution at 150–230 °C (45), as well as in the gas phase at 600 °C (FVP) (46). For synthetic purposes, this method has been used for the preparation of sterically hindered olefins and represents an extension of the *twofold extrusion methodology* [cf. (47,48)].

5.1.2. 1,3-Elimination Reactions

Silylated thioethers undergo thermal- or fluoride-catalyzed elimination to yield thiocarbonyl ylides as reactive intermediates. The generation of the parent species **1a** using CsF as the catalyst is presented in Scheme 5.4 (31,32). Under similar conditions, imino-substituted thiocarbonyl ylides **23** ($X = R_1N$) were generated starting from N-substituted dithioiminocarbonates **22** ($X = R_1N$, Scheme 5.7). They undergo [3 + 2] cycloaddition with aromatic aldehydes to give 1,3-oxathiolane-2-imines (**24**) ($X = R_1N$) (49) or enter secondary conversions leading to thiiranes as the final products (50). Starting with a silylated dithioketenacetal (**22**) (X = indane-1,3-dion-2-ylidene), thioketene (*S*)-methylide was formed and trapped *in situ* using C=O and C=C dipolarophiles (51).

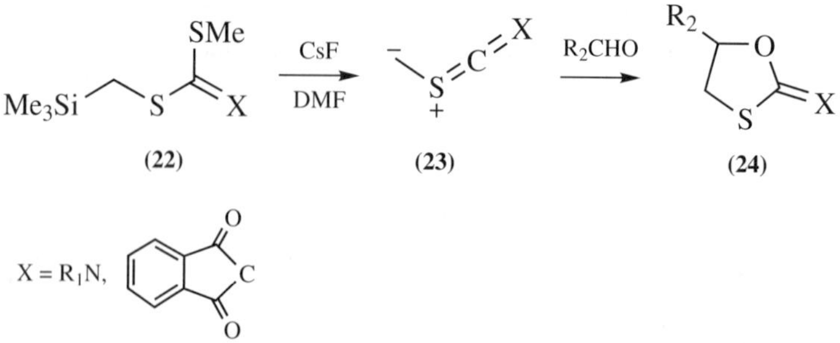

Scheme 5.7

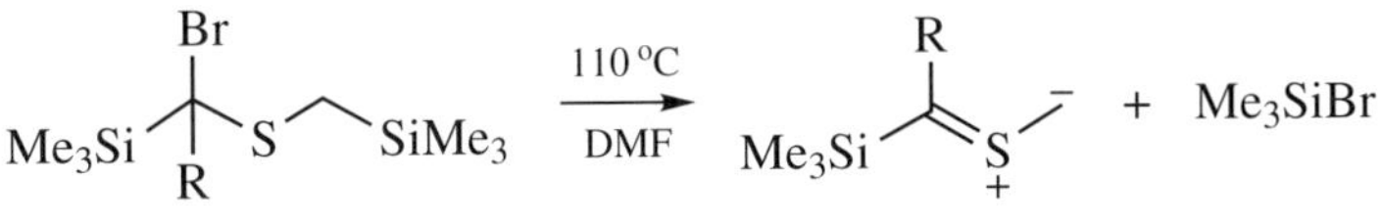

(25) (26)

R = H, Ph, Me_3Si

Scheme 5.8

α-Halogenated bis[(trimethylsilyl)methyl]thioethers (**25**) eliminate halotrimethylsilane at 110 °C in dimethylformamide (DMF) to give silylated thiocarbonyl ylides of type **26** (Scheme 5.8). In this case, no fluoride catalyst was necessary (52).

The thermolysis of bis[(trimethylsilyl)methyl]sulfoxides (**27**) at 100 °C in hexamethylphosphoramide (HMPA) results in the elimination of hexamethyldisiloxane, and thiocarbonyl ylides appear as reactive intermediates (53). This protocol allows the generation of the parent species **1a**, as well as monosubstituted and alkylidene-substituted representatives such as **1b** and **1c**, which are difficult to obtain by other methods (Scheme 5.9).

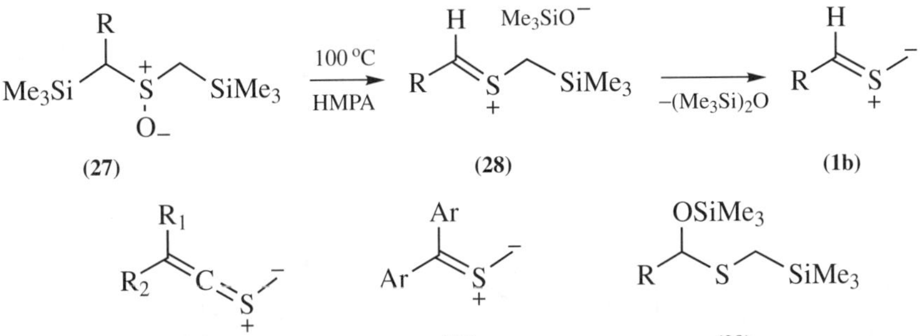

(27) (28) (1b) (1c) (1d) (29)

Scheme 5.9

The formation of **29**, the product of a sila-Pummerer rearrangement, as a minor reaction product points to the intermediacy of the ion pair **28**.

Starting with diarylthioketones and (trimethylsilyl)methyl triflate, thiocarbonyl ylides of type **1d** are easily accessible (54). The key intermediate in this reaction is believed to be an ion pair of type **28**.

Another approach to thiocarbonyl ylides involves the 1,3-elimination of HCl from α-chlorothioethers of type **30**, which are prepared by the reaction of α-chlorosulfenyl chlorides with carbanions bearing electron-withdrawing groups. Subsequent treatment with *tert*-butanolate leads to **31** (55) (Scheme 5.10).

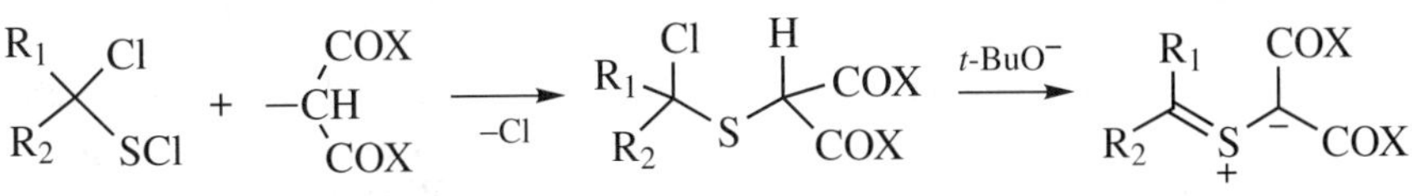

X = Ph, OR (30) (31)

Scheme 5.10

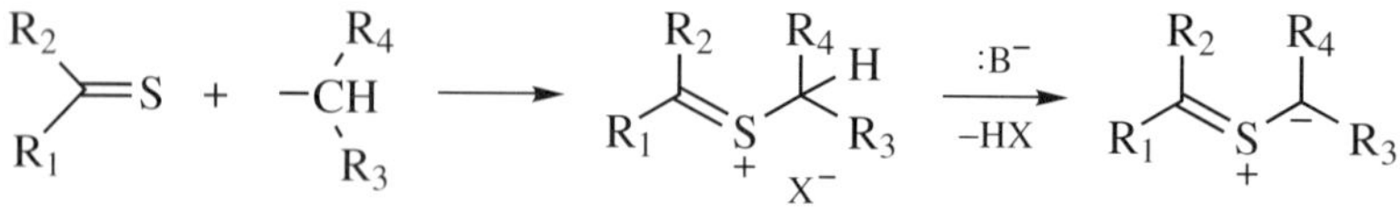

(32) (1)

Scheme 5.11

5.1.3. Deprotonation of Thioxonium Salts

As mentioned earlier, the first example of the generation of a thiocarbonyl ylide by deprotonation of a thioxonium salt was reported by Knott (18) and is presented in Scheme 5.1. This method is frequently used since the starting materials **32** are easily available *via* alkylation of C=S functionalized compounds such as thioketones, thioamides, thiourea derivatives, and dithioesters (Scheme 5.11).

Typically, thioxonium salts (**32**) are stable compounds, and the deprotonation is performed at low temperatures. This method has been used to synthesize reactive thiocarbonyl ylides as well as stable and isolable ones (56–60). Arduengo and Burgess (3) prepared differently substituted thiocarbonyl ylides from thiourea derivatives, and in the case of **33**, the structure has been established by X-ray crystallography.

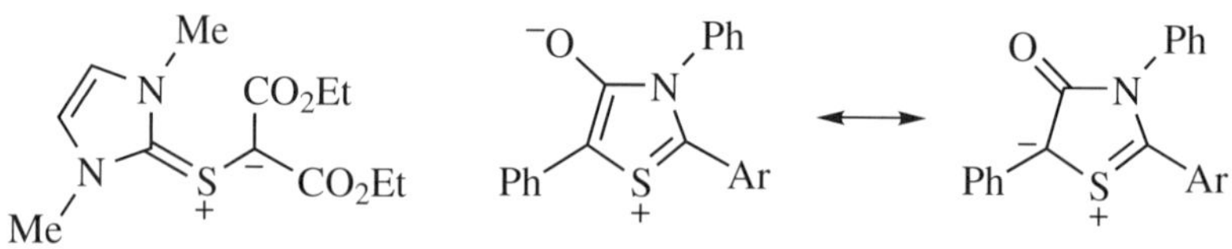

(33) (34)

Mesoionic 1,3-thiazole-4-ones of type **34** are known as thioisomünchnones. As one of the mesomeric structures demonstrates, these species contain the structural fragment characteristic of thiocarbonyl ylides (61). A convenient access to thioisomünchnones involves the reaction of *N*-arylthiobenzamides with α-bromophenylacetyl chloride (62).

5.1.4. Addition of Carbenes and Carbenoids to C=S Groups

Aliphatic diazo compounds are convenient starting materials for the generation of carbenes by photolytic or thermal methods. Decomposition of diazo compounds can be catalyzed by some metals or metal salts (63,64). Whereas copper salts were first used, rhodium(II) acetate has recently become recognized as one of the most efficient catalysts and is now used almost exlusively (65).

Addition of electrophilic carbenes and carbenoids to heteromultiple bonds has been explored as a method for the generation of 1,3-dipoles, such as nitrile ylides, azomethine ylides, and carbonyl ylides (65). Extension of this methodology to thiocarbonyl compounds opens another convenient access to reactive thiocarbonyl ylides. As previously discussed, alkyl- and aryl-substituted diazomethanes undergo a smooth [3 + 2] cycloaddition with C=S groups to yield 2,5-dihydro-1,3,4-thiadiazoles that are good precursors for thiocarbonyl ylides (cf. Schemes 5.3 and 5.5). Due to the instability of many thiocarbonyl compounds, thermally mediated carbene generation is of no practical importance. Photolytical generation of carbenes from diazo compounds in the presence of a thiocarbonyl compound is not a particularly common procedure. In an earlier mechanistic study, a mixture of diazomethane and the sterically crowded 1,1,3,3-tetramethylindane-2-thione was irradiated at 10 K in an argon matrix. Formation of an intermediate thiocarbonyl ylide was established by ultraviolet (UV) spectroscopy (66). A similar investigation was performed using diphenyldiazomethane and sterically congested thioketones (67).

The reaction of unreactive diazo compounds with thiocarbonyl groups need to be catalyzed, and involves the addition of a carbene or carbenoid to the S atom forming the thiocarbonyl ylide directly. A representative example involves the reaction of thiobenzophenone with diazomalonate, which occurs smoothly in the presence of catalytic amounts of $Rh_2(OAc)_4$. Formation of thiirane (**35**) and diphenylmethylidene malonate (**36**) is the result of ring closure of the intermediate thiocarbonyl ylide (**1e**) followed by partial desulfurization (68) (Scheme 5.12). Interestingly, on irradiation of diazo malonate and thiobenzophenone, the only product obtained was **36** (69). Other efficient protocols for the formation of thiocarbonyl ylides via carbene addition consist of the thermal decomposition of Seyferth reagents (70) and the generation of dichlorocarbene using *two-phase methodology* (71). There are also examples reported for the formal transfer of carbenes to C=S groups using substituted phenyliodonium methylides (72) and cyano-substituted oxiranes (19,20) as precursors of bivalent carbon species.

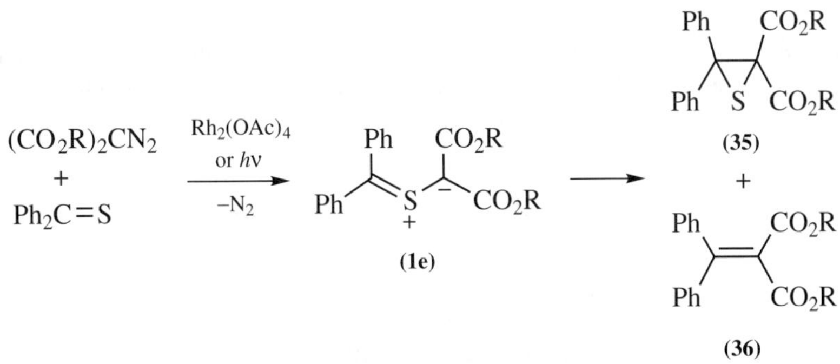

Scheme 5.12

5.1.5. Photochemical Rearrangements of Thioethers

As early as 1969, Block and Corey (73) observed the photochemical transformation of divinyl sulfide **37** to a mixture of isomeric cyclization products (**39** and **40**) (Scheme 5.13). A pathway involving the cyclic thiocarbonyl ylide **38** was proposed. A theoretical study concerning the nature of intermediates in the photochemical cyclization of divinyl sulfides was recently carried out (74). Several examples are known in which the *vinyl group* is part of an aromatic system.

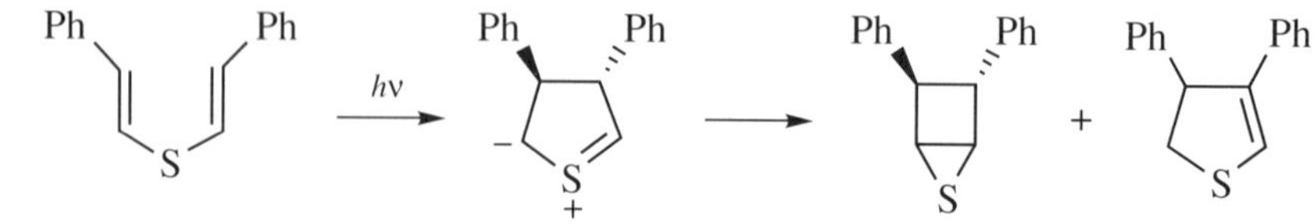

(37) **(38)** **(39)** **(40)**

Scheme 5.13

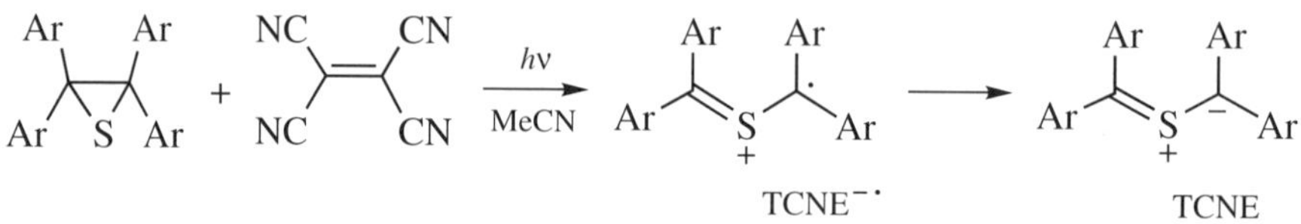

Scheme 5.14

Whereas the thermal ring-opening reaction of oxiranes and aziridines is frequently used for generation of carbonyl ylides and azomethine ylides, the analogous procedure starting with thiiranes does not produce the expected thiocarbonyl ylides (8). However, in the case of tetraaryl-substituted thiiranes, the photolytically mediated reaction with tetracyanoethylene (TCNE) is believed to occur via a single electron transfer (SET) mechanism, also involving a thiocarbonyl ylide as a likely intermediate (75,76) (Scheme 5.14).

5.2. CHEMICAL BEHAVIOR

Like other 1,3-dipolar species, thiocarbonyl ylides are able to enter intramolecular as well as intermolecular cycloaddition reactions. In this chapter, selected examples of both types will be illustrated.

5.2.1. Intramolecular Reactions

5.2.1.1. Electrocyclizations

Thiocarbonyl ylides without *push–pull* stabilization, in the absence of intercepting reagents, undergo 1,3-dipolar electrocyclization to give thiiranes. In accordance

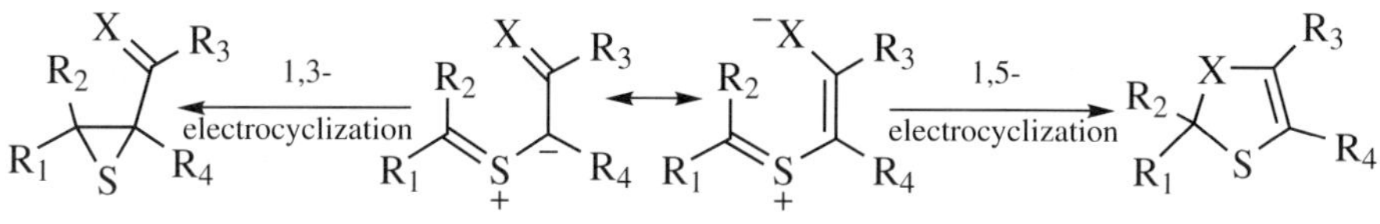

Scheme 5.15

with the Woodward–Hoffmann rules, this ring closure occurs in a conrotatory manner (8,25,46). In the case of thiocarbonyl ylides bearing an additional π-functionality, a competitive 1,5-dipolar electrocyclization is possible. In many cases, the 1,5-ring closure is the dominant reaction. However, a strong dependence on the type of substituent group has been observed (77–83) (Scheme 5.15).

5.2.1.2. 1,4-Hydrogen Shift

In some cases involving aliphatic thiocarbonyl ylides, a 1,4-H shift occurs to afford vinyl thioethers. As an example, thiocarbonyl ylides (**41**), generated by the addition of isopropylidene carbene to aliphatic thioketones, are converted to divinylthioethers (**42**) (84) (Scheme 5.16).

Attempts to generate thiocamphor (*S*)-methylide (**44**) by the addition of diazomethane to thiocamphor and subsequent N_2-elimination from the [3 + 2]-cycloadduct **43** led to enethiol ether **45** via a 1,4-H shift (Scheme 5.17). The formation of an unstable intermediate **43** was proposed on the basis of the proton nuclear magnetic resonance (^{1}H NMR) spectrum of the crude mixture. The postulated intermediate **44** could not be intercepted by dipolarophiles or methanol, and did not undergo electrocyclization to give the corresponding thiirane (41).

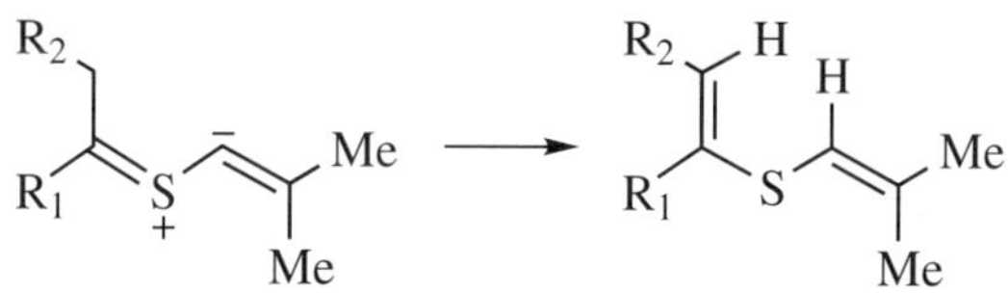

(41) **(42)**

Scheme 5.16

On the other hand, the analogous reaction of 2,2,4,4-tetramethyl-3-thioxocyclobutanone and 2-diazopropane afforded a stable cycloadduct, which upon heating eliminated N_2 and yielded a mixture of thiirane (**46**) and vinyl thioether **47** (85). A similar competition involving both an electrocyclization and a 1,4-H shift was observed in the case of diisopropylthioketone (*S*)-methylide (39).

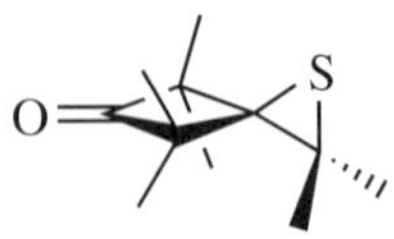

(46)

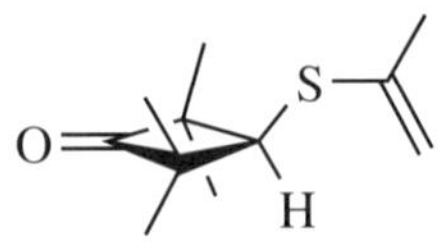

(47)

5.2.1.3. Skeleton Rearrangements

Thiocarbonyl ylides are both nucleophilic and basic compounds (40,41,86). For example, adamantanethione (*S*)-methylide (**52**) is able to deprotonate its precursor, the corresponding 2,5-dihydro-1,3,4-thiadiazole, and a 1:1 adduct is formed in a multistep reaction (40,86). Thioxonium ion (**56**) (Scheme 5.22) was proposed as a reactive intermediate. On the other hand, thiofenchone (*S*)-methylide (**48**) is not able to deprotonate its precursor but instead undergoes electrocyclization to give a mixture of diastereoisomeric thiiranes (41,87,88). The addition of a trace of acetic acid changes the reaction course remarkably, and instead of an electrocyclization product, the new isomer **51** was isolated (41,87) (Scheme 5.18). The formation of **51** is the result of a Wagner–Meerwein rearrangement of thioxonium ion **49**.

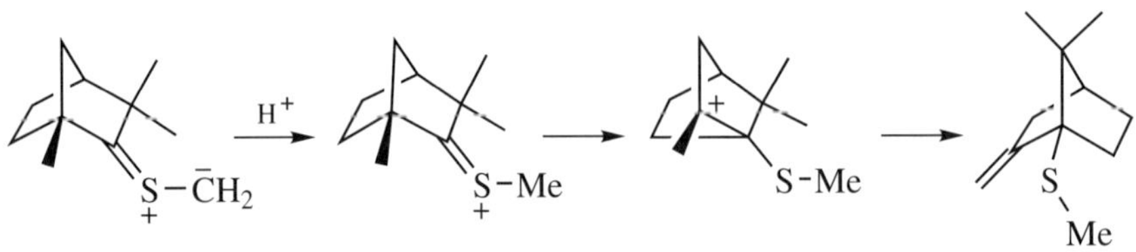

(48) **(49)** **(50)** **(51)**

Scheme 5.18

5.2.2. Intermolecular Reactions

5.2.2.1. [3 + 2] Cycloadditions

As is the case with other 1,3-dipoles, thiocarbonyl ylides undergo [3 + 2]-cycloaddition reactions producing five-membered sulfur heterocycles [cf. (8)]. These ylides belong to the class of electron-rich 1,3-dipoles (89) and, according

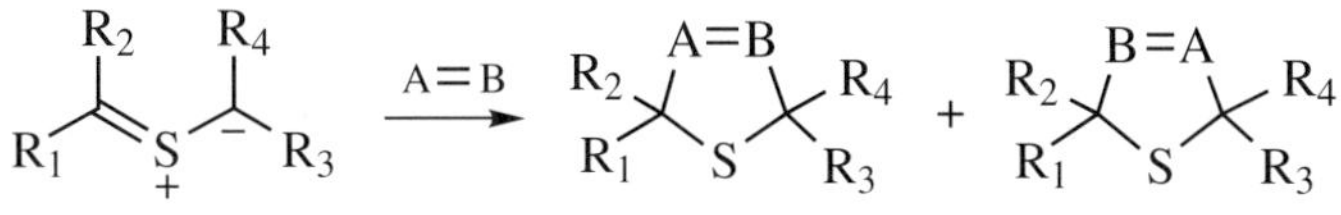

(1)

Scheme 5.19

to the classification of Sustmann (90), react preferably with electron-poor dipolarophiles. Kinetic studies using thiobenzophenone (*S*)-methylide (**16**) showed that aromatic thioketones are very efficient intercepting agents and, for this reason they were named by Huisgen as *superdipolarophiles* (91–93). Along with electron-poor C=C dipolarophiles, a variety of hetero-dipolarophiles were successfully exploited in [3 + 2] cycloadditions with differently substituted thiocarbonyl ylides (Scheme 5.19).

The formation of five-membered tetrahydrothiophene rings using nonsymmetrical reaction partners usually occurs with high regioselectively. In most cases, a stereospecific reaction was observed (9,27,94). Recently, a diastereoselective cycloaddition of the parent species **1a** to various chiral α,β-unsaturated amides was also reported (32,95).

Cycloaddition of sterically crowded 2,2,4,4-tetramethyl-3-thioxocyclobutanone (*S*)-methylide (**69**) (Scheme 5.25) with extremely electron-poor dipolarophiles are of special interest with respect to the theory of dipolar cycloaddition chemistry. In these reactions, a nonconcerted, two-step mechanism was proposed (96,97).

5.2.2.2. 1,3-Addition Reactions

Acidic compounds of type R—XH, which are able to protonate thiocarbonyl ylides, also undergo 1,3-addition leading to products of *S,S*-, *S,O*-, or *S,N*-acetal type (Scheme 5.20). Thiophenols and thiols add smoothly to thiocarbonyl ylides generated from 2,5-dihydro-1,3,4-thiadiazoles (36,38,86,98,99). Thiocamphor, which exists in solution in equilibrium with its enethiol form, undergoes a similar reaction with adamantanethione (*S*)-methylide (**52**) to give dithioacetal **53** (40) (Scheme 5.21). Formation of analogous products was observed with some thiocarbonyl functionalized NH-heterocycles (100).

Phenols and alcohols also react with substituted thiocarbonyl ylides, although for the reaction with alcohols, acid catalysis is usually recommended (36,38,41,99). Some NH-azoles are sufficiently acidic to give 1,3-adducts without the addition of a

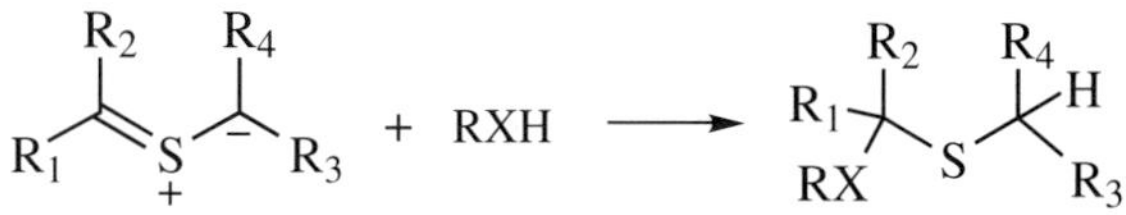

(1) X = S, O, NR

Scheme 5.20

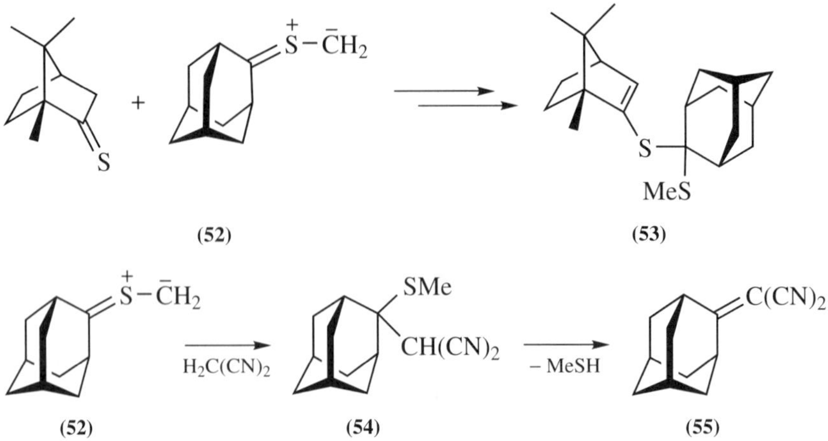

Scheme 5.21

catalyst (101,102). Malonodinitrile reacts in a similar manner with **52**, but in this case the initial adduct **54** eliminates methanethiol to give dicyanomethylidene adamantane (**55**) (40,86) (Scheme 5.21).

In some cases, aliphatic thiocarbonyl (*S*)-methylides can be protonated by 2,5-dihydro-1,3,4-thiadiazoles, which are their precursors, to give a thioxonium ion (e.g., **56**). Carbanion **57** undergoes a ring opening to thiolate **58**, which subsequently combines with **56** to give (*S,S*)-acetal **59** (40,103) (Scheme 5.22).

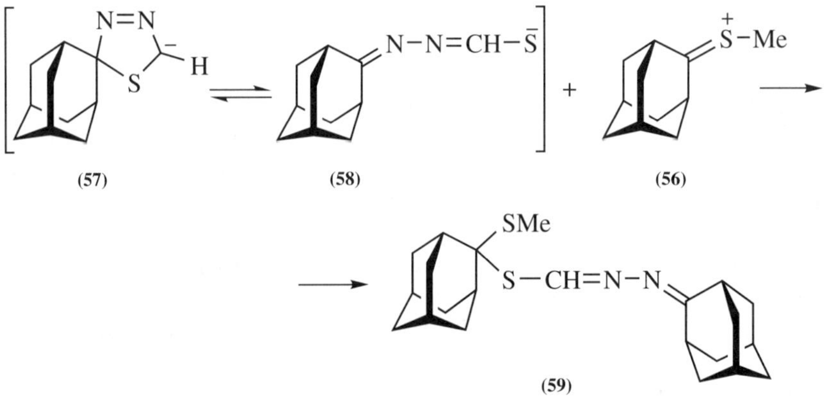

Scheme 5.22

5.2.2.3. *Dimerization Reactions*

As mentioned previously, 2,5-dihydro-1,3,4-thiadiazoles obtained from aromatic thioketones and diazomethane readily eliminate N_2 at −45 °C. (*S*)-Methylides generated under these conditions undergo a head-to-head dimerization to produce 2,2,3,3-tetraaryl-1,4-dithianes (17,28) (cf. Scheme 5.3). Formation of analogous products was also observed with some thiocarbonyl (*S*)-methylides obtained from dithioesters (60,104) or 1,3-thiazole-5(4*H*)-thiones (105).

5.3. APPLICATION IN THE SYNTHESIS OF HETEROCYCLES

5.3.1. Three-Membered Rings: Thiiranes

As mentioned on pages 317 and 324, the 1,3-dipolar electrocyclization of thiocarbonyl ylides leads to thiirane derivatives, which represents an excellent method for the preparation of those three-membered rings. Typically, thiiranes are isolated as the final products, but in some instances they are produced as intermediate compounds which spontaneously desulfurize to give alkenes [*twofold extrusion* (47,48)].

The preparation of thiiranes is most conveniently performed in solution. However, there are also protocols reported for reaction in the gas and solid phase. By using diazo and thiocarbonyl compounds in ether as solvent, both alkyl and aryl substituted thiiranes are accessible. As indicated earlier, aryl substituents destabilize the initially formed 2,5-dihydro-1,3,4-thiadiazole ring and, in general, thiiranes are readily obtained at low temperature (13,15,35). On the other hand, alkyl substituents, especially bulky ones, enhance the stability of the initial cycloadduct, and the formation of thiiranes requires elevated temperatures (36–41,88). Some examples of sterically crowded thiiranes prepared from thioketones and a macrocyclic diazo compound have been published by Atzmüller and Vögtle (106). Diphenyldiazomethane reacts with (arylsulfonyl)isothiocyanates and this is followed by spontaneous N_2 elimination to give thiirane-2-imines (**60**) (107,108). Under similar conditions, acyl-substituted isothiocyanates afforded 2:1-adducts **61** (109) (Scheme 5.23). It seems likely that the formation of **61** involves a thiirane intermediate analogous to **60**, which subsequently reacts with a second equivalent

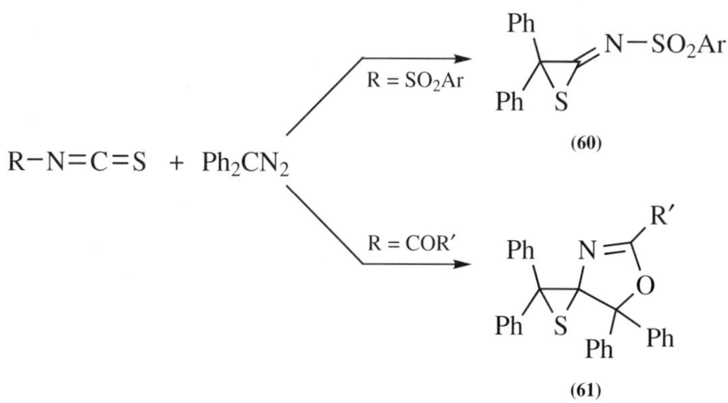

Scheme 5.23

of diphenyldiazomethane at the C=O group, and then undergoes ring closure via a 1,5-dipolar electrocyclization reaction.

Catalyst-mediated decomposition of diazo compounds in the presence of C=S compounds has found application for the preparation of thiiranes in homogeneous systems (68,110,111). Recently, a convenient procedure for the preparation of geminal dichlorothiiranes from nonenolizable thioketones and chloroform under Makosza conditions was reported (112). Another approach to 2,2-dihalogenated thiiranes (e.g., 2,2-difluoro derivatives) involves the thermolysis of *Seyferth reagents* in the presence of thioketones (113,114a). Nucleophilic dimethoxycarbene was shown to add smoothly to adamantanethione to provide a unique approach to a thiiranone (*S,S*)-dimethylacetal (114b).

The formation of tungsten complexes of thiiranes from diarylthioketones and $W(CO)_5$–benzylidene complex is believed to occur via a thiocarbonyl ylide-like intermediate (115).

The synthesis of thiiranes with subsequent elimination of sulfur is an important procedure for the creation of C=C bonds, especially for sterically crowded systems (47,48), in analogy to the *Eschenmoser-sulfide-contraction reaction* (116). The spontaneous elimination of sulfur was observed in the rhodium-catalyzed reaction of diazo compound **62**, which gave rise to the formation of cyclopentenone derivative **63** (117) (Scheme 5.24). A synthesis of indolizomycin was published by Danishefsky and co-workers (118) and involved a similar annulation step. In this case, however, the desulfurization reaction was achieved by treatment with Raney Ni.

The same mechanistic pathway was proposed to rationalize the formation of furan-2-ylidene malonate **64**, a key intermediate in the synthesis of nonactic acid (119). Other examples that involve the replacement of a C=S group by an

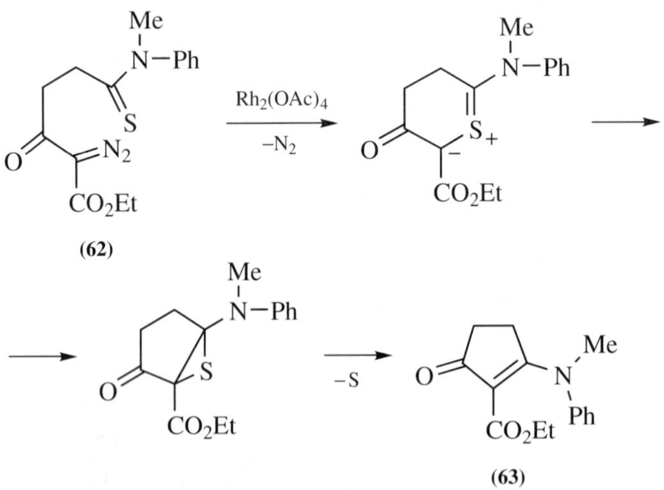

Scheme 5.24

alkylidene group (olefination), are the synthesis of β-lactam **65** (120), the steroid derivative **66** (121) [cf. also (122)], as well as the carbohydrate derivative **67** (123).

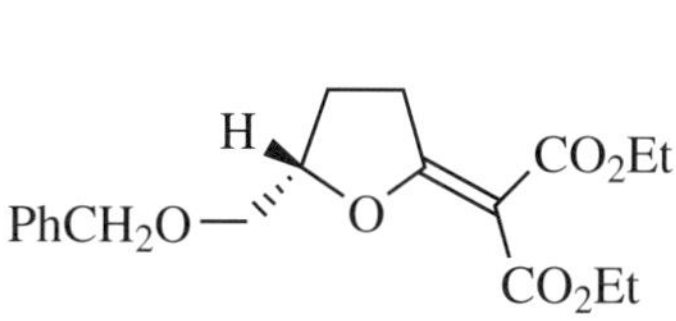

(64)

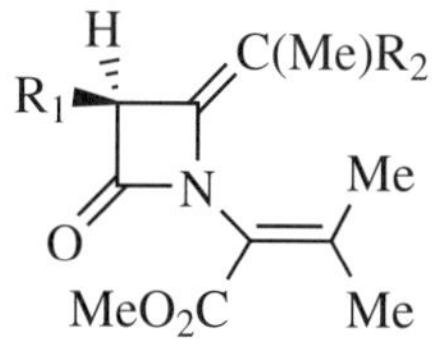

(65)

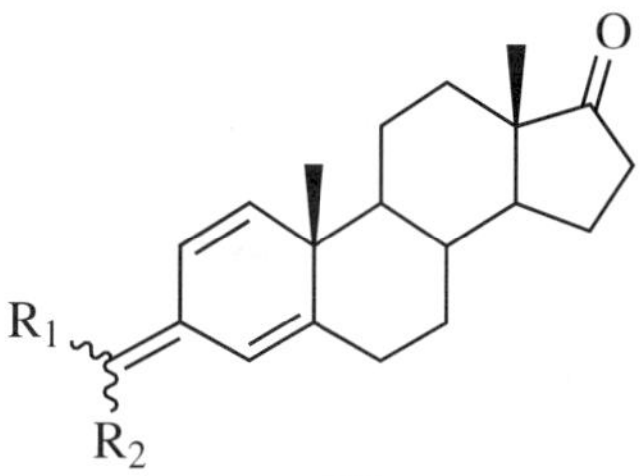

(66)

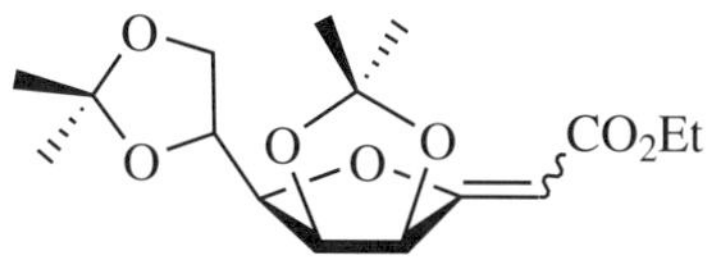

(67)

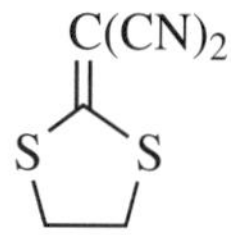

(68)

In some cases involving dicyano-substituted thiocarbonyl ylides of type **9**, ring closure to a thiirane and spontaneous desulfurization results in the formation of dicyano alkenes of type **68** (19,20). As a rule, the presence of electron-withdrawing substituents facilitates sulfur elimination. On the other hand, with alkyl and aryl substituents, desulfurization requires elevated temperature or the use of phosphanes (42,99,105,109,124–127).

5.3.2. Five-Membered Rings

5.3.2.1. Thiophene Derivatives

Numerous examples involving the preparation of tetrahydrothiophenes via [3 + 2] cycloaddition of thiocarbonyl ylides with electron-poor alkenes have been reported. Thiobenzophenone (*S*)-methylide (**16**), generated from 2,5-dihydro-1,3,4-thiadiazole (**15**) and analogous compounds, react with maleic anhydride, N-substituted maleic imide, maleates, fumarates, and fumaronitrile at −45 °C (28,91,93,98,128,129). Similar reactions with adamantanethione (*S*)-methylide (**52**) and 2,2,4,4-tetramethyl-3-thioxocyclobutanone (*S*)-methylide (**69**) occur at ca. +45°C and, generally, the products of type **70** were obtained in high yield (36,94,97,130) (Scheme 5.25). Reaction with (*E*)- and (*Z*)-configured dipolarophiles stereospecifically afford trans and cis configured adducts.

Nonactivated alkenes do not undergo reactions with thiocarbonyl ylides. However, the strained (*E*)-cyclooctene reacts easily with **16**, **52**, and **69** to yield the corresponding trans-fused tetrahydrothiophenes (94).

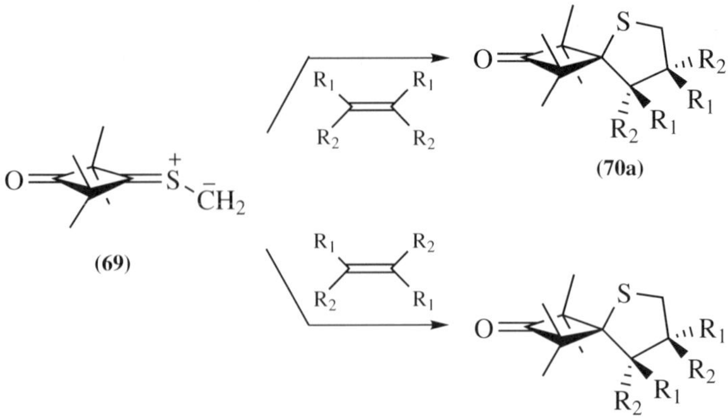

(69) (70a) (70b)

(70a) $R_1{-}R_1 = {-}CO{-}O{-}CO{-}$, $R_2 = H$
$R_1{-}R_1 = {-}CO{-}NR{-}CO{-}$, $R_2 = H$
$R_1 = CO_2R$, $R_2 = H$

(70b) $R_1 = CO_2R$, $R_2 = H$
$R_1 = CN$, $R_2 = H$

Scheme 5.25

Simply substituted thiocarbonyl ylides, including the parent system, react with activated olefins to form tetrahydrothiophene derivatives (30,31,52,53,131). Recently, the thermal desilylation method was applied toward the preparation of the C_{60}-fullerene-fused tetrahydrothiophene **71** (132) (Scheme 5.26).

An attempted synthesis of biotin using thiocarbonyl ylide cycloaddition was carried out (131,133,134). The crucial step involves the formation of the tetrahydrothiophene ring by [3 + 2] cycloaddition of a properly substituted thiocarbonyl ylide with a maleic or fumaric acid derivative (Scheme 5.27). As precursors of the thiocarbonyl ylides, compounds **25a**, **72**, and **73** were used. Further conversion of cycloadducts **74** into biotin (**75**) required several additional steps including a Curtius rearrangement to replace the carboxylic groups at C(3) and C(4) by amino moieties.

The question of regioselectivity arises when nonsymmetrically substituted alkenes, such as acrylates, acrylonitrile, (methyl)vinylketone, and cinnamic acid

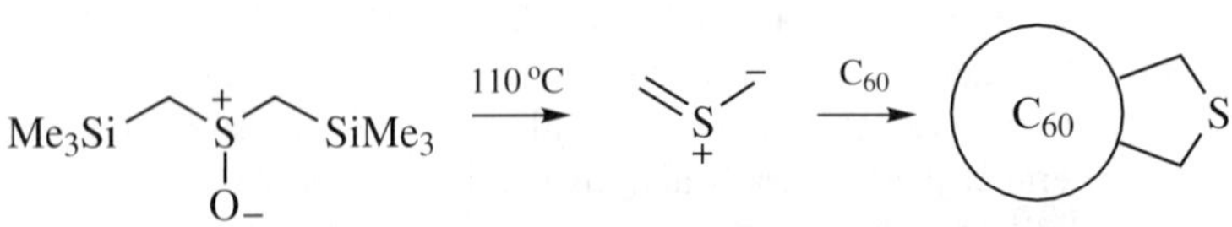

(1a) (71)

Scheme 5.26

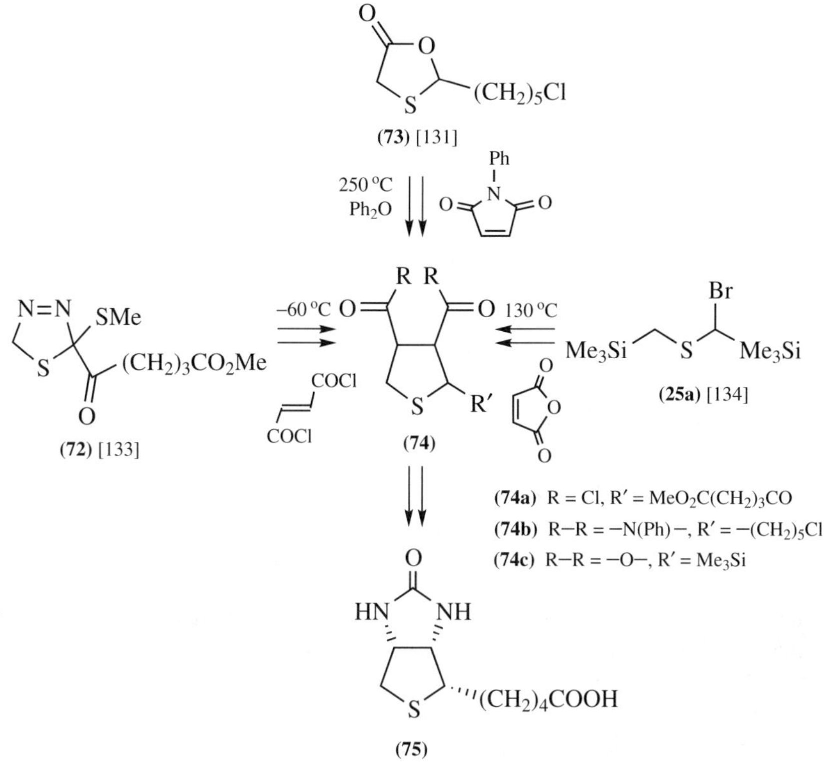

Scheme 5.27

derivatives, are used (30,31,36,52,53,94,135). Thiocarbonyl ylides **52** and **69** combine with these reagents to give tetrahydrothiophenes of type **76** and **77**, with 100% regioselectivity (36,94). On the other hand, silylated representative **26** (R = H) resulted in mixtures of regioisomers **78a,b** (52).

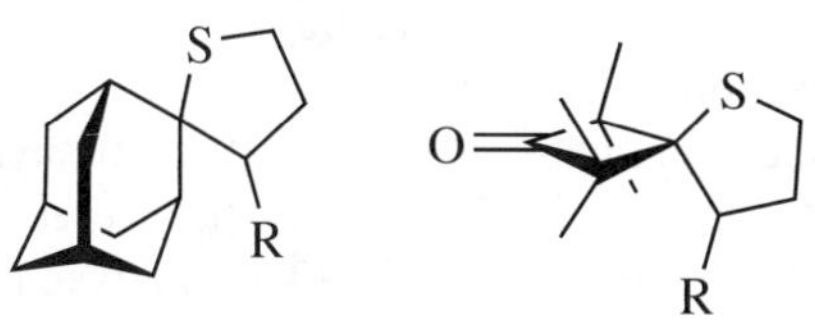

(76) R = CO_2R', CN **(77)**

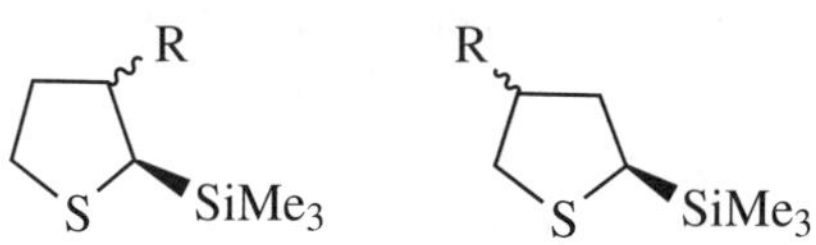

(78a) R = CO_2Me, CN **(78b)** R = CO_2Me, CN

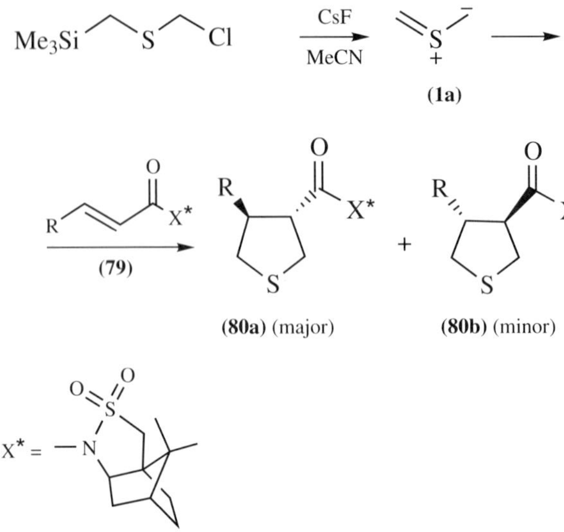

Scheme 5.28

The first diastereoselective synthesis of a tetrahydrothiophene derivative was reported by Karlsson and Högberg (32,95). The parent ylide **1a** was added to a variety of C,C-dipolarophiles (**79**) bearing (−)-(1*S*)-2,10-camphorsultam as the chiral auxiliary group to exclusively give trans-cycloadducts **80a,b** with high diastereoselectivity [diastereomeric ratio (dr) ~9:1], (Scheme 5.28).

The reaction of the sterically crowded thiocarbonyl ylide **69** with highly electron-deficient alkenes such as 2,3-dicyano fumarate and maleate, tetracyanoethene, α-cyano cinnamates, and 1,2-bis(trifluoromethyl)ethene-1,2-dicarbonitrile occurred in a nonstereospecific manner (27,89,96,97,136–138). The formation of a mixture of cis/trans tetrahydrothiophenes of type **82** is the result of a stepwise reaction involving zwitterionic intermediates of type **81** (Scheme 5.29). Ylide **69** fulfills the fundamental requirements for a two-step reaction with electron-deficient alkenes. This species corresponds to an electron-rich 1,3-dipole that also contains a bulky substituent at one terminus (89).

Interception of photochemically generated thiocarbonyl ylides with reactive dipolarophiles (e.g., *N*-phenylmaleinimide) has been used for the preparation of polycyclic tetrahydrothiophenes of type **83** (139). An example is shown in Scheme 5.30.

In the absence of a dipolarophile, thiocarbonyl ylide **84** undergoes a 1,4-hydrogen shift to give the naphthoannelated thiophene derivative **85**. Many examples of related syntheses have been reported (139–143). The photolysis of tetraarylthiiranes in the presence of tetracyanoethene represents an approach to tetrahydrothiophenes via a SET mechanism (75,76).

The mesoionic compound **86** corresponds to a thioisomünchnone (61) and was generated by the *carbenoid strategy* approach. This dipole reacts with *N*-phenylmaleinimide to give the polycyclic adduct **87** (144) (Scheme 5.31). This method,

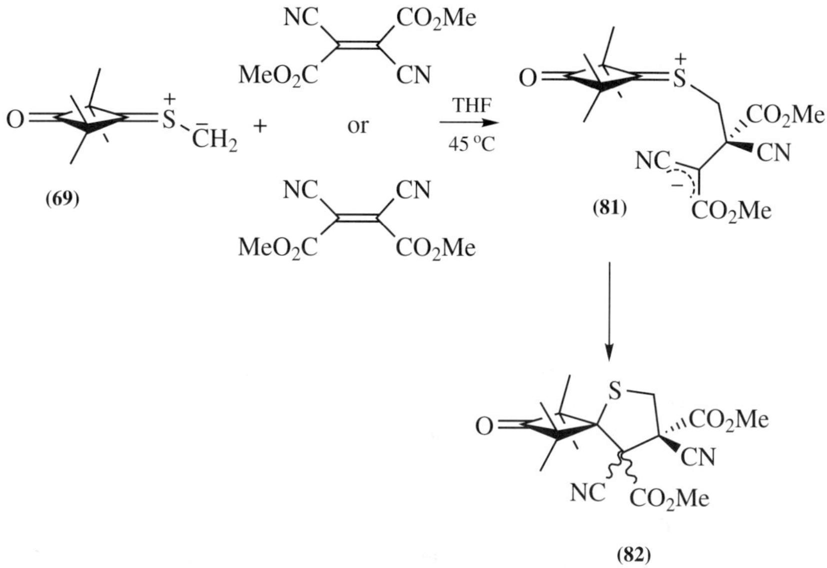

Scheme 5.29

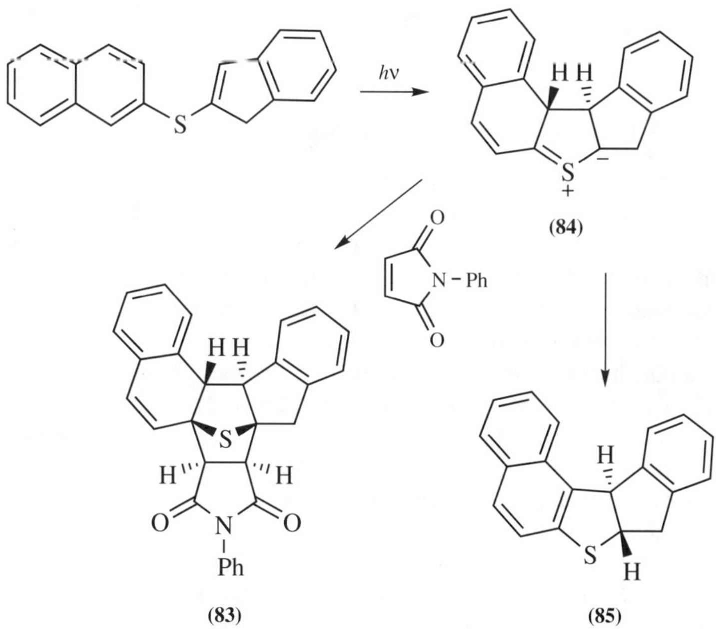

Scheme 5.30

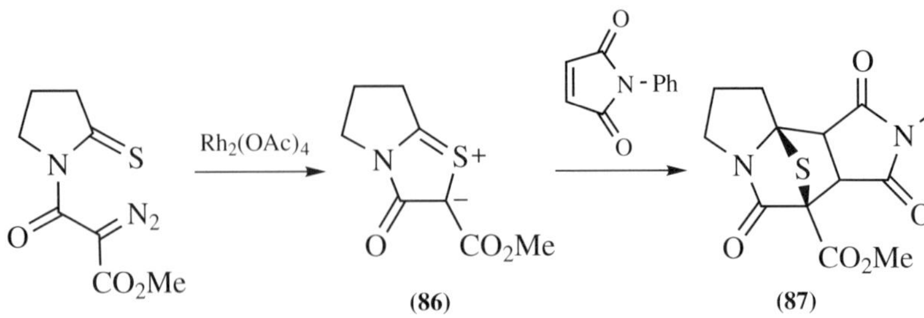

(86) (87)

Scheme 5.31

developed by Padwa and co-workers (61,117), was applied toward other annelated thiophene derivatives.

Thioisomünchnones prepared by classical methods (Section 5.1.3) have also been extensively used in [3 + 2] cycloadditions with olefinic dipolarophiles (145–148). In general, the initially formed cycloadducts are isolable. In some instances (Scheme 5.32), they extrude H_2S to afford 2-pyridone derivatives of type **88** (148,149). In another example, the double extrusion of H_2S and CO occurred to give a pyrrole derivative (150).

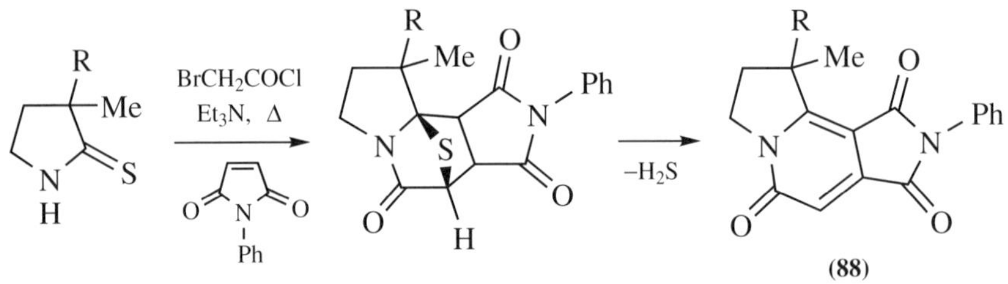

(88)

Scheme 5.32

An attractive approach toward the preparation of polycyclic systems containing a thiophene ring involves the intramolecular [3 + 2] cycloaddition of thiocarbonyl ylides. A number of representative examples were reported using mesoionic compounds. Gotthardt et al. (151) used 1,3-dithiolium-4-olates such as **89** bearing an olefinic side chain. Upon heating to 120 °C in xylene, the polycyclic tetrahydrothiophene **90** was formed (Scheme 5.33).

Thioisomünchnones are known to undergo a range of intramolecular [3 + 2] cycloadditions. One example reported by Potts et al. (62) is shown in Scheme 5.34. Heating compound **91** ($n = 1$) resulted in a 7:1 mixture of diastereoisomers **92a** and **92b**. Elongation of the side chain ($n = 2$) led to the selective formation of isomer **92a**. On the other hand, the cyclohexen-3-yl derivative produced cycloadduct **92b** as the exclusive product.

This methodology was further developed in an elegant manner by Padwa and co-workers (61). Combined with a subsequent desulfurization reaction, the method was successfully applied toward the synthesis of complex polyheterocyclic systems related to alkaloid skeletons. For example, the reaction of **94**, obtained from

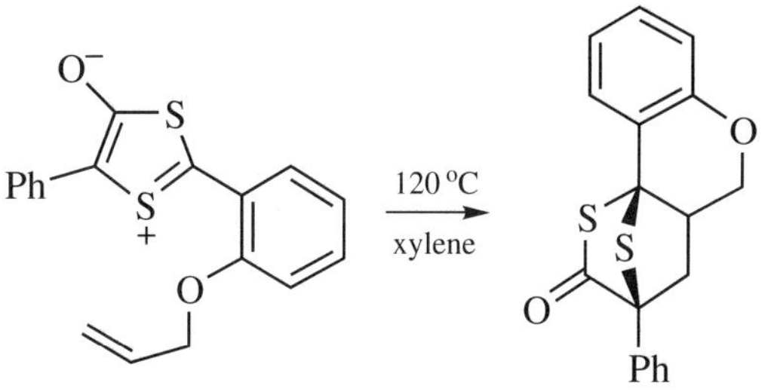

(89) (90)

Scheme 5.33

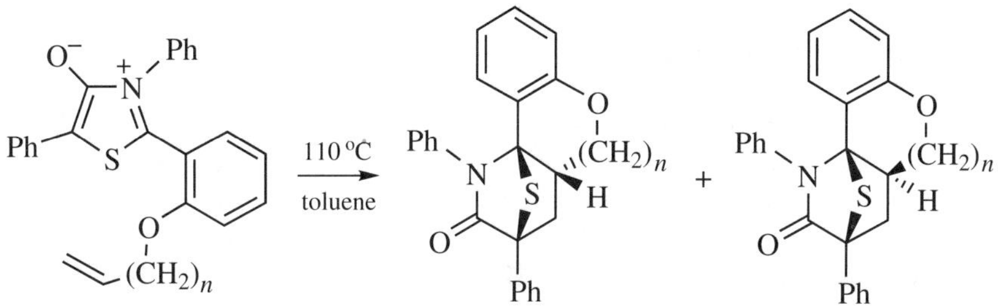

(91) (92a) (92b)

Scheme 5.34

thioamide **93** via the corresponding thioisomünchnone, with molybdenum hexacarbonyl resulted in the isolation of enamide **95**. In contrast, desulfurization using *meta*-chloroperbenzoic acid (MCPBA) produced the corresponding pyridone system (Scheme 5.35).

It is also worthy to note that the intramolecular [3 + 2] cycloaddition of thiocarbonyl ylides occurs easily with nonactivated C=C bonds, whereas the corresponding intermolecular process does not occur.

An intramolecular cycloaddition reaction was also used in the synthesis of the annelated tetrahydrothiophene (**97**), starting from 1,3-oxathiolan-5-one (**96**) (131) (Scheme 5.36). Thiocarbonyl ylide formation occurred by thermal extrusion of CO_2 at 250 °C, yielding **97** in 62% yield.

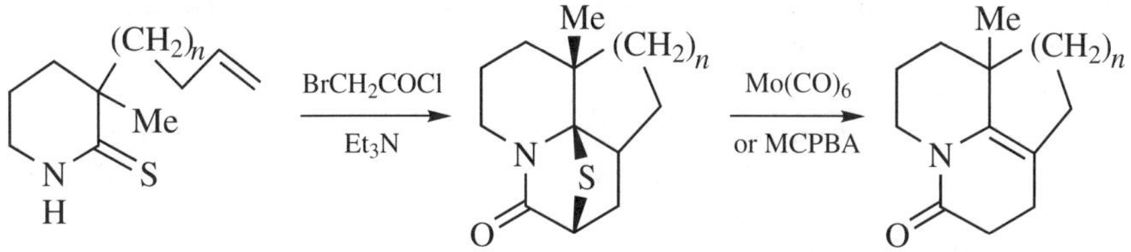

(93) (94) (95)

Scheme 5.35

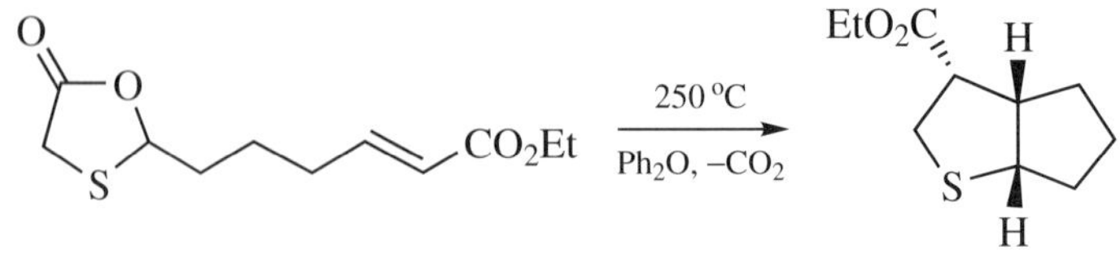

(96) **(97)**

Scheme 5.36

Reaction with acetylenic dipolarophiles represents an efficient method for the preparation of 2,5-dihydrothiophenes. These products can be either isolated or directly converted to thiophene derivatives by dehydration procedures. The most frequently used dipolarophile is dimethyl acetylenedicarboxylate (DMAD), which easily combines with thiocarbonyl ylides generated by the extrusion of nitrogen from 2,5-dihydro-1,3,4-thiadiazoles (8,25,28,36,41,92,94,152). Other methods involve the desilylation (31,53,129) protocol as well as the reaction with 1,3-dithiolium-4-olates and 1,3-thiazolium-4-olates (153–158). Cycloaddition of (*S*)-methylides formed by the N_2-extrusion or desilylation method leads to stable 2,5-dihydrothiophenes of type **98** and **99**. In contrast, bicyclic cycloadducts of type **100** usually decompose to give thiophene (**101**) or pyridine derivatives (**102**) (Scheme 5.37).

Cycloadditions with other symmetrical acetylenes were carried out by using thiocarbonyl (*S*)-methylide (**69**) (159). Interestingly, no reaction was observed when acetylene dicarboxamide was used. The reaction of **69** with cyclooctyne resulted in the formation of cycloadduct **103** (Scheme 5.38). Interestingly, the spirocyclic 2,5-dihydrothiophenes of type **103** or **104** undergo acid-catalyzed ring opening upon treatment with silica gel or trifluoroacetic acid to give thiophenes **105** and **106**, respectively.

Recently, a new approach to 3,4-disubstituted thiophenes via the 3,4-disilylated thiophene **108** was reported (160). The synthesis includes the cycloaddition of **1a** to bis(trimethylsilyl)acetylene and subsequent dehydration of cycloadduct **107** with DDQ to give **108** (Scheme 5.39). The trimethylsilyl groups can be substituted in a stepwise manner to give unsymmetrical 3,4-disubstituted thiophenes (**109**).

The reaction of thiocarbonyl ylides with propiolates affords a mixture of regioisomeric cycloadducts. Thus, 2,5-dihydrothiophenes obtained from the reaction of adamantanethione (*S*)-methylide (**52**) and methyl propiolate were produced in a 1:1 ratio (95). In the case of ylide **69**, the ratio was 1:2 in favor of the sterically less hindered isomer (160).

Intramolecular [3 + 2]-cycloadditions of thiocarbonyl ylides with nonactivated acetylenes have also been described. Most representative examples involved the use of mesoionic substrates. The initially formed polycyclic adducts of type **110** undergo spontaneous elimination of phenyl isocyanate (24,62,151). A typical example leading to compound **111** is shown in Scheme 5.40.

The thermal decomposition of 1,3-oxathiolane-5-one (**112**) produces the fused thiophene **113** via a transient thiocarbonyl ylide followed by intramolecular [3 + 2]

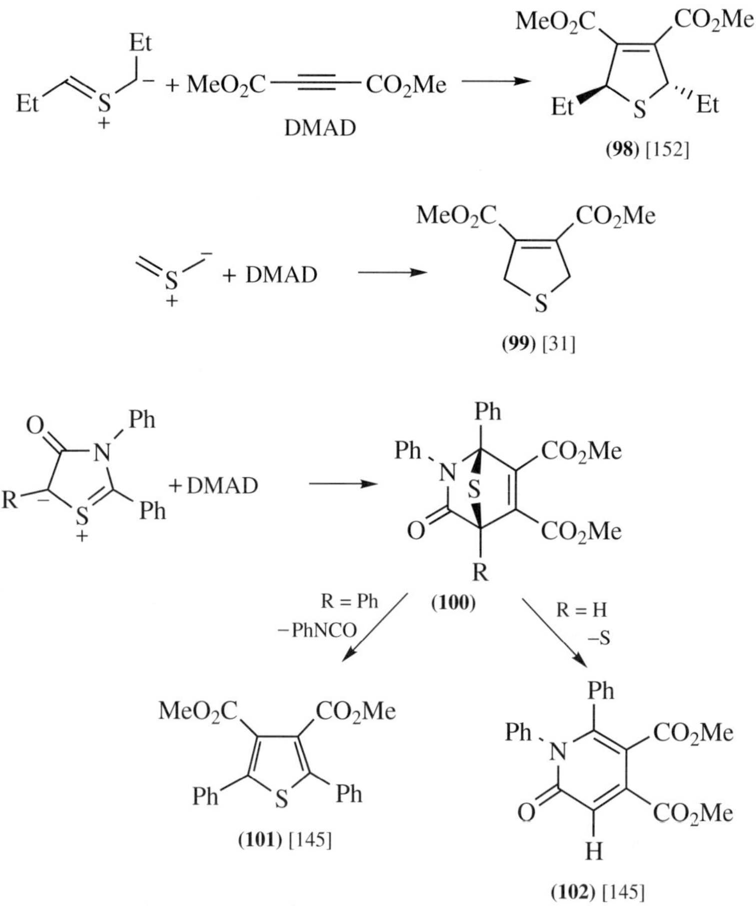

Scheme 5.37

cycloaddition. The initially formed 2,5-dihydrothiophene undergoes spontaneous dehydrogenation to give the aromatic thiophene ring present in **113** (Scheme 5.41) (131).

5.3.2.2. 1,3-Dithiolanes, 1,3-Oxathiolanes, and 1,3-Thiazole Derivatives

Based on a series of kinetic studies, Huisgen et al. (91–93) established that thiocarbonyl compounds, especially aromatic thioketones, function as very active dipolarophiles (*superdipolarophiles*) toward thiocarbonyl ylides. In fact, the trapping reaction of thiocarbonyl ylides with thiocarbonyl compounds represents an excellent method for the preparation of 1,3-dithiolanes.

Aromatic thioketone (*S*)-methylides react with aromatic and nonenolizable aliphatic thioketones to regiospecifically give 1,3-dithiolanes of type **114** and **115**

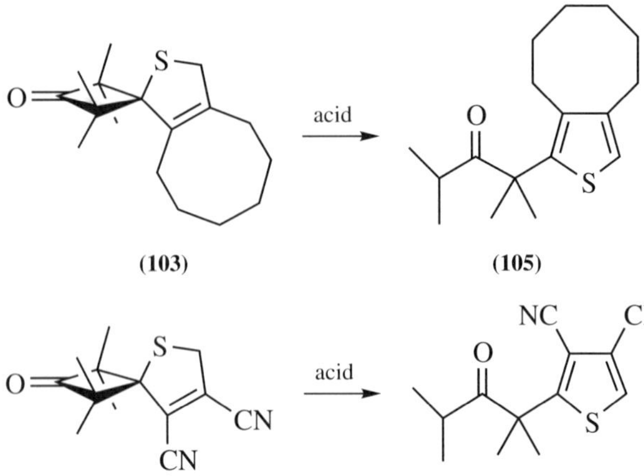

(103) (105)

(104) (106)

Scheme 5.38

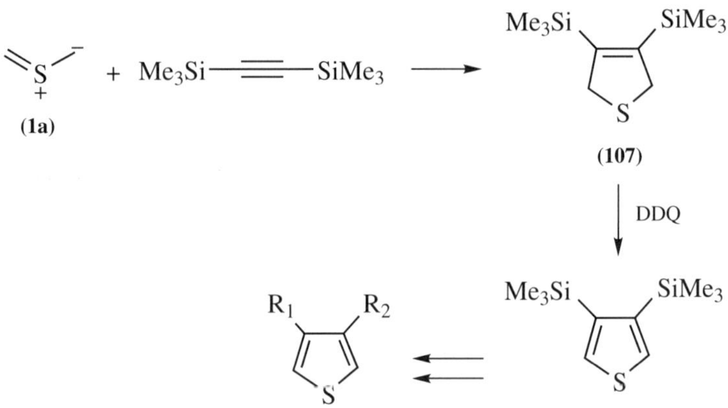

(1a) (107) (108) (109)

Scheme 5.39

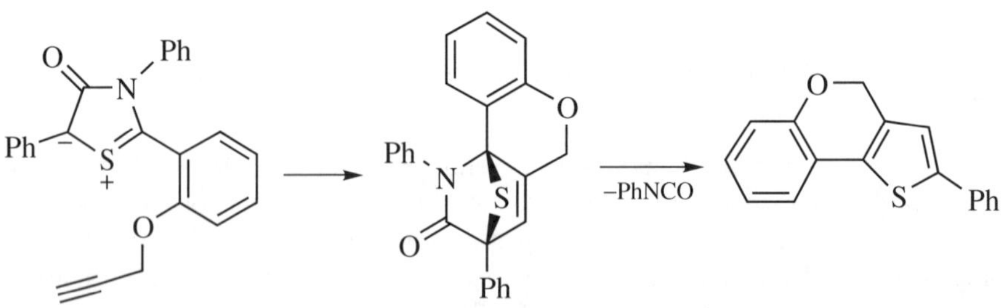

(110) (111)

Scheme 5.40

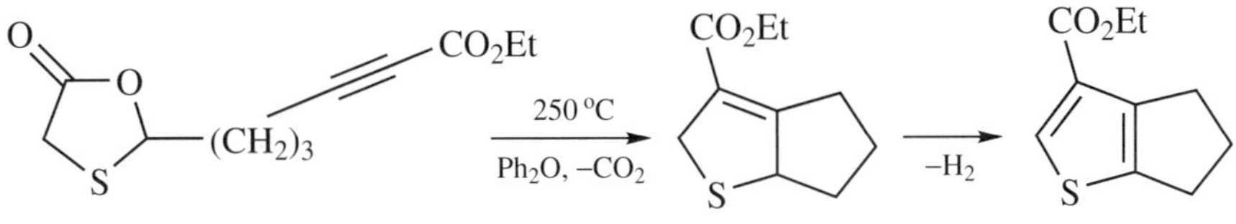

(112) (113)

Scheme 5.41

[i.e., the sterically less favored products (17,28,54,161)]. On the other hand, a mixture of regioisomeric 1,3-dithiolanes (e.g., **115** and **116**) were obtained when adamantanethione (*S*)-methylide (**52**) or 2,2,4,4-tetramethyl-3-thioxocyclobutanone (*S*)-methylide (**69**) was generated in the presence of thiobenzophenone (36,162,163).

Finally, ylides **52** and **69** were found to react regioselectively with aliphatic thioketones to give the sterically less crowded products of type **117** (36,162,163).

Other C=S compounds that have proven to be useful for the preparation of 1,3-dithiolanes are dithioesters, (161–163) trithiocarbonates, (161) and 1,3-thiazole-5-(4*H*)-thiones (164).

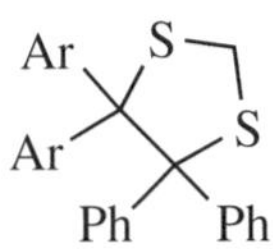

(114)

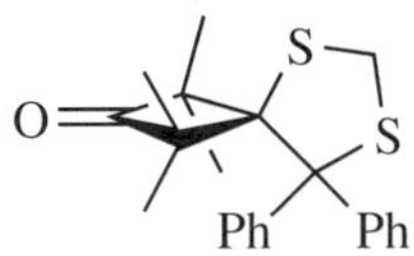

(115)

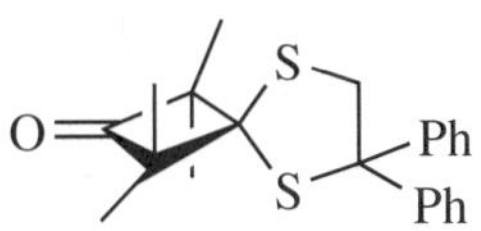

(116)

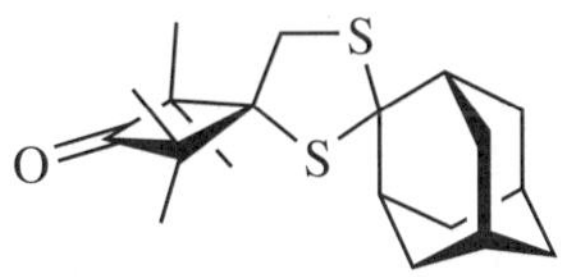

(117)

Heterocumulenes containing C=S groups have also been used as dipolarophiles for cycloaddition with thiocarbonyl ylides. Thus, the reaction of (*S*)-methylides **52** and **69**, with carbondisulfide, not only give 1:1 cycloadducts of type **118**, but also produce the corresponding 1:2 adducts **119** (162,163,165). 1,3-Dithiolane imine (**120**) was isolated as the sole product when ylide **52** was generated in the presence of phenyl isothiocyanate (163). A study carried out using sulfines (thiocarbonyl S-oxides) as trapping agents showed that ylide **69** produces 1,3-dithiolane S-oxides **121** and **122** (166) in good yield. The only known aliphatic sulfine that undergoes an analogous reaction to give product **121** was 1-oxo-2-thioxoindane S-oxide (167). Nonactivated aliphatic sulfines do not undergo cycloaddition.

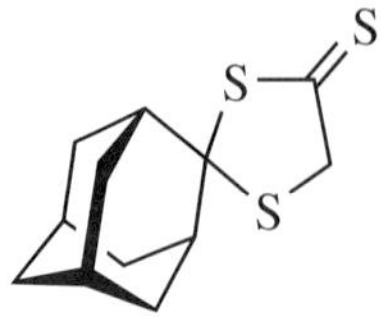

(118)

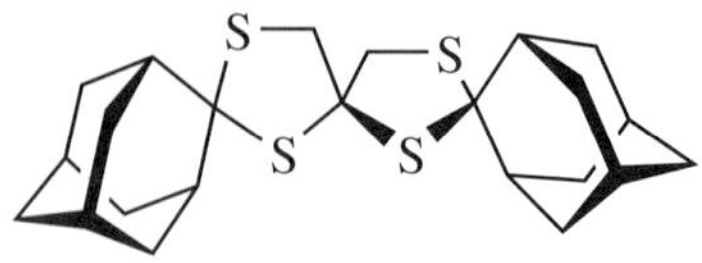

(119)

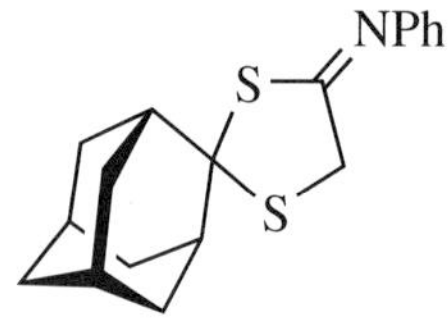

(120)

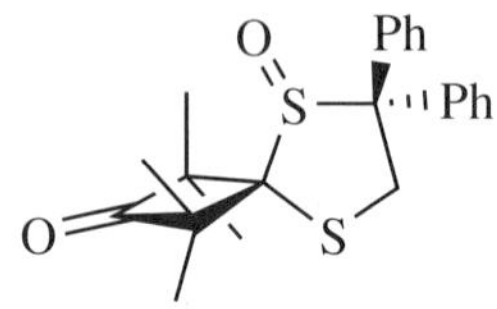

(121)

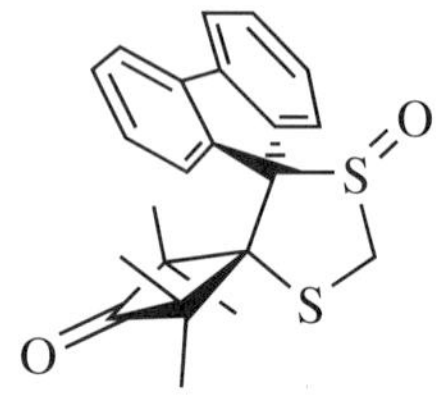

(122)

For preparative purposes, the reaction of thiocarbonyl ylides with carbonyl compounds can be considered as an alternative method for the synthesis of 1,3-oxathiolanes. Aromatic aldehydes, chloral, glyoxalates, mesoxalates, pyruvates as well as their 3,3,3-trifluoro analogues are good intercepting reagents for thioketone (*S*)-methylides (36,111,130,163). All of these [3 + 2] cycloadditions occur in a regioselective manner to produce products of type **123** and **124**.

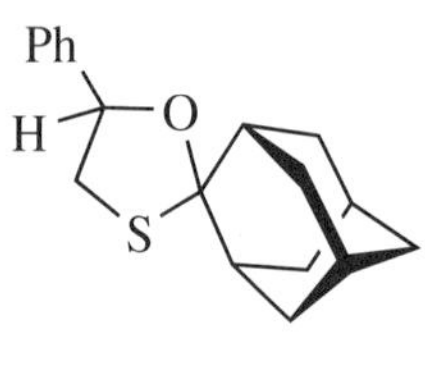

(123)

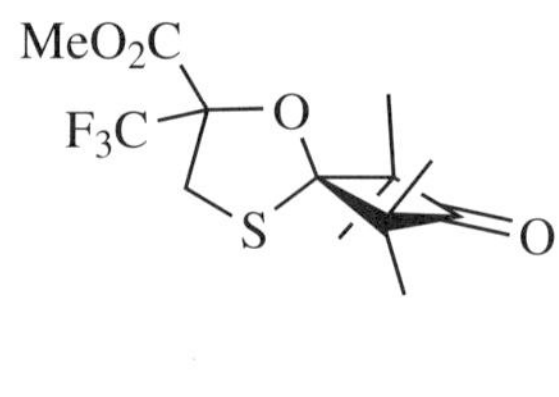

(124)

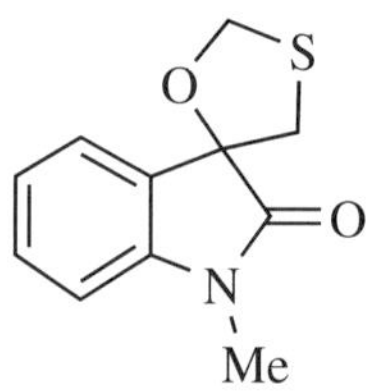

(125)

The desilylation strategy has been used for the cycloaddition of the parent thiocarbonyl ylide **1a** with aldehydes and reactive ketones. The product obtained using *N*-methyl-3-oxoindolinone as the trapping agent corresponds to the spirocyclic compound **125** (168). Thioketene (*S*)-methylide (**127**) was reported to react with aromatic aldehydes and some ketones to furnish 2-methylene-substituted 1,3-oxathiolanes (**128**) (51) (Scheme 5.42).

Other examples of functionalized thiocarbonyl ylides that have been generated by the desilylation method are those bearing an imino group (49) (see Scheme 5.7). These ylides readily undergo [3 + 2] cycloaddition with aromatic aldehydes to afford 1,3-thioxolane-2-imines of type **24** ($X = R_1N$). The reaction with ketones is sluggish, however, and the cycloadducts are obtained in very low yield.

The only reported reaction using phenylisocyanate as a trapping agent is one that involves adamantanethione (*S*)-methylide (**52**) as the dipole (163). Although a 1:1 adduct was isolated, its spectral data did not fully establish the structure of the

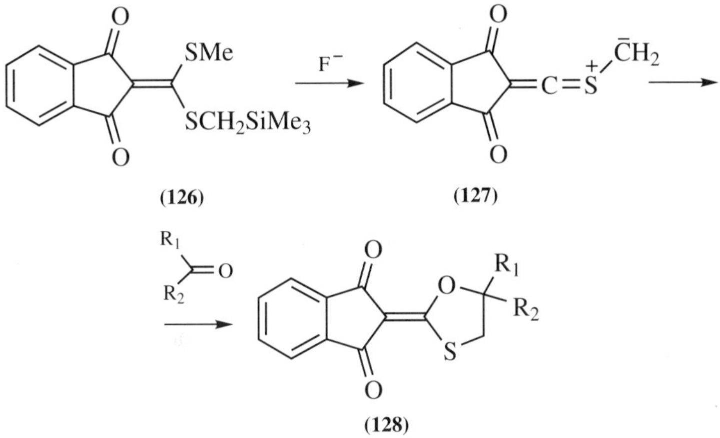

Scheme 5.42

labile product. Some contradictory results involving the reaction of isocyanates with thioisomünchnones were also published (163).

Only a few examples of the [3 + 2] cycloaddition of thiocarbonyl ylides with C=N compounds have been reported so far. By comparison with aldehydes, imines are poor dipolarophiles. For example, *N*-benzylidene methylamine and adamantanethione (*S*)-methylide (**52**) produce 1,3-thiazolidine (**129**) in only 13% yield (163). An alternative and efficient approach to 1,3-thiazolidines involves the [3 + 2] cycloaddition of azomethine ylides with thiocarbonyl compounds [cf. (169)].

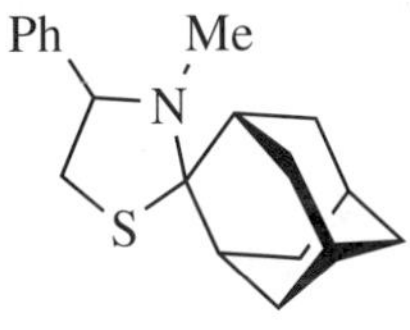

(129)

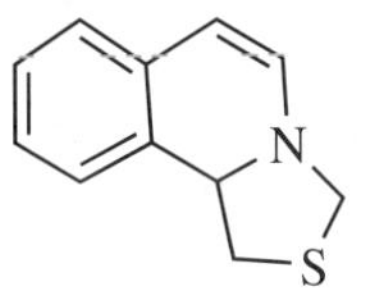

(130)

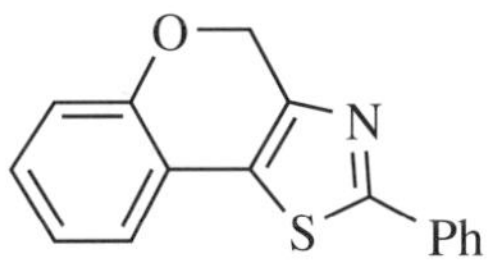

(131)

Nitrogen-containing heteroaromatic compounds react with (chloromethyl)[(trimethylsilyl)methyl]sulfide in the presence of CsF to afford fused 1,3-thiazolidines of type **130**. These compounds are the result of a formal [3 + 2] cycloaddition of the parent thiocarbonyl ylide **1a** across the C=N bond (170). In these cases, the formation of the five-membered cycloadduct is believed to occur in two steps via an intermediate onium ion.

Reactions of thiocarbonyl ylides with nitriles are scarce. Simple nitriles do not undergo bimolecular cycloaddition (171). There is, however, a single example of an intramolecular case that was reported by Potts and Dery (24c,62). By analogy to the intramolecular cycloaddition with acetylenic dipolarophiles (Scheme 5.40), the primary product derived from the reaction of a thiocarbonyl ylide with a nitrile group undergoes a subsequent elimination of phenylisocyanate to give the fused 1,3-thiazole (**131**).

5.3.2.3. *Five-Membered Rings with Three Heteroatoms*

The following types of dipolarophiles have been used successfully to synthesize five-membered heterocycles containing three heteroatoms by [3 + 2]-cycloaddition of thiocarbonyl ylides: azo compounds, nitroso compounds, sulfur dioxide, and *N*-sulfinylamines. As was reported by Huisgen and co-workers (91), azodicarboxylates were noted to be superior dipolarophiles in reactions with thiocarbonyl ylides. Differently substituted 1,3,4-thiadiazolidine-3,4-dicarboxylates of type **132** have been prepared using aromatic and aliphatic thioketone (*S*)-methylides (172). Bicyclic products (**133**) were also obtained using *N*-phenyl 1,2,4-triazoline-3,5-dione (173,174).

Nitroso compounds are seldom used as dipolarophiles for trapping reactions with thiocarbonyl ylides. However, Sheradsky and Itzhak (175) did report one example where nitrosobenzene reacts with a thioisomünchnone to give **134** as the major product.

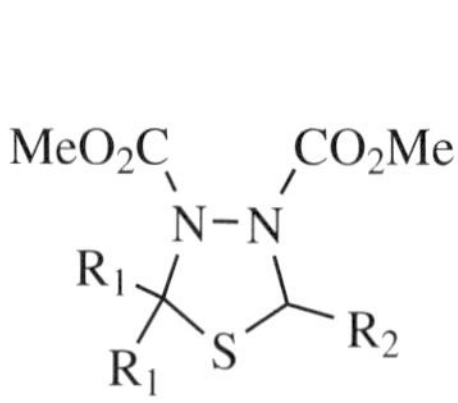

(132)

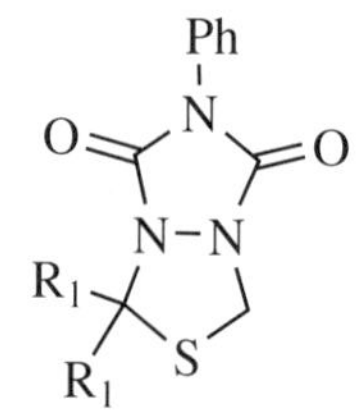

(133)

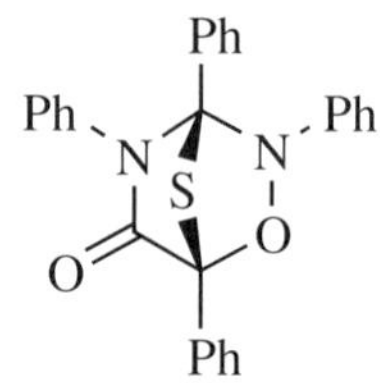

(134)

N-Sulfinylamines (R—N=S=O) are known to function as reactive dienophiles and dipolarophiles, and some examples of [3 + 2] cycloaddition with thiocarbonyl ylides have been reported (176). For example, the reaction of thiobenzophenone (*S*)-methylide (**16**) with both *N*-phenyl and *N*-tosylsulfinylamines occurs regioselectively to give 1,3,4-dithiazolidine 3-oxides (**135**). In the case of thiocarbonyl ylide **69**, reaction with *N*-phenyl sulfinylamine selectively afforded the analogous product **136** (R = Ph). However, the corresponding reaction with *N*-tosyl sulfinylamine resulted in a mixture of the N,S-adduct (**136**) (R = Tos) and the O,S-adduct **137**. Formation of a mixture of products is compatible with a stepwise reaction via a zwitterionic intermediate.

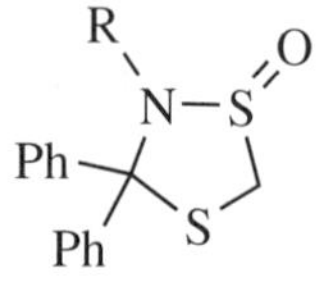

(135) R = Ph, Tos

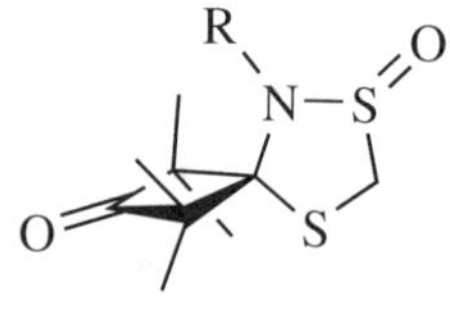

(136) R = Ph, Tos

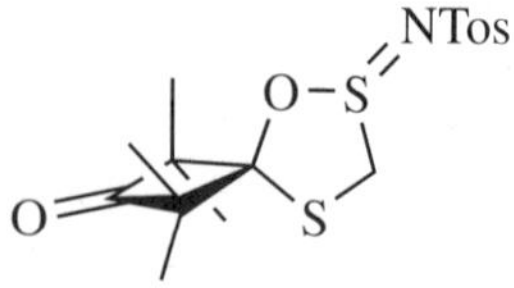

(137)

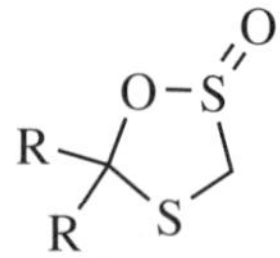

(138)

Tos = tosyl

Sulfur dioxide reacts with aliphatic thioketone (*S*)-methylides in a sealed tube at 100 °C and 1,2,4-oxadithiolane-2-oxides (**138**) are obtained. None of the regioisomeric cycloadduct was formed (177).

5.3.2.4. Five-Membered Heterocycles via 1,5-Dipolar Electrocyclization

The reaction of a thiocarbonyl and α-oxodiazo compound that leads to 1,3-oxathioles has been rationalized by a 1,5-dipolar electrocyclization reaction (178). It was suggested that an intermediate thiocarbonyl ylide bearing a C=O function at the α-position (*extended dipole*) was first formed. Due to the low reactivity of α-oxodiazo compounds, these reactions were carried out at elevated temperatures or in the presence of rhodium acetate as the catalyst. In some cases, catalysis by $LiClO_4$ was also reported (77–80).

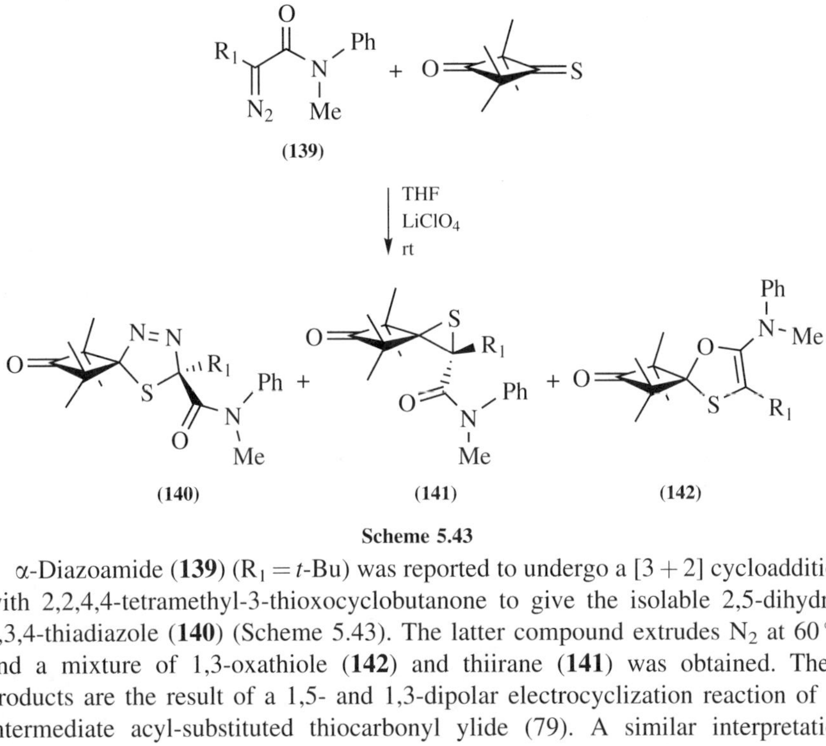

Scheme 5.43

α-Diazoamide (**139**) ($R_1 = t$-Bu) was reported to undergo a [3 + 2] cycloaddition with 2,2,4,4-tetramethyl-3-thioxocyclobutanone to give the isolable 2,5-dihydro-1,3,4-thiadiazole (**140**) (Scheme 5.43). The latter compound extrudes N_2 at 60 °C and a mixture of 1,3-oxathiole (**142**) and thiirane (**141**) was obtained. These products are the result of a 1,5- and 1,3-dipolar electrocyclization reaction of an intermediate acyl-substituted thiocarbonyl ylide (79). A similar interpretation nicely explains the formation of 1,3-oxathiole **143** starting from 1,3-thiazole-5(4*H*)-thiones and **139** ($R_1 = t$-Bu) as well as those of **144** and **145** (77,79,80).

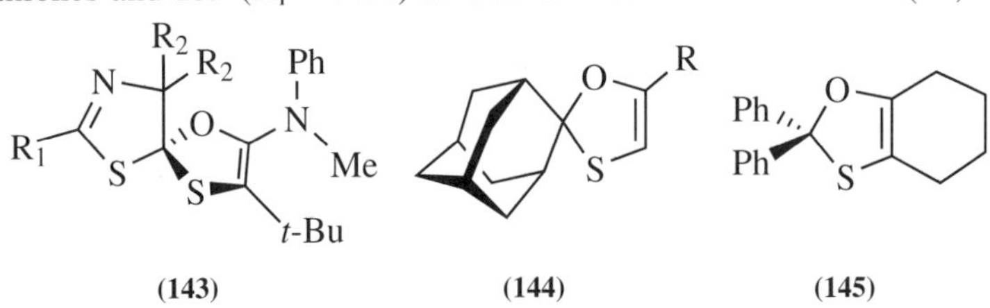

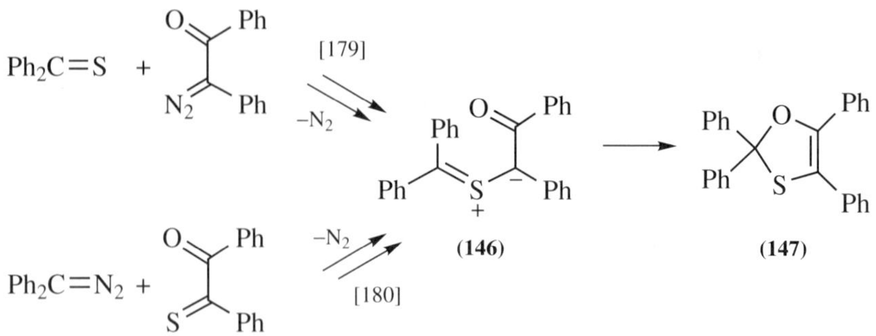

(146) **(147)**

Scheme 5.44

Acyl-substituted thiocarbonyl ylides have been suggested to undergo 1,5-dipolar cyclization to give 1,3-oxathioles (55,179–183). Convincing evidence for the intermediacy of thiocarbonyl ylide **146** involves the formation of structure **147** starting from either two different pairs of reagents (179,180) (Scheme 5.44).

Ueno and Okawara (184) were the first to explicitly formulate a *conjugated thiocarbonyl ylide* as an intermediate in the reaction of 1,3-dithiolane-2-thione with 4-bromophenacyl bromide. The initially formed thiocarbonyl ylide undergoes deprotonation with sodium hydride to give 2-(4-bromophenyl)-1-oxa-4,6,9-trithiaspiro[4.4]non-2-ene. 1,3-Diacylated thiocarbonyl ylides of type **149** (Scheme 5.45) have also been proposed as intermediates in the reaction of 1,3-diphenylpropane-1,3-dione with thionyl chloride. This reaction leads to 2,2,4-tribenzoyl-5-phenyl-

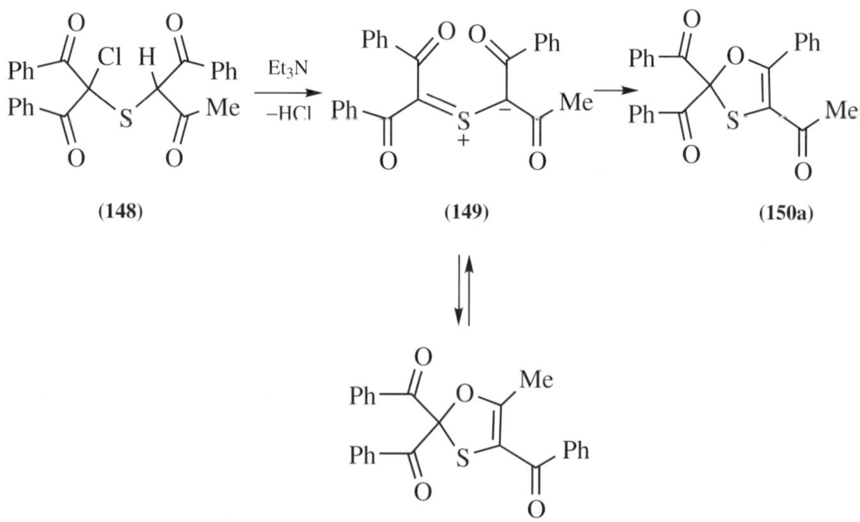

(148) **(149)** **(150a)**

(150b)

Scheme 5.45

1,3-oxathiole (181). A similar transformation involving a thallium complex of 1,3-diketone with $SOCl_2$ (185) or with an α-chlorosulfenyl chloride (55) has also been reported. In the latter case, the initially formed product **148** is deprotonated on treatment with triethylamine. Isomerization of **150a** ⇌ **150b**, which occurs already at room temperature, undoubtedly involves the intermediacy of ylide **149** (55).

The reaction of isothiocyanates with α-diazoketones only occurs in the presence of a rhodium catalyst [$Rh_2(OAc)_4$)] (81,186). Under these conditions, 1,3-oxathiol-2-imines (**151**) (Scheme 5.46) are formed by a similar pathway. An analogous reaction was encountered on treatment of phenyl-dimedonyliodone with phenylisothiocyanate (182). Another interesting example involves the reaction of diazodiketone **152** with methylisothiocyanate (Scheme 5.46). The main product isolated is 1,3-oxathiol-2-imine (**153**), which is also accompanied by some 1,3-oxazol-2(1*H*)-thione (**154**) and 2-thioxo-1,3-oxazin-4(3*H*)-one (**155**) (81). Apparently, the carbenoid intermediate derived from **152** either adds to the sulfur or nitrogen atom of the isothiocyanate giving rise to an imino-substituted thiocarbonyl ylide or a thioxo-substituted azomethine ylide. 1,5-Dipolar electrocyclization of the former produces **153**, whereas the latter cyclizes to give **154**. Compound **155** corresponds to the product of a Wolff rearrangement of the intermediate carbene that first gives an acyl ketene that is intercepted by methylisothiocyanate in a hetero-Diels–Alder reaction.

In a similar manner, the reaction of di(*tert*-butyl)ketene and dimethyl diazomalonate with $Rh_2(OAc)_4$ in benzene at 50 °C leads to the formation of 2-alkylidene-1,3-oxathiole-4-carboxylate (**157**) (R = Me) (187) (Scheme 5.47). The reversibility of the 1,5-dipolar electrocyclization was demonstrated by the thermal isomerization of **157** into allene episulfide **158**, which is obtained by heating in carbon

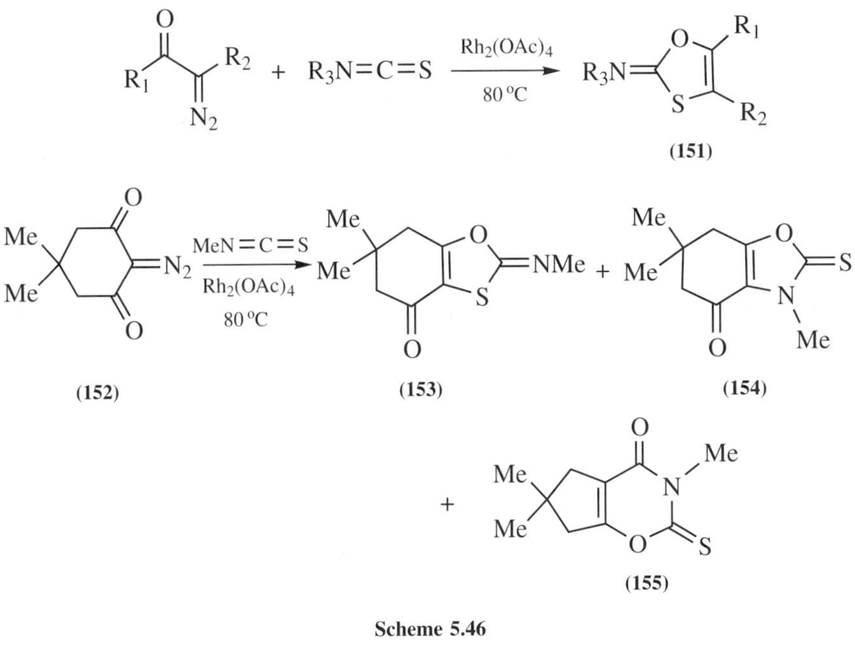

(151) (152) (153) (154) (155)

Scheme 5.46

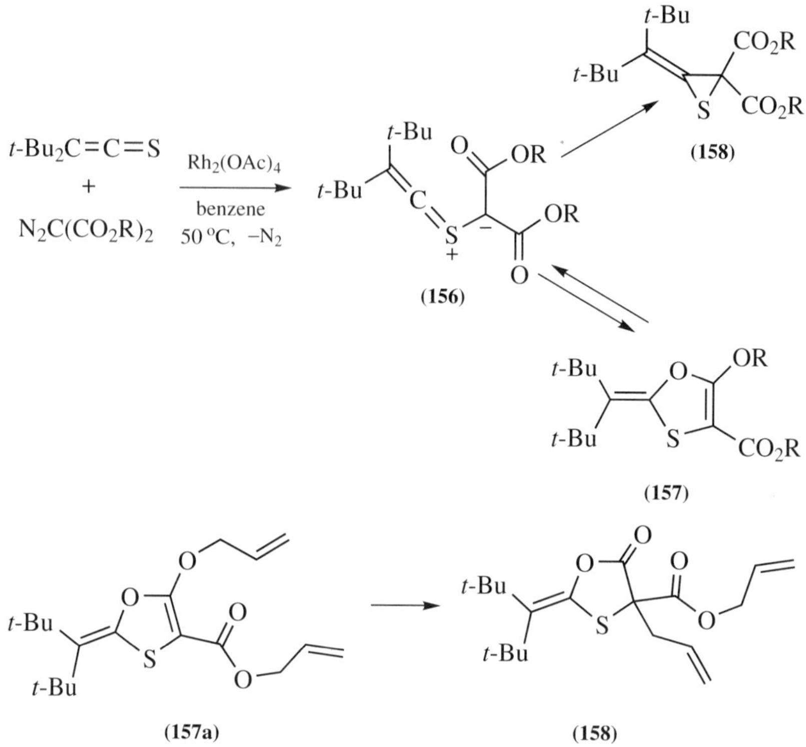

Scheme 5.47

tetrachloride at 70 °C. The intermediate thiocarbonyl ylide **156** could be intercepted with 4-phenyl-1,2,4-triazoline-3,5-dione in a 1,3-dipolar cycloaddition reaction. The equilibrium between 1,3-oxathioles and 2-acyl thiiranes that proceeds via an acyl-substituted thiocarbonyl ylide intermediate has also been described by other groups (55,82,188).

When diallyl diazomalonate or allyl aryldiazoacetates are used as substrates in the rhodium-catalyzed reaction with di(*tert*-butyl)thioketene, 4-allyl-1,3-oxathiolan-5-one derivatives of type **158** are formed (82). After the initial 1,5-dipolar electrocyclization, **157a** undergoes a subsequent Claisen rearrangement to give the thermodynamically more stable compound **158** (Scheme 5.47).

The metal-catalyzed formation of 2,3-dihydrothiophene derivatives via a 1,5-dipolar electrocyclization has been reported by Hamaguchi et al. (124). For example, the $Rh_2(OAc)_4$-catalyzed reaction of vinyldiazo compound **159** (R = Ph) with xanthione (**160**) produced the spirocyclic dihydrothiophene **161**. In contrast, when **159** containing a methyl group (R = Me) was used, thiirane **162** was the sole product (Scheme 5.48). This result was rationalized by the selective formation of an intermediate thiocarbonyl ylide **163** with (*Z*)- and (*E*)-configuration, respectively.

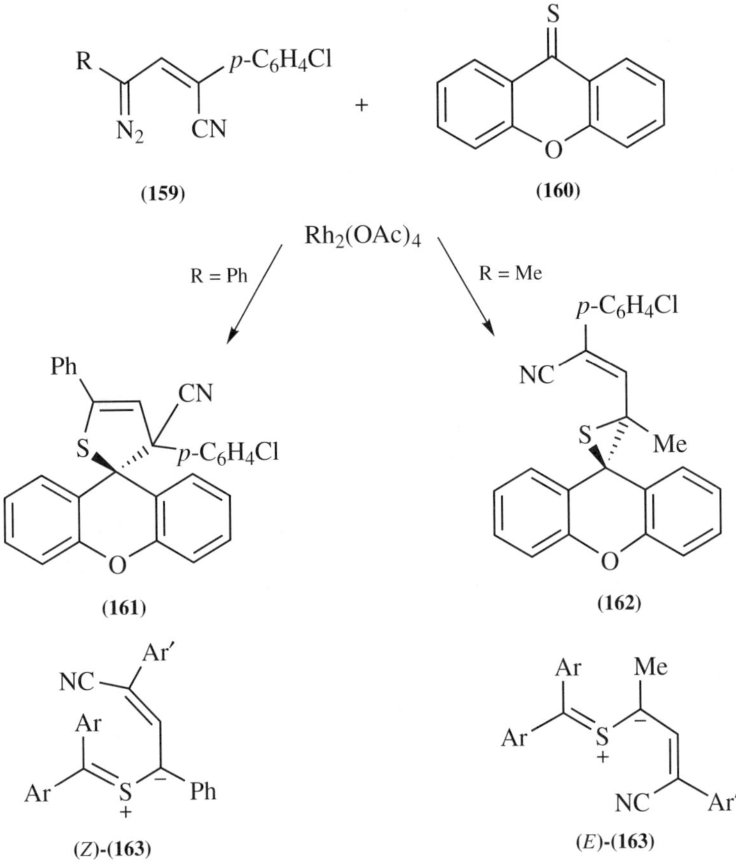

Scheme 5.48

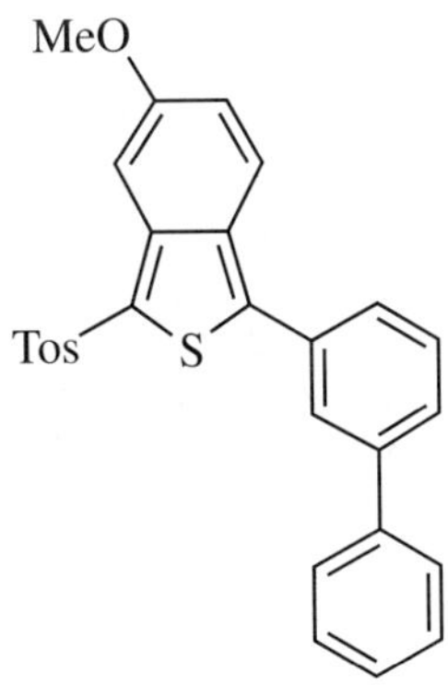

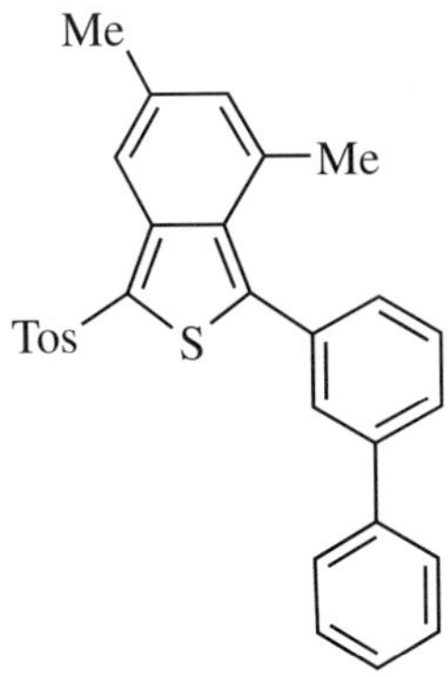

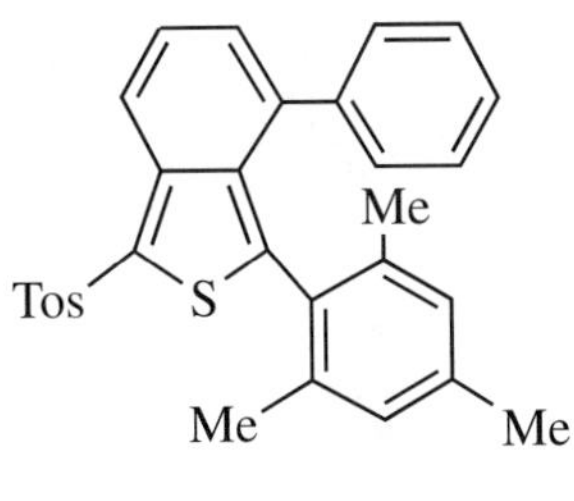

(164) **(165)** **(166)**

Tos = tosyl

Cyclization of a thiocarbonyl ylide with the C=C-bond of an aromatic ring was observed in the reaction of aryl biphenyl-2-yl ketones with di(tosyl)diazomethane in the presence of $Rh_2(OAc)_4$ (189). In the case where the aryl ring contains a 4-methoxy group, benzo[*c*]thiophene (**164**) was the only product formed. In contrast, when the aryl ring consists of a 2,4,6-trimethylphenyl group, compounds **165** and **166** were produced. It would seem that after 1,5-dipolar electrocyclization of the intermediate thiocarbonyl ylide occurs, aromatization then takes place by elimination of toluenesulfinic acid or methyl toluenesulfinate.

Another type of cyclization of thiocarbonyl ylides, which ultimately leads to 1,3-thiazole derivatives, was observed with *push–pull*-substituted dipoles bearing amino groups. For example, 1,1-diamino-3,3-dicyanothiocarbonyl ylide **9** undergoes a facile cyclization reaction when heated in water resulting in the formation of 2,4-diamino-1,3-thiazole-5-carbonitrile (19). In an analogous reaction, 1,3-thiazole-5-carbonitrile (**168**) was formed from the reaction of thiobenzamide and 3,3-bis(trifluoromethyl)oxirane-2,2-dicarbonitrile, apparently proceeding via thiocarbonyl ylide **167** (Scheme 5.49). Fused 1,3-thiazoles (**171**) were obtained from the reaction of phenyliodonium ylides of cyclic 1,3-diketones (e.g., **169**) and thioureas (72c). In both cases, ring closure occurs by the intramolecular nucleophilic addition of the amino functionality at the cyano- and carbonyl groups, respectively.

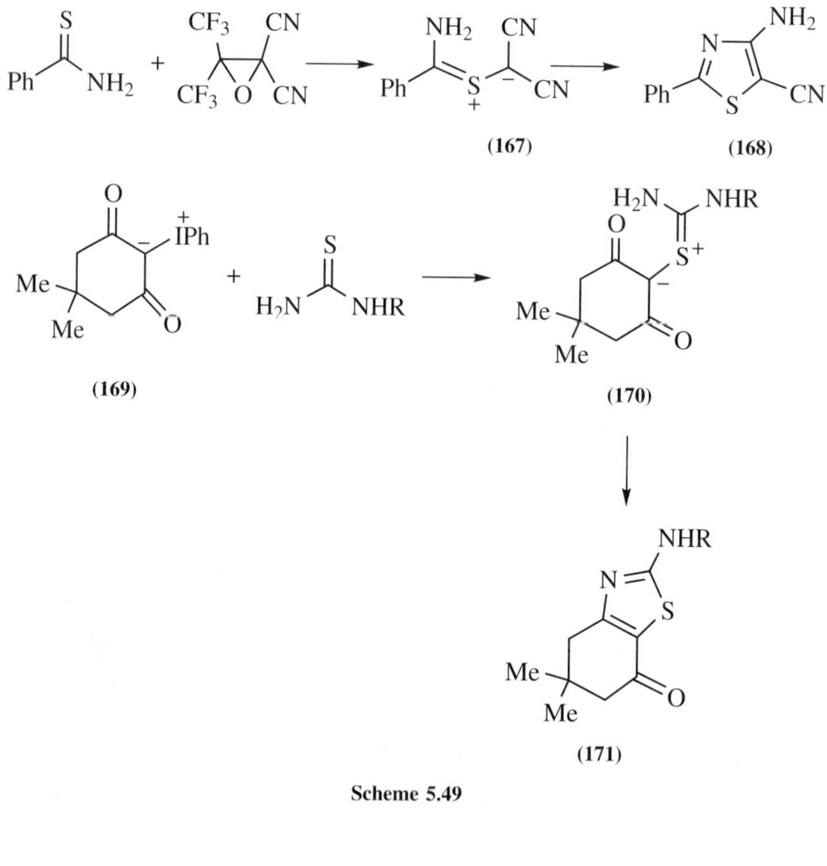

Scheme 5.49

5.3.3. Six-Membered Rings

Like many other 1,3-dipoles (e.g., nitrile ylides, imines, and oxides) (7), thiocarbonyl ylides undergo *head-to-head* dimerization to give sterically crowded 1,4-dithianes. The first reported example involves the formation of 2,2,3,3-tetraphenyl-1,4-dithiane (**18**) from thiobenzophenone (*S*)-methylide (**16**) (17,28) (cf. Scheme 5.3). Other (*S*)-methylides are known to form analogous 1,4-dithianes (e.g., thiofluorenone (*S*)-methylide yields **172**) (17). The (*S*)-methylides of 4,4-dimethyl-2-phenyl-1,3-thiazole-5(4*H*)-thione (105) and methyl dithiobenzoate (60,104) dimerize to give compounds **173** and **174**, respectively.

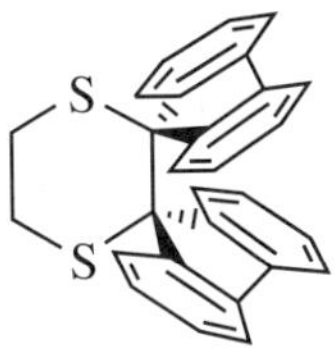

(172)

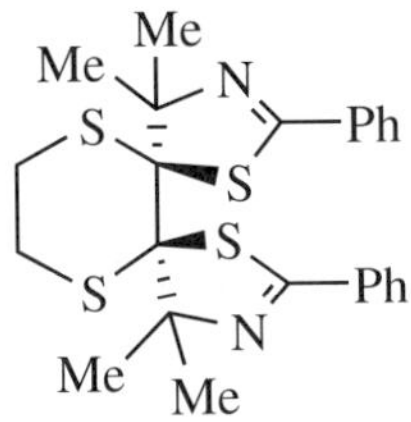

(173)

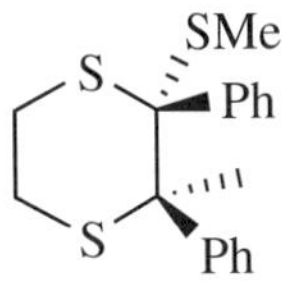

(174)

In none of these experiments was the formation of the sterically more favored *head-to-tail* dimer observed. Though the mechanism of this dimerization is not completely understood, the regiochemistry observed may be related to the biradical character of the thiocarbonyl ylide dipole.

In some reactions with thiocarbonyl ylides, 1,3-thiazine derivatives are formed by a series of consecutive reactions. For example, the interception of 3-thioxocyclobutanone (*S*)-methylide (**69**) with thiobenzamide results in the formation of the bicyclic 1,3-thiazine (**176**) (100a) (Scheme 5.50). A conceivable intermediate is the 1,3-adduct **175** as shown in Scheme 5.50.

Treatment of 4,4-dimethyl-2-phenyl-1,3-thiazole-5(4*H*)-thione with ethyl diazoacetate gives, among other products, ethyl 1,3-thiazine carboxylate (**179**) (99). The formation of **179** has been rationalized by an acid-catalyzed addition of ethyl diazoacetate to the thiocarbonyl ylide **177** to first give intermediate **178**, which undergoes a subsequent ring enlargement reaction via a Tiffeneau–Demjanov rearrangement.

5.3.4. Seven-Membered Rings

In accordance with theoretical predictions (90), the concerted pathway for 1,3-dipolar cycloaddition is replaced by a two-step mechanism when two requirements are satisfied. One of the criteria involves an extremely large difference in the highest occupied molecular orbital–lowest unoccupied molecular orbital (HOMO–LUMO) energies of the reaction partners. The other factor involves a pronounced steric hindrance at one termini of the 1,3-dipole (190). The first case of a stepwise

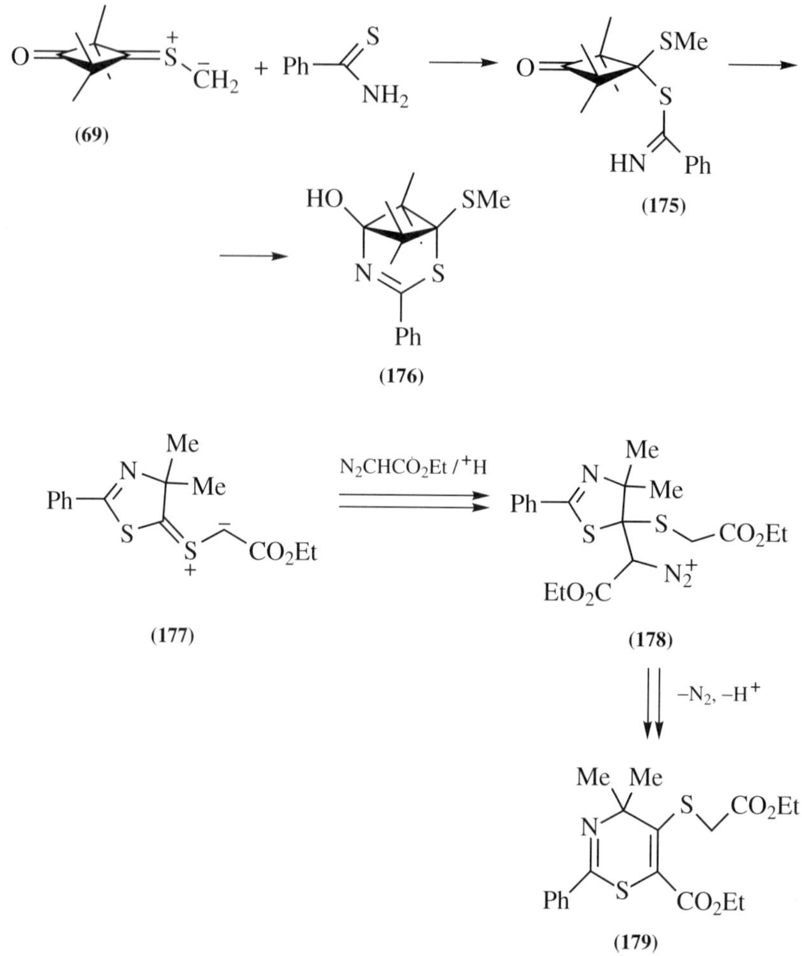

Scheme 5.50

cycloaddition reaction was reported several years ago for thiocarbonyl (*S*)-methylide (**69**) and the dimethyl fumarate–dimethyl maleate pair (96a), as well as for tetracyanoethene (TCNE) (Scheme 5.51) (96c). In the latter case, the zwitterionic intermediate **180** has two competitive reaction routes available. One of these involves the reversible formation of ketene imine **181** while the other path proceeds by an irreversible ring closure to give the thiolane **182**. This interpretation was based on the isolation of products **183** and **184**, which were obtained in reactions carried out in tetrahydrofuran (THF), which contained some methanol and water (96c,136).

Similar results were encountered in the reaction of **69** (generated thermally from the corresponding 2,5-dihydro-1,3,4-thiadiazole) with benzylidene malonodinitrile or α-cyano-substituted cinnamates (97). In these cases, seven-membered lactams or

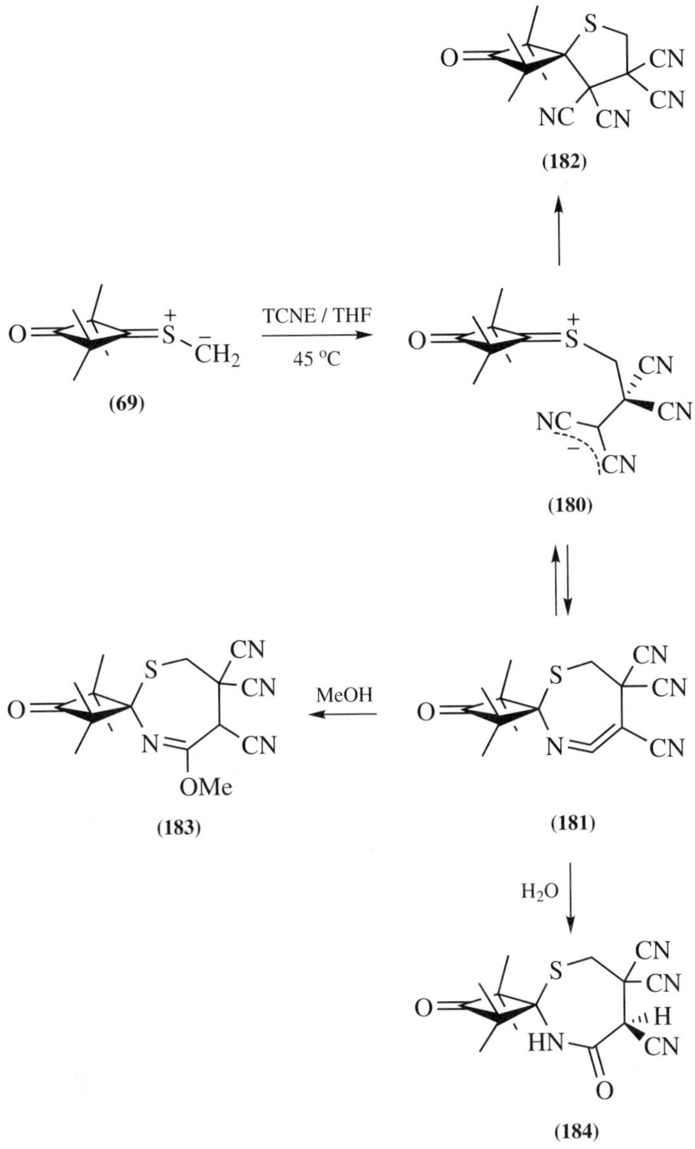

Scheme 5.51

lactim ethers were obtained in 60–70% yield and it was suggested that ketene imine **185** was a reactive intermediate.

Cycloaddition reactions were also carried out using 1,2-bis(trifluoromethyl)-fumaronitrile (137,138,191,192). Seven-membered cyclic ketene imines (**186–188**) could actually be isolated in good yields from the reactions with **69**, di(*tert*-butyl)thioketone (*S*)-methylide, and 1,1,3,3-tetramethylindane-2-thione (*S*)-methylide, respectively.

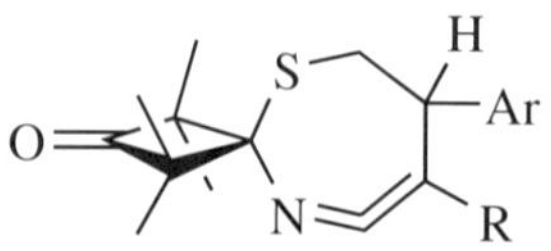

(185)

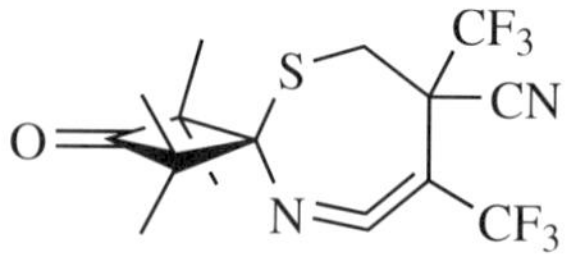

(186)

Ar = C_6H_5; 4-$NO_2C_6H_4$
R = CN; CO_2Me

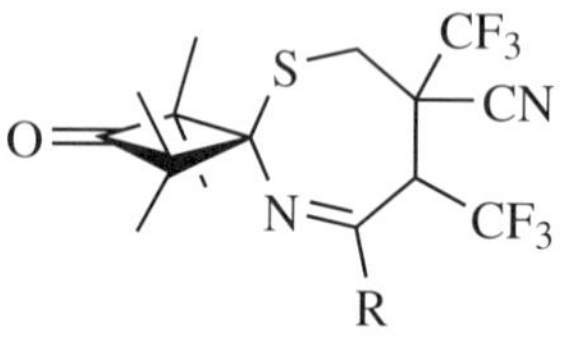

(187)

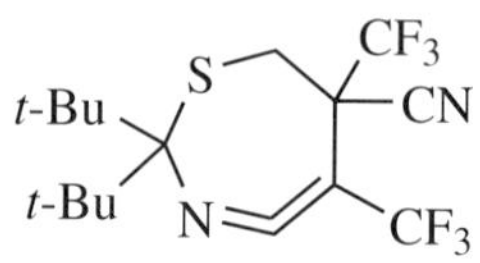

(188)

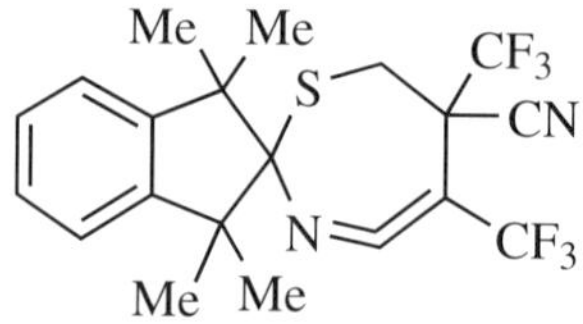

(189)

R = **a**: OMe; **b**: NH-Ph

The structure of ketene imine **188** was elucidated by means of a single-crystal X-ray diffraction analysis. Surprisingly, bond lengths of 133 and 120 pm for the cumulated system (C=C=N) hardly deviate from those found in a linear analogue (i.e., diphenylketene *p*-toluylimine) (193). Apparently, the cumulated bond system of **188** is not linear but rather is bent to an angle of 163.8°.

In accordance with expectations, the isolable trifluoromethyl substituted ketene imines (**186–188**) were found to react readily with nucleophilic agents. In the case of **186**, the reaction with methanol and aniline led to lactim ether **187a** and amidine **187b**, respectively (137).

5.4. CONCLUDING REMARKS

As a consequence of the pioneering work of both the Kellogg and Huisgen groups, thiocarbonyl ylides have become recognized as valuable synthetic intermediates for 1,3-dipolar cycloaddition chemistry. Although their history dates back to ~1920, it is only over the past two decades that these dipoles have been studied in some detail. A very important mechanistic observation was recently made by Huisgen and involves a two-step [3 + 2]-cycloaddition pathway of thiocarbonyl ylides. This reaction represents a milestone for the general understanding of 1,3-dipolar cycloaddition chemistry. Despite the progress made over the past two decades, the importance of thiocarbonyl ylides has not been fully recognized nor utilized for organic synthesis. However, the perspective for future use is very attractive. Clearly, thiocarbonyl ylides have great potential for the construction of many sulfur-containing systems in a sterecontrolled manner. Recent contributions by a host of investigators point out new possible applications of thiocarbonyl ylide cycloadditions in the field of organic synthesis.

REFERENCES

1. This is the traditional and convenient nomenclature of this class of compounds, which will be used in this chapter. According to IUPAC recommendations it should be named as (*S*)-methylenesulfonium methanide. *Glossary of Class Names of Organic Compounds and Reactive Intermediates*, IUPAC Recommendation 1995, *Pure Appl. Chem.*, **1995**, *67*, 1375.
2. J. P. Snyder, *J. Am. Chem. Soc.*, **1974**, *96*, 5005.
3. (a) A. J. Arduengo and E. M. Burgess, *J. Am. Chem. Soc.*, **1976**, *98*, 5020; (b) A. J. Arduengo and E. M. Burgess, *J. Am. Chem. Soc.*, **1976**, *98*, 5021.
4. (a) J. Fabian, U. Eckart, and A. B. Hess, Jr., *Wiss. Z. Techn. Universität Dresden*, **1995**, *44*, 99; (b) J. Fabian and A. B. Hess, Jr., *J. Org. Chem.*, **1997**, *62*, 1766.
5. (a) J. Fabian and A. Senning, *Sulfur Reports*, **1998**, *21*, 1; (b) J. Fabian and G. Mloston, *Polish J. Chem.*, **1999**, *73*, 669.
6. (a) R. Huisgen, *Angew. Chem. Int. Ed. Engl.*, **1963**, *2*, 565; (b) R. Huisgen, in *1,3-Dipolar Cycloaddition Chemistry*, A. Padwa, ed., Vol. 1, Wiley, New York, **1984**, pp. 1–176.
7. A. Padwa, ed., *1,3-Dipolar Cycloaddition Chemistry*, Wiley, New York, **1984**.
8. R. M. Kellogg, *Tetrahedron*, **1976**, *32*, 2165.
9. R. Huisgen, C. Fulka, I. Kalwinsch, X. Li, G. Mloston, J. Rodriquez Moran, and A. Pröbstl, *Bull. Soc. Chim. Belg.*, **1984**, *93*, 511.
10. P. K. Claus, in *Methoden der organischen Chemie (Houben-Weyl)*, D. Klamann, ed., Band E11/2, Thieme Verlag, Stuttgart, **1985**, pp. 1344–1359.
11. G. Mloston and H. Heimgartner, *Polish J. Chem.*, **2000**, *74*, 1503.
12. H. Staudinger and A. Gaule, *Ber. Dtsch. Chem. Ges.*, **1916**, *49*, 1897 and 1951.
13. H. Staudinger and J. Siegwart, *Helv. Chim. Acta*, **1920**, *3*, 833.
14. (a) A. Schönberg and L. Vargha, *Liebigs Ann. Chem.*, **1930**, *483*, 176; (b) A. Schönberg and L. Vargha, *Ber. Dtsch. Chem. Ges.*, **1931**, *64*, 1390; (c) A. Schönberg, D. Cernik, and W. Urban, *Ber. Dtsch. Chem. Ges.*, **1931**, *64*, 2577; (d) A. Schönberg, S. Nickel, and D. Cernik, *Ber. Dtsch. Chem. Ges.*, **1932**, *65*, 289.
15. A. Schönberg, B. König, and E. Singer, *Chem. Ber.*, **1967**, *100*, 767.
16. A. Schönberg, E. Frese, W. Knöfel, and K. Praefcke, *Chem. Ber.*, **1970**, *103*, 938 and 949.
17. R. Huisgen, I. Kalwinsch, X. Li, and G. Mloston, *Eur. J. Org. Chem.*, **2000**, 1685.
18. (a) E. B. Knott, *J. Chem. Soc.*, **1955**, 916; (b) E. B. Knott and R. A. Jeffreys, *J. Chem. Soc.*, **1955**, 927.
19. W. J. Middleton, *J. Org. Chem.*, **1966**, *31*, 3731.
20. W. J. Linn and E. Ciganek, *J. Org. Chem.*, **1969**, *34*, 2146.
21. S. Tamagaki, S. Oae, *Tetrahedron Lett.*, **1972**, 1159.
22. P. Gronski and K. Hartke, *Tetrahedron Lett.*, **1976**, 4139.
23. H. Gotthardt and B. Christl, *Tetrahedron Lett.*, **1968**, 4743, 4747, and 4751.
24. (a) K. T. Potts and U. P. Singh, *J. Chem. Soc., Chem. Commun.*, **1969**, 569; (b) K. T. Potts and P. Murphy, *J. Chem. Soc., Chem. Commun.*, **1984**, 1348; (c) K. T. Potts and M. O. Dery, *J. Chem. Soc., Chem. Commun.*, **1986**, 561.
25. J. Buter, S. Wassenaar, and R. M. Kellogg, *J. Org. Chem.*, **1972**, *37*, 4045.
26. R. B. Woodward and R. Hoffmann, *Angew. Chem. Int. Ed. Engl.*, **1969**, *8*, 781.

27. R. Huisgen, in *The Adventure Playground of Mechanisms and Novel Reactions*, J. I. Seeman, ed., American Chemical Society, Washington, D.C., **1994**, pp. 117–123.
28. I. Kalwinsch, X. Li, J. Gottstein, and R. Huisgen, *J. Am. Chem. Soc.*, **1981**, *103*, 7032.
29. (a) E. Vedejs and F. G. West, *Chem. Rev.*, **1986**, *86*, 941; (b) E. Vedejs, in: *Advances in Cycloaddition*, D. P. Curran, ed., Vol. 1, Jai Press Inc., Greenwich, **1988**, pp. 33–51.
30. Y. Terao, M. Aono, and K. Achiwa, *Heterocycles*, **1988**, *27*, 981.
31. A. Hosomi, Y. Matsuyama, and H. Sakurai, *J. Chem. Soc., Chem. Commun.*, **1986**, 1073.
32. S. Karlsson and H.-E. Högberg, *Org. Lett.*, **1999**, *1*, 1667.
33. K. Rühlmann, *J. Prakt. Chem.*, **1959**, *8*, 285.
34. (a) R. M. Kellogg and S. Wassenaar, *Tetrahedron Lett.*, **1970**, 1987; (b) R. M. Kellogg, S. Wassenaar, and J. Buter, *Tetrahedron Lett.*, **1970**, 4689.
35. R. Huisgen and X. Li, *Heterocycles*, **1983**, *20*, 2363.
36. R. Huisgen, G. Mloston, and C. Fulka, *Heterocycles*, **1985**, *23*, 2207.
37. C. E. Diebert, *J. Org. Chem.*, **1970**, *35*, 1501.
38. R. Huisgen and G. Mloston, *Polish J. Chem.*, **1999**, *73*, 635.
39. G. Mloston and R. Huisgen, *Tetrahedron Lett.*, **1989**, *30*, 7045.
40. G. Mloston and R. Huisgen, *Tetrahedron*, **2001**, *57*, 145.
41. R. Huisgen, G. Mloston, and A. Pröbstl, *Heteroatom Chem.*, **2001**, *3*, 136.
42. G. Mloston, M. Petit, A. Linden, and H. Heimgartner, *Helv. Chim. Acta*, **1994**, *77*, 435.
43. K.-T. Kang, J.-S. U, I.-N. Yoon, and C. H. Park, *J. Korean Chem. Soc.*, **1991**, *35*, 292.
44. C.Th. Pedersen, *Acta Chem. Scand.*, **1968**, *22*, 247.
45. D. H. R. Barton and B. J. Willis, *J. Chem. Soc., Perkin Trans. 1*, **1972**, 305.
46. (a) T. B. Cameron and H. W. Pinnick, *J. Am. Chem. Soc.*, **1979**, *101*, 4755; (b) T. B. Cameron and H. W. Pinnick, *J. Am. Chem. Soc.*, **1980**, *102*, 744.
47. D. H. R. Barton, F. S. Guziec, and I. Shahak, *J. Chem. Soc., Perkin Trans. 1*, **1974**, 1794.
48. F. S. Guziec and L. J. Sanfilippo, *Tetrahedron*, **1988**, *44*, 6241.
49. M. Oba, M. Yoshihara, C. Roppongi, and K. Nishiyama, *Heterocycles*, **1999**, *50*, 195.
50. Y. Tominaga, H. Ueda, K. Ogata, S. Kohra, M. Hojo, M. Ohkuma, K. Tomita, and A. Hosomi, *Tetrahedron Lett.*, **1992**, *33*, 85.
51. Y. Tominaga, S. Takada, and S. Kohra, *Tetrahedron Lett.*, **1994**, *35*, 3555.
52. (a) Y. Terao, M. Aono, and K. Achiwa, *Heterocycles*, **1986**, *24*, 1571; (b) Y. Terao, M. Tanaka, N. Imai, and K. Achiwa, *Tetrahedron Lett.*, **1985**, *26*, 3011; (c) N. Imai, H. Tokiwa, M. Aono, Y. Terao, Y. Akahori, and K. Achiwa, *Heterocycles*, **1986**, *24*, 2423; (d) Y. Terao, M. Aono, N. Imai, and K. Achiwa, *Chem. Pharm. Bull.*, **1987**, *35*, 1734.
53. (a) M. Aono, C. Hyodo, Y. Terao, and K. Achiwa, *Tetrahedron Lett.*, **1986**, *27*, 4039; (b) Y. Terao, M. Aono, I. Takahashi, and K. Achiwa, *Chem. Lett.*, **1986**, 2089.
54. M. Aono, Y. Terao, and K. Achiwa, *Chem. Lett.*, **1987**, 1851.
55. (a) K. Oka, H. Dobashi, and S. Hara, *Tetrahedron Lett.*, **1980**, *21*, 3579; (b) K. Oka, A. Dobashi, and S. Hara, *J. Am. Chem. Soc.*, **1981**, *103*, 2757.
56. E. J. Smutny, *J. Am. Chem. Soc.*, **1969**, *91*, 208.
57. (a) M. Takaku, S. Mitamura, and H. Nozaki, *Tetrahedron Lett.*, **1969**, 3651; (b) S. Mitamura, M. Takaku, and H. Nozaki, *Bull. Chem. Soc. Jpn.*, **1974**, *47*, 3152.
58. P. Gronski, K. Hartke, H. Burzlaff, R. Böhme, and A. Shaukat, *Chem. Ber.*, **1977**, *110*, 3689.
59. E. M. Burgess and M. C. Pulcrano, *J. Am. Chem. Soc.*, **1978**, *100*, 6538.
60. K. Hartke, N. Bien, W. Massa, and S. Wocadlo, *Liebigs Ann. Chem.*, **1996**, 1159.
61. (a) M. H. Osterhout, W. R. Nadler, and A. Padwa, *Synthesis*, **1994**, 123; (b) A. Padwa, S. R. Harring, D. L. Hertzog, and W. R. Nadler, *Synthesis*, **1994**, 993.

62. K. T. Potts, M. O. Dery, and W. A. Juzukonis, *J. Org. Chem.*, **1989**, *54*, 1077.

63. W. Kirmse, in *Carbene Chemistry*, A. T. Blomquist, H. H. Wasserman, eds., Academic Press, New York, **1971**.

64. M. P. Doyle, M. A. McKervey, and T. Ye, *Modern Catalytic Methods for Organic Synthesis with Diazo Compounds—from Cyclopropanes to Ylides*, Wiley, New York, **1998**.

65. (a) A. Padwa and S. F. Hornbuckle, *Chem. Rev.*, **1991**, *91*, 263; (b) A. Padwa and M. D. Weingarten, *Chem. Rev.*, **1996**, *96*, 223.

66. G. Mloston, J. Romanski, C. Schmidt, H. P. Reisenauer, and G. Maier, *Chem. Ber.*, **1994**, *127*, 2527.

67. W. G. McGimpsey and J. C. Scaiano, *Tetrahedron Lett.*, **1986**, *27*, 547.

68. G. Mloston and H. Heimgartner, *Helv. Chim. Acta*, **1996**, *79*, 1785.

69. J. A. Kaufman and S. J. Weininger, *J. Chem. Soc., Chem. Commun.*, **1969**, 593.

70. (a) K.-P. Zeller and H. Straub, in *Methoden der Organischen Chemie (Houben-Weyl)*, E. Müller, ed., Band XIII/2b, Thieme Verlag, Stuttgart, **1974**, pp. 351–384; (b) D. Seyferth, *Acc. Chem. Res.*, **1972**, *5*, 65; (c) D. Seyferth, in *Carbenes*, R. A. Moss and M. Jones, Jr., eds., Vol. 2, Wiley, New York, **1975**, pp. 101–158.

71. (a) E.V. Dehmlov, in *Methoden der Organischen Chemie (Houben-Weyl)*, M. Regitz, ed., Band E19b, Thieme Verlag, Stuttgart, **1989**, pp. 1521–1589; (b) M. Makosza and M. Fedorynski, *Polish J. Chem.*, **1996**, *70*, 1093.

72. (a) A. Varvoglis, *The Organic Chemistry of Polycoordinated Iodine*, VCH, New York, **1992**; (b) A. Varvoglis, *Hypervalent Iodine in Organic Synthesis*, Academic Press, London, **1997**; (c) H. Kamproudi, S. Spyroudis, and P. Tarantili, *J. Heterocyclic Chem.*, **1996**, *33*, 575.

73. E. Block and E. J. Corey, *J. Org. Chem.*, **1969**, *34*, 896.

74. J. Fabian and T. Wolff, *J. Photochem. Photobiol. A: Chemistry*, **1996**, *96*, 1.

75. M. Kamata and T. Miyashi, *J. Chem. Soc., Chem. Commun.*, **1989**, 557.

76. (a) R. S. Glass and W. Jung, *J. Am. Chem. Soc.*, **1994**, *116*, 1137; (b) R. S. Glass, *Main Group Chem. News*, **1994**, *2*, 4.

77. M. Kägi, A. Linden, G. Mloston, and H. Heimgartner, *Helv. Chim. Acta*, **1996**, *79*, 855.

78. G. Mloston, J. Romanski, M. Kägi, and H. Heimgartner, *Polish J. Appl. Chem.*, **1997**, *41*, 361.

79. M. Kägi, A. Linden, G. Mloston, and H. Heimgartner, *Helv. Chim. Acta*, **1998**, *81*, 285.

80. B. Kelmendi, G. Mloston, and H. Heimgartner, *Heterocycles*, **2000**, *52*, 475.

81. H. Nakano and T. Ibata, *Bull. Chem. Soc. Jpn.*, **1993**, *66*, 238.

82. H. Nakano and T. Ibata, *Bull. Chem. Soc. Jpn.*, **1995**, *68*, 1393.

83. M. Hamaguchi, N. Funakoshi, and T. Oshima, *Tetrahedron Lett.*, **1999**, *40*, 8117.

84. P. J. Stang and S. B. Christensen, *J. Org. Chem.*, **1981**, *46*, 823.

85. R. Huisgen and G. Mloston, *Heterocycles*, **1990**, *30*, 737.

86. G. Mloston and R. Huisgen, *Tetrahedron Lett.*, **1985**, *26*, 1053.

87. R. Huisgen, G. Mloston, and A. Pröbstl, *Tetrahedron Lett.*, **1985**, *26*, 4431.

88. J. M. Beiner, D. Lecadet, D. Paquer, and A. Thuillier, *Bull. Soc. Chim. France*, **1973**, 1983.

89. R. Huisgen, E. Langhals, G. Mloston, T. Oshima, and J. Rapp, *Lectures Heterocycl. Chem.*, **1987**, *9*, 1.

90. R. Sustmann, *Pure Appl. Chem.*, **1974**, *40*, 569.

91. X. Li and R. Huisgen, *Tetrahedron Lett.*, **1983**, *24*, 4181.

92. R. Huisgen and E. Langhals, *Tetrahedron Lett.*, **1989**, *30*, 5369.

93. R. Huisgen, E. Langhals, and X. Li, *Helv. Chim. Acta*, **2001**, *84*, 981.

94. G. Mloston, R. Huisgen, H. Huber, and D. S. Stephenson, *J. Heterocyclic Chem.*, **1999**, *36*, 959.

95. S. Karlsson and H.-E. Högberg, *Synthesis*, **2000**, 1863.

96. (a) R. Huisgen, G. Mloston, and E. Langhals, *J. Am. Chem. Soc.*, **1986**, *108*, 6401; (b) G. Mloston, E. Langhals, and R. Huisgen, *Tetrahedron Lett.*, **1989**, *30*, 5373; R. Huisgen, G. Mloston, and E. Langhals, *J. Org. Chem.*, **1986**, *51*, 4085.
97. R. Huisgen, J. Panelle, G. Mloston, A. Buyle Padias, and H. K. Hall, Jr., *J. Am. Chem. Soc.*, **1992**, *114*, 266.
98. R. Huisgen and X. Li, *Tetrahedron Lett.*, **1983**, *24*, 4185.
99. M. Kägi, G. Mloston, A. Linden, and H. Heimgartner, *Helv. Chim. Acta*, **1994**, *77*, 1299.
100. (a) G. Mloston, T. Gendek, A. Linden, and H. Heimgartner, *Helv. Chim. Acta*, **1999**, *82*, 290; (b) G. Mloston, T. Gendek, A. Linden, and H. Heimgartner, *Polish J. Chem.*, **1999**, *73*, 1219.
101. G. Mloston, T. Gendek, A. Linden, and H. Heimgartner, *Helv. Chim. Acta*, **1998**, *81*, 66.
102. G. Mloston, T. Gendek, and H. Heimgartner, *Polish J. Chem.*, **1998**, *72*, 66.
103. G. Mloston, A. Linden, and H. Heimgartner, *Helv. Chim. Acta*, **1996**, *79*, 31.
104. G. Mloston, J. Romanski, E. B. Rusanov, A. N. Tshernega, and Y. G. Shermolovich, *Russ. J. Org. Chem.*, **1995**, *31*, 952.
105. M. Kägi, A. Linden, H. Heimgartner, and G. Mloston, *Helv. Chim. Acta*, **1993**, *76*, 1715.
106. M. Atzmüller and F. Vögtle, *Chem. Ber.*, **1979**, *112*, 138.
107. G. L'abbé, J.-P. Dekerk, C. Martens, and S. Toppet, *J. Org. Chem.*, **1980**, *45*, 4366.
108. G. L'abbé, *Lectures Heterocycl. Chem.*, **1987**, *9*, 51.
109. G. L'abbé, A. Francis, W. Dehaen, and J. Bosman, *Bull. Soc. Chim. Belg.*, **1996**, *105*, 253.
110. T. Kawashima, S. Watanabe, and R. Okazaki, *Chem. Lett.*, **1992**, 1603.
111. G. Mloston, T. Gendek, and H. Heimgartner, *Helv. Chim. Acta*, **1996**, *79*, 1537.
112. G. Mloston, J. Romanski, J. Swiatek, and H. Heimgartner, *Helv. Chim. Acta*, **1999**, *82*, 946.
113. G. Mloston, J. Romanski, and H. Heimgartner, *Heterocycles*, **1999**, *50*, 403.
114. (a) D. Seyferth and G. J. Murphy, *J. Organometal. Chem.*, **1973**, *49*, 117; (b) M. Dawid, G. Mloston, and J. Warkentin, *Org. Lett.*, **2001**, *3*, 2455.
115. H. Fischer, K. Treier, and U. Gerbing, *J. Organometal. Chem.*, **1992**, *433*, 127.
116. M. Roth, P. Dubs, E. Götschi, and A. Eschenmoser, *Helv. Chim. Acta*, **1971**, *54*, 710.
117. A. Padwa, F. R. Kinder, W. R. Nadler, and L. Zhi, *Heterocycles*, **1993**, *35*, 367.
118. (a) G. Kim, M. Y. Chu-Moyer, and S. J. Danishefsky, *J. Am. Chem. Soc.*, **1990**, *112*, 2003; (b) G. Kim, M. Y. Chu-Moyer, S. J. Danishefsky, and G. K. Schulte, *J. Am. Chem. Soc.*, **1993**, *115*, 30.
119. (a) S. Takano, S. Tomita, M. Takahashi, and K. Ogasawara, *Synthesis*, **1987**, 1116; (b) T. Honda, H. Ishige, J. Araki, S. Akimoto, K. Hirayama, and M. Tsubuki, *Tetrahedron*, **1992**, *48*, 79.
120. M. D. Bachi, O. Goldberg, and A. Gross, *Tetrahedron Lett.*, **1978**, 4167.
121. B. Fiedler, D. Weiss, and R. Beckert, *Liebigs Ann. Rec.*, **1997**, 613.
122. D. J. Humphreys, C. E. Newall, G. H. Phillipps, and G. A. Smith, *J. Chem. Soc., Perkin Trans. 1*, **1978**, 45.
123. M. Hürzeler, B. Bernet, T. Mäder, and A. Vasella, *Helv. Chim. Acta*, **1993**, *76*, 1779; see also B. Bernet, T. Mäder, and A. Vasella, *Helv. Chim. Acta*, **1997**, *80*, 1260.
124. M. Hamaguchi, N. Funakoshi, and T. Oshima, *Tetrahedron Lett.*, **1999**, *40*, 8117.
125. A. M. Schoevaars, W. Kruizinga, R. W. J. Zijlstra, N. Veldman, A. L. Spek, and B. L. Feringa, *J. Org. Chem.*, **1997**, *62*, 4943.
126. W. Freund and S. Hünig, *J. Org. Chem.*, **1987**, *52*, 2154.
127. M. Petit, A. Linden, G. Mloston, and H. Heimgartner, *Helv. Chim. Acta*, **1994**, *77*, 1076.
128. R. Huisgen, E. Langhals, and H. Nöth, *Tetrahedron Lett.*, **1986**, *27*, 5475.
129. M. Aono, Y. Terao, and K. Achiwa, *Heterocycles*, **1995**, *40*, 249.
130. R. Huisgen and G. Mloston, *Tetrahedron Lett.*, **1985**, *26*, 1049.
131. K. C. Hong, Ph.D. Thesis, University of Michigan, **1994**.

132. H. Ishida and M. Ohno, *Tetrahedron Lett.*, **1999**, *40*, 1543.

133. V. Alcázar, I. Tapia, and J. R. Morán, *Tetrahedron*, **1990**, *46*, 1057.

134. T. Yamano, M. Tanaka, and K. Takanohashi, *Heterocycles*, **1993**, *36*, 2397.

135. V. Alcazar Montero, I. Tapia Hernandez, J. de Pascual Teresa, J. R. Moran, and R. Olabarrieta, *J. Org. Chem.*, **1989**, *54*, 3664.

136. R. Huisgen, G. Mloston, and E. Langhals, *Helv. Chim. Acta*, **2001**, *84*, 1805.

137. R. Huisgen, E. Langhals, G. Mloston, and T. Oshima, *Heterocycles*, **1989**, *29*, 2069.

138. R. Huisgen, E. Langhals, and H. Nöth, *J. Org. Chem.*, **1990**, *55*, 1412.

139. A. G. Schultz, W. Y. Fu, R. D. Lucci, B. G. Kurr, K. Ming Lo, and M. Boxer, *J. Am. Chem. Soc.*, **1978**, *100*, 2140.

140. W. G. Herkstroeter and A. G. Schultz, *J. Am. Chem. Soc.*, **1984**, *106*, 5553.

141. (a) A. G. Schultz, *Acc. Chem. Res.*, **1983**, *16*, 210; (b) A. G. Schultz and L. Motyka, in *Organic Photochemistry*, A. Padwa, ed., Vol. 6, Marcel Dekker, New York, **1983**, pp. 1–119.

142. (a) A. G. Schultz and M. B. DeTar, *J. Am. Chem. Soc.*, **1974**, *96*, 296; (b) A. G. Schultz and M. B. DeTar, *J. Am. Chem. Soc.*, **1976**, *98*, 3564.

143. T. Wolff, *J. Am. Chem. Soc.*, **1978**, *100*, 6157.

144. A. Padwa, F. R. Kindler, and L. Zhi, *Synlett*, **1991**, 287.

145. (a) K. T. Potts, E. Houghton, and U. P. Singh, *J. Org. Chem.*, **1974**, *39*, 3627; (b) K. T. Potts, J. Baum, and E. Houghton, *J. Org. Chem.*, **1974**, *39*, 3631.

146. (a) K. T. Potts, J. Baum, S. K. Datta, and E. Houghton, *J. Org. Chem.*, **1976**, *41*, 813; (b) K. T. Potts, J. Baum, and E. Houghton, *J. Org. Chem.*, **1976**, *41*, 818.

147. P. Cavalleri, F. Clerici, E. Erba, and P. Trimarco, *Chem. Ber.*, **1992**, *125*, 883.

148. D. L. Hertzog, W. R. Nadler, Z. J. Zhang, and A. Padwa, *Tetrahedron Lett.*, **1992**, *33*, 5877.

149. K. T. Potts, K. G. Bordeaux, W. R. Kuehnling, and R. L. Salesbury, *J. Org. Chem.*, **1985**, *50*, 1666.

150. K. T. Potts, K. G. Bordeaux, W. R. Kuehnling, and R. L. Salesbury, *J. Org. Chem.*, **1985**, *50*, 1677.

151. H. Gotthardt and O. M. Huss, *Liebigs Ann. Chem.*, **1981**, 347.

152. M. G. Pettett and A. B. Holmes, *J. Chem. Soc., Perkin Trans. 1*, **1985**, 1161.

153. H. Gotthardt and B. Christl, *Chem. Ber.*, **1978**, *111*, 3029.

154. (a) H. Gotthardt, M. C. Weisshuhn, and B. Christl, *Chem. Ber.*, **1976**, *109*, 740; (b) H. Gotthardt, O. M. Huss, and C. M. Weisshuhn, *Chem. Ber.*, **1979**, *112*, 1650; (c) H. Gotthardt, M. C. Weisshuhn, and B. Christl, *Chem. Ber.*, **1976**, *109*, 753.

155. H. Gotthardt and C. M. Weisshuhn, *Chem. Ber.*, **1978**, *111*, 2021 and 2028.

156. (a) H. Gotthardt, M. C. Weisshuhn, and B. Christl, *Liebigs Ann. Chem.*, **1979**, 360.

157. K. T. Potts, D. R. Choudhury, A. J. Elliott, and U. P. Singh, *J. Org. Chem.*, **1976**, *41*, 1724.

158. K. T. Potts, in *1,3-Dipolar Cycloaddition Chemistry*, A. Padwa, ed., Vol. 2, Wiley, New York, **1984**, pp. 60–72.

159. T. Gendek, G. Mloston, A. Linden, and H. Heimgartner, *Helv. Chim. Acta*, in press.

160. X.-S. Ye and H. N. C. Wong, *J. Org. Chem.*, **1997**, *62*, 1940.

161. R. Huisgen, X. Li, G. Mloston, and C. Fulka, *Eur. J. Org. Chem.*, **2000**, 1695.

162. G. Mloston and R. Huisgen, *Heterocycles*, **1985**, *23*, 2201.

163. G. Mloston, R. Huisgen, and K. Polborn, *Tetrahedron*, **1999**, *55*, 11475.

164. G. Mloston, A. Linden, and H. Heimgartner, *Helv. Chim. Acta*, **1991**, *74*, 1386.

165. G. Mloston, unpublished results.

166. G. Mloston, A. Linden, and H. Heimgartner, *Helv. Chim. Acta*, **1996**, *79*, 31.

167. G. Mloston and H. Heimgartner, *Polish J. Chem.*, **1995**, *69*, 1649.

168. A. Hosomi, S. Hayashi, K. Hoashi, S. Kohra, and Y. Tominaga, *J. Chem. Soc., Chem. Commun.*, **1987**, 1442.

169. (a) A. Padwa, Y.-Y. Chen, W. Dent, and H. Nimmesgern, *J. Org. Chem.*, **1985**, *50*, 4006; (b) R. Beugelmans, J. Chastanet, and G. Roussi, *Heterocycles*, **1987**, *26*, 3197; (c) G. Mloston and Z. Skrzypek, *Bull. Soc. Chim. Belg.*, **1990**, *99*, 167; (d) D. Planchenault, R. Wisedale, T. Gallagher, and N. J. Hales, *J. Org. Chem.*, **1997**, *62*, 3438; (e) G. Mloston, A. Linden, and H. Heimgartner, *Helv. Chim. Acta*, **1998**, *81*, 558; (f) A. Gebert, A. Linden, and H. Heimgartner, *Heterocycles*, **2001**, *54*, 691.

170. A. Hosomi, S. Hayashi, K. Hoashi, S. Kohra, and Y. Tominaga, *J. Org. Chem.*, **1987**, *52*, 4423.

171. R. Huisgen and C. Fulka, unpublished results.

172. R. Huisgen, X. Li, G. Mloston, R. Knorr, H. Huber, and D. S. Stephenson, *Tetrahedron*, **1999**, *55*, 12783.

173. (a) K. T. Potts and S. Husain, *J. Org. Chem.*, **1972**, *37*, 2049; (b) K. T. Potts, J. Baum, E. Houghton, D. N. Roy, and U. P. Singh, *J. Org. Chem.*, **1974**, *39*, 3619.

174. M. Hamaguchi and T. Nagai, *J. Chem. Soc., Chem. Commun.*, **1992**, 726.

175. T. Sheradsky and N. Itzhak, *J. Chem. Soc., Perkin Trans. 1*, **1992**, 2389.

176. R. Huisgen, G. Mloston, and K. Polborn, *Heteroatom. Chem.*, **1999**, *10*, 662.

177. G. Mloston, *Bull. Soc. Chim. Belg.*, **1990**, *99*, 265.

178. (a) E. C. Taylor and I. J. Turchi, *Chem. Rev.*, **1979**, *79*, 181; (b) R. Huisgen, *Angew. Chem. Int. Ed. Engl.*, **1980**, *19*, 947.

179. S. Mataka, S. Ishi-i, and M. Tashiro, *J. Org. Chem.*, **1978**, *43*, 3730.

180. C. Bak and K. Praefcke, *Chem. Ber.*, **1979**, *112*, 2744; C. Bak and H. Praefcke, *Z. Naturforsch. B*, **1980**, *35*, 372.

181. A. Senning, *Bull. Soc. Chim. Belg.*, **1977**, *86*, 675.

182. G. F. Koser and S.-M. Yu, *J. Org. Chem.*, **1976**, *41*, 125.

183. G. Höhne, F. Marschner, K. Praefcke, and P. Weyerstahl, *Chem. Ber.*, **1975**, *108*, 673.

184. Y. Ueno and M. Okawara, *Bull. Chem. Soc. Jpn.*, **1972**, *45*, 1797.

185. H. Ruben, D. Kaplan, A. Zalkin, and D. H. Templeton, *Acta Crystallogr., Sect. B*, **1974**, *30*, 547.

186. T. Ibata and H. Nakano, *Bull. Chem. Soc. Jpn.*, **1992**, *65*, 3088.

187. N. Tokitoh, T. Suzuki, A. Itami, M. Goto, and W. Ando, *Tetrahedron Lett.*, **1989**, *30*, 1249.

188. U. Jacobsson, T. Kempe, and T. Norin, *J. Org. Chem.*, **1974**, *39*, 2722; cf. also U. Jacobsson Kempe, T. Kempe, and T. Norin, *J. Chem. Soc., Perkin Trans. 1*, **1978**, 1547.

189. T. Saito, H. Ayukawa, N. Sumizawa, T. Shizuta, S. Motoki, and K. Kobayashi, *J. Chem. Soc., Perkin Trans. 1*, **1991**, 1405.

190. S. Vivanco, R. Lecca, A. Arrieta, P. Prieto, I. Morao, A. Linden, and F. P. Cossio, *J. Am. Chem. Soc.*, **2000**, *122*, 6078.

191. J. Greenberg and J. F. Liebman, *Strained Organic Molecules*, Academic Press, New York, **1978**, pp. 333–336.

192. R. Huisgen and G. Mloston, *Tetrahedron Lett.*, **1989**, *30*, 7041.

193. R. R. Naqvi and P. J. Wheatley, *J. Chem. Soc. A*, **1970**, 2053.

CHAPTER 6

Nitrile Oxides

Volker Jäger

Institut für Organische Chemie, Universität Stuttgart, Stuttgart, Germany

Pedro A. Colinas

Laboratorio de Estudio de Compuestos Orgánicos, Universidad Nacional de La Plata, La Plata, Argentina

6.1. Methods for the Generation of Nitrile Oxides 363
6.1.1. Dehydrohalogenation of Hydroximoyl Chlorides 364
6.1.1.1. Halogenation of Oximes 365
6.1.1.2. Dehydrohalogenation for *In Situ* Generation of Nitrile Oxides 367
6.1.1.3. One-Pot Operations for Nitrile Oxide Formation From Oximes and Subsequent Cycloaddition 367
6.1.2. Dehydration of Nitroalkenes 368
6.1.3. Oxidation of Aldoximes 370
6.1.4. Thermolysis of Furoxans 371
6.2. Cycloadditions of Nitrile Oxides 374
6.2.1. Mechanism and Relative Reactivity of Dipoles and Dipolarophiles 374
6.2.1.1. Mechanistic Studies and Calculations 374
6.2.1.2. Relative Reactivity of Dipolarophiles and Dipoles 376
6.2.2. Regioselectivity of Nitrile Oxide Cycloadditions to Olefins 380
6.2.3. Diastereoselectivity of Nitrile Oxide Cycloadditions 385
6.2.3.1. Nitrile Oxide Cycloadditions with Chiral Dipolarophiles 386
6.2.3.2. Cycloaddition of Nitrile Oxides with Achiral Olefins Bearing Chiral Auxiliaries 393
6.2.3.3. Face Selectivity of Nitrile Oxide Cycloadditions with Chiral Cycloalkenes and the Like 399
6.2.3.4. Cycloadditions of Chiral Nitrile Oxides to Olefins 400
6.3. Regio- and Stereoselectivity in Intramolecular Nitrile Oxide Cycloadditions 407
6.3.1. Regioselectivity in the Intramolecular Nitrile Oxide Cycloadditions 407
6.3.2. Stereoselectivity of Intramolecular Nitrile Oxide Cycloadditions 410
6.4. Uses of Intermolecular Nitrile Oxide Cycloadditions 416
6.4.1. Uses of Isoxazolines in Syntheses of Natural Products and Related Structures . . . 416
6.4.1.1. Natural Products Containing an Isoxazoline Group 416

The Chemistry of Heterocyclic Compounds, Volume 59: Synthetic Applications of 1,3-Dipolar Cycloaddition Chemistry Toward Heterocycles and Natural Products, Edited by Albert Padwa and William H. Pearson.
ISBN 0-471-38726-6

6.4.2. Products from "Aldol" Ring Cleavage of Isoxazoline Intermediates 417
6.4.2.1. Use of Isoxazoline Intermediates from Achiral Nitrile Oxides—Achiral Olefins 418
6.4.2.2. Use of Intermediates from Cycloadditions of Achiral Nitrile Oxides to Chiral Olefins 419
6.4.2.3. Use of Intermediates from Cycloadditions of Chiral Nitrile Oxides with Achiral Olefins 424
6.4.2.4. Use of Intermediates from Cycloadditions of Chiral Nitrile Oxides to Chiral Olefins 425
6.4.3. Aminoalcohol Ring Cleavage of Nitrile Oxide Cycloadducts; Synthesis of Amino Polyols, Amino Sugars, and Amino Acids 425
6.4.4. Uses of Nitrile Oxide Cycloadditions in the Isoxazoline Route Toward Mono- and Bicyclic Imino Acids and Imino Polyols. 431
6.5. Intramolecular Nitrile Oxide Cycloadditions: Applications in Synthesis 437
6.5.1. Intramolecular Cycloadditions to Polycyclic Isoxazolines and Furoxans 437
6.5.2. Intramolecular Cycloadditions with Ensuing Ring Cleavage of Intermediate Isoxazolines 439
6.5.2.1. Isoxazoline Ring Cleavage Reactions of the "Aldol" Type (Reduction–Hydrolysis) 440
6.5.2.2. Syntheses with Intramolecular Nitrile Oxide Cycloaddition and Isoxazoline Reduction to the γ-Aminoalcohol 454
6.5.2.3. Transformations of Isoxazolines from Intramolecular Cycloadditions via Isoxazolinium Salts 458
6.6. Conclusion 461

The chemistry of nitrile oxides, in particular their application in organic synthesis, has been continuously developed over the past two decades and represents the main theme of this chapter. The parent compound, fulminic acid (formonitrile oxide), has been known for two centuries, and many derivatives of this dipole have been prepared since that time. Several simple and convenient methods for the preparation of nitrile oxides have evolved over the years. Dehydrochlorination of hydroximoyl chlorides was first introduced by Werner and Buss in 1894 (1). A convenient synthesis of isoxazoles was reported by Quilico et al. (2–4), and then the discovery of nitrile oxide cycloadditions to alkenes was subsequently noted by the same group (5).

Most important for easy access and preparative use of nitrile oxides is the *in situ* technique, introduced by Huisgen and Mack (6) with hydroximoyl chlorides, and in a complementary way, by Mukaiyama and Hoshino (7), where nitroalkanes were submitted to dehydration by means of phenyl isocyanate and triethylamine. These discoveries were followed by the formularization of the general concept of 1,3-dipoles and their cycloadditions by Huisgen and co-workers (8). A second landmark in this arena was the application of orbital symmetry toward 1,3-dipolar cycloaddition chemistry (9). Still another important phase in the development of this field occurred in the late 1960s and 1970s and dealt with the rationalization of reactivity and selectivity of these cycloadditions. In particular, notable efforts to understand regioselectivity and stereoselectivity from *cis/trans* olefins (10) were successful and helped to establish the mechanistic course of the reaction as that of a concerted, nonsynchronous cycloaddition (11). In the 1970s, seminal concepts of *latent functionality* (12) and the *use of heterocycles in organic synthesis* (13) led to a rapid increase in methods for converting nitrile oxide cycloadducts, notably isoxazolines, into acyclic structures (14–22). Transformations to aldols (16–18), enones (16–19), and γ-aminoalcohols (20–23) were added to the standard repertoire of the chemist pursuing organic synthesis. Parallel to these findings, and often in combination, *intramolecular nitrile oxide cycloadditions* (INOC) were put to practice, first for forming heterocyclic (24), and then carbocyclic structures (25), with beautiful and imaginative applications to syntheses of natural products (17,18).

Structural features of olefin cycloadducts (i.e., isoxazolines) were extended when regioselective 4-endo- and 3′-exo-deprotonations were discovered, with the broad range of ensuing reactions of these carbanions that are common with electrophiles (19–22, 26–28). Thus, sequences consisting of *cycloaddition–modification–ring cleavage* were utilized to build up complex structures (17, 20–23).

From the 1980s on, many efforts were directed toward asymmetric induction of nitrile oxide cycloadditions to give pure (dia)stereoisomeric isoxazolines, and acyclic products derived from them (17,18,20–23). The need to obtain optically active cycloaddition products for use in the synthesis of natural products was first served by using chiral olefins, relying on 1,2-asymmetric induction, and then with optically active aldehydes or nitro compounds for the nitrile oxide part. In the latter case, insufficient induction occurs using chiral nitrile oxides, a problem still unsolved today. Finally, in the last 5 years, the first cases of successful asymmetric catalysis were found (29), which will certainly constitute a major area of study in the coming decade.

This chapter aims to outline the progress in nitrile oxide cycloadditions and synthetic uses of the cycloadducts over the past 20 years, updating the earlier chapter of Caramella and Grünanger (15) in the first edition of this series. While some duplication of the basic principles presented earlier is unavoidable, a liberal choice of examples involving important new developments in this field is offered. A number of the earlier published reviews are referred to (30–35), including some information on physiological activities of various isoxazoles (and benzo derivatives) (14). Note that a comprehensive and authoritative treatise on nitrile oxide cycloadditions leading to the 1,2-oxazole family was published in 1991 by Grünanger and Vita-Finzi (30), and represents a continuation of the seminal monograph of Grundmann and Grünanger that was put forth earlier in 1971 (10).

6.1. METHODS FOR THE GENERATION OF NITRILE OXIDES

Nitrile oxides are generally not isolable dipoles but are prepared *in situ* in the presence of a dipolarophile. However, some stable derivatives are known (see below). A common source of nitrile oxides (**1**) are aldehydes (**2**) (making it very convenient to obtain chiral, optically active derivatives) that are converted into the respective oximes (**5**). From these, there is a choice concerning the actual precursor. A hydroximoyl halide (**4**), or a nitroalkane (**6**) can be used, the latter also being available from primary alkyl halides (**3**) (36). If left alone, nitrile oxides at high temperature (110–140 °C) often rearrange to form an isocyanate. This situation has been encountered in the case of stable, nondimerizing nitrile oxides (15). Otherwise, the nitrile oxide (**1**) tends to dimerize to produce a furoxan (**8**) or, under special conditions, give rise to isomeric 1,2,4-oxadiazole-4-oxides or symmetrical 1,4,2,5-dioxadiazines (15). In order to avoid this problem, generation of nitrile oxides is performed slowly in the presence of an olefin or a dipolarophile such as **9**. The most common methods used for the preparation of nitrile oxides are the dehydrochlorination of hydroximoyl chlorides (Huisgen's *in situ* method), the oxidation of aldoximes, the dehydration of primary nitroalkanes (Mukaiyama's method) and, less importantly, the thermolysis of furoxans (Fig. 6.1).

6.1.1. Dehydrohalogenation of Hydroximoyl Chlorides

Nitrile oxides are normally obtained from oximes in two steps: halogenation of the aldoxime (**5**) to give a hydroximoyl halide (**4**) and subsequent dehydrohalogenation

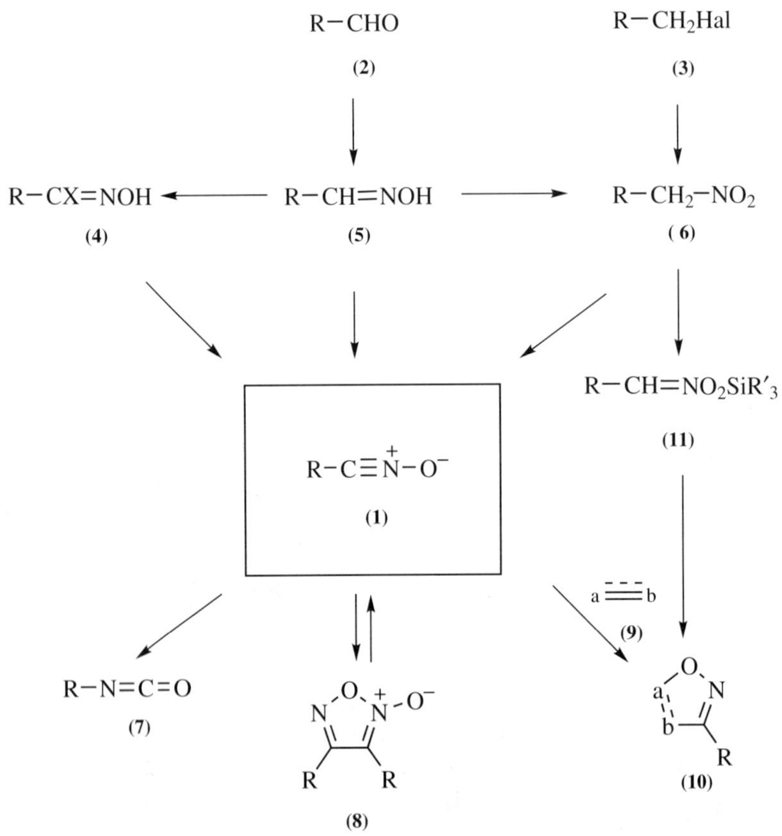

Figure 6.1

by base. It is also possible to combine both steps in one operation, although isolation and use of the pure hydroximoyl halide may have advantages such as simpler analysis of reaction progress or better reaction control. **Caution**: It must be underlined here that hydroximoyl halides are strong skin irritants and may cause abscess at the area of contact; in some cases very strong reaction with rashes on hands or arms may result and operators may develop allergenic reactions prohibiting continuation of work with these substances (37).

6.1.1.1. Halogenation of Oximes

Hydroximoyl halides (**4**) are conveniently prepared by halogenation of the respective aldoximes (**5**), for which a number of halogenating agents such as chlorine (38), *tert*-butyl hypochlorite, *N*-chlorosuccinimide (NCS) (39), or *N*-bromosuccinimide (NBS) (40) have been employed. A plausible mechanistic course involves Hal^+ addition and proton loss to give an α-halonitroso compound, often observed as a transient blue-green color, as shown for the chloro case (Scheme 6.1).

Scheme 6.1

Since this halogenation represents an electrophilic substitution (of the oxime's methine hydrogen), oximes with an alkenyl, activated aryl, or hetaryl group such as *p*-anisyl or α-furyl do not react cleanly when treated with chlorine or hypochlorite (10,20,21,41–44). In the case of *p*-anisyl (**12**), the problem was overcome (42) by employing bleach (45) or NCS (46) [near quantitative (39)], but this did not work well with the 2,4- or 3,4-dimethoxy compounds (42). Here, the 5-chloro derivative **13** was isolated (~60%) (46) in one case, and low yields (~25%) of the hydroximoyl chloride were registered with the 3,4-dimethoxy compound (42). On occasion, the above aryl group serves as a latent carboxy group (i.e., for syntheses of α-amino-γ-hydroxy acids) that becomes unmasked by oxidative degradation later-on (after cycloaddition–hydride reduction–amino protection steps) (20,21,42). α-Furyl derivatives can be obtained in good yield (86%) and purity (42) from furfuraldoxime when nitrosyl chloride (47) is used (for *in situ* preparation, with subsequent dehydrohalogenation and cycloaddition). The *bleach method* (45) also worked well for this system (Scheme 6.2).

A method that avoids the conventional chlorination of aldoximes corresponds to the reaction of nitroalkanes or conjugate nitroalkenes with titanium tetrachloride (48) (Scheme 6.3). α-Chloro (a), alkyl (a, b, c), aryl (c), α-cyano (d), and α-azido (d) hydroximoyl chlorides have been prepared in good yield in this way (48). Concerning overall efficiency, however, this would necessitate that the nitro

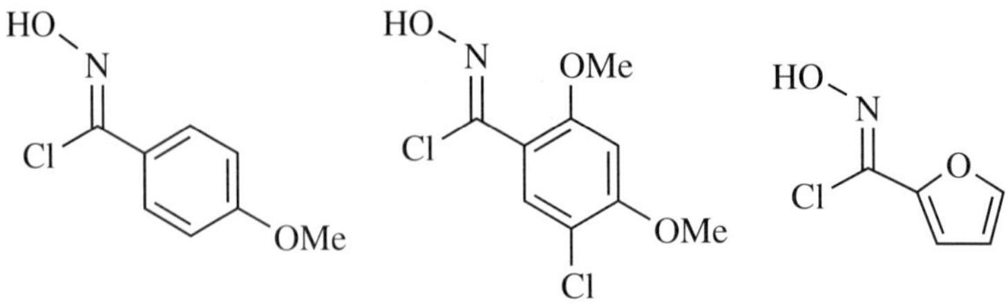

(12) (13) (14)

Scheme 6.2

(a)

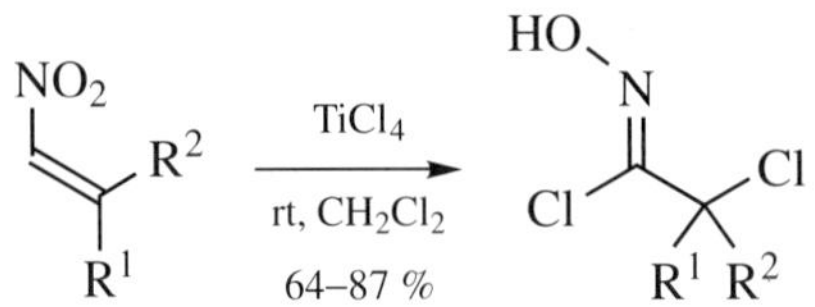

R^1 = alkyl, aryl
R^2 = H, Me
rt = Room temperature

(b)

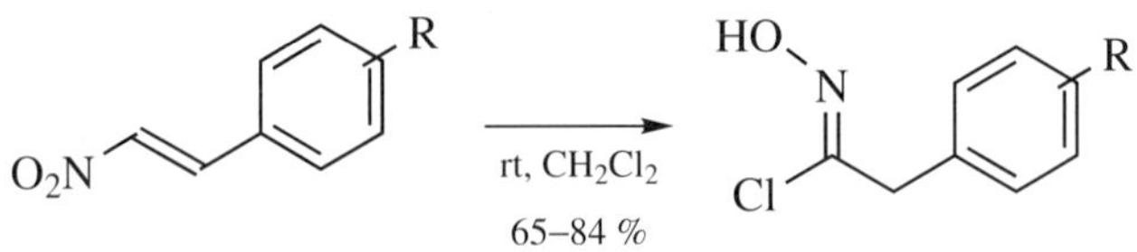

R = H, F, Cl, Br, NO_2, EtO, Me

(c)

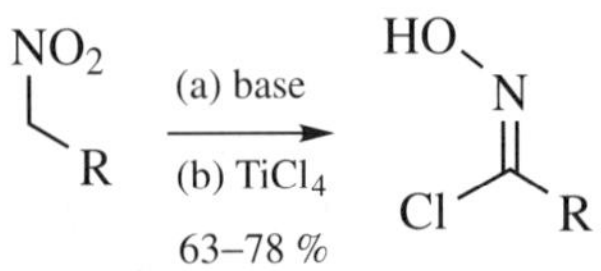

R = aryl, $CH_3OC(O)$; base = NaOMe, NaH, KH

(d)

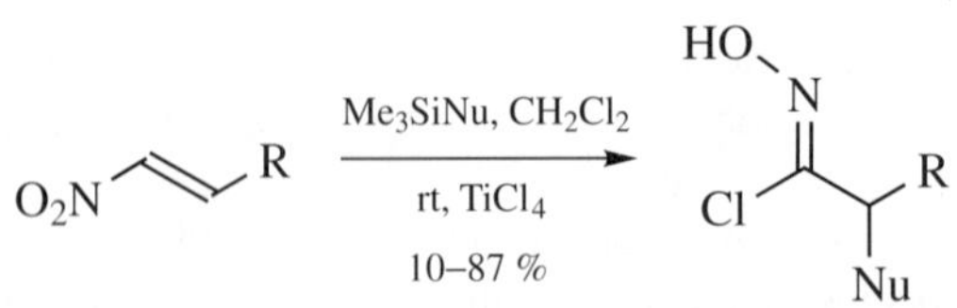

R = isopropyl, aryl; Nu = N_3, CN

Scheme 6.3

compounds be more readily available as starting materials than the oxime precursors, which is rarely the case with nitroalkenes. Tellurium tetrachloride has also been used to prepare hydroximoyl chlorides from β-nitroketones (49). The reaction of β-nitrostyrenes with Grignard or organolithium reagents was likewise reported to afford hydroximoyl chlorides (50–53). This method also provided nitrile oxides suitable for subsequent inter- or intramolecular cycloadditions.

6.1.1.2. Dehydrohalogenation for in situ Generation of Nitrile Oxides

Hydroximoyl halides can be isolated and subjected to dehydrohalogenation by slow addition of a base such as triethylamine or pyridine. Heterogeneous reactions using potassium hydrogencarbonate (38,54,55), alkali metal fluorides (56,57), or basic alumina (58) have also been employed. The heterogeneous medium allows for the slow generation of the nitrile oxide, thereby minimizing dimerization. This method was successfully used to generate nitrile oxides under high pressure conditions (10 kbar) using potassium carbonate as base (59).

Organometallics such as *n*-butyllithium or ethylmagnesium bromide have also been used for the generation of nitrile oxides, through initial O-metalation of hydroximoyl chlorides (60). Also, organotin compounds such as hexabutyldistannane, combined with sunlamp irradiation, have been employed as dehydrohalogenating agents (61). As a consequence, vinylstannanes may serve both as a dehydrochlorinating reagent and dipolarophile (62). Here, the cycloaddition occurred preferentially between the nitrile oxide and the tin-free vinylic compound to afford isoxazolines in moderate yields (62).

Several methods are available for the generation of nitrile oxides from hydroximoyl halides when base-sensitive functional groups are present. For example, the thermolysis of hydroximoyl chlorides in refluxing toluene affords nitrile oxides and subsequent cycloadducts without the formation of furoxans, allowing for easier purification of products (63). Better results were obtained by applying a stream of nitrogen, which serves to remove the hydrogen chloride liberated (64). Also, Ag(I) salts have served as dehydrohalogenating agent (65). The use of molecular sieves was reported to allow the slow formation of nitrile oxides under mild conditions (66). Although long reaction times were required, this method usually minimizes the formation of furoxans and consequently, the products are obtained in good to excellent yield (66).

6.1.1.3. One-Pot Operations for Nitrile Oxide Formation From Oximes and Subsequent Cycloaddition

Halogenation and dehydrohalogenation have been performed as a one-step operation from aldoximes using NCS in the presence of base such as pyridine (67), NBS and triethylamine (68), or a heterogeneous base such as basic alumina or Florisil (69). Solvent-free reaction conditions also proved useful for the preparation of nitrile oxides (70). A mixture of the aldoxime, NCS, alumina, and olefin was subjected to microwave irradiation; the reaction gave nitrile oxides that cycloadd to olefins or acetylenes, respectively, to produce isoxazolines or isoxazoles in good yield (70).

The most widely used, and often most convenient reagents for such one-pot reactions are sodium hypochlorite (45) or hypobromite (16). These reactions are performed in the presence of an organic base (generally triethylamine) that normally enhances the yield of cycloaddition products (45). This method was employed for many intermolecular reactions (71) and also seems especially suited for intramolecular ones (72–77) as well as for the solid-phase synthesis (78) of 2-isoxazolines. Hypohalite can also be replaced by sodium bromite in combination with a catalytic amount of tri-*n*-butyltin chloride (79). In a related method, O-tributylstannyl oximes were treated with *tert*-butyl hypochlorite to produce nitrile oxides that were trapped with alkenes or alkynes to afford the corresponding isoxazolines or isoxazoles in moderate to good yield (80).

Chloramine-T has also been employed, both as a halogenating reagent and base; the reaction proceeds in good yield with aromatic as well as with aliphatic aldoximes (81). The role of chloramine-T probably involves an initial chlorination of the aldoxime to give the hydroximoyl chloride, followed by base-catalyzed hydrogen chloride elimination to afford the nitrile oxide (81).

6.1.2. Dehydration of Nitroalkanes

The most widely used method for the dehydration of primary nitroalkanes involves their treatment with phenyl isocyanate and triethylamine, introduced in 1960 by Hoshino and Mukaiyama (7). A probable mechanism for the formation of the nitrile oxide is shown in Scheme 6.4. This method is known to be very effective for the preparation of aliphatic or aromatic nitrile oxides. In some cases, the separation of the byproduct *N,N'*-diphenylurea from the reaction mixture may be troublesome. In order to circumvent this problem, 1,4-phenylene diisocyanate was introduced (82,83). The polymeric urea that is formed as a byproduct is largely insoluble in the reaction mixture and can easily be removed.

Shimizu et al. also used ethyl chloroformate or benzenesulfonyl chloride with triethylamine as a dehydrating agent (84). The use of the former usually requires heating, in some cases resulting in low yields. In order to avoid this, Hassner and co-workers developed a new method that uses 4-dimethylaminopyridine (DMAP) instead of triethylamine (85), which they find leads to nitrile oxides under milder conditions. When ethyl chloroformate is replaced by di-*tert*-butyl carbonate, substrates containing free amino or hydroxy groups can be employed without prior protection, and the reaction leads to the respective N- or O-*tert*-butoxycarbonyl (Boc) protected products (85). Another advantage of the latter procedure is that the byproducts are readily separable from the nitrile oxide cycloadducts. This procedure also serves to generate unsaturated nitrile oxides that are useful for intramolecular reactions (86,87). In a less general version, *p*-toluenesulfonic acid

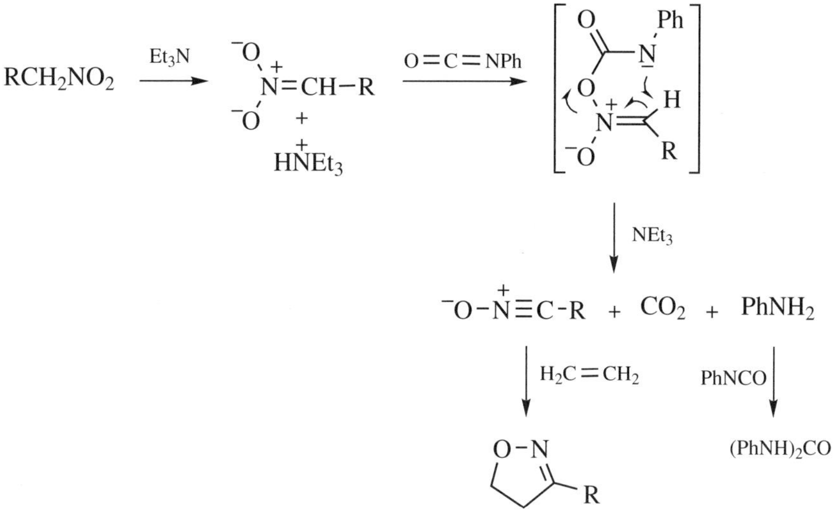

Scheme 6.4

monohydrate was used as a dehydrating agent, but the high reaction temperatures needed (reflux in mesitylene or xylene) preclude the use of nitro compounds with thermolabile functional groups (88).

Other useful dehydrating agents are dimethylaminosulfur trifluoride (DAST), methyl *N*-(triethylammoniosulfonyl)carbamate (Burgess salt), acetic anhydride, oxalyl chloride, and phosphorous oxychloride, each one in combination with triethylamine (89). Dehydration of O-silylated hydroxamic acids using trifluoromethanesulfonic anhydride and triethylamine under mild conditions also gave nitrile oxides, which in the presence of olefins led to the formation of 2-isoxazolines in moderate to good yields (90). In view of the less readily available starting materials, this method probably will be of limited use.

The generation of a nitrile oxide bearing a carbamoyl group (i.e., **16**) was effected by treating 4-nitro-3-isoxazoline-5-one (**15**) with a mixture of acetonitrile and water (Scheme 6.5). Although the mechanism of this reaction is not clear, the method allows for the formation of a functionalized nitrile oxide (**16**) and subsequent cycloaddition under mild conditions (91).

Nitrile oxides can also be generated from the thermolysis of nitro and related compounds. Thus, heating 2-nitroalkanoates in tridecane at ~230 °C produced nitrile oxides, which were trapped *in situ* with several alkene or alkyne dipolarophiles to afford 3-alkyl-2-isoxazolines or 3-alkylisoxazoles in moderate yield (92). Formonitrile oxide [usually obtained from mercury(II) fulminate (10,15)] has been prepared from the thermolysis of ethyl nitroacetate and was added *in situ* to several olefins to give cycloadducts in low to moderate yields (93). A more general approach involves the thermolysis of nitrolic acids, producing nitrile oxides under *neutral* conditions at room temperature or in refluxing tetrahydrofuran (THF) (94, 95). The starting acids are prepared from nitro or bromo precursors by nitrosation,

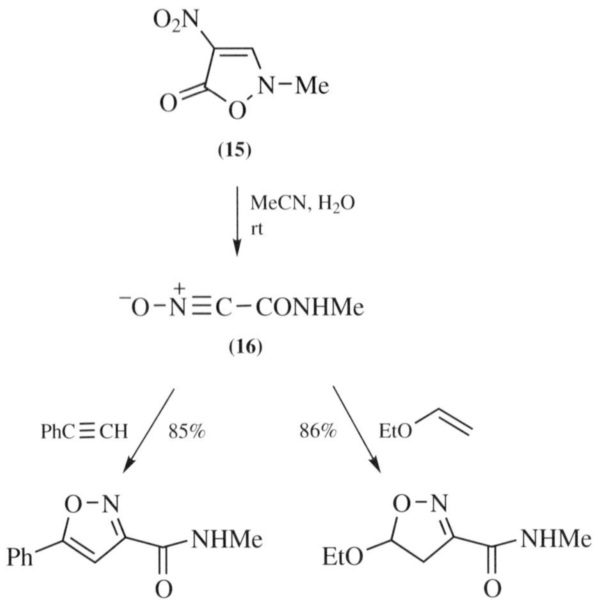

Scheme 6.5

thus limiting the range of tolerable functional groups. This reaction was used for the preparation of cycloadducts from several alkenes (including allyl alcohol) using alkyl-, alkoxycarbonyl-, or aryl nitrile oxides. The best yields were obtained with 3-phenylpropane nitrolic acid (95–97%), while the phenyl and glyoxalate derivatives gave only moderate results (40–55%) (94,95).

6.1.3. Oxidation of Aldoximes

Oxidation of aldoximes bearing electron-withdrawing groups as a method to generate nitrile oxides has been carried out using manganese dioxide (96). The reaction can be performed under mild conditions using the (*E*) or (*Z*) isomers of aldoximes. When lead tetraacetate (15,30) was used, only the latter react (97). The lead reagent has also been used for the oxidation of β-stannyl oximes (Scheme 6.6) (98). When the oximes have a β-stannyl group oriented away from the iminoyl skeleton, nitrile oxides have actually been detected from cyclohexanone oximes. In contrast, oximes bearing a primary stannyl group close to the iminoxyl moiety appear to react by a cyclization mode not involving a nitrile oxide (98).

Diammonium hexanitratocerate (IV) (CAN) has also been used as an oxidizing agent for aldoximes. This approach is especially useful for the preparation of aliphatic 2-oxo-carbonitrile oxides (99). Another useful reagent for the generation of nitrile oxides from aldoximes is 1-chlorobenzotriazole (100). Due to its reactivity toward both aldoximes and olefins, this reagent can only be used to prepare stable

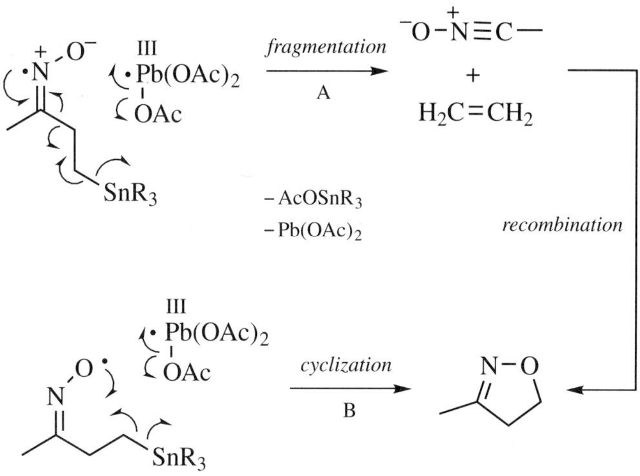

Scheme 6.6

nitrile oxides. However, in a case of intramolecular cycloaddition, isoxazoline cycloadducts were obtained in good yield employing 1-chlorobenzotriazole (100).

6.1.4. Thermolysis of Furoxans

In the absence of trapping agents, nitrile oxides normally dimerize to give furoxans (1,2,5-oxadiazole 2-oxides) (101). The furoxans have been regarded as stable, *dead-end* side products with regard to cycloaddition. However, it has been discovered that nitrile oxides can be regenerated from furoxans by thermolytic cycloreversion (101). This cycloreversion appears to be ideal for application to nitrile oxide cycloadditions since dimerization is no longer a problem. However, there are some shortcomings that detract from the generality of this approach; namely, (a) rearrangement of the nitrile oxides to isocyanates at high temperature (16), and (b) limited types of functional groups. Three types of reactions are found when furoxans are submitted to thermolysis, and these depend on the 3,4-substitution pattern (Scheme 6.7):

1. Cycloreversion into 2 mol of nitrile oxide.
2. Interconversion into a rearranged nitrile oxide.
3. Ring cleavage with concomitant fragmentation, giving nitrile oxides.

Reaction (a) occurs when the R group is alkyl, aryl, or arylsulfonyl (101–103), but only strained furoxans and furoxans bearing bulky substituents revert to nitrile oxides at moderate temperatures (101b) (Table 6.1). The nitrile oxides thus produced were trapped *in situ* with 1-hexene to afford the respective 1,3-cycloadducts in high yield (101b).

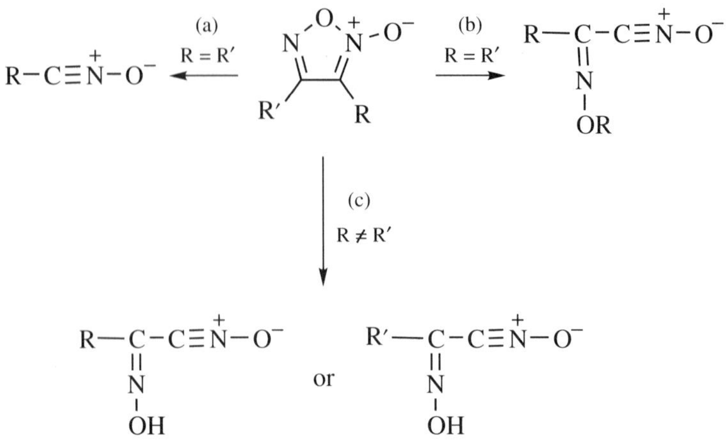

Scheme 6.7

A special case involves the thermal decomposition of 3,4-dinitrofuroxan (104). The cycloreversion is already observed at room temperature and the nitroformonitrile oxide could be trapped with electron-deficient nitriles. The cycloadditions with styrene, phenylacetylene, *trans*-stilbene, and cyclohexene, however, led to complex mixtures of products that could not be separated (104). In the related case of a furoxan with an α-hydrogen adjacent to the sulfonyl group, the reaction was proposed to proceed according to course (b) (Scheme 6.7).

Unhindered 3,4-diacylfuroxans prefer ring opening by N—O bond fission of the oxadiazole ring with concomitant migration of an acyl group to give an α-(acyloximino)-β-ketonitrile oxide [type (b)] (103) (Scheme 6.8).

The course of thermolysis of such furoxans also depends on the nature of the acyl group. For example, heating the cage-functionalized diacylfuroxan depicted in

TABLE 6.1. CYCLOADDITIONS TO 1-HEXENE USING FUROXANS AS NITRILE OXIDE PRECURSORS[a]

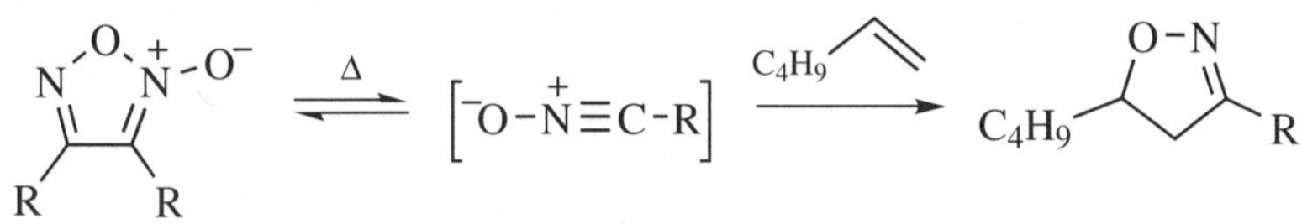

R	Temperature (°C)	Yield of Cycloadduct (%)
Me_3SiOMe_2C	165	97
$(H_3C)_3$	135	78
Me_3SiO	135	90

[a] See (101b).

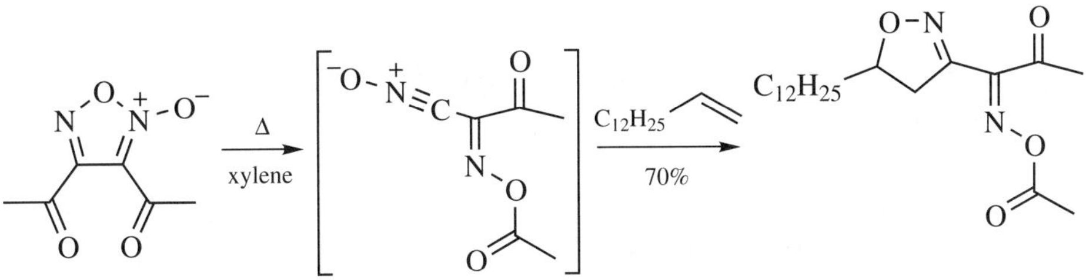

Scheme 6.8

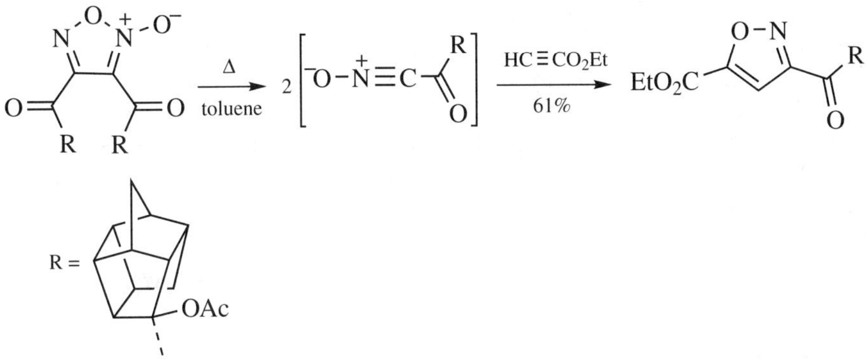

Scheme 6.9

Scheme 6.9 afforded the corresponding acylnitrile oxide [type (a)], which when trapped *in situ* with ethyl propiolate, produced the substituted isoxazole (105).

Another mode of cleavage [type (c)] was found when unsymmetrical 3,4-disubstituted furoxans were subjected to thermolysis. Thus, 3-methyl-furoxan-4-carboxylic acid was found to afford the α-oximinonitrile oxide with concurrent

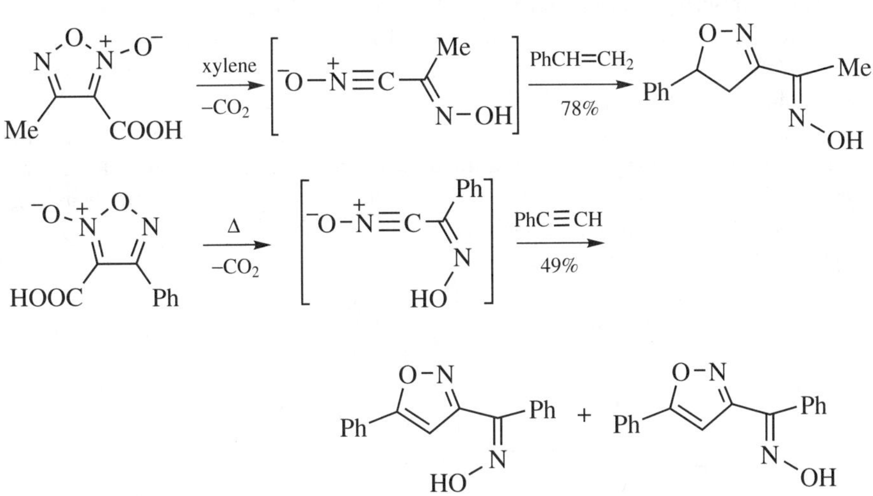

Scheme 6.10

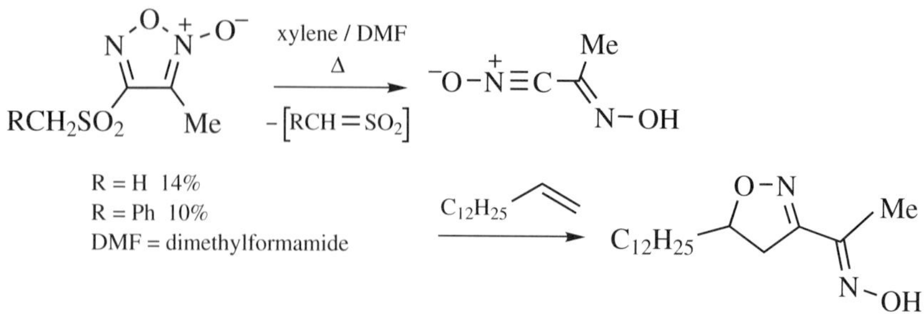

Scheme 6.11

decarboxylation (106). Thermolysis of the related 4-phenylfuroxan-3-carboxylic acid gave the α-oximino-phenylnitrile oxide (Scheme 6.10), which was trapped with phenylacetylene to give the expected isoxazole as a (*E*/*Z*) mixture (107).

Similar reactions were found when the carboxyl group was replaced by a benzylsulfonyl group (102). Here, besides the nitrile oxide (trapped as usual), the benzylidene-sulfene fragment was also formed (Scheme 6.11).

6.2. CYCLOADDITIONS OF NITRILE OXIDES

6.2.1. Mechanism and Relative Reactivity of Dipoles and Dipolarophiles

6.2.1.1. Mechanistic Studies and Calculations

In principle, cycloadditions of 1,3-dipoles with dipolarophiles may occur in a concerted or stepwise manner (108). On the basis of intensive, meticulous experimental work, Huisgen and his group unequivocally established the concerted course for this reaction (15,109,110). A model was proposed for the transition state, whereby the 4π-electron system present in 1,3-dipoles interacts with the π-bond of the dipolarophile (Scheme 6.12).

As an alternative, Firestone proposed a stepwise reaction course for the cycloaddition involving biradical intermediates (Scheme 6.13). This reaction course was based on a number of experimental facts such as small solvent effects as well as the formation of byproducts such as oximes (15,109,110).

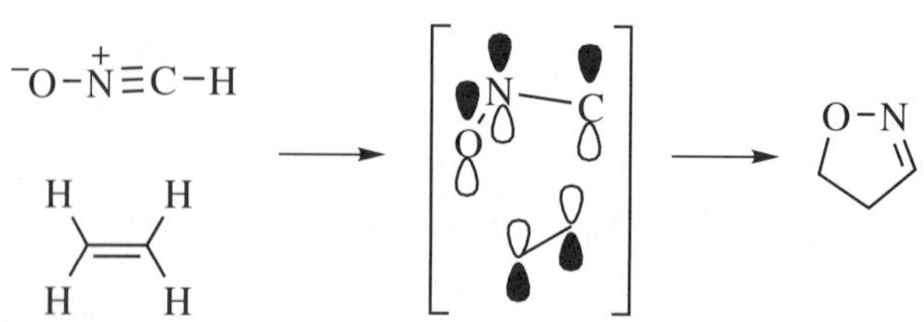

Scheme 6.12

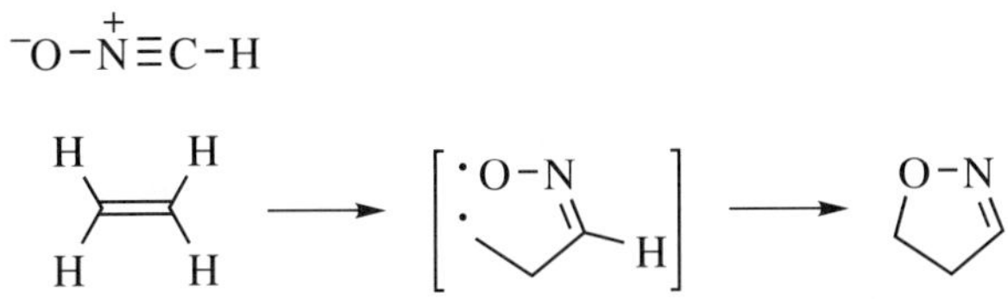

Scheme 6.13

The complete stereoselectivity of the reaction, however, is difficult to reconcile with a two-step process. This earlier controversy, however, has long since been resolved. For example, when considering results of the cycloaddition of *p*-nitrobenzonitrile oxide with *cis*- and *trans*-1,2-dideuterioethylene (111), the experiments clearly established that, within experimental limits of detection, the reaction is $\geq$98% stereoselective. If diradical intermediates were operative, significant scrambling of configuration should be observed in the products. These and other results confirm a concerted mechanism for the 1,3-dipolar cycloaddition reaction (15).

Several theoretical models have been advanced in recent years to rationalize the observed regio- and stereoselectivity of the 1,3-dipolar cycloadditions (112–114). Frontier molecular orbital (FMO) theory has had a formidable impact in helping to explain the experimental results (115). It should be pointed out that there is a tendency in this area to avoid *a priori* predictions in favor of *a posteriori* explanations (116). In the FMO approach, the energies of the highest occupied and lowest unoccupied molecular orbitals (HOMO and LUMO) of the 1,3-dipoles and the dipolarophiles are calculated, together with the MO coefficients of the reacting centers. Then, orbital energy separation of the interactions [HOMO(1,3-dipole)/LUMO(dipolarophile) and LUMO(1,3-dipole)/HOMO(dipolarophile)] are compared. When these energy differences exhibit different values, the FMO interaction corresponding to the lower value is presumed to play the dominant role (109). Regiochemical orientation is then established by the rule of *preferential facing* where both the large and small coefficients of the terminal atoms of the 1,3-dipole and dipolarophile are involved. It has to be cautioned, however, that this treatment sometimes leads to wrong predictions, in particular when steric or electrostatic effects are involved, and further when effects of interactions of other high-energy filled orbitals (*n*HOMO or OMO) with other low-energy empty ones (*n*LUMO or UMO) become operative (see Section 6.2.2) (109). For example, for cycloadditions to the acetal and thioacetal derivatives of α,β-unsaturated aldehydes, the FMO treatment correctly predicts the regiochemical result of the first case, but fails for the latter example (117).

Ab initio calculations have also been used to predict selectivity in such cycloadditions (112,113,118). These studies show that the reactants approach each other in a plane with the two new bonds forming at the same time, which appears to be concerted although not necessarily synchronous. Only one energetically favored transition state was found, and the average lengths of the newly forming bonds are scarcely affected by the presence of substituents on the alkene moiety. This result comes from the fact that the two forming bonds have, in general, small opposite charges (114). The distances of the developing C—C and C—O bonds also depend on the method used for the calculations (114). Thus, the transition state for the cycloaddition of nitrile oxides with ethene, propene, butadiene, acrylonitrile, and methyl vinyl ether, as derived from *ab initio* calculations, show that the C—C distance is shorter than that of the developing C—O bond (114,119). This asynchronicity is particularly pronounced in the transition state for the cycloaddition of fulminic acid to butadiene and formonitrile oxide to methyl vinyl ether (for discussion of some other cases see Section 6.2.2) (114,119).

Another concept that has recently been introduced is the *aromaticity* of the transition state (119). It was suggested that this aromaticity is compatible with a ring current circulating within the molecular plane of the five atoms in the transition state for cycloaddition (119). According to these calculations, such transition state aromaticity does not, however, impact upon the regioselectivity of the cycloaddition (119).

6.2.1.2. Relative Reactivity of Dipolarophiles and Dipoles

Huisgen and co-workers (120) evaluated the influence of different olefin substituents on the cycloaddition of benzonitrile oxide in diethyl ether. The relative reactivity (k_r) of ethene was set as 1. These competition studies allowed one to assess the relative rate constants of a wide range of alkenes and other dipolarophiles (Table 6.2) (15,109). The relative reactivity of several allyl-substituted olefins was also determined (121) in conjunction with studies on the diastereoselectivity of these cycloadditions (see Section 6.2.3.1). Examination of the results gathered in Table 6.2 shows that the reactivity of the dipolarophile is increased by the presence of both electron-donating and -withdrawing substituents, and that the effect of conjugation appears to be stronger than the inductive one (109). trans-Alkenes are more reactive dipolarophiles than the corresponding cis isomer as can be seen from the case of fumarate (entry 12), which has k_r 70 times higher than the maleate ester (entry 13). 1,2-Disubstitution (entry 11) decreases the reactivity more than 1,1-disubstitution does (entry 10). With cyclic compounds, it is noteworthy that ring strain and conformational effects (yet to be sorted out) strongly affect the reactivity of dipolarophiles [i.e., k_r of cyclohexene is about nine times less than that of cyclopentene (entries 3,4)]. With allyl ethers and related systems, the change to other O-substituents barely affects the reactivity (entries 17–21). Olefins with one, two, and three alkoxy groups at the allylic position were evaluated in the context of the stereochemical model (*inside alkoxy*) proposed by Houk and Jäger and their co-workers for diastereodifferentiation with α-alkoxyolefins (20,21,122–124). While there was a small change on passing from 1-alkene (k_r 0.31) to the ethyl ether (0.32) and the acetal (0.23), introduction of a third alkoxy group (as done with an orthoester) caused a dramatic decrease (0.0094) in the rate of reaction, corresponding to a raise of this transition state by ~2.9 kcal/mol (entries 1,18,22,23). With the orthoester (case 23), the anti position can no longer avoid electron withdrawal

TABLE 6.2. RELATIVE REACTIVITIES (k_r) OF ACHIRAL DIPOLAROPHILES[a]

Entry	Olefin	k_r	Reference
1		1.0	
2	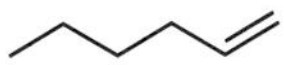	0.31	120
3		0.21	120
4		0.025	120
5	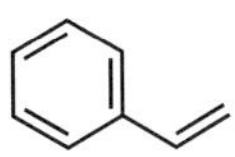	1.15	120
6	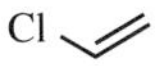	0.081	120
7		2.1	120
8	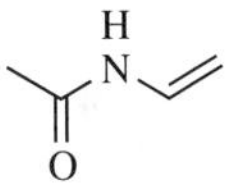	0.52	121
9		8.3	120
10	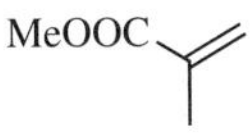	3.6	120
11		0.082	120
12	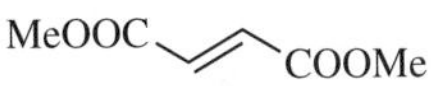	6.1	120
13	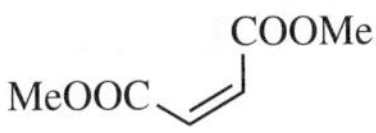	0.082	120
14	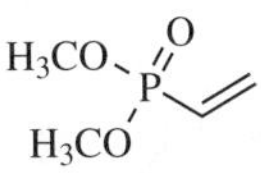	1.11	121
15	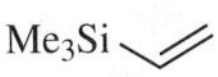	5.46	121
16	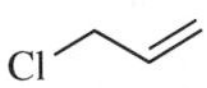	0.64	121
17	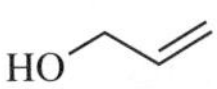	0.50	121
18	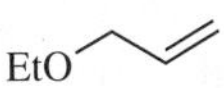	0.32	121
19	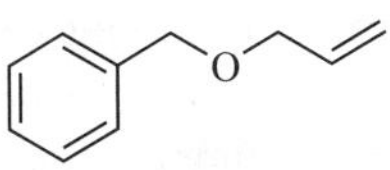	0.38	121
20		0.47	121

TABLE 6.2. (*continued*)

Entry	Olefin	k_r	Reference
21		0.46	121
22	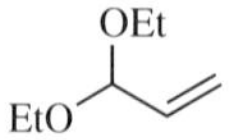	0.23	121
23	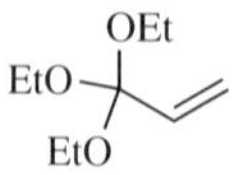	0.0094	121
24	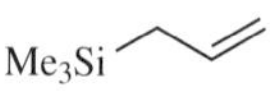	0.42	121
25	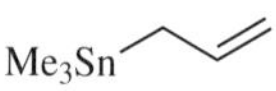	0.94	121
26	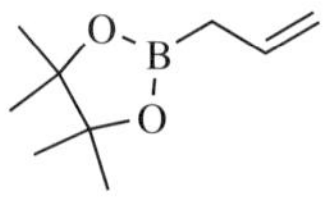	0.59	121
27		0.40	120
28	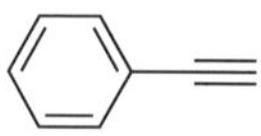	0.112	120
29	n-C_4H_9—≡	0.066	120
30	MeOOC—≡	1.24	120
31	MeOOC—≡—COOMe	3.1	120
32	MeOOC—≡—	0.030	120

[a] Reaction with benzonitrile oxide at 0 °C.

from the π-bond by σ*,π-interaction and, consequently, reduced reactivity of the olefin toward the mildly electrophilic nitrile oxide dipole occurs (11,15,109). Alkynes display lower reactivity (cases 27–32), with the influence of the substituents being similar to the ones observed with alkenes.

The relative rate constants (k_r) do not account for the fact that approach of the nitrile oxide to the π-bond can occur from both olefinic diastereofaces with two regioisomeric modes of reaction (Scheme 6.14). In the case of achiral 1-alkenes, only one regioisomer is formed. With chiral dipolarophiles, preference for one of the two is usually found (diastereodifferentiation). The relative diastereofacial reactivity ($k_r\pi$) is used to evaluate this effect (121). With ethylene, there are four possibilities of attack (two for each face corresponding to the different regioisomers), and the $k_r\pi$ of each is set as 0.25. In diastereodifferentiating cycloadditions, such as those with "α-chiral" alkenes, the major isomer generally results

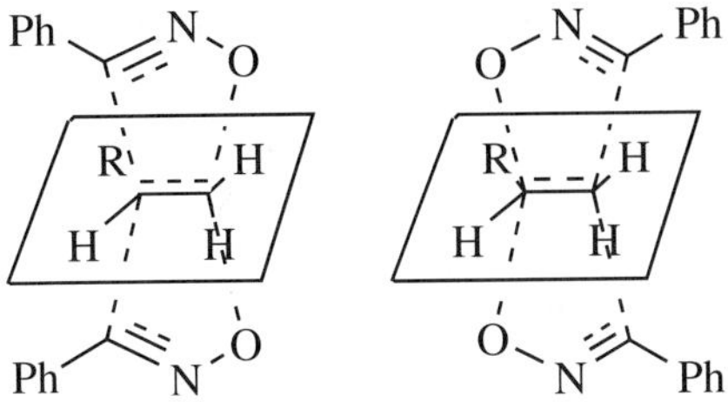

Scheme 6.14

from approach to the favored π face, which shows higher reactivity due to stereoelectronic or steric effects. The values of $k_r\pi$ for several chiral dipolarophiles in cycloadditions with benzonitrile oxide are listed in Table 6.3 (121).

The different reactivity of the two π faces observed with chiral allyl ethers has been rationalized in terms of the earlier model, with the large group positioned anti (*inside alkoxy* model) (see Section 6.2.3.1). Increased hindrance by the O-substituent's size and restricted approach/conformational mobility may account for the increasing difference of reactivity between the two faces for cases 2, 3, and 4 (see Section 6.2.3.1). With the allylic alcohol (case1), the reversed facial preference

TABLE 6.3. RELATIVE REACTIVITIES (k_r) AND RELATIVE DIASTEREOFACIAL RATES ($k_r\pi$) OF CHIRAL DIPOLAROPHILES[a,b]

Case 1	Olefin	k_r	$k_r(\pi_a)$	$k_r(\pi_b)$
1	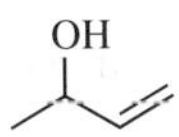	0.41	0.18	0.23
2	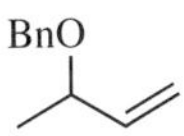	0.14	0.10	0.04
3	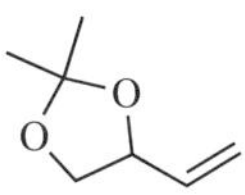	0.51	0.43	0.08
4	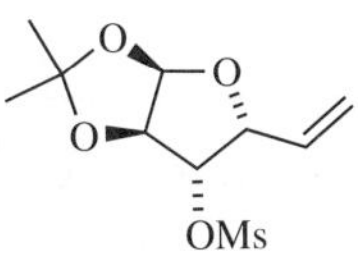	0.68	0.65	<0.03
5	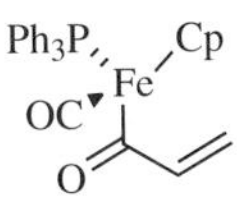	9.58	8.91	0.67
6	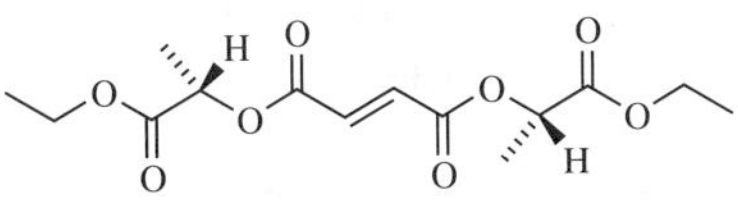	6.44	2.32/2.32	0.90/0.90

[a] Relative rate constants refer to ethylene with $k_r \equiv 1$, $k_r(\pi)$ 0.25.
[b] Reactions with benzonitrile oxide at 0 °C[121] (cf. Ref. 125).

was attributed to hydrogen bonding between the hydroxy group and the oxygen of the nitrile oxide in the transition state (123). The acryloyl–iron complex (121,125) showed particularly high reactivity for one π face and high stereodifferentiation (case 5), which was interpreted by a favored transition state in which the metal occupies the anti position with optimum σ^*, π-interaction (121). The relative reactivity of the chiral fumarate (case 6) is close to that of the ethyl ester (entry 12, Table 6.2), which shows $k_r(\pi)$ of 1.5 for each of the four directions of approach. A distinct difference in the reactivity of the two π faces was observed due to the presence of the chiral auxiliaries (see Section 6.2.3.2) (121). As far as we are aware of, comparable studies on the relative reactivity of nitrile oxides are lacking, as well as calculational studies bearing on the relative reactivities of such dipoles and dipolarophiles.

6.2.2. Regioselectivity of Nitrile Oxide Cycloadditions to Olefins

Cycloaddition of a nitrile oxide to a substituted olefin can lead to two regioisomers, the 4- and/or 5-substituted 2-isoxazoline. Reactions of monosubstituted alkenes give the 5-substituted isomers **18** with almost complete regioselectivity (10,15,30,109). This result is also supported by *ab initio* and FMO calculations (114,119). Change of substituents in the dipole has little effect on the regioselectivity of such reactions when monosubstituted alkenes are used (Table 6.4).

TABLE 6.4. RATIOS OF 4- AND 5-REGIOISOMERS **17** AND **18** IN CYCLOADDITIONS WITH METHYL ACRYLATE

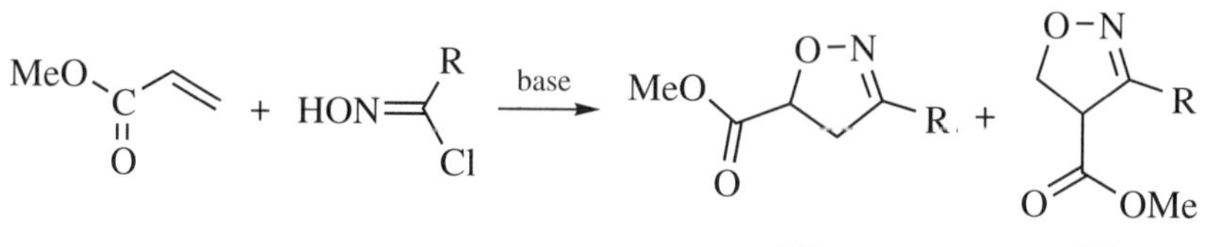

(17) **(18)**

R	Yield (%)[a]	**17**:**18**[b,c]	Reference
2,6-$Cl_2C_6H_3$	93	93.3:6.7	56
CO_2Et	94	>99:1	56
C_6H_5	99	>99:1	56
COMe	97	>99:1	56
Br	89	94.9:5.1	56
COPh	76	>99:1[d]	65
2,4,6-$Me_3C_6H_2$	99	94.5:5.5	66

[a] Yield of isolated products.
[b] The ratio was determined by gas chromatography (GC).
[c] The ratio was determined by nuclear magnetic resonance (NMR).
[d] The nitrile oxides were generated using KF[56] or AgOAc[65] or 4 Å molecular sieves (66).

The normal selectivity of some nitrile oxide cycloadditions can be reversed by tethering the dipolarophile to β-cyclodextrin (126,127). Preassociation of the modified cyclodextrins with aromatic nitrile oxides, in the form of host–guest complexes, apparently controls the relative orientations of the dipole and the dipolarophile, making it possible to reverse the normal regioselectivity of the cycloaddition. For example, the reaction of propenamide (acrylamide) with *tert*-butylphenylnitrile oxide in water afforded only the 5-regioisomer of the corresponding isoxazoline (126,127). With the propenamide-β-cyclodextrin derivative in aqueous solution, it was found that the cycloaddition gave both the 4- and 5-substituted isoxazolines in a 70:30 ratio. When the reaction was performed in DMF, the product ratio was 20:80, the interpretation being that the extent of host–guest complexation was significantly reduced. With benzonitrile oxide, only the 5-substituted isoxazoline was found to be formed, both in H_2O or DMF. The conclusion reached was that only nitrile oxides that can form a thermodynamically favorable association complex within the cyclodextrin are amenable to such changes of regioselectivity in cycloadditions to olefins (126,127).

Cycloaddition of nitrile oxides to 1,2-disubstituted alkenes may give mixtures of two regioisomers (Scheme 6.15); the product ratio (rr) will depend on the substituents present on the olefin.

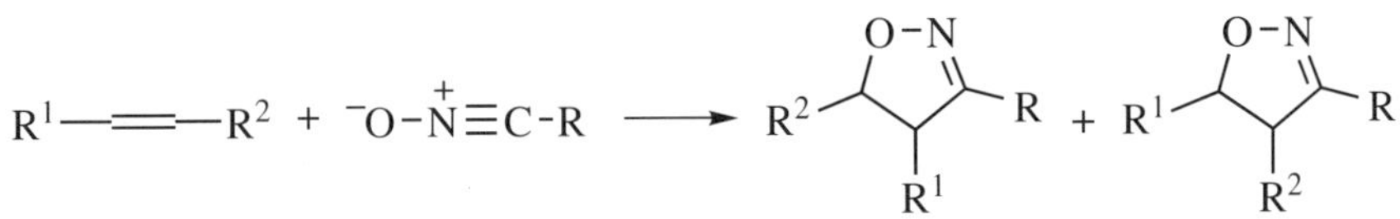

Scheme 6.15

The reaction of α,β-unsaturated acetals predominantly affords regioisomer **19** with the acetal group attached to the C(4) position of the isoxazoline (127). With the analogous thioacetal, the regioselectivity was reversed (117) (Table 6.5).

From an FMO analysis, it was found that FOs govern the regiochemical course of the cycloaddition with the acetal derivative. This treatment, however, failed to predict the major product from the thioacetal derivative (117,128). The authors' interpretation is that the regioselectivity of the cycloaddition is due to a steric interaction between the nitrile oxide substituent and the 1,3-dithiolane ring. The MNDO calculations indicate that the steric interaction between a 3-phenyl group and the dithiolane ring raises the energy of transition state **22**, thus favoring **21** (Scheme 6.16). These calculations fail, however, to correctly predict the regioselectivity of the acetal case, even though the steric bias from the acetal and thioacetal groups are similar (117).

The above example clearly demonstrates the present-day limitations of the theoretical methods for predicting regioselectivity. Note that results of most of these cycloaddition studies cite product ratios, not actual energy differences (in terms of differences of free reaction enthalpies $\Delta\Delta G^*$). Deriving energy differences from product ratios {readily done by $\Delta\Delta G^{\ddagger}_{\mathrm{T}} = -2.3\ RT \log$ [(isomer 1)/(isomer 2)]}

TABLE 6.5. RATIOS OF 4- AND 5-REGIOISOMERS **19** AND **20** IN CYCLOADDITIONS TO α,β-UNSATURATED ACETALS AND THIOACETALS[a]

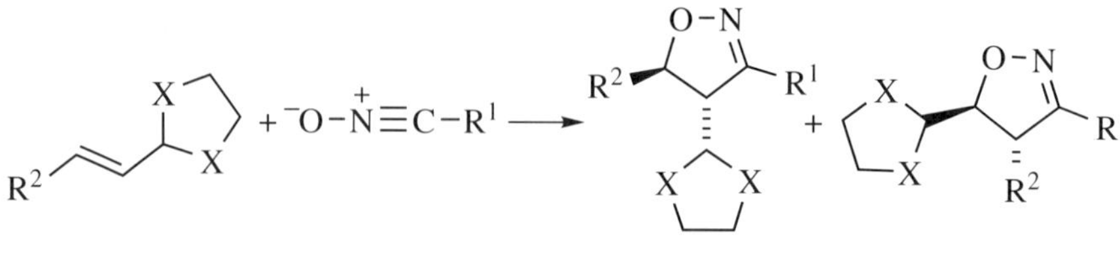

X = O or S (**19**) (**20**)

		Yield (%)[b]		**19:20**[c] ($\Delta\Delta G^{\ddagger}$ in kcal/mol)	
R^1	R^2	Acetal	Thioacetal	Acetal	Thioacetal
Ph	Me	39	32	68:32 (−0.45)	17:83 (0.94)
$PhCH_2$	Me	71	26	90:10 (−1.30)	30:70 (0.50)
p-$MeOC_6H_4$	Me	41	45	65:35 (−0.37)	19:81 (0.86)
p-$MeOC_6H_4$	Ph	62	43	85:15 (−1.03)	27:73 (0.59)

[a] See (117).
[b] Yield of product mixture.
[c] The rr was determined from crude products by high-pressure liquid chromatography (HPLC) (117).

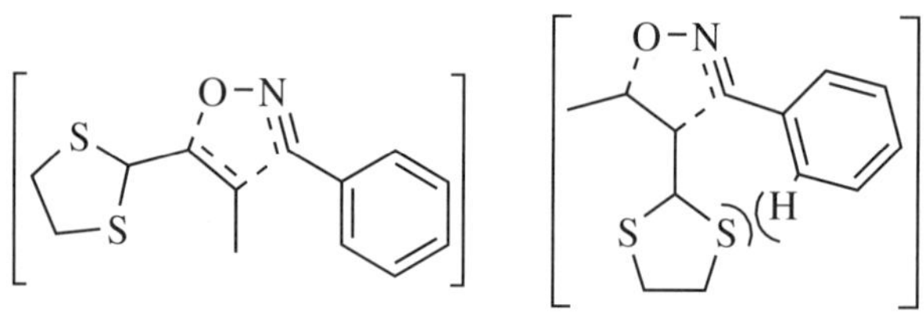

(**21**) (**22**)

Scheme 6.16

often lead to very small differences, which in many cases look tiny when the precision and limits of calculations are borne in mind.

Further examples in Table 6.6 show that, depending on the substitution pattern, the cycloaddition of nitrile oxides to various α,β-unsaturated compounds can lead to different ratios of regioisomeric adducts.

When acrylamides are used as dipolarophiles, FMO theory predicts that the 4-amido isomer should be preferred, which is contrary to the results found with tertiary amides (129). Semiempirical, *ab initio*, and density functional theory (DFT) calculations were applied to the regioisomeric transition state structures of benzonitrile oxide cycloadditions (129–131). The results suggest that there is an unfavorable steric repulsion between the phenyl ring of the nitrile oxide and the methyl group of the ester (or amide) functionalities of the dipolarophile in the transition state leading to the 4-acyl regioisomer (Scheme 6.17).

TABLE 6.6. RATIOS OF 4-ACYL/5-ACYL REGIOISOMERS IN CYCLOADDITIONS TO α,β-UNSATURATED CARBONYL COMPOUNDS[a,b]

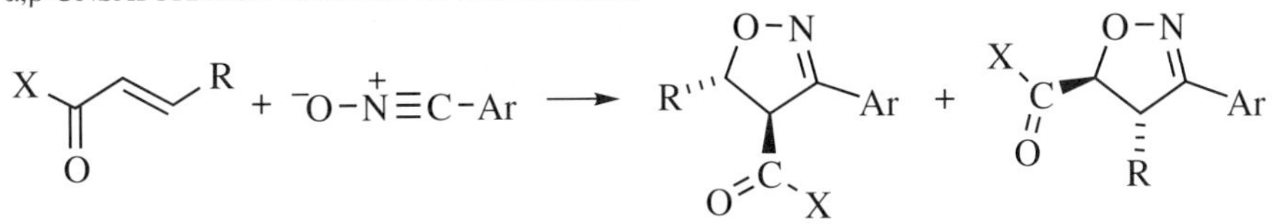

X	R = Ph 4-ClC$_6$H$_4$CNO[c]	R = Ph PhCNO	R = Ph MstCNO[d]	R = Me 4-ClC$_6$H$_4$CNO[c]	R = Me PhCNO	R = Me MstCNO
H		85:15 (73)[e]	57:43 (65)[e]		81:19 (78)[e]	89:11 (70)[e]
OMe	78:22 (85)	80:20 (84)[c]	64:36 (93)[f]	66:34 (65)	66:34 (84)[f]	73:27 (97)[f]
NH$_2$	78:22 (80)	72:28 (52)[g]	35:65 (50)[g]	59:41 (65)	56:44 (49)[g]	69:31 (42)[g]
NEt$_2$	23:77 (95)	28:72 (83)[c]	12:88 (68)[g]	16:84 (86)	12:88 (61)[g]	24:76 (88)[g]

[a] Ratios determined from crude products by NMR except when noted otherwise.
[b] Yield of the product mixture is in parentheses.
[c] Reference 129.
[d] Mesityl = Mst.
[e] Reference 131. Ratios determined by GC of the corresponding alcohols after NaBH$_4$ reduction of the reaction mixtures.
[f] Reference 132.
[g] Reference 130.

These interactions are thought to be partially overcome by the asynchronicity of the transition states, with the forming C—O bond being shorter than the C—C bond. With bulky amides or esters, these interactions favor the formation of the 5-acyl isomer (129,130). With sterically less demanding esters and amides, the interaction described above is less important than that between the β-substituent and the phenyl ring of the nitrile oxide, thus leading to the 4-acyl regioisomer as the major product (129,130).

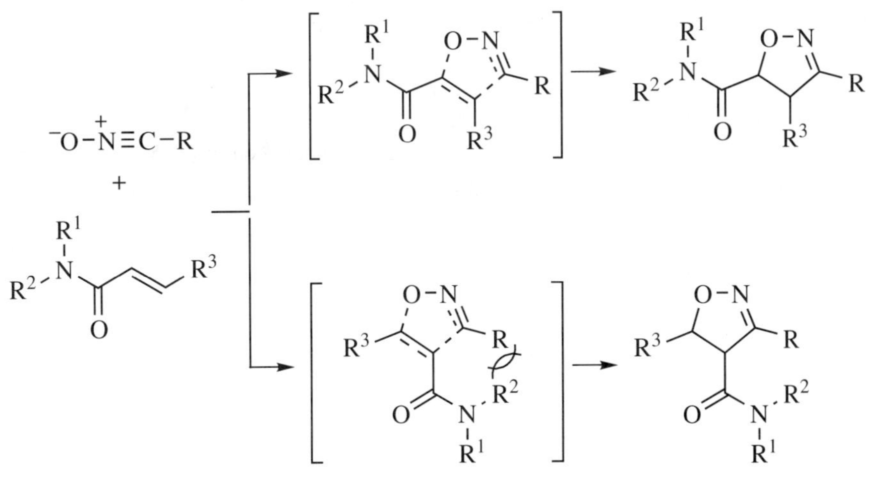

Scheme 6.17

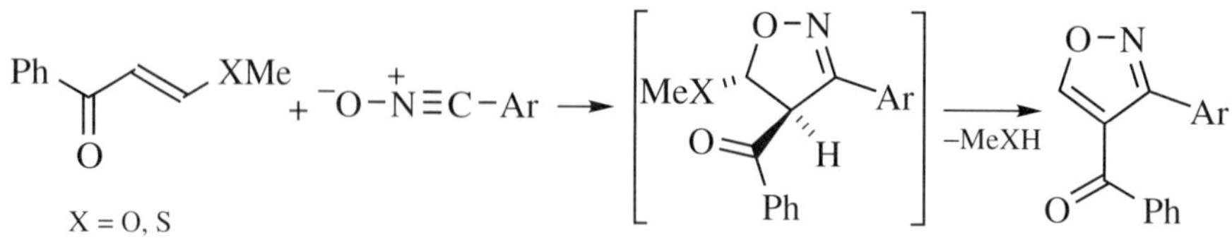

Scheme 6.18

Another cycloaddition to an α,β-unsaturated compound involves the reaction of nitrile oxides with 3-methoxy- or 3-methylthio-1-phenyl-2-propene-1-one (Scheme 6.18) (133,134). The isoxazoles that are isolated are considered to arise from the respective intermediate isoxazolines by subsequent elimination of methanol or methanethiol. The regioselectivity observed was attributed to the presence of substituents with strong electron-donating ability, and this was accommodated in terms of the FMO theory (133,134).

Hydrogen bonding between the oxygen atom of the nitrile oxide and a suitable hydrogen donor in the dipolarophile may also alter the regiochemical outcome of the cycloaddition. Thus, the cycloaddition of benzonitrile oxide to 3-methylcyclopentene in diethyl ether mainly afforded regioisomer **24** (regioisomeric mixture **23:24** rr=4:96) (135) (Scheme 6.19). When the methyl is replaced by a hydroxy group, the cycloaddition gave regioisomers **23:24** in a 35:65 ratio. The amount of regioisomer **23** could be increased to 50% using benzene as solvent, thereby suggesting that some *intramolecular* hydrogen bonding is operating in the transition state. Cycloaddition of 3-benzamidocyclopentene with benzonitrile oxide produced regioisomers **23:24** in a ratio of 90:10. The predominant formation of **23** supports the view that a hydrogen bond from the amide directs the nitrile oxide approach.

Similar to the effects of hydrogen bonding, directing effects have also been encountered with Grignard reagents (136–138). In these cycloadditions, the Grignard reagent may play a dual role, one being to act as a base to generate the nitrile oxide from the hydroximoyl halide. The other effect being that of a cation (Lewis acid) so as to effectively gather the two reactants (nitrile oxide and allylic alkoxide) in the complex (Scheme 6.20).

It was further demonstrated, both by means of experimental (136) and *ab initio* (138) studies, that the use of an allylic alkoxide instead of an alcohol could be

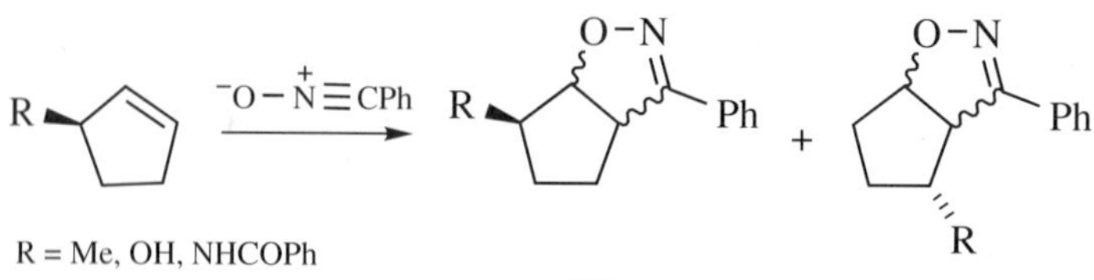

(23) (24)

Scheme 6.19

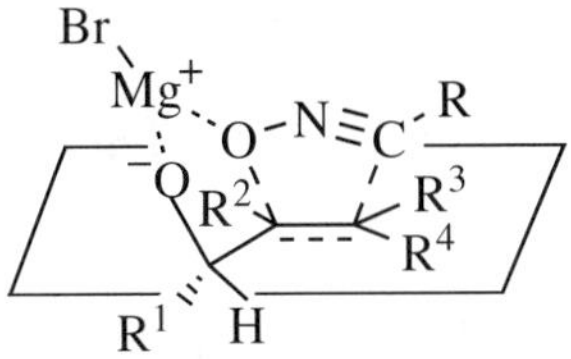

Scheme 6.20

employed to obtain high or complete regioselectivity. This concept was successfully extended to the use of homoallylic alcohols as dipolarophiles, which resulted in excellent regioselectivity (136). The results were again explained in terms of *ab initio* calculations (138). Since these reactions also invoke diastereodifferentiation, this respective study will be discussed in Section 6.2.3.1.

The cycloaddition of some nitrile oxides to *trans*-1,2-disubstituted alkeneboronates afforded the respective isoxazoline-4-boronates with high regioselectivity. These products were then used to prepare the 4-hydroxy derivatives (139) complementing earlier approaches that took advantage of 4-modification (140,141) of furan cycloaddition products (20,21) (see Section 6.4.4).

Reactions of nitrile oxides with 1,1-disubstituted alkenes afford products in which the oxygen atom of the nitrile oxide gets attached to the most crowded carbon atom of the dipolarophile. This high regioselectivity does not seem to depend on the type of substituent present on the alkene (142–152). Some of the results cannot be satisfactorily interpreted on the basis of FMO theory (149,151). Both steric and electrostatic effects often counteract each other and contribute to the regioselectivity actually observed. With trisubstituted alkenes, the orientation of cycloaddition is apparently dominated by this phenomenon. The preference is for the more substituted carbon atom to end up at the 5-position of the heterocyclic product (153,154). However, cases of opposite regiodifferentiation are also found, in particular with donor-substituted alkenes (42,155–157) (Scheme 6.21).

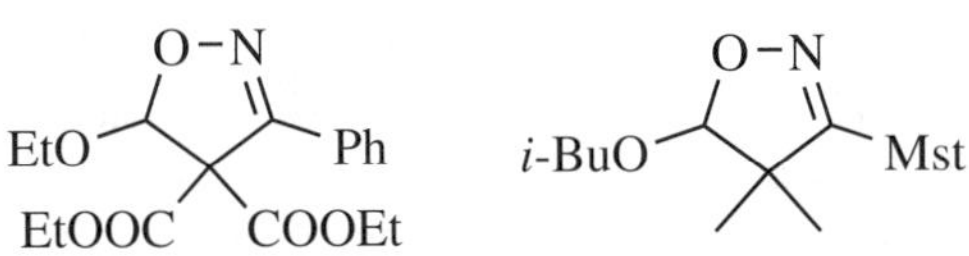

rr >95:5, 22% (157)   rr >95:5, 30% (7 kbar) (157)

Scheme 6.21

6.2.3. Diastereoselectivity of Nitrile Oxide Cycloadditions

The origin of diastereoselectivity in 1,3-dipolar cycloadditions has been intensely studied, and models useful both for rationalization and prediction have been put forward (15,109). The most obvious stereochemical consequence of a concerted cycloaddition is that the configuration about the alkene is retained in the product. This feature continues to be exploited in stereoselective synthesis. Complex problems arise when the alkene (dipolarophile) or the nitrile oxide, or both, carry one or more stereocenters on the side chain. For cases like alkenes, which are asymmetrically substituted at the allylic position, 1,2-induction occurs. This case is easily rationalized when cycloalkenes are used, often leading to high diastereomeric ratios (dr). The results are not easily predicted using acyclic chiral alkenes, where induction may vary to a large extent. Many examples of useful 1,2-induction have been reported over the past 20 years and have increasingly been exploited in organic synthesis.

When substrates such as "α-chiral" allylic alcohols are used, reactions with achiral nitrile oxides are affected both by alkoxide formation and the use of *well-coordinating* cations (136–138). In some cases, hydrogen bonding with the nitrile oxide's oxygen atom can also play an important role (135).

Another approach that relies on asymmetric induction from the alkene part, uses chiral auxiliaries of various types, thereby leading to enantiomerically enriched or pure isoxazoline products. The complexity of some of these auxiliaries is high, and more economical solutions are desirable since the *competition* is the resolution of racemic cycloadducts with an overall efficiency up to 50% yield. With chiral nitrile oxides, the situation is much less satisfactory since asymmetric induction of the 1,4-type (with 1-alkenes) is minimal, and hardly better with a 1,3-relationship of inducing–forming stereocenters, when 1,2-disubstituted alkenes are employed (Scheme 6.22). Upon separation of the two diastereomers, however, another entry to pure optically active isoxazolines is available.

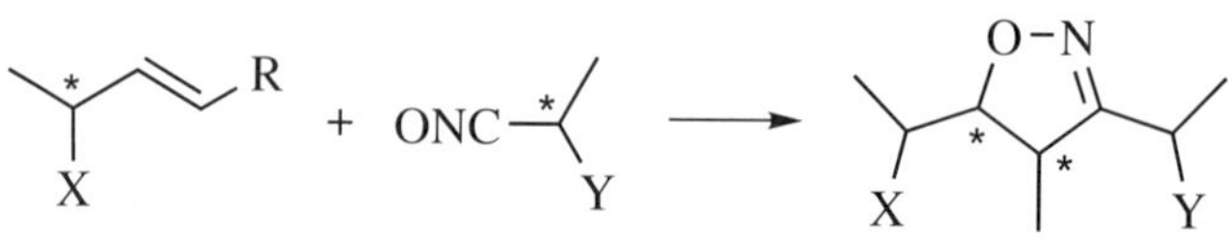

Scheme 6.22

Unlike the impressive progress that has been reported with *asymmetric catalysis* in other additions to alkenes (i.e., the Diels–Alder cycloaddition, epoxidation, dihydroxylation, aminohydroxylation, and hydrogenation) so far this is *terra incognita* with nitrile oxide cycloadditions. It is easy to predict that this will become a major topic in the years to come.

6.2.3.1. Nitrile Oxide Cycloadditions with Chiral Dipolarophiles

A large number of nitrile oxide cycloadditions to chiral olefins were reported from several groups in the early 1980s (16,18,20). A clear rationale of regioselectivity was developed only when the traditional view, which focused on ground-state conformers, was abandoned and calculational approaches to identify transition states and transition state conformations were refined (158,159). Calculations dealing with the case of formonitrile oxide (fulminic acid)–propene clearly showed the importance of both pyramidalization of the alkene part, and the preference of a pseudo-staggered conformation of the methyl group, with a distinct ($\sim 30°$) distortion from the ground-state conformation having one C—H coplanar to the $H_2C{=}CH$ moiety (158,159). Cycloaddition to monosubstituted alkenes preferentially occurs on the face of the alkene that is less sterically shielded in the transition state, and the stereoselectivity increases as the size of the R group attached to the chiral center is increased (Table 6.7).

On the basis of these results, Houk et al. (123) proposed a transition state model comprising an *inside-alkoxy effect*, in order to account for the stereoselectivities of nitrile oxide cycloaddition to a variety of alkenes that contain allylic substitution bearing hydroxy, alkoxy, or related units. The chiral allylic ethers preferably produce the anti (erythro) product, regardless of the nature of the substituent on the allylic oxygen (17,18,20). Allylic alcohols tend to favor the syn (threo) product, but with low stereoselectivity. Calculations on the relative energies of different transition state conformations indicate that an alkyl substituent at the allylic stereocenter prefers the sterically least crowded anti position. The hydroxy group slightly favors the *outside* over the *inside* position so as to maximize hydrogen bonding with the oxygen of the nitrile oxide (Scheme 6.23). Allylic alkoxy and siloxy groups tend to prefer the *inside* position. The preference of the allylic ether for the conformation with an inside-alkoxy group was rationalized in terms of secondary orbital interactions. When the allylic alkoxy group is aligned anti, the CHROR′ group becomes electron withdrawing, since the $\sigma^*_{(CO)}$ orbital overlaps with, and withdraws electron density from the alkene π orbital. With the C—O bond placed inside, it is close to the plane of the C=C-framework and overlap of the $\sigma^*_{(CO)}$ with the $\pi_{C=C}$ is minimized (123). Overlap of the electron-donating σ^*_{CH} and σ^*_{CR} orbitals with the $\pi^*(C{=}C)$ orbital is maximized, and the transition state with the nitrile oxide, which takes on the role of the more electrophilic partner (109,163, 164a), is stabilized. This effect was used to explain the outcome of cycloadditions with chiral allyl ethers derived from 2-hydroxy-3-butenal dithioacetals (165). Here, the stereoselectivity was virtually independent of the bulkiness of the oxygen protecting group or the nature of the sulfur substituent (Table 6.7). Successful application of this concept to other types of stereoselective addition to C=C systems are known (164b).

The stereochemical outcome of the cycloaddition to 3-butene-1,2-diol derivatives, cyclic acetals, or to related alkenes that possess an allylic nitrogen substituent such as 4-vinyl-oxazolines or -oxazolidines was also rationalized by this model (162) (Table 6.7). In the latter cases, the *N*-Boc group instead of the α-oxygen prefers the inside position (Scheme 6.24).

The *alkoxy-inside* model was further adapted in order to rationalize the stereoselectivities of nitrile oxide cycloadditions to alkenes that possess other allylic substituents. In the reaction of "α-chiral" alkenes (124) or allylic diphenylphosphane oxides (161) (Table 6.7), it was suggested that the largest group (L, diphenylphosphinoyl substituent) was anti, the medium sized group (M, alkyl or alkoxy substituent) was on the *inside* and the smallest group (S, hydrogen) was

TABLE 6.7. EXPERIMENTAL RATIOS OF DIASTEREOMERS FOR NITRILE OXIDE CYCLOADDITIONS TO CHIRAL ALLYL-SUBSTITUTED ALKENES[a]

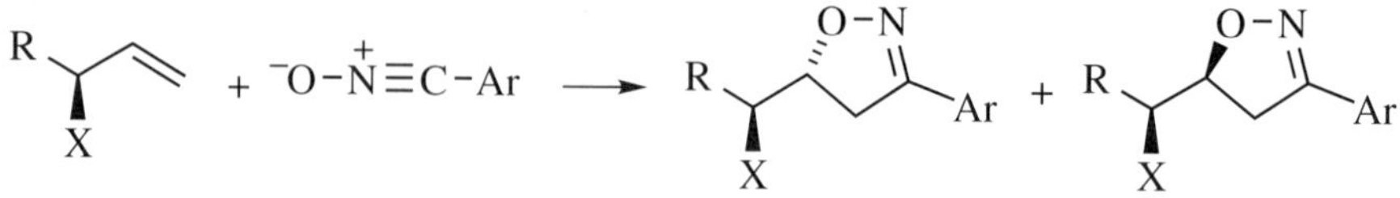

anti (erythro) syn (threo)

Ar	R	X	anti/syn (erythro/threo)
Ph	Me	OH	40:60
p-$O_2NC_6H_4$		OMe	64:36
Ph		OCH_2Ph	64:36
Ph		OTHP	63:37
Ph		$OSiMe_3$	71:29
Ph		$OSiMe_2t$-Bu	72:28
p-$O_2NC_6H_4$		$OSiMe_2Ph$	65:35
Ph		$P(O)Ph_2$	78:22[b]
p-$O_2NC_6H_4$	Et	Me	50:50[c]
		$OSiMe_2Ph$	80:20
$H_{23}C_{11}$		$P(O)Ph_2$	78:22[b]
p-$O_2NC_6H_4$	i-Pr	Me	65:35[c]
		$OSiMe_2Ph$	91:9
p-$O_2NC_6H_4$	t-Bu	Me	77:23[c]
		OH	35:65
		OMe	>95:5
		$OSiMe_3$	>95:5
p-$O_2NC_6H_4$	Ph	OH	44:56
		$OSiMe_3$	69:31
Ph	$CH(St\text{-}Bu)_2$	$OSiMe_2t$-Bu	92:8[d]
	$CH(SPh)_2$	OCH_2Ph	90:10[d]
	$CH(SPh)_2$	$OSiMe_2t$-Bu	91:9[d]
	$CH(SCH_2CH_2CH_2S)$	$OSiMe_2t$-Bu	91:9[d]
Ph	CH_2OAc	OAc	53:47
	CH_2O		69:31
	CH_2OCMe_2O		85:15
	$CH_2OC(CH_2)_5O$		81:19
	CH_2OCOO		82:18
Ph	CH_2OCMe_2BocN		66:34[e]
	CH_2OCPhN		76:24[e,f]

[a] Reference 123 except where noted otherwise.
[d] Reference 160.
[c] Reference 124.
[b] Reference 161.
[e] Reference 162.
[f] Racemic mixture of alkene.

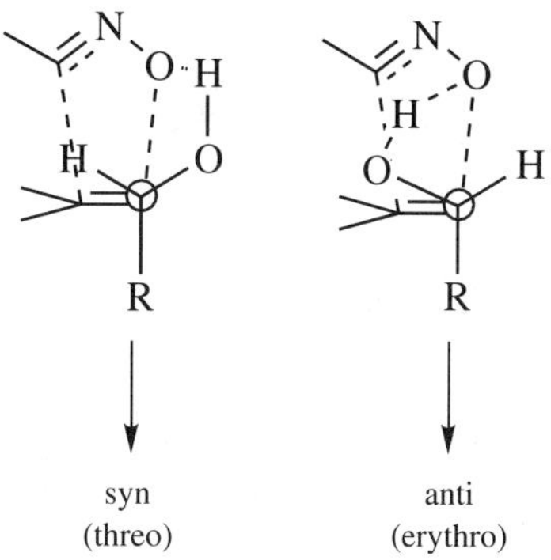

Scheme 6.23

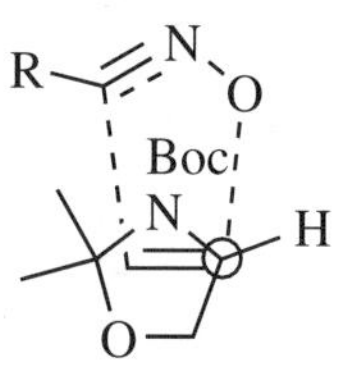

Scheme 6.24

located on the *outside* position, as a result of steric interactions (Scheme 6.25) (see also Section 6.2.1.1). These observations suggest that the *inside-alkoxy* effect is actually a combination of steric repulsion and secondary orbital interactions. The selectivity observed in the cycloaddition to unsaturated sugars was rationalized in terms of this model (22,165).

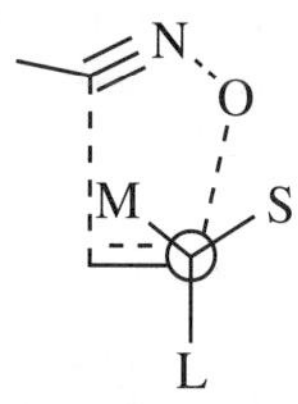

Scheme 6.25

The influence of the size and configuration of various cyclic vinyl carriers (Scheme 6.26) on the stereoselectivity of cycloaddition was studied, and included epoxides, β-lactams, dioxaborolanes, and dioxans (22). Although the anti preference was maintained in all cases, the conformation of the carrier ring must also be taken into account in order to rationalize the stereoselections observed. The highest diastereomeric ratio was observed with the vinyl-tetrahydrofuran derived from glucose, where the conformational mobility of the carrier ring is substantially locked by an acetonide clamp and one face of the C=C bond is effectively shielded (22,165,215).

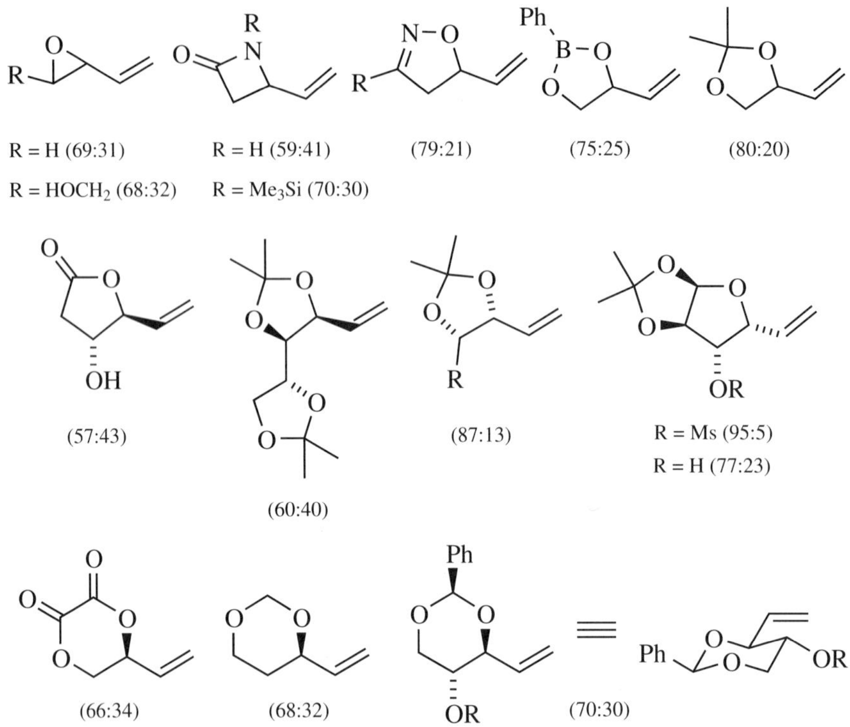

Scheme 6.26

Some additional examples, where the stereochemical outcome of the cycloaddition to chiral alkenes has been explained in terms of the Houk–Jäger model, should also be mentioned. The diastereomer ratio found in the reaction of γ-oxy-α,β-unsaturated sulfones (166), with Morita–Baylis–Hillman adducts [i.e., α-(α′-hydroxyalkyl)-acrylates (167)] (Scheme 6.27), with dispiroketal-protected 3-butene-1,2-diol (168), and with α,β-unsaturated carbonyl sugar and sugar nitroolefin (169) derivatives, all agree well with this model.

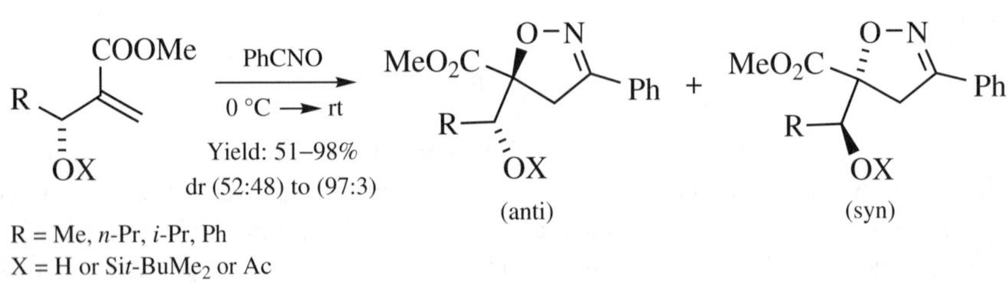

Scheme 6.27

This model of diastereoselectivity, schematized as to show the preferred approach toward erythro products, will surely be of use in stereoselective organic synthesis (Scheme 6.28) (21,22).

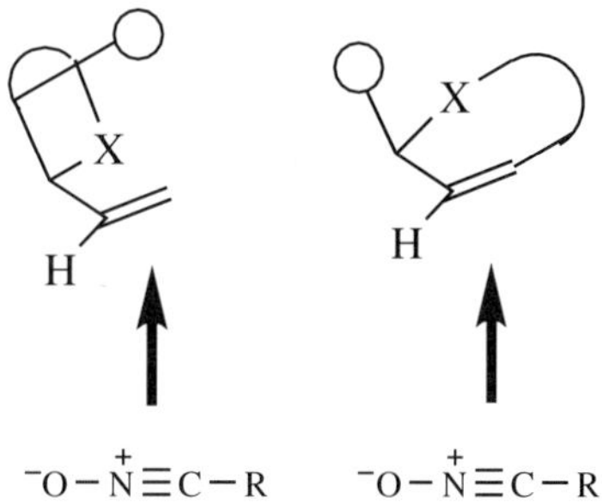

Scheme 6.28

Although the free hydroxyl group may affect the sterochemical outcome of the nitrile oxide cycloaddition by allowing for hydrogen bonding between the olefinic alcohol and nitrile oxide, the stereoselectivity found is normally poor. Hydrogen bonding to the oxygen atom of nitrile oxides [in HCNO the partial charge was found as −0.214 (170)] was inferred from the cycloaddition with allylic alcohols (see above) (170). There is some solvent dependency with cycloalkenols on the syn/anti ratio of products (171,172). A vinyl carbamate cycloadduct was reported to take up a second nitrile oxide at the C=N bond with remarkable ease (173). In line with this observation, Curran et al. (134,174) found that the use of *N*-cyclopentenyl allylic or homoallylic secondary amides allows for good regio- and stereoselection when the N—H bonds are directed toward the alkene (Scheme 6.29). This effect was not encountered, however, with cyclohexene structures or in freely rotating acyclic systems.

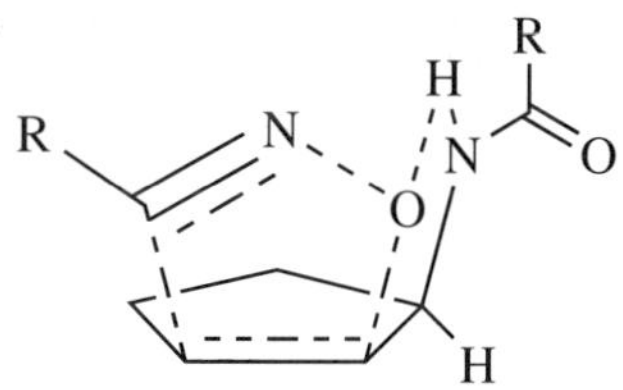

Scheme 6.29

Hydrogen bonding also accounts for the stereoselectivity found in the cycloaddition to 4-substituted 4-hydroxy-2-cyclopentenones (175). The importance of this effect is evident by comparing the results from **26** with those of unprotected **25** as illustrated in Scheme 6.30. The acetylation of the hydroxy group causes a drop in the ratio of diastereoisomers from 85:15 to 57:43 (175).

A recent breakthrough in this field was made by Kanemasa and co-workers (136–138), who outlined a method to overcome the low selectivity of the cycloaddition to allylic alcohols by converting the hydroxyl group to a magnesium alkoxide. Coordination of both reactants to the metal ion accelerates the cycloaddition, affording the isoxazoline product with good to excellent stereoselectivity (136). Two transition states were proposed as pathways to the anti (erythro) and syn

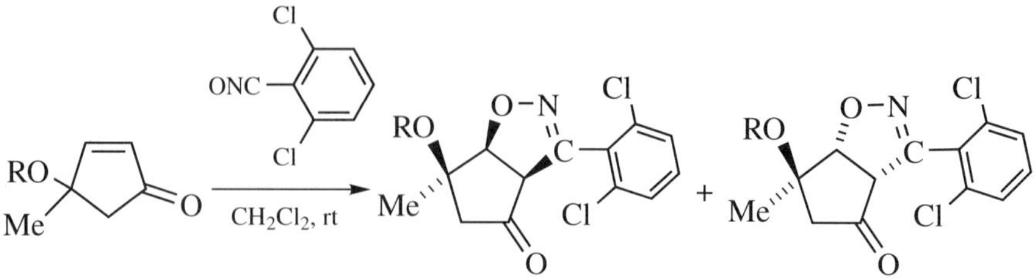

(25)	R = H	(rac)	87%	85	:	15
(26)	R = Ac	(rac)	74%	57	:	43

Scheme 6.30

(threo) cycloadducts, respectively. In both cases, the nitrile oxide and the allylic alkoxide coordinate to the magnesium ion (Scheme 6.31). The transition state leading to the erythro isomer is less favored than the other one due to repulsive interactions between the olefinic substituent R^4 and the α-substituent R^1 placed inside (136) (due to convenience, this is shown with the two enantiomers of the allylic alkoxide).

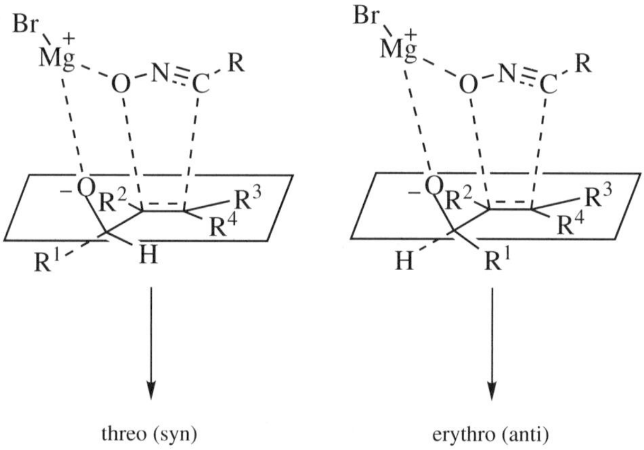

Scheme 6.31

Ab initio calculations also confirm that the use of an allyl magnesium alkoxide in place of the alcohol functionality will lead to high or complete stereoselectivity (138). When homoallylic alcohols are used, the Kanemasa protocol afforded the respective isoxazolines with poor stereoselectivity (~55:45) in the case of terminal alkenes, but with very high diastereoselectivity (up to 96:4) in the reaction of cis-1,2-disubstituted olefins (136). Extension of this concept to the reaction of α-silyl allyl alcohols also proved feasible and produced the syn (threo) adducts as nearly pure diastereomers (≥94:6) (137). Thus, the normal stereoselectivity of the cycloaddition to the Morita-Baylis–Hillman adducts (anti > syn, see above) can be reversed by prior addition of a Grignard reagent (176,177). Both this reversal and the excellent syn (threo) stereoselectivity are nicely accounted for in terms of the transition state model postulated by Kanemasa (176,177).

The use of other metal cations such as those derived from zinc, lithium, or aluminium proved less effective (136). Treatment of allyl alcohol with diethyl zinc in the presence of a catalytic amount of diisopropyl (R,R)-(+)-tartrate (DIPT) in 1,4-dioxane, however, afforded the corresponding (5*R*)-2-isoxazolines with excellent selectivity (*er* $\geq$92:8) (178). Addition of dioxane was necessary in order to avoid precipitation of the complex of zinc salts containing the DIPT moiety. Without this solvent, lower stereoselectivity was found, probably due to the precipitation mentioned above, which prevents the favorable catalytic cycle proposed (Scheme 6.32) (178).

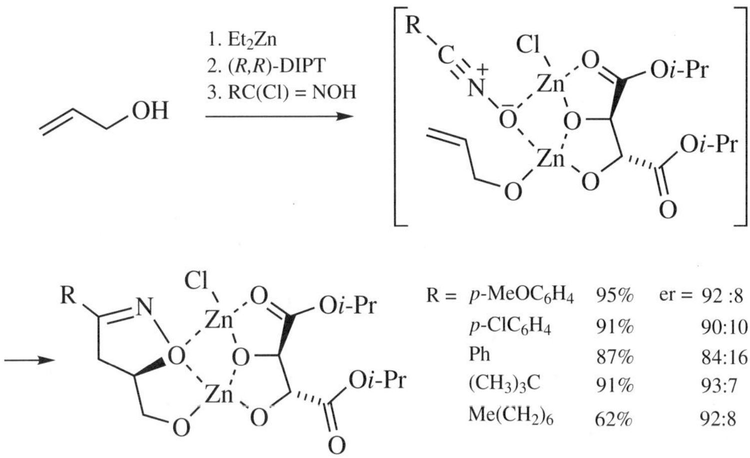

Scheme 6.32

6.2.3.2. *Cycloaddition of Nitrile Oxides with Achiral Olefins Bearing Chiral Auxiliaries*

The use of chiral auxiliaries to induce (or even control) diastereoselectivity in the cycloaddition of nitrile oxides with achiral alkenes to give 5-substituted isoxazolines has been investigated by a number of groups. With chiral acrylates, this led mostly to low or modest diastereoselectivity, which was explained in terms of the conformational flexibility of the vinyl–CO linkage of the ester (Scheme 6.33) (179). In cycloadditions to chiral acrylates (or acrylamides), both the direction of the facial attack of the dipole as well as the conformational preference of the rotamers need to be controlled in order to achieve high diastereoselection. Although the attack from one sector of space may well be directed or hindered by the chiral auxiliary, a low diastereomer ratio would result due to competing attack to the respective π-faces of both the s-cis and s-trans rotamers of the acrylate or amide.

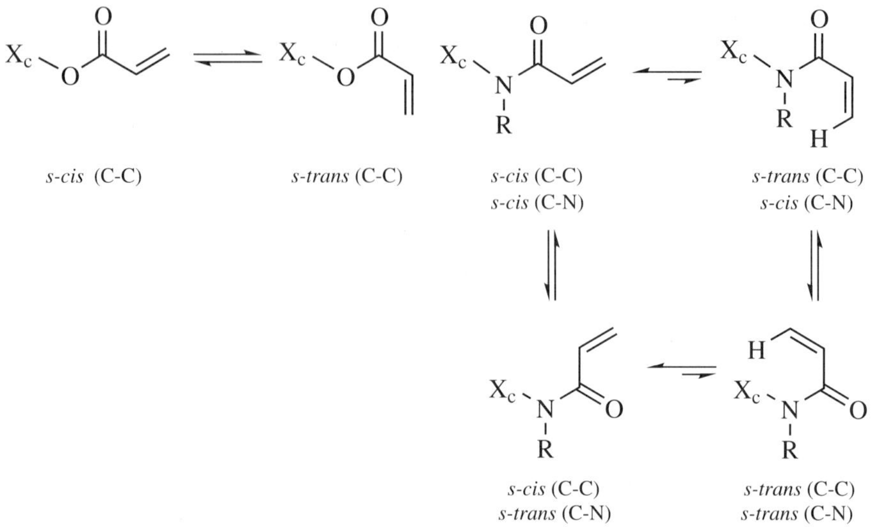

Scheme 6.33

Akiyama et al. (180) overcame this problem by employing *chiro*-inositol derivatives as chiral auxiliaries for the acrylic ester, which afforded dipolar cycloadducts with a high degree of stereoselectivity (Scheme 6.34). Formation of the major products [(5*S*)-isoxazoline-5-esters] was suggested to arise from the s-cis conformer of acrylate **27**, the minor product being derived from the s-trans conformer **28**. The bulky protective group (in this case *tert*-butyldiphenylsilyl) would effectively shield the *Re* face of the olefinic double bond and destabilize the s-trans conformer **28**.

Acrylates with other chiral auxiliaries such as 2-bornyl (181,182), or 2- or 3-menthyl (183) groups afforded the isoxazoline diastereomers with poor-to-good diastereoselectivity. Rotamers about the C—O bonds were not considered in these cases.

A solution to the O—CO—vinyl rotamer problem encountered with esters was found by using chiral amines in the form of tertiary acrylamides. In this case, the (planar) s-trans rotamers should be strongly disfavored (Scheme 6.33, right part). Since the amides supposedly have two low-energy rotamers about the C—N bond, this would again produce diastereomeric products if attack by a reagent occurred from the same direction (space sector). One such case is encountered in the cycloaddition to acrylamide **29**, bearing Oppolzer's chiral sultam auxiliary. This reaction afforded the isoxazoline-5-carboxamide with good selectivity of ($\sim$90:10) (Scheme 6.35) (179,184). The stereochemical outcome was explained in terms of the preferred s-cis conformation for both the N—C and C—C bonds. Of the four planar conformations involving the amide and C=C part, only one lacks the dipole–dipole or steric interactions that destabilize the others. On the basis of a *conventional* steric rationale, the favored approach should then be from the bottom to the *Si* face, the top face apparently being more congested due to the methyl group

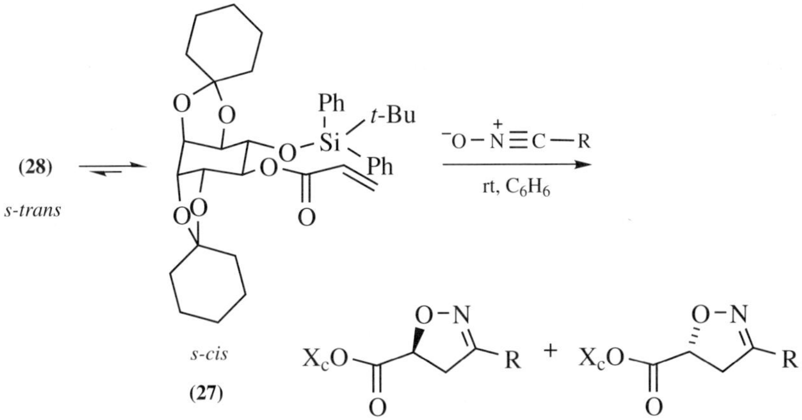

R =			
Me	71%	91	9
Et	80%	90	10
t-Bu	95%	91	9
n-C_5H_{11}	54%	90	10
Ph	81%	95	5

Scheme 6.34

at the bridge (179,184). Since the experimental results show that the dominant product results from *Re* attack, the major approach of the incoming nitrile oxide would have occurred from the top face (see Scheme 6.35). Theoretical studies by Kim et al. (185) suggest that this diastereofacial selectivity arises mainly from Coulomb repulsion between the (pseudo-axial) sultam oxygen and the nitrile oxide oxygen atom (Scheme 6.35). This picture was further supported by results with the analogous isothiazolidine system which lacks the two sulfone oxygen atoms. Here, no stereoselectivity in nitrile oxide cycloadditions was observed (185).

Curran and co-workers (186–188) also employed acrylamides with chiral auxiliaries that are based on Kemp's triacid. Benzonitrile oxide cycloaddition to these derivatives afforded isoxazoline-5-carboxamides with very high selectivity (dr up to >95:5), especially using the *bis*(lactam) **32** shown in Scheme 6.36. The selectivity in the last case was explained in terms of a repulsive interaction between the oxygen of the ring imide carbonyl group in the *bis*(lactam) with the incoming nitrile oxide. In the other cases, the stereoselectivity was thought to arise from steric effects. The inside (bottom) face of the alkene is almost completely blocked by the respective arene appendage (188).

The methyl and benzyl esters of proline were also used as chiral auxiliaries in respective acrylamides, but the isoxazoline cycloadducts were obtained with only poor to modest stereoselectivity (189,190). The related indoline-2-carboxylic acid derivative **33**, however, showed excellent ability to direct nitrile oxide attack, favoring one rotamer (Scheme 6.37), and thereby leading to 3-phenylisoxazoline-5-carboxamide

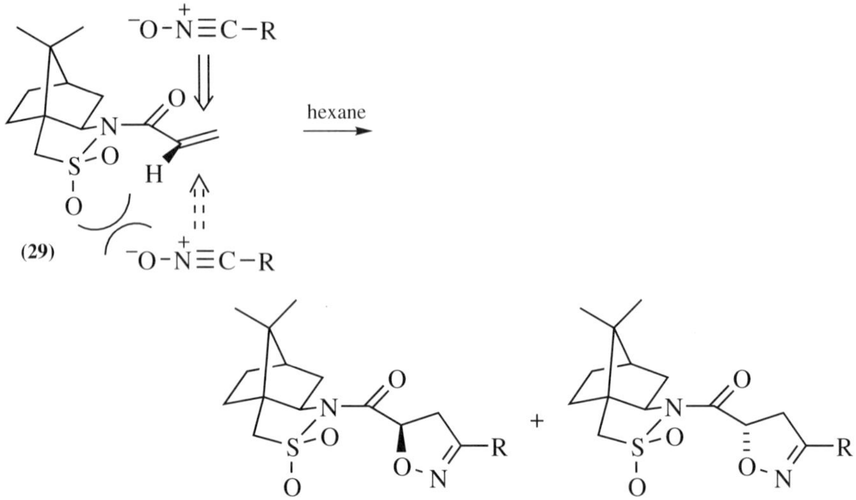

(29)

R =		
Me	90 (73%)	10
Et	90 (74%)	10
t-Bu	95 (73%)	5
Ph	95 (59%)	5

Scheme 6.35

34, with good to excellent stereoselectivity (190). Surprisingly, this selectivity varied with the nitrile oxide substituent (so far unaccounted for).

Cycloaddition to 3-acryloyl-2,2-dialkyloxazolidines (**35**) proceeded in a highly stereoselective manner (Scheme 6.38) (191), but poorly so when 4-benzyl-5,5-dimethyl-2-oxazolidinone (**36**) was used as a chiral auxiliary (Scheme 6.39).

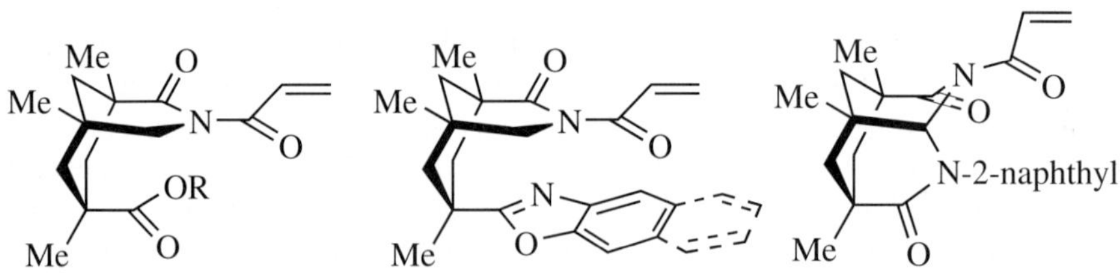

(30) (31) (32)

(30)		(31)		(32)
R = 2-phenylethyl	87:13	benzoxazolyl	96:4	95:5
R = 2-naphthyl	>95:5	naphthoxazolyl	98:2	

Scheme 6.36

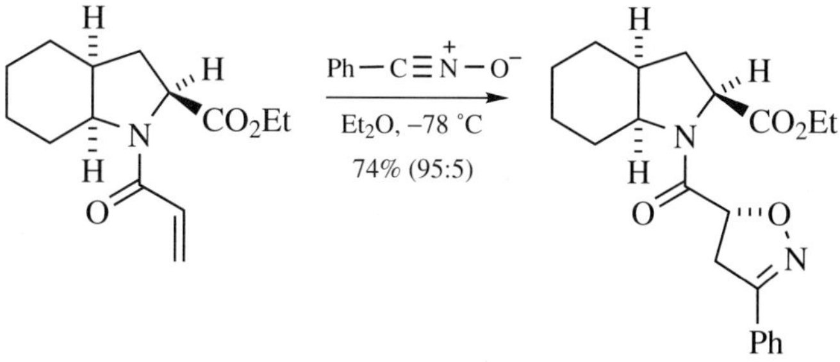

(33) (34)

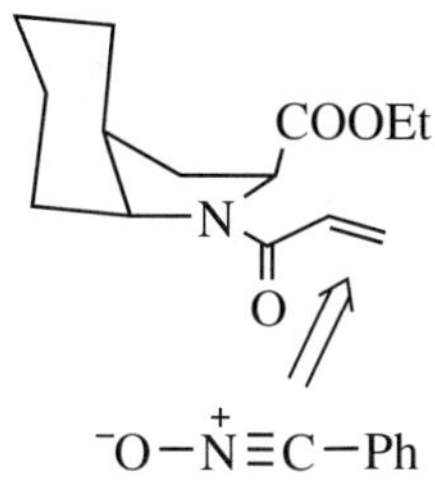

Scheme 6.37

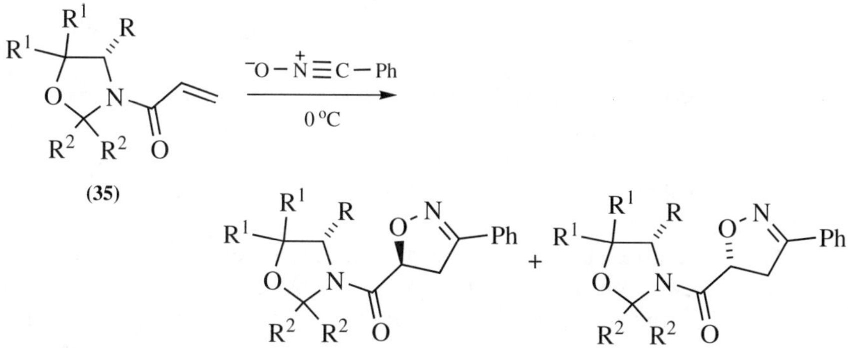

(35)

R = Ph_2CH, R^1 = H, R^2 = Me	quant	85	15
R = $PhCH_2$, R^1 = H, R^2–R^2 = $(CH_2)_5$	87%	93	7
R = R^2 = $PhCH_2$, R^1 = Me	99%	>99	1
R = $PhCH_2$, R^1 = Me, R^2–R^2 = $(CH_2)_5$	quant	>99	1

Scheme 6.38

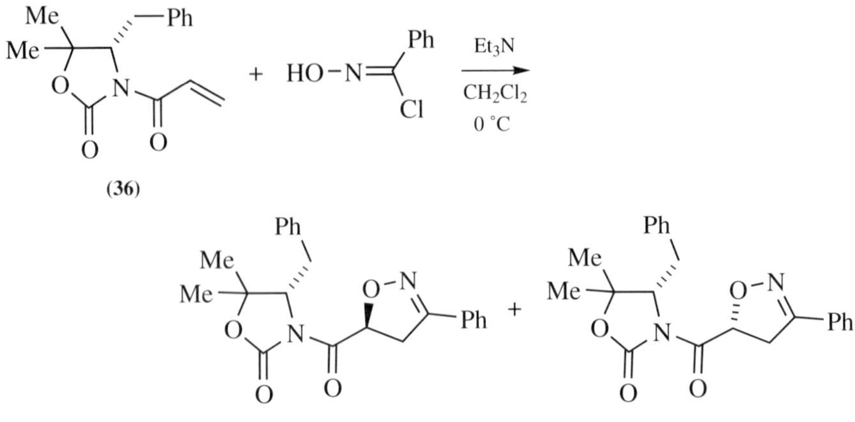

(36)

no additive	95%	43	57
$MgBr_2$	86%	96	4

Scheme 6.39

For the latter case, the stereoselectivity could be improved by the addition of magnesium bromide (192) (Scheme 6.39). The magnesium ion is thought to fix the two carbonyl groups through chelation, thus effecting blockage of the *Re* face of the vinyl group by the benzyl substituent. It was also found that cycloaddition with more nucleophilic nitrile oxides led to higher diastereomer ratios. For example, a dr of 96:4 was found with benzonitrile oxide, and a value of 73:27 with the 4-nitro derivative (192).

The cycloaddition to *N,N*-dimethylacrylamide could also be catalyzed by monoclonal antibodies. The cycloadducts were isolated with excellent selectivity (er 99:1) (193). Unfortunately, a definitive explanation both for the high regio- and stereochemical outcome of this reaction is lacking, and the preparative assets of this approach remain to be seen.

Another approach to obtain pure enantiomers of isoxazolines involves the use of chiral acrolein aminals, formed with *N,N′*-substituted diaminodiphenylethanes (194). Thus, with this chiral imidazolidinyl auxiliary in the β-position, and with unsaturated esters serving as the dipolarophile, benzonitrile oxide afforded only one regioisomeric cycloadduct with good stereoselectivity (194) (Scheme 6.40). When the analogous *N,N′*-dimethyl auxiliary was chosen, excellent stereoselectivity was accompanied by poor regioselectivity (194).

In a related case, the use of dioxazaborocines as the chiral auxiliary with benzonitrile oxide gave dipolar cycloadducts with a poor diastereomeric ratio (*dr* 68:32) (195). Similarly, the cycloaddition of benzonitrile oxide to vinyl ethers with a chiral appendage also proceeded with poor stereoselectivity (dr of 65:35) (196).

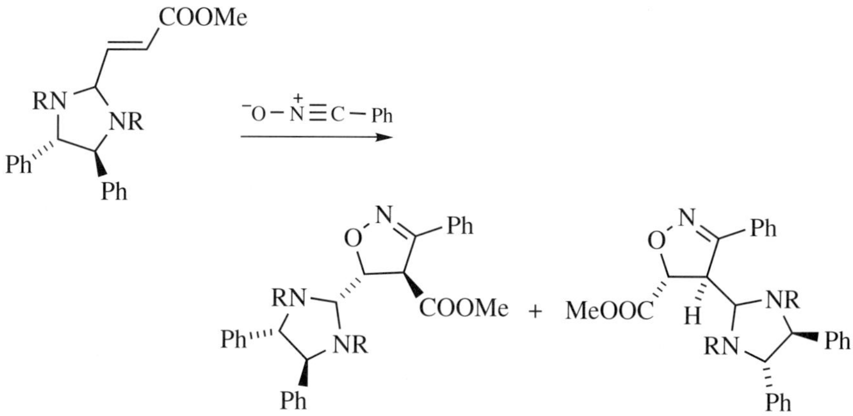

R = Me	34% (> 95:5)	16% (> 95:5)
R = Ph	71% (76:24)	

Scheme 6.40

6.2.3.3. *Face Selectivity of Nitrile Oxide Cycloadditions with Chiral Cycloalkenes and the Like*

A number of examples involving nitrile oxide cycloadditions to cyclic cis-disubstituted olefinic dipolarophiles was presented in the first edition of this treatise, notably to cyclobutene, cyclopentene, and to 2,5-dihydrofuran derivatives (15). The more recent examples discussed here also show, that the selectivity of the cycloaddition to 1,2-cis-disubstituted cyclobutenes depends on the type of substituent group present (Table 6.8; Scheme 6.41). The differences found can be explained in terms of the nonplanarity (i. e., pyramidalization) of the double bond in the transition state (15) and steric effects. In the cycloaddition to cis-3,4-diacetyl- (197) and cis-3,4-dichlorocyclobutene (198), the syn-pyramidalization of the carbon atoms of the double bond and the more facile anti deformability of the olefinic hydrogens have been invoked to rationalize the anti selectivity observed.

The opposite result was found in the cycloaddition to bicyclo[3.2.0]hept-2-ene and bicyclo[4.2.0]oct-2-ene (200). In these cases, the syn-bending of the olefinic hydrogens in the ground state and the more stable inward conformation of the cyclopentane (199) ring were assumed to be responsible for the observed syn selectivity of the cycloaddition. Surprisingly, introduction of an ester group instead of a hydrogen did not result in any distinct changes in the stereochemical course of the reaction (200). Pyramidalization of the carbon atoms of the double bond was assumed to be responsible for the selectivity in the cycloaddition to 3-acyl-2-oxa-3-azanorborn-5-ene (201). As expected, steric effects could also overcome factors related to ground-state pyramidalization and lead to the *normal*, expected outcome (202).

TABLE 6.8. CYCLOADDITIONS TO CYCLOBUTENES

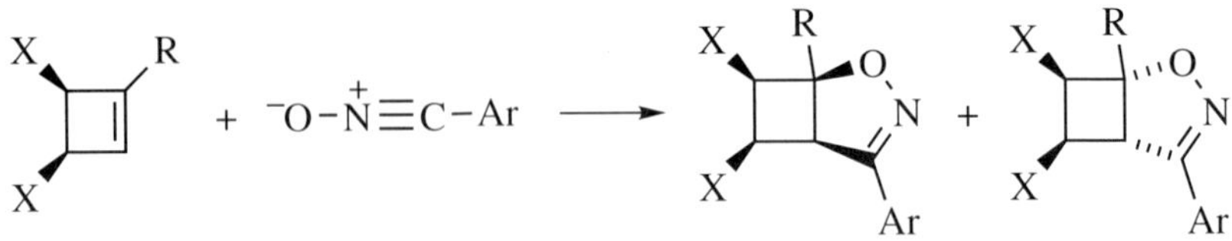

Scheme 6.41

Ar	X	R	syn/anti	Reference
Me	Cl	H	70:30	198
Ph	OAc	H	90:10	197
Ph	O–O	H	36:64	202
Ph	OCOO	H	17:83	197
Ph	$(CH_2)_3$	H	"0:100"	200
		CO_2Me	"0:100"	
Ph	$(CH_2)_4$	H	12:88	200
		CO_2Me	11:89	
Ph	Me	CO_2Me	55:45	200

More remote directing substituents are less effective. For example, the cycloaddition to 5-substituted 2-methyleneadamantanes showed poor face selectivity (203).

6.2.3.4. *Cycloadditions of Chiral Nitrile Oxides to Olefins*

Optically active aldehydes are available in abundance from amino and hydroxy acids or from carbohydrates, thereby providing a great variety of optically active nitrile oxides via the corresponding oximes. Unfortunately, sufficient 1,4- or 1,3-asymmetric induction in cycloaddition to 1-alkenes or 1,2-disubstituted alkenes has still not been achieved. This represents an interesting problem that will surely be tackled in the years to come. On the other hand, cycloadditions with achiral olefins lead to 1:1 mixtures of diastereoisomers, that on separation furnish pure enantiomers with two or more stereocenters. This process is, of course, related to the separation of racemic mixtures, also leading to both enantiomers with 50% maximum yield for each. There has been a number of applications of this principle in synthesis. Chiral nitrile oxides are stereochemically neutral, and consequently 1,2-induction from achiral alkenes can fully be exploited (see Table 6.10).

Examples of optically active aldehydes or nitroalkanes that have been used for the generation of nitrile oxides (mostly via hydroximoyl chlorides) and subsequent cycloadditions to olefins are collected in Table 6.9.

As mentioned earlier, the cycloaddition of chiral nitrile oxides to achiral alkenes generally results in poor stereoselection. The cycloaddition of glyceronitrile oxide acetonide and 2-*O*-benzyl-glyceronitrile oxide to mono-, 1,1-di- and 1,2-disubstituted olefins have been studied most extensively (18,23,121,207,215,221,225,234).

TABLE 6.9. PRECURSORS OF CHIRAL NITRILE OXIDES

Entry	R^*CHO or $R^*CH_2NO_2$	Source	Reference (R^*CHO) or ($R^*CH_2NO_2$)	References (R^*CNO)
1	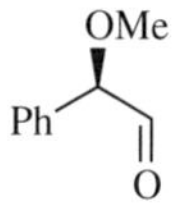	mandelic acid	204	204
2	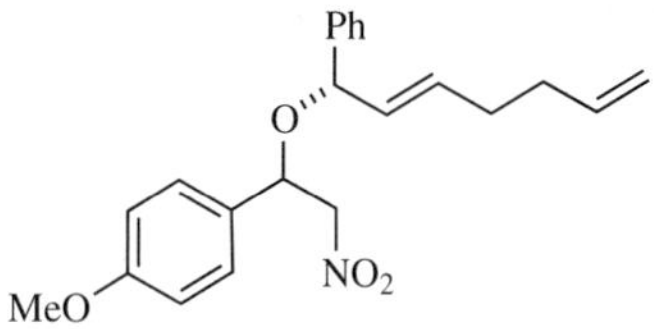	mandelic acid	205	205
3	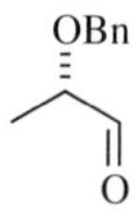	(*S*) : ethyl L-lactate	206	207
4	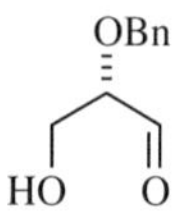	(*S*) : diethyl L-tartrate (also D)	208,209	23,42,59
5	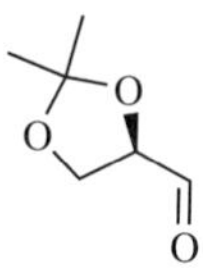	(*R*): D-mannose (*S*) : L-gulonolactone	210–213	23,59, 214–217
6	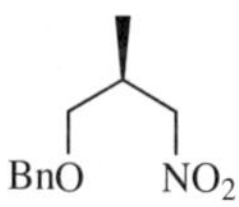	(*S*)-(+)-3-hydroxy-2-methylpropionic acid	218	218
7	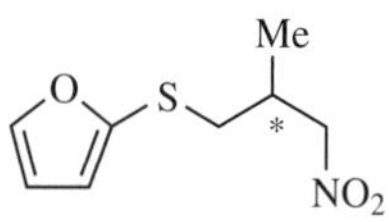	as above	219	219
8	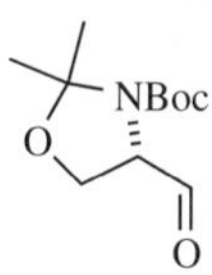	L-serinal	213,220	221
9	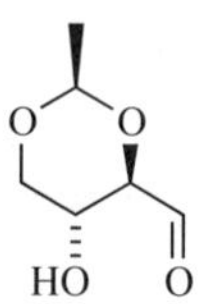	D-glucose	59,222	23,34,59

TABLE 6.9. (*continued*)

Entry	R^*CHO or $R^*CH_2NO_2$	Source	Reference (R^*CHO) or ($R^*CH_2NO_2$)	References (R^*CNO)
10	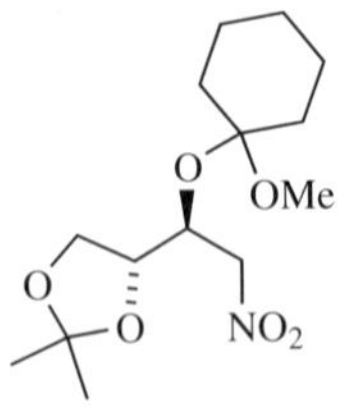	D-glyceraldehyde acetonide (D-mannose)	223	223
11	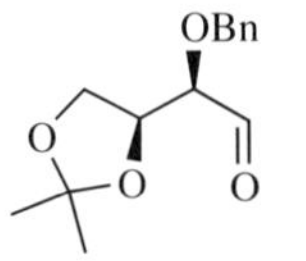	diethyl L-tartrate (also D-series)	224	221,225
12	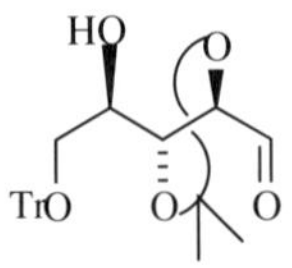	D-ribose	226	227
13	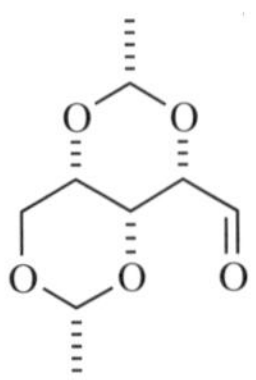	D-sorbitol	228	221
14	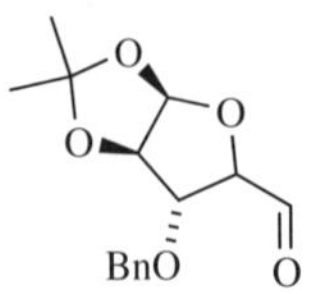	D-glucose	229	221
15	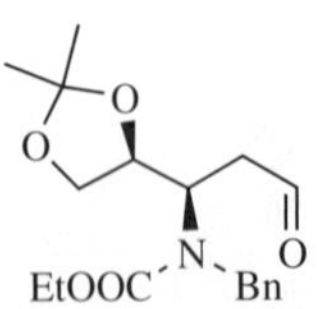	D-glyceraldehyde	221,230	221

TABLE 6.9. (*continued*)

Entry	R^*CHO or $R^*CH_2NO_2$	Source	Reference (R^*CHO) or ($R^*CH_2NO_2$)	References (R^*CNO)
16	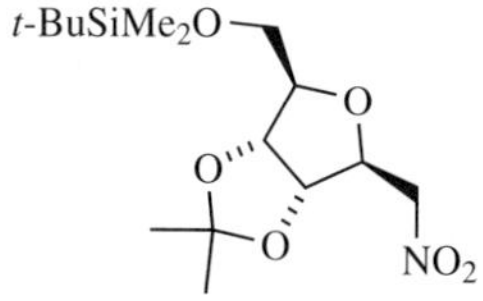	D-ribose	231	231
17	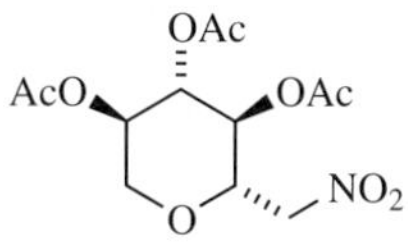	D-xylose	232	233

TABLE 6.10. CYCLOADDITIONS OF CHIRAL NITRILE OXIDES (PRECURSORS SEE TABLE 6.9) WITH ACHIRAL ALKENES

Cycloadduct		dr (rr)	Yield (%)	References
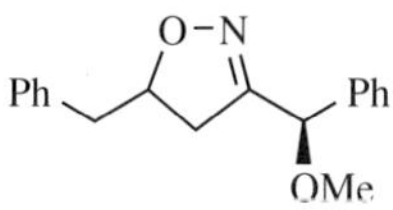		57:43	43	204
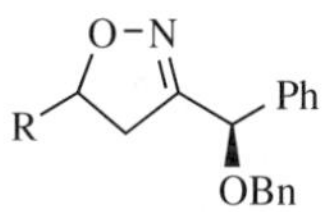	R:CO_2Me	50:50	82	207
	R:CH_2OH	50:50	78	
	R:$ClCH_2$	50:50	79	
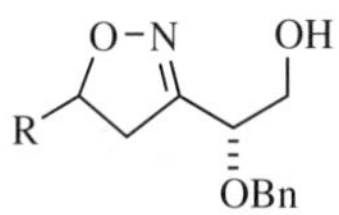	R:H		90	225
	R:Ph	53:47	39 + 41 (syn,anti)	225
	R:$ClCH_2$	50:50	46 + 46	225,234
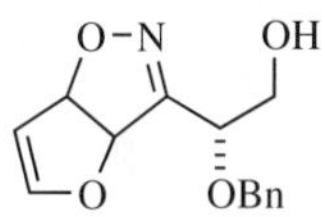		53:47	21 + 20	225,235
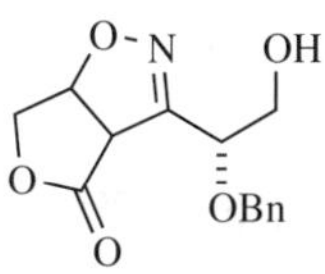		53:47 (82:18)	69	121

TABLE 6.10. (*continued*)

Cycloadduct		dr (rr)	Yield (%)	References
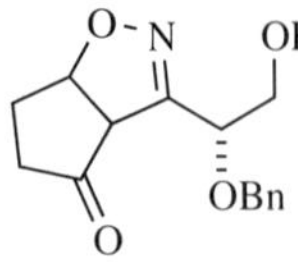		59:41 (97:3)	77	121
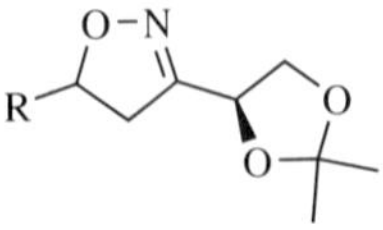	R:ClCH$_2$	50:50	39 + 35	225,234
	R:Ph	50:50	52	236
	R:MeOOC	50:50	80	216
		50:50	22 + 20	234
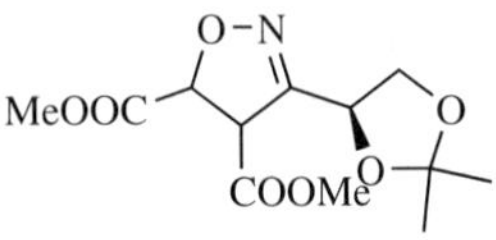	cis	70:30	33	236
	trans	50:50	62	236
		56:44	95	215
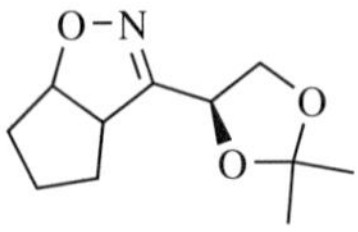		65:35	49	236
		57:43	73	214
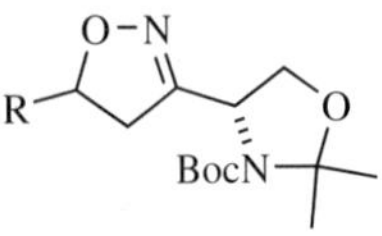	R:Ph	59:41	66	237
	R:C1CH$_2$	50:50	30 + 30	221
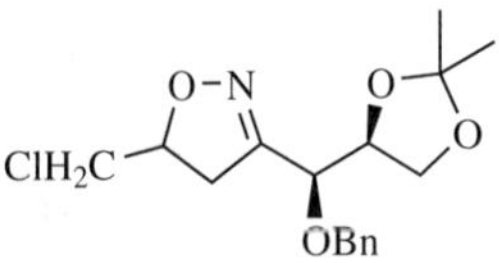		52:48	32 + 31	225
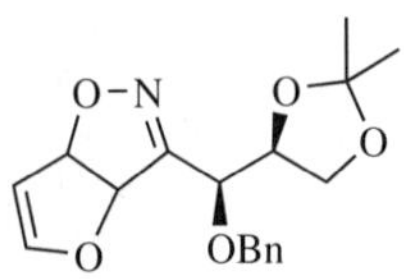		52:48	30 + 29	221,225
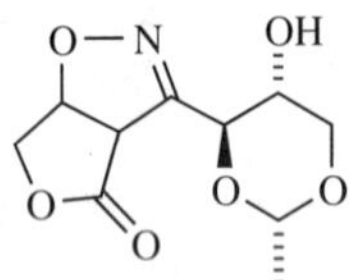		75:25 (81:19)	77	59
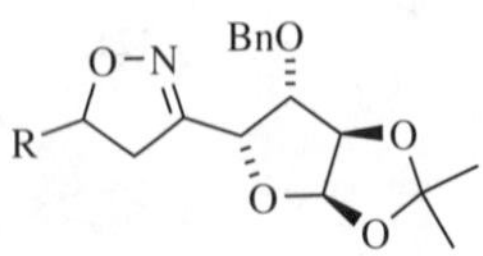	R:Ph	50:50	65	229
	R:ClCH$_2$	54:46	39 + 36	214

TABLE 6.10. (*continued*)

Cycloadduct		dr (rr)	Yield (%)	References
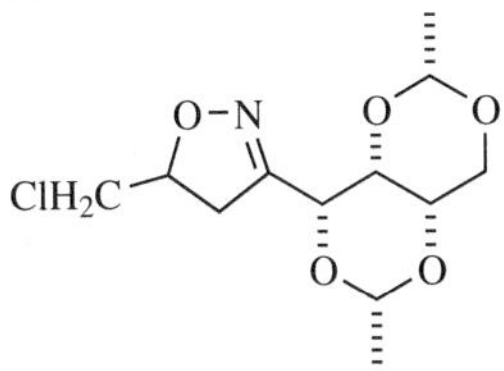		51:49	39 + 37	221
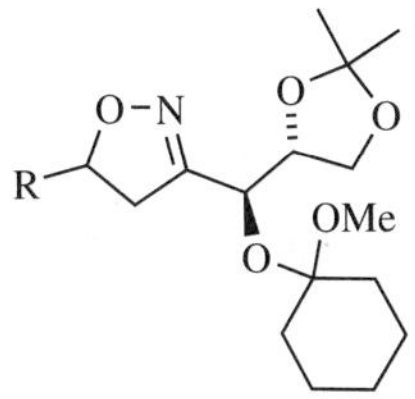	R:Ph R:C_5H_{11}	50:50 50:50	50–90%	223
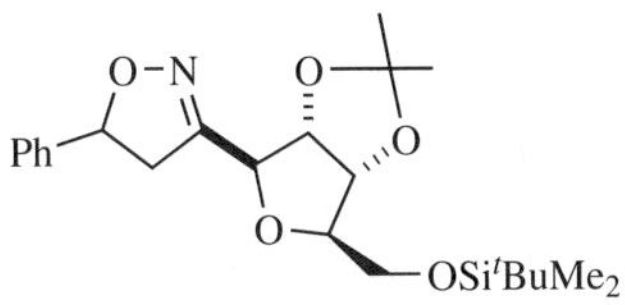		67:33 (α-anomer 50:50)	no yields are reported	231

The reactions generally produce a 1:1 mixture of diastereomeric isoxazoline adducts (Table 6.10), except for a few cases with diastereoselectivity of 70:30 (221). A selection of results concerning diastereomer ratios and yields is gathered in Table 6.10.

Cycloadditions to racemic mixtures of chiral alkenes also show that there is virtually no induction from the nitrile oxide part, meaning that there is no effect of matching–mismatching of the partners in the transition state. In the reaction with glyceraldehyde, and threose- and glucose-derived nitrile oxides, only the "usual"

TABLE 6.11. CYCLOADDITIONS OF CHIRAL NITRILE OXIDES TO CHIRAL ALKENES

Cycloadduct	dr	Yield (%)	Remarks	References
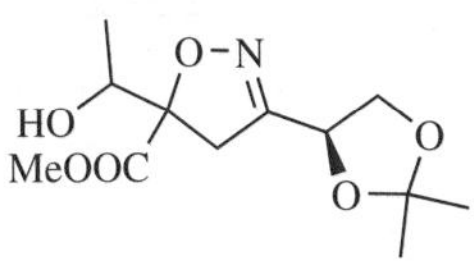	37:37:13:13	45	racemic olefin	215
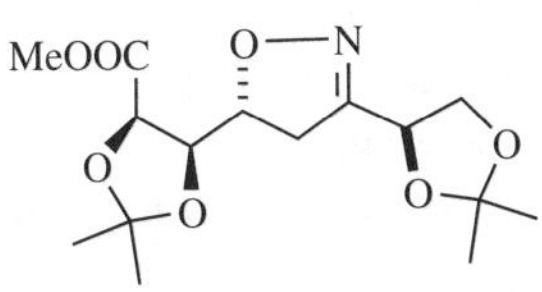	88:12	47	same dr as with achiral RCNO (22)	215

TABLE 6.11. *(continued)*

Cycloadduct	dr	Yield (%)	Remarks	References
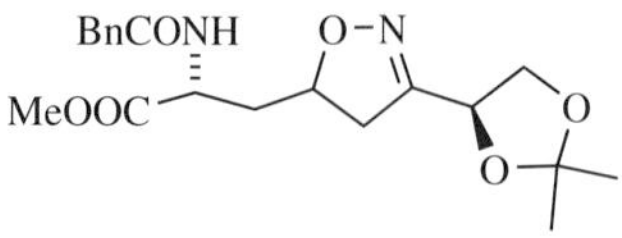	55:45	62		215
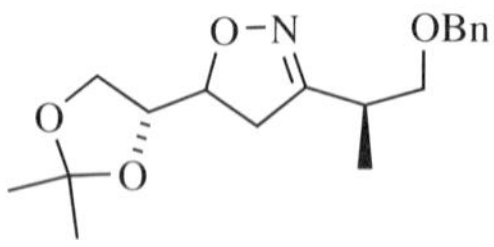	80:20	78	same dr as with achiral RCNO (20, 22, 122)	218
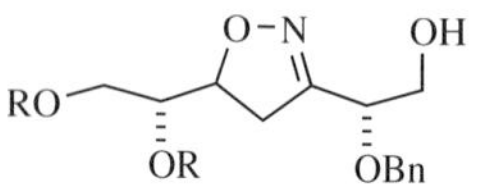 R: Ac R–R: CMe_2	29:29:21:21 39:39:11:11	65 73 + 22	Racemic olefin	225
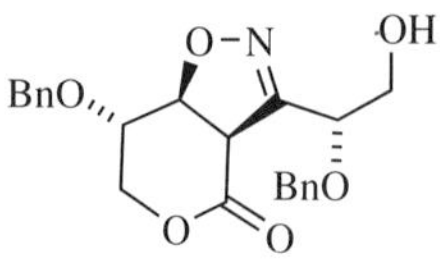	>95:5 (rr > 95:5)	52	5-regioisomer may have been formed	121
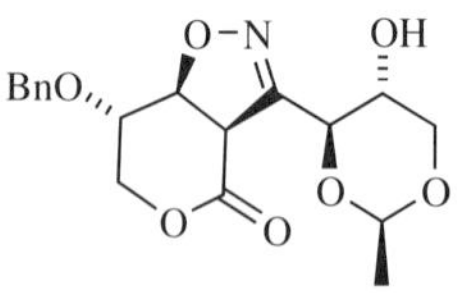	>95:5 (rr 67:33)	89	minor 5-regioisomer also with dr > 95:5	59
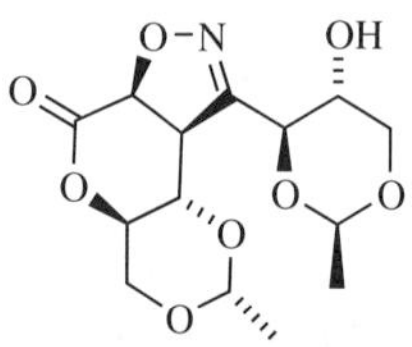	>95:5 (rr < 5:95)	58	usual regioselectivity reversed	59
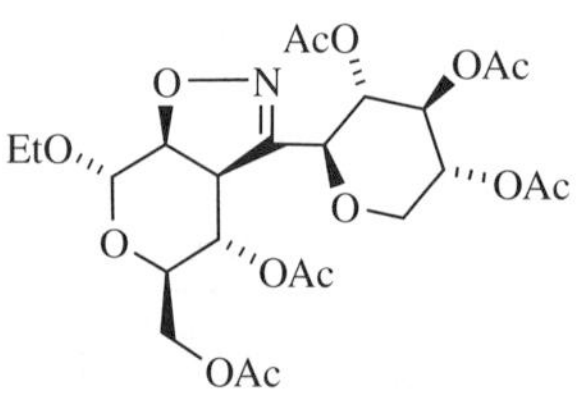	50:50	66		233

induction from the alkene was observed (Table 6.11). Of particular value for synthesis are cycloadditions to pure enantiomers of alkenes with good diastereo-differentiating ability. These reactions afford 3,5- or 3,4,5-substituted isoxazolines with high selectivities, and are similar to the reactions observed with achiral nitrile oxides. Some characteristic examples are presented in Table 6.11.

6.3. REGIO- AND STEREOSELECTIVITY IN INTRAMOLECULAR NITRILE OXIDE CYCLOADDITIONS

6.3.1. Regioselectivity in Intramolecular Nitrile Oxide Cycloadditions

Intramolecular nitrile oxide cycloadditions were first studied by Garanti and co-workers (24) in 1975, employing O-allyl derivatives of salicylic aldehyde. The first example of a carbocycle-forming process was reported in 1977 (25). This process (sometimes referred to as INOC) has seen many extensions and applications for the synthesis of natural and unnatural products alike, notably by the groups of Kozikowski, Curran, Fukumoto, and Shishido (see Section 6.4).

Many aspects of intramolecular nitrile oxide cycloadditions are similar to those of the intermolecular ones. Due to the proximity of the reacting groups, however, there are also several items that differ significantly. While HOMO–LUMO interactions and steric effects direct the intermolecular nitrile oxide cycloaddition to 1-alkenes to produce 5-substituted isoxazolines, the intramolecular cases often show a different behavior. With most of them, regioselectivity is determined by geometric constraints and cycloadditions occur in the *exo* mode to furnish the annulated bicycle (Scheme 6.42).

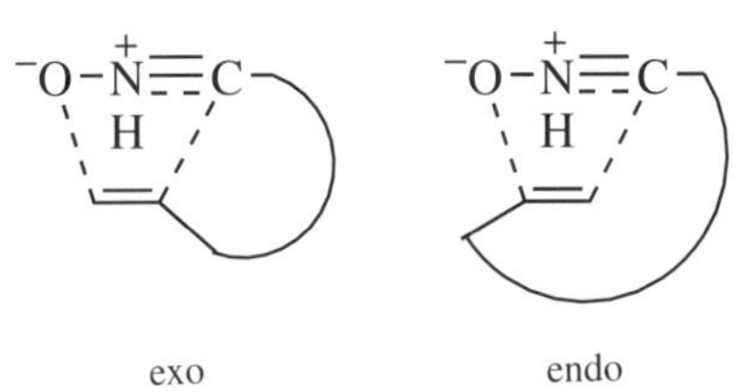

Scheme 6.42

Examples with newly formed five- and six-membered rings (cyclopentanes or cyclohexanes) abound, but altogether ring sizes of 4–19 have been observed. With long connecting tethers, the endo product (bridged bicycle) prevails, as was encountered in the intermolecular reaction and the intramolecular version has found a number of uses for the synthesis of macrocycles (204,238,246). Although there has been only one detailed study dealing with the effect of chain length on exo/endo competition and the minimum chain length required for endo cycloaddition (238), examination of the results gathered in Table 6.12 shows that the endo product is increasingly found when the link between the dipole and the dipolarophile has nine or more atoms. In contrast to the regioselectivity observed in the intermolecular cases, the regiochemical outcome of the intramolecular nitrile oxide cycloaddition is insignificantly affected by the substitution pattern of the olefin (entries 2–5 and 10–13). Also changes in the alkyl chain did not seem to affect the regioselectivity. Similar results were found in the cycloadditions of norbornadiene-tethered nitrile oxides (86,87).

TABLE 6.12. EFFECT OF CHAIN LENGTH IN INTRAMOLECULAR NITRILE OXIDE–ALKENE CYCLOADDITIONS[a]

Entry	Nitro Compound or Oxime		$n+2$[b]	Yield (%) exo	Yield (%) endo	Reference
1	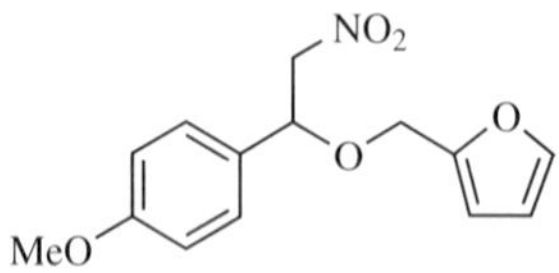		5	84		239
2	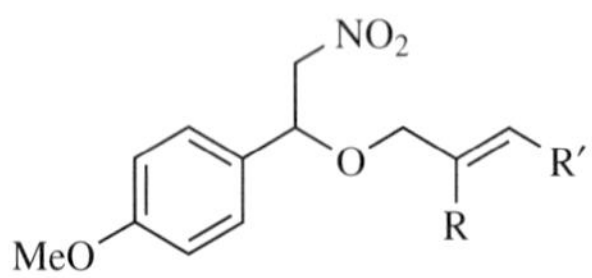	R = R′ = H	5	87		240
3		R = H; R′ = Me		79		
4		R = H; R′ = Me		78		
5		R = Me; R′ = Ph		84		
6	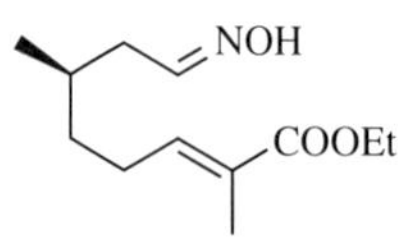		6	84		241
7	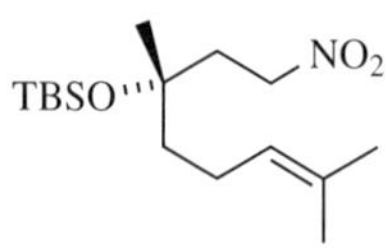		6	100		242
			6			243
8	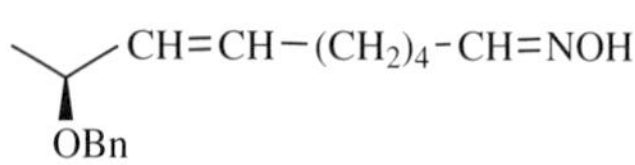	cis		62		
9		trans		70		
10	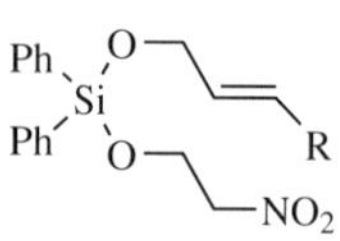	R = H	7	60		244
11		R = Ph		65		
12		R = Me		70		
13		R = CO_2Et		67		
14	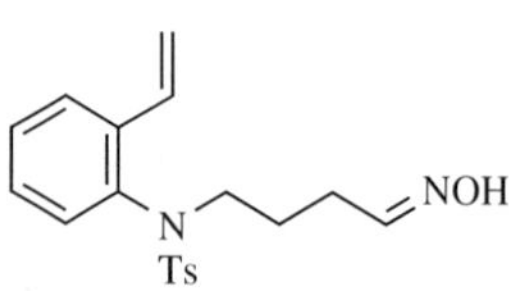		8	45		245
15	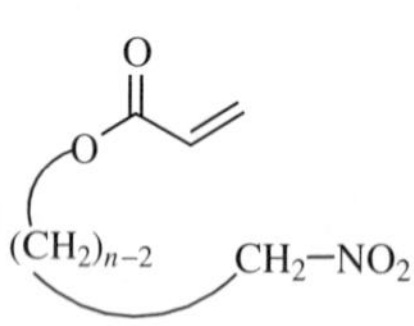		9	44		238
16			12	14	68	
17			14	10	64	
18			15		67	
19	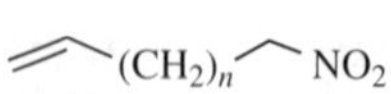		9		21	246
20			11		44	
21			12		63	
22			13		60	

TABLE 6.12. (*continued*)

Entry	Nitro Compound or Oxime		$n+2^b$	Yield (%) exo	Yield (%) endo	Reference
23	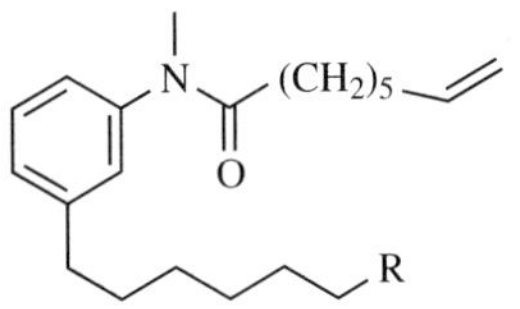	R = HC=NOH	18		51	204
24		R = CH_2NO_2			82	
25	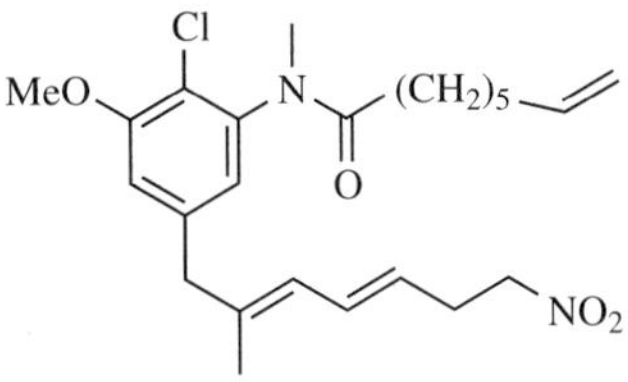		18		50	204

[a] Reaction conditions: For ω-nitroalkenyl educts treatment with PhCNO (or the like)/Et_3N, benzene (or the like), heating; for oximes halogenation with NCS (or the like)/Et_3N. Intermediate hydroximoyl chlorides usually are not observed.
[b] n = Number of chain atoms between the dipole and the dipolarophile; $(n+2)$: size of the new ring.

Side reactions that occur with intramolecular cycloaddition, such as linear oligomerization or dimerization of the nitrile oxide, are not very common when shorter chain lengths ($n \leq 7$) are used due to the entropically favored intramolecular process. A rather unusual result in this regard involves the formation of a fused cyclooctane instead of the less-strained six-membered ring (also fused) in the cycloaddition of the nitrile oxide derived from *p*-naphthoquinone (Scheme 6.43). This result is consistent with the effect of electron-withdrawal in the enedione part, leading to increased reactivity (247), and also reflects the known sluggishness of cyclohexenes towards nitrile oxides (cf. Section 6.2.1.2).

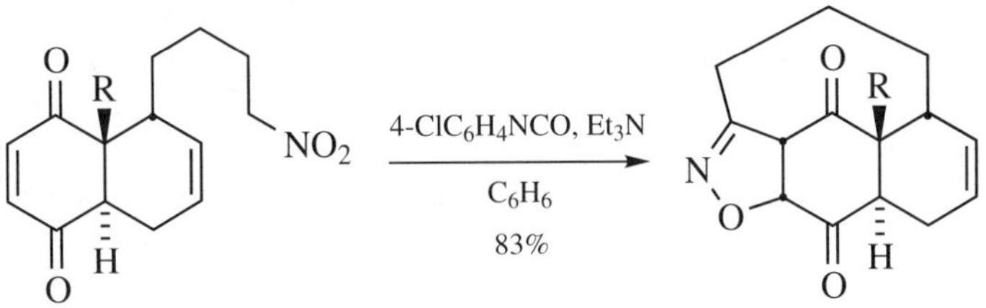

R = CO_2Me

Scheme 6.43

Nagaoka and co-worker (248) sought to apply an intramolecular cycloaddition for the synthesis of the taxane A/B rings, starting with the nitro derivative **37**

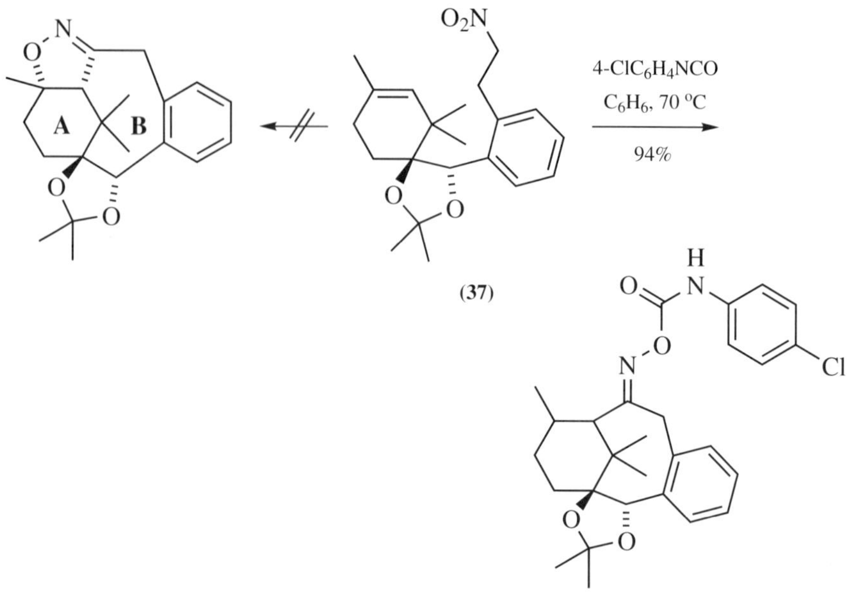

(37)

(38)

Scheme 6.44

(Scheme 6.44). The reaction gave the oxime derivative **38**, however, and the expected isoxazoline was not detected, which was attributed to geometric factors that only permit a vertical orientation, not the normal in-plane approach of the nitrile oxide unit with respect to the double bond of ring A. The preferred orientation only allows the cyclization to take place (248).

6.3.2. Stereoselectivity of Intramolecular Nitrile Oxide Cycloadditions

The stereoselectivity of intramolecular nitrile oxide cycloadditions has thoroughly been studied by several research groups. Normally, closure to a six-membered (annulated) ring proceeds with much higher stereoselectivity than that leading to a five- or seven-membered one (249–251). This result was explained by the preference of a chair over a boat transition state when forming six-membered rings. In the more flexible transition states leading to five- or seven-membered rings, the difference in energy between the competing transition state conformations was assumed to be much smaller, thereby leading to a mixture of stereoisomers (252).

The stereochemical outcome of such cycloadditions may be altered by substituents attached to the nitrile oxide–olefin linker. Hassner and co-workers (75,240,253–255) and Kurth and co-workers (256) examined the influence of a stereogenic center α to the dipole in the cycloaddition of alkene-tethered nitrile oxides that feature a sulfur or oxygen atom within the connecting chain (Table 6.13). As expected, the diastereofacial selectivity is increased in the presence of fragments with increasing steric demand. Cycloadditions of thioethers show lower selectivity than those of the oxygen analogues (256), which was attributed to an increased bond length of the C—S vs C—O single bond, thus offering a less-constrained transition state for the thioether case (256). The effect of the oxygen or sulfur atom on face selectivity was demonstrated by using a carbon analogue, which underwent dipolar cycloaddition with low selectivity (256). The rationale was that an allylic heteroatom causes a significant stereoelectronic effect as is known from the many intermolecular cases. Cycloadditions leading to 5,6-annulated heterobicycles ($n = 2$) demonstrated that the reaction proceeded with complete face selectivity in the ether cases. Moreover, the preference for the trans-isomer was now reversed in favor of formation of the cis-isomer (Table 6.13). This result was explained by a preferred chair-like transition state conformation, placing the

TABLE 6.13. STEREOSELECTIVITY IN INTRAMOLECULAR CYCLOADDITIONS WITH "α-CHIRAL" NITRILE OXIDES AND C-, O-, OR S-CONTAINING ALKENYL CHAIN

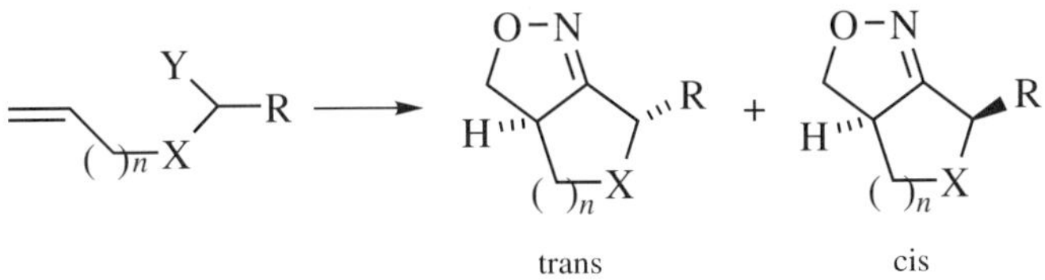

Y = CHNOH or CH_2NO_2

X	n	R	trans/cis[a]	Reference
		Me	50:50	254
S	1	Et	50:50	254
		i-Pr	50:50	254
			55:45	256
		t-Bu	67:33	254
			62:38	256
		Ph	60:40	254
S	2	*i*-Pr	22:78	254
		Ph	40:60	254
O	1	Me	71:29[b]	75
		i-Pr	80:20	256
		t-Bu	90:10	256
		Ph	90:10	256
			80:20[b]	75
O	2	Me	only cis[b]	269
		i-Pr	only cis	256
		Ph	14:86	75
			only cis	256
C	1	Me	60:40[b]	256

[a] Cycloadditions performed with the ω-nitroalkene except noted otherwise.
[b] Starting from oxime.

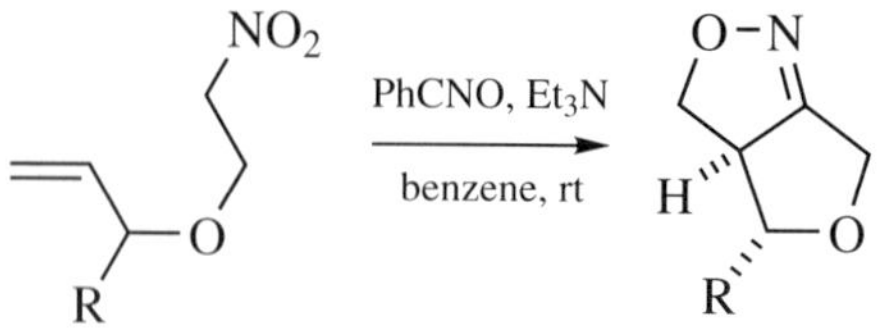

R :	Yield
Me	62%
i-Pr	79%
Ph	68%

Scheme 6.45

substituent at the nitrile oxide carbon in a quasiequatorial position (256). Also, a force-field model was employed to account for the observed product ratios (257).

When the stereogenic center is situated at the allylic position, as in nitroethyl ethers derived from 5,6-unsaturated nitro compounds, only the endo-isomer (trans) of the bicyclic structure was formed (Scheme 6.45) (256).

A directing effect of a methyl group at the allylic stereocenter, located between the nitrile oxide and the alkene moieties, on the stereochemistry of cycloaddition was found with the carbon analogues (Scheme 6.46 and Table 6.14) (18,256).

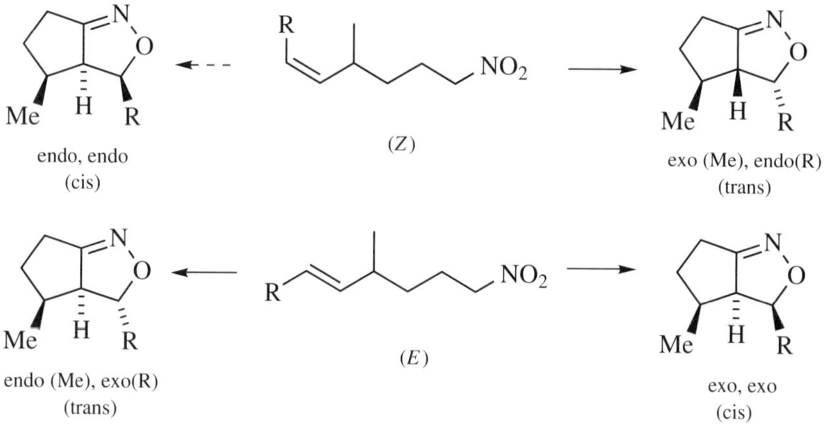

Scheme 6.46

This difference in the stereochemical outcome was explained by a CH_3/R interaction in the transition state, which favors the endo,exo-disubstituted isomer (trans) for the case of the (*Z*)-olefin (18) (Scheme 6.46). When the (*E*)-alkene or the monosubstituted (terminal) olefin was used, the steric interactions in the two transition states differ less and consequently a mixture of isomers resulted (18).

Intramolecular cycloadditions of substrates possessing several stereocenters, such as those obtained from carbohydrate derivatives, have enjoyed some recent popularity. The carbohydrate skeleton provides a highly effective means for inducing stereoselectivity in intramolecular 1,3-dipolar cycloadditions. Tatsuta and co-workers (258) reported on the intramolecular cycloaddition of the D-*xylo*

TABLE 6.14

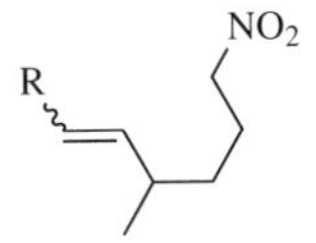

R	Products trans/cis[a]	Reference
H	71:29	256
Me (Z)	only trans	18
Me (E)	75:25	

[a] cis/trans refers to configuration of cyclopentane part, cis corresponding to *endo*-Me, trans to *exo*-Me.

derivative **39** (obtained from D-glucosamine) to give the cyclopentano-isoxazoline **40** as the sole product. This cycloadduct served as a key intermediate in the total synthesis of (−)-allosamizoline **41** (Scheme 6.47, see also Section 6.5.2.1). This strategy was also used for the synthesis of various other aminocyclopentanols (72,259–261) and as a key intermediate for calcitrol (262). The stereochemical outcome of these and related reactions has been rationalized in terms of the preferred chair-like transition model state (73,77,241) as described above.

Cozzi and co-workers (243,263) studied the influence of the double-bond configuration on the stereochemical course of the intramolecular cycloaddition of chiral alkenes, where the stereocenter is located outside the isoxazoline ring (Table 6.15). On the basis of experimental results as well as theoretical calculations, two models were proposed for the reaction with (*Z*)- and (*E*)-alkenes, in accord with the model proposed for α-X-substituted alkenes (see Section 6.2.3.1).

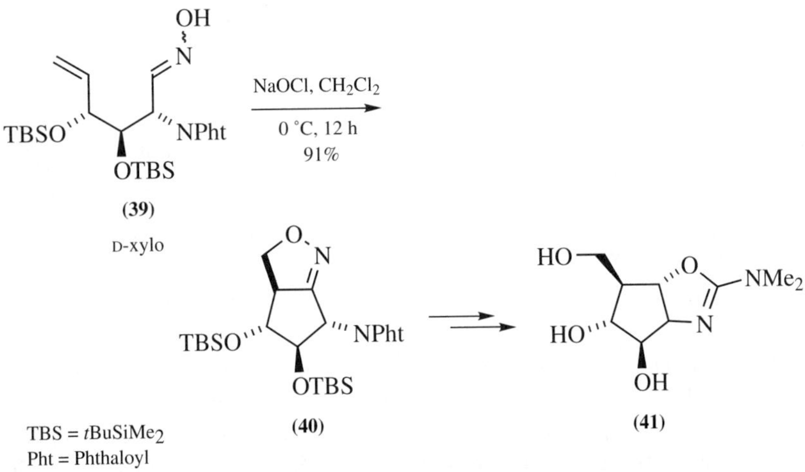

Scheme 6.47

TABLE 6.15

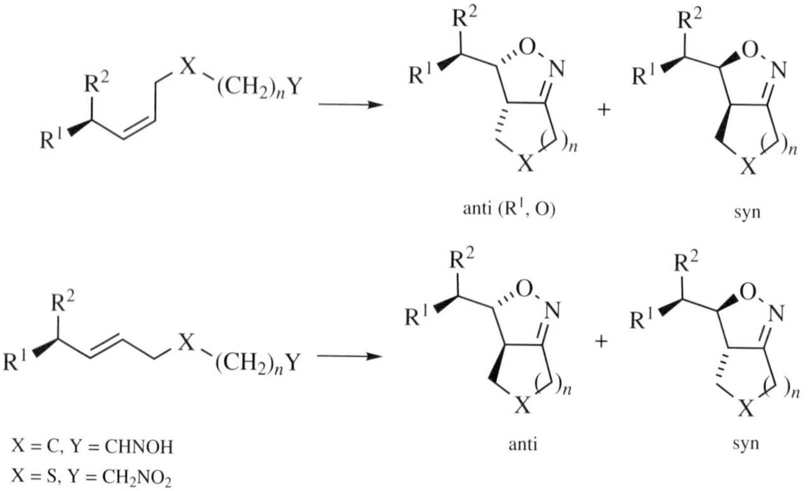

R^1	R^2	X	n		Diastereomeric Ratio[a]
OBn	Me	C	2	(*Z*)	80:20
				(*E*)	60:40
		C	1	(*Z*)	75:25
				(*E*)	58:42
		S	1	(*Z*)	66:34
				(*E*)	66:34
H_2COBn	OBn	C	2	(*Z*)	83:17
				(*E*)	77:23
		S	1	(*Z*)	63:37
				(*E*)	63:37
$H_2COC(CH_2)_5O$		C	2	(*Z*)	86:14
				(*E*)	86:14
		C	1	(*Z*)	81:19
				(*E*)	78:22
		S	1	(*Z*)	95:5
				(*E*)	83:17
i-Pr	Me	C	2	(*Z*)	66:34
				(*E*)	66:34
Me	*i*-Pr	S	1	(*Z*)	66:34
				(*E*)	66:34

[a] With X=C starting from the oxime, (243), with X=S the nitroalkene is used (264).

With (Z)-alkenes, the most crowded *inside* position should preferentially be occupied by the medium group (OR′) and the small group (H), respectively, the large group (R) being placed *anti* (Scheme 6.48). In the case of (*E*)-alkenes, the results were rationalized in terms of the *inside*-alkoxy model. *Ab initio* and force field calculations provide further support for this proposal (265). The transition

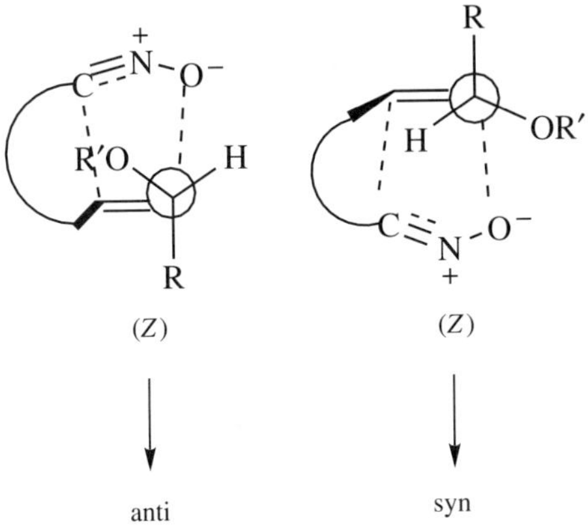

Scheme 6.48

states postulated above also account for the results found with intramolecular cycloadditions of chiral olefins bearing a sulfur atom on the carbon chain (264).

A number of intramolecular cycloadditions of alkene-tethered nitrile oxides, where the double bond forms part of a ring, have been used for the synthesis of fused carbocyclic structures (18,74,266–271). The cycloadditions afford the cis-fused bicyclic products, and this stereochemical outcome does not depend on the substituents on the alkene or on the carbon chain. When cyclic olefins were used, the configuration of the products found could be rationalized in terms of the transition states described in Scheme 6.49 (18,74,266–271). In the transition state leading to the cis-fused heterocycle, the dipole is more easily aligned with the dipolarophile if the nitrile oxide adds to the face of the cycloolefin in which the tethering chain resides. In the trans transition state, considerable nonbonded interactions and strain would have to be overcome in order to achieve good parallel alignment of the dipole and dipolarophile (74,266).

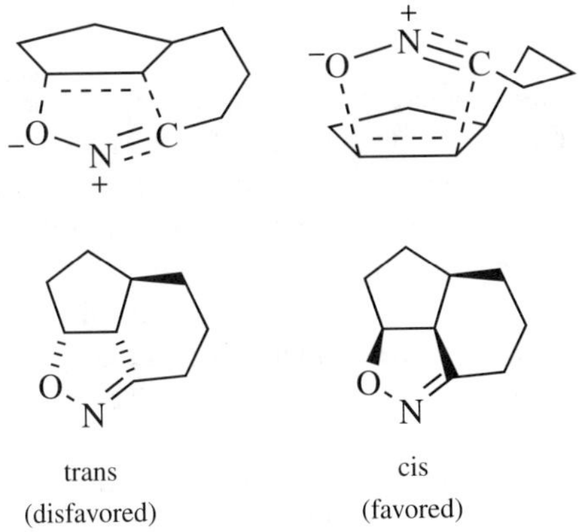

Scheme 6.49

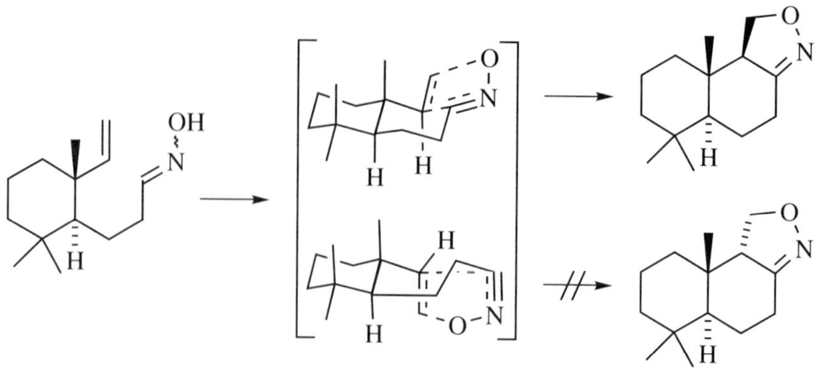

Scheme 6.50

When the alkene moiety is attached to a cyclic structure, the stereochemical outcome can satisfactorily be explained in terms of the chair and boat transition states described above (272,273) (Scheme 6.50).

The intramolecular cycloaddition of nitrile oxides to substituted furans was reported to occur with low stereoselectivity (274). Inserting a stereogenic unit within the chain connecting the dipole and dipolarophile did not increase the stereoselectivity (274).

6.4. USES OF INTERMOLECULAR NITRILE OXIDE CYCLOADDITIONS

6.4.1. Uses of Isoxazolines in Syntheses of Natural Products and Related Structures

6.4.1.1. Natural Products Containing an Isoxazoline Group

There are a few cases where isoxazolines have attracted interest as target structures in their own right. Acivicin (AT-125) **45** is an amino acid possessing a 3-chloroisoxazoline moiety, that was isolated from fermentation broths of *Streptomyces sviceus*, and displays significant activity against a number of tumors (275). This amino acid has been prepared by the cycloaddition of chloroformonitrile oxide, generated *in situ* from phosgene oxime (dichloroformaldoxime), to the phthalimide-protected vinylglycine dipolarophile **42** to produce cycloadducts **43** and **44** in a ratio of 30:70 in 52% yield (275,276) (Scheme 6.51). After separation, deprotection of the 5-(*S*)-cycloadduct **43** with hydrazine afforded acivicin in 72% yield. Another approach involved the cycloaddition of bromoformonitrile oxide, generated *in situ* from dibromoformaldoxime, to the vinylglycine derivative **46**, leading to a separable 3:2 mixture of the isoxazoline diastereomers **47** and **48**

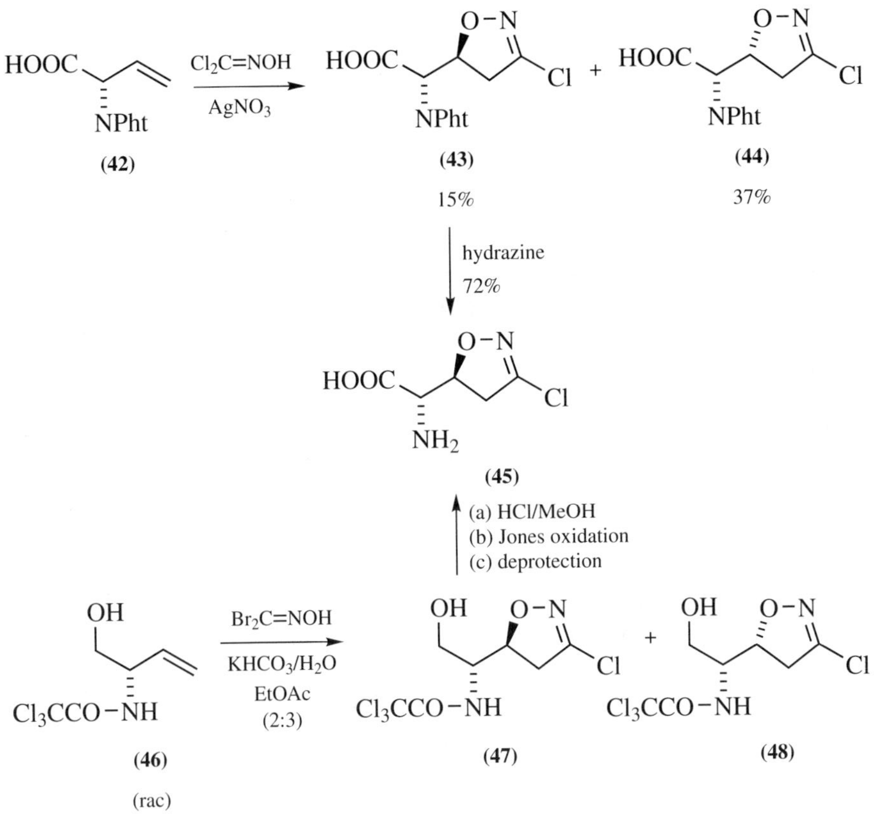

Scheme 6.51

(Scheme 6.51) (277). The erythro isomer **47** was treated with methanolic HCl followed by Jones oxidation and deprotection, to afford acivicin in racemic form.

Dihydromuscimol (**49**) is a conformationally restricted analogue of the physiologically important neurotransmitter γ-aminobutyric acid (GABA) and has been prepared using the cycloaddition of dibromoformaldoxime to *N*-Boc-allylamine followed by N-deprotection with sodium hydroxide (Scheme 6.52) (278). The individual enantiomers of dihydromuscimol were obtained by reaction of the bromonitrile oxide with (*S*)-(+)-1,2-O-isopropylidene-3-butene-1,2-diol, followed by separation of the diastereoisomeric mixture (erythro/threo 76:24), hydrolysis of respective isomers, and transformation of the glycol moiety into an amino group (279).

6.4.2. Products from "Aldol" Ring Cleavage of Isoxazoline Intermediates

A great deal of the interest in isoxazolines stems from their use in the synthesis of acyclic compounds (19). The approach to β-hydroxy carbonyl compounds via

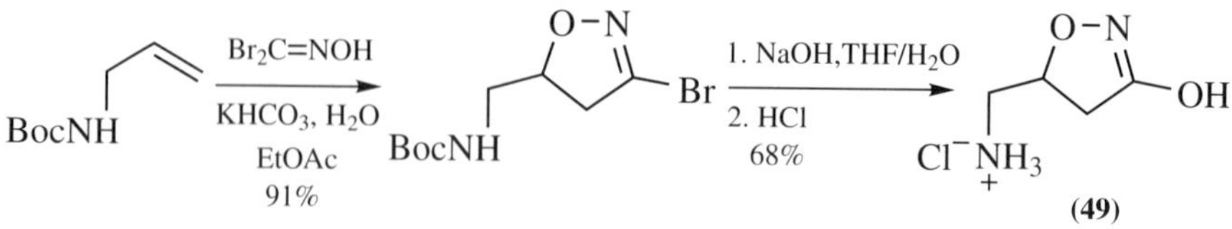

(49)

Scheme 6.52

isoxazolines by N—O bond cleavage and hydrolysis of the intermediate imine represents an alternative strategy to the aldol route (Scheme 6.53) (17,18). The first example of this transformation was reported by Torssell and Zeuthen (280) in 1978. Soon extensions and many applications, notably from the groups of Curran (17) and Kozikowski (18), both at Pittsburgh, followed. Curran (281–283) carried out a detailed investigation outlining the optimum conditions for this reduction–hydrolysis sequence. Efforts in this area have been reviewed by both by Kozikowski (18) and Curran (17). The examples given below are arranged according to the degree of complexity of the cycloaddition partners.

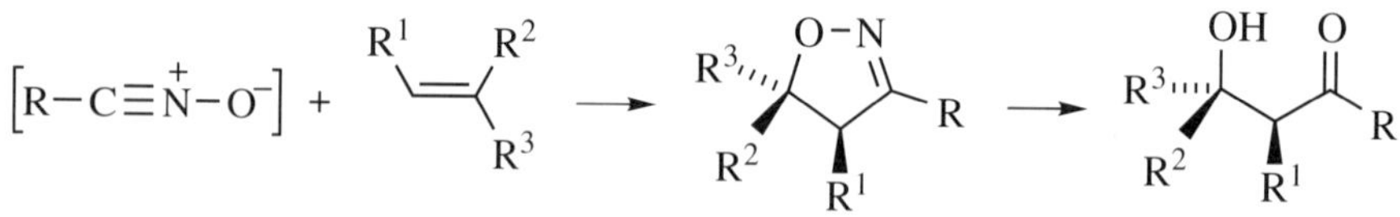

Scheme 6.53

6.4.2.1. *Use of Isoxazoline Intermediates from Achiral Nitrile Oxides—Achiral Olefins*

Talaromycin B is a spiro-acetal produced by the fungus *Talaromyces stipitatus*, the toxicity of which may be due to its ability to block outward potassium fluxes. In an elegant synthesis, the requisite open-chain polyol with hydroxy groups in the γ- and γ'-positions was assembled from nitrile oxide and olefin building blocks **50** and **51**, both of which carry a *bis*(hydroxyethyl) moiety protected as a cyclohexanone acetal (284). Hydrogenolysis of the N—O bond of isoxazoline **52** using Raney nickel, followed by treatment with aqueous acid, gave the spiroketal **53**, which was further transformed into racemic talaromycin B (**54**) (Scheme 6.54) (284).

This *cycloaddition–reduction–hydrolysis* sequence was also used in an approach to butyrolactones related to ribonolactone (71). These compounds are inducing agents of hunger and satiety in mammalians. Here, a subsequent aldol$\rightarrow$1,3-diol reduction was used, and the required carboxy function was established by oxidation of the aromatic ring with ruthenium tetroxide. Cycloaddition of benzonitrile oxide to allyl alcohol afforded an enantiomeric mixture of isoxazolines **55** and **56**, which were treated with sodium hydride and methyl iodide to achieve separation by chromatography on cellulose triacetate (71). O-Demethylation, followed by

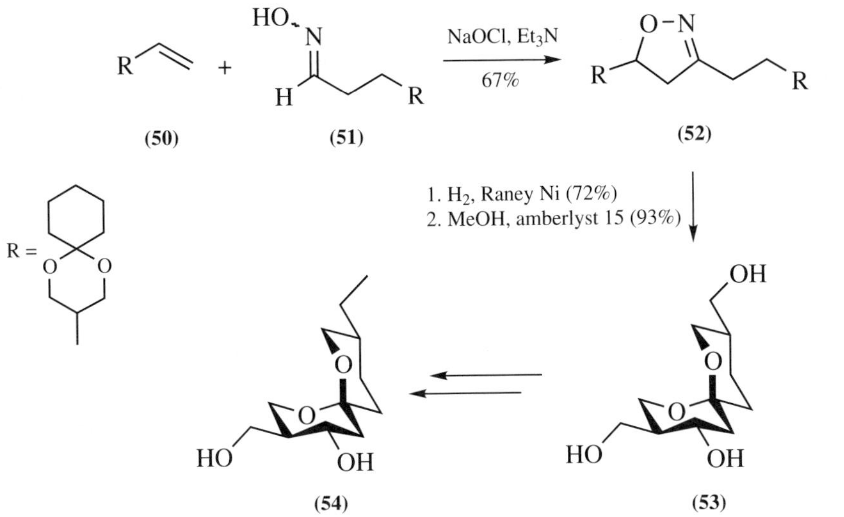

Scheme 6.54

hydrogenolysis of the isoxazoline ring gave the hydroxy ketones **57** and **58**, which were protected to allow for the oxidative degradation of the phenyl ring. Treatment with base, and then acid finally gave the corresponding 3-deoxypentonolactones **59** and **60** (Scheme 6.55) (71).

In another variation, the intermediate aldol product **64**, with an extra hydroxy group in the γ-position, was used to construct the furan ring of rosefuran (**65**), a trace component of rose oil (Scheme 6.56) (285). Here, the reaction of the nitropentene derivative **61** with crotyl acetate (**62**) afforded the 3,4,5-trisubstituted isoxazoline (**63**) in moderate yield. Removal of the acetyl group by saponification of the cycloadduct, subsequent demasking of the aldol moiety using $Mo(CO)_6$, and exposure of the ketodiol (**64**) to acid gave the target compound **65** (285).

With an α',γ-ketodiol, cyclization to produce a 3-furanone derivative is feasible, as is shown for the synthesis of ascofuranone (**71**) and geiparvarin (**72**) (Scheme 6.57) (286). The precursor for **71** was prepared by the cycloaddition of diene **66** to nitroalcohol **67**. In this case, regioselective attack occurred only on the terminal double bond. Reductive cleavage–hydrolysis of the isoxazoline adduct **68** with $Mo(CO)_6$ followed by acid-induced cyclization led to the furanone intermediate (286). A similar strategy was used for the synthesis of geiparvarin (**72**) (Scheme 6.58) (286).

6.4.2.2. Use of Intermediates from Cycloadditions of Achiral Nitrile Oxides to Chiral Olefin

In the synthesis of the compactin lactone precursor **76**, an aldol→1,3-diol conversion was included in the usual sequence, and the required aldehyde function

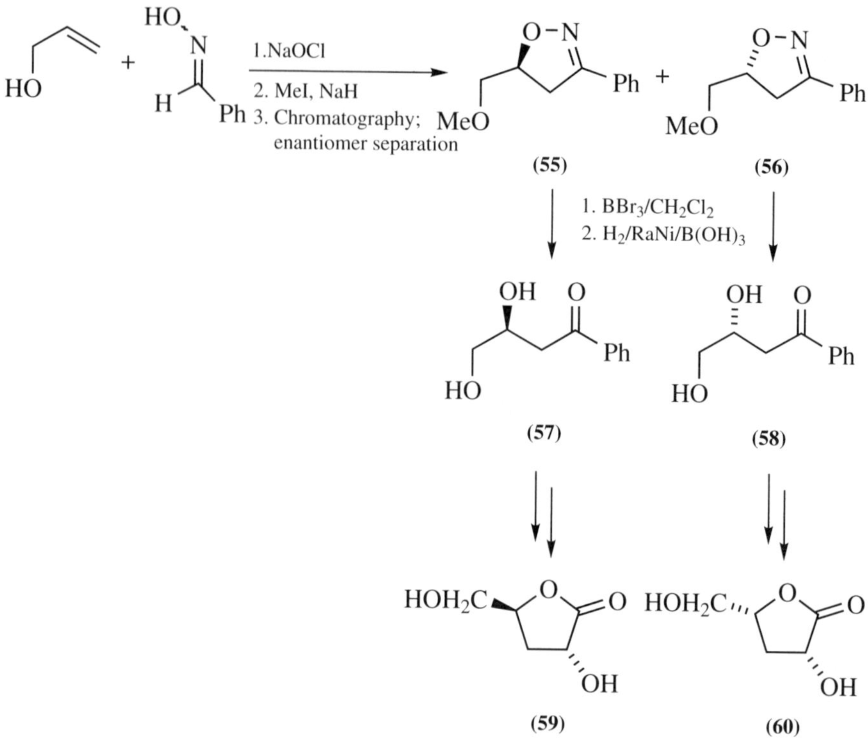

Scheme 6.55

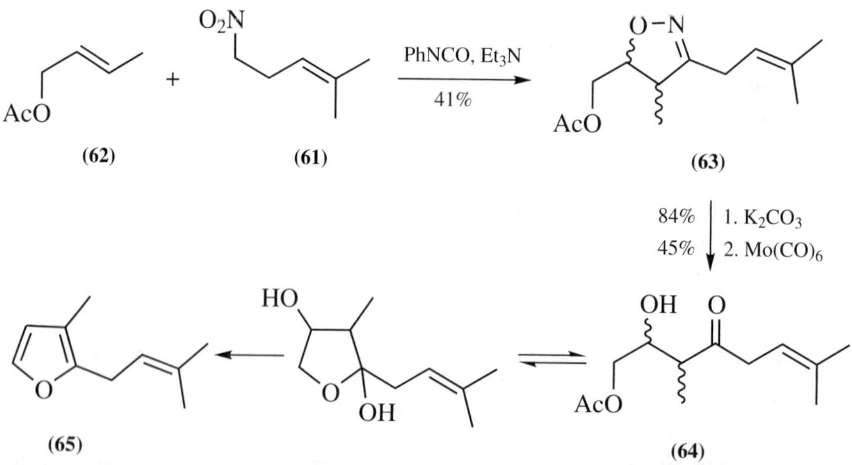

Scheme 6.56

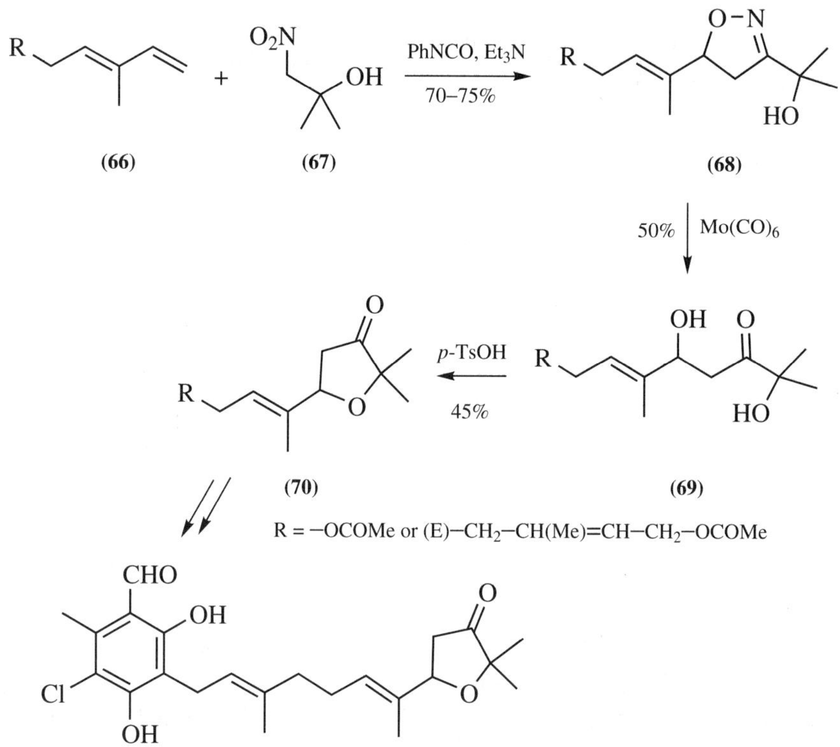

(66) (67) (68) (69) (70) (71)

R = −OCOMe or (E)−CH_2−CH(Me)=CH−CH_2−OCOMe

Scheme 6.57

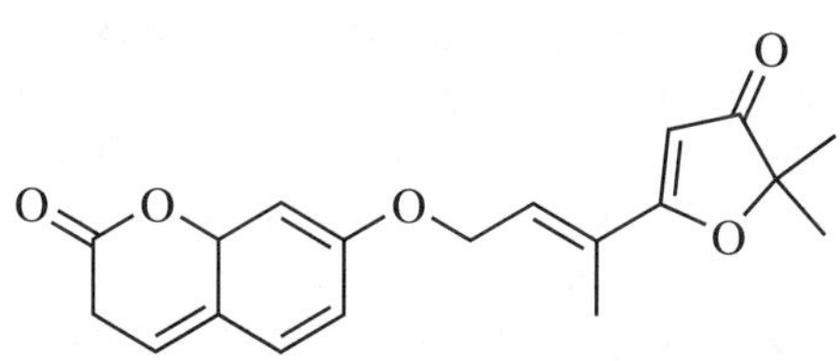

(72)

Scheme 6.58

was introduced from the nitrile oxide component (287). Thus, cycloaddition of 3-nitropropanal acetal (**73**) to dioxolane (**74**) afforded a 4:1 mixture of isomeric isozaxolines. The anti (erythro) isomer was converted to the optically active aldol which was then subjected to carbonyl reduction and acid-catalyzed cyclization to furnish the protected pyranoside (**76**), related to compactin lactone (Scheme 6.59) (287).

Nitroacetal **77** was used as the nitrile oxide precursor in a synthesis of (+)-hepialone (**82**). Curran's group carried out a 1,3-dipolar cycloaddition to

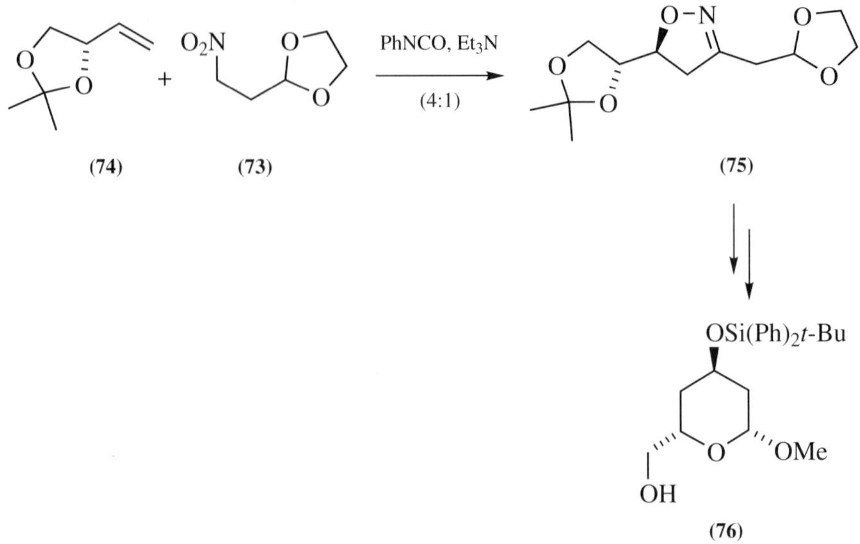

Scheme 6.59

acrylamide (**78**), derived from Oppolzer's sultam, and this afforded a 88:12 mixture of isoxazolines **79** (184). The chiral auxiliary was removed by reduction with L-Selectride, which was followed by carboxyl reduction and coupling of the derived mesylate with lithium dimethylcuprate. Raney Ni reduction–hydrolysis of the 5-ethylisoxazoline **80** then gave the β-hydroxyketone **81**, which on treatment with acid afforded (+)-hepialone **82** (Scheme 6.60) (184).

The same strategy was applied toward (−)-pestalotin (**87**) (Scheme 6.61) (288). In this case, the necessary α-hydroxypentyl side chain was introduced from the 5-carboxy group by butylmagnesium bromide addition (syn/anti dr 15:85) to the derived isoxazoline-5-carbaldehyde. Thus, cycloaddition of acetonitrile oxide (**83**) to the acrylamide, as described above, gave the 5-amide **84** in virtually quantitative yield in a ratio of 94:6. After separation, the major isomer was cleaved with L-Selectride, and this was followed by several additional steps, as outlined above, to give isoxazoline (**85**). This key intermediate was subjected to the hydrogenation–hydrolysis sequence using Raney Ni, and then further subjected to hydrochloric acid to produce keto ester **86**, a key intermediate on the way to (−)-pestalotin (**87**) (288).

Martin et al. (289) utilized the chiral bicyclic lactone **88** for a total synthesis of (+)-phyllanthocin (Scheme 6.62). Cycloaddition of **88** to acetonitrile oxide, generated *in situ* from hydroximoyl chloride (**89**), furnished cycloadduct **90** in 45% yield together with other regio- and stereoisomers. After several steps, methyl glycoside (**91**) could be obtained. From this, reduction–hydrolysis gave the aldol that was subjected to acid-catalyzed spiroacetalization to produce spiroketal (**92**), and eventually (+)-phyllanthocin (**93**) after two additional steps (289). The

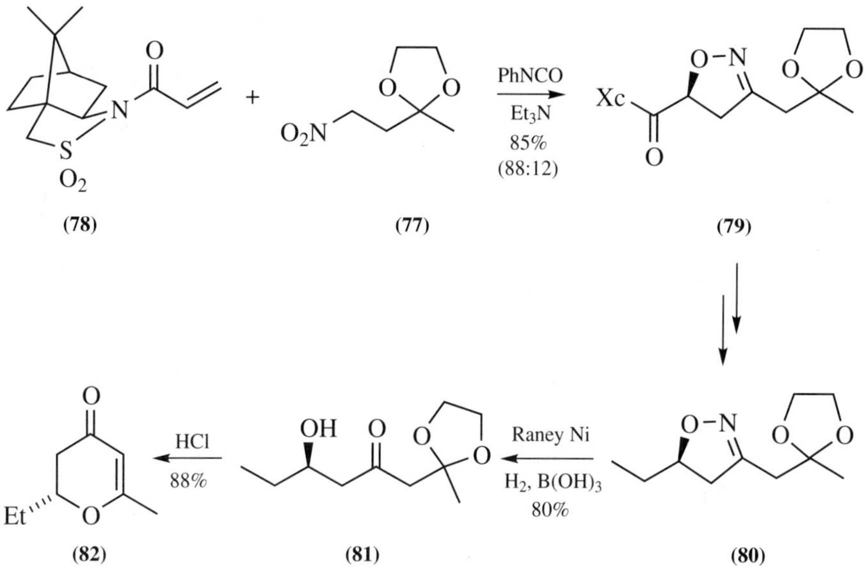

Scheme 6.60

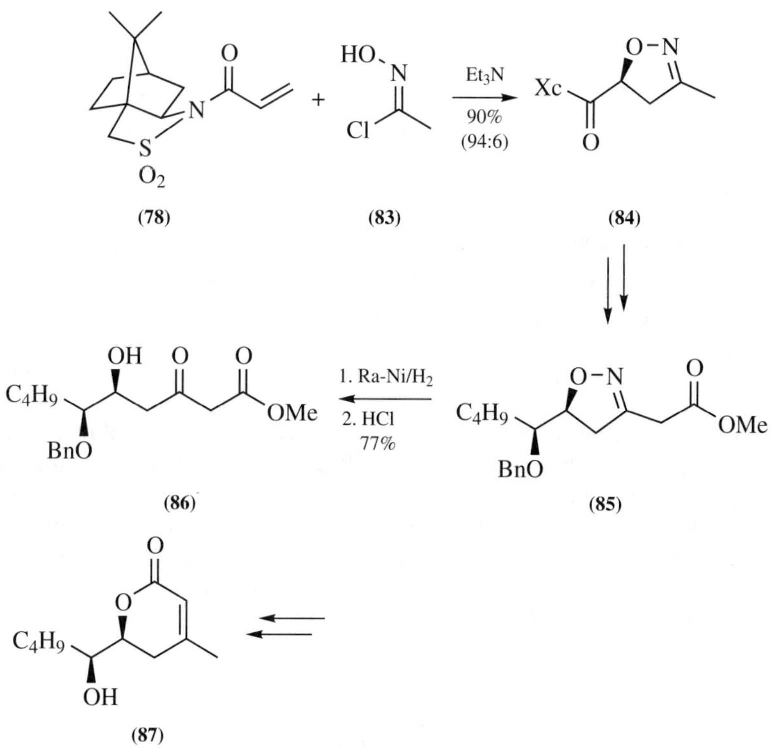

Scheme 6.61

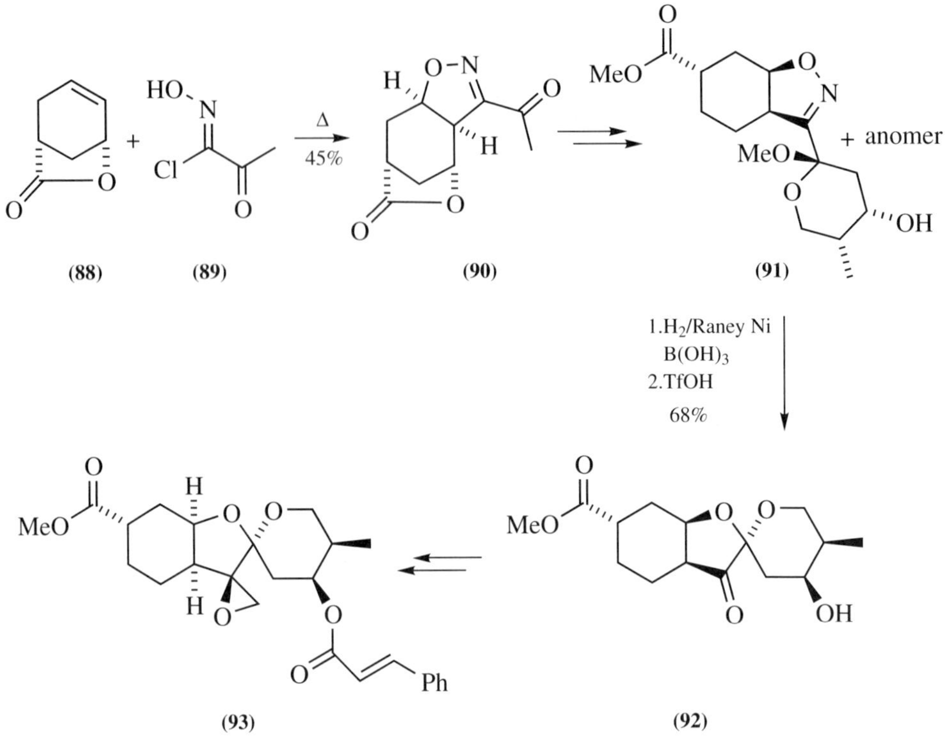

Scheme 6.62

3-acetyl and hydroxy groups of the aldol product were designed to generate the 3-furanone ring as seen in (**92**).

6.4.2.3. *Use of Intermediates from Cycloadditions of Chiral Nitrile Oxides with Achiral Olefins*

Isoxazolines can be transformed into α,β-enones by several methods from the initial aldol product. This strategy was applied by Barco et al. (285) toward the synthesis of (−)-pyrenophorin (**98**), a macrocyclic *bis*(enone-lactone) with antifungal properties. The hydroxy group was introduced from the nitrile oxide component (**95**), while the carboxy function was derived from the acrylate dipolarophile. Thus, cycloaddition of the optically active nitropentyl acetate **94** to methyl acrylate **95** afforded isoxazoline **96** as a mixture of optically active diastereomers. Reductive hydrolysis using Raney nickel/acetic acid gave β-hydroxyketone (**97**), which was subsequently utilized for the synthesis of (−)-pyrenophorin (**98**) (Scheme 6.63) (285).

A dipolar cycloaddition of a steroidal nitrile oxide to trifluoropropene, followed by reductive hydrolysis, was used to prepare the trifluoromethyl cholestanone

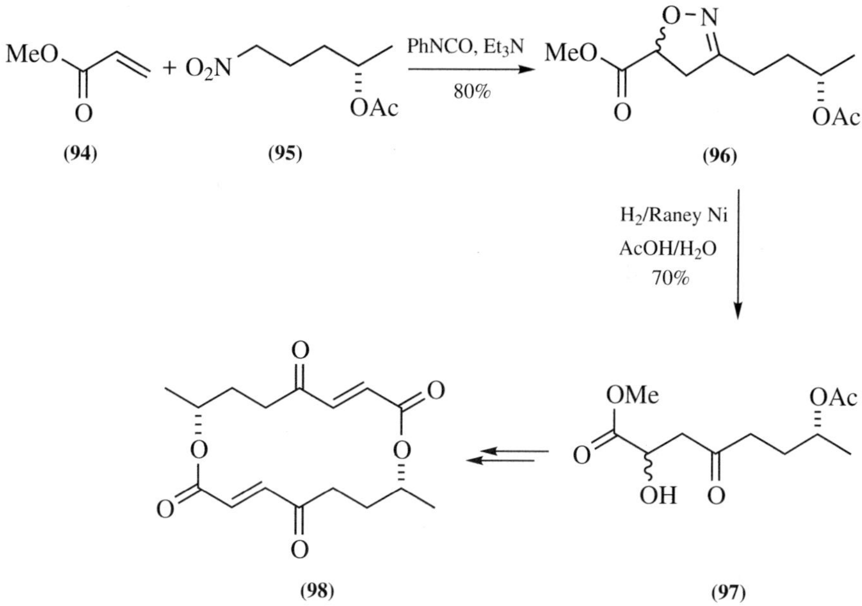

Scheme 6.63

analogue **102**. This compound was converted to fluorinated analogues of naturally occurring crinosterol and typhasterol (Scheme 6.64) (290).

6.4.2.4. Use of Intermediates from Cycloadditions of Chiral Nitrile Oxides to Chiral Olefins

Cycloaddition using a chiral dipole and dipolarophile, together with the generation of a nitrile oxide making use of an O-stannyl oxime, was recently utilized in an approach toward the synthesis of amphotericin B (**106**) (291). Thus, the nitrile oxide generated by treating O-stannyl oxime (**103**) with *tert*-butyl hypochlorite was trapped with dipolarophile **104** to afford the complex cycloadduct **105** as a 88:12 mixture of diastereomers. Reduction–hydrolysis of the major isoxazoline with $Mo(CO)_6$, followed by acid-catalyzed cyclization–acetalization using trimethyl orthoacetate, gave the protected C(1)–C(19) fragment of amphotericin B (**106**) (291), a polyene-polyol natural product with outstanding antibiotic properties (Scheme 6.65).

6.4.3. Aminoalcohol Ring Cleavage of Nitrile Oxide Cycloadducts; Synthesis of Amino Polyols, Amino Sugars, and Amino Acids

Reduction of the isoxazoline ring can lead to different mixtures of amino alcohols depending on the reducing reagent used. Catalytic hydrogenation as well

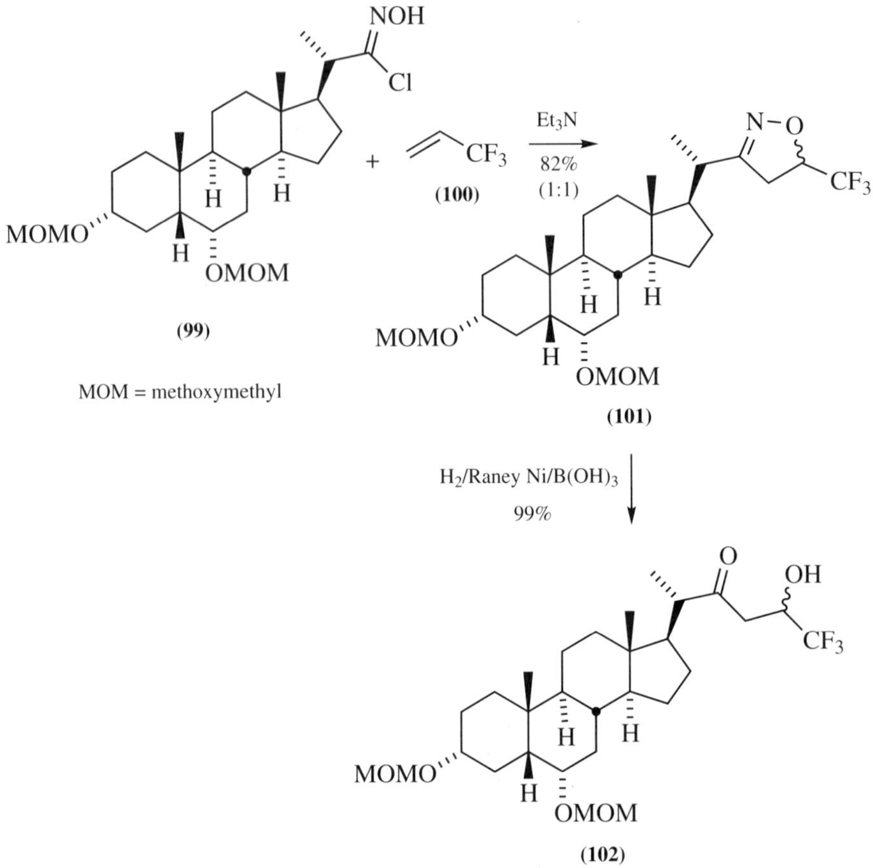

MOM = methoxymethyl

Scheme 6.64

as metal reduction are assumed to first cleave the N–O bond. For example, in alcoholic solution, the acyclic β-hydroxyimine formed from **107** undergoes further reduction to afford erythro/threo mixtures of the aminoalcohol **110** with low diastereoselectivity (20,21,292) (Scheme 6.66). Reduction of the C=N bond can be suppressed in favor of hydrolysis and formation of the aldol product **109** as described above. In contrast to this, different results are obtained with LiAlH. Under these conditions, the reduction proceeds by initial addition of hydride to the C=N bond of the heterocycle and then N–O bond cleavage of the resulting isoxazolidine **108**. This sequence leads to γ-amino alcohols with good to excellent stereoselectivity. The Jäger group has systematically studied such reductions and has demonstrated that this procedure represents a viable strategy for the stereocontrolled assembly of amino sugars, amino acids, amino polyols, and related structures (20–23).

The stereochemical outcome of most reductions using $LiAlH_4$ can be accounted for by the following assumptions: In substrates containing alkyl, phenyl, or related

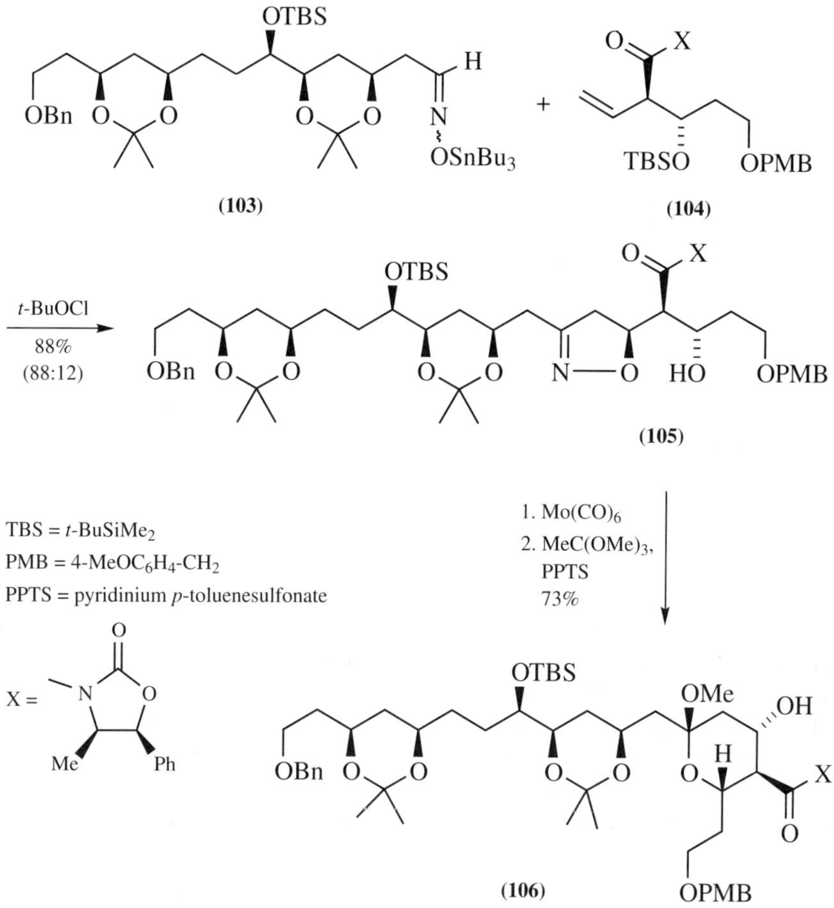

Scheme 6.65

substituents, where reduction is controlled by steric size (i.e., anti-directing) (Scheme 6.67), solvated lithium cations [Li^+S_n] are coordinated to the electron pair of the isoxazoline oxygen atom on both ring faces, but to a different degree. Hence, energetically different transition states for hydride transfer from the associated aluminate counterion are operative (20–23,27,38).

In the case of substrates containing OR or related heteroatom functionality, additional coordination complexes and transition states are involved. If a hydroxy group is present, hydride transfer preferably occurs from the corresponding alkoxyaluminate species (i.e., **113**) (Scheme 6.68). Substituents of this kind tend to be syn-directing (20–23,140,141,297).

Since the stereochemical outcome of $LiAlH_4$ reduction of isoxazolines is predictable according to the parameters described above (140,293), this concept was exploited in efficient syntheses of various amino polyols and sugars as represented

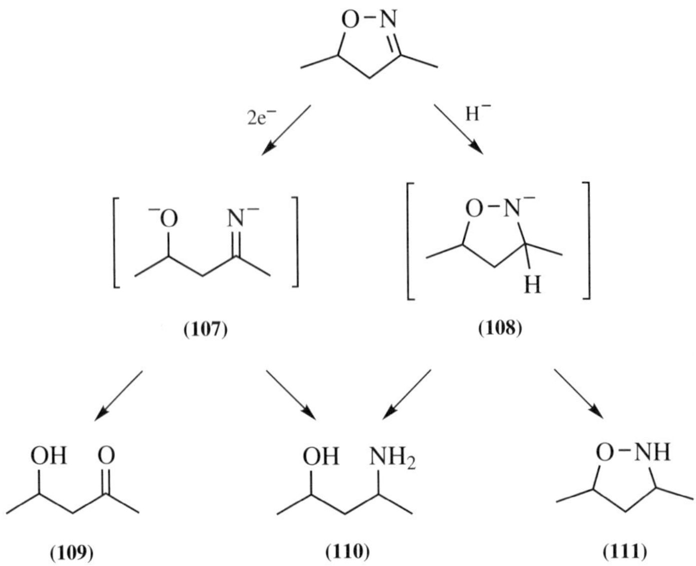

Scheme 6.66

by formula **A** in Scheme 6.69. This reductive procedure was also used to prepare hydroxy amino acids (20–22,42,293a) from isoxazoline precursors **B**–**E**.

The most straightforward approach to the 4-oxygenated isoxazolines (**B** and **C**) would involve cycloaddition to a suitable enol derivative. All attempts, however, failed to reverse (157) the normal regiochemical outcome of this cycloaddition that furnished 5-oxygenated derivatives. It proved possible to optimize the cycloaddition

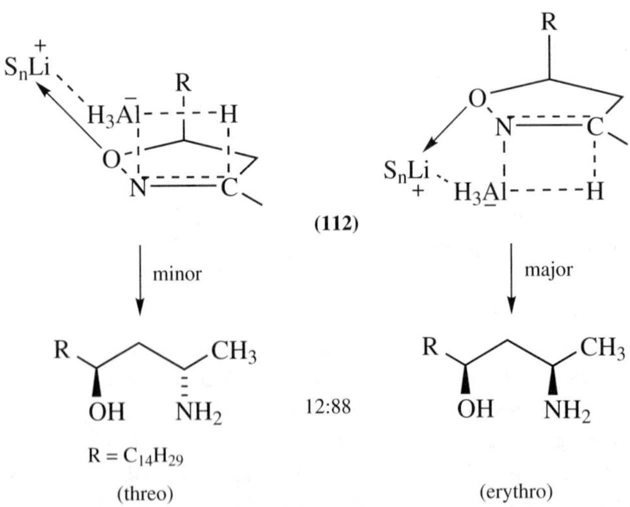

Scheme 6.67

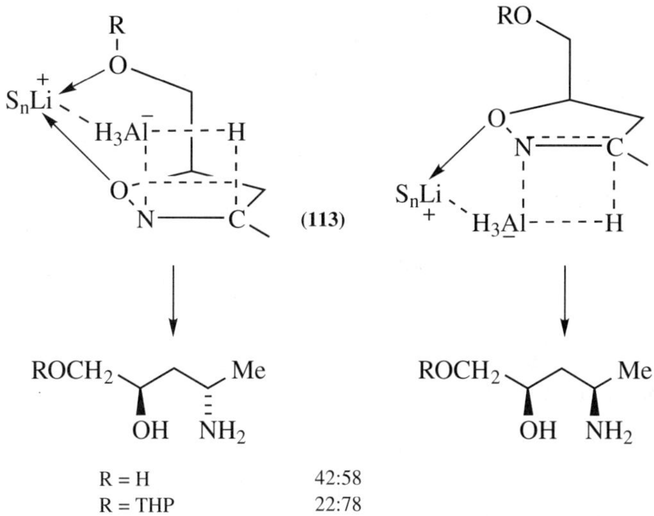

Scheme 6.68

of nitrile oxides to furan, a notorously sluggish dipolarophile, in order to secure the 4-oxygenated isomers of the cis series (20,21,82,157,294). Several furoisoxazolines were prepared in this manner, providing highly advanced, yet still versatile amino sugar precursors. A highly stereoselective addition of LiAlH to the C=N bond was anticipated in view of the steric demands of the bicyclic bowl-shaped system. This indeed was the case as reduction of furoisoxazolines with $LiAH_4$ afforded the corresponding aminoalcohols in high yield and with excellent diastereoselection

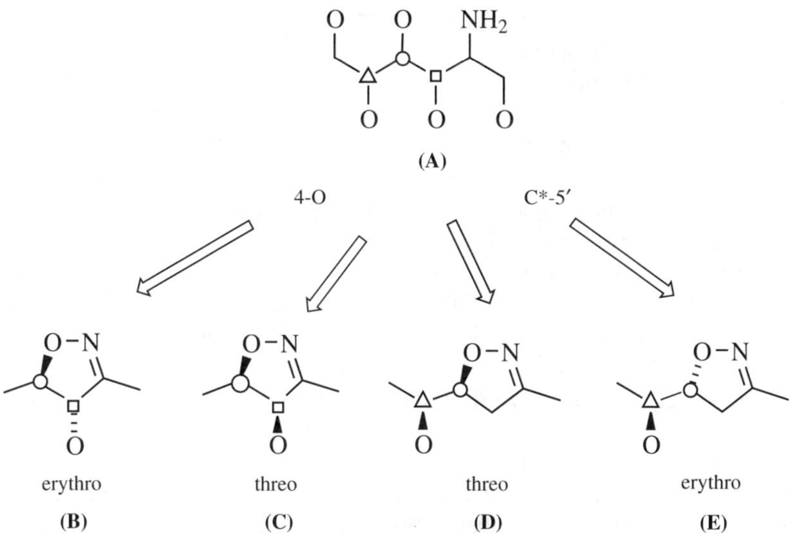

Scheme 6.69

(82,157,294). This strategy was applied to the synthesis of L-(+)-furanomycin, a naturally occurring antibiotic (295,296). Cycloaddition of the nitrile oxide generated from hydroximoyl acid chloride **114** to 2-methylfuran (**115**) afforded cycloadduct **116** as the major product. Epoxidation of **116** gave the stable, isolable enol ether epoxide **118** with high stereoselectivity. Reductive opening of the epoxide ring of **118** led to alcohol **119**, which upon reduction with $LiAlH_4$ afforded 1,3-aminoalcohol **120**. Diol removal by mesylation and sodium naphthalenide reduction introduced a double bond present in the 2,5-dihydrofuran derivative **121**. Deprotection of the side-chain diol, oxidative cleavage with sodium periodate, and further oxidation of the intermediate *N*-Boc-aminoaldehyde furnished the *N*-protected amino acid **122**. Finally, deprotection using acidic conditions followed by ion-exchange chromatography provided natural L-(+)-furanomycin **123** (Scheme 6.70). In a similar manner, the reaction of epoxide **118** with other nucleophiles led to a series of analogues (295,296). Furoisoxazolines such as **118** are further useful compounds for the synthesis of imino polyols, as a highly stereoselective hydrogenolysis was found to occur upon subjection to catalytic hydrogenation conditions (see Section 6.4.4).

The furoisoxazoline approach toward the synthesis of amino polyols represents a viable solution for the synthesis of two of the four key intermediates (**C** and **D**) outlined in Scheme 6.69. trans-4-Hydroxyisoxazolines (cf. **B**) were prepared by 4-deprotonation followed by carbanion hydroxylation. An early application of this strategy resulted in a short, highly stereoselective synthesis of racemic C_{18}-phytosphingosine (297), the long-chain base characteristic of plant sphingolipids (Scheme 6.71). Thus, cycloaddition of the nitrile oxide dipole derived from the 2-nitroethanol derivative **124** to 1-hexadecene (**125**) produced the isoxazoline **126** (140,297). This compound was regioselectively deprotonated with lithium diisopropylamide–hexamethylphosphoric triamide and the resulting 4-carbanion was quenched with trimethyl borate. The intermediate 4-boronate was oxidatively converted to the corresponding alcohol. Subsequent reduction of 4-hydroxyisoxazoline **127** with lithium aluminium hydride proceeded with high selectivity (140) and this was followed by removal of the protecting group to produce the hydrochloride of phytosphingosine (**128**) (ribo isomer) (297).

A related strategy was used to prepare D-allosamine (**134**). Cycloaddition of the dipole derived from nitroacetal (**129**) to (*S*)-vinyl dioxolane (**74**) afforded a mixture of erythro/threo isoxazolines **130:131** (Scheme 6.72). The erythro isoxazoline was subjected to hydroxylation as described above, to give 4-hydroxyisoxazoline **132** with high diastereoselectivity. Lithium aluminium hydride reduction furnished a single diastereomer of aminodiol **133**, which could be deprotected to give the hydrochloride salt of D-allosamine (**134**) (141).

Precursors of type **E** (erythro fragment) (Scheme 6.69) were obtained by cycloaddition of a nitrile oxide dipole to α-alkoxyalkenes. This strategy was used in the syntheses of DL- and D-lividosamine (298) (Scheme 6.73). Lithium aluminum hydride reduction of the erythro adduct **130** produced aminoalcohols **135** in a 78:22 ratio (2,4-erythro/threo) in high yield. The mixture was subjected to HCl hydrolysis to give the hexosamine hydrochlorides **136**, which after several steps, produced D-lividosamine **137** possessing the D-ribo configuration (298).

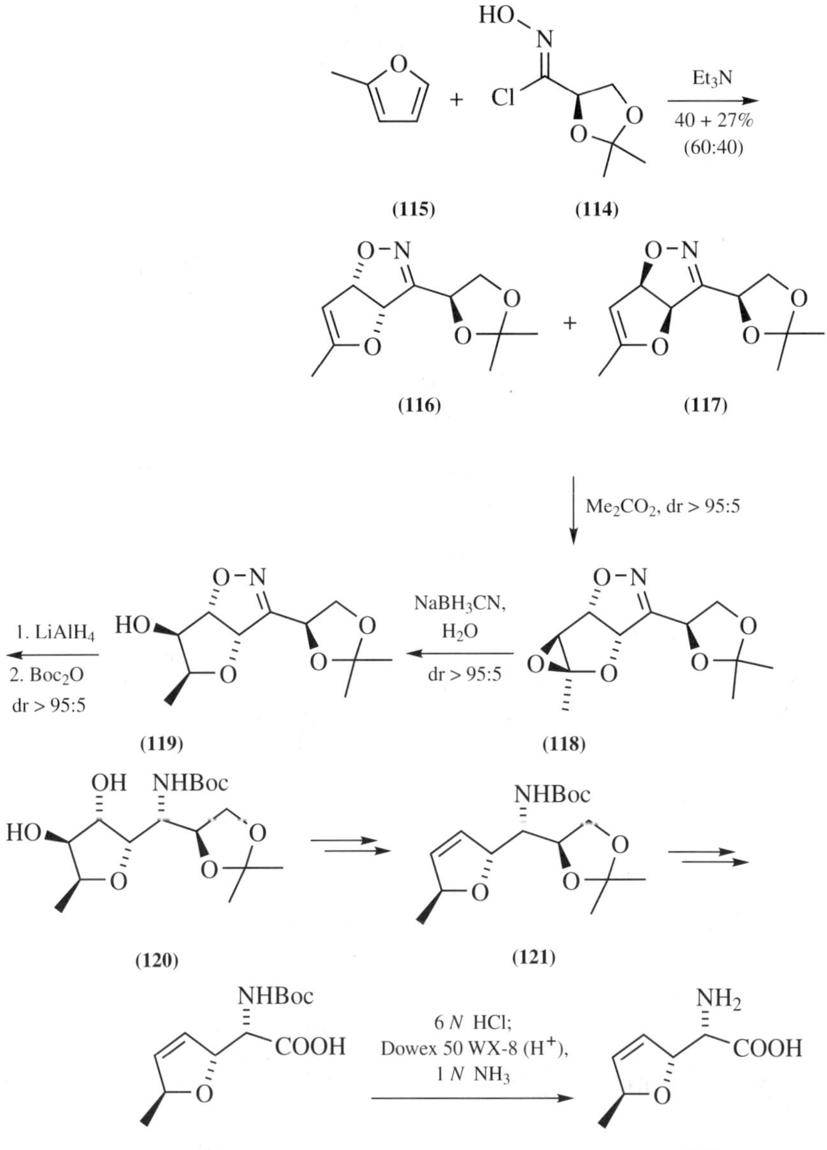

Scheme 6.70

6.4.4. Uses of Nitrile Oxide Cycloadditions in the Isoxazoline Route Toward Mono- and Bicyclic Imino Acids and Imino Polyols

Since isoxazoline formation, subsequent modification at the ring or adjacent centers and reductive cleavage can be carried out in a highly diastereoselective manner, a scheme was devised to synthesize various iminopolyols by this strategy.

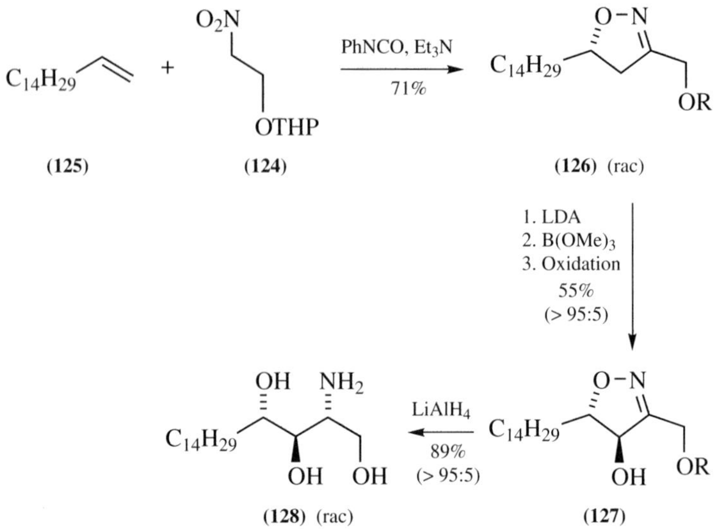

Scheme 6.71

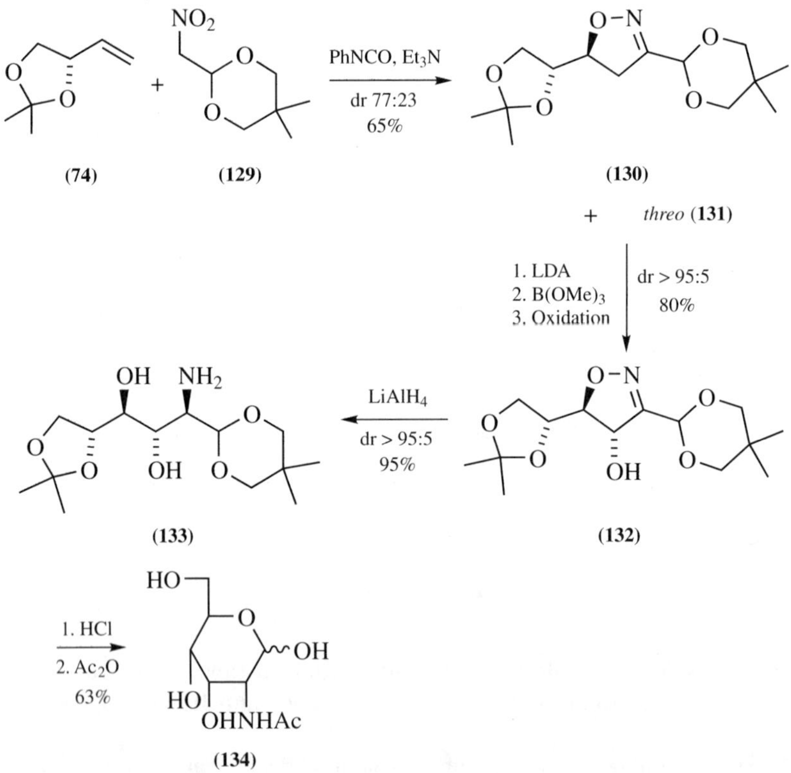

Scheme 6.72

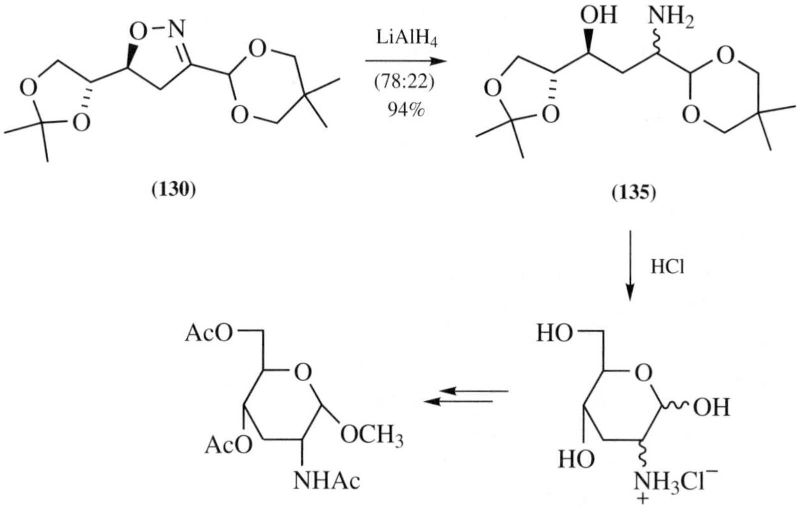

(130) (135) (137) (136)

Scheme 6.73

Iminopolyols often exhibit interesting physiological properties. Many mono and bicyclic representatives are known that contain five- and/or six-membered rings (pyrrolidines, piperidines, pyrrolizidines, indolizidines, quinolizidines) (299–302).

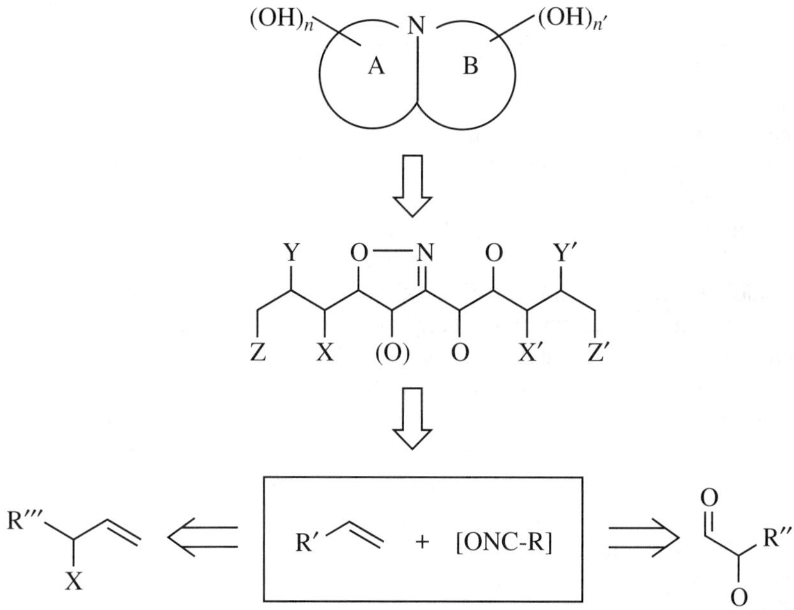

Scheme 6.74

Most noteworthy are their glycosidase-inhibiting activities that are displayed by 1,4-dideoxy-iminopentitols (303), deoxynojirimycin (304), isofagomine (305–308), and by indolizidines such as slaframine, swainsonine, or castanospermine (309). One of the respective routes devised for the synthesis of these compounds features isoxazoline intermediates. This approach furnishes the "central" nitrogen atom, together with polyhydroxy-substituted side chains containing a suitable (terminal) leaving group to effect C—N cyclization (23,221,225,234,296). Either one or both of the two building blocks would be enantiomerically pure, and are derived from the chiral pool of optically active aldehydes (or from compounds prepared from enantioselective reactions) (Scheme 6.74). A related approach involves nitroaldol additions (Henry reaction) (224).

An early example of this type of transformation was described in 1964 by Drehfahl and Hörhold (310). These authors prepared racemic 4-hydroxyproline, albeit with low diastereoselectivity for the isoxazoline reduction step [Scheme 6.75, (1)] (310). Much higher selectivity was achieved using 5-halomethylisoxazolines bearing a 3-(1-oxyalkyl) side chain, which was introduced from the nitrile oxide portion. The examples presented in Scheme 6.75 outline the synthesis of 4-hydroxy-pyrrolidines, which contain a dioxyethyl or trioxypropyl side chain. Both of these substrates were converted into the corresponding imino acids of 4-hydroxyproline (23,225,234) and 4, 1′-dihydroxyhomoproline, respectively (23,207,225) (Eq. 2, 3). The latter compound is part of an N-Val dipeptide structure, that was mistakenly proposed for the antibiotic Tü 1718B (311,312).

It is worthy to note that four of the eight possible diastereomeric pyrrolidine-polyols are accessible by this method. The configuration at C(1)′ (nitrile oxide part) depends on the enantiomeric aldehyde precursor chosen. The 5-epimers were obtained as a separable 1:1 mixture and were derived from the nonselective cycloaddition (see Section 6.2.3.4).

Fully hydroxylated 1,4-iminopolyols have been obtained using cis-4-oxyisoxa-zolines. This synthesis involves nitrile oxide cycloaddition to furan followed by various oxidative transformations of the C_2 enol ether portion of the resulting furoisoxazoline cycloadducts (21,82,293,294) (see Section 6.4.3).

1,4-Iminoheptitols were prepared from cis-4-hydroxy-5-tosyloxymethyl-isoxa-zolines. This procedure consisted of an initial bis(hydroxylation) of the dihydro-furan moiety present in **146** to give **147**, a periodate-mediated chain-shortening reaction, boranate reduction, and finally tosylation (157,225,293,294). The crucial step involves the catalytic hydrogenation of **148**, which proceeded with high stereoselectivity. The sequence of steps supposedly involves (a) N—O cleavage, (b) C=N reduction, and (c) cyclization thereby providing novel 1,4-iminoglycitols such as **149**, which possess an extended side chain (see Scheme 6.76).

The isoxazoline-furanose intermediate derived from *bis*(hydroxylation) was also submitted to direct hydrogenation. The reduction proceeded in a highly diaster-eoselective manner at the C=N double bond. In the *gluco* series starting with **150**, several deoxynojirimycin analogues with extended side chains at C-5 (carbohydrate numbering) such as **151** were obtained by this method (23,313,314). Further conversion of **151** led to indolizidine polyols (1,4,8-trideoxyimino polyols), which

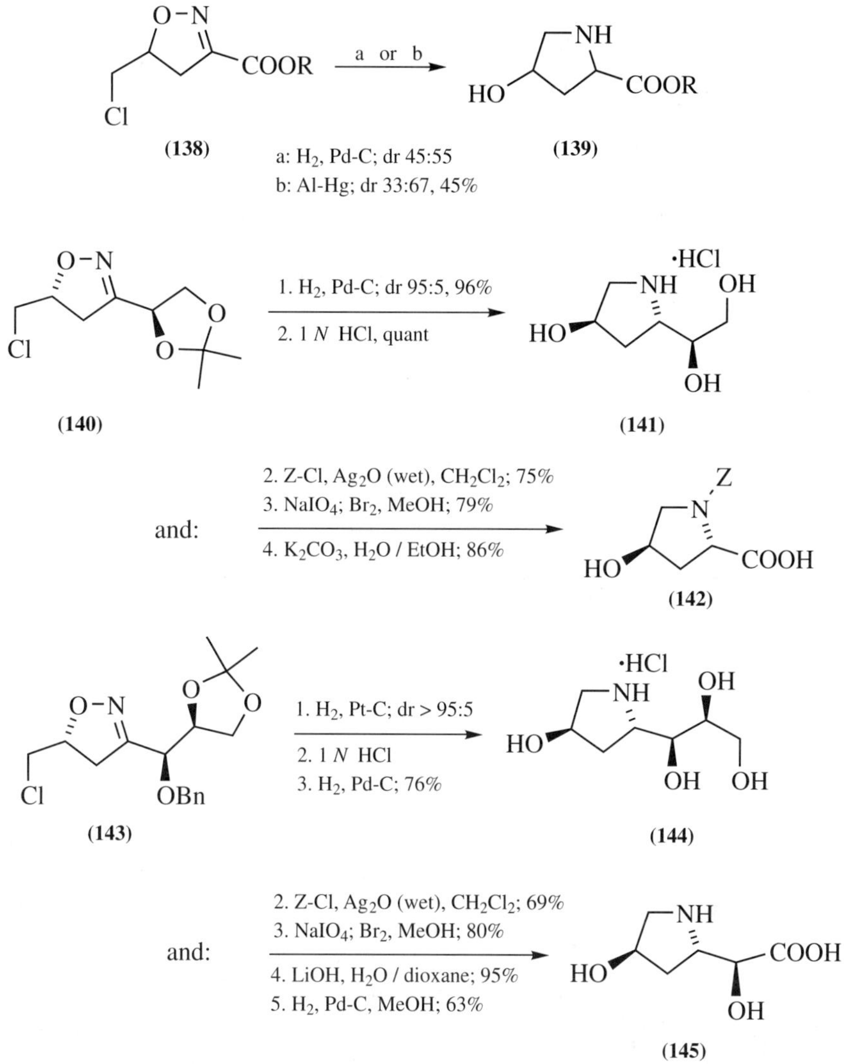

Scheme 6.75

involved a cyclization to afford novel castanospermine analogues such as **152** (see Scheme 6.77) (221,314).

This strategy was further developed so as to prepare the quinolizidine polyol **155** (315), by separating the racemic furoisoxazoline acetals (21,82,293,294) derived from the imidazolidine derivatives, which in turn, were obtained from (1*S*, 2*S*)-diphenylethylenediamine (316) (see Scheme 6.78). Aminal hydrolysis led to aldehyde **153**, and this was followed by aldehyde reduction and catalytic hydrogenation (221,313) to give (+)-deoxynojirimycin (natural). Its enantiomer was

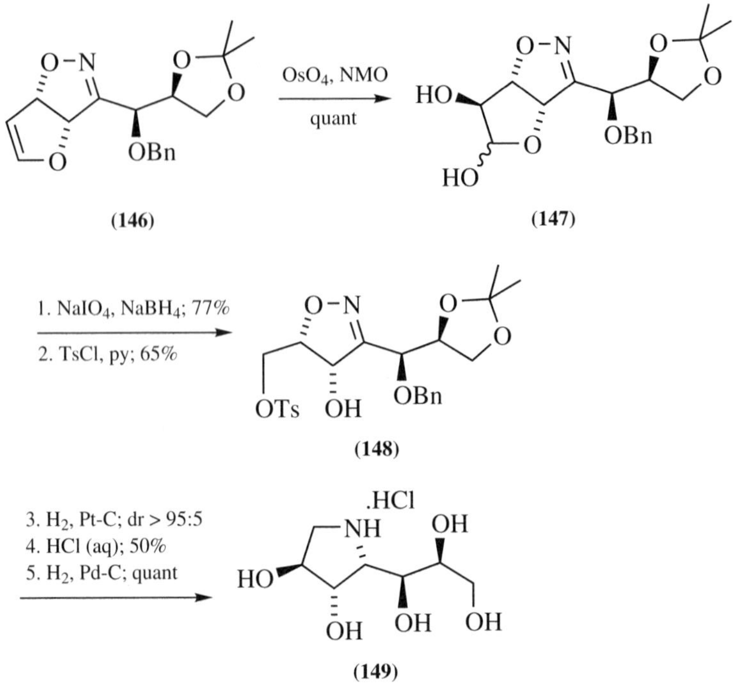

Scheme 6.76

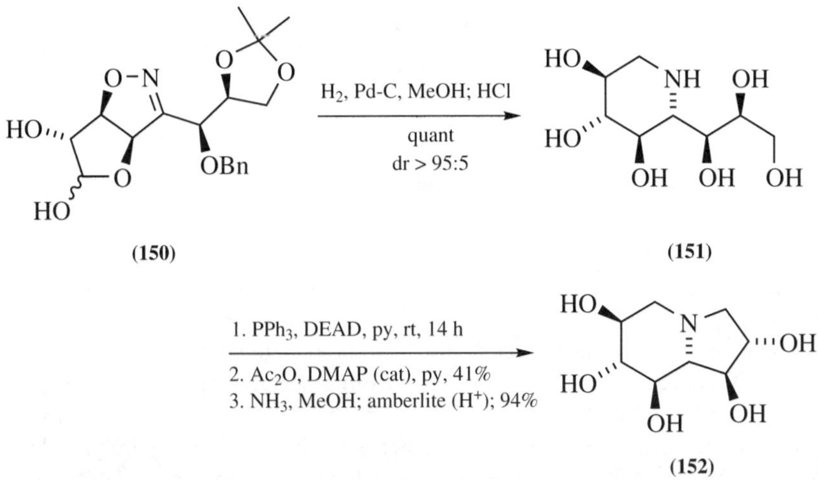

Scheme 6.77

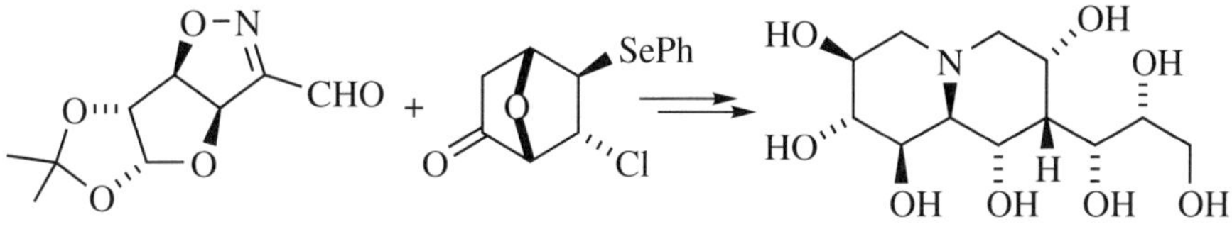

(153) (154) (155)

Scheme 6.78

obtained from the other diastereomeric furoisoxazoline (315). When the complementary hydride reduction sequence of (furo)isoxazolines was used (21,82,293, 294), deoxy-*ido*-nojirimycins was obtained (315). Combination of furoisoxazoline (**153**) with 7-oxabicyclo[2.2.1]heptane-2-one (**154**) by means of cross-aldolization led to the eventual formation of quinolizidine polyol (**155**) (317).

6.5. INTRAMOLECULAR NITRILE OXIDE CYCLOADDITIONS: APPLICATIONS IN SYNTHESIS

6.5.1. Intramolecular Cycloadditions to Polycyclic Isoxazolines and Furoxans

The intramolecular nitrile oxide–alkene cycloaddition sequence has been used for the assembly of a great variety of natural products. A target that has received special attention is that of taxol (**156**), undoubtedly due to its unique structural features, its potent anticancer activity, and its limited availability from natural sources (318,319). In 1984 Kozikowski et al. found that the treatment of nitro dienone **158** (obtained from the *p*-benzoquinone derivative **157**) with *p*-chlorophenyl isocyanate under Mukaiyama conditions afforded the unexpected eight-membered ring **159**, which is related to ring B of taxol (**156**) (Scheme 6.79).

In later work, Mioskowski and co-workers (320) used cyclohexenone **160** to prepare oxime **161** as part of a twofold nitrile oxide strategy to synthesize the basic taxol ring system. Cycloaddition of **161** was effected by means of sodium hypochlorite and gave tricyclic isoxazoline **162**, which features rings A and C of taxol (320) (Scheme 6.79). Nagaoka and co-worker tried to apply a related intramolecular cycloaddition toward the synthesis of the taxane A/B ring but this approach failed, producing only the oxime derivative (248) (see Scheme 6.44, Section 6.3.1).

Recently, the intramolecular nitrile oxide–alkene cycloaddition sequence was used to prepare spiro-*bis*(isoxazolines), which are considered useful as chiral ligands for asymmetric synthesis (321). Reaction of the dibutenyl-dioxime (**164**) (derived from the diester **163**) with sodium hypochlorite afforded a mixture of diastereomeric isoxazolines **165–167** in 74% combined yield (Scheme 6.80) (321). It was discovered that a catalytic amount of the Cu(II) complex **165**-Cu(acac)$_2$, where acac = acetylacetonate, significantly accelerated the reaction of diisopropylzinc

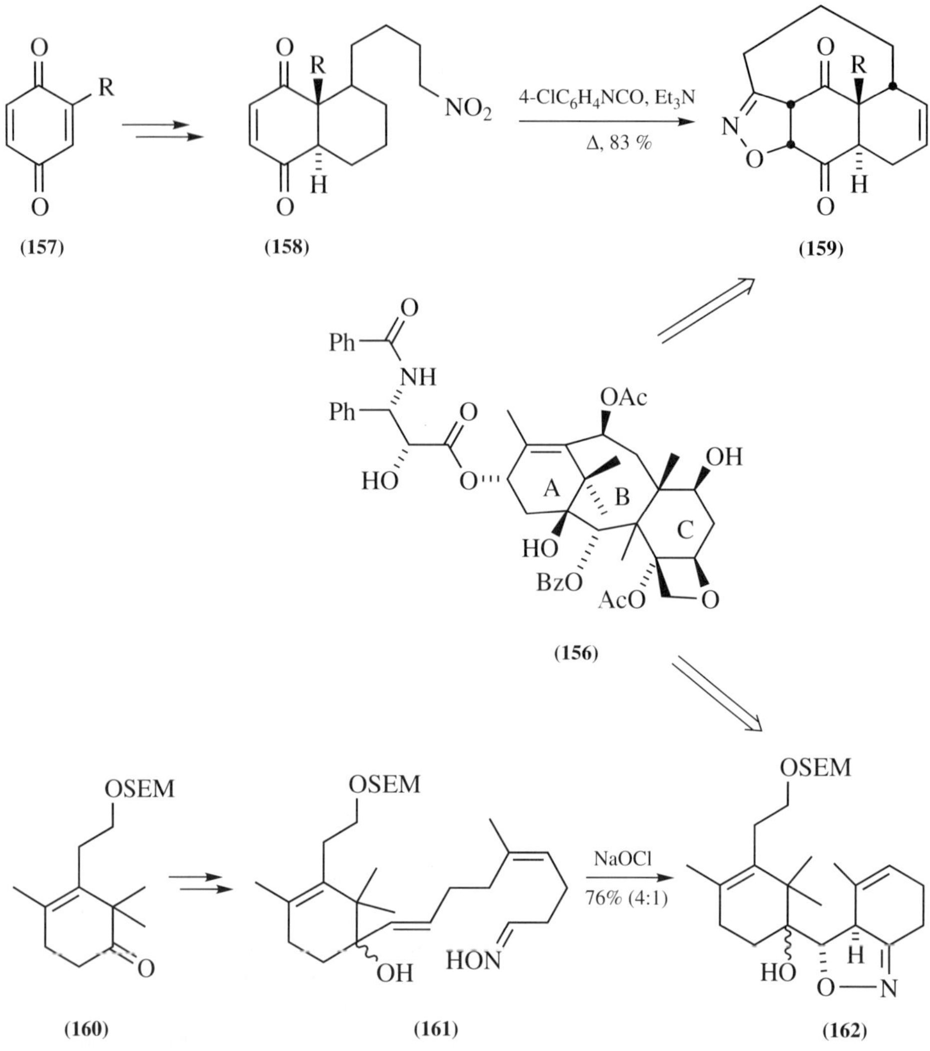

Scheme 6.79

with cyclohexenone producing the Michael adduct product with 49% enantiomeric excess (ee) (321).

As was mentioned earlier, furoxans are often encountered as unwanted byproducts in nitrile oxide cycloadditions. There are, however, some efforts to exploit this facile C–C forming dimerization for synthesis. In one case, an intramolecular *bis*(nitrile oxide) cycloaddition was used for a synthesis of biotin (322a). More recently, the intramolecular dimerization was employed for the construction of medium- and large-size rings. This was feasible if one of the two nitrile oxide functionalities was relatively hindered and stable (see Section 6.1.4). Unsymmetrical

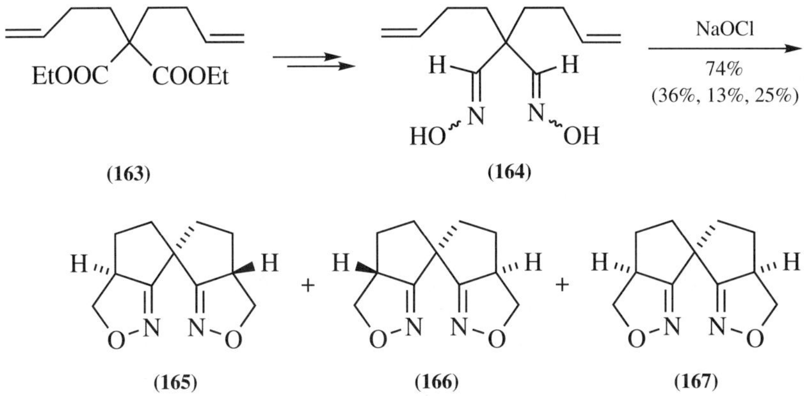

Scheme 6.80

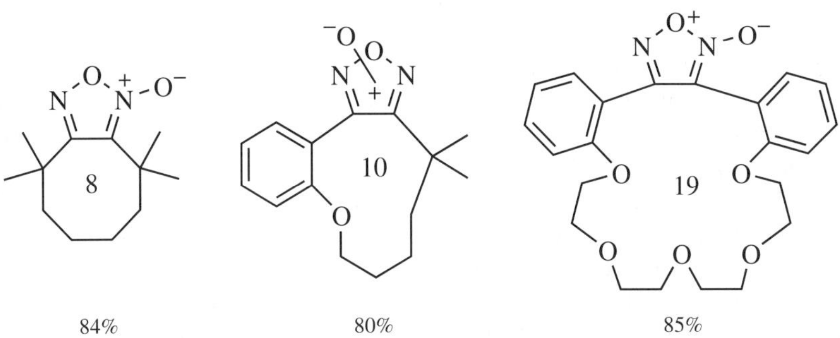

Scheme 6.80a

bicyclic furoxans could also be obtained in remarkably good yield (89,322b). Examples of 8-membered ring formation, as well as examples leading to 19-membered rings were given (322b) (Scheme 6.80a).

6.5.2. Intramolecular Cycloadditions with Ensuing Ring Cleavage of Intermediate Isoxazolines

In the great majoritiy of applications that use the intramolecular nitrile oxide–alkene cycloaddition, the intention is to prepare intermediates for the synthesis of natural products or related compounds. The most popular transformations of these isoxazolines are the following ring cleavage modes:

1. β-Hydroxyketones (or α,β-unsaturated ketones), *aldol route*.
2. β-Aminoalcohols.
3. Isoxazolinium salts, with ensuing reduction and N—O cleavage.
4. Other eliminations by base or zinc-mediated fragmentation (19,20–22,26, 292).

In this section, the different examples selected are classified according to these types and are arranged with increasing order of complexity of the isoxazoline or final product.

6.5.2.1. Isoxazoline Ring Cleavage Reactions of the "Aldol" Type (Reduction–Hydrolysis)

Isoxazolines with an Annulated Five-Membered Ring

Kozikowski and Stein (281) used the INOC strategy to prepare the 2-methylenecyclopentanone derivative **172**, which in turn was converted to sarkomycin (**173**), an antitumor agent (Scheme 6.81). The key step involved the treatment of nitroalkene **169** (obtained from bromide **168**) with *p*-chlorophenyl isocyanate–triethylamine, which furnished a single diastereomeric isoxazoline **170** in 55% yield. This compound was transformed to the aldol product **171** by Raney nickel hydrogenation using wet acetic or boric acid, followed by dehydration to the α,β-enone **172** (281), a precursor of **173**.

A variation of this route was applied to the preparation of α-methylenecyclopentane **179**, an intermediate that was employed for the synthesis of prostaglandin $PGF_{2\alpha}$ (**180**) (Scheme 6.82). The acetonide-protected oxime-diol **175** [derived from propanal (**174**)] was treated with sodium hypochlorite without the addition of base. This led to the tricyclic adduct **176** with high stereoselectivity. One of the side chains was subsequently elaborated and the fully protected cyclopentano-isoxazoline (**177**), when exposed to Raney Ni/boron trichloride, gave the 2-hydroxymethylcyclopentanone (**178**). This compound was dehydrated using mesyl chloride/pyridine to furnish enone (**179**) (324). In another related synthesis of $PGF_{2\alpha}$ the β-side-chain (3-hydroxyoctenyl) was introduced prior to the cycloaddition (325).

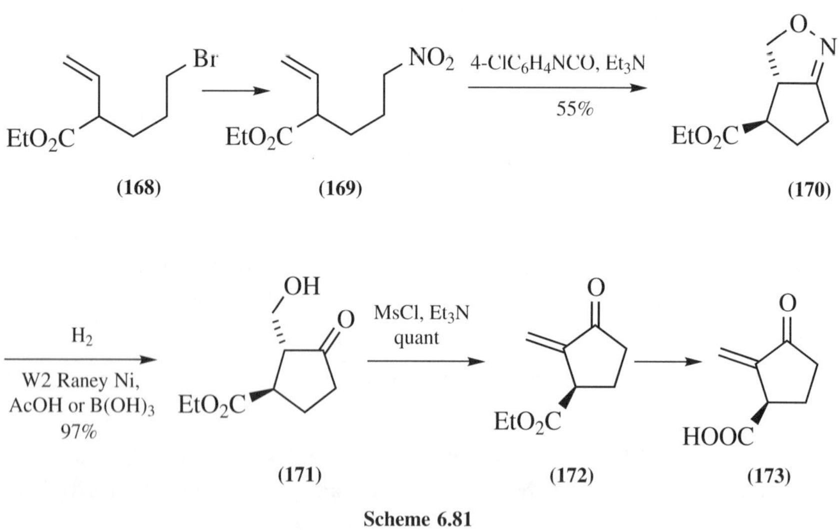

Scheme 6.81

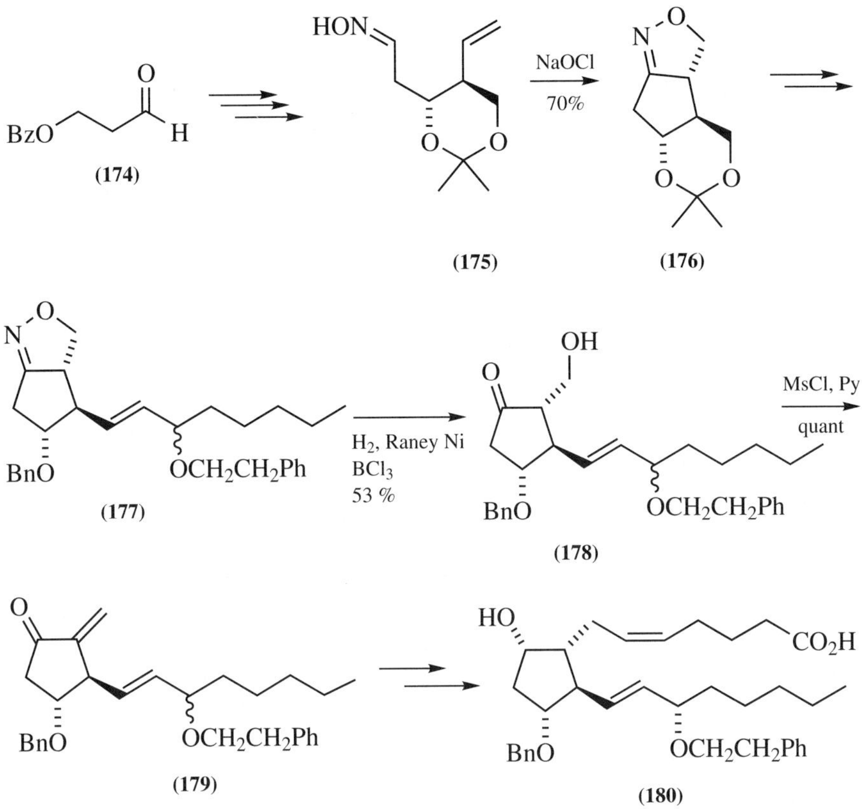

Scheme 6.82

The intramolecular cycloaddition of substrates containing several functional groups and stereocenters such as carbohydrate derivatives can be used to prepare complex cyclopentane products that are useful for the synthesis of natural products. For example, (−)-allosamizoline (**41**) [the aglycon of the pseudo-trisaccharide allosamidin (**184**), a strong chitinase inhibitor] was prepared using an intramolecular cycloaddition of D-*xylo*-hexenononitrile oxide **39** (258) (Scheme 6.83). The aldehyde precursor **181** was obtained from D-glucosamine. The resulting cyclopentano-isoxazoline **40** was subjected to reductive cleavage/hydrolysis to afford β-hydroxyketone **182**. Subsequent carbonyl reduction gave diol **183**, from which (−)-allosamizoline (**41**) was elaborated (258).

In a closely related case, D-glucose was used as the precursor for D-*xylo*-5-hexenose **185**, which in turn, was used in the synthesis of trehazolin (**189**), a strong inhibitor of trehalose (325) (Scheme 6.84). Thus, treatment of oxime **186** (derived from **185**) with sodium hypochlorite gave cycloadduct **187** in good yield. Reductive cleavage of the N—O bond of **187** followed by hydrolysis of the intermediate imine group led to the spontaneous elimination of the benzoyloxy group producing

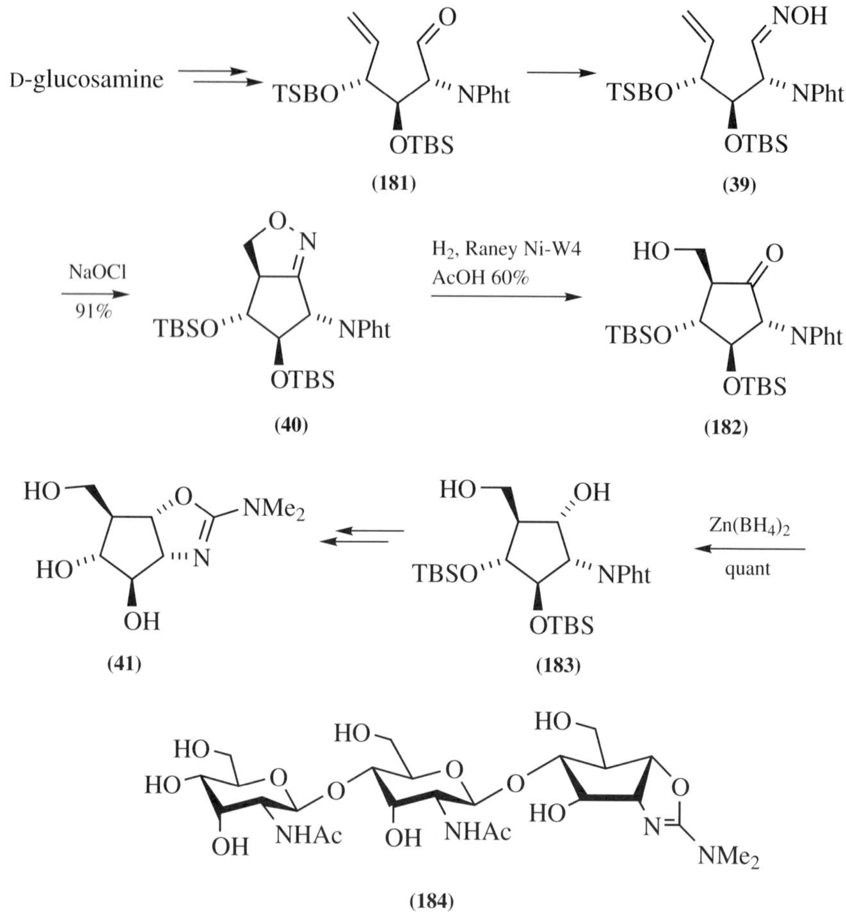

Scheme 6.83

cyclopentenone **188**. After a number of additional steps, the target compound trehazolin (**189**) was reached (325).

Intramolecular cycloadditions of alkenyl-substituted nitrile oxides produce bicyclic isoxazolines. When monocyclic olefins are used, tricyclic structures are obtained. This approach was pioneered by both Kozikowski's and Curran's groups. A typical case involves the cycloaddition of nitro compound **191** [mixture of diastereomers derived from pentenose pyranoside **190**], which produced a diastereomeric mixture of isoxazolines that contain cis-fused rings (i.e., **192**) in near quantitative yield (326) (Scheme 6.85). Further elaboration of this mixture led to epoxycyclopentano-isoxazoline **193**, which was then converted to the aldol product in the usual manner. The hydrogenation proceeded well only when rhodium on alumina was used as the catalyst, giving the required β-hydroxyketone **194**. This

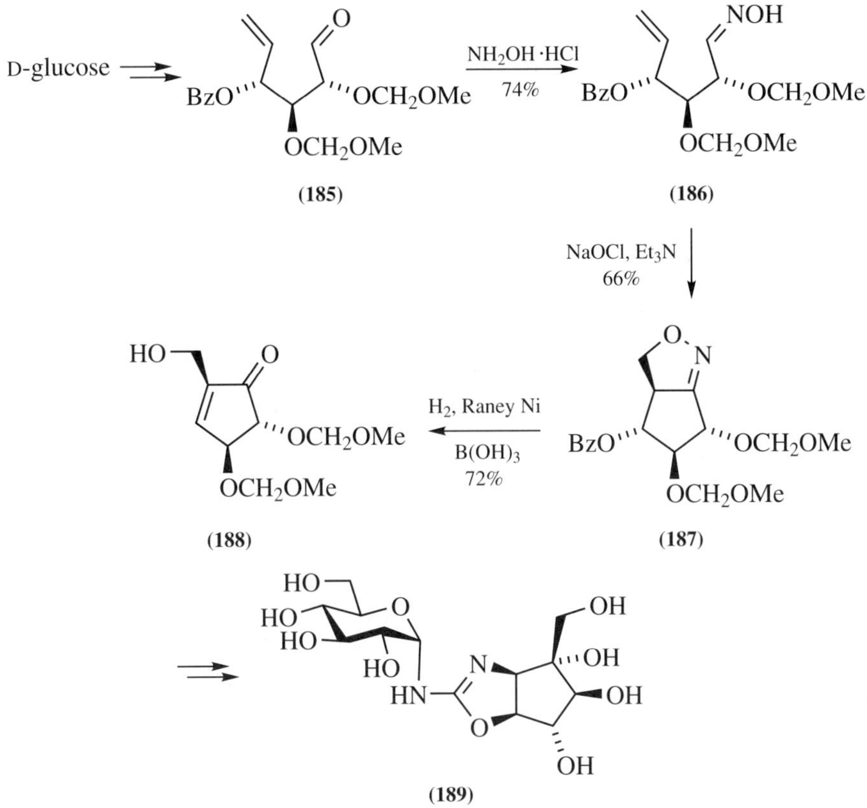

Scheme 6.84

compound was then converted to (−)-specionin (**195**), a potent antifeedant of a common pest in North American forests (326).

A related sequence was used by Kozikowski and Park (74) to prepare the ring skeleton of streptazolin (**200**), a compound that exhibits antibacterial and antifungal effects. In this approach, the tricyclic isoxazoline intermediate **198** was formed in the key cycloaddition step (Scheme 6.86). Thus, the reaction of oxime **197** (obtained from 4-piperidone) with sodium hypochlorite–triethylamine afforded tricyclic isoxazoline **198** in very good yield. This cycloadduct was converted to β-hydroxyketone **199** by reduction/hydrolysis using Raney Ni in the presence of acetic acid. Racemic streptazolin (**200**) was obtained from **199** in several additional steps (74).

An intramolecular nitrile oxide cycloaddition also served as the key step in the stereoselective assemblage of the skeleton of angular triquinane sesquiterpenes of the isocomene series. Tetracyclic isoxazoline **203** was obtained from oxime **202** [derived from tetrahydroindandione **201**] and on treatment with sodium hypochlorite

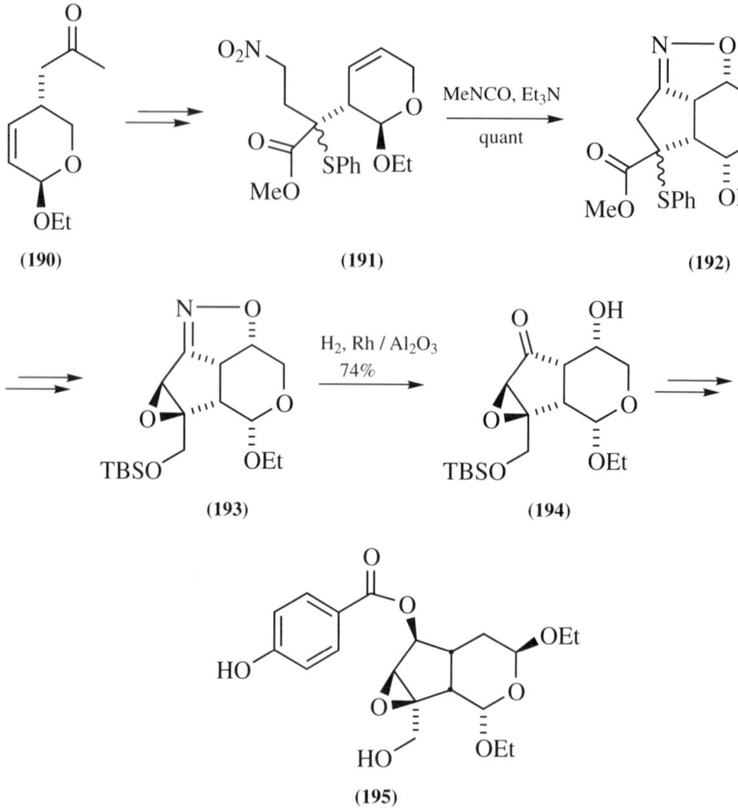

Scheme 6.85

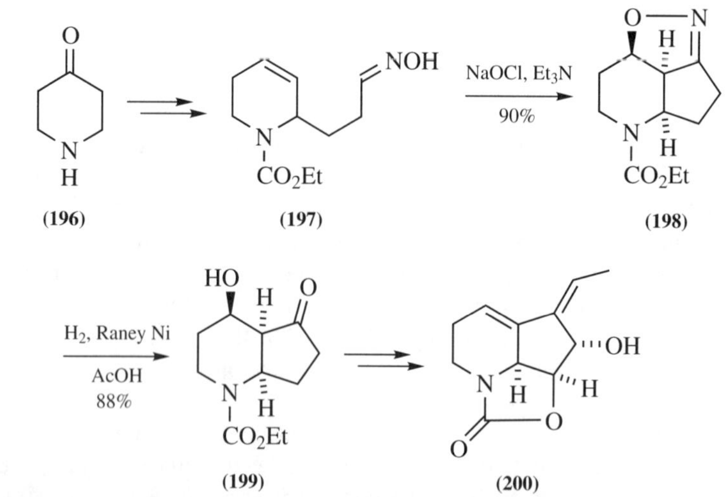

Scheme 6.86

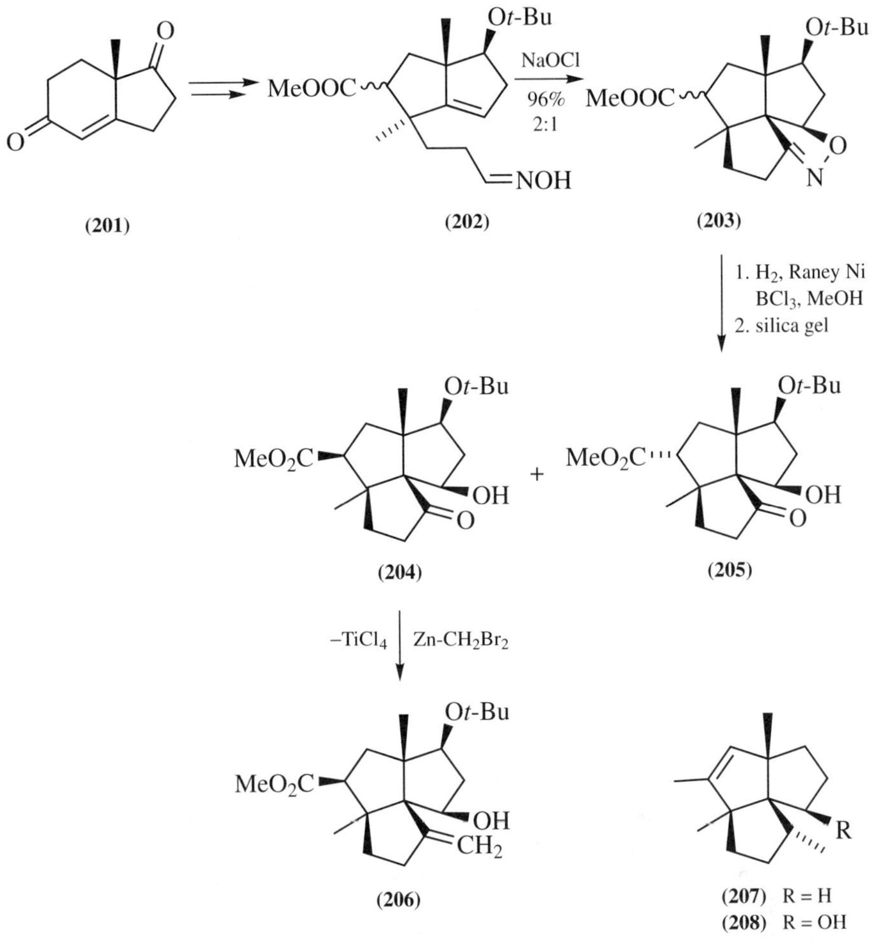

Scheme 6.87

gave **203** in excellent yield (96%) as a 2:1 diastereomeric mixture (Scheme 6.87). The mixture was subjected to Raney Ni–trimethyl borate reduction and, after hydrolysis, gave β-hydroxyketones **204** and **205**. α-Ketone **204** was then carried on to methylene-triquinane **206**, which possesses all the carbon atoms present in isocomene (**207**) and 1-hydroxyisocomene (**208**) (327).

Isoxazoline Intermediates with an Annulated Six-Membered Ring

Intramolecular nitrile oxide-alkene cycloadditions also provide efficient access to six-membered rings such as cyclohexanes or decalins that are heavily adorned with functional groups and side chains. For example, this strategy was used to prepare racemic hernaldulcin (**213**), which is a 3,6-disubstituted cyclohexenone found in a Mexican plant that possesses a strong sweet taste. Starting from (2*Z*,6*E*)-farnesal (**209**) (328) (Scheme 6.88), the aldehyde was treated with hydroxylamine

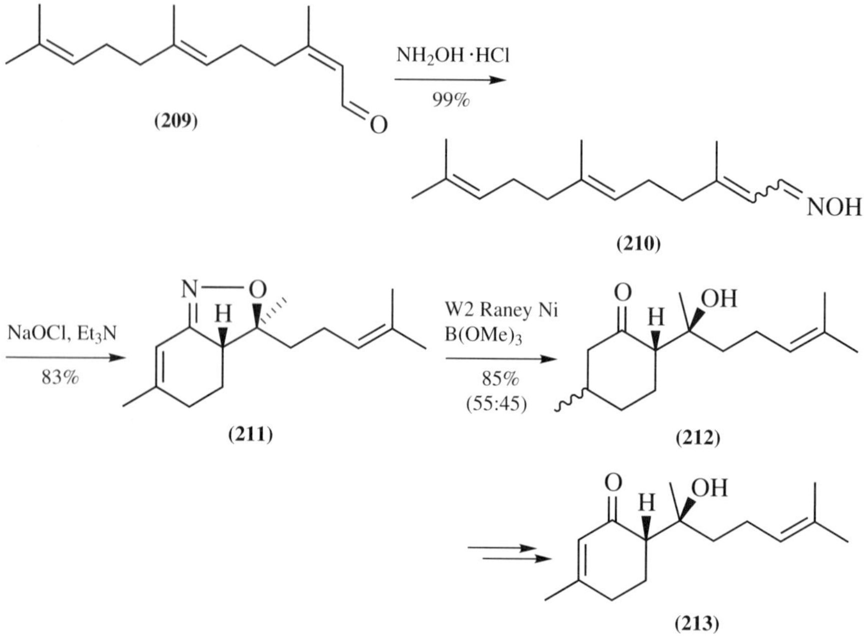

Scheme 6.88

in pyridine to give a mixture of the unsaturated oximes **210**. Treatment of **210** with aqueous sodium hypochlorite–triethylamine led to the formation of bicycle **211**. The usual catalytic hydrogenation sequence using Raney Ni afforded racemic dihydrohernalducin (**212**), which could be converted into **213** in four additional steps (328).

In a similar approach, Shishido et al. (241) used oxime **215** [derived from the monoterpene (+)-citronellal (**214**)] for the synthesis of (−)-mintlactone (**218**) and (+)-isomintlactone (**219**), sweet compounds isolated from some *Mentha* species (Scheme 6.89). Bicyclic isoxazoline **216** was obtained in good yield from the cycloaddition. As expected, the product possessing *trans*-1,4-substitution prevailed. Reductive hydrolysis of the major isomer of **216** using hydrogen–Raney Ni–trimethyl borate provided β-hydroxyketone **217**. This compound was dehydrated to give an enone and this was followed by carbonyl reduction–lactonization to complete the synthesis of both lactones **218** and **219** (241).

Intramolecular cycloaddition of the nitrile oxide intermediate generated from the unsaturated oxime **221** was used for an eventual synthesis of 1α,2β,25-trihydroxy-vitamin D_3 (262) (Scheme 6.90). Oxime **221**, prepared from tri-*O*-isopropylidene-d-mannitol (**220**), was processed as usual to give isoxazoline **222** in good yield and with excellent stereoselectivity. Conversion of **222** to the aldol **223** proceeded in the normal manner and further elaboration gave the desired diene intermediate **224** (262).

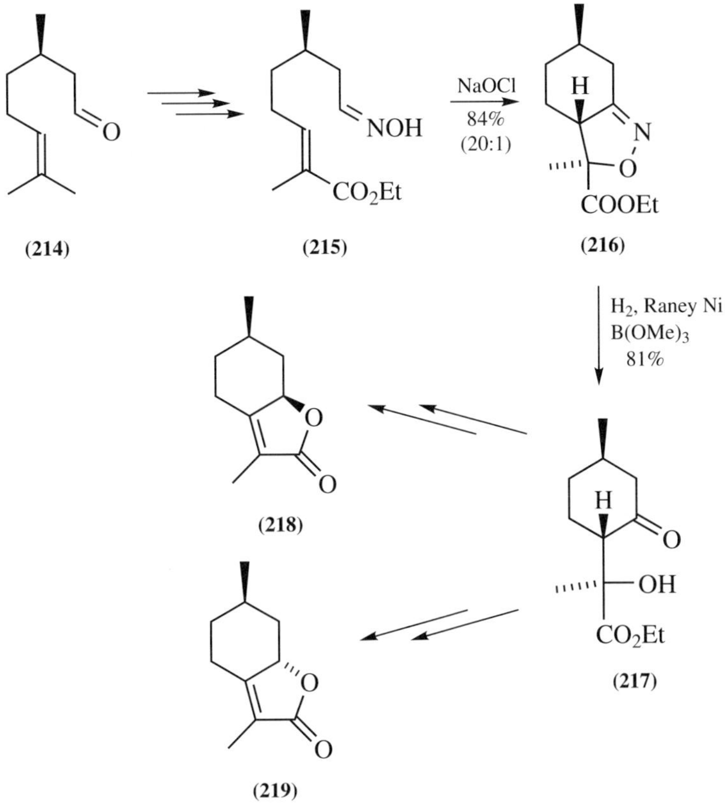

Scheme 6.89

A mixture of stereoisomers of 4-phenylsulfonyl-6-heptenal oxime **226** (assembled from the alkene **225** using sulfone carbanion chemistry) was treated with sodium hypochlorite and produced isoxazoline **227** with high stereoselectivity for each stereoisomer of **226** (77) (Scheme 6.91). Removal of the MOM-protecting group in **227** followed by replacement with a 2-bromo-2-propenyl group afforded isoxazoline **228**, which was subsequently cleaved using $TiCl_3$ (77). The hydroxy group of the aldol moiety was next protected by silylation. Cyclization of **229** occurred using di-*n*-butylcopper lithium to produce the bicyclic methylene compound **230**. This structure corresponds to an analogue of the oxahydrindene portion of the milbemycins (see **231**) and the avermectins, which are known as powerful antiparasitics (77). Another approach to the southern zone of these macrolides was initiated by Fraser-Reid and co-workers (329), who made use of the ω-nitroalkene derivative **233**, available from diacetone glucose **232**. Cycloaddition of **233** under Mukaiyama conditions afforded the tetracyclic isoxazoline **234** as a single stereoisomer (Scheme 6.92), which was subsequently converted to β-hydroxyketone **235**

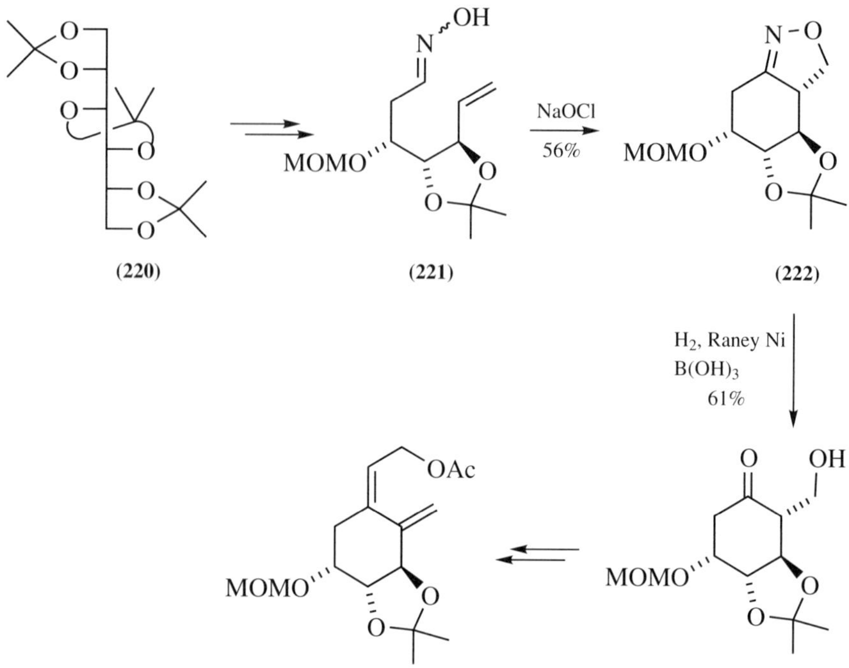

(220) (221) (222) (223) (224)

Scheme 6.90

with Raney Ni in good yield. From **235**, oxahydrindene **236** was eventually obtained (329).

The sesquiterpene skeleton has also been assembled by the intramolecular nitrile oxide cycloaddition sequence. Oxime **238** (obtained from epoxy silyl ether **237**), on treatment with sodium hypochlorite gave isoxazoline **239**, which was sequentially hydrolyzed and then subjected to the reductive hydrolysis conditions–cyclization sequence to give the furan derivative **240** (330) (Scheme 6.93). In three additional steps, compound **240** was converted to **241**. This structure contains the C_{11}–C_{21} segment of the furanoterpene *ent*-**242**, that could be obtained after several more steps (330).

Another related synthesis made use of the intramolecular cycloaddition of ω-nitroalkene **243**, also derived from geraniol epoxide **237**. Generation of the expected nitrile oxide dipole using *p*-chlorophenyl isocyanate and triethylamine quantitatively gave the annulated isoxazoline **244** as a 2:1 mixture of diastereoisomers (Scheme 6.94). Reductive hydrolysis of the cycloadduct to the aldol product followed by dehydration provided enone **245**, which was used to prepare the sesquiterpene nanaimoal **246** (242).

The intramolecular nitrile oxide–alkene cycloaddition has further been used for the construction of a tricyclic isoxazoline intermediate containing a decaline ring.

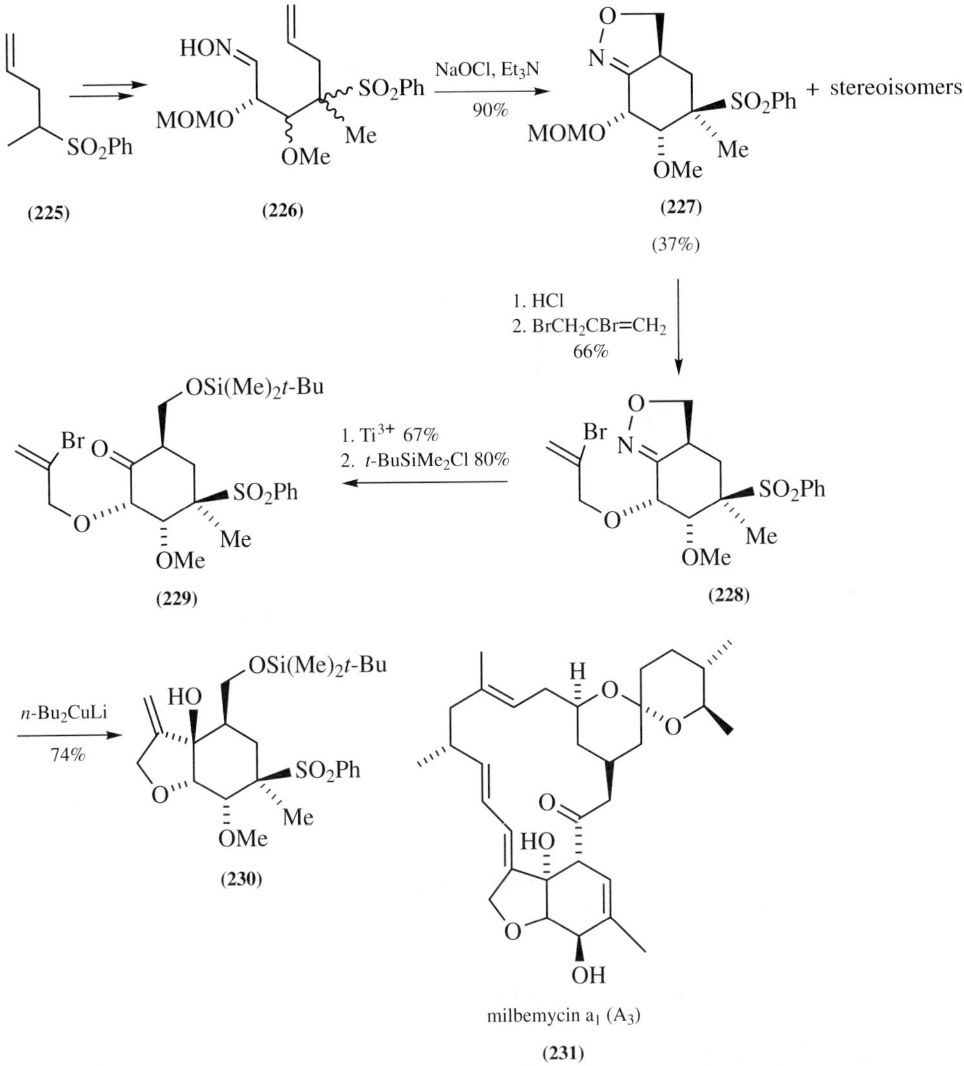

Scheme 6.91

This was demonstrated by Fukumoto and co-workers in a synthesis of (+)-albicanol (**251**), a sesquiterpene with potent fish antifeedant properties (272,273). Oxime **248** [prepared from the (+)-Wieland–Miescher ketone **247**] was subjected to cycloaddition using sodium hypochlorite and gave isoxazoline **249** in very good yield (Scheme 6.95). Conversion of **249** into β-hydroxyketone **250** was again accomplished by the reductive hydrolysis sequence using Raney Ni and trimethyl

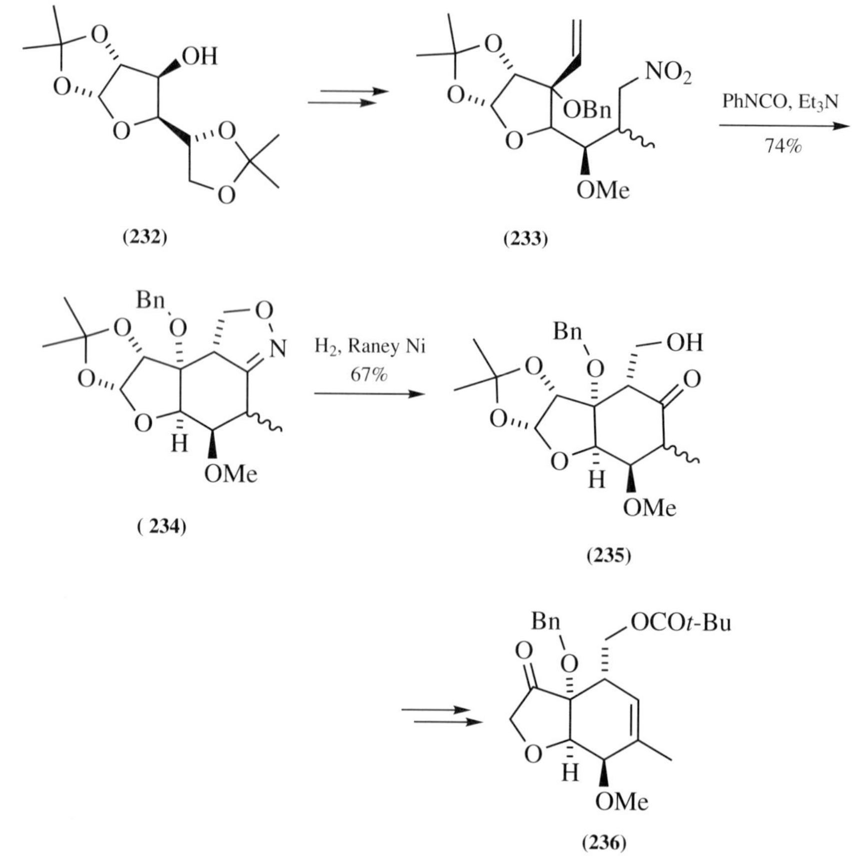

Scheme 6.92

borate. Reaction of **250** under Nozaki–Lombardo conditions for methylenation furnished (+)-albicanol **251** (272,273).

Kozikowski and Li (268) also made use of this protocol for the construction of the hexahydronaphthalene portion of the hypocholesteremic agent compactin (**256**) (see Section 6.4.2.2). The oxime derived by from alcohol **253** (via γ-lactone **252**) was heated with aqueous sodium hypochlorite in the presence of triethylamine to give the tricyclic isoxazoline adduct **254** (Scheme 6.96). Reductive hydrolysis and dehydration afforded enone **255**, which in several further steps led to compactin (**256**) (268).

A stereoselective synthesis of testosterone (**261**) was advanced by Fukumoto and co-workers (331), where ring B was joined to the C/D part by an intramolecular nitrile oxide cycloaddition. The key nitrile oxide dipole was generated *in situ* from oxime **258**, which in turn was derived from the optically active tetrahydroindanone **257**. Tetracyclic isoxazoline (**259**) was obtained as a single stereoisomer

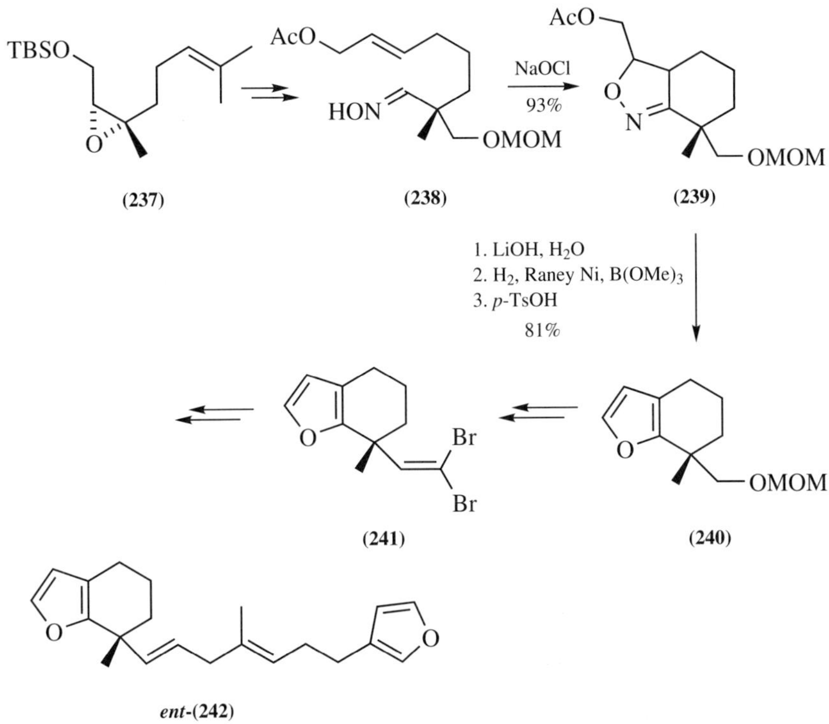

Scheme 6.93

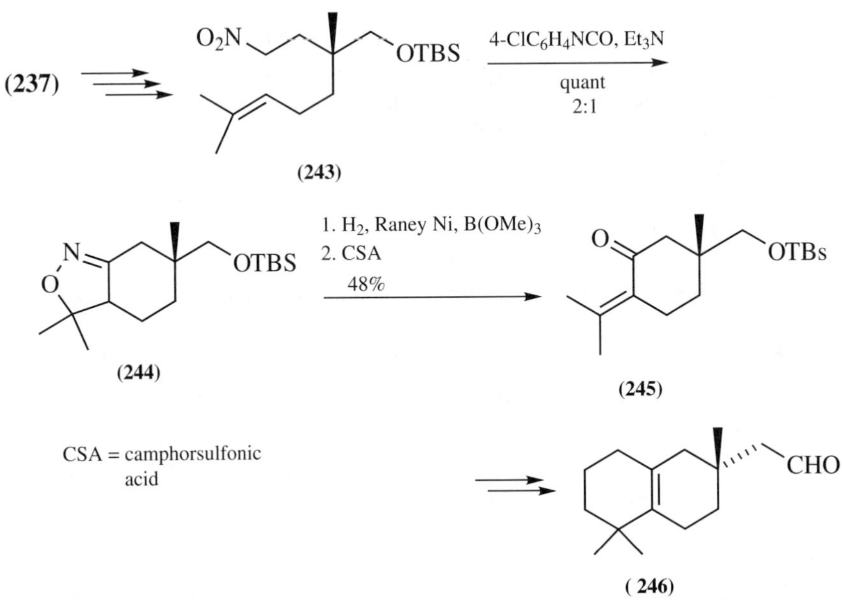

Scheme 6.94

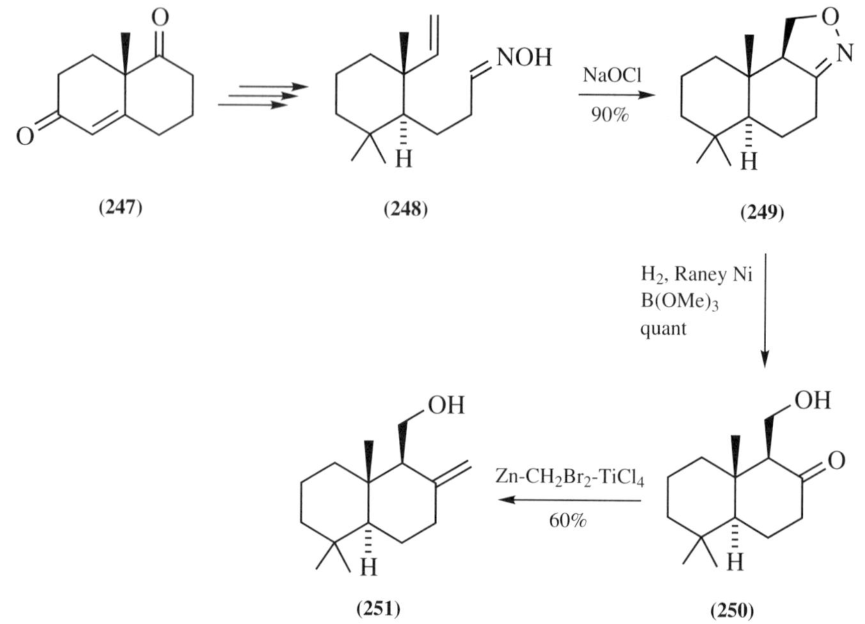

Scheme 6.95

(Scheme 6.97). Transformation of this compound to hydroxyketone **260** proceeded as usual in excellent yield. From compound **260**, ring A was elaborated resulting in a complete synthesis of testosterone **261** (331).

An intramolecular cycloaddition of the tetradecatrienyl nitroethyl ether **263** was used in the synthesis of the 14-membered bicyclic precursor **265** of crassin acetate **266**, a cembrane lactone possessing antibiotic and antineoplastic activity (332). Nitro compound **263** was obtained from farnesyl acetate (**262**) in several steps and was then treated with phenyl isocyanate and triethylamine to give the tricyclic isoxazoline **264** (Scheme 6.98). Conversion to ketone **265** was accomplished by hydrogenation of the cycloadduct with Raney Ni and boric acid followed by acetylation (332). In this case, the isoxazoline derived from a 3-butenyl nitroethyl ether moiety served to produce a 3-methylenetetrahydropyran moiety (332).

Isoxazoline Intermediates with an Annulated Seven-Membered Ring

Shibasaki and co-workers used an intramolecular nitrile oxide cycloaddition to prepare the skeleton of phorbol (**272**) (Scheme 6.99), a tumor promoter that activates protein kinase C (PKC) (333). Nitroalkene **268** was elaborated in several steps from (+)-3-carene (**267**) and was subjected to cycloaddition by means of *p*-chlorophenyl isocyanate–triethylamine to give cycloadduct **269** in 88% yield. Reductive hydrolysis employing Raney Ni and boric acid afforded hydroxyketone **270**, that was subsequently used for the construction of the optically active derivative **271**, which contains the phorbol skeleton (333).

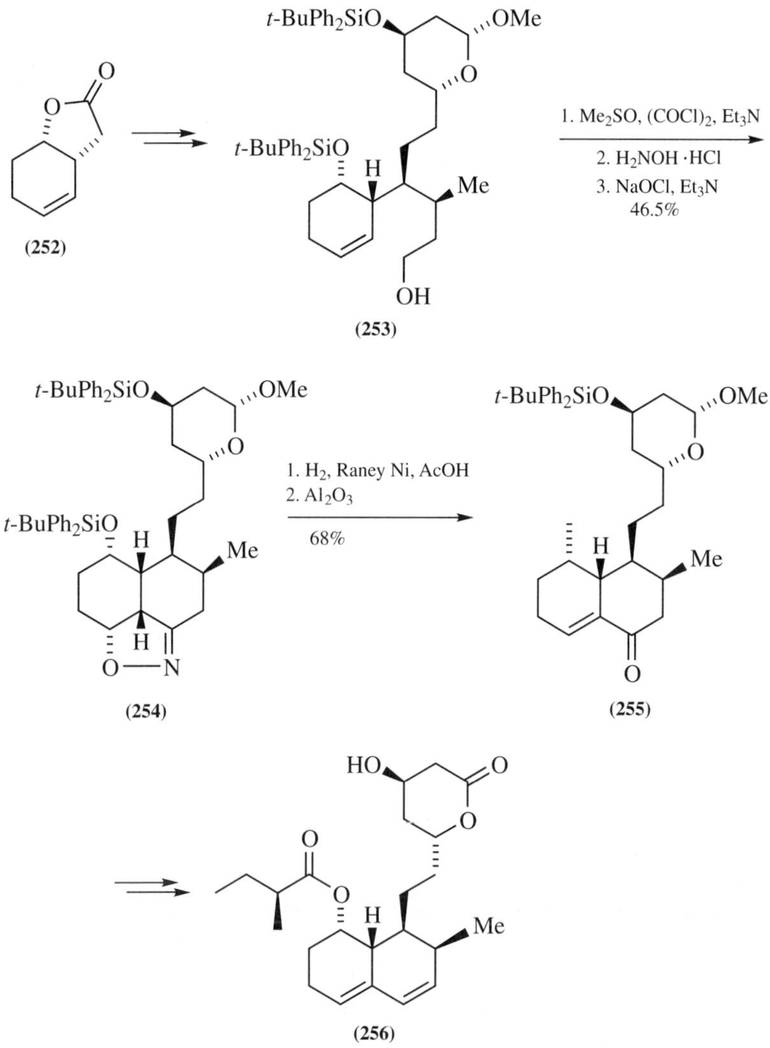

Scheme 6.96

Synthesis of Macrocycles Employing Intramolecular Nitrile Oxide Cycloaddition and Aldol Cleavage

One of the very first uses of the intramolecular nitrile oxide cycloaddition involved the synthesis of macrocyclic lactones. Asaoka et al. (238) conceived this approach to the 16-membered ring antibiotic A26771B (**277**). Nitro compound **274** [obtained from 11-acetoxydodecanal (**273**)] was dehydrated with 4-chlorophenyl isocyanate–triethylamine and this was followed by dipolar cycloaddition, which gave isoxazoline **275** as a 4:1 mixture of diastereomers (Scheme 6.100).

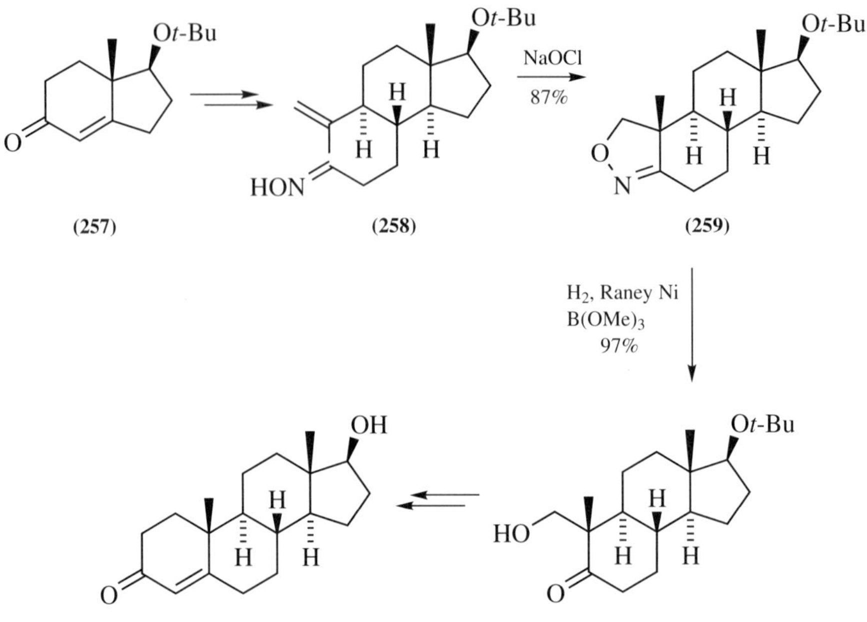

Scheme 6.97

Hydrogenation under acidic conditions followed by dehydration afforded lactone **276** that was easily converted to the target compound **277** (238).

A similar strategy was used for the synthesis of muscone (**282**) (246). Here, 1-nitro-14-pentadecene (**279**) [readily available from 12-tridecenal (**278**)] was subjected to the cycloaddition sequence under conditions similar to that described above, producing the bicyclic compound **280** in good yield (Scheme 6.101). Hydrogenolysis of the isoxazoline ring in the presence of acetic acid–water was followed by dehydration of the resulting aldol product to provide a moderate yield of the unsaturated ketone **281**. This compound was further converted to racemic muscone (**282**) by conjugate addition of a methyl group to the C=C bond (246).

6.5.2.2. Syntheses with Intramolecular Nitrile Oxide Cycloaddition and Isoxazoline Reduction to the γ-Aminoalcohol

Few examples of total syntheses have been reported that involve an intramolecular nitrile oxide cycloaddition and ensuing reduction to an aminoalcohol. The very first example was reported by Confalone et al. (334) and involved a synthesis of the naturally occurring vitamin biotin (**287**). The nitro precursor **284** was easily prepared from cycloheptene. When treated with phenyl isocyanate–triethylamine, cycloaddition led to the all-cis-fused tricyclic isoxazoline **285** with high stereoselectivity (Scheme 6.102). Reduction with $LiAlH_4$ afforded aminoalcohol **286** as a

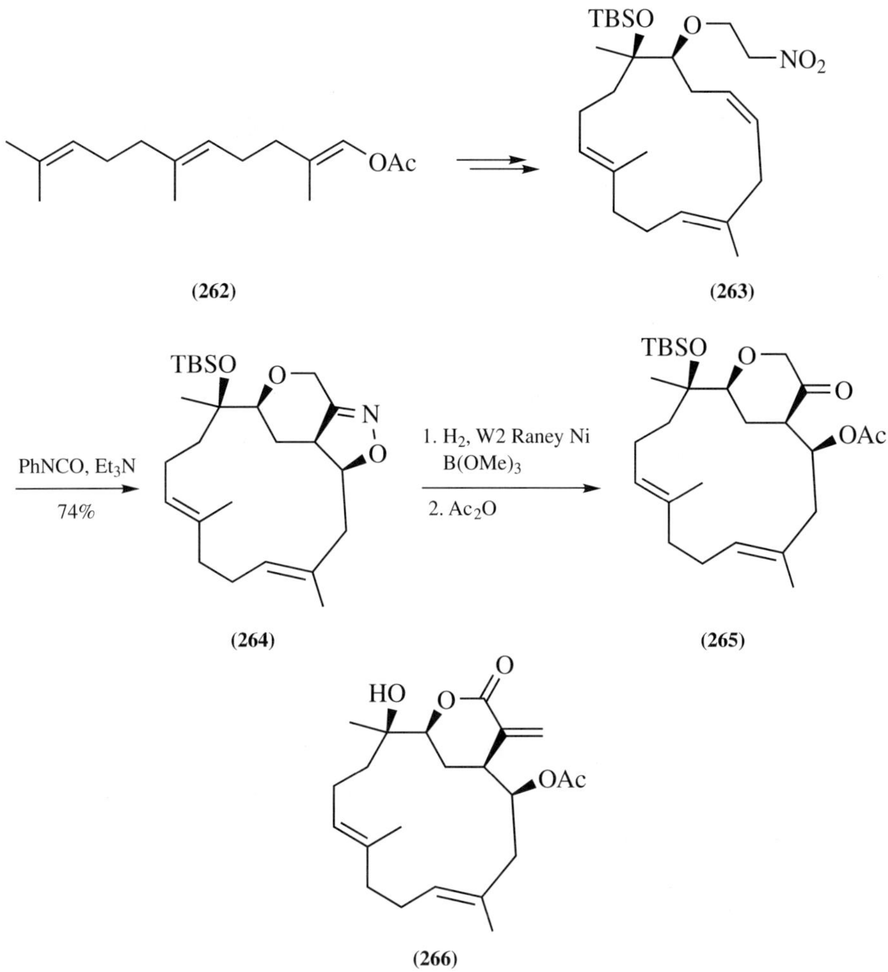

Scheme 6.98

single isomer. A number of subsequent transformations were carried out in order to complete the synthesis of racemic biotin (**287**) (334).

An interesting example has been reported where two five-membered rings were constructed in the key cycloaddition step and then used in the synthesis of papuamine (**292**) (Scheme 6.103), an alkaloid isolated from a marine sponge (335). Thus, treatment of the optically active 1-nitroethyl-2-vinylcyclohexane **289** with phenyl isocyanate resulted in the formation of tricyclic isoxazoline **290** in good yield. A single diastereoisomer, with trans-substitution at the newly formed stereocenter in the central cyclopentane ring, was obtained. Subsequent reduction and protection afforded the *N*-Boc-amino alcohol **291**, whose N-deprotected and O-benzylated derivative was taken onto the target structure of papuamine (**292**), with its intriguing C_2-symmetrical pentacyclic structure (335).

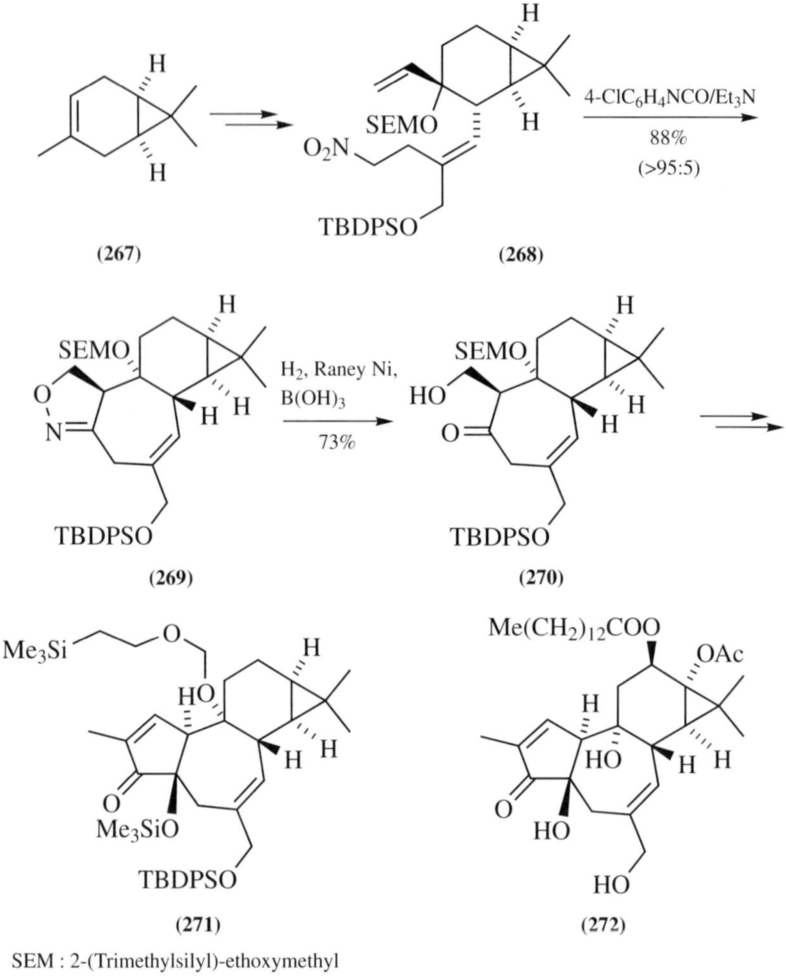

Scheme 6.99

Jäger and co-workers synthesized several aminocyclopentane polyols (336) using the above mentioned strategy (259,260,261,336). Oxime **294**, easily prepared from mannoside **293**, was treated with sodium hypochlorite under microwave irradiation to give isoxazolines **295** and **296** in moderate yield (Scheme 6.104). The isomers were separated and reduced with $LiAlH_4$ to afford the respective amino alcohols **297** and **298** in high yield as single diastereoisomers. Hydrogenolytic debenzylation of **297** using methanol–hydrobromic acid afforded the hydrobromide **299**. This compound was found to be a very strong inhibitor of α-mannosidase (259,260,336–339). Analogous routes were likewise applied to hexenoses (338) derived from glucose, galactose, and *N*-acetylglucosamine. This

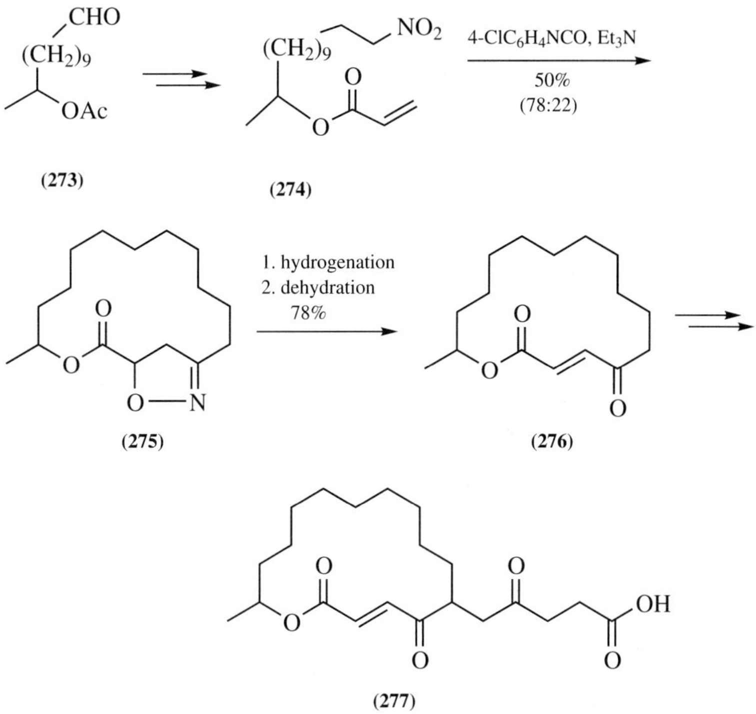

Scheme 6.100

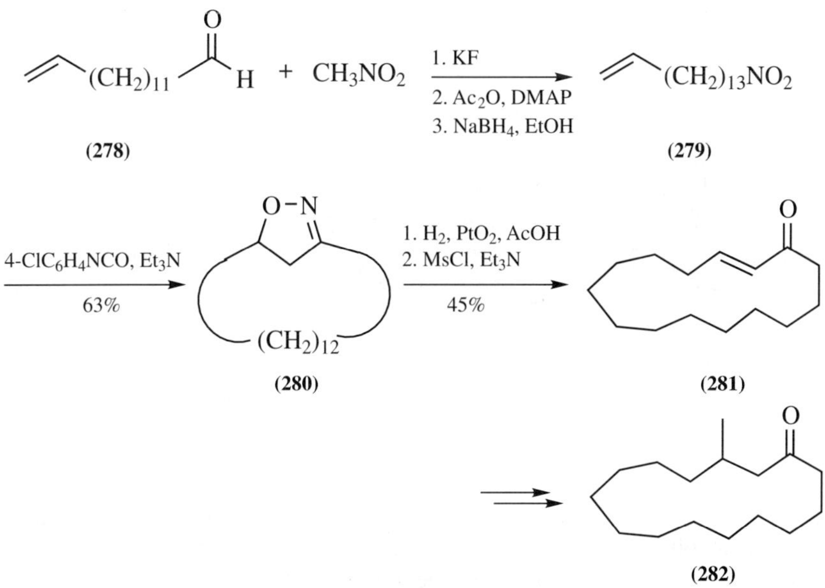

Scheme 6.101

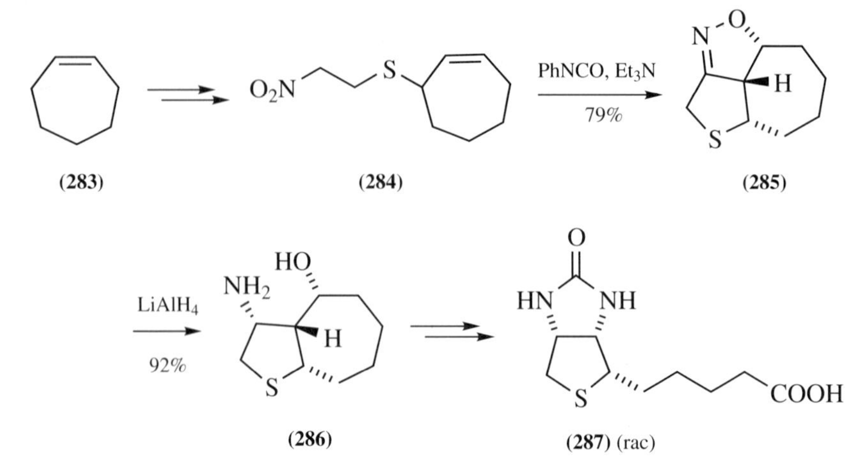

Scheme 6.102

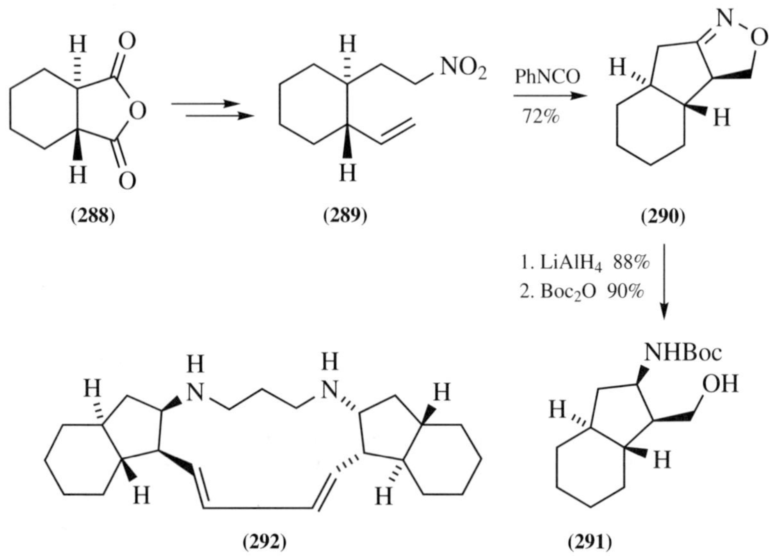

Scheme 6.103

approach resulted in the synthesis of several active inhibitors since these amino-cyclopentanepolyols effectively act as ring-contracted (deoxa) pyranoside mimics (259–261,339). An even more versatile route to these structures employed the respective hexenose nitrones (259–261,339–346).

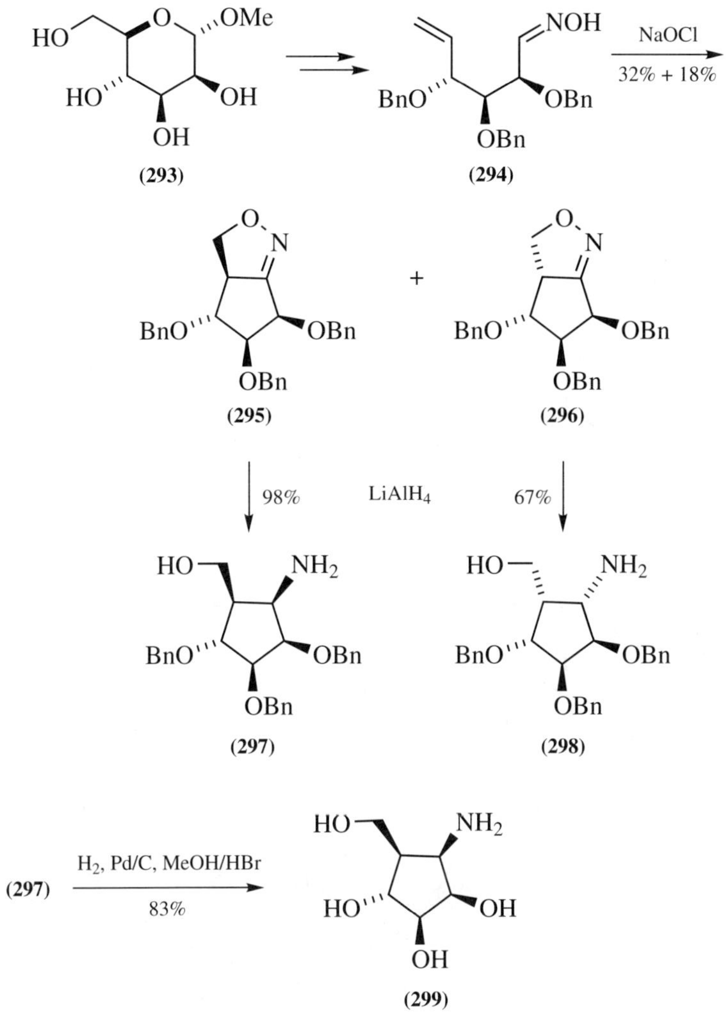

Scheme 6.104

6.5.2.3. *Transformations of Isoxazolines from Intramolecular Cycloadditions via Isoxazolinium Salts*

Isoxazolinium salts represent potentially versatile intermediates for synthesis, even though relatively few transformations of these salts are known (347–350) and synthetic uses so far have been scarce (349,350). A notable exception involves the use of an *N*-methylisoxazolinium salt. Its reduction to an isoxazolidine, and

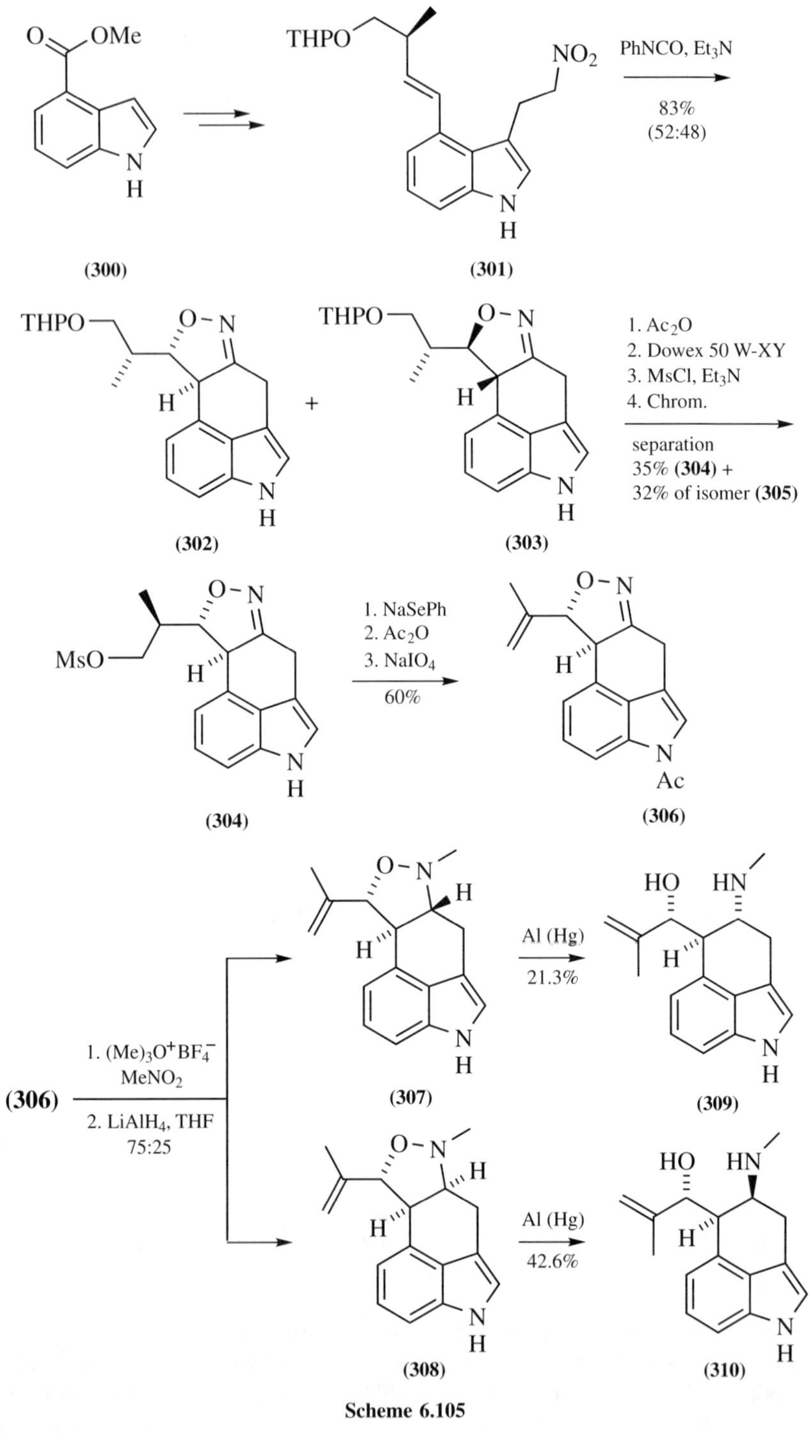
O OMe
N H
(300)
THPO
NO2
N H
(301)
PhNCO, Et3N
83%
(52:48)
THPO O–N H N H
(302)
+
THPO O–N H N H
(303)
1. Ac2O
2. Dowex 50 W-XY
3. MsCl, Et3N
4. Chrom.
separation
35% (304) +
32% of isomer (305)
MsO O–N H N H
(304)
1. NaSePh
2. Ac2O
3. NaIO4
60%
O–N H N Ac
(306)
(306)
1. (Me)3O+BF4−
MeNO2
2. LiAlH4, THF
75:25
O–N H H N H
(307)
Al (Hg)
21.3%
HO HN H N H
(309)
O–N H H N H
(308)
Al (Hg)
42.6%
HO HN H N H
(310)

Scheme 6.105

subsequent cleavage to an *N*-methylaminoalcohol was used by Kozikowski et al. (348) for a synthesis of paliclavine (**309**). The 4-alkenyl-3-nitroethylindole derivative **301** was obtained from 4-methoxycarbonylindole (**300**) (Scheme 6.105). This nitro indole derivative **301** was exposed to phenyl isocyanate–triethylamine resulting in the formation of a 1:1 diastereomeric mixture of isoxazolines **302** and **303**. The hydroxy group was liberated and mesylated and the NH-indole was acetylated to give compounds **304** and **305**, which could be separated. Indole mesylate (**304**) was subjected to elimination resulting in the formation of olefin **306**. Subsequent N-methylation of the isoxazoline followed by $LiAlH_4$ reduction afforded the *N*-methylisoxazolidines **307** and **308** as a 3:1 mixture. This mixture was subsequently treated with aluminium amalgam to give (+)-paliclavine (**309**) and epi-paliclavine (**310**) (350).

6.6. CONCLUSION

In this chapter, we described some recent advances in the generation and use of nitrile oxides as 1,3-dipoles as well as progress in rationalizing the reactivity and stereoselectivity of their cycloadditions. Nitrile oxides are readily available from aldehydes via oximes and hydroximoyl chlorides (Huisgen's *in situ* method) or from nitroalkanes (Mukaiyama method), to cite the two most versatile procedures with the broadest scope. Apart from providing heterocycles such as 1,2-oxazoles or furoxans (a class of compounds largely unexploited for the synthesis of acyclic structures), the dipolar 1,3-cycloadducts derived from alkenes (i.e., isoxazolines) have found widespread use in synthesis. This is due to several factors: (a) isoxazolines are cyclic oxime ethers constituting neutral, stable heterocycles that survive various reaction conditions and thus allow for many transformations of functional groups on the side chains; (b) treatment of isoxazolines with strong base leads to regio- and stereoselective modification at either the C(4) or $C(3)^{\alpha}$ positions via the respective carbanions and trapping with electrophiles; (c) selective modes of ring cleavage of isoxazolines are possible. One proceeds via hydrogenation–hydrolysis and leads to aldols. Another one ($LiAlH_4$ reduction) affords 1,3-aminoalcohols in a stereoselective, predictable manner. These assets help to make the 1,3-dipolar cycloaddition of nitrile oxides and the *isoxazoline route* attractive in answering some of the present-day challenges of organic synthesis. This approach also allows for the synthesis of enantiomerically pure target structures. In fact, this aspect of nitrile oxide cycloadditions has been a major focus of numerous studies over the past 20 years, and the diastereoselectivity of cycloadditions to chiral alkenes has thoroughly been examined by many groups. In recent years, a state of knowledge has been reached where the stereochemical outcome of such cycloadditions is often predictable. Using chiral nitrile oxides, or employing enantioselective catalysis, some progress has certainly been achieved. Here, however, we are still far from levelling the state of the art relative to other additions to C=C or C=O systems (i.e., hydrogenation, AE, AD, AH), which will surely constitute an increasingly important area of research for the next few decades, not only for the cycloadditions of nitrile oxides, but for other classes of 1,3-dipoles as well.

Some other aspects of nitrile oxide chemistry that will grow in importance in the future involve the use of high pressure (>10 kbar) to enforce hitherto slow or not feasible cycloadditions. Of particular relevance for nitrile oxides will be to find substantial improvements for practical, cost-efficient, large-scale reactions (351). Solid-phase techniques, adaptation for combinatorial chemistry, or improved versions for dehydrating nitroalkanes will all play an increasingly important role.

As was pointed out in Section 6.5, a major field of application for 1,3-dipolar cycloaddition of nitrile oxides over the past two decades has been the *intramolecular* version. We have greatly enjoyed selecting and presenting a sizeable number of ingenious examples of this process, which has resulted in the synthesis of a wide range of complex polycyclic target structures. We are convinced that there will be many more new and exciting applications of this protocol, to be both conceived and realized in the future. While the results summarized in this review illustrate that ring sizes from 5 to 19 have been achieved, there are also considerable gaps in between.

The many successful applications of nitrile oxide cycloadditions in synthesis are intimately linked with theory, both the simple FMO variety as well as the more sophisticated *ab initio* treatment, where the work of Sustmann and subsequently of Houk and his group has been seminal. We, the *practitioners*, have thus been supplied with a consistent view on the nature of 1,3-dipoles, their reactivity toward dipolarophiles, and the origin and interpretation of stereoselectivity of cycloaddition chemistry. It is of course desirable that our understanding of the relative reactivities of alkenes as well as of many 1,3-dipoles would be also improved, thereby leading to simple, extended recipes for the chemist practicing synthetics. We hope that this account will stimulate further advances in this field of cycloaddition chemistry and promote further uses of nitrile oxides in organic synthesis.

ACKNOWLEDGMENTS

Since 1974 in Giessen, then Würzburg, and now Stuttgart there have been ~25 associates whose dissertations have dealt with exploring the stereoselective preparation, modification, and uses of isoxazolines for synthesis. Some of this work is referred to here, and V. J. is very grateful to these co-workers for their often outstanding efforts and devotion to this field. Financial help for these studies came from DFG, AvH, DAAD, BMFT, EU-COST, the Fonds der Chemischen Industrie, the States of Hessen, Bayern, and Baden-Württemberg. Thanks are also due to DAAD for a postdoctoral fellowship for P. C. (on leave from Universidad Nacional de La Plata, La Plata, Argentina), and to Mrs. G. Kraschewski, Mrs. I. Kienzle, and Dr. D.-U. Hahn for technical help with the production of this chapter.

REFERENCES

1. A. Werner and H. Buss, *Berichte*, **1894**, *27*, 2193.
2. A. Quilico and R. Fusco, *Rend. Ist. Lombardo Sci. Lettere*, **1936**, *69*, 439.
3. A. Quilico, "Isoxazoles and Related Compounds", in *The Chemistry of Heterocyclic Compounds*, Vol. 17, A. Weissberger, ed., Interscience, New York, **1962**, pp. 1–176.
4. A. Quilico and G. Speroni, *Gazz. Chim. Ital.*, **1946**, *76*, 148.
5. A. Quilico, G. Stagno d'Alcontres, and P. Grünanger, *Gazz. Chim. Ital.*, **1950**, *80*, 479.
6. R. Huisgen and W. Mack, *Tetrahedron Lett.*, **1961**, 583; *Chem. Ber.*, **1972**, *105*, 2805.
7. T. Mukaiyama and T. Hoshino, *J. Am. Chem. Soc.*, **1960**, *82*, 5339.
8. R. Huisgen "10 Jahre Fonds der Chemischen Industrie", Düsseldorf, **1960**, p. 73; reprinted in *Naturwiss. Rundschau*, **1961**, *14*, 63.
9. R. B. Woodward and R. Hoffmann, *Angew. Chem. Int. Ed. Engl.*, **1969**, *8*, 797.
10. Ch. Grundmann and P. Grünanger, *The Nitrile Oxides*, Springer–Verlag Berlin-Heidelberg-New York **1971**.
11. M. Christl, R. Huisgen, and R. Sustmann, *Chem. Ber.*, **1973**, *106*, 3275.
12. D. Lednicer, "Latent Functionality in Organic Synthesis", *Adv. Org. Chem.*, **1972**, *8*, 17.
13. A. I. Meyers, *Heterocycles in Organic Synthesis*, Wiley, New York, **1974**.
14. S. A. Lang, Jr., and Y.-I. Lin, in *Comprehensive Heterocyclic Chemistry*, Vol. 6, A. R. Katritzky and C. W. Rees, eds., Pergamon Press, Oxford, **1984**, pp. 1–130.
15. P. Caramella and P. Grünanger, in *1,3-Dipolar Cycloaddition Chemistry*, Vol. 1, A. Padwa, ed., **1984**, Wiley-Interscience, New York, Chapter 3, p. 291.
16. K. B. G. Torssell, *Nitrile Oxides, Nitrones, and Nitronates in Organic Synthesis*, VCH, Weinheim, **1988**.
17. D. P. Curran in *Advances in Cycloaddition*, Vol. 1, D. P. Curran, ed., Jai Press Inc., Connecticut, **1988**, pp. 129–189.
18. A. P. Kozikowski, *Acc. Chem. Res.*, **1984**, *17*, 410.
19. V. Jäger and H. Grund, *Angew. Chem.*, **1976**, *88*, 27; *Angew. Chem., Int. Ed. Engl.*, **1976**, *15*, 50.
20. V. Jäger, H. Grund, V. Buss, W. Schwab, I. Müller, R. Schohe, R. Franz, and R. Ehrler, *Bull. Soc. Chim. Belg.*, **1983**, *92*, 1039.
21. V. Jäger, I. Müller, R. Schohe, M. Frey, R. Ehrler, B. Häfele, and D. Schröter, *Lect. Heterocycl. Chem.*, **1985**, *8*, 79.
22. V. Jäger, R. Franz, W. Schwab, B. Häfele, D. Schröter, D. Schäfer, W. Hümmer, E. Guntrum, and B. Seidel, in "Studies in Organic Chemistry," J. Kovac and P. Zalupsky, eds., *Chemistry of Heterocyclic Compounds*, Vol. 35, **1988**, pp. 58–75.
23. V. Jäger, R. Müller, T. Leibold, M. Hein, M. Schwarz, M. Fengler, L. Jaroskova, M. Pätzel, and P.-Y. LeRoy, *Bull. Soc. Chim. Belg.*, **1994**, *103*, 491.
24. R. Fusco, L. Garanti, and G. Zecchi, *Chim. Ind. (Milan)*, **1975**, *57*, 16.
25. V. Jäger and H. J. Günther, *Angew. Chem., Int. Ed. Engl.*, **1977**, *16*, 246.
26. H. Grund and V. Jäger, *Liebigs Ann. Chem.*, **1980**, 80.
27. V. Jäger and W. Schwab, *Tetrahedron Lett.*, **1978**, 3129.
28. (a) V. Jäger, V. Buss, and W. Schwab, *Liebigs Ann. Chem.*, **1980**, 122. (b) P. Fey, in Houben-Weyl, *Stereoselective Synthesis*, Vol. E 21b, G. Helmchen, R. W. Hoffmann, J. Mulzer, and E. Schaumann, eds., Thieme, Stuttgart, New York, **1995**, 1761 ff.; P. Fey and W. Hartwig, in Houben-Weyl, *Stereoselective Synthesis*, Vol E 21a, G. Helmchen, R. W. Hoffmann, J. Mulzer, and E. Schaumann, eds., Thieme, Stuttgart, New York, **1995**, 1030 ff.

29. K. V. Gothelf and K. A. Jorgensen, *Chem. Rev.*, **1998**, *98*, 863.
30. P. Grünanger and P. Vita-Finzi, "Isoxazoles," in *The Chemistry of Heterocyclic Compounds*, Vol. 49, Part. 1, E. C. Taylor and A. Weissberger, eds., Wiley, New York, **1991**.
31. S. Kanemasa and O. Tsuge, *Heterocycles*, **1990**, *30*, 719.
32. C. J. Easton, C. M. M. Hughes, G. P. Savage, and G. W. Simpson, *Adv. Heterocycl. Chem.*, **1994**, *60*, 261.
33. M. Cinquini and F. Cozzi, in Houben-Weyl, *Stereoselective Synthesis*, Vol. E 21c, G. Helmchen, R. W. Hoffmann, J. Mulzer, and E. Schaumann, eds., Thiem, Stuttgart, New York, **1995**, pp. 2953–2987.
34. L. Fisera, V. Ondrus, J. Kuban, P. Micuch, I. Blanarikova, and V. Jäger, *J. Heterocycl. Chem.*, **2000**, *37*, 551.
35. L. Raimondi and M. Benaglia, *Eur. J. Org. Chem.*, **2001**, 1033.
36. N. Kornblum, *Org. React.*, **1962**, *12*, 101.
37. V. Jäger and co-workers, observations over ~25 years.
38. D. Chiarino, M. Napoletano, and A. Sala, *Synth. Commun.*, **1988**, *18*, 1171.
39. K. Hallig, K. B. G. Torssell, and R. G. Hazell, *Acta Chem. Scand.*, **1991**, *45*, 736.
40. O. Tsuge, S. Kanemasa, and H. Suga, *Chem. Lett.*, **1986**, 183.
41. R. Franz, *Diplomarbeit*, Würzburg, **1981**.
42. R. Franz, *Dissertation*, Würzburg, **1984**.
43. J. Guillaumel, P. Demerseman, J.-M. Clavel, R. Royer, N. Platzer, and Ch. Brevard, *Tetrahedron*, **1980**, *36*, 2459.
44. R. Sustmann, *Pure Appl. Chem.*, **1974**, *40*, 569.
45. G. A. Lee, *Synthesis*, **1982**, 508.
46. K.-C. Liu, B. R. Shelton, and R. K. Howe, *J. Org. Chem.*, **1980**, *45*, 3916.
47. H. Rheinboldt, M. Dewald, F. Jansen, and O. Schmitz-Dumont, *Liebigs Ann. Chem.*, **1927**, *451*, 161.
48. G. Kumaran and G. H. Kulkarni, *J. Org. Chem.*, **1997**, *62*, 1516.
49. H. Suzuki, H. Shimizu, T. Inamasu, H. Tani, and R. Tamura, *Chem. Lett.*, **1990**, 559.
50. C.-F. Yao, W.-C. Chen, and Y-M. Lin, *Tetrahedron. Lett.*, **1996**, *37*, 6339.
51. C.-F. Yao, C.-S. Yang and H.-Y. Fang, *Tetrahedron Lett.*, **1997**, *38*, 6419.
52. C.-F. Yao, K-H. Kao, J.-T. Liu, C.-M. Chu, Y. Wang, W.-C. Chen, Y.-M. Lin, W.-W. Lin, M.-C. Yan, J.-Y. Liu, M.-C. Chuang, and J.-L. Shiue, *Tetrahedron*, **1998**, *54*, 791.
53. J.-Y. Liu, M.-M. Yan, W.-W. Lin, L.-Y. Wang, and C.-F. Yao, *J. Chem. Soc., Perkin Trans. 1*, **1999**, 1215.
54. D. Chiarino, M. Napoletano, and A. Sala, *Tetrahedron Lett.*, **1986**, *27*, 3181.
55. D. M. Vyas, P. M. Skonezny, T. A. Jenks, and T. W. Doyle, *Tetrahedron Lett.*, **1986**, *27*, 3099.
56. J. N. Kim, K. H. Chung, and E. K. Ryu, *Heterocycles*, **1991**, *32*, 477.
57. R. F. Cunico and L. Bedell, *J. Org. Chem.*, **1983**, *48*, 2780.
58. B. Syassi, K. Bougrin, and M. Soufiaoui, *Tetrahedron Lett.*, **1997**, *38*, 8855.
59. M. Fengler-Veith, *Dissertation*, Stuttgart, **1996**.
60. S. Kanemasa, S. Kobayashi, M. Nishiuchi, H. Yamamoto, and E. Wada, *Tetrahedron Lett.*, **1991**, *32*, 6367.
61. B. H. Kim, *Synth. Commun.*, **1987**, *17*, 1199.
62. O. Moriya, Y. Urata, and T. Endo, *J. Chem. Soc., Chem. Commun.*, **1991**, 884.
63. P. Caldirola, M. De Amici, and C. De Micheli, *Heterocycles*, **1985**, *23*, 2479.
64. J. N. Kim and E. K. Ryu, *J. Org. Chem.*, **1992**, *57*, 1088.
65. Y. Tokunaga, M. Ihara, and K. Fukumoto, *Heterocycles*, **1996**, *43*, 1771.

66. J. N. Kim and E. K. Ryu, *Heterocycles*, **1990**, *31*, 1693.

67. H.-J. Gi, Y. Xiang, R. F. Schinazi, and K. Zhao, *J. Org. Chem.*, **1997**, *62*, 88.

68. P. de la Cruz, E. Espildora, J. J. Garcia, A. de la Hoz, F. Langa, N. Martin, and L. Sanchez, *Tetrahedron Lett.*, **1999**, *40*, 4889.

69. K. B. G. Torssell, A. C. Hazell, and R. G. Hazzel, *Tetrahedron*, **1985**, *41*, 5569.

70. B. Touaux, F. Texier-Boullet, and J. Hamelin, *Heteroatom Chemistry*, **1998**, *9*, 351.

71. C. Ticozzi and A. Zanarotti, *Tetrahedron Lett.*, **1994**, *35*, 7421.

72. A. Arnone, M. Cavicchioli, A. Donadelli, and G. Resnati, *Tetrahedron Asymm.*, **1994**, *5*, 1019.

73. O. Duclos, A. Duréault, and J. C. Depezay, *Tetrahedron Lett.*, **1992**, *33*, 1059.

74. A. P. Kozikowski and P.-u. Park, *J. Org. Chem.*, **1990**, *55*, 4668.

75. A. Hassner and K. S. K. Murthy, *J. Org. Chem.*, **1989**, *54*, 5277.

76. P. J. Bhuyan, R. C. Boruah, and J. S. Sandhu, *Tetrahedron Lett.*, **1989**, *30*, 1421.

77. A. P. Kozikowski and K. E. Maloneyhuss, *Tetrahedron Lett.*, **1985**, *26*, 5759.

78. J.-F. Cheng and A. M. M. Mjalli, *Tetrahedron Lett.*, **1998**, *39*, 939.

79. O. Moriya, H. Nakamura, T. Kageyama, and Y. Urata, *Tetrahedron Lett.*, **1989**, *30*, 3987.

80. O. Moriya, H. Takenaka, Y. Urata, and T. Endo, *J. Chem. Soc., Chem. Commun.*, **1991**, 1671.

81. A. Hassner and K. M. Lokanatha Rai, *Synthesis*, **1989**, 57.

82. V. Jäger and I. Müller, *Tetrahedron*, **1985**, *41*, 3519.

83. E. J. Kantorowski, S. P. Brown, and M. J. Kurth, *J. Org. Chem.*, **1998**, *63*, 5272.

84. T. Shimizu, Y. Hayashi, H. Shibafuchi, and K. Teramura, *Bull. Chem. Soc. Jpn.*, **1986**, *59*, 2827.

85. Y. Basel and A. Hassner, *Synthesis*, **1997**, 309.

86. C. Yip, S. Handerson, R. Jordan, and W. Tam, *Org. Lett.*, **1999**, *1*, 791.

87. C. Yip, S. Handerson, G. K. Tranmer, and W. Tam, *J. Org. Chem.*, **2001**, *66*, 276.

88. T. Shimizu, Y. Hayashi, and K. Teramura, *Bull. Chem. Soc. Jpn.*, **1984**, *57*, 2531.

89. N. Maugein, A. Wagner, and C. Mioskowski, *Tetrahedron Lett.*, **1997**, *38*, 1547.

90. D. Muri, J. W. Bode, and E. M. Carreira, *Org. Lett.*, **2000**, *2*, 539.

91. N. Nishiwaki, T. Uehara, N. Asaka, Y. Tohda, M. Ariga, and S. Kanemasa, *Tetrahedron Lett.*, **1998**, *39*, 4851.

92. T. Shimizu, Y. Hayashi, H. Shibafuchi, and K. Teramura, *Bull. Chem. Soc. Jpn.*, **1987**, *60*, 1948.

93. M. G. Leslie-Smith, R. M. Paton, and N. Webb, *Tetrahedron Lett.*, **1994**, *35*, 9251.

94. D. Maiti and P. K. Bhattacharjya, *Synlett*, **1998**, 385.

95. C. Matt, A. Gissot, A. Wagner, and C. Mioskowski, *Tetrahedron Lett.*, **2000**, *41*, 1191.

96. J. Kiegiel, M. Poplawska, J. Jozwik, M. Kosior, and J. Jurczak, *Tetrahedron Lett.*, **1999**, *40*, 5605.

97. S. J. Barrow, C. J. Easton, G. P. Savage, and G. W. Simpson, *Tetrahedron Lett.*, **1997**, *38*, 2175.

98. H. Nishiyama, H. Arai, T. Ohki, and K. Itoh, *J. Am. Chem. Soc.*, **1985**, *107*, 5310.

99. N. Arai, M. Iwakoshi, K. Tanabe, and K. Narasaka, *Bull. Chem. Soc. Jpn.*, **1999**, *72*, 2277.

100. J. N. Kim and E. K. Ryu, *Synth. Commun.*, **1990**, *20*, 1373.

101. (a) A. B. Sheremetev, N. N. Makhova, and W. Friedrichsen, *Adv. Heterocycl. Chem.*, **2001**, *78*, 65. (b) D. P. Curran and C. J. Fenk, *J. Am. Chem. Soc.*, **1985**, *107*, 6023.

102. T. Shimizu, Y. Hayashi, M. Miki, and K. Teramura, *Heterocycles*, **1986**, *24*, 889.

103. T. Shimizu, Y. Hayashi, T. Taniguchi, and K. Teramura, *Tetrahedron*, **1985**, *41*, 727.

104. I. V. Ovchinnikov, N. A. Popov, N. N. Makhova, L. I. Khmel'nitskii, and V. A. Shlyapochnikov, *Mendeleev Commun.*, **1995**, 231.

105. A. P. Marchand, G. V. M: Sharma, R. Shukla, and S. G. Bott, *Heterocycles*, **1998**, *47*, 271.

106. A. Gasco, V. Mortarini, R. Calvino, and A. Serafino, *Tetrahedron Lett.*, **1974**, *8*, 627.
107. R. Fruttero, B. Ferrarotti, A. Serafino, and A. Gasco, *Liebigs Ann. Chem.*, **1990**, 335.
108. K. N. Houk, J. Gonzalez, and Y. Li, *Acc. Chem. Res.*, **1995**, *28*, 81.
109. R. Huisgen in *1,3-Dipolar Cycloaddition Chemistry*, Vol. 1, A. Padwa, ed., Wiley-Interscience, New York, **1984**, Chapter 1, p. 1.
110. K. N. Houk and K. Yamaguchi, in *1,3-Dipolar Cycloaddition Chemistry*, Vol 2, A. Padwa, ed., Wiley-Interscience, New York, **1984**, Chapter 13, p. 407.
111. K. N. Houk, R. A. Firestone, L. L. Munchausen, P. H. Mueller, B. H. Arison, and L. A. Garcia, *J. Am. Chem. Soc.*, **1985**, *107*, 7277.
112. O. Wiest and K. N. Houk, *Top. Curr. Chem.*, **1996**, *183*, 1.
113. A. Padwa in *1,3-Dipolar Cycloaddition Chemistry*, Vol 2, A. Padwa, ed., Wiley-Interscience, New York, **1984**, Chapter 12, p. 277.
114. A. Rastelli, R. Gandolfi, and M. S. Amade, *J. Org. Chem.*, **1998**, *63*, 7425, and references cited therein.
115. M. Burdisso, R. Gandolfi, S. Quartieri, and A. Rastelli, *Tetrahedron*, **1987**, *43*, 159.
116. A. Rastelli, M. Bagatti, R. Gandolfi, and M. Burdisso, *J. Chem. Soc., Faraday Trans.*, **1994**, *90*, 1077.
117. A. Kamimura and K. Hori, *Tetrahedron*, **1994**, *50*, 7969.
118. M.-D. Su, H.-Y. Liao, W.-S. Chung, and S.-Y. Chu, *J. Org. Chem.*, **1999**, *64*, 6710.
119. F. P. Cossio, I. Morao, H. Jiao, and P. von R. Schleyer, *J. Am. Chem. Soc.*, **1999**, *121*, 6737.
120. K. Bast, M. Christl, R. Huisgen, and W. Mack, *Chem. Ber.*, **1973**, *106*, 3312.
121. M. Schwarz, *Dissertation*, Würzburg, Germany, **1993**.
122. V. Jäger, R. Schohe, and E. F. Paulus, *Tetrahedron Lett.*, **1983**, *24, 5501.*
123. K. N. Houk, S. R. Moses, Y.-D. Wu, N. G. Rondan, V. Jäger, R. Schohe, and F. R. Fronczek, *J. Am. Chem. Soc.*, **1984**, *106*, 3880.
124. K. N. Houk, H.-Y. Duh, Y.-D. Wu, and S. R. Moses, *J. Am. Chem. Soc.*, **1986**, *108*, 2754.
125. W. Malisch, J. Zöller, M. Schwarz, V. Jäger, and A. M. Arif, *Chem. Ber.*, **1994**, *127*, 1243.
126. A. G. Meyer, C. J. Easton, S. F. Lincoln, and G. W. Simpson, *J. Chem. Soc., Chem. Commun.*, **1997**, 1517.
127. A. G. Meyer, C. J. Easton, S. F. Lincoln, and G. W. Simpson, *J. Org. Chem.*, **1998**, *63*, 9069.
128. T.-J. Lu, J.-F. Yang, and L.-J. Sheu, *J. Org. Chem.*, **1995**, *60*, 7701.
129. M. A. Weidner-Wells, S. A. Fraga-Spano, and I. J. Turchi, *J. Org. Chem.*, **1998**, *63*, 6319.
130. P. Caramella, D. Reami, M. Falzoni, and P. Quadrelli, *Tetrahedron*, **1999**, *55*, 7027.
131. L. Toma, P. Quadrelli, G. Perrini, R. Gandolfi, C. Di Valentin, A. Corsaro, and P. Caramella, *Tetrahedron*, **2000**, *56*, 4299.
132. M. Christl and R. Huisgen, *Chem. Ber.*, **1973**, *106*, 3345.
133. N. G. Argyropoulos, E. Coutouli-Argyropoulou, and P. Pistikopoulos, *J. Chem. Res. (S)*, **1984**, 362.
134. E. Coutouli-Argyropoulou and E. Thessalonikeos, *J. Heterocycl. Chem.*, **1991**, *28*, 429.
135. D. P. Curran, S.-M. Choi, S. A. Gothe, and F.-t. Lin, *J. Org. Chem.*, **1990**, *55*, 3710.
136. S. Kanemasa, M. Nishiuchi, A. Kamimura, and K. Hori, *J. Am. Chem. Soc.*, **1994**, *116*, 2324.
137. A. Kamimura, Y. Kaneko, A. Ohta, A. Kakehi, H. Matsuda, and S. Kanemasa, *Tetrahedron Lett.*, **1999**, *40*, 4349.
138. S. Fukuda, A. Kamimura, S. Kanemasa, and K. Hori, *Tetrahedron*, **2000**, *56*, 1637.
139. R. W. Wallace and J. Liu, *Tetrahedron Lett.*, **1994**, *35*, 7493.
140. V. Jäger, W. Schwab, and V. Buss, *Angew. Chem., Int. Ed. Engl.*, **1981**, *20*, 601.
141. V. Jäger and D. Schröter, *Synthesis*, **1990**, 556.
142. T. V. RajanBabu and G. S. Reddy, *J. Org. Chem.*, **1986**, *51*, 5458.

143. A. Guarna, A. Brandi, F. De Sarlo, A. Goti, and F. Pericciuoli, *J. Org. Chem.*, **1988**, *53*, 2426.

144. J. Baran and H. Mayr, *J. Org. Chem.*, **1989**, *54*, 5012.

145. R. K. Howe and B. R. Shelton, *J. Org. Chem.*, **1990**, *55*, 4603.

146. W. R. Dolbier, Jr., G. D. Purvis III, M. J. Seabury, G. E. Wicks, and C. R. Burkholder, *Tetrahedron*, **1990**, *46*, 7991.

147. P. Oravec, L. Fisera, and R. Gazo, *Monatsh. Chem.*, **1991**, *122*, 165.

148. P. Oravec, L. Fisera, P. Ertl, and D. Vegh, *Monatsh. Chem.*, **1991**, *122*, 821.

149. R. Jimenez, L. Perez, J. Tamariz, and H. Salgado, *Heterocycles*, **1993**, *35*, 591.

150. L. Fisera, M. Konopikova, P. Ertl, and N. Pronayova, *Monatsh. Chem.*, **1994**, *125*, 301.

151. J. Fröhlich, L. Fisera, F. Sauter, Y. Feng, and P. Ertl, *Monatsh. Chem.*, **1995**, *126*, 75.

152. H. M. Sampath Kumar, S. Anjaneyulu, and J. S. Yadav, *Synth. Commun.*, **1999**, *29*, 877.

153. J. N. Kim and E. K. Ryu, *Heterocycles*, **1990**, *31*, 663.

154. E. Coutouli-Argyropoulou, N. G. Argyropoulos, and E. Thessalonikeos, *J. Chem. Res. (S)*, **1990**, 202; *J. Chem. Res. (M)*, **1990**, 1557.

155. D. F. Corbett, *J. Chem. Soc., Perkin Trans. 1*, **1986**, 421.

156. E. Coutouli-Argyropoulou and E. Thessalonikeos, *J. Heterocycl. Chem.*, **1992**, *29*, 251.

157. I. Müller, *Dissertation*, Würzburg, **1984**.

158. Reference 15, p. 333.

159. P. Caramella, N. G. Rondan, M. N. Paddon-Row, and K. N. Houk, *J. Am. Chem. Soc.*, **1981**, *103*, 2438.

160. R. Annunziata, M. Cinquini, F. Cozzi, and L. Raimondi, *Tetrahedron*, **1988**, *44*, 4645.

161. S. K. Armstrong, E. W. Collington, J. G. Knight, A. Naylor, and S. Warren, *J. Chem. Soc., Perkin Trans. 1*, **1993**, 1433; S. K. Armstrong, S. Warren, E. W. Collington, and A. Naylor, *Tetrahedron Lett.*, **1991**, *32*, 4171.

162. E. C. Boyd and R. M. Paton, *Tetrahedron Lett.*, **1993**, *34*, 3169.

163. R. Sustmann, *Tetrahedron Lett.*, **1971**, 2717.

164. (a) R. Sustmann and H. Trill, *Angew. Chem., Int. Ed. Engl.*, **1972**, *11*, 838. (b) M. G. Finn and K. B. Sharpless, in *Asymmetric Synthesis*, J. D. Morrison, ed., Vol. 5, Chapter 8, p. 247 ff., in particular p. 281, Academic Press, Orlando, FL, **1985**; M. Eyer and D. Seebach, *J. Am. Chem. Soc.*, **1985**, *107*, 3601.

165. M. De Amici, C. De Micheli, A. Ortisi, G. Gatti, R. Gandolfi, and L. Toma, *J. Org. Chem.*, **1989**, *54*, 793.

166. J. de Blas, J. C. Carretero, and E. Dominguez, *Tetrahedron Asymm.*, **1995**, *6*, 1035.

167. R. Annunziata, M. Benaglia, M. Cinquini, F. Cozzi, and L. Raimondi, *J. Org. Chem.*, **1995**, *60*, 4697.

168. M. B. Gravestock, R. M. Paton, and C. J. Todd, *Tetrahedron Asymm.*, **1995**, *6*, 2723.

169. M. Mancera, I. Roffe, and J. A. Galbis, *Tetrahedron*, **1995**, *51*, 6349.

170. Ref. 15, p. 305.

171. C. De Micheli, A. G. Invernizzi, R. Gandolfi, and L. Scevola, *J. Chem. Soc., Chem. Commun.*, **1976**, 246.

172. P. Caramella and G. Cellerino, *Tetrahedron Lett.*, **1974**, 229.

173. F. M. Albinone, D. Vitali, R. Oberti, and P. Caramella, *J. Chem. Res. (M)*, **1980**, 4355.

174. D. P. Curran, S. A. Gothe, and S.-M. Choi, *Heterocycles*, **1993**, *35*, 1371.

175. G. Adembri, G. Giorgi, R. L. Lampiarello, M L. Paoli, and A. Sega, *J. Chem. Soc., Perkin Trans. 1*, **2000**, 2649.

176. S. Kanemasa and S. Kobayashi, *Bull. Chem. Soc. Jpn.*, **1993**, *66*, 2685.

177. P. Micuch, L. Fisera, M. K. Cyranski, T. M. Krygowski, and J. Krajcik, *Tetrahedron*, **2000**, *56*, 5465.

178. M. Shimizu, Y. Ukaji, and K. Inomata, *Chem. Lett.*, **1996**, 455.

179. D. P. Curran, B. H. Kim, J. Daugherty, and T. A. Heffner, *Tetrahedron Lett.*, **1988**, *29*, 3555.
180. T. Akiyama, K. Okada, and S. Ozaki, *Tetrahedron Lett.*, **1992**, *33*, 5763.
181. T. Olsson, K. Stern, and S. Sundell, *J. Org. Chem.*, **1988**, *53*, 2468.
182. T. Olsson, K. Stern, G. Westman, and S. Sundell, *Tetrahedron*, **1990**, *46*, 2473.
183. D. P. Curran, B. H. Kim, H. P. Piyasena, R. J. Loncharich, and K. N. Houk, *J. Org. Chem.*, **1987**, *52*, 2137.
184. D. P. Curran and T. A. Heffner, *J. Org. Chem.*, **1990**, *55*, 4585.
185. K. S. Kim, B. H. Kim, W. M. Park, S. J. Cho, and B. J. Mhin, *J. Am. Chem. Soc.*, **1993**, *115*, 7472.
186. D. P. Curran, K.-S. Jeong, T. A. Heffner, and J. Rebek, Jr., *J. Am. Chem. Soc.*, **1989**, *111*, 9238.
187. J. A. Stack, T. A. Heffner, S. J. Geib, and D. P. Curran, *Tetrahedron*, **1993**, *49*, 995.
188. D. P. Curran and M.-H. Yoon, *Tetrahedron*, **1997**, *53*, 1971.
189. H. Waldmann, *Liebigs Ann. Chem.*, **1990**, 1013.
190. Y. H. Kim, S. H. Kim, and D. H. Park, *Tetrahedron*, **1993**, *34*, 6063.
191. S. Kanemasa and K. Onimura, *Tetrahedron*, **1992**, *48*, 8645.
192. H. Yamamoto, S. Watanabe, K. Kadotani, M. Hasegawa, M. Noguchi, and S. Kanemasa, *Tetrahedron Lett.*, **2000**, *41*, 3131.
193. J. D. Toker, P. Wentworth, Jr., Y. Hu, K. N. Houk, and K. D. Janda, *J. Am. Chem. Soc.*, **2000**, *122*, 3244.
194. S. Kanemasa, T. Hayashi, H. Yamamoto, E. Wada, and T. Sakurai, *Bull. Chem. Soc. Jpn.*, **1991**, *64*, 3274.
195. C. D. Davies, S. P. Marsden, and E. S. E. Stokes, *Tetrahedron Lett.*, **1998**, *39*, 8513.
196. A. N. Boa, S. E. Booth, D. A. Dawkins, P. R. Jenkins, J. Fawcett, and D. R. Russell, *J. Chem. Soc., Perkin Trans. 1*, **1993**, 1277.
197. M. Burdisso, R. Gandolfi, M. Lucchi, and A. Rastelli, *J. Org. Chem.*, **1988**, *53*, 2123.
198. A. Rastelli, M. Bagatti, A. Ori, R. Gandolfi, and M. Burdisso, *J. Chem. Soc., Faraday Trans.*, **1993**, *89*, 29.
199. M. Burdisso, R. Gandolfi, P. Pevarello, and A. Rastelli, *J. Chem. Soc., Perkin Trans. 2*, **1988**, 753.
200. E. Cinquini, M. Freccero, R. Gandolfi, M. S. Amade, and A. Rastelli, *Tetrahedron*, **1997**, *53*, 9279.
201. P. Quadrelli, M. Mella, P. Paganoni, and P. Caramella, *Eur. J. Org. Chem.*, **2000**, 2613.
202. R. Gandolfi, G. Tonoletti, A. Rastelli, and M. Bagatti, *J. Org. Chem.*, **1993**, *58*, 6038.
203. T.-L. Tsai, W.-C. Chen, C. H. Yu, W. J. le Noble, and W.-S. Chung, *J. Org. Chem.*, **1999**, *64*, 1099.
204. S. S. Ko and P. N. Confalone, *Tetrahedron*, **1985**, *41*, 3511.
205. X. Beebe, C. L. Chiappari, M. J. Kurth, and N. E, Schore, *J. Org. Chem.*, **1993**, *58*, 7320.
206. S. K. Massad, L. D. Hawkins, and D. C. Baker, *J. Org. Chem.*, **1983**, *48*, 5180.
207. M. Hein, *Dissertation*, Stuttgart, Germany, **1996**.
208. V. Jäger and V. Wehner, *Angew. Chem. Int. Ed. Engl.*, **1989**, *28*, 469.
209. V. Wehner *Dissertation*, Würzburg, Germany, **1990**.
210. B. Häfele and V. Jäger, *Liebigs Ann. Chem.*, **1987**, 85.
211. J. Mulzer and M. Kappert, *Angew. Chem. Int. Ed. Engl.*, **1983**, *22*, 63.
212. R. Dumont and H. Pfander, *Helv. Chim. Acta.*, **1983**, *66*, 814.
213. J. Jurczak, S. Pikul, and T. Bauer, *Tetrahedron*, **1986**, *42*, 447.
214. B. Häfele, *Dissertation*, Würzburg, Germany, **1983**.
215. B. Häfele, *Dissertation*, Würzburg, Germany, **1987**
216. K. E. Larsen and K. B. G. Torssell, *Tetrahedron*, **1984**, *40*, 2985.
217. H. Yin, R. W. Franck, S.-L. Chen, G. J. Quigley, and L. Todaro, *J. Org. Chem.*, **1992**, *57*, 644.
218. A. P. Kozikowski and A. K. Ghosh, *J. Org. Chem.*, **1984**, *49*, 2762.

219. R. Annunziata, M. Cinquini, F. Cozzi, L. Raimondi, and G. Licini, *Tetrahedron*, **1991**, *47*, 3869.

220. P. Garner and J. M. Park, *J. Org. Chem.*, **1987**, *52*, 2361.

221. T. Leibold, Ph.D. Thesis, Stuttgart, Germany, **1995**.

222. M. Fengler-Veith, O. Schwardt, U. Kautz, B. Krämer, and V. Jäger , *Org. Synth.*, **2000**, *78*, 123.

223. A. P. Kozikowski, Y. Kitagawa, and J. P. Springer, *J. Chem. Soc., Chem. Commun.*, **1983**, 1460.

224. V. Jäger, R. Öhrlein, V. Wehner, P. Poggendorf, B. Steuer, J. Raczko, H. Griesser, F.-M. Kieb, and A. Menzel, *Enantiomer*, **2000**, *4*, 205.

225. R. Müller, *Dissertation*, Würzburg, **1992**.

226. A. Vasella, *Helv. Chim. Acta*, **1977**, *60*, 426.

227. M. Yoyokama and N. Yamada, *Tetrahedron Lett.*, **1989**, *30*, 3675.

228. R. C. Hockett and F. C. Schaeffer, *J. Am. Chem. Soc.*, **1947**, *69*, 849.

229. J. M. J. Tronchet, A. Jotterand, N. Le Hong, M. F. Perret, M. S. Thorndahl-Jaccard, M. J. Tronchet, J. M. Chalet, M. L. Faivre, C. Hausser, and C. Sebastian, *Helv. Chim. Acta*, **1970**, *53*, 174.

230. H. Matsunaga, T. Sakamaki, H. Nagaoka, and Y. Yamada, *Tetrahedron Lett.*, **1983**, *24*, 3009.

231. A. P. Kozikowski and S. Goldstein, *J. Org. Chem.*, **1983**, *48*, 1139.

232. A. Förtsch, H. Kogelberg, and P. Köll, *Carbohydr. Res.*, **1987**, *164*, 391.

233. I. M. Dawson, T. Johnson, R. M. Paton, and R. A. C. Rennie, *J. Chem. Soc., Chem. Commun.*, **1988**, 1339.

234. R. Müller, T. Leibold, M. Pätzel, and V. Jäger, *Angew. Chem., Int. Ed. Engl.*, **1994**, *33*, 1295.

235. V. Jäger, R. Müller, and M. Pätzel, unpublished work, 1989, 1991.

236. R. H. Jones, G. C. Robinson, and E. J. Thomas, *Tetrahedron*, **1984**, *40*, 177.

237. A. J. Blake, E. C. Boyd, R. O. Gould, and R. M. Paton, *J. Chem. Soc., Perkin Trans. 1*, **1994**, 2841.

238. M. Asaoka, M. Abe, T. Mukuta, and H. Takei, *Chem. Lett.*, **1982**, 215.

239. W. Dehaen and A. Hassner, *J. Org. Chem.*, **1991**, *56*, 896.

240. A. Hassner and W. Dehaen, *Chem. Ber.*, **1991**, *124*, 1181.

241. K. Sishido, O. Irie, and M. Shibuya, *Tetrahedron Lett.*, **1992**, *33*, 4589. For a related synthesis see: K. Sishido, T. Takata, K. Umimoto, and M. Shibuya, *Heterocycles*, **1992**, *33*, 73.

242. T. Omodani and K. Shishido, *J. Chem. Soc., Chem. Commun.*, **1994**, 2781.

243. R. Annunziata, M. Cinquini, F. Cozzi, C. Gennari, and L. Raimondi, *J. Org. Chem.*, **1987**, *52*, 4674.

244. R. Dogbéavou and L. Breau, *Synlett*, **1997**, 1208.

245. A. P. Kozikowski and B. B. Murgrage, *J. Chem. Soc., Chem. Commun.*, **1988**, 198.

246. M. Asaoka, M. Abe, and H. Takei, *Bull. Chem. Soc. Jpn.*, **1985**, *58*, 2145.

247. A. P. Kozikowski, K. Hiraga, J. P. Springer, B. C. Wang, and Z.-B. Xu, *J. Am. Chem. Soc.*, **1984**, *106*, 1845.

248. Y. Hirai and H. Nagaoka, *Tetrahedron Lett.*, **1997**, *38*, 1969.

249. A. Hassner and K. S. K. Murthy, *Tetrahedron Lett.*, **1987**, *28*, 4097.

250. A. Hassner, in *Heterocycles in Bio-organic Chemistry*, Royal Society of Chemistry, London **1991**, pp. 130–143 (there are no references given in this review).

251. A. Hassner, K. S. K. Murthy, W. Dehaen, and O. Friedman, *J. Heterocycl. Chem.*, **1994**, *31*, 687 (there are no references given in this review).

252. A. Hassner, K. S. K. Murthy, A. Padwa, W. H. Bullock, and P. D. Stull, *J. Org. Chem.*, **1988**, *53*, 5063.

253. K. S. K. Murthy and A. Hassner, *Tetrahedron Lett.*, **1987**, *28*, 97.

254. A. Hassner and W. Dehaen, *J. Org. Chem.*, **1990**, *55*, 5505.

255. K. M. Lokanatha Rhai and A. Hassner, *Heterocycles*, **1990**, *30*, 817.

256. H. R. Kim, H. J. K. Kim, J. L. Duffy, M. M. Olmstead, K. Ruhlandt-Senge, and M. J. Kurth, *Tetrahedron Lett.*, **1991**, *32*, 4259.

257. F. K. Brown, L. Raimondi, Y.-D. Wu, and K. N. Houk, *Tetrahedron Lett.*, **1992**, *33*, 4405.

258. M. Nakata, S. Akazawa, S. Kitamura, and K. Tatsuta, *Tetrahedron Lett.*, **1991**, *32*, 5363. For a related approach to (−)-allosamizoline, see T. Kitahara, N. Suzuki., K. Koseki, and K. Mori, *Biosci. Biotech. Biochem.*, **1993**, *57*, 1906.

259. M. Kleban, *Dissertation*, Stuttgart, Germany, **1996**.

260. M. Kleban, P. Hilgers, J. N. Greul, R. D. Kugler, J. Li, S. Picasso, P. Vogel, and V. Jäger, *Chem BioChem.*, **2001**, 365.

261. P. Hilgers (née Pruschek), *Dissertation*, Stuttgart, Germany, **2000**.

262. T. Takahashi, M. Nakazawa, Y. Sakamoto, and K. N. Houk, *Tetrahedron Lett.*, **1993**, *34*, 4075. For other related syntheses see K. Tatsuta, Y. Niwata, K. Umezawa, K. Toshima, and M. Nakata, *Tetrahedron Lett.*, **1990**, *31*, 1171; B. Lygo, M. Swiatyj, H. Trabsa, and M. Voyle, *Tetrahedron Lett.*, **1994**, *35*, 4197.

263. R. Annunziata, M. Cinquini, F. Cozzi, and L. Raimondi, *J. Chem. Soc., Chem. Commun.*, **1987**, 529.

264. R. Annunziata, M. Cinquini, F. Cozzi, G. Dondio, and L. Raimondi, *Tetrahedron*, **1987**, *43*, 2369.

265. L. Raimondi, Y.-D. Wu, F. K. Brown, and K. N. Houk, *Tetrahedron Lett.*, **1992**, *33*, 4409.

266. A. P. Kozikowski, B. B. Mugrage, B. C. Wang, and Z.-b. Xu, *Tetrahedron Lett.*, **1983**, *24*, 3705.

267. D. L. Comins and A. H. Abdullah, *Tetrahedron Lett.*, **1985**, *26*, 43.

268. A. P. Kozikowski and C.-S. Li, *J. Org. Chem.*, **1987**, *52*, 3541.

269. K. Bock and T. Skrydstrup, *J. Chem. Soc., Perkin Trans. 1*, **1991**, 1181.

270. J. H. Rigby and T. W. McGuire, *Tetrahedron Lett.*, **1993**, *34*. 3017.

271. A. Fernandez-Mateos, G. P. Coca, R. R. Gonzalez, and C. T. Hernandez, *J. Org. Chem.*, **1996**, *61*, 9097.

272. K. Shishido, Y. Tokunaga, N. Omachi, K. Hiroya, K. Fukumoto, and T. Kametani, *J. Chem. Soc., Chem. Commun.*, **1989**, 1093.

273. K. Shishido, Y. Tokunaga, N. Omachi, K. Hiroya, K. Fukumoto, and T. Kametani, *J. Chem. Soc. Perkin Trans. 1*, **1990**, 2481.

274. R. Annunziata, M. Cinquini, F. Cozzi, and L. Raimondi, *Tetrahedron Lett.*, **1989**, *30*, 5013.

275. P. A. Wade, M. K. Pillay, and S. M. Singh, *Tetrahedron Lett.*, **1982**, *23*, 4563.

276. P. A. Wade, S. M. Singh, and M. K. Pillay, *Tetrahedron*, **1984**, *40*, 601.

277. D. M. Vyas, Y. Chiang, and T. W. Doyle, *Tetrahedron Lett.*, **1984**, *25*, 487.

278. P. Caldirola, M. De Amici, and C. De Micheli, *Tetrahedron Lett.*, **1986**, *27*, 4651.

279. M. De Amici, C. De Micheli, and V. Misani, *Tetrahedron*, **1990**, *46*, 1975.

280. K. Torssell and O. Zeuthen, *Acta Chem. Scand.*, **1978**, *B32*, 118.

281. A. P. Kozikowski and P. D. Stein, *J. Am. Chem. Soc.*, **1982**, *104*, 4023.

282. D. P. Curran and D. H. Singleton, *Tetrahedron Lett.*, **1983**, *24*, 2079.

283. D. P. Curran, *J. Am. Chem. Soc.*, **1983**, *105*, 5826.

284. A. P. Kozikowski and J. G. Scripko, *J. Am. Chem. Soc.*, **1984**, *106*, 353.

285. A. Barco, S. Benetti, C. De Risi, G. P. Pollini, and V. Zanirato, *Tetrahedron*, **1995**, *51*, 7721.

286. M. Aghazade Tabrizi, P. G. Baraldi, M. Guarneri, S. Manfredini, G. P. Pollini, and D. Simoni, *Tetrahedron Lett.*, **1991**, *32*, 683.

287. A. P. Kozikowski and C.-S- Li, *J. Org. Chem.*, **1985**, *50*, 778.

288. J. Zhang and D. P. Curran, *J. Chem. Soc., Perkin Trans. 1*, **1991**, 2627.

289. S. F. Martin, M. S. Dappen, B. Dupre, C. J. Murphy, and J. A. Colapret, *J. Org. Chem.*, **1989**, *54*, 2209.

290. B. Jiang, Y. Liu and W.-s. Zhou, *J. Org. Chem.*, **2000**, *65*, 6231.

291. G. J. McGarvey, J. A. Mathys, and K. J. Wilson, *Tetrahedron Lett.*, **2000**, *41*, 6011.

292. V. Jäger, "Synthesen mit Isoxazolinen", *Habilitationsschrift*, Giessen, **1979**.

293. (a) V. Jäger and I. Müller, *Tetrahedron Lett.*, **1985**, *26*, 2997. (b). B. J. Banks, A. G. M. Barrett, M. A. Russell, and D. J. Williams, *J. Chem. Soc., Chem. Commun.*, **1983**, 873; G. Zimmermann, W. Hass, H. Faasch, H. Schmalle, and W. A. König, *Liebigs Ann. Chem.*, **1985**, 2165; A. G. Barrett and S. A. Leibold, *J. Org. Chem.*, **1990**, *55*, 5818.

294. I. Müller and V. Jäger, *Tetrahedron Lett.*, **1982**, *23*, 4777.

295. P. J. Zimmermann, I. Blanarikova, and V. Jäger, *Angew. Chem. Int. Ed.*, **2000**, *39*, 910.

296. P. J. Zimmermann, *Dissertation*, Stuttgart, Germany, **2000**.

297. W. Schwab and V. Jäger, *Angew. Chem., Int. Ed. Engl.*, **1981**, *20*, 603.

298. V. Jäger and R. Schohe, *Tetrahedron*, **1984**, *40*, 2199.

299. A. E. Stütz, *Iminosugars as Glycosidase Inhibitors*, Wiley-VCH, Weinheim, **1999**.

300. Y. Chapleur, *Carbohydrate Mimics*, Wiley-VCH, Weinheim, **1998**.

301. J. Cossy and P. Vogel, *Stud. Nat. Prod. Chem.*, **1993**, *12*, 275.

302. A. B. Hughes and A. J. Rudge, *Nat. Prod. Rep.*, **1994**, *11*, 135.

303. G. W. J. Fleet, S. J. Nicholas, P. W. Smith, S. V. Evans, L. E. Fellows, and R. J. Nash, *Tetrahedron Lett.*, **1985**, *26*, 3127; N. Asano, R. J. Nash, R. J. Molyneux, and G. W. J. Fleet, *Tetrahedron Asymm.*, **2000**, *11*, 1645.

304. G. Legler, *Adv. Carbohydr. Chem. Biochem.*, **1990**, *48*, 319.

305. H. Liu, X. Liang, H. Sohoel, A. Bülow, and M. Bols, *J. Am. Chem. Soc.*, **2001**, *123*, 5116.

306. T. M. Jespersen, W. Dong, M. R. Sierks, T. Skrydstrup, I. Lundt, and M Bols, *Angew. Chem., Int. Ed. Engl.*, **1994**, *33*, 1778.

307. Y. Ichikawa and Y. Igaraschi, *Tetrahedron Lett.*, **1995**, *36*, 4585; Y. Ichikawa, Y. Igaraschi, M. Ichikawa, and Y. Suhura, *J. Am. Chem. Soc.*, **1998**, *120*, 3007.

308. M. Bols, *Acc. Chem. Res.*, **1998**, *31*, 1.

309. J. P. Michael, *Nat. Prod. Rep.*, **1994**, *11*, 639.

310. G. Drehfahl and H.-H. Hörhold, *Chem. Ber.*, **1964**, *97*, 159.

311. A. Kern, G. Bovermann, G. Jung, M. Wanning, and M. Zähner, *Liebigs Ann. Chem.*, **1989**, 361.

312. T. Postels, *Diplomarbeit*, Hamburg, **1990**.

313. V. Jäger, R. Müller, T. Leibold, U. Stahl, J. Raczko, B. Steuer, P. Poggendorf, and T. Gracza, *Abstr. Papers XVIth Int. Carbohydr. Symp. IUPAC*, Paris, July 5–10, **1992**, IL-A9.

314. J. Raczko, T. Leibold, R. Müller, and V. Jäger, *Abstract Papers VIIth European Carbohydr. Symposium*, Krakow, Poland, 22–27. August, **1993**.

315. C. Schaller, P. Vogel, and V. Jäger, *Carbohydr. Res.*, **1998**, *314*, 25.

316. P. Mangeney, A. Alexakis, and J. F. Normant, *Tetrahedron Lett.*, **1988**, *29*, 2677.

317. C. Schaller and P. Vogel, *Synlett*, **1999**, 1219.

318. K. C. Nicolaou, W.-M. Dai, and, R. K. Guy, *Angew. Chem., Int. Ed. Engl.*, **1994**, *33*, 15.

319. K. C. Nicolaou and, E. J. Sorensen, *Classics in Total Synthesis*, VCH, Weinheim, **1996**, pp. 655–672.

320. L. Alcaraz, J. J. Harnett, C. Mioskowski, T. Le Gall, D.-S. Shin, and J. R. Falck, *J. Org. Chem.*, **1995**, *60*, 7209.

321. M. A. Arai, T. Arai, and H. Sasai, *Org. Lett.*, **1999**, *1*, 1795; cf. Review: R. Noyori and M. Kitamura, *Angew. Chem., Int. Ed. Engl.*, **1991**, *30*, 49.

322. (a) M. Marx, F. Marli, J. H. Reisdorff, S. Sandemeier, and S. Clark, *J. Am. Chem. Soc.*, **1977**, *99*, 6754. (b) N. Maugein, A. Wagner, and C. Mioskowski, *J. Org. Chem.*, **1999**, *64*, 8428.

323. T. Takahashi, T. Shimayama, M. Miyazawa, M. Nakazawa, H. Yamada, K. Takatori, and M. Kajiwara, *Tetrahedron Lett.*, **1992**, *33*, 5973.

324. A. P. Kozikowski and P. D. Stein, *J. Org. Chem.*, **1984**, *49*, 2301.

325. Y. Kobayashi, H. Miyazaki, and M. Shiozaki, *J. Am. Chem. Soc.*, **1992**, *114*, 10065. Y. Kobayashi, H. Miyazaki, and M. Shiozaki, *J. Org. Chem.*, **1994**, *59*, 813.

326. B. H. Kim, P. B. Jacobs, R. L. Elliot, and D. P. Curran, *Tetrahedron*, **1988**, *44*, 3079. For other intramolecular cycloadditions that afford tricyclic isoxazolines, see: M. Toyota, M. Matsuura, and K. Fukumoto, *Heterocycles*, **1995**, *41*, 661.
327. M. Ihara, Y. Tokunaga, N. Taniguchi, and K. Fukumoto, *J. Org. Chem.*, **1991**, *56*, 5281.
328. G.-C. Zheng and H. Kakisawa, *Bull. Chem. Soc. Jpn.*, **1989**, *62*, 602.
329. M. Prashad and B. Fraser-Reid, *J. Org. Chem.*, **1985**, *50*, 1566.
330. T. Bando and K. Shishido, *J. Chem. Soc., Chem. Commun.*, **1996**, 1357. For other related syntheses of bicyclic furans see: K. Shishido, K. Umimoto, and M. Sibuya, *Heterocycles*, **1994**, *38*, 641; K. Shishido, K. Umimoto, M. Ouchii, O. Irie, T. Omodani, T. Takata, and M. Shibuya, *J. Chem. Res. (S)*, **1993**, 58, *J. Chem. Res. (M)*, **1993**, 328. K. Shishido, T. Takata, T. Omodani, and M. Shibuya, *Chem. Lett.*, **1993**, 557.
331. M. Ihara, Y. Tokunaga, and K. Fukumoto, *J. Org. Chem.*, **1990**, *55*, 4497. M. Ihara, Y. Tokunaga, N. Taniguchi, and K. Fukumoto, *Tetrahedron*, **1991**, *47*, 6635.
332. M. A. Tius and K. Reddy, *Tetrahedron Lett.*, **1991**, 3605.
333. K. Sugita, K. Shigeno, C. F. Neville, H. Sasai, and M. Shibasaki, *Synlett*, **1994**, 325; K. Shigeno, K. Ohne, T. Yamaguchi, H. Sasai, and M. Shibasaki, *Heterocycles*, **1992**, *33*, 161.
334. P. N. Confalone, G. Pizzolato, D. L. Confalone, and M. R. Uskokovic, *J. Am. Chem. Soc.*, **1980**, *102*, 1954.
335. A. Saha and A. Bhattacharjya, *J. Chem. Soc., Chem. Commun.*, **1997**, 495.
336. For a review on aminocyclopentitol glycosidase inhibitors see: A. Berecibar, C. Grandjean, and A. Siriwardena, *Chem. Rev.*, **1999**, *99*, 779.
337. V. Jäger and S. Picasso, *The 16th ICHC* (Montana State University, Bozeman), **1997**, Abstr. IL-VI-2.; M. Kleban, U. Kautz, J. Greul, P. Hilgers, R. Kugler, H.-Q. Dong, and V. Jäger, *Synthesis*, **2000**, 1027.
338. J. N. Greul, M. Kleban, B. Schneider, S. Picasso, and V. Jäger, *ChemBioChem.*, **2000**, *2*, 368 and Ref. 260.
339. R. A. Farr, N. P. Peet and M. S. Kang, *Tetrahedron Lett.*, **1990**, *31*, 7109; N. P. Peet, E. W. Huber, and R. A. Farr, *Tetrahedron Lett.*, **1991**, *32*, 7537; R. A. Farr, A. K. Holland, E. W. Huber, and N. P. Peet, *Tetrahedron*, **1994**, *50*, 1033. R. A. Farr (Merrell-Dow) US-Pat. 5382709 **1995**, [*Chem. Abstr.*, *116*, 59902 (**1995**)].
340. J. Greul, *Dissertation*, Stuttgart, Germany, **2000**.
341. B. Bernet and A. Vasella, *Helv. Chim. Acta*, **1979**, *62*, 1990; B. Bernet and A. Vasella, *Helv. Chim. Acta* **1979**, *62*, 2401; B. Bernet and A. Vasella, *Helv. Chim. Acta*, **1979**, *62*, 2411.
342. T. K. M. Shing, D. A. Eisley, and J. G. Gilhouly, *J. Chem. Soc., Chem. Commun.*, **1989**, 1280.
343. S. Moutel and M. Shipman, *Synlett*, **1998**, 1333.
344. P. J. Dransfield, S. Moutel, M. Shipman, and V. Sik, *J. Chem. Soc., Perkin Trans. 1*, **1999**, 3349.
345. R. J. Ferrier and P. Prasit, *J. Chem. Soc., Chem. Commun.*, **1981**, 983.
346. R. J. Ferrier and S. Middleton, *Chem. Rev.*, **1993**, *93*, 2779.
347. A. Cerri, C. De Micheli, and R. Gandolfi, *Synthesis*, **1974**, 710.
348. A. P. Kozikowski, Y.-Y. Chen, B. C. Wang, and Z.-B. Xu, *Tetrahedron*, **1984**, *40*, 2345.
349. P.-Y. Le Roy, *Dissertation*, Stuttgart, **1997**.
350. M. Henneböhle, *Dissertation* (planned), Stuttgart,
351. A. Studer and *Dissertation*, *Tetrahedron*, **1997**, *53*, 6681.

CHAPTER 7

Nitrile Ylides and Nitrile Imines

John T. Sharp

Department of Chemistry, University of Edinburgh, Edinburgh, Scotland

7.1. Physical and Spectroscopic Properties 474
7.1.1. Of Nitrile Ylides 474
7.1.2. Of Nitrile Imines 475
7.2. Generation 478
7.2.1. Of Nitrile Ylides 478
7.2.1.1. Nitrile Ylide Synthons 479
7.2.1.2. Reactions of Carbenes with Nitriles 485
7.2.1.3. Photolysis of Azirines 488
7.2.1.4. Other Routes to Nitrile Ylides 490
7.2.2. Of Nitrile Imines 492
7.2.2.1. Transient Nitrile Imines 492
7.2.2.2. Stabilized Nitrile Imines 495
7.3. [2 + 3]-Cycloaddition Reactions 496
7.3.1. Of Nitrile Ylides 496
7.3.2. Of Nitrile Imines 502
7.3.2.1. Of Stabilized Nitrile Imines 502
7.3.2.2. Selectivity in Nitrile Imine Cycloaddition Reactions 505
7.3.2.3. Of Nitrile Imines to Heterocycles 507
7.3.2.4. Intramolecular Cycloaddition Reactions of Nitrile Imines 511
7.4. Electrocyclization Reactions 517
7.4.1. Of Nitrile Ylides 517
7.4.1.1. Alkene-Conjugated Nitrile Ylides 518
7.4.1.2. Diene-Conjugated Nitrile Ylides 520
7.4.1.3. Triene-Conjugated Nitrile Ylides 524
7.4.2. Of Nitrile Imines 525
7.5. Reactions with Nucleophiles 526
7.5.1. Reactions of Nitrile Ylides 526
7.5.2. Reactions of Nitrile Imines 527
7.6. Conclusion 530

The Chemistry of Heterocyclic Compounds, Volume 59: Synthetic Applications of 1,3-Dipolar Cycloaddition Chemistry Toward Heterocycles and Natural Products, Edited by Albert Padwa and William H. Pearson.
ISBN 0-471-38726-6

7.1. PHYSICAL AND SPECTROSCOPIC PROPERTIES

7.1.1. Physical and Spectroscopic Properties of Nitrile Ylides

The first nitrile ylide stable enough to be isolated (i.e., **1**) has been prepared by the carbene/nitrile method (1). For this dipole, the anionic component is stabilized by electron delocalization and the nitrilium component by the steric bulk of the adamantyl group to such an effect that it has a melting point of 230 °C. The X-ray structure showed that the nitrile ylide moiety is close to linear and much like the resonance structure shown below.

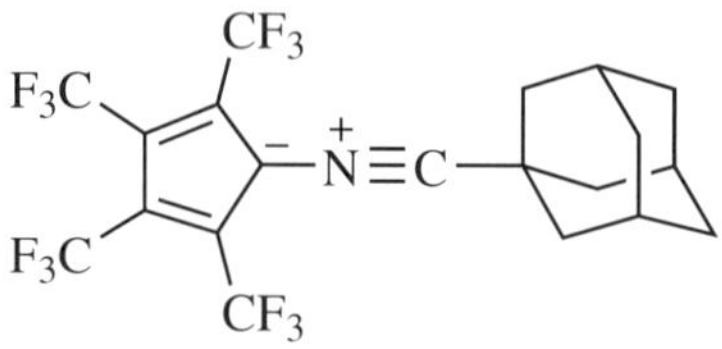

(1)

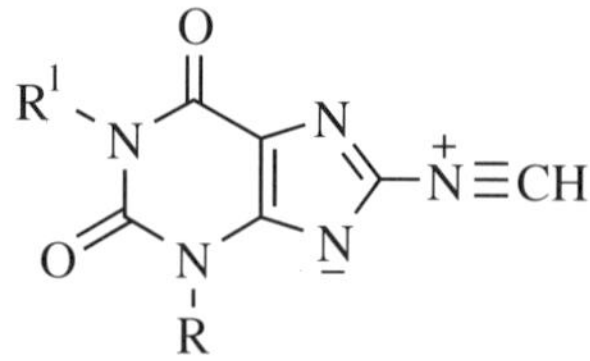

(2)

Another example of a stable nitrile ylide (**2**; R,R^1 = alkyl) was generated by the thermal decomposition of 7-azido-1,3-disubstituted lumazines in xylene (2). The product is formed *via* a complex multistep process. As in the case of **1**, the stability is attributed to strong resonance stabilization of the anionic moiety but it is notable that this is the first isolable nitrile ylide that does not also rely on the presence of a bulky substituent at the nitrilium carbon.

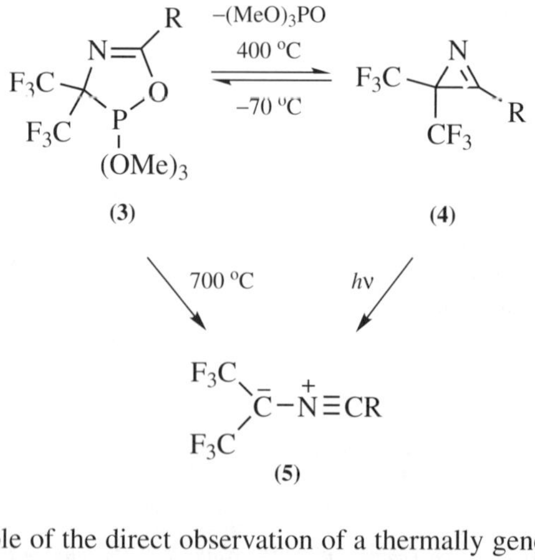

The first example of the direct observation of a thermally generated nitrile ylide has been reported (3). This was achieved by studying the thermal flash vacuum pyrolysis (FVP) decomposition of the oxazaphospholes **3** (R=*t*-Bu, Ph) *via* the condensation of the products on the KBr window of an infrared (IR) spectrometer at −196 °C. The reaction path was found to be strongly temperature dependent. Thus, pyrolysis at 400 °C gave quantitative conversion into trimethyl phosphate and the azirine **4**, which recombined quantitatively when the window was warmed up to −70 °C. However, on pyrolysis at 700 °C, the azirine was not detected and instead a species thought to be the nitrile ylide **5**, absorbing strongly at 2250 cm^{-1}, was formed. This formulation was supported by the fact that the same species was produced by photolysis of the matrix isolated azirine.

Work on the molecular structure of benzonitrilio methylide **8/9** has been carried out *via* Fourier transform infrared (FTIR) studies on it and five isotope-labeled variants. The nitrile ylides were generated in nitrogen matrices at 12 K either directly, by photolysis of the azirine **7**, or indirectly from the azidostyrenes **6** (4).

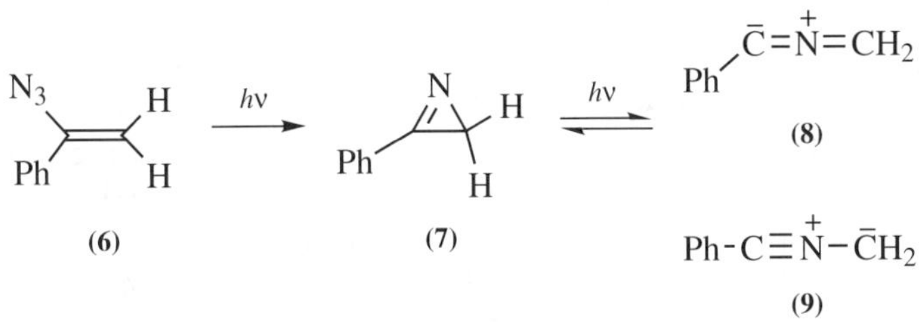

It was found that the azirine-nitrile ylide isomerization was a completely reversible process. The unlabeled nitrile ylide showed a prominent band at 1926 cm^{-1} that underwent a 66-cm^{-1} shift with ^{15}N substitution. This shift was interpreted as being consistent with an *allene-like* skeleton (**8**) rather than the alternative *propargyl-like* structure (**9**). This conclusion was supported by the spectra from the ^{13}C- and ^{2}H-labeled variants. Warming the nitrile ylide in a xenon matrix from 12 to 82 K provided no new absorptions suggesting that the *allene-like* structure may also be adopted in solution. Some absorption spectra for benzonitrilio benzylide (DPNY) and some substituted benzonitrilio methylides obtained *via* pulsed-laser photolysis of azirines are given in Table 7.1 (5).

7.1.2. Physical and Spectroscopic Properties of Nitrile Imines

The long-sought parent nitrile imine **10/11** was produced for the first time in 1994 from 1,2,4-triazole in a beam experiment and was characterized by

TABLE 7.1. ABSORPTION SPECTRA FOR SUBSTITUTED BENZONITRILE YLIDES

		4-X-Benzonitrilio methylide				
	DPNY	MeO	Me	H	Cl	CN
λ_{max}nm	235	260	236	230	240	238
	345	288	284	280	283	282

neutralization–reionization mass spectrometry (NRMS) (6). It was subsequently generated by photolysis of the tetrazole **12** in an argon matrix at 12 K and identified by comparison of its observed IR spectrum with the *ab initio* calculated spectrum (QCISD/6-31G**) (7). The same species was also found to be present in the matrix-isolated products from FVP of the same substrate at 800 °C. Further irradiation of the nitrile ylide resulted in its conversion into a previously unknown species considered to be a HCN—NH complex.

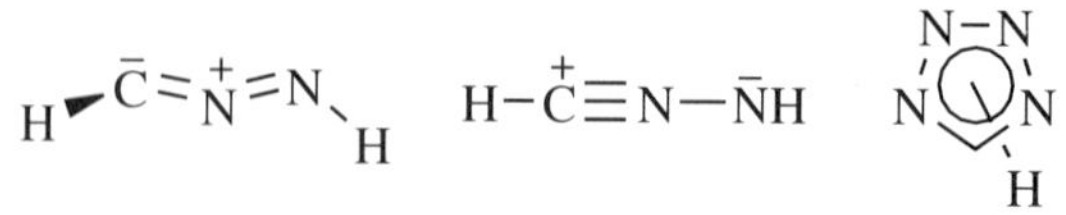

(10) **(11)** **(12)**

Early theoretical calculations indicated that the lowest energy structure for this species was the planar propargyl form **11** but also that the molecule was *floppy* and the nonplanar allenic structure **10** of similar energy was also accessible. More recently, it has been shown that the calculated energy difference between the two structures is very dependent on the theoretical method used, and that high-level calculations show the allenic structure **10** to be the only stable minimum, favored over **11** by ~ 14 kJ mol^{-1} (8). Other calculations report similar conclusions in respect of the preferred geometry (7,9), but the fact that the terminal nitrogen atom is negatively polarized indicates some contribution from **11** to the wave function (6). The triplet lies ~ 161 kJmol^{-1} above the singlet ground state (10).

The silyl substituted nitrile imine **14**, generated by flash vacuum pyrolysis of the tetrazole **13**, has been directly observed by mass spectrometry (MS); and by IR (2955, 2230 cm^{-1}) and ultraviolet (UV) (ν_{max} 303 nm) via its condensation at 77 K (11). The silyl group was selected in order to prevent isomerization into azines and carbodiimides that are known reactions of N-alkyl analogues. The nitrile imine proved to be stable indefinitely at 77 K but disappeared on warming to 170–180 K. Cocondensation at 77 K with dipolarophiles gave the expected cycloadducts (e.g., **15**) from ethyl propiolate. Other N-silylated analogues were also prepared. The flash pyrolysis process was later optimized using real-time photoelectron spectroscopic gas analysis (12). The bands were assigned on the basis of geometry-optimized MINDO calculations.

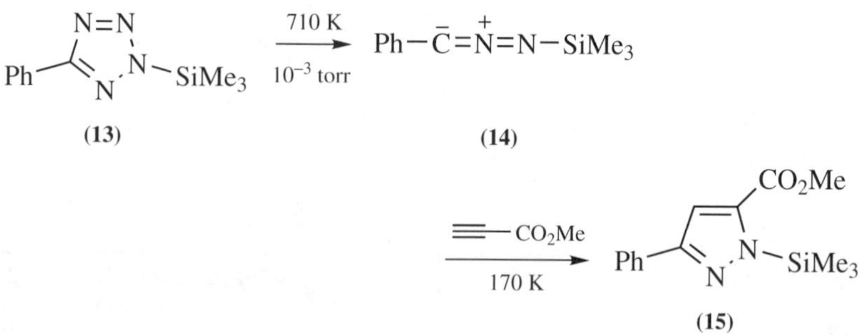

(13) **(14)**

(15)

Laser flash photolysis (LFP) studies have been carried out on the generation of N,C-diaryl nitrile imines from sydnones and from tetrazoles in solution at 77 K. They were found to have lifetimes of milliseconds and were quenched by dimethyl acetylenedicarboxylate (DMAD) ($k_q = 5\text{–}9 \times 10^3 M^{-1} s^{-1}$) and by carboxylic acids ($k_q = 10^4\text{–}10^9\ M^{-1}\ s^{-1}$) (13). The strong dependence of ν_{max} on the nature of the aromatic substituents in N,C-diaryl nitrile imines was interpreted by a linear free energy relationship as due to intramolecular charge transfer (14).

In the late 1980s, accustomed thinking about the nature of nitrile imines was upturned by the discovery that these intermediates could be stabilized by the use of appropriate C and N substituents to such a degree that they could be obtained as stable isolable compounds (15,16). Some examples are shown below **16–24**. Most of them are quite thermally stable and are insensitive to air and water (e.g., **16**). The first such nitrile imine (**16**) to be prepared has a mp of 100 °C. Even example **23** with carbon-based substituents has a mp of 60 °C. It appears that the major factor in the kinetic stability of these compounds is the steric bulk of the C and N substituents.

$$(i\text{-Pr}_2N)_2P(=S)-C\equiv\overset{+}{N}-\overset{-}{N}-P(Ni\text{-Pr}_2)_2 \qquad \textbf{(16)}$$

$$(i\text{-Pr}_2N)_2P(=S)-C\equiv\overset{+}{N}-\overset{-}{N}-SiR_3 \qquad \textbf{(17)}$$

$$(i\text{-Pr}_2N)_2\ddot{P}-C\equiv\overset{+}{N}-\overset{-}{N}-SiR_3 \qquad \textbf{(18)}$$

$$(i\text{-Pr}_2N)_2P(=S)-C\equiv\overset{+}{N}-\overset{-}{N}-B(Ni\text{-Pr}_2)_2 \qquad \textbf{(19)}$$

$$(i\text{-Pr}_2N)_2\ddot{P}-C\equiv\overset{+}{N}-\overset{-}{N}-B(Ni\text{-Pr}_2)_2 \qquad \textbf{(20)}$$

$$i\text{-Pr}_3Si-C\equiv\overset{+}{N}-\overset{-}{N}-Si(i\text{-Pr}_3) \qquad \textbf{(21)}$$

$$(i\text{-Pr}_2N)_2B-C\equiv\overset{+}{N}-\overset{-}{N}-B(Ni\text{-Pr}_2)_2 \qquad \textbf{(22)}$$

$$Ph_3C-C\equiv\overset{+}{N}-\overset{-}{N}-CPh_3 \qquad \textbf{(23)}$$

$$(i\text{-Pr}_2N)_2P(=S)-C\equiv\overset{+}{N}-\overset{-}{N}-\overset{+}{P}(CH_3)(Ni\text{-Pr}_2)_2\ \ TfO^- \qquad \textbf{(24)}$$

Thus in the N-silyl substituted series, **17** and **18**, which rearrange thermally to the corresponding diazo compounds, the stability increases through the series R=Me, Ph, *i*-Pr. As discussed below, these compound undergo the usual cycloaddition and electrocyclization reactions of nitrile imines and are not simply over-stabilized curiosities. The usefulness in synthesis of those with P–C bonds is probably limited since these bonds are not easily broken, but products derived from those with C–Si and C–B bonds (e.g., **21** and **22**) should be capable of further elaboration (16). Structural data from X-ray crystallography have been obtained for several of these compounds (16). They show a range that runs from compound **23**, which is of the allenic type **10**, to the phosphonio derivative **24**, which is closer to the propargyl type structure **11**. The IR, UV–visible, and nuclear magnetic resonance (NMR) data for these compounds have been collated and discussed (16). The IR absorption frequency for the C—N—N system correlated with the variation in structure between *allenic* (2010–2100 cm^{-1}) and *propargylic* (2140–2170 cm^{-1}) as revealed by the X-ray data. In respect of this absorption, these stable nitrile imines seem to constitute a separate class to the *transient* C-aryl species and the N-silyl species **14** whose IR spectra were obtained in matrices, all of which have more nitrile-like absorptions >2200 cm^{-1} (16). NMR spectroscopy is particularly useful in distinguishing stable nitrile imines from their isomeric diazo compounds, ^{13}C spectra can be used but ^{14}N spectra are better since they show a ~70 ppm difference in the chemical shift of the α-N (17). Low-temperature NMR studies using derivatives **25** and **26** (R=*i*-Pr, R′=cyclohexyl), which have chiral substituents, have been used to demonstrate that these compounds retain their chiral nonplanar *allenic* structures **10** in solution (18). The activation free energy for racemization was found to be ~30 kJ mol^{-1} and, on the basis of *ab initio* calculations, it was suggested that the most likely mechanism was inversion at carbon followed by followed by rotation of the $P(S)(NR_2)_2$ moiety.

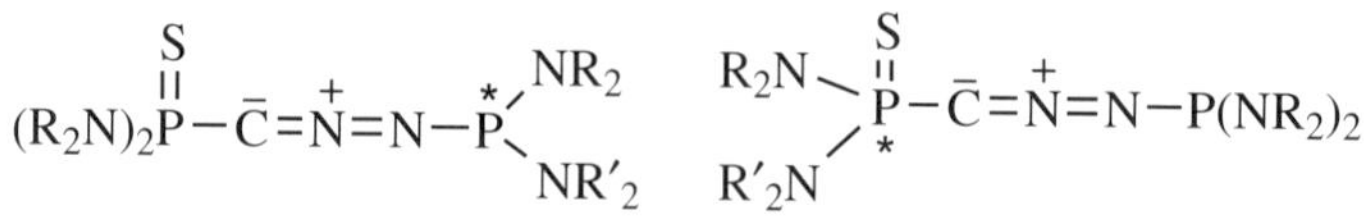

(25) **(26)**

7.2. GENERATION

7.2.1. Generation of Nitrile Ylides

The most significant development in the generation of *true* nitrile ylides to emerge since the last review (19) is concerned with the direct formation of these species by the reactions of carbenes and carbenoids with nitriles. The classical methods, however, via the elimination of hydrogen chloride from imidoyl chlorides and the photochemical ring opening of azirines, still continue to provide the major *synthetic* access routes to these species. The former has undergone some useful methodological development and has been extended to new substituted species, and the latter has been much utilized in pulsed-laser photolysis work on the study of nitrile ylide spectra and the determination of kinetic parameters in their reactions. Much ingenuity and effort have also been devoted to finding an easy general route to the simple alkyl substituted nitrile ylides not easily available by either of these methods. Most of this work was based on desilylation chemistry and although it did not lead to the formation of *true* nitrile ylides it has produced much fascinating chemistry and led to the development of a range of nitrile ylide *synthons*, which are of much interest to the synthetic chemist.

7.2.1.1. *Nitrile Ylide Synthons*

Much work has been reported in the last 15 years on the development of *nitrile ylide synthons* (i.e., species that are not true nitrile ylides but that achieve the same overall synthetic result when used in pericyclic reactions). The aims motivating this work were to improve access to nitrile ylides with a wider range of substituents. In the event, although it did not lead to *true* nitrile ylides, the work did provide easy routes to species that are not only their synthetic equivalents but that also circumvent limitations in *true* nitrile ylide reactivity or regioselectivity. Much of this chemistry has been based on the generation of azomethine ylides that have a leaving group (X) on the *imine* carbon (e.g., **27**). These species add easily to a wide range of dipolarophiles and the product **28** then reacts further, usually spontaneously, via an elimination reaction to give a final product **29**, which is equivalent to that of a nitrile ylide cycloaddition.

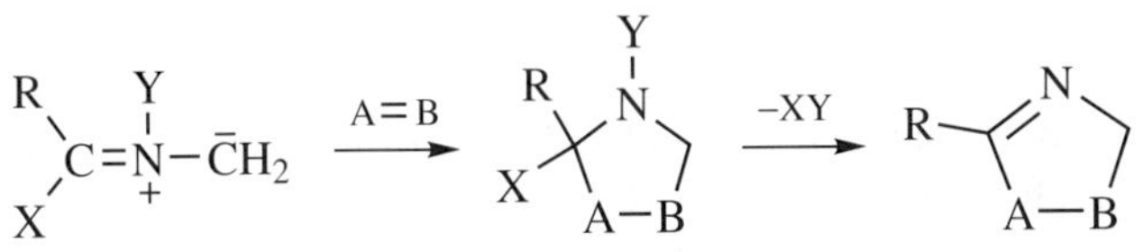

(27) **(28)** **(29)**

The highly effective desilylation routes to nonstabilized azomethinc ylides have provided the basis for much of this chemistry. Thus, the reaction of *N*-(silylmethyl)-thioimidates (**30**) with AgF in the presence of a range of dipolarophiles (electron-deficient alkenes and alkynes, and aldehydes) led to the isolation of *nitrile ylide* adducts in generally high yields (20,21). Differences in reactivity and regioselectivity

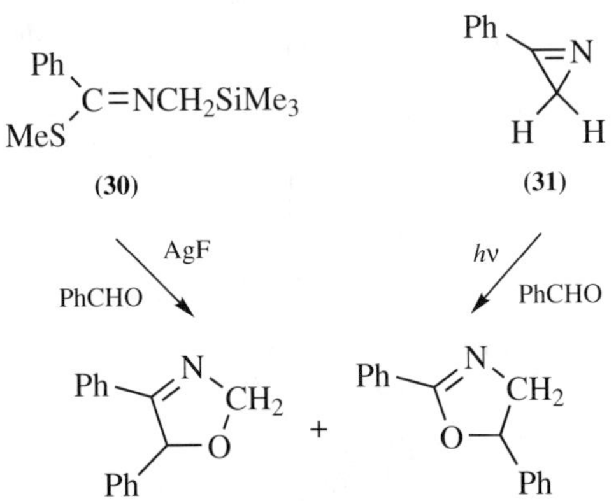

(32) **(33)**

from that shown by *authentic* nitrile ylides generated from azirines indicated that the reacting species were different. For example, in the reaction with benzaldehyde, photolysis of the azirine **31** gave **32** only whereas the desilylation route from **30** gave **32** as the minor product (16%) together with **33** (84%).

The mechanism proposed involves desilyation of the silver complexed imidate and cycloaddition by the azomethine ylide **35** to give **37** followed by elimination (22).

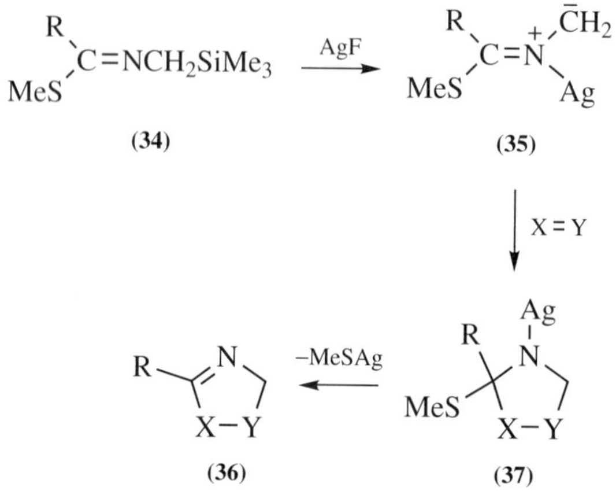

N-(Silylmethyl)thioimidates (**34**) also undergo water-induced desilylation leading to the N-protonated azomethine ylides (**38**). These ylides react with a range of electron-deficient alkenes and alkynes, aldehydes, and ketones followed by elimination of methane thiol to give formal nitrile ylide adducts (e.g., **40**) (23,24). The reactivity of these species is rather dependent on the nature of R (e.g., good for R=Ph but less so for R=Et or *i*-Pr), which may be due to competition from tautomerization to give the *N*-methylthioimidate (**39**).

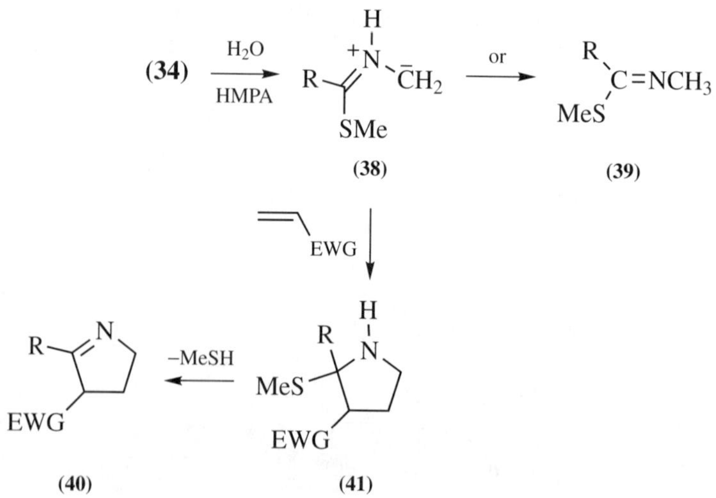

HMPA = hexamethylphosphoramide

N-Silylmethyl-amidines and -thioamides (**42**) (X=NR′ or S) undergo alkylation at X with, for example methyl triflate, and then fluorodesilylation to give the azomethine ylides **43** (identical with **38** for the thioamides) (25,26). Cycloaddition followed by elimination of an amine or thiol, respectively, again leads to formal nitrile ylide adducts. These species again showed the opposite regioselectivity in reaction with aldehydes to that of true nitrile ylides. The thioamides were generally thought to be better for use in synthesis than the amidines and this route leads to better yields and less substituent dependence than the water-induced desilylation discussed above.

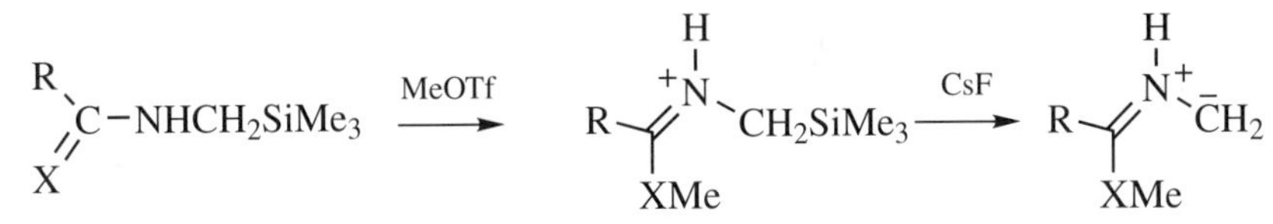

(42) **(43)** **(44)**

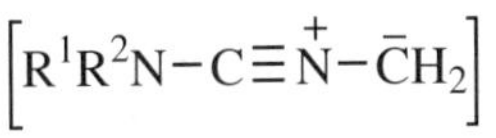

R^1R^2 = Ph,H; $(CH_2)_4$; $(CH_2)_5$; $(CH_2)_2O(CH_2)_2$

(45)

This work has been extended from aryl and alkyl substituted systems (**42**) (R=aryl, alkyl) to analogues where R is an amino group, so giving access to synthetic equivalents of the nonstabilized amino nitrile ylides (**45**). Adducts were obtained in good-to-moderate yield with *N*-methylmaleimide (NMMA), DMAD, electron-deficient alkenes and aromatic aldehydes (27,28), and with sulfonylimines and diethyl azodicarboxylate (29). Similarly the *N*-[(trimethylsilyl)methyl]-thiocarbamates (**46**) undergo selective S-methylation with methyl triflate and subsequent fluorodesilylation in a one-pot process at room temperature to generate the azomethine ylides **47**.

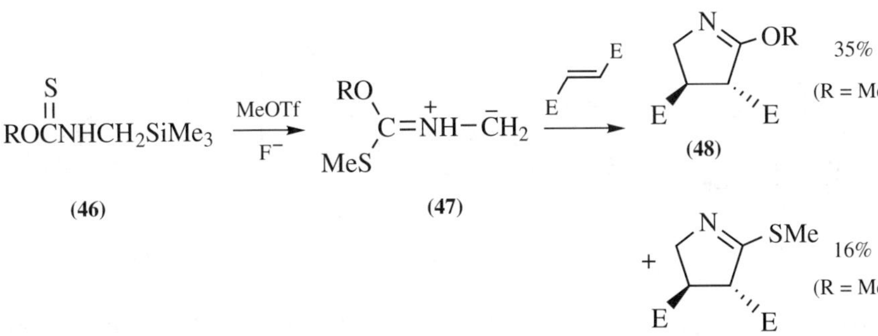

(46) **(47)** **(48)** 35% (R = Me) **(49)** 16% (R = Me)

The latter added to a range of electron-deficient alkenes to give adducts that spontaneously underwent elimination reactions to give a mixture of 2-alkoxy- and 2-methylthiopyrrolines (**48** and **49**) (30). Reaction with DMAD gave only the corrresponding 2-alkoxypyrrole (61%). The same basic theme has been extended to the *N*-silylmethylimidates (**50**) [R=Ph, tetramethylsilyl (TMS)], which lead to the azomethine ylides (**52**) containing methoxy as the leaving group. Interestingly, the reactants were converted into **52** by reaction with phenyltrifluorosilane that served as alkylating agent and desilylating agent as shown (31). In the case where R=TMS, the elimination was accompanied by TMS migration leading to **53** (R=TMS), which hydrolyzed during work up to give the N-unsubstituted product **53** (R=H).

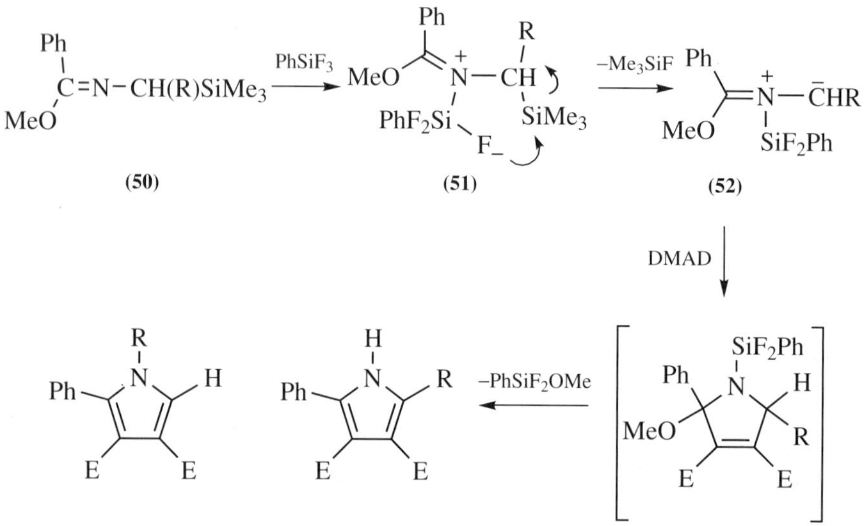

(50) (51) (52) (53) (54) (55)

Analogues with a siloxy leaving group (**58**) (R^1=Ar, R=H, Me) have been generated from *N*-(silylbenzyl)-benzamides (**56**) by an intramolecular silicon migration (32).

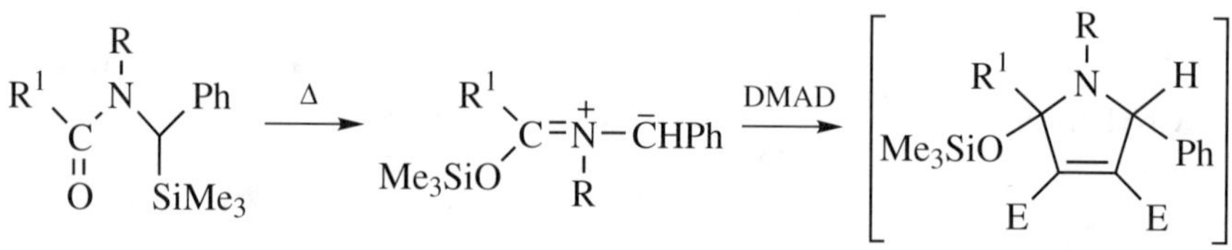

(56) (57) (58)

Azomethine ylides containing a leaving group can also be produced by the standard tautomerization route (e.g., species **60** with a cyano group on the ylide carbon). These species (e.g., **60** [R^1=Ph, PhCO,PhCH=CH; R^2=H,Ph]) reacted with maleimides and with fumarate, maleate, and acrylate esters to give adducts **61**, which readily eliminated hydrogen cyanide (33).

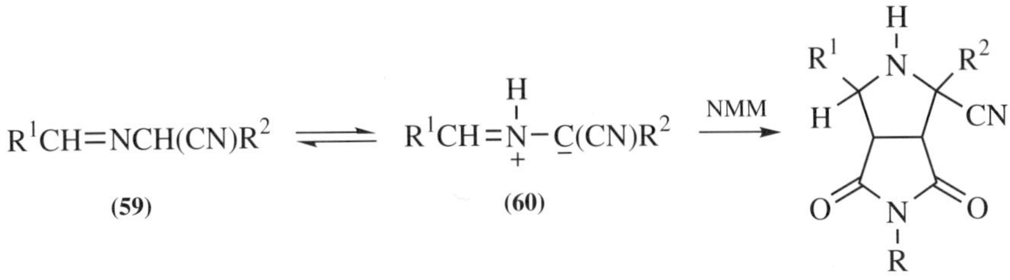

(59) (60) (61)

The imidates (**62**) also serve as nitrile ylide synthons via cycloaddition and subsequent spontaneous elimination of ethanol (34,35). Cycloadditions were carried out to aldehydes, leading to 2-oxazolines (e.g., **64**) and to isocyanates and isothiocyanates. In the preparation of the 2-oxazolines, a solvent-less mixture of the imidate and the required aldehyde were heated at 70 °C and the cycloadducts **64** (R = Ph, 2-furyl, Me_2CH, 2-HO—C_6H_4, 2-pyridyl, cinnamyl) were isolated in yields of 64–91%.

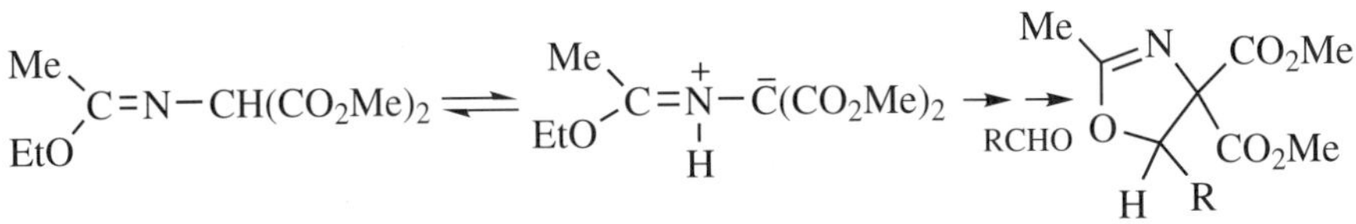

(62) (63) (64)

An alternative route to nitrile ylide synthons, based on the 2-azaallyl anion **65**, is also accessible from the *N*-(silylmethyl)thioimidates (**34**) (24). Desilylation without prior quaternization at nitrogen generates the 2-azaallyl anion **65**, which reacts with monosubstituted electron-deficient alkenes and with aromatic aldehydes and ketones to give the formal nitrile ylide adducts **67** and **69**. Reaction with fumarate and maleate esters gave Michael adducts as the isolated products but, with monosubstituted analogues, the reaction continues by cyclization as shown to give pyrrolines (e.g., **67**). Interestingly, the latter are formed by attack at the γ-carbon, and hence show the opposite regioselectivity to the adducts formed from the species **35** or **38**. Reaction with aromatic aldehydes, however, occurs solely via the α-carbon with concomitant cyclization to give the 2-oxazolines (**69**) in moderate to good yield.

A similar reaction of **70** leads to an *amino nitrile ylide* synthon (36,37), which reacts with a range of aromatic and heteoaromatic aldehydes to give the 2-oxazolines (**71**), but which fails to react with aliphatic aldehydes, simple ketones, or activated alkenes.

Similarly, the iminodithiocarbonate **72** serves as a *thio nitrile ylide* synthon (38). When treated with fluoride ion in the presence of aldehydes and ketones, it gave the alcohols **73** as isolable products that were converted into **74** on treatment with silical gel. Best results were obtained with aromatic aldehydes containing electron-withdrawing groups; yields were poorer when electron-donating groups were present and for ketones.

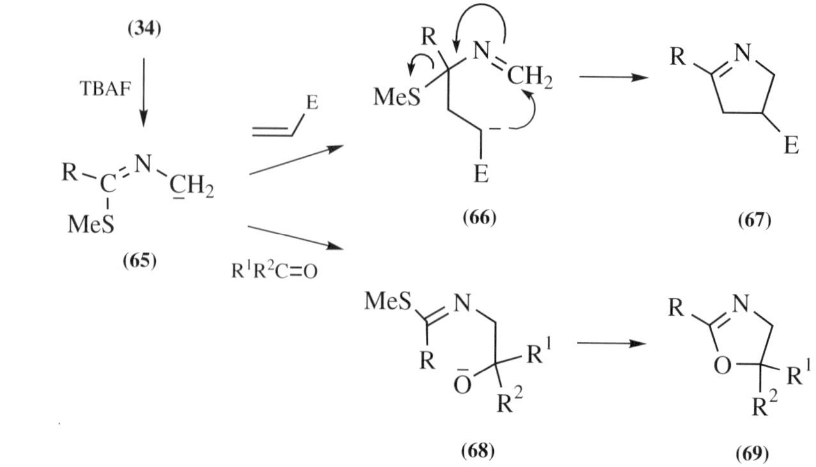

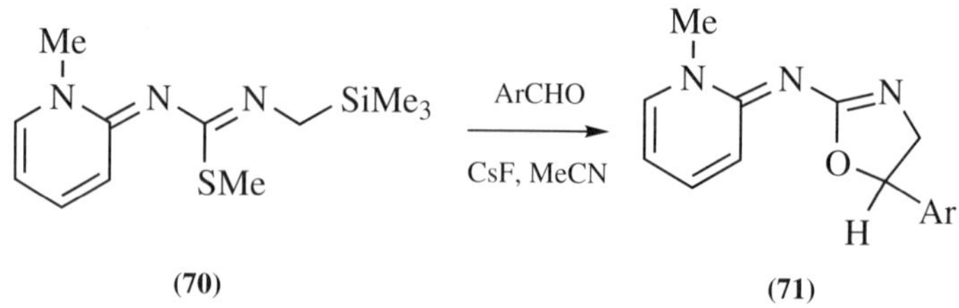

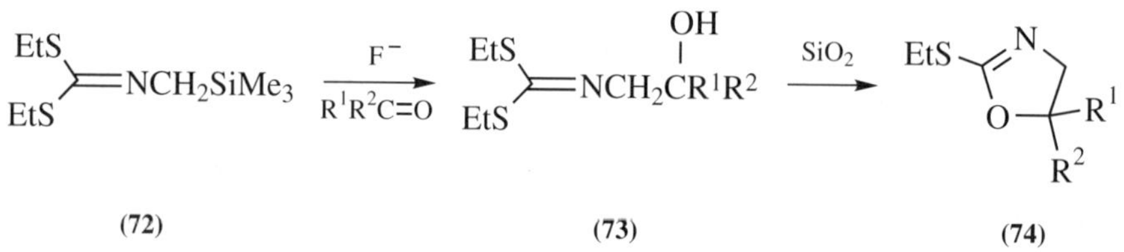

The 4-phospha-1,3-butadiene **77/80** serves as an effective synthon for the unknown H-substituted nitrile ylide **79** in [3 + 2]-cycloaddition reactions with a range of electron-poor dipolarophiles (e.g., reaction with DMAD gave **78** in 80% yield). Similar yields were also obtained using methyl propiolate, azodicaboxylic esters, ethyl acrylate, and acrylonitrile (39). The reactant was generated under very mild conditions from **75** as shown below.

The imino carbene complexes of tungsten and chromium (e.g., **81**) also serve as nitrile ylide synthons (40). The tungsten complexes gave higher yields of the adducts and were strongly regioselective for product **82**. For example, for **81** (M = W; R^1 = Me, Ph; R^3 = Pr, Ph; R^4 = H), the pyrrole **82** was produced in yields of 65–75% with <1% of **83**. This route to pyrroles thus has clear advantages over

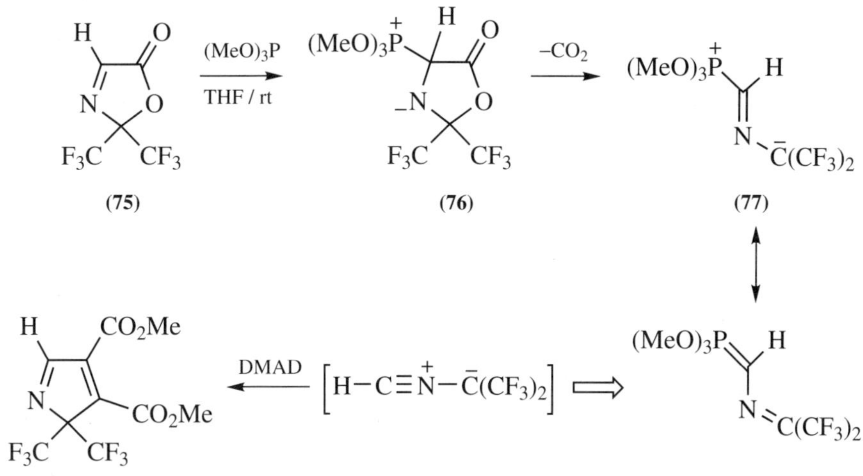

(75) (76) (77) (78) (79) (80)

THF = tetrahydrofuran
rt = room temperature

analogous routes using nitrile ylides themselves in terms of both reactivity and regioselectivity. The tungsten complexes also reacted in a similar way with benzonitrile, styrene, methyl acrylate, and benzaldehyde giving cycloadducts in moderate to good yields.

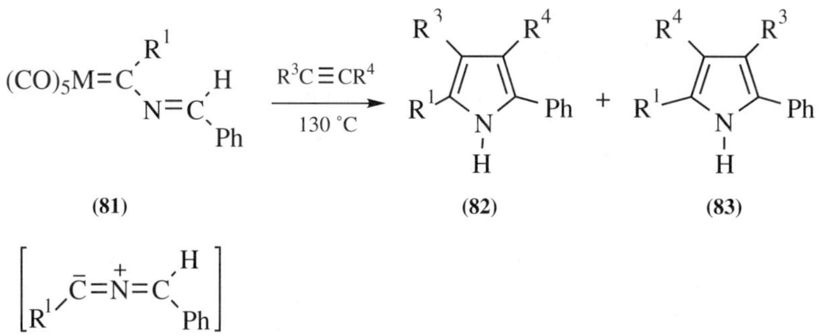

(81) (82) (83)

7.2.1.2. Reactions of Carbenes with Nitriles

The direct assembly of nitrile ylides by the reaction of carbenes (e.g., **84**) with nitriles has, in principle, great potential for structural variation as both components are easily accessible. The first reports of such reactions appeared in the early 1980s and were concerned with the reaction of nitriles with stabilized carbenes such as **84** and **86** (41–43).

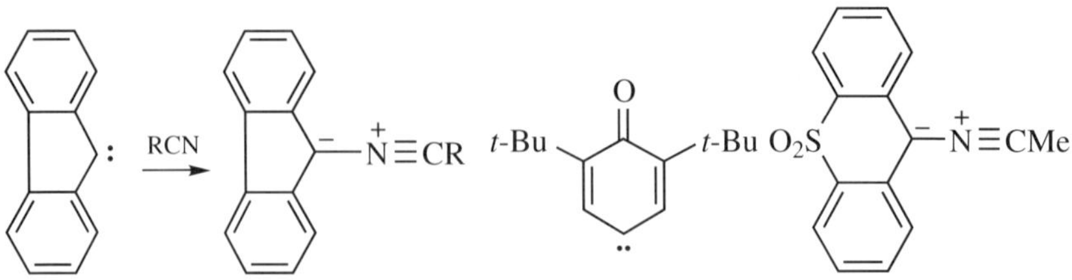

(84) **(85)** **(86)** **(87)**

More recent developments, as discussed in detail below, have shown that the reaction is not limited to stabilized carbenes but appears to be a general route although the equilibrium does not always lie in favor of the nitrile ylide. For example, early work showed that, although the reaction is well established for fluorenylidene (**84**), a nitrile ylide spectrum was not observed when diphenylcarbene precursors were photolysed in acetonitrile (44). Much of the work in this area has utilized LFP to generate the carbenes from diazo compound or diazirine precursors and has been directed at the observation of the spectroscopic properties of the nitrile ylides or the determination of the kinetic parameters of their reactions with dipolarophiles. However, the method does not yet seem to have found significant application in new *synthetic* reactions apart from its use in oxazole synthesis via rhodium acetate generated carbenoids (see Section 7.4.1.1). Strong chemical, kinetic, and spectroscopic evidence has been obtained for the formation of the nitrile ylide (**91**) from 1-napthylcarbene (**90**) and acetonitrile (45,46). The chemical evidence was provided by a trapping experiment in which the diazirine **88** was decomposed by flash photolysis in acetonitrile containing acrylonitrile as a dipolarophile. This gave a mixture of the (*E*) and (*Z*) isomers of **93** identical with that produced using **92** as a source of the *authentic* nitrile ylide. The nitrile ylide was found to have a lifetime in excess of 100 μs and a build up time of 100 ns when acetonitrile was used as a solvent at ambient temperature.

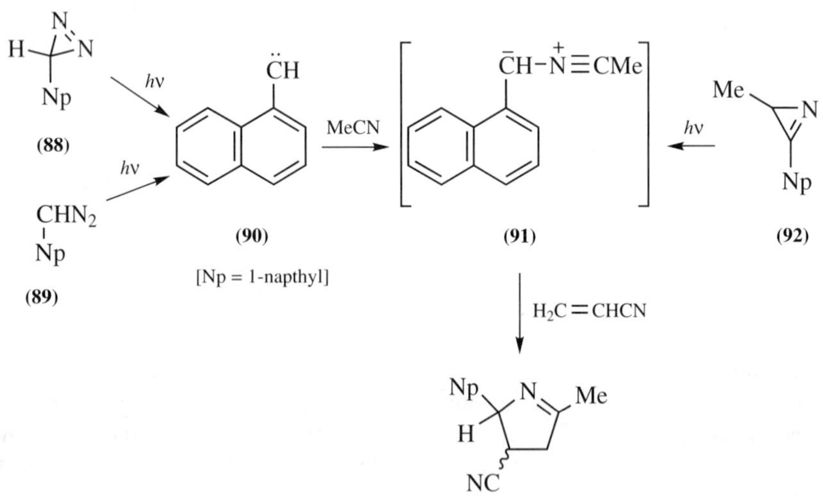

(93)

Confirmation was provided by the observation that the species produced by the photolysis of two different carbene sources (**88** and **89**) in acetonitrile and by photolysis of the azirine **92** all had the same strong absorption band at 390 nm and all reacted with acrylonitrile at the same rate ($k = 4.6 \times 10^5\ M^{-1}s^{-1}$). Rate constants were also measured for its reaction with a range of substituted alkenes, methanol and *tert*-butanol. Laser flash photolysis work on the photolysis of 9-diazothioxanthrene in acetonitrile also produced a new band attributed the nitrile ylide **87** (47). The first alkyl-substituted example, acetonitrilio methylide (**95**), was produced in a similar way by the photolysis of diazomethane or diazirine in acetonitrile (20,21). This species showed a strong absorption at 280 nm and was trapped with a variety of electron-deficient olefinic and acetylenic dipolarophiles to give the expected cycloadducts (e.g., **96** and **97**) in high yields. When diazomethane was used as the precursor, the reaction was carried out at −40 °C to minimize the rate of its cycloaddition to the dipolarophile. In the reactions with unsymmetrical dipolarophiles such as acrylonitrile, methyl acrylate, or methyl propiolate, the ratio of regioisomers was found to be ~1:1.

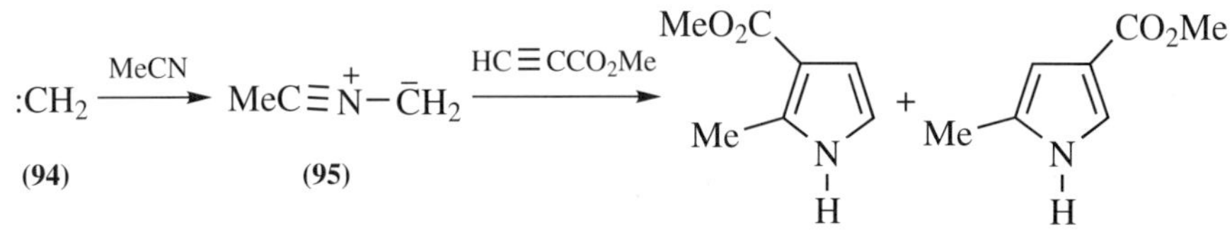

(94) **(95)** **(96)** **(97)**

The heat of formation of **95**, formed by this route, has been determined experimentally using photoacoustic calorimetry (PAC) (48). The reaction was found to be highly exothermic with a $\Delta H_f = 70.8$ kcal mol^{-1}. LFP studies have also been carried out on the generation of phenylcarbene and its perfluoro analogue from diazo compound precursors in the presence of acetonitrile (49). In both cases, the carbenes were found to react with rate constants of $2.4 \times 10^6\ M^{-1}\ s^{-1}$ (in Freon as solvent) to give the expected nitrile ylides that had intense transient absorption peaks (λ_{max} 350 nm). It was concluded that the two carbenes have similar properties (i.e., a triplet ground state in equilibrium with a low-lying singlet) by which the reaction with the nitrile occurs. It was calculated that the lifetimes of the spin-equilibrated species are, respectively, 190 and 500 ns and that both would have a lifetime of 22 ns in neat acetonitrile. In contrast, it was reported in the early 1980s that chloro(phenyl)carbene, a typical singlet carbene, does not produce a nitrile ylide absorption spectrum when generated in acetonitrile. Further studies on the generation of this type of carbene support the conclusion that the reaction does occur, but the equilibrium does not lie in favor of the nitrile ylide. Thus, in the photolysis of the diazirine **98** (biph = 4-Ph–C_6H_4) in the presence of acetonitrile at low temperature in a solid argon matrix (50), new IR bands attributed to the nitrile ylide were observed and it was found that these bands predominated over the carbene absorptions at high acetonitrile/diazirine ratios. The attribution of these new bands to the nitrile ylide was supported by their correspondence with values calculated for the allenyl structure **100** using the RHF/6-31G* basis set.

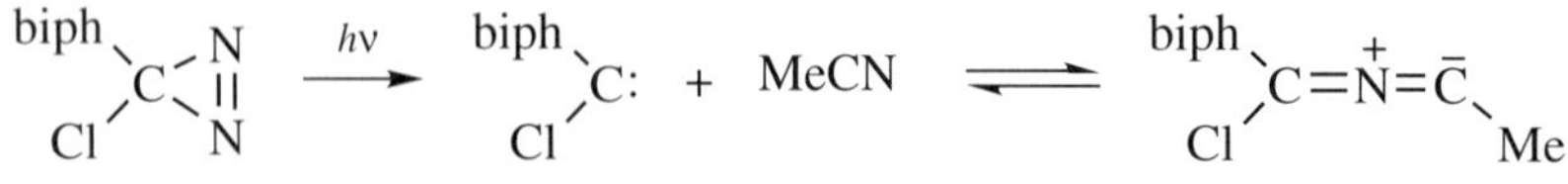

(98) (99) (100)

The interaction of this carbene with a range of nitriles was also studied by LFP (51). In only one case, that of the nitrile **101**, was the concentration of the nitrile ylide high enough to give a measurable absorption spectrum. In the presence of **101** it was found that the carbene absorption at 370 nm decayed with the appearance of a weak absorption in the 420–490-nm range (λ_{max} 440 nm), which was attributed to the nitrile ylide **102**.

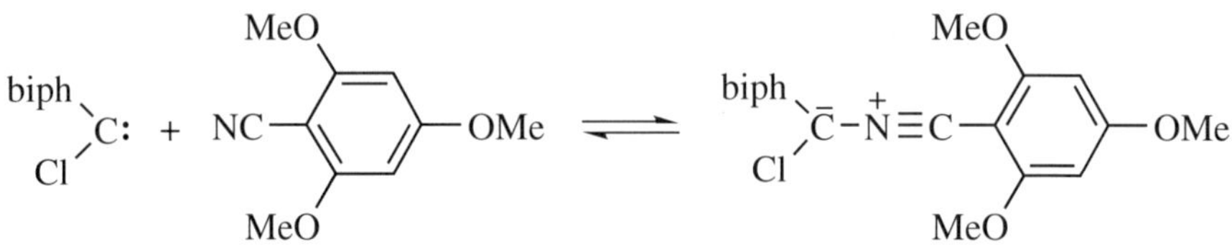

(99) (101) (102)

7.2.1.3. Photolysis of Azirines

Much work has been done since the early 1980s on the detailed investigation of the azirine–nitrile ylide interconversion using pulsed-laser photolysis. Thus the azirines **103** ($R^1 = R^2 =$ Ph, $R^3 =$ H; $R^1 =$ Me, $R^2 = R^3 =$ Ph; $R^1 =$ β-napthyl, $R^2 =$ Me, $R^3 =$ H), on irradiation in isooctane, gave intense long-lived absorptions (250–400 nm) attributed to the nitrile ylides **104** (44). Quenching studies with electon-deficient alkenes led to the determination of absolute rate constants that were similar to those reported earlier for steady-state trapping experiments. The nitrile ylide–olefin reactions are discussed in more detail in Section 7.3.1.

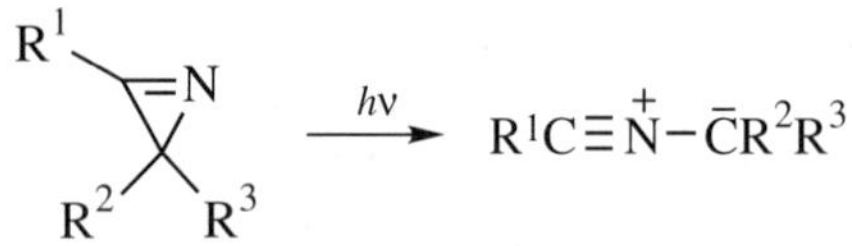

(103) (104)

Theoretical work on the unsubstituted azirine–nitrile ylide system indicates that the latter is formed directly from the $n\pi^*$ excited S_1 state of the azirine via an S_1/S_0 conical intersection (52). Wavelength-dependent experiments using 3-phenylazirine (**103**) ($R^1 =$ Ph; R^2, $R^3 =$ H) showed that the nitrile ylide is formed only via ($n-\pi^*$) excitation (248.5 nm) while excitation at 308 nm ($\pi-\pi^*$) leads to a different transient attributed to a triplet state (5). In the case of 2,3-diphenylazirine, the $n\pi^*$ and $\pi\pi^*$ states were found to be nearly degenerate and no wavelength-dependent photochemistry was observed. Experimental studies on the mechanism have also been carried out via LFP of 3-(4-biphenylyl)-2*H*-azirine (**105**) in cyclohexane (53,54). Studies of triplet sensitization and of reactions with oxygen and with propanone indicate that the excited singlet state $^{1}AZ^{*}$ decays rapidly into either the nitrile ylide (**106**) or the triplet state of the azirine $^{3}AZ^{*}$, which has a lifetime of 1.5 μs. Since the stational photoirradiation of the azirine in benzophenone did not give any adducts, it was concluded that the structure of $^{3}AZ^{*}$ was the closed form and decayed to the ground state. The nitrile ylide reacted with propanone and acrylonitrile with rate constants of 2.5×10^{3} and 2.3×10^{6} dm^{3} mol^{-1} s^{-1}, respectively.

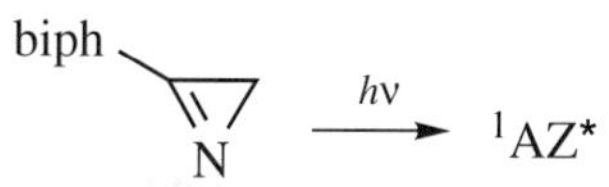

(105) (AZ)

$$^{1}AZ^{*} \longrightarrow {}^{3}AZ^{*}$$

$$^{1}AZ^{*} \longrightarrow \text{biph-C}{\equiv}\overset{+}{N}{-}\bar{C}H_2$$

(106)

The photochemical conversion of 3-phenylazirine into benzonitrilio methylide in a nitrogen matrix at 12 K was found to be a completely reversible process (4). In contrast, however, it is reported that acetonitrilio fluorenylide **85** (R=Me) is photostable at 77 K and does not convert into the corresponding azirine (55). This stability was attributed to a steric limitation on bond rotation.

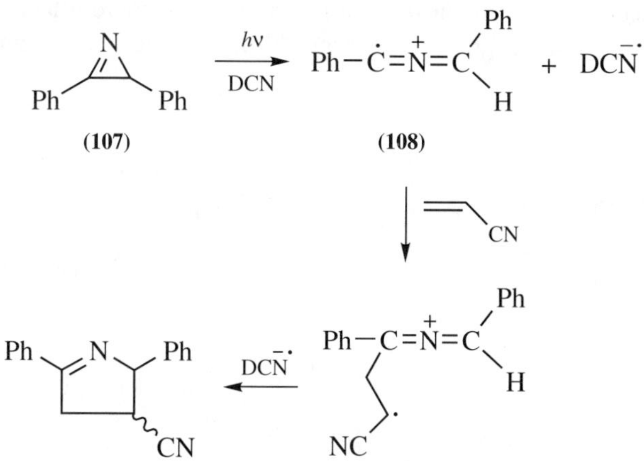

(107) **(108)** **(109)** **(110)**

The ring opening of azirines under conditions of photoinduced electron transfer (PET) gives, not a nitrile ylide, but the short-lived 2-azaallenyl radical cation **108**

(56). Irradiation with 350-nm light in the presence of the electron acceptor, 1,4-napthalenedicarbonitrile (DCN) effected electron transfer from the azirine to the excited sensitizer. After losing one electron, the azirine ring opened to give **108**, which was found to react with acrylonitrile with a complete lack of diastereoselectivity to give **109**. It was suggested that the reaction followed the stepwise process shown. The mechanism was supported by trapping experiments with trifluoroethanol.

In contrast, photolysis followed by γ-irradiation of 3-(4-biphenylyl)-2*H*-azirine **105** resulted in the formation of the nitrile ylide radical anion **111** (57). Two mechanisms were suggested, either the one shown in the scheme, or *via* the photochemical opening of the azirine radical anion.

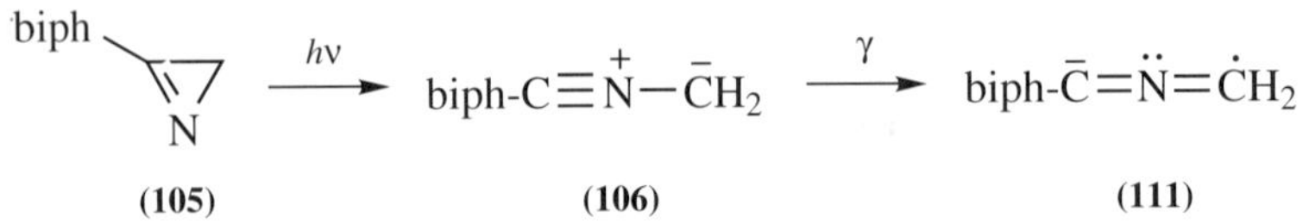

7.2.1.4. *Other Routes to Nitrile Ylides*

The Dehydrochlorination of Imidoyl Chlorides

This method has been used extensively for the generation of diene- and triene-conjugated nitrile ylides (see Section 7.4.1.2) using strong bases. Potasium *tert*-butoxide (58) was used for the most part but in recent work (59,60) it was reported that lithium bis(trimethylsilyl)amide is both more effective and more convenient. It has also been shown that thermally unstable imidoyl chlorides for use in these reactions can be prepared by reaction of the corresponding amides with chlorodimethylformiminium chloride at 0 °C, a reaction that is more effective than using either thionyl chloride or phosphorus pentachloride at higher temperatures (61).

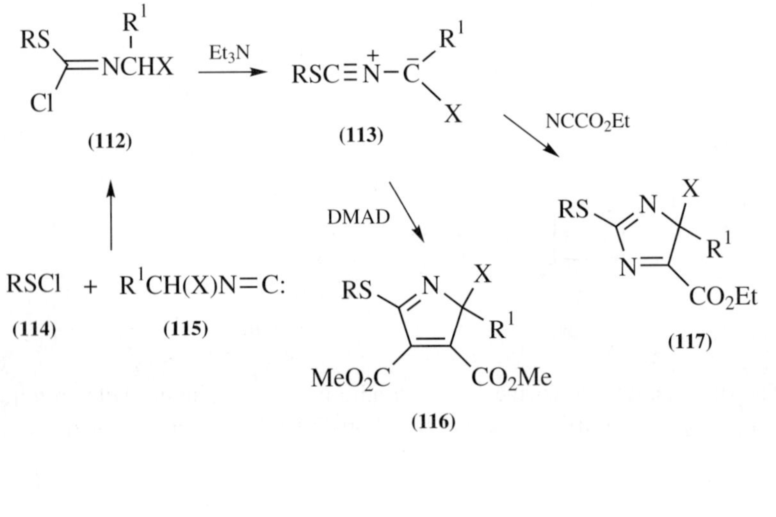

The thioalkyl substituted nitrile ylides (**113**) (*thiocyanate* ylides) have been generated from imidoyl chlorides (**112**) that contain an electron-withdrawing group X [X=CN, 4-NO_2—C_6H_5, Tos, $P(O)(OEt)_2$; R^1= H, Me] by treatment with triethylamine (62–64). They were trapped in moderate yields using DMAD and ethyl cyanoformate to give **116** and **117**, respectively, and dimethyl fumarate. The precursors **112** were easily prepared by reaction of the appropriate sulfenyl chloride **114** (R=Me, Ph, 2,4-di-NO_2—C_6H_3, 2-NO_2—C_6H_4, 4-Cl-2-NO_2—C_6H_3, 4-Cl—C_6H_4), with the appropriate isonitrile (**115**). In the cases where X=Tos or $P(O)(OEt)_2$, it was found that the reaction was best carried out in heterogeneous medium (Al_2O_3/KOH) where the intermediates could be trapped to give adducts in moderate to good yield (64). The reactions with diethyl fumarate led, in cases where X=Tos, to pyrroles *via* cycloaddition and subsequent elimination of *p*-toluenesulfinic acid and/or thiophenols.

Phosphoryl substituted nitrile ylides have also been generated via the imidoyl chloride–base route using precursors **118** (R=Et, *c*-C_6H_{11}, *t*-Bu) prepared by the addition of an acid chloride to diethyl isocyanomethylphosphonate (**120**) (65). Treatment of the imidoyl chloride with triethylamine at −10 to 0 °C in the presence of dipolarophiles gave adducts in yields of up to 55% (e.g., **119** and **121**) in ratios 1:4, 1:3 for R=Et and *c*=C_6H_{11}, respectively.

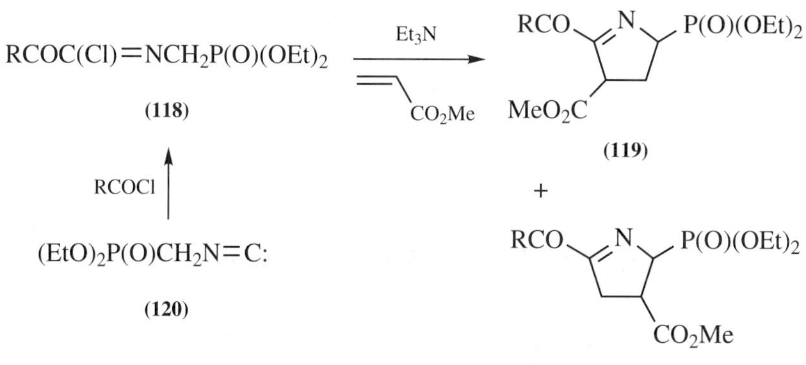

Miscellaneous Routes to Nitrile Ylids

The cyano-substituted nitrile ylides **123** have been generated via 1,1-elimination reactions. For example, the benzylidene derivative **122** (R=Ph) eliminated benzene on vapor phase pyrolysis to give **123** (R=Ph), which reacted via 1,5-electrocyclization [see also (66)] to give the isoindole **124** (41%) (67). In a similar way, **122** [R=$(CH_2)_3CH{=}CH_2$] gave the corresponding nitrile ylide that reacted *via* intramolecular cycloaddition to give the pyrroline derivative **126**.

Felhammer and co-workers (68–71) (and references cited therein) has shown that metal coordinated α-deprotonated isocyanides (e.g., **127** and **128**) are genuine 1,3-dipoles of the nitrile ylide type that react with various dipolarophiles by [3 + 2]

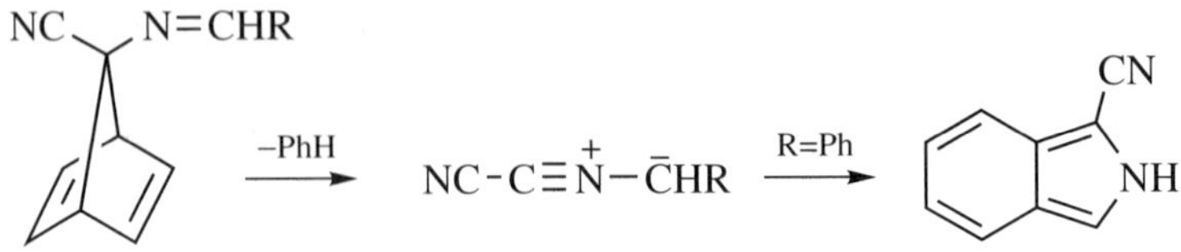

(122) **(123)** **(124)**

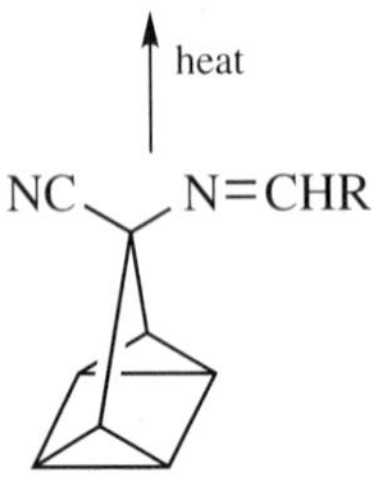

(125)

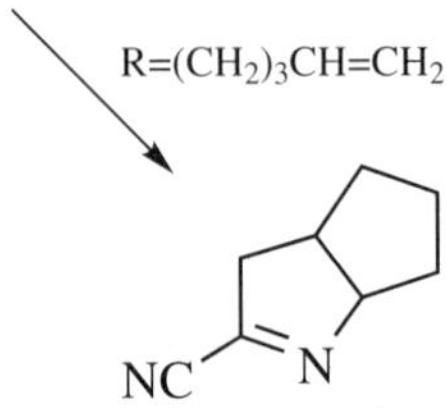

(126)

cycloaddition to give C-metalated five-membered heterocycles including pyrroles, pyrrolines, imidazoles, oxazoles, oxazolines, and thiazoles carrying metal complex substituents.

$$[(CO)_5MC{\equiv}N{-}CHR]^- \quad [Ph_3BC{\equiv}N{-}CHR]^-$$

(127) **(128)**

(M = Cr,W; R = CO_2Et, Tos)

7.2.2. Generation of Nitrile Imines

7.2.2.1. Transient Nitrile Imines

Of the general methods for the generation of transient nitrile imines for use in synthesis (19), perhaps the most convenient are the base-induced dehydrochlorination of hydrazonyl chlorides and the oxidation of hydrazones. Developments in both of these areas have either increased the convenience of the method or given a deeper insight into the reaction mechanism.

Generation from Hydrazonyl Chlorides

$$R^1{-}NH{-}N{=}\underset{\displaystyle Cl}{C}{-}R^2 \xrightarrow[-HCl]{\text{base}} R^1{-}\bar{N}{-}\overset{+}{N}{\equiv}C{-}R^2$$

(129) **(130)**

This reaction is one of the most versatile and, hence, most extensively used routes to nitrile imines (**130**). It does, however, have the disadvantage that some hydrazonyl halides (**129**) are skin-active and exposure can lead to sensitization and allergy. Greater convenience and a solution to this problem can be achieved by using *in situ* preparation in a *one-pot* chlorination–dehydrochlorination procedure (72). Thus the treatment of *N*-acylhydrazines with triphenylphosphine–hexachloroethane (Ph_3PCl_2) in the presence of triethylamine and the dipolarophile leads to the expected cycloadducts in high yields (e.g., **133** from **131**). The method is effective for cycloaddition to both olefinic and acetylenic dipolarophiles and for 1,5-electrocyclization reactions of conjugated nitrile imines.

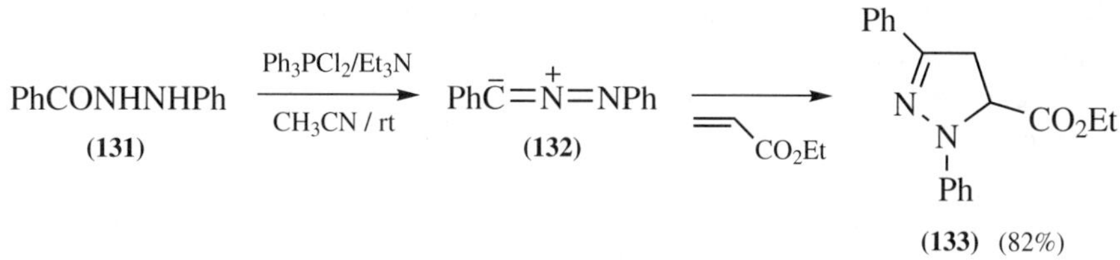

Silver carbonate, when used as the base in the dehydrochlorination of hydrazonyl chlorides, has a unique effect in enhancing yields and favoring particular modes of reaction, for example, in the promotion of the tandem intermolecular–intramolecular synthesis of macrocycles **302** discussed in Section 7.3.2.4 (73). Such differences have led to some investigations into the mechanism of its reaction. In this work (74), the hydrazonyl chloride **134** was reacted with 2 *M* equiv of silver carbonate in the presence of various allylic alcohols.

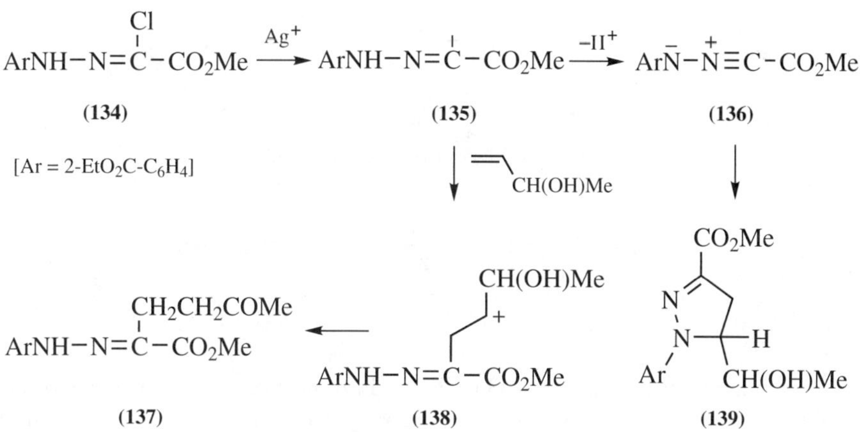

In all cases, the generation of a true nitrile imine was confirmed by the formation of the expected cycloadducts (e.g., **139**) with the expected regioselectivity. However, in several cases, carbonyl derivatives (e.g., **137**) were also formed in relatively low yield. These compounds were not formed when triethylamine was used as the base. From these data, it was suggested that the dehydrochlorination is a two-step process, involving first silver-ion promoted dehalogenation to give the carbocation

135, followed by deprotonation to give the nitrile ylide. Compound **137** is derived by interception of the carbocation **135** by reaction with the alkene to give **138**, followed by a pinacol rearrangement. These conclusions were supported by further work using homoallylic and homopropagylic alcohols (75).

The use of water as a medium for nitrile imine cycloaddition reactions using the imidoyl chloride–base generation method has been investigated. It was found that satisfactory yields of cycloadducts could be obtained provided that a cationic surfactant was used and that sufficient of an organic cosolvent (e.g., THF) was added to obtain a homogeneous medium (76). It has also been reported that microwave heating can give enhanced yields of cycloadducts and shorter reaction times in the reaction of hydrazonyl chlorides with *N*-methylmorpholine in the presence of alkenes, and when using alumina as the base under solvent-free conditions (77).

The use of organomagnesium reagents as bases leads to complexation of the nitrile imines (e.g., **141**), which has been found to have a strong effect in promoting syn selectivity in reactions with methyl 2-(1-hydroxyalkyl)acrylates via coordination of the metal atom with the alcoholic oxygen (e.g., leading to the formation of **142**). Lithium complexation had little effect (78).

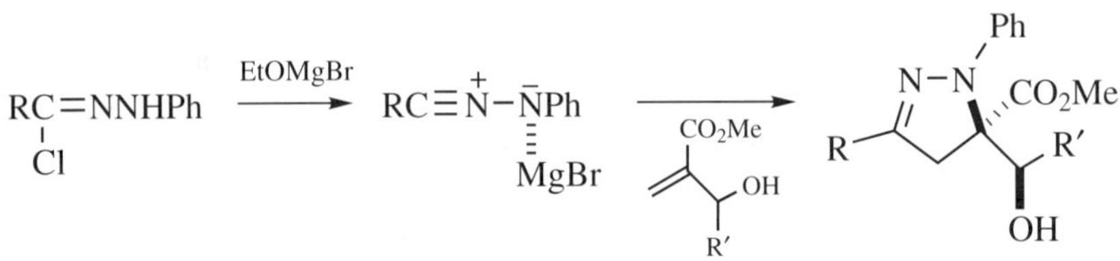

(140) **(141)** **(142)**

The Oxidation of Hydrazones

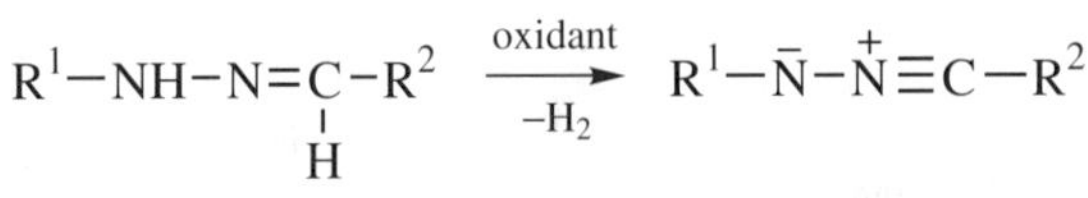

(143) **(144)**

The oxidation of hydrazones **143** provides, in principle, a very convenient route to nitrile imines from easily accessible starting materials. However, the earliest reagent used, lead tetraacetate, was of limited effectiveness as yields were low and the reaction often gave high yields of diacylhydrazides as byproducts. Work has been done on the application of several other oxidants to this process to produce a more effective general route. The one that has proved most popular is chloramine T (*N*-chloro-*N*-sodio-*p*-toluenesulfonamide, CAT) which is used under mild conditions and has been shown to work well for both cycloaddition (79) (e.g., in the preparation of **146** from **145**) and electrocyclization (80) reactions.

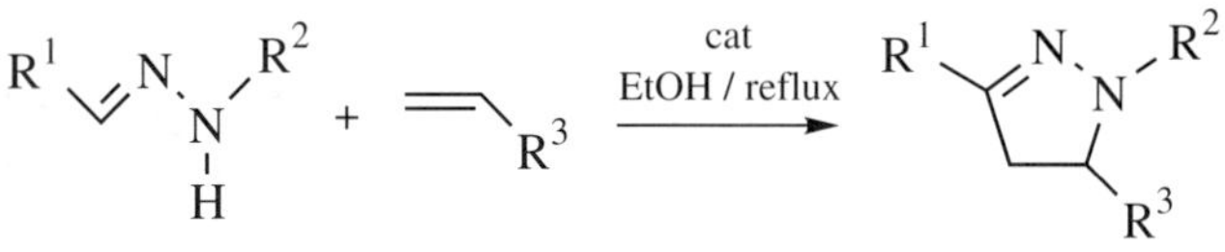

(145) **(146)** (70–90%)

R^1 = Ar, *i*-Pr, C_5H_{11}; R^2 = Ph, 4-$NO_2C_6H_4$; R^3 = CN, CO_2Et, Ph

2,3-dichloro-5,6-dicyno-1,4-benzoquinone mercuric acetate (81), phenyliodine diacetate (82), and (DDQ) 2,3-dichloro-5,6-dicyano-1,4,-benzoquinone (83) have also been used.

7.2.2.2. Stabilized Nitrile Imines

Two general methods are available for the assembly of the sterically stabilized species **16–24**, both starting from metallic derivatives of diazo compounds (**147**) (16). The latter have two nucleophilic centers and can, in principle, react with electrophiles at C giving the functionalized diazo compounds (**148**), or at N, which leads to the nitrile imines (**149**).

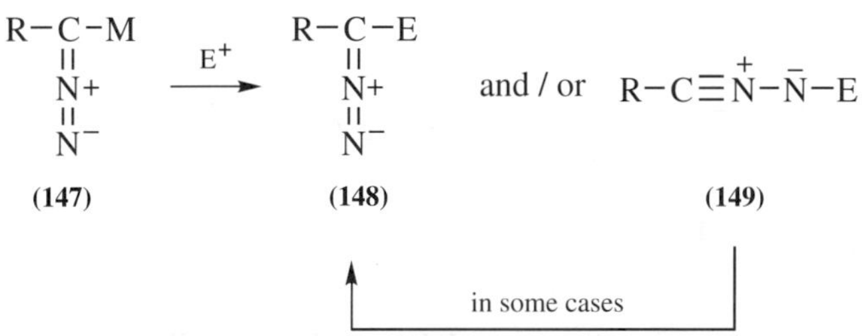

(147) **(148)** **(149)**

In principle, it would appear that the presence of bulky substituents R and E should kinetically favor the formation of nitrile imines. Lithium salts have been used for much of this work and it has been shown that this is generally the case [e.g., for **151** (X=S)] bulky electrophiles favor the nitrile imine (e.g., **153**) while smaller analogues favor diazo compounds (e.g., **152**). Reducing the size of the substituent on the diazo compound has a similar effect (e.g., **150** from **151**) (X=lone pair).

Stannyl diazo compounds are less reactive than their lithium analogues and, containing a bigger metal atom, strongly favor reaction at *N* to give nitrile imines (e.g., **155** from **154**). The use of bis(trimethylstannyl)diazomethane (**156**) provides an easy one-step route to symmetrically substituted systems (e.g., **157**).

Both experiment and calculations have shown that the diazo compounds are the more thermodynamically stable isomers and in some cases, notably for *N*-silyl substituents, thermally induced isomerizations from **149** to give **148** can be an important factor.

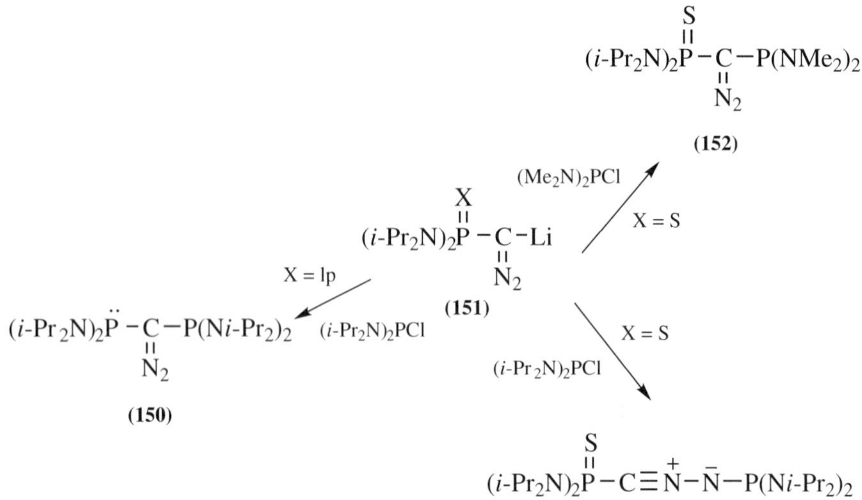

(150) (151) (152) (153)

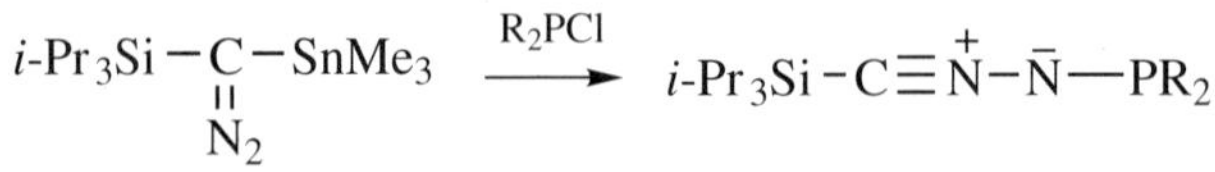

(154) (155)

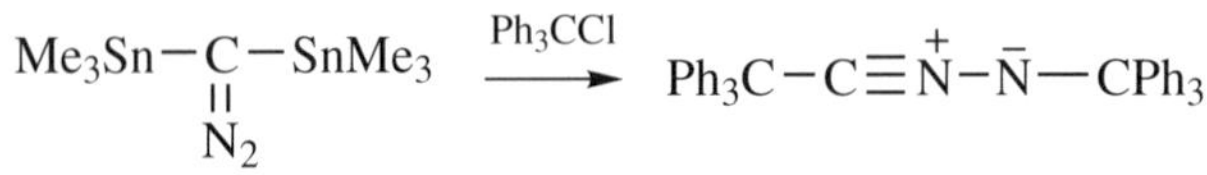

(156) (157)

7.3. [2 + 3]-CYCLOADDITION REACTIONS

7.3.1. Cycloaddition Reactions of Nitrile Ylides

Pulsed-laser photolysis of the azirines **158a–c** in the presence of electron-deficient alkenes (44,66) allowed the determination of the bimolecular quenching rate constants (k_q) for reactions with acrylonitrile (1.0–5.4 $\times$ 10^6 M^{-1} s^{-1}) and fumaronitrile (0.5–7.4 $\times$ 10^9 M^{-1} s^{-1}). The relative value for azirine **158a** was close to that obtained in earlier steady-state trapping experiments. Temperature studies for the reactions of the nitrile ylide **158a** showed that k_q for methyl acrylate is insensitive to temperature while that for diethyl fumarate initially increases with decreasing temperature and then begins to decrease at temperatures below −65 °C. This type of Arrhenius behavior, previously observed for the quenching of phosphorescence from ketones, carbenes, and singlet oxygen, was explained as being consistent with a kinetic scheme involving the reversible formation of a nitrile ylide–alkene complex **160**.

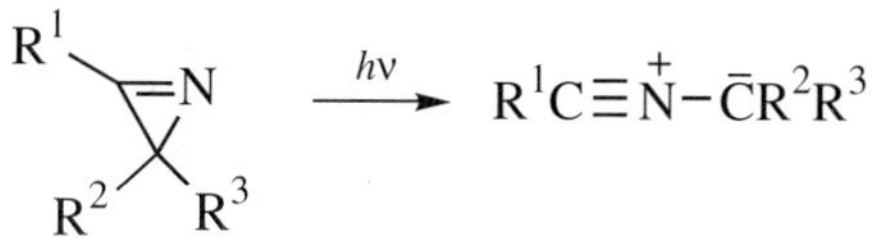

(158) **(159)**

(a) $R^1, R^2 = Ph, R^3 = H$
(b) $R^1 = Me, R^2, R^3 = Ph$
(c) $R^1 = \beta$-napthyl, $R^2 = Me, R^3 = H$

$$\text{NY} + \text{alkene} \underset{k_{-1}}{\overset{k_1}{\rightleftharpoons}} \underset{\textbf{(160)}}{\text{NY / alkene}} \xrightarrow{k_2} \text{product}$$

Some other data showing substituent effects on the rate constants for cycloadditions with electron-deficient alkenes is given in Table 7.2 (5).

The mechanism of the reaction of nitrile ylides with the carbonyl group has been studied via LFP of 3-biphenyl-4-yl-2*H*-azirine (**105**) (84). The nitrile ylide **106** was found to decay according to pseudo-first-order kinetics and the rate constants for cycloaddition with a range of aldehydes, ketones, esters, and carbon dioxide were measured. The results were correlated with the ionization potentials of the carbonyl compounds in a Sustmann plot. From this it was concluded that the major frontier molecular orbital (FMO) interaction was that between the nitrile ylide highest occupied molecular orbital (HOMO) and the dipolarophile lowest unoccupied molecular orbital (LUMO) and that the mechanism was the same as that (above) suggested by Padwa for alkene cycloaddition. The linear relationship between the log k value and the π ionization potential (IP) was taken to indicate that the reaction proceeds through a complex of configuration **161** rather than **162**. It was suggested that Padwa's "intermediate complex" **160** was also of this type.

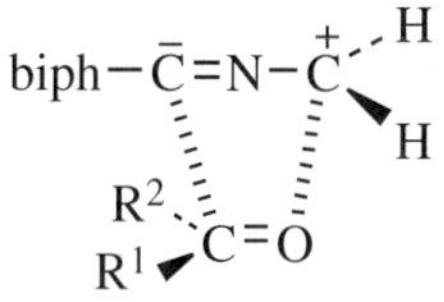

(161)

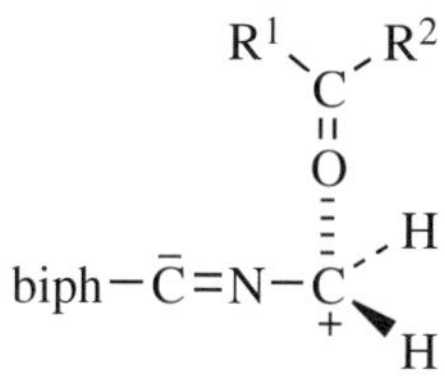

(162)

TABLE 7.2. RATE CONSTANTS (s^{-1} M^{-1}) FOR CYCLOADDITION REACTIONS OF SUBSTITUTED BENZONITRILE YLIDES

		4-X-Benzonitrilio methylide			
	DPNY	Cl	H	Me	MeO
Acrylonitrile	8.9×10^5	9.4×10^5	1.4×10^6	1.9×10^6	2.2×10^6
DMAD	1.2×10^7	1×10^7	1.7×10^7	7.4×10^7	
TCNE		3.5×10^9	5×10^9	5.9×10^5	6.9×10^9

(TCNE = tetracyanoethene)

Theoretical calculations on the cycloaddition reactions of a range of 1,3-dipoles to ethene in the gas phase have been carried out (85) with optimization of the structures of these precursor complexes and the transition states for the reactions at the B3LYP/6-31G* level. Calculated vibration frequencies for the orientation complexes revealed that they are true minima on the potential energy surface. The dipole–alkene bond lengths in the complexes were found to be about twice that in the final products and binding was relatively weak with energies <2 kcal mol^{-1}. Calculations on the cycloaddition reactions of nitrilium and diazonium betaines to ethene indicate that the former have smaller activation energies and are more exothermic.

Some interesting work has been carried out on the generation and reactions of nitrile ylides (e.g., **164** and **171**) with different substituents at the *nitrile* carbon. The effect of such substituents on regioselectivity in cycloaddition reactions throws

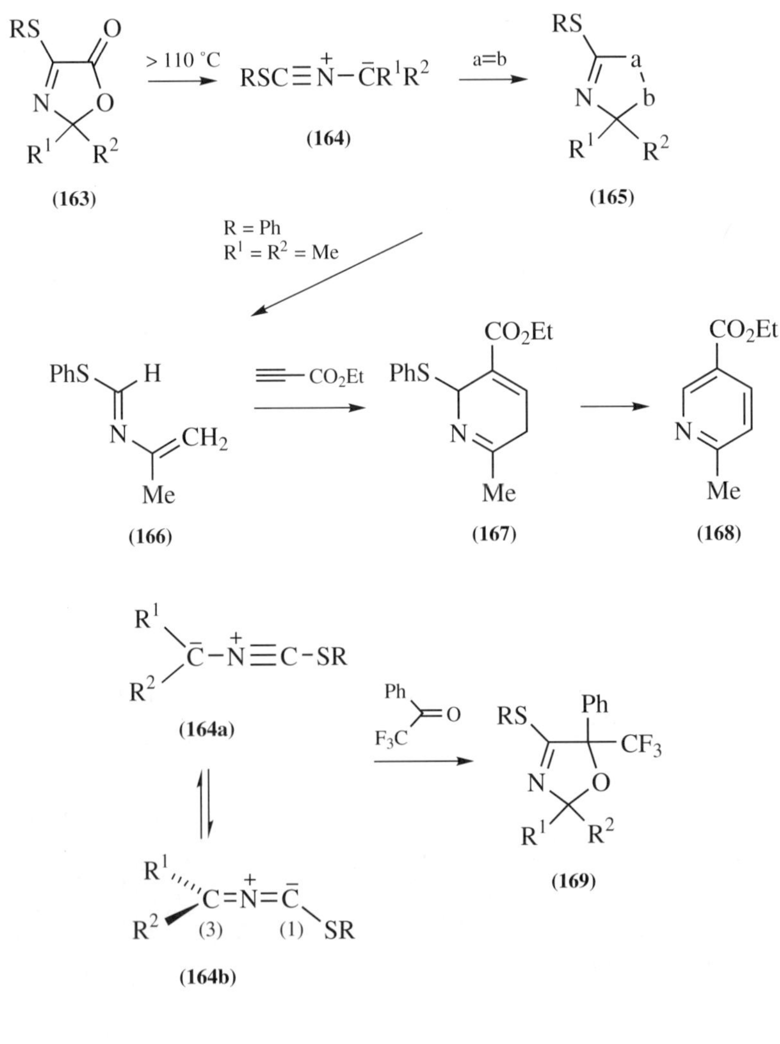

some light on their effect on the geometry of the nitrile ylides. The thio-substituted nitrile ylides (**164**) (R = Ar, $PhCH_2$, and R^1, $R^2 = (CH_2)_{3\ \text{or}\ 5}$; Me, Me; Ph, Me) were generated by the thermolysis (>110 °C) of 4-substituted 3-oxazolin-5-ones (**163**) (86,87). They were trapped unexceptionally in moderate to good yield (35–97%) by reactive dipolarophiles such as dimethyl acetylenedicarboxylate, diethyl azodicarboxylate, diethyl fumarate, and 4,4-dimethyl-2-phenyl-1,3-thiazol-5-thione (88). In the absence of dipolarophiles, the examples where [R^1, $R^2 = (CH_2)_5$, and Me, Me] reacted by isomerization to give 2-azabutadienes (e.g., **166**), which were subsequently trapped by reaction with ethyl propiolate to give, initially, **167**, which eliminated thiophenol to give **168**. The most interesting feature of this chemistry, however, is that the cycloadditions with trifluoroacetophenone proceeded regiospecifically to give the isomer **169**. This result was expected for the examples where R^1 and R^2 were electron-donating groups since these are predicted by the Houk calculations to favor the bent allene-type structure **164b**, which has the greatest HOMO coefficient on the nitrile C [(C1)]. However, the observation of the same regioselectivity for **164** (R^1 = Ph, $R^2 = CF_3$) was unexpected since electron-withdrawing groups would be expected to favor the linear propargyl geometry **164a**, where the largest HOMO coefficient is on [(C3)]. Thus it would appear that **164** (R^1 = Ph, $R^2 = CF_3$) favors the allenyl geometry, thus leading to speculation by the authors that the presence of the thio substituent on [(C1)] may be overwhelming the effect of the electron-withdrawing CF_3 group.

In an attempt to investigate this theory further, the analogous alkoxy-substituted nitrile ylides **171** were investigated. Species of this type were previously unknown but were successfully generated for this work by the thermolysis of the 4-alkoxy substituted 1,3-oxazol-5-ones **170** (89). In the absence of a dipolarophile, the nitrile ylides reacted *via* 1,5-electrocyclization to give the isoindoles **173** [see also

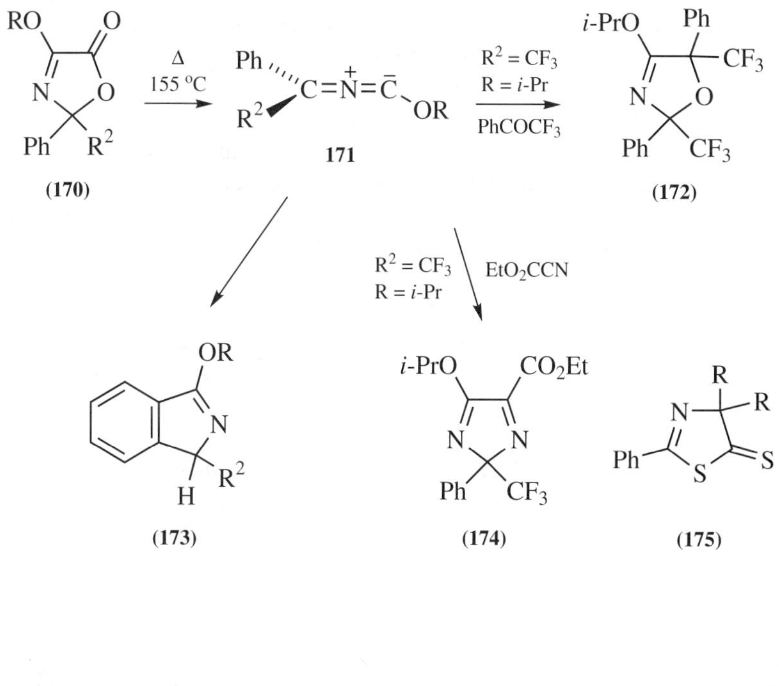

TABLE 7.3. BOND LENGTHS AND CALCULATED GEOMETRY FOR **176/177**

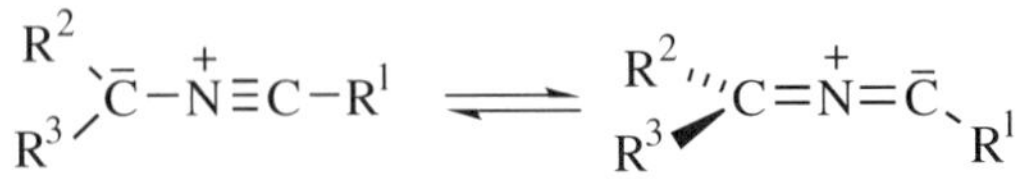

(176) **(177)**

	$R^2, R^3 = Me$				$R^2, R^3 = CF_3$			
R^1	R^1–C–N	C–N–C	C(1)–N	N–C(3)	R^1–C–N	C–N–C	C(1)–N	N–C(3)
Me	145.8	166.4	1.209	1.296	178.3	179.0	1.170	1.316
OH	117.1	163.8	1.288	1.270	121.4	159.9	1.284	1.262
SH	141.5	168.0	1.218	1.289	175.1	178.4	1.182	1.301

[a] R^1 = Me, OH, SH WITH R^2, R^3 = Me and CF_3 and bond angles are in degrees and bond lengths are in angstroms.

(66,67)]. However, when generated in trifluoacetophenone, **171** (R = *i*-Pr, $R^2 = CF_3$) reacted by a regiospecific but not stereospecific route to give the adducts **172**. Thus it was shown that the same strong regioselectivity is obtained for the 1-alkoxy substituted nitrile ylides as for their 1-phenyl, -arylthio, and -amino analogues. This nitrile ylide also gave adducts with ethyl cyanoformate (36%) (only the regioisomer **174**) and with DMAD (48%), diethyl azodicarboxylate (19%), and dimethyl fumarate (23%) (90). Only in the case of cycloaddition to 2-phenyl-3-thia-1-azaspiro[4,4]non-1-ene-4-thione (**175**) [RR = $(CH_2)_4$] did the reaction prove to be nonregioselective (90,91). In order to rationalize the regioselectivity results, theoretical calculations were carried out to determine how the geometry of nitrile ylides is affected by both the nature of the [(C1)] substituent and of the electron-withdrawing–electron-donating properties of the substituents at [(C3)] (90). The results (Table 7.3) show that the effect of Me and SH at the nitrile carbon are similar but that the dipoles with the OH group are significantly more bent in both cases and, even with $R^1R^2 = CF_3$, seem to prefer the bent allenyl structure.

In the first addition of nitrile ylides to α,β-unsaturated lactones (92), it was found that the reaction of benzonitrilio 4-nitrobenzylide with the lactones **178–180** were strongly regioselective. Compounds **178** and **179** reacted to give [e.g., **181** from **178** (51%)] and its methyl analogue from **179** (54%). The exocyclic double bond in **180** was, however, the most reactive and gave **182** (69%). The six-membered analogue of **178** was less reactive (34%) and the seven-membered analogue failed to react.

It has also been shown (93) that the additions of benzonitrilio benzylide to the substituted analogues **183** and its (*E*) isomer were not only highly regioselective but also highly stereoselective. Thus in both cases, NMR examination of the product showed that only a single diastereoisomer (e.g., **184**) was obtained.

Nitrile ylides, generated by the imidoyl chloride–base route, have been added to 1-azetines (**185**) (X = O, S) to give the adducts **186** in moderate to good yields (42–68%) (94,95). These examples are among the first cycloadditions to 1-azetines. In the case where Ar = Ph, the NMR spectrum of the product showed only one set

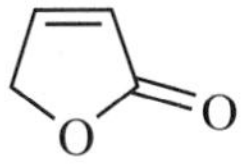
(178)

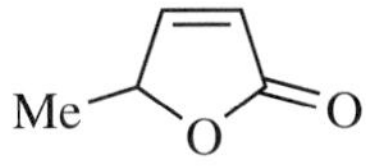
(179)

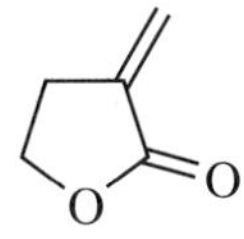
(180)

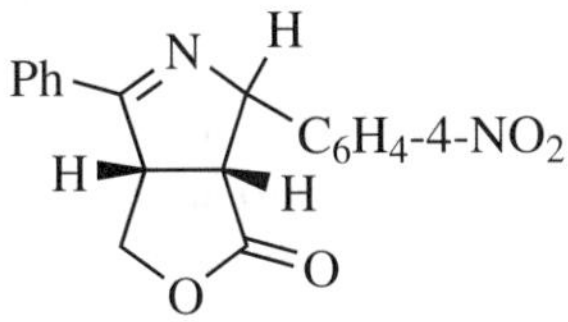
(181)

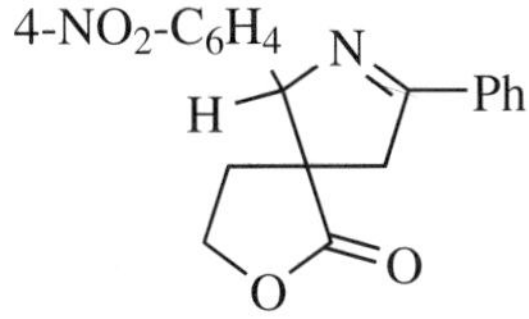
(182)

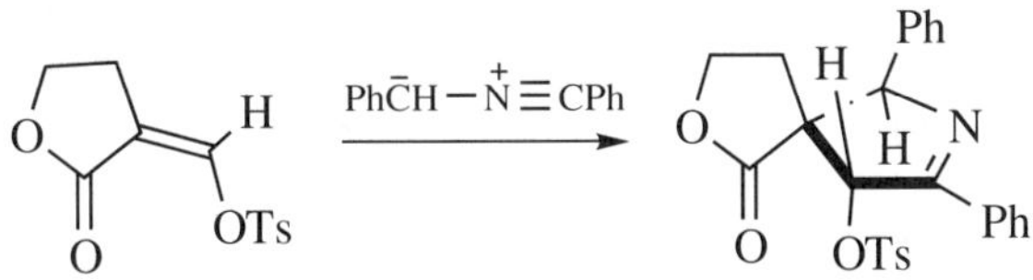
(183) (184)

of diastereoisomers and X-ray crystallography showed that the ethoxy and the 4-nitrophenyl groups were *trans* related.

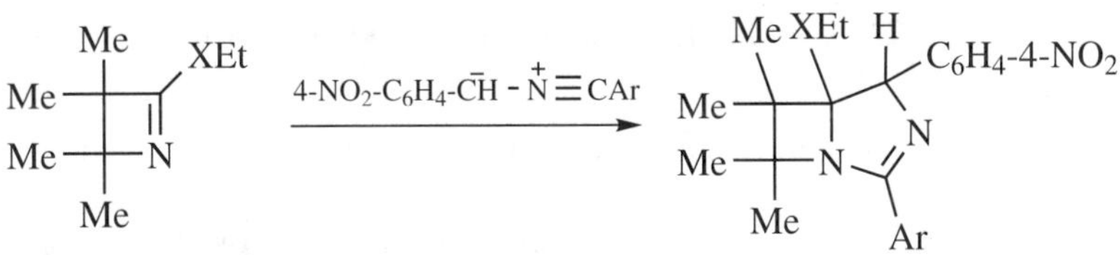
(185) (186)

Extensive work has been done to determine and understand the factors controlling diastereoselectivity in the cycloaddition of nitrile oxides to alkenes but very little is known about nitrile ylides in this regard. Work on their reactions with alkenes that are geminally disubstituted with electron-withdrawing groups (e.g., **187**) has illustrated some of the difficulties in such studies. When the imidoyl chloride–base route was used to generate the nitrile ylides it was found that the products **188** epimerized under the reaction conditions. When the azirine route was used, the reaction was complicated by the photochemical isomerization of the dipolarophiles (96,97). Thus, in both cases, it proved impossible to determine the *kinetic* product ratio.

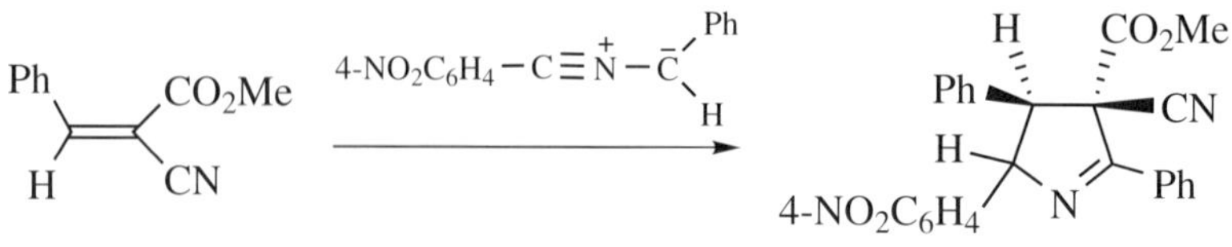

(187) **(188)**

Irradiation of 2,3-diphenyl-2*H*-azirine in the presence of C_{60} fullerene leads to the formation of mono- and oligo adducts (98,99). A monoadduct, 1,9-(3,4-dihydro-2,5-diphenyl-2*H*-pyrrolo)fullerene-60 was isolated and characterized. Mechanistic studies showed that under conditions of direct irradiation it was formed by a classic nitrile ylide cycloaddition but in the presence of 1,4-napthalenedicarbonitrile (DCA) it resulted from reaction of the radical cation intermediate **108**. Cycloaddition reactions have also been carried out with diazaphospholes and diazaarsoles (100) to give adducts of the type **189** (A = As,P) and with cyanogen to give **190** and with aryldiazocyanides where addition to both the azo moiety and the cyano group were observed (101).

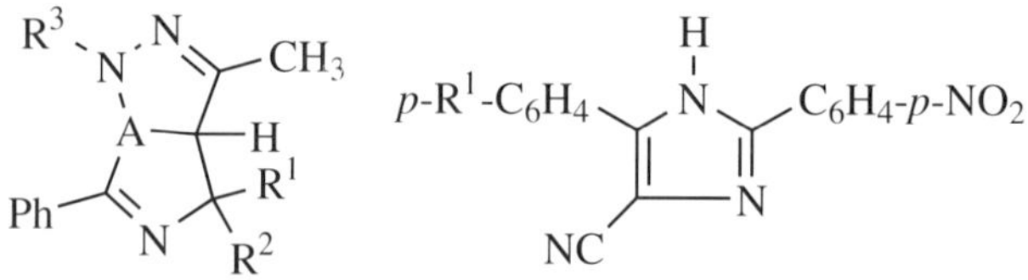

(189) **(190)**

7.3.2. Cycloaddition Reactions of Nitrile Imines

Variously substituted nitrile imines are easily available and react readily with a wide range of double and triple bonds. Intermolecular cycloaddition is therefore an area of major interest, and a large proportion of the papers on the use of nitrile ylides in synthesis is concerned with the exploitation of this reaction. Space limitation means, regrettably, that work leading to results that were predictable on the basis of known chemistry (19) has generally not been included.

7.3.2.1. Cycloaddition of Stabilized Nitrile Imines

One of the most remarkable recent developments in nitrile imine chemistry was the discovery that these species could be stabilized by the incorporation of appropriately large substituents at the C and N termini. This finding resulted in the preparation of a range of compounds (e.g., **16–24**) with high enough levels of thermal and air stability to allow them to exist as isolable compounds with quite high melting points (see Section 7.1.2). It is interesting therefore to find that these compounds, in spite of their stability, still retain useful levels of typical 1,3-dipole reactivity in cycloaddition reactions (16). For example, **16** gave the expected cycloadducts on reaction at room temperature (102,103) with methyl acrylate (55%), methyl propiolate (50%), dimethyl fumarate (70%), methyl isocyanate (65%), *N*-phenylmaleimide (**191**, 85%), methyl vinyl ketone (82%), and 1,4-naphthaquinone (72%). Example **23** without heteroatoms gave rather higher yields under milder reaction conditions [e.g., methyl acrylate (**192**, 90%), methyl propiolate (85%), and dimethyl fumarate (91%)] (104). However, their reactivity characteristics are somewhat different from those of transient nitrile imines. The latter are Sustmann Class II 1,3-dipoles (19), which means that in their reactions with alkenes and alkynes, they are reactive to those that are electron rich and electron poor, but least reactive those without electron-withdrawing or -donating groups. The stabilized nitrile imines are less ambiphilic in nature and are moved toward either dipole (HO) control or dipole (LU) control depending on the nature of their substituents (103).

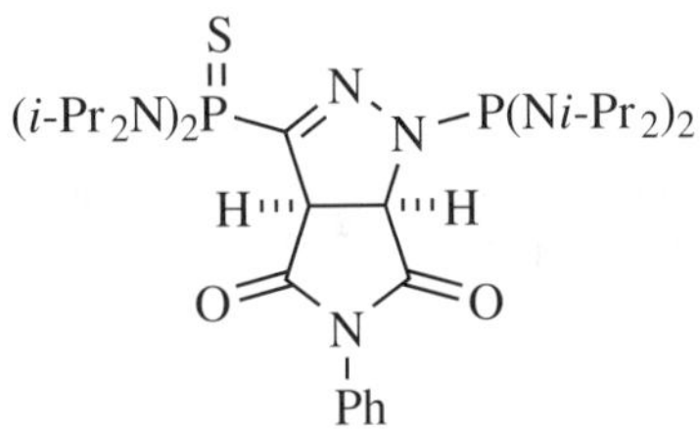

(191)

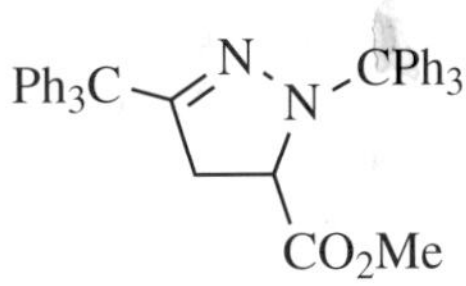

(192)

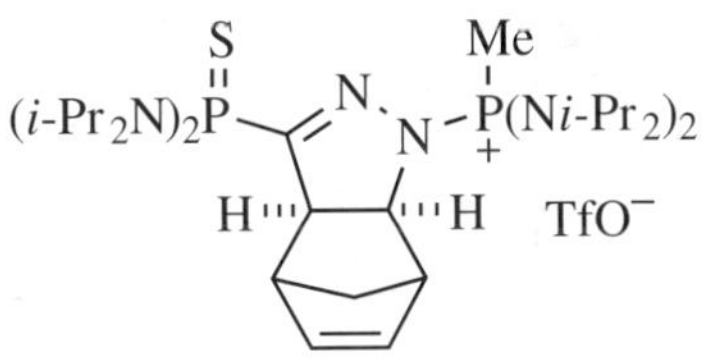

193

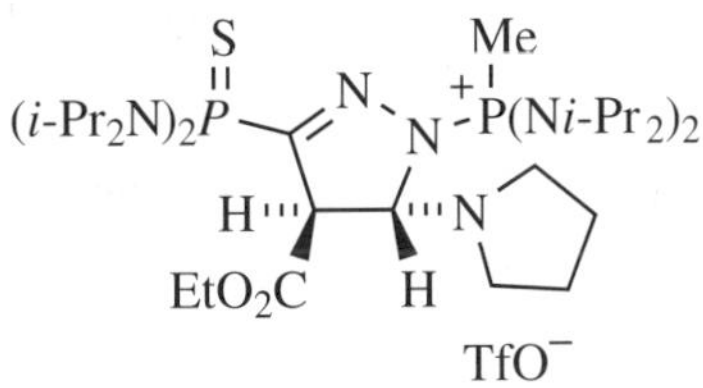

(194)

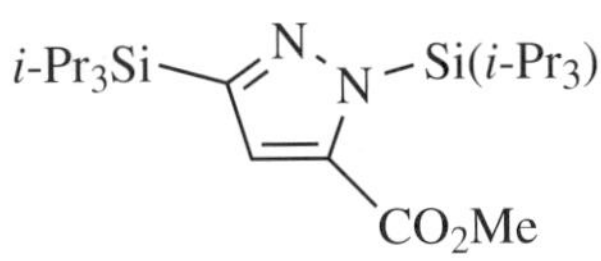

(195)

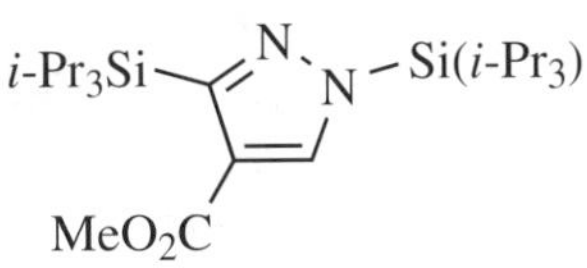

(196)

Thus examples such as **16** that have a phosphino substituent on the N atom are more nucleophilic in character and react readily with electron-poor alkenes and alkynes (see above) but fail to react with those that are electron rich. In contrast, the ones with a phosphonio substituent (e.g., **24**) were found to react readily with electron-rich double bonds such as those in norbornadiene and the enamine ethyl trans-2-pyrrolidineacrylate to give, respectively **193** (80%) and **194** (78%). The regioselectivity for all the reactions with monosubstituted dipolarophiles discussed above was the same as that for transient nitrile imines (i.e., a very high level of selectivity favoring the formation of the 5-substituted products). However, the bis-(silyl) derivative **21** was different and gave mixtures of the 4- and 5-substituted cycloadducts with methyl acrylate and with methyl propiolate (e.g., **195** and **196** from the latter in a 52:48 ratio). This change was related to the raising of the dipole FMO energies by the presence of the two silyl groups (105,106). Species **21** has also been added to the C=S bonds of thio-aldehydes and thio -ketones generated *in situ* by a retro-Diels–Alder reaction (107), and to the P=S bond in Lawesson's reagent (108).

The N-phosphino derivative **16** shows an interesting duality of reactivity. It can give typical reactions of the 1,3-dipolar moiety or can react with more electrophilic reactants via nucleophilic attack by the phosphine center (109,110). Thus, with dimethyl acetylenenedicarboxylate it does not give a [3 + 2] cycloadduct as with methyl propiolate (see above) but instead gives the 1,2,4λ^5-diazaphosphinine (**198**) (90%) via **197**. The analogous reaction with TCNE rather surprisingly goes at one of the cyanide groups to give **199**. The reaction of **16** with methyl triflate gives the stable phosphonio derivative **24**, and reactions with sulfur, selenium, and phenyl azide give the heterocycles **200** (X=S, Se, NPh) in high yield.

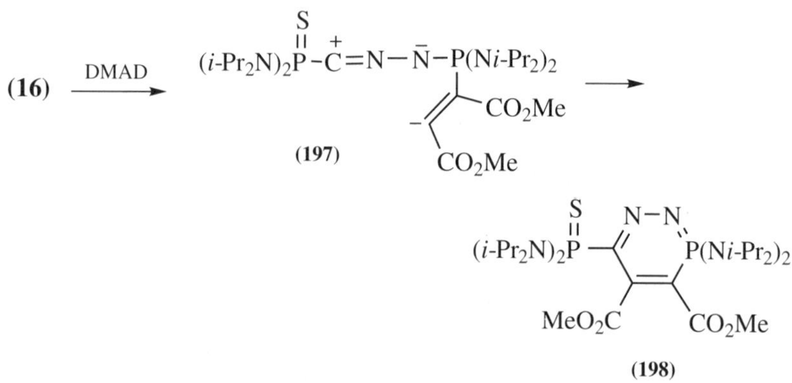

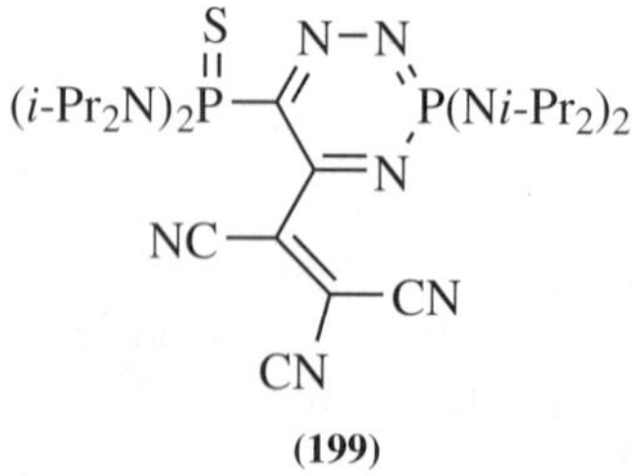

(199)

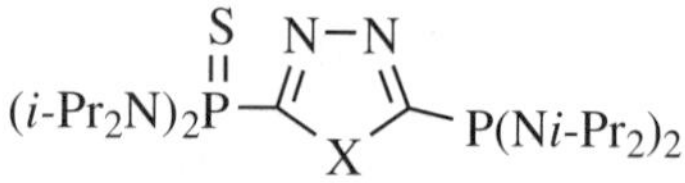

(200)

7.3.2.2. *Selectivity in Nitrile Imine Cycloaddition Reactions*

Extensive studies on diastereoselectivity in the reactions of 1,3-dipoles such as nitrile oxides and nitrones have been carried out over the last 10 years. In contrast, very little work was done on the reactions of nitrile imines with chiral alkenes until the end of the 1990s and very few enantiomerically pure nitrile imines were generated. The greatest degree of selectivity so far has been achieved in cycloadditions to the Fischer chromium carbene complexes (**201**) to give, initially, the pyrazoline complexes **202** and **203** (111,112). These products proved to be rather unstable and were oxidized *in situ* with pyridine N-oxide to give predominantly the (4*R*,5*S*) product **204** in moderate yield (35–73%).

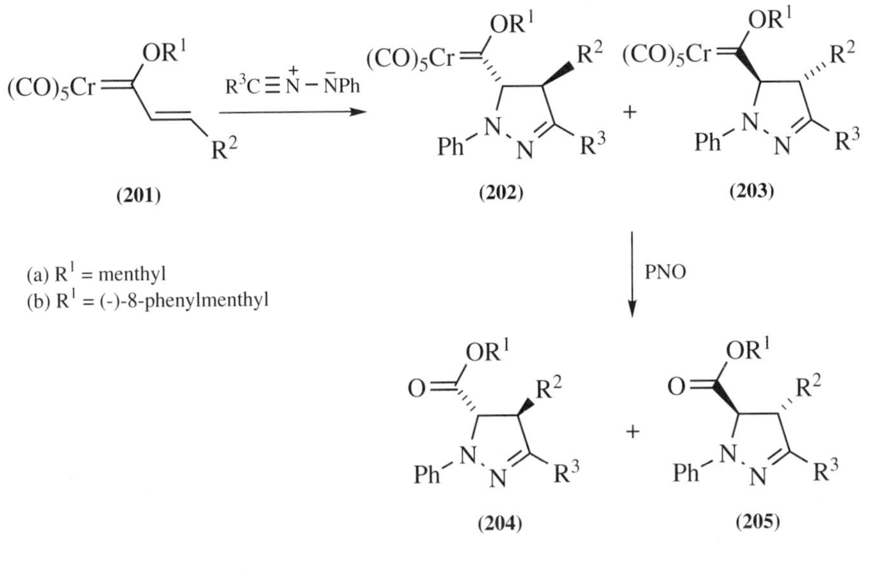

(201) (202) (203)

(a) R[1] = menthyl
(b) R[1] = (-)-8-phenylmenthyl

(204) (205)

High levels of regioselectivity were achieved for both **201a** and **201b** but only the latter gave useful levels of diastereoselectivity (>95:5 for a wide range of substituent variations). This contrasts with the corresponding reaction of ethyl cinnamate, which gave both regioisomers. A very high level of stereoselectivity has also been achieved by the complementary use of chiral substituents on both reactants. Thus stereoselectivity in the reaction of the nitrile imine **206** derived from D-galactose phenylhydrazone was enhanced by the use of chiral acrylic esters and amides as dipolarophiles. Only low diastereoisomeric excesses (de) (8–23%) were obtained with (+) and (−) menthyl acrylates and (1*R*)-*endo*-fenchyl acrylate but the use of Oppozer's *N*-acryloyl-(−)-camphor sultam (**207**) gave **208** only, a de of 100% (113). The (+) isomer, however, constituting a mismatched pair, gave a de of only 59%.

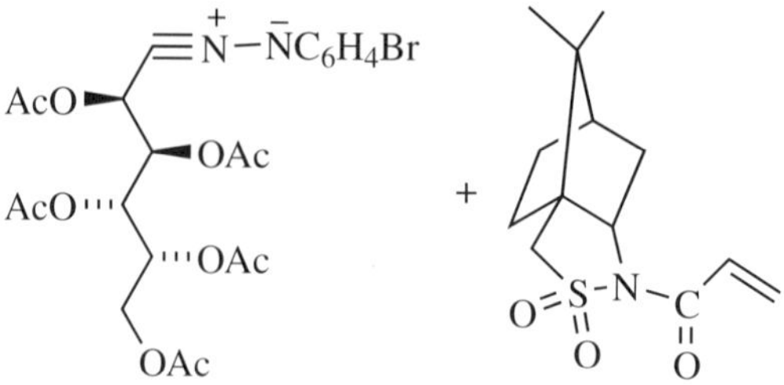

(206) **(207)**

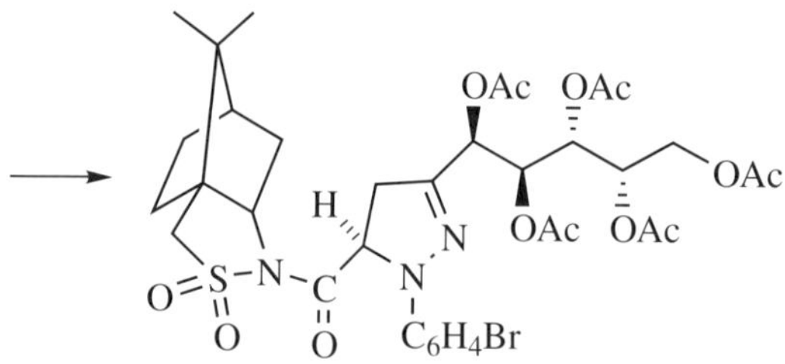

(208)

Pairs of diastereoisomeric pyrazolines have been formed with modest stereoselectivity in a number of reactions. The reaction of the chiral acrylic ester **209** with the stable nitrile imine **23** gave the products in a $\sim$ 3:1 ratio in 60% overall yield (18). In the addition of nitrile imines to the double bond in 2-vinylcarbapenem (**212**), the products, separable by chromatography, were formed in a $\sim$ 2:1 ratio (114). Similar selectivity was achieved using the chiral alkenone **210** and related reactants; the best being a 70:30 ratio for the reaction with **213** ($R^2{=}CO_2Et$) (115). The complementary effect of the presence of chirality in the nitrile imines was also investigated using the novel species **213** [$R^2{=}CO_2$-menthyl, $CO_2CH_2CH(MeEt)$], derived from enantiomerically pure hydrazonyl chlorides, but it was found that this had little influence on stereoselectivity.

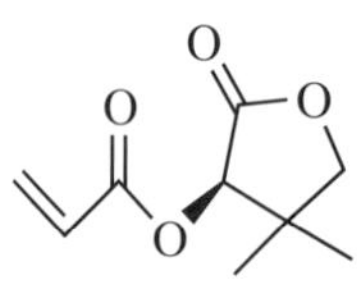

(209)

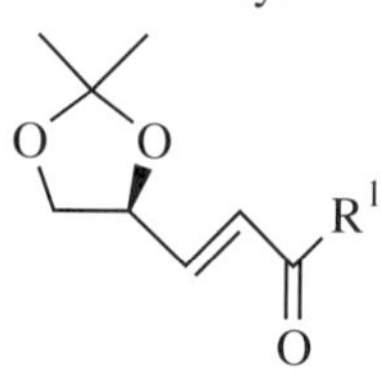

(210)

$\text{Glyc-C}{\equiv}\overset{+}{N}{-}\overset{-}{N}{-}C_6H_4NO_2$

(211)

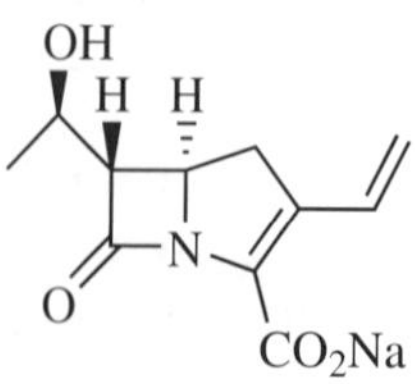

(212)

$R^2C{\equiv}\overset{+}{N}{-}\overset{-}{N}Ph$

(213)

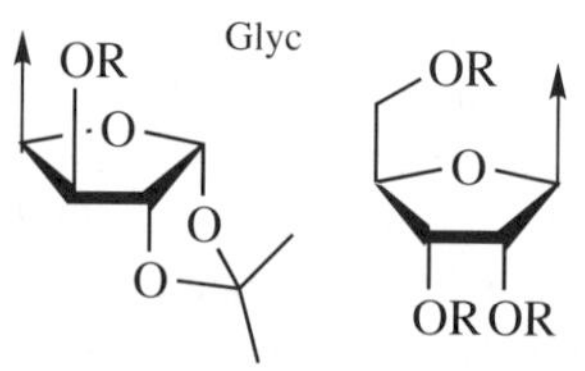

R = Me, Bn R = Bz, H

Enantiomerically pure nitrile imines (**211**) have also been generated by the lead tetraacetate oxidation of aldehydo sugar *p*-nitrophenyl hydrazones. Reaction with methyl acrylate gave the pyrazolines as a 1:1 mixture of the (5*S*) and (5*R*) epimers, which were resolvable in some cases (116).

Complexation has been utilized to control site selectivity. The iron tricarbonyl complex **216** reacts with nitrile imines only at the cyclobutene double bond, whereas the uncomplexed analogue reacts at all three sites (117).

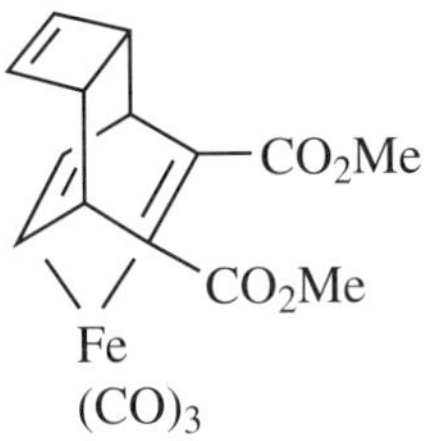

(216)

$Ph_3SiN{=}SO$

(214)

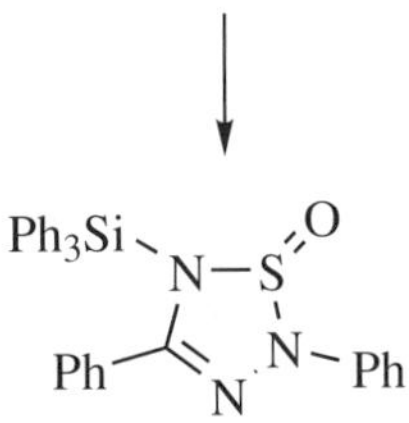

(217)

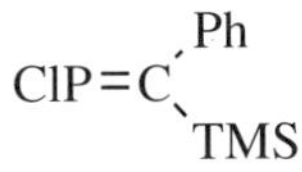

(215)

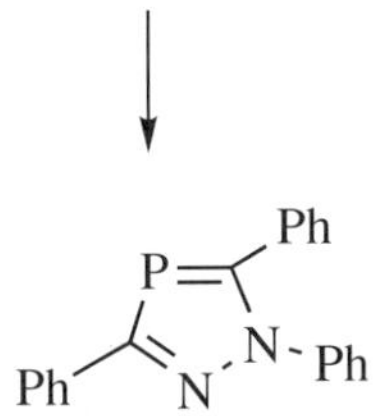

(218)

Nitrile imines have been added to C_{60} fullerene (118–120); the N=S bond in the triphenylsilylsulfinylamine **214** to give **217** (75%) (121), and the P=C bond in **215** to give **218** (84%) after the elimination of trimethylsilyl chloride (122).

7.3.2.3. Cycloaddition of Nitrile Imines to Heterocycles

In most cases, the cycloaddition to double bonds in heterocyclic rings is uncomplicated by side reactions and often goes with a high degree of peri- and regioselectivity. Where this is the case the chemistry is not discussed in detail but the structures of representative dipolarophiles are given with the reaction site indicated by arrows (full arrow = C-terminus, broken arrow = N-terminus). Reactions occurring with endocyclic double bonds are discussed for the various ring sizes in order, and then those at exocyclic bonds. The area was reviewed in 1994 (123).

Cycloaddition to Four-Membered Heterocycles

Nitrile imines generated by the benzaldehyde phenylhydrazone–chloramine-T route have been added to 1-azetines (**219**) (X=O, S) (95,124). However, the adducts **220** (X=O, S) proved to be less stable than those derived from nitrile ylides (**186**) (94) and in many cases fragmented spontaneously to give the 5-butenyl-1,2,4-triazoles (**221**) as the isolated products. It was found, however, that the use of an electron-withdrawing group in the Ar substituent had a stabilizing effect allowing, for example, **220** (X=O, S; Ar=4-NO_2—C_6H_4) to be isolated in high yield.

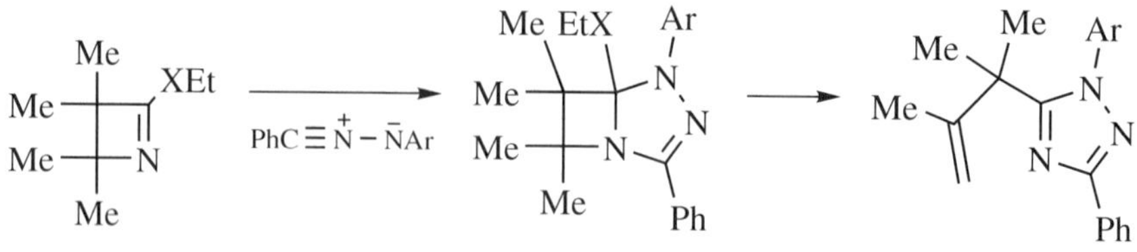

(219) (220) (221)

Cycloaddition to Five-Membered Heterocycles

Cycloadditions have been carried out to 3*H*-indoles (**222**, **223**) (125,126), *N*-arylmaleimides (**224**) (127,128), 1,2λ^5-azaphospholes (**225**) (129), 5(4*H*)-oxazolones (**226**) (130), and 4,5-dihydrooxazoles (**230**) (131). The primary cycloadducts from the reaction of oxazolones (e.g., **226** with diaryl nitrile imines), derived from tetrazoles in refluxing anisole, do not survive. They appear to lose carbon dioxide and undergo a dimerization–fragmentation sequence to give the triazole **228** and the diarylethene **229** as the isolated products (130). In cases where the two aryl substituents on the oxazole are not the same, then, due to tautomerism, isomeric mixtures of products are obtained.

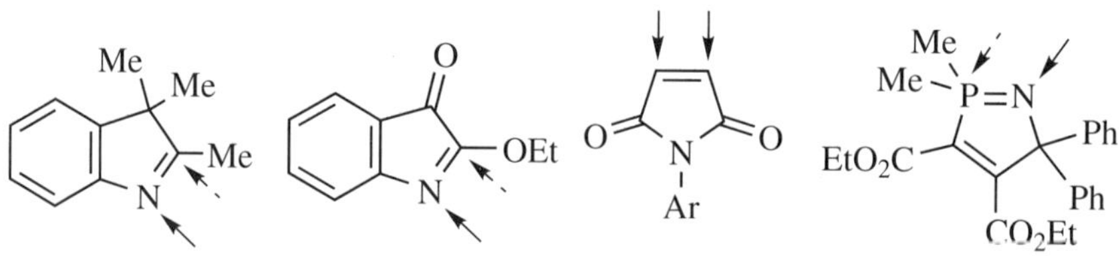

(222) (223) (224) (225)

The adducts **231** from cycloaddition to the 4,5-dihydrooxazoles **230** are also prone to undergo ring opening to give **232**, depending on the nature of the substituents R and R[1](131).

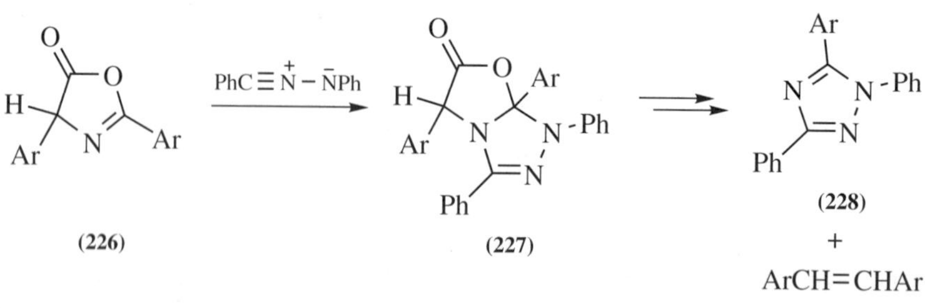

(226) (227) (228) + ArCH=CHAr (229)

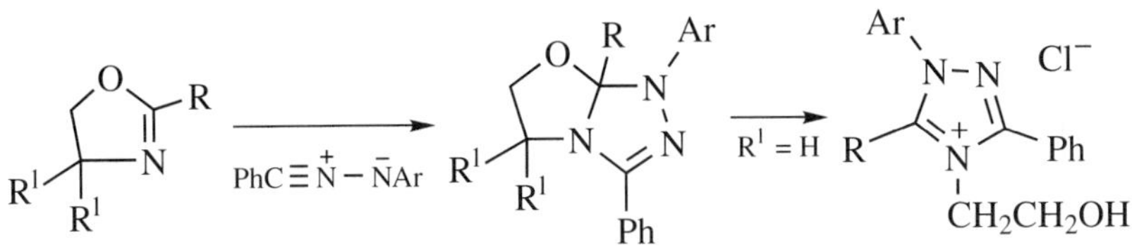

(230) (231) (232)

Cycloaddition to Six-Membered Heterocycles

Pyridine (**233**), quinoline (**235**), and isoquinoline (**236**) all react at their C=N bonds with complete regioselectivity to give mono adducts in high yield (132). The bis(adducts) (e.g., **234**) are not significant byproducts under normal conditions but can be formed as the major product when a fourfold excess of nitrile imine is used (133). The mono- and bis-adducts of pyridine undergo interesting rearrangements on heating (133).

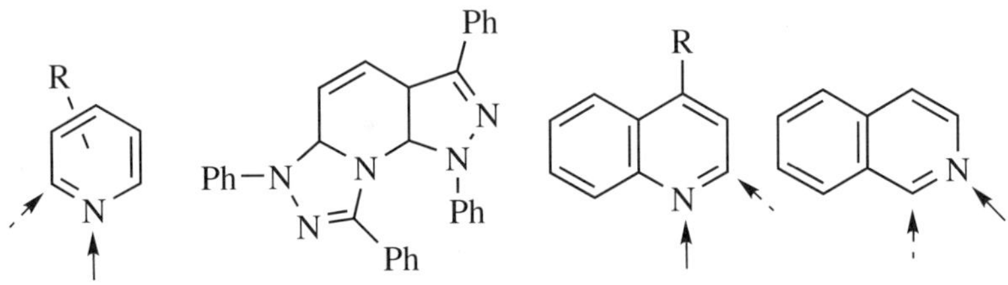

(233) (234) (235) (236)

Pyrimidines (**237**) react to give monoadducts but in the case of pyrazines (**238**) the monoadducts proved to be highly reactive to further cycloaddition (134). The 5-substituted pyridazinones (**239**) (X=$EtSO_2$, I, pyrrolidino, Cl) undergo cycloaddition to the C=C bond as shown followed by elimination of HX to give fused pyrazoles (135).

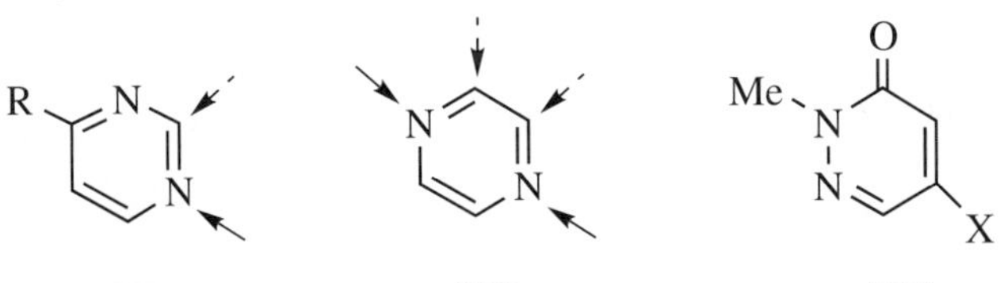

(237) (238) (239)

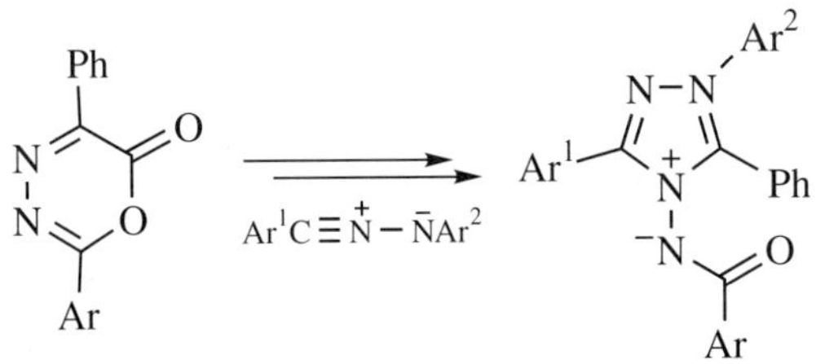

(240) (241)

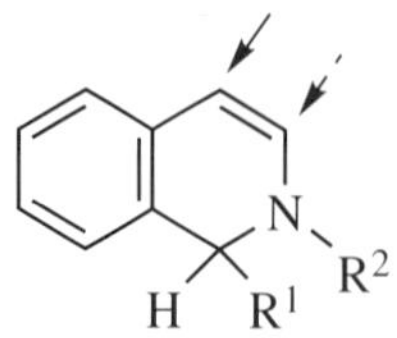 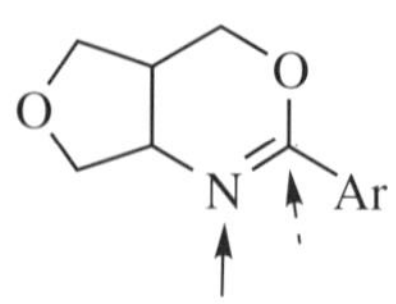 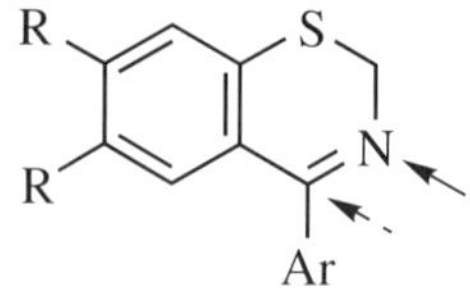 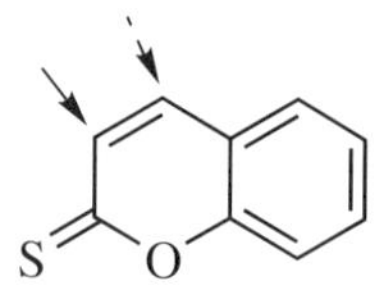

(242) **(243)** **(244)** **(245)**

The 1,3,4-oxadiazin-6-one (**240**) undergoes cycloaddition followed by a remarkable rearrangement to give the triazole *N*-imine **241** and an open-chain product (136). Cycloadditions have also been carried out with the following ring systems: 1,2-dihydroisoquinoline (**242**) (137); dihydro-1,3-oxazine (**243**) (138,139), 2*H*-1, 3-benzothiazine (**244**) (140,141), and 2*H*-1-pyran-2-thione (**245**) (142).

Cycloaddition to Seven-Membered Heterocycles

Nitrile imines have been added to 1,4-diazepines (**246**) (143,144), 1,2, 4-triazepines (**247**) (145), 1-benzazepines (**248**) (146), 1,4-benzodiazepines (**249**–**251**) (147–149), 1,5-benzodiazepines (**252**, **253**) (X=NH) (143,150–153), 1,5-benzothiazepines (**253**) (X=S) (153). Interest in this area has been stimulated by the known pharmacological activity of many compounds with five-membered heterocyclic rings fused to a benzodiazepine skeleton.

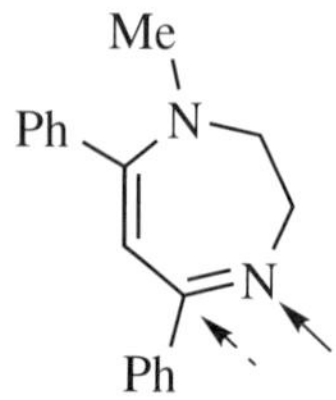 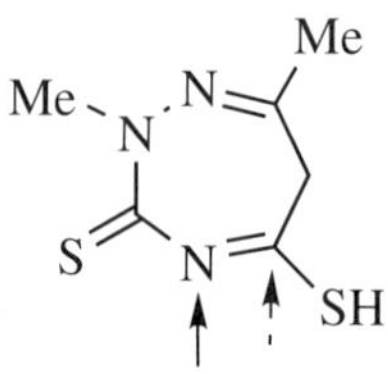 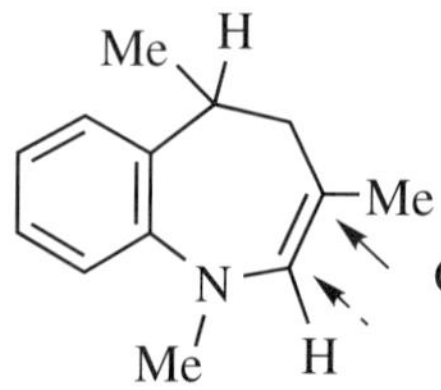 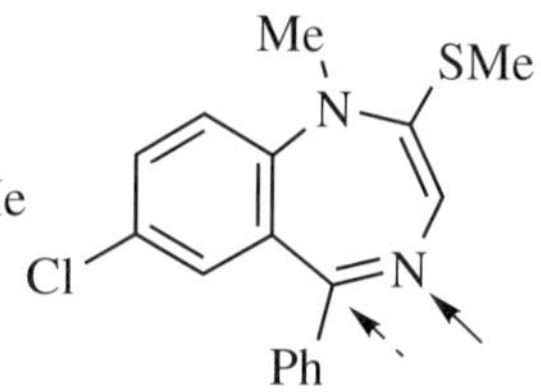

(246) **(247)** **(248)** **(249)**

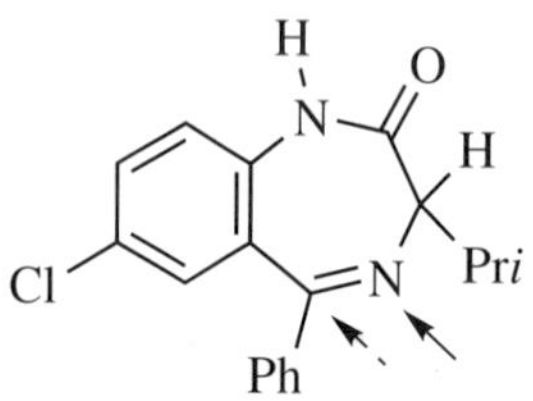 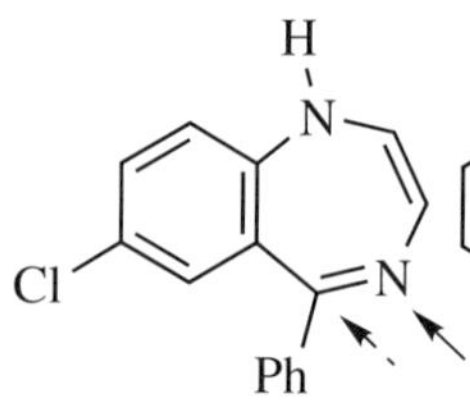 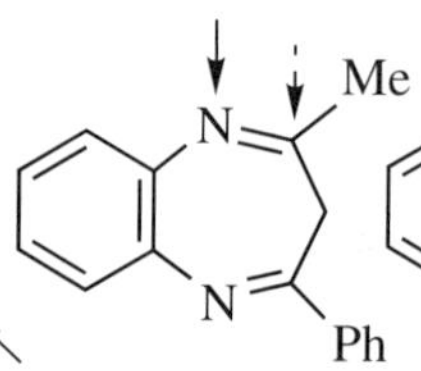 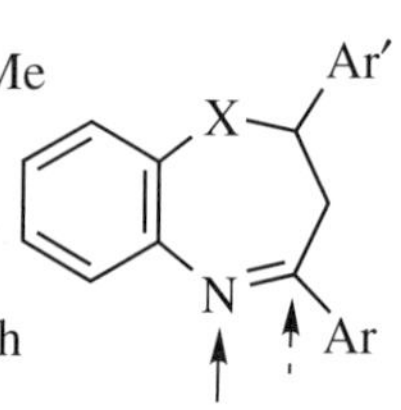

(250) **(251)** **(252)** **(253)**

Cycloaddition Reactions with Heterocyclic Exocyclic Double Bonds

Cycloadditions to exocyclic C=C bonds have been carried out for α-methylenebutyrolactones (**254**) (154,155), 4-methylenepyrazol-3-ones (**255**) (156), 4,5-dihydro-5-methylene-1*H*-pyrazoles (**256**) (157), 4,5-dihydroisoxazoles (**257**) (158),

4-methyleneoxazole-5(4*H*)-ones (**258**) (159), and 4-methylene-5*H*-thiazolones (**259**) (160).

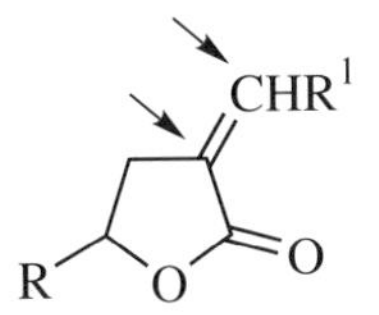

(254)

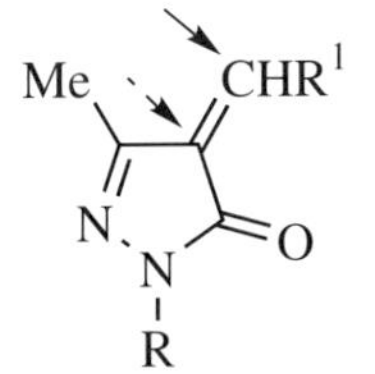

(255)

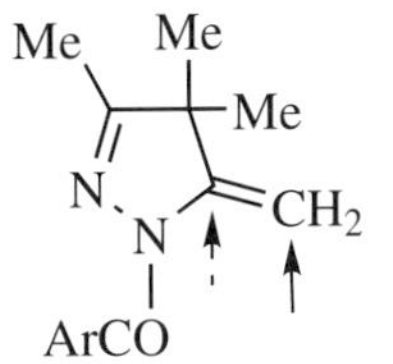

(256)

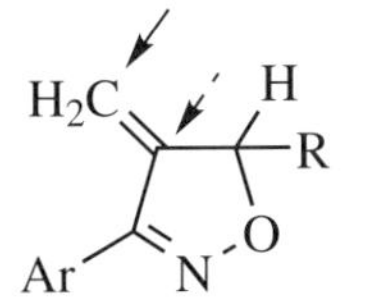

(257)

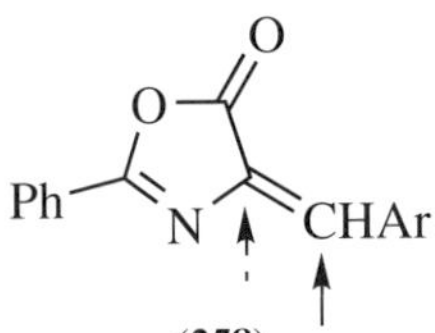

(258)

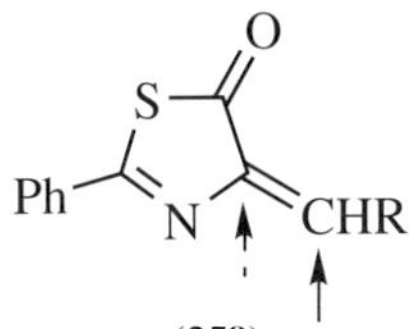

(259)

The thiocarbonyl group is a highly reactive dipolarophile and in general this group dominates the reactivity of nonenolisable exocyclic thioketones as illustrated for the systems shown: 5-methylene-2-thioxo-1,3-thiazolinin-4-one (**260**) (161), pyrimidone-2- and -4-thiones (**261**, **262**) (134), pyrazolo[1,5,4-*ef*][1,5]benzodiazepin-6-thione (**263**) (162). 2-Thiono-4-imidazolidinone (**264**) also gave a C=S cycloadduct as expected but, in the case of the analogue **265** with an additional exocyclic methylene group, the latter proved to be more reactive (163).

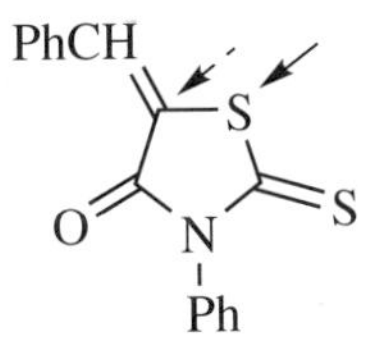

(260)

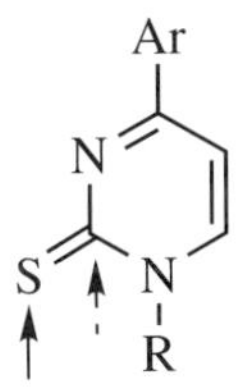

(261)

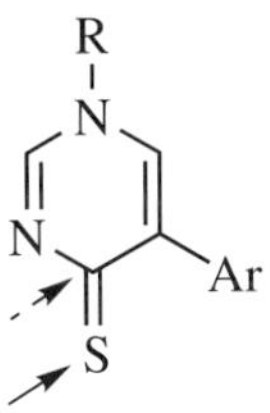

(262)

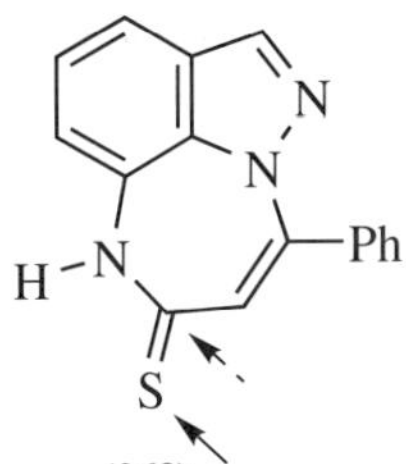

(263)

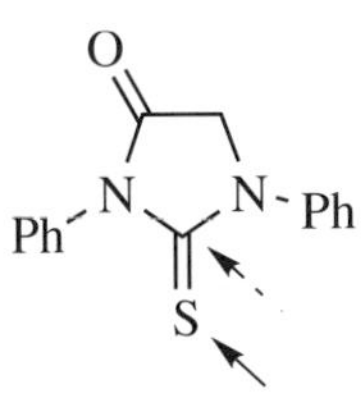

(264)

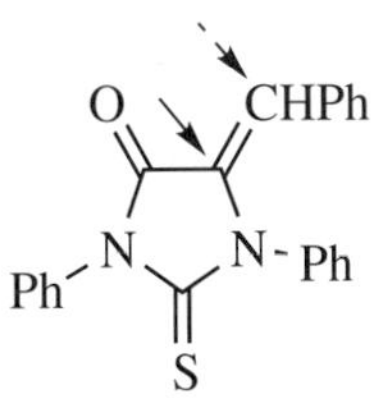

(265)

7.3.2.4. *Intramolecular Cycloaddition Reactions of Nitrile Imines*

The work described here has been partitioned into four sections, depending on the length of the link between the reacting groups, and a final one on tandem reactions. Recent work has focussed on diastereoselectivity and the reactions of chiral nitrile imines. The area was reviewed in 1998 (164).

Three-Atom Tether

Intramolecular cycloaddition of nitrile ylides to olefinic dipolarophiles linked to the dipole by a three-atom chain leads to pyrazoles fused to five-membered rings. Work on stereoselectivity in such reactions has been carried out using the reactant **266** in which the alkene moiety is linked to the C-terminus via a tether that incorporates an enantiomerically pure (*R*) stereogenic group (165). Both diastereoisomers **267** and **268** were isolated and it was found that the reaction showed moderate stereoselectivity favoring **267**.

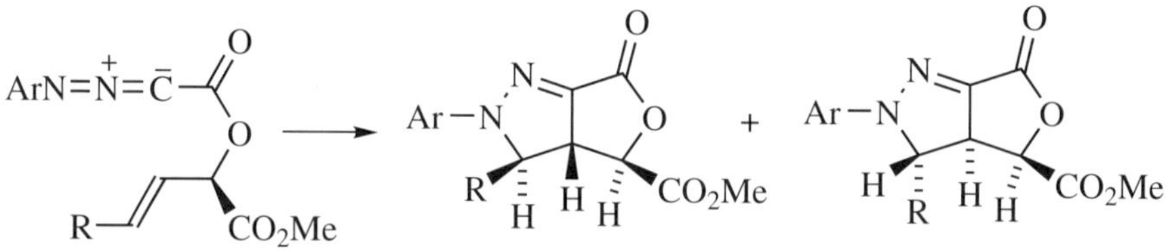

(266) **(267)** **(268)**

The analogue **269** reacted via a remarkable intramolecular cycloaddition to the thiophene ring to give **270** as the primary product, which was then partially consumed by intermolecular cycloadditions of **269** to both the pyrazole C=N and the residual thiophene C=C to give **271** and **272**, respectively (166,167).

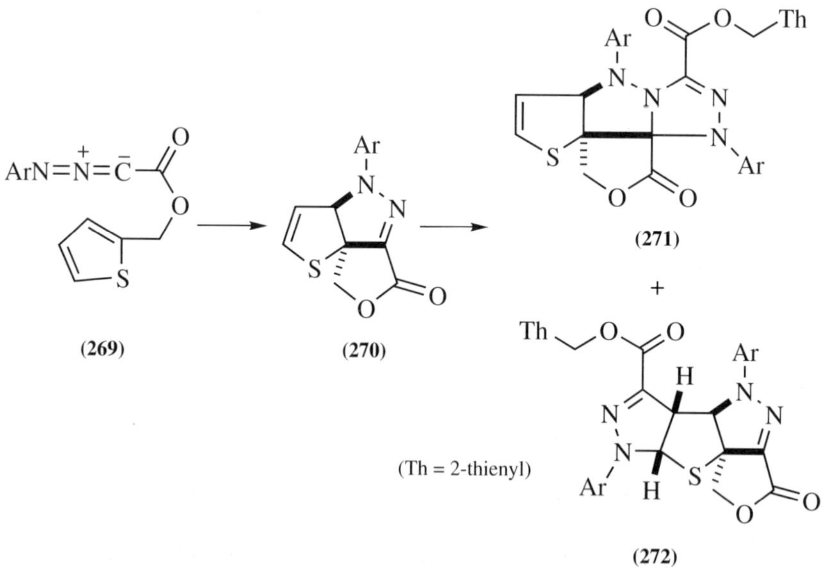

(269) **(270)** **(271)** **(272)**

The isoxazole analogue **273** underwent a similar process via a completely periselective intramolecular reaction with the C=C of the isoxazole (Isox) (168).

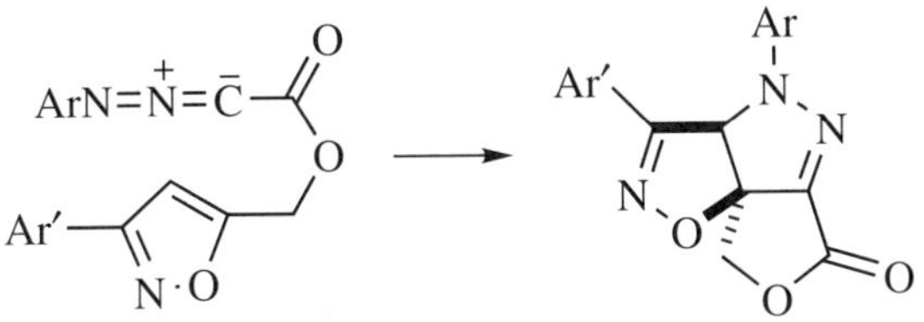

(273) **(274)**

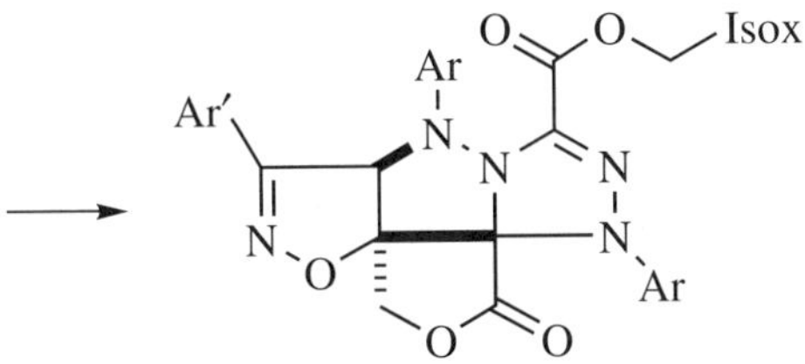

(275)

However, in this case none of the primary product **274** was isolated but reacted further by cycloaddition exclusively to the pyrazole C=N to give **275**.

Four-Atom Tether

Reactants in which the N terminus of the dipole is linked by a four-atom chain to a dipolarophile (e.g., **276**) lead to pyrazolines or pyrazoles **277**; or 1,2,4-triazoles (**278**) fused to a six-membered ring. Examples have been reported where X=S (169) [and for a pyrazolo fused analogue (170)], X=SO (171), X=SO_2 (172), X=N (173), and X=O (174–176). The case where X=S (169) is interesting in that reaction of **279** at room temperature led to the benzothiadiazines **281** via nucleophilic attack of the S atom, followed by a [2,3]-sigmatropic shift. However, this process was reversed on heating to give **282** as the thermodynamic product.

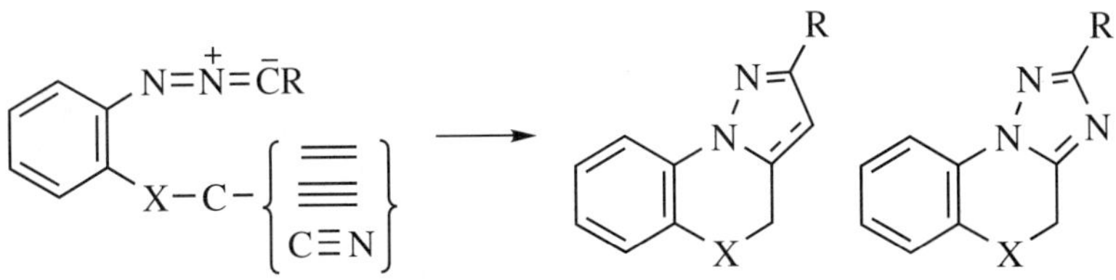

(276) **(277)** **(278)**

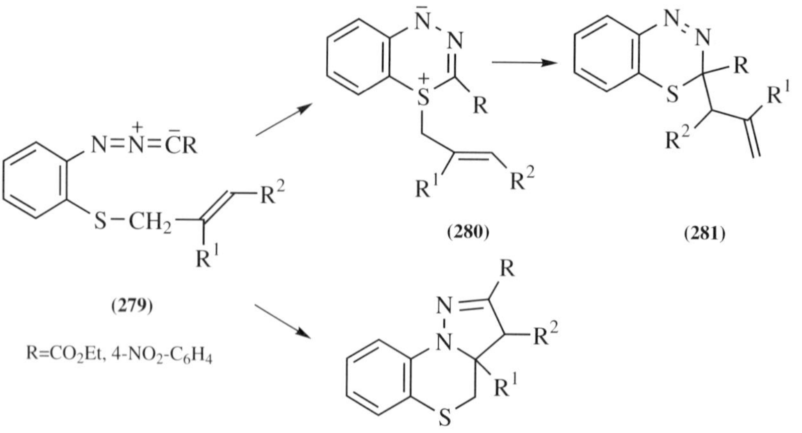

(279) **(280)** **(281)** **(282)**

The sulfinyl analogues **283**, however, did not follow a similar path but gave the cycloadducts **284** directly (171). This reaction showed surprisingly strong stereoselectivity in the case where R=Me to give only the *anti* product **284** (R=Me).

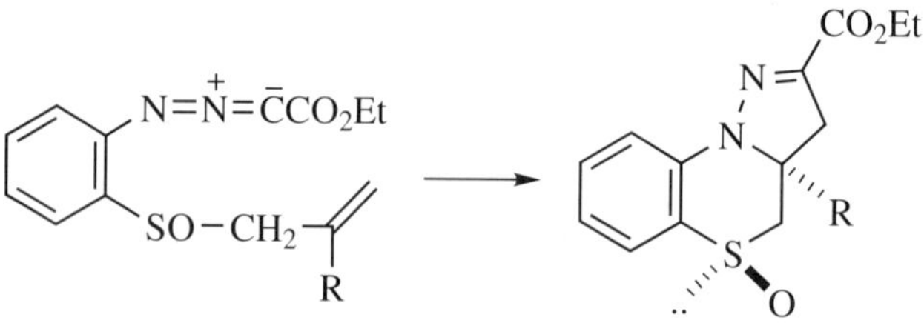

(283) **(284)**

Five-Atom Tether

Extension of the linkage to five atoms as in **285** provides routes to pyrazolines or pyrazoles **286**, or 1,2,4-triazoles **287**, fused to a seven-membered ring. The products are potentially biologically active and examples have been reported for X=N (177–181), X=O (181–185) and for a pyrazolo fused analogue (186) and X=S (187). In some cases, [e.g., (183)], these reactions are accompanied by tandem intramolecular–intermolecular reactions leading to the formation of macrocycles (see the section Tandem Intermolecular–Intramolecular Cycloaddition Reactions).

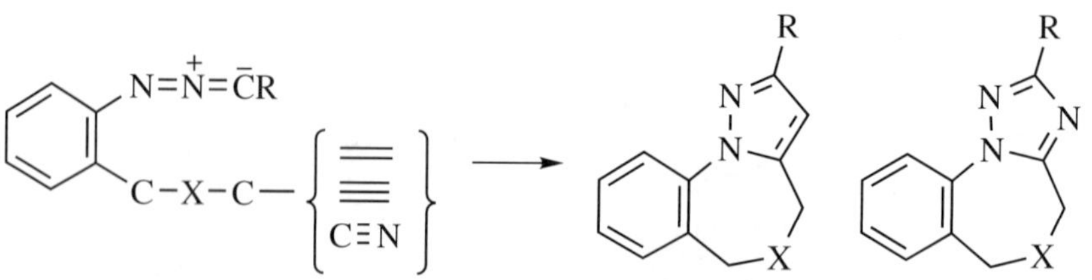

(285) **(286)** **(287)**

In recent work, a homochiral substituent has been incorporated into the reactant to allow the separation of enantiomerically pure products. Thus, the homochiral reactant **288**, prepared from (*S*)-1-phenylethylamine, gave a pair of diastereoisomers (**289**) and (**290**) that were separated by chromatography and identified via X-ray crystallography (178). The nitrile imine was generated by the hydrazonyl chloride–base route. The reaction showed only modest stereoselectivity that favored **289** when silver carbonate was used as the base but it was found that this was reversed when triethylamine was used. However, this was not the case for a related reaction (179).

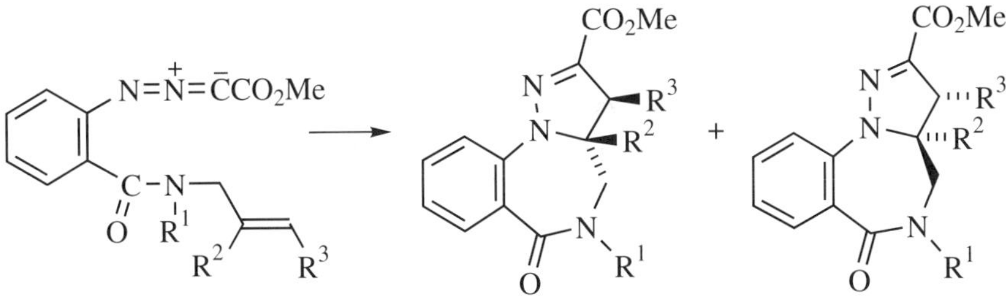

(288) **(289)** **(290)**

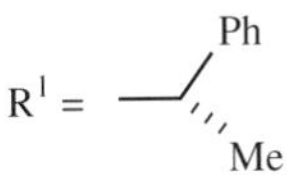

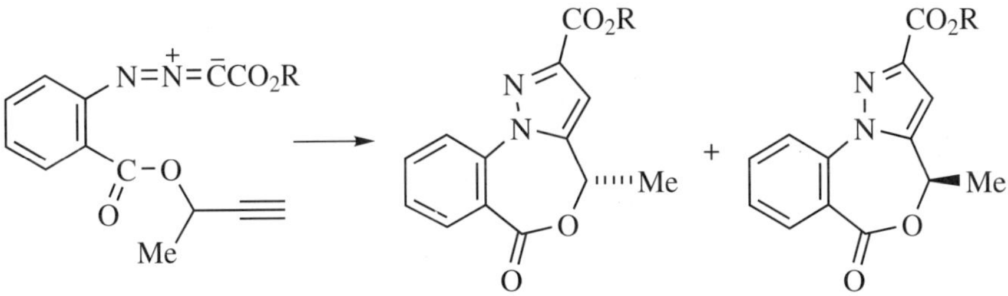

(291) **(292)** **(293)**

Similarly, **291**, which has a homochiral menthyl ester group (R), gave the cycloadducts **292** and **293** (181).

Tether Greater than or Equal to Six Atoms

Reactants **294** with a six-atom linkage have been used as a route to pyrazolo [1,5-*a*][5,1]benzoxocines (**295** and **296**). Using alkenes as dipolarophiles gave **295** in moderate-to-good yields (21–61%) (183), terminal alkynes gave **296** (21–47%) but only 7% was obtained for an example with a terminal methyl group (184).

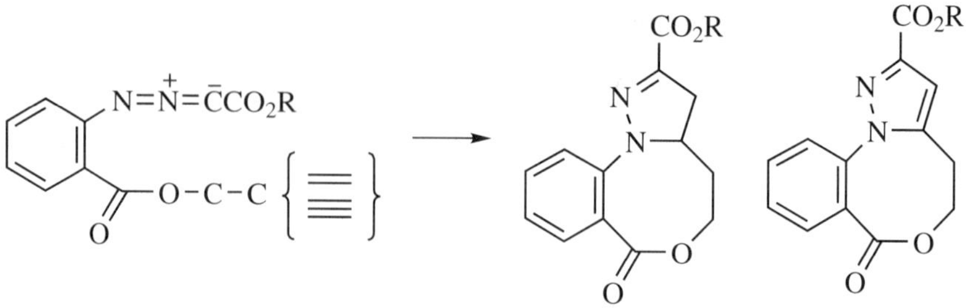

(294) (295) (296)

Longer tethers, as shown in structures **297a,b** have yielded pyrazoline fused macrocycles with 9–13-membered rings (188). Reactants of type **297a** gave **298** (11–28%) while those of type **297b** gave **299** in much higher yields (39–56%), possibly due to higher conformational flexibility in the side chain. The acetylenic analogue of **297a** (Z=CH_2CH_2) gave the pyrazole corresponding to **298** (26%).

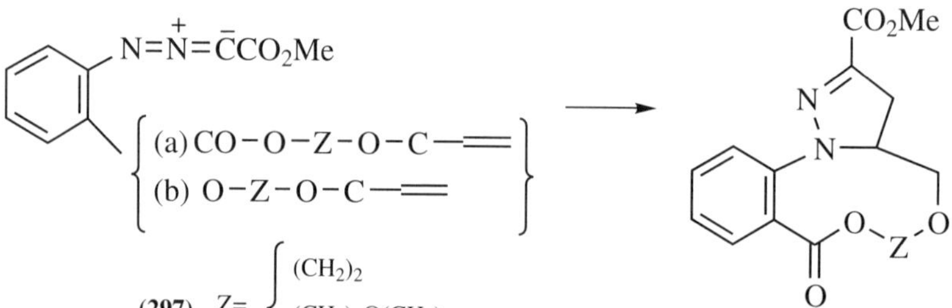

(297) Z= $(CH_2)_2$, $(CH_2)_2O(CH_2)_2$, o-C_6H_4

(298)

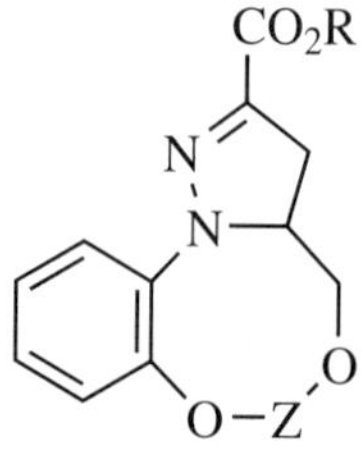

(299)

Tandem Intermolecular–Intramolecular Cycloaddition Reactions

Reactants with long tethers (e.g., **300**) can also take an alternative path involving first an intermolecular cycloaddition with the hydrazonyl chloride precursor to give **301** followed by an intramolecular cycloaddition to give the *bis*(pyrazolocyclophanes) (**302**) (73). Byproducts of this type were also been reported in earlier work (183,188).

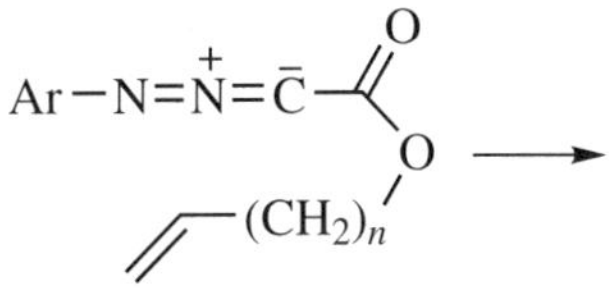

(300)

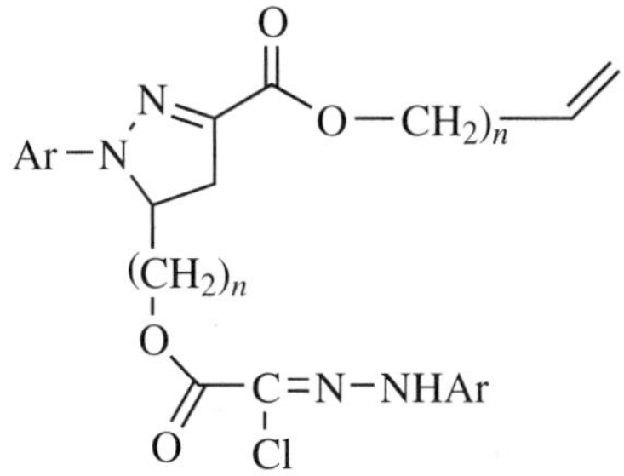

(301)

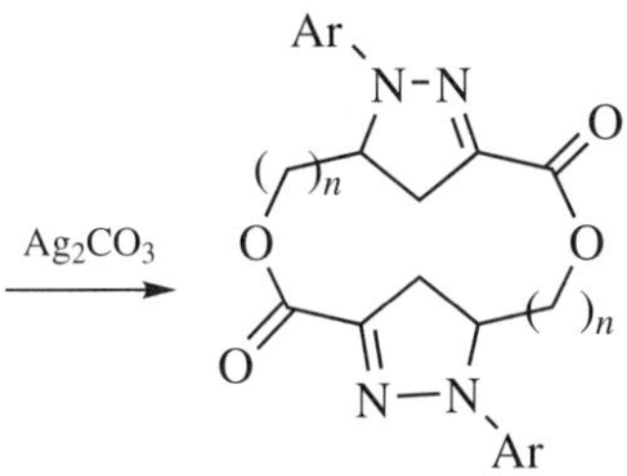

(302)

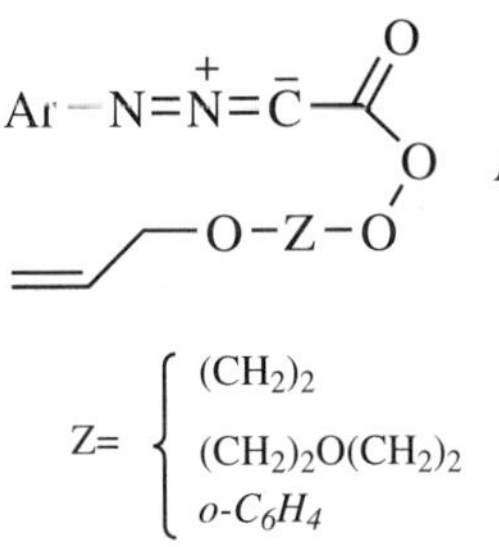

(303)

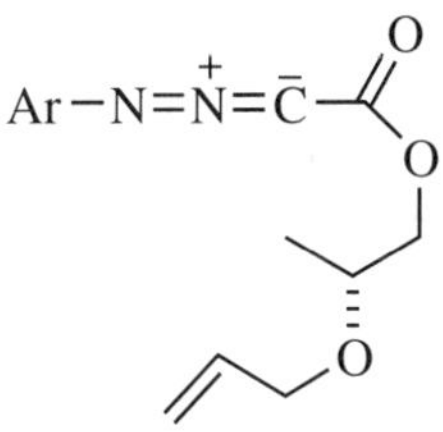

(304)

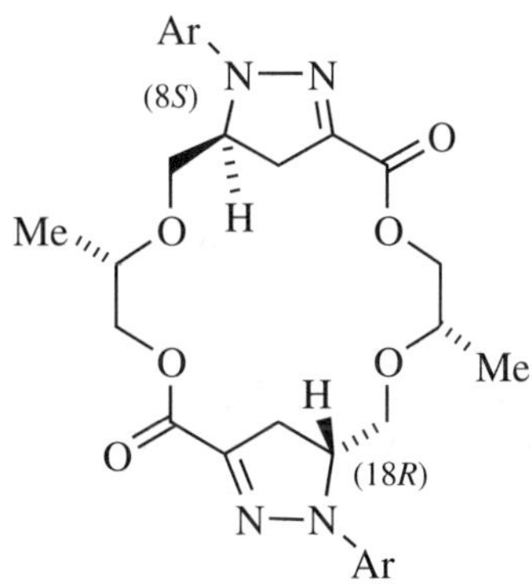

(305)

Reactants of type **303** gave similar reactions and, interestingly, the enantiomerically pure example **304** reacted via a fully diastereoselective cycloaddition in the first step leading to **305** (66%) (189).

7.4. ELECTROCYCLIZATION REACTIONS

7.4.1. Electrocyclization Reactions of Nitrile Ylides

Some interesting new chemistry has been produced on the well-known 1,5-electrocyclization reaction of alkene-conjugated nitrile ylides but the greatest volume of work has been concerned with the reactions of systems with extended conjugation that provide good routes to fused azepines and other heterocycles.

7.4.1.1. *Alkene-Conjugated Nitrile Ylides*

Ab initio and density functional calculations have been carried out on the mechanism of the 1,5-electrocyclization reactions of conjugated nitrile ylides (190). The results indicate that vinyl-conjugated systems (**306**) (X=CH_2) cyclize via the classical electrocyclization pathway—a pericyclic, monorotatary process with a relatively early transition state in which there is substantial torsion of the vinyl group as well as pyramidalization at C(5). In contrast, systems with a heteroatom at the cyclization site **306** (X=NH, O) react via a pseudo-pericyclic process that is characterized by the in-plane attack of the lone pair of the heteroatom on the nitrile ylide. Such reactions have a lower activation energy.

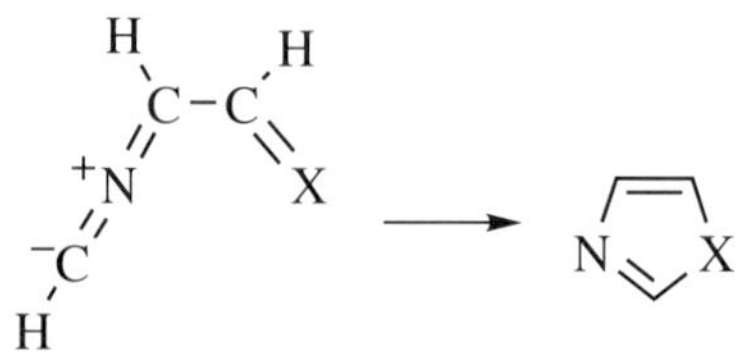

(306) **(307)**

Reactions of this type have been used in the construction of the pyrrolylfuran derivatives (**309**) (R=H, 57% and R=Cl, 24%) using nitrile ylides generated by the imidoyl chloride–KO*t*-Bu method (191). This reaction was a key step *en route* to simplified analogues of roseophilin, however, the reaction failed for the derivative with a methoxy group at the 4 position of the furan ring.

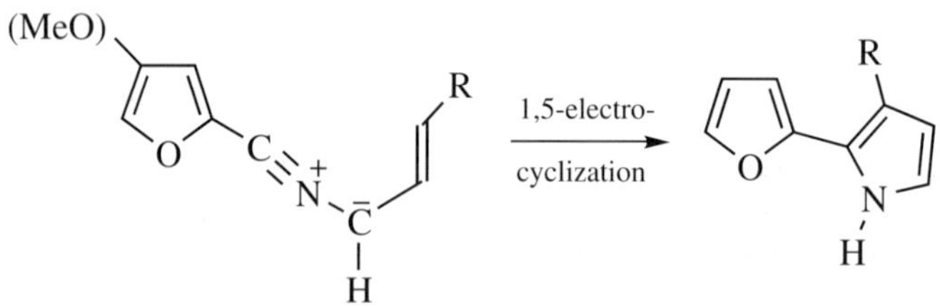

(308) **(309)**

A similar 1,5-electrocyclization involving an aromatic ring was observed for the nitrile ylide **310** (R^1=Ph, R^2=Me) (66). This reaction gave **311** as the initial product, which then rearranged via a 1,3-hydrogen shift to give the isoindole **312**.

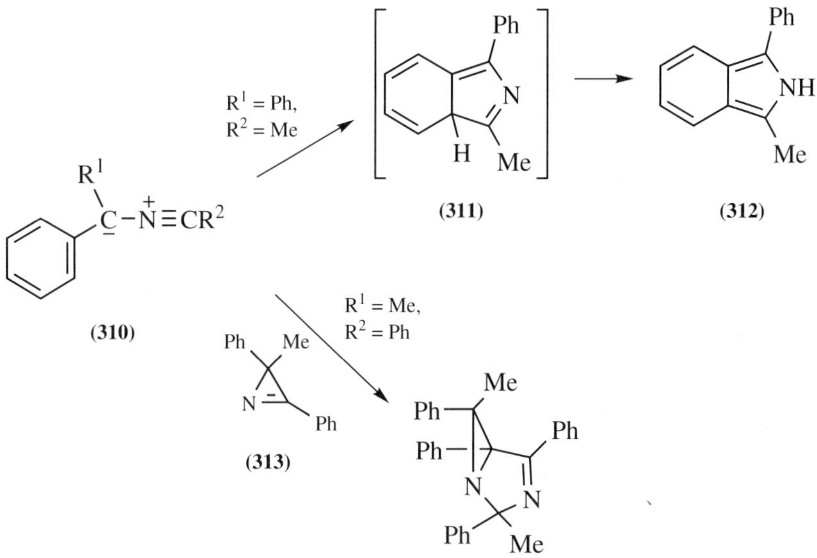

(310) (311) (312) (313) (314)

The intermediacy of the nitrile ylide was demonstrated by a trapping experiment using methyl acrylate in which it was found that the formation of **312** was entirely suppressed. Interestingly and unexpectedly, the nitrile ylide **310** (R^1=Me, R^2=Ph) failed to follow a similar path and reacted only via cycloaddition of the nitrile ylide to the double bond of the precursor azirine **313** to give **314**. Its failure to undergo 1,5-electrocyclization was attributed to steric destabilization of the required syn isomer.

The synthesis of oxazoles via the reaction of α-diazoketones **316** and esters (e.g., **315**) with nitriles in the presence of a copper catalyst has been modified by the replacement of copper with rhodium(II) acetate. This is a further example of the generation of nitrile ylides by the reaction of nitriles with carbenes–carbenoids. Dimethyl diazomalonate (**315**) was found to react with a wide range of nitriles to give the oxazoles (**317**) (R^1=OMe, R^2=CO_2Me) in generally moderate to good yield, formed by 1,5-electrocyclization of the carbonyl-conjugated nitrile ylide intermediates **318** (R^1=OMe, R^2=CO_2Me) (192). The analogues **318** (R^1=Ar, R^2=H), formed from the diazo-ketones (**316**), reacted in a similar way (193,194) but could also be partially intercepted by reaction with DMAD to give the adduct **319** (R^1=Ar) in low yield. In neat benzonitrile, the reaction gave **317** (R=Ph, R^1=Ar, R^2=H) (62%) and **319** (E=CO_2Me, R=Ph, R^1=Ar) (11%). The interception provides convincing experimental evidence for the intermediacy of the acyl-substituted nitrile ylide in these reactions. A higher cycloadduct–oxazole ratio was obtained using diazoacetates but the overall yields were lower. The use of N,N-disubstituted cyanamides R^3R^4NCN (R^3R^4=H, Me, Et, Ph, *i*-Pr in various combinations) in place of benzonitrile similarly gave 2-aminooxazoles (**317**) (R=R^3R^4N) in generally high yields but, cyanamide itself and monosubstituted derivatives, gave only low yields (195).

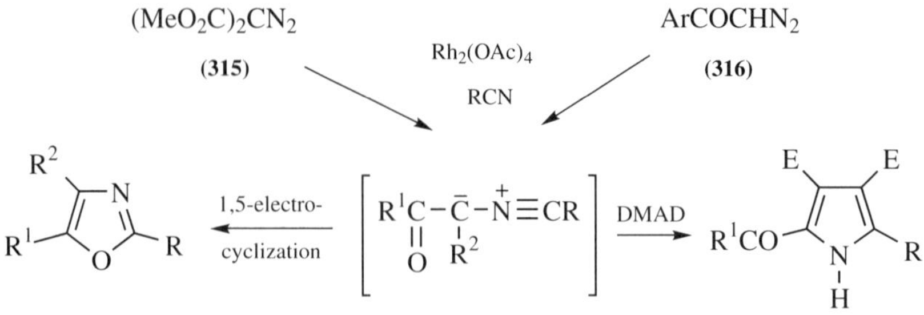

(317) (318) (319)

Similar products (e.g., **320**) (R=Et), formed in the same way from ethyl diazoacetate, were detected in NMR spectra of the reaction mixtures but did not survive chromatographic workup. However, the addition of DMAD to the crude product gave rise to the adduct **322** in 58% yield, most likely formed via an equilibrium between the oxazole and the nitrile ylide as shown. Further work showed that the presence of a bulky ester group had a strong stabilizing effect on the oxazole ring [e.g., when R=*t*-Bu, adamantyl or 2,6-di-*tert*-butyl-4-methylphenyl the corresponding heterocycles **320** were isolated in 49, 33, and 25% yields, respectively (196)]. On heating in the presence of DMAD, they gave adducts **322** and on reaction with methanol gave, for example, **323** in virtually quantitative yield. Kinetic studies supported a mechanism in which the methanol attacks the nitrile ylide rather than the oxazole itself.

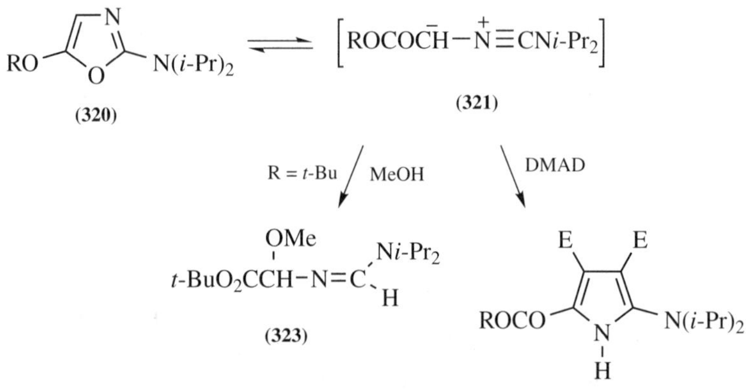

(320) (321) (323) (322)

7.4.1.2. *Diene-Conjugated Nitrile Ylides*

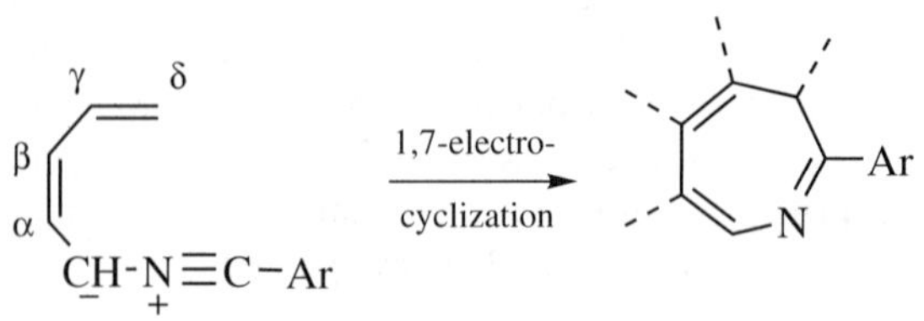

(324) (325)

Extensive work has been carried out on the 1,7-electrocyclization of diene-conjugated nitrile ylides (**324**) leading to fused heterocyclic systems containing the azepine ring (**325**). Reactions of this type for all 1,3-dipoles have been reviewed (197,198).

Reactants in which both the α,β and the γ,δ bonds were aromatic in character [e.g., **326** (R=H; R′=OMe, Cl; Ar=Ph, 2-ClC_6H_4, 4-MeC_6H_4, 2-FC_6H_4, 3,4-di-MeOC_6H_3)] cyclized readily in a two-step process to give dibenz[*c*,*e*]azepines (**328**) in high yields (67–90%) (58,199). This result contrasts with the reactions of the analogous diazo compounds that did not cyclize but instead reacted via loss of nitrogen. Interestingly, the presence of an ortho methyl group (e.g., **326**) (R=Me) prevented cyclization by its steric inhibition of conjugation in the transition state, with the result that the nitrile ylide reacted only via dimerization.

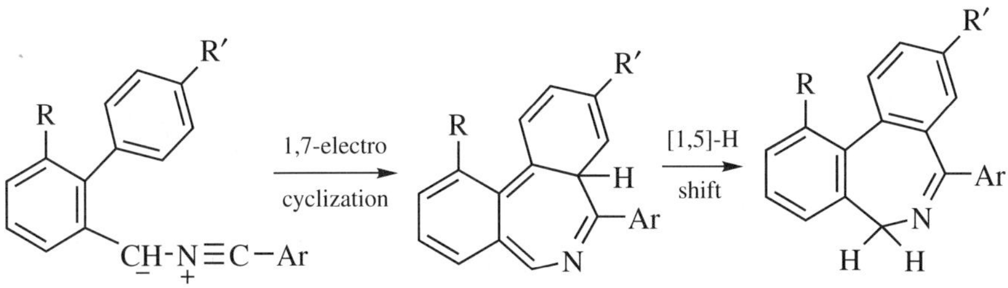

(326) **(327)** **(328)**

The nitrile ylides were generated from amides via the imidoyl chloride–base method and hence the reaction is, overall, the electrocyclic equivalent of a Bischler–Napieralski type of process. However, it has the advantage that it is effective for cyclization on to both electron-rich and electron-poor aromatic rings, unlike the Bischler–Napieralski reaction itself, which is an electrophilic process and only works well for electron-rich rings.

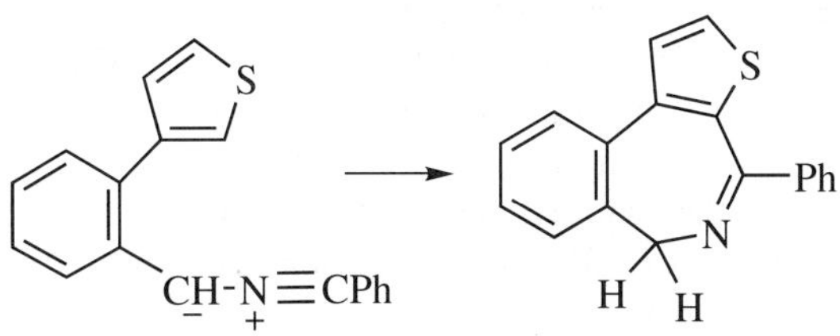

(329) **(330)**

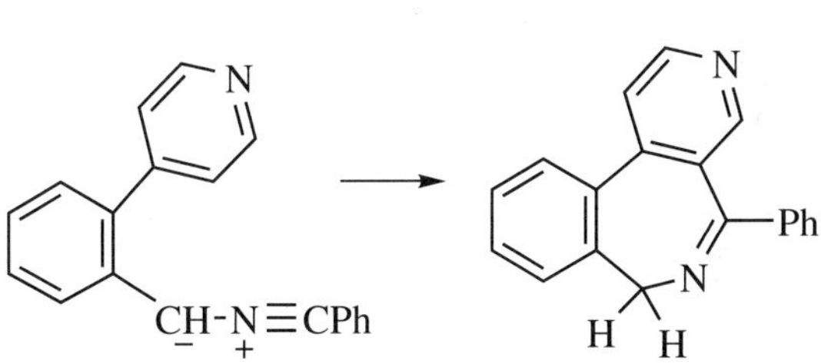

(331) **(332)**

The method was extended to systems in which either of the benzene rings were replaced by a five- or six-membered heterocycle (e.g., **329** and **331**) thus giving effective routes to fully unsaturated heterocyclo[*d*][2]benzazepine systems (61) [e.g., **330** (65%) and **332** (81%)]. The combination of this cyclization process with the use of Pd(0) catalyzed cross-coupling to prepare the required biaryl starting materials provides an easy general route to such systems. The reactions of **329** and **331** again illustrate the capability for cyclizing at both electron-rich and electron-poor sites. This facility was further examined and quantified by carrying out intramolecular competition experiments using reactants of the type **334** in which the rate of reaction of a substituent A (alkenes, substituted aryl rings and various heterocycles) was measured relative to that of an unsubstituted phenyl group via determination of the product ratio **333:335** (200,201).

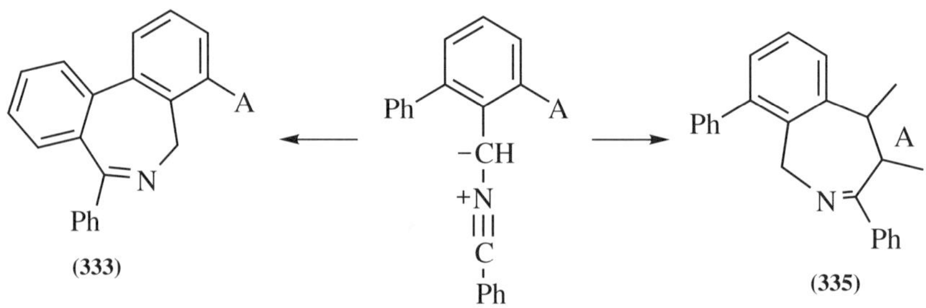

(333) **(334)** **(335)**

Alkenyl groups and the thiophene ring were, predictably, found to be >100× more reactive than the phenyl group since the rate of cyclization might be expected to be related to the *double-bond* character of the γ,δ-bond. However, more surprisingly, in cases where the A substituent was a substituted aryl group it was found that *all* aromatic substituents at the 3′- and 4′- positions, irrespective of their electronic nature, increased reactivity compared to the unsubstituted phenyl group. The effect was strongest for 3′ substituents (NO_2 > 100, OMe 5.6, bis-CF_3 32, bis-CH_3 8.3) and weak for 4′ substituents [e.g., 2.8 (CF_3), 1.5 (CH_3)]. Carbonyl ylides did not show the same pattern of reactivity (202).

The related system **336** with an olefinic α,β-bond cyclized in a similar way to give the 3*H*-2-benazepines (**337**) (203). Deuterium labeling studies showed that the cyclization step is irreversible.

(336) **(337)**

Interestingly, the 1,7 cyclization path was completely dominant in these reactions whereas the analogous diazo- compounds (**336**; N for CPh) reacted only via 1,5-electrocyclization to give pyrazoles. This difference in periselectivity may be due in part to the presence of the bulky phenyl group on the attacking carbon of the nitrile ylide. However, it may also reflect a difference in the ease of in-plane bending for the two dipoles that is related to the relative strengths of the orthogonal π bonds. Some work has also been done on systems related to **336** but with the dipole moiety inverted. It was found that they do not undergo cyclization but react by what is formally a dimerization reaction to give an imidazole (204).

Extensive work has also been carried out on the system **338**, which has an olefinic γ,δ-bond, again using the imidoyl chloride generation route (205). This result confirmed an early report by Padwa that the unsubstituted system **338** (R^2, R^3=H), generated by azirine photolysis, cyclized by 1,1-cycloaddition to give the corresponding cyclopropa[*c*]isoquinoline **340**. By the use of a variety of substituents R^2 and R^3, it was shown unambiguously that the reaction is entirely stereospecific. This parallels earlier results for the 1,5-electrocyclization of alkene-conjugated nitrile ylides. The products **340** were found to be quite stable at room temperature but rearranged on heating. Thus, for example, the exo and endo isomers **340** and **342** underwent a relatively rapid equilibration via **341** at 60 °C and, when either R^2 or R^3=H, both decayed slowly to give the 1*H*-2-benzazepine (**339**) in quantitative yield via a [1,5]-hydrogen migration.

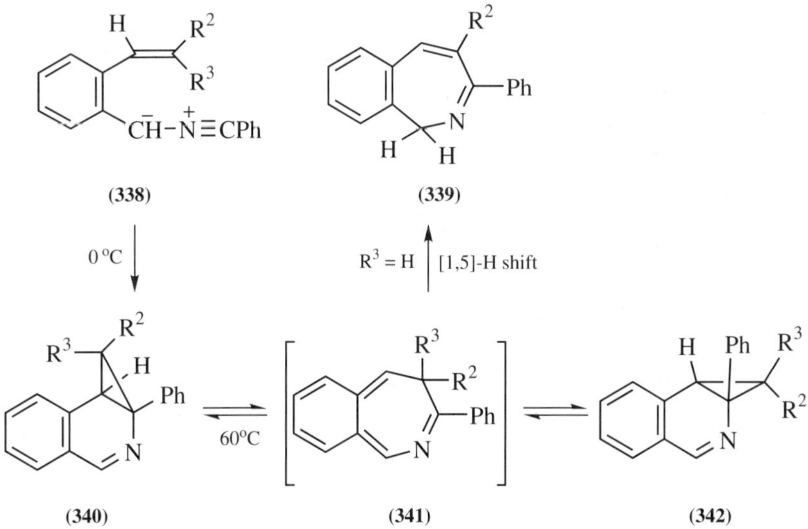

Analogues **343** with R^1=H but without a migratable hydrogen (R^2, $R^3 \neq$ H) gave 5*H*-2-benzazepines (**345**) *via* a walk rearrangement followed by ring expansion. However, in cases where either R^2 or R^3 was a methyl group, then the presence of a group larger than hydrogen in the R^1 position (Me or Et) served to divert the reaction completely into another path leading to the 4-alkenyl-1,4-dihydroisoquinolines (**346**) *via* a homo[1,5]-sigmatropic hydrogen shift (59).

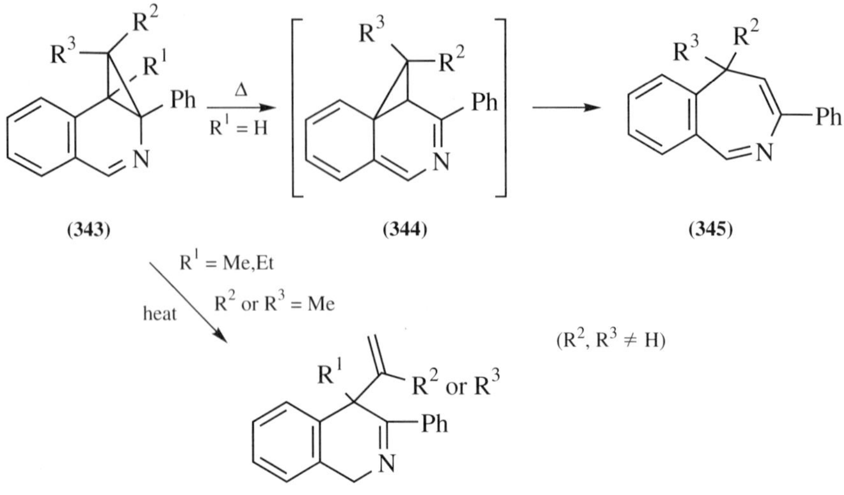

(343) (344) (345)

(346)

7.4.1.3. *Triene-Conjugated Nitrile Ylides*

When the conjugation is further extended, as in the triene-conjugated nitrile ylides (**347**), it is interesting that 1,1-cycloaddition still appears to be the favored primary process. In the case of the cis isomers **347** [R^1, R^2=Me, Ph or $(CH_2)_3$; R^3=Ph, Me, CO_2Me] the proximate products (**348**) were not isolated but rearranged spontaneously *via* an aza-Cope process to give the bridged isoquinolines (**349**) in moderate yield (20–65%) (60,206).

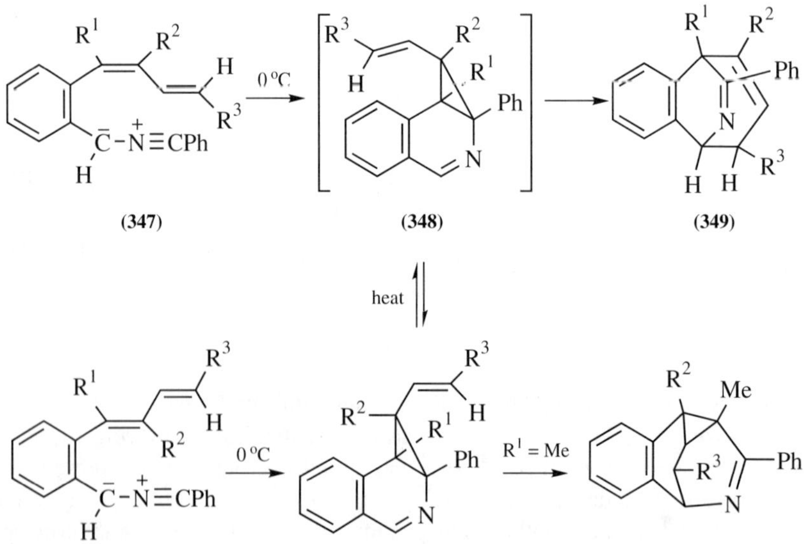

(347) (348) (349)

(350) (351) (352)

However, the trans reactants **350** gave the exo isomers **351** as isolable products. On heating, the latter reacted by two paths, either *via exo/endo* equilibration (cf. **340/342**) forming **348**, which rearranged to give the bridged isoquinolines (**349**), or via a more complex path leading to the azabenzo[3,4]barbaralanes (**352**). The factors controlling the partitioning between these two pathways are not fully understood but it appears that the latter predominates in cases where the endo isomer **348** is sterically disfavored by the presence of two non-hydrogen substituents R^1 and R^2.

7.4.2. Electrocyclization Reactions of Nitrile Imines

Virtually all the basic work on 1,5- and 1,7-electrocyclization reactions of conjugated nitrile imines had been done before the mid-1980s but it is covered effectively in general reviews on 1,7-electrocyclization reactions (197,198) and in a review on intramolecular reactions of nitrile imines (164). Some of this chemistry is illustrated by the reactants **353** in the following scheme (207). The examples that have a trans cyano group were converted into the 1,2-benzodiazepines (**355**) while the ones with a cis cyano group provided an internal competition between 1,1-cycloaddition to the γ,δ-double bond to give **356** and intramolecular cycloaddition to the cyano group to give **357**. In cases where $R^2{=}H$ the competition favored **357** but this was reversed when $R^2{=}Ph$.

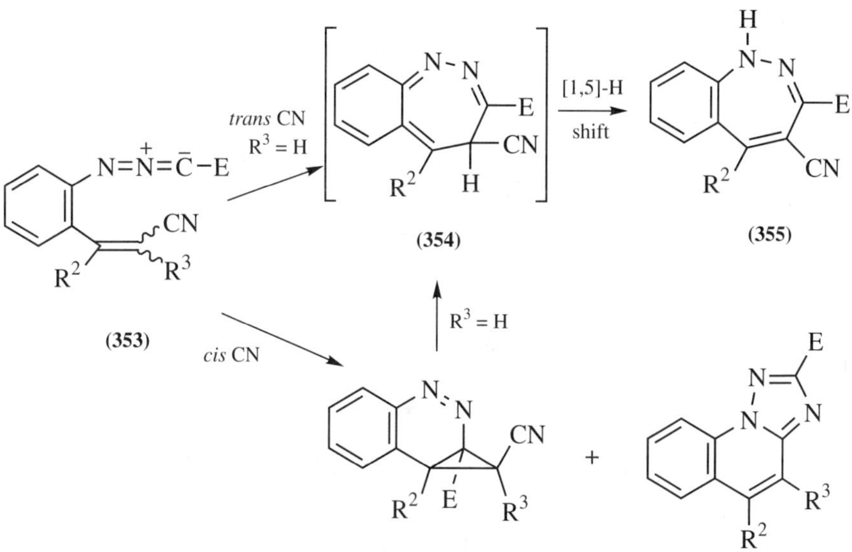

(353) (354) (355) (356) (357)

Similarly the nitrile imines **359** and **362**, formed by thermal rearrangements from **358** and **361**, respectively, cyclized *via* 1,7-electrocyclization to give **360** (55–88%) (208) and **363** (30–68%) (209).

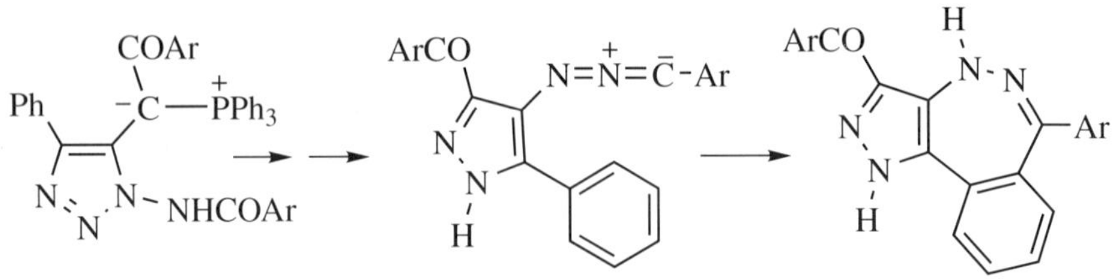

(358) **(359)** **(360)**

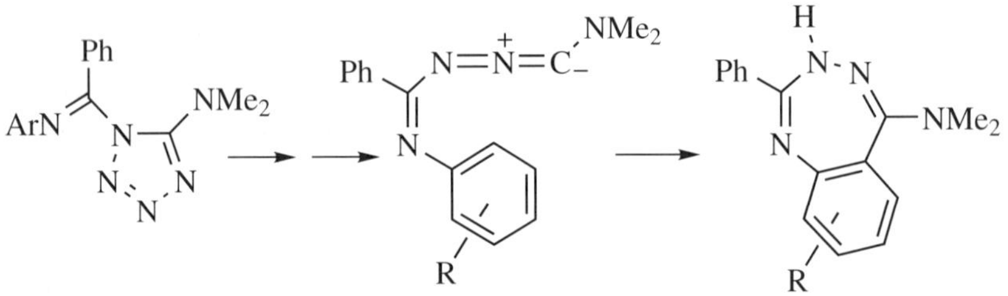

(361) **(362)** **(363)**

Reaction of dichorobenzaldoxazine with sodium azide followed by 1-propanethiol–triethylamine gave the tetrazine imide **367** as the major product. Its formation is thought to involve the reaction sequence shown which includes an unprecedented electrocyclization of the nitrile imine **366** as the final step (210).

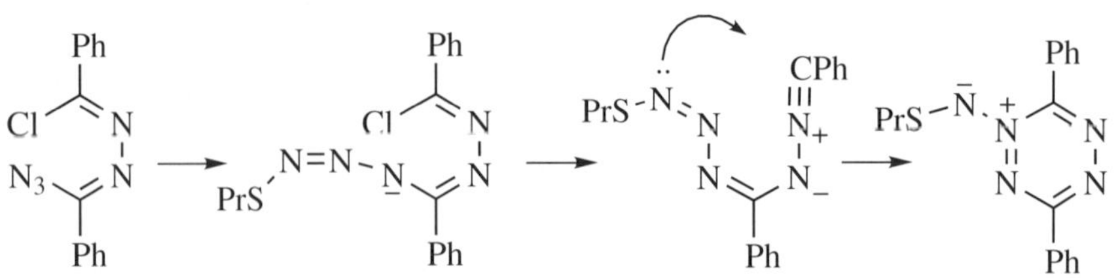

(364) **(365)** **(366)** **(367)**

7.5. REACTIONS WITH NUCLEOPHILES

7.5.1. Reactions of Nitrile Ylides with Nucleophiles

It has been known for a long time that nitrile ylides react with alcohols to give alkoxyimines (e.g., **372**). Recent LFP studies (211,212) have confirmed that the azaallyl cation **370** is an intermediate in this reaction.

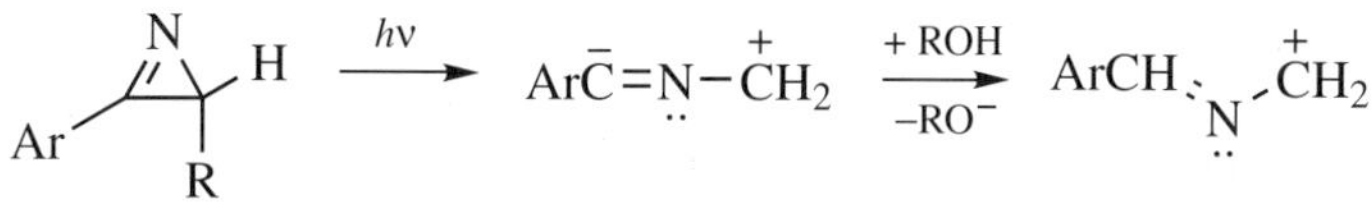

(368) (369) (370)

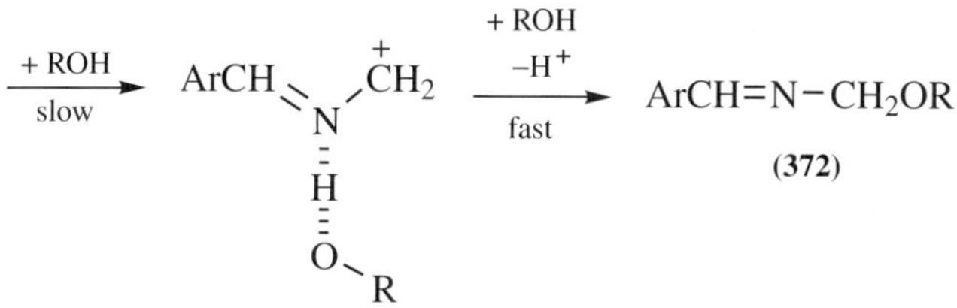

(371) (372)

It was found that the rate of the protonation step depended on both the acidity of the alcohol and the nature of the substituent on the nitrile ylide. After protonation, the azaallyl cation **370** might have been expected to react rapidly with the alkoxy anion but, instead, it was found that it decayed according to pseudo-first-order kinetics *via* reaction with an alcohol molecule at a rate which depended on the alcohol pK_s. The overall mechanism proposed (212) is shown in the scheme.

7.5.2. Reactions of Nitrile Imines with Nucleophiles

The reactions of nitrile imines with nucleophilic reagents in which the addition is followed by cyclization of the primary product, provide a useful route to heterocycles. In most of these reactions, the experiments were carried out by the treatment of hydrazonyl chlorides, as nitrile imine precursors, with triethylamine in the presence of the nucleophilic reagent. The originators generally made the assumption that it was nitrile imines rather than the hydrazonyl chlorides that were the reacting species.

This general process is illustrated by the addition of a range of α-aminoacid esters (**374**) to the nitrile imine **373** (R=Ac, CO_2Me, COPh; R^1=*p*-X–C_6H_4), which proceed with no detectable racemization to give the chiral 1,2,4-triazin-6-ones (**376**) in high yields (213,214).

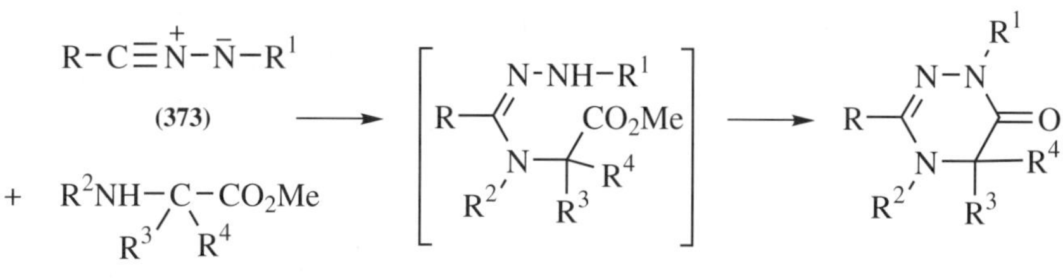

(373) (374) (375) (376)

On the basis of earlier work on the addition of primary and secondary amines, it was suggested that the primary products are the amidrazone adducts **375**, which cyclize under the reaction conditions to give **376**. β-Aminoesters gave acyclic adducts analogous to **375**. Alkoxycarbonylhydrazides (e.g., **377**) undergo a similar addition step to give **378**. These products, however, failed to cyclize *via* nucleophilic substitution on the ester group but, remarkably, when R^3=Me, underwent thermal oxidative closure on heating with charcoal to give **379** (215).

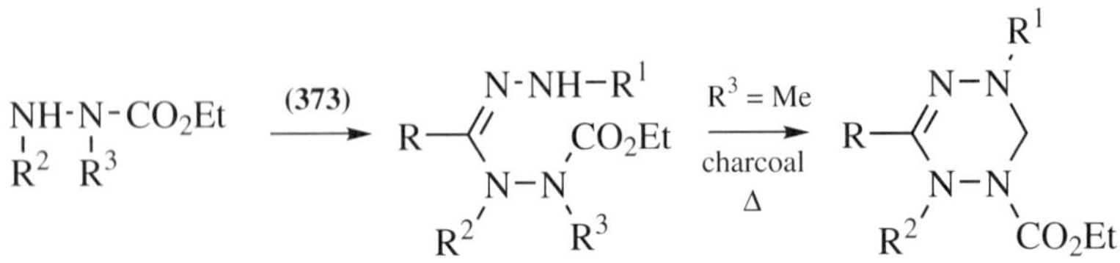

(377) **(378)** **(379)**

Methyl hydrazones of a wide range of aldehydes and ketones (**380**) undergo addition–cyclization to give 1,2,4,5-tetrazines (**382**) *via* **381** (216–219). Substituent effects on the ring-chain tautomerism between **381** and **382** were studied by NMR spectroscopy.

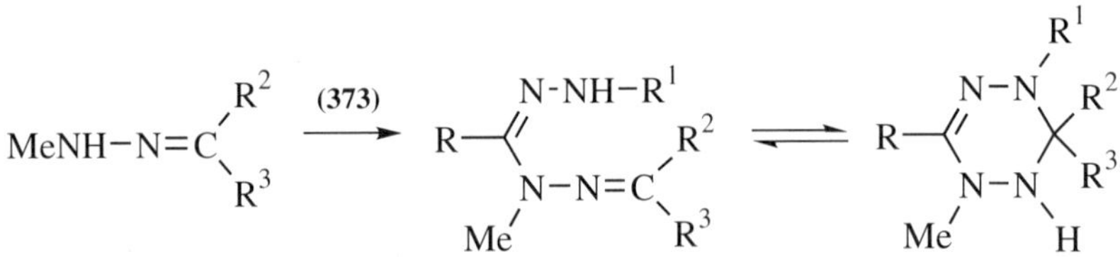

(380) **(381)** **(382)**

The analogous hydrazones (**383**), however, gave only the acyclic adducts (**384**) (220). The latter failed to cyclize when heated in ethanol but did undergo oxidative cyclization when treated with Pd on carbon to give **385**.

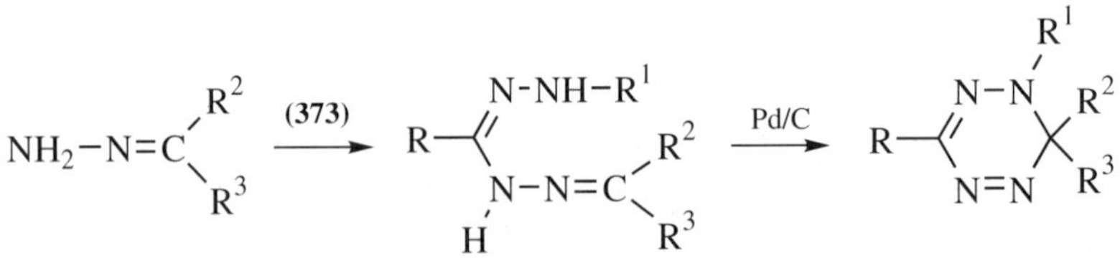

(383) **(384)** **(385)**

Nitrile imines also undergo nucleophilic attack by enamines [e.g., the ketene aminals (**386**) that react to give the pyrazoles **388** (221)]. The intermediate adduct

387 was isolated when Ar=2,4-di-NO_2—C_6H_3. The alternative concerted cycloaddition pathway was ruled out by using an analogue of **386** (R^2, $R^3 \neq H$) in which tautomerism was prevented.

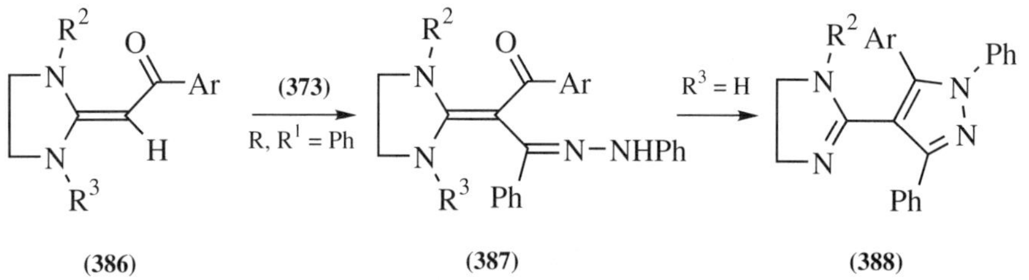

Enamine-type addition has also been observed in 1,2-diazepines (222) and in the intramolecular sense in the reaction of the tetrazole-derived nitrile imine **389** (223).

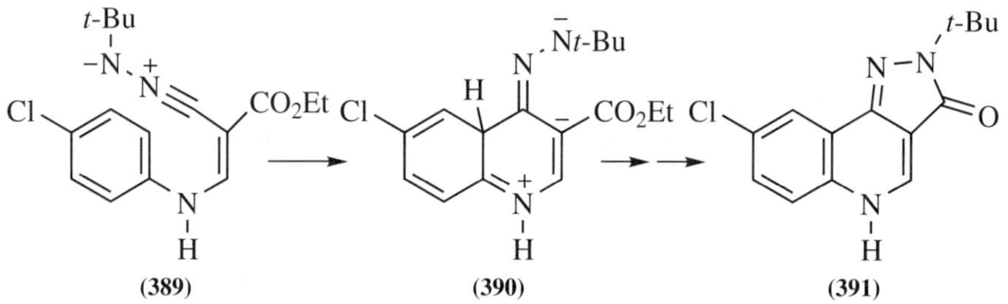

Some earlier contradictory results concerning the reactions of N-unsubstituted azoles have been clarified (224,225) and it has been shown that pyrazole, imidazole, 1,2,4-triazole, and benzotriazole all undergo nucleophilic addition to give (e.g., **392**) from pyrazole and **373** (R, R^1=Ph).

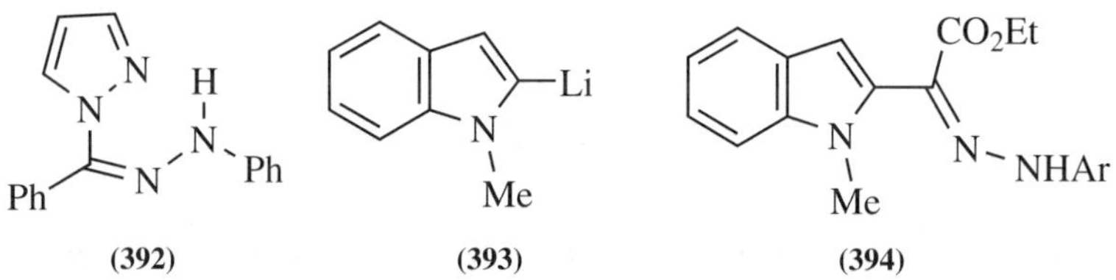

When nitrile imines **373** (R=CO_2Et, R^1=Ar) are generated in the presence of N-methylindole from hydrazonyl chlorides using an excess of butyllithium, then the organolithium derivative **393** reacts exclusively via nucleophilic addition to give **394** (226). However, when a Grignard reagent is used as the base then cycloadducts are also formed in a rapid process involving addition to a complex of the indole and the Grignard reagent [see also (78)]. These additions showed the opposite regioselectivity to that expected but the yields of the adducts were rather low.

Trifluoroacetylacetonitrile (**396**) can also react with nitrile imines **395** via two reaction paths depending on their structure. Those with an electron donating N substituent reacted via cycloaddition to the enolic double bond to give **397** while those containing an electron-withdrawing group reacted *via* nucleophilic addition of the enolate anion to give **398**, which subsequently cyclized to give **399** (227).

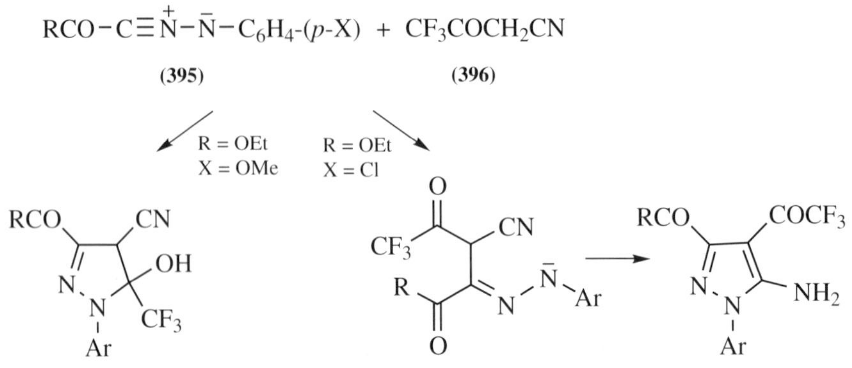

(395) **(396)** **(397)** **(398)** **(399)**

The long-established multistep [3 + 3] reaction of α,β-unsaturated P(III) compounds with nitrile imines leading to phosphorodiazo heterocycles has been reviewed (228) and further extend to the reactions of **400**, which lead to **402**, (229) and references cited therein.

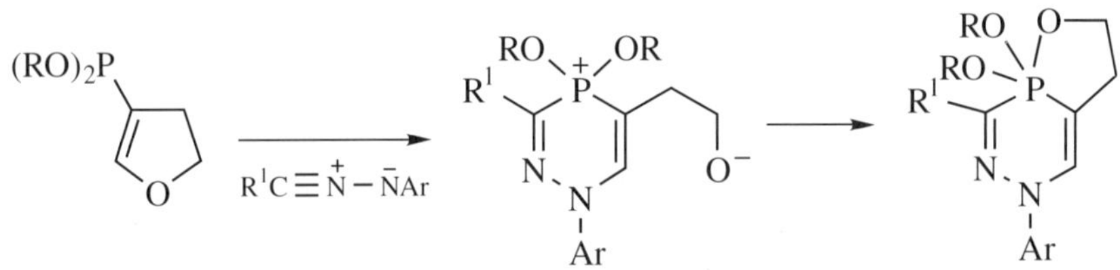

(400) **(401)** **(402)**

7.6. CONCLUSION

Over the last 25 years both nitrile ylides and nitrile imines have continued to provide fascinating and synthetically useful chemistry. In both cases, the exploitation of [3 + 2]-cycloaddition chemistry with an increasing range of dipolarophiles has continued as a key route to five-membered heterocycles. The major development of new chemistry, however, has been in the extensive exploration of intramolecular reactions both in cycloaddition chemistry and in the electrocyclization of 1,3-dipoles with extended conjugation. Such chemistry harnesses the unique reactivity of 1,3-dipoles in the synthesis of relatively elaborate structures but does require the design and preparation of quite complex reactants containing both the 1,3-dipole precursor and the dipolarophilic component. However, access to this chemistry is becoming much easier via the application of new synthetic procedures (e.g., Pd(0) catalyzed coupling reactions) and this will increasingly extend its application to a wider range of synthetic targets.

The extent to which the two 1,3-dipoles have been utilized in synthesis is clearly related to the ease with which they can be generated and the range of substituents available. Thus, the use of nitrile imines in cycloaddition reactions has far exceeded that of nitrile ylides. The main routes to the latter via the dehydohalogenation of imidoyl chlorides and the photolysis of azirines have continued to provide good service, but they have well-known limitations. Attempts to find new and better routes have led to much ingenious and interesting chemistry and have produced some highly effective routes to *nitrile ylide synthons*. However, the only really new route to true nitrile ylides to emerge has been the direct route via the reaction of carbenes or carbenoids with nitriles. The potential of this route is obvious but, while much mechanistic work has been done, the exploitation of this reaction in a synthetic sense and in the invention of new chemistry has been very limited thus far.

REFERENCE

1. E. P. Janulis, S. R. Wilson, and A. J. Arduengo, *Tetrahedron Lett.*, **1984**, *25*, 405.
2. W. Pfleiderer, *Heterocycles*, **1989**, *28*, 203.
3. C. Wentrup, S. Fischer, H. M. Berstermann, M. Kuzaj, H. Luerssen, and K. Burger, *Angew. Chem., Int. Ed. Engl.*, **1986**, *25*, 85.
4. E. Orton, S. T. Collins, and G. C. Pimentel, *J. Phys. Chem.*, **1986**, *90*, 6139.
5. J. Mattay, E. Albrecht, M. Fagnoni, A. Heidbreder, S. Hintz, T. Kirschberg, M. Klessinger, J. Mahlmann, I. Schlachter, and S. Steenken, *J. Inf. Recording*, **1996**, *23*, 23.
6. N. Goldberg, A. Fiedler, and H. Schwarz, *Helv. Chim. Acta*, **1994**, *77*, 2354.
7. G. Maier, J. Eckwert, A. Bothur, H. P. Reisenauer, and C. Schmidt, *Liebigs Ann. Chem.*, **1996**, 1041.
8. M. W. Wong and C. Wentrup, *J. Am. Chem. Soc.*, **1993**, *115*, 7743.
9. S. Kawauchi, T. Tachibana, M. Mori, Y. Shibusa, and T. Yamabe, *J. Mol. Struct. (Theochem)*, **1994**, *310*, 255.
10. K. Kuhler, R. T. Palmer, B. L. Wittkamp, and M. R. Hoffmann, *J. Mol. Struct. (Theochem)*, **1996**, *360*, 41.
11. C. Wentrup, S. Fischer, A. Maquestiau, and R. Flammang, *Angew. Chem., Int. Ed. Engl.*, **1985**, *24*, 56.
12. H. Bock, R. Dammel, S. Fischer, and C. Wentrup, *Tetrahedron Lett.*, **1987**, *28*, 617.
13. K. Bhattacharyya, D. Ramaiah, P. K. Das, and M. V. George, *J. Photochem.*, **1987**, *36*, 63.
14. C. Csongar, P. Leihkauf, V. Lohse, and G. Tomaschewski, *J. Prakt. Chem.*, **1985**, *327*, 96.
15. G. Sicard, A. Baceiredo, and G. Bertrand, *J. Am. Chem. Soc.*, **1988**, *110*, 2663.
16. G. Bertrand and C. Wentrup, *Angew. Chem., Int. Ed. Engl.*, **1994**, *33*, 527.
17. K. H. Vonlocquenghien, R. Reau, and G. Bertrand, *J. Chem. Soc., Chem. Commun.*, **1991**, 1192.
18. J. L. Faure, R. Reau, M. W. Wong, R. Koch, C. Wentrup, and G. Bertrand, *J. Am. Chem. Soc.*, **1997**, *119*, 2819.
19. A. Padwa ed., *1,3-Dipolar Cycloaddition Chemistry*, Wiley, New York, 1984.
20. N. J. Turro, Y. Cha, I. R. Gould, A. Padwa, J. R. Gasdaska, and M. Tomas, *J. Org. Chem.*, **1985**, *50*, 4415.
21. A. Padwa, J. R. Gasdaska, M. Tomas, N. J. Turro, Y. Cha, and I. R. Gould, *J. Am. Chem. Soc.*, **1986**, *108*, 6739.

22. A. Padwa, J. R. Gasdaska, G. Haffmanns, and H. Rebello, *J. Org. Chem.*, **1987**, *52*, 1027.
23. O. Tsuge, S. Kanemasa, T. Yamada, and K. Matsuda, *Heterocycles*, **1985**, *23*, 2489.
24. O. Tsuge, S. Kanemasa, T. Yamada, and K. Matsuda, *J. Org. Chem.*, **1987**, *52*, 2523.
25. O. Tsuge, S. Kanemasa, and K. Matsuda, *Chem. Lett.*, **1985**, 1411.
26. O. Tsuge, S. Kanemasa, and K. Matsuda, *J. Org. Chem.*, **1986**, *51*, 1997.
27. O. Tsuge, T. Hatta, Y. Kakura, H. Tashiro, H. Maeda, and A. Kakehi, *Chem. Lett.*, **1997**, 945.
28. O. Tsuge, T. Hatta, H. Tashiro, Y. Kakura, H. Maeda, and A. Kakehi, *Tetrahedron*, **2000**, *56*, 7723.
29. O. Tsuge, T. Hatta, H. Tashiro, and H. Maeda, *Heterocycles*, **2001**, *55*, 243.
30. O. Tsuge, T. Hatta, M. Shinozuka, and H. Tashiro, *Heterocycles*, **2001**, *55*, 249.
31. K. Washizuka, S. Minakata, I. Ryu, and M. Komatsu, *Tetrahedron*, **1999**, *55*, 12969.
32. M. Ohno, M. Komatsu, H. Miyata, and Y. Ohshiro, *Tetrahedron Lett.*, **1991**, *32*, 5813.
33. O. Tsuge, K. Ueno, S. Kanemasa, and K. Yorozu, *Bull. Chem. Soc. Jpn.*, **1986**, *59*, 1809.
34. J. M. Lerestif, J. P. Bazureau, and J. Hamelin, *Synlett*, **1995**, 647.
35. J. M. Lerestif, L. Toupet, S. Sinbandhit, F. Tonnard, J. P. Bazureau, and J. Hamelin, *Tetrahedron*, **1997**, *53*, 6351.
36. S. Kohra and Y. Tominaga, *Heterocycles*, **1994**, *38*, 1217.
37. S. Kohra, K. Ueda, and Y. Tominaga, *Chem. Pharm. Bull.*, **1995**, *43*, 204.
38. M. Oba, M. Yoshihara, J. Nagatsuka, and K. Nishiyama, *Heterocycles*, **1997**, *45*, 1913.
39. K. Burger, H. Neuhauser, and M. Eggersdorfer, *Synthesis*, **1987**, 924.
40. V. Dragisich, W. D. Wulff, and K. Hoogsteen, *Organometallics*, **1990**, *9*, 2867.
41. D. Griller, L. Hadel, A. S. Nazram, M. S. Platz, P. C. Wong, T. G. Savino, and J. C. Scaiano, *J. Am. Chem. Soc.*, **1984**, *106*, 2227.
42. P. B. Grasse, B.-E. Brauer, J. J. Zupanic, K. J. Kaufmann, and G. B. Schuster, *J. Am. Chem. Soc.*, **1983**, *105*, 6833.
43. A. S. Kende, P. Hebeisen, P. J. Sanfilippo, and B. H. Toder, *J. Am. Chem. Soc.*, **1982**, *104*, 4244.
44. N. J. Turro, D. A. Hrovat, I. R. Gould, A. Padwa, W. Dent, and R. J. Rosenthal, *Angew. Chem., Int. Ed. Engl.*, **1983**, *22*, 625.
45. R. L. Barcus, B. B. Wright, M. S. Platz, and J. C. Scaiano, *Tetrahedron Lett.*, **1983**, *24*, 3955.
46. L. M. Hadel, M. S. Platz, and J. C. Scaiano, *Chem. Phys. Lett.*, **1983**, *97*, 446.
47. A. A. Abdel-Wahab, S. H. Doss, H. Durr, N. J. Turro, and I. R. Gould, *J. Org. Chem.*, **1987**, *52*, 429.
48. J. A. Lavilla and J. L. Goodman, *Tetrahedron Lett.*, **1988**, *29*, 2623.
49. A. Admasu, A. D. Gudmundsdottir, and M. S. Platz, *J. Phys. Chem. A*, **1997**, *101*, 3832.
50. I. Naito, K. Nakamura, T. Kumagai, A. Oku, K. Hori, K. Matsuda, and H. Iwamura, *J. Phys. Chem. A*, **1999**, *103*, 8187.
51. I. Naito, A. Oku, Y. Fijiwara, and Y. Tanimoto, *J. Chem. Soc., Perkin Trans. 2*, **1999**, 1051.
52. C. Bornemann and M. Klessinger, *Chem. Phys.*, **2000**, *259*, 263.
53. I. Naito, A. Ishida, K. Yamamoto, S. Takamuku, K. Isomura, and H. Taniguchi, *Chem. Lett.*, **1989**, 1615.
54. I. Naito, H. Morihara, A. Ishida, S. Takamuku, K. Isomura, and H. Taniguchi, *Bull. Chem. Soc. Jpn.*, **1991**, *64*, 2757.
55. J. C. Scaiano, W. G. Mcgimpsey, and H. L. Casal, *J. Am. Chem. Soc.*, **1985**, *107*, 7204.
56. F. Muller, J. Mattay, and S. Steenken, *J. Org. Chem.*, **1993**, *58*, 4462.
57. A. Ishida, I. Naito, and S. Takamuku, *Chem. Lett.*, **1992**, 1913.
58. K. E. Cullen and J. T. Sharp, *J. Chem. Soc., Perkin Trans. 1*, **1993**, 2961.
59. J. P. Strachan, J. T. Sharp, and S. Parsons, *J. Chem. Soc., Perkin Trans. 1*, **1998**, 807.

60. J. P. Strachan, J. T. Sharp, and M. J. Crawshaw, *J. Chem. Soc., Perkin Trans. 1*, **1999**, 443.
61. H. Finch, D. H. Reece, and J. T. Sharp, *J. Chem. Soc., Perkin Trans. 1*, **1994**, 1193.
62. R. Bossio, S. Marcaccini, and R. Pepino, *Tetrahedron Lett.*, **1986**, *27*, 4643.
63. R. Bossio, S. Maraccini, P. Paoli, R. Pepino, and C. Polo, *Heterocycles*, **1990**, *31*, 1855.
64. F. Berree, E. Marchand, and G. Morel, *Tetrahedron Lett.*, **1992**, *33*, 6155.
65. W. S. Huang, Y. X. Zhang, and C. Y. Yuan, *J. Chem. Soc., Perkin Trans. 1*, **1996**, 1893.
66. A. Padwa, R. J. Rosenthal, W. Dent, P. Filho, N. J. Turro, D. A. Hrovat, and I. R. Gould, *J. Org. Chem.*, **1984**, *49*, 3174.
67. R. W. Hoffmann and W. Barth, *Chem. Ber.*, **1985**, *118*, 634.
68. W. P. Fehlhammer, D. Achatz, U. Plaia, and A. Volkl, *Z. Naturforsch., B: Chem. Sci.*, **1987**, *42*, 720.
69. W. P. Fehlhammer, A. Volkl, U. Plaia, and G. Beck, *Chem. Ber.*, **1987**, *120*, 2031.
70. U. Eckert, C. Robl, and W. P. Fehlhammer, *Organometallics*, **1993**, *12*, 3241.
71. W. P. Fehlhammer, H. Hoffmeister, U. Eckert, and C. Robl, *Z. Naturforsch., B: Chem. Sci.*, **1993**, *48*, 1448.
72. H. Wamhoff and M. Zahran, *Synthesis*, **1987**, 876.
73. G. Broggini, L. Garanti, G. Molteni, and G. Zecchi, *Tetrahedron*, **1998**, *54*, 2843.
74. L. Garanti, G. Molteni, and G. Zecchi, *Heterocycles*, **1995**, *40*, 777.
75. G. Broggini, L. Garanti, G. Molteni, and G. Zecchi, *Heterocycles*, **1997**, *45*, 1945.
76. G. Molteni, M. Orlandi, and G. Broggini, *J. Chem. Soc., Perkin Trans. 1*, **2000**, 3742.
77. H. Kaddar, J. Hamelin, and H. Benhaoua, *J. Chem. Res., Synop.*, **1999**, 718.
78. S. Kanemasa and S. Kobayashi, *Bull. Chem. Soc. Jpn.*, **1993**, *66*, 2685.
79. K. M. L. Rai and A. Hassner, *Synth. Commun.*, **1989**, *19*, 2799.
80. P. Bourgeois, R. Cantegril, A. Chene, J. Gelin, J. Mortier, and J. Moyroud, *Synth. Commun.*, **1993**, *23*, 3195.
81. K. M. L. Rai and N. Linganna, *Synth. Commun.*, **1997**, *27*, 3737.
82. D. W. Chen and Z. C. Chen, *Synth. Commun.*, **1995**, *25*, 1617.
83. J. E. Bishop, K. A. Flaxman, B. S. Orlek, P. G. Sammes, and D. J. Weller, *J. Chem. Soc., Perkin Trans. 1*, **1995**, 2551.
84. I. Naito, Y. Fujiwara, Y. Tanimoto, A. Ishida, and S. Takamuku, *Bull. Chem. Soc. Jpn.*, **1995**, *68*, 2905.
85. M. D. Su, H. L. Liao, W. S. Chung, and S. Y. Chu, *J. Org. Chem.*, **1999**, *64*, 6710.
86. P. Wipf and H. Heimgartner, *Chimia*, **1984**, *38*, 357.
87. P. Wipf, R. Prewo, J. H. Bieri, G. Germain, and H. Heimgartner, *Helv. Chim. Acta*, **1988**, *71*, 1177.
88. H. Heimgartner, *Phosphorus Sulfur Silicon Relat. Elem.*, **1991**, *58*, 281.
89. N. Bozhkova and H. Heimgartner, *Helv. Chim. Acta*, **1989**, *72*, 825.
90. N. Bojkova and H. Heimgartner, *Heterocycles*, **1998**, *47*, 781.
91. A. Linden, N. Bojkova, and H. Heimgartner, *Acta Crystallogr., Sect. C: Cryst. Struct. Commun.*, **1998**, *54*, 373.
92. P. de March, M. el Arrad, M. Figueredo, and J. Font, *Tetrahedron*, **1998**, *54*, 11613.
93. J. Castulik, J. Jonas, and C. Mazal, *Collect. Czech. Chem. Commun.*, **2000**, *65*, 708.
94. A. B. N. Luheshi, R. K. Smalley, P. D. Kennewell, and R. Westwood, *Tetrahedron Lett.*, **1990**, *31*, 123.
95. K. Hemming, A. B. N. Luheshi, A. D. Redhouse, R. K. Smalley, J. R. Thompson, P. D. Kennewell, and R. Westwood, *Tetrahedron*, **1993**, *49*, 4383.
96. J. Perrocheau, R. Danionbougot, and R. Carrie, *Bull. Soc. Chim. Fr.*, **1994**, *131*, 973.

97. J. Perrocheau, D. Danion, and R. Carrie, *Bull. Soc. Chim. Belg.*, **1995**, *104*, 39.
98. J. Averdung, E. Albrecht, J. Lauterwein, H. Luftmann, J. Mattay, H. Mohn, W. H. Muller, and H. U. Termeer, *Chem. Ber.*, **1994**, *127*, 787.
99. J. Averdung and J. Mattay, *Tetrahedron*, **1996**, *52*, 5407.
100. Y. Y. C. Y. Ko, R. Carrie, F. Desarlo, and A. Brandi, *Can. J. Chem.*, **1983**, *61*, 1105.
101. W. Ried and M. Fulde, *Helv. Chim. Acta*, **1988**, *71*, 1681.
102. M. Granier, A. Baceiredo, Y. Dartiguenave, M. Dartiguenave, M. J. Menu, and G. Bertrand, *J. Am. Chem. Soc.*, **1990**, *112*, 6277.
103. F. Palacios, J. Pagalday, V. Piquet, F. Dahan, A. Baceiredo, and G. Bertrand, *J. Org. Chem.*, **1997**, *62*, 292.
104. R. Reau, G. Veneziani, F. Dahan, and G. Bertrand, *Angew. Chem., Int. Ed. Engl.*, **1992**, *31*, 439.
105. F. Castan, A. Baceiredo, and G. Bertrand, *Angew. Chem., Int. Ed. Engl.*, **1989**, *28*, 1250.
106. F. Castan, A. Baceiredo, D. Bigg, and G. Bertrand, *J. Org. Chem.*, **1991**, *56*, 1801.
107. J. Hasserodt, H. Pritzkow, and W. Sundermeyer, *Liebigs Ann. Chem.*, **1995**, 95.
108. N. Dubau-Assibat, A. Baceiredo, and G. Bertrand, *J. Org. Chem.*, **1995**, *60*, 3904.
109. M. Granier, A. Baceiredo, M. Nieger, and G. Bertrand, *Angew. Chem., Int. Ed. Engl.*, **1990**, *29*, 1123.
110. M. Granier, A. Baceiredo, G. Bertrand, V. Huch, and M. Veith, *Inorg. Chem.*, **1991**, *30*, 1161.
111. J. Barluenga, F. Fernandez-Mari, E. Aguilar, A. L. Viado, and B. Olano, *Tetrahedron Lett.*, **1998**, *39*, 4887.
112. J. Barluenga, F. Fernandez-Mari, R. Gonzalez, E. Aguilar, G. A. Revelli, A. L. Viado, F. J. Fananas, and B. Olano, *Eur. J. Org. Chem.*, **2000**, 1773.
113. A. Agocs, A. Benyei, L. Somogyi, and P. Herczegh, *Tetrahedron-Asymmetry*, **1998**, *9*, 3359.
114. G. Burton, G. J. Clarke, J. D. Douglas, A. J. Eglington, C. H. Frydrych, J. D. Hinks, N. W. Hird, E. Hunt, S. F. Moss, A. Naylor, N. H. Nicholson, and M. J. Pearson, *J. Antibiot.*, **1996**, *49*, 1266.
115. L. Grubert, G. Galley, and M. Patzel, *Tetrahedron-Asymmetry*, **1996**, *7*, 1137.
116. J. M. J. Tronchet, J. F. Tronchet, and F. Barbalatrey, *Heterocycles*, **1993**, *36*, 833.
117. M. Freccero, R. Gandolfi, and M. Sarzi-Amade, *Heterocycles*, **1998**, *49*, 415.
118. S. Muthu, P. Maruthamuthu, R. Ragunathan, P. R. V. Rao, and C. K. Mathews, *Tetrahedron Lett.*, **1994**, *35*, 1763.
119. U. Hartmann and G. Tomaschewski, *J. Inform. Recording*, **1996**, *22*, 573.
120. Y. Matsubara, H. Muraoka, H. Tada, and Z. Yoshida, *Chem. Lett.*, **1996**, 373.
121. G. A. Ahmed, N. A. Ismail, and F. A. Yassin, *Phosphorus Sulfur Silicon Relat. Elem.*, **1992**, *69*, 253.
122. G. Markl, I. Troetschschaller, and W. Holzl, *Tetrahedron Lett.*, **1988**, *29*, 785.
123. A. S. Shawali, *Chem. Rev.*, **1993**, *93*, 2731.
124. A. B. N. Luheshi, R. K. Smalley, P. D. Kennewell, and R. Westwood, *Tetrahedron Lett.*, **1990**, *31*, 127.
125. E. Coutouliargyropoulou and E. Malamidouxenikaki, *J. Chem. Res., Synop.*, **1991**, 106.
126. D. J. Miller, R. M. Scrowston, P. D. Kennewell, and R. Westwood, *Tetrahedron*, **1994**, *50*, 5159.
127. H. A. Ead, H. M. Hassaneen, M. A. Abdallah, and H. A. H. Mousa, *Arch. Pharm. (Weinheim, Ger.)*, **1991**, *324*, 35.
128. K. K. Singal and M. R. Giare, *Indian J. Heterocycl. Chem.*, **1998**, *8*, 121.
129. K. Polborn and A. Schmidpeter, *Heteroatom Chemistry*, **1999**, *10*, 243.
130. M. L. Gelmi, D. Pocar, and R. Riva, *Heterocycles*, **1992**, *34*, 315.
131. P. D. Kennewell, D. J. Miller, R. M. Scrowston, and R. Westwood, *J. Chem. Soc., Perkin Trans. 1*, **1994**, 3563.
132. L. Grubert, W. Jugelt, H. J. Bress, H. Koppel, U. Strietzel, and A. Dombrowski, *Liebigs Ann. Chem.*, **1992**, 885.

133. P. Caramella, A. G. Invernizzi, E. Pastormerlo, P. Quadrelli, and A. Corsaro, *Heterocycles*, **1995**, *40*, 515.

134. L. Grubert, M. Patzel, W. Jugelt, B. Riemer, and J. Liebscher, *Liebigs Ann. Chem.*, **1994**, 1005.

135. G. Krajsovszky, A. Gaal, N. Haider, and P. Matyus, *J. Mol. Struct. (Theochem)*, **2000**, *528*, 13.

136. N. G. Argyropoulos, E. Corobili, and E. P. Raptopoulou, *J. Heterocycl. Chem.*, **1996**, *33*, 591.

137. S. Kitane, A. Sebban, J. Vebrel, and B. Laude, *Bull. Soc. Chim. Belg.*, **1989**, *98*, 105.

138. G. Stajer, G. Bernath, A. E. Szabo, P. Sohar, G. Argay, and A. Kalman, *Tetrahedron*, **1987**, *43*, 5461.

139. G. Stajer, A. E. Szabo, G. Bernath, and P. Sohar, *Tetrahedron*, **1987**, *43*, 1931.

140. J. Szabo, L. Fodor, E. Baniakoto, G. S. Talpas, G. Bernath, and P. Sohar, *Tetrahedron*, **1992**, *48*, 4957.

141. L. Fodor, M. S. Elgharib, J. Szabo, G. Bernath, and P. Sohar, *Heterocycles*, **1984**, *22*, 537.

142. A. K. Baruah, D. Prajapati, and J. S. Sandhu, *Heterocycles*, **1986**, *24*, 1527.

143. A. Baouid, A. Benharref, A. Hasnaoui, and J. P. Lavergne, *Bull. Soc. Chim. Belg.*, **1994**, *103*, 743.

144. A. Hasnaoui, A. Baouid, and J. P. Lavergne, *J. Heterocycl. Chem.*, **1991**, *28*, 73.

145. M. Elmessaoudi, A. Hasnaoui, M. Elmouhtadi, and J. P. Lavergne, *Bull. Soc. Chim. Belg.*, **1992**, *101*, 977.

146. C. Hamon, J. Lorans, J. C. Corbel, J. M. Jego, L. Toupet, and P. Uriac, *Bull. Soc. Chim. Fr.*, **1997**, *134*, 525.

147. R. Benelbaghdadi, A. Benharref, A. Hasnaoui, J. P. Lavergne, M. Giorgi, and M. Pierrot, *Acta Crystallogr., Sect. C: Cryst. Struct. Commun.*, **1998**, *54*, 1343.

148. R. Benelbaghdadi, A. Hasnaoui, and J. P. Lavergne, *Bull. Soc. Chim. Belg.*, **1997**, *106*, 813.

149. R. Benelbaghdadi, A. Hasnaoui, J. P. Lavergne, M. Giorgi, and M. Pierrot, *Synth. Commun.*, **1998**, *28*, 4221.

150. M. Essaber, A. Baouid, A. Hasnaoui, A. Benharref, and J. P. Lavergne, *Synth. Commun.*, **1998**, *28*, 4097.

151. A. Chiaroni, C. Riche, A. Baouid, A. Benharref, A. Hasnaoui, and J. P. Lavergne, *Acta Crystallogr., Sect. C: Cryst. Struct. Commun.*, **1995**, *51*, 961.

152. M. Essaber, A. Baouid, A. Hasnaoui, M. Giorgi, and M. Pierrot, *Acta Crystallogr., Sect. C: Cryst. Struct. Commun.*, **1998**, *54*, 519.

153. A. Aatif, A. Baouid, H. Hasnaoui, A. Benharref, and M. Pierrot, *Acta Crystallogr., Sect. C: Cryst. Struct. Commun.*, **1999**, *55*, 1528.

154. S. Stverkova, Z. Zak, and J. Jonas, *Liebigs Ann. Chem.*, **1995**, 477.

155. J. B. F. Dunstan, G. M. Elsey, R. A. Russell, G. P. Savage, G. W. Simpson, and E. R. T. Tiekink, *Aust. J. Chem.*, **1998**, *51*, 499.

156. D. Singh and D. Singh, *J. Indian Chem. Soc.*, **1991**, *68*, 580.

157. S. Papadopoulos and J. Stephanidoustephanatou, *J. Heterocycl. Chem.*, **1987**, *24*, 309.

158. M. Ballabio, P. D. Croce, V. Massari, D. Pocar, A. Riva, and P. Trimarco, *J. Chem. Res., Synop.*, **1983**, 140.

159. E. Coutouliargyropoulou, N. G. Argyropoulos, and E. Thessalonikeos, *J. Chem. Res., Synop.*, **1990**, 202.

160. E. Coutouliargyropoulou and E. Thessalonikeos, *Liebigs Ann. Chem.*, **1990**, 1097.

161. A. Linden, E. M. A. H. Awad, and H. Heimgartner, *Acta Crystallogr., Sect. C: Cryst. Struct. Commun.*, **1999**, *55*, 1877.

162. E. M. Rakib, M. Benchidmi, E. M. Essassi, A. El Bouadili, M. Khouili, J. M. Barbe, and M. D. Pujol, *Heterocycles*, **2000**, *53*, 571.

163. H. M. Hassaneen, H. A. Daboun, H. A. Abdelhadi, and N. A. Abdelreheim, *Phosphorus Sulfur Silicon Relat. Elem.*, **1995**, *107*, 269.

164. G. Broggini, G. Molteni, and G. Zecchi, *Heterocycles*, **1998**, *47*, 541.
165. G. Broggini, L. Garanti, G. Molteni, and G. Zecchi, *Tetrahedron-Asymmetry*, **1999**, *10*, 487.
166. G. Broggini, L. Garanti, G. Molteni, and G. Zecchi, *J. Chem. Soc., Perkin Trans. 1*, **1998**, 4103.
167. G. Broggini, L. Garanti, G. Molteni, and G. Zecchi, *Heterocycles*, **2000**, *53*, 917.
168. G. Broggini, L. Garanti, G. Molteni, and G. Zecchi, *Heterocycles*, **2000**, *53*, 831.
169. L. Bruche, L. Garanti, and G. Zecchi, *J. Chem. Soc., Perkin Trans. 1*, **1984**, 2535.
170. G. Labbe, S. Emmers, W. Dehaen, and L. K. Dyall, *J. Chem. Soc., Perkin Trans. 1*, **1994**, 2553.
171. G. Broggini, L. Bruche, A. Farina, S. V. Meille, G. Molteni, and G. Zecchi, *J. Chem. Res., Synop.*, **2000**, 390.
172. L. Bruche and G. Zecchi, *J. Chem. Res., Synop.*, **1986**, 210.
173. G. Broggini, L. Garanti, G. Molteni, and G. Zecchi, *Synthesis*, **1996**, 1076.
174. H. Meier and H. Heimgartner, *Helv. Chim. Acta*, **1985**, *68*, 1283.
175. D. Janietz, K. Khoudary, and W. D. Rudorf, *J. Prakt. Chem.*, **1987**, *329*, 343.
176. D. Janietz and W. D. Rudorf, *Tetrahedron*, **1989**, *45*, 1661.
177. L. Bruche and G. Zecchi, *Tetrahedron*, **1989**, *45*, 7427.
178. G. Broggini, L. Garanti, G. Molteni, T. Pilati, A. Ponti, and G. Zecchi, *Tetrahedron-Asymmetry*, **1999**, *10*, 2203.
179. G. Broggini, G. Casalone, L. Garanti, G. Molteni, T. Pilati, and G. Zecchi, *Tetrahedron-Asymmetry*, **1999**, *10*, 4447.
180. G. Broggini, L. Garanti, G. Molteni, and G. Zecchi, *Tetrahedron*, **1998**, *54*, 14859.
181. G. Molteni and T. Pilati, *Tetrahedron-Asymmetry*, **1999**, *10*, 3873.
182. L. Garanti and G. Zecchi, *J. Heterocycl. Chem.*, **1993**, *30*, 559.
183. G. Broggini, L. Bruche, L. Garanti, and G. Zecchi, *J. Chem. Soc., Perkin Trans. 1*, **1994**, 433.
184. G. Broggini, L. Garanti, G. Molteni, and G. Zecchi, *Heterocycles*, **1994**, *38*, 1601.
185. G. Broggini, L. Garanti, G. Molteni, and G. Zecchi, *Synthesis*, **1995**, 1483.
186. S. M. Sayed, *J. Chin. Chem. Soc.*, **2000**, *47*, 933.
187. G. Broggini, L. Bruche, and G. Zecchi, *Org. Prep. Proced. Int.*, **1992**, *24*, 352.
188. G. Broggini, L. Garanti, G. Molteni, and G. Zecchi, *Tetrahedron*, **1997**, *53*, 3005.
189. G. Broggini, G. Molteni, and T. Pilati, *Tetrahedron-Asymmetry*, **2000**, *11*, 1975.
190. W. M. F. Fabian, C. O. Kappe, and V. A. Bakulev, *J. Org. Chem.*, **2000**, *65*, 47.
191. A. Furstner, T. Gastner, and H. Weintritt, *J. Org. Chem.*, **1999**, *64*, 2361.
192. R. Connell, P. Scavo, P. Helquist, and B. Akermark, *Tetrahedron Lett.*, **1986**, *27*, 5559.
193. T. Ibata and K. Fukushima, *Chem. Lett.*, **1992**, 2197.
194. K. Fukushima and T. Ibata, *Bull. Chem. Soc. Jpn.*, **1995**, *68*, 3469.
195. K. Fukushima and T. Ibata, *Heterocycles*, **1995**, *40*, 149.
196. K. Fukushima, Y. Q. Lu, and T. Ibata, *Bull. Chem. Soc. Jpn.*, **1996**, *69*, 3289.
197. G. Zecchi, *Synthesis*, **1991**, 181.
198. P. W. Groundwater and M. Nyerges, *Adv. Heterocycl. Chem.*, **1999**, *73*, 97.
199. P. W. Groundwater, C. Struthers-Semple, and J. T. Sharp, *J. Chem. Soc., Chem. Commun.*, **1987**, 1367.
200. K. E. Cullen and J. T. Sharp, *J. Chem. Soc., Chem. Commun.*, **1991**, 658.
201. K. E. Cullen and J. T. Sharp, *J. Chem. Soc., Perkin Trans. 1*, **1995**, 2565.
202. D. F. OShea and J. T. Sharp, *J. Chem. Soc., Perkin Trans. 1*, **1997**, 3025.
203. P. W. Groundwater and J. T. Sharp, *Tetrahedron*, **1992**, *48*, 7951.
204. P. W. Groundwater, A. J. Morton, and R. Salter, *J. Chem. Soc., Chem. Commun.*, **1993**, 1789.

205. K. R. Motion, I. R. Robertson, J. T. Sharp, and M. D. Walkinshaw, *J. Chem. Soc., Perkin Trans. 1*, **1992**, 1709.

206. J. P. Strachan, J. T. Sharp, and S. Parsons, *J. Chem. Soc., Chem. Commun.*, **1996**, 2739.

207. L. Bruche, L. Garanti, and G. Zecchi, *J. Chem. Soc., Perkin Trans. 1*, **1983**, 539.

208. E. Laskos, P. S. Lianis, N. A. Rodios, A. Terzis, and C. P. Raptopoulou, *Tetrahedron Lett.*, **1995**, *36*, 5637.

209. G. V. Boyd, J. Cobb, P. F. Lindley, J. C. Mitchell, and G. A. Nicolaou, *J. Chem. Soc., Chem. Commun.*, **1987**, 99.

210. P. Kaszynski and V. G. Young, *J. Am. Chem. Soc.*, **2000**, *122*, 2087.

211. I. Naito, A. Ishida, S. Takamuku, K. Isomura, and H. Taniguchi, *J. Chem. Soc., Perkin Trans. 2*, **1992**, 1985.

212. E. Albrecht, J. Mattay, and S. Steenken, *J. Am. Chem. Soc.*, **1997**, *119*, 11605.

213. M. M. Elabadelah, A. Q. Hussein, and B. A. Thaher, *Heterocycles*, **1991**, *32*, 1879.

214. D. J. Collins, T. C. Hughes, and W. M. Johnson, *Aust. J. Chem.*, **1996**, *49*, 463.

215. M. R. El Haddad, A. E. S. Ferwanah, and A. M. Awadallah, *J. Prakt. Chem.*, **1998**, *340*, 623.

216. M. M. Elabadelah, A. Q. Hussein, M. R. Kamal, and K. H. Aladhami, *Heterocycles*, **1988**, *27*, 917.

217. A. Q. Hussein, M. M. Elabadelah, K. A. Adhami, and A. S. Abushamleh, *Heterocycles*, **1989**, *29*, 1163.

218. M. M. Elabadelah, A. Q. Hussein, and A. S. Abushamleh, *J. Prakt. Chem.*, **1991**, *333*, 61.

219. M. M. Elabadelah, A. Q. Hussein, and H. A. Saadeh, *Heterocycles*, **1991**, *32*, 1063.

220. A. Q. Hussein, *J. Chem. Res., Synop.*, **1996**, 174.

221. B. Liu, M. X. Wang and Z. T. Huang, *Tetrahedron Lett.*, **1999**, *40*, 7399.

222. M. Elmessaoudi, A. Hasnaoui, and J. P. Lavergne, *Synth. Commun.*, **1994**, *24*, 513.

223. S. Isoda and H. Kanno, *Heterocycles*, **1992**, *33*, 273.

224. A. Delahoz, C. Pardo, A. Sanchez, and J. Elguero, *J. Chem. Res., Synop.*, **1990**, 294.

225. L. Grubert, W. Jugelt, H. J. Bress, and U. Strietzel, *Z. Chem.*, **1990**, *30*, 286.

226. T. Benincori, F. Sannicolo, L. Trimarco, L. Bonati, S. Grandi, D. Pitea, and C. Gatti, *J. Phys. Org. Chem.*, **1998**, *11*, 455.

227. H. F. Zohdi, *J. Chem. Res., Synop.*, **1992**, 82.

228. Y. G. Trishin, V. N. Chistokletov, and A. A. Petrov, *Sov. Sci. Rev. B. Chem.*, **1991**, *15*, 1.

229. Y. G. Trishin, V. I. Namestnikov, and V. K. Bel'skii, *Russ. J. Gen. Chem.*, **1999**, *69*, 736.

CHAPTER 8

Diazoalkanes

Gerhard Maas

Division of Organic Chemistry I, University of Ulm, Ulm, Germany

Dedicated to Manfred Regitz

8.1. Cycloadditions with C=C Bonds 541
8.1.1. Reactivity and Regioselectivity 541
8.1.2. Diazo Dipoles 543
8.1.3. Alkenes 549
8.1.4. Diastereofacial Selectivity 552
8.2. Cycloadditions with Carbon–Heteroatom Double Bonds 557
8.2.1. C=N Bonds 557
8.2.2. C=P Bonds 559
8.2.3. C=As Bonds 567
8.2.4. C=S Bonds 569
8.2.5. C=Se and C=Te Bonds 575
8.3. Cycloadditions with Heteroatom–Heteroatom Double Bonds 576
8.3.1. P=X Bonds 576
8.3.2. Other Heteroatom–Heteroatom Double Bonds 578
8.4. Cycloadditions with Heterocumulenes 578
8.5. Cycloadditions with Triple-Bond Systems 581
8.5.1. Alkynes 581
8.5.2. Nitriles 586
8.5.3. Heteroatom–Heteroatom Triple Bonds 586
8.5.4. Phosphaalkynes 587
8.6. Intramolecular Cycloaddition Reactions 592
8.6.1. Intramolecular Cycloadditions Involving C=C Bonds 592
8.6.2. Intramolecular Cycloadditions Involving C≡C Bonds 601
8.6.3. Intramolecular Cycloadditions Involving Other Multiple Bonds 604
8.7 "Ketocarbene Dipoles" as Building Blocks in Heterocyclic Synthesis 604
Conclusions 609

The Chemistry of Heterocyclic Compounds, Volume 59: Synthetic Applications of 1,3-Dipolar Cycloaddition Chemistry Toward Heterocycles and Natural Products, Edited by Albert Padwa and William H. Pearson.
ISBN 0-471-38726-6

The history of cycloaddition chemistry using aliphatic diazo compounds began in the 1890s when Buchner (1) and von Pechmann (2) reported that ethyl diazoacetate and diazomethane underwent cycloaddition across carbon–carbon multiple bonds. Ever since that time, diazo compounds have occupied a major place in [3 + 2]-cycloaddition chemistry (3,4). For a long time, diazo compounds, as well as organic azides, have been one of the more synthetically useful classes of 1,3-dipoles. No doubt this was because many different mono- and disubstituted diazo compounds could be prepared (Scheme 8.1) and isolated in pure form, in contrast to other 1,3-dipoles that are typically generated as transient species.

The achievements dealing with the cycloaddition chemistry of aliphatic diazo compounds up to the year 1982 have been summarized in an in-depth review (5) that contains more than 700 references. This chapter covers the literature from 1983 to the spring of 2001. A few earlier references are included where it seemed appropriate to put things in context.

During the past two decades, diazo compounds have been extensively used as substrates for transition metal catalyzed reactions. These include both intra- and intermolecular reactions with multiple-bond systems (6–8). In addition, substantial progress with 1,3-dipolar cycloaddition chemistry of the intact diazo function with π-bonds of alkenes has been made as well. Major developments include their reaction with novel dipolarophiles, in particular of functionalized alkenes, alkynes, phosphaalkenes, and phosphaalkynes. Among the diazo compounds that have emerged as versatile cycloaddition partners, diazo(trimethylsilyl)methane ($TMSCHN_2$) and its lithio derivative [$TMSC(Li)N_2$] (9–11) as well as the structurally novel diazocumulenes (12) of the types $R_2C{=}C{=}N_2$ and $R_3P{=}C{=}N_2$ deserve special mention. The commercially available $TMSCHN_2$ is a safe, nonexplosive and nonmutagenic substitute of diazomethane, and its cycloaddition abilities can be greatly expanded by metalation with butyl lithium.

Due to space limitations, it is not possible to provide a comprehensive coverage of all 1,3-dipolar cycloaddition chemistry carried out using diazo compounds over the past two decades. Rather, attention will be given to the most significant developments, including the synthesis of novel heterocyclic systems, the preparation of well-established heterocycles (such as pyrazoles and pyrazolines) with novel functionalities, as well as stereoselective cycloadditions. A discussion of the theoretical, mechanistic, and kinetic aspects of these 1,3-dipolar cycloaddition reactions will be kept to a minimum, but references to important work in these areas will be given at appropriate places. Authoritative reviews dealing with the

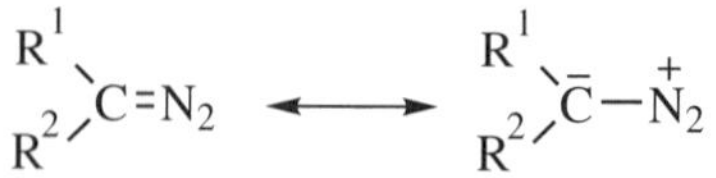

$R^1 = R^2 = H$, or
R^1, R^2 = almost any combination of the following:
alkyl, aryl, hetaryl, alkenyl, alkynyl,
COR, COOR, CN, CF_3, NO_2, MR_3 (M = Si, Ge, Sn)
PR_2, POR_2, $PO(OR)_2$, SO_2R, etc.

Scheme 8.1

classification, mechanism (13) and theory (14) of 1,3-dipolar cycloaddition reactions are available in the literature.

Numerous methods to prepare individual classes of aliphatic diazo compounds have been extensively developed. The major strategies for their synthesis involve the alkaline cleavage of *N*-alkyl-*N*-nitroso-ureas, -carboxamides and -sulfonamides, dehydrogenation of hydrazones, as well as diazo group transfer from sulfonyl and related azides to active methylene compounds, and electrophilic diazoalkane substitution reactions. These synthetic methods have been comprehensively reviewed (15,16). Useful information on the preparation of selected diazo compounds can be found elsewhere (6,17).

8.1. CYCLOADDITIONS WITH C=C BONDS

Diazo compounds combine with a broad range of alkenes to form 3*H*-4,5-dihydropyrazoles (Δ^1-pyrazolines). Depending on the substitution pattern, some of these cycloadducts are stable enough to be isolated. Others undergo rapid tautomerization to give Δ^2-pyrazolines or are transformed into pyrazoles by a 1,2-elimination reaction, or extrude molecular nitrogen to form cyclopropanes. A large number of examples of these reactions have been presented in an earlier review (5). Olefinic π-bonds continue to be used as substrates for the cycloaddition chemistry of diazo compounds. Since it is not possible to review the results comprehensively, we will limit this survey to interesting developments using novel diazo compounds and novel olefinic substrates, as well as examples of stereoselective cycloaddition reactions. Some general remarks on the reactivity of diazo dipoles toward alkenes will be given first.

8.1.1. Reactivity and Regioselectivity

The reactivity of diazo compounds toward olefins is nicely rationalized by frontier molecular orbital (FMO) theory (14,18). According to this concept, cycloadditions of simple diazoalkanes are highest occupied molecular orbital (HOMO, dipole)–lowest unoccupied molecular orbital (LUMO, dipolarophile) controlled. Diazomethane cycloadditions are generally accelerated with respect to ethylene by electron-attracting substituents on the olefin and decelerated by electron-donating ones. Cycloadditions of diazoacetic esters and diazomalonates are accelerated by both types of substituents on the olefinic π-bond due to the control by frontier orbital interactions (i.e., HOMO(dipole)–LUMO(dipolarophile) and HOMO(dipolarophile)–LUMO(dipole). With diazomalonates and diazo(phenylsulfonyl)acetates, the latter interaction is of greater importance, so that the rate-increasing effect is more pronounced for electron-rich than for electron-deficient C=C bonds. Electron-releasing substituents at the diazo function raise the HOMO energy of the dipole thereby accelerating the cycloaddition (19). Dipolar cycloadditions of diazo(phenyl)methane, diazo(diphenyl)methane, and their substituted derivatives with both alkenes and alkynes are also controlled by the HOMO (dipole)–LUMO(dipolarophile) interaction (20).

In this context, note that the cycloaddition rate of a crown ether-annulated diazo(diphenyl)methane such as **1** with maleic anhydride is ion-selectively retarded in the presence of alkali perchlorates (21). This observation was attributed to a lowering of the HOMO (dipole) level, due to the electron-withdrawing electrostatic effect of the complexed cation. A parallel to the negative Hammett ρ values for cycloadditions of ring-substituted diazo(diphenyl)methanes with TCNE (22) ($\rho = -2.67$) and chloranil (23) ($\rho = -1.67$) was drawn.

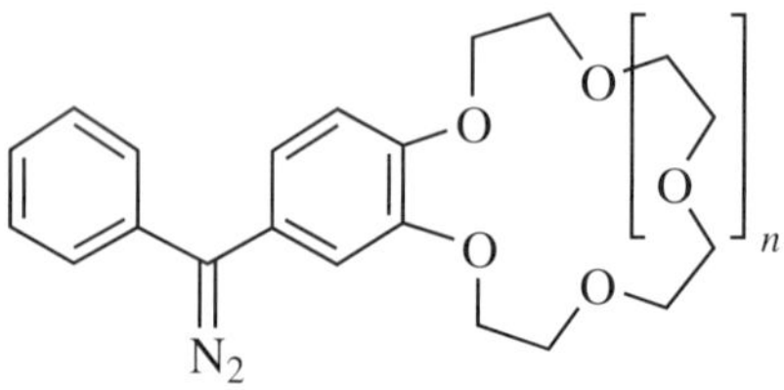

(1) ($n = 0$–3)

Diazo compounds add to many types of C=C bonds with (nearly) exclusive or at least predominant formation of one of the two regioisomeric cycloadducts. As a typical example, monosubstituted olefins react with diazomethane to preferentially give 3-substituted Δ^1-pyrazolines, regardless of the electronic nature of the substituent. Based on second-order perturbation theory, the results can be rationalized by consideration of the magnitude of the orbital coefficients on the frontier molecular orbitals of the reactants (13). There are cases, however, where this approach does not correctly predict the regiochemistry (e.g., the formation of 3-ethoxypyrazoline from ethyl vinyl ether and diazomethane). Using a perturbational treatment that included polar interactions, covalent stabilization and noncovalent repulsion terms, Sustmann et al. were able to show that closed-shell repulsions have an important effect on the regioselectivity of 1,3-dipolar cycloadditions (24,25). Distortions of the dipolarophile in the regioisomeric transition structures, caused by these noncovalent interactions, result in FMO interactions that favor the formation of 3-ethoxypyrazoline. In an alternative approach, based on calculations carried out mainly at the HF/3-21G level, the regioselective formation of 3-methoxypyrazoline was traced to the heteroallyl-type 4π conjugation in the O–C=C unit of methyl vinyl ether and its different perturbation by the two regioisomeric approaches of the dipole. It was concluded that this perturbation begins well in advance of the transition state and is more favorable when a C–N rather than a C–C bond is formed at the α-C atom of the enol ether (26).

Not unexpectedly, predictions on reactivity and regiochemistry based on a FMO treatment can be overruled by steric effects (13). The change in reactivity may be illustrated by the following example: 2-Phenylsulfonyl-norbornadiene reacts with 2-diazopropane at the electron-deficient C=C bond, as expected. However, 2-phenylsulfonyl-3-trimethylsilyl-norbornadiene reacts with the same dipole at the unsubstituted double bond, probably as a result of the steric bulk of the trimethylsilyl group (27).

The potential energy surface involving the gas-phase cycloaddition of diazomethane with ethylene has recently been investigated using density functional theory and coupled cluster [CCSD(T)] calculations (28,29). Combining these results with the configuration mixing model of Pross and Shaik, it was concluded (29) that the singlet–triplet splitting of diazomethane (and of other 16-electron dipoles as well) correlates strongly with the reactivity of the dipole and the exothermicity of the reaction. Electropositive substituents tend to lower the singlet–triplet gap and facilitate the cycloaddition.

8.1.2. Diazo Dipoles

Diazomethane continues to be widely used in cycloaddition chemistry with alkenes, especially with a view to subsequent transformation of the resulting Δ^1-pyrazoline into a cyclopropane. Diazo(trimethylsilyl)methane has gained importance in recent years as a commercially available and less problematic substitute of diazomethane. It undergoes 1,3-dipolar cycloaddition with ethylene (30), acrylonitrile (31), α,β-unsaturated carbonyl compounds (32) (including a cyclobutenone) (33), 1-acetyl-1-(4-nitrobenzoyloxy)ethene (34), 1,4-quinones (35), but not with 1,2-quinones that give benzodioxoles (35). Scheme 8.2 demonstrates that the initially formed Δ^1-pyrazolines are usually not isolated due to fast subsequent reactions such as tautomerization and oxidative aromatization. Protodesilylation often occurs during the actual reaction or during workup. Further cycloadditions using this diazo compound are mentioned in various sections of this chapter.

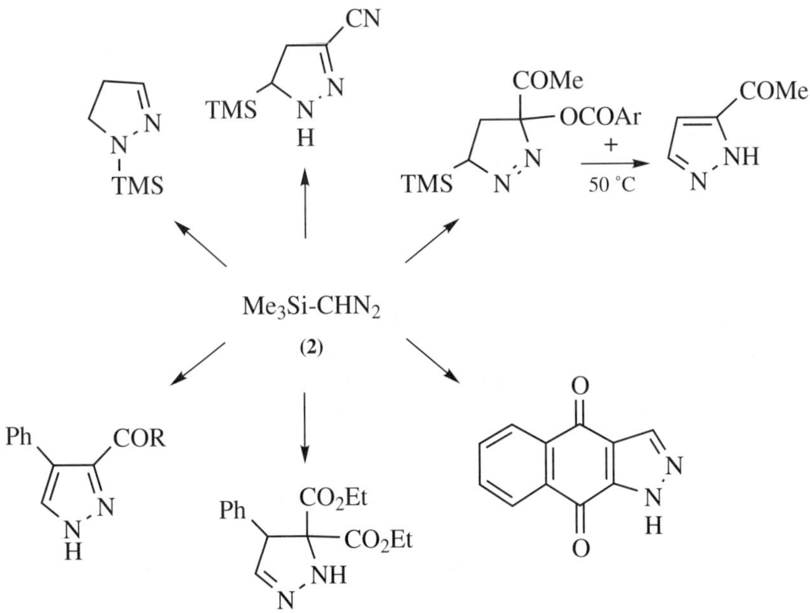

Scheme 8.2

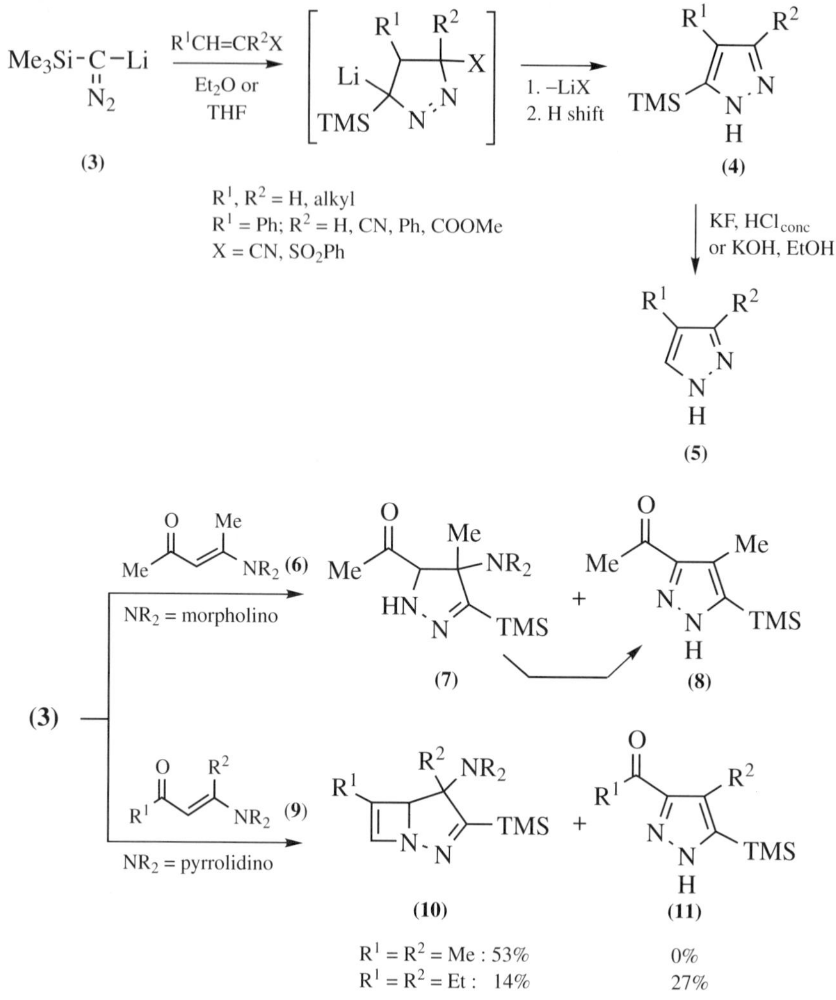

Scheme 8.3

Metalation of **2** with lithium diisopropylamide (LDA) generates diazo(trimethylsilyl)methyl lithium (**3**), which reacts with α,β-unsaturated nitrile 36 and phenylsulfones (37) to form 3(or 5)-trimethylsilyl-1*H*-substituted pyrazole **4** that can be desilylated to furnish pyrazoles **5** (Scheme 8.3).

The use of **3** also allows for [3 + 2] cycloaddition with enaminoketones (38) (Scheme 8.3). When the morpholine-derived enaminoketon **6** was used, a mixture of Δ^2-pyrazoline **7** and pyrazole **8** was obtained. Complete transformation of **7** into **8** was achieved by treatment with water. In the case of the pyrrolidine-derived enaminoketone **9**, pyrazole **11** and diazabicycloheptadiene **10** are formed competitively. In the formation of the latter compound, reaction of a second equivalent of **3** with the carbonyl group of **9** is involved. Reaction of uracil as

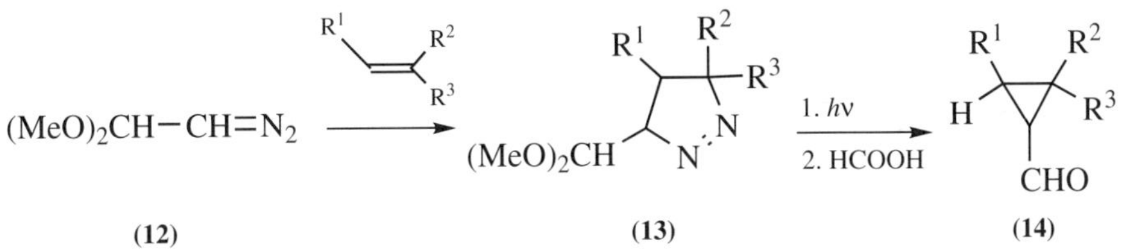

$R^1 = R^2 = H$, $R^3 = Ph$; $R^1 = H$, $R^2 = Me$, $R^3 = CO_2Me$
R^1--$R^2 = (CH_2)_3$, $R^3 = CN$
R^1--$R^2 = (CH_2)_4$, $R^3 = CO_2Me$

Scheme 8.4

well as of 5-substituted uracils and uridines with **3** were also investigated (39). Formation of the expected bicyclic 4,5-dihydro-3-silyl-1*H*-pyrazoles was observed in the 5-H, 5-F, and 5-NO_2 cases, which is analogous to their reactions with diazomethane. 5-Bromouracil reacted with **3** to give 3-trimethylsilyl-1 *H*-pyrazolo [4,3-*d*]uracil.

Diazoacetaldehyde dimethylacetal (**12**) has been used as a substitute for diazoacetaldehyde in 1,3-dipolar cycloadditions with 1-benzopyran-2(*H*)-ones (40), styrene, methyl methacrylate, 1-cyanocyclopentene, and methyl cyclohexene-1-carboxylate (41). The resulting Δ^1-pyrazolines were readily transformed in two steps into cyclopropanecarbaldehydes [e.g., **13** → **14** (Scheme 8.4)]. In a similar manner, 3-phenylcyclopropane-1,2-dicarbaldehyde was obtained from the reaction of **12** with dimethyleneketal of cinnamic aldehyde.

[3 + 2] Cycloaddition using the unusually functionalized 2-diazo-1,1,1-trifluoro-3-nitropropane (**15**) could be achieved with methyl acrylate, methacrylic acid chloride, and esters (Scheme 8.5), but not with the 1,2-disubstituted C=C bonds of β-nitrostyrene, ethyl cinnamate, and 4-methyl-3-penten-2-one (42). In these cycloadditions, **15** is considerably less reactive than 2-diazo-1,1,1-trifluoroethane

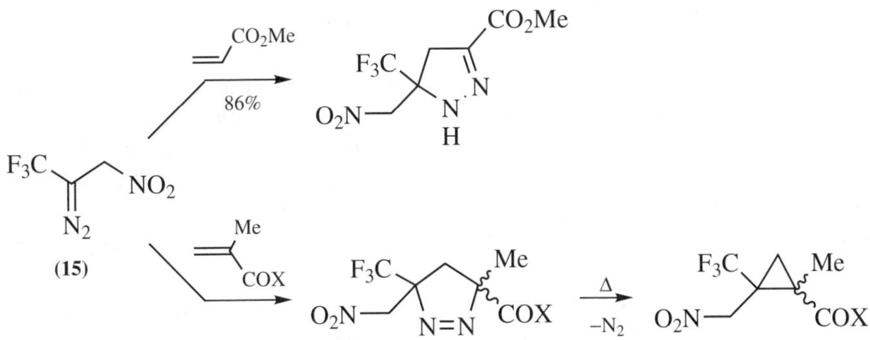

R = Cl: 24 h	90% (*E*/*Z*) = (89:11)	75% (85:15)
R = OMe: 168 h	95% (86:14)	84% (89:11)
R = OPh: 96 h	70% (100:0)	100% (100:0)

Scheme 8.5

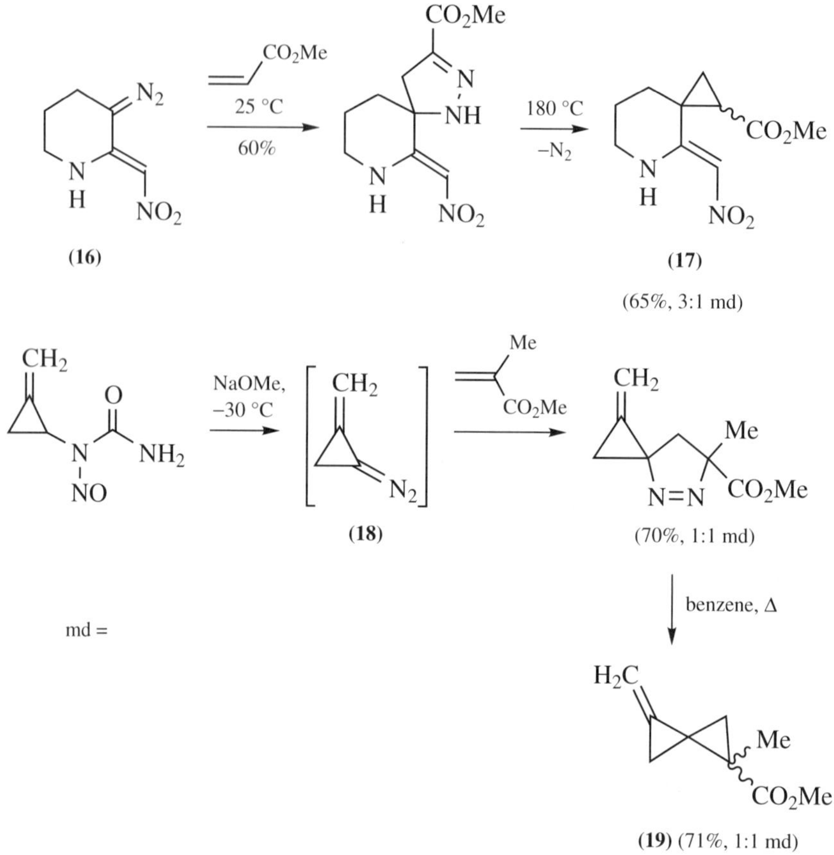

Scheme 8.6

and 2-diazo-1,1,1-trifluoro-2-(*p*-tolyl)ethane. With enaminoketones (43) and ethyl 3-morpholinocrotonate (44), other reaction pathways leading to nitrogen-free products are encountered.

Cyclic diazoalkanes have been used to generate spirocyclopropanes *via* ring contraction of pyrazolines formed in the initial cycloaddition step. Two novel examples are the transformation of 3-diazo-2-nitromethylenepiperidine (**16**) into 5-aza-spiro[2.5]octane **17** (45) and the conversion of 1-diazo-2-methylenecyclopropane (**18**) into methylene-spiro[2.2]pentane **19** (46) (Scheme 8.6). Related reactions have also been reported for diazocyclopropane (47) and diazospiropentane (48).

Other novel diazo compounds that have been subjected to 1,3-dipolar cycloaddition with activated alkenes, and that give unusually functionalized pyrazolines (Scheme 8.7), include 1-diazo-3-trimethylsilylpropan-2-one (**20**) (49), 2-diazomethyl-4(5*H*)-furanones (**21**) (50), methyl 2-diazo-5-methylanilino-5-oxopentanoate (**22**) (51), 2-(acylamino)-2-diazoacetates (**23**) (51), ethyl 2-diazo-4,4,4-trichloro-3-(ethoxycarbonylamino)butyrate (**24**) (52), and diazopropyne (53).

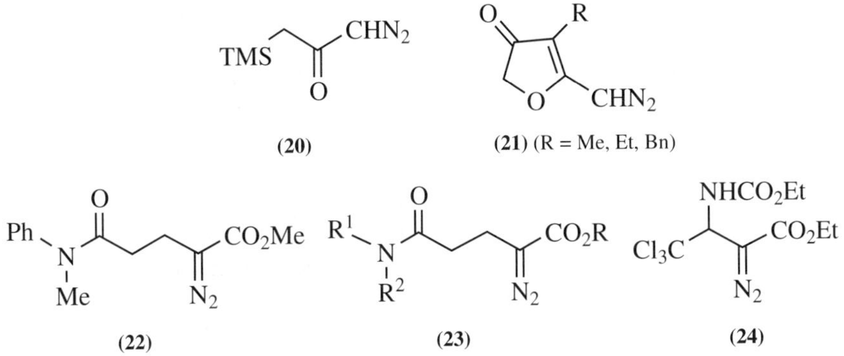

(20) **(21)** (R = Me, Et, Bn) **(22)** **(23)** **(24)**

Scheme 8.7

An interesting preparation of aliphatic diazoalkanes ($R^1R^2C{=}N_2$; R^1, $R^2{=}$ alkyl) involves the photolysis of 2-alkoxy-2,5-dihydro-1,3,4-oxadiazoles (see Scheme 8.49). When the photolysis is carried out in the presence of an appropriate dipolarophile, the diazo compounds can be intercepted (prior to their further photolysis) by a [3 + 2] cycloaddition reaction (54). As an example, 2-diazopropane was intercepted with *N*-phenylmaleimide (54) and norbornenes (55) to give the corresponding Δ^1-pyrazolines.

The (η^4-diene tricarbonyliron)-substituted diazocarbonyl compounds **25** have been found to undergo 1,3-dipolar cycloaddition with methyl acrylate in high yield, but with little or no diastereoselectivity (56). Nevertheless, the facile chromatographic separation of the diastereomeric products **26a,b** and **27a,b** (Scheme 8.8), permits the synthesis of pure enantiomers when optically active diazo compounds (**25**) [enantiomeric excess (ee) >96%] are employed. When the reaction of **25** ($R{=}CO_2Et$) with methyl acrylate was carried out at 70 °C, cyclopropanes instead of Δ^2-pyrazolines were formed. The enantiomerically pure

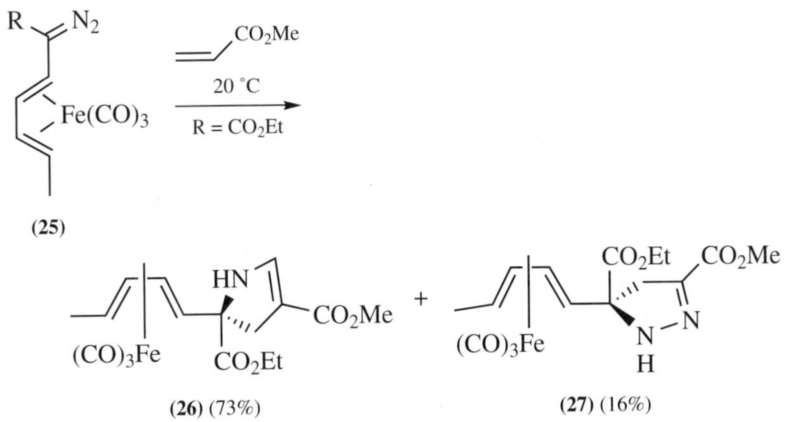

(25) **(26)** (73%) **(27)** (16%)

Scheme 8.8

cis- and *trans*-1-(penta-1,3-dienyl)cyclopropane-1,2-dicarboxylic esters could be obtained after separation of the diastereoisomers and metal decomplexation.

Diazocumulenes represent a structurally unusual class of diazo compounds (12). Diazocumulenes of the type $R_2C{=}C{=}N_2$ have not been isolated so far, undoubtedly because of the high propensity to form a vinylidenecarbene by elimination of N_2. Nevertheless, when $Me_2C{=}C{=}N_2$ was generated in a Horner–Emmons reaction from dimethyl diazomethylphosphonate and acetone in the presence of 3,3-dimethylcyclopropene, 4,4-dimethyl-2-(methylvinyl)-1,4-dihydropyridazine could be isolated (57) (see Scheme 8.65). This product is most likely formed by isomerization of the initial 1,3-dipolar cycloaddition product. On the other hand, the reaction of (1-diazo-2-oxoalkyl)silanes **29** with activated alkenes (i.e., *N*-phenylmaleimide, maleic anhydride, norbornene, and norbornadiene) gave cycloadducts such as **31** and **32** (Scheme 8.9) (58). These compounds are considered to be the direct trapping products of 1-diazo-1-alkenes **30**, which are present in minor but nondetectable quantities as a result of its equilibrium with diazoketones **29**. Kinetic experiments in the presence of a large excess of *N*-phenylmaleimide support this assumption. The results are definitely not compatible with an alternate pathway involving cycloaddition of **29** followed by a 1,3(C $\rightarrow$ O) silyl shift. Cycloaddition of **29:30** with 3,3-dimethylcyclopropene gave 4-acyl-2-silyl-2,3-diazabicyclo[3.1.0]hex-3-enes in equilibrium with 6-acyl-1-silyl-1,4-dihydropyridazines. In these cases, no direct evidence concerning the reacting dipole is available (59).

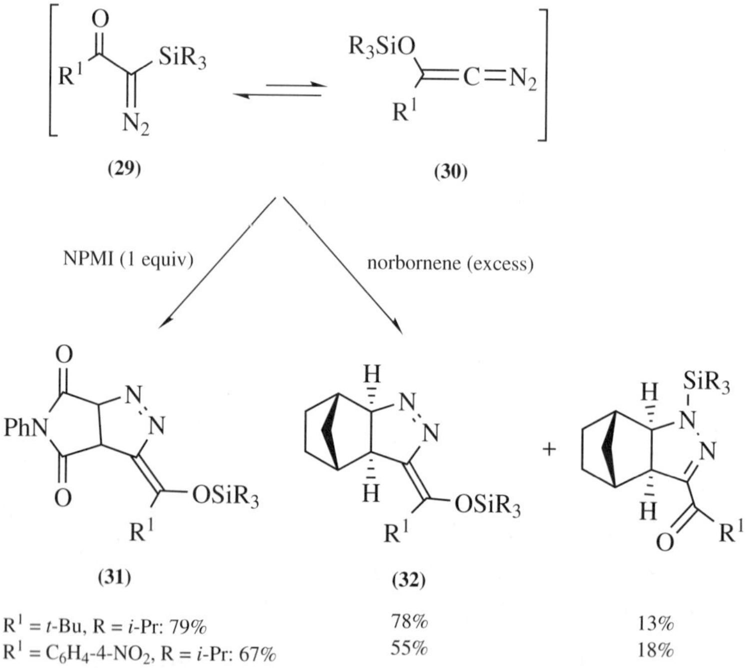

Scheme 8.9

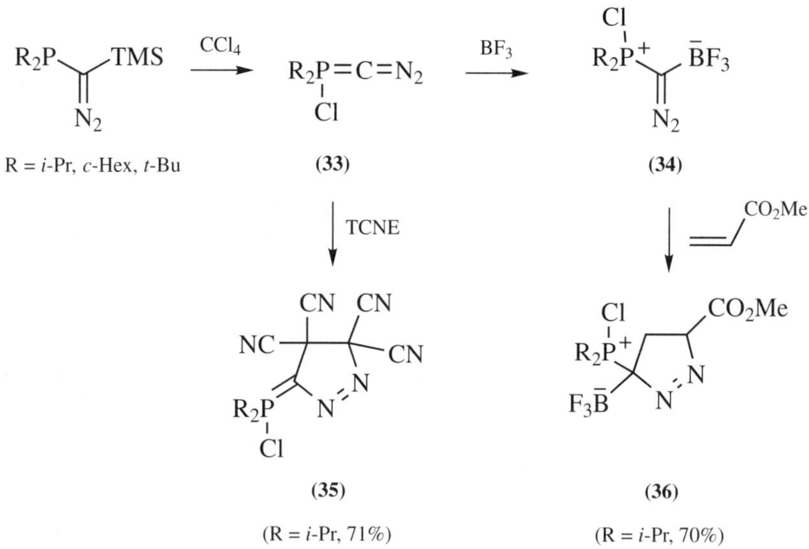

Scheme 8.10

(Diazomethylene)phosphoranes **33** (Scheme 8.10), which represent another type of diazocumulenes (12) are easily obtained by the oxidative ylidation of the corresponding phosphanyl(trimethylsilyl)diazomethane with CCl_4. The increased stability of these compounds as compared with diazocumulenes ($R_2C{=}C{=}N_2$) is probably due to the ylidic character of the P=C bond. These diazo compounds exhibit the expected dipolar reactivity toward electron-deficient alkenes, alkynes, phosphaalkenes, and heterocumulenes (12). Thus, **33** reacts with TCNE to form Δ^1-pyrazoline **35** (60). Furthermore, **33** could be converted into the phosphonio-borate-substituted diazo compound **34**, which underwent subsequent cycloaddition with electron-deficient alkenes (e.g., **34** → **36**) (61).

8.1.3. Alkenes

Of the many substituted and functionalized alkenes that have been combined with diazo dipoles to give Δ^1-pyrazolines or products derived from them (i.e., Δ^2-pyrazolines, pyrazoles, cyclopropanes), only a selection will be mentioned. These include α-alkylidene-cycloalkanones (62), -flavanones, -thioflavanones, -chromanones, and thiochromanones (63,64); α-arylidene-indanones and -indolones (65); diarylideneacetones (66); 1-benzopyran-2(*H*)-ones (coumarins) (67,68); 4-nitro-1,2-oxazoles (69); 2-alkylidene-2-cyanoacetates (70); dimethyl 2,3-dicyanofumarate (71); tetracyanoethylene (72); tetraethyl ethylenetetracarboxylate (72); 1,4-quinones (35,73–75); 2-X-1,1,1-trifluoro-2-propene [X = Br, (76), SPh, SOPh, SO_2Ph (77)]; nitroalkenes (78) including sugar nitroalkenes (79); 1-diethoxyphosphoryl-1-alkenyl-sulfoxides (80); methyl 2-(acetylamino)cinnamate and –acrylate

(81); methyl (*E*)-2-acylamino-3-cyanoprop-2-enoates (82); enaminones of the type $R^1COCH{=}C(Me)NHR^2$ ($R^1 = Me$, OEt) (83); vinylselenides (84); 1-alkenylboronates (85); cyclopropenes (86–90); *tert*-butyl 2,3,4-tri-*tert*-butylcyclobutadienecarboxylate (91) (with diazomethane, while Ph_2CN_2 and 2-diazo-1,3-dicarbonyl compounds gave cyclopropenyl-substituted ketazines); 1*H*-1,2-diazepines (92); allene (93); phenylsulfonyl-1,2-propadienes (94); and other allenes (95); α,β-unsaturated chromium carbene complexes (96–98) (see Scheme 8.17). The ruthenium allenylidene complex $[Ru(\eta^5\text{-}C_5H_5)(C{=}C{=}CPh_2)(CO)(P\textit{i}\text{-}Pr_3)]BF_4$ reacts with ethyl diazoacetate to provide a cyclic carbene complex $[Ru(\eta^5\text{-}C_5H_5)\{{=}CCH{=}C(OEt)OC{=}CPh_2\}(CO)(P\textit{i}\text{-}Pr_3)]BF_4$ by a formal ketocarbene addition at the C=C bond (99).

Stanovnik and co-workers (100,101) systematically investigated the cycloaddition reactions of diazoalkanes with unsaturated nitrogen heterocycles, such as azolo[1,5-*a*]pyridines, pyridazin-3(2*H*)-ones, and [*b*]-fused azolo- and azinopyridazines. The Stanovnik group have studied the further transformations of the products and reviews of this chemistry are available. In a typical example, the reaction of 6-chlorotetrazolo[1,5-*b*]pyridazine (**37**) with 2-diazopropane yields the *NH*,*NH*-dihydro-pyrazolo[4,3-*d*]tetrazolo[1,5-*b*]pyridazine **38** (102) (Scheme 8.11). The latter substrate reacts with acetone to produce an azomethine imine **39** that thermally rearranges to give the fused dihydro-1,2-diazepine **40**. The azomethine imine obtained with glucose can be trapped with methyl acrylate to furnish the *C*-nucleoside **41** (103).

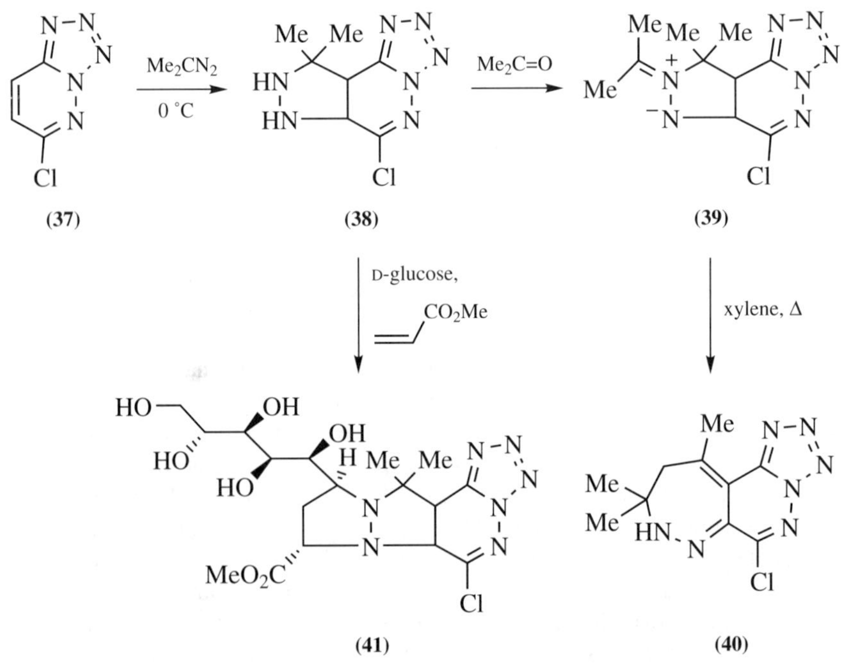

Scheme 8.11

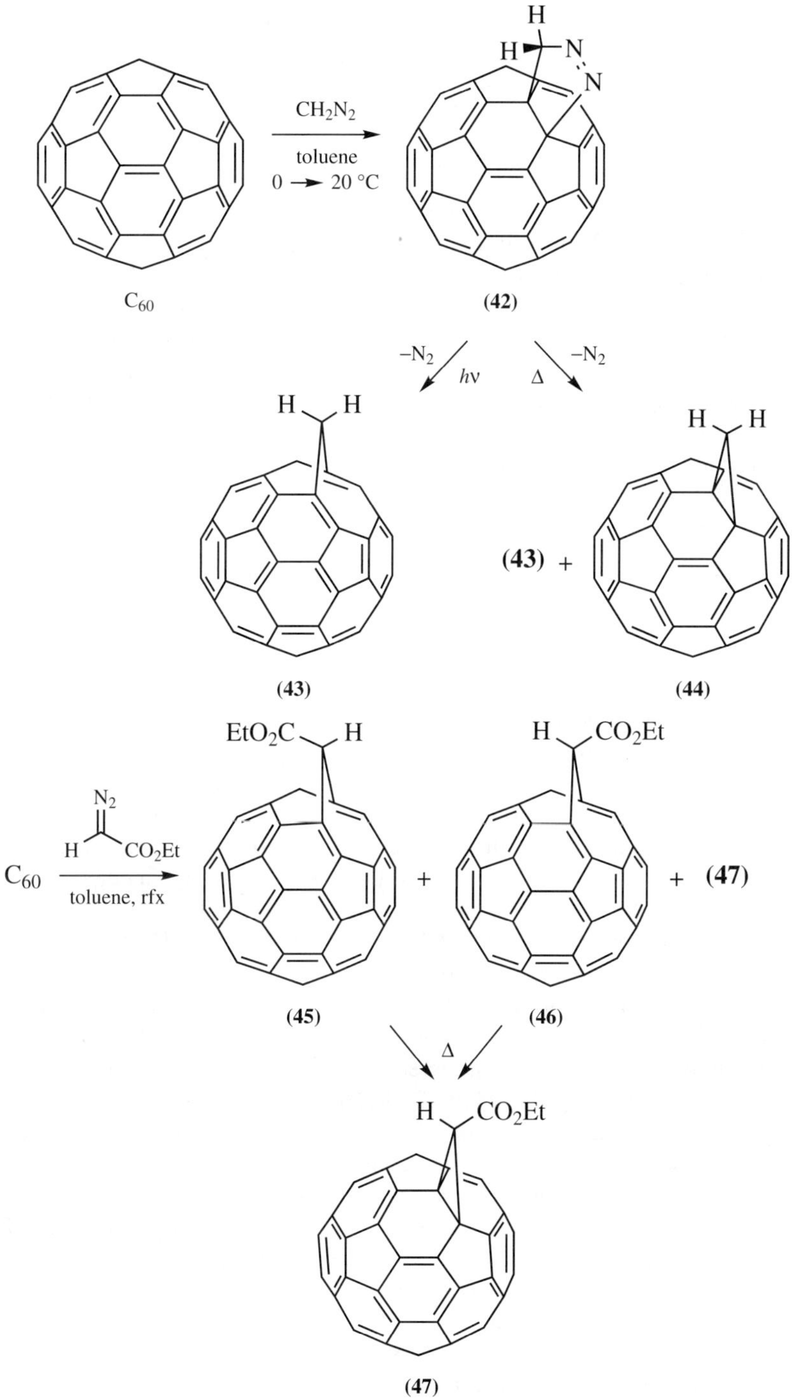

CH_2N_2
toluene
0 → 20 °C
C_{60}
(42)
$-N_2$
hν
Δ
$-N_2$
(43)
(43) +
(44)
N_2
H
CO_2Et
C_{60}
toluene, rfx
EtO_2C
H
H
CO_2Et
+
+ (47)
(45)
(46)
Δ
H
CO_2Et
(47)

Scheme 8.12

Diazo compounds also undergo cycloaddition with fullerenes [for reviews, see (104),(105)]. These reactions are HOMO(dipole)–LUMO(fullerene) controlled. The initial Δ^1-pyrazoline **42** can only be isolated from the reaction of diazomethane with [60]fullerene (106) (Scheme 8.12) or higher substituted derivatives of C_{60} (107). Loss of N_2 from the thermally labile **42** resulted in the formation of the "6,5-open" 1,2-methanofullerene (**43**) (106). On the other hand, photolysis produced a 4:3 mixture of **43** and the "6,6-closed" methanofullerene (**44**) (108). The three isomeric pyrazolines obtained from the reaction of [70]fullerene and diazomethane behaved analogously (109). With all other diazo compounds so far explored, no pyrazoline ring was isolated and instead the methanofullerenes were obtained directly. As a typical example, the reaction of C_{60} with ethyl diazoacetate yielded a mixture of two "6,5-open" diastereoisomers **45** and **46** as well as the "6,6-closed" adduct **47** (110). In contrast to the parent compound **43**, the ester-substituted structures **45** and **46**, which are formed under kinetic control, could be thermally isomerized into **47**. The fomation of multiple CPh_2 adducts from the reaction of C_{60} and diazodiphenylmethane was also observed (111). The mechanistic pathway that involves the extrusion of N_2 from pyrazolino-fused [60]fullerenes has been investigated using theoretical methods (112).

8.1.4. Diastereofacial Selectivity

In the context of stereoselective organic synthesis, diastereofacial-selective cycloadditions of diazoalkanes and diazoacetates with functionalized alkenes has attracted some attention. 3,4-Disubstituted cyclobutenes were studied as dipolarophiles by the groups of Gandolfi and co-workers (113) and Martin and co-workers (114). The transition state structures of the cycloaddition of diazomethane with *cis*-3,4-dimethylcyclobutene was investigated theoretically by DFT methods (113a).

5-Monosubstituted 2(5*H*)-furanones (γ-butenolides) **48** [Scheme 8.13, R = CH_2OH (115), CH_2OBn (115), CH_2OSiPh_2t-Bu (116–118), CH_2OTs (119), OMe (120–122), menthyloxy (120,123), SR, SOR, and SO_2Ph (124)] gave cycloadducts with both diazomethane and ethyl diazoacetate with excellent anti-π-facial selectivity and with high regiospecifity [i.e., with the diazo carbon attached to C(4) of the lactone]. On the other hand, low diastereoselectivities were found in the reaction of diazomethane with **48** [R = OMe (120,122)] and its 3-Me, 4-Me, and 3-Hal derivatives (122). The addition of diazomethane to the two C(5) epimers of furanone sulfoxide **49** occurred with complete π-facial diastereoselectivity. In this case, approach of the dipole is controlled by the chiral sulfoxide group rather than the ethoxy substituent (125). In some cases, the pyrazolines obtained from optically active γ-butenolides were used to synthesize cyclopropanes with high enantioselectively (see below). An example involves the synthesis of methyl *cis*-chrysanthemate from optically active **48** (R = *i*-Pr) (126).

The diastereofacial-directing effect of a chiral center attached directly to the C=C bond was also demonstrated using other enantiopure dipolarophiles, such as vinyl sulfoxide **50** (80) (Scheme 8.14; formation a single diastereoisomer of a

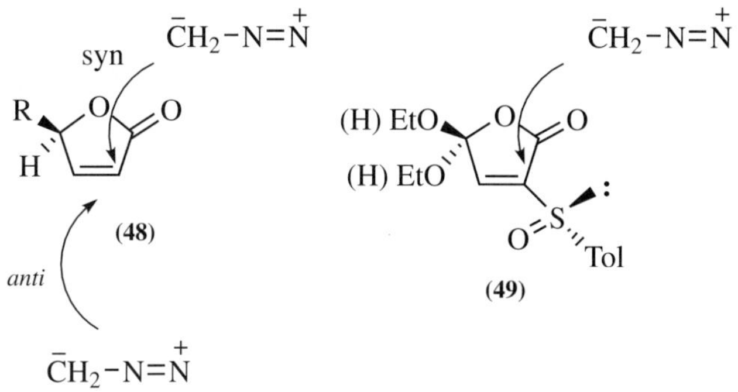

Scheme 8.13

cyclopropane derived from Ph_2CN_2, and of a Δ^1-pyrazoline from Me_2CN_2). Other examples include 5-alkylidene-1,3-dioxan-4-ones **51** [(R^1 = Me, Et, *i*-Pr, Ph; R^2 = *t*-Bu, *c*-Hex; reaction with $RCHN_2$, R = H, Me, $SiMe_3$ (127,128)], α-benzylidene-β-lactone **52** [with CH_2N_2 (128)], azlactone **53** [(*E*)- and (*Z*)-isomer, with CH_2N_2 (129)], α-alkylidene-γ-lactones **54** and **55** (130), acyclic γ-alkoxy- and γ-amino-α,β-unsaturated ketones such as **56** and **57** [with diazomethane, diazoethane,

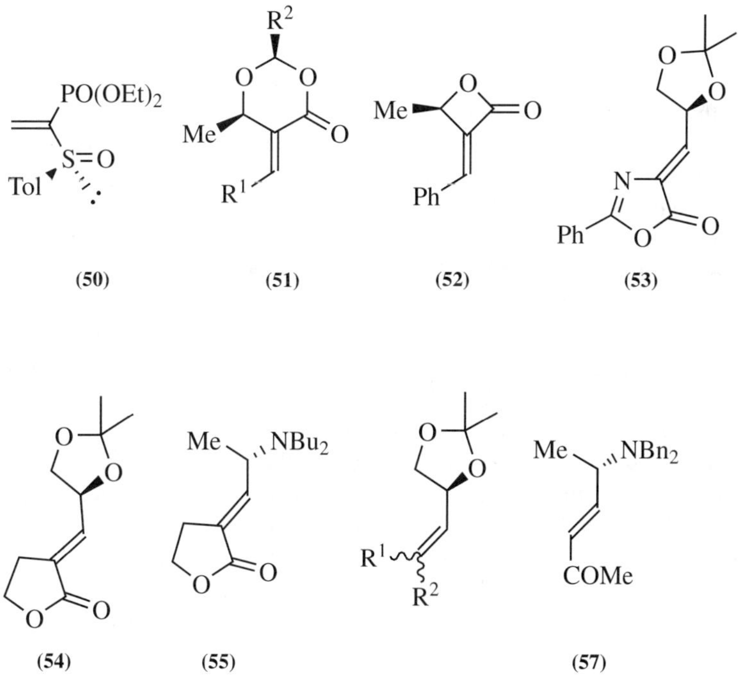

Tol = tolyl

(**56**): R^1 = H, R^2 = COMe (*E*)
(**58**): R^1 = CO_2R, SO_2Ph, Ph, NO_2

Scheme 8.14

diazodiphenylmethane (131)], and γ-alkoxy-α,β-unsaturated esters **58** and related compounds (118,131–133) [e.g., $R^1 = CO_2Me$, $R^2 = H$; $R^1 = CO_2Et$, $R^2 = Me$; $R^1 = CO_2Me$, $R^2 = NHCbz$ or NHAc; $R^1 = NO_2$, $R^2 = Me$; $R^1 = SO_2Ph$, $R^2 = H$; reactions with diazomethane (133)]. Cycloaddition of diazomethane occurs with high π-facial syn-selectivity in the case of **51**, **52**, **54**, **56**, **57**, and **58**, but with anti-selectivity for **52**. Density functional calculations for the cycloaddition of CH_2N_2 with **58** ($R^1 = NO_2$, $R^2 = Me$ and $R^1 = NHAc$, $R^2 = CO_2Me$) predict a preference for the formation of the syn-product and suggest that the chirality at the dioxolane stereogenic center is mainly responsible for this stereochemistry (133). For the case of γ-dibenzylamino-α,β-didehydro-α-aminoacid esters (134), the diastereoselectivity of diazomethane cycloaddition is opposite to that predicted by the principle of 1,3-allylic strain.

The spiropyrazolines obtained from **51** were converted into enantiopure Δ^2-pyrazoline-3-carboxylates and 1-(hydroxyethyl)cyclopropane-1-carboxylates (128). Those obtained from **54** and **55** were transformed into optically active α-spirocyclopropyllactones and 3-amino-3-(hydroxyethyl)pyrrolidin-2-ones (130). The spiropyrazoline obtained from a chiral propylidene-diketopiperazine and diazomethane was converted into (+)-(1*R*,2*S*)-1-amino-2-ethyl)cyclopropane-1-carboxylic acid (allocoronamic acid) (135).

Since the π-facial diastereoselectivity of furanone **48** is opposite to that of open-chain α,β-unsaturated ester **58** (R^1 or $R^2 = CO_2R$), it was possible to transform the resulting pyrazolines into cyclopropanes with complementary absolute configuration. As an example, the reaction of (5*S*)-2(5*H*)-furanone (**59**) (Scheme 8.15) with diazomethane proceeded by anti-attack to give pyrazoline **60**, which was converted into chiral 2-formylcyclopropane-1-carboxylate **61**, and then further transformed into 4,5-methano-2,3-didehydroaminoacid ester **62** by a Wittig–Horner olefination (118). On the other hand, diazomethane reacted with syn-selectivity using the unsaturated ester **63** to give Δ^2-pyrazoline **64**, which was readily transformed into *ent*-**61** and further into *ent*-**62**. Both the starting materials **59** and **63** derive their chirality from D-glyceraldehyde acetonide.

The formation of spirocyclopropanes from the reaction of diazodiphenylmethane and (−)-8-phenylmenthyl esters of acrylic acid and methyl fumarate occurred with a modest level of diastereofacial selectivity (136). In contrast, diastereoselectivities of 90:10 were achieved in the cycloadditions of diazo(trimethylsilyl)methane with acrylamides **65** derived from camphor sultam as the chiral auxiliary (137) (Scheme 8.16). Interestingly, the initial cycloadducts **66** afforded the nonconjugated Δ^2-pyrazolines **67** on protodesilylation; the latter were converted into optically active azaproline derivatives **68**. In a related manner, acrylamide **69** was converted into Δ^2-pyrazolines **70a,b** (138). The major diastereoisomer **70a** was used to synthesize indolizidine **71**. The key step in this synthesis involves the hydrogenolytic cleavage of the pyrazoline ring.

The cycloaddition of several diazoalkanes with (2-arylvinyl)-[(−)-(8-phenylmenthyloxy]methylene chromium complexes **72** gave the Δ^2-pyrazolines **73** with high diastereoselectivity. These compounds were converted into pyrazolinecarboxylates **74** by *N*-protection and metal decomplexation (98) (Scheme 8.17). It is

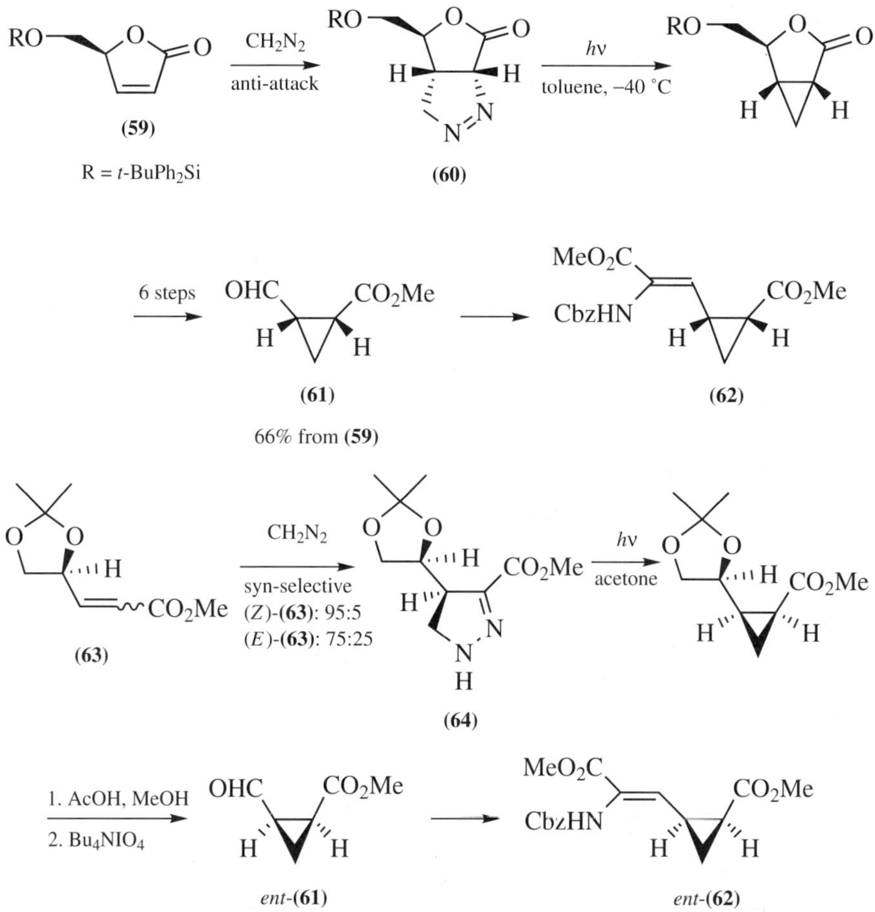

Scheme 8.15

interesting to note that in the above cases, the chiral auxiliary completely shields the (*re*,*re*) face of the olefinic C=C bond of **72**, in contrast to the only moderate π-facial selectivity observed with acrylates (136) and the cinnamate bearing the same chiral ester residue (98). Conversion of **73** into the optically active pyrazolidine-3-carboxylic acids (azaprolines) and protected 1,3-diamines was achieved (98).

The first effective enantioselective 1,3-dipolar cycloaddition of diazoalkanes catalyzed by chiral Lewis acids was reported in the year 2000 (139). Under catalysis using zinc or magnesium complexes and the chiral ligand (*R*,*R*)-DBFOX/Ph, the reaction of diazo(trimethylsilyl)methane with 3-alkenoyl-2-oxazolidin-2-one **75** ($R^2 = H$) gave the desilylated Δ^2-pyrazolines (4*S*,5*R*)-**76** ($R^1 = Me$: 87% yield, 99% ee at −40 °C) (Scheme 8.18). Simple replacement of the oxazolidinone with the 4,4-dimethyloxazolidinone ring resulted in the formation of (4*R*,5*S*)-**77** ($R^1 = Me$: 75% yield, 97% ee at −78 °C).

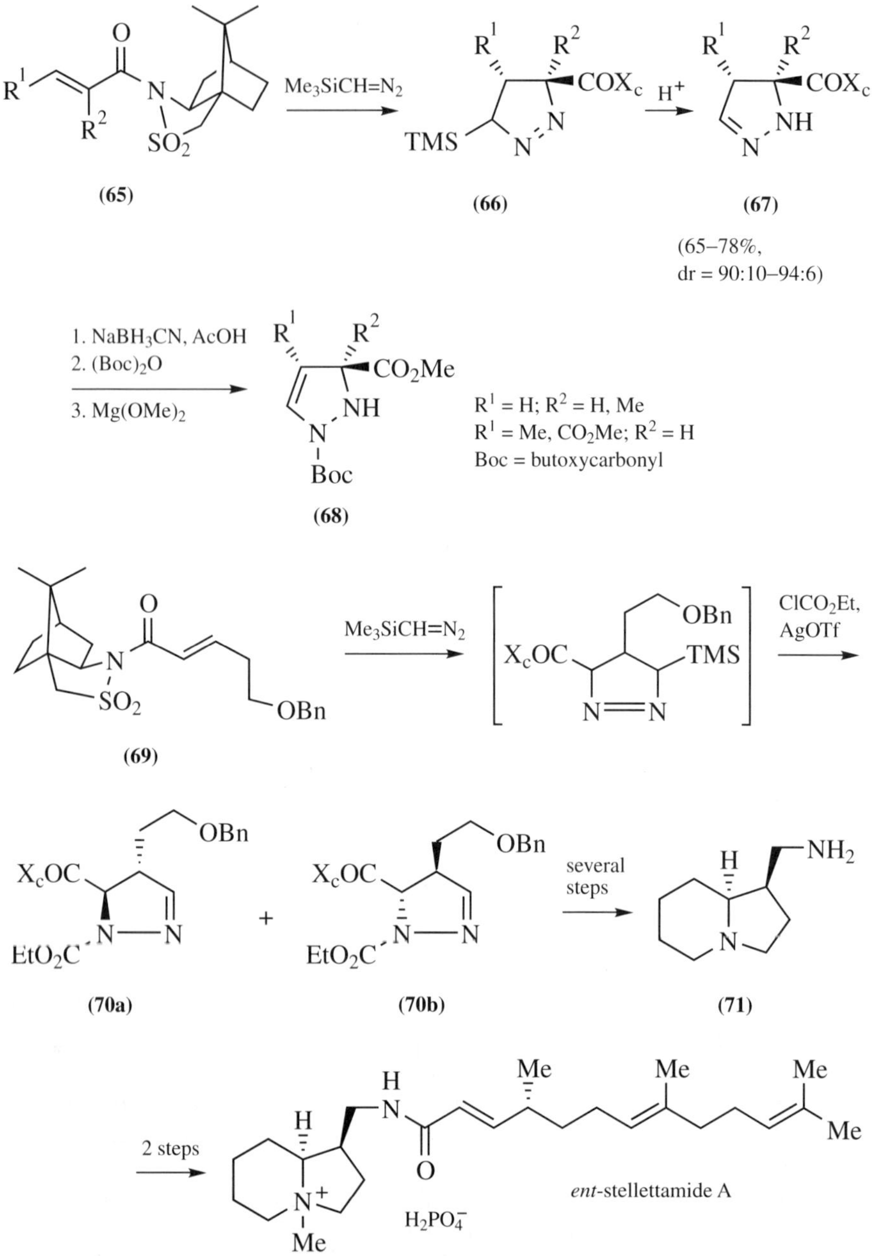
Me3SiCH=N2
COXc
TMS
H+
(65)
(66)
(67)
(65–78%,
dr = 90:10–94:6)
1. NaBH3CN, AcOH
2. (Boc)2O
3. Mg(OMe)2
CO2Me
NH
Boc
(68)
R1 = H; R2 = H, Me
R1 = Me, CO2Me; R2 = H
Boc = butoxycarbonyl
(69)
OBn
XcOC
ClCO2Et,
AgOTf
EtO2C
(70a)
(70b)
several
steps
NH2
(71)
2 steps
H2PO4−
ent-stellettamide A

Scheme 8.16

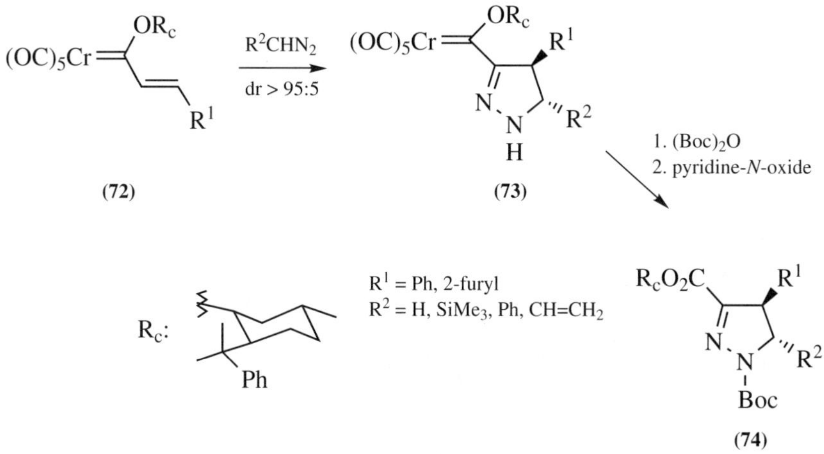

Scheme 8.17

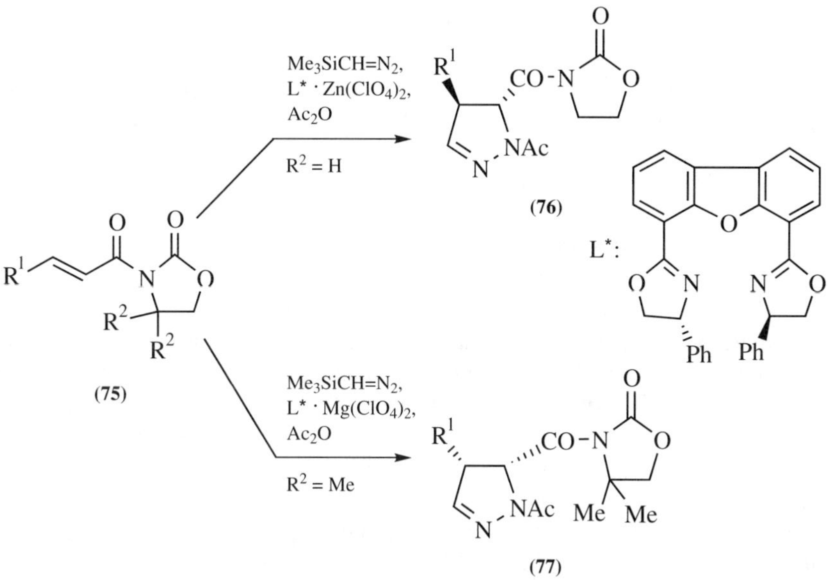

Scheme 8.18

8.2. CYCLOADDITIONS WITH CARBON–HETEROATOM DOUBLE BONDS

8.2.1. C=N Bonds

In an earlier survey of the cycloaddition chemistry of diazo compounds with imines (5), it was noted that diazoalkanes react particularly well with both electron-deficient acyclic and strained cyclic imines, such as 2*H*-azirines, producing

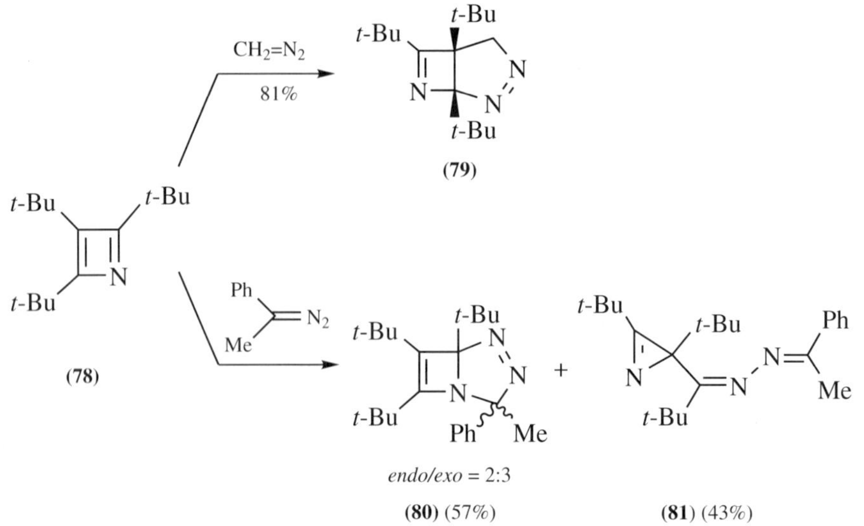

Scheme 8.19

1,2,3-triazolines or derived products. During the past two decades, this particular subject has not received much attention, and only a few examples will be mentioned.

The kinetically stabilized azacyclobutadiene **78** (Scheme 8.19) reacts with diazomethane at the C=C bond with formation of **79**, while 1-diazo-1-phenylethane adds across the sterically less hindered C=N bond to furnish the bicyclic 1,2,4-triazoline **80** (140a). The latter product is also accompanied by 2*H*-azirinylketazine **81**. It was shown that the isomerization **80** → **81**, for which a dipolar intermediate was suggested, can be achieved both thermally and photochemically. With α-diazocarbonyl compounds (e.g., 2-diazoindane-1,3-dione), ring-contraction products analogous to **81** are obtained as the exclusive products (140b).

Carrié and co-workers studied the cycloaddition of oxime esters derived from methyl cyanoacetate and malonate esters **82** (Scheme 8.20) with diazomethane and some monosubstituted derivatives. Thermally labile 1,2,3-triazolines **83** were obtained when tosyloxy- and benzoyloxyimines were used (141), while methyl acetoxyimino-cyanoacetate (**82**, X = CN, Y = CO_2Me, R^1 = Ac) gave products derived from both a 1,2,3- and a 1,2,4-triazoline, depending on the structure of the diazo compound (142). Not unexpectedly, diazomethane reacted with the corresponding imino-malononitrile (**82**, X = Y = CN) system at the nitrile function rather than at the C=N bond (143).

Elimination of N_2 from triazolines **83** occurred thermally (⩽20 °C) as well as photochemically. Depending on the substituent pattern and the method used, different products were obtained, such as aziridines **84a** and **84b**, oximines **85**, and 1,3-oxazoles **86** (143).

Diazo(trimethylsilyl)methane reacts with *N*-sulfonylimines derived from aldehydes to give 2-(trimethylsilyl)aziridines in good yield and with high cis-stereo-

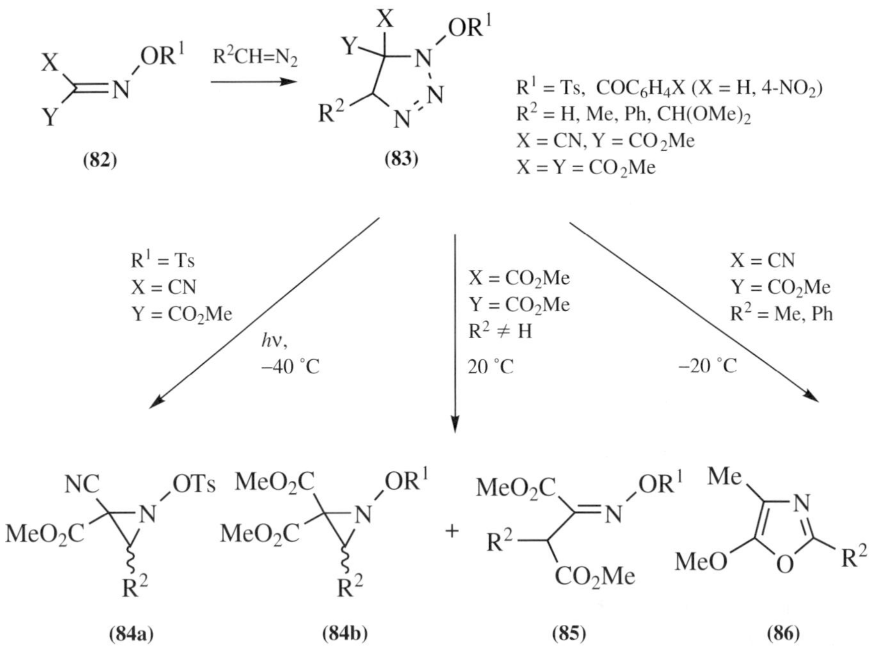

Scheme 8.20

selectivity (144). Ethyl diazoacetate and diazo(phenyl)methane do not react in an analogous manner. Diazo(trimethylsilyl)methane underwent cycloaddition with the iminium salt derived from cyclohexanone and pyrrolidine at the $C{=}N^+$ bond to afford, after loss of N_2 and the $SiMe_3$ group, a dispiro-aziridinium salt (145). The same transformation has been reported to occur with diazomethane. It is not clear whether the aziridines are formed *via* a transient triazoline intermediate or by nucleophilic addition of the diazo compound at the imine or iminium function, followed by loss of nitrogen from the resulting diazonium betaine, and then ring closure. It should be mentioned that the isolation of aziridines from diazo compounds and imines under the action of typical Lewis acids (146,147), copper (148,149), and rhodium (145,150) catalysts is a subject of intense research activity.

8.2.2. C=P Bonds

Phosphaalkenes that possess a $\lambda^3\sigma^2$-phosphorus atom can be isolated when appropriately substituted (151). These systems exhibit a much more expressed dipolarophilic than dienophilic reactivity, probably as a consequence of the polarity of the P=C bond. The [3 + 2] cycloaddition of diazo compounds with phosphaalkenes **87** leads to 4,5-dihydro-3*H*-1,2,4-diazaphospholes **88** (Scheme 8.21) that are not always isolated. Quite often, elimination of molecular nitrogen occurs during the cycloaddition at or below 20 °C. In other cases, N_2 extrusion is achieved at

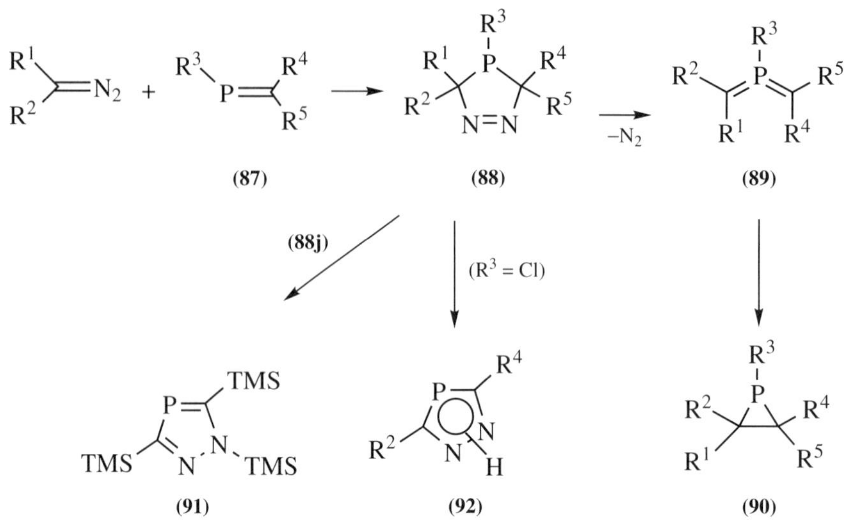

Scheme 8.21

elevated temperature or by photochemical activation. The resulting bis(alkylidene)phosphorane system **89** typically undergoes a thermally induced, conrotatory 4π-electrocyclic ring closure to give phosphiranes **90**. Depending on the substituent pattern, this reaction occurs somewhere between −80 and + 120 °C. These general remarks are illustrated by the following examples (Scheme 8.21 and Table 8.1): The *P*-bis(trimethylsilyl)amino-4-phosphapyrazolines **88a** (152), **88b** (153), and **88c**

TABLE 8.1. CONVERSION OF 4-PHOSPHAPYRAZOLINES (**88**) INTO PHOSPHIRANES (**90**) OR 1,2,4-DIAZAPHOSPHOLES (**92**) (SEE SCHEME 8.21)[a]

88 → 90	R^1	R^2	R^3	R^4	R^5	Reference
a	H	TMS	$N(TMS)_2$	H	TMS	152
b	H	*t*-Bu	$N(TMS)_2$	H	TMS	153
c	H	TMS	$N(i\text{-}Pr)_2$	H	TMS	154
d	Ph	Ph	PhC≡C	TMS	TMS	155
e	Ph	Ph	Ph	TMS	TMS	156
f	Ph	Ph	mesityl	TMS	TMS	156
g	H	Ph	$N(i\text{-}Pr)_2$, *t*-Bu	Ph	TMS	157
h	aryl, Me[b]	aryl, Me[b]	Cl	Ph	TMS	157
i	H	*t*-Bu	Cl	TMS	TMS	158
j	H	TMS	Cl	TMS	TMS	158
k	H	mesityl	Cl	TMS	TMS	158
88 → 92						
l	H	*c*	Cl	Ph	TMS	159
m	H	CO_2Me	Cl	TMS	TMS	160
n	H	CO_2Me	Cl	CO_2Et	TMS	161

[a] Tetramethylsilane = TMS = $SiMe_3$.
[b] Ph_2CN_2, 9-diazofluorene, bis(2-naphthyl)diazomethane, Me_2CN_2.
[c] H, Ph, COPh, CO_2Et, $PO(OMe)_2$, $POPh_2$.

(154) are thermally rather stable; thermolysis of **88b,c** at 80–120 °C gives the phosphiranes **90b,c**. The photolysis of **88a** was used to generate the bis(alkylidene)phosphorane **89a**, which cyclized to phosphirane **90a** only at 190 °C over a 3-day period.

Phosphapyrazolines **88d** (155), **88e** (156), and **88g** (157) lose N_2 at 0 °C and yield phosphiranes **90** *via* spectroscopically detectable phosphacumulenes **89**. In contrast to **88e**, the *P*-mesityl-phosphapyrazoline system **88f** (156) can be isolated. *P*-Chloro-phosphapyrazolines **88h** (157) and **88i–k** (158) also lose N_2 under the cycloaddition conditions or during work up and yield phosphiranes **90h–k**. The cycloaddition products **88i,j,l–n**, obtained from monosubstituted diazo compounds and *P*-chloro-*C*-trimethylsilylphosphaalkenes, undergo a fast reaction with conservation of the ring nitrogen atoms. In the case of **88i,j**, these reactions compete with the formation of a phosphirone. 1,2-Elimination of HCl from **88j** followed by a 1,3(C → N) silyl shift yields 1,3,5-tris(trimethylsilyl)-1,2,4-diazaphosphole **91** (158), whereas 1,2-elimination of Me_3SiCl from **88** followed by a 1,3(C → N) H shift generates 1*H*-1,2,4-diazaphospholes **92l** (159), **92m** (160), and **92n** (161). Tautomerization of Δ^1-phosphapyrazolines **88** to form Δ^2-phosphapyrazolines has also been observed [e.g., **88i**: $SiMe_3$ shift (158); **88**, $R^1 = H$, $R^2 = t$-Bu, $R^3 = SiMe_3$, $R^4 = OSiMe_3$, $R^5 = t$-Bu: H shift (162)].

The cycloadduct obtained from ethyl diazoacetate and the cyclic phosphaalkene 9-*tert*-butyl-1,3-diphenyl-10-phospha-1,3-etheno-1*H*-benzopyran-4(3*H*)-one underwent spontaneous [3 + 2] cycloreversion and produced ethyl 5-*tert*-butyl-1,2,4-diazaphosphole-3-carboxylate (163). Still another transformation was found for *P*-trimethylsilyl-substituted diazaphospholes system **94**, which suffered dediazoniation under the cycloaddition conditions and yielded phosphaalkene **95** (162) (Scheme 8.22). It was proposed that N_2 extrusion and $SiMe_3$ migration occur in concert. On the other hand, the cycloaddition products derived from phosphaalkene **93** and 2,2-dimethyl-1-diazopropane or diazo(trimethylsilyl)methane simply underwent tautomerization to the corresponding Δ^2-phosphapyrazoline (162) (**94**, R = *t*-Bu: H shift; R = $SiMe_3$: $SiMe_3$ shift).

In certain cases, 4,5-dihydro-1,2,3-diazaphospholes rather than 3,5-dihydro-1,2,4-diazaphospholes are formed from the [3 + 2] cycloaddition reaction of diazo compounds with phosphaalkenes. This regiochemistry was encountered in the reaction of (mesityl)P=CPh_2 with diazodiphenylmethane and was attributed to steric factors (164). Electronic factors may explain the orientation found in the

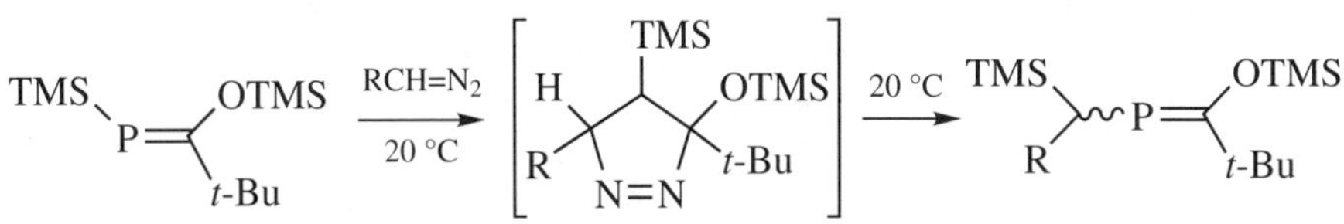

(93) **(94)** **(95)**

R = H, Me, CO_2t-Bu

Scheme 8.22

cycloaddition of diazoacetates with the electron-rich double bond of (η^5-C_5Me_5)(CO_2)Fe—P=C(NMe)$_2$ from which the N^2-metalated 4-dimethylamino-1,2,3-diazaphosphole-5-carboxylic ester was isolated (165). The latter reaction is also noteworthy because it was discovered that other electron-rich phosphaalkenes such as *P*-phenyl-*C*-(dialkylamino)phosphaalkenes do not react with diazo compounds (such as diazodiphenylmethane, 1-diazo-1-phenylethane, diazoacetates, and diazomalonates) as 1,3-dipoles, but rather react with them at the nucleophilic phosphorus atom *via* a Staudinger-type reaction to give phosphazine-derived products (166). Diazocumulenes also undergo 1,3-dipolar cycloaddition with phosphaalkenes. The first example reported was that of (diazomethylene)phosphorane **33**, which reacted with [bis(trimethylsilyl)methylene]chlorophosphane or the related bis[bis(trimethylsilyl)methylene]chlorophosphorane to give 3-phosphoranylidene-1,2,4-diazaphospholes **96** and **97** by cycloaddition followed by elimination of Me_3SiCl (167) (Scheme 8.23). According to their NMR data and X-ray diffraction analysis, these heterocycles have a significant bonding contribution from a heterophospholyl betaine structure.

α-Silyl-α-diazoketones **29** react with various phosphaalkenes to form 5-alkylidene-3*H*-1,2,4-diazaphospholes (**98–101**, Scheme 8.23) or products derived from them (**102**, **103**) (168). By analogy to their reactions with alkenes (see Section 8.1), it was assumed that 1-diazoalkenes **30**, which are minor equilibrium partners of diazoketones **29**, are selectively trapped from the equilibrium mixture. It is interesting to note that these diazo compounds also afford a [3 + 2] cycloaddition product **103** when the electron-rich phosphaalkene mesityl—P=CH—NMe$_2$ group was used. Diazodiphenylmethane and dimethyl diazomalonate underwent a Staudinger reaction rather than a cycloaddition with PhP=C(R)NMe$_2$ (R = H, Me), as was mentioned above (166).

Alkylidene-phosphapyrazolines **98–101** are much more thermally stable than their relatives **88**, which do not possess the *exo*-methylene substitution. Dediazoniation of **98** required heating in toluene at 110 °C and gave one or more of the following products, probably *via* intermediate diphenylmethylene(vinylidene)phosphoranes: methylenephosphiranes, (2-siloxyvinyl)phosphanes, 2*H*-1,3-oxaphospholes, and 1-alkylidene-2,3-dihydro-1*H*-benzo[*c*]phospholes (169). Thermolysis of **100** (R = *t*-Bu, 1-adamantyl) afforded isolable 2-phosphabutadienes (169). The photochemical elimination of N_2 from **98** generated cyclic azomethine imine dipoles **104** (Scheme 8.24), which rearrange to compounds **105** and **106** that could be further trapped with DMAD to form **107** (170).

Some highly reactive phosphaalkenes, generated by base-assisted β-elimination at low temperature, have been trapped with appropriate 1,3-dipoles (171). In this manner, dihydro-1,2,4-diazaphospholes were obtained from phosphaalkenes $R^1R^2C{=}PR^3$ [$R^1 = H$, $R^2 = Me$, $R^3 = H$ (172); $R^1 =$ alkyl, $R^2 = Cl$, $R^3 = H$ (173); $R^1 = CO_2Et$, COPh, CONMe$_2$, $R^2 = H$, SiMe$_3$, $R^3 = Cl$ (174)] and ethyl diazoacetate or diazo(trimethylsilyl)methane. By the same strategy, the transient 1-phospha-1,3-butadiene **108** was trapped to give diazaphosphole **109** after elimination of HCl (175) (Scheme 8.25). It is interesting to note that the cycloaddition reaction occurred exclusively at the hetero-double bond.

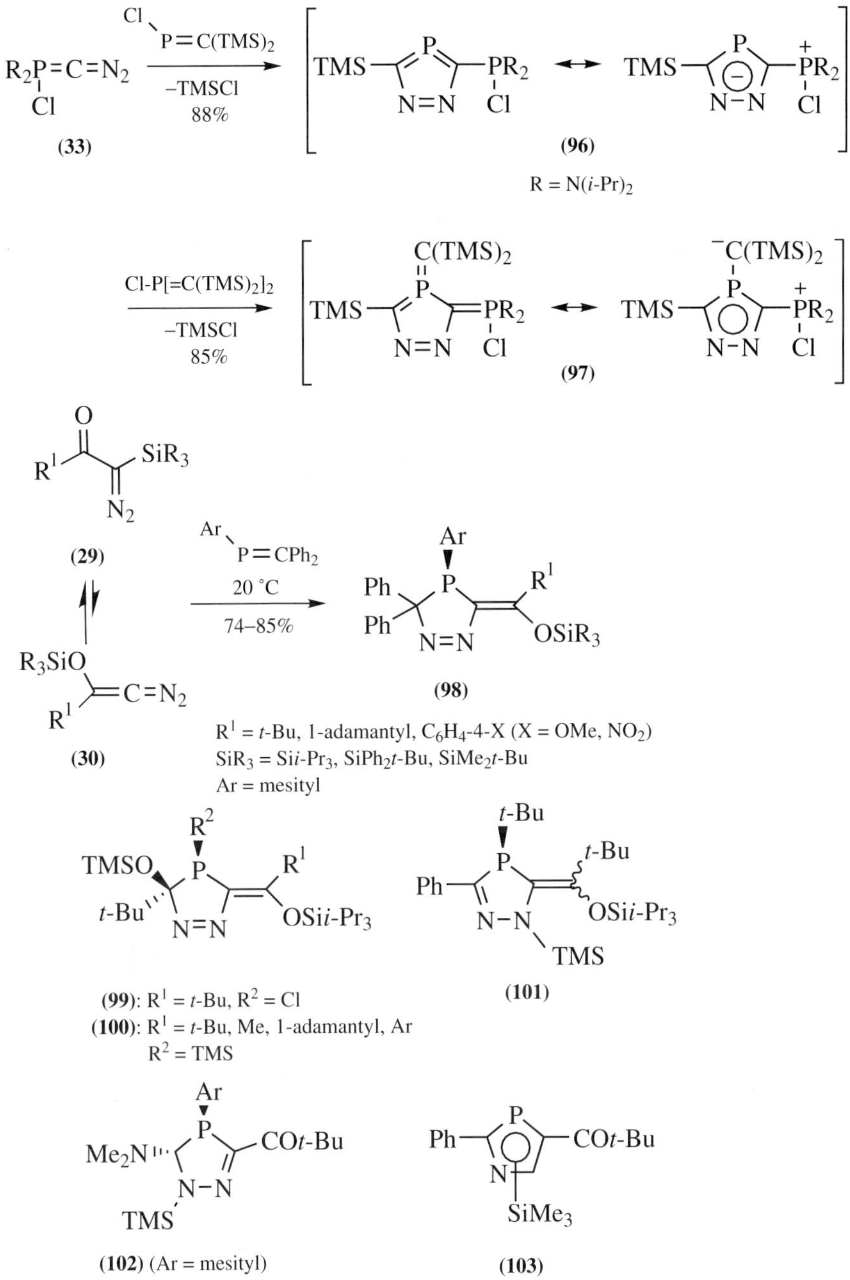

(33)
–TMSCl
88%
(96)
R = N(i-Pr)2
Cl-P[=C(TMS)2]2
–TMSCl
85%
(97)
(29)
(30)
20 °C
74–85%
(98)
R1 = t-Bu, 1-adamantyl, C6H4-4-X (X = OMe, NO2)
SiR3 = Sii-Pr3, SiPh2t-Bu, SiMe2t-Bu
Ar = mesityl
(99): R1 = t-Bu, R2 = Cl
(100): R1 = t-Bu, Me, 1-adamantyl, Ar
R2 = TMS
(101)
(102) (Ar = mesityl)
(103)

Scheme 8.23

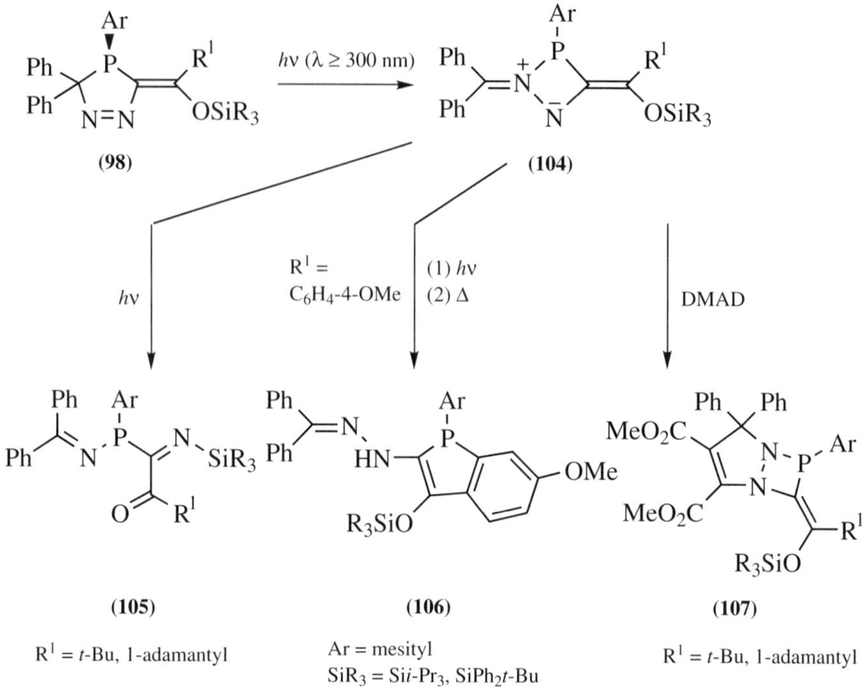

Scheme 8.24

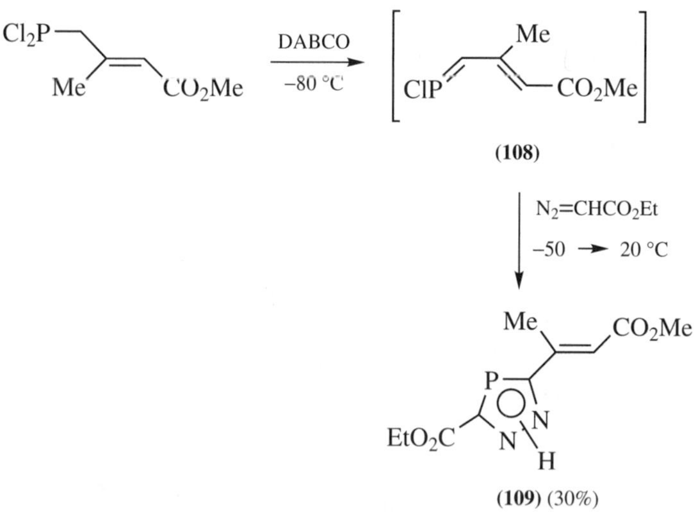

Scheme 8.25

1,3-Dipolar cycloaddition chemistry has also been achieved with P=C bonds incorporated in the ring system of some heterophospholes, in spite of the heteroaromatic character of the latter. This chemistry was developed by Arbuzov and co-workers (176) who reviewed the use of 1,2,3-diazaphospholes **110** as dipolarophiles (Scheme 8.26). The reaction rate was facilitated by increasing the electron density in the diazo dipole ($Ph_2CN_2 < PhMeCN_2 < Me_2CN_2$) (177).

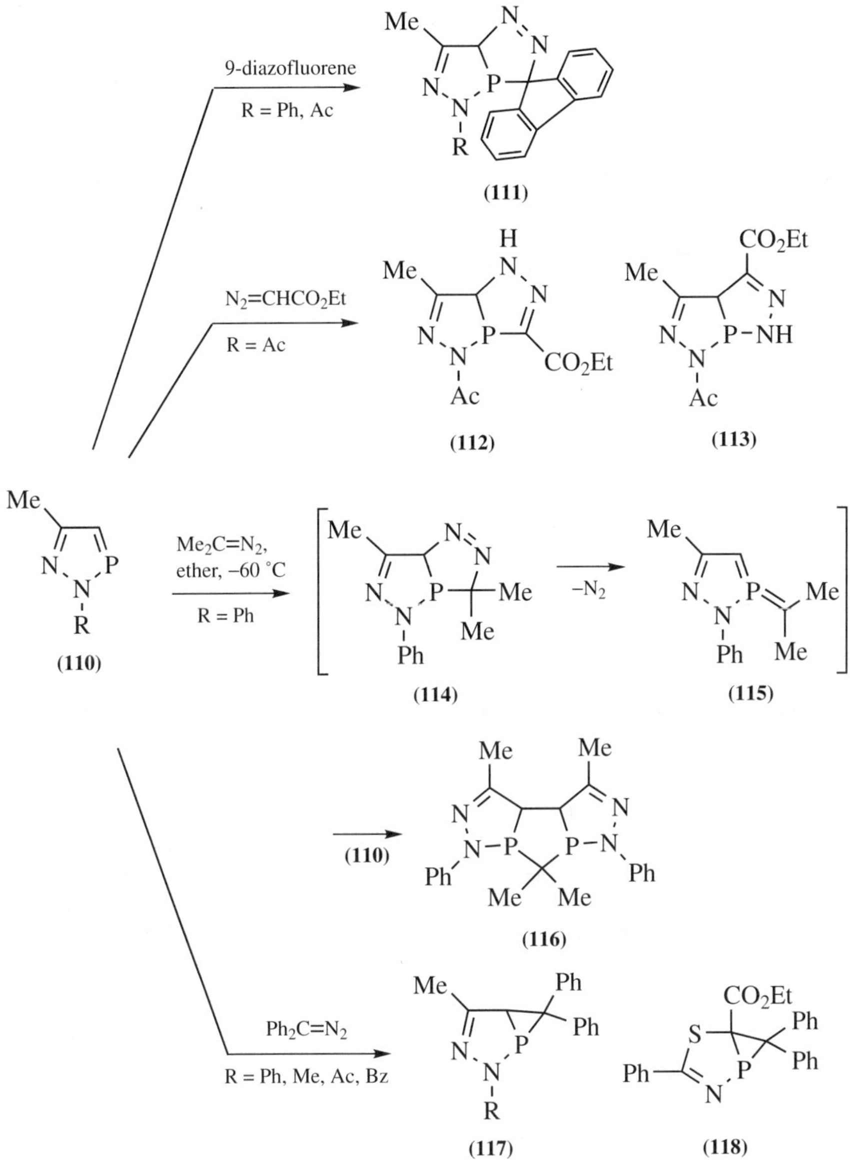

Scheme 8.26

The course of the reaction of **110** with diazo compounds is highly dependent on the nature of the diazo substrate as well as the reaction conditions (solvent, temperature) as well.

With 9-diazofluorene, [3 + 2]cycloaddition products of type **111** were obtained in pentane or hexane, while N_2 loss and subsequent cyclotrimerization occurred in CH_2Cl_2, CCl_4, and benzene (178). With acceptor-substituted diazo compounds of the type $RCH{=}N_2$ [$R = CO_2Et$ (179), COMe (180,181), and $PO(OEt)_2$ (182)], the initially formed Δ^1-phosphapyrazolines rapidly tautomerized and bicyclic Δ^2-phosphapyrazolines such as **112** and **113** were formed.

With 2-diazopropane, two stereoisomers of the resulting tricyclic product **116** (cis, anti, cis and cis, syn, cis) were obtained (183). Formation of **116** can be rationalized by N_2 extrusion from the cycloaddition product **114** and a subsequent $[3_{4\pi} + 2_{2\pi}]$ cycloaddition of the resulting 3-alkylidene-1,2,3-diazaphosphole **115** with the remaining heterophosphole **110**. When an excess of 2-diazopropane was used, **115** was trapped by 1,3-dipolar cycloaddition across the exocyclic P=C bond.

The reaction of 1,2,3-diazaphospholes **110** with diazodiphenylmethane leads directly to bicyclic phosphiranes **117** (184). The analogous product **118** was obtained using a 1,3,4-thiaazaphosphole (185). It is reasonable to assume that these phosphiranes result from a 4π-electrocyclization of the bis(methylene)phosphorane unit of a reaction intermediate analogous to **115**.

Similar to their reaction with phosphaalkenes, 1-diazo-2-(oxoalkyl)silanes **29** react with various heterophospholes by [3 + 2] cycloaddition of the diazocumulene system **30** (which is in equilibrium with **29**) across the P=C bond. With 2-acyl-1,2,3-diazaphospholes **119** ($R^2 = Ac$, Bz; no reaction with $R^2 = Me$, Ph up to 60 °C), the expected cycloaddition products **120** (Scheme 8.27) could be isolated (186). Elimination of N_2 from these bicyclic Δ^1-pyrazolines occurred upon heating at 100 °C and furnished the tricyclic systems **122** when SiR_3 was a trialkylsilyl group. Apparently, the thermolysis of **120** generates the 5-alkenylidene-1,2,5-diazaphosphole **121** (by N_2 extrusion) as well as diazaphosphole **119** (by a [3 + 2] cycloreversion process), which recombine in an intermolecular cycloaddition to furnish **122**. When $SiR_3 = SiPh_2t$-Bu, a formal intramolecular [3 + 2] cycloaddition of the C=P=C unit with an aromatic C=C bond occurs and the polycyclic compound **123** is obtained (187).

In contrast to **120**, the cycloaddition products derived from diazocumulenes **30** and 1,2-thiaphospholes **124** (as well as 1,2,3,4-triazaphospholes **127**) cannot be isolated, since molecular nitrogen is readily lost under the reaction conditions (Scheme 8.28). Phosphorus heterocycles **125** and **126** are produced from **124** (188). The major product is formed by a path analogous to that outlined in Scheme 8.27. Compound **126** is probably formed by a 4π-electrocyclic ring closure of an intermediate 2-alkenylidene-1,2-thiaphosphole. In the 1,2,3,4-triazaphosphole case, N_2 extrusion from the initial cycloaddition product **128** takes place and generates a 4-imino-1,2,4-diazaphosphole derivative **129**. This transient intermediate undergoes further reaction to form either the dispiroannulated 1,3-diaza-2,4-diphosphetidine **130** or dihydro-1,2,4(λ^5)-diazaphosphole **131**, depending on the substituent at the imino group (189).

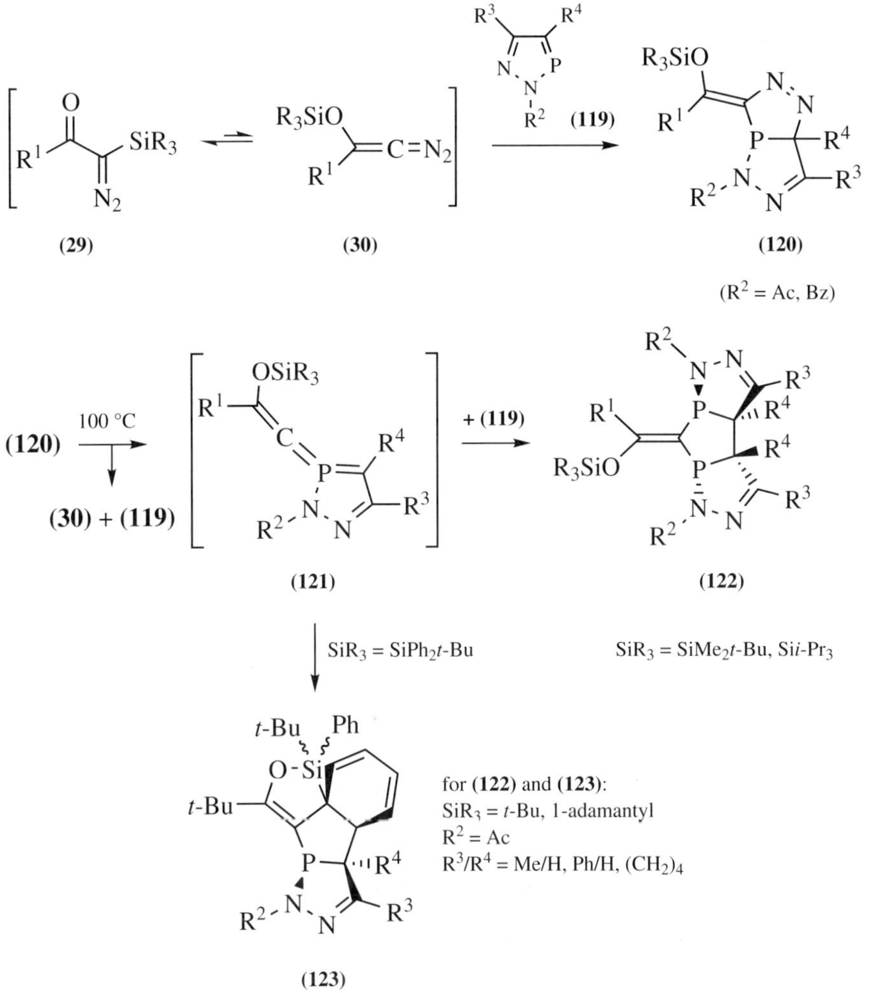

Scheme 8.27

The P=C bond of $\lambda^5\sigma^3$-phosphoranes is also accessible to a 1,3-dipolar cycloaddition reaction. The formation of **97** (Scheme 8.23), of 3,5-dihydro-1,2,4-diazaphosphole sulfide **132** from an amino(methylene)thioxophosphorane (190) and of phosphirane imine **133** from an amino(imino)alkylidenephosphorane (191) (Scheme 8.29) illustrate the chemical behavior of phosphaalkenes containing a $\lambda^3\sigma^2$-phosphorus atom.

8.2.3. C=As Bonds

Although compounds containing a C=As bond have attracted much less attention than those with a C=P bond, the available results show that the

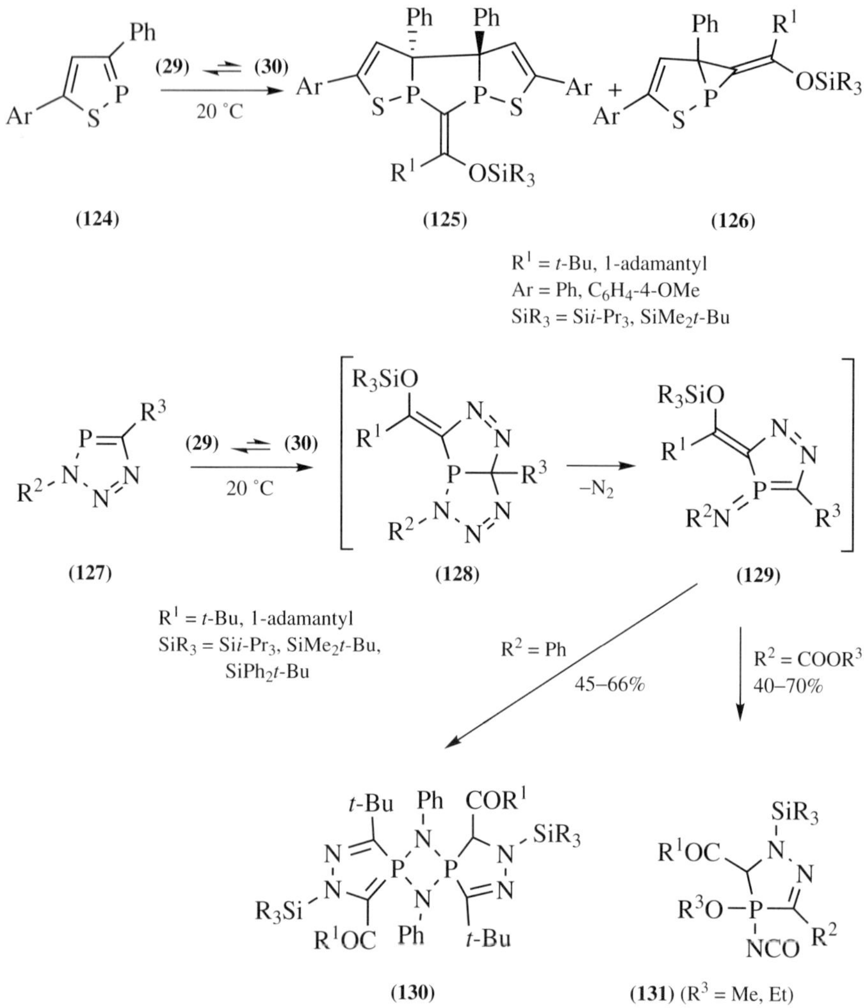

Scheme 8.28

dipolarophilic character of these two hetero-double bonds toward diazo compounds is rather similar. Thus, the short-lived arsaalkenes **134** were trapped *in situ* by cycloaddition with ethyl diazoacetate to give 1,2,4-diazaarsoles **135** (192) (Scheme 8.30). A transient 1-chloro-1-arsa-1,3-butadiene has been generated and trapped in the same manner as that shown in Scheme 8.25 [**108** → **109**, As instead of P (175)]. On the other hand, the *in-situ* generated arsaalkene $MeO_2C(R)C{=}AsCl$ (R=Me, CN) reacted with ethyl diazoacetate to give **135** ($R^1 = CO_2Et$), perhaps *via* a cycloaddition–cycloreversion reaction of an intermediate 3*H*-1,2,4-diazaarsole (193).

By close analogy to heterophospholes (see Scheme 8.26), the C=As bond incorporated in heteroarsoles is also amenable to 1,3-dipolar cycloaddition. Thus,

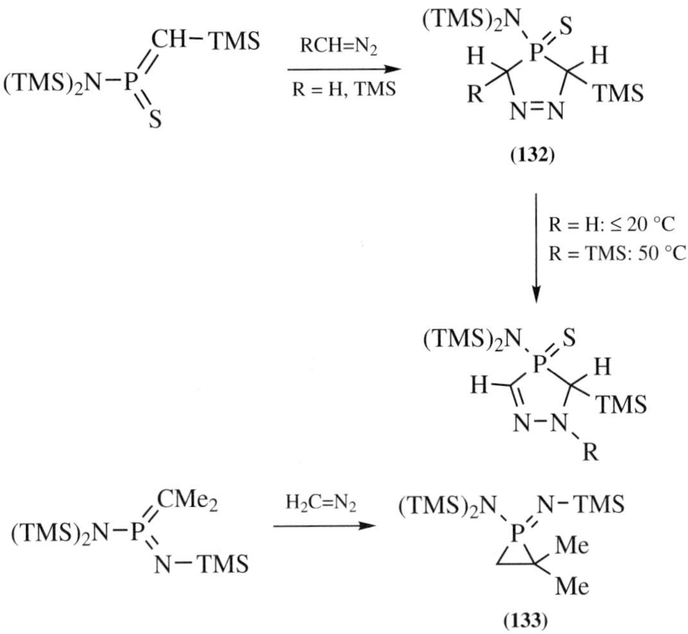

Scheme 8.29

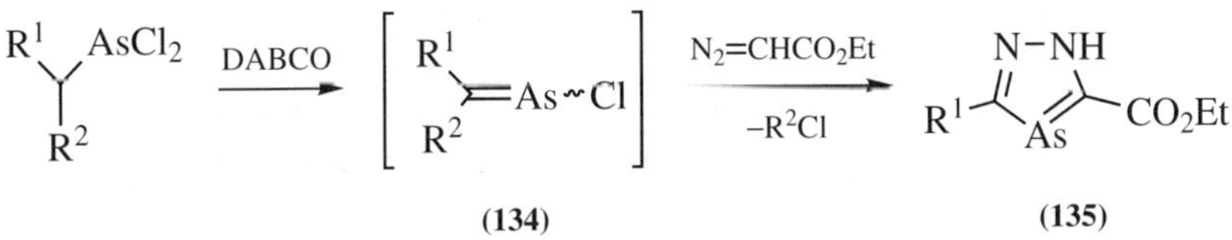

$R^1 = CO_2Et, CONMe_2$; $R^2 = H, TMS$

Scheme 8.30

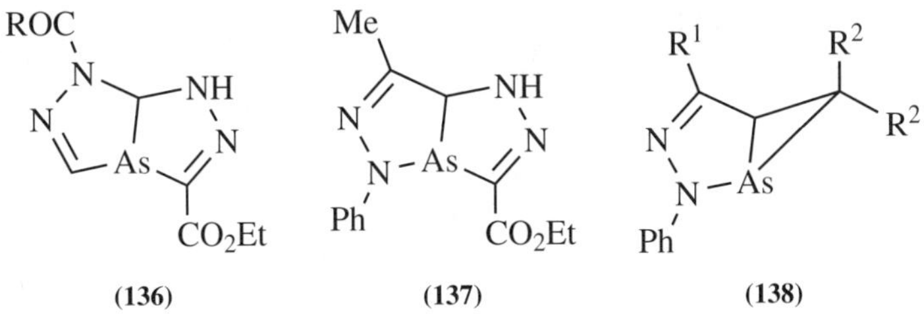

R = Me, Ph

$R^1 = Me, Ph$; $R^2 = Me, Ph$

Scheme 8.31

products **136** (194) and **137** (181,195) (Scheme 8.31) were obtained from the corresponding 1,2,4- and 1,2,3-diazaarsoles, respectively. When 2-diazopropane or diazodiphenylmethane rather than ethyl diazoacetate was allowed to react with 1,2,3-diazaarsoles, the bicyclic arsiranes **138** were obtained (183,196–198). With 2-diazopropane, the expected [3 + 2]-cycloaddition product could be detected and was identified as a precursor to the bicyclic arsirane (197).

8.2.4. C=S Bonds

Reactions of diazo compounds with thioketones leading to 1,3-dithiolanes or thiiranes have been known since 1920, but the mechanism of the so-called Schönberg reaction was only clarified by Huisgen and co-workers in 1981 (199). For example, diazomethane reacts with thiobenzophenone at −78 °C in a [3 + 2] manner to give 2,5-dihydro-1,2,4-thiadiazole **139** (Scheme 8.32). This product rapidly eliminates N_2 at −45 °C providing thiocarbonyl ylide **140** that can be trapped by a broad range of dipolarophiles [electrophilic alkenes, alkynes, N=N bonds (199–201)] to give various five-membered sulfur heterocycles **141**. When excess thiobenzophenone is present, the 1,3-dithiolane **142** is formed. In the absence of trapping reagents, **140** produces thiirane **143** by 1,3-cyclization or 1,4-dithian **144** by cyclodimerization. An analogous study was carried out for the reaction of thiobenzophenone with diazoethane, diazophenylmethane, and diazodiphenylmethane (202). Remarkably, thiobenzophenone *S*-diphenylmethylide could not be trapped intermolecularly and thiirane formation was the only reaction observed.

Related investigations of the reaction of diazo compounds with alkyl-substituted thioketones [$R_2C{=}S$, R = Et, Pr, *i*-Pr, *t*-Bu (203); 2,2,4,4-tetramethylcyclobutan-1-one-3-thione (204), and adamantanethione (205,206)] showed that the 3,3-dialkyl-1,2,4-thiazolines are thermally more stable than the 3,3-diphenyl analogue **139** and

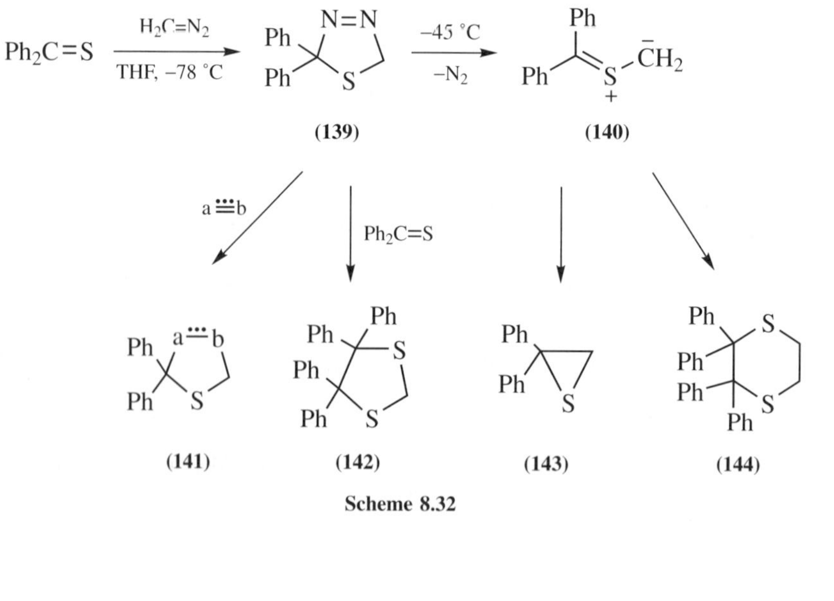

Scheme 8.32

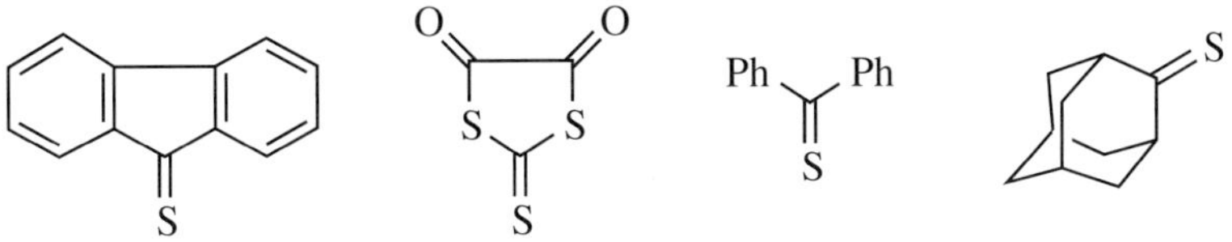

450,000[a] 66,800[a] 10,200 101

$(NC)_2C{=}C(CN)_2$ 3180

$RC{\equiv}C{-}CO_2Et$ R = H: 10.2; R = CO_2Et: 76.4

$H_2C{=}CHCO_2Et$ 8.1

Scheme 8.33. Rate Constants for the 1,3-Dipolar Cycloaddition of $Ph_2C{=}N_2$ with C=S and Other Dipolarophiles; $10^3\ k_2\ [M^{-1}\ s^{-1}]$, in DMF ([a] in $CHCl_3$) at 40 °C (DMF = dimethylformamide).

can be isolated. The mechanistic picture outlined in Scheme 8.32 was also helpful to rationalize the formation of a tricyclic dithiepane derivative from the reaction of thiotropone and diazomethane. In this case, a 4,5-dispiroannulated 1,3-dithiolane was the only reaction intermediate that could be isolated (207). Huisgen and co-workers (209,209) performed kinetic measurements on the cycloaddition reactions of thiocarbonyl compounds with diazoalkanes and other nucleophilic 1,3-dipoles. The rate constants for some selected thioketones (209) are given in Scheme 8.33 and demonstrate that thioketones are superior to other commonly used dipolarophiles. This exceptionally high reactivity was explained by perturbation MO theory and was based on the fact that thiocarbonyl compounds have a remarkably small HOMO–LUMO energy separation compared with C=O compounds.

The high reactivity of the C=S bond toward diazo dipoles was also helpful in the characterization of thiopivaldehyde, a monomeric thioaldehyde persistent in inert organic solvents for 16 h at 20 °C. This thioaldehyde underwent ready cycloaddition with ethyl diazoacetate (210). Similarly, the diazocumulene 9-(diazomethylene)fluorene could be generated by diazotization of 9-(aminomethylene) fluorene and was trapped by the cycloaddition with thiobenzophenone (211).

Cycloaddition of diazo dipoles at C=S bonds is not always orientation specific. Mixtures of 1,3,4-thiadiazolines and 1,2,3-thiadiazolines were formed when diazomethane was combined with open-chain thiones $R_2C{=}S$ (203) (R = Et, Pr, *i*-Pr, *t*-Bu) or with adamantanethione (205,206) (Scheme 8.34). Also, the products obtained from thiopivaldehyde and ethyl diazoacetate are derived from the two regioisomeric cycloadducts (210). The generally observed dependence of the regioisomer ratio on the solvent polarity (203,206) [e.g., **145:146** from adamantanethione: 91:9 in pentane at −20 °C, 10:90 in methanol at −30 °C (206)] was attributed to the large dipole moment of the transition state leading to the 1,2,3-thiadiazoline (203,212). *Ab initio* calculations of the cycloaddition of thioformaldehyde with diazomethane gave activation enthalpies (3-21G* basis set), which favor the 1,2,3-isomer by 2.4 kcal mol^{-1} (only 0.5 kcal mol^{-1} by a CASSCF single-point calculation of the 3-21G* transition structure). This energy difference, however, vanishes for diethyl thioketone (212). This result suggests that increasing

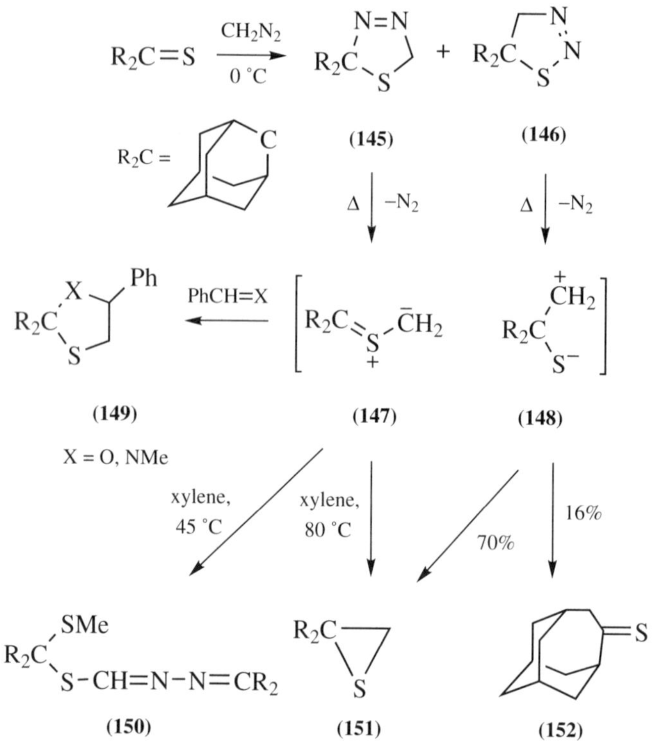

Scheme 8.34

the steric demand of the thione substituents directs the cycloaddition towards the 1,3,4-isomer, in agreement with the experimental findings.

The isomeric adamantane-spirothiadiazolines (**145** and **146**) (Scheme 8.34) exhibit different thermal stability [**145**: $\tau_{1/2} = 33$ min at 45 °C; **146**: $\tau_{1/2} = 25.6$ min at 110 °C (206a)]. Elimination of N_2 from **145** generates thiocarbonyl ylide **147** that was trapped not only with the dipolarophiles mentioned above for **140**, but also with aldehydes and imines (206a) (**147** → **149**). Without a trapping reagent, thiirane **151** was formed from both **145** (at 80 °C) and **146**. In the latter case, the extrusion probably proceeds *via* intermediate **148** and is accompanied by homo-adamantanethione **152** and a trace of methyleneadamantane. When **145** was decomposed at 45°C rather than at 80 °C, dimer **150** was also obtained. The isolation of **150** suggests that ylide **147** is also able to act as a base toward its precursor **145** (213). In fact, **147** can also be trapped with other protic nucleophiles.

The chemistry outlined in Schemes 8.32 and 8.34 illustrates the complexity of reactions that occur between thiocarbonyl compounds and diazo compounds. Heimgartner and co-workers (214–217) observed a similar reactivity pattern when they combined 1,3-thiazol-5(4*H*)-thiones (**153**) with diazoalkanes. When ethyl diazoacetate was used, additional reaction pathways occurred giving rise to a complex mixture of products (218). An interesting aspect of this chemistry involves

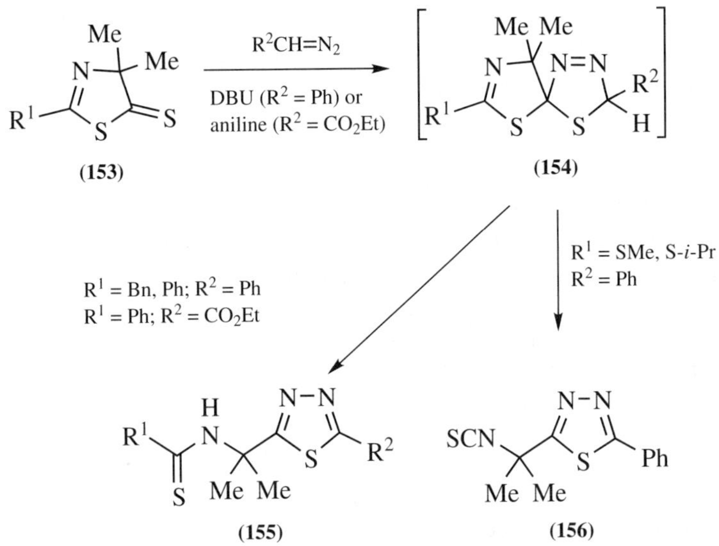

Scheme 8.35

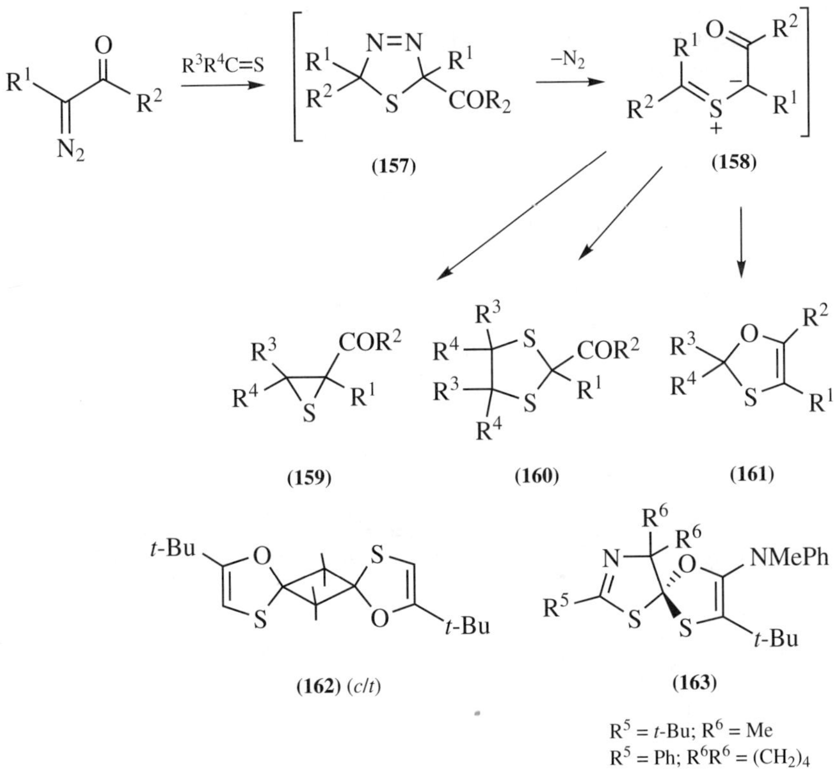

Scheme 8.36

the conversion of the initially formed 1,3,4-thiazoline ring **154** into functionalized 1,3,4-thiadiazoles **155** or **156** by a base-assisted 1,4-elimination reaction (217,218) (Scheme 8.35). Since the thiadiazolines cannot be isolated due to their thermal lability, the base must be present prior to the cycloaddition step.

When thiocarbonyl and α-diazocarbonyl compounds are combined, acyl-substituted thiocarbonyl ylides **158** are generated from a nonisolable 3-acyl-1,2,4-thiadiazoline **157** (Scheme 8.36). In addition to giving acylthiiranes **159** and 1,3-dithiolanes **160**, dipoles **158** can also 1,5-cyclize to produce 1,3-oxathioles **161**. Acyl-thiocarbonyl ylides derived from diazoketones [e.g., $HC(O)C(N_2)R$, R = Ph, *t*-Bu (219,220); 2-diazocyclohexanone (221)] produce 1,3-oxathioles [e.g., **162** (220), Scheme 8.36], while those derived from diazoesters (218,222,223) lead to thiiranes by 1,3-cyclization. Ylides derived from α-diazocarboxamides form 1,3-oxathioles (e.g., **163**) and thiiranes (e.g., **159**, $R^1 = t$-Bu, $R^2 =$ NMePh, $R^3 = R^4 =$ Ph), depending on the nature of the substituents (220). A related 1,5-cyclization of an aminomethyl-thiocarbonyl ylide formed from dimethyl 3-anilino-2-diazobutanedioate was also reported (224).

Thiiranes that are obtained from the reaction of diazo dipoles with C=S bonds can be transformed into alkenes by desulfurization. This reaction sometimes occurs spontaneously, but more often is achieved by treatment with phosphanes (225). This important methodology represents an alternative for the Wittig reaction and has high merit for the preparation of sterically hindered (226–229) and uncommonly functionalized alkenes (214,216,217,230,231). Some examples are given in

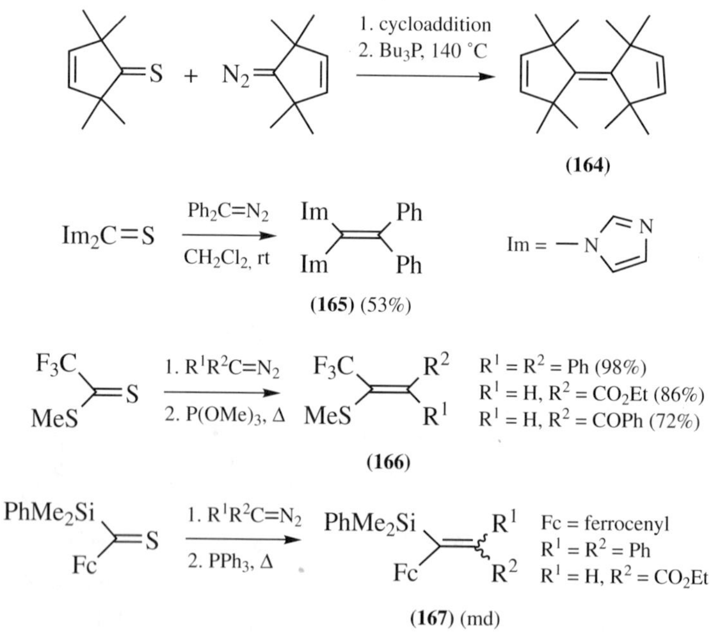

Scheme 8.37

Scheme 8.37 [**164** (228), **165** (231) (but formation of 2-R-1,3,4-thiadiazole with $RCH{=}N_2$, R = H, Me, Ph, CO_2R'), **166** (230) (accompanied by 16% of 1,3-oxathiole if $R^1 = H$, $R^2 = PhCO$), **167** (232)].

8.2.5. C=Se and C=Te Bonds

Reactions of diazo compounds with selenoketones have been much less investigated than thioketones. It was found that di(*tert*-butyl)selenoketone (**168a**) reacted with diazodiphenylmethane to regioselectively give 2,5-dihydro-1,3,4-selenadiazole **169a**, which on heating lost N_2 and selenium to furnish the tetrasubstituted alkene **170a** (Scheme 8.38) (233). This methodology is quite useful for the synthesis of sterically hindered alkenes (229). In fact, olefins **170b** (233,234), **170c** (235a), **170d** (227,235b), and some related alkenes (227) were prepared by this method. Since the cycloaddition of diazodiphenylmethane with selenoketone (**168a**) was faster than with the corresponding thioketone, it was originally hoped that the selenadiazoline route to sterically hindered alkenes would work in those cases where the thiadiazoline route (see Section 8.2.4) failed. However, the selenadiazoline route tolerates only a bit more of steric hindrance and the elusive tetra(*tert*-butyl)ethene was not accessible by this method, since di(*tert*-butyl)selenoketone failed to react with di(*tert*-butyl)diazomethane (233). The reaction of 4,4-dimethyl- (or dimethoxy-)selenobenzophenone with $R^3R^4C{=}N_2$ ($R^3 = H$, $R^4 = SiMe_3$; R^3, $R^4 = Ph$, 4-Tol, 4-MeO-C_6H_4) at temperatures $\leqslant 20\,°C$ furnished not only the unsymmetrical alkenes **170**, but also one or both of the symmetrical alkenes, $R^1R^2C{=}R^1R^2$ and $R^3R^4C{=}CR^3R^4$ (236). This indicates that the selenadiazoline intermediate **169** is highly unstable toward both N_2 extrusion and the two possible [3 + 2] cycloreversion pathways.

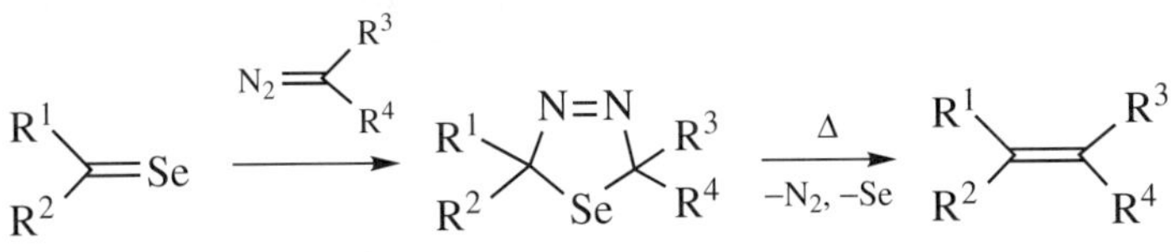

(168) **(169)** **(170)**

(169), (170)	**a**	**b**	**c**	**d**
R^1R^2C	$(t\text{-Bu})_2C$	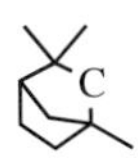	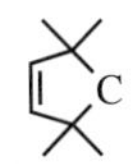	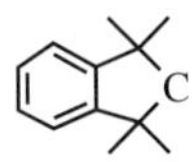
R^3R^4C	Ph_2C		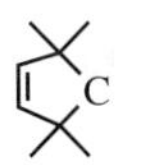	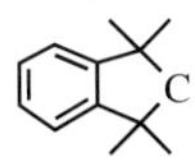

Scheme 8.38

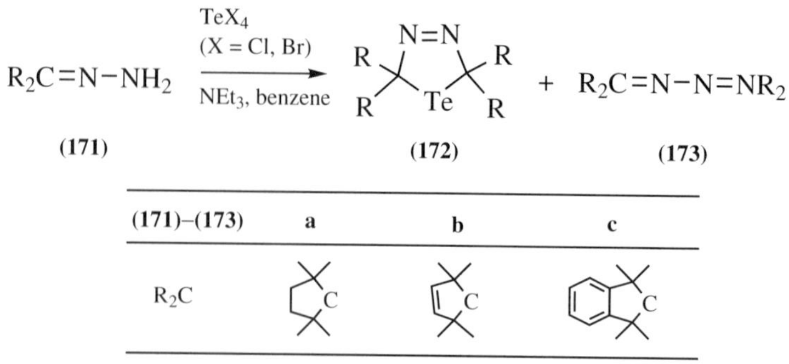

Scheme 8.39

Stable representatives of telluroketones were not available until recently. 2,5-Dihydro-1,3,4-telluradiazole **172** can be prepared from hydrazones **171** and tellurium tetrachloride or tetrabromide in yields of 19-48 and 42–55%, respectively (Scheme 8.39) (237). Ketazines **173** were obtained as byproducts of this reaction in 32–45% yield. It was proposed that TeX_4 is first reduced to TeX_2 in the reaction medium. The latter species probably generates both the diazo compound and the telluroketone derived from hydrazone **3** and then a [3 + 2] cycloaddition occurs. In fact, both **172** and **173** were obtained when hydrazones **3a–c** were treated with the difficult-to-handle tellurium dichloride in the presence of triethylamine.

Telluradiazolines **172** are extremely light sensitive and brief ultraviolet UV-irradiation of **172a** in solution or in the solid state afforded ketazine **173a** quantitatively. On the other hand, **172a** was completely stable at 80 °C. Thermolysis at 160 °C gave ketazine **173a**, alkene **170d**, and some other products which suggests that a [3 + 2] cycloreversion process is operating at this temperature.

8.3. CYCLOADDITIONS WITH HETEROATOM–HETEROATOM DOUBLE BONDS

8.3.1. P=X Bonds

In contrast to the P=C bond of phosphaalkenes (Section 8.2.2), double bonds between phosphorus and a heteroatom have not been used much as dipolarophiles. Most of the studies reported so far were devoted to the reactivity of the (λ^3)P=N bond of iminophosphanes. Amino(iminophosphanes) react with diazoalkanes to form 4,5-dihydro-3*H*-1,2,3,4-triazaphospholes or, by N_2 loss from the latter, to imino(alkylidene)-λ^5-phosphoranes (5,238). With *P*-halogeno-(arylimino)phosphanes **174** and the appropriate diazo compounds, 3*H*-1,2,3,4-triazaphospholes **175** (167) and **176** (239) (Scheme 8.40) were obtained as the major products after cycloaddition and eliminative aromatization.

The P=N bond of amino(imino)thioxophosphorane **177** is readily cyclopropanated even at 0 °C, with formation of phosphaaziridine **178**. An intermediate [3 + 2]

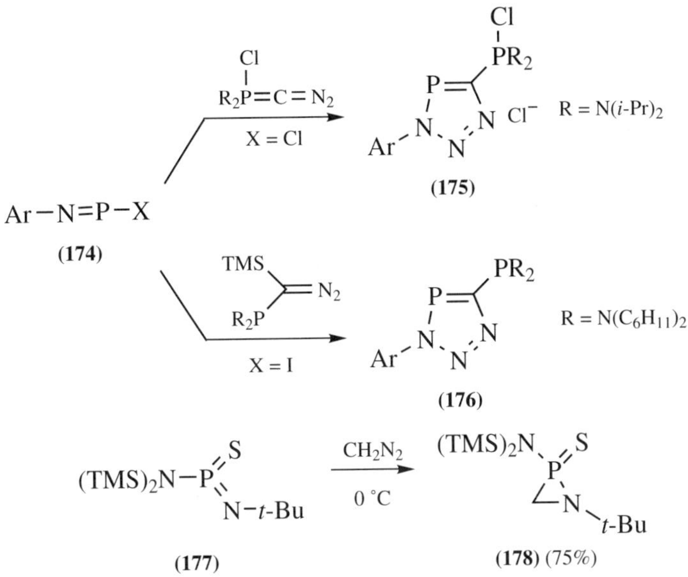

Scheme 8.40

cycloaddition product is likely (190). Bis(trimethylsilylimino)[bis- (trimethylsilyl)amino]phosphorane reacts in an analogous manner (240). On the other hand, the P=N bond present in alkylidene(amino)iminophosphoranes does not compete with the P=C bond for a diazo dipole (see Scheme 8.29).

Little is known about using a P—S bond as a dipolarophilic unit. Indirect evidence of a 1,3-dipolar cycloaddition in the case of 2,2-dimethyl-1-diazopropane with the short-lived amino(thioxo)phosphane R_2N–P=S (R = $SiMe_3$) exists (190). More remarkable is the formation of 1,3,4,2-thiadiazaphospholine **182** from diazo compound **180** and 0.5 equiv of Lawesson's reagent (**179**) (Scheme 8.41) (241). This and similar results with nitrones and nitrilimines suggest that the monomeric dithiometaphosphate form of **179** can be trapped in a dipolar cycloaddition across the P=S bond. A spontaneous 1,3-R shift in cycloadduct **181** would then lead to the final product.

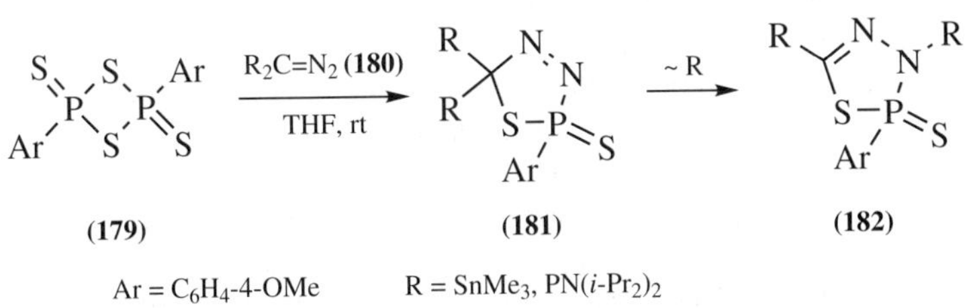

Scheme 8.41

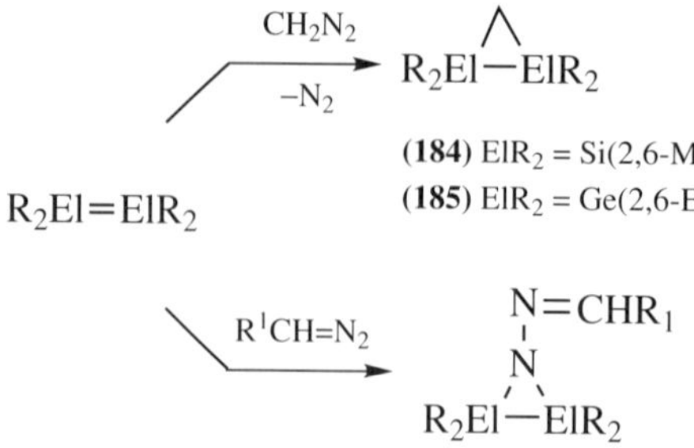

(**184**) $ElR_2 = Si(2,6\text{-}Me_2\text{-}C_6H_3)_2$
(**185**) $ElR_2 = Ge(2,6\text{-}Et_2\text{-}C_6H_3)_2$

(**186**) $ElR_2 = Si(2,4,6\text{-}Me_3\text{-}C_6H_2)$; R^1 = Ph
(**187**) $ElR_2 = Ge(2,4,6\text{-}i\text{-}Pr_3\text{-}C_6H_2)_2$; R^1 = H, TMS

Scheme 8.42

8.3.2. Other Heteroatom–Heteroatom Double Bonds

Compounds with a Si=Si or Ge=Ge bond (i.e., disilenes and digermenes) can be isolated when the double bond contains bulky substituents. Only a few cycloaddition reactions with diazo compounds are known, and two reaction modes have been observed. One of the paths leads to the formation of disiliranes **184** (242) and digermiranes **185** (243) (Scheme 8.42), probably by ring contraction of an initially formed [3 + 2] cycloaddition product. The other path involves a 1,1-cycloaddition of the diazoalkane to give disilaaziridine **186** (244) and digermaaziridine **187** (245). This nitrene-like reactivity is rather uncommon although some intramolecular examples are known (see Section 8.6.1).

8.4. CYCLOADDITIONS WITH HETEROCUMULENES

A statement made in an earlier review (5), *that the synthetic potential of cycloadditions of diazo compounds with heterocumulenes does not appear to have been extensively probed*, is still valid. This may be due in part to the complexity of the reaction. Instead of a concerted [3 + 2] cycloaddition, nucleophilic addition of the diazo compound at the heterocumulene can lead to a diazonium betaine from which several different products can arise with or without conservation of the azo moiety.

Diazo(trimethylsilyl)methyl lithium (**3**) was found to be the reagent of choice for the synthesis of azoles from heterocumulenes (Scheme 8.43). The reaction is typically carried out in ether at 0–20 °C. Thus, alkyl- (or aryl-)substituted ketenimines are transformed into 1,2,3-triazoles **188** (246), while C-acceptor-substituted ketenimines yield either 4-aminopyrazoles **189** or 1,2,3-triazoles, depending on the substituents (247). Isocyanates are converted into 5-hydroxy-1,2,3-triazoles **190** (248). Reaction of **3** with isothiocyanates are strongly solvent dependent.

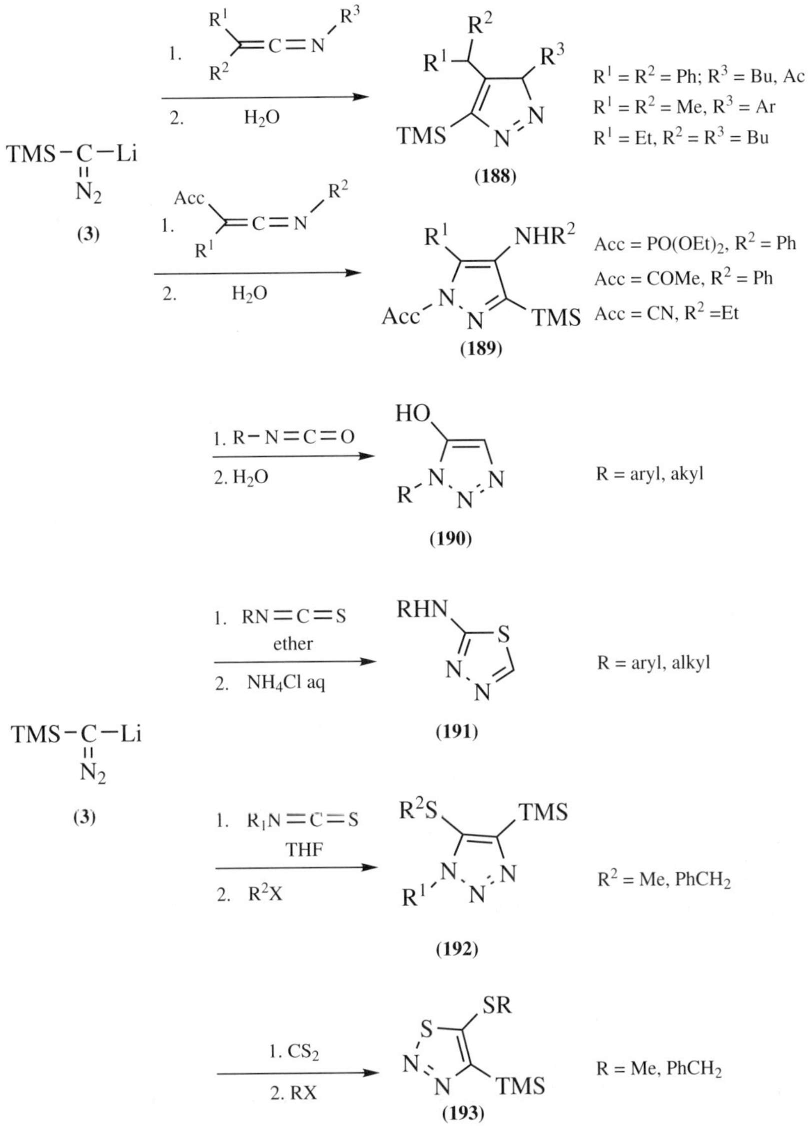

Scheme 8.43

2-Amino-1,2,4-thiadiazoles **191** are obtained when ether is used (249), while 5-alkylthio-1,2,3-triazoles **192** result when the reaction is carried out in THF (250). Reaction of **3** with carbon disulfide leads to 5-alkylthio-1,2,3-thiadiazoles **193** (251). While **3** can act as a synthetic equivalent of the RC—N—N synthon (R = H, $SiMe_3$) in all these reactions, it should be emphasized that it does not react by a concerted 1,3-dipolar cycloaddition but rather by a stepwise polar mechanism. The highly nucleophilic character of **3** can account for why diazomethane and

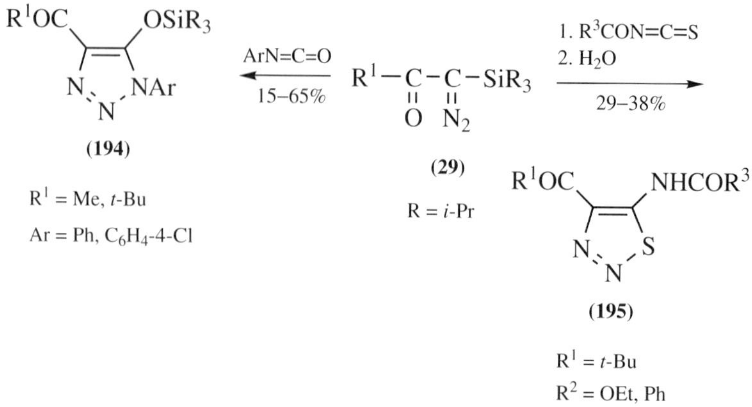

Scheme 8.44

diazo(trimethylsilyl)methane either do not react or react differently with the same heterocumulenes.

Silyl-substituted diazoketones **29** cycloadd with aryl isocyanates to form 1,2,3-triazoles **194** (252) (Scheme 8.44). This reaction, which resembles the formation of 5-hydroxy-1,2,3-triazoles **190** in Scheme 8.43, has no analogy with other diazocarbonyl compounds. The beneficial effect of the silyl group in **29** can be seen from the fact that related diazomethyl-ketones do not react with phenyl isocyanate at 70 °C (252). Although the exact mechanistic details are unknown, one can speculate that the 2-siloxy-1-diazo-1-alkene isomer **30** [rather than **29** (see Section 8.1)] is involved in the cycloaddition step. With acyl isocyanates, diazoketones **29** cycloadd to give 5-acylamino-1,2,3-thiadiazoles **195** by addition across the C=S bond (252), in analogy with the behavior of diazomethyl-ketones and diazoacetates (5).

Only a few examples of the reaction of diazo compounds with isoselenocyanates are available and this is probably due to the difficulty of preparing these heterocumulenes. Although aroylisoselenocyanates **196** (Scheme 8.45) are readily formed from the reaction of acid chlorides and KSeCN, they were never isolated in pure form. Nevertheless, it was possible to trap the *in situ* generated heterocumulenes **196** by [3 + 2] cycloaddition with diazo compounds. Thus, 1,2,3-selenadiazole **197** was obtained from **196** and diazomethane (253). Treatment of benzoylisoselenocyanate with diazodiphenylmethane furnished benzo[*b*]selenophene **198** in 27% yield (254). The initial [3 + 2] cycloaddition product rapidly lost N_2 and the intermediate dipolar species recyclized with participation of the phenyl ring to give **198**. Generation of **196** in the presence of ethyl diazoacetate resulted in the formation of 2-aroyl-5-aroylimino-1,2,3-selenadiazoles in moderate yield (255). It was assumed that the initial cycloaddition product underwent tautomerization to form the 2*H*-1,2,3-selenadiazole that was rapidly *N*-acylated by excess acid chloride. The acyl group could be removed by treatment with morpholine to give 5-aroylamino-1,2,3-selenadiazoles **199**. The formation of **197** and **199** indicates that the 1,3-dipolar cycloaddition of diazo compounds across the C=Se bond of

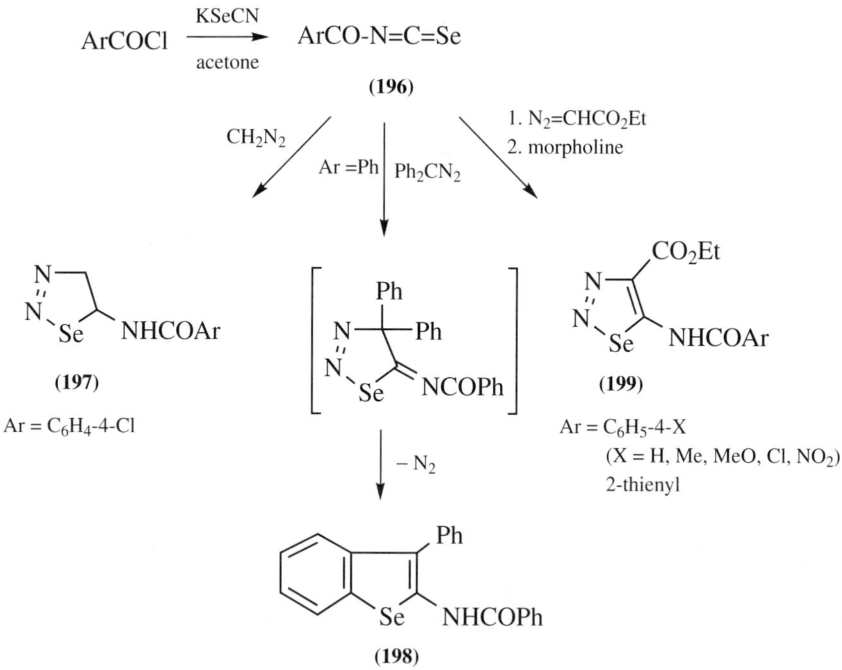

Scheme 8.45

isoselenocyanates and selenoketones (Section 8.2.5) occurs with different regioselectivity.

Ketenes rarely produce [3 + 2]-cycloaddition products with diazo compounds. The reaction possibilities are complex, and nitrogen-free products are often obtained (5). Formation of a cyclopropanone represents one possibility. Along these lines, the synthesis of (*Z*)-2,3-bis(trialkylsilyl)cyclopropanones and (*Z*)-2-trialkylsilyl-3-(triethylgermyl)cyclopropanones from diazo(trialkylsilyl)methanes and appropriate silyl- or germylketenes has been reported (256,257). It was found that subsequent reaction of the cyclopropanone with the diazoalkane was not a problem, in contrast to the reaction of diazomethane with the same ketenes. The high cycloaddition reactivity of diazomethylenephosphoranes also extends to heterocumulenes. The compound $R_2P(Cl){=}C{=}N_2$ ($R = N(i\text{-}Pr)_2$) reacts with CS_2, PhNCO and PhNCS to give the corresponding 1,2,3-triazole derivative (60).

8.5. CYCLOADDITIONS WITH TRIPLE-BOND SYSTEMS

8.5.1. Alkynes

The 1,3-dipolar cycloaddition of diazo compounds with alkynes represents a standard method for the preparation of pyrazoles. The initially formed

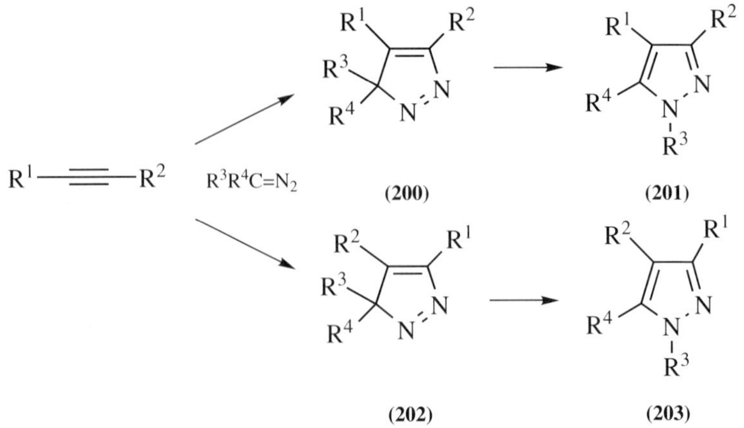

Scheme 8.46

3*H*-pyrazoles **200** and/or their regioisomers **202** (Scheme 8.46) can be isolated if the substituent group is a poor migrator (i.e., H, $SiMe_3$, acyl, phosphoryl). These substrates are interesting precursors of cyclopropenes and of vinyldiazoalkanes. A 1,5-(C $\rightarrow$ N) substituent migration converts the 3*H*-pyrazoles **200** into 1*H*-pyrazoles **201** (and **202** into **203**). For the case where $R^1 = H$, an equilibrium of the two NH-tautomeric forms is often found or postulated. Isomerization of **200** by two sequential 1,5-sigmatropic shifts followed by 1,5 (C $\rightarrow$ N) is also known. For example, the isomerization of dimethyl 3,3-dialkyl-3*H*-pyrazole-4,5-dicarboxylates into dimethyl 4,5-dialkylpyrazole-1,3-dicarboxylates represents such a case (258).

In terms of reactivity, the same general principles as observed with olefinic dipolarophiles (see Section 8.1) apply. For simple diazoalkanes, the cycloaddition is HOMO(dipole)–LUMO(dipolarophile) controlled. Conjugating and electron-withdrawing substituents at the triple bond accelerate the reaction. The cycloaddition of diazomethane with the more electron-rich triple bond of ethoxyacetylene is also known. This reaction was recently reinvestigated and it was found that it produces not only 4-ethoxypyrazole but also a small amount of the 3-ethoxy isomer (96:4 mixture) (25). PM3 calculations of the transition structures as well as an analysis of perturbation energies suggest that FMO interactions which result from distortions in the transition structures due to closed-shell repulsions between the reactants account for the observed regiochemistry. By using the same theoretical treatment, the opposite regiochemistry encountered when ethyl vinyl ether was used as a dipolarophile could also be explained (25).

[3 + 2] Cycloadditions have been reported for many combinations of diazo compounds and alkynes (5). A few recent examples are given in Table 8.2. An inspection of entries 2–4 shows that the regiochemical behavior of internal sulfonylalkynes is totally reversed when the second substituent (R^1) is changed from Me or Ph to $SiMe_3$. This difference was explained in terms of a steric effect of

TABLE 8.2. PYRAZOLES **200–203** FROM ALKYNES $R^1C{\equiv}CR^2$ AND DIAZOCOMPOUNDS $R^3R^4CN_2$

Entry	R^1	R^2	R^3	R^4	Pyrazole	Reference
1	TMS	TMS	H	CO_2Me	**201a**	259
2	Me, Ph	SO_2Ph	Me	Me	**200b** (95%)	260
3	TMS	SO_2Tol	Me	Me	**202c** (>98%)	260
4	TMS	COMe	Me	Me	**202e**(~100%)	260
5	PhSe	SO_2Tol	H	H	**203d** (81%)	261
6	TMS	SO_2Tol	Li	$P(N\textit{i}\text{-}Pr_2)_2$	**201f**[a] ($R^3 = H$)	262
7	TMS	SO_2Tol	$P(N\textit{i}\text{-}Pr_2)_2$	H	**201g**[b]	262
8	Ph	$C{\equiv}CPh$	Me	Me	**200h**(61%),[c] **202f** (15%)	263
9	CF_3	CH(Me)OAc	H	H	**201i** (63%)[d]	264

[a] 3-Bis(diisopropylamino)thiophosphoryl-5-tosyl-4-trimethylsilyl-1*H*-pyrazole (65%) was isolated after treatment with S_8.
[b] 1-Bis(diisopropylamino)thiophosphoryl-3-tosyl-4-trimethylsilyl-1*H*-pyrazole was isolated after treatment with S_8 (93%).
[c] A small amount of bipyrazole was also formed.
[d] With excess diazomethane, **201i** is transformed into the 1-Me—4-CF_3—3-R^2 and 1-Me—4-CF_3—5-R^2 derivatives.

the $SiMe_3$ group that overrules the orientation expected on the basis of FMO interactions (260). A similar effect was invoked to rationalize the opposite regiochemistry of diazocarbonyl compounds with terminal and trimethylsilyl-substituted alkynyliodonium salts **204** (265) (Scheme 8.47). The formation of (pyrazol-4-yl)phenyliodonium salts from **204**, $R^1 = COt$-Bu (265) or CN (266),

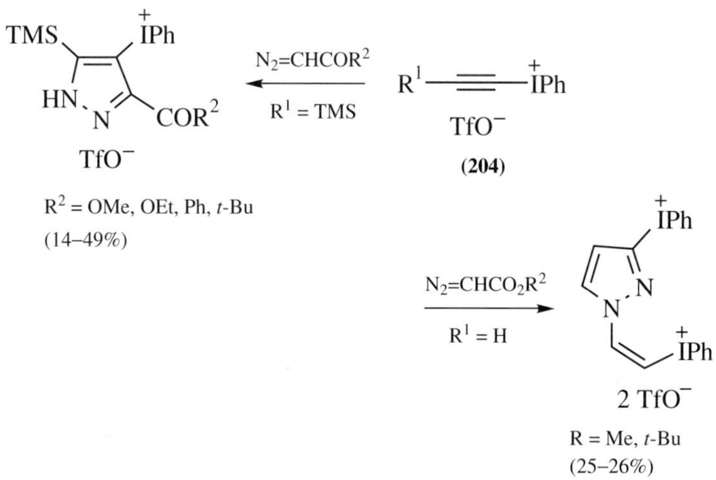

Scheme 8.47

with diazoacetates, however, is in agreement with expectations based on FMO theory.

Trifluoromethyl-substituted pyrazoles are easily obtained using trifluoromethylalkynes as dipolarophiles (Table 8.2, entry 9). Thus, treatment of 4,4,4-trifluorobut-2-ynoic acid with excess diazomethane gave methyl 4-(trifluoromethyl)pyrazole-4-carboxylate (45%) accompanied by its N^2- (32%) and N^1-methylated (6.5%) derivatives (267). Another convenient route to CF_3-substituted pyrazoles involves dipolar cycloaddition of appropriately CF_3-substituted alkenes followed by eliminative aromatization (76,77,268). For example, the reaction of alkenes such as $(CF_3)_2C{=}C(H)COAr$ with ethyl diazoacetate gave 4-aroyl-5-trifluoromethylpyrazole-3-carboxylates (268).

Cycloaddition of diazo(trimethylsilyl)methane across the triple bond of alkynylcarbene tungsten or chromium complexes **205** proceeds regioselectively and produces pyrazoles **206** (Scheme 8.48) (269). These compounds were demetalated to give pyrazolecarboxylates **207**. Thus, the carbene complex **205** can act as the synthetic equivalent of the corresponding propiolic acid ester and offers the advantage of regiochemical control. The cycloaddition of Me_3SiCHN_2 with methyl but-2-ynoate afforded a 35:65 mixture of pyrazole **207** and its regioisomer (269). The conversion of pyrazole **206** (R = Me) into pyrazoloquinones **208** illustrates additional synthetic opportunities of these pyrazolylcarbene complexes.

Acetylenedicarboxylic acid esters and related activated alkynes are routinely used as dipolarophiles for diazo dipoles. Recent examples include the use of diazo compounds **20** (49), **23** (51), and **24** (52) (Scheme 8.7), **25** (56) (Scheme 8.8), diazoacetaldehyde dimethylacetal (41) (which after cycloaddition and deprotection gave the corresponding pyrazole-3-carbaldehyde), ethyl 3-diazopyruvate (270), *p*-tolyl-trifluoromethyldiazomethane (271), bis(trifluoromethyl)diazomethane (272), and diazomethylenephosphoranes (60).

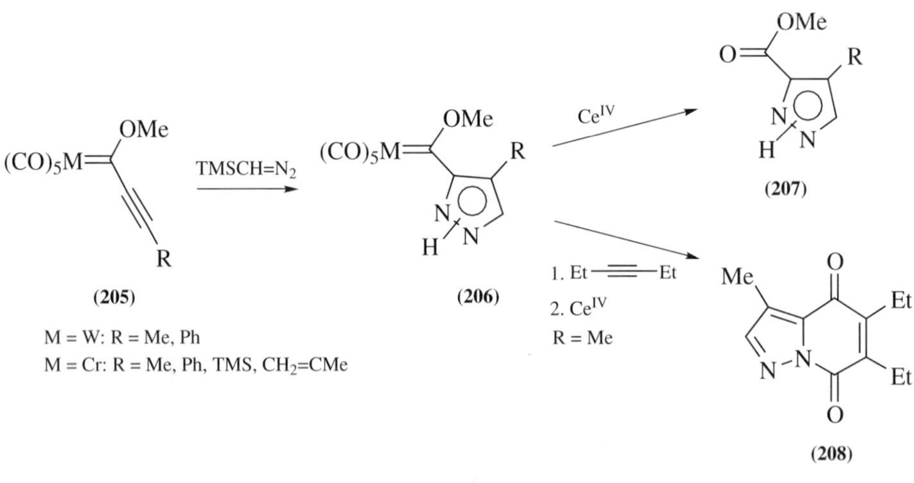

Scheme 8.48

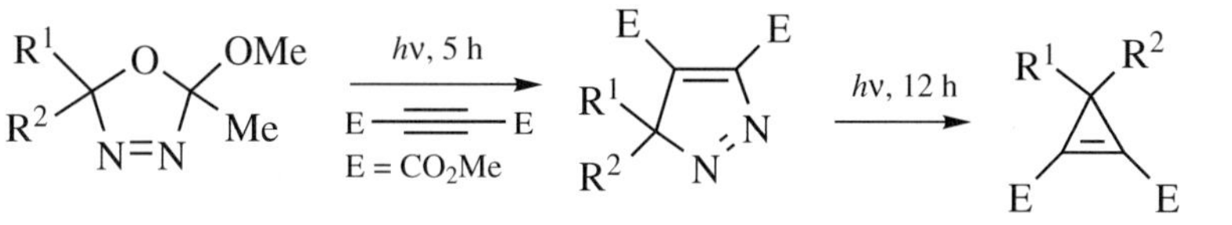

(209) (210) (70–87 %) (211)

$R^1 = R^2 = Me, Et$

$R^1 = R^2 = (CH_2)_4$

$R^1 =$ cyclopropyl, $R^2 = Me$

$R^1 = R^2 = Me, Et$

Scheme 8.49

Another method used to prepare dialkyl-substituted diazomethanes involves the photolysis of 2-alkoxy-2,5-dihydro-1,3,4-oxadiazoles (**209**), which can be prepared by the oxidative cyclization of *N*-acetylhydrazones. The diazoalkanes are trapped *in situ* by cycloaddition with dimethyl acetylenedicarboxylate (54) (Scheme 8.49). The resulting pyrazoles **210** are converted into cyclopropenes **211** by continued irradiation.

There are several reports in the literature dealing with the bimolecular $[3+2]$ cycloaddition reactions of alkynyl-substituted diazo compounds. Propargyl diazoacetate **212**, when stored for 2 weeks at 0 °C, was transformed into an oligomer to which the constitution **213** was assigned (273) (Scheme 8.50). The alkynyl-diazoketone **214** requires a much higher temperature and is transformed into pyrazole **215**, which probably arises from intermolecular cycloaddition, pyrazole tautomerization, and carbenic N/H insertion (274). The inter-intramolecular

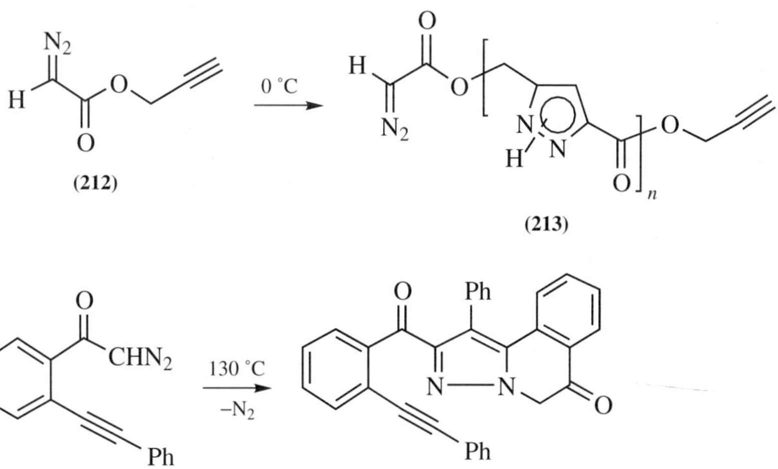

(212) (213) (214) (215)

Scheme 8.50

cycloaddition cascade of (alkynyldiisopropylsilyl)oxy-diazoacetates is described in more detail in Section 8.6.2.

8.5.2. Nitriles

Diazo compounds generally do not undergo [3 + 2] cycloaddition with unactivated nitriles under purely thermal, noncatalyzed conditions. The formation of 4-R-5-trimethylsilyl-1*H*-1,2,3-triazoles from the reaction of diazo(trimethylsilyl)-methyl lithium and a broad range of nitriles [RCN; R = alkyl, aryl, SEt, OPh, $PO(OEt)_2$] appears to be an exception, but this reaction most likely occurs in a stepwise manner with initial nucleophilic attack at the nitrile (275).

If the C≡N function is attached to an electron-withdrawing group, 1,3-dipolar cycloaddition with diazoalkanes occurs leading to 1,2,3-triazoles (5, 276). When diazomethane is used, the initially formed NH-triazole is not isolated due to a rapid subsequent NH deprotonation followed by N-methylation. Consequently, a mixture of the three *N*-methyltriazoles is formed when methyl cyanoformate (71) (**216**) or trichloroacetonitrile (276) (**217**) is treated with excess diazomethane (Scheme 8.51). Huisgen and co-workers found that methyl diazoacetate reacts with TCNE by a 1,3-dipolar cycloaddition at the C=C bond and not, as published earlier by other authors, at one of the nitrile functions (72).

8.5.3. Heteroatom–Heteroatom Triple Bonds

Diazomethane reacts with 4-nitrobenzenediazonium chloride to give the 1-aryltetrazole **218** (Scheme 8.52) in addition to other products (277). This long-known reaction was revisited when the dipolarophilic character of the arenediazonium salt was realized (278). The tetrazole probably arises by a concerted 1,3-dipolar cycloaddition rather than by a two-step process. However, the observed regiochemistry is difficult to reconcile with either mechanism.

A concerted [3 + 2] cycloaddition (or a two-step process) followed by deprotonation, can also explain the formation of tetrazole **220** (279). In this case, the amidinio-substituted diazophosphoryl compound **219**, which has a resonance

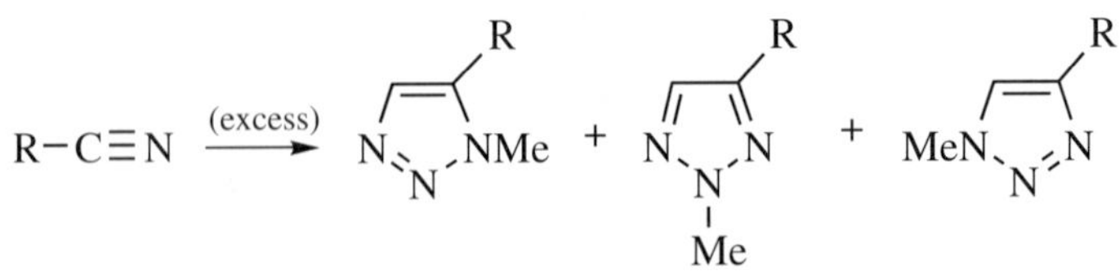

(**216**): R = CO_2Me yield: 77% (62.3: 31.2: 6.5)

(**217**): R = CCl_3 – (9.8: 88.6: 1.6)

Scheme 8.51

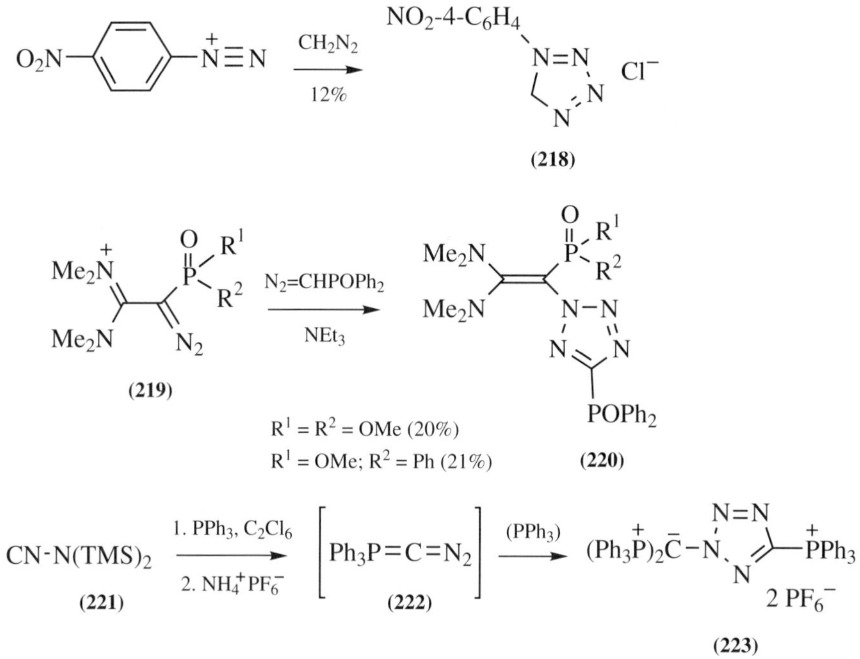

Scheme 8.52

contribution from the 2,2-bis(dimethylamino)ethenediazonium structure, can also function as the dipolarophile. Cycloaddition of a diazo dipole with a diazonium function may also be involved in the formation of tetrazole **222** (280). It was suggested that isocyanoamine **221** is first converted into the transient diazomethylenephosphorane **222**, which then cyclodimerizes with the incorporation of a PPh_3 molecule. Diazocumulenes $R_3SiO(R^1)C{=}C{=}N_2$ may also function as N≡N dipolarophiles towards diazo compounds (281).

Diazo(trimethylsilyl)methyl lithium reacts with white phosphorus (P_4) in THF to form the 5-trimethylsilyl-1,2,3,4-diazadiphospholide lithium salt (282). This novel heteroaromatic system formally represents the cycloaddition product of a diazo dipole and a P≡P bond. However, the true mechanism of this reaction is not known.

8.5.4. Phosphaalkynes

Phosphaalkynes of the type RC≡P, featuring a three-valent phosphorus atom with coordination number 1 ($\lambda^3\sigma^1$-P), represent novel organophosphorus compounds. Their chemistry has been extensively investigated since 1981, when the first synthesis of a kinetically stabilized phosphaalkyne (*t*-BuC≡P) was reported (283). Several reviews on the cycloaddition chemistry of these compounds with diazo compounds have been published (284–286).

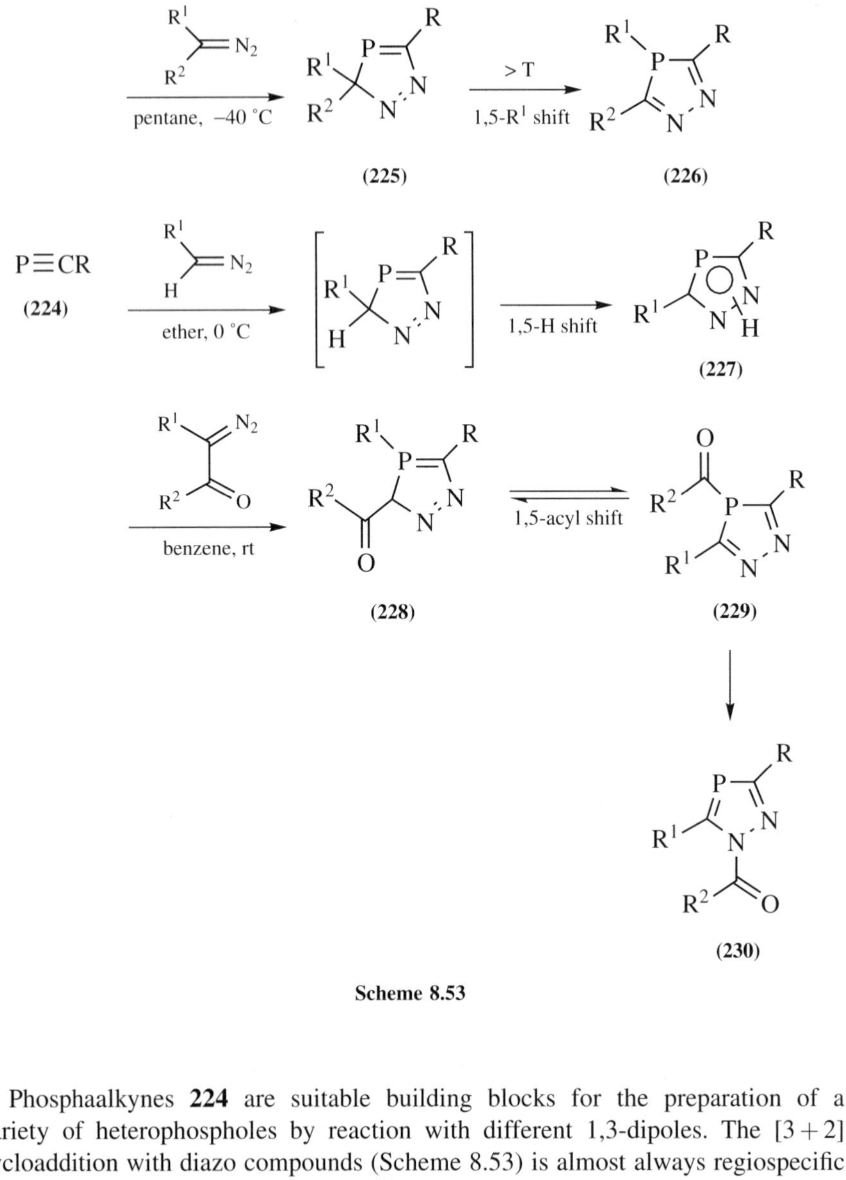

Scheme 8.53

Phosphaalkynes **224** are suitable building blocks for the preparation of a variety of heterophospholes by reaction with different 1,3-dipoles. The [3 + 2] cycloaddition with diazo compounds (Scheme 8.53) is almost always regiospecific and affords 3*H*-1,2,4-diazaphospholes **225** as the initial products. These structures can be isolated in those cases where the diazo carbon atom bears two substituents that are not capable of undergoing a further 1,5-sigmatropic shift. If a 1,5-shift occurs, the group usually migrates to the phosphorus atom rather than to the nitrogen atom. Interestingly, for the 3-trimethylsilyl-substituted derivative **225** [$R = t$-Bu, $R^1 = SiMe_3$, $R^2 = P{=}C(SiMe_3)Ph$], a silyl shift to nitrogen was found (287). With some 3,3-dialkyl-substituted derivatives of **225** (R^1,R^2 = alkyl), a proton-catalyzed isomerization to **226** occurred upon storing in $CHCl_3$ (288).

The 3*H*-1,2,4-diazaphospholes formed from the reaction of diazomethane and its monosubstituted derivatives ($R^1CH{=}N_2$; $R^1 = H$, alkyl, aryl, acyl, phosphoryl) could not be isolated due to a rapid 1,5-H shift leading to 2*H*-1,2,4-diazaphospholes **227**. When diazo(trimethylsilyl)methane or [bis(diisopropylamino)phosphino]diazomethane was used, the 1,5-$SiMe_3$ [or PR_2, $R = N(i\text{-}Pr)_2$] shift completely dominates over the H shift (289,290). In the case of open-chain or cyclic α-diazoketones, cycloadducts **228** cannot be isolated due to rapid acyl shifts giving **229** and ultimately **230** (289). This transformation offers a versatile method to prepare [*b*]-fused 1,2,4-diazaphospholes from cyclic α-diazoketones and phosphaalkynes (289).

Tert-butylphosphaacetylene (**224**, R = *t*-Bu) has played a major role in understanding the chemistry of kinetically stabilized phosphaalkynes. This alkyne has been subjected to 1,3-dipolar cycloaddition chemistry using a broad range of diazo compounds (286): [**226** (R = *t*-Bu): R^1,R^2 = alkyl, and $R^1 \cdots R^2$ = cycloalkyl (286,288); $R^1 = R^2$ = Ph (291). – **227** (R = *t*-Bu): R^1 = H, Me, Ph, *t*-Bu, COOR, COPh, $POPh_2$ (289,292); R^1 = H, $SiMe_3$ instead of H (289); R^1 = H, $P(N(i\text{-}Pr)_2)_2$ instead of H (290). – **230**: R^1 = Me, Ph, SO_2Ph, CONHPh, R^2 = Me (289); $R^1 = R^2$ = Ph (289); $R^1 \cdots R^2$ = ring systems (289)]. Other kinetically stabilized phosphaalkynes behave analogously [e.g., **224**, R = 1-adamantyl (293), *i*-Pr (294), CH_2*t*-Bu (294), 1-methylcyclopentyl and 1-methylcyclohexyl (294), $SiMe_3$ (295), mesityl (296)]. Short-lived phosphaalkynes such as methylidinephosphane (297), (HC≡P) and $H_3C{-}C{\equiv}P$ (298) were easily characterized by their 1,3-dipolar cycloaddition reactions with diazomethane and diazoacetates. From HC≡P (**231**, Scheme 8.54) and *tert*-butyl diazoacetate, a mixture of the two regioisomeric adducts **232** and **233** was obtained. This represents the only example in the literature of a nonregiospecific 1,3-dipolar cycloaddition of a diazo dipole with an electronically unaltered P≡C bond. Simple FMO theory cannot explain the observed regioselectivity with phosphaalkynes, since the orbital coefficients at P and C are nearly equal in both the HOMO and the LUMO. *Ab initio* calculations on the [3 + 2]-cycloaddition reaction of HC≡P with diazomethane were carried out at different levels of theory, including CCSD(T)/6-311 + *G** // MP2/6-311 + *G**. They show that the barrier leading to 3*H*-1,2,4-diazaphosphole is lower in energy (by 1–2 kcal mol^{-1}, depending on the theoretical level) than the one leading to 5*H*-1,2,3-diazaphosphole thereby indicating that this reaction is under kinetic control (299). However, the small difference in activation barriers does not really adequately explain the preference for the 3*H*-1,2,4-diazaphosphole that was observed experimentally.

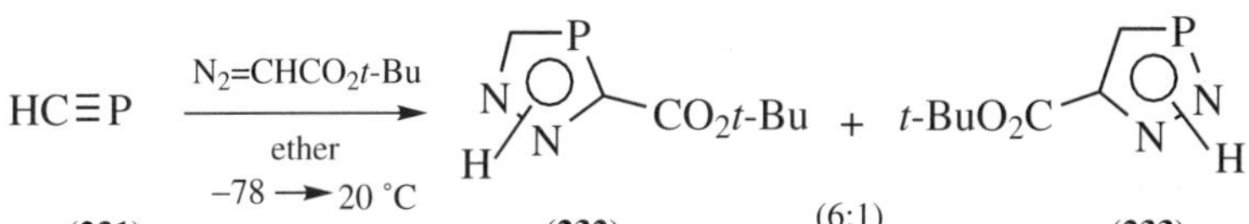

(231) **(232)** **(233)**

Scheme 8.54

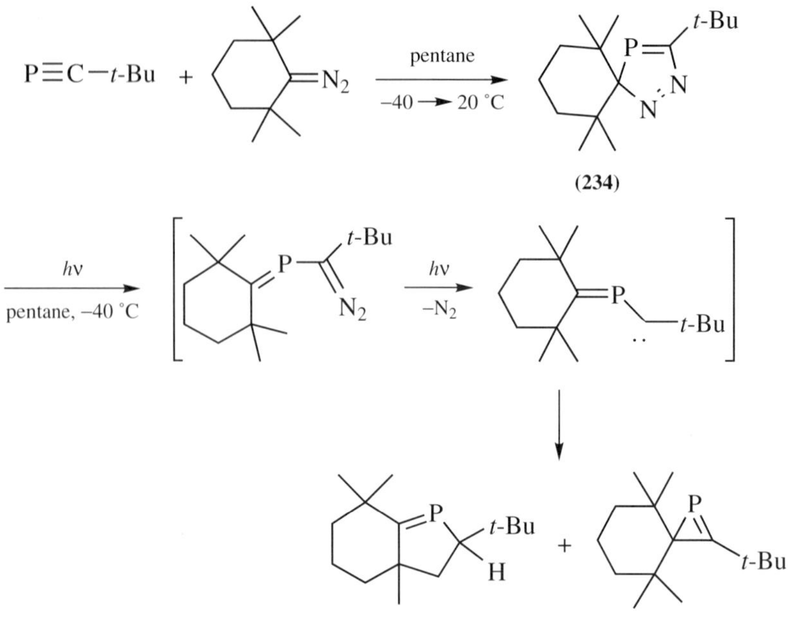

(235) (54%) **(236)** (11%)

Scheme 8.55

The potential usefulness of 3*H*-1,2,4-diazaphospholes for the synthesis of other phosphorus heterocycles was demonstrated using **234** (Scheme 8.55) which, after UV irradiation at −40 °C, gave a 5:1 mixture of 4,5-dihydro-3*H*-phospholes **235** and **236**. The latter compound is the first example of the previously unknown 2*H*-phosphirene system (288). This reaction probably proceeds *via* a photochemical ring opening of **234** and formation of a phosphavinyl carbene.

When phosphaalkynes are exposed to bis- and tris(diazo) compounds, bis- or tris(1,2,4-diazaphosphol-5-yl) compounds are formed that may be further converted into a variety of novel heterocyclic systems. For example, bis- and tris[diazo(trimethylsilyl)methyl]phosphanes **237** and **240** afforded bis- and tris(diazaphospholyl)phosphanes **238** and **241** after cycloaddition with *tert*-butylphosphaacetylene followed by a subsequent 1,5-silyl shift (Scheme 8.56) (300). Reaction with electrophilic halides at the *N*-silyl functions allows the introduction of a heteroatom bridge between the diazaphosphole ring leading to polycyclic ring systems such as **239** and **242**.

The reaction of the electron-rich phosphaalkyne **243** with monosubstituted diazo compounds ($R^1CH{=}N_2$, $R^1 = CF_3$, CO_2Me) is remarkable because it furnishes a mixture of the regioisomeric cycloaddition products (1*H*-1,2,4- and 1*H*-1,2,3-diazaphospholes) where the *normal* regioisomer predominates (301). When

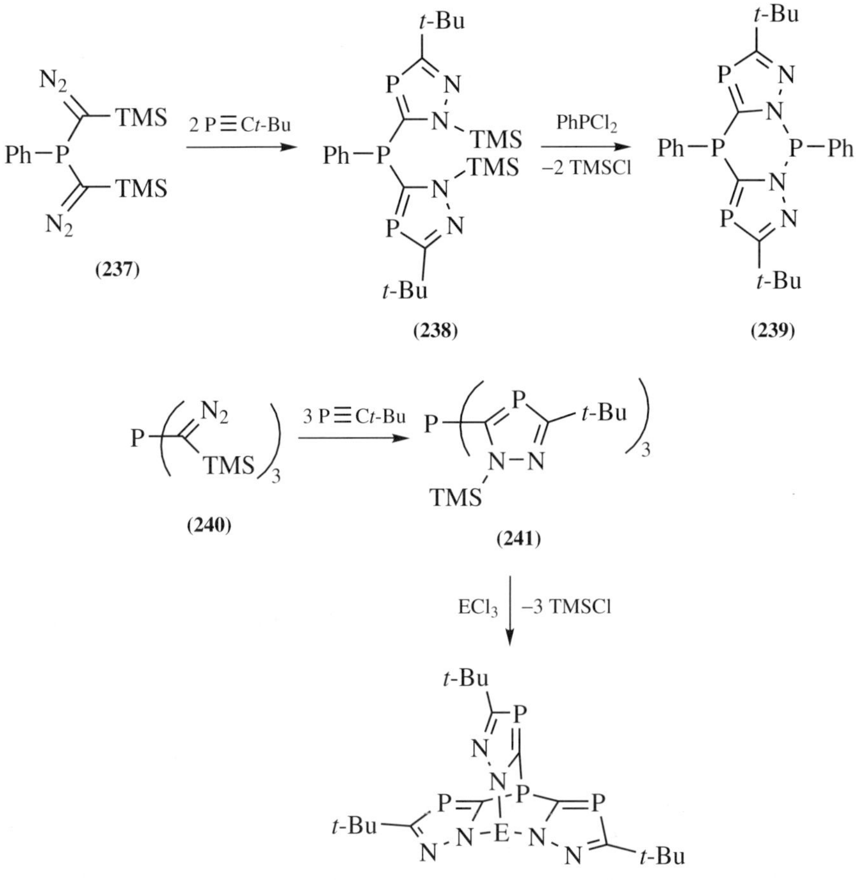

(242)

E = P, As, Sb, MeSi, MeGe

Scheme 8.56

disubstituted aryldiazoalkanes are used, the reaction does not stop at the expected 1,2,4-diazaphosphole **244** but rather gives the bicyclic products **246** (302) (Scheme 8.57). It was proposed that **244** is first transformed into **245** by a [4 + 2] cycloaddition with excess phosphaalkyne and that this is followed by cycloelimination of (i-Pr)$_2$N–CN and subsequent reaction of **245** with the diazo compound to give **246**. This mechanistic postulate is supported by the isolation of **245** ($R^1 \cdots R^1$ = CH=CH–CH=CH) from the reaction of **243** with diazocyclopentadiene.

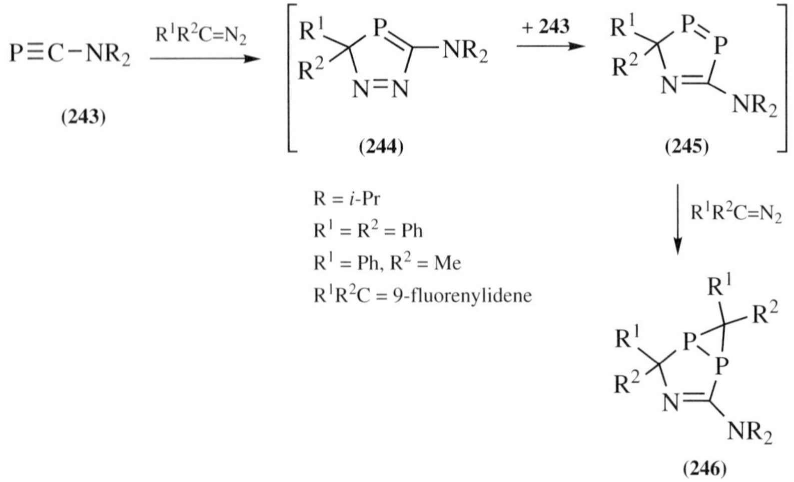

Scheme 8.57

8.6. INTRAMOLECULAR CYCLOADDITION REACTIONS

8.6.1. Intramolecular Cycloadditions Involving C=C Bonds

Earlier results dealing with the intramolecular cycloaddition reactions of olefinic diazo compounds have been briefly summarized (303,304). The [3 + 2] cycloaddition reactions can be achieved when the diazo carbon and the C=C bond are separated by three or more atoms. Conformational and steric effects are clearly important in these reactions. Open-chain diazoalkenes **247a** (305) and **247b** (306), generated from either a tosylhydrazone salt or by thermolysis of an iminoaziridine precursor, undergo [3 + 2] cycloaddition with formation of a 2,3-diazabicyclooctene (Scheme 8.58). Interestingly, the diazo-compound analogous to **247b**, but with one CH_2 group less in the chain, did not cyclize (306). Intramolecular 1,3-dipolar cycloaddition was also observed for 2-alkenyl-1-diazocyclobutanes (**248**) (307) (Scheme 8.58). *o*-Allyl, *o*-[(*Z*)- and (*E*)-2-butenyl], and *o*-(3-butenyl)-substituted diazotoluenes also undergo intramolecular pyrazoline formation when generated from the corresponding tosylhydrazones or aziridinylimines (305). As demonstrated in Scheme 8.59, the cycloaddition of **249** is stereospecific and proceeds with retention of configuration about the olefinic double bond. However, this is not the case for the subsequent Δ^1-pyrazoline → cyclopropane ring contraction that occurs at elevated temperatures. α-Diazoketone **250** isomerizes into pyrazoline **251** at $-10\,°C$ (Scheme 8.59) and the loss of N_2 from the latter proceeds at 25 °C to give a 1:1 mixture of a cyclopropanoindene and an indene (308).

The intramolecular [3 + 2] cycloaddition reaction of diazoalkane **252** (Scheme 8.60) is remarkable for its high asymmetric induction. Due to the presence of the

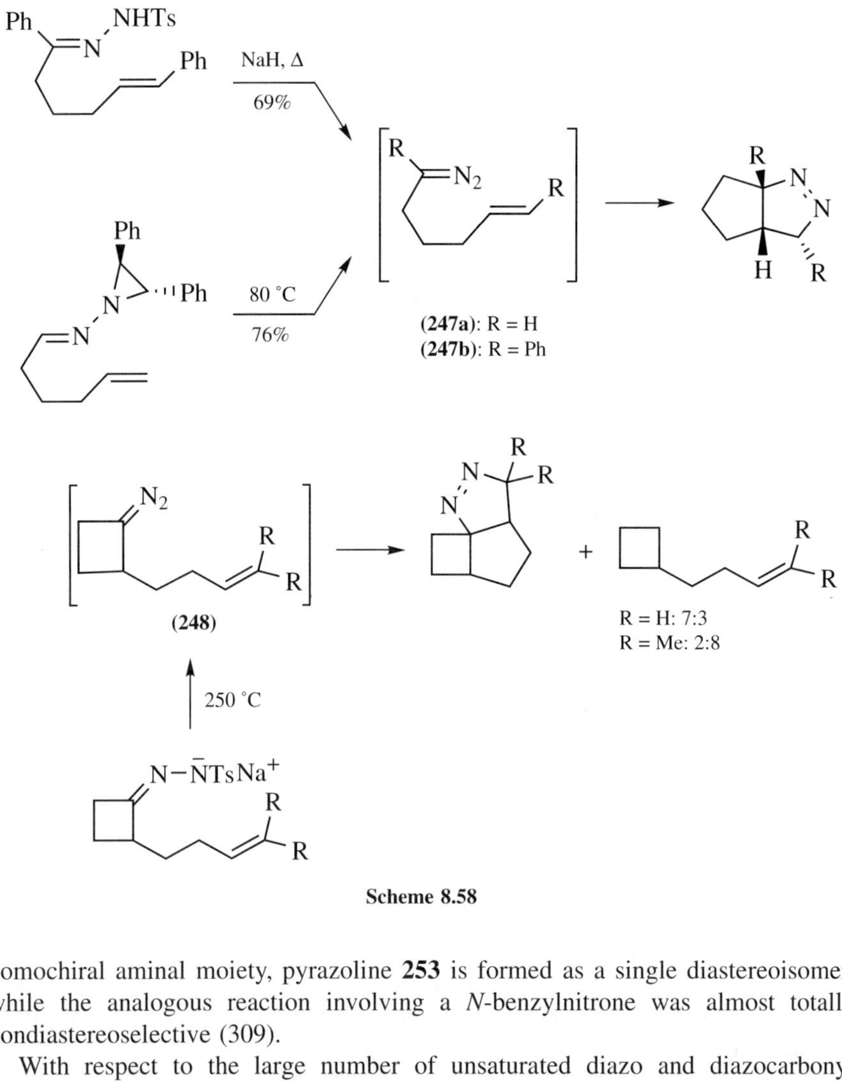

Scheme 8.58

homochiral aminal moiety, pyrazoline **253** is formed as a single diastereoisomer, while the analogous reaction involving a *N*-benzylnitrone was almost totally nondiastereoselective (309).

With respect to the large number of unsaturated diazo and diazocarbonyl compounds that have recently been used for intramolecular transition metal catalyzed cyclopropanation reactions (6–8), it is remarkable that 1,3-dipolar cycloadditions with retention of the azo moiety have only been occasionally observed. This finding is probably due to the fact that these [3 + 2]-cycloaddition reactions require thermal activation while the catalytic reactions are carried out at ambient temperature. *N*-Allyl carboxamides appear to be rather amenable to intramolecular cycloaddition. Compounds **254–256** (Scheme 8.61) cyclize intramolecularly even at room temperature. The faster reaction of **254c** (310) and diethoxyphosphoryl-substituted diazoamides **255** (311) as compared with diazoacetamides **254a** (312) ($\tau_{1/2} \approx 25$ h at 22 °C) and **254b** (310), points to a LUMO (dipole) – HOMO(dipolarophile) controlled process. The Δ^1-pyrazolines expected

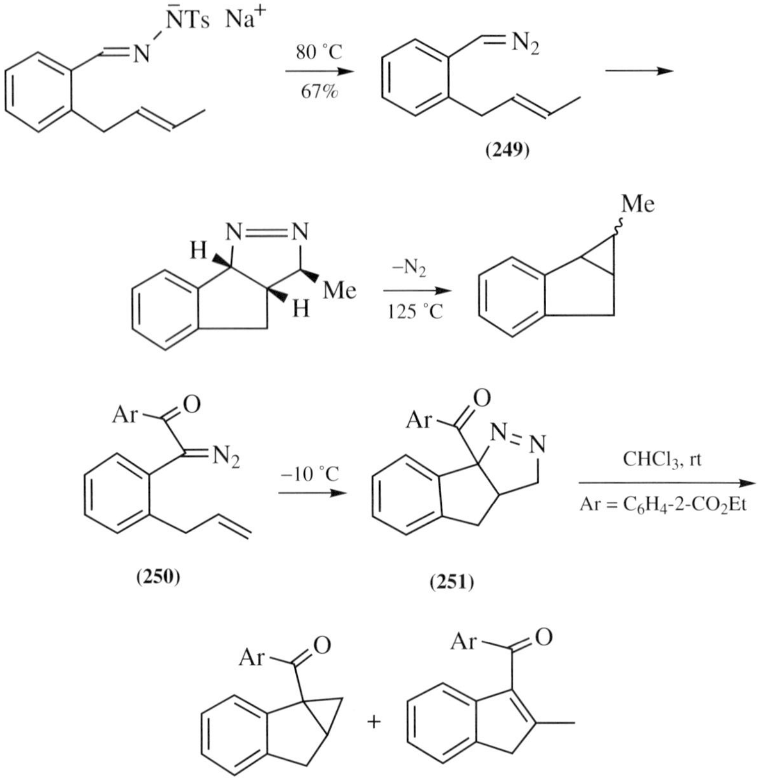

(249)

(250) (251)

Scheme 8.59

from **255** and **256** were not isolated, as they readily isomerize into Δ^2-pyrazolines **257** or form cyclopropanes **258** by spontaneous extrusion of N_2 (311).

(Allyloxysilyl)diazoacetates **259** are transformed into oxasiloles **260** upon heating (313) (Scheme 8.62). It is assumed that a Δ^1-pyrazoline is formed as a transient intermediate that loses N_2 and generates a 1,3-diradical from which the product is formed by transposition of a H atom.

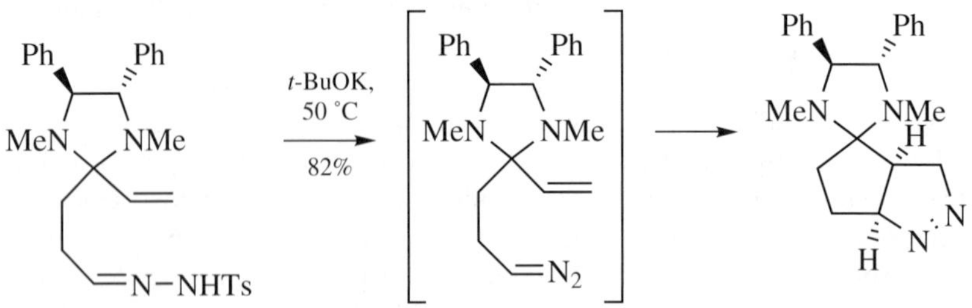

(252) (253)

Scheme 8.60

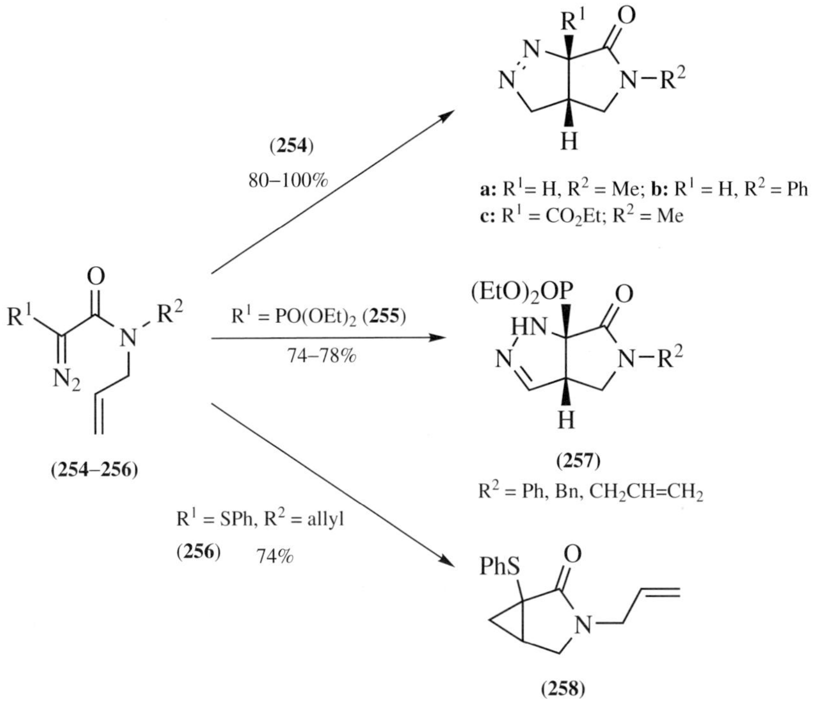

Scheme 8.61

In a novel total synthesis of the tricyclic sesquiterpene (−)-longifolene, an intramolecular diazoalkane cycloaddition to a cyclohexadienone ring followed by thermal ring contraction of the resulting pyrazoline gave the tricyclic vinylcyclopropane **261** and this constitutes the key steps in this synthesis (314) (Scheme 8.63). The interesting features of this sequence are the separation of dipole and dipolarophile by five atoms and the formation of a seven-membered ring in the cycloaddition step.

The reaction (315) shown in Scheme 8.64 is unusual in two respects. First, tosylhydrazone sodium salts usually do not fragment at room temperature

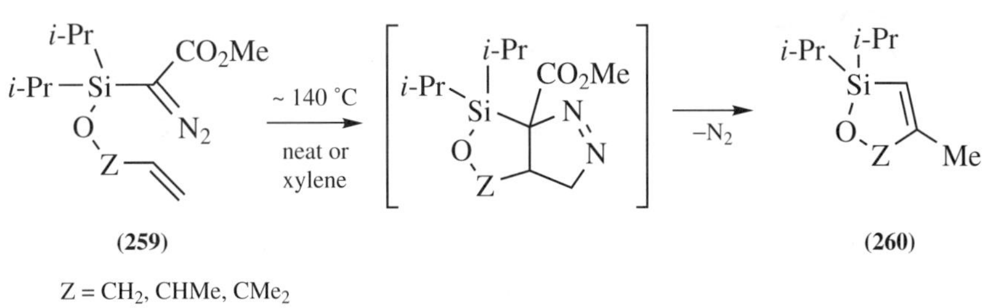

Scheme 8.62

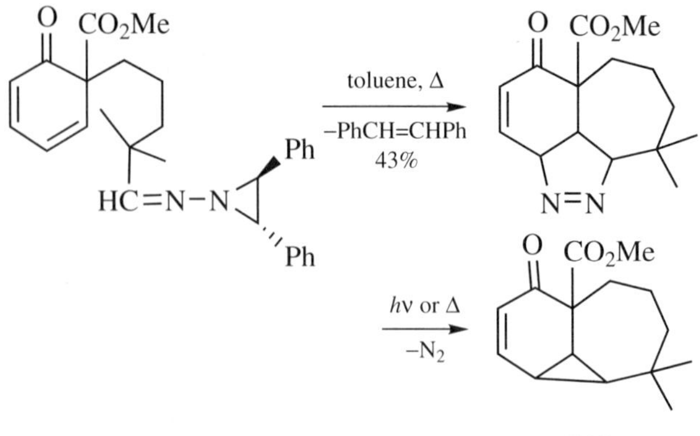

(261)

Scheme 8.63

(**262**$\rightarrow$**263**). Second, the formation of the diazobenzazocine derivatives **264a–e** represents an unprecedented reaction for intramolecular 1,3-dipolar cycloaddition reactions of diazo compounds. Note that diazo compounds such as **247a** (305) and **248** (307) also give bridged diazabicyclo[n.2.1]alkenes rather than fused diazabicyclo[n.3.0]alkenes upon treatment with BF_3–etherate, but these transformations

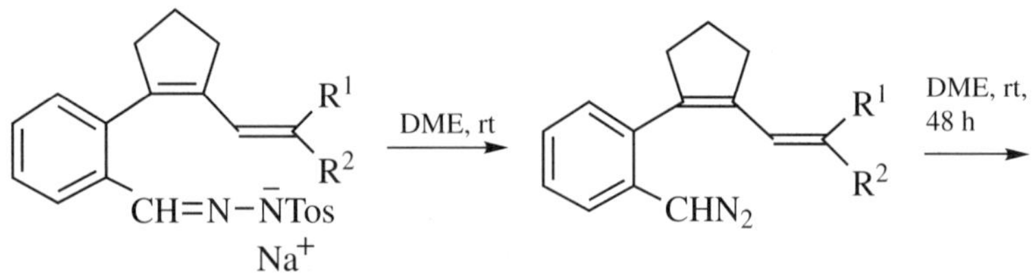

(262) **(263)**

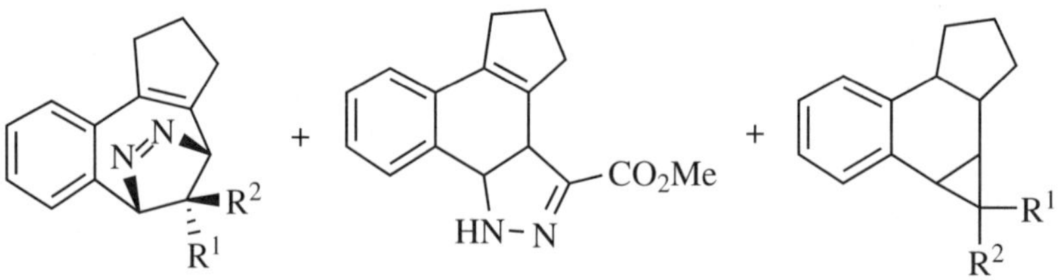

(264) **(265)** **(266)**

a: $R^1 = R^2 = H$ (**264**, 86%); **b**: $R^1 = H$, $R^2 = Me$ (**264**, 75%); **c**: $R^1 = H$, $R^2 = Ph$ (**264**, 67%; **266**, 19%); **d**: $R^1 = R^2 = Me$ (**264**, 88%); **e**: $R^1 = H$, $R^2 = CO_2Me$ (**264**, 28%; **265**, 59%)

Scheme 8.64

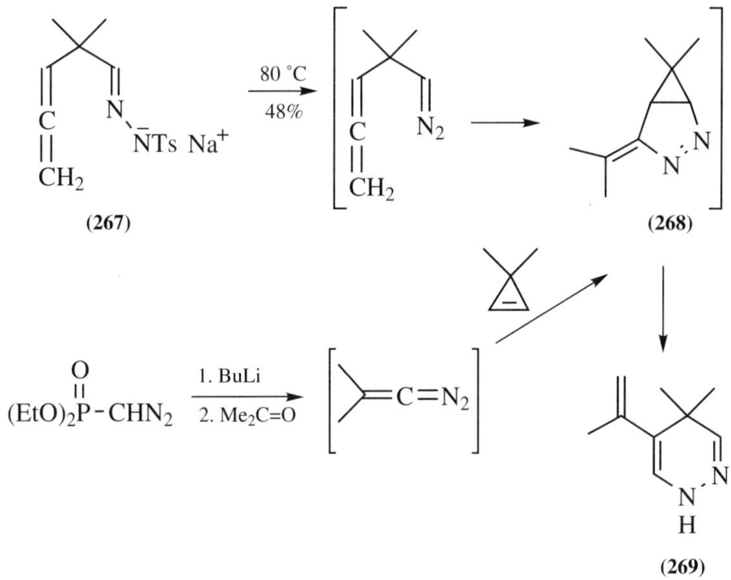

Scheme 8.65

proceed by a stepwise, polar mechanism. Only when the C=C bond is activated by an ester group does the [3 + 2] cycloaddition proceed with the expected regioselectivity (**263** → **265e**), but even in this case only to a minor extent. In line with the resulting stereochemistry of the product, a helical transition state for the cycloaddition leading to **264** was proposed. Replacement of the cyclopentene by a cyclohexene ring (or a CMe=CPh moiety) in **263** favors the formation of cyclopropane **266**. In the case of the CMe=CPh unit (see **283** in Scheme 8.69), a 1,10b-dihydropyrrolo[2,1-*a*]phthalazine is also formed and probably arises form an initial intramolecular 1,1-cycloaddition of the diazo functionality.

An exceptionally interesting example of an intramolecular [3 + 2] cycloaddition, in which the diazo dipole and the olefinic C=C bond are separated by only one carbon atom, is outlined in Scheme 8.65. The thermal decomposition of the allenic tosylhydrazone sodium salt **267** produced 1,4-dihydropyridazine **269** (57). It is assumed that diazabicyclohexene **268** is a short lived reaction intermediate. This suggestion is supported by the observation that the generation of the diazocumulene 1-diazo-2-methyl-1-propene in the presence of 3,3-dimethylcyclopropene also leads to **269**.

7-Alkoxy-5-diazomethyl-5*H*-benzocyclopentenes of type **270** undergo an unusual isomerization reaction leading to tetracyclic azo compounds **271** (316,317) (Scheme 8.66). The reaction readily occurs upon chromatographic workup of the diazo compounds that are prepared by electrophilic diazoalkane substitution using benzotropylium salts. The isolated diazo compounds are thermally converted into **271**. The isomerization reaction was interpreted as a formal [4 + 3] cycloaddition. Since [4π + 4π] cycloaddition reactions are thermally disallowed processes, a

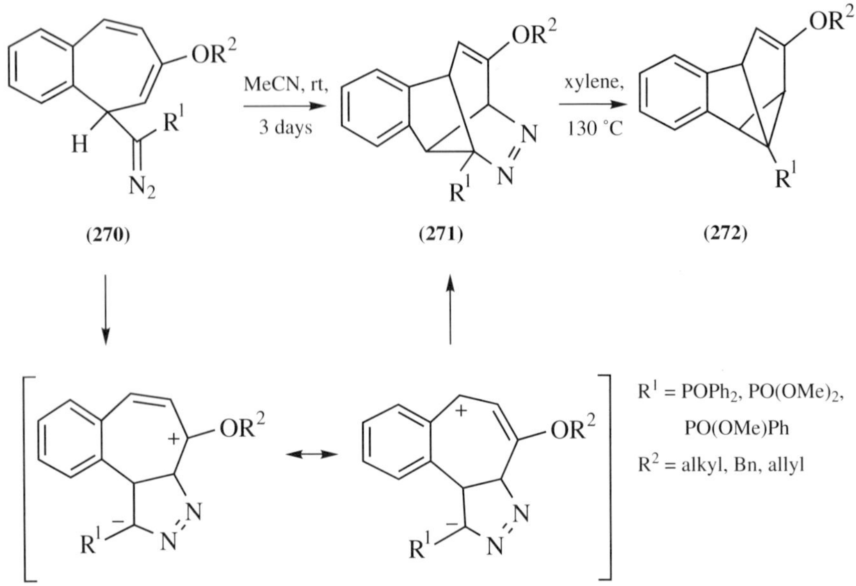

Scheme 8.66

stepwise mechanism was proposed (Scheme 8.66). The fact that the 7-alkoxy substituent is necessary for the isomerization of **270** to occur lends support to this mechanistic proposal and renders less likely an alternative 1,3-dipolar cycloaddition across the 8,9-double bond followed by a formal 1,3-shift of a nitrogen atom. Compounds **271** were used to synthesize benzosemibullvalenes (**272**) by thermal extrusion of nitrogen.

Intramolecular reactions that differ from the 1,3-dipolar type are also known for olefinic diazo compounds. α,β-Unsaturated diazo compounds are known to undergo 1,5-cyclization to give pyrazoles. This reaction type may be considered as a variant of an intramolecular [3 + 2] cycloaddition. A recent *ab initio* and DFT study classified the cyclization of vinyldiazomethane to 3*H*-pyrazole as a monorotatory pericyclic process (318).

As shown in Scheme 8.67, the cyclization of diazoalkenes **273** requires thermal activation and not only affords 3*H*-pyrazoles **274**, but also cyclopropenes **275** that are formed from carbene intermediates (319). The activation parameters for cyclopropene formation (i.e., N_2 elimination from **273**) have been determined (320). A novel example involves the cyclization of the 3-nitro-1-diazoprop-2-ene derivative **276** into pyrazolopyridine derivative **277** (45).

The 1,5-cyclization reaction also occurs with (phosphavinyl)diazoalkanes. Thus, (methylenephosphanyl)diazoalkanes **278**, generated by electrophilic diazoalkane substitution, readily cyclizes at low temperatures (321,322) (Scheme 8.68). The expected 3*H*-1,2,4-diazaphospholes **279** were not detected due to their rapid conversion into 1*H*-1,2,4-diazaphospholes **280** by a silyl shift and concomitant aromatization.

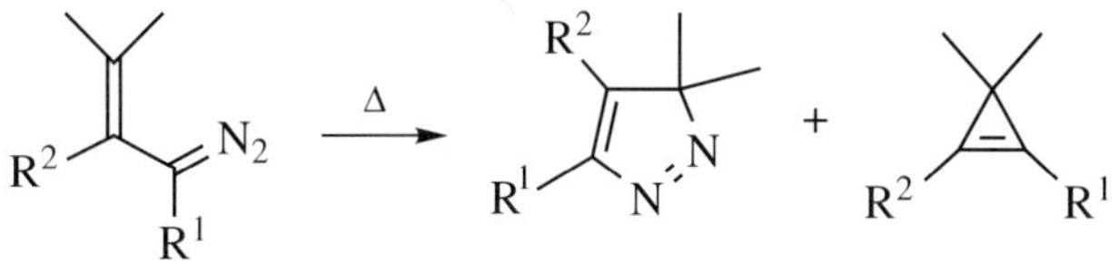

(273) (274) (275)

	(274)	(275)
$R^1 = R^2 = Me$ (30–50 °C):	98%	< 2%
$R^1 = Ph, R^2 = Me$ (70–90 °C):	16–25%	75–84%
$R^1 = Me, R^2 = Ph$ (40–60 °C):	82–92%	8–18%

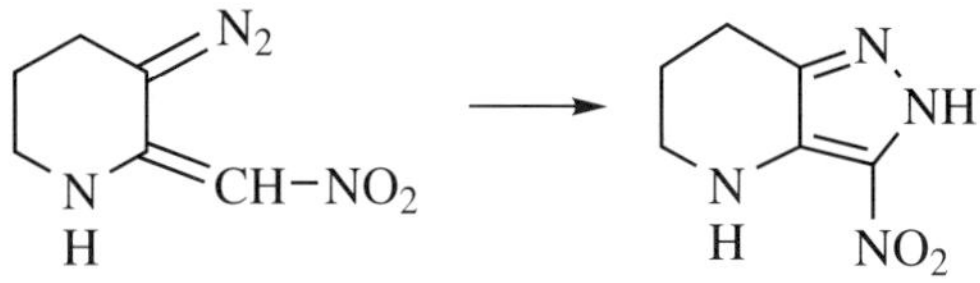

(276) (277)

Scheme 8.67

The terminal nitrogen atom of a diazo dipole is also able to undergo an intramolecular, nitrene-like, [2 + 1] cycloaddition across a C=C bond. This aziridination reaction, which is in principle reversible, has been termed a 1,1-cycloaddition. It was first encountered for a number of allyl-diazomethane derivatives, all of which were phenyl substituted at the diazo carbon (323–327). Theoretical studies (327,328) suggest that the nitrene-like character of the terminal nitrogen atom of the diazo group is enhanced if the attached substituent is π-attracting and σ-withdrawing, properties fullfilled by a phenyl group. Examples **281** and **282** (324), shown in Scheme 8.69, illustrate that the cycloaddition is

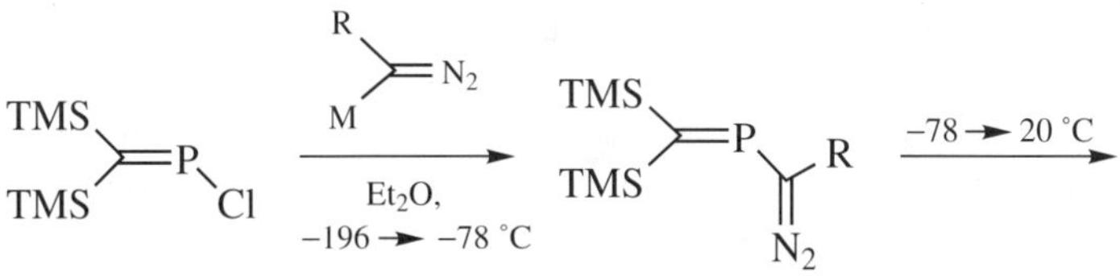

(278)

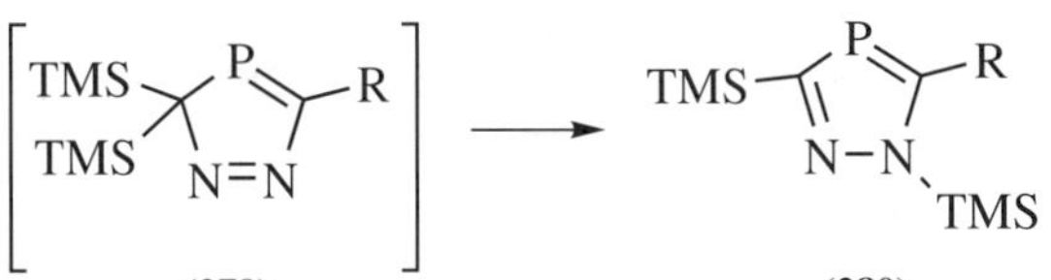

R = TMS, mesityl (M = Li)
R = CO_2t-Bu (M = HgR′)
R = $POPh_2$ (M = Ag)

(279) (280)

(82–86%)

Scheme 8.68

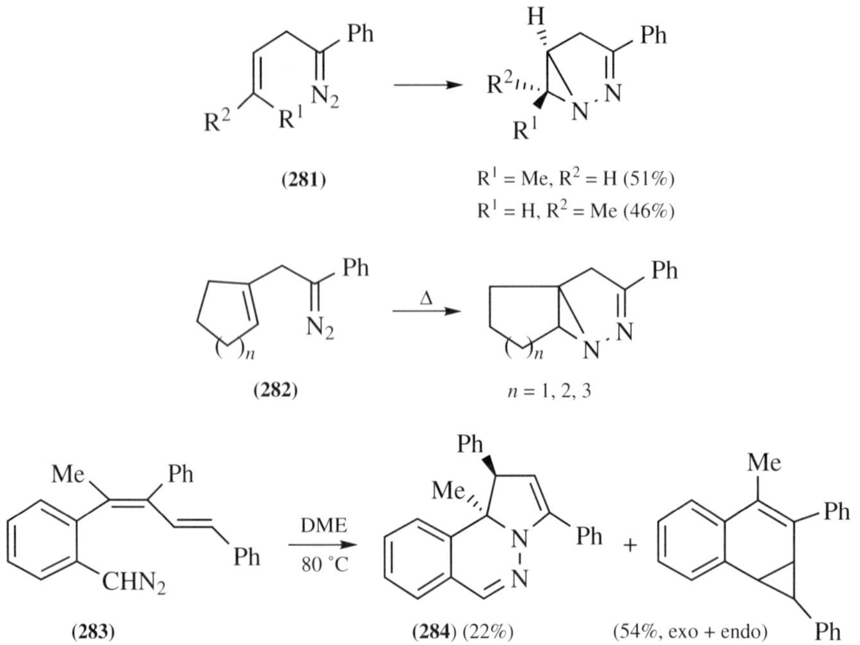

Scheme 8.69

stereospecific and occurs with retention of stereochemistry about the double bond. An intramolecular 1,1-cycloaddition at the disubstituted C=C bond of **283** may account for the formation of pyrrolo[2,1-*a*]phthalazine derivative **284**. It was assumed that the resulting vinylaziridine undergoes fast ring expansion under the reaction conditions (315).

Various α,β;γ,δ-unsaturated 1,3-dipoles are known to undergo 1,7-cyclization by a 8π-electrocyclization process (329,330), and the corresponding diazo compounds behave similarly. 5-Diazopenta-1,3-diene derivatives such as **285** (Scheme 8.70), generated *in situ* by thermolysis of the corresponding tosylhydrazone sodium salts, cyclize to form 1,2-diazepines (**286**) (331). Sharp and co-workers studied the mechanism, scope, and limitations of this transformation. It was found that cis-substitution about the γ,δ-double bond prevents the 1,7-cyclization and directs the system toward 1,5-cyclization (332,333) (i.e., formation of a 3*H*-pyrazole), and that the α,β-double bond can be part of a phenyl ring (334). In special cases, the γ,δ-double bond can be incorporated as part of an aromatic [**287** → **288** (335)] or 2- or 3-thienyl ring as well (336).

A related 1,7-cyclization has been invoked to account for the formation of diazepine **292** from the electrophilic diazoalkane substitution reaction of ethyl diazoacetate and dimethyl diazomethylphosphonate, with the 2,4,6-trimethylpyrylium salt **289** (337) (Scheme 8.70). While the expected 4-(diazomethyl)-4*H*-pyran **290** could be isolated (20–22%), the 2-substituted isomer **291** was not. It was proposed that this latter species underwent 6π-electrocyclic ring opening followed

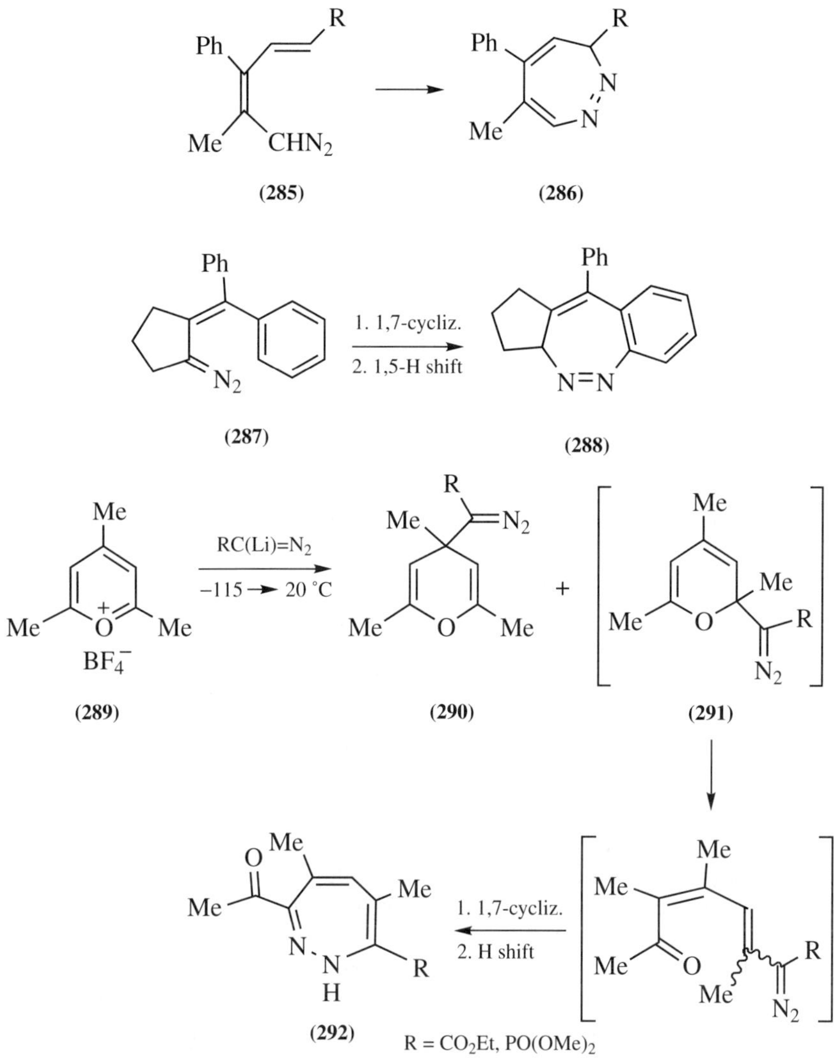

Scheme 8.70

by a 1,7-cyclization of the resulting $\alpha,\beta;\gamma,\delta$-unsaturated diazo compound and a subsequent tautomeric H-shift to give **292** (19–21% yield). If this interpretation is correct, it should be noted that the 1,7-cyclization reaction occurs in spite of the 1,2-cis-substitution about the γ,δ-double bond.

8.6.2. Intramolecular Cycloadditions Involving C≡C Bonds

Intramolecular [3 + 2] cycloaddition reactions of diazo dipoles across carbon–carbon triple bonds are rare. The first reported example involves the quantitative

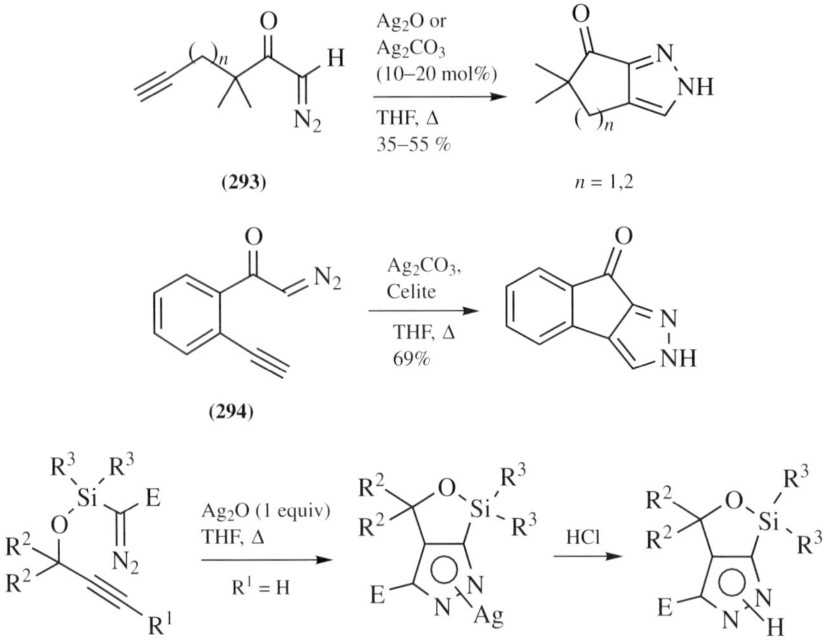

(293) n = 1,2

(294)

(295) (296)

E = CO_2Me, R^3 = *i*-Pr

Scheme 8.71

formation of phenanthro[9,10-*d*]-1*H*-pyrazole from 2-diazo-2′-ethynylbiphenyl (338). Acetylenic diazo ketones **293** and **294** undergo an intramolecular cycloaddition under Ag(I) catalysis (Scheme 8.71). This reaction proceeds at the complete expense of the expected Wolff rearrangement (339). It appears that silver ion activates the triple bond as a dipolarophile. Whether this occurs by side-on (339) or end-on (340) coordination is not clear. Intramolecular cycloadditions of acetylenic diazo compounds are highly sensitive toward structural conditions. Thus, the α′,α′-dimethyl substitution of α-diazoketones **293** is seemingly a prerequisite, but cannot be easily rationalized by a simple Thorpe–Ingold effect (339).

Alkynyl(diisopropylsilyl)oxy-diazoacetates (**295**) undergo intramolecular 1,3-dipolar cycloaddition in good yield when $R^1 = H$ (isolation of silver pyrazolide **296** was possible) and R^2,R^2 = Me,Me or $(CH_2)_5$, but no reaction occurred when $R^1 = H$, $R^2 = H$ or R^1 = Me (340). The silicon substitution is apparently crucial. Replacement of the Si(*i*-Pr)$_2$ in **295** ($R^1 = R^2 = H$) by Si(*t*-Bu)$_2$ allowed an uncatalyzed intramolecular [3 + 2] cycloaddition to take place [xylene, 140–160 °C, 11% yield (340)], while Ag(I) catalysis led to decomposition. A diazoacetic acid (2-propyn-1-yl)oxysilyl ester also produced a bicyclic pyrazole, but in low yield. On the other hand, the same diazo compound **295**, which reacted *intra*-molecularly under silver ion catalysis, underwent dimerization by an *inter*molecular

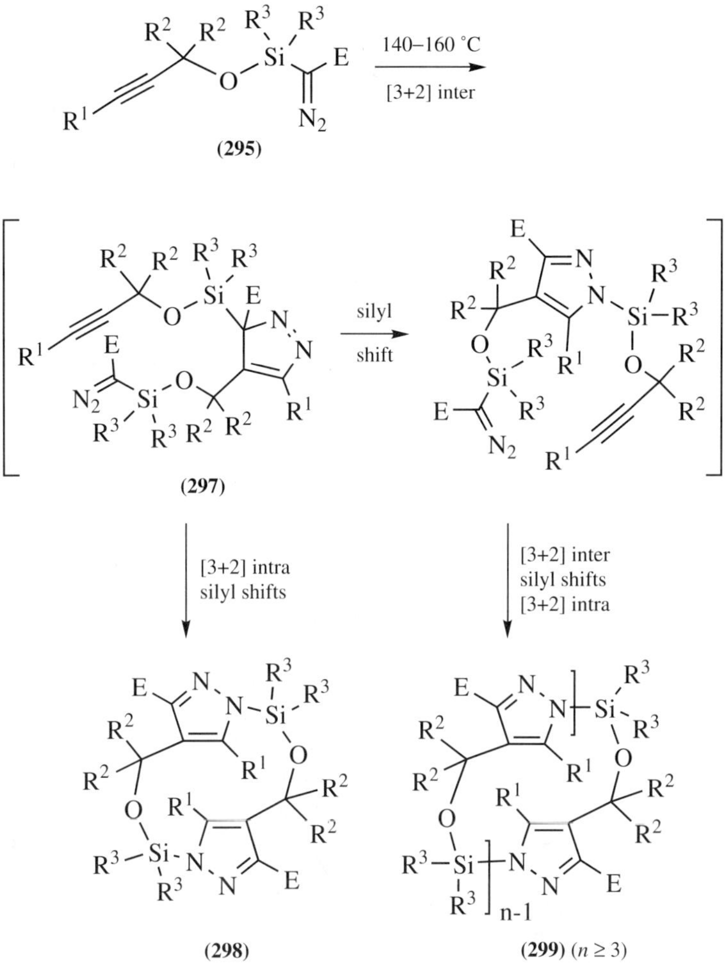

E = CO_2Me

R^1	H	H	H	H	Me
R^2	Me	Me	$-(CH_2)_5-$	H	H
R^3	Me	*i*-Pr	*i*-Pr	*i*-Pr	*i*-Pr

Scheme 8.72

1,3-dipolar cycloaddition reaction under thermal conditions (341) (Scheme 8.72). The resulting pyrazoles **297** can either give [3.3](1,4)pyrazolophanes **298** by intramolecular [3 + 2] cycloaddition followed by a silyl shift, or proceed to give higher cyclooligomers **299** by a sequence of a silyl shift, inter-, and then intramolecular cycloaddition. Cyclooligomers up to $n = 5$ were detected by FD

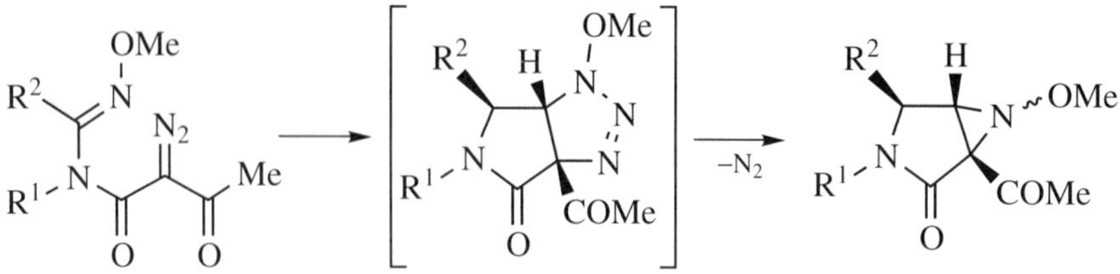

(300) **(301)** **(302)**

Scheme 8.73

mass spectrometry, and those with $n = 2\text{–}4$ were characterized by **X**-ray diffraction analysis.

8.6.3. Intramolecular Cycloadditions Involving Other Multiple Bonds

Diazoamides of type **300** rapidly cyclize to form aziridines **302** (342) (Scheme 8.73). It is conceivable that this reaction proceeds through a 1,2,3-triazoline intermediate **301**, which is the consequence of a LUMO(dipole)–HOMO(dipolarophile) controlled intramolecular [3 + 2] cycloaddition. Some remarkable steric effects were encountered for this cyclization. While the piperidine derivative [**300**, $R^1,R^2 = (CH_2)_4$] readily cyclized by diazo group transfer at 0 °C in 88% yield, the pyrrolidine analogue [**300**, $R^1,R^2 = (CH_2)_3$] had to be heated for 1–2 days in polar solvents. The corresponding acyclic diazoamide (**300**, $R^1 = R^2 = H$) possessed a half-life of >10 days at ambient temperature. The intramolecular aziridination reaction, however, could be readily achieved under catalysis using $Rh_2(OAc)_4$.

The photolysis or thermolysis of certain 1,3-bis(diazomethyl)trisilanes furnished products that are derived from bicyclic pyrazolines formed by intramolecular 1,3-dipolar cycloaddition at the Si=C bond of a diazosilene intermediate (343,344).

8.7. "KETOCARBENE DIPOLES" AS BUILDING BLOCKS IN HETEROCYCLIC SYNTHESIS

Photochemical or thermal extrusion of molecular nitrogen from α-diazocarbonyl compounds generates α-carbonylcarbenes. These transient species possess a resonance contribution from a 1,3-dipolar (**303**, Scheme 8.74) or 1,3-diradical form, depending on their spin state. The three-atom moiety has been trapped in a [3 + 2] cycloaddition fashion, but this reaction is rare because of the predominance of a fast rearrangement of the ketocarbene into a ketene intermediate. There are a steadily increasing number of transition metal catalyzed reactions of diazocarbonyl compounds with carbon–carbon and carbon–heteroatom double bonds, that, instead of affording three-membered rings, furnish five-membered heterocycles which

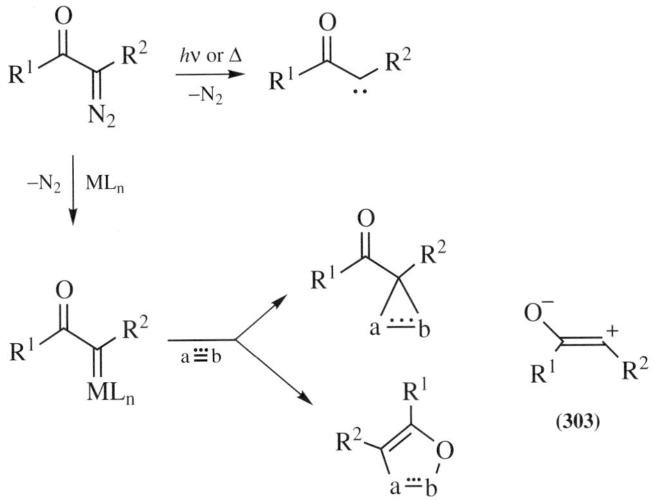

(303)

Scheme 8.74

structurally correspond to [3 + 2] cycloaddition products of ketocarbene dipoles **303**.

Certain acyclic diazoketones react with electron-rich alkenes (such as enol ethers) to form dihydrofurans. The catalyst of choice is $Rh_2(OAc)_4$ for these and related transformations, although copper catalysts have been used as well. A variety of α-diazoketones and α-diazoaldehydes can be used, including $PhCOCHN_2$, $N_2CHCOCOOEt$, $MeCOC(N_2)COMe$, and $MeCOC(N_2)COOR$ (345,346). Wenkert (347) and Alonso (348) studied the scope of this dihydrofuran synthesis [Scheme 8.75; e.g., **304** → **305**; R = H (349), COOEt (350)].

More recently, Pirrung and co-workers established the facility with which the $Rh_2(OAc)_4$ catalyzed reaction of 2-diazocyclohexane-1,3-dione (**306**) and its substituted derivatives (Scheme 8.76) occurs with dihydrofuran and dihydropyran (351–353), vinyl acetates (354) (**306** → **307**), terminal alkynes (355) (**306** → **308**), methoxyallene (355), trimethylsilylketene (serving as a synthetic equivalent for ketene) (355), and heteroaromatic compounds (353). This reaction is quite useful

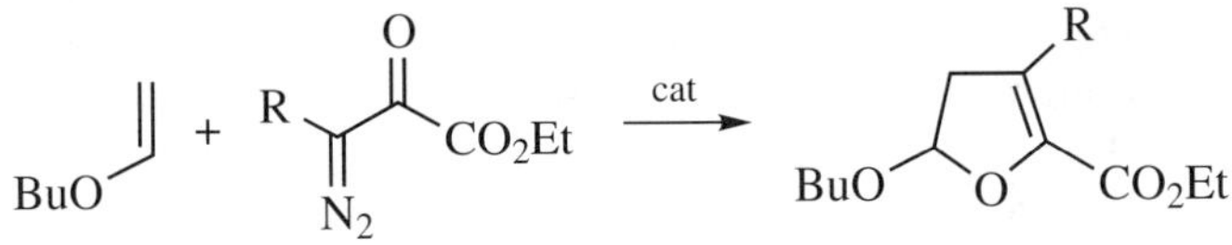

(304) (305)

R = H: $Cu(acac)_2$, benzene, 80 °C; 62%
R = CO_2Et: $Rh_2(OAc)_4$; 95%
acac = acetylacetonate

Scheme 8.75

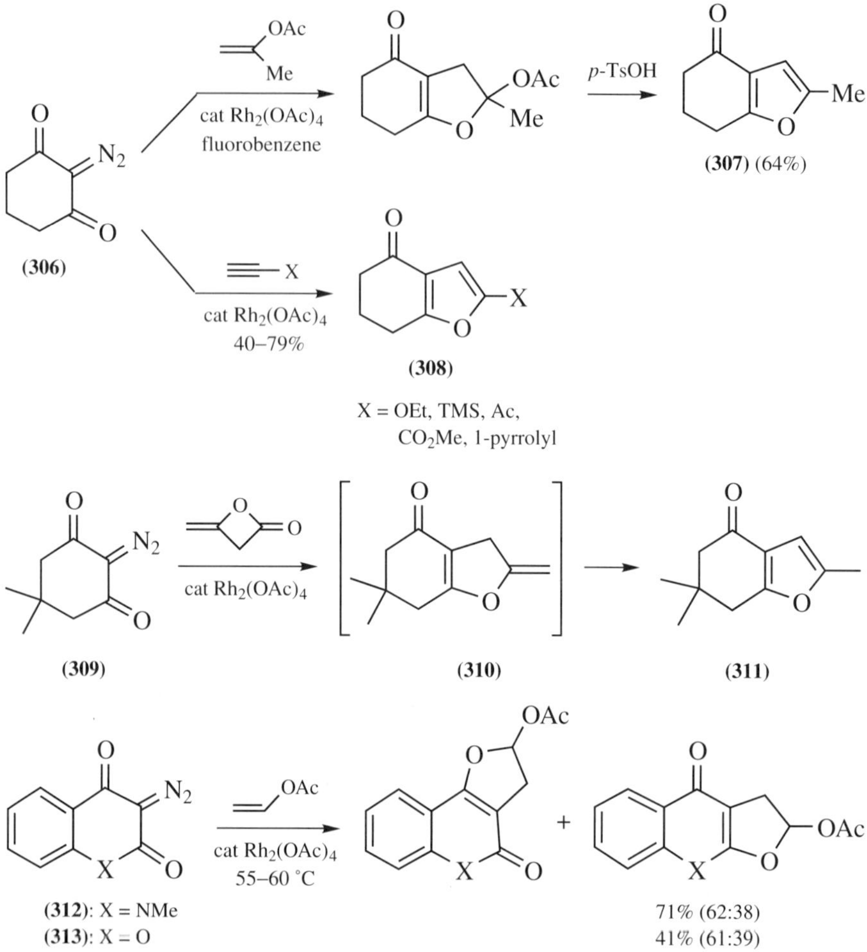

Scheme 8.76

for the synthesis of annulated dihydrofurans and furans. Diketene can also be used as a substrate (356) (e.g, **309** → **311**). In this case, a formal [3 + 2] cycloaddition followed by CO_2 extrusion was proposed to account for the formation of the 3-methylenedihydrofuran derivative **310** that subsequently tautomerizes to give furan **311**. 2-Diazoindane-1,3-dione and diketene afforded a product analogous to **310** in 28% yield (356). 3-Diazo-1-methylquinoline-2,4-dione (357) (**312**) and 3-diazochromene-2,4-dione (358) (**313**), due to their unsymmetrical substitution, led to mixtures of linearly and angularly fused products.

When the chiral catalyst tetrakis(binaphtholphosphate)dirhodium was used, moderate enantiomeric excess (50%) was obtained in cycloaddition reactions with furans and dihydrofurans (359). This method has been applied to the synthesis of several natural products that contain fused furan rings, e.g., the human platelet

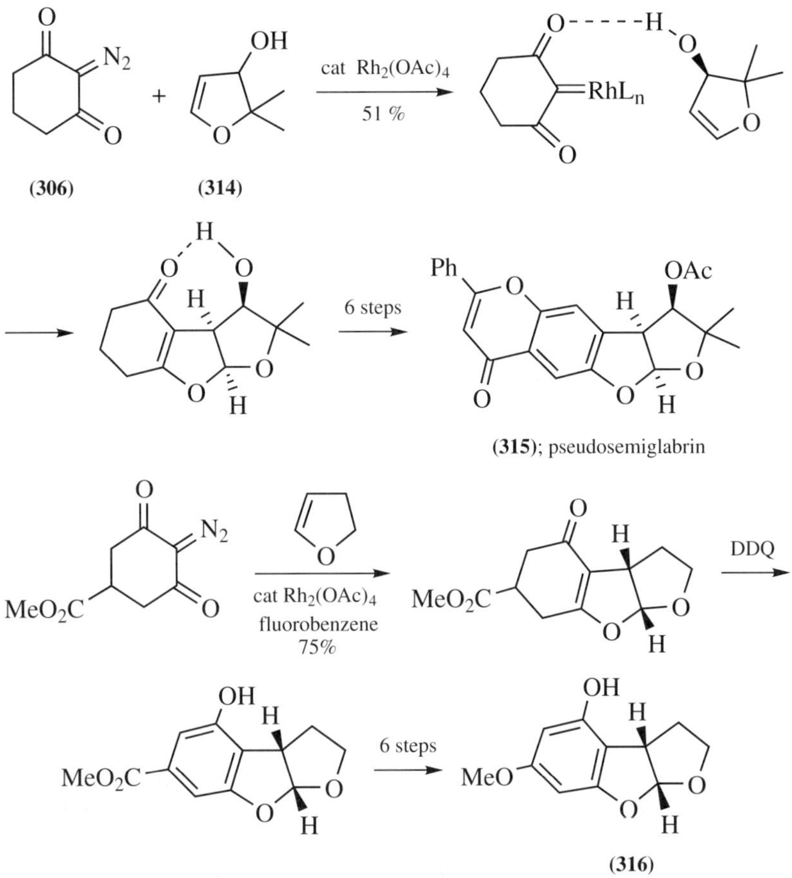

Scheme 8.77

aggregation antagonist pseudosemiglabrin (360) (**315**, Scheme 8.77) [in racemic form and, by the use of the chiral 3-hydroxy-2,3-dihydrofuran **314**, as the (+)-enantiomer), of the (±)-aflatoxin B_2 precursor **316** (361), and of furoquinoline alkaloids (357)].

The mechanistic pathways for the metal-catalyzed [3+2] cycloaddition of ketocarbenes described in this section is not yet completely understood. These reactions seem to depend on the diazo substrate and also on the nature of the diazocarbonyl group. Reaction with electron-rich alkenes have been explained by addition of the electrophilic metal–carbene intermediate (see Scheme 8.74) to generate a dipolar species that undergoes 1,5-cyclization either prior to or after elimination of the metal fragment (345,346,348). For cycloaddition reactions involving furans, thiophenes, and pyrroles as substrates, a cyclopropanation–ring opening–1,5-cyclization cascade sequence has been proposed (353) (Scheme 8.78). This sequence occurs by the initial formation of the cyclopropane intermediate **317**, which forms **318** by ring-opening. Betaine **318** can undergo 1,5-cyclization or

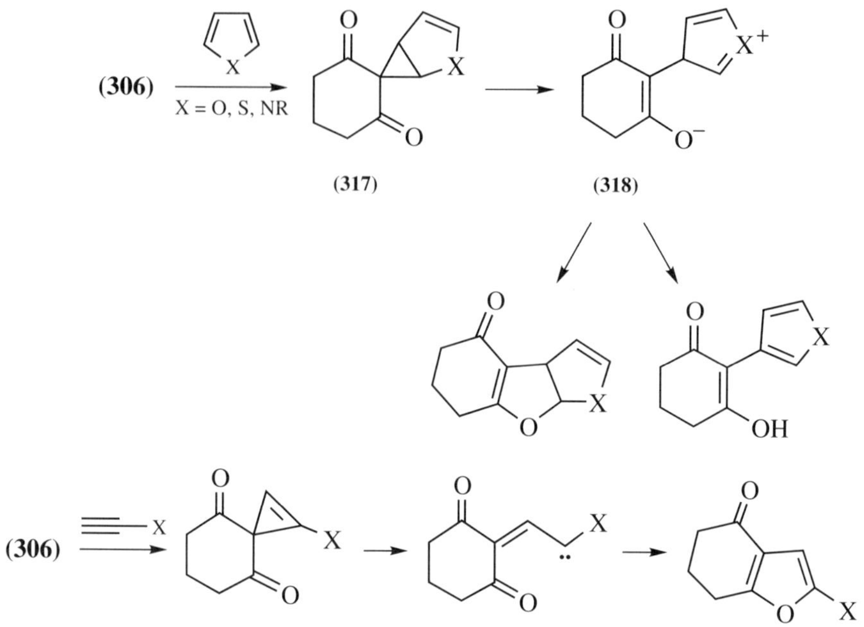

Scheme 8.78

generate a 2-hetaryl-cyclohexane-1,3-dione by a proton shift. Since ring-opening of **317** can occur in two directions, the formation of regioisomeric products is possible. To account for the cycloaddition products obtained with alkynes, a pathway involving cyclopropenation, ring opening to form a vinylcarbene, and 1,5-cyclization has been suggested (355) (Scheme 8.78). This mechanism nicely accounts for the formation of the same regioisomer from **306** and both electron-rich and electron-poor alkynes.

The reaction of α-diazocarbonyl compounds with nitriles produces 1,3-oxazoles under thermal (362,363) and photochemical (363) conditions. Catalysis by Lewis acids (364,365), or copper salts (366), and rhodium complexes (367) is usually much more effective. This latter transformation can be regarded as a formal $[3+2]$ cycloaddition of the ketocarbene dipole across the C≡N bond. More than likely, the reaction occurs in a stepwise manner. A nitrilium ylide (**319**) (Scheme 8.79) that undergoes 1,5-cyclization to form the 1,3-oxazole ring has been proposed as the key intermediate.

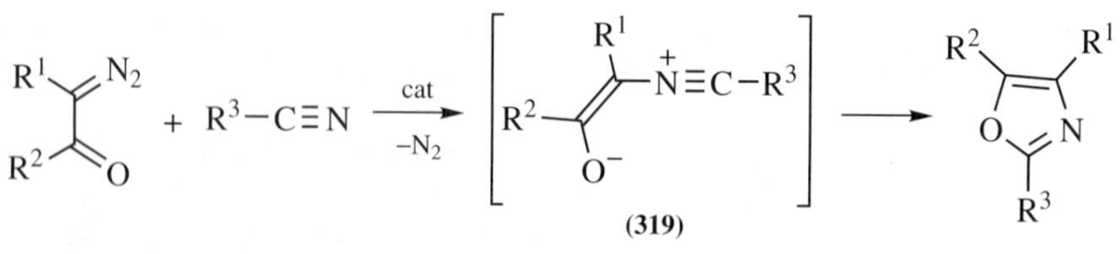

Scheme 8.79

TABLE 8.3. SYNTHESIS OF 1,3-OXAZOLES FROM α-DIAZOCARBONYL COMPOUNDS AND NITRILES

R^1	R^2	R^3	Catalyst	Yield (%)	Reference
H	Ph	Et	$BF_3 \cdot Et_2O$	99	365
H	4-(2-diazo acetyl)phenyl	Me	$BF_3 \cdot Et_2O$	98[a]	365
H	4-X–C_6H_4 (X=H, CN, NO_2, OMe)	(i-Pr)$_2$N	$Rh_2(OAc)_4$	76–95	372
CO_2Me	OMe	EtOCH=CH	$Rh_2(OAc)_4$	97	373
$SiMe_3$	OMe	CO_2Me	$Rh_2(pfb)_4$[b]	51	374
$SiEt_3$	OEt	H_2C=CH	$Rh_2(OAc)_4$	53	375
CO–O-o-phenylene[c]		$ClCH_2$	$Rh_2(OAc)_4$	95	358
CO_2Et	CF_3, C_5F_{11}, $(CF_2)_3Cl$	Ph, Bn, $ClCH_2$, CH=CHMe	$Rh_2(OAc)_4$		376

[a] 1,3-Bis(2-methyl-1,3-oxazol-5-yl)benzene is formed.
[b] Perfluorobutyrate=pf b.
[c] 3-Diazochromene-2,4-dione.

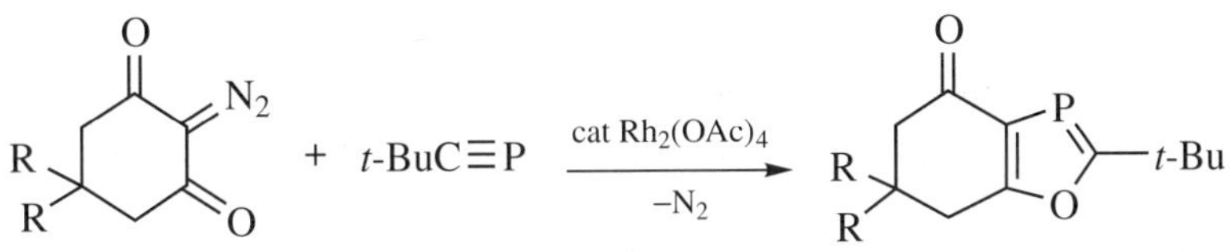

(320)

R = H (39%), Me (52%)

Scheme 8.80

Some examples that illustrate the scope of this transformation are given in Table 8.3 [for further examples, see (368,369)]. Note that this reaction, in contrast to the synthesis of dihydrofurans and furans, is not limited to α-diazoketones as the α-carbonylcarbene moiety concealed in diazoesters also works.

This oxazole synthesis has been used to prepare some important biologically active compounds such as 3-[2-R-1,3-oxazol-5-yl)indoles (370) (R = Me: pimprinine; R = Et: pimprinethine; R = Pr: WS-30581A] and 1-Boc-2-chloro-3-(2-methyl-1,3-oxazol-5-yl)indole, a structural fragment of the cytotoxic marine peptide diazonamide A (371).

The first example of the [3 + 2] cycloaddition of ketocarbenes with a phosphorus–carbon triple bond has also been reported. Thus, the rhodium-catalyzed reaction of 2-diazocyclohexane-1,3-diones with *tert*-butylphosphaethyne gave 1,3-oxaphosphole **320** (377) (Scheme 8.80) in moderate yield.

CONCLUSION

More than a hundred years after their discovery, aliphatic diazo compounds continue to be widely used in dipolar cycloaddition chemistry. During the past two decades, we have witnessed many applications of novel diazo compounds as 1,3-dipoles. Perhaps even more important is the cycloaddition with carbon–heteroatom and heteroatom–heteroatom multiple bonds, which gives rise to novel heterocyclic ring systems. These heterocycles can be synthetic targets *per se* (*e.g.* 4-phospha-pyrazolines or 1,2,4-diazaphospholes) but can also be used as intermediates for further synthetic transformations. A typical example involves the conversion of 2,5-dihydro-1,3,4-thiadiazoles, obtained from diazo compounds and thiones, into sterically hindered olefins or into thiocarbonyl ylides. The synthetic potential of *intramolecular* 1,3-dipolar cycloadditions of unsaturated diazo compounds, in particular of acetylenic ones, has not yet been fully appreciated. A large number of such diazo compounds have recently been used recently for metal-catalyzed intramolecular cyclopropanation (or cyclopropenation) or other carbenoid reactions. The ability of these diazo compounds to form bicyclic pyrazoles and pyrazolines has often gone unnoticed due to the need for thermal activation of this type of internal cycloaddition. Finally, much information is now available on the diastereofacial selectivity of the [3 + 2] cycloaddition of diazo compounds with alkenes bearing chiral centers.

It should be noted, however, that the 1,3-dipolar cycloaddition chemistry of diazo compounds has been used much less frequently for the synthesis of natural products than that of other 1,3-dipoles. On the other hand, several recent syntheses of complex molecules using diazo substrates have utilized asymmetric induction in the cycloaddition step coupled with some known diazo transformation, such as the photochemical ring contraction of Δ^1-pyrazolines into cyclopropanes. This latter process often occurs with high retention of stereochemistry. Another useful transformation involves the conversion of Δ^2-pyrazolines into 1,3-diamines by reductive ring-opening. These and other results show that the 1,3-dipolar cycloaddition chemistry of diazo compounds can be extremely useful for stereoselective target-oriented syntheses and presumably we will see more applications of this type in the near future.

ACKNOWLEDGMENTS

Thanks are due to the Deutsche Forschungsgemeinschaft and the Fonds der Chemischen Industrie for financial support of our research on diazo compounds including their cycloaddition chemistry. I am grateful to Joachim Nikolai for his help with the drawing of the formula schemes.

REFERENCES

1. E. Buchner, *Ber. Dtsch. Chem. Ges.*, **1888**, *21*, 2637.
2. H. von Pechmann, *Ber. Dtsch. Chem. Ges.*, **1895**, *28*, 855.
3. Review: R. Huisgen, *Angew. Chem.*, **1963**, *75*, 604; *Angew. Chem. Int. Ed. Engl.*, **1963**, *2*, 565.
4. Review: R. Huisgen, *Angew. Chem.*, **1963**, *75*, 742; *Angew. Chem. Int. Ed. Engl.*, **1963**, *2*, 633.

5. M. Regitz and H. Heydt, in *1,3-Dipolar Cycloaddition Chemistry*, A. Padwa, ed., Wiley, New York, 1984, Chapter 4.
6. M. P. Doyle, M. A. McKervey, and T. Ye, *Modern Catalytic Methods for Organic Synthesis with Diazo Compounds—From Cyclopropanes to Ylides*, Wiley, New York, 1998.
7. M. P. Doyle, in *Comprehensive Organometallic Chemistry II*, Vol. 12, L. S. Hegedus, ed., Elsevier, Oxford, 1995, Chapters 5.1 and 5.2.
8. Review: T. Ye and M. A. McKervey, *Chem. Rev.*, **1994**, *94*, 1091.
9. Review: T. Shioiri and T. Aoyama, in *Advances in the Use of Synthons in Organic Chemistry*, Vol. 1, A. Dondoni, ed., JAI Press, London, **1993**, pp. 51–101.
10. Review: T. Shioiri, and T. Aoyama, *J. Synth. Org. Chem. Jpn.*, **1986**, *44*, 149.
11. Review: J. Podlech, *J. Prakt. Chem.*, **1998**, *340*, 679.
12. Review: D. Bourissou and G. Bertrand, *C. R. Acad. Sci. Paris, Série IIb*, **1996**, *322*, 489.
13. R. Huisgen, in *1,3-Dipolar Cycloaddition Chemistry*, A. Padwa, ed., Wiley, New York, **1984**, Chapter 1.
14. K. D. Houk and K. Yamaguchi, in *1,3-Dipolar Cycloaddition Chemistry*, A. Padwa, ed., Wiley, New York, **1984**, Chapter 13.
15. M. Regitz and G. Maas, *Diazo Compounds—Properties and Synthesis*, Academic Press, Orlando, FL, **1986**.
16. M. Böhshar, J. Fink, H. Heydt, O. Wagner, and M. Regitz, in *Houben-Weyl, Methoden der organischen Chemie*, Vol. E14b, D. Klamann and H. Hagemann, eds., Thieme, Stuttgart, **1990**, pp. 961–1371.
17. H. Zollinger, *Diazo Chemistry II—Aliphatic, Inorganic and Organometallic Compounds*, VCH, Weinheim, **1995**, pp 11–94.
18. W. Bihlmaier, R. Huisgen, H.-U. Reissig, and S. Voss, *Tetrahedron Lett.*, **1979**, *28*, 2621.
19. R. Huisgen and J. Geittner, *Heterocycles*, **1978**, *11*, 105.
20. (a) L. Fišera, J. Geittner, R. Huisgen, and H.-U. Reissig, *Heterocycles*, **1978**, *10*, 1532. (b) Y. D. Samuilov, A. I. Movchan, S. E. Solov'eva, and A. I. Konovalov, *Zh. Org. Khim.*, **1984**, *20*, 2179.
21. K. Kokubo, Y. Kanematsu, and T. Oshima, *Chem. Lett.*, **2000**, 1283.
22. T. Oshima, A. Yoshioka and T. Nagai, *J. Chem. Soc., Perkin Trans. 2*, **1979**, 1283.
23. T. Oshima and T. Nagai, *Bull. Chem. Soc. Jpn.*, **1980**, *53*, 3284.
24. R. Sustmann and W. Sicking, *Chem. Ber.*, **1987**, *120*, 1653.
25. R. Sustmann, W. Sicking and M. Felderhoff, *Tetrahedron*, **1990**, *46*, 783.
26. A. Rastelli, M. Bagatti, R. Gandolfi, and M. Burdisso, *J. Chem. Soc., Faraday Trans.*, **1994**, *90*, 1077.
27. A. Padwa and M. Woods Wannamaker, *J. Chem. Soc., Chem. Commun.*, **1987**, 1742.
28. V. Branchadell, E. Muray, A. Oliva, R. M. Ortuno, and C. Rodriguez-Garcia, *J. Phys. Chem. A.*, **1998**, *102*, 10106.
29. M.-D. Su, H.-Y. Liao, W.-S. Chung, and S.-Y. Chu, *J. Org. Chem.*, **1999**, *64*, 6710.
30. R. T. Conlin and Y.-W. Kwak, *J. Organomet. Chem.*, **1985**, *293*, 177.
31. D. Seyferth, H. Menzel, A. W. Dow, and T. C. Flood, *J. Organomet. Chem.*, **1972**, *44*, 279.
32. T. Aoyama, Y. Iwamoto, S. Nishigaki, and T. Shioiri, *Chem. Pharm. Bull.*, **1989**, *37*, 253.
33. H.-D. Martin, R. Iden, F.-J. Mais, G. Kleefeld, A. Steigel, B. Fuhr, O. Rümmele, A. Oftring, and E. Schwichtenberg, *Tetrahedron Lett.*, **1983**, *24*, 5469.
34. P. Herrera, A. Nagarajan, M. A. Morales, F. Méndez, H. A. Jiménez-Vázquez, L. G. Zepeda, and J. Tamariz, *J. Org. Chem.*, **2001**, *66*, 1252.
35. T. Aoyama, T. Nakano, S. Nishigaki, and T. Shioiri, *Heterocycles*, **1990**, *30*, 375.
36. S. Inoue, K. Sudo, and T. Aoyama, *Heterocycles*, **1984**, *21*, 620. T. Aoyama, S. Inoue, and T. Shioiri, *Tetrahedron Lett.*, **1984**, *25*, 433.

37. T. Asaki, T. Aoyama, and T. Shioiri, *Heterocycles*, **1988**, *27*, 343.
38. T. Ito, K. Hatano, Y. Kurono, T. Aoyama, and T. Shioiri, *Heterocycles*, **1993**, *35*, 41.
39. R. Saladino, L. Stasi, C. Crestini, R. Nicoletti, and M. Botta, *Tetrahedron*, **1997**, *53*, 7045.
40. H. Abdallah, R. Grée and R. Carrié, *Bull. Soc. Chim. Fr.*, **1984**, 338.
41. H. Abdallah, R. Grée, and R. Carrié, *Bull. Soc. Chim. Fr.*, **1985**, 794.
42. A. Y. Aizikovich, V. Y. Korotaev, M. I. Kodess, and A. Y. Barkov, *Zh. Org. Khim.*, **1998**, *34*, 383.
43. A. Y. Aizikovich, V. Y. Korotaev, and L. E. Yaroslavtseva, *Zh. Org. Khim.*, **1994**, *30*, 989.
44. A. Y. Aizikovich, V. Y. Korotaev, M. I. Kodess, and A. Y. Barkov, *Zh. Org. Khim.*, **1998**, *34*, 1149.
45. I. S. Hutchinson, S. A. Matlin, and A. Mete, *Tetrahedron Lett.*, **2001**, *42*, 1773.
46. Y. V. Tomilov, I. P. Klimenko, E. V. Shulishov, and O. M. Nefedov, *Russ. Chem. Bull.*, **2000**, *49*, 1207.
47. (a) Y. V. Tomilov, I. V. Kostyuchenko, E. V. Shulishov, and O. M. Nefedov, *Izv. Akad. Nauk, Ser. Khim.*, **1997**, 532. (b) Y. V. Tomilov, E. V. Shulishov, G. P. Okonnishnikova and O. M. Nefedov, *Izv. Akad. Nauk, Ser. Khim.*, **1995**, 2199.
48. Y. V. Tomilov, E. V. Shulishov, G. P. Okonnishnikova, and O. M. Nefedov, *Mendeleev Commun.*, **1997**, 200.
49. O. Tsuge, S. Kanemasa, T. Suzuki, and K. Matsuda, *Bull. Chem. Soc. Jpn.*, **1986**, *59*, 2851.
50. L. A. Tolochko, A. M. Sipyagin, V. G. Kartsev, Z. G. Aliev, L. O. Atovmyan, and A. T. Lebedev, *Khim. Geterotsikl. Soedin.*, **1988**, 454.
51. A. Padwa, Z. J. Zhang, and L. Zhi, *J. Org. Chem.*, **2000**, *65*, 5223.
52. A. Belaissaoui, S. Jacquot, C. Morpain, G. Schmitt, J. Vebrel, and B. Laude, *Can. J. Chem.*, **1997**, *75*, 523.
53. Y. V. Tomilov, G. P. Okonnishnikova, E. V. Shulishov, K. N. Shavrin, and O. M. Nefedov, *Izv. Akad. Nauk SSSR, Ser. Khim.*, **1998**, 2278; engl. transl. 2208.
54. M. W. Majchrzak, M. Békhazi, I. Tse-Sheepy, and J. Warkentin, *J. Org. Chem.*, **1989**, *54*, 1842.
55. M. W. Majchrzak and J. Warkentin, *J. Phys. Org. Chem.*, **1990**, *3*, 339.
56. M. Franck-Neumann, P. Geoffroy, and A. Winling, *Tetrahedron Lett.*, **1998**, *39*, 5643.
57. P. M. Lahti and J. A. Berson, *J. Am. Chem. Soc.*, **1981**, *103*, 7011.
58. (a) R. Munschauer and G. Maas, *Angew. Chem.*, **1991**, *103*, 312; *Angew. Chem. Int. Ed. Engl.*, **1991**, *30*, 306. (b) R. Munschauer, Doctoral Thesis, Universität Kaiserslautern, **1991**.
59. R. Munschauer and G. Maas, *Chem. Ber.*, **1992**, *125*, 1227.
60. J.-M. Sotiropoulos, A. Baceiredo, and G. Bertrand, *J. Am. Chem. Soc.*, **1987**, *109*, 4711; *Bull. Soc. Chim. Fr.*, **1992**, *129*, 367.
61. J.-M. Sotiropoulos, A. Baceiredo, K. Horchler von Locquenghien, F. Dahan, and G. Bertrand, *Angew. Chem.*, **1991**, *103*, 1171; *Angew. Chem. Int. Ed. Engl.*, **1991**, *30*, 1154.
62. A. Levai, A. M. S. Silva, T. Patonay, and J. A. S. Cavaleiro, *J. Heterocycl. Chem.*, **1999**, *36*, 1215.
63. G. Toth, A. Szollosy, A. Levai, and G. Kotovych, *J. Chem. Soc., Perkin Trans. 2*, **1986**, 1895.
64. G. Toth, A. Levai, A. Szollosy, and H. Duddeck, *Tetrahedron*, **1993**, *49*, 863.
65. A. Levai and T. Patonay, *J. Heterocycl. Chem.*, **1999**, *36*, 747.
66. N. Boukamcha, R. Gharbi, M. T. Martin, A. Chiaroni, Z. Mighri, and A. Khemiss, *Tetrahedron*, **1999**, *55*, 449.
67. A. Bojilova, I. Videnova, C. Ivanov, N. A. Rodios, A. Terzis, and C. P. Raptopoulou, *Tetrahedron*, **1994**, *50*, 13023.
68. Ă. Cerček, B. Stanovnik, A. Stimac, and M. Tišler, *Heterocycles*, **1987**, *26*, 2425.
69. R. Nesi, D. Giomi, S. Turchi, P. Tedeschi, and F. Ponticelli, *Gazz. Chim. Ital.*, **1993**, *123*, 633.
70. W. Nagai, Y. Hirata, K. Hosomi, and T. Higuchi, *J. Heterocycl. Chem.*, **1994**, *31*, 225.
71. R. Huisgen, A. Mitra, and J. R. Moran, *Heterocycles*, **1986**, *24*, 2429; *Chem. Ber.*, **1987**, *120*, 159.

72. U. Eichenauer, R. Huisgen, A. Mitra, and J. R. Moran, *Heterocycles*, **1987**, *25*, 129; *Chem. Ber.*, **1987**, *120*, 153.

73. T. Oshima, S. Fujii, T. Takatani, K. Kokubo, and T. Kawamoto, *J. Chem. Soc., Perkin Trans. 2*, **1999**, 1783; and references cited therein.

74. V. N. Kovtonyuk, L. S. Kobrina, I. Y. Bagryanskaya, and Y. V. Gatilov, *Russ. J. Org. Chem.*, **1999**, *35*, 67.

75. M. F. Aldersley, F. M. Dean, and B. E. Mann, *J. Chem. Soc., Perkin Trans. 1*, **1986**, 2217.

76. B. Jian, Y. Xu, and J. Yang, *J. Fluorine Chem.*, **1994**, *67*, 83.

77. M.-A. Plancquaert, M. Redon, Z. Janousek, and H. G. Viehe, *Tetrahedron*, **1996**, *52*, 4383.

78. A. Baranski and V. I. Kelarev, *Khim. Geterotsikl. Soedin.*, **1990**, 435.

79. (a) M. Mancera, E. Rodriguez, I. Roffe, and J. A. Galbis, *J. Org. Chem.*, **1988**, *53*, 5648. (b) M. Mancera, E. Rodriguez, I. Roffe, J. A. Galbis, C. F. Conde, and A. Conde, *Carbohydr. Res.*, **1991**, *210*, 327.

80. W. H. Midura, J. A. Krysiak, and M. Mikolajczyk, *Tetrahedron*, **1999**, *55*, 14791.

81. (a) A. N. Anisimova, G. A. Berkova, and L. I. Deiko, *Zh. Org. Khim.*, **1998**, *34*, 1263. (b) N. Anisimova, G. A. Berkova, and L. I. Deiko, *Zh. Obshch. Khim.*, **1998**, *68*, 1165.

82. L. Pizzioli, B. Ornik, J. Svete, and B. Stanovnik, *Helv. Chim. Acta*, **1998**, *81*, 231.

83. R. Augusti and C. Kascheres, *J. Org. Chem.*, **1993**, *58*, 7079.

84. M. Hamaguchi and T. Nagai, *J. Org. Chem.*, **1989**, *54*, 3957.

85. M. Jazouli, B. Carboni and R. Carrié, *Heteroatom Chem.*, **1994**, *5*, 513.

86. R. J. Bushby and M. V. Jesudason, *J. Chem. Res. (S)*, **1986**, 126.

87. G. Maier and B. Wolf, *Synthesis*, **1985**, 871.

88. M. S. Baird and H. H. Hussain, *Tetrahedron*, **1987**, *43*, 215.

89. H. Heydt, H.-W. Breiner, V. Hell, and M. Regitz, *Z. Naturforsch.*, **1987**, *42b*, 210.

90. A. R. Dulayymi and M. S. Baird, *Tetrahedron*, **1998**, *54*, 12897.

91. U.-J. Vogelbacher, P. Eisenbarth, and M. Regitz, *Angew. Chem.*, **1984**, *96*, 691; *Angew. Chem. Int. Ed. Engl.*, **1984**, *23*, 708.

92. P. Gesche, F. Klinger, W. Müller, J. Streith, H. Strub, and R. Sustmann, *Chem. Ber.*, **1985**, *118*, 4682.

93. A. Rastelli, M. Bagatti, and R. Gandolfi, *Tetrahedron*, **1994**, *50*, 5561.

94. A. Padwa, M. A. Filipkowski, D. N. Kline, S. S. Murphree, and P. E. Yeske, *J. Org. Chem.*, **1993**, *58*, 2061.

95. Review: G. Broggini, and G. Zecchi, *Gazz. Chim. Ital.*, **1996**, *126*, 479.

96. C. Baldoli, P. Del Buttero, E. Licandro, S. Maiorana, A. Papagni, and A. Zanotti-Gerosa, *J. Organomet. Chem.*, **1994**, *476*, C27.

97. E. Licandro, S. Maiorana, A. Papagni, A. Zanotti-Gerosa, F. Cariati, S. Bruni, M. Moret, and A. Chiesi-Villa, *Inorg. Chim. Acta*, **1994**, *220*, 233.

98. J. Barluenga, F. Fernández-Marí, A. L. Viado, E. Aguilar, B. Olano, S. García-Granda, and C. Moya-Rubiera, *Chem. Eur. J.*, **1999**, *5*, 883.

99. M. A. Esteruelas, A. V. Gómez, A. M. López, M. C. Puerta, and P. Valerga, *Organometallics*, **1998**, *17*, 4959.

100. Review: B. Stanovnik, *Tetrahedron*, **1991**, *47*, 2925.

101. Review: B. Stanovnik, B. Jelen, C. Turk, M. Žličar, and J. Svete, *J. Heterocycl. Chem.*, **1998**, *35*, 1187.

102. M. Žlicar, B. Huč, B. Stanovnik, and M. Tišler, *Synlett*, **1990**, 707.

103. M. Žlicar, B. Stanovnik, and M. Tišler, *J. Heterocycl. Chem.*, **1993**, *30*, 1209.

104. Review: F. Diederich, and C. Thilgen, *Science*, **1996**, *271*, 317.

105. Review: F. Diederich, L. Isaacs, and D. Philp, *Chem. Soc. Rev.*, **1994**, 243.

106. T. Suzuki, Q. Li, K. C. Khemani, and F. Wudl, *J. Am. Chem. Soc.*, **1992**, *114*, 7301.

107. R. F. Haldimann, F.-G. Klärner, and F. Diederich, *Chem. Commun.*, **1997**, 237.

108. A. B. Smith III, R. M. Strongin, L. Brard, G. T. Furst, and W. J. Romanow, *J. Am. Chem. Soc.*, **1993**, *115*, 5829.

109. A. B. Smith III, R. M. Strongin, L. Brard, G. T. Furst, W. J. Romanow, K. G. Owens, R. J. Goldschmidt, and R. C. King, *J. Am. Chem. Soc.*, **1995**, *117*, 5492.

110. L. Isaacs, A. Wehrsig, and F. Diederich, *Helv. Chim. Acta*, **1993**, *76*, 1231.

111. T. Suzuki, Q. Li, K. C. Khemani, F. Wudl, and Ö. Almarsson, *Science*, **1991**, *254*, 1186.

112. E.-U. Wallenborn, R. F. Haldimann, F.-G. Klärner, and F. Diederich, *Chem. Eur. J.*, **1998**, *4*, 2258.

113. (a) M. Freccero, R. Gandolfi, M. Sarzi-Amade, and A. Rastelli, *J. Chem. Soc., Perkin Trans.*, *2*, **1998**, 2413. (b) E. Cinquini, M. Freccero, R. Gandolfi, M. Sarzi-Amade, and A. Rastelli, *Tetrahedron*, **1997**, *53*, 9279. (c) M. Burdisso, A. Gamba, R. Gandolfi, L. Toma, A. Rastelli, and E. Schiatti, *J. Org. Chem.*, **1990**, *55*, 3311. See also references cited therein.

114. (a) H. Hake, H. Landen, H. D. Martin, B. Mayer, A. Steigel, G. Distefano, and A. Modelli, *Tetrahedron Lett.*, **1988**, *29*, 6601. (b) H. Landen, B. Margraf, and H. D. Martin, *Tetrahedron Lett.*, **1988**, *29*, 6593. (c) H. Landen, B. Margraf, H. D. Martin, and A. Steigel, *Tetrahedron Lett.*, **1988**, *29*, 6597.

115. R. M. Ortuño, J. Bigorra, and J. Font, *Tetrahedron*, **1987**, *43*, 2199.

116. S. Hanessian and P. J. Murray, *Tetrahedron*, **1987**, *43*, 5055.

117. N. Hanafi and R. M. Ortuño, *Tetrahedron: Asymmetry*, **1994**, *5*, 1657.

118. M. Martín-Vilà, N. Hanafi, J. M. Jiménez, A. Alvarez-Larena, J. F. Piniella, V. Branchadell, A. Oliva, and R. M. Ortuño, *J. Org. Chem.*, **1998**, *63*, 3581.

119. P. I. Butler, T. Clarke, C. Dell, and J. Mann, *J. Chem. Soc., Perkin Trans. 1*, **1994**, 1503.

120. B. de Lange and B. L. Feringa, *Tetrahedron Lett.*, **1988**, *29*, 5317.

121. E. Keller, B. de Lange, M. T. Rispens, and B. L. Feringa, *Tetrahedron*, *49*, **1993**, 8899.

122. F. Fariña, M. V. Martín, and F. Sánchez, *Heterocycles*, **1986**, *24*, 2587.

123. M. T. Rispens, E. Keller, B. de Lange, R. W. J. Zijlstra, and B. L. Feringa, *Tetrahedron: Asymmetry*, **1994**, *5*, 607.

124. R. Alguacil, F. Fariña, M. V. Martín, and M. C. Paredes, *Tetrahedron*, **1999**, *55*, 229.

125. J. L. García Ruano, A. Fraile, and M. R. Martín, *Tetrahedron: Asymmetry*, **1996**, *7*, 1943.

126. M. Franck-Neumann, M. Sedrati, J.-P. Vigneron, and V. Bloy, *Angew. Chem.*, **1985**, *97*, 995; *Angew. Chem. Int. Ed. Engl.*, **1985**, *24*, 996.

127. A. Bartels and J. Liebscher, *Tetrahedron: Asymmetry*, **1994**, *5*, 1451.

128. A. Bartels, P. G. Jones, and J. Liebscher, *Synthesis*, **1998**, 1645.

129. (a) C. Cativiela, M. D. Díaz-de-Villegas, A. I. Jiménez, and F. Lahoz, *Tetrahedron Lett.*, **1994**, *35*, 617. (b) C. Cativiela, M. D. Díaz-de-Villegas, J. I. García, and A. I. Jiménez, *Tetrahedron*, **1997**, *53*, 4479.

130. A. Otto, B. Ziemer, and J. Liebscher, *Synthesis*, **1999**, 965.

131. G. Galley, M. Pätzel, and P. G. Jones, *Tetrahedron*, **1995**, *51*, 1631.

132. S. Baskaran, J. Vasu, R. Prasad, K. Kodukulla, G. K. Trivedi, and J. Chandrasekhar, *Tetrahedron*, **1996**, *52*, 4515.

133. E. Muray, A. Alvarez-Larena, J. F. Piniella, V. Branchadell, and R. M. Ortuño, *J. Org. Chem.*, **2000**, *65*, 388.

134. M. T. Reetz, F. Kayser, and K. Harms, *Tetrahedron Lett.*, **1992**, *33*, 3453.

135. C. Alcaraz, A. Herrero, J. L. Marco, E. Fernández-Alvarez, and M. Bernabé, *Tetrahedron Lett.*, **1992**, *33*, 5605.

136. K. Okada, F. Samizo, and M. Oda, *Chem. Lett.*, **1987**, 93.

137. M. R. Mish, F. M. Guerra, and E. M. Carreira, *J. Am. Chem. Soc.*, **1997**, *119*, 8379.

138. G. A. Whitlock and E. M. Carreira, *J. Org. Chem.*, **1997**, *62*, 7916.

139. S. Kanemasa and T. Kanai, *J. Am. Chem. Soc.*, **2000**, *122*, 10710.

140. (a) U. J. Vogelbacher, M. Ledermann, T. Schach, G. Michels, U. Hees, and M. Regitz, *Angew. Chem.*, **1988**, *100*, 304; *Angew. Chem. Int. Ed. Engl.*, **1988**, *27*, 272. (b) T. Schach and M. Regitz, unpublished results; cited in M. Regitz, *Nachr. Chem. Tech. Lab.*, **1991**, *39*, 9.

141. J. Perrocheau and R. Carrié, *Bull. Soc. Chim. Belges*, **1993**, *102*, 749.

142. J. Perrocheau and R. Carrié, *J. Chem. Res. (S)*, **1995**, 303; *(M)*, **1995**, 1927.

143. J. Perrocheau, R. Carrié, and J.-P. Fleury, *Can. J. Chem.*, **1994**, *72*, 2458.

144. V. K. Aggarwal and M. Ferrara, *Org. Lett.*, **2000**, *2*, 4107.

145. M. Moran, G. Bernardinelli, and P. Müller, *Helv. Chim. Acta*, **1995**, *78*, 2048.

146. K. Juhl, R. G. Hazell and K. A. Jørgensen, *J. Chem. Soc., Perkin Trans. 1*, **1999**, 2293.

147. J. C. Antilla and W. D. Wulff, *J. Am. Chem. Soc.*, **1999**, *121*, 5099.

148. K. B. Hansen, N. S. Finney, and E. N. Jacobsen, *Angew. Chem.*, **1995**, *107*, 750; *Angew. Chem. Int. Ed. Engl.*, **1995**, *34*, 676.

149. K. G. Rasmussen and K. A. Jørgensen, *J. Chem. Soc., Chem. Commun.*, **1995**, 1401.

150. M. P. Doyle, W. Hu, and D. J. Timmons, *Org. Lett.*, **2001**, *3*, 933.

151. R. Appel, in *Multiple Bonds and Low Coordination in Phosphorus Chemistry*, M. Regitz and O. J. Scherer, eds. Thieme, Stuttgart, **1990**, pp. 157–219.

152. E. Niecke, M. Leuer, D.-A. Wildbredt, and W. W. Schoeller, *J. Chem. Soc., Chem. Commun.*, **1983**, 1171.

153. E. Niecke, W. W. Schoeller and D. A. Wildbredt, *Angew. Chem.*, **1981**, *93*, 119; *Angew. Chem. Int. Ed. Engl.*, **1981**, *20*, 131.

154. P. Becker, H. Brombach, G. David, M. Leuer, H.-J. Metternich, and E. Niecke, *Chem. Ber.*, **1992**, *125*, 771.

155. R. Appel and C. Casser, *Chem. Ber.*, **1985**, *118*, 3419.

156. R. Appel, T. Gaitzsch, F. Knoch, and G. Lenz, *Chem. Ber.*, **1986**, *119*, 1977.

157. G. Märkl, W. Hölzl, and I. Trötsch-Schaller, *Tetrahedron Lett.*, **1987**, *28*, 2693.

158. W. Schnurr and M. Regitz, *Z. Naturforsch.*, **1998**, *43b*, 1285.

159. G. Märkl and I. Trötsch, *Angew. Chem.*, **1984**, *96*, 899; *Angew. Chem. Int. Ed. Engl.*, **1984**, *23*, 901.

160. Y. Y. C. Yeung Lam Ko, and R. Carrié, *J. Chem. Soc., Chem. Commun.*, **1984**, 1640.

161. P. Pellon and J. Hamelin, *Tetrahedron Lett.*, **1986**, *27*, 5611.

162. F. Zurmühlen, W. Rösch, and M. Regitz, *Z. Naturforsch.*, **1985**, *40b*, 1077.

163. S. G. Ruf, U. Bergsträsser, and M. Regitz, *Eur. J. Org. Chem.*, **2000**, 2219.

164. T. A. van der Knaap, T. C. Klebach, F. Visser, R. Lourens, and F. Bickelhaupt, *Tetrahedron*, **1984**, *40*, 991.

165. L. Weber, O. Kaminski, H.-G. Stammler, and B. Neumann, *Organometallics*, **1995**, *14*, 581.

166. M. Rahmouni, M. Abbari, R. Carrié, and M. Soufiaoui, *Bull. Soc. Chim. Belg.*, **1993**, *102*, 719.

167. A. Baceiredo, M. Nieger, E. Niecke, and G. Bertrand, *Bull. Soc. Chim Fr.*, **1993**, *130*, 757.

168. B. Manz and G. Maas, *Tetrahedron*, **1996**, *52*, 10053.

169. B. Manz, U. Bergsträsser, J. Kerth, and G. Maas, *Chem. Ber./Recueil*, **1997**, *130*, 779.

170. B. Manz, J. Kerth, and G. Maas, *Chem. Eur. J.*, **1998**, *4*, 903.

171. Review: A. C. Gaumont and J. M. Denis, *Chem. Rev.*, **1994**, *94*, 1413.

172. J. L. Cabioch, Ph.D. Thèse, Université de Rennes 1, **1989**; see Ref. 105 in (439).

173. C. Grandin, E. About-Jaudet, N. Collignon, J. M. Denis, and P. Savignac, *Heteroat. Chem.*, **1992**, *3*, 337.

174. (a) P. Pellon and J. Hamelin, *Tetrahedron Lett.*, **1986**, *27*, 5611. (b) T. A. Riley, S. B. Larson, T. L. Avery, R. A. Finch, and R. K. Robins, *J. Med. Chem.*, **1990**, *33*, 572.
175. P. Pellon, S. Himdi-Kabbab, I. Rault, F. Tonnard, and J. Hamelin, *Tetrahedron Lett.*, **1990**, *31*, 1147.
176. Review: B. A. Arbuzov, and E. N. Dianova, *Phosphorus Sulfur*, **1986**, *26*, 203.
177. B. A. Arbuzov, E. N. Dianova, I. Z. Galeeva, I. A. Litvinov, Y. T. Struchkov, A. N. Chernov, and A. V. Ilyasov, *Zh. Obshch. Khim.*, **1985**, *55*, 3.
178. B. A. Arbuzov, E. N. Dianova, and S. M. Sharipova, *Izv. Akad. Nauk SSSR, Ser. Khim.*, **1981**, 1113.
179. B. A. Arbuzov, A. F. Lisin, and E. N. Dianova, *Izv. Akad. Nauk SSSR, Ser. Khim.*, **1983**, 2395.
180. B. A. Arbuzov, E. N. Dianova, and E. Y. Zabotina, *Zh. Obshch. Khim.*, **1987**, *57*, 1699.
181. E. Zabotina and E. Dianova, *Phosphorus, Sulfur*, **1993**, *77*, 246.
182. B. A. Arbuzov, E. N. Dianova, and E. Y. Zabotina, *Zh. Obshch. Khim.*, **1985**, *55*, 1471.
183. (a) I. A. Litvinov, Y. T. Struchkov, B. A. Arbuzov, E. N. Dianova, and I. Z. Galeeva, *Izv. Akad. Nauk SSSR, Ser. Khim.*, **1982**, 2718. (b) I. A. Litvinov, Y. T. Struchkov, B. A. Arbuzov, E. N. Dianova, and I. Z. Galeeva, *Izv. Akad. Nauk SSR, Ser. Khim.*, **1984**, 2023.
184. B. A. Arbuzov and E. N. Dianova, *Izv. Akad. Nauk SSSR, Ser. Khim.*, **1977**, 1453.
185. B. A. Arbuzov, E. N. Dianova, R. T. Galiaskarova, and A. Schmidpeter, *Chem. Ber.*, **1987**, *120*, 597.
186. J. Kerth and G. Maas, *Eur. J. Org. Chem.*, **1999**, 2633.
187. J. Kerth and G. Maas, *Eur. J. Org. Chem.*, **2001**, 1581.
188. J. Kerth, Ph.D. Thesis, University of Ulm, **1998**.
189. J. Kerth, U. Werz, and G. Maas, *Tetrahedron*, **2000**, *56*, 35.
190. E. Niecke, J. Böske, B. Krebs, and M. Dartmann, *Chem. Ber.*, **1985**, *118*, 3227.
191. E. Niecke and D. A. Wildbredt, *Chem. Ber.*, **1980**, *113*, 1549.
192. S. Himdi-Kabbab, P. Pellon, and J. Hamelin, *Tetrahedron Lett.*, **1989**, *30*, 349.
193. S. Himdi-Kabbab and J. Hamelin, *Tetrahedron Lett.*, **1991**, *32*, 2755.
194. G. Märkl and H. Seitz, *Tetrahedron Lett.*, **1986**, *27*, 2957.
195. B. A. Arbuzov, E. Y. Zabotina, E. N. Dianova, I. A. Litvinov, V. A. Naumov, S. K. Latypov, and A. V. Ilyasov, *Heteroatom Chem.*, **1992**, *3*, 151.
196. B. A. Arbuzov, E. N. Dianova, E. Y. Zabotina, and N. A. Chadaeva, *Zh. Obshch. Khim.*, **1983**, *53*, 835.
197. B. A. Arbuzov, E. N. Dianova, I. Z. Galeeva, R. R. Shagidullin, Y. Y. Efremov, P. P. Chernov, *Izv. Akad. Nauk SSSR, Ser. Khim.*, **1986**, 230.
198. B. A. Arbuzov, E. N. Dianova, and E. Y. Zabotina, *Izv. Akad. Nauk SSSR, Ser. Khim.*, **1987**, 2819.
199. I. Kalwinsch, L. Xingya, J. Gottstein, and R. Huisgen, *J. Am. Chem. Soc.*, **1981**, *23*, 7032.
200. L. Xingya and R. Huisgen, *Tetrahedron Lett.*, **1983**, *24*, 4181.
201. R. Huisgen, X. Li, G. Mlostoń, and C. Fulka, *Eur. J. Org. Chem.*, **2000**, 1695.
202. R. Huisgen and L. Xingya, *Heterocycles*, **1983**, *20*, 2363.
203. G. Mlostoń and R. Huisgen, *Tetrahedron Lett.*, **1989**, *30*, 7045.
204. (a) R. Huisgen, G. Mlostoń, and C. Fulka, *Heterocycles*, **1985**, *23*, 2207. (b) R. Huisgen, G. Mlostoń, and E. Langhals, *J. Am. Chem. Soc.*, **1986**, *108*, 6401.
205. A. P. Krapcho, D. R. Rao, M. P. Silvon, and B. Abegaz, *J. Org. Chem.*, **1971**, *36*, 3885.
206. (a) R. Huisgen and G. Mlostoń, *Tetrahedron Lett.*, **1985**, *26*, 1049. (b) R. Huisgen and G. Mlostoń, *Polish J. Chem.*, **1999**, *73*, 635.
207. R. Huisgen and J. R. Moran, *Tetrahedron Lett.*, **1985**, *26*, 1057.
208. Review: L. Fišera, R. Huisgen, I. Kalwinsch, E. Langhals, X. Li, G. Mlostoń, K. Polborn, J. Rapp, W. Sicking, and R. Sustmann, *Pure Appl. Chem.*, **1996**, *68*, 789.
209. R. Huisgen and E. Langhals, *Tetrahedron Lett.*, **1989**, *30*, 5369.
210. E. Vedejs and R. G. Wilde, *J. Org. Chem.*, **1986**, *51*, 117.

211. W. Ando, T. Furukata, and T. Takata, *Tetrahedron Lett.*, **1985**, *26*, 4499.

212. R. Sustmann, W. Sicking, and R. Huisgen, *J. Org. Chem.*, **1993**, *58*, 82.

213. G. Mlostoń and R. Huisgen, *Tetrahedron Lett.*, **1985**, *26*, 1053.

214. G. Mlostoń and H. Heimgartner, *Helv. Chim. Acta*, **1992**, *75*, 1825.

215. M. Kägi, A. Linden, H. Heimgartner, and G. Mlostoń, *Helv. Chim. Acta*, **1993**, *76*, 1715.

216. G. Mlostoń, M. Petit, A. Linden, and H. Heimgartner, *Helv. Chim. Acta*, **1994**, *77*, 435.

217. M. Petit, A. Linden, G. Mlostoń, and H. Heimgartner, *Helv. Chim. Acta*, **1994**, *77*, 1076.

218. M. Kägi, G. Mlostoń, A. Linden, and H. Heimgartner, *Helv. Chim. Acta*, **1994**, *77*, 1299.

219. M. Kägi, A. Linden, G. Mlostoń, and H. Heimgartner, *Helv. Chim. Acta*, **1996**, *79*, 8955.

220. M. Kägi, A. Linden, G. Mlostoń, and H. Heimgartner, *Helv. Chim. Acta*, **1998**, *81*, 285.

221. B. Kelmendi, G. Mlostoń, and H. Heimgartner, *Heterocycles*, **2000**, *52*, 475.

222. G. Mlostoń, J. Romanski, A. Linden, and H. Heimgartner, *Helv. Chim. Acta*, **1997**, *80*, 1992.

223. M. Kägi, G. Mlostoń, and H. Heimgartner, *Polish J. Chem.*, **1998**, *72*, 678.

224. J. Romański, G. Mlostoń, A. Linden, and H. Heimgartner, *Pol. J. Chem.*, **1999**, *73*, 475.

225. D. H. R. Barton and B. J. Willis, *J. Chem. Soc., Perkin Trans. 1*, **1972**, 305.

226. D. H. R. Barton, F. S. Guziec, Jr., and I. Shanak, *J. Chem. Soc., Perkin Trans. 1*, **1974**, 1794.

227. A. Krebs, W. Rüger, B. Ziegenhagen, M. Hebold, I. Hardtke, R. Müller, M. Schütz, M. Wietzke, and M. Wilke, *Chem. Ber.*, **1984**, *117*, 277.

228. A. Krebs, B. Kaletta, W.-U. Nickel, W. Rüger, and L. Tikwe, *Tetrahedron*, **1986**, *42*, 1693.

229. Review: F. S. Guziec, Jr., and L. J. SanFilippo, *Tetrahedron*, **1988**, *44*, 6241.

230. F. Laduron, C. Nyns, Z. Janousek, and H. G. Viehe, *J. Prakt. Chem.*, **1997**, *339*, 697.

231. (a) A. Martvoń, L. Floch, S. Sekretár, *Tetrahedron,*, **1978**, *34*, 453. (b) G. Mlostoń, T. Gendek, A. Linden, and H. Heimgartner, *Helv. Chim. Acta*, **1998**, *81*, 66.

232. B. F. Bonini, M. Comes-Franchini, M. Fochi, G. Mazzanti, A. Ricci, and G. Varchi, *Tetrahedron Lett.*, **1999**, *40*, 6473.

233. T. G. Back, D. H. R. Barton, M. R. Britten-Kelly, and F. S. Guziec, Jr., *J. Chem. Soc., Chem. Commun.*, **1975**, 539; *J. Chem. Soc., Perkin Trans. 1*, **1976**, 2079.

234. F. S. Guziec, Jr., and C. J. Murphy, *J. Org. Chem.*, **1980**, *45*, 2890.

235. E. R. Cullen, F. S. Guziec, Jr., and C. J. Murphy, *J. Org. Chem.*, **1982**, *47*, 3563. (b) F. S. Guziec Jr., L. J. SanFilippo, C. J. Murphy, C. A. Moustakis, and E. R. Cullen, *Tetrahedron*, **1985**, *41*, 4843.

236. K. Okuma, K. Kojima, and H. Ohta, *Phosphorus, Sulfur, Silicon*, **1993**, *80*, 259.

237. (a) M. Minoura, T. Kawashima, N. Tokitoh, and R. Okazaki, *Tetrahedron*, **1997**, *53*, 8137. (b) R. Okazaki, M. Minoura, and T. Kawashima, *Chem. Lett.*, **1993**, 1047.

238. E. Niecke, in *Multiple Bonds and Low Coordination in Phosphorus Chemistry*, M. Regitz and O. J. Scherer, eds., Thieme, Stuttgart, **1990**, pp. 293–320.

239. V. Romanenko, A. O. Gudima, A. N. Chernega, J.-M. Sotiropoulos, G. Alcaraz, and G. Bertrand, *Bull. Soc. Chim. Fr.*, **1994**, *131*, 748.

240. E. Niecke and W. Flick, *Angew. Chem.*, **1975**, *87*, 355; *Angew. Chem. Int. Ed. Engl.*, **1975**, *14*, 363.

241. N. Dubau-Assibat, A. Baceiredo, and G. Bertrand, *J. Org. Chem.*, **1995**, *60*, 3904.

242. S. Masamune, S. Murakami, H. Tobita, and D. J. Williams, *J. Am. Chem. Soc.*, **1983**, *105*, 7776.

243. W. Ando and T. Tsumuraya, *Organometallics*, **1988**, *7*, 1882. (b) T. Tsumuraya, S. Sato and W. Ando, *Organometallics*, **1990**, *9*, 2061.

244. H. Piana and U. Schubert, *J. Organomet. Chem.*, **1988**, *348*, C19.

245. H. Schäfer, W. Saak, and M. Weidenbruch, *Organometallics*, **1999**, *18*, 3159.

246. T. Aoyama, S. Katsuta, and T. Shioiri, *Heterocycles*, **1989**, *28*, 133.

247. T. Aoyama, T. Nakano, K. Marumo, Y. Uno, and T. Shioiri, *Synthesis*, **1991**, 1163.

248. T. Aoyama, M. Kabeya, A. Fukushima, and T. Shioiri, *Heterocycles*, **1985**, *23*, 2363.
249. T. Aoyama, M. Kabeya, A. Fukushima, and T. Shioiri, *Heterocycles*, **1985**, *23*, 2367.
250. T. Aoyama, M. Kabeya, and T. Shioiri, *Heterocycles*, **1985**, *23*, 2371.
251. T. Aoyama, Y. Iwamoto, and T. Shioiri, *Heterocycles*, **1986**, *24*, 589.
252. D. Mayer and G. Maas, *Zh. Org. Khim.*, **1994**, *30*, 1391.
253. G. Suchar and P. Kristian, *Chem. Zvesti*, **1985**, *29*, 422; *Chem. Abstr.*, **1975**, *83*, 97149e.
254. G. L'abbé, J.-P. Dekerk, C. Martens, and S. Toppet, *J. Org. Chem.*, **1980**, *45*, 4371.
255. Y. Zhou and H. Heimgartner, *Helv. Chim. Acta*, **2000**, *83*, 539.
256. G. S. Zaitseva, A. N. Kisin, E. N. Fedorenko, V. M. Nosova, L. I. Livantsova, and Y. I. Baukov, *Zh. Obshch. Khim.*, **1987**, *57*, 22049.
257. G. S. Zaitseva, I. F. Lutsenko, A. V. Kisin, Y. I. Baukov, and J. Lorberth, *J. Organomet. Chem.*, **1988**, *345*, 253.
258. C. S. Frampton, M. W. Majchrzak, and J. Warkentin, *Can. J. Chem.*, **1991**, *69*, 373.
259. G. Maier, D. Volz, and J. Neudert, *Synthesis*, **1992**, 561.
260. A. Padwa, M. Woods Wannamaker, *Tetrahedron*, **1990**, *46*, 1145.
261. T. G. Back, R. J. Bethell, M. Parvez, J. A. Taylor, and D. Wehrli, *J. Org. Chem.*, **1999**, *64*, 7426.
262. D. Buorissou and G. Bertrand, *Tetrahedron Lett.*, **1999**, *40*, 883.
263. M. A. Kuznetsov, Y. V. Dorofeeva, V. V. Semenovskii, V. A. Gindin, and A. N. Studenikov, *Tetrahedron*, **1992**, *48*, 1269.
264. L. Sibous and A. E. Tipping, *J. Fluorine Chem.*, **1993**, *62*, 39.
265. G. Maas, M. Regitz, U. Moll, R. Rahm, F. Krebs, R. Hector, P. J. Stang, C. M. Crittell, and B. L. Williamson, *Tetrahedron*, **1992**, *48*, 3527.
266. P. J. Stang and P. Murch, *Tetrahedron Lett.*, **1997**, *38*, 8793.
267. S. Tajammal and A. E. Tipping, *J. Fluorine Chem.*, **1990**, *47*, 45.
268. J. Cyrener, C. Lauterbach, and K. Burger, *J. Fluorine Chem.*, **1996**, *78*, 55.
269. K. S. Chan and W. D. Wulff, *J. Am. Chem. Soc.*, **1986**, *108*, 5229.
270. A. V. Sukhotin, V. G. Kartsev, K. I. Kobrakov, and A. B. Serov, *Khim. Geterotsikl. Soedin*, **1995**, 176.
271. V. V. Linev, A. F. Kolomiets, and A. V. Fokin, *Izv. Akad. Nauk SSSR, Ser. Khim.*, **1991**, 1589.
272. V. V. Linev, A. F. Kolomiets, and A. V. Fokin, *Izv. Akad. Nauk SSSR, Ser. Khim.*, **1992**, 452.
273. E. A. Shapiro, A. B. Dyatkin, and O. M. Nefedov, *Izv. Akad. Nauk SSSR, Ser. Khim.*, **1991**, 1051.
274. A. Padwa, U. Chiacchio, D. J. Fairfax, J. M. Kassir, A. Litrico, M. A. Semones, and S. L. Xu, *J. Org. Chem.*, **1993**, *58*, 6429.
275. T. Aoyama, K. Sudo, and T. Shioiri, *Chem. Pharm. Bull.*, **1982**, *30*, 3849.
276. S. Danoun, G. Baziard-Mouysset, J.-L. Stigliani, G. Commenges, A. Carpy, and M. Payard, *Bull. Soc. Chim. Fr.*, **1995**, *132*, 943.
277. R. Huisgen and H.-J. Koch, *Liebigs Ann. Chem.*, **1955**, *591*, 200.
278. F. Bronberger and R. Huisgen, *Tetrahedron Lett.*, **1984**, *25*, 65.
279. G. Maas, H. Gümbel, G. Weise, and M. Regitz, *Chem. Ber.*, **1985**, *118*, 2105.
280. G. Zinner, G. Beck, W. P. Fehlhammer, and N. Wiberg, *J. Organomet. Chem.*, **1989**, *368*, 23.
281. A. Fronda, F. Krebs, B. Daucher, T. Werle, and G. Maas, *J. Organomet. Chem.*, **1992**, *424*, 253.
282. C. Charrier, N. Maigrot, L. Ricard, P. le Floch, and F. Mathey, *Angew. Chem.*, **1996**, *108*, 2282; *Angew. Chem. Int. Ed. Engl.*, **1996**, *35*, 2133.
283. G. Becker, G. Gresser, and W. Uhl, *Z. Naturforsch.*, **1981**, *36b*, 16.
284. Review: M. Regitz, and P. Binger, *Angew. Chem.*, **1988**, *100*, 1541; *Angew. Chem. Int. Ed. Engl.*, **1988**, *27*, 1484.

285. M. Regitz and P. Binger, in *Multiple Bonds and Low Coordination in Phosphorus Chemistry*, M. Regitz and O. J. Scherer, eds., Thieme, Stuttgart, **1990**, pp. 58–111.

286. Review: M. Regitz, *Chem. Rev.*, **1990**, *90*, 191.

287. W. Schnurr and M. Regitz, *Synthesis*, **1989**, 511.

288. O. Wagner, G. Maas, and M. Regitz, *Angew. Chem.*, **1987**, *99*, 1328; *Angew. Chem. Int. Ed. Engl.*, **1987**, *26*, 1257.

289. W. Rösch, U. Hees, and M. Regitz, *Chem. Ber.*, **1987**, *120*, 1645.

290. M. Sanchez, R. Réau, C. J. Marsden, M. Regitz, and G. Bertrand, *Chem. Eur. J.*, **1999**, 274.

291. A. H. Cowley, S. W. Hall, R. A. Jones, and C. M. Nunn, *J. Chem. Soc., Chem. Commun.*, **1988**, 867.

292. W. Rösch and M. Regitz, *Angew. Chem.*, **1984**, *96*, 898; *Angew. Chem. Int. Ed. Engl.*, **1984**, *23*, 900.

293. T. Allspach, M. Regitz, G. Becker, and W. Becker, *Synthesis*, **1986**, 31.

294. W. Rösch, U. Vogelbacher, T. Allspach, and M. Regitz, *J. Organomet. Chem.*, **1986**, *306*, 39.

295. W. Fiedler, M. Regitz, and G. Bertrand, *Bull. Soc. Chim. Belg.*, **1997**, *106*, 455.

296. A. Mack, E. Pierron, T. Allspach, U. Bergsträsser, and M. Regitz, *Synthesis*, **1998**, 1305.

297. E. P. O. Fuchs, M. Hermesdorf, W. Schnurr, W. Rösch, H. Heydt, and M. Regitz, *J. Organomet. Chem.*, **1988**, *338*, 329.

298. M. Hermesdorf, M. Birkel, H. Heydt, M. Regitz, and P. Binger, *Phosphorus, Sulfur, Silicon*, **1989**, *46*, 31.

299. L. Nyulászi, P. Várnai, W. Eisfeld, and M. Regitz, *J. Comput. Chem.*, **1997**, *18*, 609.

300. H. Keller and M. Regitz, *Tetrahedron Lett.*, **1988**, *29*, 925.

301. J. Grobe, D. Le Van, M. Hegemann, B. Krebs, and M. Läge, *Chem. Ber.*, **1992**, *125*, 411.

302. J. Grobe, D. Le Van, F. Immel, B. Krebs, and M. Läge, *Chem. Ber.*, **1996**, *129*, 1271.

303. P. A. Wade, in *Comprehensive Organic Synthesis*, Vol. 4, B. M. Trost, I. Fleming, and M. F. Semmelhack, eds., Pergamon Press, Oxford, **1991**, pp. 1151–1157.

304. A. Padwa, in *1,3-Dipolar Cycloaddition Chemistry*, A. Padwa, ed., Wiley, New York, **1984**, pp. 304–316.

305. A. Padwa and H. Ku, *J. Org. Chem.*, **1980**, *45*, 3756.

306. T. Miyashi, Y. Nishizawa, Y. Fuji, K. Yamakawa, M. Kamata, S. Akao, and T. Mukai, *J. Am. Chem. Soc.*, **1986**, *108*, 1617.

307. U. H. Brinker, T. Schrievers, and L. Xu, *J. Am. Chem. Soc.*, **1990**, *23*, 8609.

308. A. Padwa, L. Precedo, and M. A. Semones, *J. Org. Chem.*, **1999**, *64*, 4079.

309. M. E. Jung and A. Huang, *Org. Lett.*, **2000**, *2*, 2659.

310. H. Sturm, K. H. Ongania, J. J. Daly, and W. Klötzer, *Chem. Ber.*, **1981**, *114*, 190.

311. T. Minami, M. Kamitamari, T. Utsunomiya, T. Tanaka, and J. Ichikawa, *Bull. Chem. Soc. Jpn.*, **1993**, *66*, 1496.

312. M. P. Doyle and A. V. Kalinin, *J. Org. Chem.*, **1996**, *61*, 2179.

313. G. Maas, F. Krebs, T. Werle, V. Gettwert, and R. Striegler, *Eur. J. Org. Chem.*, **1999**, 1939.

314. A. G. Schultz and S. Puig, *J. Org. Chem.*, **1985**, *50*, 915.

315. J. T. Sharp, P. Wilson, S. Parsons, and R. O. Gould, *J. Chem. Soc., Perkin Trans. 1*, **2000**, 1139.

316. M. Böhshar, H. Heydt, G. Maas, H. Gümbel, and M. Regitz, *Angew. Chem.*, **1985**, *97*, 571; *Angew. Chem. Int. Ed. Engl.*, **1985**, *24*, 597.

317. M. Regitz, M. Böhshar, S. Arenz, and H. Heydt, *Chem. Ber.*, **1989**, *122*, 565.

318. W. M. F. Fabian, V. A. Bakulev, and C. O. Kappe, *J. Org. Chem.*, **1998**, *63*, 5801.

319. J. A. Pincock and N. C. Mathur, *J. Org. Chem.*, **1982**, *47*, 3699.

320. J. A. Pincock and K. P. Murray, *Canad. J. Chem.*, **1979**, *57*, 1403.

321. W. Schnurr and M. Regitz, *Synthesis*, **1989**, 511.

322. A. Igau, A. Baceiredo, and G. Bertrand, *New. J. Chem.*, **1989**, *13*, 359.

323. Y. Nishizawa, T. Miyashi, and T. Mukai, *J. Am. Chem. Soc.*, **1980**, *102*, 1176.

324. T. Miyashi, Y. Nishizawa, Y. Fujii, and T. Mukai, *Heterocycles*, **1981**, *16*, 147.

325. T. Miyashi, Y. Nishizawa, Y. Fujii, K. Yamakawa, M. Kamata, S. Akao, and T. Mukai, *J. Am. Chem. Soc.*, **1986**, *108*, 1617.

326. A. Padwa and H. Ku, *Tetrahedron Lett.*, **1980**, *21*, 1009.

327. A. Padwa, A. Rodriguez, M. Tohidi, and T. Fukunaga, *J. Am. Chem. Soc.*, **1983**, *108*, 933.

328. A. Padwa and M. Tohidi, *J. Chem. Soc., Chem. Commun.*, **1984**, 295.

329. Review: G. Zecchi, *Synthesis*, **1991**, 181.

330. Review: P. W. Groundwater and M. Nyerges, *Adv. Heterocycl. Chem.*, **1999**, *73*, 97.

331. I. R. Robertson and J. T. Sharp, *J. Chem. Soc., Chem. Commun.*, **1983**, 1003.

332. I. R. Robertson and J. T. Sharp, *Tetrahedron*, **1984**, *40*, 3095.

333. D. P. Munro and J. T. Sharp, *Tetrahedron Lett.*, **1980**, *21*, 4109.

334. K. L. M. Stanley, J. Dingwall, J. T. Sharp, and T. W. Naisby, *J. Chem. Soc., Perkin Trans. 1*, **1979**, 1433.

335. T. K. Miller, J. T. Sharp, G. J. Thomas, and I. Thompson, *Tetrahedron Lett.*, **1981**, *22*, 1537.

336. T. K. Miller and J. T. Sharp, *J. Chem. Soc., Perkin Trans. 1*, **1984**, 223.

337. K.-L. Hoffmann, G. Maas, and M. Regitz, *J. Org. Chem.*, **1987**, *52*, 3851.

338. J. P. Mykytka and W. M. Jones, *J. Am. Chem. Soc.*, **1975**, *97*, 5933.

339. A. S. Kende and M. Journet, *Tetrahedron Lett.*, **1995**, *36*, 3087.

340. G. Maas and V. Gettwert, *Tetrahedron*, **2000**, *56*, 4139.

341. G. Maas, V. Gettwert, F. Krebs, and G. Schmidtberg, *Chem. Eur. J.*, **2000**, *6*, 1646.

342. D. L. Wright and M. C. McMills, *Org. Lett.*, **1999**, *1*, 667.

343. W. Ando, H. Yoshida, K. Kurishima, and M. Sugiyama, *J. Am. Chem. Soc.*, **1991**, *113*, 7790.

344. W. Ando, M. Sugiyama, T. Suzuki, C. Kato, Y. Arakawa, and Y. Kabe, *J. Organomet. Chem.*, **1995**, *499*, 99.

345. Review: M. P. Doyle, *Chem. Rev.*, **1986**, *86*, 919.

346. Review: G. Maas, *Top. Curr. Chem.*, **1987**, *137*, 75.

347. Reviews: (a) E. Wenkert, *Acc. Chem. Res.*, **1980**, *13*, 27. (b) E. Wenkert, *Heterocycles*, **1980**, *14*, 1703.

348. M. E. Alonso, A. Morales, and A. W. Chitty, *J. Org. Chem.*, **1982**, *47*, 3747.

349. M. E. Alonso, P. Jano, M. I. Hernandez, A. S. Greenberg, and E. Wenkert, *J. Org. Chem.*, **1983**, *48*, 3047.

350. E. Wenkert, T. P. Ananthanarayan, V. F. Ferreira, M. G. Hoffmann, and H. S. Kim, *J. Org. Chem.*, **1990**, *55*, 4975.

351. M. C. Pirrung, J. Zhang, and A. T. McPhail, *J. Org. Chem.*, **1991**, *56*, 6269.

352. M. C. Pirrung and Y. R. Lee, *J. Chem. Soc., Chem. Commun.*, **1995**, 673.

353. M. C. Pirrung, J. Zhang, K. Lackey, D. D. Sternbach, and F. Brown, *J. Org. Chem.*, **1995**, *60*, 2112.

354. M. C. Pirrung and Y. R. Lee, *Tetrahedron Lett.*, **1994**, *35*, 6231.

355. M. C. Pirrung, J. Zhang, and A. T. Morehead, *Tetrahedron Lett.*, **1994**, *35*, 6229.

356. P. V. Murphy, T. J. O'Sullivan, B. D. Kennedy, and N. W. A. Geraghty, *Perkin Trans. 1*, **2000**, 2126.

357. M. C. Pirrung and F. Blume, *J. Org. Chem.*, **1999**, *64*, 3642.

358. Y. R. Lee, J. Y. Suk, and B. S. Kim, *Tetrahedron Lett.*, **1999**, *40*, 6603.

359. M. C. Pirrung and J. Zhang, *Tetrahedron Lett.*, **1992**, *40*, 5987.

360. M. C. Pirrung and Y. R. Lee, *J. Am. Chem. Soc.*, **1995**, *117*, 4814.

361. M. C. Pirrung and Y. R. Lee, *Tetrahedron Lett.*, **1996**, *37*, 2391.

362. R. Huisgen, H. S. Sturm, and G. Binsch, *Chem. Ber.*, **1964**, *97*, 2864.

363. M. I. Komendantov, V. N. Novinskii, and R. R. Bekmukhametov, *Zh. Org. Khim.*, **1973**, *9*, 427 engl. transl. 431.

364. M. P. Doyle, M. Oppenhuizen, R. C. Elliott, and M. R. Boelkins, *Tetrahedron Lett.*, **1978**, *26*, 2247.

365. T. Ibata and R. Sato, *Bull. Chem. Soc. Jpn.*, **1979**, *52*, 3597.

366. P. G. Moniotte, A. J. Hubert, and P. Teyssié, *J. Organomet. Chem.*, **1975**, *88*, 115.

367. R. Connell, F. Scavo, P. Helquist, and B. Åkermark, *Tetrahedron Lett.*, **1986**, *27*, 5559.

368. Reference 6, pp. 593–594 and 630–633.

369. G. Maas, in *Houben-Weyl, Methoden der Organischen Chemie*, Vol. E19b, M. Regitz, ed., Part 2 Thieme, Germany, **1989**, pp. 1119–1122.

370. K. J. Doyle and C. J. Moody, *Synthesis*, **1994**, 1021.

371. J. P. Konopelski, J. M. Hottenroth, H. Monzó Ottra, E. A. Véliz, and Z.-C. Yang, *Synlett*, **1996**, 609.

372. K. Fukushima and T. Ibata, *Heterocycles*, **1995**, *40*, 149.

373. R. D. Connell, M. Tebbe, P. Helquist, and B. Åkermark, *Tetrahedron Lett.*, **1991**, *32*, 17.

374. M. Alt and G. Maas, *Tetrahedron*, **1994**, *50*, 7435.

375. P. C. Ducept and S. P. Marsden, *Synlett*, **2000**, 692.

376. Y. Wang and S. Zhu, *J. Fluorine Chem.*, **2000**, *103*, 139.

377. S. G. Ruf, A. Mack, J. Steinbach, U. Bergsträsser, and M. Regitz, *Synthesis*, **2000**, 360.

CHAPTER 9

Azides

Chin-Kang Sha and A. K. Mohanakrishnan

Department of Chemistry, National Tsing Hua University, Taiwan, Republic of China

9.1. Synthesis of Heterocycles via 1,3-Dipolar Cycloadditions of Azides . . . 624
9.1.1. Reaction with Alkenes . . . 624
9.1.2. Reaction with the Double Bond of Dienes and Conjugated Systems . . . 637
9.1.3. Reaction with Alkynes and Nitriles . . . 639
9.1.4. Reaction with Allylic Carbocations . . . 644
9.1.5. Reaction with Phosphorus Ylides . . . 645
9.1.6. Reaction with Fullerenes . . . 646
9.2. Synthesis of Natural Products, Intermediates, and Their Analogues via 1,3-Dipolar Cycloaddition of Azides . . . 649
9.2.1. Reaction with Alkenes . . . 649
9.2.2. Reaction with the Double Bond of Dienes . . . 664
9.2.3. Reaction with Alkynes and Nitriles . . . 672
9.3. Conclusion . . . 676

Organic azides belong to the propargyl–allenyl category of dipoles, and are popular for synthetic transformations because of their ready availability (Scheme 9.1) (1).

Since the discovery of triazole formation from phenyl azide and dimethyl acetylenedicarboxylate in 1893, synthetic applications of azides as 1,3-dipoles for the construction of heterocyclic frameworks and core structures of natural products have progressed steadily. As the 1,3-dipolar cycloaddition of azides was comprehensively reviewed in the 1984 edition of this book (2), in this chapter we recount developments of 1,3-dipolar cycloaddition reactions of azides from 1984 to 2000, with an emphasis on the synthesis of not only heterocycles but also complex natural products, intermediates, and analogues.

The Chemistry of Heterocyclic Compounds, Volume 59: Synthetic Applications of 1,3-Dipolar Cycloaddition Chemistry Toward Heterocycles and Natural Products, Edited by Albert Padwa and William H. Pearson.
ISBN 0-471-38726-6

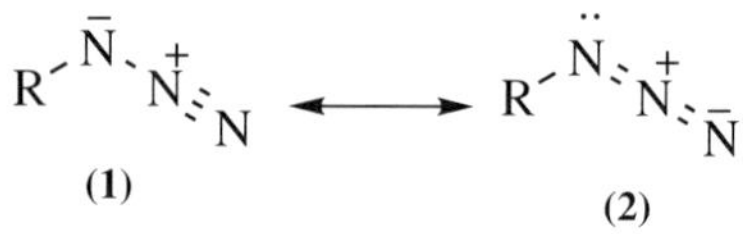

Scheme 9.1

9.1. SYNTHESIS OF HETEROCYCLES VIA 1,3-DIPOLAR CYCLOADDITIONS OF AZIDES

9.1.1. Reaction with Alkenes

In 1984, a facile synthesis of pyrrolo[3,4-*b*]indole (**5**) as a stable indole-2,3-quinodimethane analogue using an intramolecular azide–alkene cycloaddition–cycloreversion strategy was reported (Scheme 9.2) (3). Treatment of bromo compound **3** with NaN_3 in aqueous tetrahydrofuran (THF) produced the triazoline **4** via an intramolecular 1,3-dipolar cycloaddition of an intermediate azide. Treatment of the triazoline **4** with *p*-toluenesulfonic acid (*p*-TSA) effected 1,3-dipolar cycloreversion of **4** to give pyrroloindole **5** in 82% yield along with diethyl diazomalonate.

This method was extended to the synthesis of not only pyrrolo[3,4-*c*]pyridine (**8a**) and pyrrolo[3,4-*b*]pyridine (**8b**) (4a), but also the salt of several labile heteroaromatic pyrroles **9–11** (4b) (Scheme 9.3).

This method was also applied to preparation of thieno[3,4-*c*]pyrroles, including the first preparation of the parent compound (Scheme 9.4) (5a,b). When treated with NaN_3, the bromo compounds (**12**) underwent 1,3-dipolar cycloaddition to yield the triazolines **13**. 1,3-Dipolar cycloreversion gave the salts of the thieno[3,4-*c*]pyrroles

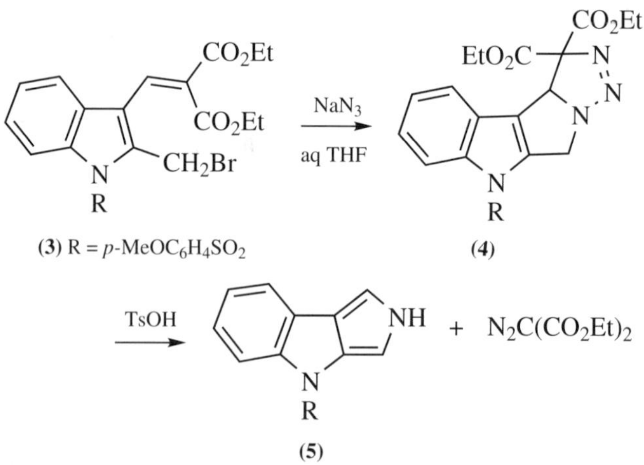

Scheme 9.2

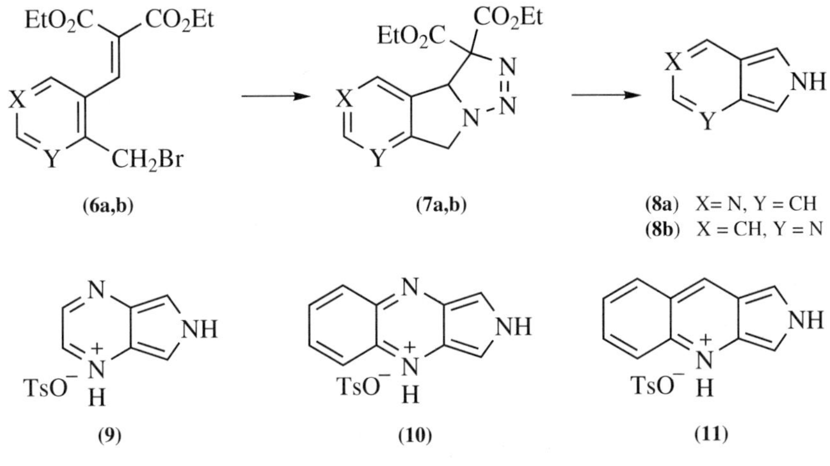

(6a,b) **(7a,b)** **(8a)** X = N, Y = CH **(8b)** X = CH, Y = N

(9) **(10)** **(11)**

Ts = tosyl

Scheme 9.3

14. Neutralization with sodium carbonate in an nuclear magnetic resonance (NMR) tube allowed detection of the highly labile parent compound **15** ($R^1 = R^2 = H$).

Synthesis of the parent system of benz[*de*]isoquinolinium-1-ide (**21**) was achieved from the corresponding bromo compound **17** via the triazoline intermediate **18** using the same strategy (Scheme 9.5) (5c).

Goldsmith and Soria (6b) reported a novel approach to a synthesis of the cyclopentanoid ring system **25** based on the 1,3-dipolar cycloaddition of *p*-bromobenzenesulfonyl azides with the electron-rich 1,4-cyclohexadienol ether **22** and subsequent ring contraction at moderate temperature (Scheme 9.6) (6a).

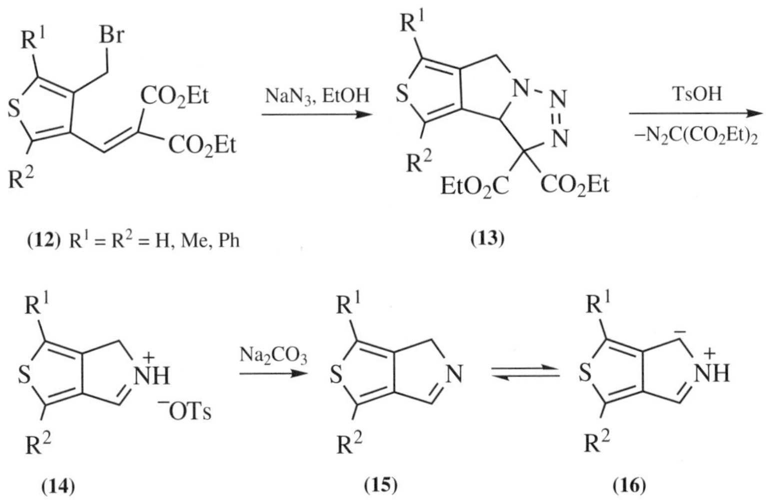

(12) $R^1 = R^2 =$ H, Me, Ph **(13)**

(14) **(15)** **(16)**

Scheme 9.4

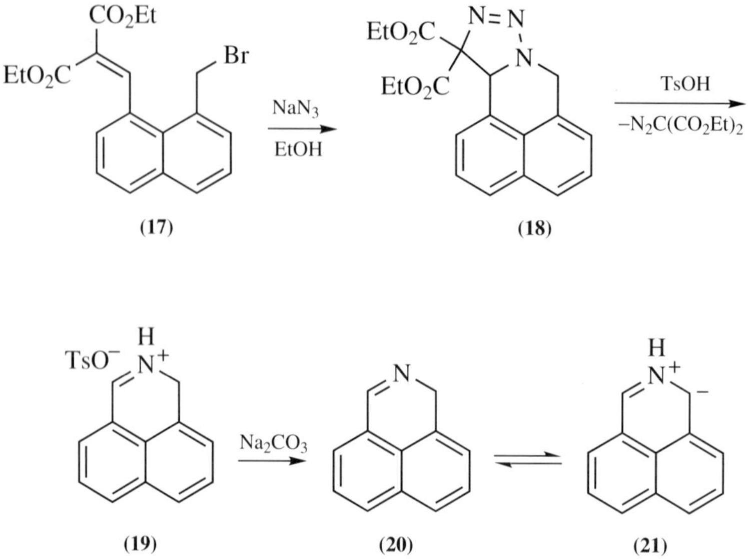

Scheme 9.5

1,3-Dipolar cycloaddition occurred preferentially at the electron-rich double bond of **22** to give the unstable triazoline **23**, which on thermolysis led to extrusion of nitrogen and rearrangement to give the cyclopentenoid compound **25**. The 1,3-dipolar cycloaddition–rearrangement sequence was subsequently extended to ultrasonic conditions.

A synthesis of the 11-aryl-11-aza[5.3.1]propellan-2-one **29** was accomplished by the intermolecular cycloaddition of the cycloheptenone **26** with an aryl azide. The

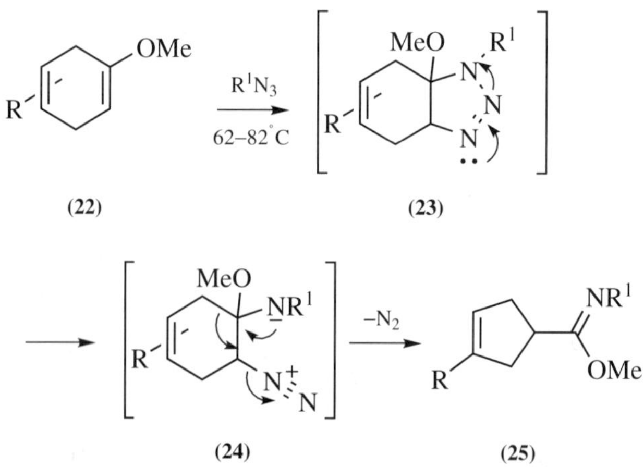

Scheme 9.6

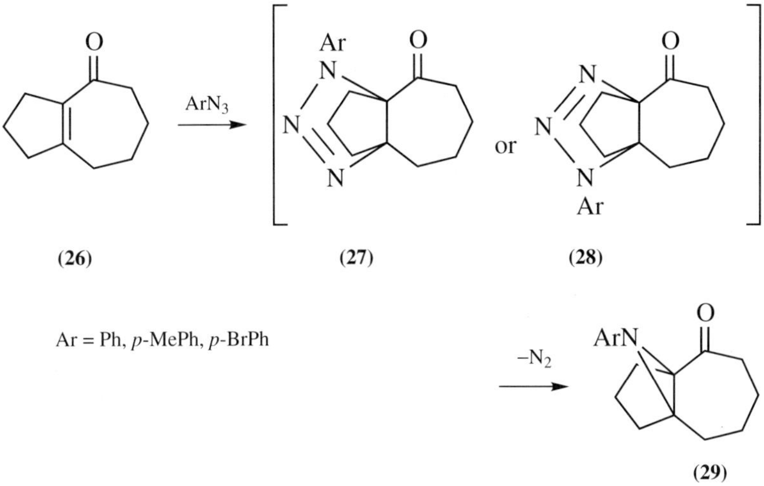

(26) (27) (28) (29)

Scheme 9.7

aziridine **29** was obtained through the intermediacy of a triazoline, either **27** or **28** (Scheme 9.7) (7).

Hassner et al. (8) reported a novel synthesis of 2,5-dihydroxyoxazoles (**32**) using an intramolecular azide–alkene cycloaddition. The ratio of reagents (aldehyde/allyl alcohol/hydrazoic acid, 1:3:9) was critical for the preparation of azide **30** (Scheme 9.8).

Junjappa and co-workers (9) reported the cycloaddition of sodium azide to the polarized ketene-(*S*,*S*)-acetal **33** to give the triazole **35**; they also reported an intermolecular cycloaddition of tosyl azide **37** with the enamine **36** to give an unstable triazoline intermediate **38**. Ring opening **38** followed by a Dimroth rearrangement afforded the triazole **41** (Scheme 9.9).

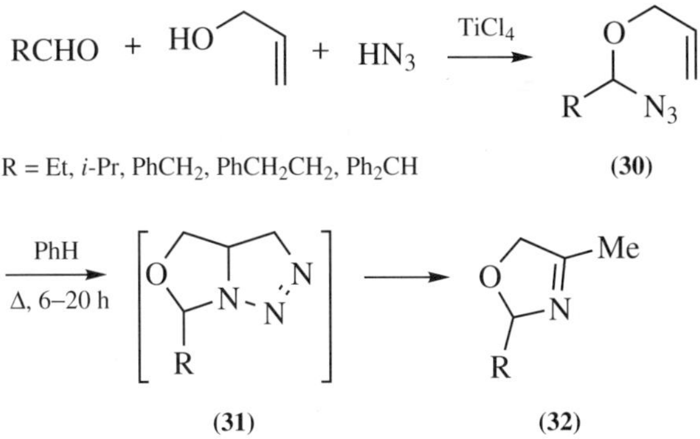

(30) (31) (32)

Scheme 9.8

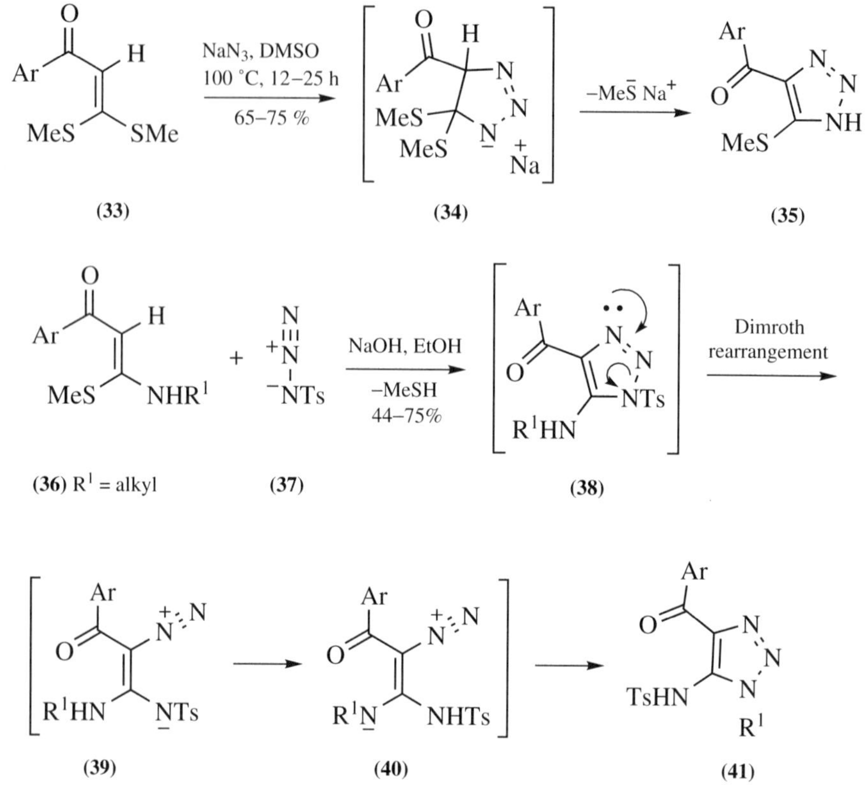

Scheme 9.9

Keshava et al. (10) reported a facile synthesis of the fused tricyclic β-lactams **44** and **45** via an intramolecular 1,3-dipolar cycloaddition of an azide with an alkene (Scheme 9.10). 1,3-Dipolar cycloaddition of the azides **43** in benzene at reflux gave a mixture of cis and trans tricyclic β-lactams **44** and **45**. As the ring size increased

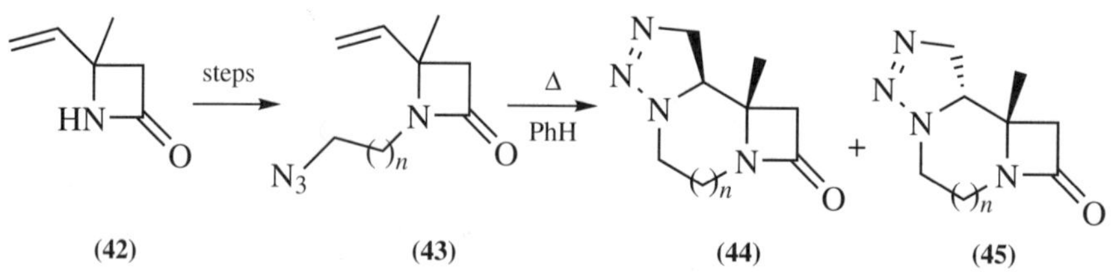

$n = 1$ only cis
$n = 2$ cis/trans (9:1)
$n = 3$ cis/trans (6:4)

Scheme 9.10

(n = 1–3), the rate and stereoselectivity of the intramolecular cycloaddition decreased.

Gallagher and co-workers (11) reported the cycloadditions of electron-deficient azides with the ketene-(*S*,*S*)-acetals (**46**) (Scheme 9.11). Reaction of *p*-toluenesulfonyl azide with **46** gave the unstable cycloadducts **47**, which underwent rearrangement to afford compounds **48** in 24–89% yield. In the case of ethyl azidoformate,

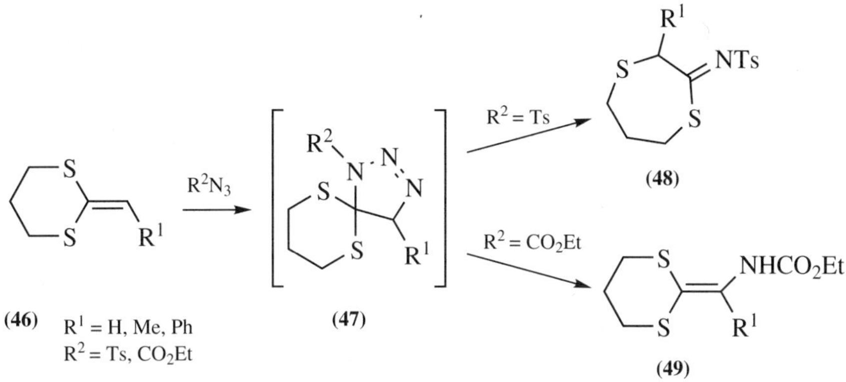

Scheme 9.11

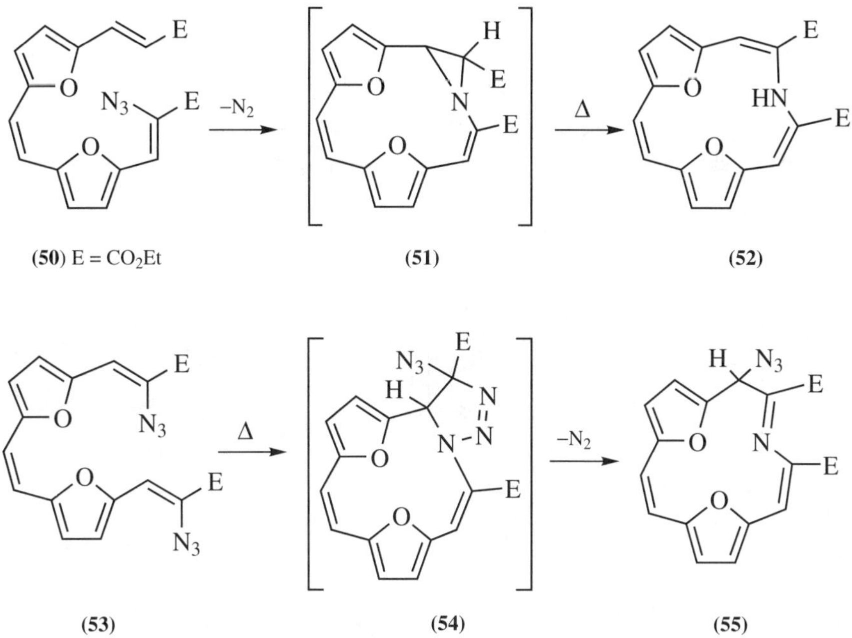

Scheme 9.12

the corresponding cycloadducts **47** led to the formation of the amino ketene-(*S*,*S*)-acetals (**49**) in 5–33% yield.

Ogawa et al. (12) used an intramolecular azide–alkene cycloaddition strategy to synthesize the oxygen-bridged aza[15]annulene **52** and the aza[15]annulene dicarboxylate **55** (Scheme 9.12). 1,3-Dipolar cycloaddition of vinyl azide to the acrylate moiety followed by extrusion of nitrogen gave the aziridine **51**. Rearrangement of **51** afforded the aza[15]annulene **52**. The same approach was used to synthesize the aza[15]annulene **55**.

An unexpected rearrangement involving an intramolecular azide–alkene cycloaddition in the morphinane series was reported by d'Angelo and co-workers (13) (Scheme 9.13). Treatment of the mesylate (**56**) with sodium azide in dimethylformamide (DMF) at 60 °C generated compound **59**. Intramolecular cycloaddition of the corresponding azide, formed from **56**, gave the intermediate triazoline **57**, which on rearrangement and extrusion of nitrogen afforded compound **59**.

Kadaba (14) reported the first example of an intermolecular cycloaddition of sodium azide with a vinyl azide (**60**) to give the tetrazole **62** in 25% yield (Scheme 9.14).

Hassner et al. (15) explored the synthetic utility of the polymeric azidation reagent **63** to prepare di- and tri-azidomethanes **64** and **66** at ambient temperature. They also studied the intermolecular cycloaddition of these azides with dimethyl acetylenedicarboxylate (DMAD) to yield the corresponding cycloadducts

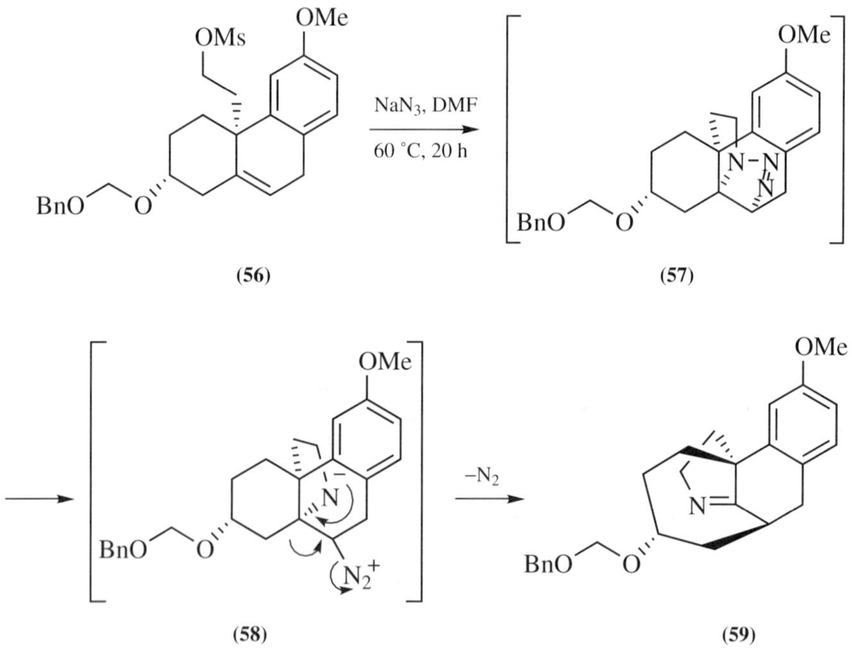

Bn = benzyl

Scheme 9.13

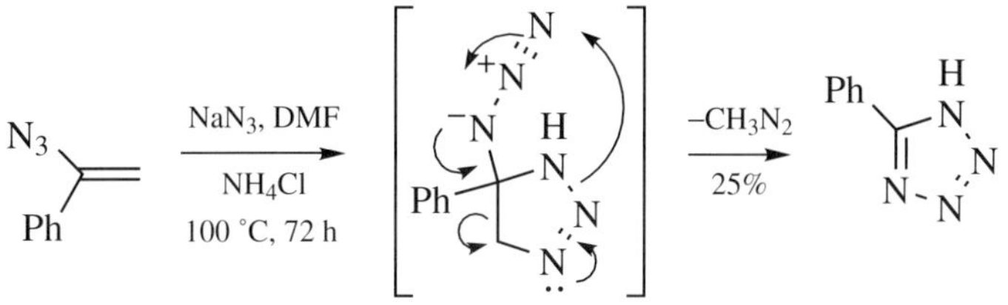

(60) **(61)** **(62)**

Scheme 9.14

65 and **68**, respectively (Scheme 9.15). In the case of triazidomethane **66**, the initial cycloadduct **67** underwent rearrangement to give compound **68**. X-ray crystallographic analyses confirmed the structures of bis(triazoles) **65** and the rearranged tristriazoles **68**.

Weinreb and co-workers (16) reported a high-pressure-induced 1,3-dipolar cycloaddition of alkyl and phenyl azides with electron-deficient alkenes at ambient temperature. As a representative example, phenyl azide underwent cycloaddition with methyl crotonate (**69**) at 12 kbar to give the triazoline **70** (43%) and the β-amino diazoester **71** (53%). The high-pressure conditions resulted in high yield and a shorter reaction time (Scheme 9.16).

Benati et al. (17) reported intermolecular cycloadditions of aryl azides with 1,4-naphthoquinone (**72**) at ambient temperature. The triazoline intermediate **73** was unstable even at room temperature, leading to the formation of a mixture of products **74–77** (Scheme 9.17).

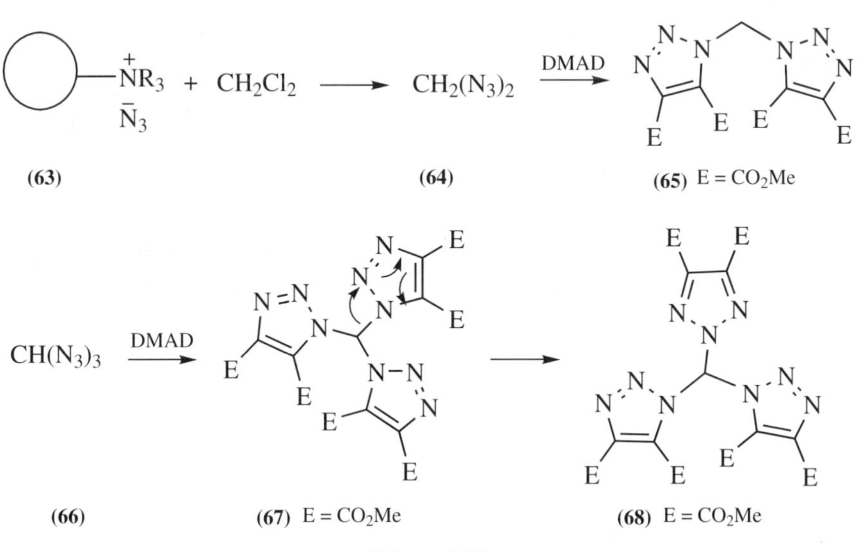

(63) **(64)** **(65)** E = CO_2Me

(66) **(67)** E = CO_2Me **(68)** E = CO_2Me

Scheme 9.15

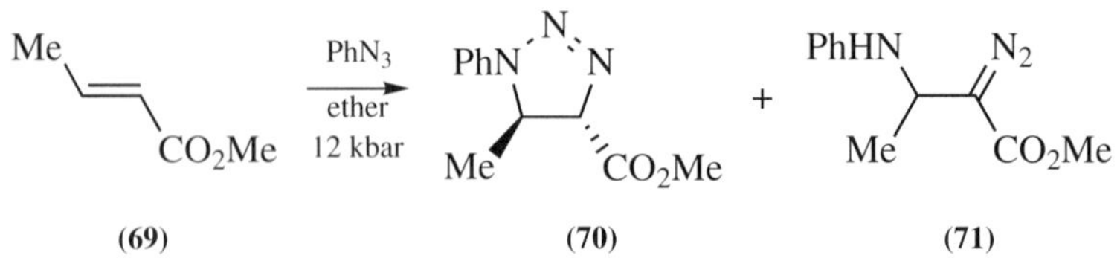

Scheme 9.16

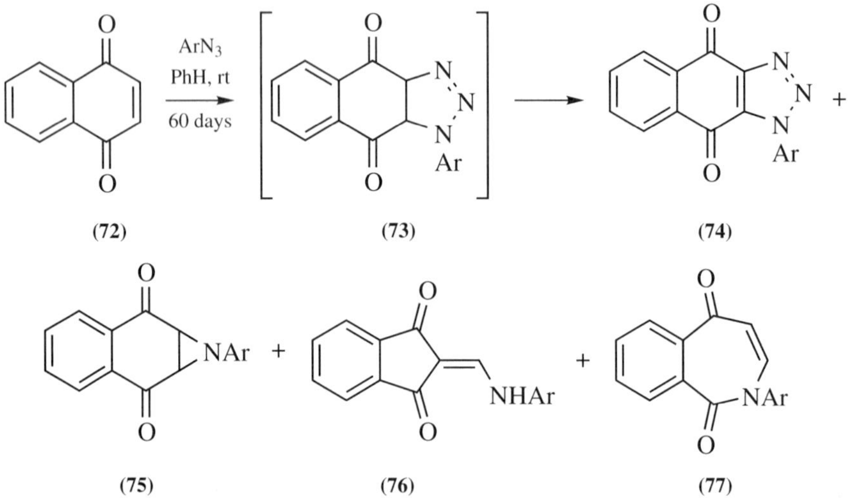

Scheme 9.17

Kadaba (18) reported that the intermolecular 1,3-dipolar cycloaddition of the aryl azides **79** with the enamides **78** in refluxing ethanol gave the triazoles **80** (Scheme 9.18).

Soufiaoui and co-workers (19) reported an intermolecular 1,3-dipolar cycloaddition of tosyl azide with the 1,2-dihydroisoquinoline **81** to give the triazoline **82**, which, on extrusion of nitrogen followed by aromatization via elimination of H_2 or

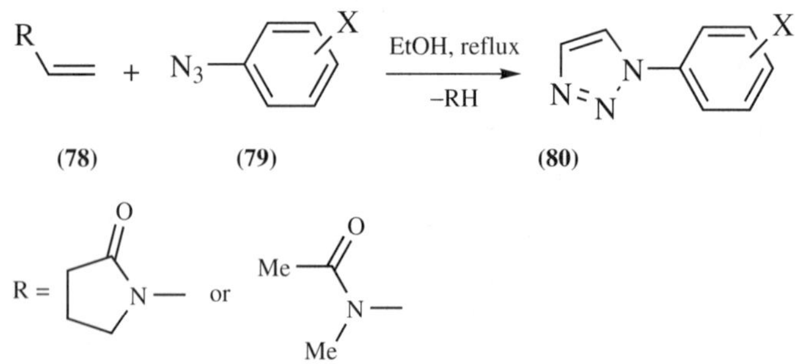

Scheme 9.18

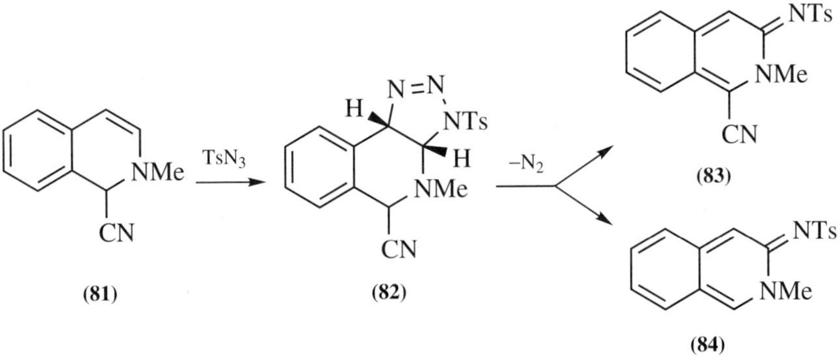

Scheme 9.19

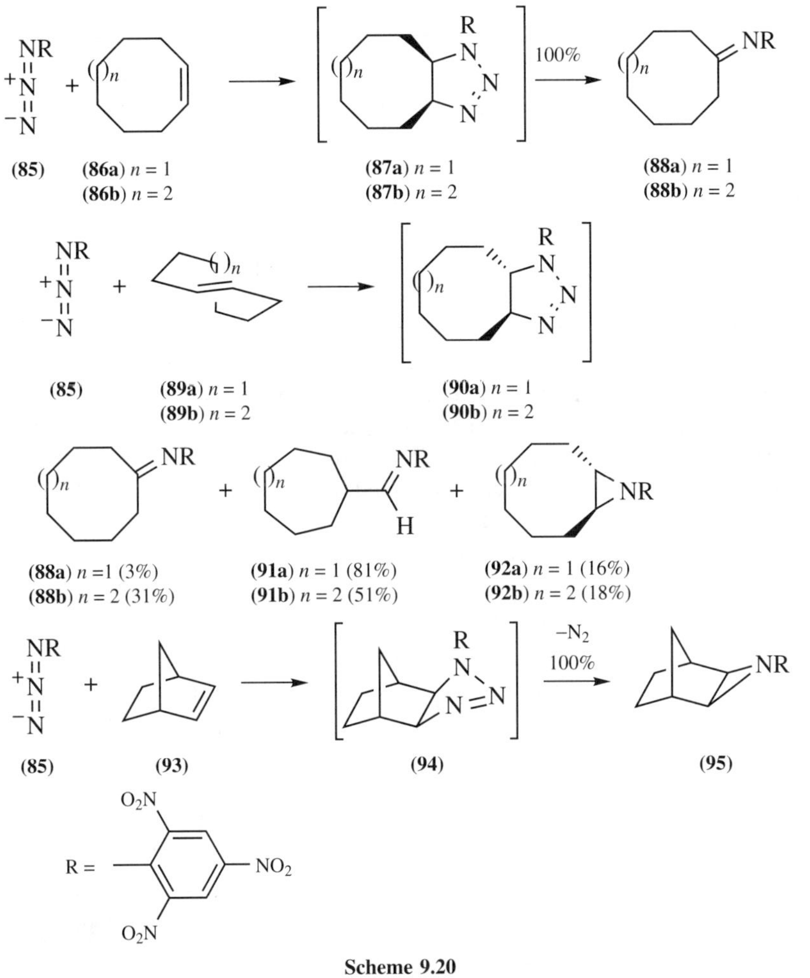

Scheme 9.20

HCN, produced a mixture of the *o*-quinodimethane analogues **83** and **84** (Scheme 9.19).

Shea and Kim (20) investigated in detail the intermolecular 1,3-dipolar cycloaddition of picryl azide (**85**) with a series of mono- and bicyclic alkenes including *trans*-cycloalkenes and bridgehead alkenes (Scheme 9.20). In the cases of *cis*-cyclooctene (**86a**) and *cis*-cyclononene (**86b**), decomposition of the initially formed cycloadducts **87a** and **87b** followed by 1,2-hydride shift gave cyclooctylidene-2,4,6-trinitroaniline (**88a**) and cyclononylidene-2,4,6-trinitroaniline (**88b**), respectively. When the cycloadditions were performed with *trans*-cylooctene (**89a**) and *trans*-cyclononene (**89b**), the corresponding cycloadducts **90a** and **90b** underwent both 1,2-hydride migration and 1,2-alkyl migration to give the ketimines **88** and aldimines **91**, respectively; aziridines **92** were also isolated in small proportions. 1,3-Dipolar cycloaddition of picryl azide to norbornene (**93**) followed by extrusion of nitrogen gave the exo aziridine **95** in good yield. Based on their results, they confirmed that strain involved in these alkenes affects both the reactivity and regiochemistry of the cycloaddition reactions.

Conformational constraints induced by various ortho-substitutents in 1-allyloxy-2-azidomethylbenzenes (**97**) were used to accelerate intramolecular cycloadditions of the azide group to alkenes (21) (Scheme 9.21). For the unsubstituted azide **96**, high temperature was required for the cycloaddition and the yield of the cycloadduct **100** was low. The monosubstituted azide **97** underwent cycloaddition in refluxing benzene in 10 h to give the cycloadduct **101** in good yield. Disubstituted azides **98** and **99** underwent 1,3-dipolar cycloaddition in 5–7 h to give the triazolines **102** and **103**.

De Kimpe and Boeykens (22) reported synthesis of the β-lactam derivatives **107** via cycloaddition of azides with 2-methyleneazetidines (**104**) (Scheme 9.22). Because of electronic control, the intermolecular cycloaddition of the azide with the enamine double bond resulted in the formation of the triazoline intermediate **105**, ring opening and rearrangement of which gave the imino lactam **107**. Although all attempts to convert compound **107** to the corresponding β-lactam **108** under acidic conditions were unsuccessful, under basic conditions compound **107** was converted into the β-amino amides **109**.

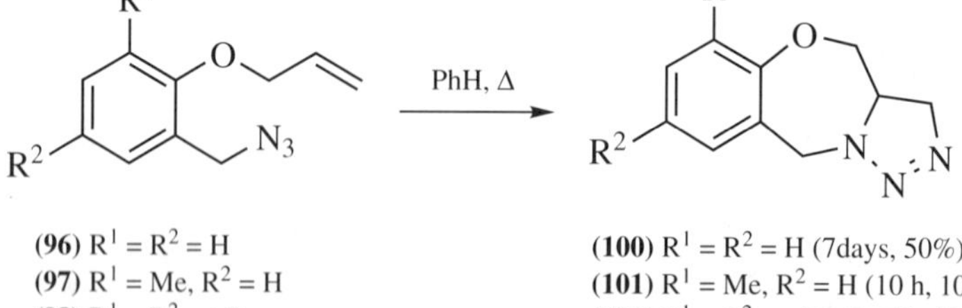

(**96**) $R^1 = R^2 = H$
(**97**) $R^1 = Me, R^2 = H$
(**98**) $R^1 = R^2 = t\text{-Bu}$
(**99**) $R^1 = R^2 = Br$

(**100**) $R^1 = R^2 = H$ (7days, 50%)
(**101**) $R^1 = Me, R^2 = H$ (10 h, 100%)
(**102**) $R^1 = R^2 = t\text{-Bu}$ (5–7 h, 100%)
(**103**) $R^1 = R^2 = Br$ (5–7 h, 100%)

Scheme 9.21

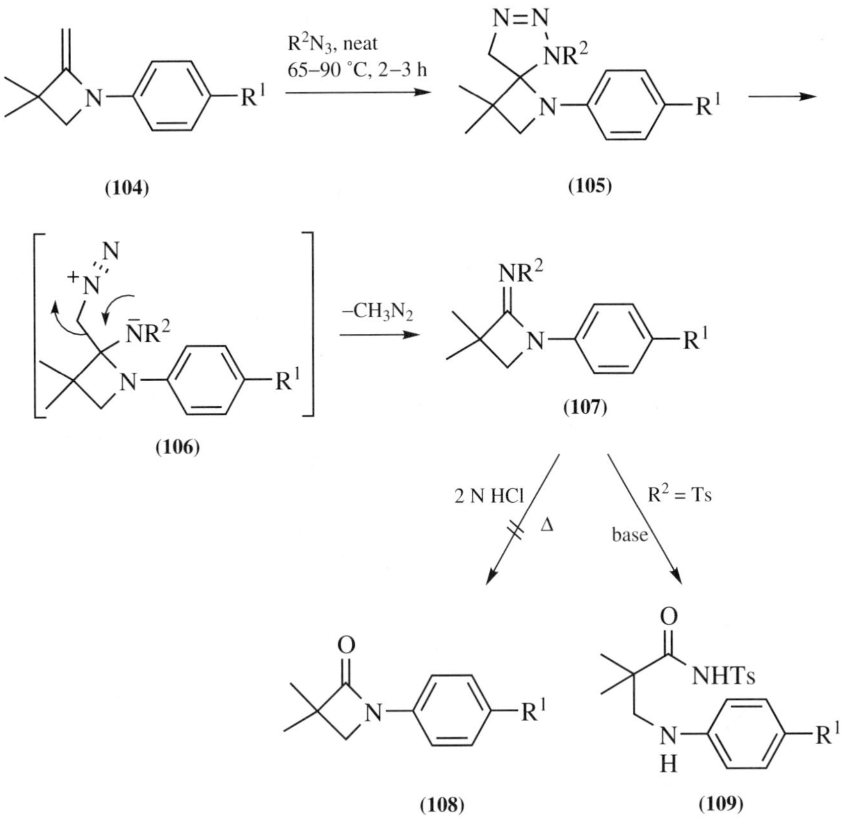

Scheme 9.22

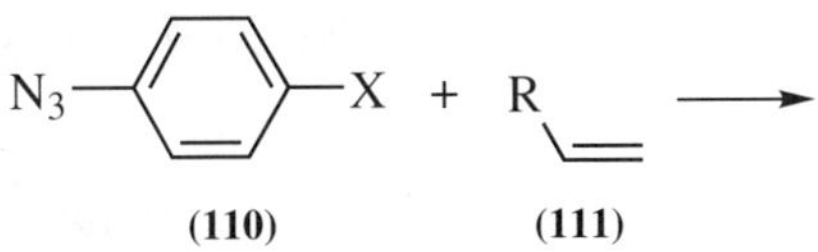

X = OMe, Me, Cl, CF_3, NO_2
R = 2-pyridyl, 4-pyridyl

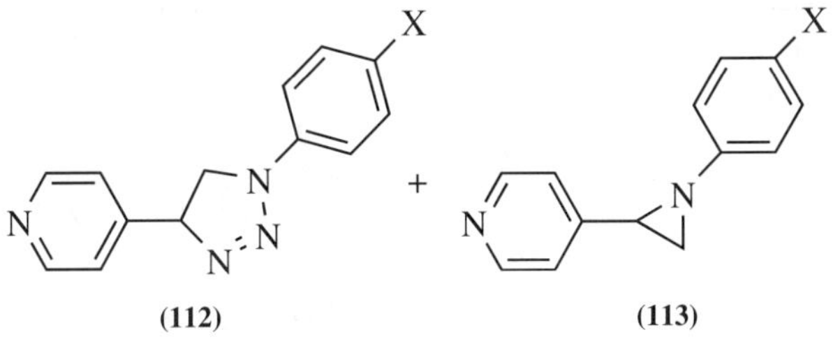

Scheme 9.23

Lin and Kadaba (23) reported the intermolecular 1,3-dipolar cycloadditions of aryl azides (**110**) with vinyl pyridines (**111**) to give a mixture of pyridyltriazolines (**112**) and aziridines (**113**) (Scheme 9.23).

The stereoselective intermolecular cycloaddition of azides to chiral cyclopentanone enamines was reported, but both product yields and enantiomeric excesses (ee) were low (24) (Scheme 9.24). Ethyl azidoformate (**115**) and *N*-mesyl azidoformamimidate (**116**) underwent 1,3-dipolar cycloaddition with the monosubstituted chiral enamine **114** to give products **117** and **118** in low yields with ee of 24 and 18%, respectively. Intermolecular cycloaddition of the *N*-mesyl azidoformamimidate **116** with the disubstituted C_2-symmetric chiral enamine **119** proceeded with good diastereoselectivity to give compound **120** in 18% yield. On cleavage of the enamine unit, compound **120** afforded **118** with low ee.

Warrener and co-workers (25) exploited a 1,3-dipolar cycloaddition reaction to synthesize a 7-azanorbornane **124** (Scheme 9.25). The cyclobutene-1,2-diester **121** underwent smooth cycloaddition with benzyl azide to give the triazoline **122**, which

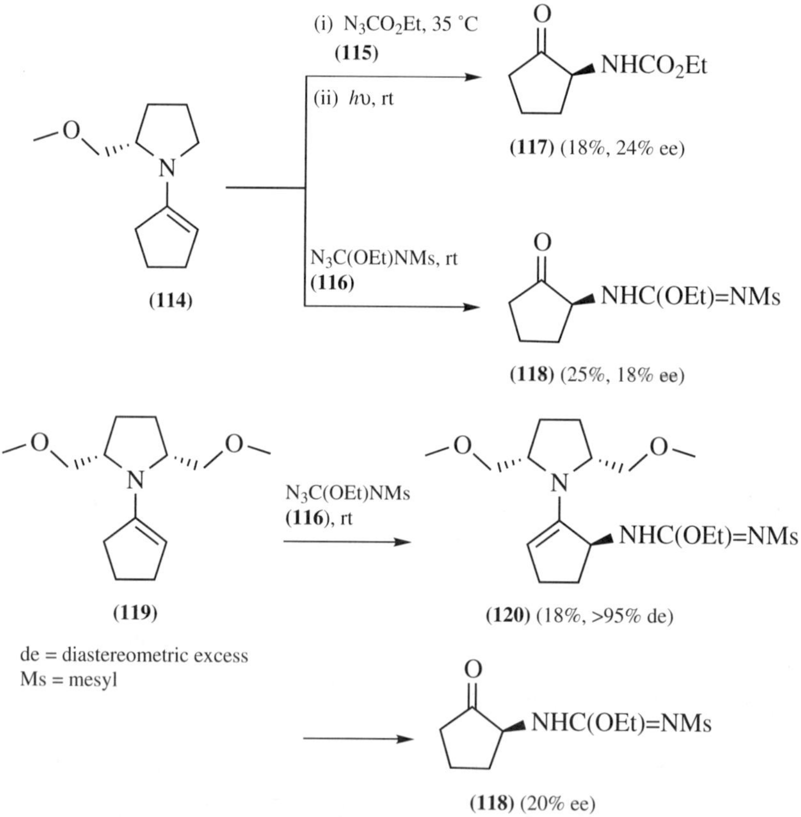

Scheme 9.24

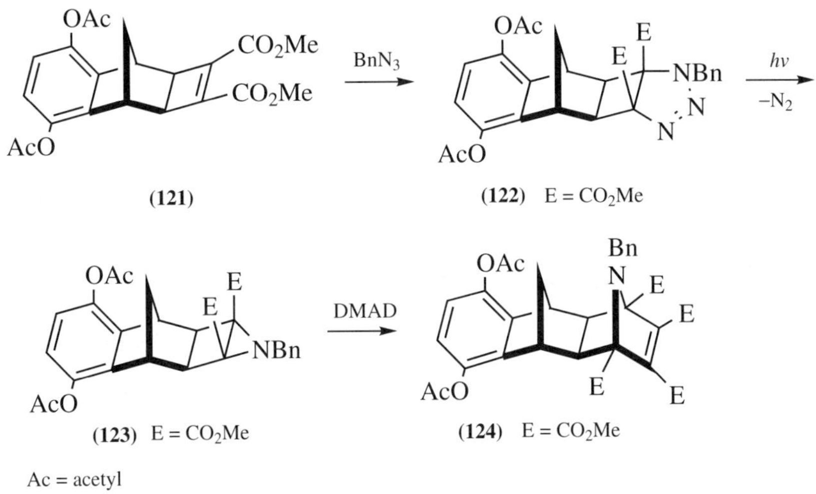

Scheme 9.25

upon extrusion of nitrogen under photochemical conditions gave the aziridine **123**. The latter rearranged thermally into an azomethine ylide, which underwent cycloaddition with DMAD to give the 7-azanorbornane **124**.

9.1.2. Reaction with the Double Bond of Dienes and Conjugated Systems

Vogel and Delavier (26) reported a synthesis of the 6-azabicyclo[3.2.2]nonane skeleton **130** using an intramolecular azide–alkene cycloaddition strategy (Scheme 9.26). When refluxed in xylene, the azide **126** underwent an intramolecular 1,3-dipolar cycloaddition with the internal alkene. Nitrogen extrusion and subsequent rearrangement led to a mixture of compounds **128**, **129**, and **130**. Reactions of azides with the double bond of dienes were also used in various total syntheses of alkaloids, and will be discussed later in Section 9.2.2.

Wedegaertner and Kattak (27) reported the intermolecular cycloaddition of aryl azides with allenes (Scheme 9.27). Cycloaddition of an aryl azide with the 1,2-propanediene **131** produced a mixture of the isomeric triazolines **132** and **133**, whereas when the cycloaddition was conducted with cyclonona-1,2-diene **134**, the triazoline **135** was the sole product. X-ray crystallographic analysis confirmed the structure of **135**.

Clerici and co-workers (28) reported an intermolecular cycloaddition of azides with the isothiazole dioxides **136** to give the triazolines **137**; further heating of cycloadduct **137**, just above its melting point, resulted in the extrusion of nitrogen to give the aziridine **138** (Scheme 9.28).

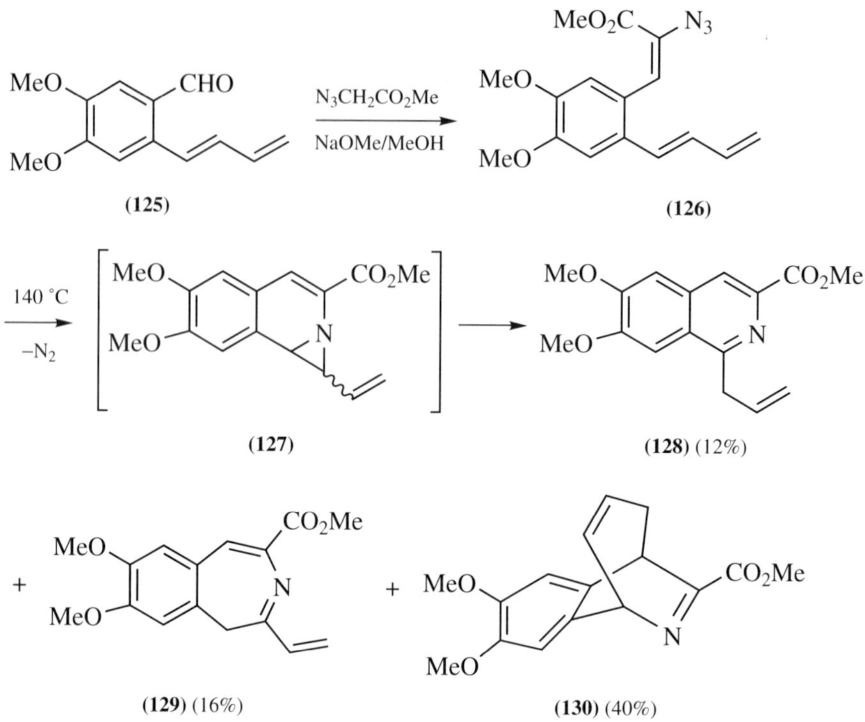

Scheme 9.26

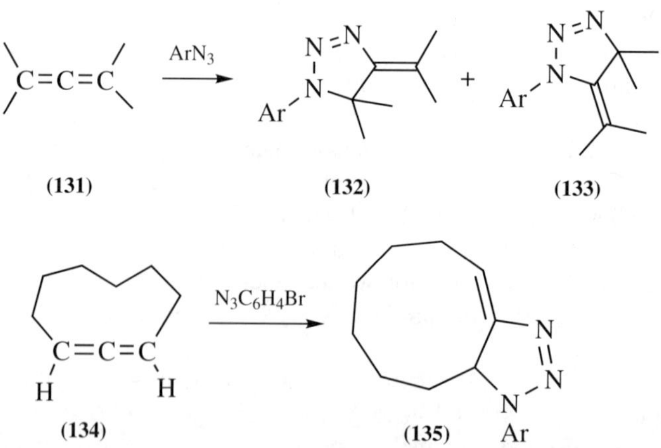

Scheme 9.27

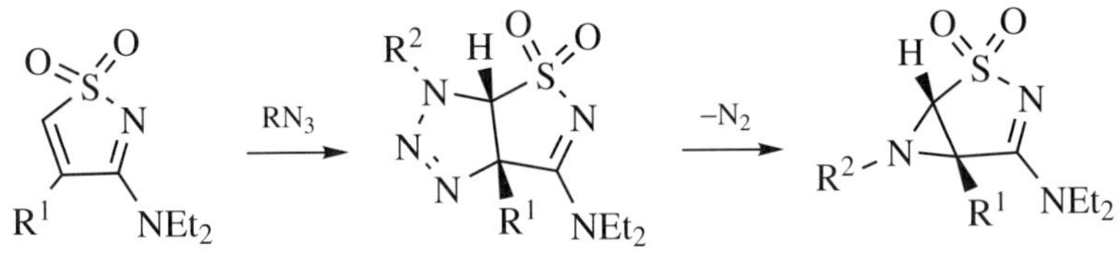

(136) (137) (138)

Scheme 9.28

9.1.3. Reaction with Alkynes and Nitriles

Pearson et al. (29) described a synthesis of mono- and disubstituted 5,6-dihydro-4*H*-pyrrolo[1,2-*c*]-1,2,3-triazole analogues of antitumor dehydropyrrolizidine alkaloids using an intramolecular azide–alkyne cycloaddition (Scheme 9.29). On treatment with sodium azide in dimethyl sulfoxide (DMSO), the chloro alcohol **139** produced the azide **140**, which on heating in refluxing toluene gave the triazoline **141**. On treatment with acetic anhydride in pyridine, triazoline (**141**) furnished the target compound **142** in 96% yield. By a similar procedure, the triazole analogue **144** was prepared in 69% yield from the corresponding azide **143**.

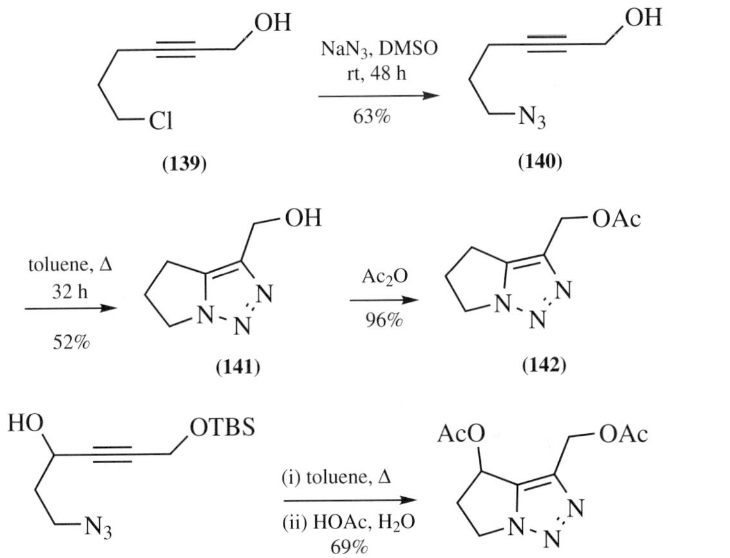

(139) (140) (141) (142) (143) (144)

Scheme 9.29

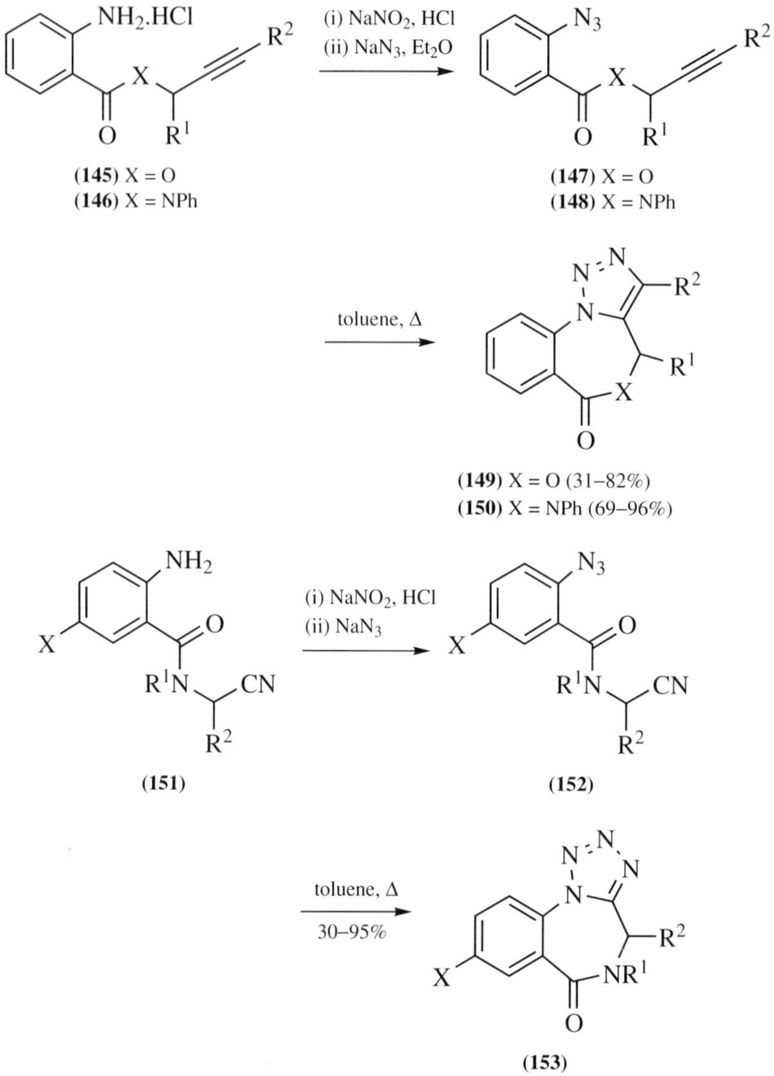

Scheme 9.30

Garanti et al. (30a) reported a synthesis of the 1,2,3-triazolo[1,5-*a*][4.1]benzoxazepine **149** via an intramolecular cycloaddition of an aryl azide with an acetylene (Scheme 9.30). By using a similar strategy, the 1,2,3-triazolo[1,5-*a*][1,4-*b*]benzodiazepine **150**, an analogue of Flumazenil, was also reported (30b,c). As an extension of this method, the 1,2,3-triazolo[1,5-*a*][1,4]benzodiazepine-6-one **153** was synthesized using an intramolecular 1,3-dipolar cycloaddition of an azide with a cyano group (30d).

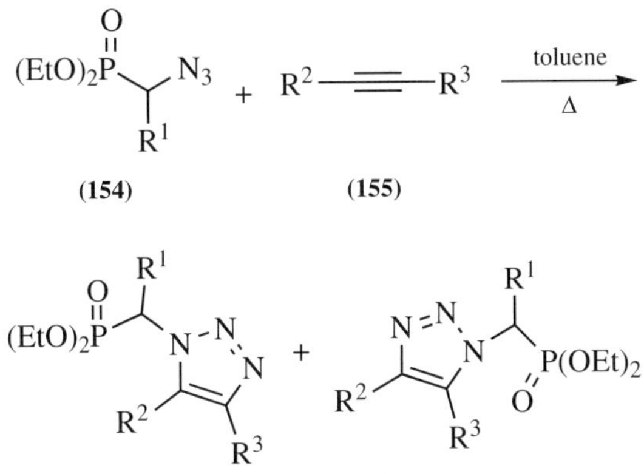

(154) (155) (156) (157)

Scheme 9.31

Palacios et al. (31) reported the synthesis of the isomeric 4,5-disubstituted 1-(diethoxyphosphorylmethyl)-1,2,3-triazoles **156** and **157** via thermal intermolecular 1,3-dipolar cycloadditions of the azidoalkylphosphonates **154** with the disubstituted acetylenes **155** (Scheme 9.31).

A general synthesis of functionalized 1,2,3-triazolyl acylsilanes (**160**) was based on the intermolecular cycloaddition of azides **159** with the alkynyl acylsilane **158** (Scheme 9.32) (32). The resulting triazolyl acylsilanes (**160**) were smoothly converted into their corresponding aldehydes **161** upon treatment with sodium hydroxide in ethanol.

Brillante and co-workers (33) conducted an intermolecular 1,3-dipolar cycloaddition of the aryl azide **162** with (trimethylsilyl)acetylene under high-pressure conditions (Scheme 9.33). The rate of cycloaddition increased logarithmically with pressure, and the yield of cycloadduct **163** was almost quantitative.

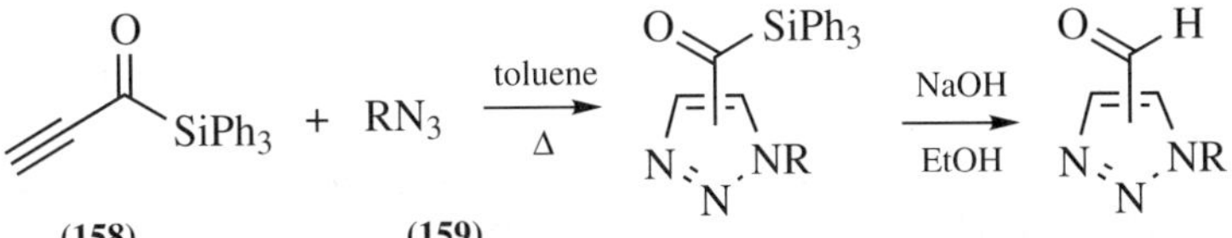

(158) (159) (160) (161)

Scheme 9.32

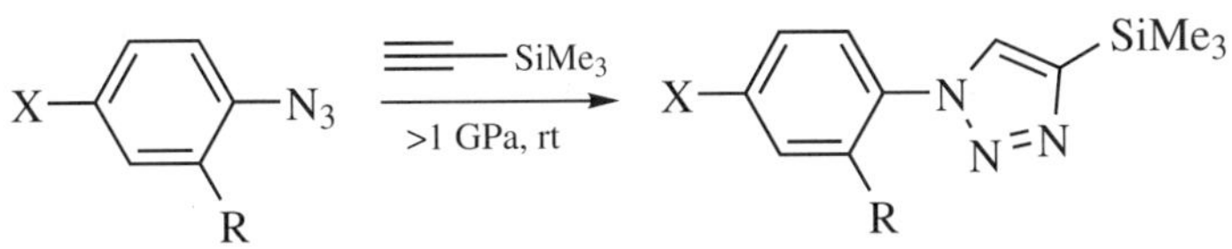

(162) (163)

Scheme 9.33

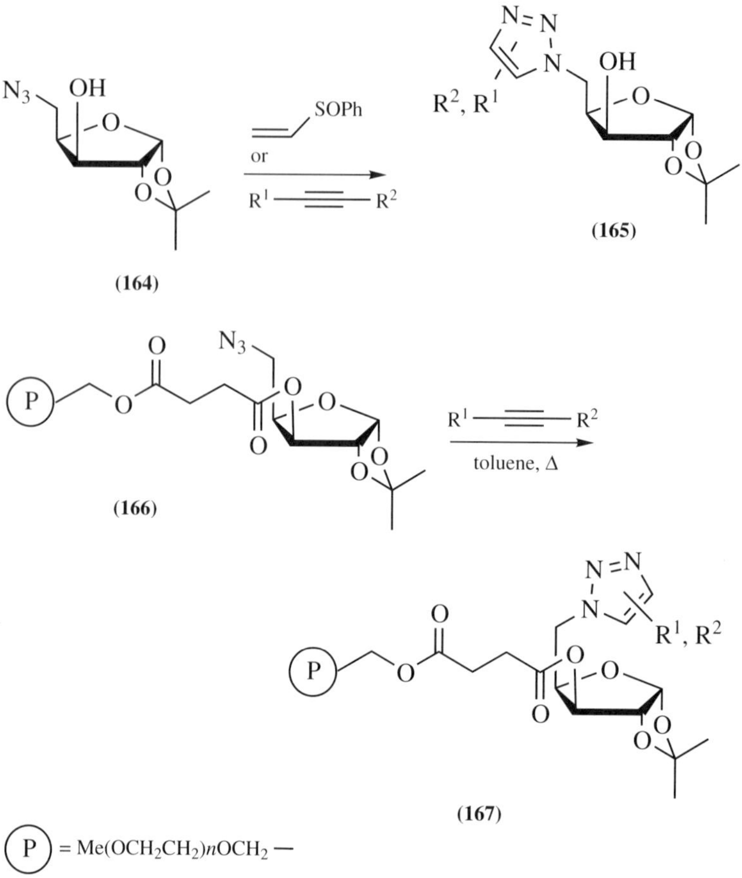

(164) (165) (166) (167)

Scheme 9.34

Freeze and Norris (34) reported the 1,3-dipolar cycloaddition of 5-azido-5-deoxy-1,2-*O*-isopropylidene-D-xylofuranose (**164**) with acetylenic dipolarophiles to give the triazoles **165** (Scheme 9.34). This process was subsequently extended using the soluble polymer-supported azide (**166**) to produce the corresponding triazoles **167** in 50–95% yield. Dipolarophiles present in large excess facilitated the cycloaddition of the polymer-supported azide **166**. Purification of the triazole **167** was achieved by filtration.

An efficient synthesis of the 1-allyl-6-(1′,2′,3′-triazolyl) analogue **170** of 1-[2-hydroxyethoxy)methyl]-6-(phenylthio)thymine (HEPT), an anti-human immunodeficiency virus (HIV) reverse transcriptase inhibitor, was reported using an intermolecular 1,3-dipolar cycloaddition of the azide **169** with acetylenes (35) (Scheme 9.35). Azidouracil (**169**), when refluxed with an acetylene in equimolar proportions in toluene, gave the corresponding triazoles (**170**) in excellent yield.

The synthesis of the 2-triazolylpyrimido[1,2,3-*cd*]purine-8,10-diones **172** and **173** was achieved using the 1,3-dipolar cycloaddition of azides with the terminal

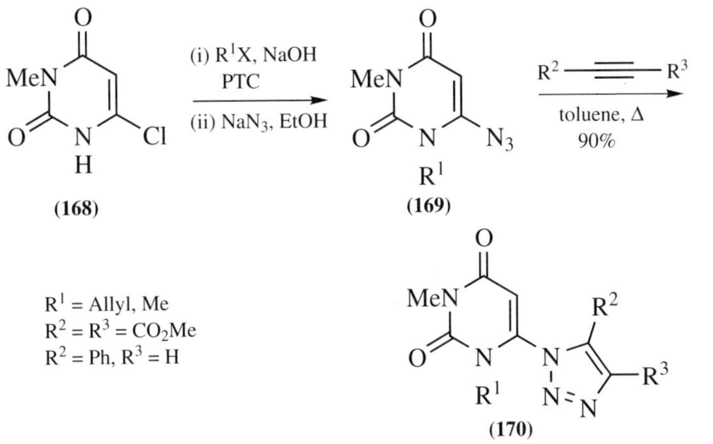

Scheme 9.35

triple bond of 5,6-dihydro-2-ethynyl-9-methyl-4*H*,8*H*-pyrimido[1,2,3-*cd*]purine-8,10(9*H*)dione (**171**) (36) (Scheme 9.36).

The preparation of compound **175**, a structurally diverse analogue of the carbocyclic nucleoside ribavarine **176**, was reported using an intermolecular 1,3-dipolar cycloaddition of the cyclopentyl azide **174** with methyl propiolate (37) (Scheme 9.37).

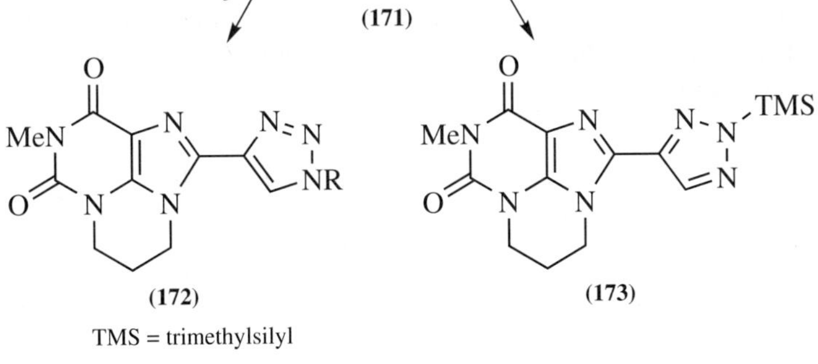

Scheme 9.36

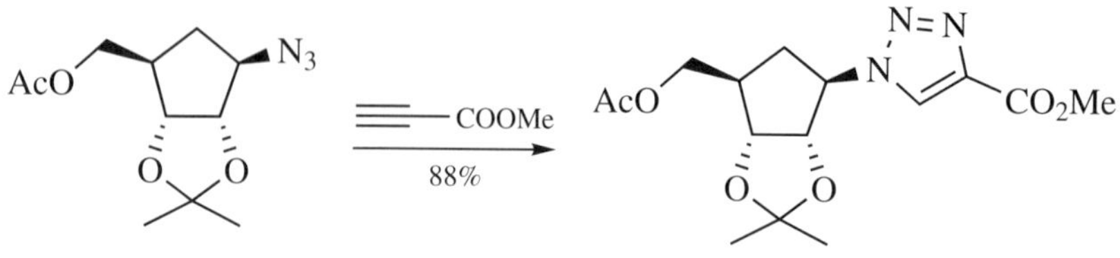

(174) (175)

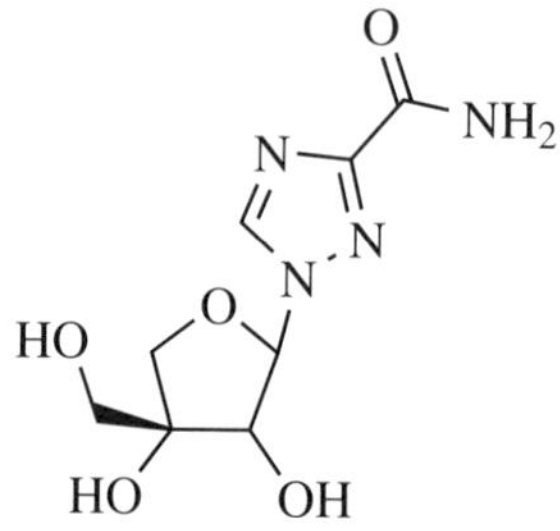

(176) ribavarine

Scheme 9.37

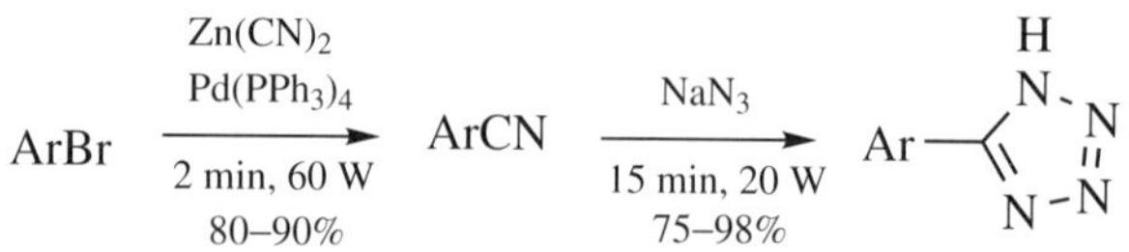

(177) (178) (179)

Ar = aryl, naphthyl, pyridyl, thienyl

Scheme 9.38

The microwave-assisted preparation of aryl tetrazoles **179** was reported using the intermolecular 1,3-dipolar cycloaddition of aryl nitriles **178** with sodium azide (38) (Scheme 9.38).

The synthesis of the heterocyclic dendrimer **181** was based on the intermolecular 1,3-dipolar cycloaddition of the azide **180** with acetylenedicarboxylic acid and its esters (39) (Scheme 9.39).

9.1.4. Reaction with Allylic Carbocations

Pearson et al. (40) observed an unprecedented low-temperature [3+2] or [3+3] cycloaddition of azides with allylic carbocations, yielding triazolines or dihydrotriazolines (Scheme 9.40). When the hydroxy azide **182** was treated with $SnCl_4$ at $-78\,°C$, a diastereomeric mixture of crystalline triazolines **183** was obtained. A

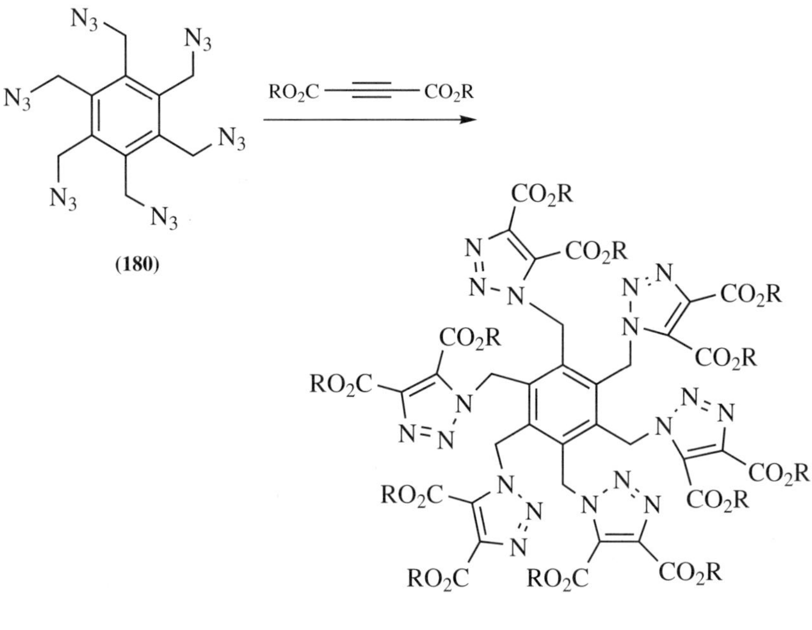

(181) R = H, Me, *t*-Bu

Scheme 9.39

mechanism involving the initial formation of an allylic carbocation intermediate **184** was proposed. The intermediate **184** might undergo a low-temperature 1,3-dipolar cycloaddition of the azide with the allylic cation segment of the indole in either a concerted or stepwise manner via the aminodiazonium ion **185** to give the cationic intermediate **186**, which on abstraction of a chloride ion may afford compound **183**. A closely related azide **187** underwent smooth [3+2] cycloaddition to yield the triazoline **188**. The cyclization of indoles **189** and **190**, lacking an electron-withdrawing sulfonyl group at the indole NH, underwent [3+3] cycloaddition to give the dihydrotriazolines **191** and **192** in 44 and 85% yields, respectively.

9.1.5. Reaction with Phosphorus Ylides

The synthesis of the 1-(D-apio-D-furanosyl)-1,2,3-triazoles **196**, structurally related analogues of the antibiotic ribavarin A **176**, was achieved through an intermolecular 1,3-dipolar cycloaddition of the azide **193** with the 2-oxoalkylidenetriphenylphosphorane (41) (Scheme 9.41).

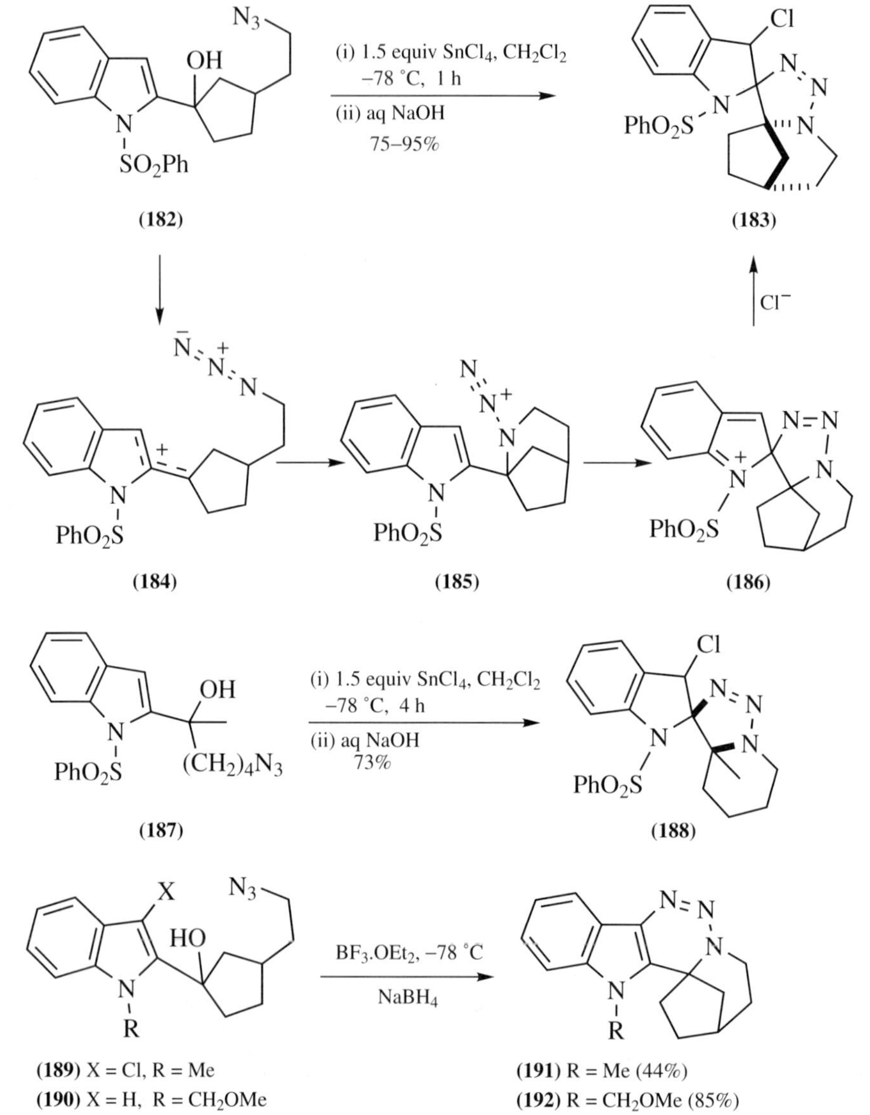

Scheme 9.40

9.1.6. Reaction with Fullerenes

Kobayashi and co-workers (42) reported a general synthetic pathway for the incorporation of oligosaccharides into [60]fullerene via cycloadditions between glycosyl azides and the fullerene (Scheme 9.42). The cycloaddition of 2,3,4,6-tetra-*O*-acetyl-D-glucopyranosyl azide **197** with [60]fullerene in refluxing chlorobenzene

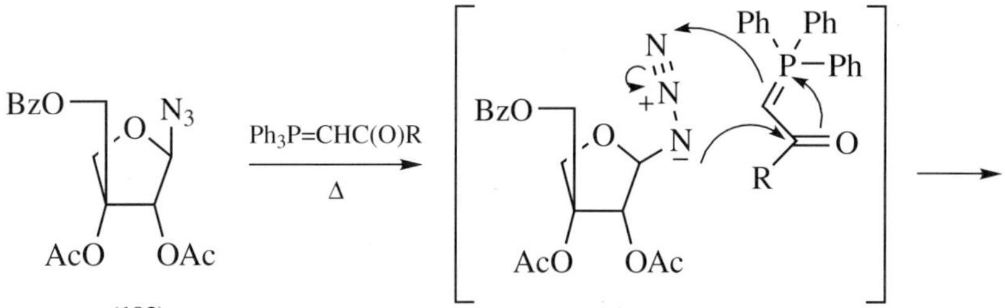

(193) (194)

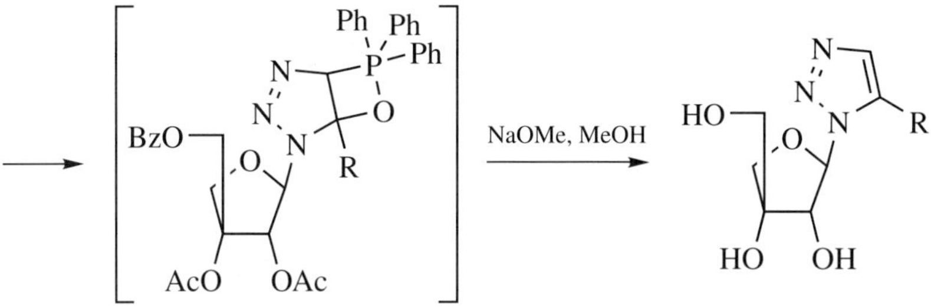

(195) (196) R = Me, Et, CO_2Me, $CONH_2$

Bz = benzoyl

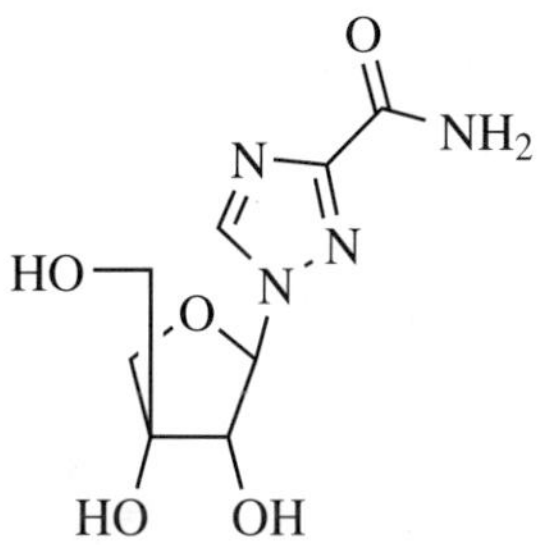

(176) ribavarine

Scheme 9.41

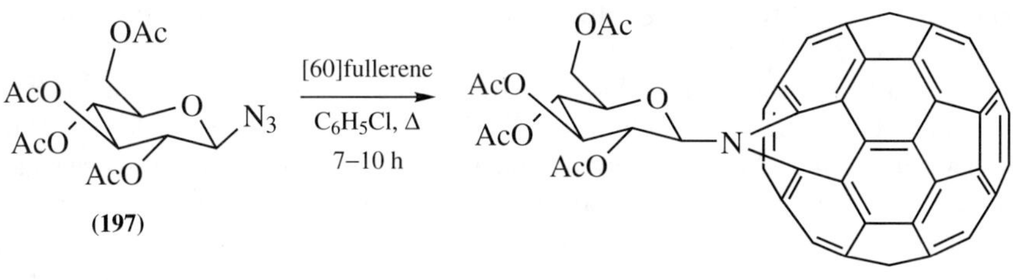

(197) (198)

Scheme 9.42

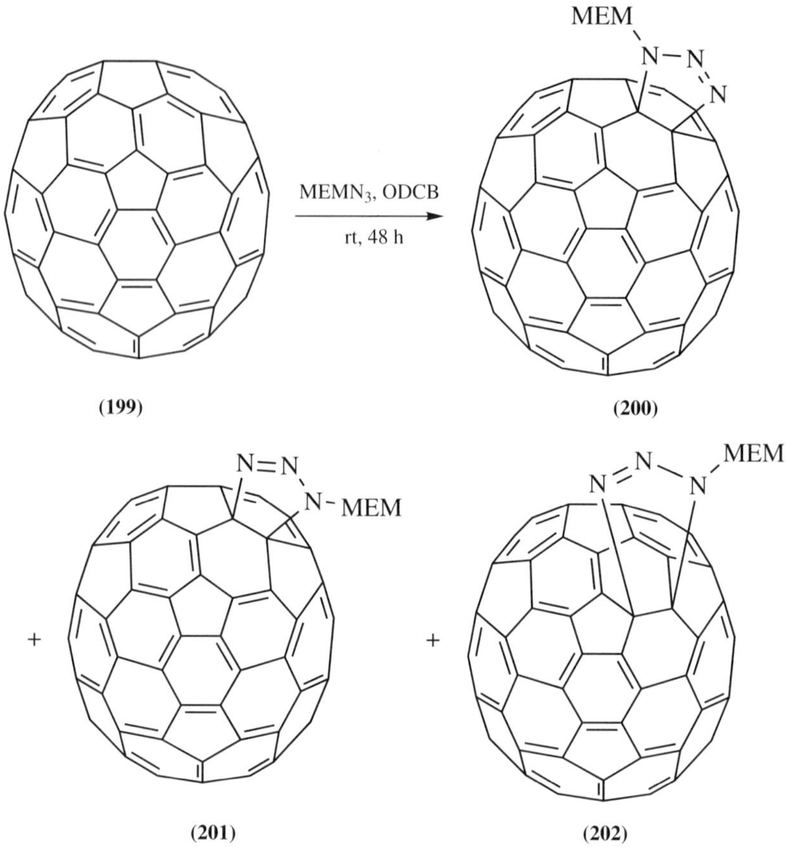

MEM = $MeOCH_2CH_2OCH_2$–

ODCB = o-dichlorobenzene

Scheme 9.43

followed by extrusion of nitrogen gave the azafulleroid **198** in 28% yield. Extension of this cycloaddition reaction to di- and trisaccharide azides confirmed the generality of the reaction for the production of the corresponding fullerene conjugates in 27–50% yields.

Bellavia-Lund and Wudl (43) investigated the 1,3-dipolar cycloaddition of 2-(methoxyethoxy)methyl azide with [70]fullerene (**199**) (Scheme 9.43). Three isomeric triazolines **200–202** were obtained. Thermolysis of these triazolines gave the corresponding azafulleroids and fulleroaziridines, as a mixture, respectively.

The synthesis of the triazolino[4′,5′:1,2][60]fullerene **204**, a novel donor-acceptor dyad exhibiting efficient electron-transfer dynamics, was reported by Guldi et al. (44) (Scheme 9.44). The azido tetrathiafulvalene **203**, on heating with [60]fullerene in *o*-dichlorobenzene at 60 °C, gave the triazoline **204** in 24%

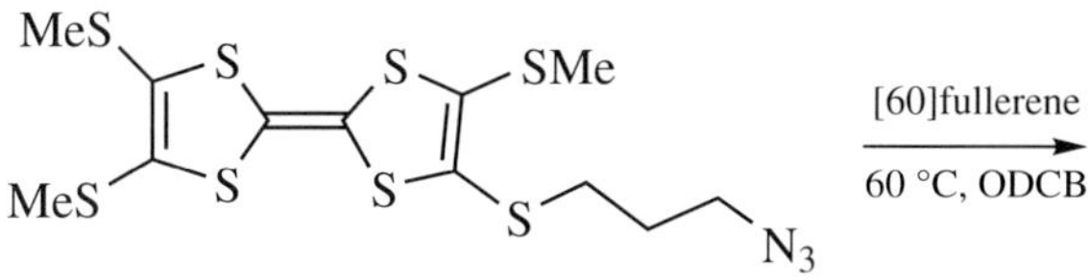

(203)

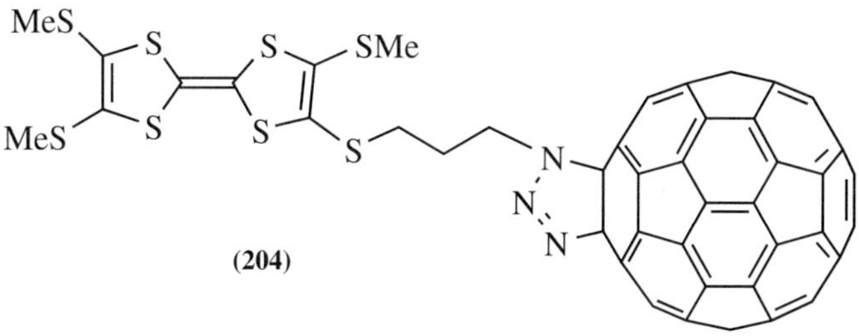

(204)

Scheme 9.44

yield, or 54% yield based on consumed [60]fullerene. The temperature of the cycloaddition must not be >60 °C to avoid formation of the corresponding azafulleroids and fulleroaziridines. The electrochemical and photophysical properties of **204** were also reported.

9.2. SYNTHESIS OF NATURAL PRODUCTS, INTERMEDIATES, AND THEIR ANALOGUES VIA 1,3-DIPOLAR CYCLOADDITION OF AZIDES

9.2.1. Reaction with Alkenes

Sha et al. (45) reported an intramolecular cycloaddition of an alkyl azide with an enone in an approach to a cephalotaxine analogue (Scheme 9.45). Treatment of the bromide **205** with NaN_3 in refluxing methanol enabled the isolation of compounds **213** and **214** in 24 and 63% yields, respectively. The azide intermediate **206** underwent 1,3-dipolar cycloaddition to produce the unstable triazoline **207**. On thermolysis of **207** coupled with rearrangement and extrusion of nitrogen, compounds **213** and **214** were formed. The lactam **214** was subsequently converted to the *tert*-butoxycarbonyl (*t*-Boc)-protected spirocyclic amine **215**. The exocyclic double bond in compound **215** was cleaved by ozonolysis to give the spirocyclic ketone **216**, which was used for the synthesis of the cephalotaxine analogue **217**.

When the enone **218**, possessing a methyl group at the C(2) position, was treated with NaN_3 in DMF, aziridine **219** was isolated as the only product (46) (Scheme 9.46). This reaction was adopted for the total synthesis of (±)-desamylperhydrohistrionicotoxin (**224**) from the azido enone **220**.

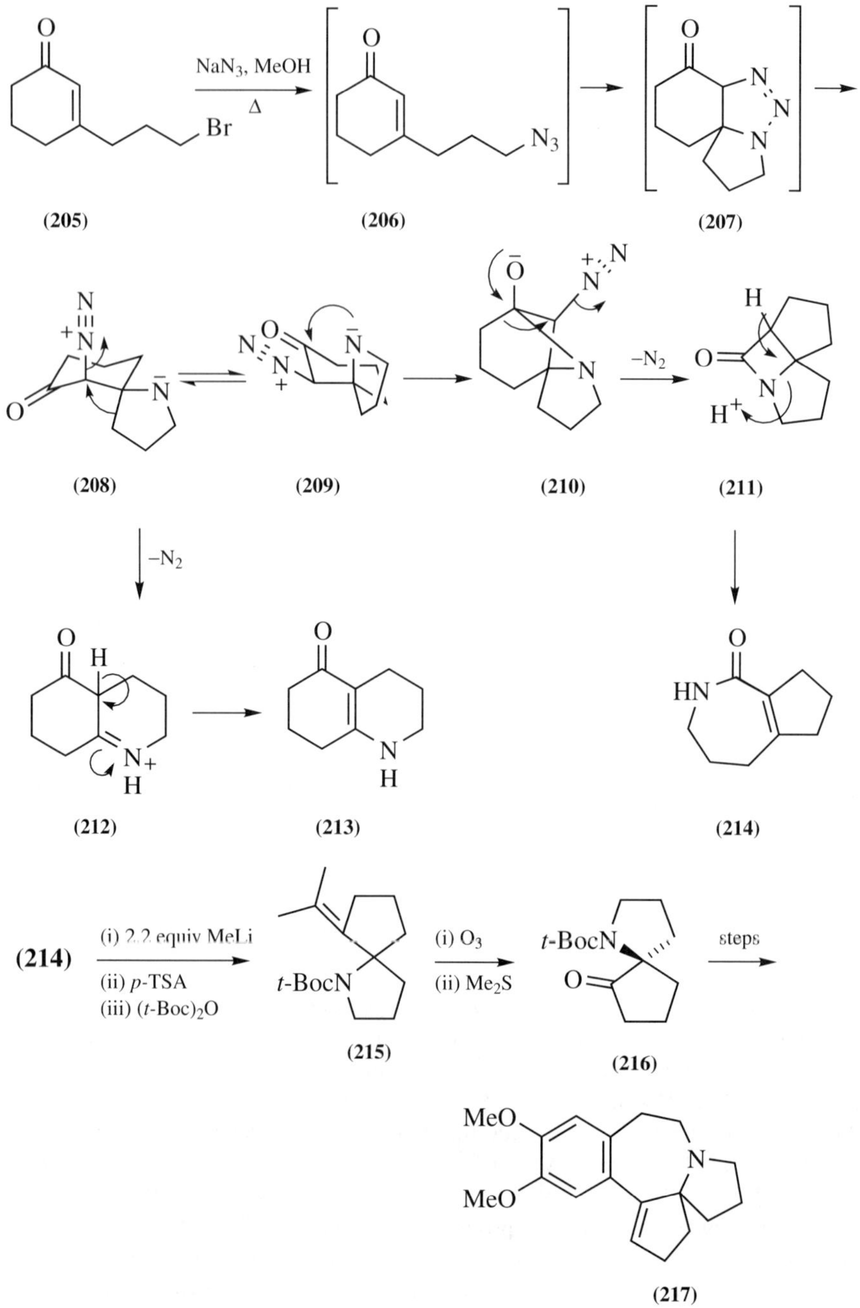

NaN3, MeOH
Δ
(205)
(206)
(207)
(208)
(209)
−N2
(210)
(211)
−N2
(212)
(213)
(214)
(214)
(i) 2.2 equiv MeLi
(ii) p-TSA
(iii) (t-Boc)2O
(215)
(i) O3
(ii) Me2S
(216)
steps
(217)

Scheme 9.45

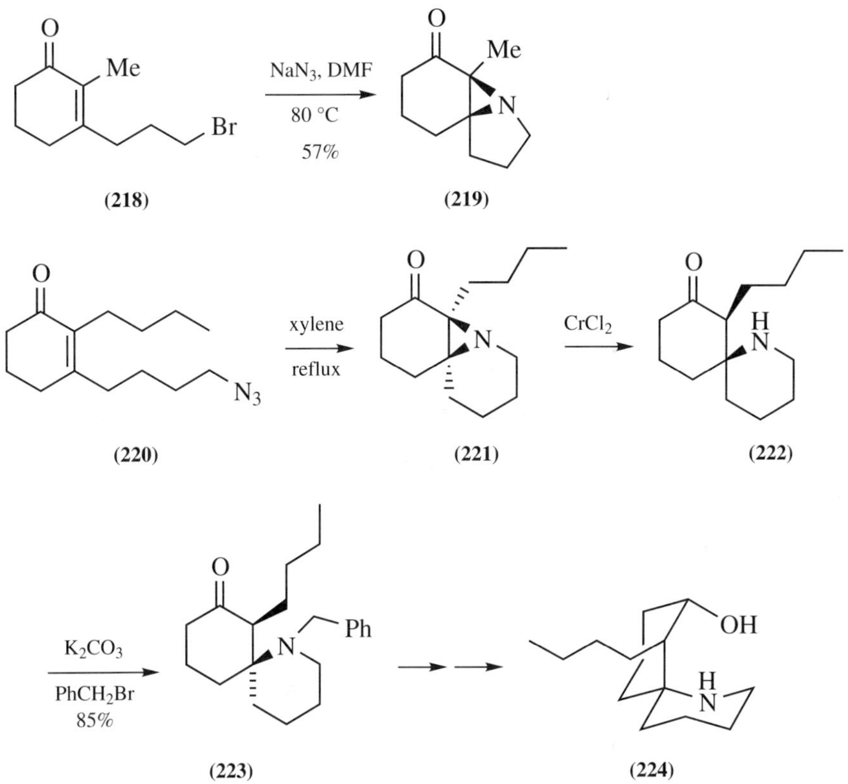

Scheme 9.46

Kozikowski and Greco (47) reported a direct approach to a total synthesis of (±)-clavicipitic acid (**229**) involving an intramolecular azide 1,3-dipolar cycloaddition (Scheme 9.47). Treatment of the malonate ester **225** with NaH and TsN_3 gave the required azide **226**. On heating at 190–195 °C in *o*-dichlorobenzene for 8 h, the azide **226** gave the imine **228** via an unstable triazoline intermediate **227**. The fact that the intramolecular cycloaddition of azide **226** proceeded readily in a polar solvent, *o*-dichlorobenzene, but poorly in toluene, provided indirect evidence for the formation of a polar intermediate during the course of the formation of the seven-membered ring. The imine **228** was transformed smoothly to the target molecule **229** in three steps.

Buchanan et al. (48) reported a new route to the synthesis of the chiral hydroxypyrrolidines **234** and **238** from D-erythrose (**230**) via an intramolecular cycloaddition of an azide with an alkene (Scheme 9.48). Wittig reaction of the acetonide **230** with (carbethoxyethylene)triphenylphosphorane gave the (*E*) and (*Z*) alkenes **231** and **232**. On conversion into the triflate followed by its reaction with KN_3, the (*E*) isomer **231** allowed the isolation of the triazoline **234** in 68% overall yield, which on treatment with sodium ethoxide afforded the diazo ester **235** in 86% yield.

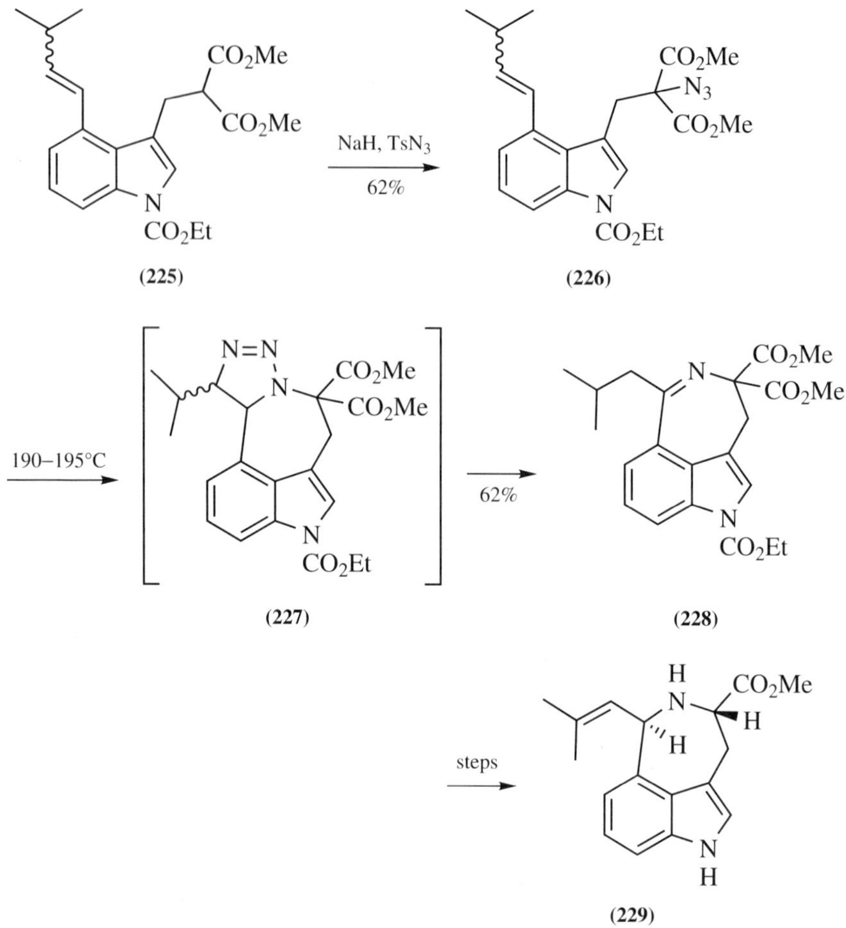

Scheme 9.47

Catalytic reduction of **235** produced the chiral hydroxypyrrolidine **236** in 89% yield. When the (*Z*) alkene **232** was subjected to a similar sequence, the diazo ester **237** was obtained directly in 65% yield. Catalytic hydrogenation of the diazo ester **237** afforded the corresponding chiral hydroxypyrrolidine **238** in 91% yield.

Fukuyama and Yang (49) developed a highly efficient synthesis of the tetracyclic intermediate **241**, used in a total synthesis of mitomycin A (Scheme 9.49). The required azide **240** was produced from **239** in several steps. Upon heating in refluxing toluene, the azide **240** underwent smooth intramolecular cycloaddition with the unsaturated lactone followed by extrusion of nitrogen to give aziridine **241** in 85% yield.

Cha and co-workers (50) reported a short enantioselective synthesis of the indolizidine alkaloid, (−)-swainsonine (**247**) involving an intramolecular cycloaddition

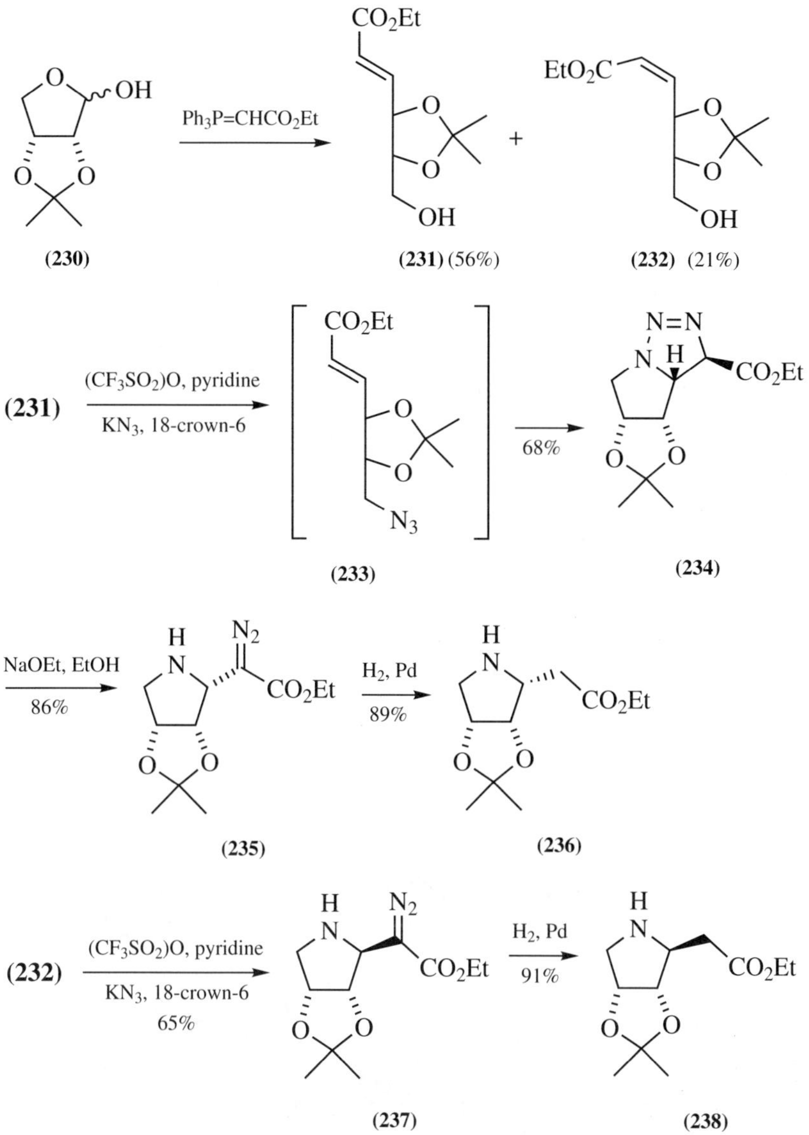

Scheme 9.48

of an azide to an alkene (Scheme 9.50). Wittig reaction of 2,3-*O*-isopropylidene-D-erythrose (**242**) followed by tosylation afforded compound **243**. Treatment of the tosylate **243** with NaN_3 generated the imino ester **245** in 81% yield via the triazoline intermediate **244**. Isomerization of the imino ester **245** to an enamine followed by thermal cyclization in refluxing toluene gave the lactam **246**. Hydroboration of **246** followed by oxidation and subsequent cleavage of the acetonide unit afforded (−)-swainsonine (**247**) in good yield.

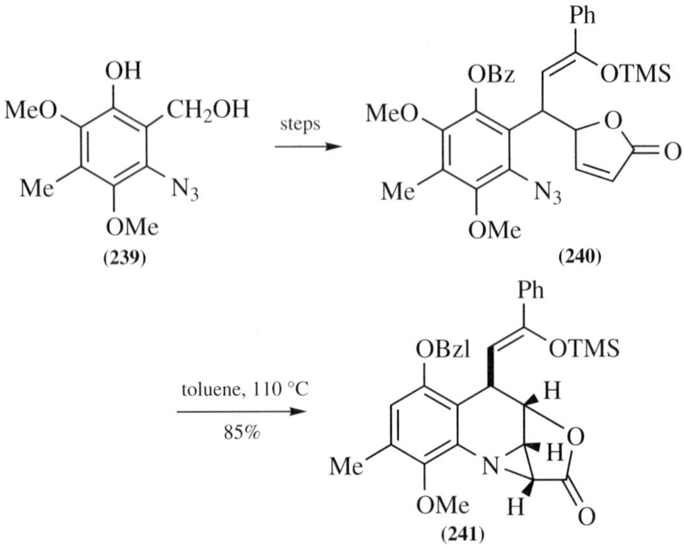

Scheme 9.49

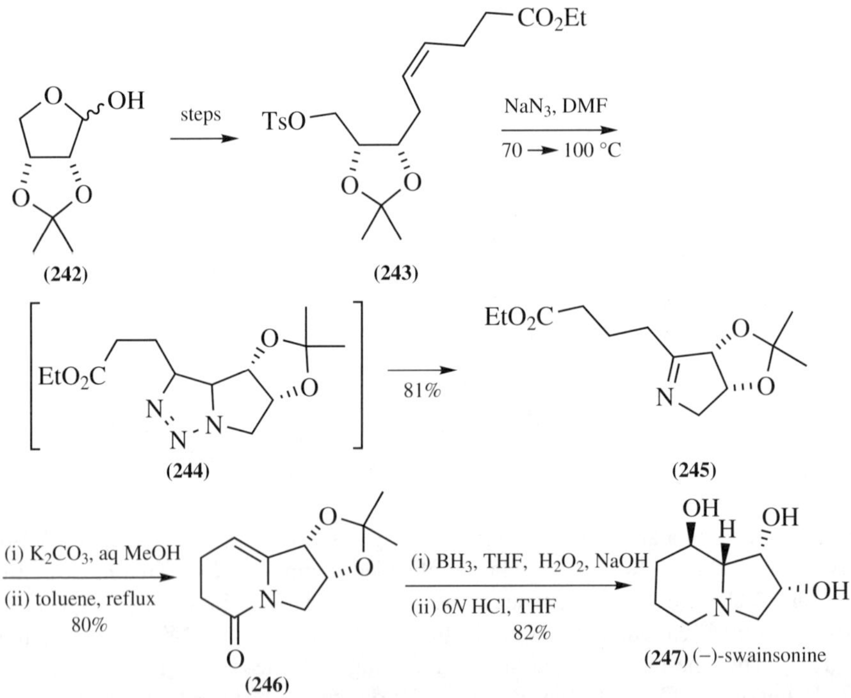

Scheme 9.50

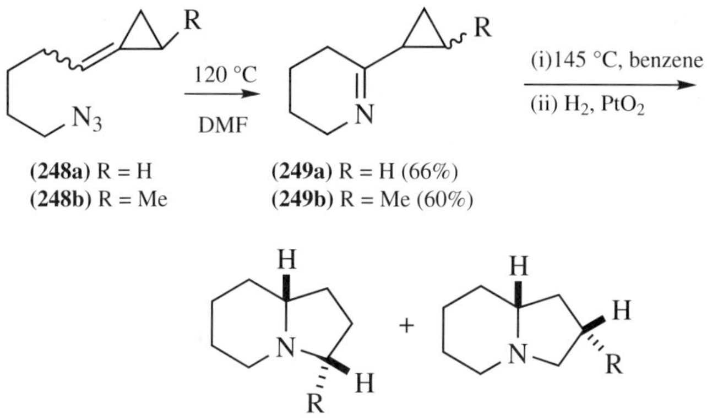

(248a) R = H
(248b) R = Me
(249a) R = H (66%)
(249b) R = Me (60%)
(250a) R = H (42%)
(250b) R = Me
(251b) R = Me (47%)

Scheme 9.51

Pearson and co-workers (51) reported a simple synthesis of the indolizidines **250** and **251** based on the intramolecular cycloaddition of azides with the exocyclic double bond of cyclopropanes (Scheme 9.51). The intramolecular cycloaddition of azides **248a** and **248b** at 120 °C in DMF afforded the corresponding cyclopropyl

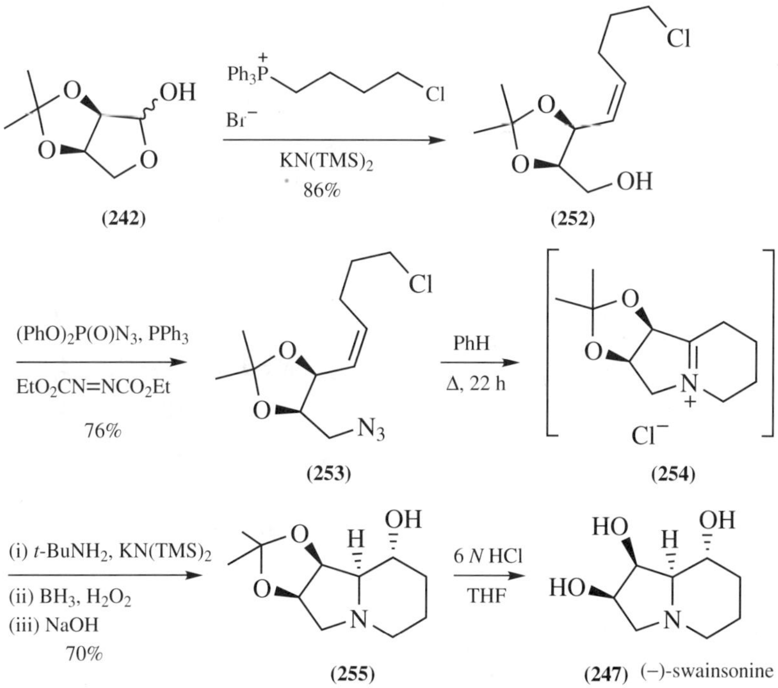

(242) **(252)** **(253)** **(254)** **(255)** **(247)** (−)-swainsonine

Scheme 9.52

imines **249a,b** in 66 and 60% yields, respectively. On thermal rearrangement followed by catalytic hydrogenation, compounds **249a,b** afforded not only the respective indolizidines **250a,b** but also **251b**.

Pearson and Lin (52) developed an elegant approach to the synthesis of optically active (−)-swainsonine (**247**) from isopropylidene-D-erythrose (**242**) (Scheme 9.52). Wittig reaction of the acetonide **242** led to the (*Z*) alkene **252** in 86% yield. The chloro alcohol **252** was converted to the azide **253** in 76% yield, which subsequently underwent 1,3-dipolar cycloaddition, isomerization and hydroboration–oxidation to give the indolizidine **255** in 70% overall yield. Cleavage of the acetonide unit in **255** using 6 *N* HCl gave the target molecule **247** in 85% yield.

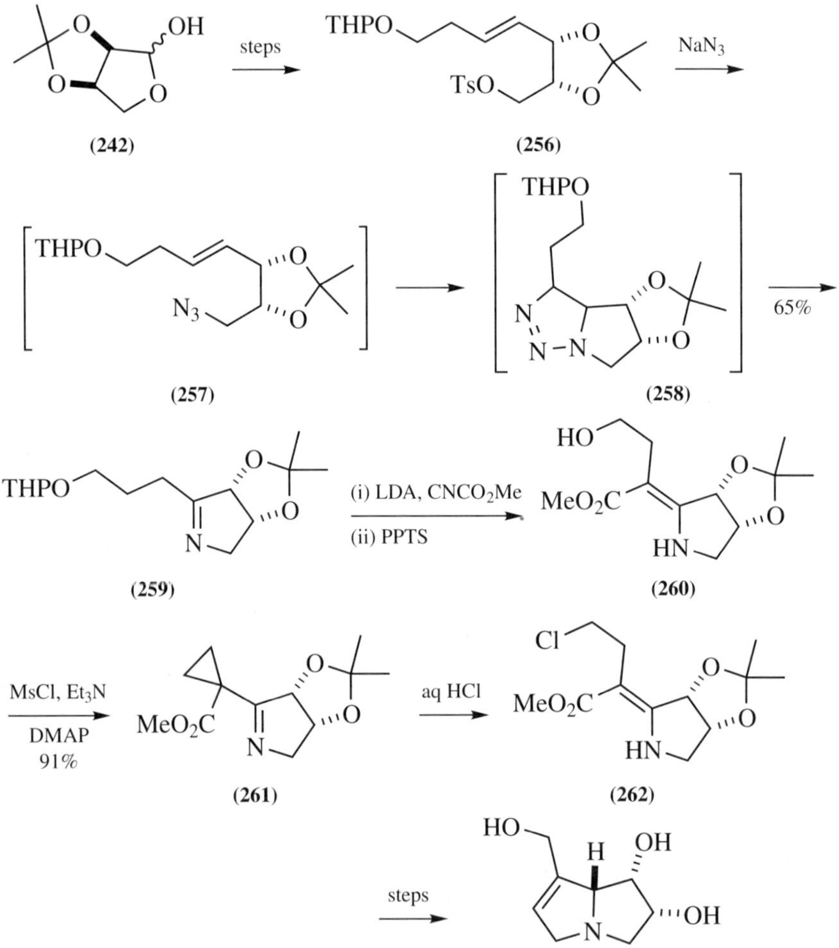

(242) **(256)** **(257)** **(258)** **(259)** **(260)** **(261)** **(262)** **(263)** (+)-crotanecine

LDA = lithium diisopropylamide
DMAP = 4-N,N-dimethylaminopyridine
PPTS = pyridinium p-toluenesulfonate

Scheme 9.53

Starting also with compound **242**, Bennett III and Cha (53) synthesized (+)-crotanecine (**263**) employing a 1,3-dipolar cycloaddition strategy (Scheme 9.53). Wittig reaction of the acetonide **242** with a THP-protected phosphonium salt followed by tosylation produced the tosylate **256**. Displacement of the tosyl group with NaN_3 in DMF followed by smooth 1,3-dipolar cycloaddition and nitrogen extrusion gave the imine **259** in 65% overall yield from acetonide **242**. Carbomethoxylation of the imine **259** followed by acid-catalyzed isomerization generated enamine **260**, which on treatment with mesyl chloride and Et_3N followed by aqueous HCl gave the chloro enamine **262** via the intermediate **261**. The chloro enamine **262** was then smoothly transformed into (+)-crotanecine (**263**) by conventional methods.

Cha and co-workers (54) described an enantioselective total synthesis of (−)-slaframine (**269**) based on an intramolecular cycloaddition of an azide (Scheme 9.54). On reaction with NaN_3 in DMF at 60 °C followed by intramolecular

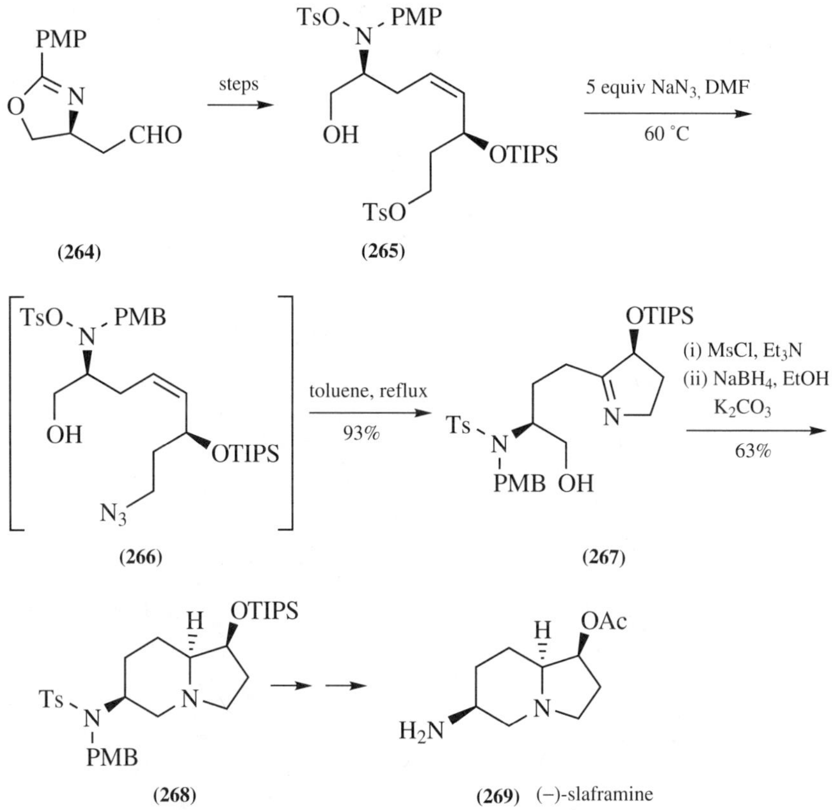

PMP = p-methoxyphenyl
PMB = p-methoxybenzyl
TIPS = $Si(iPr)_3$

Scheme 9.54

1,3-dipolar cycloaddition in toluene at reflux, tosylate **265** produced the imine **267** in 93% yield. Mesylation followed by reduction of the iminium ion gave compound **268** in 63% yield. Conventional reaction conditions to remove the protecting groups afforded (−)-slaframine (**269**) in 43% overall yield from the aldehyde **264**.

Garner et al. (55) developed an asymmetric approach to the 3,8-diazabicylco[3.2.1]octane moiety of the antitumor antibiotic quinocarcin (**275**) via an intermolecular 1,3-dipolar cycloaddition (Scheme 9.55). 1,3-Dipolar cycloaddition of methyl azide with the chiral dipolarophile **270** gave the corresponding triazoline **271** in 73–88% yields. Nitrogen was extruded from the triazoline **271** under photochemical conditions to give the aziridines **272** in 78–92% yields. Transformation of aziridines **272** into the chiral azomethine ylide **273**, followed by its cycloaddition with alkenes, gave **274**. Compound **274** was then converted into quinocarcin (**275**) via a Pictet–Spengler cyclization.

Pearson and Schkeryantz (56) developed a novel approach for synthesis of (±)-lycorane (**280**) using an intramolecular cycloaddition of an azide with an ω-chloro alkene (Scheme 9.56). The bromide **276** was smoothly converted into the required chloro azide **277** in several steps. 1,3-Dipolar cycloaddition of the azide **277** in benzene at 140 °C followed by extrusion of nitrogen gave the unstable

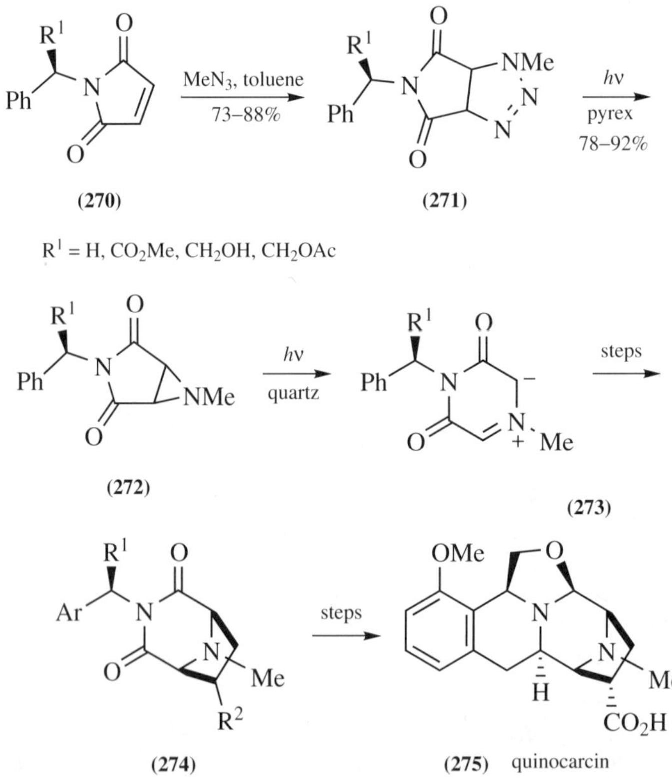

Scheme 9.55

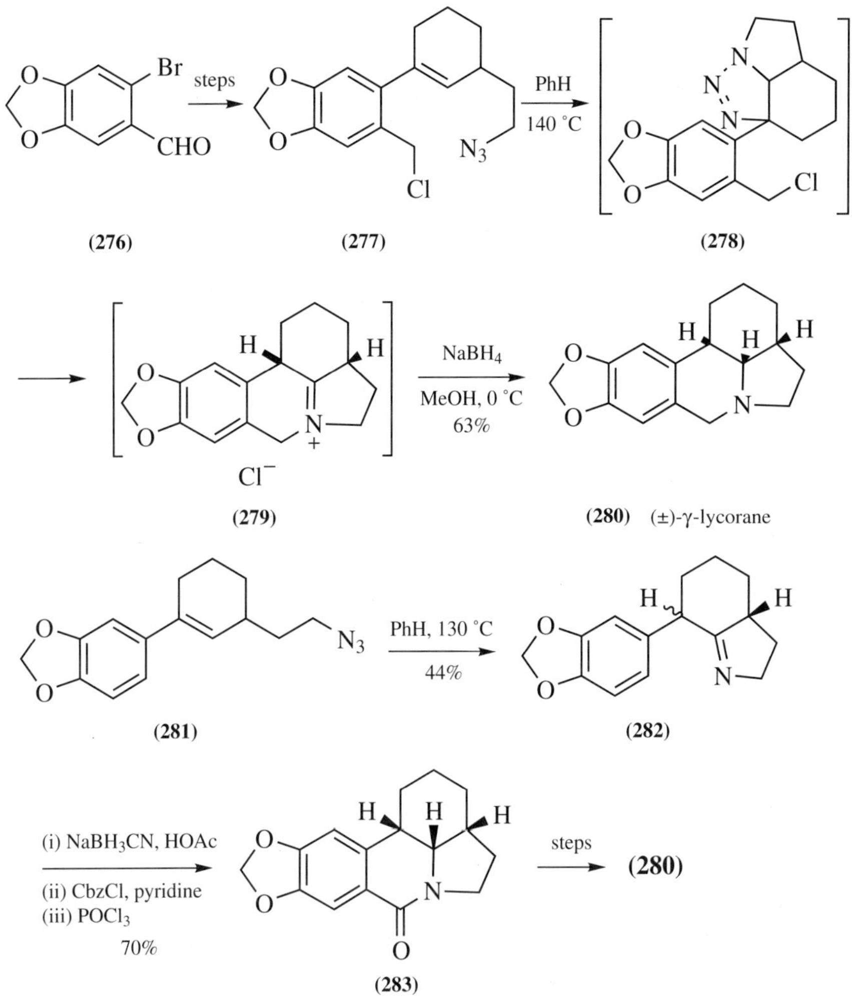

Scheme 9.56

iminium salt **279**. Compound **279** was reduced with sodium borohydride to afford (±)-γ-lycorane (**280**) in 63% yield from the azide **277**. An alternative strategy for (±)-γ-lycorane was developed via the intramolecular cycloaddition of the azide **281**. 1,3-Dipolar cycloaddition of **281** followed by nitrogen extrusion gave the imine **282**. Reduction of **282** with sodium cyanoborohydride followed by Bischler–Napieralski cyclization gave **283**, which was converted into (±)-γ-lycorane (**280**) in several steps.

Pearson and Walavalkar (57) reported a facile approach to the synthesis of (±)-tyloporine (**288**) based on an intramolecular cycloaddition of an azide with an ω-chloroalkene (Scheme 9.57). The required (*Z*) alkene **285** was prepared from homoveratric acid (**284**). Treatment of the chloro alkene **285** with sodium azide

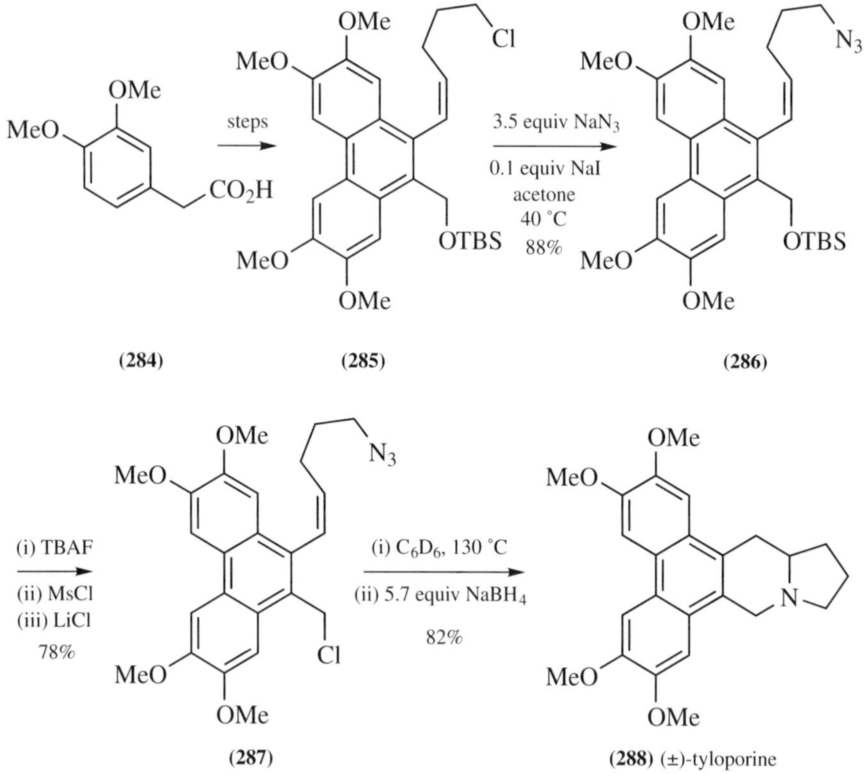

Scheme 9.57

gave the azide **286** in 88% yield. Deprotection followed by mesylation and displacement by chloride afforded chloro azide **287** in 78% overall yield. Heating of **287** at 130 °C in deuterated benzene followed by reduction with sodium borohydride gave the target molecule **288** in 82% yield.

Two enantioselective syntheses of (+)-biotin (**293**) from L-cysteine were reported based upon the intramolecular 1,3-dipolar cycloadditions of carbamoyl azides **289** and **291** by Deroose and De Clercq (58) (Scheme 9.58). Thermolysis of the carbamoyl azides **289** and **291** gave the triazolines **290** and **292**, respectively. Both **290** and **292** were converted into (+)-biotin (**293**) in several steps.

Schkeryantz and Pearson (59) reported a total synthesis of (±)-crinane (**298**) using an intramolecular azide–alkene cycloaddition (Scheme 9.59). The allylic acetate **294** was first subjected to an Ireland–Claisen rearrangement followed by reduction to give alcohol **295**, which was then converted into the azide **296** using Mitsunobu conditions. Intramolecular cycloaddition of the azide **296** in refluxing toluene followed by extrusion of nitrogen gave the imine **297** in quantitative yield. On reduction with sodium cyanoborohydride and subsequent reaction with

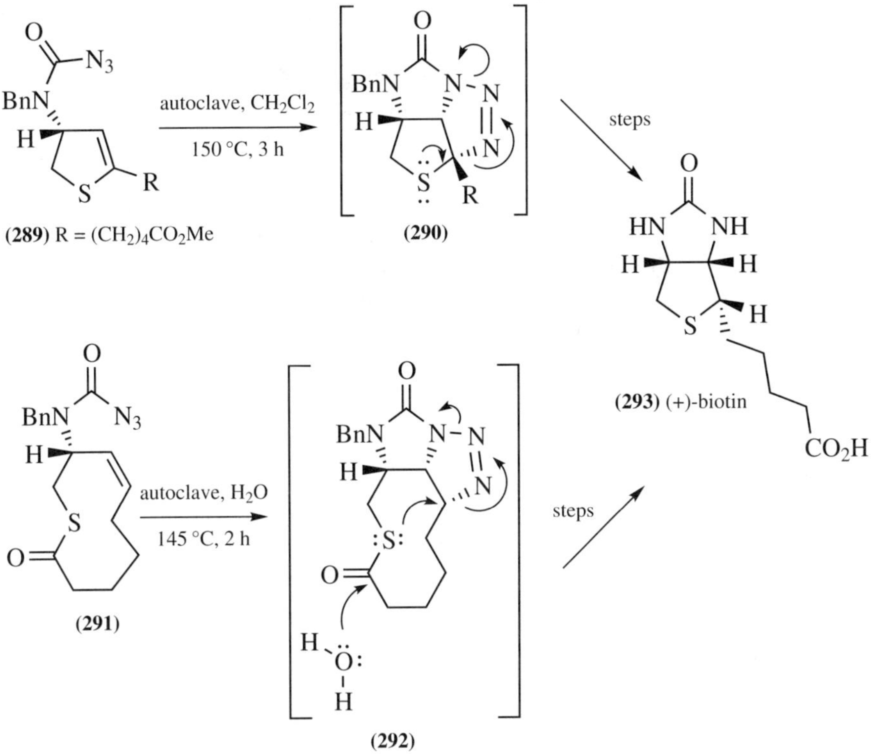

Scheme 9.58

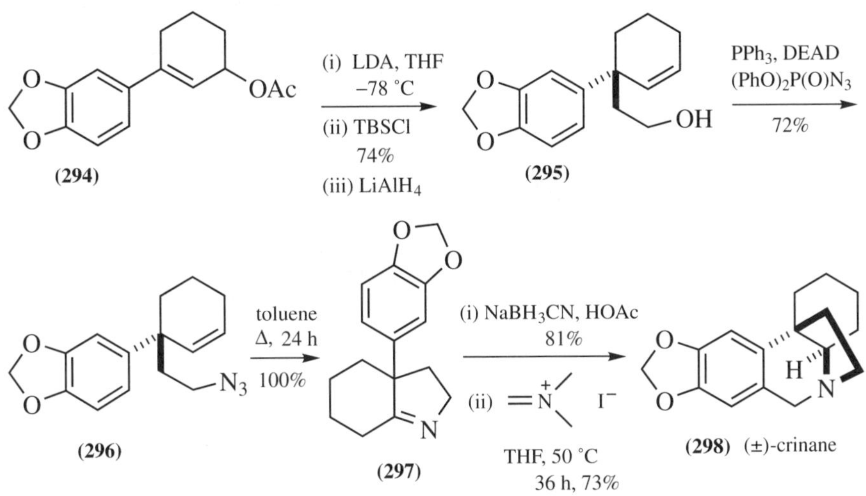

Scheme 9.59

Eschenmoser's salt at 50 °C in THF, the imine **297** gave the target molecule **298** in 77% overall yield.

Molander and Hiersemann (60) reported the preparation of the spirocyclic keto aziridine intermediate **302** in an approach to the total synthesis of (±)-cephalotaxine (**304**) via an intramolecular 1,3-dipolar cycloaddition of an azide with an electron-deficient alkene (Scheme 9.60). The required azide **301** was prepared by coupling the vinyl iodide **299** and the aryl zinc chloride **300** using a Pd(0) catalyst in the presence of *tris*-2-furylphosphine. Intramolecular 1,3-dipolar cycloaddition of the azido enone **301** in boiling xylene afforded the desired keto aziridine **302** in 76% yield. Hydroxylation of **302** according to Davis's procedure followed by oxidation with Dess–Martin periodinane delivered the compound **303**, which was converted to the target molecule (±)-cephalotaxine (**304**).

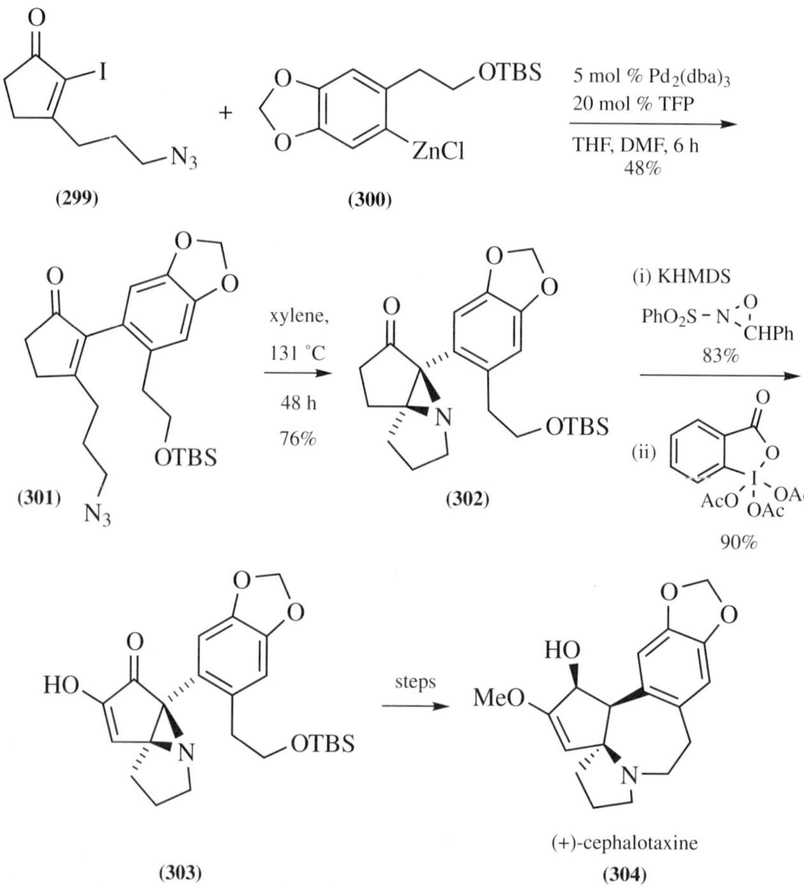

dba = dibenzylideneacetone

Scheme 9.60

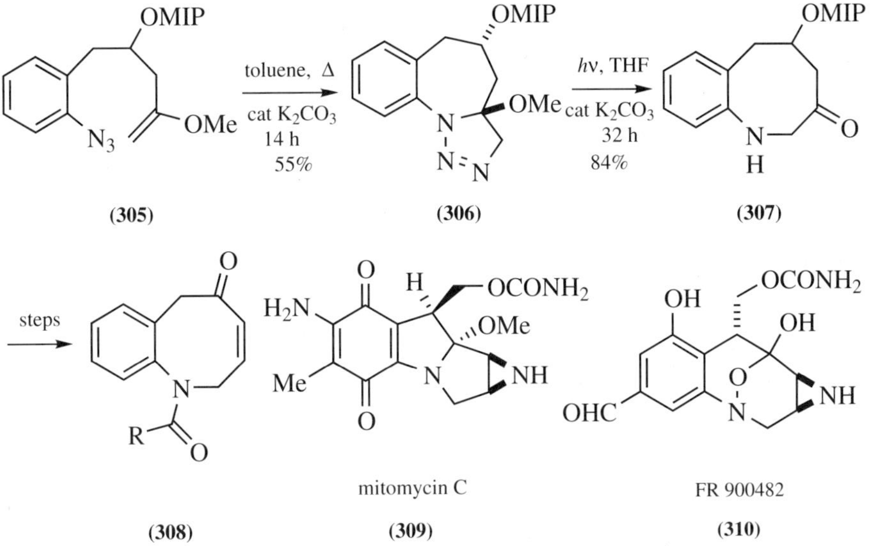

Scheme 9.61

Ciufolini et al. (61) reported a facile assembly of the benzazocenone **307** as a part of the total synthesis of the antitumor alkaloids mitomycin C (**309**) and FR 900482 (**310**) based on intramolecular 1,3-dipolar cycloadditions of aryl azides with electron-rich alkenes (Scheme 9.61). Azide **305** was heated in refluxing toluene with a catalytic amount of K_2CO_3 to give the triazoline **306** in 55% yield. Irradiation of a solution of the triazoline **306** in wet THF with a sun lamp gave an 84% yield of the required benzazocene **308**, which was converted to the target molecules **309** and **310**.

Fleet and co-workers (62) reported the preparation of D-glucose- and D-galactose related triazole carboxylic acids **315** and **318** as glycosidase inhibitors by the intramolecular 1,3-dipolar cycloaddition of an azide with an electron–deficient alkene (Scheme 9.62). Wittig reaction of the azide **312** with (carbomethoxymethylene)triphenylphosphorane in toluene at 120 °C furnished the triazoline **314** in 54% yield. Oxidation of **314** with bromine followed by cleavage of the silyl protecting group and subsequent hydrolysis of the ester afforded the D-*gluco*-triazole carboxylic acid **315** in 90% overall yield. The D-*galacto*-triazole carboxylic acid **318** was similarly prepared from the corresponding azide **316**.

An efficient stereoselective synthesis of the (pyrrolidin-2-ylidene)glycinate intermediate **325** was reported in a total synthesis of carzinophilin (**326**), employing an intramolecular cycloaddition of an azide with an alkene (63) (Scheme 9.63). The arabinose derivative **319** was converted into the required azide **321** via the triflate **320**. Thermolysis of the azide **321** at 50 °C in THF produced the unstable triazoline **322**, which on rearrangement gave the (pyrrolidin-2-ylidene)glycinate **325** in 60–72% overall yield from the triflate **320**.

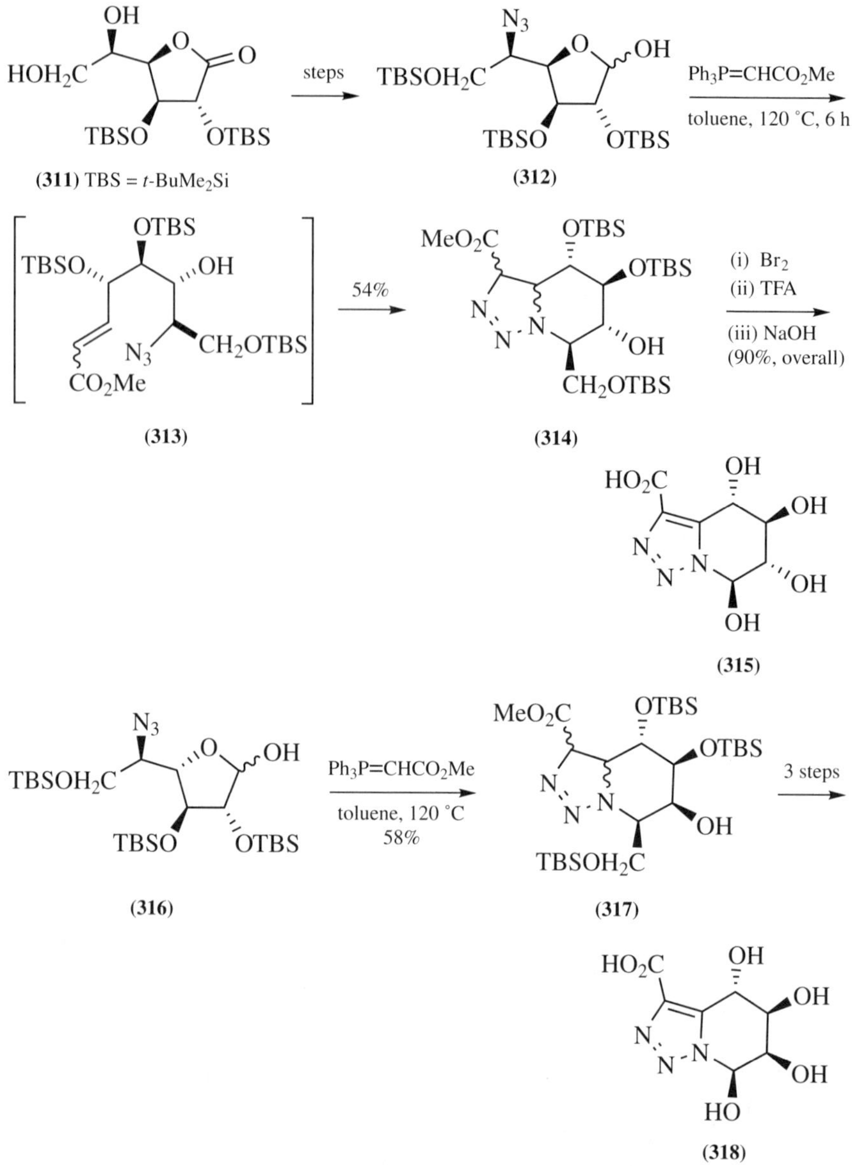

Scheme 9.62

9.2.2. Reaction with the Double Bond of Dienes

Pearson et al. (64) developed an approach to the fused bicyclic 3-pyrrolines **328** based on an intramolecular azide–alkene cycloaddition (Scheme 9.64). Azides (**327**) were heated at various temperatures between 70 and 110 °C to afford the

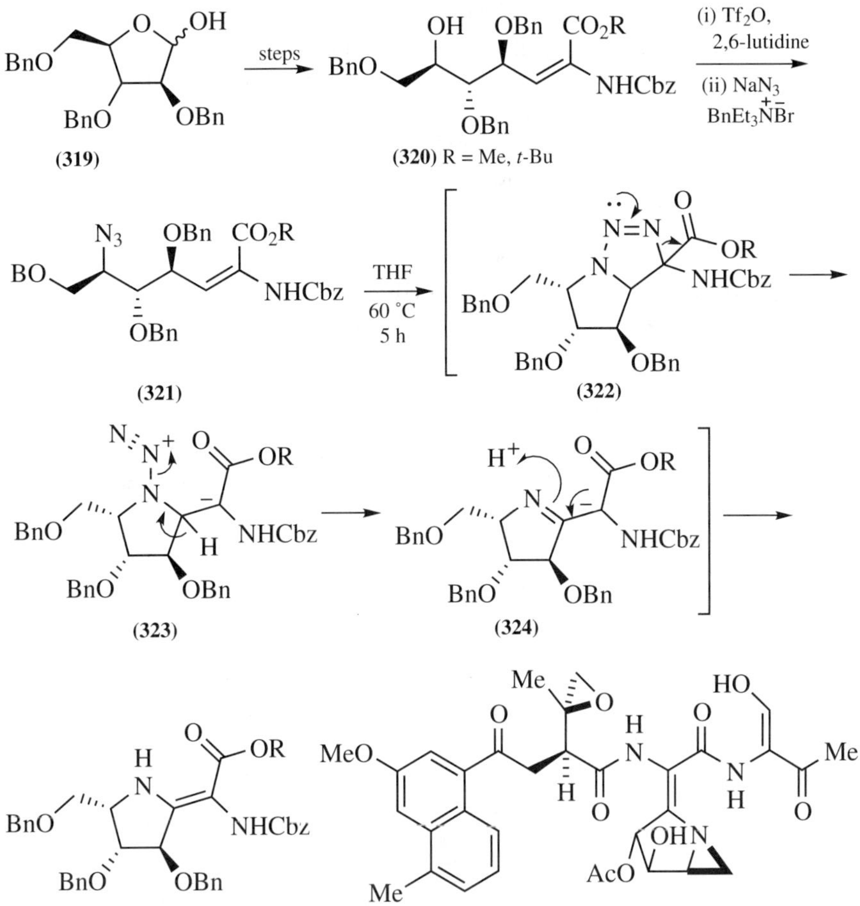

(319) (320) R = Me, *t*-Bu (321) (322) (323) (324) (325) (326) carzinophilin

(325) R = Me (72% from **320**)
R = *t*-Bu (60% from **320**)
Cbz = carbobenzoxy

Scheme 9.63

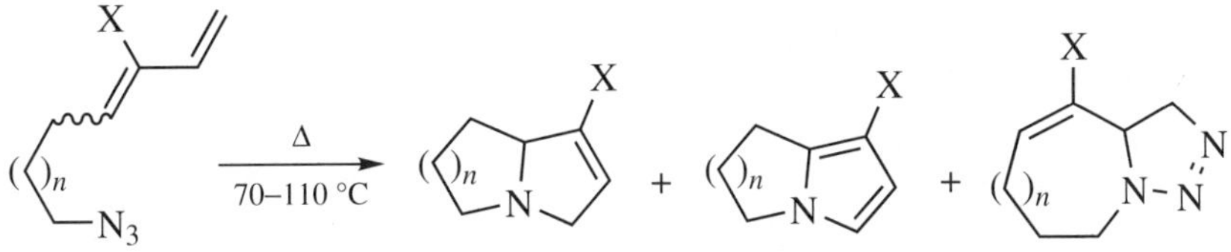

(327) (328) (329) (trace) (330)

n = 1 or 2
X = SPh, OTBS, $OCH(CH_3)OEt$

Scheme 9.64

pyrroline **328**, traces of the pyrrole **329**, and compound **330**, the result of cycloaddition of the azide at the distal monosubstituted olefin. The relative amounts of **328–330** were dependent on the tether length and the nature of the group X.

Hudlicky et al. (65) reported a formal stereoselective total synthesis of the oxygenated pyrrolizidine alkaloids platynecine (**336**), dihydroxyheliotridane (**337**), hastanecine (**341**), and tumeforcidine (**342**), involving an intramolecular azide–diene cycloadditions (Scheme 9.65). Intramolecular 1,3-dipolar cycloaddition of

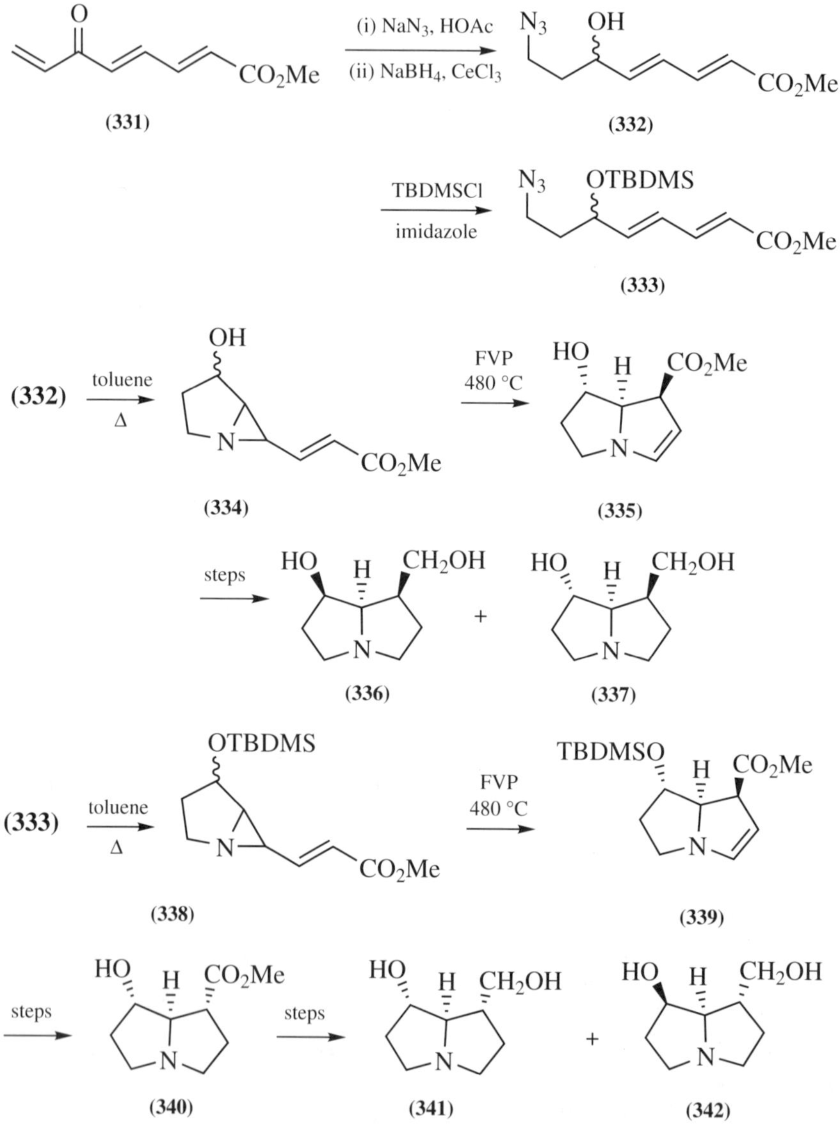

Scheme 9.65

azides **332** and **333** was highly regioselective for the more electron-rich double bond of the diene to yield the corresponding aziridines **334** and **338**. Flash vacuum pyrolysis (FVP) of these aziridines at 480 °C gave the respective pyrrolidines **335** and **339**. These pyrrolidines were converted into the pyrrolizidine natural products using well-established procedures.

When this strategy was applied to the chiral azide **343**, a severe loss of optical activity occurred, but the synthetic sequence was continued to afford the target molecule **347** in almost racemic form (66) (Scheme 9.66).

Trihydroxyheliotridane-6,7-*O*-acetonide (**351**) was synthesized using the same method (67) (Scheme 9.67). When refluxed in benzene, the azide **348** afforded the diastereomeric vinyl aziridines **349** in 44% yield. On FVP of vinyl aziridine (**349**) followed by catalytic reduction and subsequent $LiAlH_4$ reduction of the ester group, trihydroxyheliotridane-6,7-*O*-acetonide (**351**) was formed in 34% yield.

Pearson et al. (68) reported a versatile approach to pyrrolizidine and indolizidine alkaloids such as **355**, **247**, and **362** using intramolecular cycloadditions of azides with electron-rich dienes (Scheme 9.68). Azido dienes **353**, **357**, and **360** that possess a electron-donating group on the diene were prepared from the respective compounds **352**, **356**, and **359**. On heating at 100 °C, the azido diene **353** underwent smooth intramolecular 1,3-dipolar cycloaddition in a stereoselective

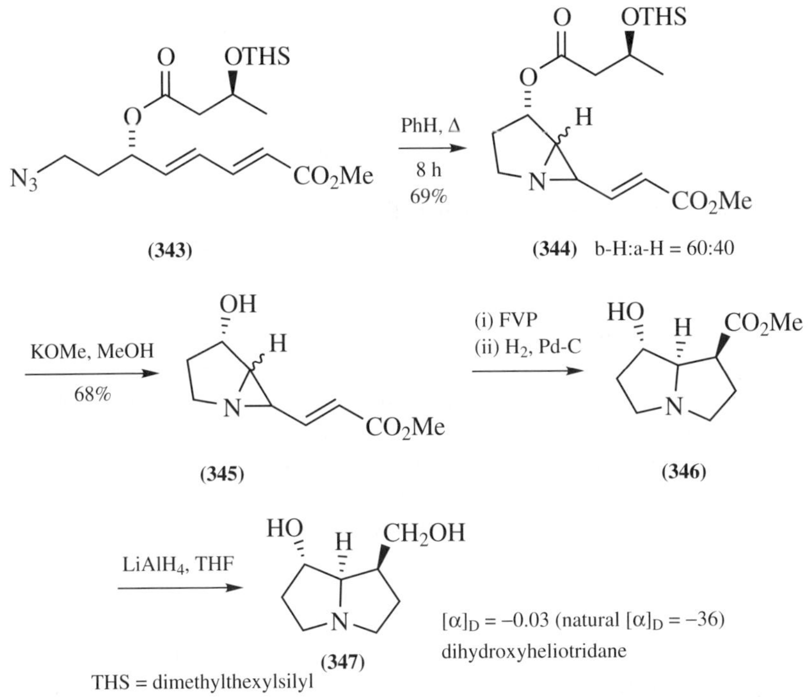

Scheme 9.66

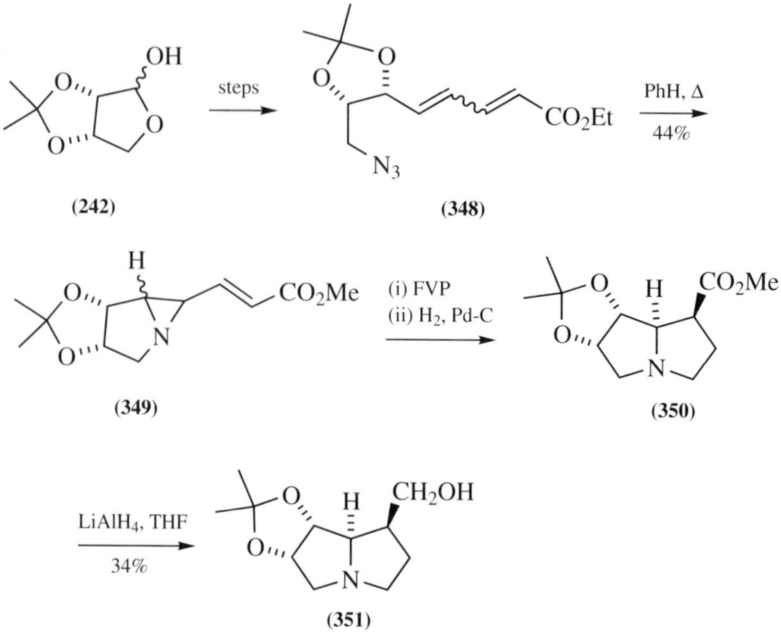

Scheme 9.67

manner to give pyrrolizidine **354** in 74% yield. Azido dienes **357** and **360** similarly afforded the corresponding indolizidines **358** and **361** in 62 and 55% yields, respectively.

The core skeleton of the indolizidine alkaloid gephyrotoxin (**365**) was prepared via the 1,3-dipolar cycloaddition of azido diene **363** (69) (Scheme 9.69).

By using the same method, an attempt toward the synthesis of 3-epiaustraline (**371**) was unsuccessful (70) (Scheme 9.70). Thermolysis of the azido–diene **367** afforded the dehydropyrrolizidine **368** and the triazoline **369** in equal amounts. All attempts to hydrolyze the vinyl sulfide unit of **368** to the ketone **370** were futile, although a more conventional route to these alkaloids proved to be successful.

Cha and co-workers (71) reported an enantioselective synthesis of the amyloglucosidase inhibitor 6,7-diepicastanospermine (**376**) based on an intramolecular 1,3-dipolar cycloaddition of an azide with a diene (Scheme 9.71). Intramolecular 1,3-dipolar cycloaddition of the azide **372** proceeded with complete diastereoselectivity to provide the vinyl aziridine **373** in 52% yield. Ring opening of **373** followed by protection of the alcohol gave the pyrrolidine **374** in 65% yield. Sharpless asymmetric dihydroxylation of **374** followed by cyclization and acetylation afforded the amide **375** in 70% yield. Reduction of the amide group in **375** with borane followed by removal of the acetyl units produced the target molecule **376** in 71% yield.

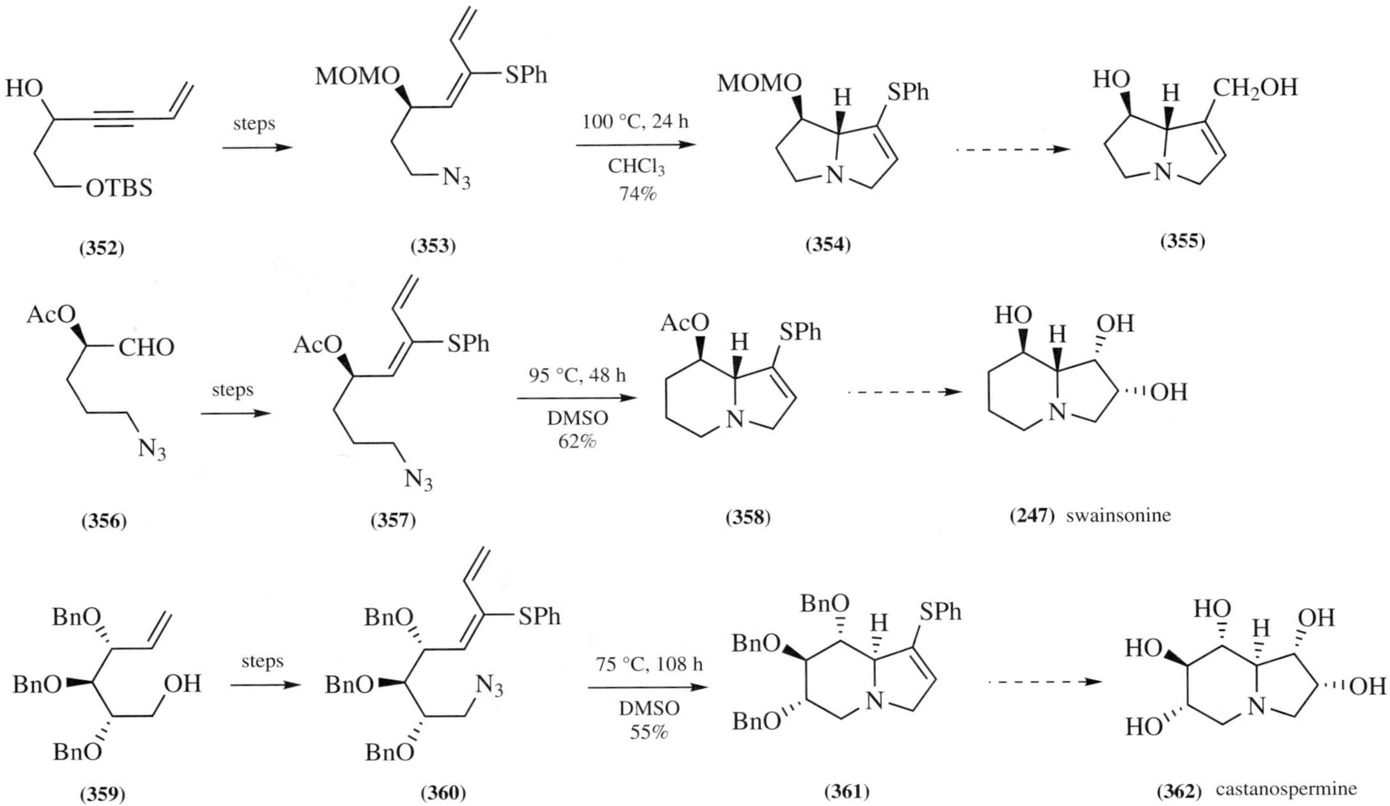

(352)
steps
(353)
100 °C, 24 h
$CHCl_3$
74%
(354)
(355)
(356)
steps
(357)
95 °C, 48 h
DMSO
62%
(358)
(247) swainsonine
(359)
steps
(360)
75 °C, 108 h
DMSO
55%
(361)
(362) castanospermine

Scheme 9.68

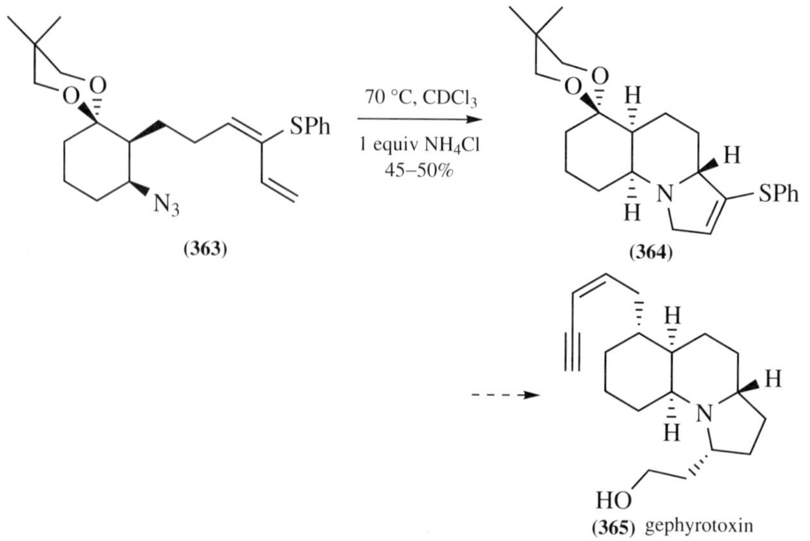

Scheme 9.69

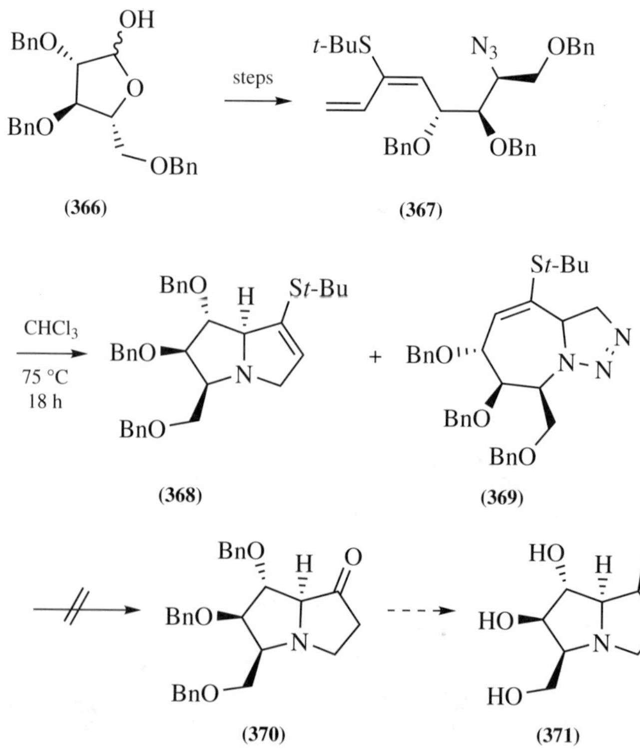

Scheme 9.70

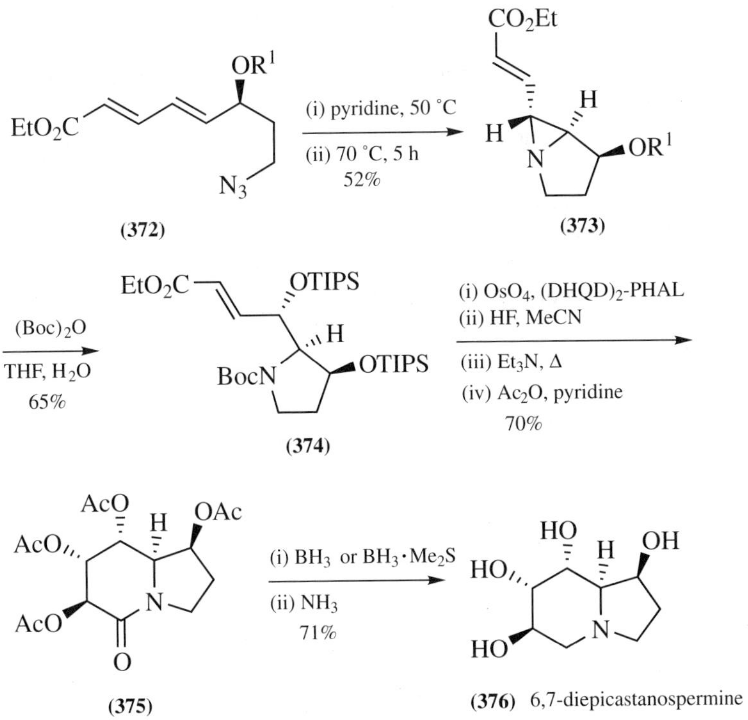

Scheme 9.71

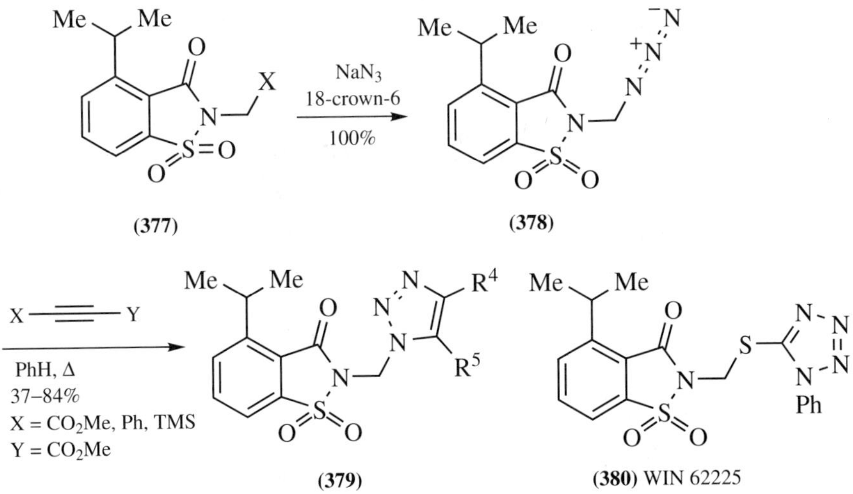

Scheme 9.72

9.2.3. Reaction with Alkynes and Nitriles

Hlasta and Ackerman (72) reported a synthesis of the triazoles **379**, related to the human leuokocyte elastase inhibitor WIN 62225 (**380**), based on an intermolecular 1,3-dipolar cycloaddition of the azide **378** with alkynes (Scheme 9.72). They also investigated in detail the effect of steric and electronic factors on the regioselectivity of the cycloaddition reaction. (Azidomethyl)benzisothiazolone (**378**) underwent smooth 1,3-dipolar cycloaddition with various disubstituted acetylenes to give the corresponding triazoles (**379**) in 37–84% yields. Electron-deficient acetylenic dipolarophiles reacted more rapidly with the azide to give the respective triazoles.

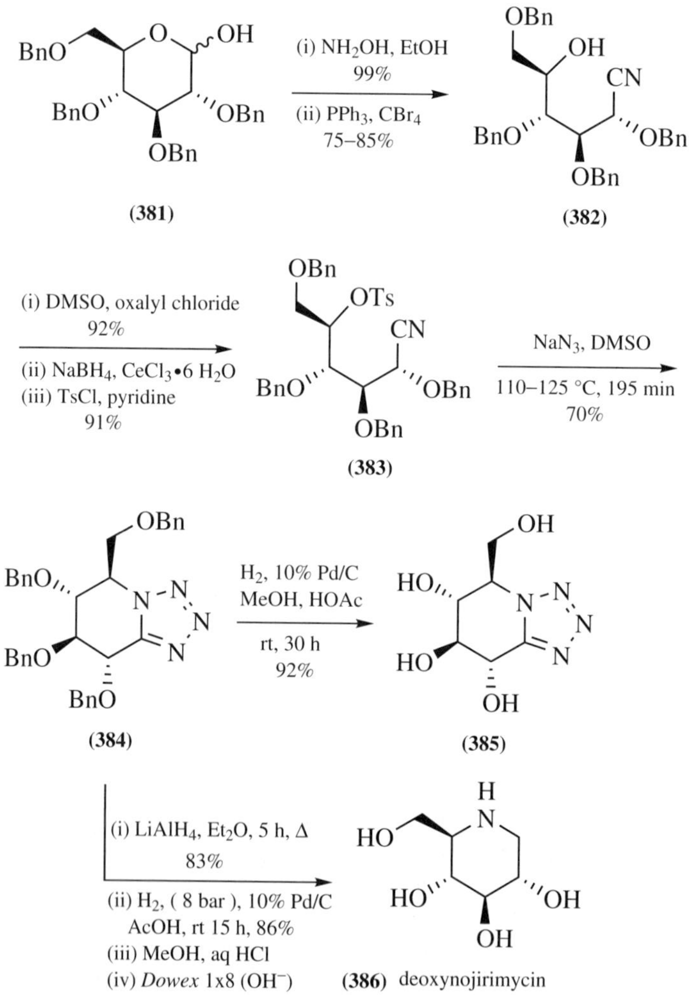

Scheme 9.73

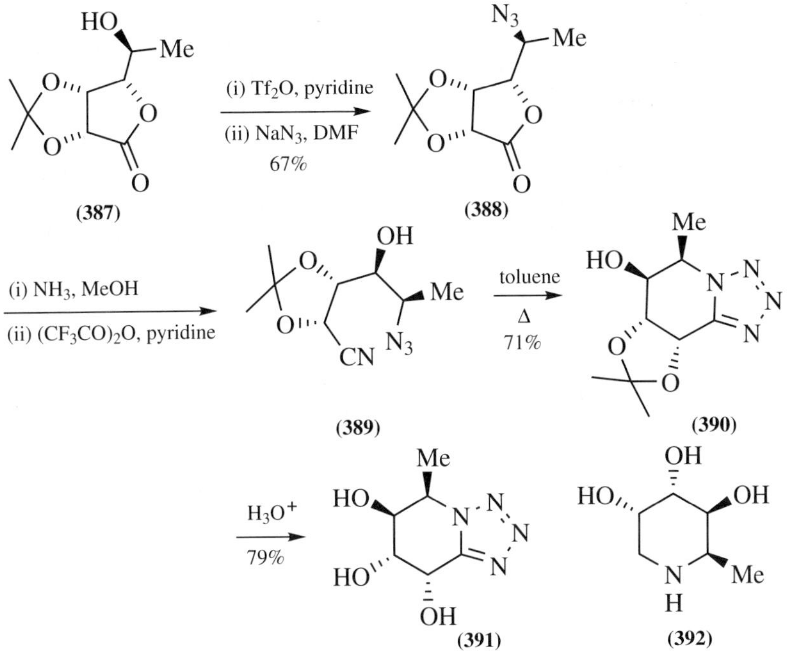

Scheme 9.74

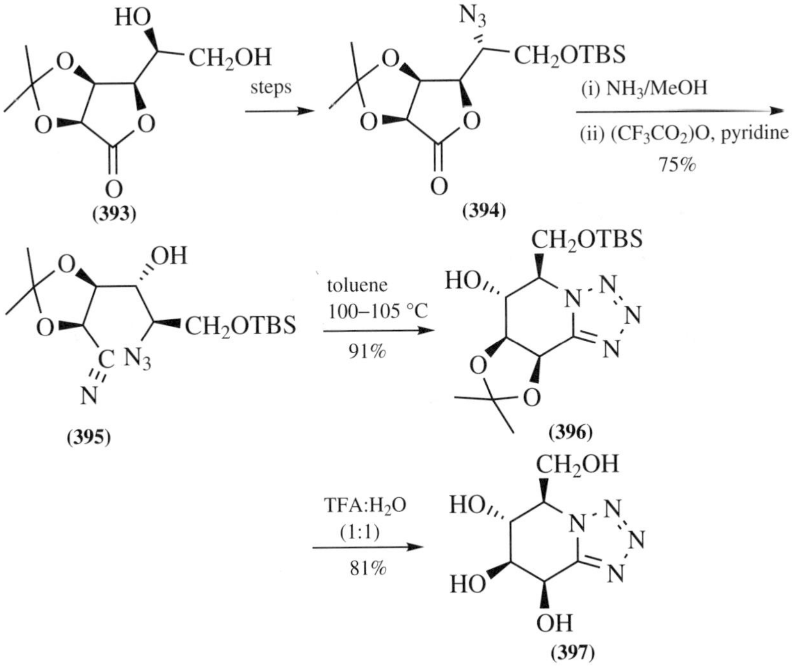

Scheme 9.75

Ermert and Vasella (73) reported a novel synthesis of the glucose-derived tetrazole **385** as a new glucosidase inhibitor via the intramolecular cycloaddition of an azide with a nitrile (Scheme 9.73). Ring opening of *tetra-O*-benzylglucose (**381**) with hydroxylamine followed by dehydration afforded the nitrile **382** in good yield. Compound **382** was then smoothly converted into the tosylate **383**, which on reaction with NaN_3 in DMSO at 110–125 °C gave the tetrazole **384** in 70% yield. Removal of the benzyl protecting groups produced the glucosidase inhibitor **385** in 92% yield. Tetrazole **384** was also converted into deoxynojirimycin (**386**) in a few steps.

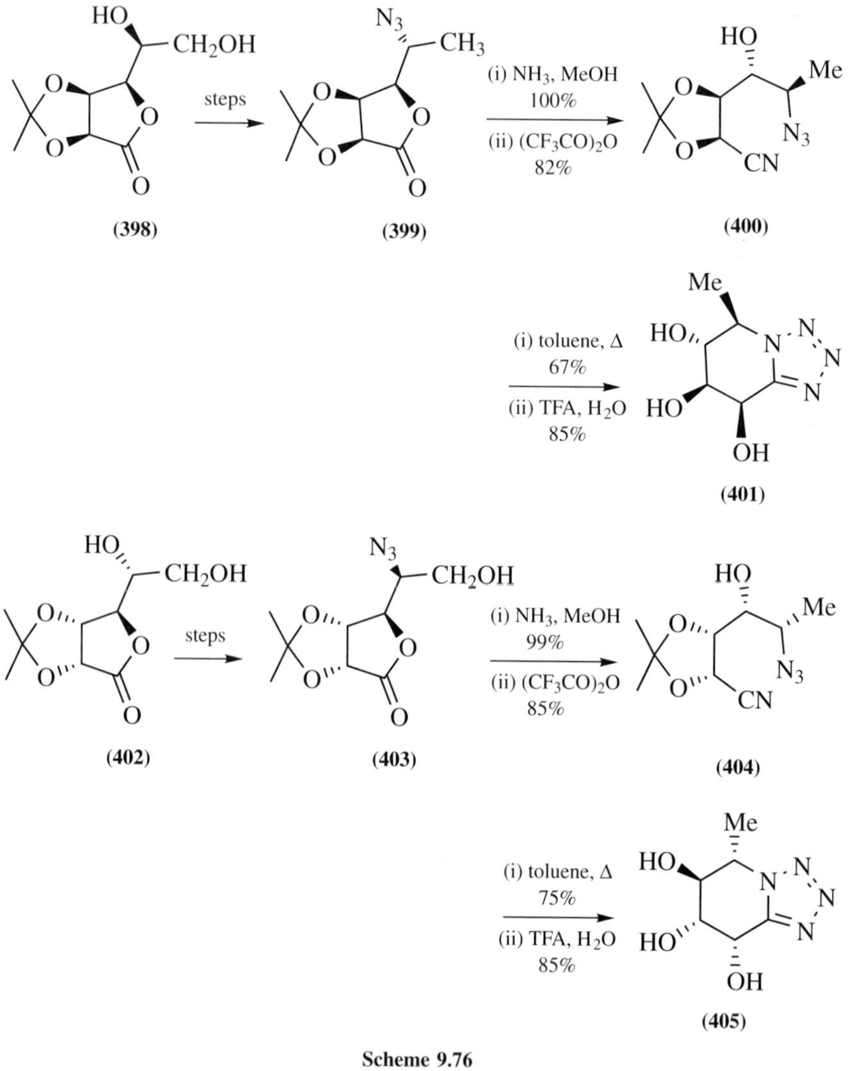

Scheme 9.76

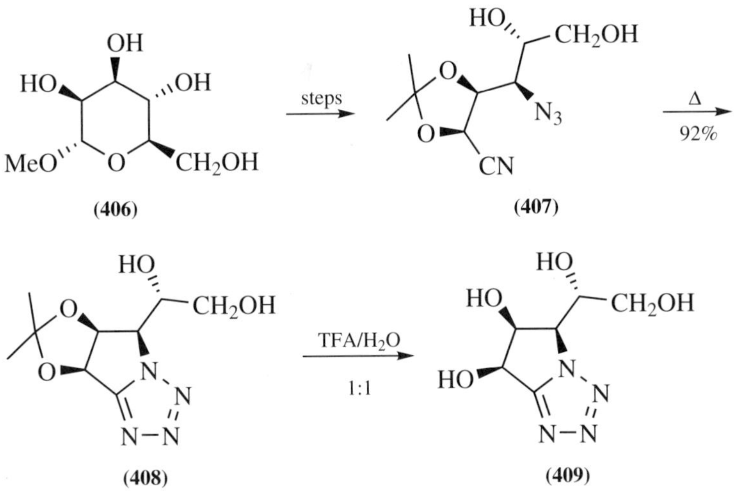

Scheme 9.77

Fleet and co-workers (74) reported a synthesis of a tetrazole analogue of 5-*epi*-L-deoxyrhamnojirimycin (**392**), a potent inhibitor of naringinase, based on an intramolecular cycloaddition of an azide with a nitrile (Scheme 9.74). Esterification of the lactone **387** with triflic anhydride followed by treatment with NaN_3 gave the azido lactone **388** in 67% overall yield. Ring opening of **388** with ammonia followed by dehydration yielded the azido nitrile **389**. Heating of azide **389** in refluxing toluene followed by acid hydrolysis of the ketal unit delivered the tetrazole **391** in 75% overall yield.

Fleet and co-workers (75a) synthesized various tetrazoles from manno- and rhamnopyranoses, as well as furanoses, based on the intramolecular 1,3-dipolar cycloadditions of azides with nitriles (Scheme 9.75). All of these tetrazoles were tested for their inhibitory activities toward both glycosidases and other sugar-processing enzymes. D-Mannopyranotetrazole (**397**) was prepared from L-gluono-lactone (**393**). Azide **394** on ring opening with ammonia followed by dehydration with trifluoroacetic anhydride gave the azido nitrile **395**. Intramolecular 1,3-dipolar cycloaddition of **395** in refluxing toluene followed by deprotection produced the D-mannopyranotetrazole **397** in 86% overall yield.

A similar synthetic approach was used to prepare D-rhamnopyranotetrazole (**401**) and its enantiomer L-rhamnopyranotetrazole (**405**) from the corresponding azides **399** and **403** (Scheme 9.76).

D-Mannofuranotetrazole (**409**) was prepared analogously from methyl pyranoside (**406**) (75b) (Scheme 9.77). Methyl pyranoside (**406**) was smoothly converted into the azido nitrile **407**. On heating in DMSO, the azido nitrile **407** gave the tetrazole **408** in 92% yield. The ketal unit in **408** was cleaved using 1:1 TFA/H_2O to give the target molecule D-mannofuranotetrazole (**409**) in 85% yield.

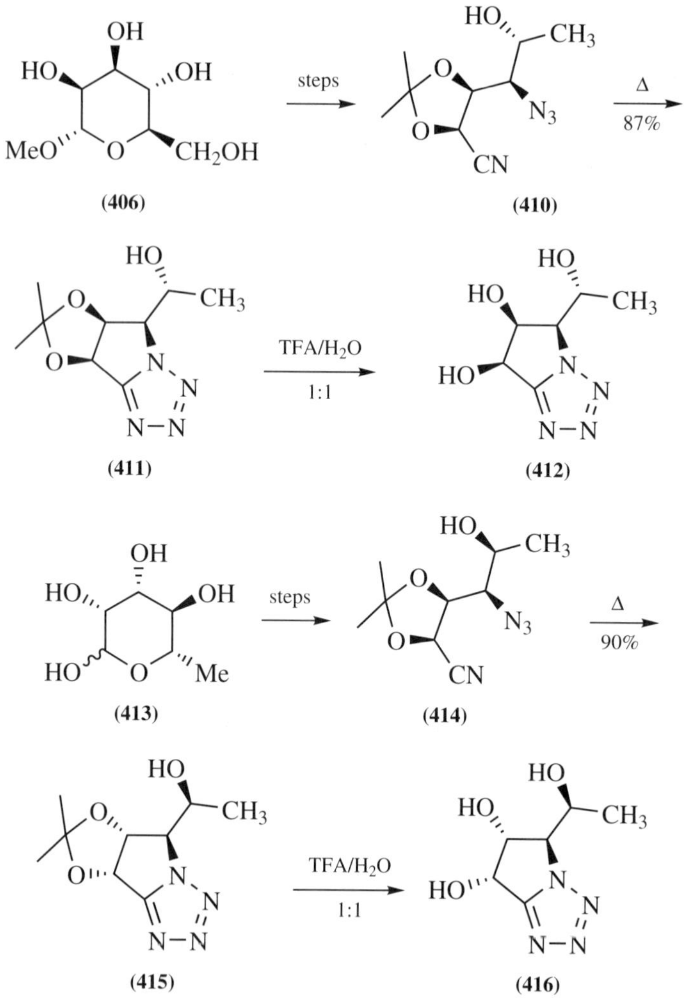

Scheme 9.78

D-Rhamnofuranotetrazole (**412**) and L-rhamnofuranotetrazole (**416**) were similarly prepared from methyl pyranoside (**406**) and L-rhamnose (**413**) in 8 and 15% overall yields, respectively (Scheme 9.78).

9.3. CONCLUSION

The 1,3-dipolar cycloaddition of azides combined with further synthetic transformations is a highly useful reaction for the synthesis of heterocycles and natural products. Even though the chemistry of azide cycloadditions has been known for

>100 years, applications of this strategy for the synthesis of natural products have been realized only toward the end of the twentieth century. We believe that the synthetic potential and new applications of this useful class of 1,3-dipoles will be continue to be developed in the future.

REFERENCES

1. (a) E. F. V. Scriven, *Chem. Rev.*, **1988**, *88*, 297. (b) E. F. V. Scriven, ed., *Azides and Nitrenes, Reactivity and Utility*, Academic Press, Inc., New York, 1984.
2. W. Lwowski, in *1,3-Dipolar Cycloaddition Chemistry*, A. Padwa, ed., Wiley-Interscience, New York, 1984, Chapter 5.
3. (a) C.-K. Sha, K.-S. Chuang, and J.-J. Young, *J. Chem. Soc., Chem. Commun.*, **1984**, 1552. (b) C.-K. Sha, K.-S. Chuang, and S.-J. Wey, *J. Chem. Soc., Perkin Trans. 1.*, **1987**, 977.
4. (a) C.-Y. Tsai and C.-K. Sha, *Tetrahedron Lett.*, **1987**, *28*, 1419. (b) Y.-C. Li, M. S. Thesis, National Tsing Hua University, Taiwan,1987.
5. (a) C.-K. Sha, C.-P. Tsou, and S.-L. Wang, *J. Chem. Soc., Chem. Commun.*, **1988**, 320. (b) C.-K. Sha, C.-P. Tsou, C.-Y. Tsai, J.-M. Liu, Y.-C. Li, F.-Y. Tsai, S.-J. Way, J.-J. Young, K.-S. Chuan, and R.-H. Yeh, *J. Chin. Chem. Soc.*, **1992**, *39*, 635. (c) C.-K. Sha and D.-C. Wang, *Tetrahedron*, **1994**, *50*, 7495.
6. (a) D. J. Goldsmith and J. J. Soria, *Tetrahedron Lett.*, **1986**, *27*, 4701. (b) D. Goldsmith and J. J. Soria, *Tetrahedron Lett.*, **1991**, *32*, 2457.
7. E. Coutouli-Argyropoulou and E. Malamidou-Xenikaki, *J. Heterocyclic Chem.*, **1986**, *23*, 1681.
8. A. Hassner, A. S. Amarasekara, and D. Andisik, *J. Org. Chem.*, **1988**, *53*, 27.
9. R. T. Chakrasali, H. Ila, and H. Junjappa, *Synthesis*, **1988**, 453 and 851.
10. K. S. Keshava Murthy and A. Hassner, *Tetrahedron Lett.*, **1987**, *28*, 97.
11. W. O. Moss, R. H. Bradbury, N. J. Hales, and T. Gallagher, *Tetrahedron Lett.*, **1988**, *29*, 6475.
12. (a) H. Ogawa, M. Kumemura, T. Imoto, I. Miyamoto, H. Kato, and Y. Taniguchi, *J. Chem. Soc., Chem. Commun.*, **1984**, 423. (b) H. Ogawa, M. Kumemura, and T. Imoto, *Tetrahedron Lett.*, **1988**, *29*, 219.
13. H. Sdassi, G. Revial, M. Pfau, and J. d'Angelo, *Tetrahedron Lett.*, **1989**, *30*, 3427.
14. P. K. Kadaba, *Synlett*, **1990**, 349.
15. A. Hassner, M. Stern, H. E. Gottlieb, and F. Frolow, *J. Org. Chem.*, **1990**, *55*, 2304.
16. G. T. Anderson, J. R. Henry, and S. M. Weinreb, *J. Org. Chem.*, **1991**, *56*, 6946.
17. L. Benati, P. C. Montevecchi, and P. Spagnolo, *J. Chem. Soc., Perkin Trans. 1*, **1991**, 71.
18. P. K. Kadaba, *J. Org. Chem.*, **1992**, *57*, 3075.
19. S. Kitane, L. Chraïbi, and M. Soufiaoui, *Tetrahedron*, **1992**, *48*, 8935.
20. K. J. Shea and J.-S. Kim, *J. Am. Chem. Soc.*, **1992**, *114*, 4846.
21. (a) B. S. Orlek, P. G. Sammes, and D. J. Weller, *J. Chem. Soc., Chem. Commun.*, **1993**, 607. (b) B. S. Orlek, P. G. Sammes, and D. J. Weller, *Tetrahedron*, **1993**, *49*, 8179.
22. N. De Kimpe and M. Boeykens, *J. Org. Chem.*, **1994**, *59*, 5189.
23. Z. Lin and P. K. Kadaba, *J. Heterocyclic Chem.*, **1997**, *34*, 1645.
24. S. Fioravanti, L. Pellacani, D. Ricci, and P. A. Tardella, *Tetrahedron: Asymmetry.*, **1997**, *8*, 2261.
25. D. N. Butler, J. R. Malpass, D. Margetic, R. A. Russell, G. Sun, and R. N. Warrener, *Synlett*, **1998**, 588.
26. C. Vogel and P. Delavier, *Tetrahedron Lett.*, **1989**, *30*, 1789.
27. D. K. Wedegaertner, R. K. Kattak, I. Harrison, and S. K. Cristie, *J. Org. Chem.*, **1991**, *56*, 4463.

28. F. Clerici, F. Galletti, D. Pocar, and P. Roversi, *Tetrahedron*, **1996**, *52*, 7183.

29. W. H. Pearson, S. C. Bergmeier, and J. A. Chytra, *Synthesis*, **1990**, 156.

30. (a) L. Garanti, G. Molten, and G. Zecchi, *Heterocycles*, **1994**, *38*, 291. (b) G. Broggini, G. Molteni, and G. Zecchi, *Synthesis*, **1995**, 647. (c) G. Broggini, L. Garanti, G, Molteni, and G. Zecchi, *Heterocycles*, **1999**, *51*, 1295. (d) G. Broggini, G, Molteni, A. Terraneo, and G. Zecchi, *Tetrahedron*, **1999**, *55*, 14803.

31. F. Palacios, A. M. Ochoa de Retana, and J. Pagalday, *Heterocycles*, **1994**, *38*, 95.

32. A. Degl'Innocenti, P. Scafato, A. Capperucci, L. Bartoletti, A. Mordini, and G. Reginato, *Tetrahedron Lett.*, **1995**, *36*, 9031.

33. V. Melai, A. Brillante, and P. Zanirato, *J. Chem. Soc., Perkin Trans. 2*, **1998**, 2447.

34. S. Freeze and P. Norris, *Heterocycles*, **1999**, *51*, 1807.

35. P. B. Bhuyan, H. N. Borah, and J. S. Sandhu, *J. Chem. Soc., Perkin Trans. 1*, **1999**, 3083.

36. O. Simo, A. Rybar, and J. Alfoldi, *J. Heterocyclic Chem.*, **2000**, *37*, 1033.

37. R. Kuang, A. K. Ganguly, T.-M. Chan, B. N. Pramanik, D. J. Blythin, A. T. McPhail, and A. K. Saksena, *Tetrahedron Lett.*, **2000**, *41*, 9575.

38. M. Alterman and A. Hallberg, *J. Org. Chem.*, **2000**, *65*, 7984.

39. G. V. Wuytswinkel, B. Verheyde, F. Compernolle, S. Toppet, and W. Dehaen, *J. Chem. Soc., Perkin Trans. 1*, **2000**, 1337.

40. (a) W. H. Pearson, W.-K. Fang, and J. W. Kampf, *J. Org. Chem.*, **1994**, *59*, 2682. (b) W. H. Pearson, *J. Heterocyclic. Chem.*, **1996**, *33*, 1489.

41. F. Hammerschmidt, J.-P. Polsterer, and E. Zbiral, *Synthesis*, **1995**, 415.

42. A. Yashiro, Y. Nishida, M. Ohno, S. Eguchi, and K. Kobayashi, *Tetrahedron Lett.*, **1998**, *39*, 9031.

43. C. Bellavia-Lund and F. Wudl, *J. Am. Chem. Soc.*, **1997**, *119*, 943.

44. D. M. Guldi, S. Gonzalez, N. Martin, A. Anton, J. Garin, and J. Orduna, *J. Org. Chem.*, **2000**, *65*, 1978.

45. (a) C.-K. Sha, S.-L. Ouyang, D.-Y. Hsieh, and T.-H. Hseu, *J. Chem. Soc., Chem. Commun.*, **1984**, 492. (b) C.-K. Sha, J.-J. Young, C.-P. Yeh, S.-C. Chang, and S.-L. Wang. *J. Org. Chem.*, **1991**, *56*, 2694.

46. C.-K. Sha, S.-L. Ouyang, D.-Y. Hsieh, R.-C. Chang, and S.-C. Chang, *J. Org. Chem.*, **1986**, *51*, 1490.

47. P. Kozikowski and M. N. Greco, *J. Org. Chem.*, **1984**, *49*, 2310.

48. J. G. Buchanan, A. R. Edgar, and B. D. Hewitt, *J. Chem. Soc., Perkin Trans. 1*, **1987**, 2371.

49. T. Fukuyama and L. Yang, *Tetrahedron Lett.*, **1986**, *27*, 6299.

50. R. B. Bennett III, J.-R. Choi, W. D. Montgomery, and J. K. Cha, *J. Am. Chem. Soc.*, **1989**, *111*, 2580.

51. P. C. Heidt, S. C. Bergmeier, and W. H. Pearson, *Tetrahedron Lett.*, **1990**, *31*, 5441.

52. W. H. Pearson and K.-C. Lin, *Tetrahedron Lett.*, **1990**, *31*, 7571.

53. R. B. Bennett III and J. K. Cha, *Tetrahedron Lett.*, **1990**, *31*, 5437.

54. J.-R. Choi, S. Han, and J. K. Cha, *Tetrahedron Lett.*, **1991**, *32*, 6469.

55. P. Garner, W. B. Ho, S. K. Grandhee, W. J. Youngs, and V. O. Kennedy, *J. Org. Chem.*, **1991**, *56*, 5893.

56. W. H. Pearson and J. M. Schkeryantz, *J. Org. Chem.*, **1992**, *57*, 6783.

57. W. H. Pearson and R. Walavalkar, *Tetrahedron*, **1994**, *50*, 12293.

58. F. D. Deroose and P. J. De Clercq, *J. Org. Chem.*, **1995**, *60*, 321.

59. J. M. Schkeryantz and W. H. Pearson, *Tetrahedron*, **1996**, *52*, 3107.

60. G. A. Molander and M. Hiersemann, *Tetrahedron Lett.*, **1997**, *38*, 4347.

61. M. A. Ciufolini, M. Chen, D. P. Lovett, and M. V. Deaton, *Tetrahedron Lett.*, **1997**, *38*, 4355.

62. T. M. Krülle, C. de la Fuente, L. Pickering, R. T. Aplin, K. E. Tsitsanou, S. E. Zographos, N. G. Okionomakos, R. J. Nash, R. C. Griffiths, and G. W. J. Fleet, *Tetrahedron: Asymmetry*, **1997**, *8*, 3807.

63. Y. Konda, T. Sato, K. Tsushima, M. Toda, A. Kusunoki, M. Sakayanagi, N. Sato, K. Takeda, and Y. Harigaya, *Tetrahedron*, **1999**, *55*, 12723.

64. W. H. Pearson, J. E. Celebuski, Y.-F. Poon, B. R. Dixon, and J. H. Glans, *Tetrahedron Lett.*, **1986**, *27*, 6301.

65. T. Hudlicky, G. Seoane, and T. C. Lovelace, *J. Org. Chem.*, **1988**, *53*, 2094.

66. T. Hudlicky, G. Seoane, J. D. Price, and K. G. Gadamasetti, *Synlett*, **1990**, 433.

67. T. Hudlicky, H. Luna, J. D. Price, and F. Rulin, *J. Org. Chem.*, **1990**, *55*, 4683.

68. W. H. Pearson, S. C. Bergmeier, S. Degan, K.-C. Lin, Y.-F. Poon, J. M. Schkeryantz, and J. P. Williams, *J. Org. Chem.*, **1990**, *55*, 5719.

69. W. H. Pearson and Y.-F. Poon, *Tetrahedron Lett.*, **1989**, *30*, 6661.

70. W. H. Pearson and J. V. Hines, *J. Org. Chem.*, **2000**, *65*, 5785.

71. N.-S. Kim, C. H. Kang, and J. K. Cha, *Tetrahedron Lett.*, **1994**, *35*, 3489.

72. D. J. Hlasta and J. H. Ackerman, *J. Org. Chem.*, **1994**, *59*, 6184.

73. P. Ermert and A. Vasella, *Helv. Chim. Acta*, **1991**, 2043.

74. J. P. Shilvock, J. R. Wheatley, B. Davis, R. J. Nash, R. C. Griffiths, M. George Jones, M. Müller, S. Crook, D. J. Watkin, C. Smith, G. S. Besra, P. J. Brennan, and G. W. J. Fleet, *Tetrahedron Lett.*, **1996**, *37*, 8569.

75. (a) B. G. Davis, T. W. Brandstetter, L. Hackett, B. G. Winchester, R. J. Nash, A. A. Watson, R. C. Griffiths, C. Smith, and G. W. J. Fleet, *Tetrahedron*, **1999**, *55*, 4489. (b) B. G. Davis, R. J. Nash, A. A. Watson, C. Smith, and G. W. J. Fleet, *Tetrahedron*, **1999**, *55*, 4501.

CHAPTER 10

Mesoionic Ring Systems

Gordon W. Gribble

Department of Chemistry, Dartmouth College, Hanover, NH

10.1. Synthesis 682
10.1.1. Münchnones (1,3-oxazolium-5-olates) 682
10.1.2. Isomünchnones (1,3-oxazolium-4-olates) 687
10.1.3. Sydnones (1,2,3-oxadiazolium-5-olates) 696
10.1.4. Thiomünchnones (1,3-oxathiolium-5-olates) 696
10.1.5. Thioisomünchnones (1,3-thiazolium-4-olates) 697
10.1.6. 1,3-Thiazolium-4-thiolates 698
10.1.7. Oxamünchnones (1,3-dioxolium-4-olates) 699
10.2. Cycloaddition Reactions 700
10.2.1. Münchnones (1,3-oxazolium-5-olates) 700
10.2.2. Isomünchnones (1,3-oxazolium-5-olates) 725
10.2.3. Sydnones (1,2,3-oxadiazolium-5-olates) 735
10.2.4. Thiomünchnones (1,3-oxathiolium-5-olates) 738
10.2.5. Thioisomünchnones (1,3-thiazolium-4-olates) 739
10.2.6. 1,3-Thiazolium-4-thiolates 742
10.2.7. Oxamünchnones (1,3-dioxolium-4-olates) 742
10.2.8. 1,3-Dithiolium-4-olates 742
10.2.9. 1,3-Oxathiolium-4-olates 744
10.2.10. 1,3,2-Oxathiazolium-5-olates 745
10.2.11. Miscellaneous Mesoionic Heterocycles 745
10.3. Concluding Remarks 747

In the nearly 20 years since the publication of the outstanding review of mesoionic ring systems in *1,3-Dipolar Cycloaddition Chemistry* by Potts (1), the mystique surrounding these fascinating molecules has evaporated to reveal a collection of remarkably versatile compounds that are powerful 1,3-dipoles in cycloaddition chemistry. The main focus of this chapter consists of the synthetic

The Chemistry of Heterocyclic Compounds, Volume 59: Synthetic Applications of 1,3-Dipolar Cycloaddition Chemistry Toward Heterocycles and Natural Products, Edited by Albert Padwa and William H. Pearson.
ISBN 0-471-38726-6

applications of mesoionic heterocycles in 1,3-dipolar cycloaddition reactions, particularly those involving münchnones (**1**) (1,3-oxazolium-5-olates) and the newer isomünchnones (**2**) (1,3-oxazolium-4-olates), about which most of the chemistry has been reported. Lesser work has been described involving the mesoionic sydnones (**3**) (1,2,3-oxadiazolium-5-olates), thiomünchnones (**4**) (1,3-oxathiolium-5-olates), thioisomünchnones (**5**) (1,3-thiazolium-4-olates), and other mesoionic ring systems.

Since the nomenclature and classification of mesoionic heterocyclic ring systems has been adequately presented by Potts (1), and has been discussed further by Ollis et al. (2), these topics will not be repeated here.

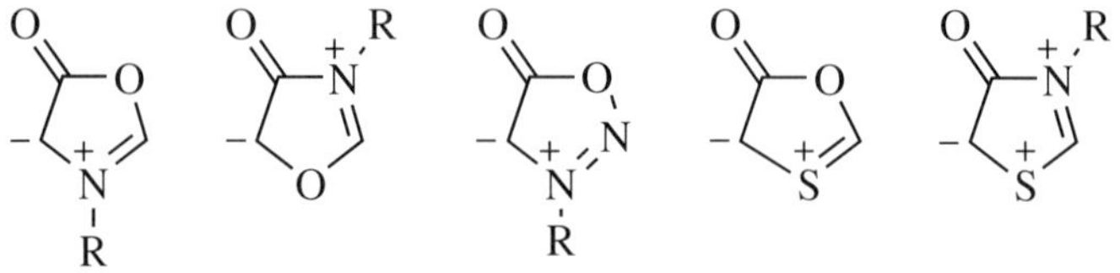

(**1**) (**2**) (**3**) (**4**) (**5**)

10.1. SYNTHESIS

While this section ostensibly deals only with the new syntheses of mesoionic heterocycles, it is appropriate in some cases to include the accompanying 1,3-dipolar cycloaddition reactions reported by the authors.

10.1.1. Münchnones (1,3-oxazolium-5-olates)

The traditional synthesis of münchnones involves the cyclodehydration of N-acylamino acids usually with acetic anhydride or another acid anhydride. Potts and Yao (3) were apparently the first to employ dicyclohexylcarbodiimide (DCC) to generate mesoionic heterocycles, including münchnones. Subsequently, Anderson and Heider (4) discovered that münchnones can be formed by the cyclodehydration of N-acylamino acids using *N*-ethyl-*N′*-dimethylaminopropylcarbodiimide (EDC) or silicon tetrachloride. The advantage of EDC over DCC is that the urea byproduct is water soluble and easily removed, in contrast to dicyclohexylurea formed from DCC. Although the authors conclude that the traditional Huisgen method of acetic anhydride is *still the method of choice*, these two newer methods are important alternatives. Some examples from the work of Anderson and Heider are shown. The *in situ* generated münchnones (not shown) were trapped either with dimethyl acetylenedicarboxylate (DMAD) or ethyl propiolate.

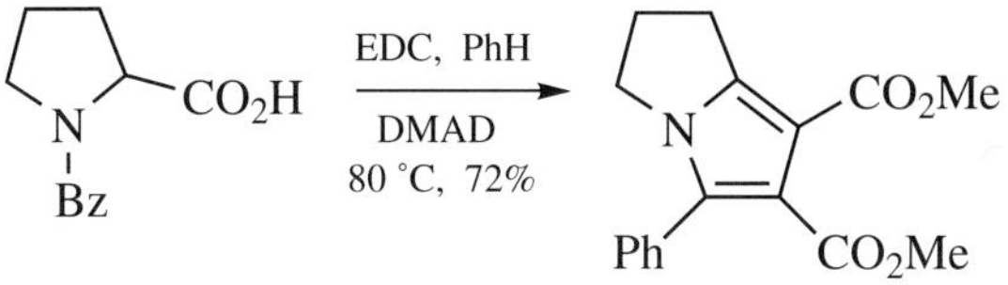

(6) **(7)**

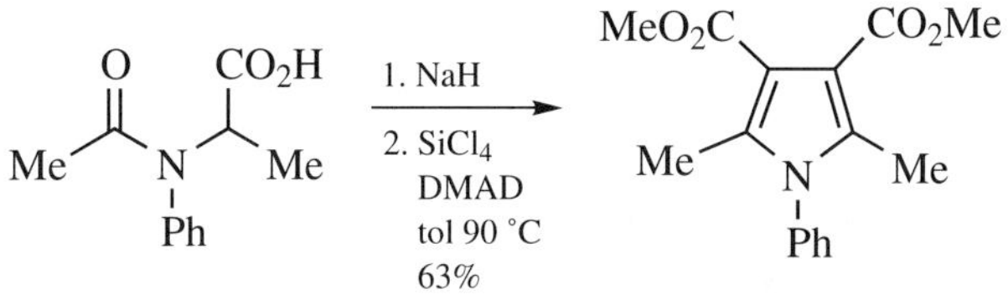

(8) **(9)**

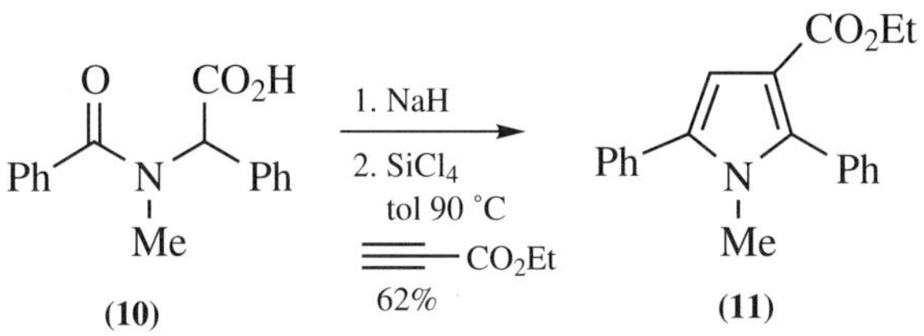

(10) **(11)**

Bz = benzoyl
tol = toluene

During a peptide synthesis study, Slebioda (5) apparently independently observed the formation of münchnone (**12**) from *N*-benzoylphenylalanine and DCC. He was able to isolate and fully characterize this crystalline münchnone along with oxazolone **13**. The tautomerization of **13** to **12** as a function of base and solvent is best effected with triethylamine in DMF (68% yield).

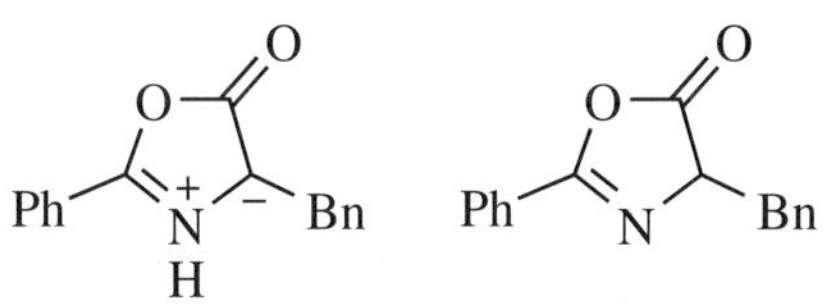

(12) **(13)**

Bn = benzyl

Another important route to münchnones is that described by Boyd and Wright (6,7) involving the cyclodehydration of N-acylamino acids with acetic anhydride in the presence of perchloric acid to give 1,3-oxazolonium salts (*N*-alkylated azlactones). Upon exposure to mild base (triethylamine or sodium carbonate) münchnones are formed. Hershenson and Pavia (8) reported a variation involving the *in situ* alkylation of azlactones (**14**), deprotonation with 2,6-di-*tert*-butylpyridine, and trapping the resulting münchnones (**15**) with DMAD to give pyrroles (**16**) in good yield.

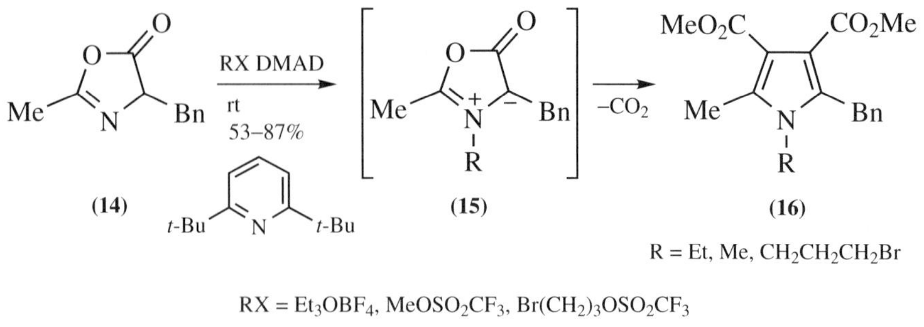

Wilde (9) generated N-acyl münchnones for the first time via the acylation and desilylation of 5-siloxyoxazoles. Thus, for example, exposure of **17** to acetyl chloride in the presence of DMAD affords *N*-acetylpyrrole (**19**) via *N*-acetylmünchnone **18**. Other trapping agents (thiocarbonyls, *N*-phenylmaleimide) failed and the analogous trimethylsiloxyoxazoles were unusually labile and gave lower yields of pyrroles (13–15%). Ethyl chloroformate was also used in this sequence and gave, for example, pyrroles **20–21**.

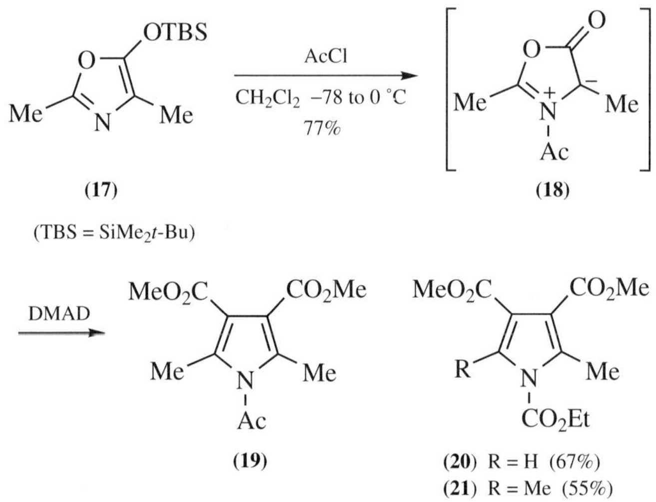

Moreover, Wilde (9) also found that N-acyl münchnones are prone to equilibrate through ring-chain tautomerism, a process well known for N-alkyl münchnones (10–13). Thus, as shown in Scheme 10.1, oxazole **22** reacts with benzoyl chloride in the presence of DMAD to give a mixture of pyrroles **24** and **27** presumably via ketene **25**, which (slowly) equilibrates münchnones **23** and **26**. Capture of münchnone **23** by DMAD is faster than equilibration to münchnone **26**.

In a protocol similar to that reported by Hershenson and Pavia (8) (see above), Wilde (9) found that *N*-methyl münchnone (**28**) could be generated using methyl triflate and trapped with DMAD and thiobenzophenone to give **29** and **31**,

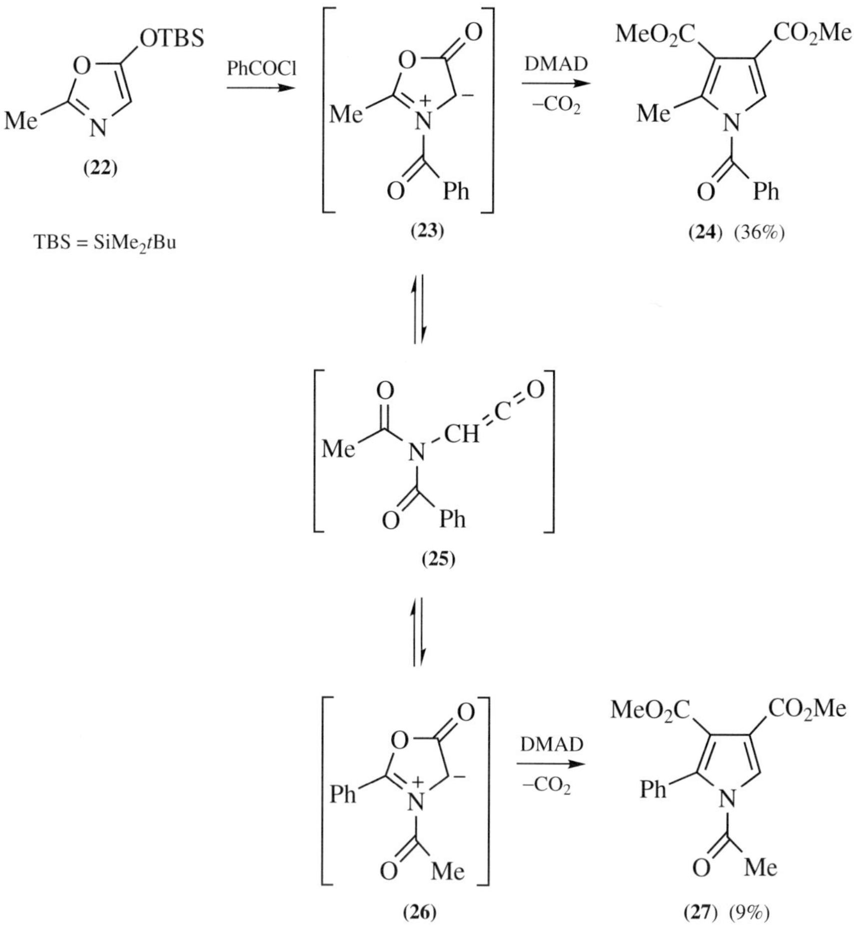

Scheme 10.1

respectively (Scheme 10.2). Other trapping agents such as trimethylsilyl chloride and *p*-toluenesulfonyl chloride afforded lower yields of cycloadducts with DMAD.

Armstrong and co-workers (14–17) employed the elegant Ugi four-component condensation (18,19) to construct münchnone precursors and, following deprotonation, münchnones. The overall sequence, which has been adapted to the solid-phase synthesis of pyrroles by Armstrong and co-workers (17) and, independently, by Mjalli et al. (20), is illustrated in Scheme 10.3 for the synthesis of pyrrole **35** (17). The Ugi condensation leads directly to **33** and münchnone (**34**) formation is induced with acid. Trapping of **34** with DMAD and cleavage from the resin yields **35**. Mjalli et al. (20) generated several other solid-supported münchnones, which were converted into pyrroles by trapping with DMAD and other alkynes.

Merlic et al. (21) reported a novel generation of münchnones from acylamino chromium carbene complexes (Scheme 10.4). Thus, exposure of complex **36** to

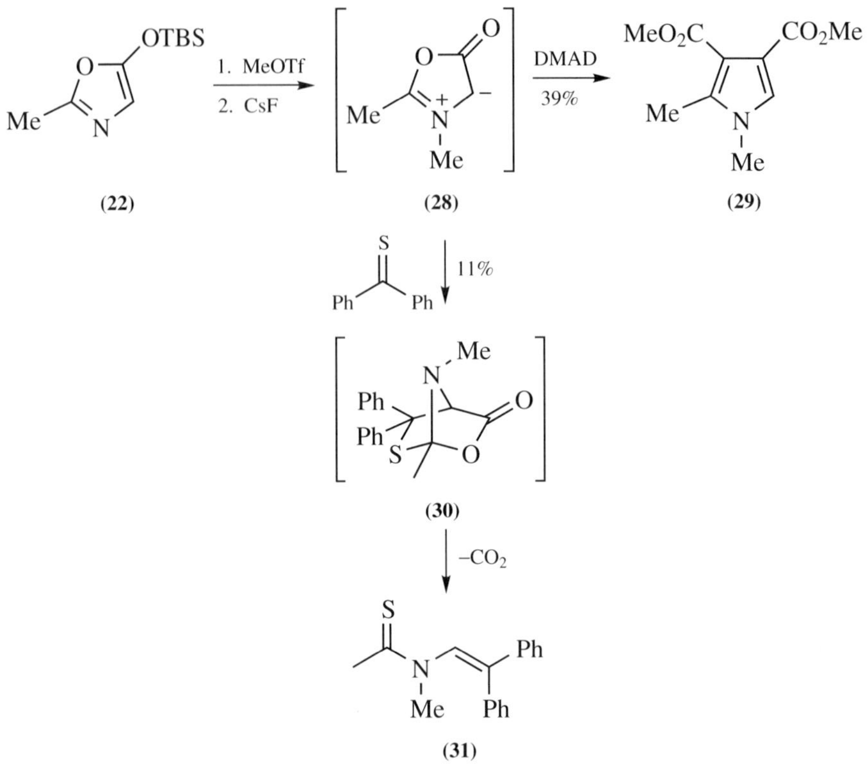

Scheme 10.2

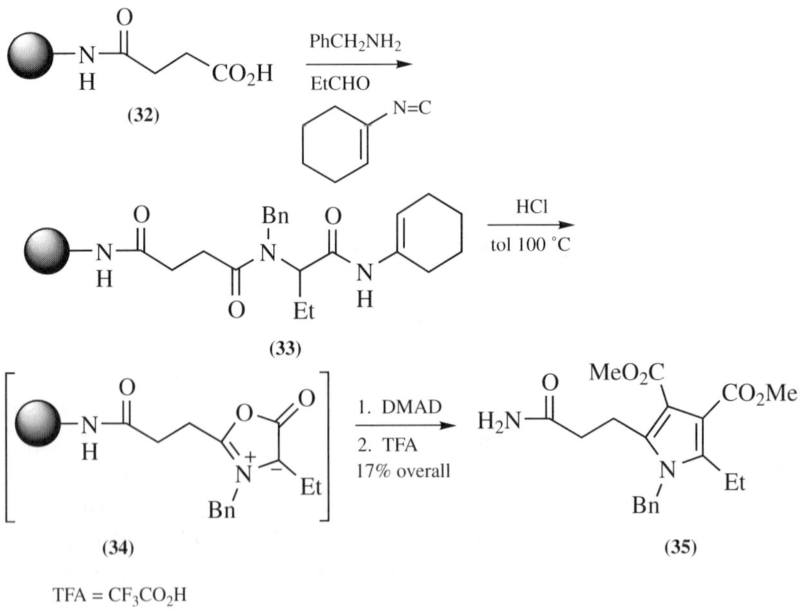

Scheme 10.3

carbon monoxide affords the chromium ketene complex **37**, which cyclizes to münchnone **38** and can be isolated in 27% yield. Performing this reaction in the presence of DMAD furnishes pyrrole **39** in 78% yield. The related but more stable chelate carbene complex **40** also reacts with pressurized carbon monoxide to give münchnones that can be trapped with alkynes to afford the expected pyrroles **41**.

The relatively sensitive acylamino chromium complexes (e.g., **43**) can be prepared *in situ* from stable amino carbene complexes (e.g., **42**) as shown for the generation of münchnone **44** and conversion to pyrrole **45** with DMAD (Scheme 10.5) (21).

10.1.2. Isomünchnones (1,3-oxazolium-5-olates)

Although the mesoionic 1,3-oxazolium-4-olates, "isomünchnones", occupied only a few pages in the reviews by Potts (1) and Gingrich and Baum (10), in the intervening years this ring system has exploded in popularity, largely due to the efforts of Padwa and co-workers. Padwa has summarized his isomünchnone work in several reviews (22–27). While isomünchnones are rarely isolable, these carbonyl

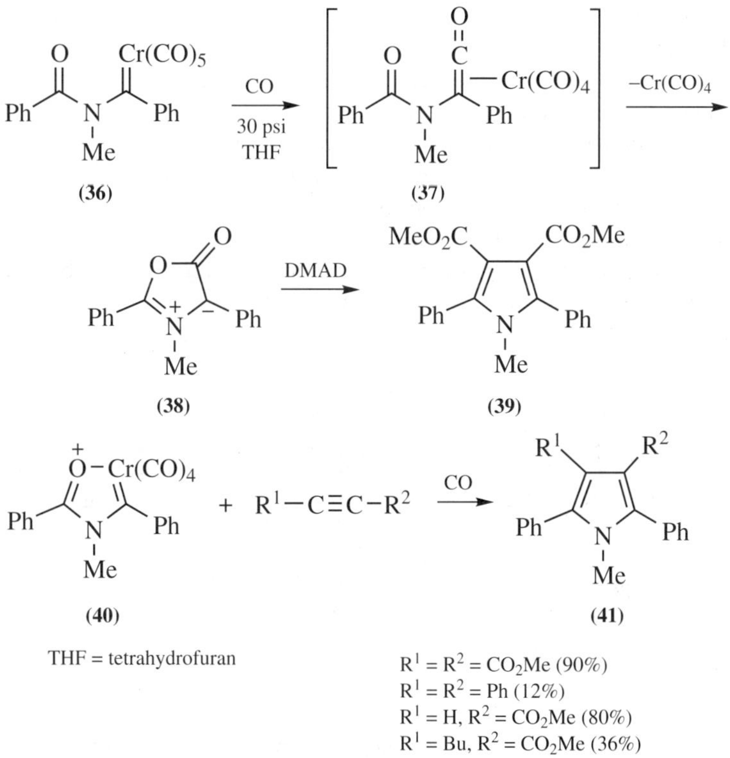

Scheme 10.4

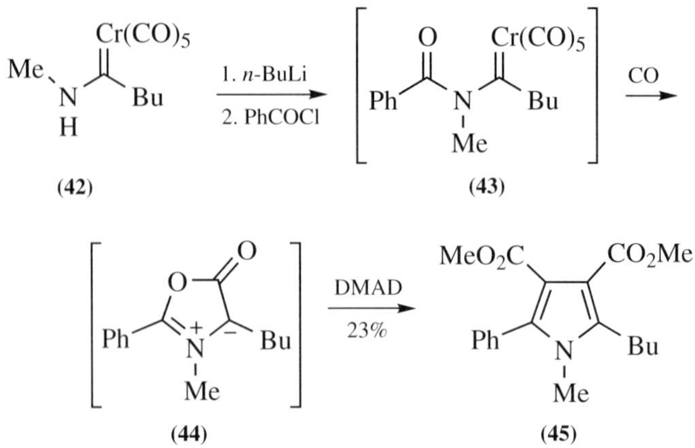

Scheme 10.5

ylide 1,3-dipoles often undergo 1,3-dipolar cycloaddition reactions extremely well as discussed in Section 10.2.2.

Only a few methods are available for the synthesis of isomünchnones, and the original techniques are summarized by Gingrich and Baum (10). Padwa and coworkers (25) provided a more recent historical account of the synthesis of isomünchnones. The major new development is the generation of isomünchnones using the rhodium(II)-catalyzed decomposition of α-diazo carbonyl compounds. In continuation of earlier work (28), Haddadin and Tannus (29) extended their new synthesis of isomünchnones **51** from *N*-benzoylphenylglyoxanilides (**50**) as illustrated in Scheme 10.6 to a series of new compounds.

In some cases these highly colored isomünchnones were quite stable for weeks (**51a–c**), moderately stable for 2–10 min (**51d–n**), or unstable within seconds (**51o**). Isomünchnones **51a–c** crystallized out of the reaction mixture.

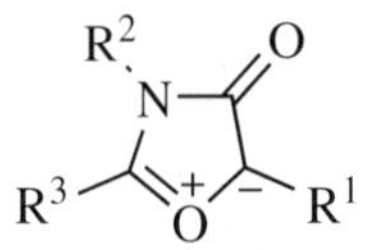

(**51a**) $R^1 = R^2 = R^3 = Ph$
(**51b**) $R^1 = R^2 = Ph$, $R^3 = 4\text{-}NO_2\text{-}Ph$
(**51c**) $R^1 = 4\text{-}Br\text{-}Ph$, $R^2 = Ph$, $R^3 = 4\text{-}NO_2\text{-}Ph$
(**51d**) $R^1 = R^2 = Ph$, $R^3 = 4\text{-}MeO\text{-}Ph$
(**51e**) $R^1 = R^2 = Ph$, $R^3 = t\text{-}Bu$
(**51f**) $R^1 = R^2 = Ph$, $R^3 = CH{=}CHPh$
(**51g**) $R^1 = R^3 = Ph$, $R^2 = 2,5\text{-}Me_2\text{-}Ph$
(**51h**) $R^1 = Ph$, $R^2 = 2,5\text{-}Me_2\text{-}Ph$, $R^3 = 4\text{-}NO_2\text{-}Ph$
(**51i**) $R^1 = R^3 = Ph$, $R^2 = 2\text{-naphthyl}$
(**51j**) $R^1 = Ph$, $R^2 = 2\text{-naphthyl}$, $R^3 = 4\text{-}NO_2\text{-}Ph$
(**51k**) $R^1 = 4\text{-}Br\text{-}Ph$, $R^2 = Ph$, $R^3 = 4\text{-}Me\text{-}Ph$
(**51l**) $R^1 = 4\text{-}Br\text{-}Ph$, $R^2 = Ph$, $R^3 = CH{=}CHPh$
(**51m**) $R^1 = 4\text{-}Me\text{-}Ph$, $R^2 = R^3 = Ph$
(**51n**) $R^1 = R^2 = Ph$, $R^3 = 1\text{-naphthyl}$
(**51o**) $R^1 = R^2 = Ph$, $R^3 = Me$

Only a few of these isomünchnones (**51**) reacted with *N*-phenylmaleimide (NPM) to give stable adducts. Thus, although **51a** gave a mixture of exo and

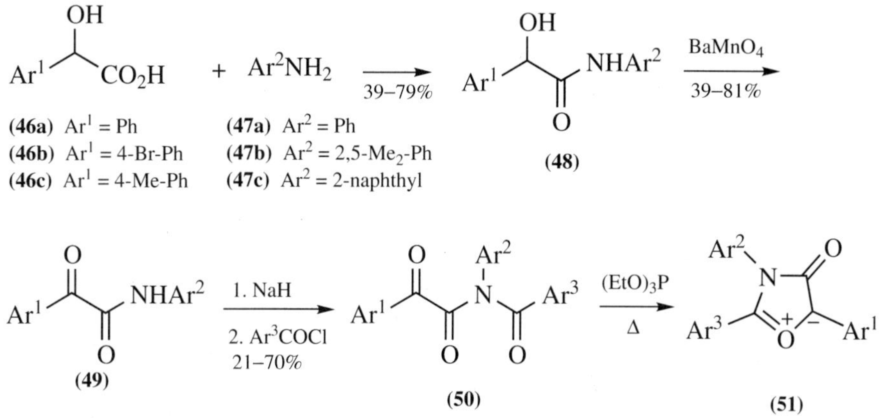

Scheme 10.6

endo adducts (28), **51b** and **51c** gave mainly endo adducts **52** and **53**, respectively. Haddadin et al. (28) were unable to isolate cycloadducts of the other isomünchnones with NPM.

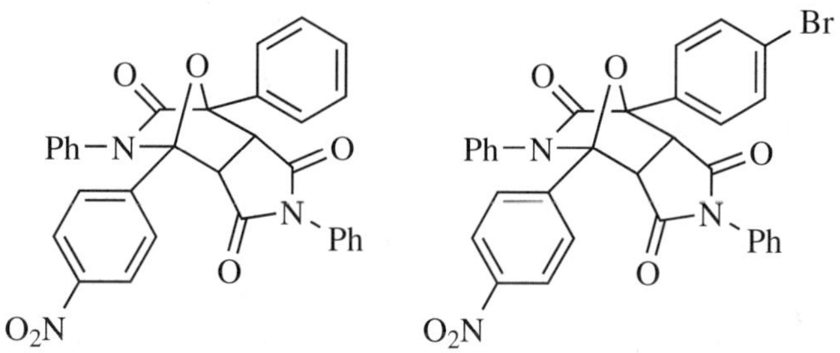

(52) (36%) **(53)** (46%)

Mathias and Moore (30–33) described a new synthesis of isomünchnones **55** via the thermal cyclization of *N*-(chloroacetyl)lactams (**54**) (Scheme 10.7). These isomünchnones can be captured by NPM to give fused 2-pyridones in moderate yields. Cycloadducts from the reaction with DMAD are produced in much lower yields ($\leq$17%), and other olefinic dipolarophiles (fumarate, maleate, acrylate, and dicyanocyclobutene) are unreactive. Reaction of *N*-(chloroacetyl)benzamide (**57**) in the presence of NPM gave **58** in low yield.

Doyle et al. (34) were the first group to generate isomünchnones from diazo imides using Rh(II) catalysis. For example, isomünchnone **60** was produced from diazo imide **59**, but attempts to trap this species with ethyl acrylate were unsuccessful. The only material identified was the isomünchnone hydrolysis product. This use of Rh(II) to generate a rhodium–carbenoid species from an α-diazo carbonyl compound is reminiscent of the first successful synthesis of

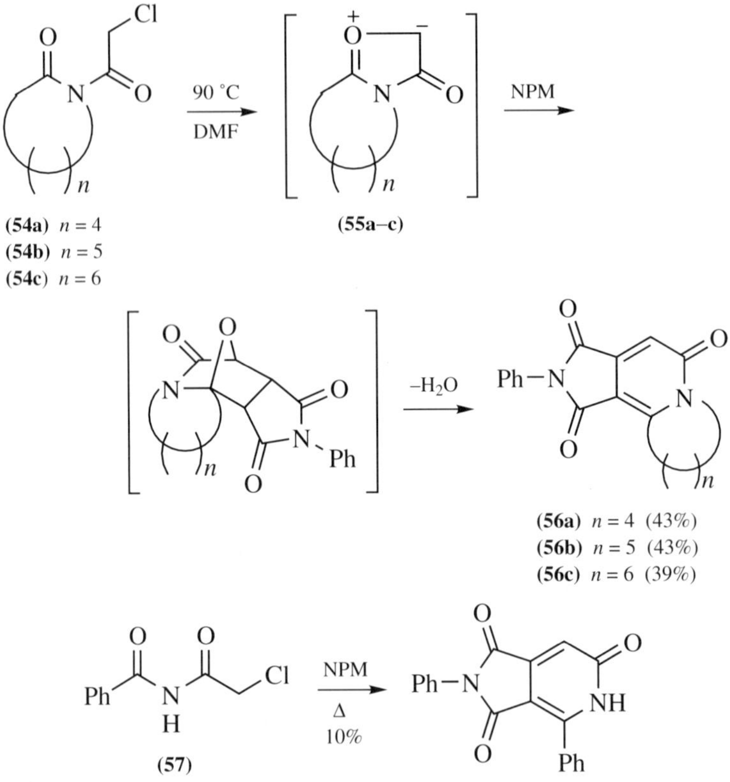

(**54a**) $n = 4$
(**54b**) $n = 5$
(**54c**) $n = 6$

(**55a–c**)

(**56a**) $n = 4$ (43%)
(**56b**) $n = 5$ (43%)
(**56c**) $n = 6$ (39%)

(**57**)

(**58**)

DMF = dimethylformamide

Scheme 10.7

isomünchnones by Hamaguchi and Ibata (35), which involved Cu(II) to generate a copper–carbenoid species from an α-diazo imide that subsequently cyclized to an isomünchnone.

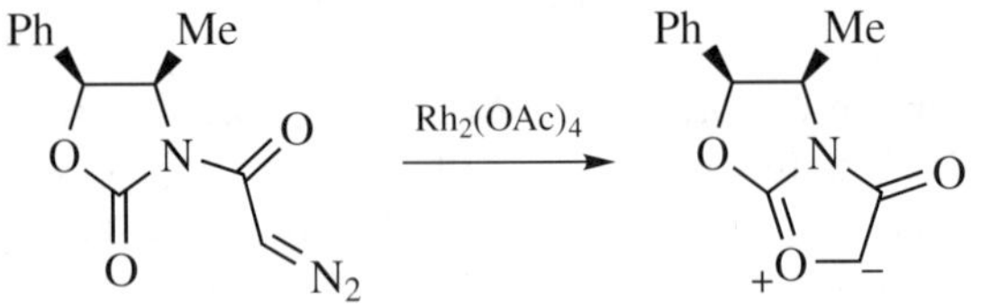

(**59**) (**60**)

The first successful generation and trapping of isomünchnones using this strategy was described independently by Maier et al. (36,37) and Padwa et al. (38,39). Maier and Evertz (36) were the first workers to report the intramolecular dipolar cycloaddition of isomünchnones to alkenes, the reaction that Padwa would later exploit so spectacularly. Thus, diazo imide **62** was readily prepared from

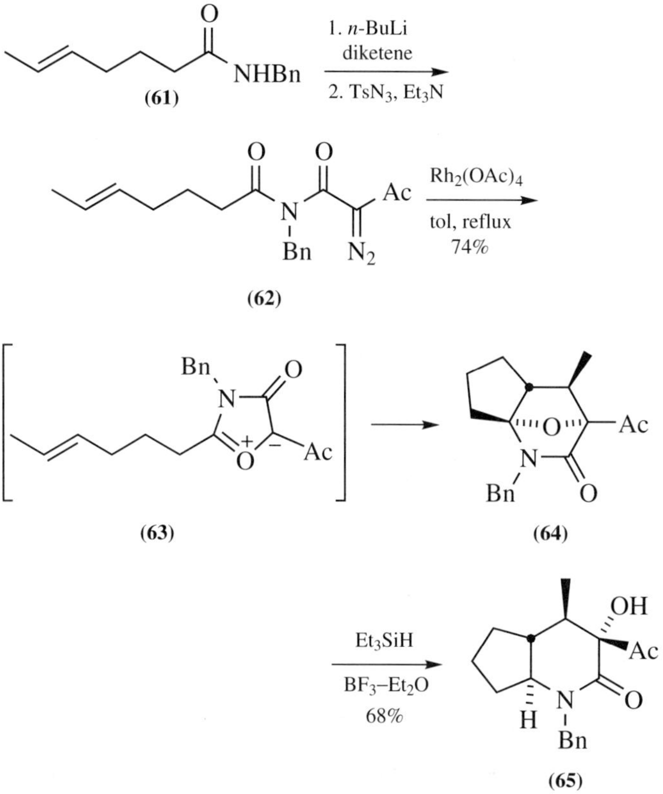

Scheme 10.8

amide **61** by acylation and diazo transfer (Scheme 10.8). Reaction of **62** with rhodium acetate generates isomünchnone **63**, which smoothly cyclizes to afford tricycle **64**. Several other examples are described in this article. Reductive ring opening of **64** led to 2-piperidone **65**.

Maier and Schöffling (37) extended this intramolecular isomünchnone cycloaddition to a synthesis of fused furans by employing an alkyne dipolarophile (Scheme 10.9). Thus, the diazo acetylenes (**66**) are smoothly converted to furans (**69**) via isomünchnones (**67**) with catalytic rhodium acetate.

Padwa et al. (38) also explored the rhodium-catalyzed reaction of diazo imides to form isomünchnones (Scheme 10.10). Thus, **70** smoothly forms isomünchnones **71** that can be intercepted in high yield with DMAD to give furans **73**, following loss of methyl isocyanate from the cycloadducts **72**.

The bicyclic isomünchnones **75** can be trapped with NPM to afford adducts **76** (38).

Trapping isomünchnone **77** with DMAD led to furan isocyanate **78**, after a retro-Diels–Alder reaction (Scheme 10.11) (38).

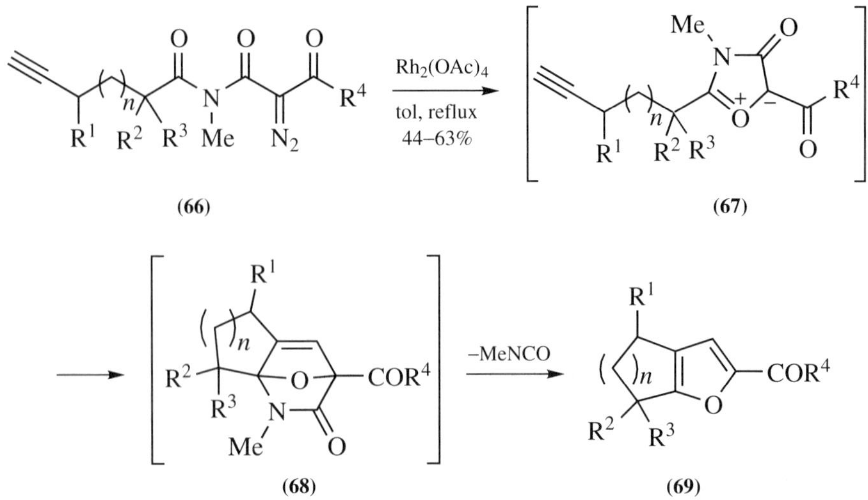

R^1 = H, OCOPh; R^2 = H, Me; R^3 = H, Me; R^4 = Me, OMe; n = 1, 2

Scheme 10.9

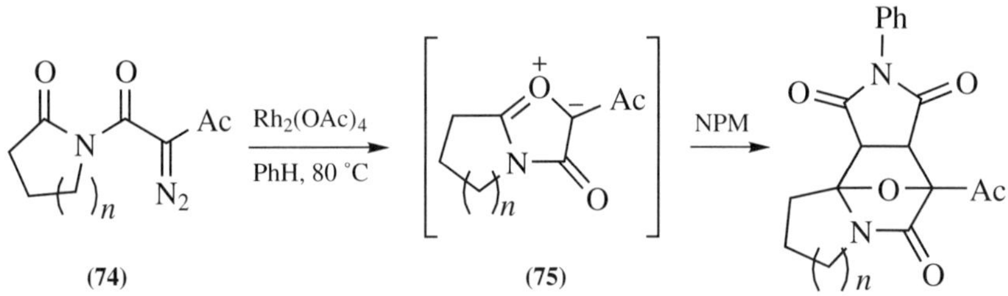

(76a) $n = 0$ (61%)
(76b) $n = 1$ (78%)
(76c) $n = 2$ (85%)
(76d) $n = 3$ (75%)

Doyle et al. (39) expanded the rhodium-catalyzed generation of isomünchnones from diazoacetacetamides and subsequent trapping with dipolarophiles (38). As shown in Scheme 10.12, in the case of diazoacetoacetyl urea (**79**) the derived isomünchnone **80** reacts with methyl propiolate to give a 2:1 mixture of cycloadducts **81**. The resulting regiochemistry is successfully rationalized using frontier molecular orbital (FMO) theory as being isomünchnone–HOMO controlled. This result represents one of the few reactions in which the cycloadducts from isomünchnones and alkynes are stable.

The newest method for generating isomünchnones was reported by Padwa and co-workers (40,41). Thus, Kuethe and Padwa developed an exciting new application of the venerable Pummerer reaction of imidosulfoxides to generate and trap isomünchnones with alkenes. For example, the readily prepared imidosulfoxide **82**

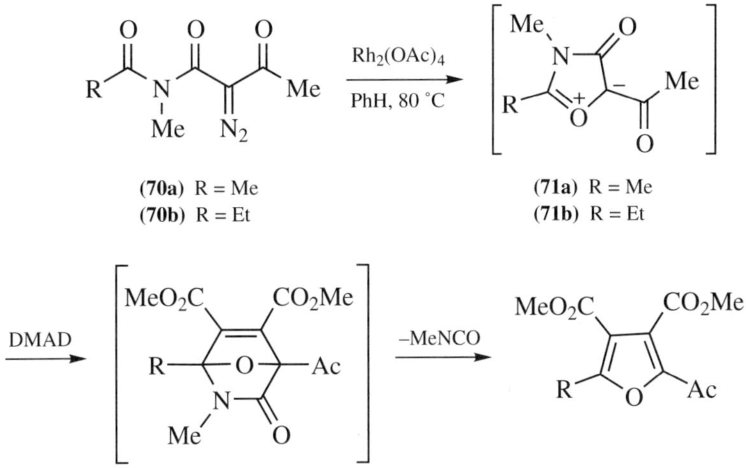

(70a) R = Me
(70b) R = Et

(71a) R = Me
(71b) R = Et

(72)

(73a) R = Me (82%)
(73b) R = Et (86%)

Scheme 10.10

upon exposure to acetic anhydride and a trace of *p*-toluenesulfonic acid affords isomünchnone **83**. Trapping with NPM or maleic anhydride yields cycloadducts **84** (Scheme 10.13).

This strategy is a powerful route to bicyclic pyridones and their transformation products. Thus, these workers (40,41) applied this methodology to formal syntheses of the lupinine alkaloids (±)-lupinine and (±)-anagyrine (**89**) (Scheme 10.14). Imidosulfoxide (**82**) is converted to the corresponding isomünchnone that is trapped with methyl acrylate to give **85**. Oxidation, ring opening, and triflate formation

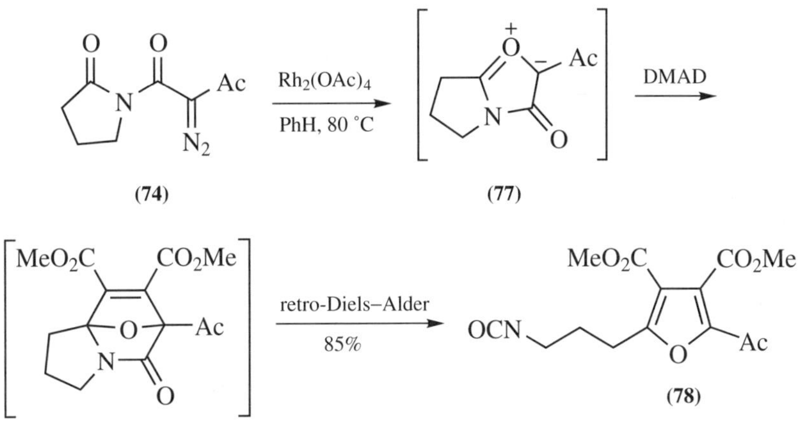

(74) **(77)** **(78)**

Scheme 10.11

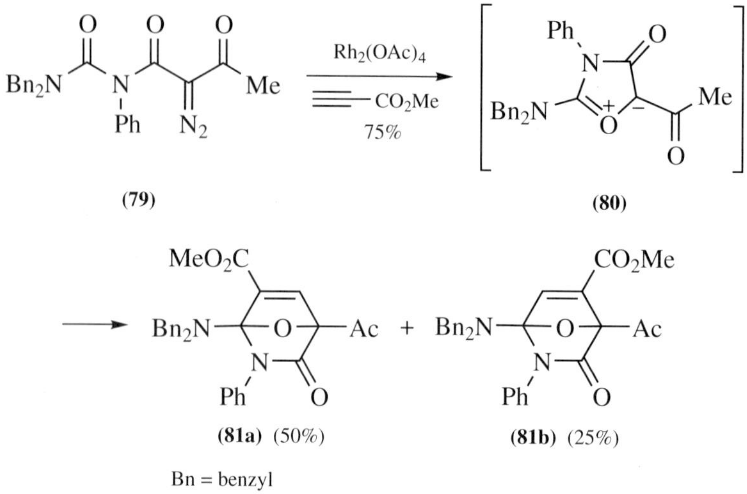

Scheme 10.12

affords **86**. A Stille cross-coupling installs the pyridine unit and further manipulation leads to **88**, which has previously been converted to (±)-anagyrine (**89**).

In the full account of this work, Padwa et al. (41) demonstrated that the 1,3-dipolar cycloaddition is an endo cycloaddition and the regiochemistry is consistent with that of a HOMO–dipole controlled process as judged from the products **91** and **92** that arise from the reaction between isomünchnone **90** and methyl propiolate and phenyl vinyl sulfone, respectively (Scheme 10.15). Isomünchnone **90** is also trapped with DMAD to give the expected furan in 41% yield.

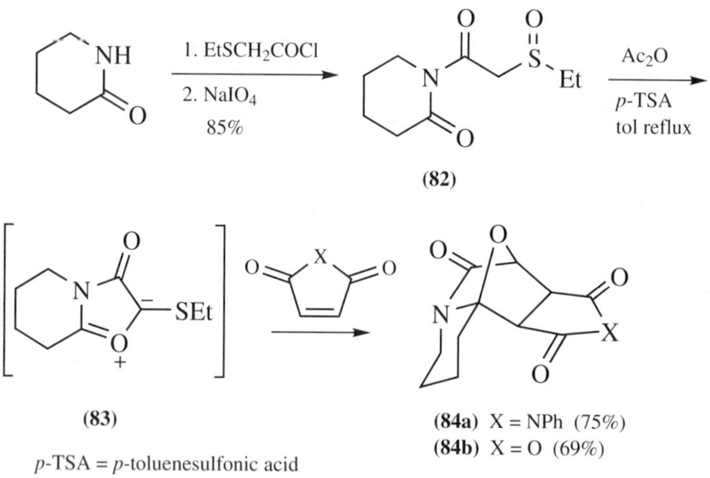

Scheme 10.13

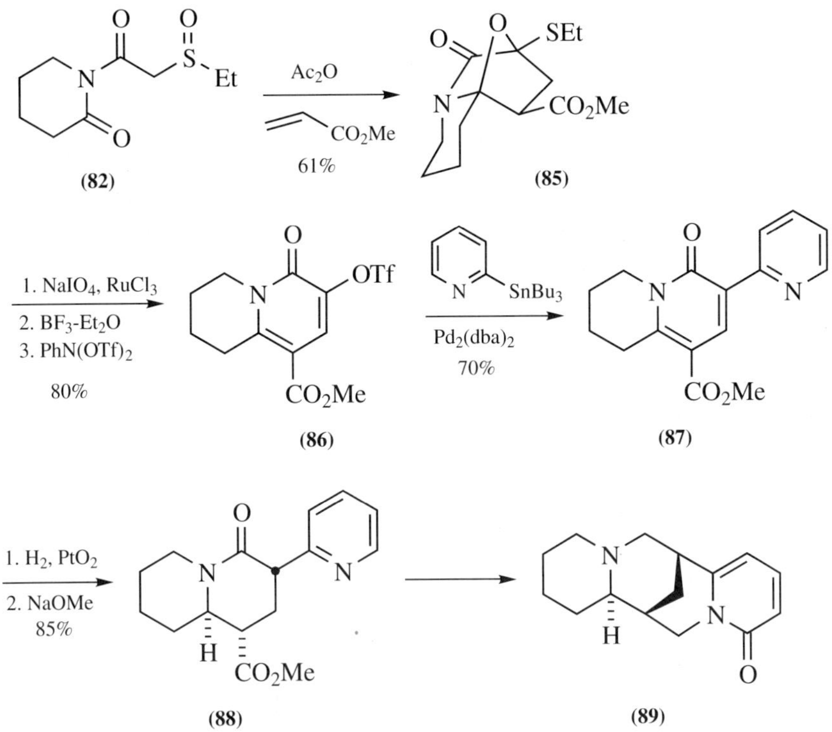

dba = dibenzylideneacetone

Scheme 10.14

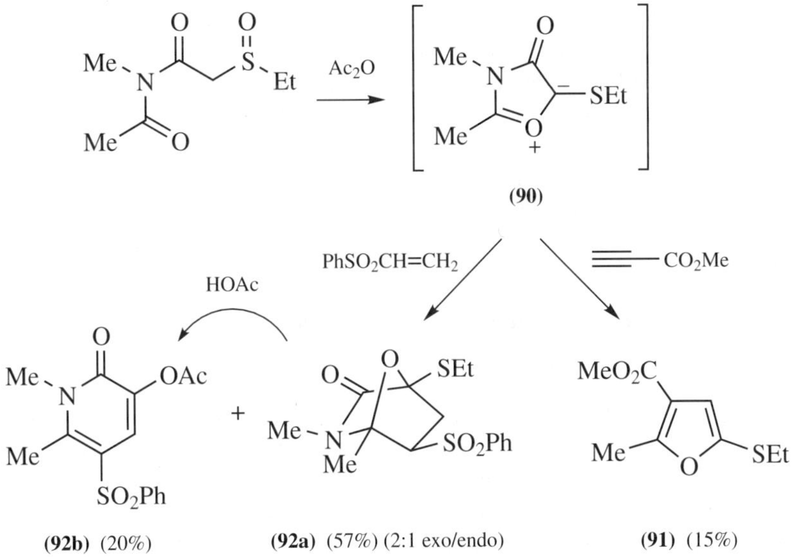

Scheme 10.15

10.1.3. Sydnones (1,2,3-oxadiazolium-5-olates)

The classical synthesis of mesoionic 1,2,3-oxadiazolium-5-olates (**3**), the so-called "sydnones", by Earl and Mackney (42), involving the cyclodehydration of an N-nitrosoglycine derivative, continues to be the only viable method for their preparation. However, Turnbull and co-worker (43) reported that the use of neutral nitrosation conditions avoids the formation of C-nitrosation and allows for the synthesis of previously inaccessible 3-arylsydnones, such as 3-(2-acetylphenyl) sydnone. The method is illustrated below.

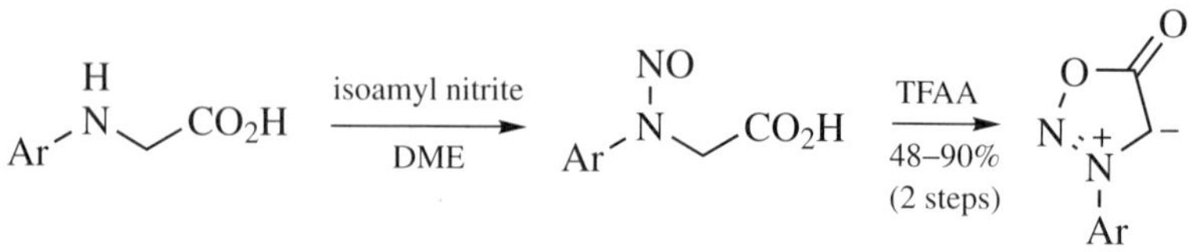

(93)

Ar = Ph, 2-CN-Ph, 2-CO_2Me-Ph, 2-CO_2H-Ph, 2-Me-Ph, 2-MeO-Ph, 2-NO_2-Ph, 2-Br-Ph, 2-Ac-Ph
DME = 1,2-dimethoxyethane

Turnbull et al., who has been a pioneer in the exploration of sydnone chemistry, has effected several derivatizations of sydnones that promise to make these compounds more attractive for 1,3-dipolar cycloaddition studies. This chemistry includes nitration of the aryl ring in 3-arylsydnones (44), halogenation studies (45,46), C(4) acylation (47), and dilithiation of 3-phenylsydnone (48,49). Kalinin and co-workers (50) described the lithiation and functionalization of the methyl group in 3-methyl-4-phenylsydnone. Petride (51) has also studied the acylation of both sydnones and münchnones, and the conformational preferences of the resulting compounds including several novel 4-unsubstituted monocyclic münchnones. The C(4) iodination of sydnones under mild conditions has been discovered by Dumitrascu et al. (52) and has led to an excellent synthesis of 5-iodopyrazoles as is described in Section 10.2.3.

10.1.4. Thiomünchnones (1,3-oxathiolium-5-olates)

The original synthesis of thiomünchnones (**4**) (1,3-oxathiolium-5-olates) (53–55), which involves the cyclodehydration of an (*S*)-acylthioglycolic acid, was used by Gribble and co-workers (56) to generate 2,4-diphenylthiomünchnone (**94**) and trap it with 1,5-cyclooctadiene in a tandem 1,3-dipolar cycloaddition sequence to afford sulfide **95** (Scheme 10.16). Apart from this one report, this mesoionic heterocycle has received no attention since the review by Potts (1).

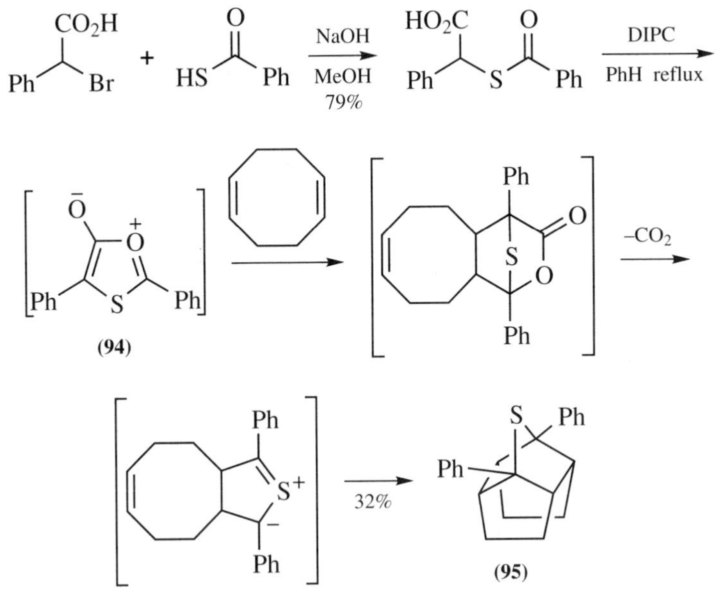

Scheme 10.16

10.1.5. Thioisomünchnones (1,3-thiazolium-4-olates)

In contrast to thiomünchnones, thioisomünchnones (**5**) (1,3-thiazolium-4-olates) have received considerable attention over the past three decades, and a rich array of 1,3-dipolar cycloaddition chemistry is described in Section 10.2.5. These heterocycles were initially constructed by Potts (1,53,57) from thioamides and α-halo acid chlorides, an example of which is shown in Scheme 10.17 (58).

Shortly thereafter, Potts and Murphy (59) found that thioisomünchnones could be synthesized by a Rh(II)-catalyzed cyclization of a diazothioamide, as illustrated for **97**. Several other thiomünchnones (**98–100**) were similarly prepared (59).

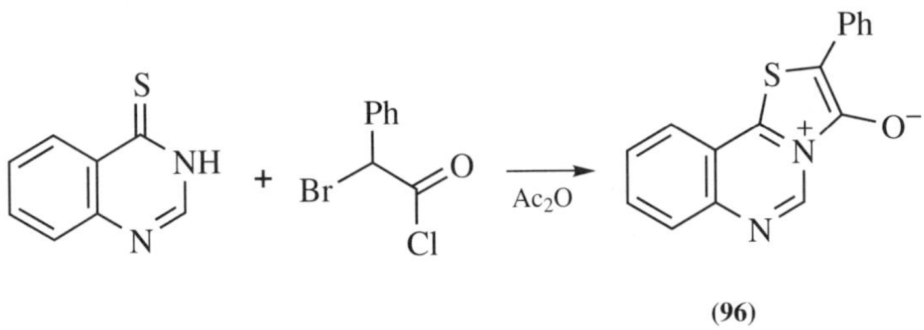

Scheme 10.17

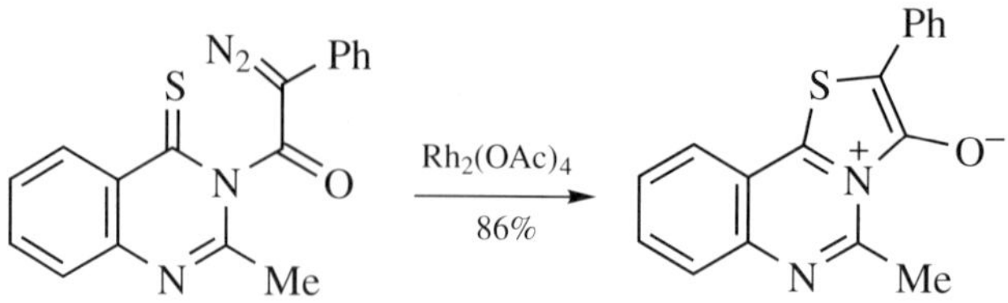

(97)

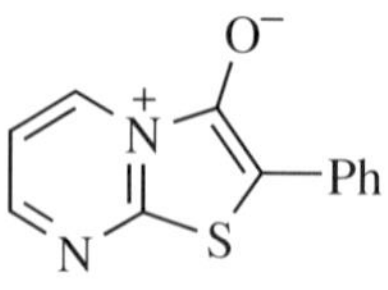

(98) (93%)

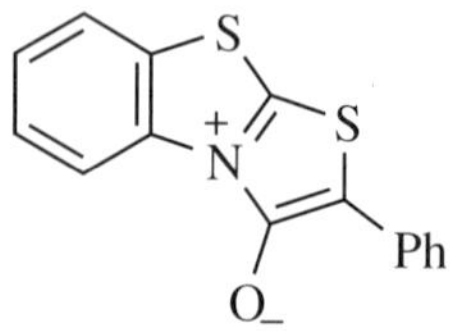

(99) (76%)

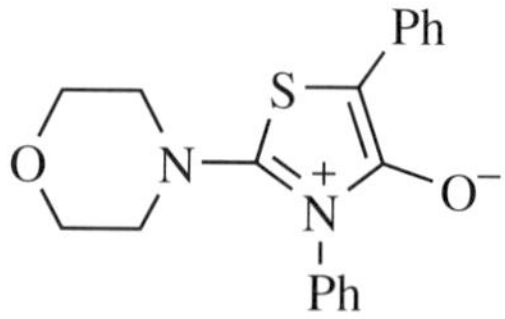

(100) (88%)

Padwa utilized this rhodium-catalyzed cyclization of diazothiocarbonyl compounds to great effect in the generation and trapping of thioisomünchnones (60,61). For example, thioisomünchnone (**101**) is readily prepared (and isolated) from the reaction of 2-thiopyrrolidinone and diketene, followed by diazo transfer and rhodium induced cyclization (Scheme 10.18) (60). Trapping with NPM afforded cycloadduct **102**.

10.1.6. 1,3-Thiazolium-4-thiolates

The sulfur analogues of thioisomünchnones are 1,3-thiazolium-4-thiolates (**104**) and two synthetic routes to these compounds have been described by Yoshii and

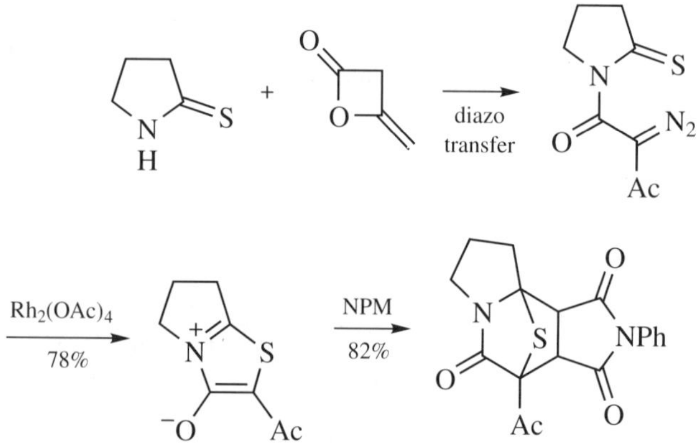

(101) (102)

Scheme 10.18

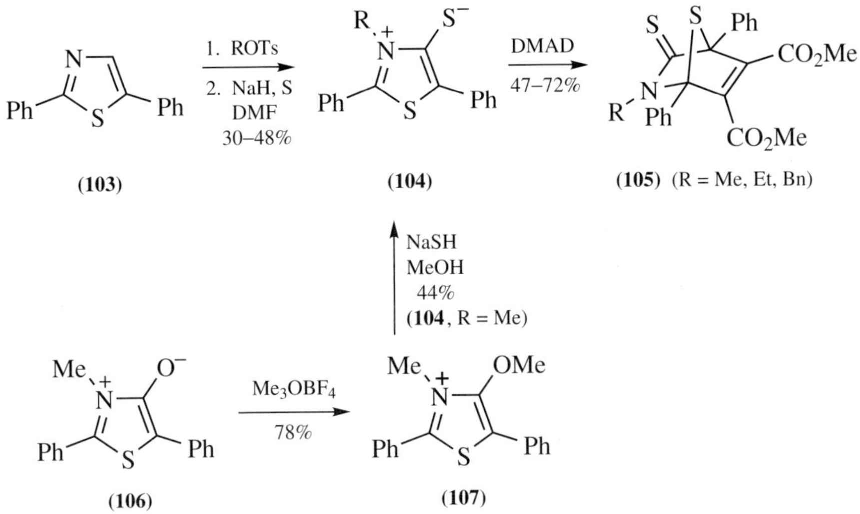

Scheme 10.19

co-workers (62) (Scheme 10.19). Alkylation of thiazole **103** followed by thiation gave mesoionics **104**. Reaction with DMAD gave the unexpectedly stable cycloadducts **105**. Alternatively, exchange of the exocyclic oxygen of thioisomünchnone (**106**) for sulfur via **107** was less efficient than direct thiation.

Somewhat earlier, Souizi and Roberts (63) reported mesoionic heterocycle interconversions leading to 1,3-thiazolium-4-thiolates, 1,3-thiazolium-4-olates, and 1,3-dithiolium-4-thiolates from 1,3-dithiolium-4-olates. This elegant chemistry, which involves cycloaddition reactions, is presented in Section 10.3.8.

10.1.7. Oxamünchnones (1,3-dioxolium-4-olates)

The first synthesis of 1,3-dioxolium-4-olates (here defined as *oxamünchnones*) was reported in 1980 by Berk et al. (64) but it was work of Hamaguchi and Nagai (65,66) that demonstrated the accessibility and utility of these new mesoionic heterocycles in cycloaddition reactions. Thus, reaction of diazoacetic benzoic anhydrides **108** with a π-allyl palladium complex affords oxamünchnones **109**.

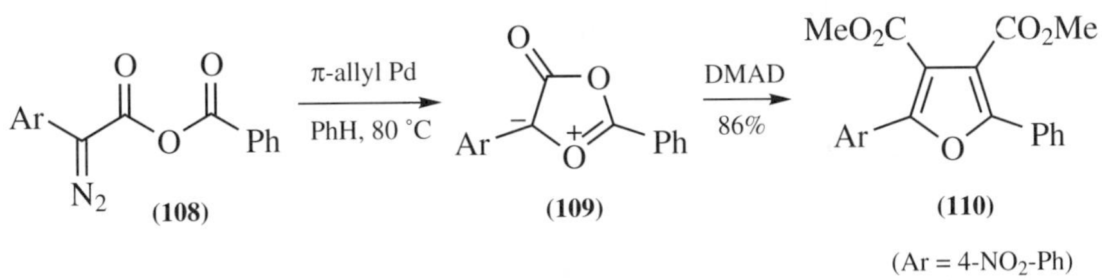

Scheme 10.20

Subsequent trapping with DMAD affords the expected furan **110** in excellent yield (65) (Scheme 10.20). Additional chemistry of this novel mesoionic heterocycle is presented in Section 10.2.7.

10.2. CYCLOADDITION REACTIONS

The statement made in 1986 by Gingrich and Baum (10), with regard to münchnones, that "the most important reactions (of münchnones) from a synthetic point of view are 1,3-dipolar cycloaddition reactions," certainly applies to **all** mesoionic heterocycles and is more true today than it was in 1986. Although the factors governing the regioselectivity of unsymmetrical mesoionic cycloadditions are not always completely understood, the synthetic utility of this chemistry is enormous and indisputable.

As we have seen in previous discussions, DMAD is a powerful and reliable dipolarophile that is routinely used to trap mesoionic heterocycles. If DMAD is unable to ambush a suspected mesoionic heterocycle, then the latter most probably has not been generated!

10.2.1. Münchnones (1,3-oxazolium-5-olates)

Anderson et al. (67,68) used the münchnone generation–DMAD trapping protocol to synthesize various pyrrolizines as potential antileukemic agents, such as **111** and the pyrrolo[1,2-*c*]thiazoles **112**.

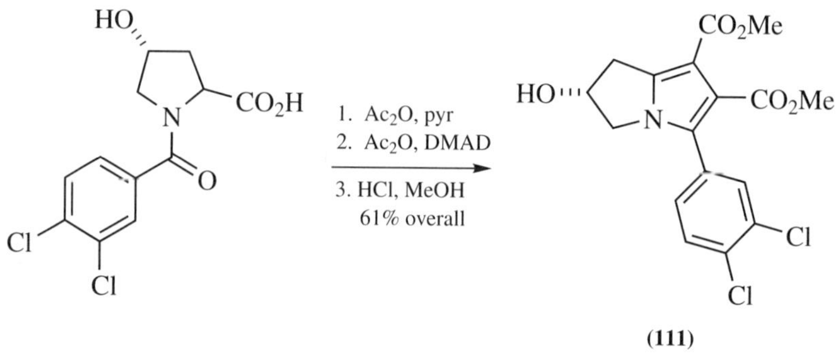

(111)

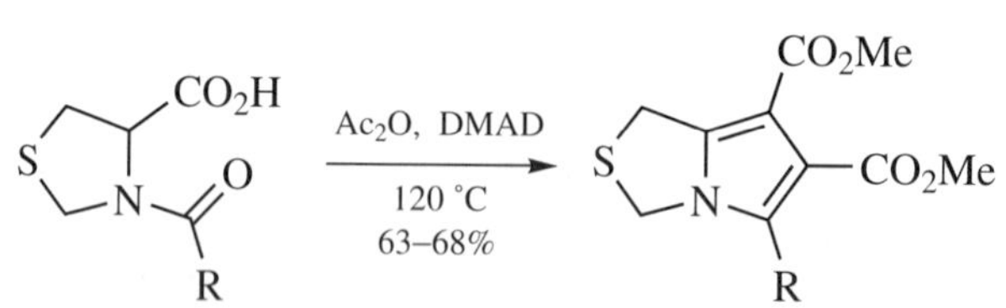

R = Me, 3,4-Cl_2-Ph

(112)

Györgydeák et al. (69) converted a series of 2-substituted 3-acyl-1,3-thiazolidine-4-carboxylic acids (**113**) into the corresponding pyrrolo[1,2-*c*]thiazoles (**115**)

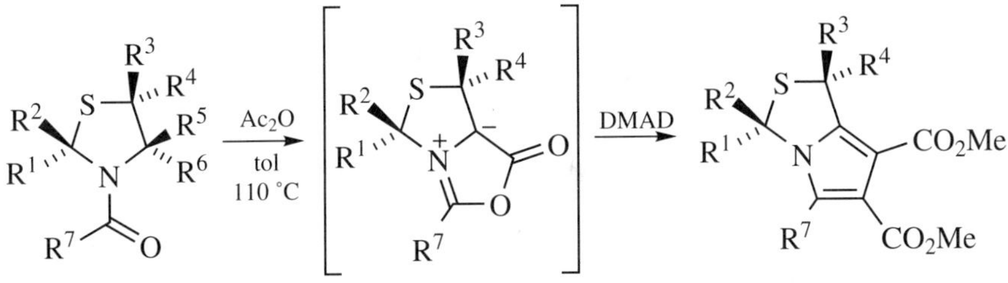

(113) (114) (115)

R^1, R^2 = H, Ph, Ar, sugar; R^3, R^4 = H, Me; R^7 = Me, Et, *t*-Bu

Scheme 10.21

via münchnones (**114**), which were trapped with DMAD (Scheme 10.21). Similarly, Pinho e Melo et al. (70) found that heating a diastereomeric mixture of (2*R*,4*R*)- and (2*S*,4*R*)-2-phenylthiazolidine-2-carboxylic acids with acetic anhydride in the presence of DMAD affords the expected pyrrolothiazole in good yield and 99% enantiometric excess (ee). Similar dipolar cycloadditions occur with methyl propiolate, methyl vinyl ketone, and acrylonitrile.

Robba and co-workers (71) synthesized the 2-azapyrrolo[1,2-*a*]indole ring system via a münchnone cycloaddition strategy (Scheme 10.22). Thus, trapping the münchnone derived from proline derivative **116** gave pyrrole **117** in 75% yield. Further elaboration yielded the desired **118** and subsequent target compounds.

Vasella and co-workers (72) employed münchnone chemistry in the synthesis of several pyrrolopyridines and imidazopyridines as novel inhibitors of β-D-glucosidases (Scheme 10.23). Thus, treatment of the lactam glycine (**119**) with acetic

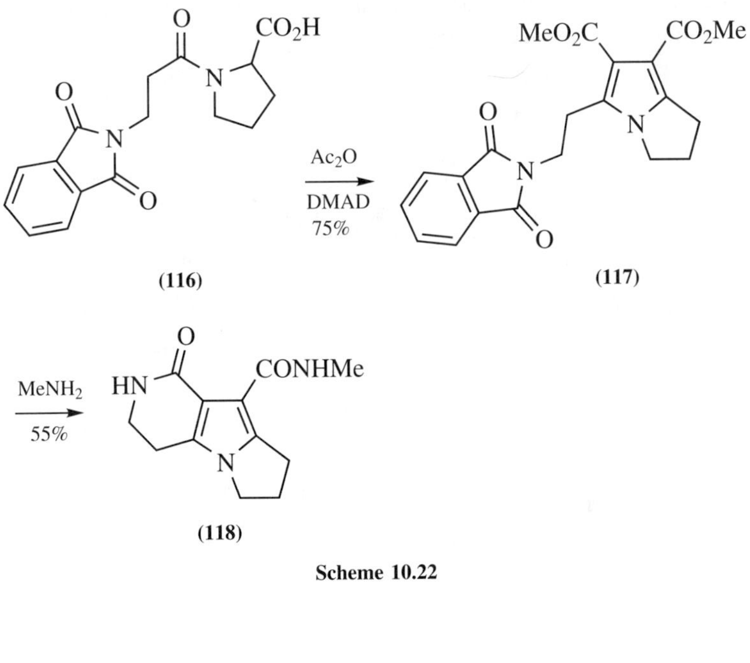

(116) (117)

(118)

Scheme 10.22

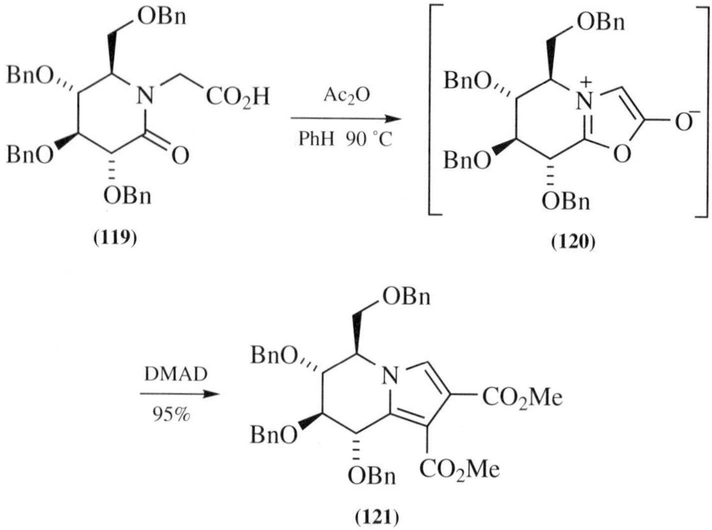

(119) (120) (121)

Scheme 10.23

anhydride in the presence of DMAD affords pyrrole (**121**) via münchnone **120** in 95% yield. These workers also generated **120** in the presence of methyl propiolate using EDC to afford a mixture of isomeric pyrroles (77% yield).

The imidazopyridine **123** was synthesized by treating amino acid **122** with *p*-toluenesulfonyl cyanide and DCC. The yield of **123** was lower (38%) when the münchnone was generated with mesyl chloride (72). The minor amide product **124** is proposed to arise from the diastereoisomeric münchnone adduct that undergoes fragmentation rather than decarboxylation.

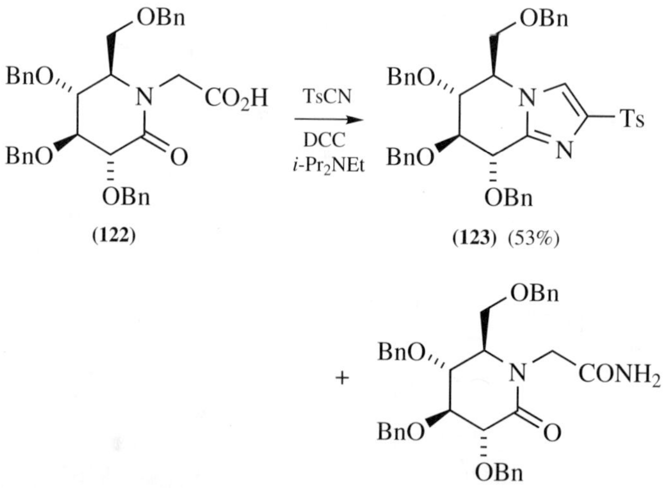

(122) (123) (53%) (124) (23%)

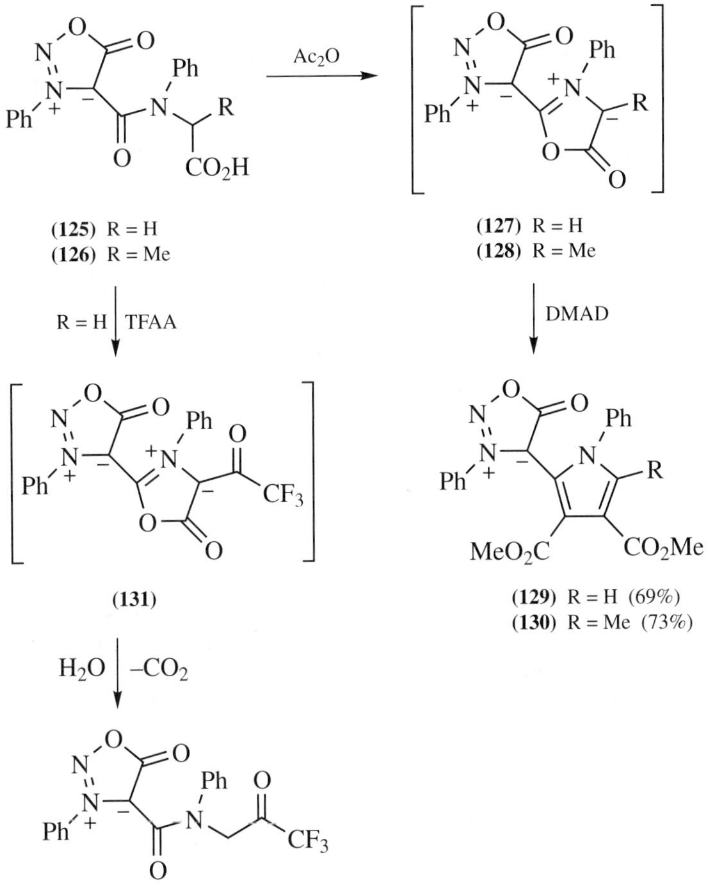

(**125**) R = H
(**126**) R = Me

(**127**) R = H
(**128**) R = Me

(**131**)

(**129**) R = H (69%)
(**130**) R = Me (73%)

(**132**)

TFAA = Trifluoroacetic acid anhydride

Scheme 10.24

Chan and co-workers (73) generated the novel münchnone–sydnone hybrids **127**–**128** by cyclodehydration of the sydnone glycine (**125**) and alanine (**126**), respectively. Trapping with DMAD gave **129** and **130**. Maleic anhydride and dimethyl maleate failed to capture these münchnone–sydnone hybrids. Exposure of **125** to trifluoroacetic anhydride gave acylated münchnone **131** [by nuclear magnetic resonance (NMR)]. Workup afforded the hydrolysis product **132** (Scheme 10.24).

Yamanaka and co-workers (74) effected the 1,3-dipolar cycloaddition between münchnones **134** and polyfluoro-2-alkynoic acid esters to afford the corresponding 4-(polyfluoroalkyl)pyrrole-3-carboxylates (**135**). The reaction proceeds rapidly at low temperature and the various münchnones (**134**) were generated from 1,3-oxazolium perchlorates **133** using the method of Boyd and Wright (6,7). Under these reaction conditions methyl 2-butynoate failed to react with **134**, but DMAD gave the corresponding pyrrole in 75% yield. Interestingly, changing the solvent from toluene to acetonitrile promoted Michael-addition products of triethylamine to the alkyne ester. The rapid reaction of these polyfluorinated esters in this 1,3-dipolar cycloaddition reaction is attributed to a lowering of the lowest occupied molecular orbital (LUMO) as a result of the powerful electron-withdrawing ability of the fluorinated alkyl group.

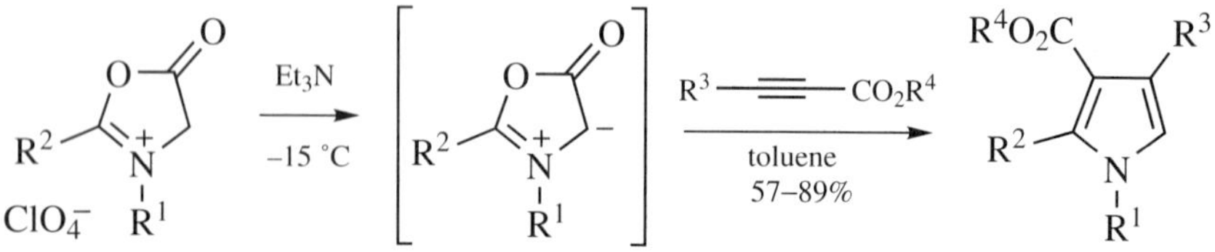

(133) **(134)** **(135)**

R^1 = Ph, Me (one example); R^2 = Ph, 4-NO_2-Ph, 4-Cl-Ph, 4-Me-Ph, 4-MeO-Ph, Me;
R^3 = CHF_2, CF_3, $H(CF_2)_3$, CO_2Me; R^4 = Me, n-C_6H_{11} (one example)

Padwa et al. (75) found that the unsymmetrical münchnone **137**, which was generated from *N*-acetyl-*N*-benzylglycine (**136**) and refluxing acetic anhydride, reacts with methyl propiolate to give an 8:1 mixture of pyrroles **138** and **139**. The same product ratio is obtained from the reaction of methyl propiolate and the azomethine ylide derived from *N*-benzyl-*N*-(α-cyanoethyl)-*N*-[(trimethylsilyl)-methyl]amine.

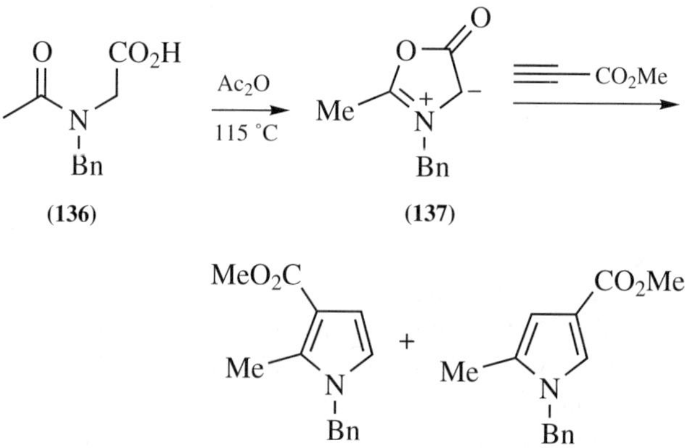

(138) (60%) **(139)** (8%)

Similarly, Toupet et al. (76) determined the structures of the pyrrole adducts from the cycloaddition of several münchnones with methyl phenylpropionate by X-ray crystallography. These highly regioselective cycloadditions are in accord with FMO predictions. This same French team (77) studied the cycloaddition reactions of münchnones with methyl propiolate and methyl 3-phenylpropiolate to

TABLE 10.1. 1,3-DIPOLAR CYCLOADDITION REACTIONS OF *IN SITU* GENERATED MÜNCHNONES AND METHYL PROPIOLATE AND METHYL 3-PHENYLPROPIOLATE[a]

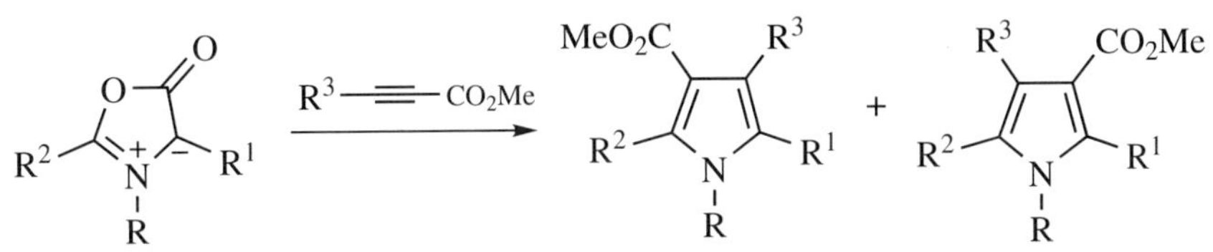

(140) **(141)**

R	R^1	R^2	R^3	Yield (%)	% Regiochemistry **140**	**141**
Me	Ph	Me	H	100	48	52
Me	Me	Ph	H	100	65	35
Ph	Ph	Me	H	100	35	65
Ph	Me	Ph	H	100	65	35
Me	Ph	4-NO_2–Ph	H	100	50	50
Me	Ph	Me	Ph	65	95	5
Me	Me	Ph	Ph	55	95	5
Ph	Ph	Me	Ph	40	90	10
Ph	Me	Ph	Ph	47	90	10
Me	Ph	4-MeO–Ph	Ph	40	100	0
Me	4-Me–Ph	Ph	Ph	60	100	0
Me	Ph	4-NO_2–Ph	Ph	70	100	0
Me	3-NO_2–Ph	Ph	Ph	82	100	0

[a] Reference (77).

give mixtures of the regioisomeric pyrroles **140** and **141** (Table 10.1). Yebdri and Texier (78) also studied cycloaddition reactions of the well-known münchnone **142**, derived from proline and acetic anhydride, with methyl propiolate and methyl 3-phenylpropiolate to give mixtures of pyrrolizines **143** and **144**, respectively.

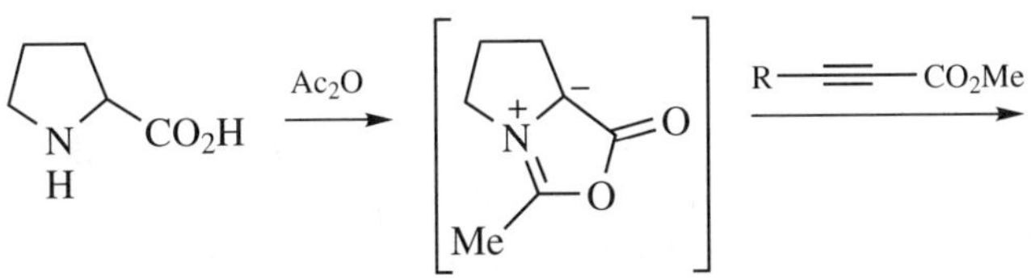

(142)

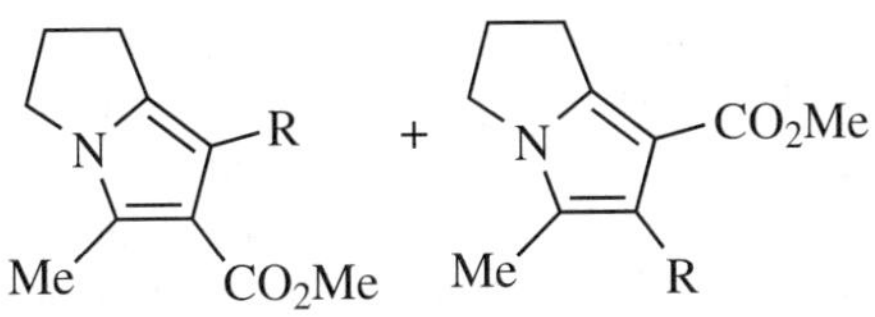

(143a) R = H (21%) **(143b)** R = H (42%)
(144a) R = Ph (40%) **(144b)** R = Ph (21%)

TABLE 10.2. 1,3-DIPOLAR CYCLOADDITION REACTIONS OF *IN SITU* GENERATED MÜNCHNONES AND TERMINAL ALKYNES[a,b]

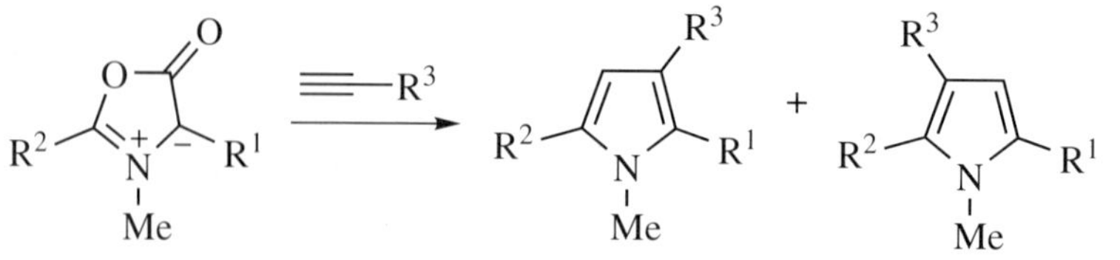

(145) **(146)**

				% Regiochemistry	
R^1	R^2	R^3	Yield (%)	% **145**	% **146**
Ph	H	CO_2Et	77	75	25
H	Ph	CO_2Et	30	14	86
Ph	H	COPh	80	75	25
H	Ph	COPh	58	0	100
Ph	H	Ph	34	100	0
Me	H	CO_2Et	51	84	16
H	Me	CO_2Et	24	25	75
Me	H	COPh	85	80	20
H	Me	COPh	11	0	100
Me	H	Ph	4	100	0
Ph	Me	CO_2Et	87	38	62
Me	Ph	CO_2Et	99	43	57
Ph	Me	COPh	29	2	98
Me	Ph	COPh	83	18	82
Ph	Me	Ph	50	98	2
Me	Ph	Ph	44	99	1

[a] Münchnones were generated from the corresponding N-acylamino acids using acetic anhydride in toluene at 80°C.
[b] Reference (79).

Dalla Croce and La Rosa (79) examined the 1,3-dipolar cycloaddition reactions of unsymmetrical münchnones with terminal alkynes to give pyrroles **145** and **146** (Table 10.2). The reaction is generally regioselective with the major pyrrole isomer from the monosubstituted münchnones having adjacent hydrogens, irrespective of the münchnone substituent. With the disubstituted münchnones the major regioisomer is derived from attachment of C(4) of the münchnone and the β-carbon of the alkyne, except for phenylacetylene, which shows the opposite regiochemistry. The authors interpret this behavior as a consequence of the electron-rich nature of phenylacetylene as a dipolarophile and having a larger LUMO coefficient on the α-carbon. Gribble et al. (80) observed the same behavior with *N*-benzylmünchnones **147** and **150** to give pyrroles **148** and **149**, respectively, in a highly regioselective fashion (Scheme 10.25).

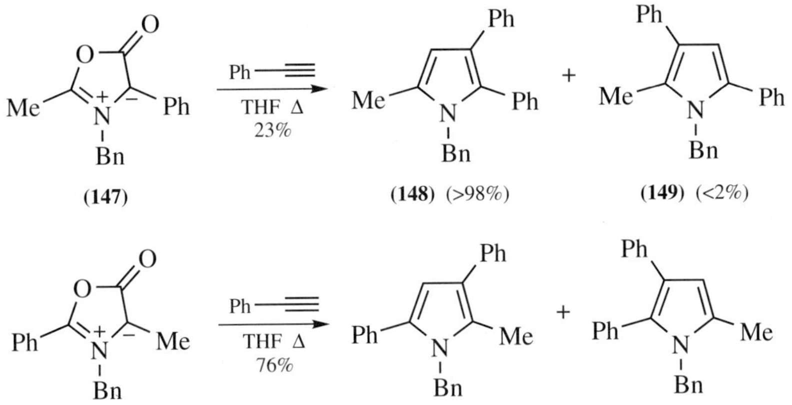

(147) (148) (>98%) (149) (<2%)

(150) (149) (100%) (148) (0%)

Scheme 10.25

Coppola et al. (81) extensively studied the dipolar cycloaddition of methyl propiolate with unsymmetrical münchnones. In addition to their own results (Table 10.3), these investigators summarized much previous data on these cycloadditions. In the authors' words: "No single criterion can successfully be used to correlate the experimental observations regarding the regioselectivity in münchnone cycloaddition reactions." Steric and electronic effects must be considered in

TABLE 10.3. 1,3-DIPOLAR CYCLOADDITION REACTIONS OF *IN SITU* GENERATED MÜNCHNONES AND METHYL PROPIOLATE[a]

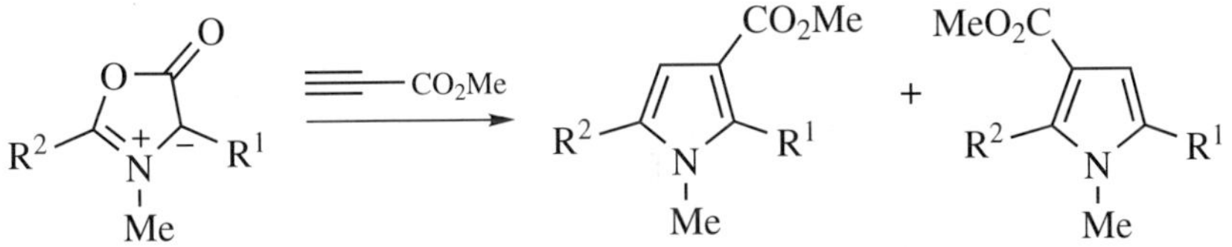

(153) (154)

R^1	R^2	Yield (%)	% Regiochemistry: **%153**	**%154**
i-Pr	H	56	57	43
H	*i*-Pr	92	25	75
Me	*i*-Pr	88	67	33
i-Pr	Me	84	23	77
Me	Et	64	55	45
t-Bu	H	53	67	33
Bn	H	95	83	17
H	Bn	91	16	84

[a] Reference (81).

analyzing these cycloadditions. One revealing result from this study is the use of C(13) labeled münchnones **151** and **152**, which essentially exhibit no regiochemical bias in reacting with methyl propiolate. These workers have also examined the influence of thio-substitution on the cycloaddition of münchnones with methyl propiolate (81). The results are consistent with an unsymmetrical transition state where bond formation between the two least encumbered positions occurs first.

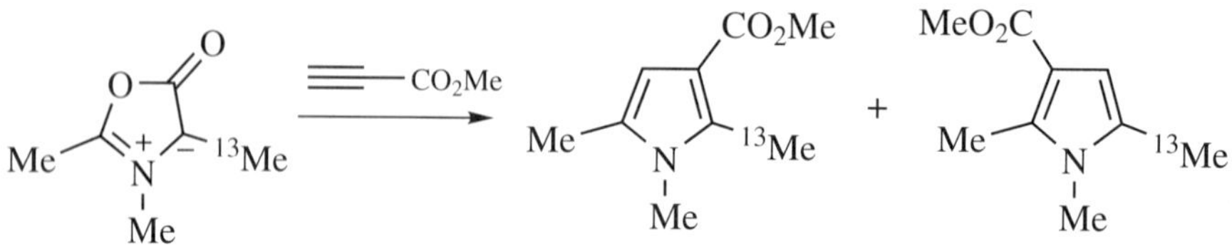

(151) 53 : 47

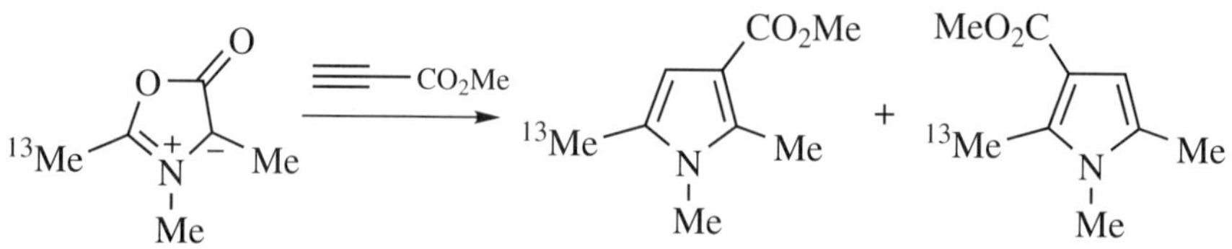

(152) 55 : 45

Coppola et al. (82) reexamined the reaction of münchnone **155**, derived from *N*-formylproline, with methyl propiolate (Scheme 10.26). In contrast to an earlier study that reported complete regioselectivity (83), the present work found both regioisomers **156** and **157** in this reaction, with **156** predominating.

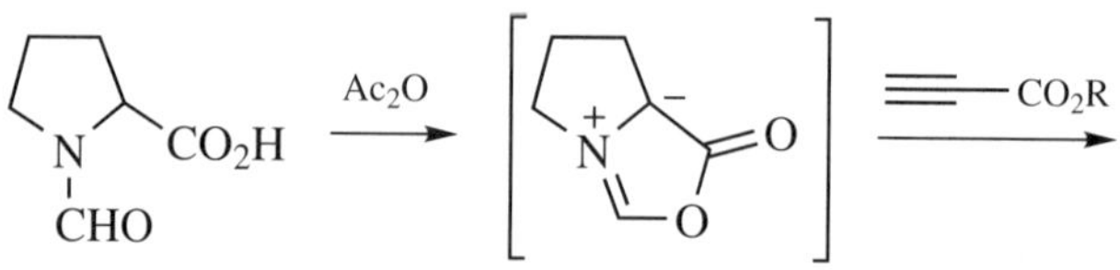

(155)

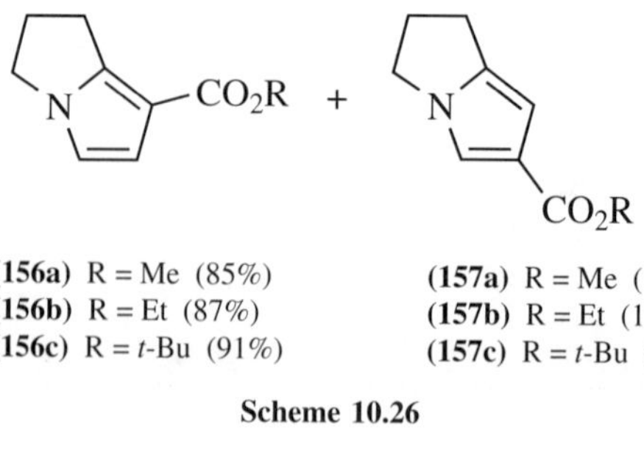

(156a) R = Me (85%)
(156b) R = Et (87%)
(156c) R = *t*-Bu (91%)

(157a) R = Me (15%)
(157b) R = Et (13%)
(157c) R = *t*-Bu (9%)

Scheme 10.26

Despite the uncertainties regarding regiochemistry, the reaction of propiolates with münchnones has found use in synthesis. Kane and co-workers (84) synthesized the calcium channel activator FPL 64176 (**161**) using a münchnone cycloaddition protocol. Thus, reaction of amino acid **158** with acetic anhydride in the presence of acetylenic dipolarophile **159** gave pyrrole **160** in 49% yield. Base-induced elimination of the 4-nitrophenethyl protecting group afforded FPL 64176 (**161**) in 85% yield.

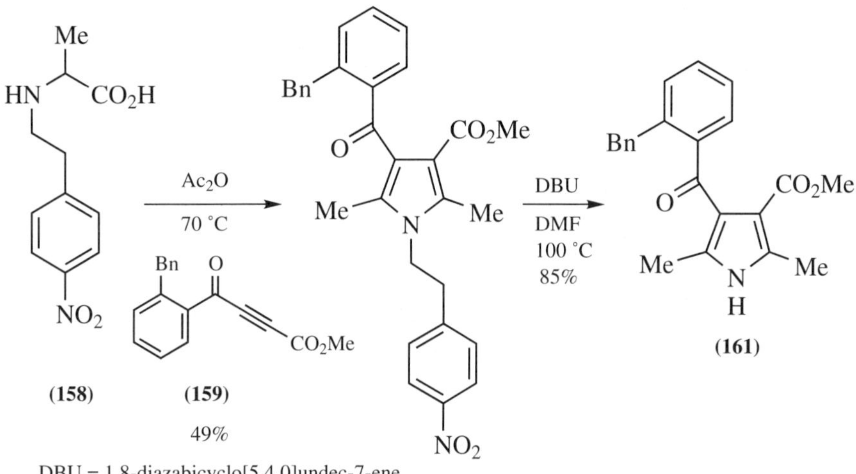

(158) **(159)** **(160)** **(161)**

DBU = 1,8-diazabicyclo[5.4.0]undec-7-ene

Similar to the work of Anderson et al. (67,68) described above and in continuation of their proline–derived münchnone generation and DMAD trapping (71) (**116** → **117**), Ladurée et al. (85) also utilized the reaction between the münchnones derived from cyclic N-acylamino acids and DMAD to prepare a series of potential new antileukemic agents derived from the pyrrole cycloadducts. La Rosa and co-workers (86) find that münchnones react with arylsulfonyl alkynes to furnish 3-arylsulfonyl pyrroles in good yields. In the case of unsymmetrical münchnones, good regioselectivity is observed. The pyrroles were identified using two-dimensional (2D) NMR techniques. The regioselectivity is explained in terms of a preferred highest occupied molecular orbital (HOMO)-münchnone–LUMO-alkyne interaction, which is one of the few cases where FMO theory is satisfactory in explaining the regiochemistry of münchnone 1,3-dipolar cycloadditions. Intramolecular 1,3-dipolar cycloadditions represent a powerful synthetic tool. Kato et al. (87) were apparently the first to report an intramolecular münchnone–alkyne cycloaddition. Thus, münchnones **163**, as generated from *N*-acylamino acids **162**, yield the corresponding benzopyrano[4,3-*b*]pyrroles **165** after extrusion of carbon dioxide from adduct **164** (Scheme 10.27). Interestingly, attempts to divert the intramolecular cycloaddition by the addition of *N*-phenylmaleimide had no effect on the reaction pathway.

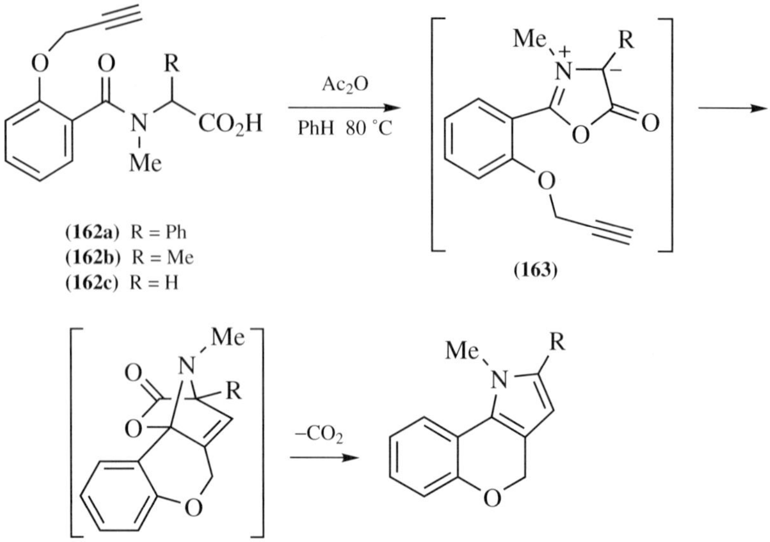

(162a) R = Ph
(162b) R = Me
(162c) R = H

(163)

(164)

(165a) R = Ph (96%)
(165b) R = Me (92%)
(165c) R = H (85%)

Scheme 10.27

Sainsbury et al. (88) utilized the intramolecular 1,3-dipolar cycloaddition of tetrahydroquinolines **166** to construct 11-acetoxycarbonyl-1,2,3,4,5,10-hexahydroindeno[2,3-*a*]isoquinoline (**167**). The presence of an ester functionality on the alkyne is essential for the cycloaddition to occur. With one less methylene group in the tether, the yield of the corresponding pentaleno[2,3-*a*]isoquinoline is 37%.

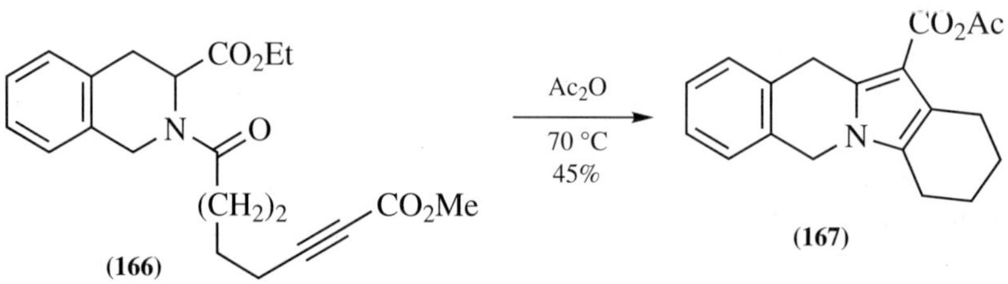

(166) **(167)**

Pinho e Melo et al. (89) employed an intramolecular münchnone cycloaddition to construct several 1*H*-pyrrolo[1,2-*c*]thiazole derivatives from N-acylthiazolidines and acetic anhydride. Martinelli and co-workers (90,91) employed an intramolecular münchnone cycloaddition to craft a series of 4-keto-4,5,6,7-tetrahydroindoles (**168–171**) in two steps. The requisite acetylenic precursors were prepared from glutaric anhydride (or 3-methylglutaric anhydride). The overall sequence is illustrated for the synthesis of **168**. An electrophilic acetylenic unit appears to be necessary for successful intramolecular 1,3-dipolar cycloaddition.

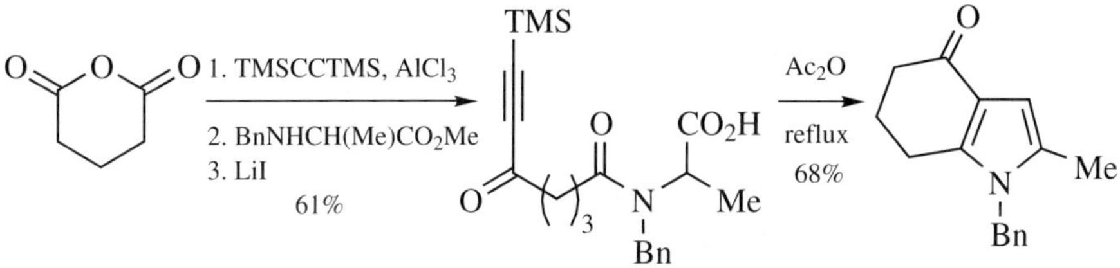

(168)

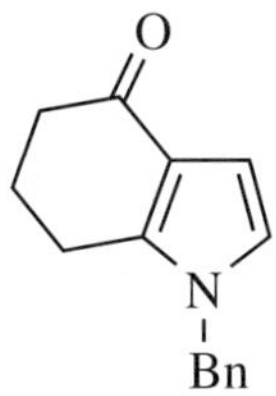

(169) (47%)

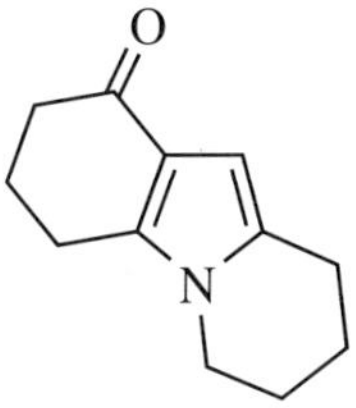

(170) (51%)

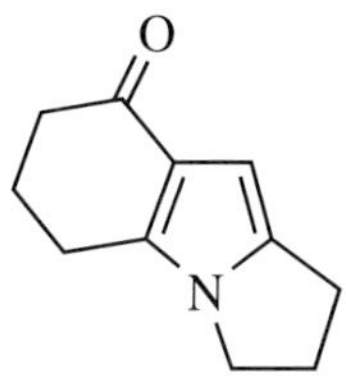

(171) (45%)

TMS = tetramethylsilyl

Jursic (92) studied the cycloaddition reaction of a münchnone with acetylene from several theoretical standpoints using density functional theory on AM1 geometries. The predicted activation energy for the 1,3-dipolar cycloaddition is 11.49 kcal/mol and the elimination of carbon dioxide from the cycloadduct to give a pyrrole is 5.82 kcal/mol. Both reactions are extremely exothermic as observed experimentally.

The 1,3-dipolar cycloaddition reactions of münchnones with olefinic dipolarophiles continues to be of enormous interest with regard to both mechanisms and synthetic applications. Unlike the comparable cycloadditions with acetylenic dipolarophiles that yield only pyrroles, reactions of münchnones with olefinic dipolarophiles can lead to a variety of interesting products. Nan'ya et al. (93,94) reexamined the double cycloaddition of münchnone **38** with 2 equiv of several maleimides (**172**) to give cycloadducts **173** and **174** (Scheme 10.28). Whereas *N*-phenylmaleimide (**172a**) gives only the exo,endo adduct **173a** (75%), *N*-methylmaleimide (**172b**) yields a mixture of exo,endo adduct **173b** (50%) and exo,exo adduct **174b** (22%). Similarly, maleimide (**172c**) itself affords exo,endo adduct **173c** (44%) and exo,exo adduct **174c** (22%). Huisgen and co-workers (95,96) had earlier reported that münchnone **38** and *N*-phenylmaleimide (**172a**) gave the exo,exo adduct **174** (R=Ph).

Nan'ya et al. (97) also reported the synthesis of isoindolediones by the reaction of münchnones with 1,4-benzoquinones. Reactions with an unsymmetrical münchnone were not regioselective. Several groups have examined the reactions of münchnones with unsaturated nitriles, including 2-chloroacrylonitrile (98), cinnamonitrile (78,99) and fumaronitrile (78) to give unexpected products in several cases. Eguchi and co-workers (100) studied the cycloaddition of several münchnones with electron-deficient trifluoromethylated olefins. Thus, münchnones **176**

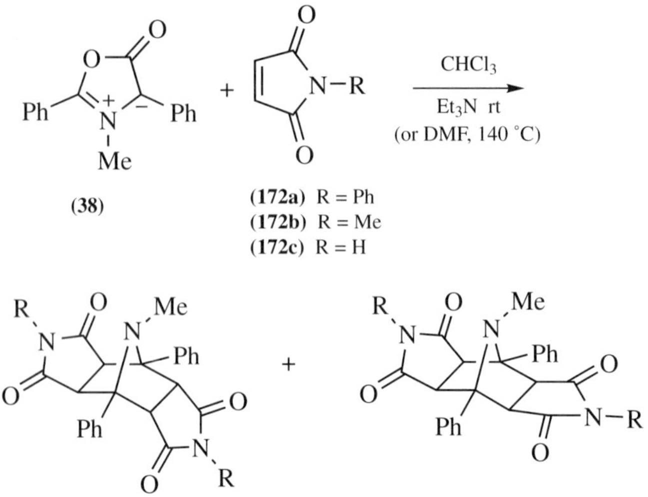

(38)

(172a) R = Ph
(172b) R = Me
(172c) R = H

(173a) R = Ph (75%)
(173b) R = Me (50%)
(173c) R = H (44%)

(174b) R = Me (22%)
(174c) R = H (22%)

Scheme 10.28

react with alkenes **177** to give the corresponding 3-(trifluoromethyl)pyrroles **178**. Yields of pyrroles from alkene **177a** are higher (56–89%) than those from alkene **177b** (9–33%). The regiochemistry in **178** was easily established by spin–spin coupling between the CF_3 group and the C-5H in the proton NMR (^{1}H NMR) spectrum.

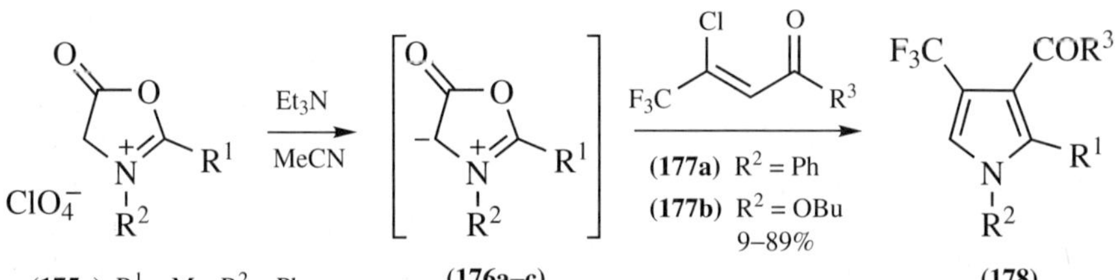

(175a) R^1 = Me, R^2 = Ph
(175b) R^1 = Ph, R^2 = Me
(175c) R^1 = R^2 = Ph

(176a–c)

(178)

Gelmi and co-workers (101) found that vinyl phosphonium salts serve as alkyne synthetic equivalents in reacting with münchnones to give pyrroles. The münchnones were generated in the usual way *in situ* by cyclodehydration of the corresponding N-acylamino acids with acetic anhydride. In the case of 1-propenyltriphenylphosphonium bromide, a single regioisomer was obtained. The regiochemistry apparently obtains from a strong interaction between the phosphonium group and the carbonyl group, which overwhelms the simple polarization of the

TABLE 10.4. 1,3-DIPOLAR CYCLOADDITION REACTIONS OF MÜNCHNONES AND METHYLENE TRIAZOLES[a]

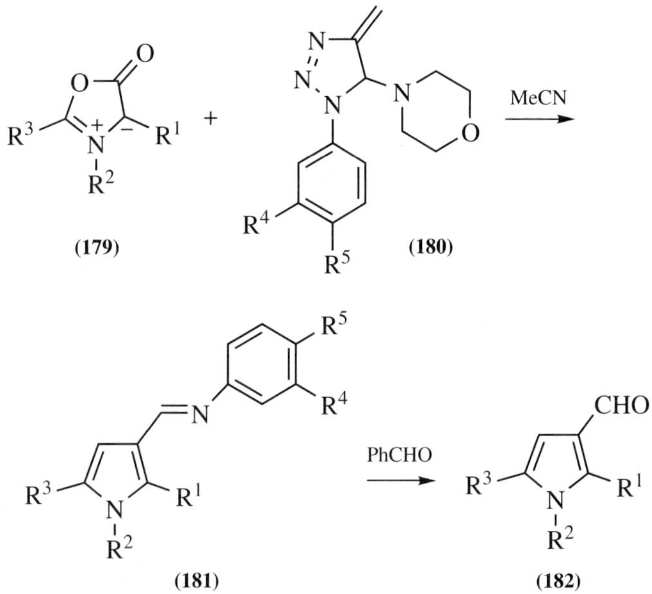

R^1	R^2	R^3	R^4	R^5	% Yield **181**[b]	% Yield **182**
Ph	Me	Ph	CF_3	H	10	65
Ph	Me	4-MeO–Ph	CF_3	H		68
Ph	Me	2-Cl–Ph	CF_3	H		81
4-MeO–Ph	Me	4-MeO–Ph	CF_3	H		69
Ph	Et	Ph	CF_3	H		68
Ph	Ph	Ph	H	NO_2		89
4-MeO–Ph	Me	Ph	CF_3	H		52
Ph	Me	Me	H	NO_2		78
Me	Me	Ph	H	NO_2	23	45

[a] Reference 102.
[b] In most cases, the imine **181** was not isolated but directly converted to **182**.

vinyl group. This group also found that azlactones react with vinyl phosphosphonium salts to give N-unsubstituted pyrroles.

Erba et al. (102) observed a novel formation of pyrrole imines **181** from the 1,3-dipolar cycloaddition of münchnones **179** and 5-amino-1-aryl-4,5-dihydro-4-methylene-1,2,3-triazoles **180** (Table 10.4). Treatment with benzaldehyde yielded 3-formylpyrroles (**182**). The reaction presumably involves loss of carbon dioxide, nitrogen, and morpholine from the initial cycloadduct. Unsymmetrical münchnones behave regioselectively and furnish products derived from bonding between C(2) of the münchnone and the methylene terminus of **180**. This study also featured the synthesis of several new münchnones.

This same Milan group (103,104) described the reactions of münchnones with 4-methylene-4,5-dihydroisoxazoles **183** to give cycloadducts **184**, and with benzylideneisoxazolones **185** to give cycloadducts **186** as a mixture of isomers (Scheme 10.29). The latter undergo facile loss of benzonitrile to afford pyrroles **187**. This transformation, which also results in the interconversion of the isomeric **186**, involves ring opening to a zwitterionic species. The spiro compounds are isomeric at the spiro center and also exist as a mixture of regioisomers.

Clerici and co-workers (105,106) investigated the behavior of münchnones with an isothiazole 1,1-dioxide and a vinylisothiazole 1,1-dioxide. The initial

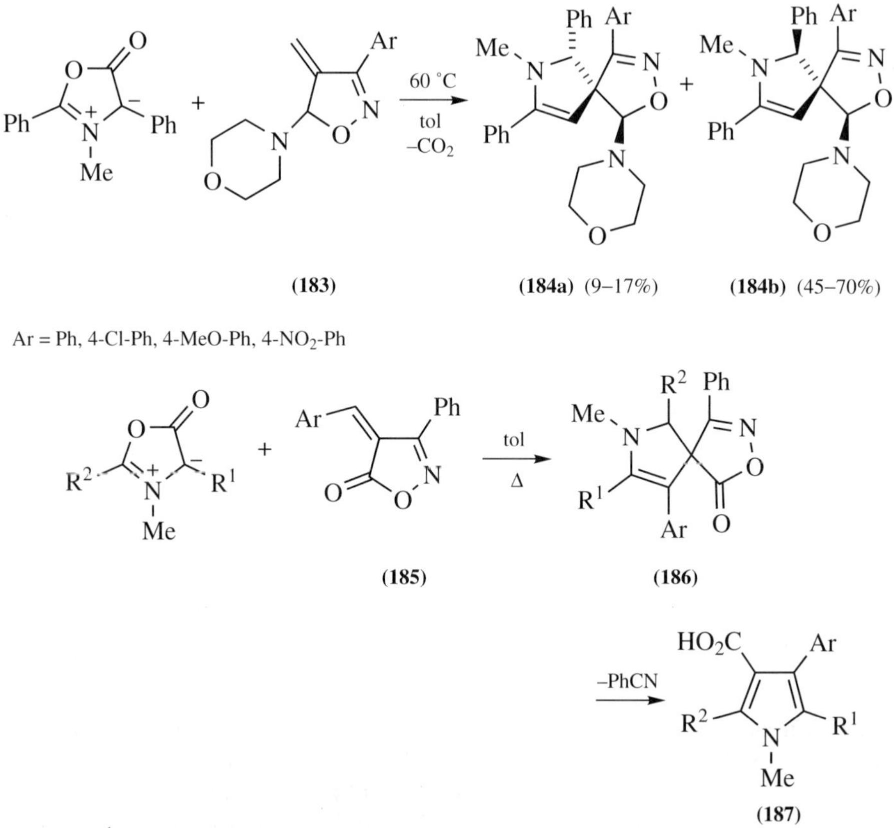

R^1 = Ph, Me, 4-Cl-Ph, 4-MeO-Ph

R^2 = Ph, Me, 4-Cl-Ph, 4-MeO-Ph, 4-Me-Ph

Ar = Ph, 4-NO_2-Ph, 4-*i*-Pr-Ph, 4-Me-Ph, 2-thienyl

Scheme 10.29

cycloadducts are converted to pyrroles on heating. An example of the latter transformation is shown in **188** → **189**.

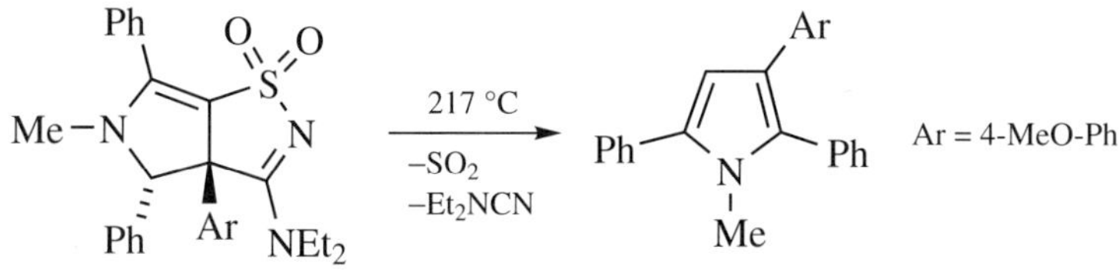

(188) **(189)**

The ability of the nitro group to serve both as a powerful electron–withdrawing and leaving group in the form of nitrous acid has led to its employment in münchnone cycloaddition chemistry. Thus, Jiménez and co-workers (107,108) studied the reaction of münchnone **190** with sugar derivatives **191** and **192**. These nitroalkenes react in a highly regioselective fashion to give pyrroles **193** and **194** in 69 and 65% yield, respectively. Deacylation and periodate cleavage afford pyrrole aldehyde **195**. A series of nuclear Overhauser effect (NOE) experiments established the structure of **195** and the regioselectivity of the cycloaddition reaction. This cycloaddition, like many involving münchnones, is opposite to that predicted by FMO theory. However, the authors carried out an *ab initio* MO (MP2/6-31B) calculation on a simpler system, the results of which do agree with experimental observation. Furthermore, from these calculations the authors conclude that this 1,3-dipolar cycloaddition of a münchnone and a nitroalkene is concerted but slightly asynchronous.

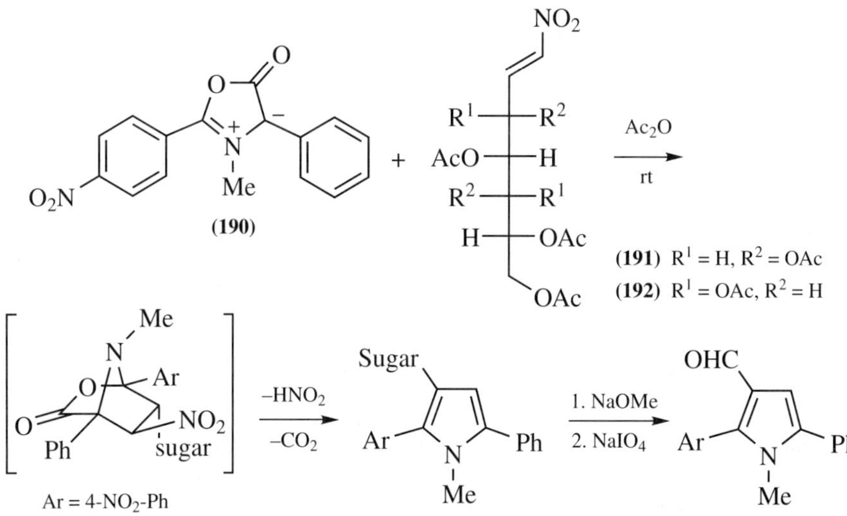

(191) $R^1 = H, R^2 = OAc$
(192) $R^1 = OAc, R^2 = H$

(193) $R^1 = H, R^2 = OAc$
(194) $R^1 = OAc, R^2 = H$

(195)

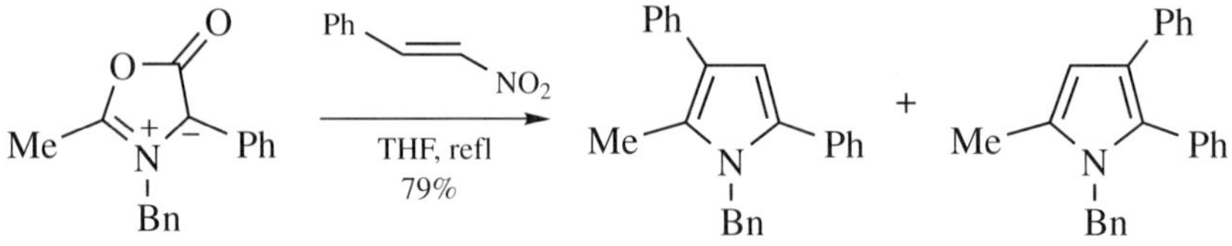

(147) **(149)** (85%) **(148)** (15%)

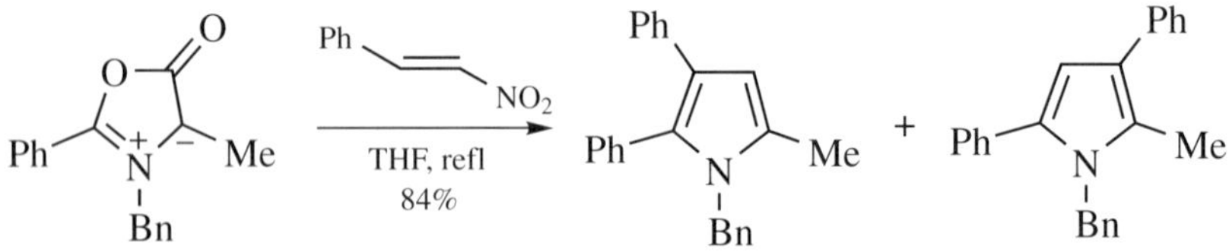

(150) **(148)** (90%) **(149)** (10%)

Scheme 10.30

Gribble et al. (80) also observed apparent anti-FMO regiochemistry in the dipolar cycloadditions of münchnones **147** and **150** with β-nitrostyrene to give pyrroles **148** and **149** (Scheme 10.30). The isomers were distinguished by NMR and NOE experiments.

The idea of employing the reaction of a nitroarene or nitroheterocycle with a münchnone to synthesize a fused pyrrole ring system has been developed by two groups. Nesi et al. (109) found that münchnone **38** reacts with 3-methyl-4-nitroisoxazole (**196**) and 4-nitro-3-phenylisoxazole (**197**) to give the corresponding 5*H*-pyrrolo[3,4-*d*]isoxazoles **198** and **199**, respectively, in good yield. Presumably loss of carbon dioxide in a retro-Diels–Alder reaction follows loss of nitrous acid.

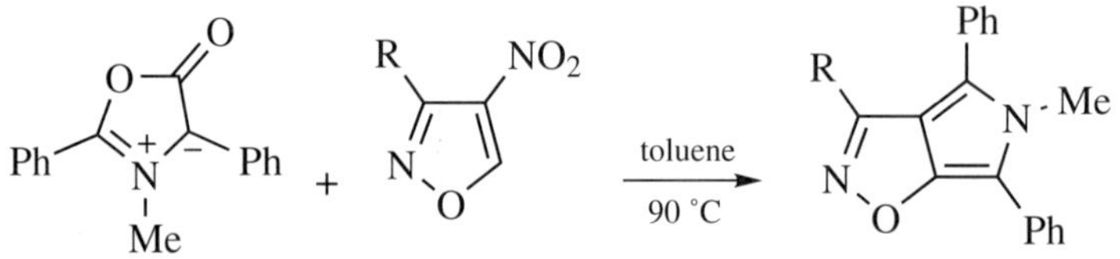

(38) **(196)** R = Me **(197)** R = Ph **(198)** R = Me (52%) **(199)** R = Ph (69%)

Gribble et al. (110,111) found that münchnones react smoothly with 2- and 3-nitroindoles to afford pyrrolo[3,4-*b*]indoles **200** and **201** (Table 10.5). In the case of unsymmetrical münchnones the regiochemistry, which can be very clean, does not follow simple FMO theory and probably results from a combination of electronic, steric, and dipole interactions. The structures of regioisomers were established

TABLE 10.5. 1,3-DIPOLAR CYCLOADDITION REACTIONS OF *IN SITU* GENERATED MÜNCHNONES AND 2- AND 3-NITROINDOLES[a,b]

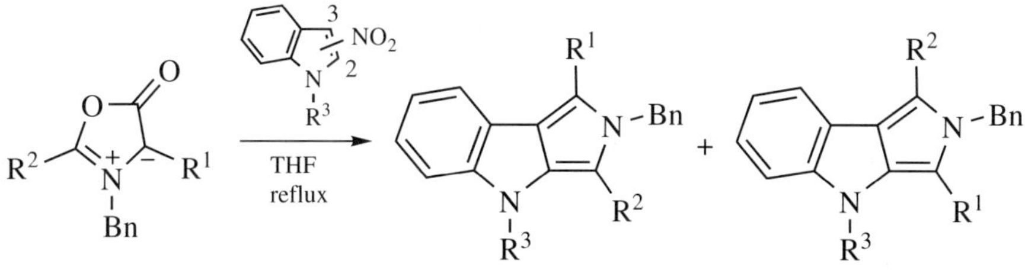

R^1	R^2	R^3	NO_2	Yield (%)	Regiochemistry % **200**	Regiochemistry % **201**	Reference
Ph	Ph	CO_2Et	2	94			110
Ph	Ph	CO_2Et	3	60			110
Me	Me	CO_2Et	2	53			110
Me	Me	CO_2Et	3	39			110
Ph	Ph	SO_2Ph	2	76			110
Ph	Ph	SO_2Ph	3	65			110
Me	Me	SO_2Ph	2	17			110
Me	Me	SO_2Ph	3	67			110
Ph	Me	CO_2Et	2	85	10	90	111
Me	Ph	CO_2Et	2	88	10	90	111
Ph	Me	CO_2Et	3	65	95–100	0–5	111
Me	Ph	CO_2Et	3	89	40–50	50–60	111
Ph	Me	SO_2Ph	3	74	94–100	0–6	111
Me	Ph	SO_2Ph	3	76	10–30	70–90	111
Me	Ph	Me	3	$<1\%$[c]			111

[a] Münchnones were generated from the corresponding N-acylamino acids using DIPC in refluxing THF.
[b] References (110) and (111).
[c] Reflux in diglyme for 24 h.

using a combination of NOE techniques, independent synthesis, and X-ray crystallography. This novel pyrrole annulation reaction was also employed to synthesize benzo[*b*]furo[2,3-*c*]pyrroles and benzo[*b*]thieno[2,3-*c*]pyrroles (110).

The higher reactivity of ring-strained olefins has been exploited by several workers in 1,3-dipolar cycloaddition reactions of münchnones. Thus, Kato and co-workers (112) reported that münchnone **38** reacts with 1,2,3-triphenyl-1*H*-phosphirene (**202**) to give 1-methyl-2,3,4,5-tetraphenylpyrrole (**203**) (45% yield). Control experiments demonstrated that phosphirene **202** does not decompose to diphenylacetylene appreciably under the reaction conditions. Moreover, the reaction of diphenylacetylene and münchnone **38** afforded only a 21% yield of pyrrole **203**.

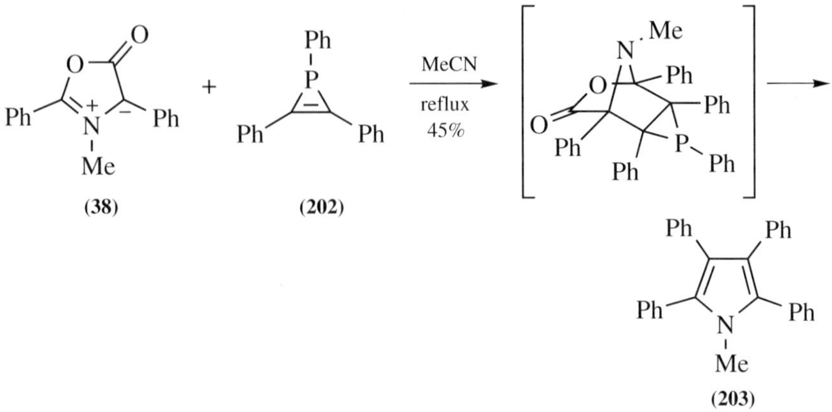

(38) (202) (203)

Unfortunately, this same group found that münchnone **38** gave only complex product mixtures upon reaction with benzocyclopropene, in an unsuccessful attempt to synthesize a methanoazonine (113). As is presented later, a similar reaction was successful with an isomünchnone. Likewise, Kato et al. (114) were unable to induce münchnones **38** and **190** to react cleanly with benzocyclobutadiene.

Martin et al. (115) found that münchnone **38** reacts with isopropylidenecyclobutenone (**204**) to form dihydroazepine **205**. At room temperature the two bis(adducts) **206** and **207** were isolated, although the regiochemistry of the cycloaddition has not been established.

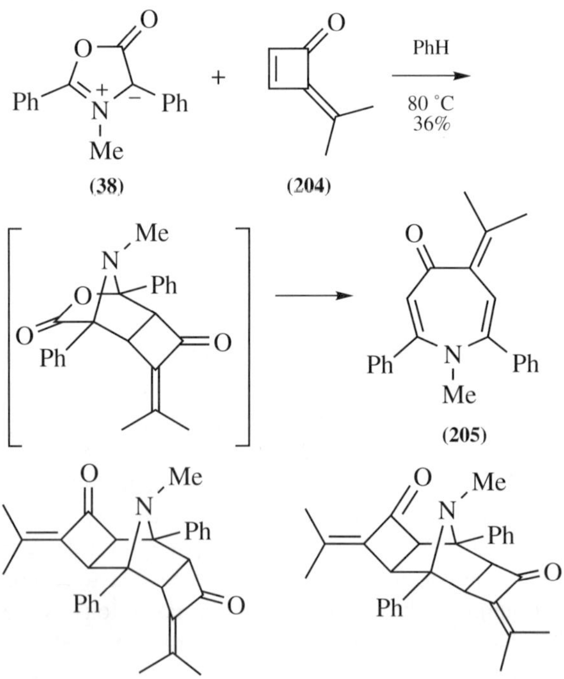

(38) (204) (205)

(206) (20%) (exo, endo) **(207)** (4%) (exo,exo)

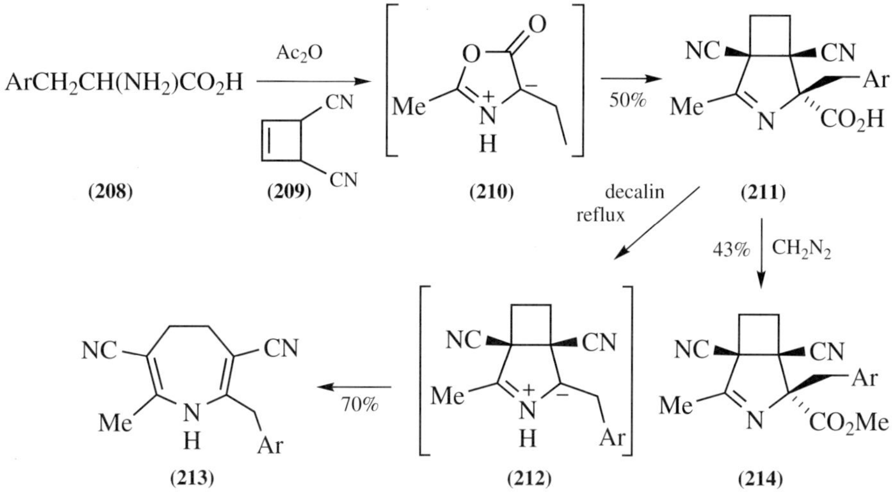

Scheme 10.31

Maryanoff et al. (116) studied of the reaction of 1,2-dicyanocyclobutene (**209**) with münchnones, and they find that 3-(4-chlorophenyl)alanine (**208**) reacts with **210** in the presence of acetic anhydride to give imino acid **211** (Scheme 10.31). Esterification of **211** afforded ester **214**, whose structure was confirmed by X-ray crystallography. Further heating of **211** yielded the originally expected dihydroazepine **213**. The isolation of acid **211** suggests that decarboxylation of the münchnone primary adducts need not be concerted. Maryanoff and Turchi (117) pursued a detailed theoretical study of the reaction between 1,2-dicyanocyclobutene (**209**) and münchnone **210**. The results from these AM1 molecular orbital calculations led to the conclusions that the transition state leading to the exo cycloadduct **211** is favored electrostatically and that azomethine ylide **212** is a discrete intermediate in the formation of dihydroazepine **213**.

Turchi (118) also reported cycloaddition reactions between münchnones **215** and **209** to afford dihydroazepine **216** in high yield. Further cyclization of **216** gave tricycle **217**. Likewise, diester **218** reacts with münchnone **38** to give dihydroazepine **219**.

Kato et al. (119) explored reactions of fulvenes with a variety of mesoionic heterocycles. Unfortunately, reactions of münchnone **38** with several fulvenes afforded complex mixtures in each case, and no identifiable products were reported, although Friedrichsen and co-workers (120–122) previously reported the reaction between münchnones and fulvenes to give cycloadducts. Kato et al. (123) also studied the cycloaddition reactions of tropone with several mesoionic heterocycles. Despite heroic efforts, the reaction of tropone with münchnone **38** was complex and could not be unraveled. However, as described later, the reaction of tropone with isomünchnones was successful. Wu et al. (124) effected the cycloaddition between a münchnone and fullerene-60 (C_{60}) to give the corresponding dihydropyrrole in excellent yield.

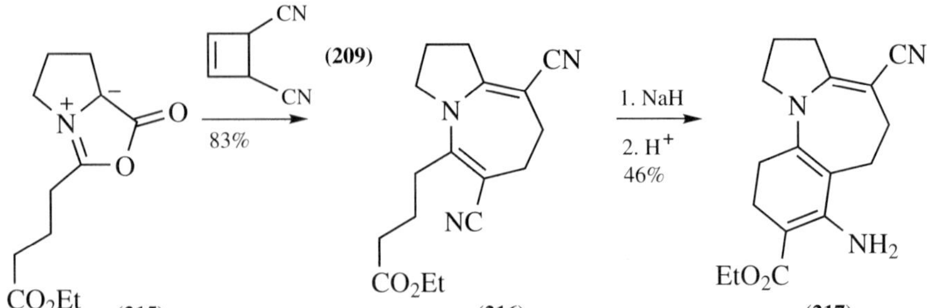

(215) (216) (217)

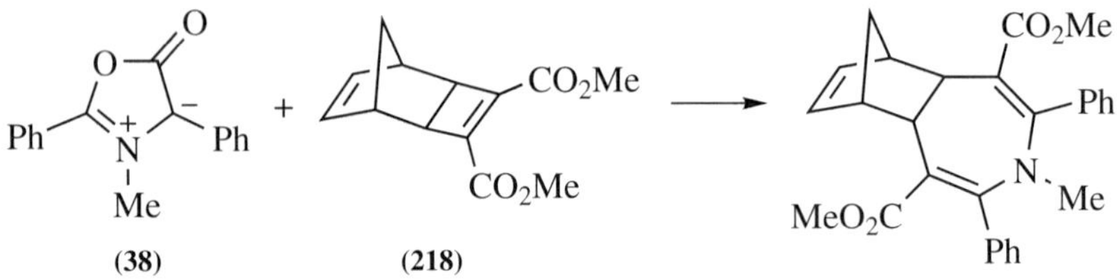

(38) (218) (219)

Gribble et al. (125) found that münchnones **38** and **220** react in a novel tandem fashion with 1,5-cyclooctadiene (**221**) to give the caged compounds, 10-methyl- (**222a**) and 10-benzyl-9,10-diphenyl-10-azatetracyclo$[6.3.0.0.^{4,11}0^{5,9}]$undecane (**222b**). The latter compound was characterized by X-ray crystallography. A similar reaction of münchnone **220** with 1,3,5,7-cyclooctatetraene afforded the corresponding cycloadduct in low yield, the photolysis of which gave azahomopentaprismane in good yield.

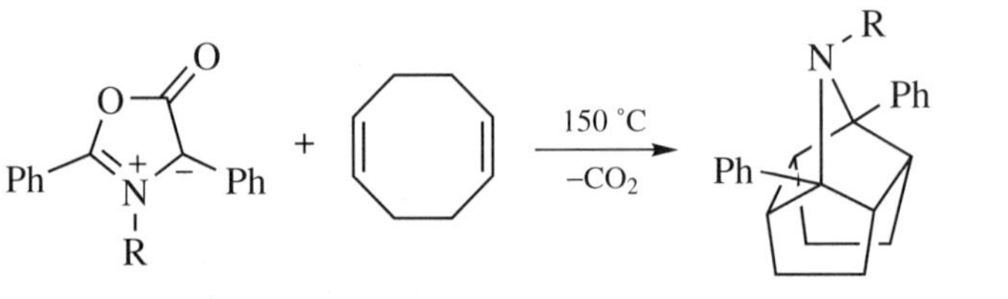

(38) R = Me (**221**) **(222a)** R = Me (57%)
(220) R = Bn **(222b)** R = Bn (74%)

Padwa et al. (126) applied the intramolecular münchnone–olefin cycloaddition reaction to craft a series of novel caged compounds **224** and **225**. Whereas the alanine–derived münchnone from **223a** affords only **224a**, the 2-phenylglycine–derived münchnone from **223b** gives **225** as the major regioisomer. Cyclization of the münchnones derived from glycine derivatives is also highly regioselective to afford the expected cycloadducts.

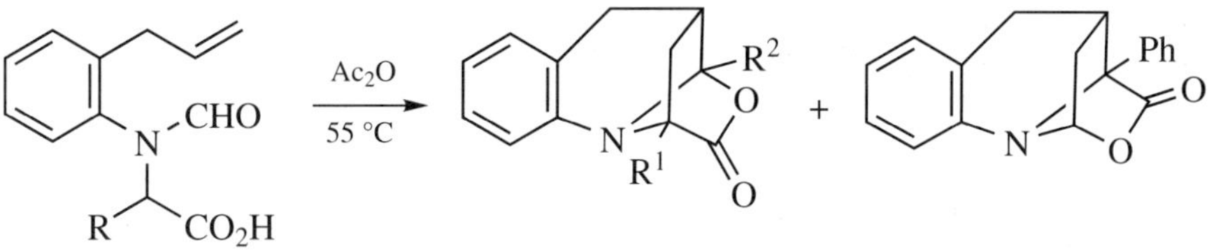

(223a) R = Me
(223b) R = Ph

(224a) R^1 = Me, R^2 = H (75%)
(224b) R^1 = Ph, R^2 = H (26%)

(225) (58%)

Kawase (127,128) reexamined the reaction of münchnones with oxygen. Not only is this a powerful method for the synthesis of imides, but, based on ^{18}O labeling experiments, the mechanism of this autoxidation is different from that originally proposed by Huisgen and co-workers (129). This reaction is particularly useful for the preparation of tetrahydroisoquinolones and tetrahydrocarbolones.

The father of münchnone chemistry, Rolf Huisgen, reported (130) the 1,3-dipolar cycloaddition of münchnone **38** with *p*-nitrophenyldiazonium salt **226** to give the triazolium salt **227**. This reaction would appear to be an important new route to these heterocycles.

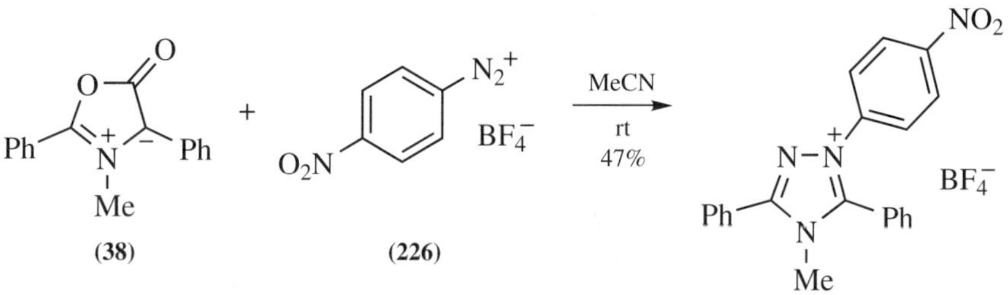

(38) **(226)** **(227)**

Ferraccioli and co-workers (131,132) employed the reaction of münchnones with N-(phenylsulfonyl)imines as a general synthesis of imidazoles (**228**) (Table 10.6). Regioselectivity is very high and was determined using NOE measurements. The authors utilize the perturbation MO treatment, which is less approximate than the FMO method, to predict correctly the observed regiochemistry.

Bilodeau and Cunningham (133) effected the solid-phase synthesis of a series of imidazoles via münchnone generation and trapping regioselectively with N-tosylimines. Hamper et al. (134) employed a solid-phase protocol for the generation of a large library of 5-(trifluoroacetyl)imidazoles (200 compounds) using the reaction between münchnones and benzamidines. Dalla Croce and co-workers (135) pursued the chemistry of unsaturated N-(phenylsulfonyl)imines and münchnone **38**. While the major products are imidazoles, pyrroles and amides also are formed in this reaction. This same group (136) generated various bicyclic münchnones and trapped them with imines to afford either imidazoles or spirocyclic β-lactams depending on conditions, although mixtures are produced and yields of the

TABLE 10.6. 1,3-DIPOLAR CYCLOADDITION REACTIONS OF *IN SITU* GENERATED MÜNCHNONES[a] AND N-(PHENYLSULFONYL)IMINES[b]

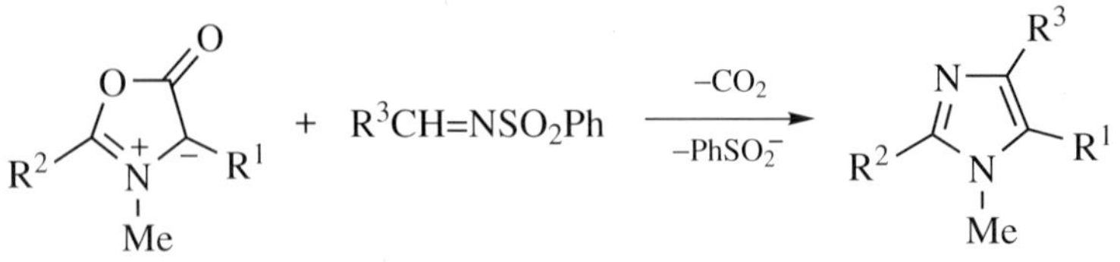

(228)

R^1	R^2	R^3	% Yield **228**
Ph	Ph	Ph	64
Me	Ph	Ph	29
Ph	Me	Ph	40
Ph	Ph	H	20
Ph	Ph	4-NO_2–Ph	65
Me	Ph	4-NO_2–Ph	50
Ph	Me	4-NO_2–Ph	30
Ph	Ph	4-MeO–Ph	45
Me	Ph	4-MeO–Ph	55
Ph	Me	4-MeO–Ph	42

[a] The münchnones were generated *in situ* by the cyclodehydration of their N-acylamino acid precursors with DCC in toluene (25–60 °C).
[b] Reference (131).

β-lactams are invariably low. The latter can arise by nucleophilic attack on the münchnone itself or on the ring-opened ketene tautomer. Storr and co-workers (137) employed the cycloaddition of münchnones **230** and **231** with 2-phenylbenzazete (**229**) to craft 3*H*-1,3-benzodiazepines **232** and **233**. Upon heating and subsequent hydrolysis these benzodiazepines undergo conversion to 2,3-diarylindoles **236** and **237**, as summarized in Scheme 10.32. The observed regiochemistry is consistent with that seen with other electron–deficient dipolarophiles.

Rodríguez et al. (138) found that münchnones react with nitrosobenzene in hot xylene to give N-benzoyl-N′-phenylbenzamidines. At room temperature the intermediate oxadiazolines can be isolated.

Apparently independently, Märkl et al. (139) and Regitz and co-workers (140–142) discovered that 1,3-dipolar cycloaddition reactions of münchnones and phosphaalkenes or phosphaalkynes provide a direct synthesis of 1,3-azaphospholes (**240**) (Table 10.7). The intermediate cycloadducts cannot be isolated. The various phosphaalkynes were generated from phosphaalkenes or, in the case of methylidynephosphane (**239**, R^4=H), by flash vacuum pyrolysis of either **239** (R^4=*t*-Bu) or dichloromethylphosphine.

Münchnones can undergo ring opening to ketene tautomers, which can then engage in chemistry. A few examples of this pathway have been described since the reviews by Potts (1) and Gingrich and Baum (10). For example, münchnones react

TABLE 10.7. 1,3-DIPOLAR CYCLOADDITION REACTIONS OF MÜNCHNONES AND PHOSPHAALKYNES

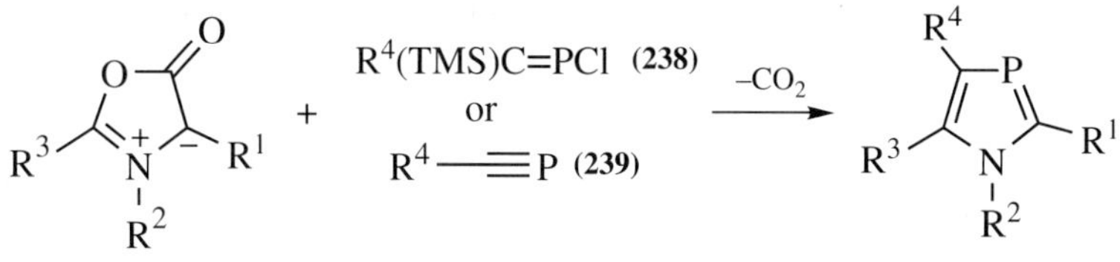

(240)

Dipolarophile	R^1	R^2	R^3	R^4	% Yield **240**	Reference
238	Ph	Ph	Ph	Ph	50	139
238	4-Cl–Ph	Ph	Ph	Ph	40	139
238	4-MeO–Ph	Ph	Ph	Ph	43	139
238	Ph	Me	Ph	Ph	33	139
238	4-MeO–Ph	Me	Ph	Ph	55	139
239	Ph	Me	Ph	*t*-Bu	63	140
239	Ph	Me	Ph	H	23	141
239	Ph	Me	Ph	Mes[a]	46	142

[a] Mesityl = Mes.

with unsaturated imines to give β-lactams (135), and Regitz and co-workers (143) found that the stable 2,3,4-tri-*tert*-butylazete reacts with münchnone **38** to afford a mixture of unseparable (*E*) and (*Z*) isomers of oxaazabicyclo[2.2.0]hexenes.

In a series of papers, Laude and co-workers (144–149) examined 1,3-dipolar cycloaddition reactions of münchnone imines derived from Reissert compounds. For example, münchnone imine **241** undergoes a smooth intramolecular 1,3-dipolar cycloaddition with the tethered alkyne unit to afford pyrrole **242** after extrusion of HNCO (144).

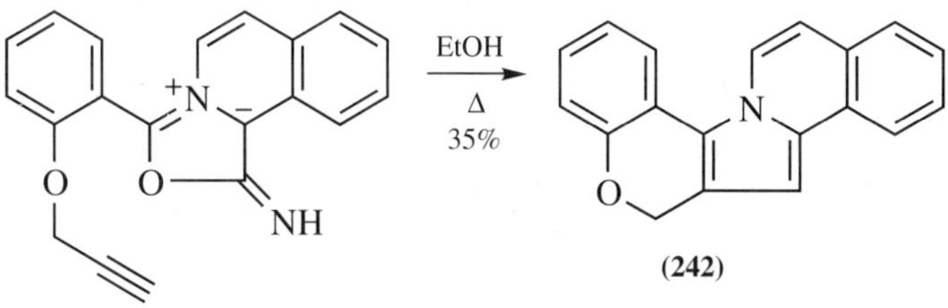

(241) (242)

This French group (145) has also been able to divert the usual Diels–Alder cycloaddition pathway of Reissert salts with olefinic esters to a 1,3-dipolar cycloaddition pathway by the addition of triethylamine. In addition, münchnone imine **243** can be trapped with dipolarophiles to furnish **244** (146). No Diels–Alder cycloadducts derived from the oxazolium salt were detected. In contrast, fumarate

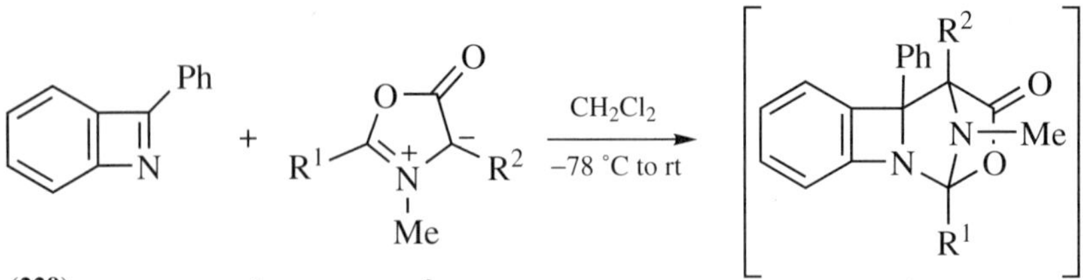

(229) **(230)** R^1 = 4-Me-Ph, R^2 = Ph
(231) R^1 = Ph, R^2 = 4-Me-Ph

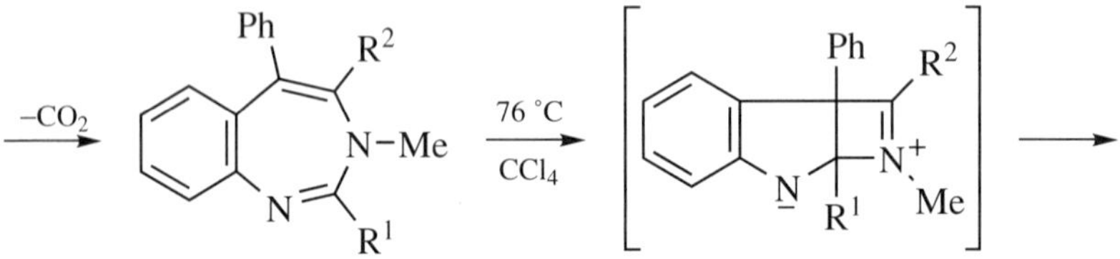

(232) R^1 = 4-Me-Ph, R^2 = Ph (47%)
(233) R^1 = Ph, R^2 = 4-Me-Ph (44%)

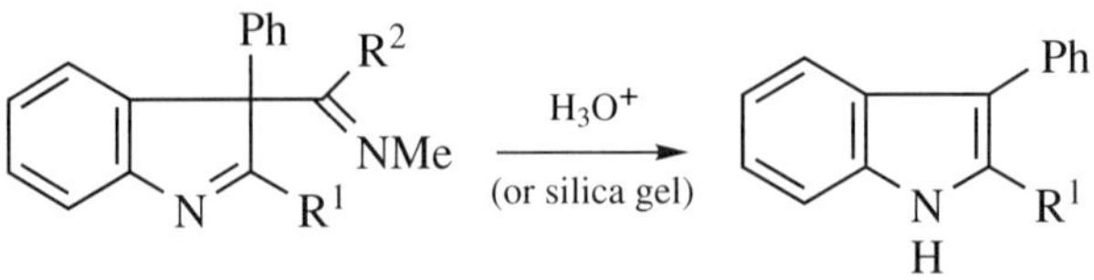

(234) R^1 = 4-Me-Ph, R^2 = Ph (92%) **(236)** R^1 = 4-Me-Ph
(235) R^1 = Ph, R^2 = 4-Me-Ph (88%) **(237)** R^1 = Ph (83%)

Scheme 10.32

and acrylate esters give only Diels–Alder cycloadducts from the tautomeric oxazolium salt. Also, benzoquinones and 1,4-naphthoquinone react in a 1,3-dipolar fashion with münchnone imines derived from Reissert compounds (147).

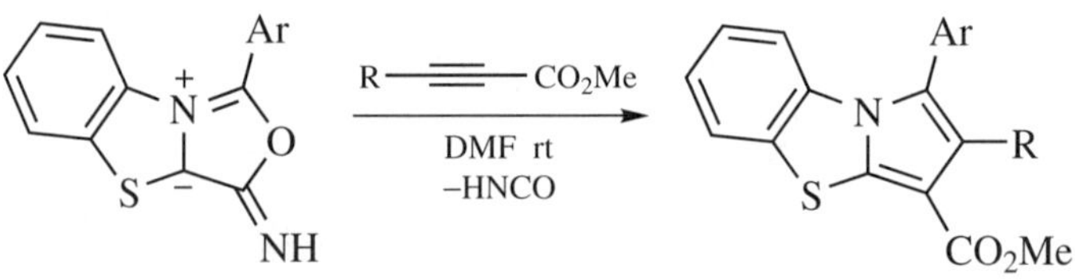

(243) **(244a)** R = CO_2Me (26%)
(244b) R = H (65%)
Ar = 4-NO_2-Ph

In a reinvestigation of earlier work by McEwen et al. (150), Laude and coworkers (148,149) showed that the product from the reaction between münchnone imine **245** and several acrylates affords 2-pyridones (e.g., **248**) and not a 3-carboethoxypyrrole as originally claimed. Thus, once again, a 1,3-dipolar

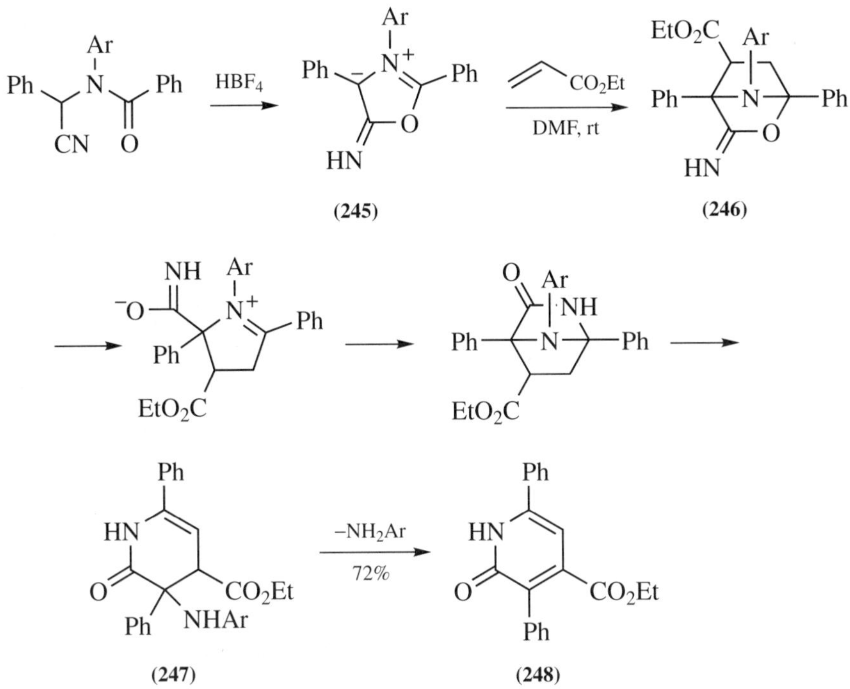

Scheme 10.33

cycloaddition reaction triumphs over a Diels–Alder cycloaddition. The process is illustrated in Scheme 10.33. In some cases, cycloadduct **246** and dihydropyridone **247** could be isolated.

10.2.2. Isomünchnones (1,3-oxazolium-5-olates)

The reader was given a taste of the power of isomünchnone dipolar cycloaddition chemistry in Section 10.2.1. As discussed by Potts (1) and Gingrich and Baum (10), the isomünchnone ring system—a "masked" carbonyl dipole—is exceptionally reactive as a 1,3-dipole in 1,3-dipolar cycloaddition reactions. In the intervening years since these two excellent reviews the major research efforts in isomünchnone chemistry have entailed synthetic applications to specific targets such as alkaloids and other natural and unnatural products.

Kato et al. (113) have had much better success in performing 1,3-dipolar cycloadditions with isomünchnones than with münchnones (see above). Thus, the room temperature union of isomünchnone **51a** with benzocyclopropene (**249**) leads to a syn-cycloadduct **250**. The latter is remarkably stable and is recovered unchanged upon heating to 300°C. It is also impervious to the action of tributylphosphine, in Kato's abortive attempt to excise the bridging oxygen, which would have led to a methanooxonine.

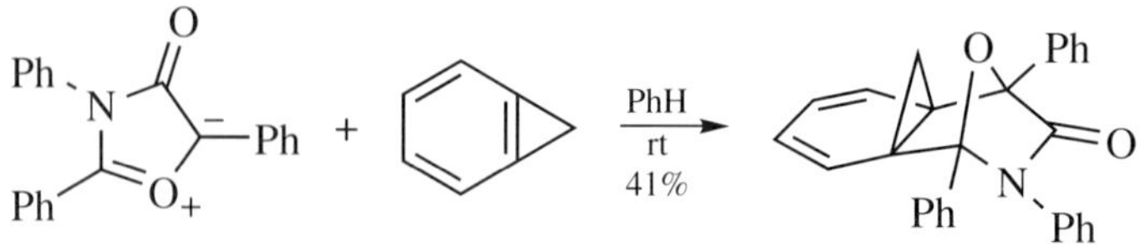

(51a) (249) (250)

Kato et al. (123) also found that isomünchnone **51a** reacts with tropone (**251**) to afford **252**, which is apparently the first example of a [4π+6π] cycloadduct involving both a mesoionic heterocycle and a carbonyl ylide (Scheme 10.34). The one-pot reaction of **253** gave **252** in somewhat higher yield. Whereas heating the latter in bromobenzene affords *o*-benzoylmandelanilide (69%), heating **252** in refluxing toluene in the presence of DMAD leads to furan **254**.

Kato et al. (151,152) explored the chemistry of 2-*tert*-butylfulvenes with isomünchnones, as well as with several other mesoionic compounds, in a novel approach to pseudo-hetero-azulenes. Thus, isomünchnone **51a**, generated as before *in situ* from *N*-benzoylphenylglyoxyanilide **253** with triethylphosphite, reacts with 2-*tert*-butyl-6-(dimethylamino)fulvene to give the [4π+6π] adduct diphenylcyclopenta[*c*]pyran in low yield. Likewise, reaction of **51a** with dimethylfulvene gave a mixture of two adducts, one of which arises from a [4π+2π] cycloaddition.

Regitz and co-workers (143) found that 2,3,4-tri-*tert*-butylazete reacts with isomünchnones to give relatively labile cycloadducts. This group (153) has employed the cycloaddition of isomünchnones **256** with phosphaalkynes **257** to prepare 1,3-oxaphospholes **258** (Scheme 10.35). This sequence is clearly the method of choice for the synthesis of the relatively little investigated 1,3-oxaphospholes. The presumed bicyclic intermediates could not be detected by NMR.

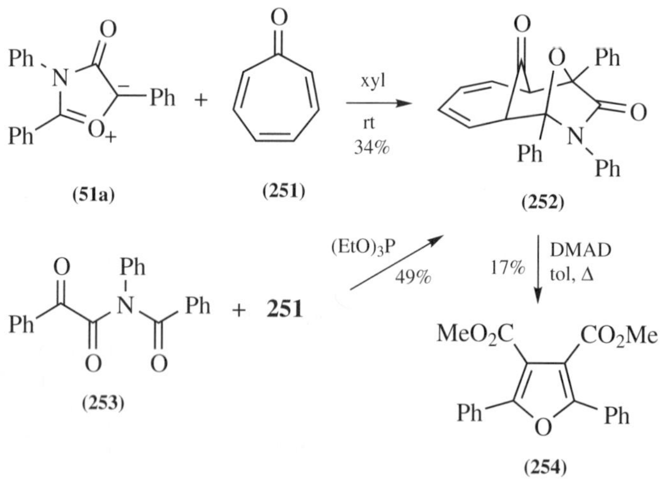

Scheme 10.34

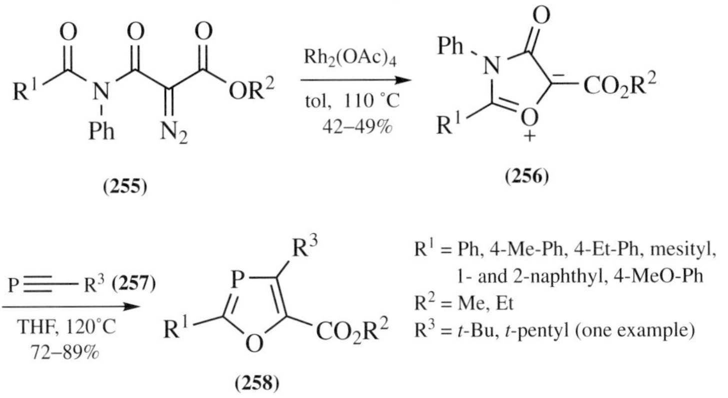

Scheme 10.35

Padwa and his co-workers have been the consummate practitioners of the rhodium(II)-catalyzed decomposition of α-diazo carbonyl compounds leading to isomünchnones and their subsequent 1,3-dipolar cycloaddition reactions. Padwa and Hertzog (154) described intermolecular cycloaddition reactions of isomünchnones with both electron-rich and -deficient dipolarophiles. The resulting regiochemistry is in accord with FMO theory. Diazo imide **259** is readily prepared from 2-pyrrolidinone and under the usual conditions it reacts via **260** with *N*-phenylmaleimide to give the expected cycloadducts (86% yield; *endo*/*exo*, 2.4:1), and also with DMAD to give a furan following ring opening. Interestingly, isomünchnone **260** could not be generated from chloroacetyl lactam **54** ($n=3$) as described earlier (Scheme 10.7).

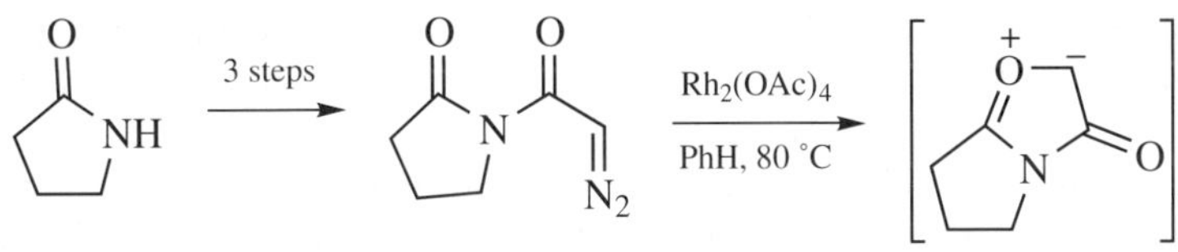

(259) **(260)**

Whereas **260** does not react with electron-rich dipolarophiles, the more delocalized isomünchnone **261** does react with both electron-rich and -deficient dipolarophiles (154). A detailed FMO analysis is consistent with these observations and with the regiochemistry exhibited by diethyl ketene acetal and methyl vinyl ketone as shown in Scheme 10.36. The reaction of **261** with the ketene acetal to give **262** is LUMO-dipole HOMO-dipolarophile controlled (so-called Type III process). In contrast, the reaction of **261** with methyl vinyl ketone to give **263** is HOMO-dipole LUMO-dipolarophile controlled (so-called Type I process). In competition experiments using a mixture of *N*-phenylmaleimide and ketene acetal only a cycloadduct from the former was isolated. This result is consistent with a smaller energy gap for

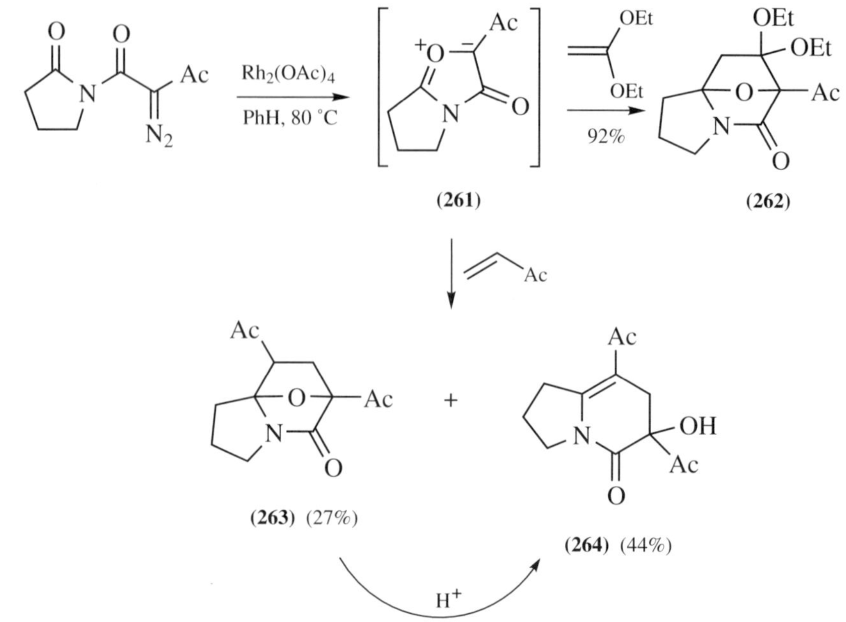

Scheme 10.36

this Type I process than for the Type III reaction ($\Delta E = 7.05$ vs 8.69 eV). The difference in reactivity between isomünchnones **260** and **261** is also manifest in their behavior with methyl propiolate.

In a careful study of rhodium catalysts for the decomposition of α-diazo imides, Padwa and co-workers (155,156) found that perfluorinated ligands greatly favor isomünchnone formation, whereas acetate leads to the generation of a six-membered carbonyl ylide. These catalysts include rhodium perfluorobutyroamidate [$Rh_2(pfm)_4$], rhodium perfluorobutyrate [$Rh_2(pfb)_4$], and rhodium trifluoroacetate [$Rh_2(tfa)_4$], but not rhodium acetate. The mechanistic basis for this dramatic ligand effect is not understood. Interestingly, the addition of the strong Lewis acid $Sc(OTf)_3$ to the $Rh_2(OAc)_4$ reaction mixture diverts the pathway to an isomünchnone mode. Padwa and Prein (157) have also presented an extensive experimental and theoretical study of the 1,3-dipolar cycloaddition reactions of isomünchnones with olefinic dipolarophiles, and Wudl, Padwa, and their co-workers (158) find that several isomünchnones react with C_{60} reversibly to form the expected cycloadducts. These workers suggest that such cycloadducts could function of isomünchnone repositories, since heating them in the presence NPM affords C_{60} and the known isomünchnone–NPM cycloadducts.

The dipolar cycloaddition chemistry of isomünchnones is a powerful and concise route to polycyclic azaheterocycles, and Padwa has been the pioneer in this effort. Sheehan and Padwa (159) employed the rhodium-catalyzed isomünchnone generation and subsequent trapping to a synthesis of 2-pyridones and the alkaloid

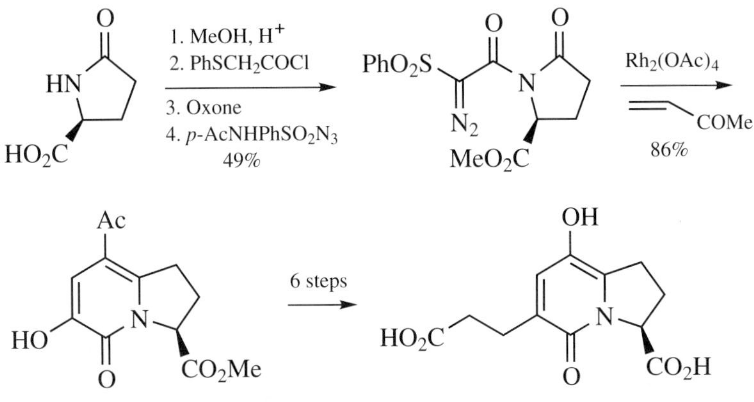

(265)

Scheme 10.37

(±)-ipalbidine. Expansion of this protocol led to the synthesis of several other highly substituted 2-pyridones (160,161). For example, the angiotensin converting enzyme inhibitor (−)-A58365A (**265**) was synthesized using this protocol (Scheme 10.37). The appropriate selection of starting materials allowed for the synthesis of the indolizidine alkaloids δ-coniceine and a formal synthesis of (±)-septicine.

Padwa and Prein (162) generated several chiral isomünchnones, using the rhodium catalyzed deamination of chiral α-diazo imides, and trapped them with various dipolarophiles.

The best results were obtained with those isomünchnones derived from phenylalanine methyl ester, apparently due to π-stacking and shielding of one face. This work and that of Harwood (see below) represent the first examples of acyclic stereocontrol in the 1,3-dipolar cycloaddition reactions of mesoionic heterocycles. Isomünchnones derived from alanine and leucine were less stereocontrolling. The results from this study reinforce the notion that the actual 1,3-dipole is an isomünchnone-rhodium carbenoid species and not the free isomünchnone. Independently, Harwood (163–165) also demonstrated the role of chiral-templated isomünchnones in 1,3-dipolar cycloaddition reactions. Thus, using the rhodium (II)-catalyzed decomposition of diazo carbonyl compounds, Harwood and co-workers (163) explored cycloadditions of isomünchnone derivatives of (5*R*)- and (5*S*)-phenyloxazin-2,3-dione. Along with the work of Padwa (see above), these reactions appear to represent the first examples of chirally templated isomünchnone 1,3-dipolar cycloadditions. For example, the isomünchnones derived from diazo imides **266** react with methyl propiolate to form adducts **267** (164). Diazo imides **266** (R=CO_2Et, Ac, H) also react with maleimides and DMAD to afford adducts with high endo/exo selectivities and moderate diastereofacial selectivities. Harwood and co-workers (165) also found that the isomünchnones derived from these diazo imides react with aromatic aldehydes with excellent diastereofacial- and exo-selectivity.

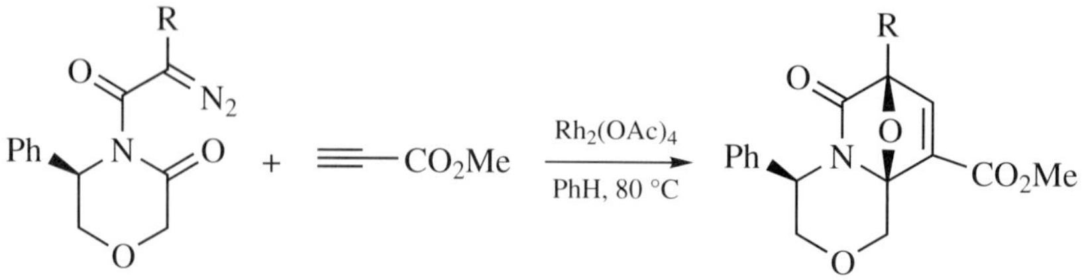

(266a) R = CO_2Et
(266b) R = Ac

(267a) R = CO_2Et (65%)
(267b) R = Ac (70%)

Kappe et al. (166) employed an isomünchnone generation-trapping sequence to access conformationally restricted dihydropyrimidine derivatives as novel calcium channel modulators. For example, the conformationally restricted analogues **269** were prepared via intramolecular cycloadditions from the isomünchnones generated from α-diazo imides **268**. The structures of these cycloadducts were established by X-ray crystallography.

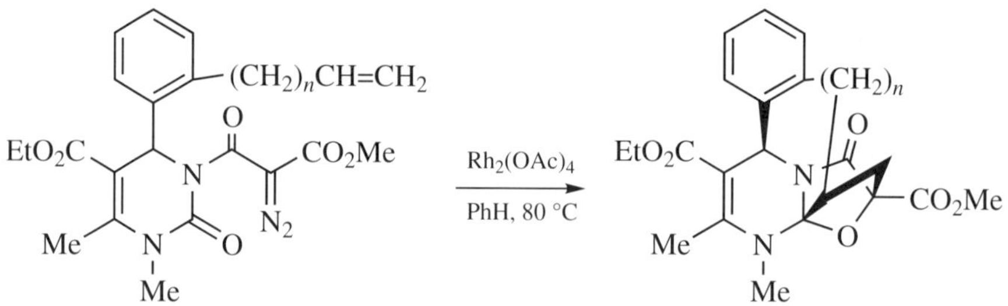

(268a) $n = 0$
(268b) $n = 1$

(269a) $n = 0$ (92%)
(269b) $n = 1$ (88%)

Kappe (167) investigated the generation and trapping of "aminoisomünchnones". Interestingly, when diazoacetylurea (**270**) is treated with rhodium acetate, the ammonium ylide **272** is obtained in good yield (Scheme 10.38). The structure of this unanticipated product was established by X-ray crystallography. However, when this reaction is performed in the presence of DMAD, then 2-aminofuran (**275**) is isolated, presumably indicating the generation of isomünchnone **274** via the rhodium-carbenoid intermediate **271**. The reversibility of the sequence was shown by the fact that ylide **272** reacts with DMAD in the presence of rhodium acetate to give a small amount (10–15%) of 2-aminofuran (**275**), presumably via an equilibrium concentration of isomünchnone **271**. In a continuation of these observations, Padwa et al. (168) examined in depth the rhodium-catalyzed chemistry of α-diazo esters leading either to ammonium or carbonyl ylides depending on the reaction conditions.

Gowravaram and Gallop (169) adapted the rhodium-catalyzed generation of isomünchnones from diazo imides to the solid-phase synthesis of furans, following a 1,3-dipolar cycloaddition reaction with alkynes. A variety of furans were prepared in this fashion. With unsymmetrical electron-deficient alkynes (e.g., methyl

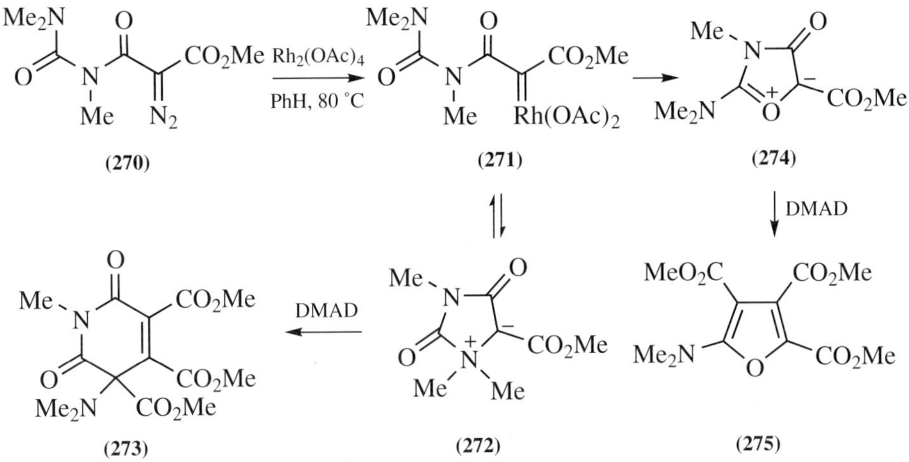

Scheme 10.38

propiolate), the anticipated regiochemistry is observed; that is, HOMO–dipole LUMO–dipolarophile, as seen previously. Independently, Austin and co-workers (170,171) also adopted the isomünchnone generation and trapping protocol to the solid-phase synthesis of furans. Model studies revealed that rhodium perfluorobutyroamidate [$Rh_2(pfm)_4$] afforded none of a heptatriene byproduct, which forms via a tandem cyclopropanation–Cope rearrangement. Moreover, this catalyst is more soluble in organic solvents than is rhodium trifluoroacetate. This same group (172) found that isomünchnones react with electron-rich enol ethers to give the corresponding cycloadducts in high yield. In each case a single diastereomer is isolated. The authors conclude that similar FMO energetics obtain for both electron-rich and -deficient dipolarophiles in their 1,3-dipolar cycloaddition reactions with isomünchnones.

As Moody and co-workers (173) discovered, isomünchnones can occasionally form even when they are not the desired product! Thus, these workers inadvertently obtained an oxazolidinedione via an isomünchnone rather than the desired oxoindoline when a diazo compound was treated with rhodium(II) perfluorobutyramide in their studies leading ultimately to a synthesis of the marine alkaloid convolutamydine C.

Hamaguchi and Nagai (174) revised the structures for the 1:1 adduct of various mesoionic heterocycles and isocyanates that were originally proposed by Potts et al. (175–177). Thus, the acylated isomünchnone structure **279** is now proposed for the reaction product of isomünchnone **276** and aryl isocyanates and aryl thioisocyanates (Scheme 10.39). Similar acylated isomünchnones were obtained with benzoylisocyanate, phenylthioisocyanate, and benzoylthioisocyanate. The NMR evidence seems compelling for this revision.

Although Maier achieved the first intramolecular 1,3-dipolar cycloaddition reaction of an isomünchnone, it has been Padwa who has unleashed the synthetic

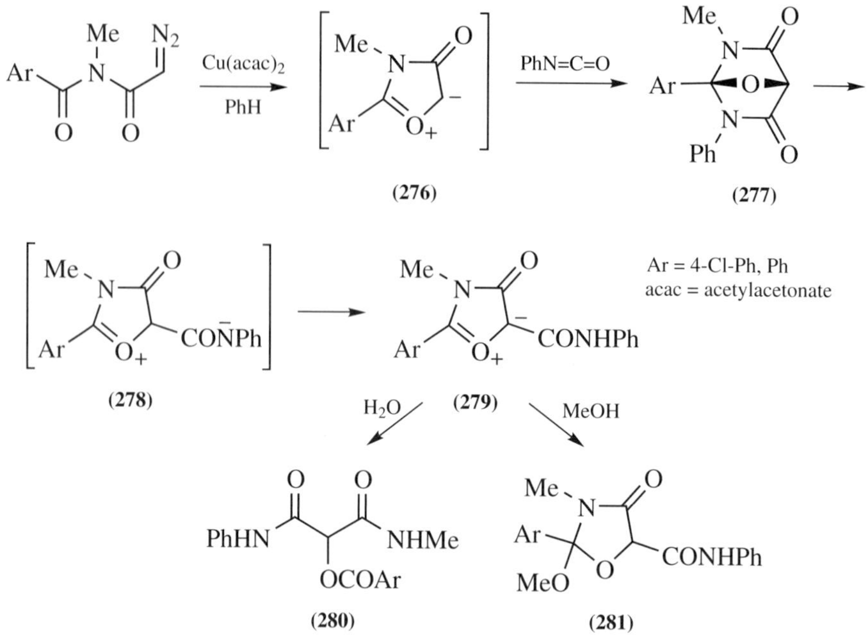

Scheme 10.39

utility of this strategy. Thus, Padwa and co-workers (178,179) found that isolated π-bonds can successfully and efficiently capture the *in situ* generated isomünchnones as shown by examples **282** → **283** (Scheme 10.40). The indole double bond in **284** intercepts an isomünchnone 1,3-dipole to give the single diastereomer **285**, the structure of which is confirmed by X-ray crystallography. The initial cycloadducts, such as **283**, can be ring opened with acid to *N*-acyliminium ions that can then either deprotonate to an enamide or be reduced to a hydroxy enamide.

Padwa et al. (180,181) effected intramolecular 1,3-dipolar cycloaddition reactions of isomünchnones tethered with other examples of π-systems. The preferred exo cycloaddition mode that is observed experimentally is supported by molecular mechanics calculations. In those cases where ring strain prevents an intramolecular cycloaddition, the isomünchnone can nevertheless be intercepted by *N*-phenylmaleimide in an intermolecular reaction. Padwa et al. (182) extended this methodology to intramolecular 1,3-dipolar cycloaddition reactions of isomünchnones to tethered heterocycles, and to the synthesis of several alkaloids. Thus, syntheses of vallesamidine (183), erythrinane homologues (184), lycopodine (185), and a clever approach to lysergic acid (186) have been described by Padwa and co-workers. Although ultimately unsuccessful, the approach to lysergic acid provided much interesting chemistry, including several examples of intramolecular isomünchnone 1,3-dipolar cycloadditions (186) (Scheme 10.41). The desired cycloaddition reaction **287** → **288** proceeded in very high yield. However, the double bond in **289** could not be isomerized as required for a synthesis of lysergic acid.

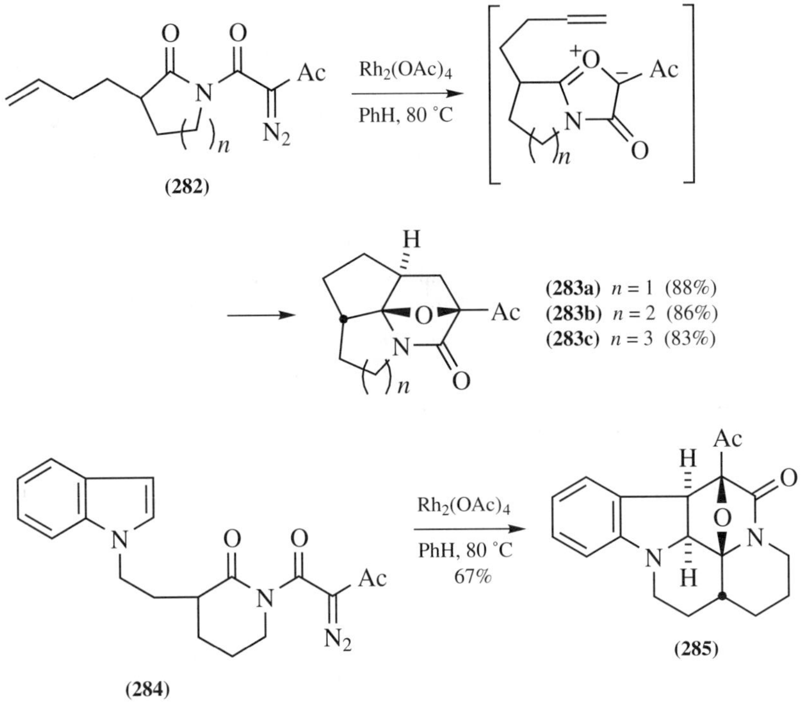

Scheme 10.40

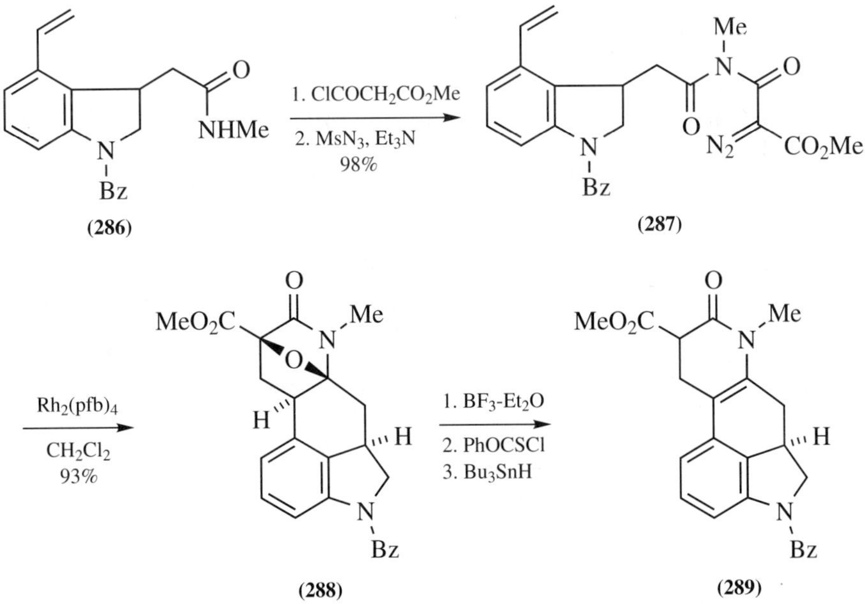

Scheme 10.41

Padwa et al. (187,188) concisely summarized his "domino cycloaddition/*N*-acyliminium ion cyclization cascade" process, which involves sequentially the generation of an isomünchnone 1,3-dipole, intramolecular 1,3-dipolar cycloaddition reaction, *N*-acyliminium ion formation, and, finally, Mannich cyclization. Kappe and co-workers (189) utilized Padwa's cyclization–cycloaddition cascade methodology to construct several rigid compounds that mimic the putative receptor-bound conformation of dihydropyridine-type calcium channel modulators.

As mentioned in Section 10.1.2, Padwa and co-workers (40,41) employed the Pummerer reaction to generate and trap isomünchnones. This group (190,191) has now adapted the intramolecular version of this tactic to the synthesis of several alkaloids of the pyridine, quinolizidine, and clavine classes. In each case, a 2-pyridone serves as the keystone intermediate. For example, Kuethe and Padwa (190) employed this Pummerer reaction of imidosulfoxides that contain tethered π-bonds in a formal synthesis of the frog alkaloid (±)-pumiliotoxin C. They also used this methodology to synthesize the azafluorenone alkaloid onychine (**295**) (Scheme 10.42) (191). Generation of the thionium ion **291** under standard

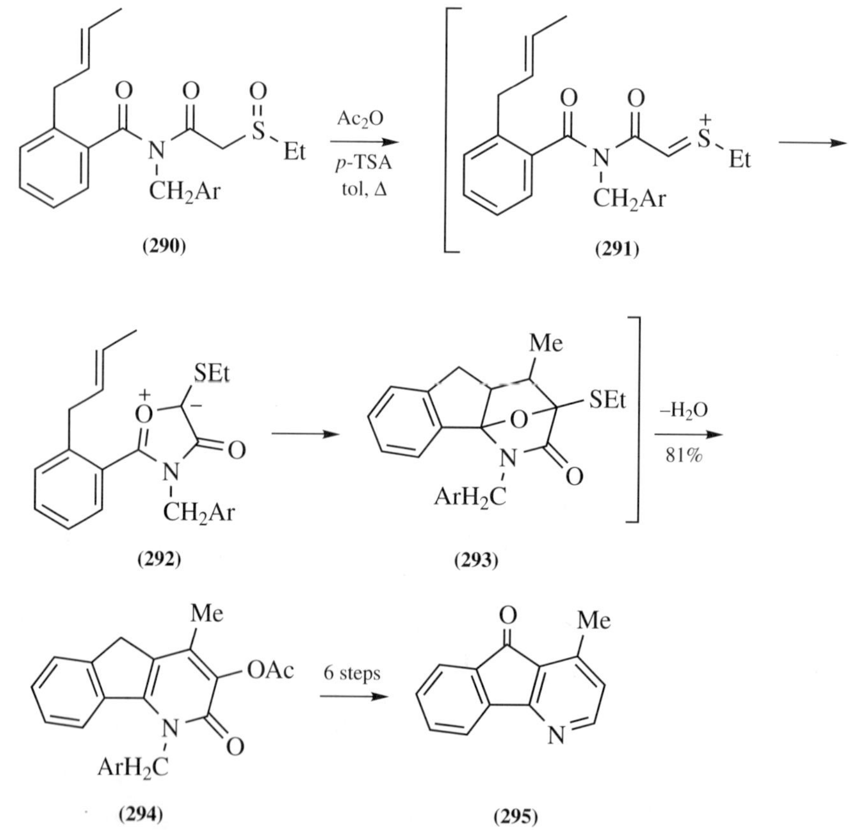

Scheme 10.42

Pummerer reaction conditions was followed by cyclization to isomünchnone **292** and hence to cycloadduct **293**, which loses water to form α-pyridone **294**. Subsequent manipulation involving deoxygenation and debenzylation completed the synthesis. In similar fashion, the azaanthraquinone alkaloid dielsiquinone was synthesized for the first time. Also, the quinolizidine alkaloids (±)-lupinine and (±)-anagyrine, and the ergot alkaloid (±)-costaclavine were synthesized using this Pummerer cyclization–cycloaddition cascade of imidosulfoxides and isomünchnones.

10.2.3. Sydnones (1,2,3-oxadiazolium-5-olates)

Since the early review on sydnones by Stewart (192) and the subsequent coverage by Potts (1), several new applications of these remarkably stable mesoionic heterocycles have been described. In particular, the synthesis of pyrazoles from sydnones has been pursued by several groups. Badachikar et al. (193) prepared several new potential antibacterials (**297**) from the appropriate sydnones **296**, which were synthesized in the standard fashion by the cyclodehydration of the corresponding *N*-nitroso-*N*-arylglycine.

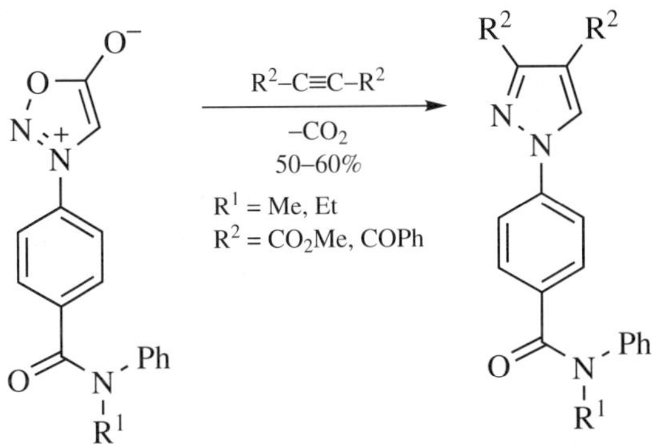

(296) (297)

Dumitrascu and co-workers (52) transformed 4-halosydnones into 5-halopyrazoles by cycloaddition with DMAD and methyl propiolate followed by retro-Diels–Alder loss of CO_2. Turnbull and co-workers (194) reported that the cycloadditions of 3-phenylsydnone with DMAD and diethyl acetylenedicarboxylate to form pyrazoles can be achieved in supercritical carbon dioxide. Nan'ya et al. (195) studied this sydnone in its reaction with 2-methylbenzoquinone to afford the expected isomeric indazole-4,7-diones. Interestingly, Sasaki et al. (196) found that 3-phenylsydnone effects the conversion of 1,4-dihydronaphthalene-1,4-imines to isoindoles, presumably by consecutive loss of carbon dioxide and *N*-phenylpyrazole from the primary cycloadduct. Ranganathan et al. (197–199) studied dipolar cycloadditions with the sydnone **298** derived from *N*-nitrosoproline (Scheme 10.43). Both acetylenic and olefinic dipolarophiles react with **298**. In

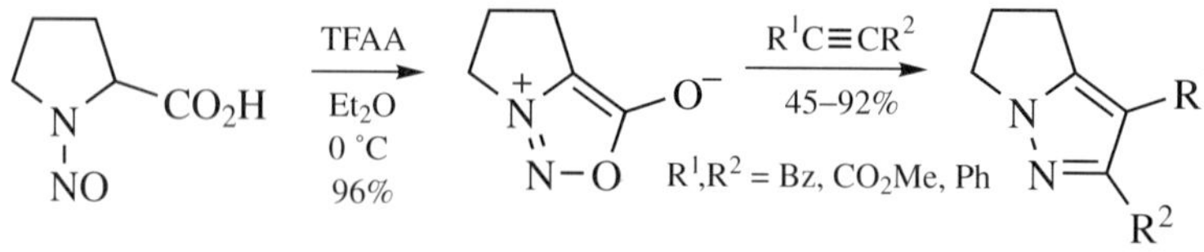

(298) **(299)**

Scheme 10.43

the reaction of **298** with phenylacetylene, the natural product withasomnine (**299**; R^1=Ph, R^2=H) was the minor regioisomer (18%) (198). The reaction with 3-phenylpropiolic acid afforded a novel ring-opened product formed via an intramolecular redox process (199).

Märkl and Pflaum (200) employed sydnones **300** to construct the unusual diazaphospholes **301**.

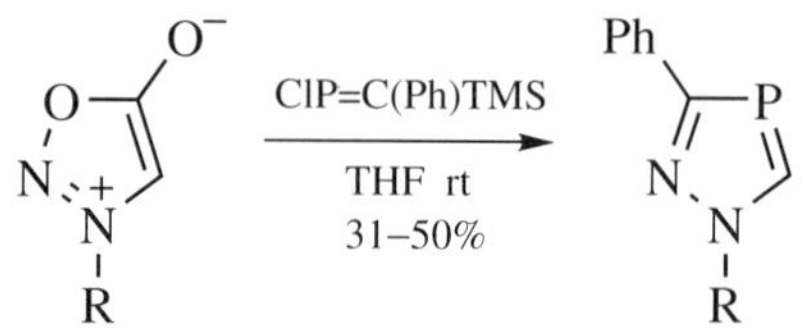

(300a) R = Ph
(300b) R = Bn
(301)

Destro et al. (201) examined the reactions of several sydnones with triazoles to give pyrazoles, and the stable bicyclic sydnone **302** has been prepared by Storr and co-workers (202). Reaction with DMAD affords the pyrazolothiazole **303**.

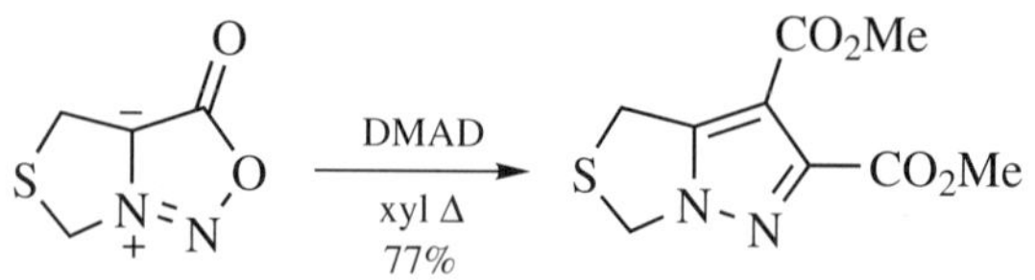

(302) **(303)**

Although in the reaction with olefinic dipolarophiles the sydnone primary cycloadduct cannot lead to an aromatic pyrazole upon loss of CO_2, it can afford a new 1,3-dipole that can be trapped in a second 1,3-dipolar cycloaddition reaction. Thus, Sun (203–205) has shown that maleimides react with sydnones to form bis(adducts) **304** in excellent yields. Both endo/exo and exo/exo isomers are obtained with the latter predominating. The reaction of sydnones with bis(maleimides) affords a novel set of polyimides having good thermal stability and solubility (205). Upon thermolysis adducts **304** undergo cycloreversion to fused pyrazoles and maleimides, and heating adduct **304** (R^1=R^2=R^3=Ph) in the presence of diphenylacetylene gave 1,3,4-triphenylpyrazole (52%) (204). Apparently independently,

Nan'ya et al. (93,94) also reported the reaction of sydnones with *N*-methyl- and *N*-phenylmaleimides. Gribble and Hirth (206) extended the work of Weintraub (207) on the tandem dipolar cycloaddition of sydnones with 1,5-cyclooctadiene to afford diazatetracycloundecanes **305**.

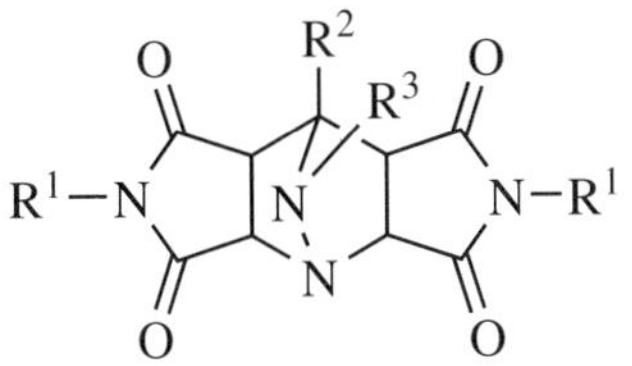

(304)

R^1 = Ph, Me, H, tol
R^2 = H, Ph, Et
R^3 = Ph, Pr

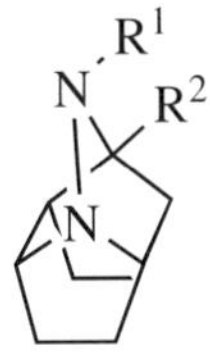

(305)

R^1 = Ph, Bn, Me, Ar, *t*-Bu, allyl
R^2 = H, Ph

Meier and Heimgartner (208) achieved an intramolecular sydnone–alkene cycloaddition to give adduct **306** in 50% yield. Other tether lengths did not so react, but photolysis of these other sydnones led to novel fused pyrazoles *via* decarboxylation and subsequent cycloadditions from the subsequent nitrile-imines.

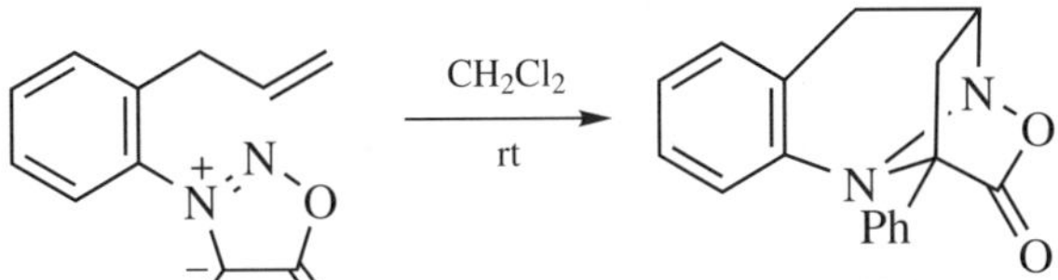

(306)

Vedejs and Wilde (209) reported the reaction of 3-phenylsydnone with thiopivaldehyde to give the ring-opened hydrazone **307**, and Mais et al. (210) studied the reactions of sydnones with isopropylidenecyclobutenone to give diazepines such as **308**. These latter workers present extensive FMO data for sydnones. Triazepines **309** are formed from the reaction of sydnones and 2,3,4-tri-*tert*-butylazete (143). Photolysis of **309** yields the corresponding 3,4-di-*tert*-butylpyrazoles.

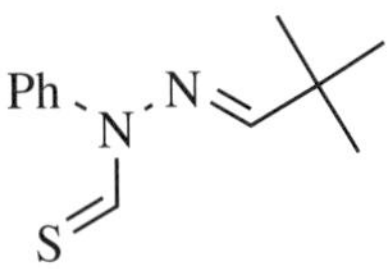

(307)

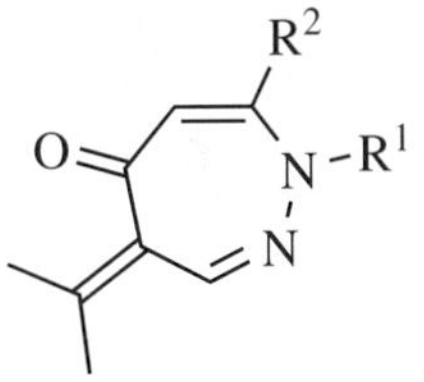

(308)

R^1 = Ph, Me, Bn; R^2 = H, Me, Cl

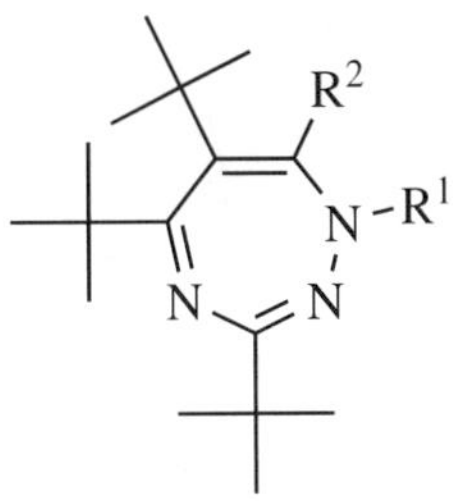

(309)

R^1 = Ph, Me, 2-Me-Ph, 4-Me-Ph; R^2 = H, Me

The versatile *tert*-butylphosphaalkyne (**310**) reacts with sydnones **311** to afford regioselectively diazaphospholes **312** (140), similar to the behavior observed by these workers with münchnones and similar to **300** → **301** (200). In the case of **311** (R^1=H, R^2=Me) a mixture of regioisomers is obtained.

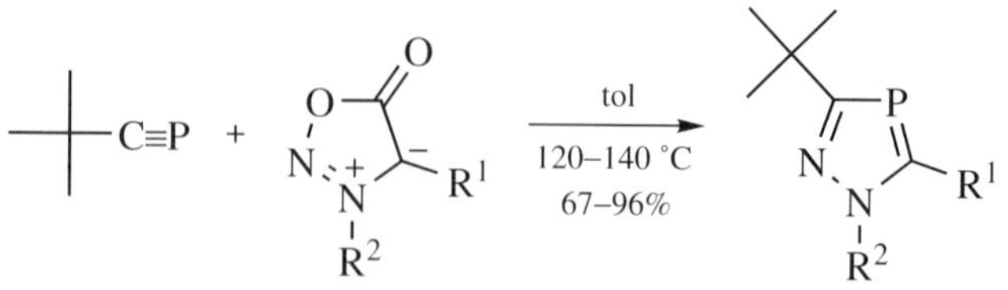

(310) **(311)** **(312)**

R^1 = H, Me, Ph; R^2 = Me, Ph

In his exhaustive survey of mesoionic cycloaddition reactions, Kato et al. (119) examined the behavior of 3-phenylsydnone (**300a**) with several potential dipolarophiles. Although the reaction of **300a** with some fulvenes was complex or did not proceed (119), reactions did occur with 2-*tert*-butyl-6,6-dimethylfulvene and 2-*tert*-butyl-6-acetoxy (or dimethylamino) fulvene to afford **313** and **314**, respectively (151,152). Heating a xylene solution of tropone with **300a** gave a low yield (10%) of cycloadduct **315** (123). The yield was increased to 33% in boiling bromobenzene in the presence of cupric acetate. The benzodiazepine **316** was formed in 30% yield from **300a** and benzocyclobutadiene (114). However, 3-methylsydnone and 3-methyl-4-phenylsydnone failed to react under these conditions.

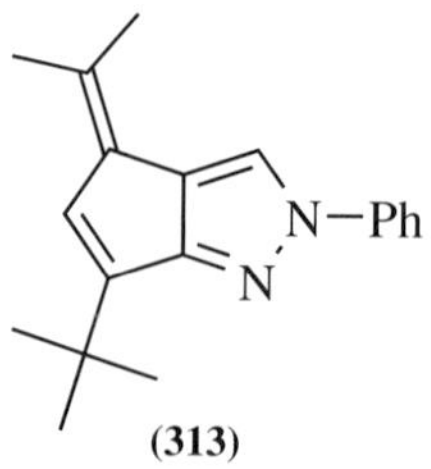

(313)

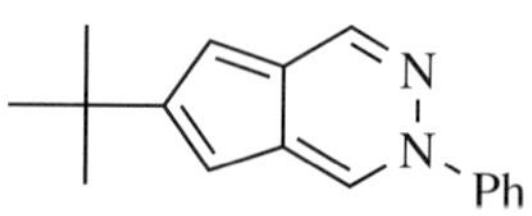

(314)

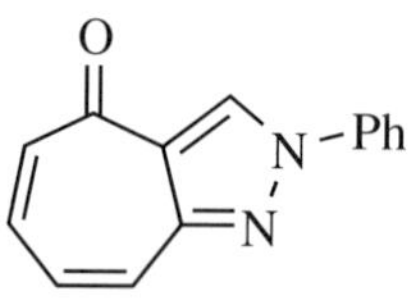

(315)

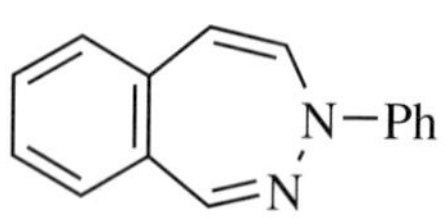

(316)

10.2.4. Thiomünchnones (1,3-oxathiolium-5-olates)

The only report involving thiomünchnones (**4**), other than that presented in Section 10.1.4 (cf. Scheme 10.16), was an unsuccessful reaction between 2-phenyl-4-(4-methylphenyl)thiomünchnone and tropone (123).

10.2.5. Thioisomünchnones (1,3-thiazolium-4-olates)

The generation and study of thioisomünchnones (**5**) has been very popular over the past decade. Padwa and co-workers (25,61) provided summaries of his work and reviews of the work of others in this area. Robert and co-workers have been pioneers in the synthesis of thioisomünchnones (211) and in exploring its cycloaddition chemistry (212,213). With olefinic dipolarophiles (maleate, fumarate, acryonitrile, ethyl vinyl ether) a variety of thioisomünchnones **317** provide the isolable expected cycloadducts (212). With acetylenic dipolarophiles the products are either 2-pyridones or thiophenes (213). The regiochemistry is discussed extensively in terms of FMO theory. Hamaguchi and Nagai (174) suggested that the thioisomünchnone–isocyanate cycloadducts described by Potts et al. (214,215) are actually C(5) acylated thioisomünchnones, although evidence per se is not given in this paper. Sheradsky and Itzhak (216) isolated cycloadducts **318** in 52–94% yield from reactions of thioisomünchnones **317** ($R^2 = R^3 = Ph$) and dipolarophiles such as dialkyl azodicarboxylates and *N*-phenyl-1,2,4-triazoline-3,5-dione (PTAD). Potts et al. (58,217,218) extensively examined both inter- and intramolecular cycloadditions of thioisomünchnones, and, more recently, Kappe et al. (166) employed fused thioisomünchnone **319** to craft novel calcium channel modulators.

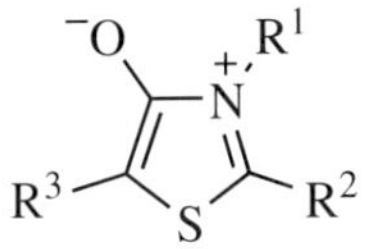

(317)

R^1 = Ph, Bn, CH=CH$_2$
R^2 = Ph, Ar, CN, SPh, Me$_2$N
R^3 = H, Ph, Ar

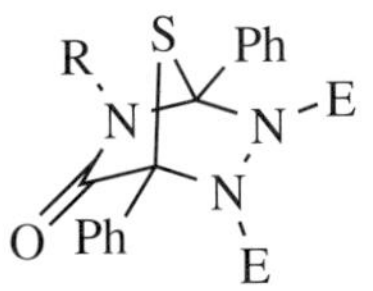

(318)

R = Ph, 4-Me-Ph, 4-CN-Ph, Bn, C_6H_{11}
E = Me, Et

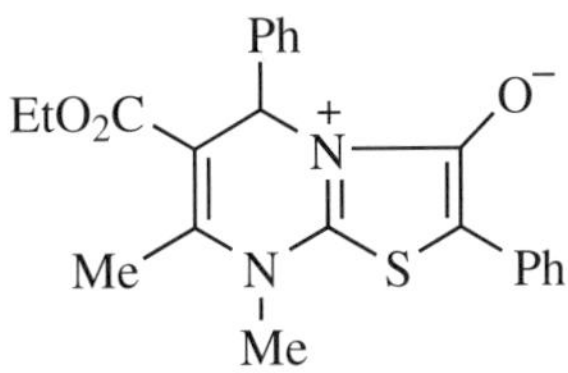

(319)

Cavalleri et al. (219) studied the cycloaddition of thioisomünchnones with 5-amino-4-methylene-1,2,3-triazolines. The resulting cycloadducts afford 2-pyridones with Raney Ni or 8-thia-6-azabicyclo[3.2.1]octanes with acid. Kato has continued his extensive exploration of cycloaddition studies of mesoionic heterocycles with atypical dipolarophiles. Thus, whereas thioisomünchnone (**320**) gave a complex mixture with 6,6-dimethylfulvene (119), it reacted with 2-*tert*-butyl-6,6-dimethylfulvene to afford **321** in 17% yield (151,152). The cycloaddition of **320** with benzocyclobutadiene was much cleaner to give **322** in 70% yield (114).

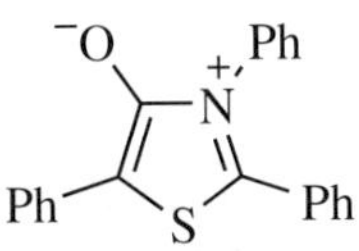

(320)

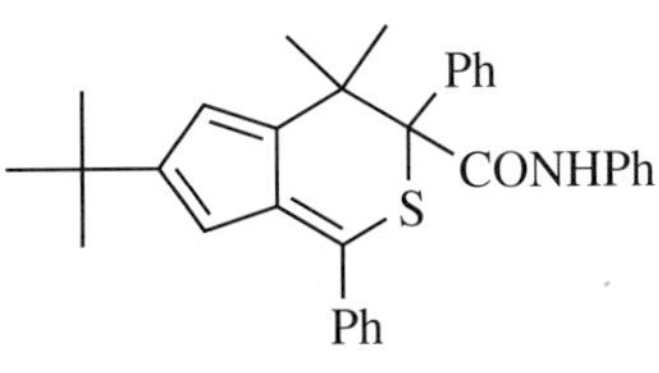

(321)

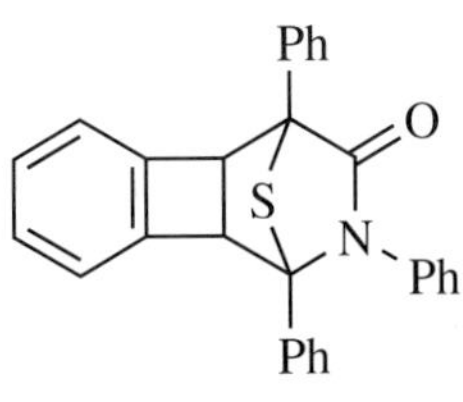

(322)

Avalos and co-workers (220–228) extensively investigated the 1,3-dipolar cycloaddition chemistry of 2-aminothioisomünchnones with both acetylenic and olefinic dipolarophiles. For example, sugar derivatives of the mesoionic imidazo[2,1-*b*]thiazolium-3-olate system react regioselectively with a variety of acetylenic dipolarophiles [DMAD, diethyl azodicarboxylate (DEAD), methyl propiolate, ethyl phenylpropiolate] to give the corresponding imidazo[1,2-*a*]pyridin-4-ones (e.g., **323**) following sulfur extrusion from the not isolable cycloadducts (220). Similarly, these thioisomünchnones react with diethyl azodicarboxylate and arylisocyanates in the expected fashion (221), and also with aryl aldehydes to form episulfides (222).

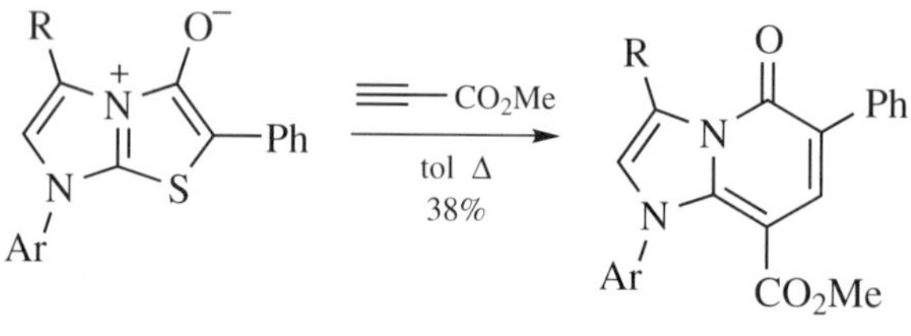

Ar = 4-EtO-Ph
R = arabinose derivative

(323)

Avalos et al. (223,224) also found that simpler 2-aminothioisomünchnones react with nitroalkenes to give dihydrothiophenes and other products, work that includes detailed MO calculations that rationalize both the reactivity of the thioisomünchnones and the observed regioselectivity (224). These same thioisomünchnones (**324**) react with aryl aldehydes to provide β-lactams **325** following fragmentation and subsequent cyclization of the primary cycloadducts (225). Novel 1,2,4-triazines are produced when **324** is exposed to diethyl azodicarboxylate (226), and detailed synthetic and mechanistic studies have been reported for the reactions of **324** with alkynes (227) and chiral 1,2-diaza-1,3-butadienes (228).

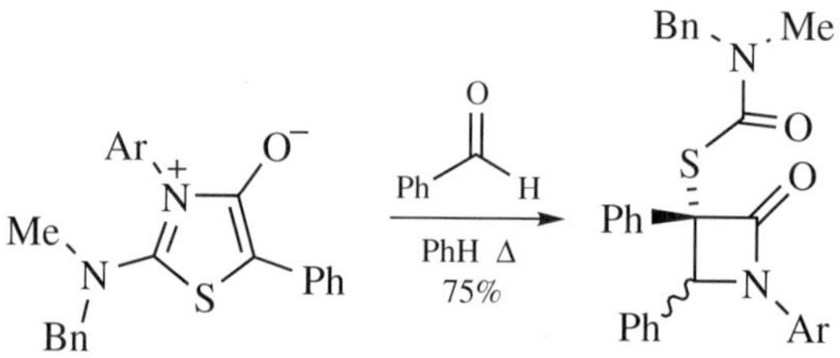

(324) **(325)**

Ar = Ph, 4-NO₂-Ph 65:35 (β/α)

Padwa and co-workers (25,61,229–233) explored 1,3-dipolar cycloaddition behavior of thioisomünchnones with the goal of establishing its utility in natural product synthesis. The Rh(II) catalyzed reaction of diazothioamide **326** with NPM gave adduct **328** in 75% yield, presumably via thioisomünchnone **327** (229). Numerous other model systems have been investigated by this group (25,61,230,

231) who have shown that this protocol is a powerful one for the construction of nitrogen heterocycles.

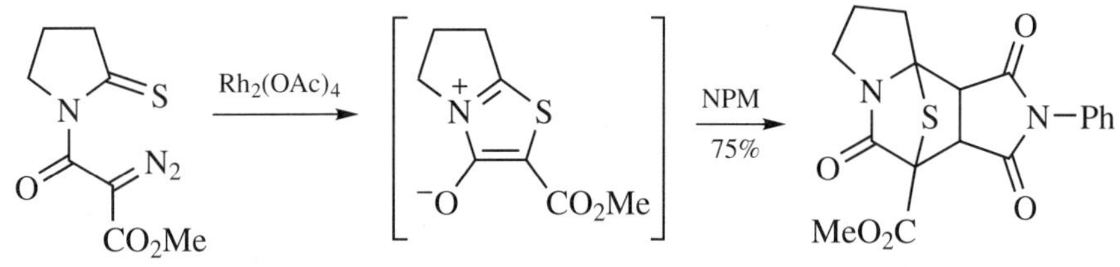

(326) **(327)** **(328)**

Padwa and co-workers (232,233) adapted their thioisomünchnone generation–cycloaddition strategy to the synthesis of several tetrahydroisoquinoline alkaloids and the indolo[2,3-*a*]quinolizidine alkaloid alloyohimbane **333** (Scheme 10.44).

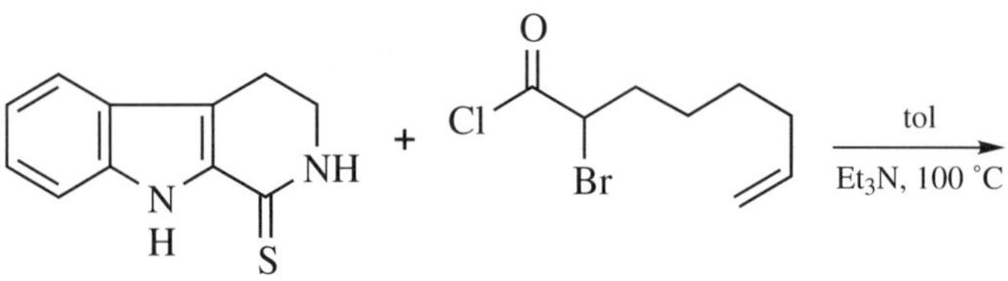

(329) **(330)**

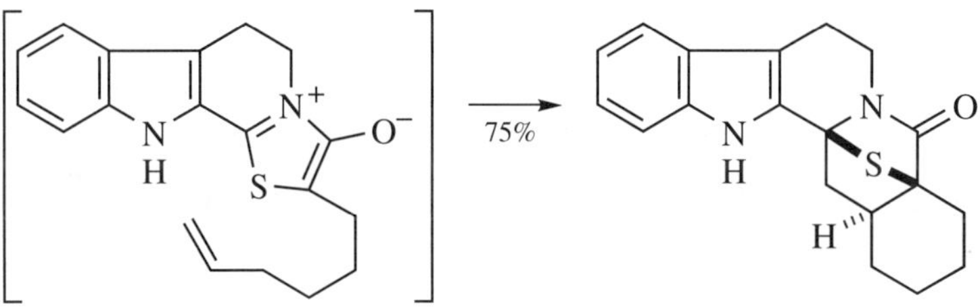

(331) **(332)**

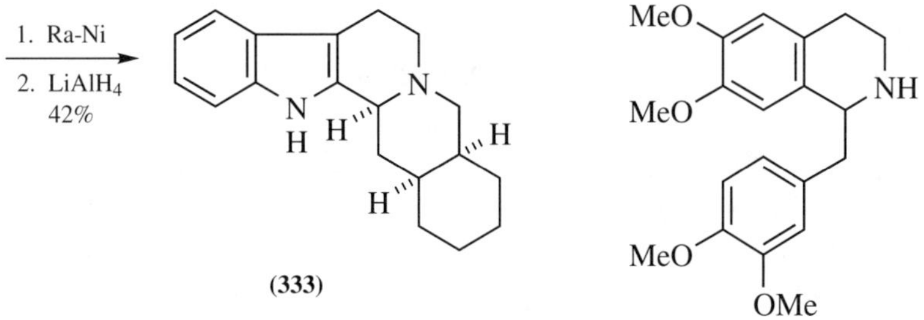

(333)

(334)

Scheme 10.44

Treatment of thiocarboline (**329**) with acid chloride **330** gave cycloadduct **332** in good yield, presumably via thioisomünchnone **331**. Reduction afforded alloyohimbane (**333**). These workers also synthesized tetrahydropapaverine (**334**).

10.2.6. 1,3-Thiazolium-4-thiolates

Apart from work of Yoshii and co-workers (62) described in Section 10.1.6 (cf. Scheme 10.19), no additional cycloaddition reactions of 1,3-thiazolium-4-thiolates have been reported.

10.2.7. Oxamünchnones (1,3-dioxolium-4-olates)

Hamaguchi and Nagai (65,66) generated oxamünchnones **109** as described in Section 10.1.7. They find that in addition to a cycloaddition with DMAD (Scheme 10.20), these oxamünchnones react with acenaphthylene, maleimides, phenylacetylene, and dibenzoylacetylene to afford **335–338**, respectively, in good yields (65,66). In the case of **335** and **336** the exo/endo ratio varies from 3:2 to 5:1 depending on the substituents.

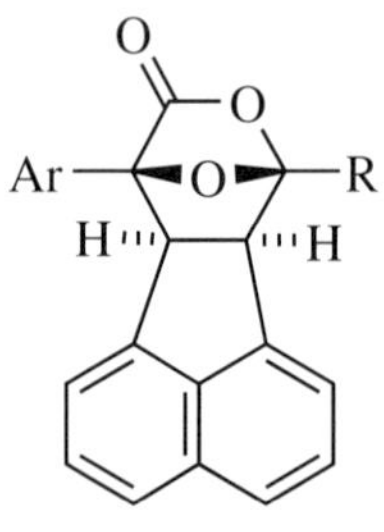

(335)

R = Me, Ph
Ar = 4-NO_2-Ph

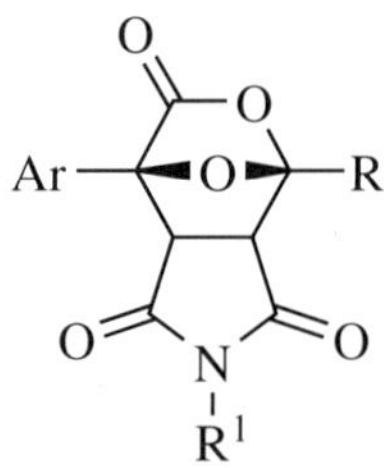

(336)

R^1 = Me, Ph

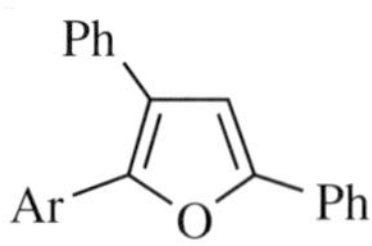

(337)

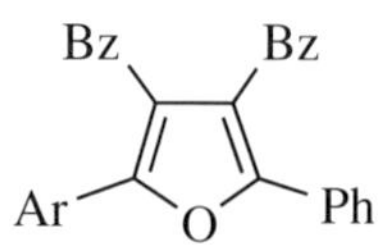

(338)

10.2.8. 1,3-Dithiolium-4-olates

Although the 1,3-dithiole mesoionic heterocycle 1,3-dithiolium-4-olate (**339**) was first synthesized in 1968 by Gotthardt and Christl (234,235) and shown to react with acetylenic dipolarophiles to give thiophenes (236), this ring system has been relatively neglected until the more recent work of Kato and Regitz. For example, Kato et al. (113,237,238) found that the benzocyclopropene adduct **340** with 2,5-diphenyl-1,3-dithiolium-4-olate (**339**, $R^1{=}R^2{=}Ph$) undergoes a range of interesting reactions (heat, light, peracid, tributylphosphine). Thus, upon heating, **340** is transformed into methanothionine **341**, and photolysis of **340** results in extrusion of carbonyl sulfide and formation of a cyclohepta[*c*]thiophene. The corresponding benzocyclopropene adduct from 5-methyl-2-phenyl-1,3-dithiolium-4-olate (**339**) ($R^1 = Me$, $R^2 = Ph$) did not undergo comparable chemistry.

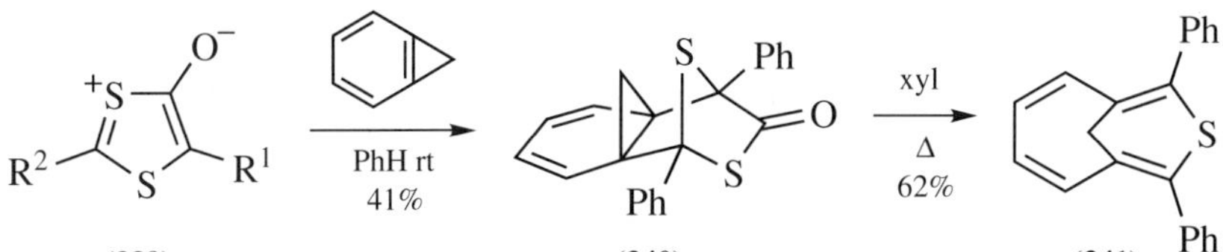

(**339**) (**340**) (**341**)

R^1 = H, Me, Ar
R^2 = Ar

Kato et al. observed that mesoionic **339** reacts with fulvenes, (119,151,152) benzocyclobutadiene (114), tropone (123), and triphenylphosphirene (112) to give cycloadducts **342–345**, respectively. Only a sampling of products is given as these reactions are quite complex. For example, 2-*tert*-butyl-6,6-dimethylfulvene gives a product type different from **342** (151,152).

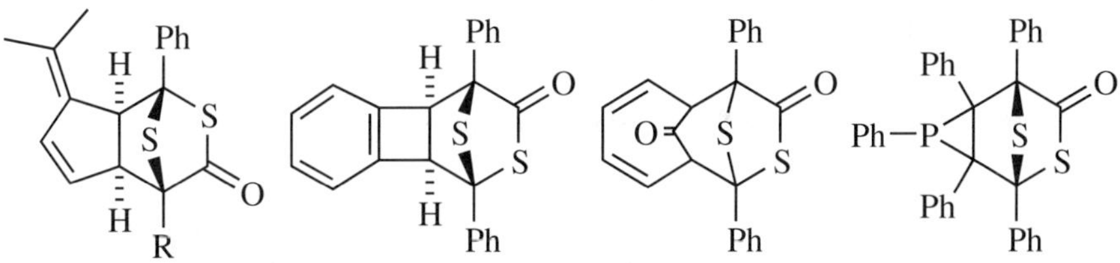

(**342a**) R = Ph (49%)
(**342b**) R = Me (32%)
(**342c**) R = H (27%)
(**343**) (63%)
(**344**) (19%)
(**345**) (77%)

Regitz and co-workers (140,142) continued their exploration of the chemistry of mesoionic heterocycles with phosphaalkynes. Thus, **339** ($R^1 = R^2 = Ph$) reacts with *tert*-butyl- and mesitylphosphaacetylene to give 1,3-thiaphospholes (**346**). This same 2,5-diphenyl-1,3-dithiolium-4-olate reacts with 2,3,4-tri-*tert*-butylazete to give **347** in low yield (143). This reaction is believed to proceed via diphenylthiirene by loss of carbonyl sulfide from **339**.

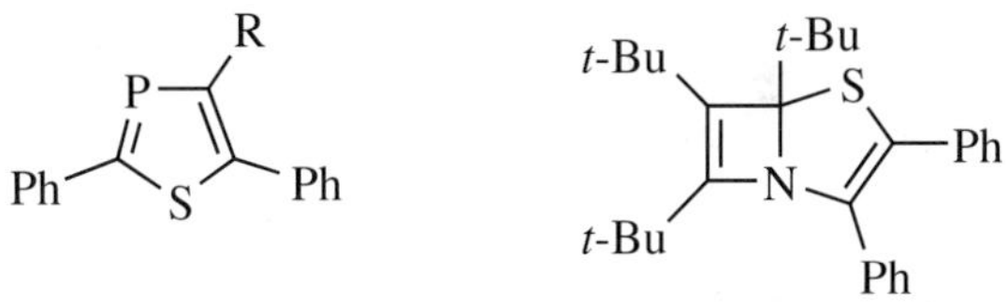

(**346a**) R = *t*-Bu (91%)
(**346b**) R = Mes (59%)
(**347**) (15%)

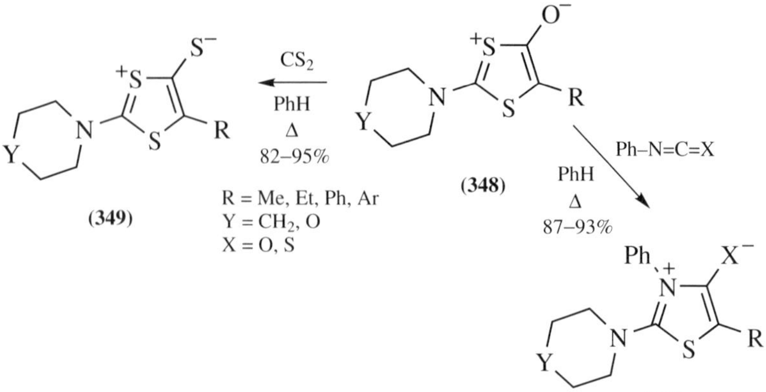

(349) (348) (350)

R = Me, Et, Ph, Ar
Y = CH_2, O
X = O, S

Y = CH_2, R = Ar, X = O, S

Robert and co-workers (239,240) discovered novel conversions of 2-amino-1,3-dithiolium-4-olates (**348**) into other mesoionic heterocycles. For example, reaction of **348** with carbon disulfide, phenyl isocyanate, or phenyl isothiocyanate affords 1,3-dithiolium-4-thiolates (**349**), 1,3-thiazolium-4-olates (**350**), and 1,3-thiazolium-4-thiolates (**351**), respectively. Some of these reactions proceed via the ring-opened ketene tautomer of **348** (240).

10.2.9. 1,3-Oxathiolium-4-olates

Gotthardt et al. (241,242) were the first to synthesize and study the cycloaddition chemistry of 1,3-oxathiolium-4-olates **351**. Although these original examples were not isolable, more recent work by this group (243–246) has led to the isolation of 5-trifluoroacetyl derivatives **352**. All of these mesoionics undergo smooth 1,3-dipolar cycloaddition reactions with alkynes to afford the expected furans after loss of carbonyl sulfide. In addition, some alkenes such as norbornadiene and 1-methylindole react with these heterocycles. For example, the latter dipolarophile reacts with the trichloroacetyl version of **352** (R=Me) to give **353** in good yield (246). Kato et al. (113) found that **351** (R=SMe) and **352** (R=Me) gave only trace amounts of cycloadducts with benzocyclopropene.

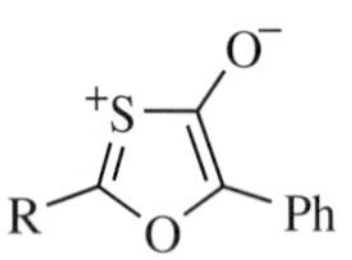

(351)

R = SMe, piperidinyl, morpholinyl, pyrrolidinyl

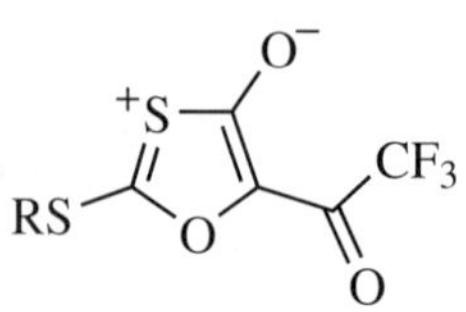

(352)

R = Me, Et, *i*-Pr, allyl

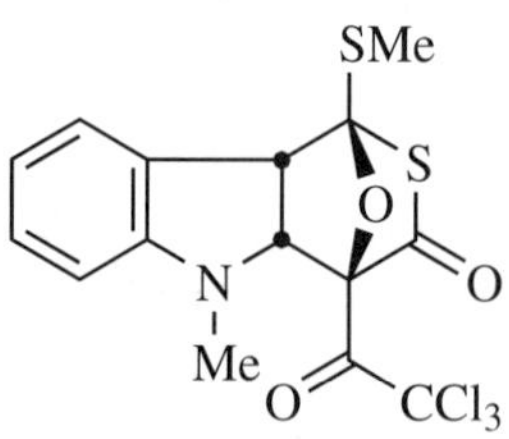

(353) (64%)

10.2.10. 1,3,2-Oxathiazolium-5-olates

Although the mesoionic 1,3,2-oxathiazolium-5-olates (**354**) were initially synthesized in 1960, it was Gotthardt (247,248) who developed a general route to these heterocycles and demonstrated their utility in isothiazole ring construction by the usual alkyne or alkene cycloaddition followed by extrusion of carbon dioxide. However, in recent years Kato and co-workers (112-114,119,151,152,237) investigated the chemistry of 4-phenyl-1,3,2-oxathiazolium-5-olate (**354**, R=Ph). Whereas **354** (R=Ph) gave tars or complex mixtures with some fulvenes (119), benzobutadiene (114), and triphenylphosphirene (112), and gave a low yield of an unstable adduct **355** with benzocyclopropene (113,237), this mesoionic heterocycle reacted with 3-*tert*-butyl-6-dimethylaminofulvene to give **356** in 29% yield plus a smaller amount of an isomer (151,152).

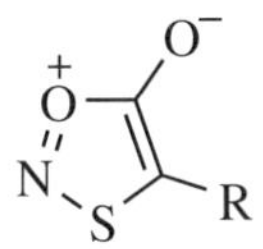

(354)

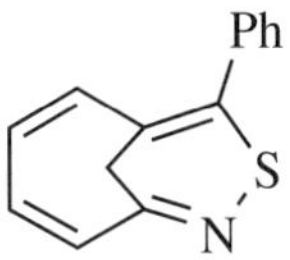

(355)

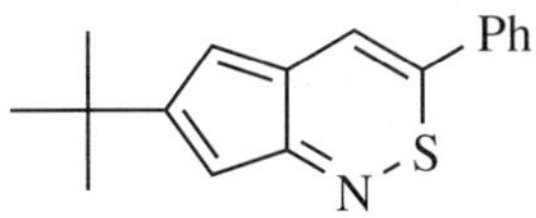

(356)

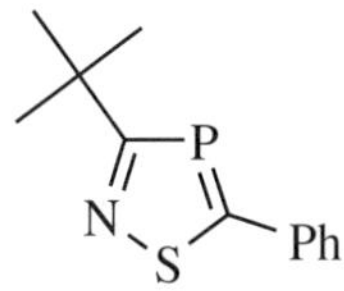

(357)

Regitz and co-workers (140) found that **354** (R=Ph) reacts with *tert*-butylphosphaacetylene to give thiazaphosphole **357** in 30% yield as the major isomer (9:1). An independent synthesis of the minor isomer, which is derived from 3-phenylthiazirene, confirmed the assignment.

10.2.11. Miscellaneous Mesoionic Heterocycles

The 1,3-dipolar cycloaddition reactions of several other mesoionic heterocycles have been investigated since Potts review (1). Kato et al. (113,114,123) found that the 1,3-thiazolium-5-olate (**358**) ring system affords low yields or complex reaction mixtures with benzocyclopropene (113), benzocyclobutadiene (114), and tropone (123). Likewise, Vedejs and Wilde (209) isolated in low yields a cycloadduct of **358d** with thiopivaldehyde, along with a ring-opened thiamide. Also, 1,3-thiazolium-5-olate **359** reacts with thiopivaldehyde to give **360** (209).

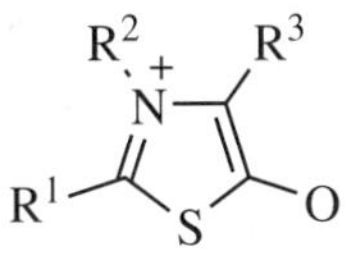

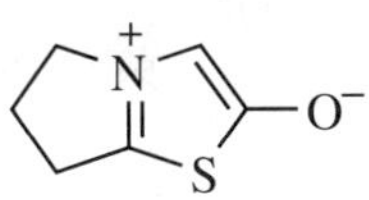

(359)

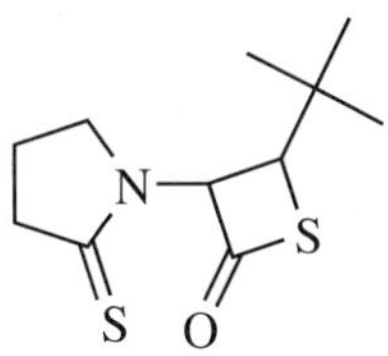

(360)

(358a) R^1 = Ph, R^2 = Me, R^3 = H
(358b) R^1 = SMe, R^2 = Me, R^3 = Ph
(358c) R^1 = R^3 = Ph, R^2 = Me
(358d) R^1 = Me, R^2 = Ph, R^3 = H

The mesoionic 1,2,3-triazolium-4-olates **361a,b** and 1,2,3-triazolium-4-thiolate **361c** did not react with benzocyclopropene (113), and the attempted reaction between 1,3-diazolium-4-olate **362** and benzocyclobutadiene failed (114). In contrast to these negative results, Ichinari et al. (249) found that the isomeric 1,3-diazolium-4-olates (1,3-imidazolium-4-olates) **363** and **364** react with DMAD to form pyridones and pyrroles, respectively.

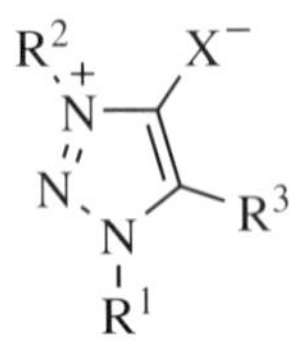

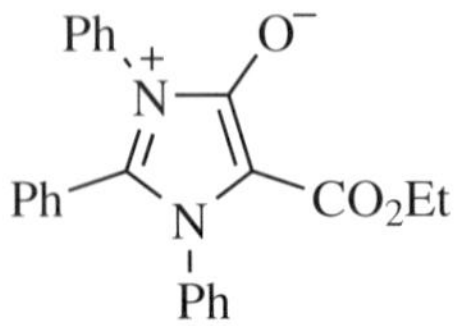

(362)

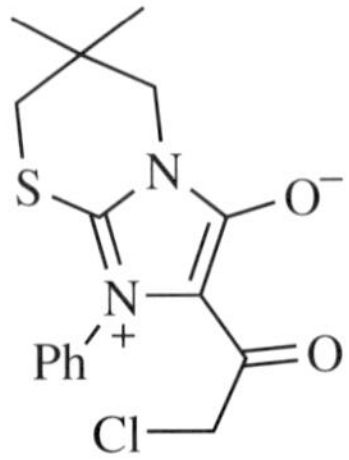

(363)

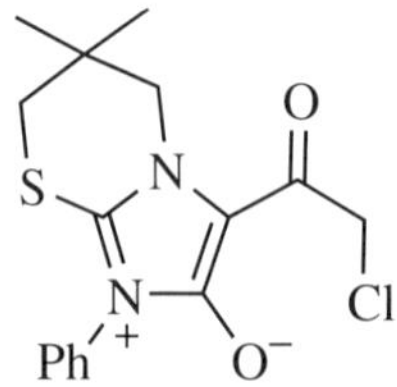

(364)

(361a) $R^1 = Me, R^2 = Bn, R^3 = H, X = O$
(361b) $R^1 = Bn, R^2 = Me, R^3 = H, X = O$
(361c) $R^1 = R^2 = R^3 = Me, X = S$

Coppola et al. (250) showed that the extent of the 1,3-dipolar cycloaddition pathway of tricyclic mesoionic 1,3-diazolium-4-olates **365** with DMAD and methyl propiolate decreases as the size of the variable ring decreases. Thus, whereas with **365c** and methyl propiolate the "normal" cycloadduct **366** (after fragmentation) is obtained in high yield, the same reaction with **365a** affords **367** as the major compound, which is the product of a stepwise pathway, along with only 8% of the dipolar cycloaddition product. Coppola et al. (250) interpret these results in terms of an unsymmetrical transition state that diverts a portion of the reaction through a stepwise pathway as the ring size decreases in **365**. Berrée and Morel (251) generated a series of 1,3-thiazolium-5-amidates (**368**) and studied their reactions with a variety of alkyne and alkene dipolarophiles. Products include pyrroles, pyrrolines, aminobutadienes, and thiopyrans.

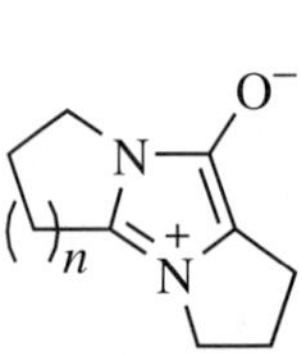

(365a) $n = 1$
(365b) $n = 2$
(365c) $n = 3$

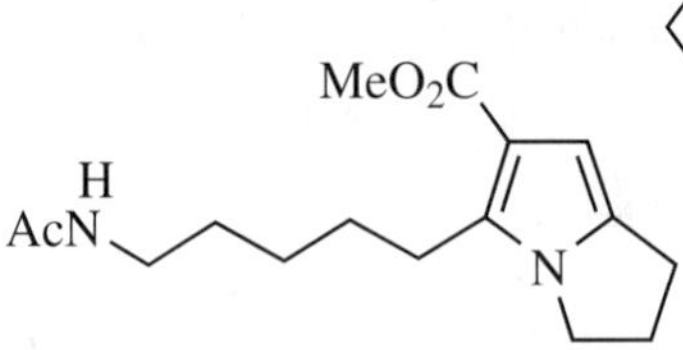

(366) (89%)

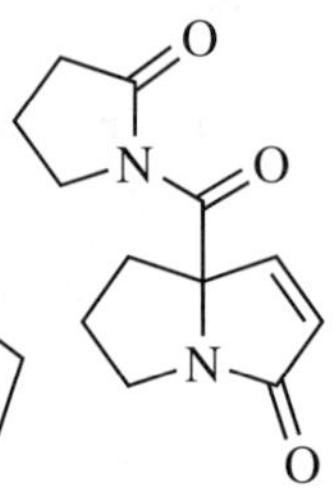

(367) (22%)

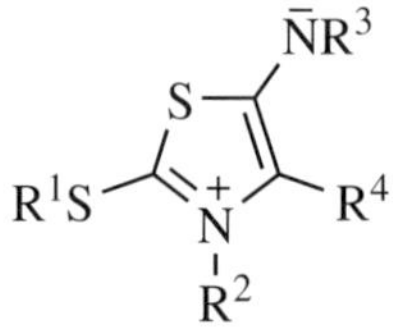

(368)

$R^1 = Me, Ph$
$R^2 = Me, Bn, CH_2CO_2Et, CH_2PO(OEt)_2, i\text{-}Pr$
$R^3 = t\text{-}Bu, Bn, CH_2CO_2Et$
$R^4 = Ph, NMe_2$

10.3. CONCLUDING REMARKS

Over the past two decades, the 1,3-dipolar cycloaddition chemistry of mesoionic heterocycles has matured to become a powerful synthetic tool for accessing a multitude of organic structures. We have seen that a variety of pyrroles, furans, thiophenes, imidazoles, thiazoles, azaphospholes, oxaphospholes, thiaphospholes, pyridones, azepines, and fused heterocycles are available from this methodology. Several complex nitrogen-containing natural products have been synthesized by utilizing a mesoionic heterocycle in a 1,3-dipolar cycloaddition as the key step. Despite this success in the synthesis arena, the factors governing the regiochemistry of these cycloadditions are not fully understood and a unified treatment is not yet possible. Whereas the 1,3-dipolar cycloadditions of some mesoionic heterocycles are rationalized in terms of simple FMO theory, many others cannot be so interpreted. It is clear that a combination of steric, electronic, dipole, and other ground- and transition state interactions play a role in determining the regiochemistry of these reactions. The success reported with *ab initio* MO calculations that have been performed on a few cycloadditions portends a greater future understanding of the regiochemistry of the 1,3-dipolar cycloaddition reactions of mesoionic heterocycles. In any event, the future of the 1,3-dipolar cycloaddition chemistry of mesoionic heterocycles in organic synthesis is assured.

REFERENCES

1. K. T. Potts, in *1,3-Dipolar Cycloaddition Chemistry*, A. Padwa, ed., Vol. 2, Wiley, New York, **1984**, pp. 1–82.
2. (a) W. D. Ollis, S. P. Stanforth, and C. A. Ramsden, *Tetrahedron*, **1985**, *41*, 2239. (b) C. G. Newton and C. A. Ramsden, *Tetrahedron*, **1982**, *38*, 2965.
3. K. T. Potts and S. Yao, *J. Org. Chem.*, **1979**, *44*, 977.
4. W. K. Anderson and A. R. Heider, *Synth. Commun.*, **1986**, *16*, 357.
5. M. Slebioda, *Polish J. Chem.*, **1997**, *71*, 1045.
6. G. V. Boyd and P. H. Wright, *J. Chem. Soc., Perkin Trans. 1*, **1972**, 909.
7. G. V. Boyd and P. H. Wright, *J. Chem. Soc., Perkin Trans. 1*, **1972**, 914.
8. F. M. Hershenson and M. R. Pavia, *Synthesis*, **1988**, 999.
9. R. G. Wilde, *Tetrahedron Lett.*, **1988**, *29*, 2027.
10. H. L. Gingrich and J. S. Baum, in *Oxazoles; The Chemistry of Heterocyclic Compounds*, I. J. Turchi, ed., A. Weissberger and E. C. Taylor, eds., Wiley, New York, **1986**, Vol. 45, pp. 731–961.
11. R. Huisgen, E. Funke, F. C. Schaefer, and R. Knorr, *Angew. Chem. Int. Ed. Engl.*, **1967**, *6*, 367.
12. H. O. Bayer, R. Huisgen, R. Knorr, and F. C. Schaefer, *Chem. Ber.*, **1970**, *103*, 2581.
13. E. Funke and R. Huisgen, *Chem. Ber.*, **1971**, *104*, 3222.
14. T. A. Keating and R. W. Armstrong, *J. Am. Chem. Soc.*, **1995**, *117*, 7842.
15. T. A. Keating and R. W. Armstrong, *J. Am. Chem. Soc.*, **1996**, *118*, 2574.
16. T. A. Keating and R. W. Armstrong, *J. Org. Chem.*, **1996**, *61*, 8935.
17. A. M. Strocker, T. A. Keating, P. A. Tempest, and R. W. Armstrong, *Tetrahedron Lett.*, **1996**, *37*, 1149.

18. A. Dömling and I. Ugi, *Angew. Chem. Int. Ed. Engl.*, **2000**, *39*, 3168.
19. I. Ugi, S. Lohberger, and R. Karl, in *Comprehensive Organic Synthesis*, C. H. Heathcock, ed., Vol. 2, Pergamon, Oxford, *1991*, Chapter 4.6.
20. A. M. M. Mjalli, S. Sarshar, and T. J. Baiga, *Tetrahedron Lett.*, **1996**, *37*, 2943.
21. C. A. Merlic, A. Baur, and C. C. Aldrich, *J. Am. Chem. Soc.*, **2000**, *122*, 7398.
22. A. Padwa, *Acc. Chem. Res.*, **1991**, *24*, 22.
23. A. Padwa and S. F. Hornbuckle, *Chem. Rev.*, **1991**, *91*, 263.
24. A. Padwa and K. E. Krumpe, *Tetrahedron*, **1992**, *48*, 5385.
25. M. H. Osterhout, W. R. Nadler, and A. Padwa, *Synthesis*, **1994**, 123.
26. A. Padwa and M. D. Weingarten, *Chem. Rev.*, **1996**, *96*, 223.
27. A. Padwa, *Top. Curr. Chem.*, **1997**, *189*, 121.
28. M. J. Haddadin, A. M. Kattan, and J. P. Freeman, *J. Org. Chem.*, **1982**, *47*, 723.
29. M. J. Haddadin and H. T. Tannus, *Heterocycles*, **1984**, *22*, 773.
30. L. J. Mathias and D. R. Moore, *J. Am. Chem. Soc.*, **1985**, *107*, 5817.
31. L. J. Mathias and D. R. Moore, *Polym. Prep.*, **1986**, *27*, 118.
32. D. R. Moore and L. J. Mathias, *Macromolecules*, **1986**, *19*, 1530.
33. D. R. Moore and L. J. Mathias, *J. Org. Chem.*, **1987**, *52*, 1599.
34. M. P. Doyle, R. L. Dorow, J. W. Terpstra, and R. A. Rodenhouse, *J. Org. Chem.*, **1985**, *50*, 1663.
35. M. Hamaguchi and T. Ibata, *Tetrahedron Lett.*, **1974**, 4475.
36. M. E. Maier and K. Evertz, *Tetrahedron Lett.*, **1988**, *29*, 1677.
37. M. E. Maier and B. Schöffling, *Chem. Ber.*, **1989**, *122*, 1081.
38. A. Padwa, D. L. Hertzog, and R. L. Chinn, *Tetrahedron Lett.*, **1989**, *30*, 4077.
39. M. P. Doyle, R. J. Pieters, J. Taunton, H. Q. Pho, A. Padwa, D. L. Hertzog, and L. Precedo, *J. Org. Chem.*, **1991**, *56*, 820.
40. J. T. Kuethe and A. Padwa, *J. Org. Chem.*, **1997**, *62*, 774.
41. A. Padwa, T. M. Heidelbaugh, and J. T. Kuethe, *J. Org. Chem.*, **1999**, *64*, 2038.
42. J. C. Earl and A. W. Mackney, *J. Chem. Soc.*, **1935**, 899.
43. J. Applegate and K. Turnbull, *Synthesis*, **1988**, 1011.
44. K. Turnbull, T. L. Blackburn, and J. J. Miller, *J. Heterocycl. Chem.*, **1996**, *33*, 485.
45. K. Turnbull, T. L. Blackburn, and D. T. Esterline, *J. Heterocycl. Chem.*, **1990**, *27*, 1259.
46. K. Turnbull, T. L. Blackburn, and D. B. McClure, *J. Heterocycl. Chem.*, **1994**, *31*, 1631.
47. K. Turnbull and J. C. George, *Synth. Commun.*, **1996**, *26*, 2757.
48. K. Turnbull and D. M. Krein, *Tetrahedron Lett.*, **1997**, *38*, 1165.
49. K. Turnbull, C. Sun, and D. M. Krein, *Tetrahedron Lett.*, **1998**, *39*, 1509.
50. I. A. Cherepanov, S. N. Lebedev, and V. N. Kalinin, *Synlett*, **1998**, 667.
51. H. Petride, A. D. Cort, C. Florea, and M. Caproiu, *Rev. Roum. Chim.*, **1999**, *44*, 249.
52. F. Dumitrascu, C. Draghici, D. Dumitrescu, L. Tarko, and D. Raileanu, *Liebigs Ann./Recueil*, **1997**, 2613.
53. K. T. Potts, S. J. Chen, J. Kane, and J. L. Marshall, *J. Org. Chem.*, **1977**, *42*, 1633.
54. K. Masuda, J. Adachi, and K. Nomura, *Chem. Pharm. Bull.*, **1977**, *25*, 1471.
55. K. T. Potts, J. Kane, E. Carnahan, and U. P. Singh, *J. Chem. Soc., Chem. Commun.*, **1975**, 417.
56. W. R. Sponholtz III, H. A. Trujillo, and G. W. Gribble, *Tetrahedron Lett.*, **2000**, *41*, 1687.
57. K. T. Potts and D. R. Choudhury, *J. Org. Chem.*, **1978**, *43*, 2697.
58. K. T. Potts, K. Bordeaux, W. Kuehnling, and R. Salsbury, *J. Chem. Soc., Chem. Commun.*, **1984**, 213.
59. K. T. Potts and P. Murphey, *J. Chem. Soc., Chem. Commun.*, **1984**, 1348.

60. A. Padwa, F. R. Kinder, W. R. Nadler, and L. Zhi, *Heterocycles*, **1993**, *35*, 367.

61. A. Padwa, S. R. Harring, D. L. Hertzog, and W. R. Nadler, *Synthesis*, **1994**, 993.

62. K. Takezawa, K. Nomura, E. Yoshii, and K. Masuda, *Chem. Pharm. Bull.*, **1984**, *32*, 4637.

63. A. Souizi and A. Robert, *Synthesis*, **1982**, 1059.

64. H. C. Berk, K. E. Zwikelmaier, and J. E. Franz, *Synth. Commun.*, **1980**, *10*, 707.

65. M. Hamaguchi and T. Nagai, *J. Chem. Soc., Chem. Commun.*, **1985**, 190.

66. M. Hamaguchi and T. Nagai, *J. Chem. Soc., Chem. Commun.*, **1985**, 1319.

67. W. K. Anderson, C.-P. Chang, and H. L. McPherson, Jr., *J. Med. Chem.*, **1983**, *26*, 1333.

68. W. K. Anderson and R. H. Mach, *J. Med. Chem.*, **1987**, *30*, 2109.

69. Z. Györgydeák, L. Szilágyi, J. Kajtár, G. Argay, and A. Kálmán, *Monatsh. Chem.*, **1994**, *125*, 189.

70. T. M. V. D. Pinho e Melo, M. I. L. Soares, D. M. Barbosa, A. M. d'A. Roche Gonsalves, A. M. Beja, J. A. Paixão, M. R. Silva, and L. Alte da Veiga, *Tetrahedron*, **2000**, *56*, 3419.

71. D. Ladurée, J.-C. Lancelot, and M. Robba, *Tetrahedron Lett.*, **1985**, *26*, 1295.

72. T. Granier, F. Gaiser, L. Hintermann, and A. Vasella, *Helv. Chim. Acta*, **1997**, *80*, 1443.

73. C. W. Lo, W. L. Chan, Y. S. Szeto, and C. W. Yip, *Chem. Lett.*, **1999**, 513.

74. K. Funabiki, T. Ishihara, and H. Yamanaka, *J. Fluorine Chem.*, **1995**, *71*, 5.

75. A. Padwa, Y.-Y. Chen, W. Dent, and H. Nimmesgern, *J. Org. Chem.*, **1985**, *50*, 4006.

76. P. L. Toupet, M. Mazari, F. Texier, and R. Carrié, *Acta Crystallogr.*, **1991**, *C47*, 1528.

77. F. Texier, M. Mazari, O. Yebdri, F. Tonnard, and R. Carrié, *Bull. Soc. Chim. Fr.*, **1991**, *128*, 962.

78. O. Yebdri and F. Texier, *J. Heterocycl. Chem.*, **1986**, *23*, 809.

79. P. Dalla Croce and C. La Rosa, *Heterocycles*, **1988**, *27*, 2825.

80. G. W. Gribble, W. M. Simon, E. T. Pelkey, and H. A. Trujillo, unpublished results.

81. B. P. Coppola, M. C. Noe, D. J. Schwartz, R. L. II Abdon, and B. M. Trost, *Tetrahedron*, **1994**, *50*, 93.

82. B. P. Coppola, M. C. Noe, and S. S.-K. Hong, *Tetrahedron Lett.*, **1997**, *38*, 7159.

83. M. T. Pizzorno and S. M. Albonico, *J. Org. Chem.*, **1974**, *39*, 731.

84. B. Santiago, C. R. Dalton, E. W. Huber, and J. M. Kane, *J. Org. Chem.*, **1995**, *60*, 4947.

85. D. Ladurée, J.-C. Lancelot, M. Robba, E. Chenu, and G. Mathé, *J. Med. Chem.*, **1989**, *32*, 456.

86. P. Dalla Croce, P. Gariboldi, and C. La Rosa, *J. Heterocycl. Chem.*, **1987**, *24*, 1783.

87. H. Kato, S.-Z. Wang, and H. Nakano, *J. Chem. Soc., Perkin Trans. 1*, **1989**, 361.

88. M. Sainsbury, R. H. Strange, P. R. Woodward, and P. A. Barsanti, *Tetrahedron*, **1993**, *49*, 2065.

89. T. M. V. D. Pinho e Melo, D. M. Barbosa, P. J. R. S. Ramos, A. M. d'A. Rocha Gonsalves, T. L. Gilchrist, A. M. Beja, J. A. Paixão, M. R. Silva, and L. Alte da Veiga, *J. Chem. Soc., Perkin Trans. 1*, **1999**, 1219.

90. D. R. Hutchison, N. K. Nayyar, and M. J. Martinelli, *Tetrahedron Lett.*, **1996**, *37*, 2887.

91. N. K. Nayyar, D. R. Hutchison, and M. J. Martinelli, *J. Org. Chem.*, **1997**, *62*, 982.

92. B. S. Jursic, *J. Mol. Struc. (Theochem)*, **1996**, *365*, 55.

93. S. Nan'ya, S. Goto, and Y. Butsugan, *J. Heterocycl. Chem.*, **1990**, *27*, 1519.

94. S. Nan'ya, Y. Kurachi, and Y. Butsugan, *J. Heterocycl. Chem.*, **1991**, *28*, 1853.

95. H. Gotthardt, R. Huisgen, and F. C. Schaefer, *Tetrahedron Lett.*, **1964**, 487.

96. H. Gotthardt and R. Huisgen, *Chem. Ber.*, **1970**, *103*, 2625.

97. S. Nan'ya, T. Tange, E. Maekawa, and Y. Ueno, *J. Heterocycl. Chem.*, **1986**, *26*, 1267.

98. J. L. Fabre, D. Farge, C. James, and D. Lavé, *Tetrahedron Lett.*, **1985**, *26*, 5447.

99. F. Texier, M. Mazari, O. Yebdri, F. Tonnard, and R. Carrié, *Tetrahedron*, **1990**, *46*, 3515.

100. T. Okano, T. Uekawa, N. Morishima, and S. Eguchi, *J. Org. Chem.*, **1991**, *56*, 5259.

101. F. Clerici, M. Luisa Gelmi, and P. Trimarco, *Tetrahedron*, **1998**, *54*, 5763.
102. E. Erba, M. L. Gelmi, D. Pocar, and P. Trimarco, *Chem. Ber.*, **1986**, *119*, 1083.
103. P. D. Croce, C. La Rosa, M. L. Gelmi, and M. Ballabio, *J. Chem. Soc., Perkin Trans. 2*, **1988**, 423.
104. E. Clerici, E. Erba, P. Mornatti, and P. Trimarco, *Chem. Ber.*, **1989**, *122*, 295.
105. P. Baggi, F. Clerici, M. L. Gelmi, and S. Mottadelli, *Tetrahedron*, **1995**, *51*, 2455.
106. F. Clerici, M. L. Gelmi, R. Soave, and M. Valle, *Tetrahedron*, **1998**, *54*, 11285.
107. M. Avalos, R. Babiano, I. Bautista, J. I. Fernandez, J. L. Jimenez, J. C. Palacios, J. Plumet, and F. Rebolledo, *Carbohydr. Res.*, **1989**, *186*, C7.
108. M. Avalos, R. Babiano, A. Cabanillas, P. Cintas, J. L. Jiménez, J. C. Palacios, M. A. Aguilar, J. C. Corchado, and J. Espinosa-García, *J. Org. Chem.*, **1996**, *61*, 7291.
109. R. Nesi, D. Giomi, S. Turchi, P. Tedeschi, and F. Ponticelli, *Gazz. Chim. Ital.*, **1993**, *123*, 633.
110. G. W. Gribble, E. T. Pelkey, and F. L. Switzer, *Synlett*, **1998**, 1061.
111. G. W. Gribble, E. T. Pelkey, W. M. Simon, and H. A. Trujillo, *Tetrahedron*, **2000**, *56*, 10133.
112. T. Kobayashi, H. Minemura, and H. Kato, *Heterocycles*, **1995**, *40*, 311.
113. H. Kato, S. Toda, Y. Arikawa, M. Masuzawa, M. Hashimoto, K. Ikoma, S.-Z. Wang, and A. Miyasaka, *J. Chem. Soc., Perkin Trans. 1*, **1990**, 2035.
114. H. Kato, T. Kobayashi, K. Horie, K. Oguri, and M. Moriwaki, *J. Chem. Soc., Perkin Trans. 1*, **1993**, 1055.
115. H.-D. Martin, F.-J. Mais, B. Mayer, H.-J. Hecht, M. Hekman, and A. Steigel, *Monatsh.*, **1983**, *114*, 1145.
116. C. A. Maryanoff, C. B. Karash, I. J. Turchi, E. R. Corey, and B. E. Maryanoff, *J. Org. Chem.*, **1989**, *54*, 3790.
117. C. A. Maryanoff and I. J. Turchi, *Heterocycles*, **1993**, *35*, 649.
118. I. J. Turchi, private communication of unpublished results.
119. H. Kato, N. Aoki, Y. Kawamura, and K. Yoshino, *J. Chem. Soc., Perkin Trans. 1*, **1985**, 1245.
120. W. Friedrichsen and W.-D. Schröer, *Liebigs Ann. Chem.*, **1981**, 476.
121. W. Friedrichsen, W.-D. Schröer, and T. Debaerdemaeker, *Liebigs Ann. Chem.*, **1981**, 491.
122. T. Debaerdemaeker, W.-D. Schröer, and W. Friedrichsen, *Liebigs Ann. Chem.*, **1981**, 502.
123. H. Kato, T. Kobayashi, K. Tokue, and S. Shirasawa, *J. Chem. Soc., Perkin Trans. 1*, **1993**, 1617.
124. S.-H. Wu, W.-Q. Sun, D.-W. Zhang, L.-H. Shu, H.-M. Wu, J.-F. Xu, and X.-F. Lao, *J. Chem. Soc., Perkin Trans. 1*, **1998**, 1733.
125. G. W. Gribble, W. R. Sponholtz III, F. L. Switzer, F. J. D'Amato, and M. P. Byrn, *Chem. Commun.*, **1997**, 993.
126. A. Padwa, R. Lim, J. G. MacDonald, H. L. Gingrich, and S. M. Kellar, *J. Org. Chem.*, **1985**, *50*, 3816.
127. M. Kawase, *J. Chem. Soc., Chem. Commun.*, **1990**, 1328.
128. M. Kawase, *Chem. Pharm. Bull.*, **1997**, *45*, 1248.
129. H. O. Bayer, R. Huisgen, R. Knorr, and F. C. Schaefer, *Chem. Ber.*, **1970**, *107*, 2581.
130. F. Bronberger and R. Huisgen, *Tetrahedron Lett.*, **1984**, *25*, 65.
131. R. Consonni, P. Dalla Croce, R. Ferraccioli, and C. La Rosa, *J. Chem. Res. (S)*, **1991**, 188.
132. L. Bonati, R. Ferraccioli, and G. Moro, *J. Phys. Org. Chem.*, **1995**, *8*, 452.
133. M. T. Bilodeau and A. M. Cunningham, *J. Org. Chem.*, **1998**, *63*, 2800.
134. B. C. Hamper, K. D. Jerome, G. Yalamanchili, D. M. Walker, R. C. Chott, and D. A. Mischke, *Biotechnol. Bioeng.*, **2000**, *71*, 28.
135. P. Dalla Croce, R. Ferraccioli, C. La Rosa, and T. Pilati, *J. Chem. Soc., Perkin Trans. 2*, **1993**, 1511.
136. P. Dalla Croce, R. Ferraccioli, and C. La Rosa, *Tetrahedron*, **1995**, *51*, 9385.
137. P. W. Manley, C. W. Rees, and R. C. Storr, *J. Chem. Soc., Chem. Commun.*, **1983**, 1007.

138. H. Rodríguez, H. Pavez, A. Márquez, and P. Navarrete, *Tetrahedron*, **1983**, *39*, 23.

139. G. Märkl and G. Dorfmeister, *Tetrahedron Lett.*, **1986**, *27*, 4419.

140. W. Rösch, H. Richter, and M. Regitz, *Chem. Ber.*, **1987**, *120*, 1809.

141. E. P. O. Fuchs, M. Hermesdorf, W. Schnurr, W. Rösch, H. Heydt, M. Regitz, and P. Binger, *J. Organomet. Chem.*, **1988**, *338*, 329.

142. A. Mack, E. Pierron, T. Allspach, U. Bergstässer, and M. Regitz, *Synthesis*, **1998**, 1305.

143. P. Bach, U. Bergsträsser, S. Leininger, and M. Regitz, *Bull. Soc. Chim. Fr.*, **1997**, *134*, 927.

144. G. Schmnitt, N. D. An, J. Verbrel, and B. Laude, *Bull. Soc. Chim. Belg.*, **1986**, *95*, 215.

145. G. Schmitt, B. Laude, J. Vebrel, N. Rodier, and F. Theobald, *Bull. Soc. Chim. Belg.*, **1989**, *98*, 113.

146. M. Berrabah, G. Schmitt, N. D. An, and B. Laude, *Bull. Soc. Chim. Belg.*, **1991**, *100*, 613.

147. K. Monnier, G. Schmitt, and B. Laude, *Tetrahedron Lett.*, **1992**, *33*, 1609.

148. S. Perrin, K. Monnier, B. Laude, M. M. Kubicki, and O. Blacque, *Tetrahedron Lett.*, **1998**, *39*, 1753.

149. S. Perrin, K. Monnier, B. Laude, M. Kubicki, and O. Blacque, *Eur. J. Org. Chem.*, **1999**, 297.

150. W. E. McEwen, A. V. Grossi, R. J. MacDonald, and A. P. Stamegna, *J. Org. Chem.*, **1980**, *45*, 130.

151. H. Kato, T. Kobayashi, M. Ciobanu, H. Iga, A. Akutsu, and A. Katehi, *Chem. Commun.*, **1996**, 1011.

152. H. Kato, T. Kobayashi, M. Ciobanu, and A. Kakehi, *Tetrahedron*, **1997**, *53*, 9921.

153. S. G. Ruf, U. Bergsträsser, and M. Regitz, *Tetrahedron*, **2000**, *56*, 63.

154. A. Padwa and D. L. Hertzog, *Tetrahedron*, **1993**, *49*, 2589.

155. M. Prein and A. Padwa, *Tetrahedron Lett.*, **1996**, *37*, 6981.

156. M. Prein, P. J. Manley, and A. Padwa, *Tetrahedron*, **1997**, *53*, 7777.

157. A. Padwa and M. Prein, *J. Org. Chem.*, **1997**, *62*, 6842.

158. R. González, B. W. Knight, F. Wudl, M. A. Semones, and A. Padwa, *J. Org. Chem.*, **1994**, *59*, 7949.

159. S. M. Sheehan and A. Padwa, *J. Org. Chem.*, **1997**, *62*, 438.

160. C. S. Straub and A. Padwa, *Org. Lett.*, **1999**, *1*, 83.

161. A. Padwa, S. M. Sheehan, and C. S. Straub, *J. Org. Chem.*, **1999**, *64*, 8648.

162. A. Padwa and M. Prein, *Tetrahedron*, **1998**, *54*, 6957.

163. R. Angell, M. G. B. Drew, M. Fengler-Veith, H. Finch, L. M. Harwood, A. W. Jahans, and T. T. Tucker, *Tetrahedron Lett.*, **1997**, *38*, 3107.

164. R. Angell, M. Fengler-Veith, H. Finch, L. M. Harwood, and T. T. Tucker, *Tetrahedron Lett.*, **1997**, *38*, 4517.

165. M. G. B. Drew, M. Fengler-Veith, L. M. Harwood, and A. W. Jahans, *Tetrahedron Lett.*, **1997**, *38*, 4521.

166. C. O. Kappe, K. Peters, and E.-M. Peters, *J. Org. Chem.*, **1997**, *62*, 3109.

167. C. O. Kappe, *Tetrahedron Lett.*, **1997**, *38*, 3323.

168. A. Padwa, J. P. Snyder, E. A. Curtis, S. M. Sheehan, K. J. Worsencroft, and C. O. Kappe, *J. Am. Chem. Soc.*, **2000**, *122*, 8155.

169. M. R. Gowravaram and M. A. Gallop, *Tetrahedron Lett.*, **1997**, *38*, 6973.

170. D. L. Whitehouse, K. H. Nelson, Jr., S. N. Savinov, and D. J. Austin, *Tetrahedron Lett.*, **1997**, *38*, 7139.

171. D. L. Whitehouse, K. H. Nelson, Jr., S. N. Savinov, R. S. Löwe, and D. J. Austin, *Bioorg. Med. Chem.*, **1998**, *6*, 1273.

172. S. N. Savinov and D. J. Austin, *Chem. Commun.*, **1999**, 1813.

173. S. Miah, C. J. Moody, I. C. Richards, and A. M. Z. Slawin, *J. Chem. Soc., Perkin Trans. 1*, **1997**, 2405.

174. M. Hamaguchi and T. Nagai, *J. Chem. Soc., Chem. Commun.*, **1985**, 726.

175. K. T. Potts and S. Husain, *J. Org. Chem.*, **1972**, *37*, 2049.

176. K. T. Potts, J. Baum, E. Houghton, D. N. Roy, and U. P. Singh, *J. Org. Chem.*, **1974**, *39*, 3619.

177. K. T. Potts, J. Baum, S. K. Datta, and E. Houghton, *J. Org. Chem.*, **1976**, *41*, 813.
178. D. L. Hertzog, D. J. Austin, W. R. Nadler, and A. Padwa, *Tetrahedron Lett.*, **1992**, *33*, 4731.
179. A. Padwa, D. L. Hertzog, W. R. Nadler, M. H. Osterhout, and A. T. Price, *J. Org. Chem.*, **1994**, *59*, 1418.
180. A. Padwa, D. J. Austin, and A. T. Price, *Tetrahedron Lett.*, **1994**, *35*, 7159.
181. A. Padwa, D. J. Austin, A. T. Price, and M. D. Weingarten, *Tetrahedron*, **1996**, *52*, 3247.
182. A. Padwa, D. L. Hertzog, and W. R. Nadler, *J. Org. Chem.*, **1994**, *59*, 7072.
183. A. Padwa, S. R. Harring, and M. A. Semones, *J. Org. Chem.*, **1998**, *63*, 44.
184. J. P. Marino, Jr., M. H. Osterhout, A. T. Price, M. A. Semones, and A. Padwa, *J. Org. Chem.*, **1994**, *59*, 5518.
185. A. Padwa, M. A. Brodney, J. P. Marino, Jr., and S. M. Sheehan, *J. Org. Chem.*, **1997**, *62*, 78.
186. J. P. Marino, Jr., M. H. Osterhout, and A. Padwa, *J. Org. Chem.*, **1995**, *60*, 2704.
187. A. Padwa, *Chem. Commun.*, **1998**, 1417.
188. A. Padwa, M. A. Brodney, J. P. Marino, Jr., M. H. Osterhout, and A. T. Price, *J. Org. Chem.*, **1997**, *62*, 67.
189. B. Jauk, F. Belaj, and C. O. Kappe, *J. Chem. Soc., Perkin Trans. 1*, **1999**, 307.
190. J. T. Kuethe and A. Padwa, *Tetrahedron Lett.*, **1997**, *38*, 1505.
191. A. Padwa, T. M. Heidelbaugh, and J. T. Kuethe, *J. Org. Chem.*, **2000**, *65*, 2368.
192. F. H. C. Stewart, *Chem. Rev.*, **1964**, *64*, 129.
193. S. V. Badachikar, R. K. Tikare, and G. S. Puranik, *Ind. J. Chem.*, **1986**, *25B*, 1079.
194. A. E. McGowin, L. Jackson, L. W. Marshall, and K. Turnbull, *Org. Prep. Proc. Int.*, **2001**, *33*, 100.
195. S. Nan'ya, K. Katsuraya, E. Maekawa, K. Kondo, and S. Eguchi, *J. Heterocycl. Chem.*, **1987**, *24*, 971.
196. T. Sasaki, T. Manabe, and S. Nishida, *J. Org. Chem.*, **1980**, *45*, 479.
197. D. Ranganathan and S. Bamezai, *Tetrahedron Lett.*, **1983**, *24*, 1067.
198. D. Ranganathan and S. Bamezai, *Synth. Commun.*, **1985**, *15*, 259.
199. D. Ranganathan, S. Bamezai, H. Cun-heng, and J. Clardy, *Tetrahedron Lett.*, **1985**, *26*, 5739.
200. G. Märkl and S. Pflaum, *Tetrahedron Lett.*, **1986**, *27*, 4415.
201. R. Destro, E. Erba, L. Fosti, D. Pocar, and D. Scarcella, *Ann.*, **1985**, 1377.
202. O. B. Sutcliffe, R. C. Storr, T. L. Gilchrist, and P. Rafferty, *Tetrahedron*, **2000**, *56*, 10011.
203. K. K. Sun, *Tetrahedron Lett.*, **1986**, *27*, 317.
204. K. K. Sun, *Tetrahedron Lett.*, **1986**, *27*, 321.
205. K. K. Sun, *Macromolecules*, **1987**, *20*, 726.
206. G. W. Gribble and B. H. Hirth, *J. Heterocycl. Chem.*, **1996**, *33*, 719.
207. P. M. Weintraub, *Chem. Commun.*, **1970**, 760.
208. H. Meier and H. Heimgartner, *Helv. Chim. Acta*, **1986**, *69*, 927.
209. E. Vedejs and R. G. Wilde, *J. Org. Chem.*, **1986**, *51*, 117.
210. F.-J. Mais, H. Dickopp, B. Middelhauve, H.-D. Martin, D. Mootz, and A. Steigel, *Ber. Chem.*, **1987**, *120*, 275.
211. M. Baudy and A. Robert, *J. Chem. Soc., Chem. Commun.*, **1976**, 23.
212. M. Baudy, A. Robert, and G. Guimon, *Tetrahedron*, **1982**, *38*, 1241.
213. M. Baudy, A. Robert, and G. Guimon, *Tetrahedron*, **1982**, *38*, 2129.
214. K. T. Potts, J. Baum, E. Houghton, D. N. Roy, and U. P. Singh, *J. Org. Chem.*, **1974**, *39*, 3619.
215. K. T. Potts, J. Baum, S. K. Datta, and E. Houghton, *J. Org. Chem.*, **1976**, *41*, 813.
216. T. Sheradsky and N. Itzhak, *J. Chem. Soc., Perkin Trans. 1*, **1987**, 1979.
217. K. T. Potts and M. O. Dery, *J, Chem. Soc., Chem. Commun.*, **1986**, 561.

218. K. T. Potts, M. O. Dery, and W. A. Juzukonis, *J. Org. Chem.*, **1989**, *54*, 1077.

219. P. Cavalleri, F. Clerici, E. Erba, and P. Trimarco, *Chem. Ber.*, **1992**, *125*, 883.

220. P. Areces, M. Avalos, R. Babiano, L. González, J. L. Jiménez, J. C. Palacios, and M. D. Pilo, *Carbohydr. Res.*, **1991**, *222*, 99.

221. M. Avalos, R. Babiano, M. J. Diánez, J. Espinosa, M. D. Estrada, J. L. Jiménez, A. López-Castro, M. M. Méndez, and J. C. Palacios, *Tetrahedron*, **1992**, *48*, 4193.

222. P. Areces, M. Avalos, R. Babiano, L. González, J. L. Jiménez, M. M. Méndez, and J. C. Palacios, *Tetrahedron Lett.*, **1993**, *34*, 2999.

223. M. Avalos, R. Babiano, A. Cabanillas, P. Cintas, M. J. Diánez, M. D. Estrada, J. L. Jiménez, A. López-Castro, J. C. Palacios, and S. P. Garrido, *J. Chem. Soc., Chem. Commun.*, **1995**, 2213.

224. M. Avalos, R. Babiano, A. Cabanillas, P. Cintas, F. J. Higes, J. L. Jiménez, and J. C. Palacios, *J. Org. Chem.*, **1996**, *61*, 3738.

225. M. Avalos, R. Babiano, P. Cintas, M. B. Hursthouse, J. L. Jiménez, M. E. Light, I. López, and J. C. Palacios, *Chem. Commun.*, **1999**, 1589.

226. M. J. Arévalo, M. Avalos, R. Babiano, P. Cintas, M. B. Hursthouse, J. L. Jiménez, M. E. Light, I. López, and J. C. Palacios, *Tetrahedron Lett.*, **1999**, *40*, 8675.

227. M. J. Arévalo, M. Avalos, R. Babiano, P. Cintas, M. B. Hursthouse, J. L. Jiménez, M. E. Light, I. López, and J. C. Palacios, *Tetrahedron*, **2000**, *56*, 1247.

228. M. Avalos, R. Babiano, P. Cintas, F. R. Clemente, R. Gordillo, J. L. Jiménez, J. C. Palacios, and P. R. Raithby, *J. Org. Chem.*, **2000**, *65*, 5089.

229. A. Padwa, F. R. Kinder, and L. Zhi, *Synlett*, **1991**, 287.

230. D. L. Hertzog, W. R. Nadler, Z. J. Zhang, and A. Padwa, *Tetrahedron Lett.*, **1992**, *33*, 5877.

231. A. Padwa and Z. J. Zhang, *Heterocycles*, **1994**, *37*, 441.

232. T. M. Heidelbaugh, B. Liu, and A. Padwa, *Tetrahedron Lett.*, **1998**, *39*, 4757.

233. A. Padwa, L. S. Beall, T. M. Heidelbaugh, B. Liu, and S. M. Sheehan, *J. Org. Chem.*, **2000**, *65*, 2684.

234. H. Gotthardt and B. Christl, *Tetrahedron Lett.*, **1968**, 4743.

235. H. Gotthardt, M. C. Weisshuhn, and B. Christl, *Chem. Ber.*, **1976**, *109*, 740.

236. H. Gotthardt, M. C. Weisshuhn, and B. Christl, *Chem. Ber.*, **1976**, *109*, 753.

237. H. Kato and S. Toda, *J. Chem. Soc., Chem. Commun.*, **1982**, 510.

238. H. Kato, Y. Arikawa, M. Hashimoto, and M. Masuzawa, *J. Chem. Soc., Chem. Commun.*, **1983**, 938.

239. A. Souizi and A. Robert, *Synthesis*, **1982**, 1059.

240. M. Bssaibis, A. Robert, and A. Souizi, *J. Chem. Soc., Chem. Commun.*, **1993**, 998.

241. H. Gotthardt, M.C. Weisshuhn, and K. Dörhöfer, *Angew. Chem. Inter. Ed. Engl.*, **1975**, *14*, 422.

242. H. Gotthardt, M.C. Weisshuhn, and K. Dörhöfer, *Chem. Ber.*, **1978**, *111*, 3336.

243. H. Gotthardt, S. Schoy-Tribbensee, and U. Feist, *Angew. Chem. Inter. Ed. Engl.*, **1982**, *21*, 779.

244. H. Gotthardt, U. Feist, and S. Schoy-Tribbensee, *Chem. Ber.*, **1985**, *118*, 774.

245. H. Gotthardt and U. Feist, *Chem. Ber.*, **1985**, *118*, 785.

246. H. Gotthardt, H.-G. Kinzelmann, U. Feist, and J. Buddrus, *Chem. Ber.*, **1986**, *119*, 2317.

247. H. Gotthardt, *Chem. Ber.*, **1972**, *105*, 188.

248. H. Gotthardt, *Chem. Ber.*, **1972**, *105*, 196.

249. M. Ichinari, T. Sato, and Y. Hayase, *Heterocycles*, **1988**, *27*, 227.

250. B. P. Coppola, M. C. Noe, R. L. Abdon, and R. G. Konsler, *J. Org. Chem.*, **1993**, *58*, 7324.

251. F. Berrée and G. Morel, *Tetrahedron*, **1995**, *51*, 7019.

CHAPTER 11

Effect of External Reagents

Shuji Kanemasa

Institute of Advanced Material Study, Kyushu University, Kasugakoen, Japan

11.1. Azomethine Ylides . . . 757
11.1.1. Introduction . . . 757
11.1.2. Metal N-Alkylideneglycinates . . . 758
11.1.3. 2-Azaallyl Anions . . . 759
11.1.4. N-Metalated Azomethine Ylides . . . 759
11.1.4.1. Irreversible Metalation of (N-Alkylideneamino)acetonitriles . . . 761
11.1.4.2. Reversible Metalation of α-(Alkylideneamino) Esters . . . 763
11.1.4.3. anti-Selective Michael Additions . . . 768
11.1.4.4. Asymmetric Azomethine Ylide Cycloadditions . . . 772
11.1.4.5. Asymmetric Michael Additions of Enantiopure Camphor Imines . . . 774
11.1.4.6. Enantioselective Reactions . . . 778
11.2. Nitrile Oxides . . . 779
11.2.1. Introduction . . . 779
11.2.2. Magnesium Ion Mediated Reactions . . . 780
11.2.2.1. Diastereocontrol of Nitrile Oxide Cycloadditions . . . 780
11.2.2.2. Regiocontrol of Nitrile Oxide Cycloadditions . . . 780
11.2.2.3. Solvent Effects . . . 782
11.2.2.4. Kinetics . . . 784
11.2.2.5. Chemoselectivity . . . 786
11.2.2.6. Catalytic Cycle . . . 787
11.2.2.7. Related Work and Applications . . . 788
11.2.3. Lewis Acid-Mediated Diastereoselective Reactions . . . 789
11.2.4. Other Mediators in Nitrile Oxide Cycloadditions . . . 790
11.2.5. Catalyzed Enantioselective Reactions . . . 793
11.3. Nitrones . . . 794
11.3.1. Introduction . . . 794
11.3.2. Titanium Ion-Mediated Reactions . . . 795
11.3.3. Magnesium Ion-Mediated Reactions . . . 797
11.3.4. Transesterification–Dipolar Cycloadditions . . . 801
11.3.5. Catalyzed Enantioselective Reactions . . . 804

The Chemistry of Heterocyclic Compounds, Volume 59: Synthetic Applications of 1,3-Dipolar Cycloaddition Chemistry Toward Heterocycles and Natural Products, Edited by Albert Padwa and William H. Pearson.
ISBN 0-471-38726-6

11.4. Reactions Involving Other 1,3-Dipoles 804
11.4.1. Carbonyl Ylides 804
11.4.2. Nitrile Ylide Equivalents 806
11.5. Conclusion 809

The stereoselective synthesis of heterocyclic compounds using 1,3-dipolar cycloadditions is now regarded to be one of the most important research subjects in modern synthetic organic chemistry. The synthetic versatility of 1,3-dipolar cycloaddition reactions arises from the fact that (a) they provide a general and direct synthetic access to the skeletons of five-membered heterocycles, and (b), they can be effectively used, after subsequent functional group manipulation, to construct heterocycles other than five-membered rings as well as other organic functional groups (1–6). One advantage is the high stereospecificity of 1,3-dipolar cycloaddition reactions under ordinary reaction conditions, where the configurations of both the 1,3-dipoles and the dipolarophiles are completely expressed in the stereostructure of cycloadducts (1,5). If high diastereo- and/or enantiocontrol is also possible using 1,3-dipolar cycloadditions, these reactions will become powerful synthetic tools much like the Diels–Alder reaction, where effective catalysis by *external reagents* such as Lewis acids has been well established. However, most 1,3-dipolar cycloaddition reactions are simple thermal processes, unaided by external reagents. The use of external reagents to control the stereo-, regio-, and enantioselectivity of 1,3-dipolar cycloadditions is a relatively new area and has been examined with a limited number of dipoles. This field is expected to grow rapidly in the near future. A common difficulty is the competition between the dipole and the dipolarophile for the external reagent when it is a Lewis acid. When one tries to activate electrophilic dipolarophiles with Lewis acidic external reagents, binding to the 1,3-dipole can predominate, resulting in a slower reaction. Such strong binding also causes decreased reactivity of the 1,3-dipoles toward nucleophilic dipolarophiles. Strong binding of 1,3-dipoles to a Lewis acid is often not avoidable because of the electronic structure of the 1,3-dipoles and the typical presence of Lewis basic sites. Accordingly, Lewis acidic reagents may combine with the basic site(s) on certain 1,3-dipoles, making relatively stable salt-like complexes. The coordinating power of external reagents must be carefully adjusted in order to be successful in attaining high rate enhancements. The nature of the ligands on the Lewis acid and the lability and exchangeability of these ligands are typical issues to be considered.

Some of the strategies for the control of regio-, diastereo-, and enantioselectivity in 1,3-dipolar cycloaddition reactions are shown below. These methods have been realized to some degree in recent years, and will form the basis of the discussion in this chapter.

Metalated 1,3-dipoles or 1,3-dipole equivalents.
Metal-tethered 1,3-dipolar cycloaddition via an internal delivery mechanism.
Metal-mediated 1,3-dipolar cycloadditions (stoichiometric).

Metal-mediated 1,3-dipolar cycloadditions (catalytic).

Other activation methods.

This chapter deals mainly with the 1,3-dipolar cycloaddition reactions of three 1,3-dipoles: azomethine ylides, nitrile oxides, and nitrones. These three have been relatively well investigated, and examples of external reagent-mediated stereocontrolled cycloadditions of other 1,3-dipoles are quite limited. Both nitrile oxides and nitrones are 1,3-dipoles whose cycloaddition reactions with alkene dipolarophiles produce 2-isoxazolines and isoxazolidines, their dihydro derivatives. These two heterocycles have long been used as intermediates in a variety of synthetic applications because their rich functionality. When subjected to reductive cleavage of the N—O bonds of these heterocycles, for example, important building blocks such as β-hydroxy ketones (aldols), α,β-unsaturated ketones, γ-amino alcohols, and so on are produced (7–12). Stereocontrolled and/or enantiocontrolled cycloadditions of nitrones are the most widely developed (6,13). Examples of enantioselective Lewis acid catalyzed 1,3-dipolar cycloadditions are summarized by Jørgensen in Chapter 12 of this book, and will not be discussed further here.

11.1. AZOMETHINE YLIDES

11.1.1. Introduction

The chemistry of azomethine ylides has changed dramatically in the last two decades. The previous review in this series (14) covered only the traditional methods for the generation of these dipoles, such as the thermal or photochemical ring opening of aziridines (15–18). Since then, there have appeared a variety of new generation methods (19), including (a) the thermal tautomerization of imines bearing an electron withdrawing group at the α-position of the N-substituent (20–24), (b) the desilylation of N-alkylidene(α-silylalkyl)amines (25–32), (c) the dehydrative condensation of α-amino esters and nitriles with carbonyl compounds (33–37), (d) the decarboxylative condensation of α-amino acids with carbonyl compounds (34,36,38–41), and (e) the deprotonation of tertiary amine-N-oxides with strong bases (42,43).

These newly discovered methods have found wide synthetic application. Examples include the generation and cycloaddition of stabilized N-unsubstituted azomethine ylides, nonstabilized N-substituted azomethine ylides, and even the parent azomethine ylides bearing no carbon substituents (19). However, these modern procedures often require severe reaction conditions such as high reaction temperatures, the use of polar solvents, and the use of strong bases, among others. The poor stereo- and regioselectivities that are often observed in the cycloadditions of nonstabilized azomethine ylides have discouraged their use in the stereocontrolled synthesis of complex molecules.

11.1.2. Metal N-Alkylideneglycinates

N-Metalated azomethine ylides (44) are probably the most synthetically useful of these dipoles, since they can be generated from readily available N-alkylideneamino esters or nitriles through a simple activation method. The resulting 1,3-dipoles show high reactivity with α,β-unsaturated carbonyl acceptors to provide excellent stereo- and regioselectivities. A variety of asymmetric versions have been reported.

The first example of N-metalated azomethine ylides was reported in 1979 by Casella et al. (45). The action of triethylamine on *N*-pyruvylideneglycinatocopper(II) in pyridine and subsequent trapping of the resulting anionic species with acrylonitrile or methyl acrylate gave the chelated cycloadducts (Scheme 11.1). Removal of the copper metal from the chelated cycloadducts with a cation exchange resin liberated 2,5-pyrrolidinedicarboxylic acids (46). Although the reactions are highly stereoselective as confirmed later by Grigg et al. (47), a ready epimerization took place during the demetalation procedure to give mixtures of several stereoisomeric cycloadducts.

Later, Grigg et al. (47) reported that metals other than Cu(II) could be incorporated in the glycine imines of pyruvic acid or phenylglyoxylic acid (47). Such isolable chelates containing Zn(II), Cd(II), and Pb(II) metals also undergo stereoselective cycloadditions to 1,2-disubstituted electron-deficient alkenes. However, the reactions with less reactive dipolarophiles such as acrylonitrile, methyl acrylate, and phenyl vinyl sulfone give mixtures of regio- and stereoisomers. The kinetic and thermodynamic nature of such cycloadditions has been discussed (47). The ready epimerization in the demetalation step is a disadvantage. A Michael addition–cyclization mechanism previously proposed by Casella et al. (45,46) was questioned by Grigg et al. (47), and a new mechanism involving N-metalated azomethine ylide 1,3-dipoles was suggested.

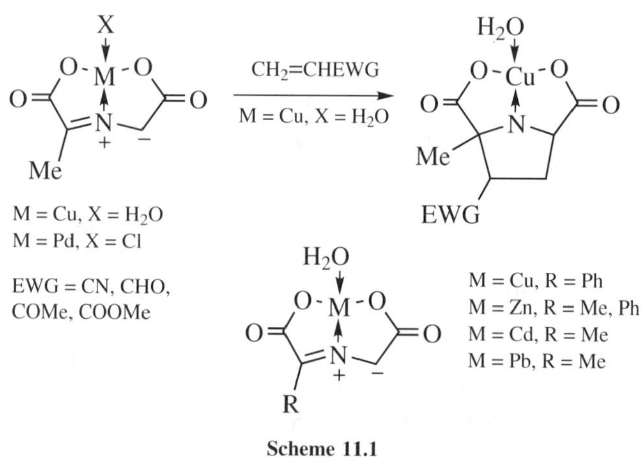

Scheme 11.1

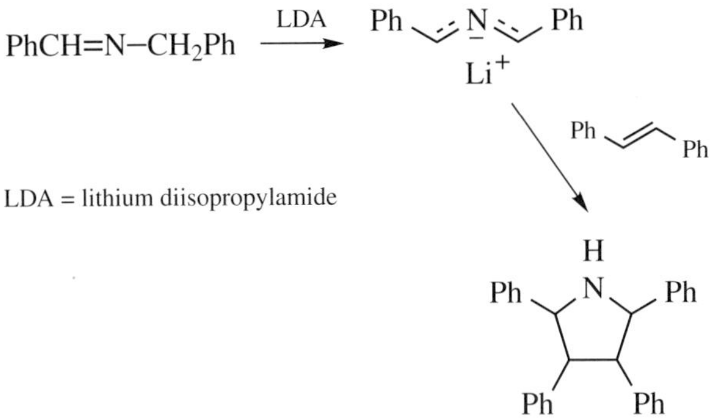

Scheme 11.2

11.1.3. 2-Azaallyl Anions

The use of lithium amides to metalate the α-position of the N-substituent of imines generates 2-azaallyl anions, typically stabilized by two or three aryl groups (Scheme 11.2) (48–62), a process pioneered by Kauffmann in 1970 (49). Although these reactive anionic species may be regarded as N-lithiated azomethine ylides if the lithium metal is covalently bonded to the imine nitrogen, they have consistently been discussed as 2-azaallyl anions. Their cyclization reactions are characterized by their enhanced reactivity toward relatively unactivated alkenes such as ethene, styrenes, stilbenes, acenaphtylene, 1,3-butadienes, diphenylacetylene, and related derivatives. Accordingly, these cycloaddition reactions are called anionic [3 + 2] cycloadditions. Reactions with the electron-poor alkenes are rare (54,57). Such reactivity makes a striking contrast with that of N-metalated azomethine ylides, which will be discussed below (Section 11.1.4).

Note that Pearson has extended the classical anionic [3 + 2] cycloadditions to allow the generation of nonstabilized 2-azaallyl anions, and has successfully applied this methodology to the field of alkaloid total synthesis. A key discovery was that (2-azaallyl)stannanes are capable of undergoing tin–lithium exchange to generate the nonstabilized anions (63–76), which can be trapped either intramolecularly or intermolecularly with unactivated alkenes to produce pyrrolidines, often in a stereoselective fashion. Thus, a variety of 2-azaallyl anions are accessible by his method. A few examples of Pearson's contributions are illustrated in Scheme 11.3 (70,76).

11.1.4. N-Metalated Azomethine Ylides

In 1980, Grigg et al. (77) reported the reactions of α-(arylideneamino) esters with methyl acrylate in the presence of a catalytic amount of sodium methoxide, producing both Michael adducts and cycloadducts in a stereoselective fashion, the

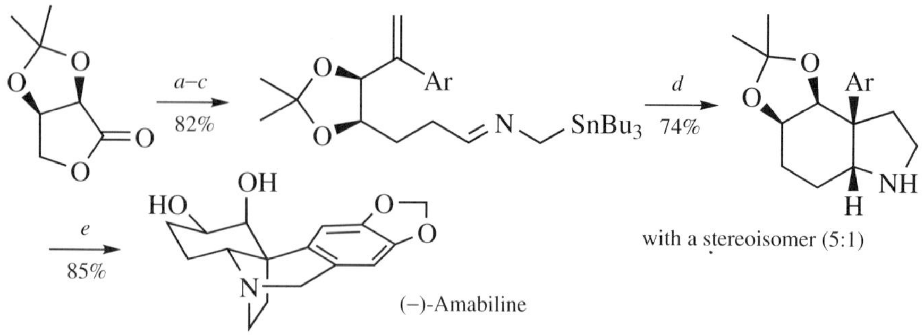

a: (1) ArLi, (2) $CH_2{=}PPh_3$. b: (1) Tf_2O, py, (2) $CyN{=}CHCH_2Li$, (3) H^+.
c: n-$Bu_3SnCH_2NH_2$ (1 equiv), 4Å MS in Et_2O. d: (1) n-BuLi (1.9 equiv) in THF, −78 °C,
(2) H_2O. e: (1) $Me_2N^+{=}CH_2\ I^-$ in MeCN, Δ (2) HCl in MeOH

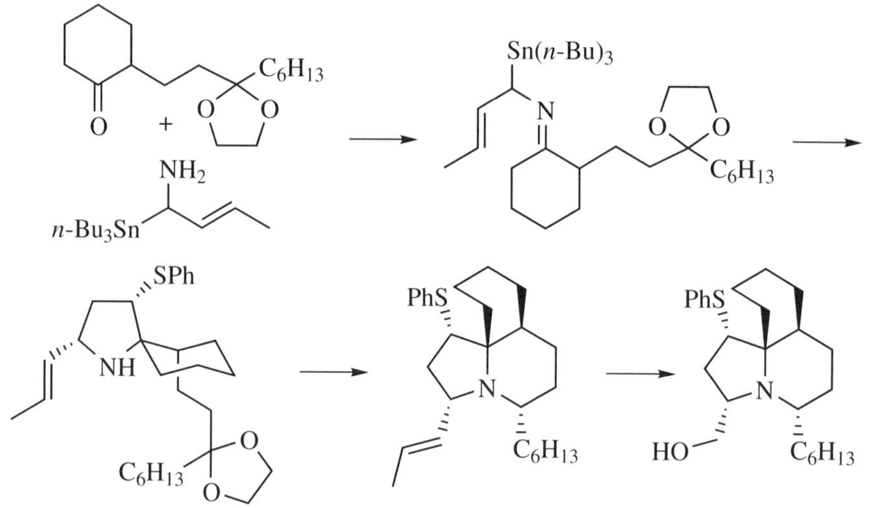

Scheme 11.3

product ratios depending on the nature of the arylidene substituents (Scheme 11.4). Although the author has not mentioned such a point, this report corresponds to the first report of the direct generation of N-metalated azomethine ylides from α-(alkylideneamino) esters.

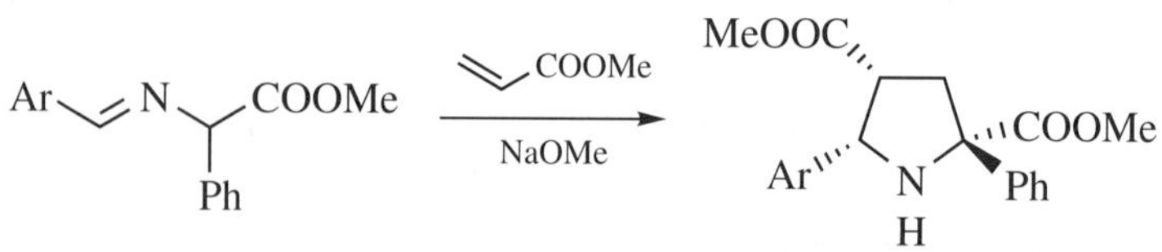

Ar = 2-Furyl

Scheme 11.4

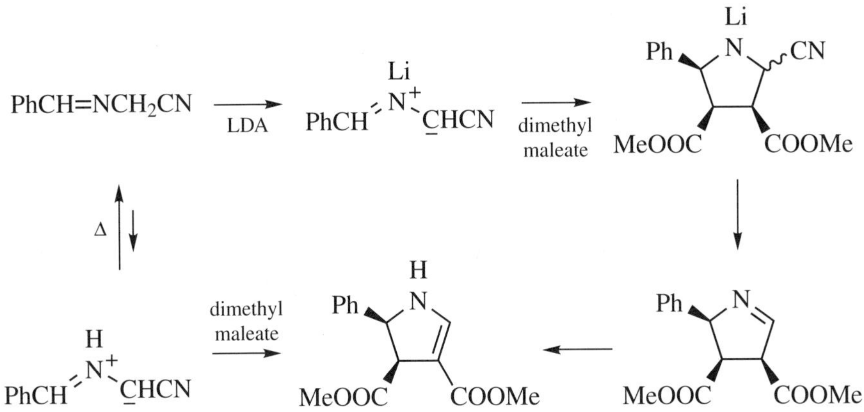

Scheme 11.5

11.1.4.1. *Irreversible Metalation of (N-Alkylideneamino)acetonitriles*

Metalation at the α-position of the N-substituent of imines is facilitated by an attaching electron-withdrawing substituent at this position, for example, a cyano group (Scheme 11.5). The successful generation of N-metalated azomethine ylides through α-metalation of the N-substituent of imines has been demonstrated with LDA and α-(benzylideneamino)acetonitrile (78). On treatment with LDA (1 equiv) at −78 °C, a colorless solution of the imine nitrile in tetrahydrofuran (THF) turns bright red, probably the color of the N-lithiated azomethine ylide. This color fades immediately as dimethyl maleate is added to the solution, and dimethyl *cis*-5-phenyl-2-pyrroline-3,4-dicarboxylate is obtained as a single stereoisomer in 71% yield (Scheme 11.5). Exclusive formation of the stereoselective cycloadduct involves (a) the lithiation of the imine with LDA generating the N-lithiated azomethine ylide of the cyano-stabilized type, (b) the endo-selective cycloaddition of the ylide to dimethyl maleate, (c) the spontaneous elimination of LiCN from the lithiated cycloadduct, and (d) the spontaneous 1,3-proton migration of the resulting 1-pyrroline intermediate producing the 2-pyrroline. The same 2-pyrroline is obtained stereoselectively in a similar reaction involving a N-unsubstituted azomethine ylide, generated thermally from the same imine nitrile, although at a much higher reaction temperature. The pyrroline is produced in 66% yield after 27 h at reflux in toluene (79). N-Unsubstituted azomethine ylides, generated thermally from imines, are less suitable than the N-lithiated ylides, and do not to react at room temperature with α,β-unsaturated esters such as acrylates, crotonates, and cinnamates. This is because an imine/azomethine ylide equilibrium is very unfavorable.

α-Substituted α-(alkylideneamino)acetonitriles are also activated by lithiation with LDA. The resulting N-lithiated ylides react with a variety of α,β-unsaturated ester dipolarophiles to give endo-cycloadducts and/or the LiCN eliminated derivatives in

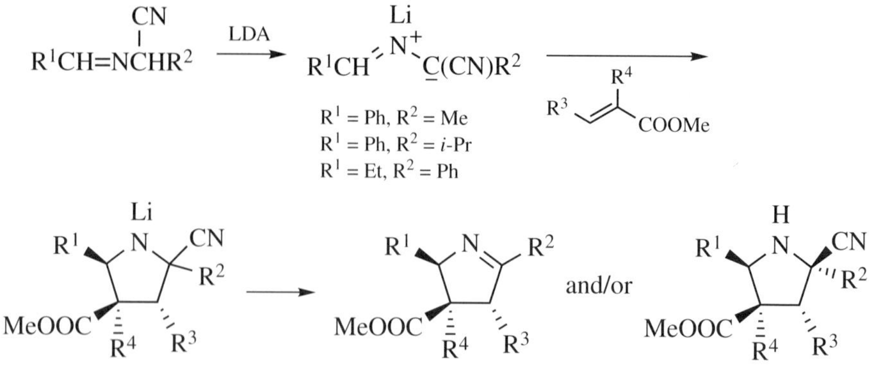

Scheme 11.6

satisfactory yields (Scheme 11.6). The imines derived form enolizable aldehydes can be successfully employed in the LDA induced cycloadditions.

N-Unsubstituted azomethine ylides may be generated thermally (79), and the N-metalated, 2-azaallyl anion versions may be generated by action of nonmetallic bases such as 1,8-diazabicyclo[5.4.0]undec-7-ene (DBU) on certain imines (80). Although they are assumed to show similar chemical properties, these two species usually show different reaction patterns, as shown in Scheme 11.7, where the regio- and stereoselectivities of the cycloadditions are quite different (24,78–80). Metalation of (alkylideneamino)acetonitriles can be performed with metallic bases other than LDA. Thus, butyllithium, ethylmagnesium bromide, and magnesium bromide-diisopropylamide are also effective (78). The N-magnesioazomethine

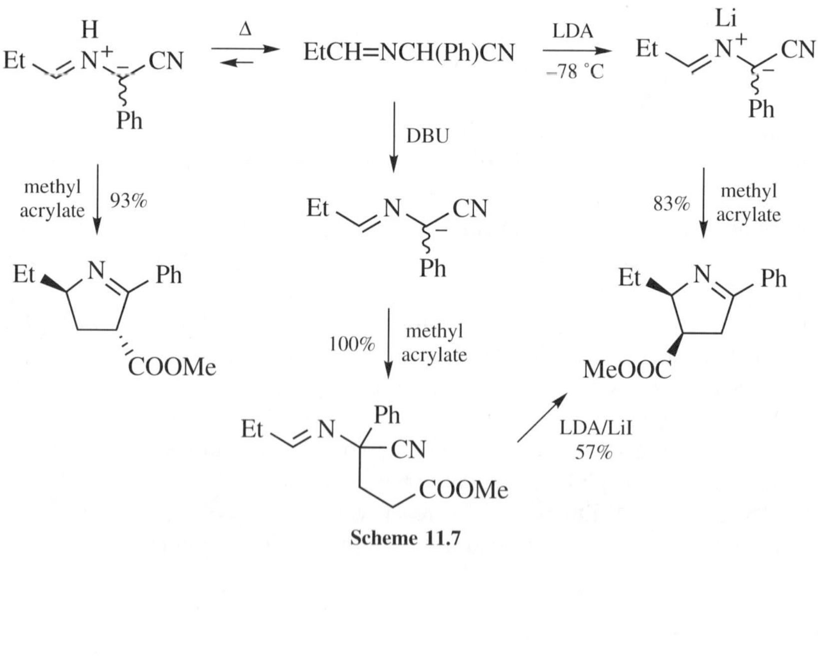

Scheme 11.7

ylide generated from (benzylideneamino)acetonitrile and ethylmagnesium bromide is exclusively the (*E,Z*) isomer, while poor (*E,Z*)-/(*E,E*)-selectivity results when butyllithium is used.

Thus, N-metalated azomethine ylides of the cyano-stabilized type are important in the stereoselective synthesis of pyrrolines. These reactive intermediates are regarded as synthetic equivalents of nonstabilized nitrile ylides.

11.1.4.2. Reversible Metalation of α-(Alkylideneamino)esters

N-Metalated azomethine ylides generated from α-(alkylideneamino) esters can exist as tautomeric forms of the chelated ester enolate (Scheme 11.8). On the basis of the reliable stereochemical and regiochemical selectivities described below, it is clear that the N-metalated tautomeric contributor of these azomethine ylides is important. Simple extension of the above irreversible lithiation method to α-(alkylideneamino) esters is not very effective, and cycloadditions of the resulting lithiated ylides to α,β-unsaturated carbonyl compounds are not always clean reactions. When the α-(alkylideneamino) esters bear a less bulky methyl ester moiety, or when α,β-unsaturated carbonyl compounds are sterically less hindered, these species suffer from nucleophilic attack by the organometallics, or the metalated cycloadducts undergo further condensation reactions (81–85).

The most convenient access to N-metalated azomethine ylides from α-(alkylideneamino) esters is their reversible generation by the combined use of lithium bromide and triethylamine (LiBr/NEt_3) (86). Thus, treatment of methyl *N*-benzylideneglycinate with LiBr/NEt_3 in THF at room temperature generates the N-lithiated ylide, which is trapped with N-methylmaleimide to give the endo-cycloadduct in 83% yield (Scheme 11.9). Lithium bromide works as a weak Lewis acid to associate to both the carbonyl oxygen and the imine nitrogen of the methyl *N*-benzylideneglycinate. Thus, the acidity of the α-hydrogen is enhanced so as to be readily deprotonated by such a weak base such as triethylamine, leading to the (*E,E*)-ylide. The lithiated ylide undergoes endo-selective cycloadditions to a variety of α,β-unsaturated esters and ketones to afford proline ester derivatives (86). In all these cycloadducts, the phenyl moiety derived from the ylide is cis to both the ester group (COOMe) derived from the ylide and the carbonyl group (EWG) derived from the acceptor. Although the reaction of the enolate with N-methylmaleimide is

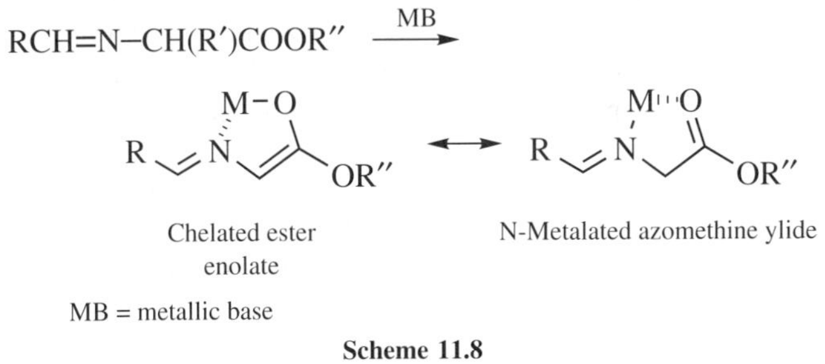

Scheme 11.8

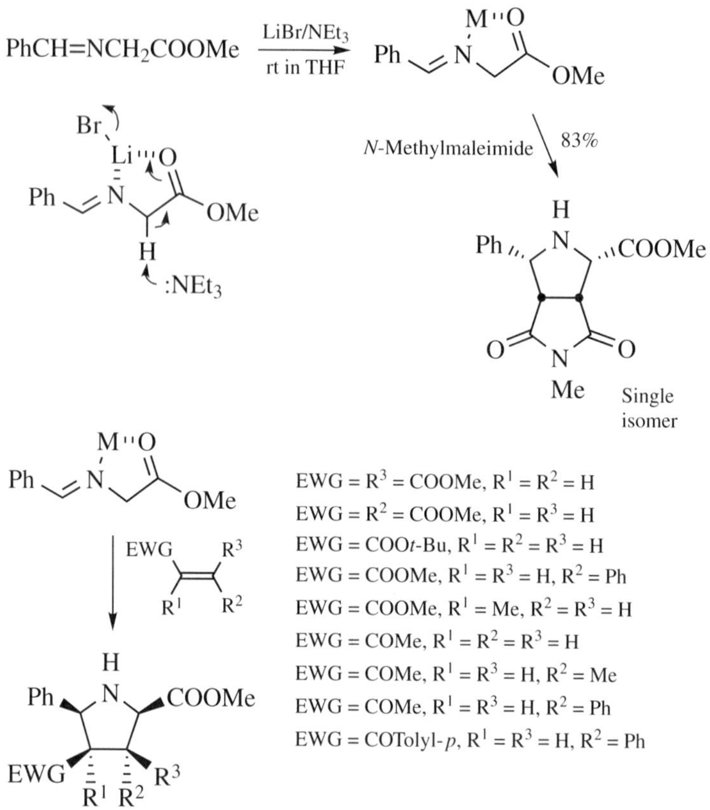

Scheme 11.9

also exclusively endo-selective as mentioned above, acrylonitrile shows poor stereoselectivity (endo/exo = 29: 71). Thus, the α,β-unsaturated carbonyl functionality is essential for the highly endo-selective cycloadditions.

In general, LiBr and NEt_3 are employed in 1.5 and 1.2 equiv, respectively. Although the reaction becomes rather slower, catalytic amounts of LiBr/NEt_3 (0.1 equiv each) are also sufficient. In reactions with the highly reactive dipolarophile N-methylmaleimide, the catalytic reaction results in a better yield. A similar lithiation is possible with α-substituted (alkylideneamino)acetates and (alkylideneamino)-acetamides to generate lithium enolates (86). Cycloadditions with a variety of α,β-unsaturated carbonyl compounds leads to endo cycloadducts. However, the reaction with acrylonitrile is again nonstereoselective.

When the α-substituent is methyl (R = Me), deprotonation occurs readily with NEt_3 at room temperature. Although the cycloaddition step of the resultant ylide with sterically less hindered or reactive dipolarophiles (*N*-methylmaleimide, methyl acrylate, methyl methacrylate, and dimethyl fumarate) proceeds faster than the ylide generation step, the cycloaddition step becomes rate determining if sterically

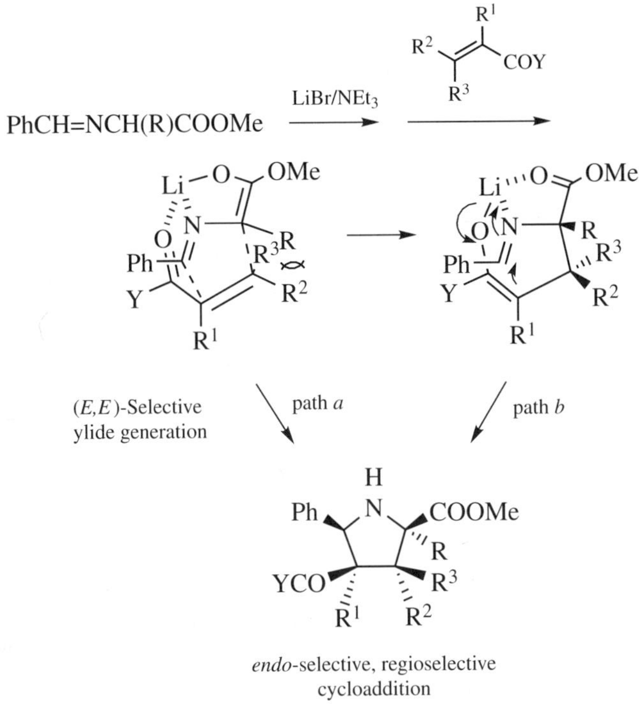

Scheme 11.10

hindered methyl crotonate is employed. A rate acceleration is observed using DBU instead of NEt_3 in the latter case (82). When the α-substituent is a bulky group such as isopropyl ($R = i$-Pr), DBU rather than NEt_3 is also needed for the ylide formation. Although the resulting ylide undergoes a smooth cycloaddition with methyl methacrylate, methyl crotonate is no longer useful as a dipolarophile. This decreased reactivity is due to steric repulsion between the β-methyl group of the crotonate and the isopropyl group of the ylide.

The reaction mechanism proposed for the $LiBr/NEt_3$ induced azomethine ylide cycloadditions to α,β-unsaturated carbonyl acceptors is illustrated in Scheme 11.10. The (*E,E*)-ylides, reversibly generated from the imine esters, interact with acceptors under frontier orbital control, and the lithium atom of ylides coordinates with the carbonyl oxygen of the acceptors. Either through a direct cycloaddition (path a) or a sequence of Michael addition–intramolecular cyclization (path b), the cycloadducts are produced with endo- and regioselectivity. Path b is more likely, since in some cases Michael adducts are isolated.

Treatment of 2-(benzylideamino)propanoate with NEt_3 and metal halides other than LiBr leads to N-metalated azomethine ylides (Scheme 11.11) (87). Although N-sodiated ylides generated by using NaI/NEt_3 afford the Michael adduct as the major product in reactions with methyl crotonate, other metalated ylides produce

Reversible Generation

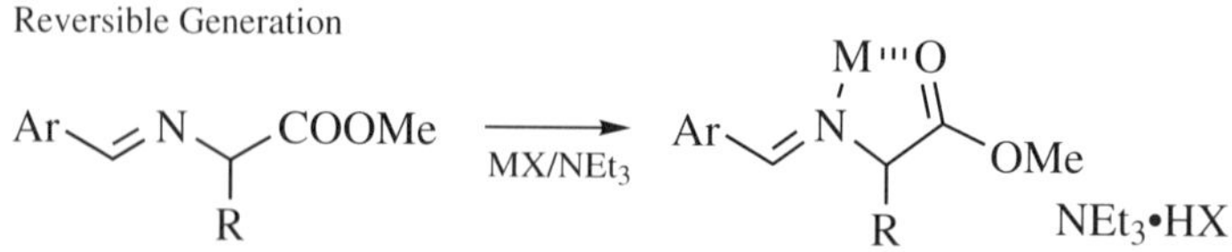

(10h): Ar = R = Ph
(27a): Ar = 2-Naph, R = H
(27b): Ar = 2-Naph, R = Me

Ar = 2-Naph, R = H, M = Li
Ar = R = Ph, M = Li
Ar = Ph, R = Me, M = ZnI
Ar = R = Ph, M = ZnBr
Ar = Ph, R = Me, M = $Al(Et)_2$
Ar = Ph, R = Me, M = MgBr
Ar = Ph, R = Me, M = Ag
Ar = 2-Naph, R = H, M = Ag
Ar = R = Ph, M = Ag
Ar = 2-Naph, R = H, M = MnBr
Ar = 2-Naph, R = Me, M = MnBr

Scheme 11.11

the endo-cycloadducts as the only products. Similar reversible generation of a variety of metal enolates using NEt_3/LiBr, LiI, LiOAc·H_2O, $ZnBr_2$, AgOAc, $AgNO_3$, Ag_2CO_3, Ag_2·tartrate, $MgBr_2 \cdot Et_2O$, and $MnBr_2$ are now known (88,89), where THF, MeCN, dimethylformamide (DMF), dimethyl sulfoxide (DMSO), and N-methylacetamide are used as the reaction solvents.

Methyl (naphthylideneamino)acetate undergoes dimerization to produce a mixture of two diastereomeric imidazolidines when treated with $Mg(ClO_4)_2$ or $CoCl_2$ (89). Other imines can also be used as acceptor molecules (Scheme 11.12). With the exception of azomethine ylides incorporating sodium and titanium ions, other N-metalated ylides undergo highly endo-selective cycloadditions with α,β-unsaturated

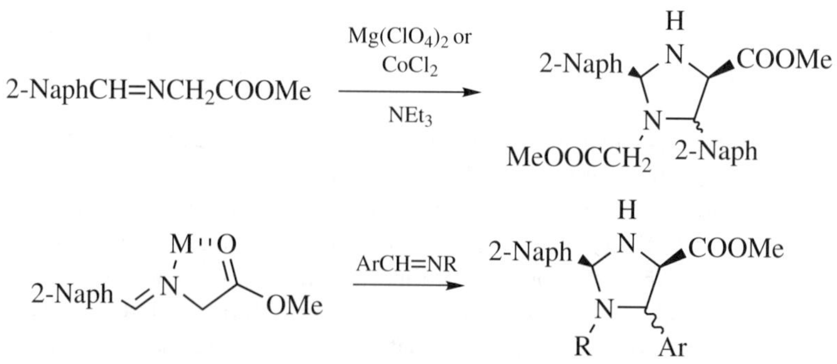

M = $MgClO_4$
M = CoCl
M = ZnBr
2-Naph = 2-Naphthyl

Ar = 2-Naph, R = Me
Ar = 2-Naph, R = $PhCH_2$
Ar = Ph, R = $PhCH_2$

Scheme 11.12

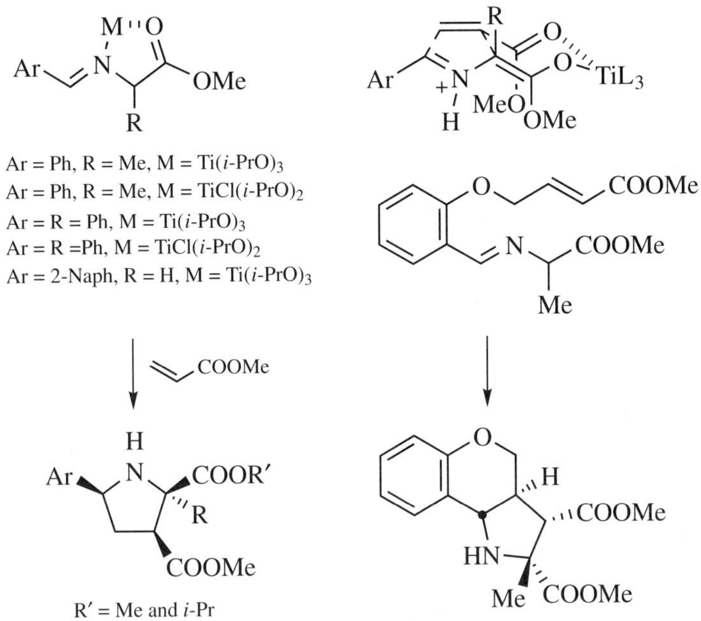

Scheme 11.13

carbonyl compounds. With the exception of titanium ylides (see Scheme 11.13 and 11.14), the stereoselectivity of the cycloadditions is independent of the metal ion.

Titanium ylides are generated from imine esters with titanium isopropoxide chlorides and amines or by transmetalation of the N-lithiated ylides (90,91). The regioselectivity of their reactions with methyl acrylate is opposite to that normally observed (90). A transition state is proposed in Scheme 11.13 to explain this alternative regioselectivity. Intramolecular cycloadditions of the titanium ylides offer a synthetic application of this regioselectivity.

With methyl acrylate and methyl methacrylate bearing no substituent at the electrophilic 3-position, cycloadditions at a higher temperature [room temperature (rt)] lead to the exo-cycloadducts having an ester group trans to the adjacent 2-phenyl group. On the other hand, at a lower temperatures (−78 to −20 °C), regioisomeric cycloadducts having the ester group cis to the adjacent 2-CO_2t-Bu group are the only products (Scheme 11.14). It is most likely that the selectivity is dependent of the stereochemistry of the reacting intermediate ylides. The structure seems to change with a change of the reaction temperature at which the metalation is carried out. Thus, the intermediate generated at −78 °C, probably a chelated ylide structure, reacts with methyl crotonate to give the endo-cycloadduct at −78 °C or a mixture of endo- and exo-cycloadducts at room temperature. The intermediate generated at room temperature, probably a nonchelated ylide like the transition state shown in Scheme 11.13, gives the regioisomeric exo-cycloadducts regardless of the reaction temperature.

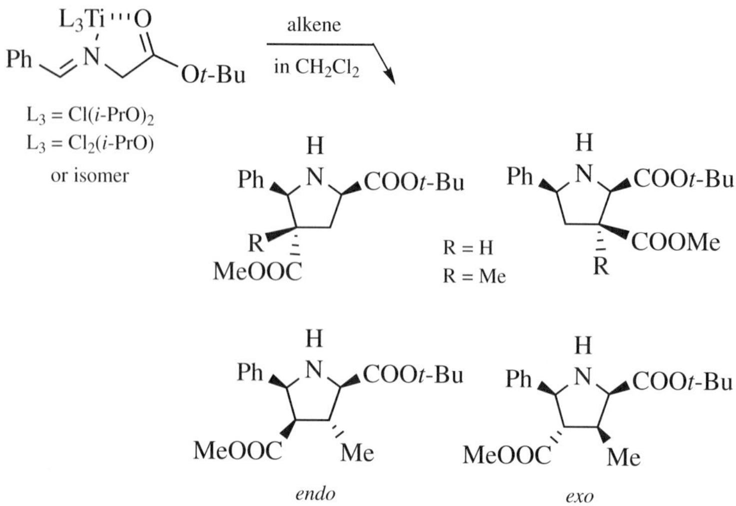

Scheme 11.14

11.1.4.3. *Anti-Selective Michael Additions*

When 2-(benzylideneamino)propanoate is treated with methyl acrylate in the presence of LiBr/NEt_3 in THF, the Michael adduct (18%) is formed as a minor product together with the cycloadduct (79%) as a major product in a stereoselective fashion (Scheme 11.15).

Based on this result as well as related investigations, the following have been unveiled (87): (a) Under equilibrating conditions (use of LiBr/NEt_3), the cycloaddition and the Michael addition path compete each other via the common intermediate **A** (Scheme 11.16); (b) the intermediate **B** is stereoselectively formed through the chelation- and frontier orbital controlled transition state **A**; (c) the

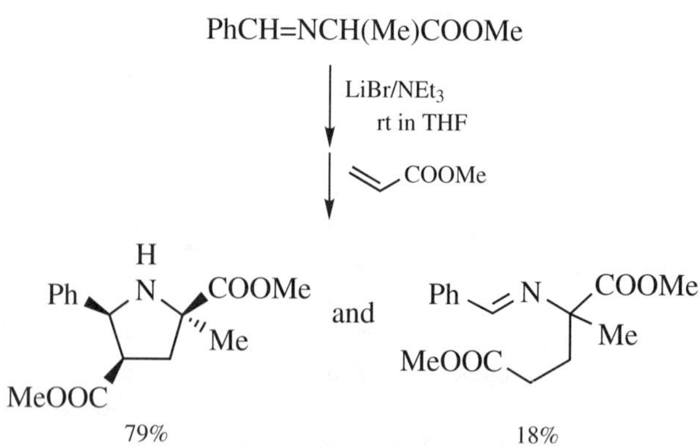

Scheme 11.15

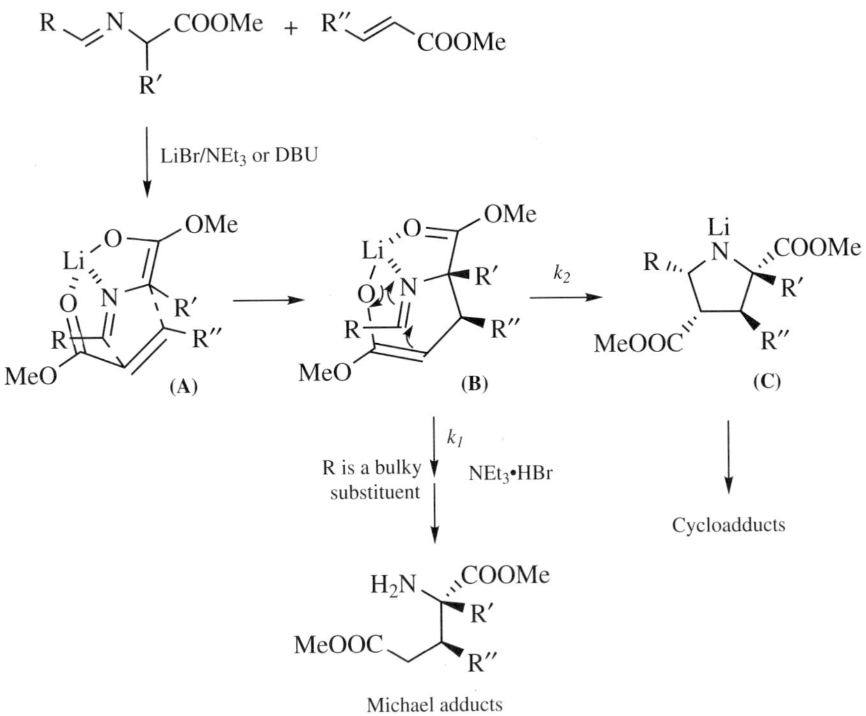

Scheme 11.16

product ratio depends on the ratio of reaction rates k_1/k_2; (d) the cycloaddition path is favored when the reaction is performed in a dilute solution because quenching of **B** with $NEt_3{\cdot}HBr$, producing Michael adducts, is an intermolecular process and cyclization of **B** (imine/aldol reaction) leading to cycloadducts is an intramolecular process; (e) use of a base stronger than NEt_3 (e.g., DBU) results in higher concentrations of both **B** and the ammonium salt quencher (DBU·HBr), so that the Michael addition path becomes more favored, and (f) no Michael adducts are formed when the enolates are generated under irreversible conditions.

The above reaction mechanism indicates that the Michael addition path should become the major one if the lithiated ylide derived from a bulky aldehyde is employed (e.g., 2,2-dimethylpropanal). The lithium or magnesium ylides of methyl (2,2-dimethylpropylideneamino)acetate, when generated by the irreversible bases LDA or *t*-BuMgCl, undergo stereoselective cycloadditions with methyl crotonate (Scheme 11.17) (82). On the other hand, under the equilibrating conditions using LiBr/DBU, a 7:6 mixture of diastereomeric Michael adducts is produced after 4 h at room temperature. The latter is a result of overreaction under the equilibrating conditions. When the same reaction is quenched at short reaction time (10 min), the anti-adduct is the only isomer isolated in 77% yield (82). Thus, reactions of (2,2-dimethylpropylideneamino)acetate with α,β-carbonyl compounds under equilibrating conditions offer a convenient method for anti-selective Michael additions.

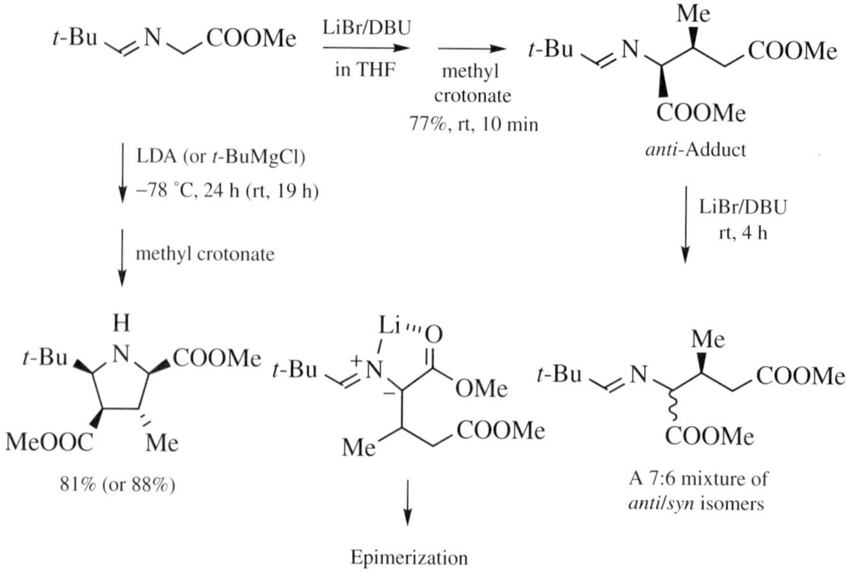

Scheme 11.17

The anionic intermediate formed in the Michael reaction, formed through a rigid transition state, undergoes a sluggish intramolecular cyclization due to serious steric interaction between the *tert*-butyl and methoxyl groups, instead giving rise to the acyclic materials upon quenching with DBU·HBr Since this product bears an another acidic hydrogen, it is again lithiated reversibly with LiBr/DBU to cause epimerization. This epimerization occurs more readily when the imine proton to be abstracted is sterically less hindered by the β-substituent derived from the acceptor molecule (Scheme 11.18). For example, the Michael adduct from methyl acrylate yields the 1:2 adduct as the sole product. The initially formed anti-adducts do not suffer from epimerization when the α-position of the N-substituent the of imines is already substituted. For example, the reaction of the alanate imine with methyl crotonate is exclusively anti-selective (82).

The imine protecting group of the anti-selective Michael adducts is readily removed by treatment with aqueous acetic acid. The resulting α-amino esters undergo a spontaneous intramolecular condensation to give the corresponding 5-oxopyrrolidine-2-carboxylates (Scheme 11.19). Michael additions to a variety of α,β-unsaturated ketones result in the exclusive formation of anti-isomers of the adducts if the reaction conditions are carefully set up to avoid epimerization (82). When a bulky *tert*-butyl substituent is attached at the β-position of acceptor, the Michael addition requires a much longer reaction time (1 h at −5 °C). No epimerization takes place under these conditions. High steric hindrance by the *tert*-butyl group effectively inhibits further lithiation. Hydrolytic removal of the alkylidene protecting group produces 4,5-*trans*-1-pyrroline-5-carboxylates in a stereospecific fashion.

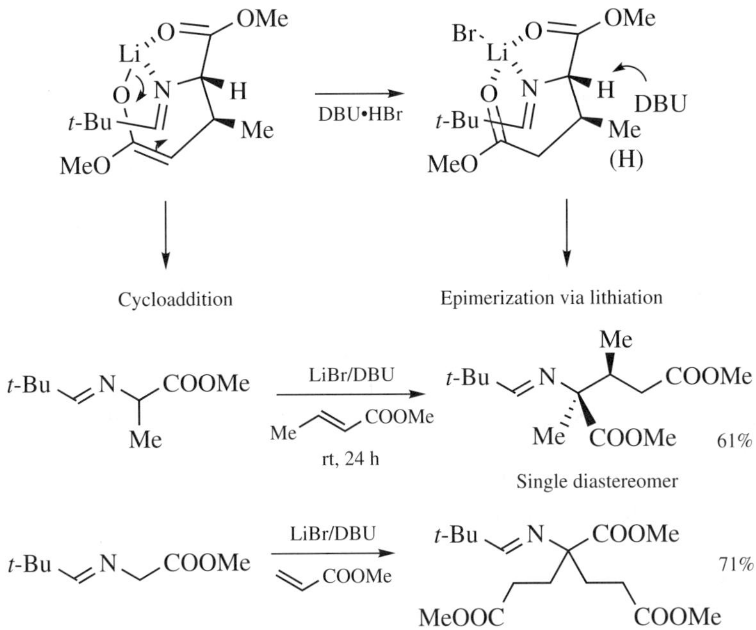

Scheme 11.18

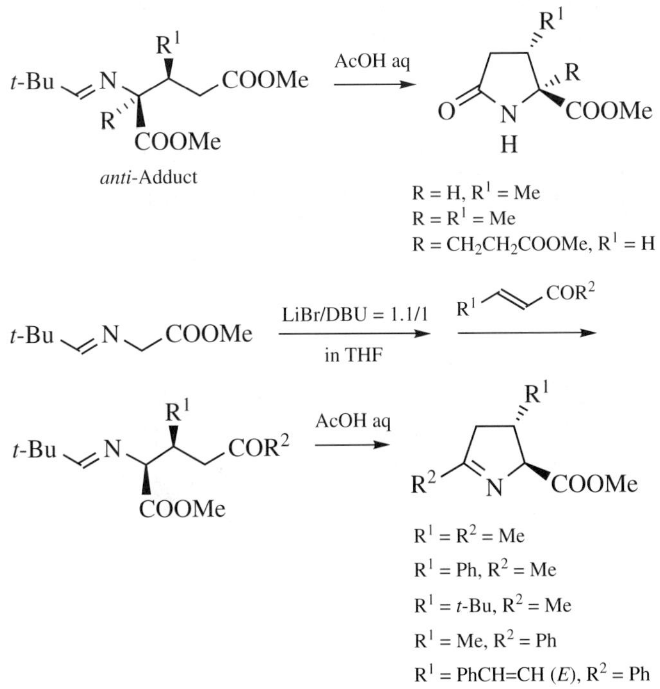

Scheme 11.19

11.1.4.4. Asymmetric Azomethine Ylide Cycloadditions

Although the first attempts at asymmetric azomethine ylide cycloadditions were reported by Padwa's group (92), the acyclic azomethine ylides chosen, bearing an α-chiral alkyl substituent on the nitrogen, showed poor diastereoselectivities (93,94). When the chiral center is fixed in a cyclic structure (95) or when chirality is introduced in an intramolecular cycloaddition system (96–98), high selectivities have been accomplished. There are only a few examples known of asymmetric cycloadditions of achiral azomethine ylides to chiral dipolarophiles where cyclic azomethine ylides (99,100) or cyclic chiral dipolarophiles (94) were used.

Enhanced reactivity as well as high endo-selectivity based on the rigid transition structure of N-metalated azomethine ylides is attractive for asymmetric 1,3-dipolar cycloaddition reactions. There are several reports known for the design of effective chiral nucleophiles in asymmetric cycloadditions.

A new dipolarophile bearing a chirality-controlling heterocyclic auxiliary at the β-position is readily accessible from (*S*)-*N*-benzylvalinol and methyl (*E*)-4-oxo-2-propenoate. However, the dipolarophile is available only as an 86:14 equilibrium mixture of trans and cis stereoisomers (Scheme 11.20) (84). When this is used without separation in the reaction with the *N*-lithiated azomethine ylide derived from methyl (benzylideneamino)acetate in THF at −78 °C for 3.5 h, a mixture of two diastereomeric cycloadducts (75:25) was obtained in 82% yield. These two cycloadducts are derived from the trans and cis isomers of acceptor, indicating that both cycloadditions were highly diastereoselective.

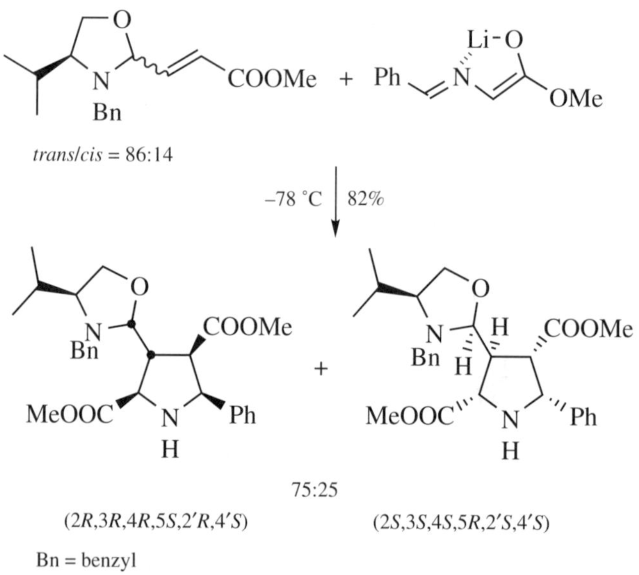

Scheme 11.20

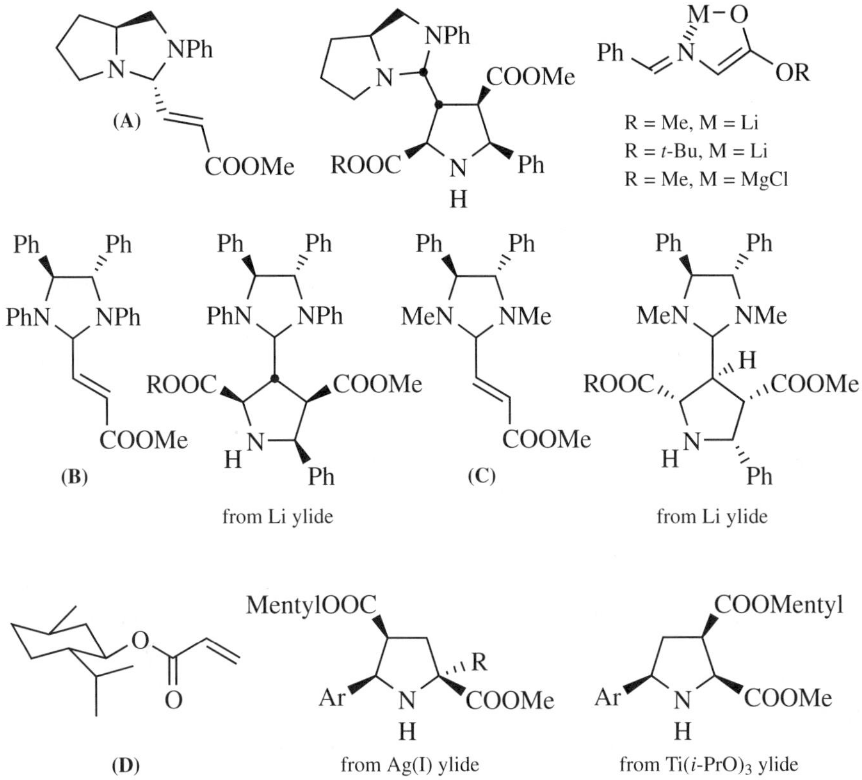

Scheme 11.21

The α,β-unsaturated ester (3*R*,7a*S*)-**A** bearing a perhydropyrrolo[1,2-*c*]imidazole chiral auxiliary at the β-position is available from (*S*)-2-(anilinomethyl)pyrrolidine and methyl (*E*)-4-oxo-2-propenoate (101). Its cycloaddition reactions with N-metalated azomethine ylides are exclusively diastereoselective, regardless of the metal atom M contained in metal enolates, the size of the alkoxy substituents, the reaction temperatures, and the methods of enolate generation (Scheme 11.21). Condensation of the enantiomers of 1,2-dianilino-1,2-diphenylethane (81) and 1,2-bis(methylamino)-1,2-diphenylethane with methyl (*E*)-4-oxo-2-propenoate gives the α,β-unsaturated esters **B** and **C** bearing a C_2-symmetric imidazolidine chiral auxiliary at the β-position. Reactions of (4*S*,5*S*)-**B** with the lithium or magnesium ylides derived from methyl (benzylideneamino)acetate show diastereoselectivities in the range of 96:4–85:15 depending on the reaction temperatures and the methods used to generate the enolates, the major products being the (2*R*,3*R*,4*R*,5*S*)-stereoisomer. On the other hand, reactions of the same ylides with (4*S*,5*S*)-**C** shows the opposite diastereoselectivity to the above case to give the (2*S*,3*S*,4*S*,5*R*)-stereoisomer as the sole products, regardless of the reaction conditions. Grigg and co-workers (102) independently reported the asymmetric cycloadditions of silver enolates derived from α-(alkylideneamino) esters and AgOAc/NEt_3. They employed both enantiomers of menthyl acrylate **D**. With (1*R*,2*S*,5*R*)-**D**, the *re* face exclusively participates in the attack by the silver enolates to produce diastereoselective pyrrolidines. In the cycloadditions of both enantiomers with the titanium enolates generated from methyl (2-naphthylideneamino)acetate and Ti(OPr-*i*)$_3$Cl/NEt_3, single diastereomers of 2,3-pyrrolidinedicarboxylate with an opposite regiochemistry are produced.

11.1.4.5. Asymmetric Michael Additions of Enantiopure Camphor Imines

Enantiocontrolled anti-selective Michael additions of the metalated ylides derived from α-(alkylideneamino)alkanoates are attractive as a new synthetic route to enantiomers of α-amino esters. Although there are a variety of chiral glycine equivalents available, only the enolates derived from 1,4-dihydropyrazine have been successfully applied to asymmetric Michael additions (Scheme 11.22) (103–105). In these reactions, the diastereoselectivities are high.

One problem in the anti-selective Michael additions of *N*-metalated azomethine ylides is ready epimerization after the stereoselective carbon–carbon bond formation. The use of the camphor imines of α-amino esters should work effectively because camphor is a readily available bulky chiral ketone. With the camphor auxiliary, high asymmetric induction as well as complete inhibition of the undesired epimerization is expected. The lithium enolates derived from the camphor imines of α-amino esters have been used by McIntosh's group for asymmetric alkylations (106–109). Their Michael additions to some α,β-unsaturated carbonyl compounds have now been examined, but no diastereoselectivity has been observed (108). It is also known that the *N*-pinanylidene-substituted α-amino esters function as excellent Michael donors in asymmetric Michael additions (110). Lithiation of the camphor

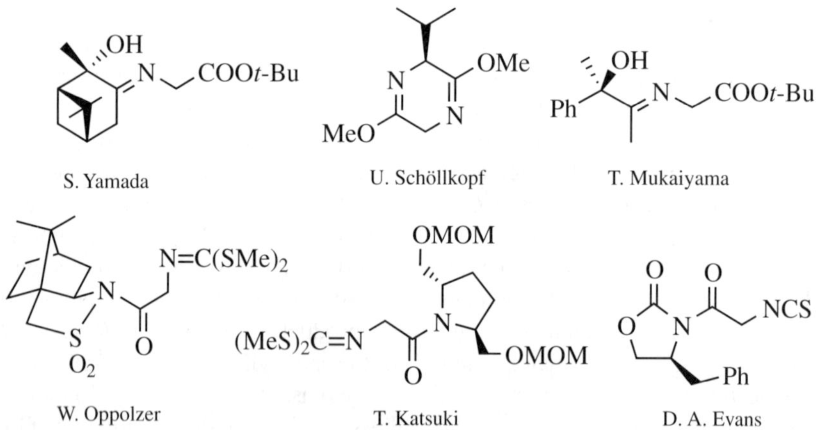

Scheme 11.22

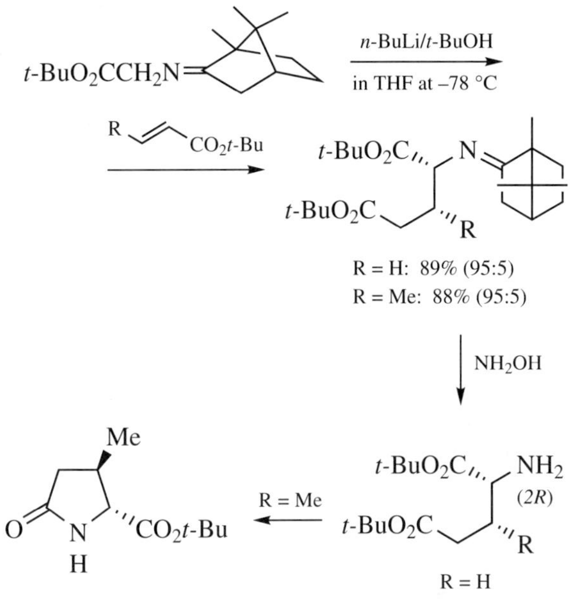

Scheme 11.23

imine of *tert*-butyl glycinate with butyllithium at $-78\,^{\circ}C$ in THF followed by treatment with *tert*-butyl acrylate or crotonate gives the Michael adducts in high diastereoselectivities of 95:5 in both cases (Scheme 11.23) (111). The camphor imine of ethyl glycinate can be reversibly lithiated with LiBr/NEt_3 or LiBr/DBU in THF and the Michael addition to methyl acrylate is completed at $0\,^{\circ}C$ in a few minute (ds = 82:18), while that to methyl crotonate is extremely sluggish (ds = 74:26). To attain high diastereoselectivity, the lithiated ylide generated with butyllithium has to be pretreated with *t*-BuOH (1 equiv) at $-78\,^{\circ}C$ in THF (111). It should be emphasized that epimerization at the 2-position of the Michael adducts is totally inhibited under these reaction conditions. The α-position is sterically protected by the bulky camphor moiety from attack by the bases employed for the enolate generation. Even a strong organometallic base such as butyllithium cannot deprotonate at this position.

Alkylidenemalonates were found to be excellent acceptor molecules (111). Reactions of lithium ylides with dimethyl alkylidenemalonates at $-78\,^{\circ}C$ in THF in the presence of *t*-BuOH were diastereoselective for all the substituents R except methyl, producing Michael adducts as single diastereomers (Scheme 11.24). The only exception was dimethyl ethylidenemalonate, which produces an 86:14 mixture of diastereomeric adducts, the minor diastereomer being syn-adduct. Since dimethyl alkylidenemalonates bear two geminal methoxycarbonyl moieties, one is cis to the terminal substituent R and the other trans, so of these ester substituents can participate in chelate formation in the transition state. When the terminal substituent R is small, there is a chance for the syn-adduct to be produced, which

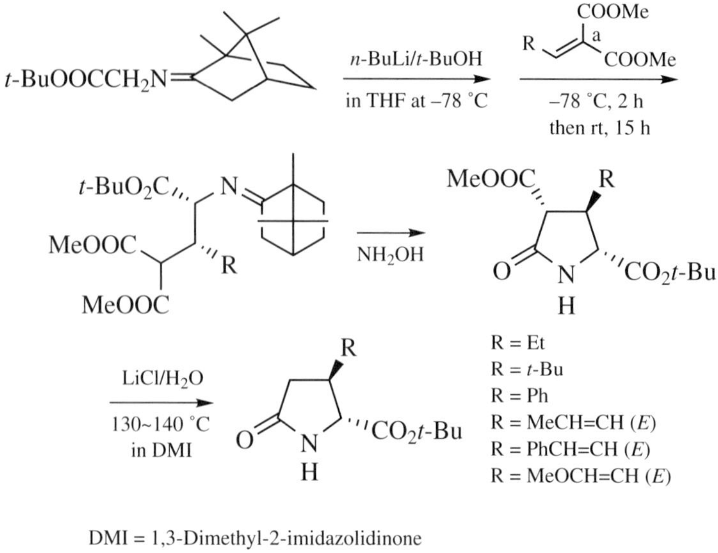

Scheme 11.24

has actually been observed. One of the ester moieties derived from the alkylidenemalonate can be removed after the Michael addition so that alkylidenemalonates can be regarded as the synthetic equivalents of α,β-unsaturated esters.

The lithium ylide generated from methyl (2,2-dimethylpropylideneamino)acetate and LiBr/DBU reacts with an α,β-unsaturated ester bearing a perhydropyrrolo[1,2-*c*]imidazole chiral auxiliary at the β-position at −78 °C in THF to give a quantitative yield of the Michael adduct as a single diastereomer (Scheme 11.25) (112). When the reaction is performed at room temperature, the diastereoselectivity decreases to 95:5 and some epimerization is observed. Accordingly, epimerization is not a serious problem as long as the reaction temperature is kept at or below −78 °C.

Ethyl (4*S*,2*E*)-3-(2,2-dimethyl-1,3-dioxolan-4-yl)acrylate (113,114) has been widely used as a chiral α,β-unsaturated ester bearing a five-membered dioxolane chiral auxiliary at the β-position in asymmetric reactions (115–117), including cycloadditions (94,118–120). Although one example of asymmetric dipolar cycloaddition using a nonstabilized azomethine ylide is known, the diastereoselectivity observed is low (94). Use of the same acceptor in reactions with the N-lithiated ylide generated from ethyl (2,2-dimethylpropylideneamino)acetate and LDA, at −78 °C in THF in the presence of *t*-BuOH gives a mixture of the Michael adduct and the cycloadduct in a stereoselective fashion (Scheme 11.25) (121). When the crude mixture is treated with NH_2OH in ethanol, the lactam (40%) and the unreacted cycloadduct (9%) are obtained after *N*-Boc protection, both as single diastereomers. Based on the absolute configurations, the Michael addition has arisen from reaction of the *si*-face of acceptor with the lithium ylide. Such

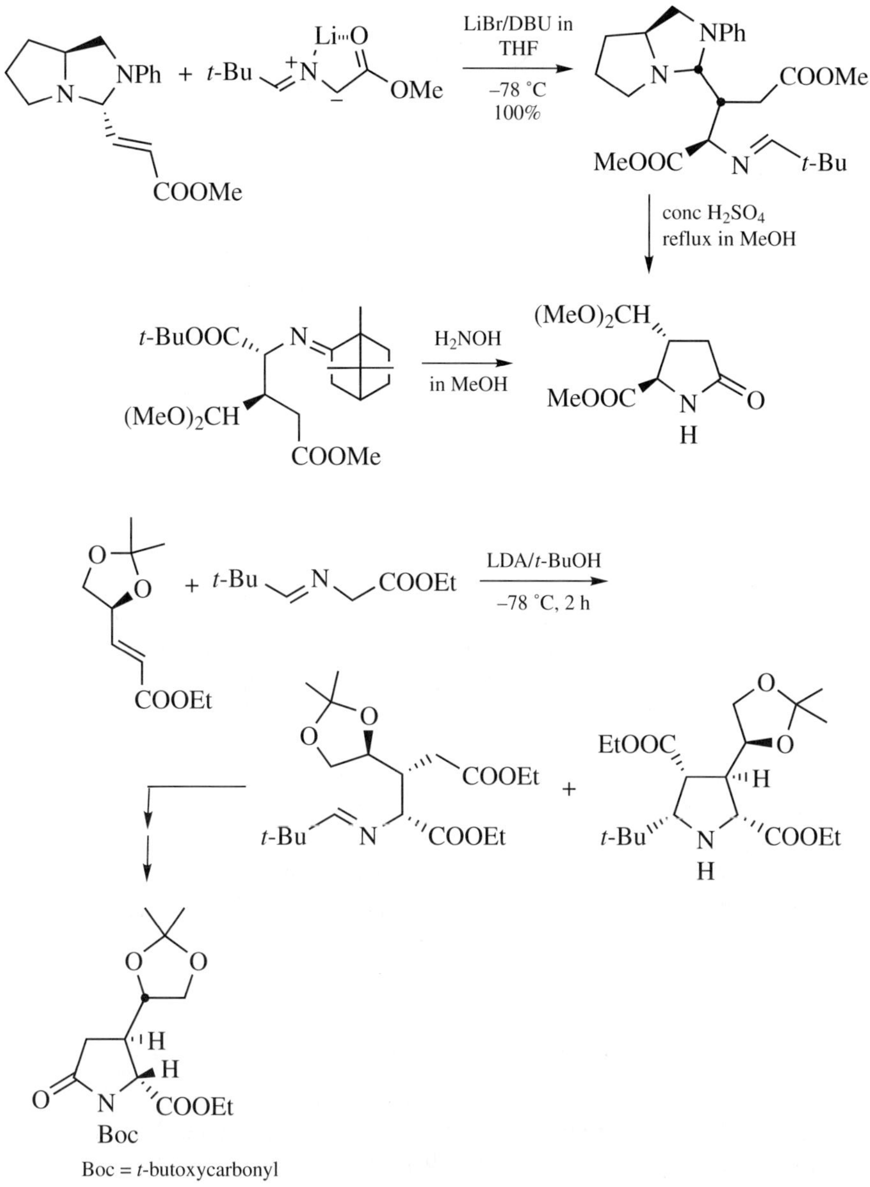

Scheme 11.25

si-face attack by a nucleophile is very unusual in asymmetric reactions using this acceptor (94).

The double chiral-inductive asymmetric Michael addition of the camphor-based lithium ylide to 3-(4*S*-2,2-dimethyl-1,3-dioxolan-4-yl)acrylate produces exclusively the

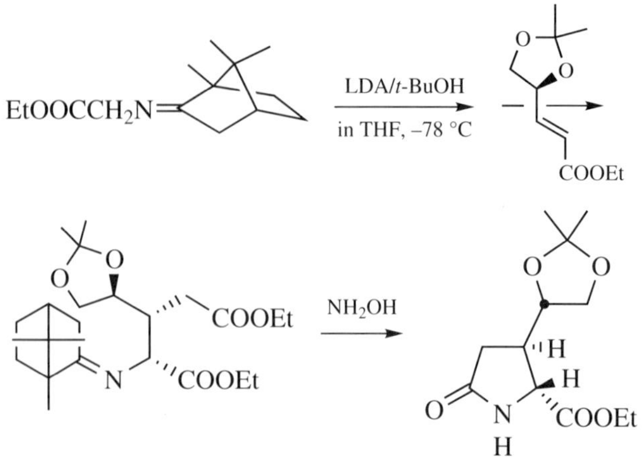

Scheme 11.26

(2*R*,3*R*)-diastereomeric product (73%), which corresponds to an adduct formed via an attack of the ylide to the *si*(Cβ)-face of the acceptor (Scheme 11.26) (122,123). Such a combination is a matching pair, since the (1*R*,4*R*)-ylide has participated predominantly in the Michael addition using the racemate of the camphor-based lithium ylide (2 equiv) to ethyl 3-(4*S*-2,2-dimethyl-1,3-dioxolan-4-yl)acrylate to give the enantiopure adduct in 81% yield together with the unreacted ylide, which mainly consists of (1*S*,4*S*)-enantiomer.

High diastereofacial selectivities are observed in cycloadditions and Michael additions with α,β-unsaturated esters having chiral heterocyclic auxiliary at the β-position, as shown in Schemes 11.20, 11.21, and 11.25, and cannot be well-explained using Kozikowski's "*anti*-periplanar model" (124,125) or Houk's "inside alkoxy model" (126,127). Both the anti-periplanar conformation and the syn-periplanar conformation of the acceptors participate in the transition structures, depending on nonbonding interactions in the dipole–chiral auxiliary pair (121).

11.1.4.6. Enantioselective Reactions

The above azomethine ylide cycloadditions have been extended to an enantio-selective version involving amino alcohols both as chiral ligands and amine bases. Thus, reactions of the N-metalated azomethine ylides derived from achiral methyl 2-(arylmethyleneamino)acetates, cobalt(II) chloride [or manganese(II) bromide], and chiral amino alcohols, 1 and 2 equiv each, with methyl acrylate as solvent have been performed to provide the enantiomer-enriched pyrrolidine-2,4-dicarboxylates with the enantioselectivities of up to 96% enantiomeric excess (ee) (128,129). However, a large excess of the metal ions and the chiral source (ligand and base) have to be employed.

11.2. NITRILE OXIDES

11.2.1. Introduction

1,3-Dipolar cycloaddition reactions between nitrile oxides and alkenes produce 2-isoxazolines. Through reductive cleavage of the N—O bond of the 2-isoxazolines, the resulting heterocycles can be readily transformed into a variety of important synthetic intermediates such as β-hydroxy ketones (aldols), β-hydroxy esters, α,β-unsaturated carbonyl compounds, γ-amino alcohols, imino ketones and so forth (7–12).

There are some other synthetic advantages in nitrile oxide cycloadditions: (a) This reaction takes place in an stereospecific manner, so that geometry of the starting alkenes is completely expressed in the cycloadducts; (b) the above building blocks can be constructed in stereospecific fashion starting from sterically defined alkenes; (c) nitrile oxides are highly reactive 1,3-dipoles that react with many kinds of monosubstituted alkenes to produce 5-substituted-2-isoxazolines in a regioselective fashion; and (d) hydroximoyl halides are common nitrile oxide precursors, and are readily available, as are the alkene reaction partners. Thus, nitrile oxide cycloadditions with alkenes provide a general route to 2-isoxazolines (9,10).

Control of reaction selectivities with external reagents has been quite difficult. Unsolved problems remaining in the field of nitrile oxide cycloadditions are (a) Nitrile oxide cycloadditions to 1,2-disubstituted alkenes are sluggish, the dipoles undergoing facile dimerization to furoxans in most cases; (b) the reactions of nitrile oxides with 1,2-disubstituted alkenes nonregioselective; (c) stereo- and regiocontrol of this reaction by use of external reagents are not yet well developed; and (d) there are few examples of catalysis by Lewis acids known, as is true for catalyzed enantioselective reactions.

No reliable methods for stereo- and regiocontrol of nitrile oxide cycloaddition reactions to alkene dipolarophiles were available before 1991. The first successful example for the control of nitrile oxide cycloadditions by the use of a metallic external reagent was reported by Kanemasa et al. in 1994 (130). His group developed a new method for the generation of nitrile oxide–Lewis acid complexes involving the treatment of hydroximoyl chlorides with organometallic bases such as butyllithium, ethylmagnesium bromide, and diethylaluminum chloride (131). Therein, metalation of the highly acidic hydroximoyl chloride hydrogen was followed by 1,3-elimination of the metal chlorides. The resulting nitrile oxides and metal chlorides, which behave as Lewis acids, coordinate to each other. Thus, the benzonitrile oxide–MgBrCl complex is formed when benzhydroximoyl chloride is treated with a ethylmagnesium bromide. Reaction of this magnesium complex with methyl 2-hydroxymethylacrylate shows reversal of the diastereoselectivity of the cycloaddition as compared to free benzonitrile oxide.

11.2.2. Magnesium Ion-Mediated Reactions

11.2.2.1. Diastereocontrol of Nitrile Oxide Cycloadditions

Kanemasa et al. (130) found that nitrile oxides show high reactivity with allylic alcohols when magnesium ions are present in the reaction. For example, without magnesium ion, the reaction of benzonitrile oxide with 1-penten-3-ol is slow and nonstereoselective, a 67:37 mixture of syn- and anti-cycloadducts being produced in a combined yield of 40%. On the other hand, when the reaction is run with the magnesium salt of 1-penten-3-ol in dichloromethane, a quantitative yield of the syn-cycloadduct is produced as the sole product (Scheme 11.27). This highly syn-selective reaction is applicable to a variety of α-substituted allylic alcohol substrates. This rate enhancement and high diastereoselectivity is specific to magnesium ions; other metal ions such as lithium, zinc, and aluminum are ineffective, showing little improvement of stereoselectivity. The presence of alcohol additives such as isopropyl alcohol, if in small excess, does not affect the reaction, except that the reaction slightly accelerated (130,131).

11.2.2.2. Regiocontrol of Nitrile Oxide Cycloadditions

Although monosubstituted alkenes usually show relatively high reactivity in nitrile oxide cycloadditions to give 5-substituted-2-isoxazolines in a regioselective

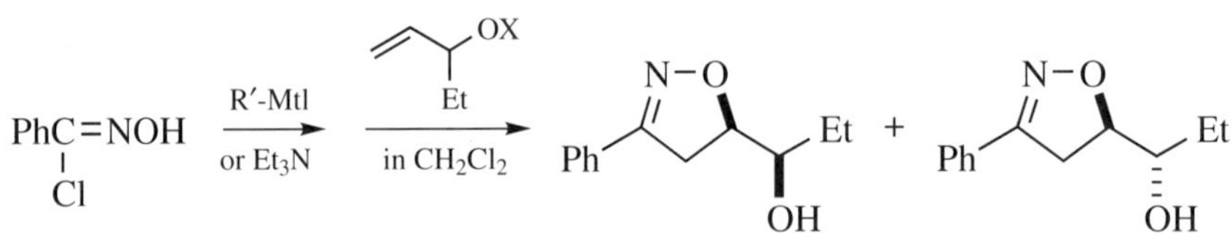

syn-Isomer *anti*-Isomer

Mtl: metal

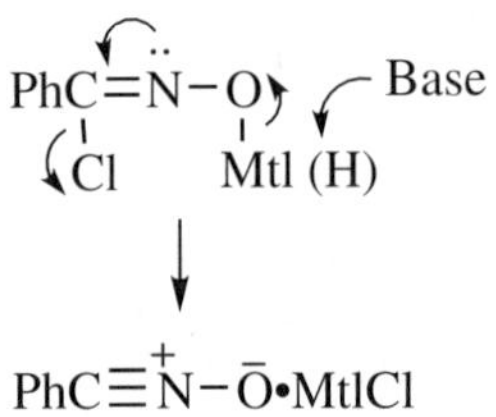

Nitrile oxide / LA complex

Base	X	Yield	Isomer ratio *syn*/*anti*
Et_3N	H	40%	67:33
n-BuLi		54	75:25
Et_3Zn		79	77:23
Et_3Al		32	71:21
EtMgBr		75	95:5
None	MgBr	92	>99:1
		99*	>99:1

**i*-PrOH (2 equiv) was added.

Scheme 11.27

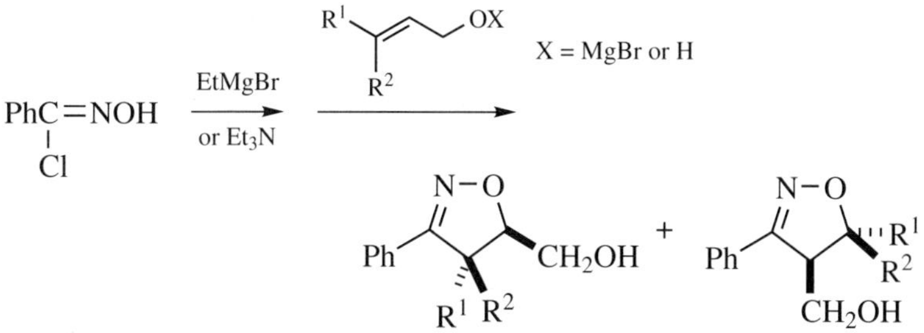

Acceptor	EtMgBr method			Et_3N method		
	X	Yield	Isomer ratio	X	Yield	Isomer ratio
R^1 = Me, R^2 = H	MgBr	82%	>99:1	H	46%	46:54
R^1 = *n*-Pr, R^2 = H		68	>99:1		44	55:45
R^1 = H, R^2 = *n*-Pr		67	96:4		0	
R^1 = Ph, R^2 = H		68	>99:1		46	20:80
R^1 = R^2 = Me		20	97:3		19	1:99

Scheme 11.28

manner, multisubstituted alkene dipolarophiles such as 1,2-disubstituted and 1,1,2-trisubstituted alkenes are much less reactive. For example, the nitrile oxide cycloaddition reaction between benzonitrile oxide and crotyl alcohol is poor, both in terms of reactivity and regioselectivity, producing a 46:54 mixture of regioisomeric cycloadducts in a combined yield of 46% (Scheme 11.28). In reactions with trisubstituted alkenes, the regioselectivity is mainly determined by a steric effect to produce the regioisomeric cycloadducts in which the more substituted olefinic carbon combines with the oxygen atom of nitrile oxides. On the other hand, nitrile oxide cycloadditions with substituted allylic alcohols as their magnesium salts produces 2-isoxazoline-5-methanol regioisomers selectively (Scheme 11.28) (130,132). This high regiocontrol is applicable to a variety of multisubstituted allylic alcohols.

Both the (*E*)- and (*Z*)-isomers of the magnesium salt of 2-hexen-1-ol show high regioselectivities in favor of 2-isoxazolin-5-methanol derivatives. Although cinnamyl alcohol gives the 2-isoxazoline-4-methanol as the major regioisomer under thermal conditions due to the electronic directing effect by the γ-phenyl substituent, the magnesium ion-mediated reaction shows reversal of the regioselectivity to produce the 2-isoxazoline-5-methanol regioisomer exclusively. The thermal reaction with 3-methyl-2-buten-1-ol is completely controlled by steric effects, giving only the 2-isoxazoline-4-methanol product, while completely reversal of the regioselectivity results in the presence of magnesium ions, albeit in low yield.

Since a magnesium alkoxide undergoes rapid ligand exchange with free alcohols even in dichloromethane at ambient temperature, the regiocontrol of nitrile oxide cycloadditions of multisubstituted allylic alcohols can be effectively attained

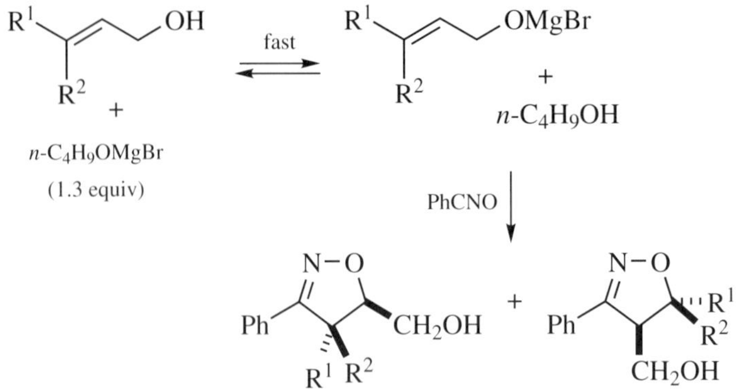

Acceptor	Yield	Isomer ratio
R^1 = Me, R^2 = H	87%	>99:1
R^1 = *n*-Pr, R^2 = H	90	>99:1
R^1 = H, R^2 = *n*-Pr	100	98:2
R^1 = Ph, R^2 = H	92	>99:1
R^1 = R^2 = Me	47	>99:1

Scheme 11.29

simply by adding the magnesium alkoxide of other alcohols (Scheme 11.29). The procedure is simple: a nitrile oxide is generated in dichloromethane from the hydroximoyl chloride precursor and triethylamine in the presence of allylic alcohols, then a slight excess of butoxy magnesium bromide is added. By this modified method, highly regioselective nitrile oxide cycloadditions can be achieved (130,132). Both the regioselectivity and yield are improved.

11.2.2.3. Solvent Effects

Proper selection of the reaction solvent is important to attain high yields and selectivities, especially in reactions using α-substituted terminal allylic alcohols. Less coordinating solvents such as dichloromethane or toluene should be employed. With THF as the solvent, magnesium ion coordination slows the rate and lowers the diastereoselectivities (Scheme 11.30), which indicates that chelation plays an important role in the transition state structure of the magnesium ion-mediated reaction.

On the other hand, the regioselectivity is not seriously affected by the choice of reaction solvent. Even when THF is used in the reaction with crotyl alcohol, the 2-isoxazoline-5-methanol regioisomer is the only product produced (Scheme 11.31). The reaction rate is decreased in THF, but is still much faster than that in the absence of magnesium ions. Kanemasa et al. (130) concluded that the magnesium-mediated nitrile oxide cycloadditions to allylic alcohols compete with the thermal

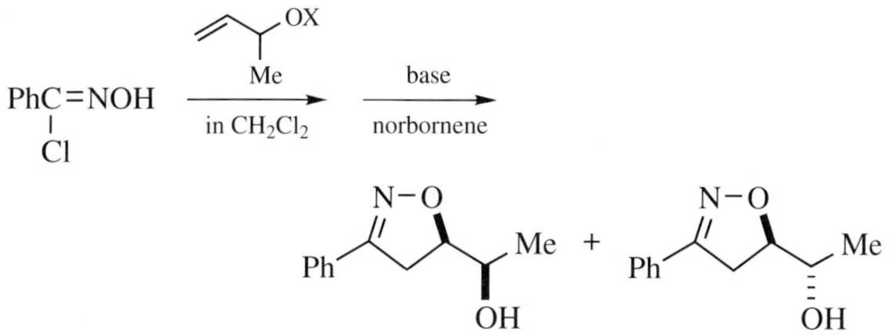

Solvent	Base	X	Relative rate	Isomer ratio *syn/anti*
CH_2Cl_2	Dipolarophile	MgBr	14	96:4
THF		MgBr	0.09	68:32
CH_2Cl_2	Et_3N	H	0.059	61:39

Five equivalent each of the dipolarophile was used.

Scheme 11.30

reactions so that use of coordinating solvents such as THF decreases the mediated reactions. As a result, selectivities are relatively lowered. The reactivity of multi-substituted allylic alcohol substrates in nitrile oxide cycloaddition reactions is much less than that of terminal allylic alcohols, so that the rate difference between the two reaction conditions is small in the cases of terminal allylic alcohols. This is why

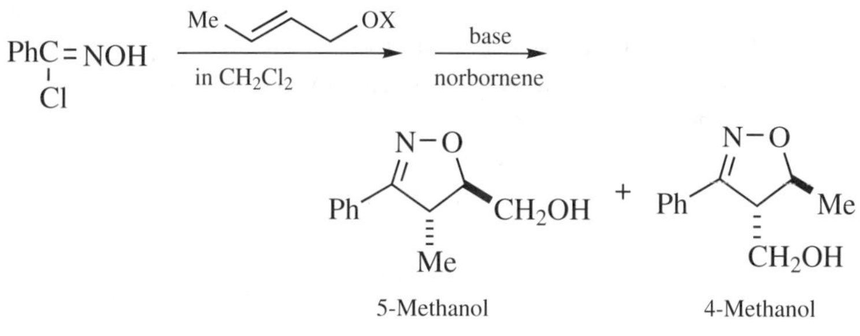

X	Base	Relative rate	Isomer ratio
MgBr	None	9	5-Methanol only
	n-BuOH (1)	13	5-Methanol only
	i-PrOH (1)	19	5-Methanol only
	t-BuOH (1)	19	5-Methanol only
	THF (CH_2Cl_2:THF = 10:1 v/v)	1	5-Methanol only
	THF (as solvent)	0.1	5-Methanol only
H	None	0.0013	46:54

Scheme 11.31

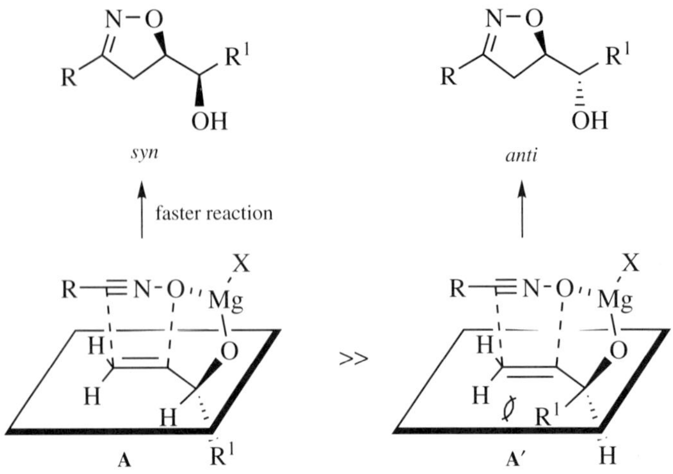

Scheme 11.32

reactions with terminal allylic alcohols are more sensitive to the coordinating ability of reaction solvent (130). It is again interesting to note that the addition of a small amount of an alcohol increases the rates of magnesium ion-mediated nitrile oxide cycloaddition reactions.

The magnesium ion-mediated nitrile oxide cycloadditions of allylic alcohols can be explained with a chelated transition state structure (Scheme 11.32). The magnesium ion of the allylic magnesium alkoxides is coordinated to the nitrile oxide. The metal ion serves to bind both the 1,3-dipole (nitrile oxide) and the dipolarophile (allylic alcohol) to activate them via an intramolecular cycloaddition reaction. Accordingly, this reaction is regarded as a metal-tethered intramolecular 1,3-dipolar cycloaddition reactions. The cycloadducts are always regioselective, giving 2-isoxazoline-5-methanol derivatives. When the allylic alcohol substrates are substituted at the α-position, the sterically more favored transition structure **A** participates in the reaction giving syn cycloadducts (130).

11.2.2.4. Kinetics

Based on the competitive cycloaddition reactions of benzonitrile oxide to a mixture of a magnesium allylic alkoxide and the reference dipolarophile norbornene, kinetic data in the form of relative reaction rates were collected and are shown in Table 11.1 (130,132,133). Reactions using the benzonitrile oxide–magnesium bromide chloride complexes and free allylic alcohols (reaction M) are only ~10 times faster than the reactions in the absence of magnesium ions (reaction L). On the other hand, reactions using benzonitrile oxide and the magnesium allylic alkoxides (reaction N) are much faster. Rate acceleration (N/L) in the reactions with magnesium alkoxides is larger for the internal allylic alcohols than for the terminal alcohols. The maximum rate acceleration factor of

TABLE 11.1. KINETIC DATA IN THE FORM OF RELATIVE REACTION RATES

Terminal Allylic Alcohols — Internal Allylic Alcohols — Homoallylic Alcohols

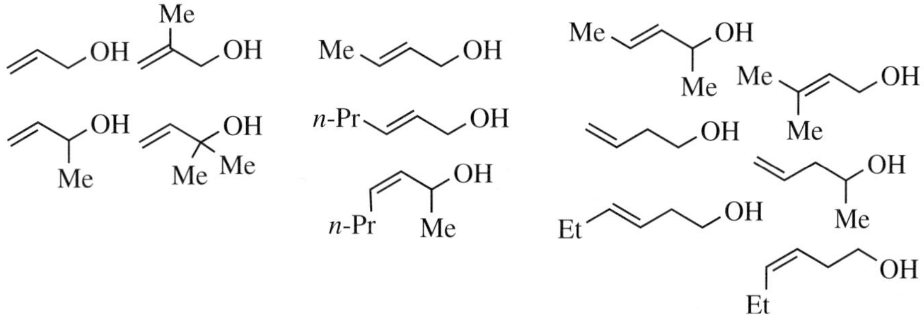

Entry	Dipolarophile	Reaction L[a]	Reaction M[b]	Reaction N[c]	M/L	N/L
1	Norbornene	1	1	1		
		Terminal Allylic Alcohols				
2	2-Propen-1-ol	0.064 (1)	0.79	130 (1)	12	2,030
3	2-Methyl-2-propen-1-ol	0.031 (0.5)	0.42	19 (0.1)	14	610
4	3-Buten-2-ol	0.059 (0.9)	0.46	14 (0.1)	8	240
5	2-Methyl-3-buten-2-ol	0.048 (0.8)	0.14	3.6 (0.03)	3	75
		Internal Allylic Alcohols				
6	(*E*)-2-Buten-1-ol	0.0013 (0.02)	0.014	9 (0.07)	11	6,900
7	(*E*)-2-Hexen-1-ol	0.0012 (0.02)	0.012	3.8 (0.03)	10	3,170
8	(*Z*)-2-Hexen-1-ol	0.00056 (0.009)	0.0063	3 (0.02)	11	5,360
9	(*E*)-3-Penten-2-ol	0.0009 (0.01)	0.0079	0.76 (0.006)	9	840
10	3-Methyl-2-buten-1-ol	0.00022 (0.003)	0.0013	3.6 (0.03)	6	16,000
		Homoallylic Alcohols				
11	3-Buten-1-ol	0.041 (0.6)	0.34	20 (0.2)	8	490
12	4-Penten-2-ol	0.034 (0.5)	0.18	1.3 (0.01)	5	40
13	(*E*)-3-Hexen-1-ol	0.00098 (0.02)	0.0072	0.1 (0.0008)	7	100
14	(*Z*)-3-Hexen-1-ol	0.00032 (0.005)	0.0024	0.2 (0.002)	8	625
		Other Olefins				
15	3-Buten-2-one	0.82	0.82	1		
16	Methyl acrylate	0.39	0.44	1		
17	3-Phenylpropene	0.012	0.018	2		

[a] Reaction L: PhCNO and free alcoholic forms of dipolarophiles were used.
[b] Reaction M: PhCNO·MgBrCl and free alcoholic forms of dipolarophiles were used.
[c] Reaction N: PhCNO and the magnesium alkoxides of dipolarophiles were used.

N/L recorded was 16,000 for the reactions using the magnesium alkoxide of 3-methyl-2-buten-1-ol as a multisubstituted internal allylic alcohol substrate, which is why regiocontrol is still effective in the reactions in a highly coordinating solvent such as THF. Rate enhancement is much lower in the nitrile oxide cycloaddition reactions using homoallylic alcohol substrates.

Chemoselectivity, regioselectivity, and diastereoselectivity in benzonitrile oxide cycloadditions

With

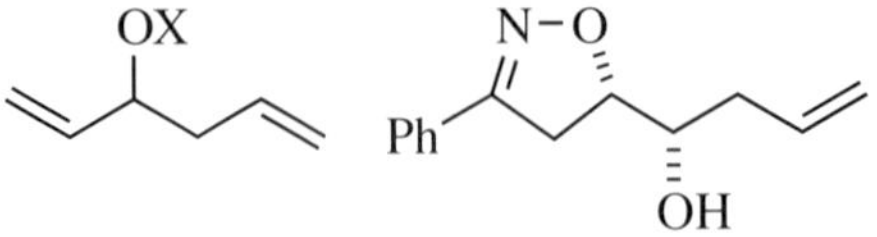

X = MgBr: 94% as the single product (>99:1)
X = H: 100% (four isomers = 67:23:17:16)

With

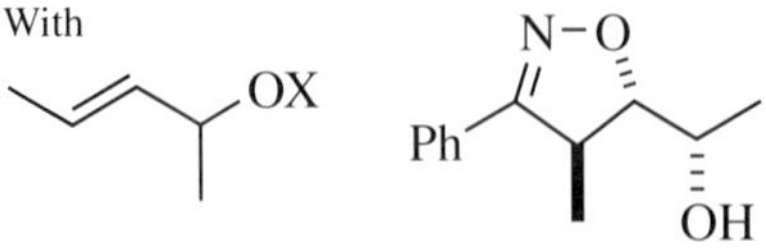

X = MgBr: 53% as the single regioisomer (94:6)
X = H: 28% (four isomers = 55:27:12:6)

With

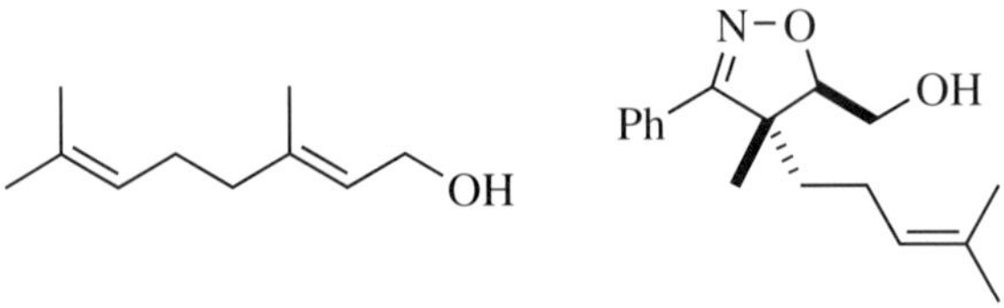

With *n*-BuMgBr (1.3 equiv): 26% as the single product
None: 67% (three isomers = 40:33:27)

With

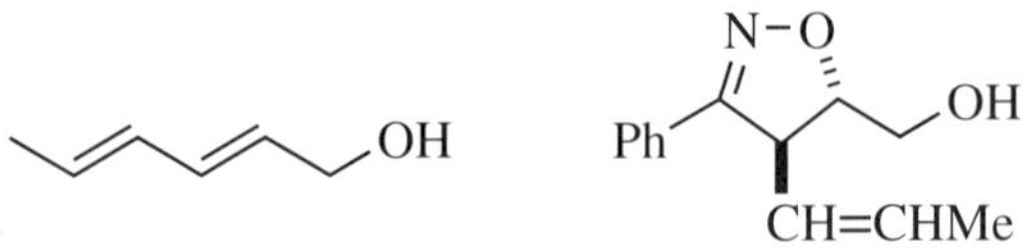

With *n*-BuMgBr (1.3 equiv): 88% as the single product
None: 67% (four isomers = 44:42:7:7)

Scheme 11.33

11.2.2.5. Chemoselectivity

The reaction of benzonitrile oxide with a substrate bearing both allylic and homoallylic alcohol functionalities in the absence or presence of magnesium ion provides very different chemo- and diastereoselectivities (Scheme 11.33) (130). Since the both alkenes are terminal, it was expected that the cycloadditions would proceed regioselectively to give 5-substituted 2-isoxazolines. The reaction mediated by magnesium ion proceeded in an entirely syn-selective manner. However, the reaction without magnesium ion gave a mixture of four isomers. 3-Penten-2-ol is an internal alkene and bears a chiral center. The benzonitrile oxide cycloaddition to this alkene was regioselective in the presence of magnesium ion, whereas four isomers were formed in the absence of magnesium ion. When dipolarophiles bear two internal alkenes, one of which is allylic, the magnesium ion-mediated reactions are exclusively chemoselective, producing single regioisomers of cycloadducts at only the allylic alkenes. These examples indicate the high reliability of the magnesium ion-mediated nitrile oxide cycloaddition methodology.

11.2.2.6. Catalytic Cycle

Information about the catalytic cycle of magnesium ion-mediated nitrile oxide cycloadditions has been gathered by Kanemasa et al. (134). This data is relevant to the possibility of using the magnesium ion mediated nitrile oxide cycloaddition with allylic alcohols in an enantioselective fashion. When stoichiometric amounts of magnesium ion are reduced to 10 mol% in competitive reactions of mesitylnitrile oxide with norbornene and 2-propen-1-ol, a turnover number of 8.3 is observed. With a 1 mol% of magnesium ion, the turnover number is improved to 34. This means that the magnesium ion can be effectively liberated from the bidentate-chelated product/magnesium ion complex (Scheme 11.34). Other important information observed is (a) a concentration effect on the reaction rate and (b) ligand-based rate acceleration. The reaction rate becomes faster under dilute conditions and the addition of a bidentate ligand of the bisether type increases the reaction rate. However, no successful examples of catalyzed asymmetric reactions of the magnesium ion-mediated nitrile oxides have yet been achieved.

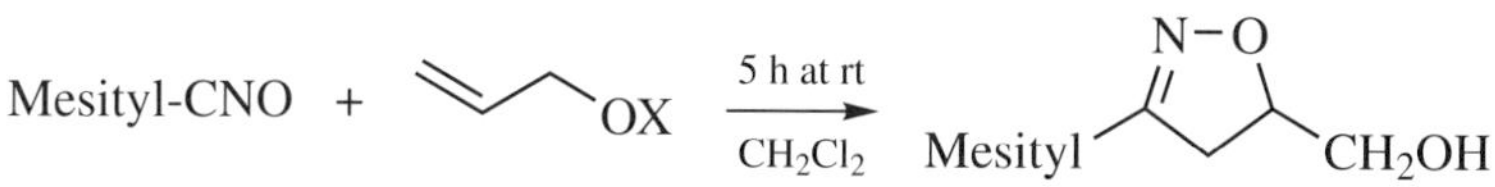

	Yield/%	Tunover No.
Uncatalyzed	6	–
X = MgBr (0.1)	89	8.3
X = MgBr (0.01)	40	34

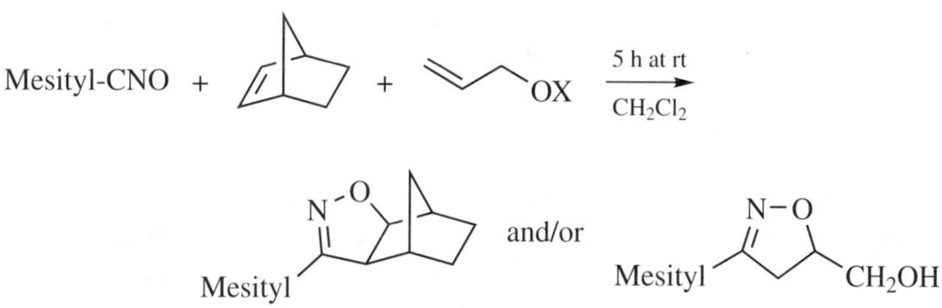

	M	Yield (%)	Product ratio	Rate acceleration
Uncatalyzed	0.05	97	1:5.9	1
H, $MgBr_2$•OEt_2 (1)	0.05	100	3.1:1	18
H, $MgBr_2$•OEt_2 (1)	0.005	95	12.1:1	72
X = MgBr (1)	0.005	99	71:1	419
X = MgBr (1)[a]	0.005	100	116:1	684

[a] *rac*-1,2-bis(benzyloxy)-1,2-diphenylethane (1 equiv).

Scheme 11.34

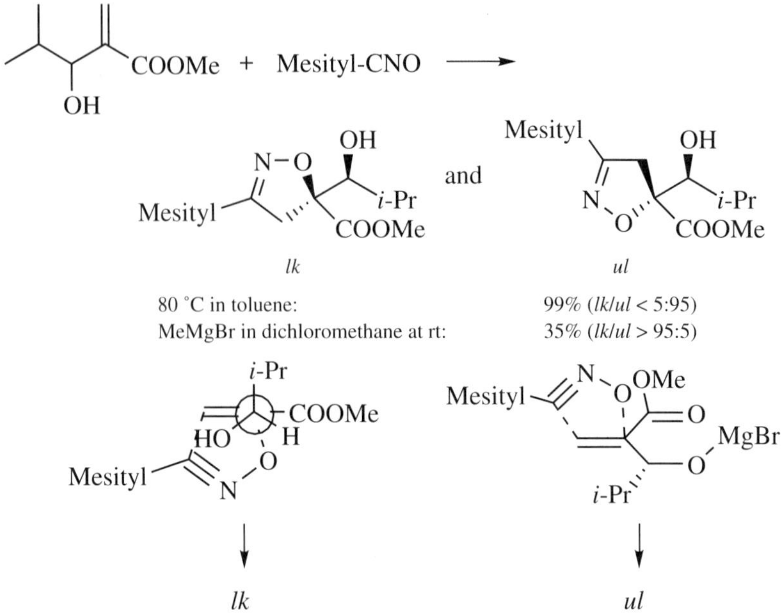

Scheme 11.35

11.2.2.7. Related Work and Applications

Fisera and co-workers (135,136) investigated the Mg(II) ion-mediated nitrile oxide method for the reaction between mesitylnitrile oxide and α-methylene-β-hydroxy esters (Baylis–Hillman adducts). This reaction is exclusively diastereoselective even at high temperatures (80 °C) in the absence of Mg(II) ion when an appropriate dipolarophile such as methyl 3-hydroxy-2-methylene-4-methylpentanoate is chosen, the *ul*-stereoisomeric cycloadduct being produced as the single product (Scheme 11.35). It is interesting that a complete reversal of diastereoselectivity is observed, giving the *lk*-isomer, when an equimolar amount of MeMgBr is present in the same reaction. It is concluded that intramolecular chelation between the hydroxyl oxygen and the ester oxygen is preferred in this case rather than intermolecular chelation. Although not so dramatic, some previous examples of the reversal of diastereoselectivity by the presence of metallic ions in nitrile oxide cycloadditions to allylic alcohol substrates are also known (131,137).

Carreira and co-workers (138) successfully applied the Mg(II) ion-mediated nitrile oxide cycloaddition method to the total synthesis of epothilones A and B (Scheme 11.36). The key step in the synthesis was a hydroxyl-directed syn-selective nitrile oxide cycloaddition using a phosphorus-functionalized aliphatic nitrile oxide. This cycloaddition step served not only to introduce a heterocycle-substituted appendage into the skeleton of the epothilones, but also to assemble two

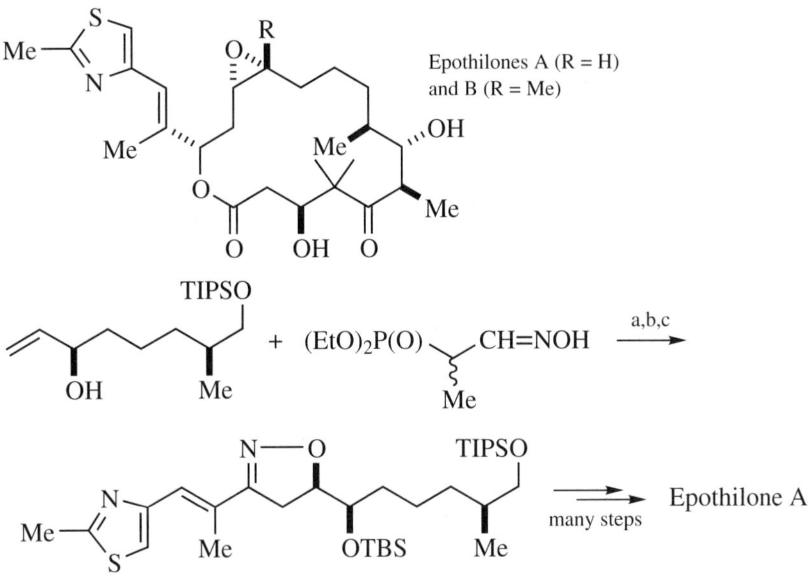

(a) t-BuOCl in CH_2Cl_2. (b) 1.3 equiv of (R)-3-buten-2-ol, 3.3 equiv of i-PrOH, 3 equiv of EtMgBr, rt. (c) LiCl, DBU, 2-methylthiazole-4-carboxaldehyde in MeCN, rt

Scheme 11.36

hydroxyl-bearing stereocenters of the skeleton in a stereoselective fashion. This hydroxyl-directed nitrile oxide cycloaddition methodology is also useful as a general solution to the synthesis of polyketide building blocks. Carreira has further utilized the regio- and diastereoselective nitrile oxide cycloaddition as key reaction in an efficient synthesis of erythronolide A seco-acid (139).

11.2.3. Lewis Acid Mediated Diastereoselective Reactions

Yamamoto and co-workers (135,135–137) recently reported a new method for stereocontrol in nitrile oxide cycloadditions. Metal ion-catalyzed diastereoselective asymmetric reactions using chiral electron-deficient dipolarophiles have remained unreported except for reactions using α-methylene-β-hydroxy esters, which were described in Section 11.2.2.6. Although synthetically very useful and, hence, attractive as an entry to the asymmetric synthesis of 2-isoxazolines, the application of Lewis acid catalysis to nitrile oxide cycloadditions with 4-chiral 3-(2-alkenoyl)-2-oxazolidinones has been unsuccessful, even when >1 equiv of Lewis acids are employed. However, as shown in the Scheme 11.37, diastereoselectivities in favor of the *lk*-cycloadducts are improved (diastereomer ratio = 96 : 4) when the reactions are performed in dichloromethane in the presence of 1 equiv of $MgBr_2$ at higher than normal concentrations (0.25 vs 0.083 M) (140). The Lewis acid

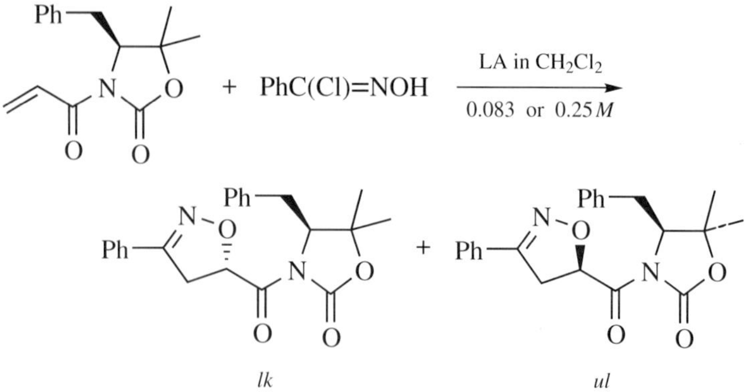

Additive	M	*lk*:*ul*
none	0.17	43:57
ZnI_2 (1.2)	0.17	45:55
$MgBr_2$ (1.0)	0.083	71:29
$MgBr_2$ (0.5)	0.083	48:52
$MgBr_2$ (1.0)	0.25	96:4

Scheme 11.37

($MgBr_2$) is incorporated between the carbonyl oxygens of the acceptor substrate, so that only one diastereoface of the activated alkene of the acceptor is susceptible to attack by the nitrile oxide.

Yamamoto et al. (141,142) succeeded in carrying out catalytic versions of the above reaction by using of a slow *in situ* generation of the nitrile oxide by the slow addition of triethylamine to a acetonitrile solution of the hydroximoyl chloride precursors. Both $MgBr_2$ and $Yb(OTf)_3$ are effective as promoters in these cases. The use of Lewis acid–alcohol mixtures such as $Mg(ClO_4$/EtOH (1:1) and $La(OTf)_3$/*i*-PrOH (1:1) is even more effective (Scheme 11.37).

11.2.4. Other Mediators in Nitrile Oxide Cycloadditions

Regiocontrol of nitrile oxide cycloadditions through inclusion complexes has been reported by Easton and co-workers (143). The 1,3-dipolar cycloaddition of 4-*tert*-butylbenzonitrile oxide with 6^A-acrylamido-6^A-deoxy-β-cyclodextrin in aqueous solution favors the formation of the 4-substituted isoxazoline (4-Amide/5-Amide = 2.3:1, Scheme 11.38). This regioselectivity is in sharp contrast to the normal predominance of the 5-substituted regioisomer using monosubstituted alkene dipolarophiles. This reversal of regioselectivity is thought to occur through self-assembly of the reactants within the cyclodextrin annulus, since the relative stability of the inclusion complexes should be dependent on the nature of the reaction solvent, which has actually been observed. The same reaction in DMF provides the opposite regioselectivity (4-Amide/5-Amide = 1:4).

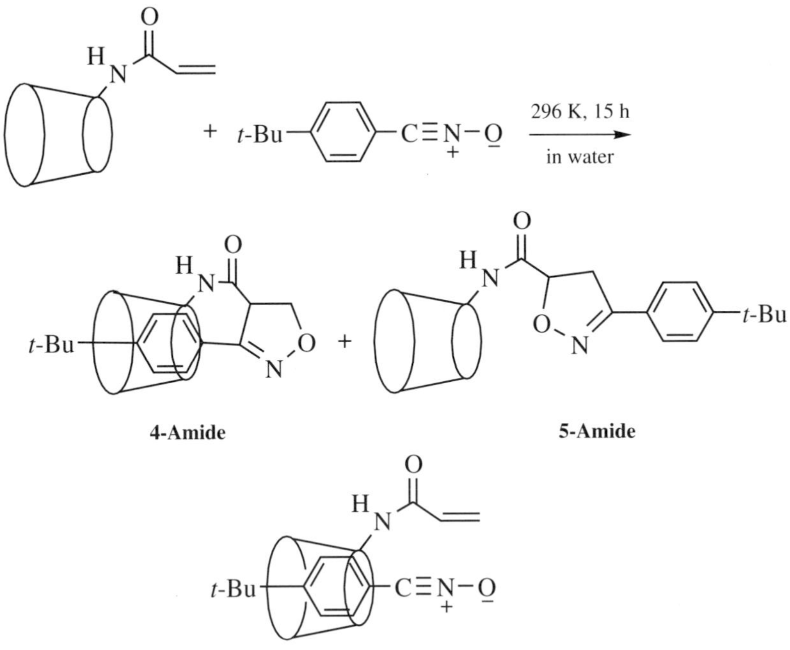

Scheme 11.38

Researchers at Scripps demonstrated the first successful example of an antibody-catalyzed regio- and enantiocontrolled 1,3-dipolar cycloaddition. In order to control the regiochemistry of a nitrile oxide cycloaddition reaction with *N,N*-dimethylacrylamide, they designed a derivative of *o*-amino-*N,N*-dimethylbenzamide **A** as a hapten (Scheme 11.39) (144). Two of fifteen monoclonal antibodies generated against hapten **A** catalyze the completely regioselective formation of 5-(*N,N*-dimethylcarbamoyl)isoxazoline with multiple turnovers (>50). The most efficient antibody, 29G12, catalyzes this dipolar cycloaddition with good enantioselectivity to give 5-(*N,N*-dimethylcarbamoyl)isoxazoline with up to 98% ee. Although the hapten utilized is achiral, the antibody has created chirality in the cycloadduct, because the immune system, by nature of its inherent chirality, has produced a stereochemical environment capable of exquisite stabilization of one enantiomeric transition state in the reaction.

Wilcox and co-workers (145) reported that the stereoselectivity of 1,3-dipolar cycloaddition reactions can be controlled in a predictable manner when ion pairs are located at a proper position near the reaction site (Scheme 11.40). He has employed an *N*-phenylmaleimide substrate having a chiral center in the substituent at ortho position of the phenyl group. Due to serious steric hindrance, this phenyl group suffers hindered rotation around the aryl–nitrogen bond (rotation barrier: 22 kcal/mol). Four diastereomeric cycloadducts are possible in the cycloaddition step with a nitrile oxide. When the cycloaddition reaction is carried out in

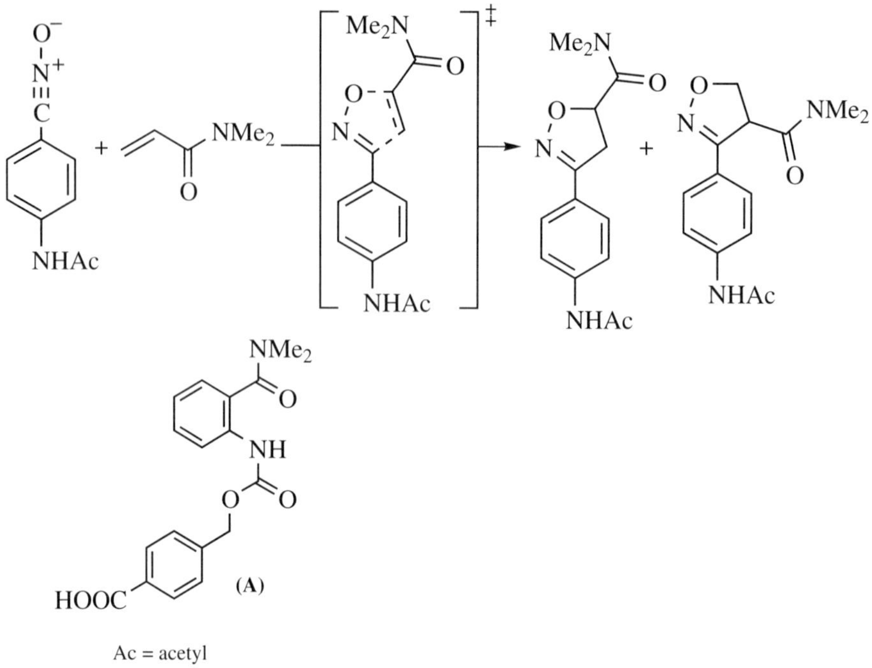

Scheme 11.39

chloroform, one diastereomer **B** was selectively produced (**A/B/C/D** = 0:20:1:2). In the transition structure producing isomer **B**, the positively charged ammonium group interacts with the negatively charged nitrile oxide oxygen to afford high regio- and diastereofacial selectivity, producing **B** as the major product.

A general method for the preparation of enantiomerically pure 4-hydroxy-2-isoxazolines has been reported by Wallace (146). The first version involved nitrile oxide cycloadditions with vinylboronic esters, leading to the regioselective formation of 2-isoxazoline-4-boronates, which were then oxidized with *tert*-butyl hydroperoxide to give 4-hydroxy-2-isoxazolines. This reaction occurred with the retention of the configuration of the starting alkenes (Scheme 11.41). Sodium percarbonate ($Na_2CO_3{\cdot}5H_2O_2$), known as a "solid form" or "dry carrier" of hydrogen peroxide, was found to work effectively both as a base for the nitrile oxide generation and as an oxidizing reagent. Thus, the chiral vinyl boronates derived from Oppolzer's camphor sultam have been applied successfully to the production of enantiopure 4-hydroxy-2-isoxazolines in good chemical yields and with high diastereoselectivities. When sodium percarbonate was used in aqueous THF, the chiral auxiliary was cleaved under the reaction conditions to give, after a one-pot esterification with diazomethine, enantiopure methyl 4-hydroxyl-2-oxazoline-5-carboxylate in good yield (147).

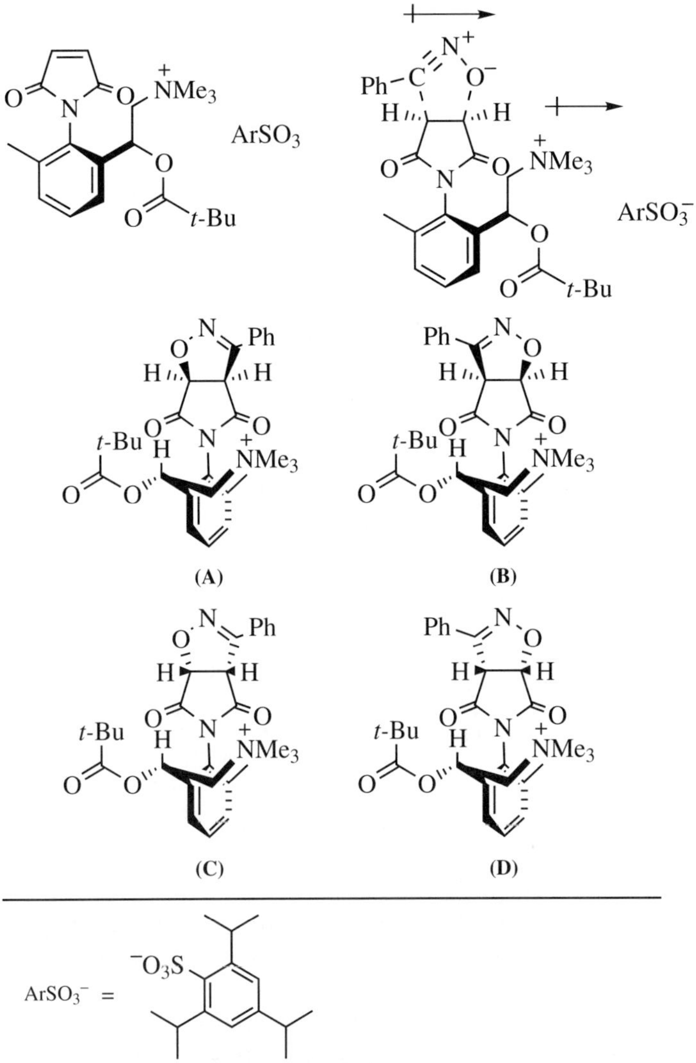

Scheme 11.40

11.2.5. Catalyzed Enantioselective Reactions

Many groups have tackled the development of catalytic asymmetric versions of nitrile oxide cycloaddition reactions using chiral Lewis acid catalysts. However, Ukaji is the first and the only chemist who has succeeded in the achievement of such processes involving nitrile oxide cycloaddition reactions. He studied reactions

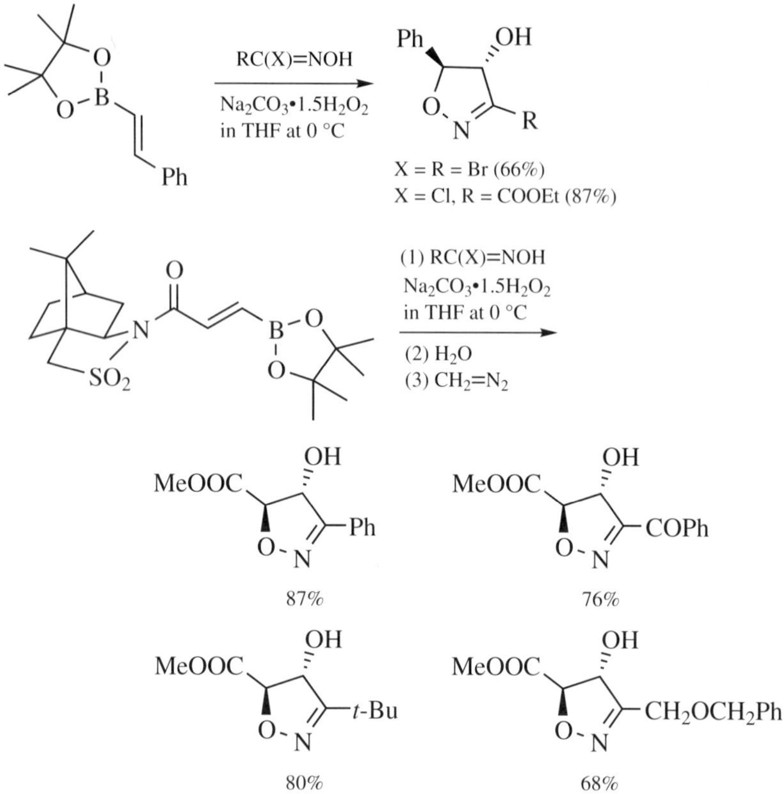

Scheme 11.41

between nitrile oxides and allylic alcohols in the presence of zinc catalysts modified by the addition of chiral ligands derived from natural tartaric acid. Although his pioneering work is a great contribution to the development of this field, the catalytic efficiency is not yet satisfactory, since the rate enhancement of the catalyzed reactions is so far insufficient. Accordingly, this research field still remains as a large challenge. Detailed results have been discussed elsewhere in this book, and therefore only the relevant references are cited here (148–155).

11.3. NITRONES

11.3.1. Introduction

Nitrones are the most widely studied of the 1,3-dipoles in the field of catalyzed enantioselective 1,3-dipolar cycloaddition reactions. Effective catalysis using a variety of chiral Lewis acid catalysts has been reported for the nitrone cycloaddition reactions using both electron-deficient and electron-rich dipolarophiles (6,156,157). The conceptual origin of this work is no doubt a short communication in 1992 from Kanemasa's group, which will be described in Section 11.3.2 (158).

The use of mediators to improve reactivity or selectivity in nitrone cycloaddition chemistry begins with the nitrone generation step. As is well known, the N-alkylation of oximes provides one of the most direct and convenient synthetic routes to N-alkylated nitrones from readily available aldehydes and ketones. Electrophilic mediators have been employed to activate alkenes for N-alkylation, both in intramolecular and intermolecular reactions. They include activation of the internal alkene function by the action of (a) strong nonmetallic electrophiles such as phenylselenenyl sulfate (159), and (b) metallic catalysts such as Ag(I) (160) and Pd(II) ions (161). The use of (c) external electrophilic alkenes such as α,β-unsaturated ketones (162), esters (163), sulfones (164,165), and nitriles (163) leads to N-alkylation of oximes in the presence of mediators, providing an important route to N-alkyl nitrones. However, it is not clear that these promoters work effectively to improve reactivity and/or selectivity of the cycloaddition steps of the resulting nitrones. Accordingly, the author will not describe these generation methods in detail.

One exception is the reaction of acetone oxime with divinyl ketone in the presence of an equimolar amount of zinc(II) bromide (162). Acetone oxime reacts with divinyl ketone on heating in THF at reflux, leading to both conjugate addition and nitrone cycloaddition, producing a 5:1 mixture of regioisomers with 8-oxa-1-azabicyclo[3.2.1]octan-4-one as the major isomer (Scheme 11.42). On the other hand, in the presence of an equimolar amount of zinc(II) bromide, 7-oxa-1-azabicyclo[3.2.1]octan-4-one is the major isomer (97:3) in a total yield of 97%, indicating that the Lewis acid has controlled the regioselectivity of the second step, namely, the cycloaddition.

11.3.2. Titanium Ion-Mediated Reactions

A chelating enone employed in a Lewis acid catalyzed nitrone cycloaddition reaction should result in a rate enhancement. The first example of such catalysis

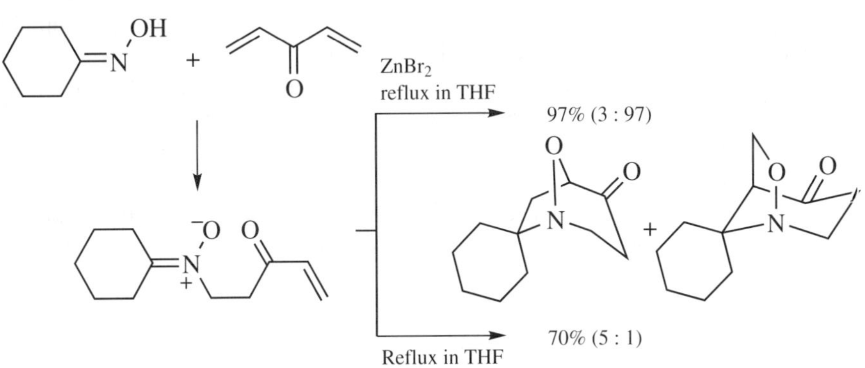

Scheme 11.42

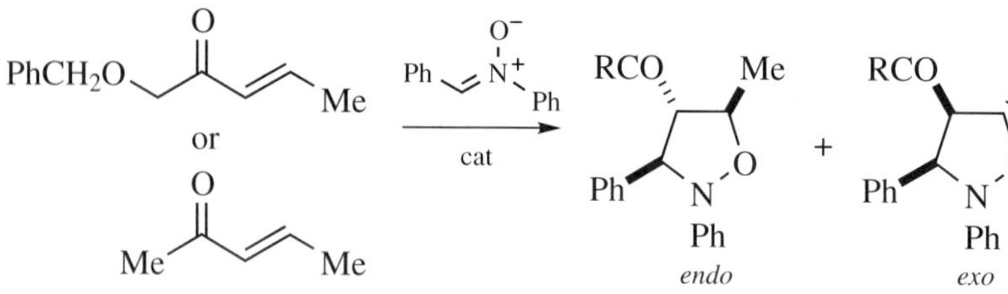

Nitrone	Dipolarophile	Catalyst	Temp (°C)	Yield (%)	*endo/exo*
R′ = Me	R = Me	None	80	70	73:27
		$Ti(Oi\text{-}Pr)_2Cl_2$	rt	49	77:23
	R = $PhCH_2OCH_2$	None	80	76	40:60
		$Ti(Oi\text{-}Pr)_2Cl_2$	0	50	>99:1
R′ = Ph		None	80	89	67:33
		$Ti(Oi\text{-}Pr)_2Cl_2$	0	74	>99:1

Scheme 11.43

was reported in the nitrone cycloadditions to (*E*)-1-benzyloxy-3-penten-2-one (158). The reaction proceeds in the presence $Ti(Oi\text{-}Pr)_2Cl_2$ at 0 °C, whereas it requires 80 °C without the Lewis acid catalyst (Scheme 11.43). Although the reaction of nonchelating (*E*)-3-penten-2-one as a reference dipolarophile is accelerated by the catalyst, the stereoselectivity is not improved over the purely thermal process. On the other hand, the $Ti(Oi\text{-}Pr)_2Cl_2$ mediated reaction of chelating (*E*)-1-benzyloxy-3-penten-2-one leads to the exclusive formation of the endo-cycloadduct.

The regioselectivity of nitrone cycloadditions is usually effected by both steric and electronic factors. In nitrone cycloadditions of monosubstituted electron-deficient alkenes under thermal conditions, electronic control steers the nitrone oxygen to the β-carbon of acceptor, whereas steric factors direct it to the α-carbon. Accordingly, a mixture of two regioisomers is produced under thermal conditions, as shown in Scheme 11.44 for dimethyl 2-oxo-3-butenylphosphonate. With $Ti(O\ i\text{-}Pr)_2Cl_2$, which coordinates in a bidentate chelated fashion with this dipolarophile, only the 4-acyl regioisomer is produced (158).

Scheme 11.45 shows a proposal for the transition state involved in a related Lewis acid catalyzed cycloaddition where a disubstituted dipolarophile is used and endo/exo issues are examined. By coordination of the bulky titanium Lewis acid catalyst, the α-carbon of the acceptor becomes sterically more hindered, disfavoring exo-approach, which involves a serious steric interaction of the benzylidene phenyl moiety with the ligands on the titanium ion. Accordingly, the endo-cycloadduct is the only product observed.

High regiocontrol is based on the predominant contribution of electronic factors. Since the nitrone carbon and the α-carbon of the acceptor are the sterically more hindered termini in each reagent, the formation of the electronically controlled regioisomer (e.g., the 4-acylisoxazolidines) should be much less favored.

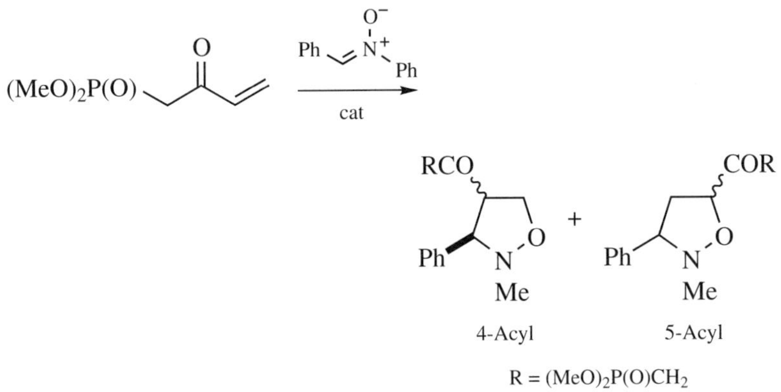

No catalyst	110 °C	Three isomers [4-Acyl:5-Acyl = 34:66 (1:1)]
Ti(O*i*-Pr)$_3$Cl	rt	Single isomer (*endo*-Isomer of 4-acyl derivative)

Scheme 11.44

Nevertheless, this regioisomer is the only one produced in the catalyzed reaction, indicating that electronic factors have become dominant as a result of coordination to the Lewis acid.

11.3.3. Magnesium Ion-Mediated Reactions

High stereo- and regiocontrol in nitrile oxide cycloadditions to allylic alcohols in the presence of magnesium ions (see Section 11.2.2) makes extension of this

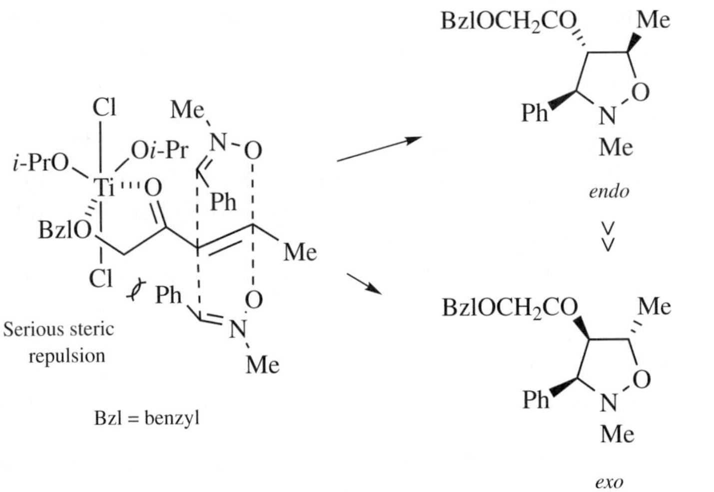

Scheme 11.45

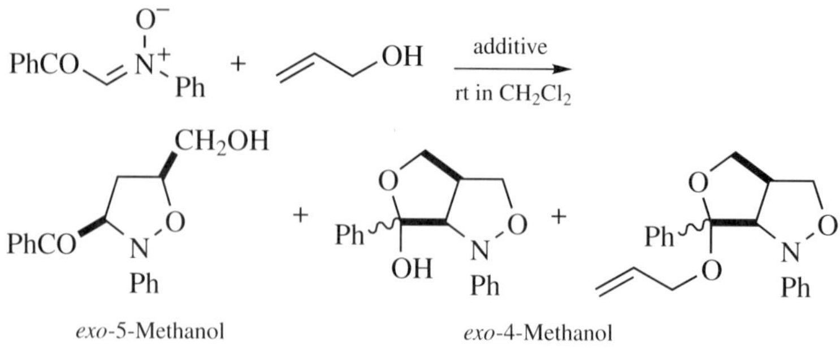

Additive	Equiv	Product	Yield (%)	Ratio
None		*exo*-5-Methanol	78	97:3
EtMgBr	1	*exo*-5-Methanol	93	98:2
Et_2Zn	1	*exo*-4-Methanol/hemiacetal	57	single
$MgBr_2$•Et_2O	1	*exo*-5-Methanol	81	single
$ZnBr_2$	1	*exo*-4-Methanol/acetal	65	single

Scheme 11.46

methodology to nitrones an attractive endeavor. Cycloaddition of *N*-benzoylmethyleneaniline N-oxide with either the magnesium alkoxide of 2-propen-1-ol or 2-propen-1-ol in the presence of $MgBr_2 \cdot Et_2O$ produces the exo-isomer of the isoxazolidine-5-methanol cycloadduct as the sole product (Scheme 11.46). However, similar reactions using the zinc alkoxide or the allyl alcohol in the presence of $ZnBr_2$ gives the ring-fused isoxazolidine as either hemiacetal or acetal derivatives. It is interesting that the regioselectivities of nitrone cycloadditions with allyl alcohol show such divergent behavior depending on the metal ions (166).

The magnesium ion-mediated reaction of crotyl alcohol leads to the exclusive formation of the exo-isomer of the isoxazolidine-5-methanol cycloadduct, showing regioselectivity opposite to that found in the uncatalyzed reaction (Scheme 11.47).

The magnesium ion-mediated nitrone cycloadditions of an α,γ-disubstituted allyl alcohol are stereoselective, and show opposite regioselectivity to that observed when zinc-mediated reactions are examined (Scheme 11.48) (167). The exo–syn-isomer of the isoxazolidine-5-alcohol regioisomer and the exo–syn-isomer of the isoxazolidine-4-alcohol regioisomer are the exclusive cycloadducts in the magnesium- and zinc-mediated reactions, respectively.

The above dramatic dependence of regio- and stereoselectivity on the nature of the metal can be explained by the reaction mechanism shown in Scheme 11.49 (167). The nitrone cycloadditions of allylic alcohols are again magnesium-specific just like the nitrile oxide reactions described in Section 11.2.2. Magnesium ions accelerate the reaction through a metal ion-bound intramolecular cycloaddition path. On the other hand, zinc ions afford no such rate acceleration, but these ions catalyze the acetalization at the benzoyl carbonyl moiety of the nitrone to provide a hemiacetal intermediate. The subsequent intramolecular regio- and stereoselective cycloaddition reaction gives the observed products.

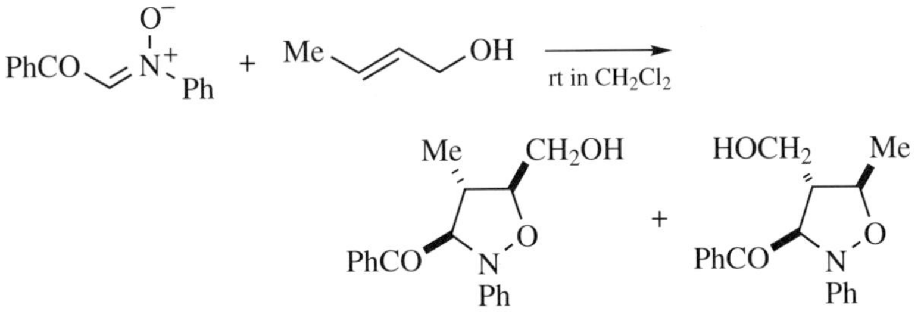

exo-5-Methanol *endo*-4-Methanol

Additive	Equiv	Product	Yield (%)	Ratio
None		*exo*-5-Methanol + *endo*-4-Methanol	94	18:82
EtMgBr	1	*exo*-5-Methanol + *endo*-4-Methanol	97	98:2
$MgBr_2$•Et O	1	*exo*-5-Methanol	77	single

Scheme 11.47

Carbonyl-substituted nitrones are formed as mixture of (*E*)- and (*Z*)-stereoisomers. By coordination to a Lewis acid, the (*Z*)-isomers are expected to be more stabilized due to tight complexation. Thus, a 2.8:1 (*E*/*Z*)-mixture of *N*-(methoxycarbonylmethylene)methylamine *N*-oxide isomerizes to the (*Z*)-isomer in the presence of $MgBr_2 \cdot Et_2O$ and undergoes a regio- and exo-selective cycloaddition reaction to 2-propen-1-ol to give the isoxazolidine-5-methanol as a single product

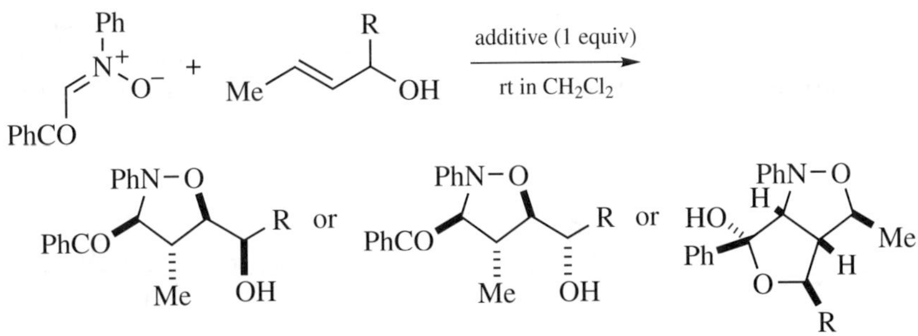

5-Alcohol/*exo-syn* 5-Alcohol/*exo-anti* 4-Alcohol/*exo-syn*

	Additive	Time (h)	Yield (%)	Products				
R = Me	$MgBr_2$•Et_2O	5	85	96	:	4	:	--
	$ZnBr_2$	5	46	--	:	--	:	only
R = *i*-Pr	$MgBr_2$•Et_2O	5	47	only	:	--	:	--
	$ZnBr_2$	20	33	--	:	--	:	only

Scheme 11.48

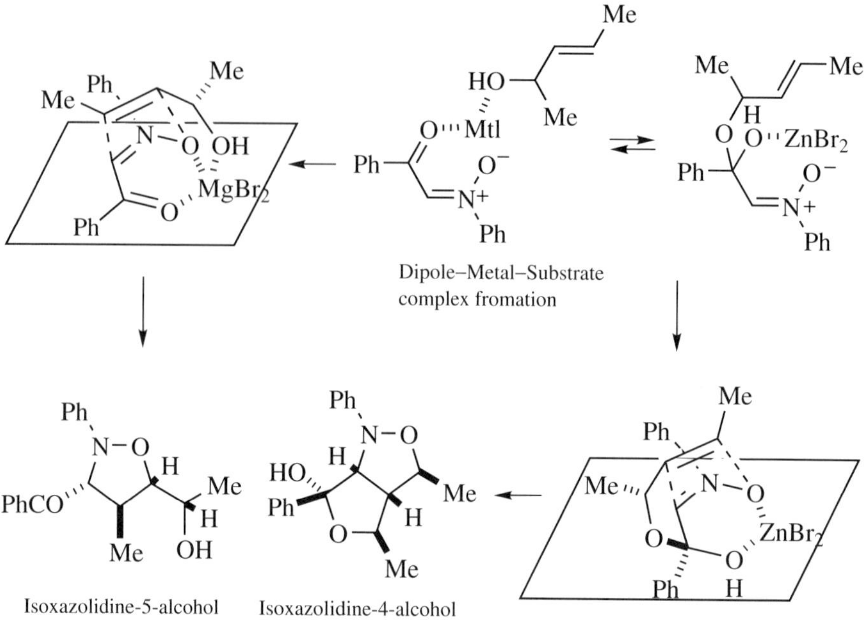

Scheme 11.49

(Scheme 11.50) (168). On the other hand, a 1:1 stereoisomeric mixture of the cycloadducts is formed in the absence of $MgBr_2 \cdot Et_2O$.

The magnesium ion-mediated nitrone cycloaddition methodology can be successfully applied to the asymmetric reactions of a chiral cyclic nitrone of the lactone type (Scheme 11.51) (169). Although (*R*)-2(*H*)-oxo-5-phenyl-5,6-dihydro-1,4-oxazine *N*-oxide is an (*E*)-nitrone, all of its reactions are highly diastereoselective, producing isoxazolidine 5-alcohol cycloadducts. The chelated transition

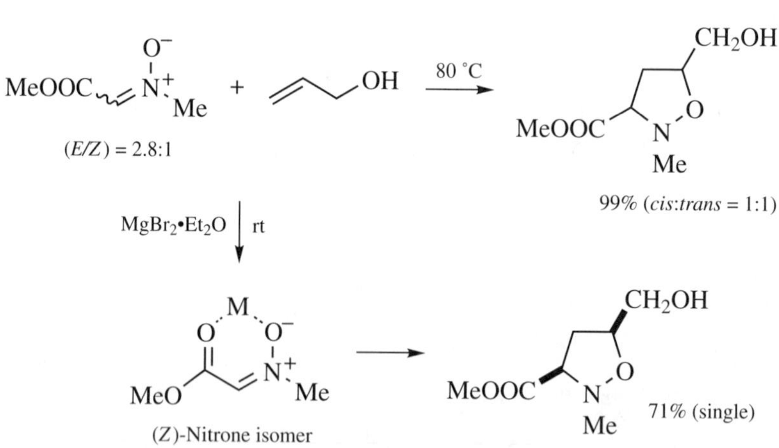

Scheme 11.50

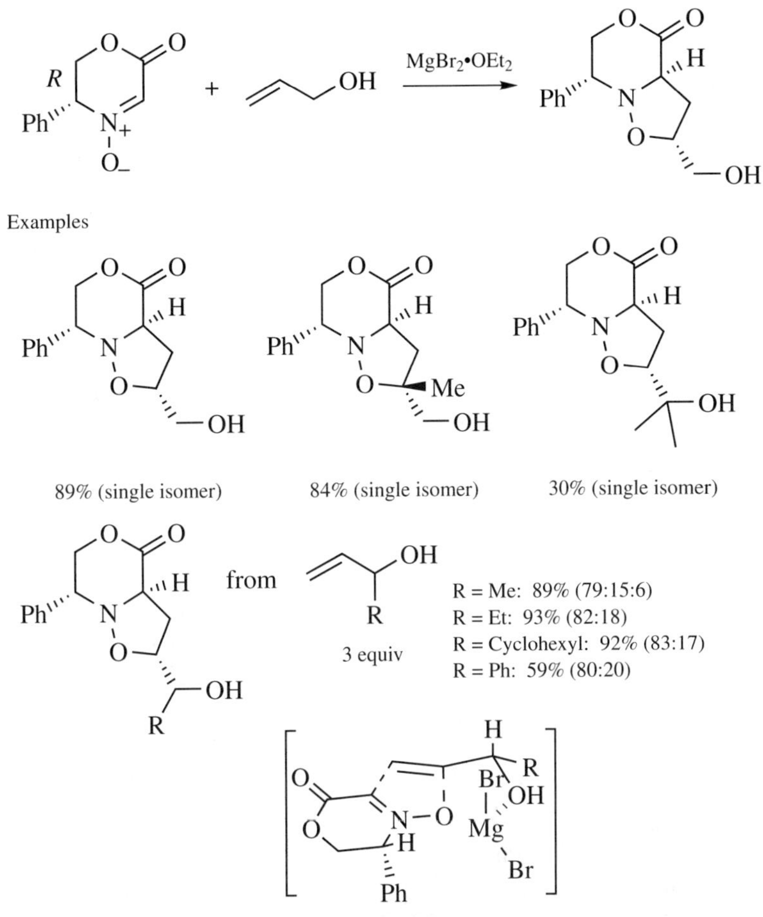

Scheme 11.51

state structure shown is responsible for the high diastereofacial selectivity. When allylic alcohols having α-chirality are employed in a large excess, the *si*-face selectively participates in the reaction, albeit the diastereoselectivities are modest.

11.3.4. Transesterification–Dipolar Cycloadditions

Tamura et al. (170–172) discovered that, when reactions of ester-substituted nitrones with allylic alcohols are performed in the presence of an equimolar amount of titanium tetraisopropoxide under heating or at room temperature, transesterification takes place to form new nitrones bearing an inner alkene dipolarophile. The resulting nitrone substrates undergo regio- and stereoselective intramolecular cycloaddition reactions to give the ring-fused isoxazolidines (Scheme 11.52). This tandem transesterification/[3 + 2]-cycloaddition method leads to the selective

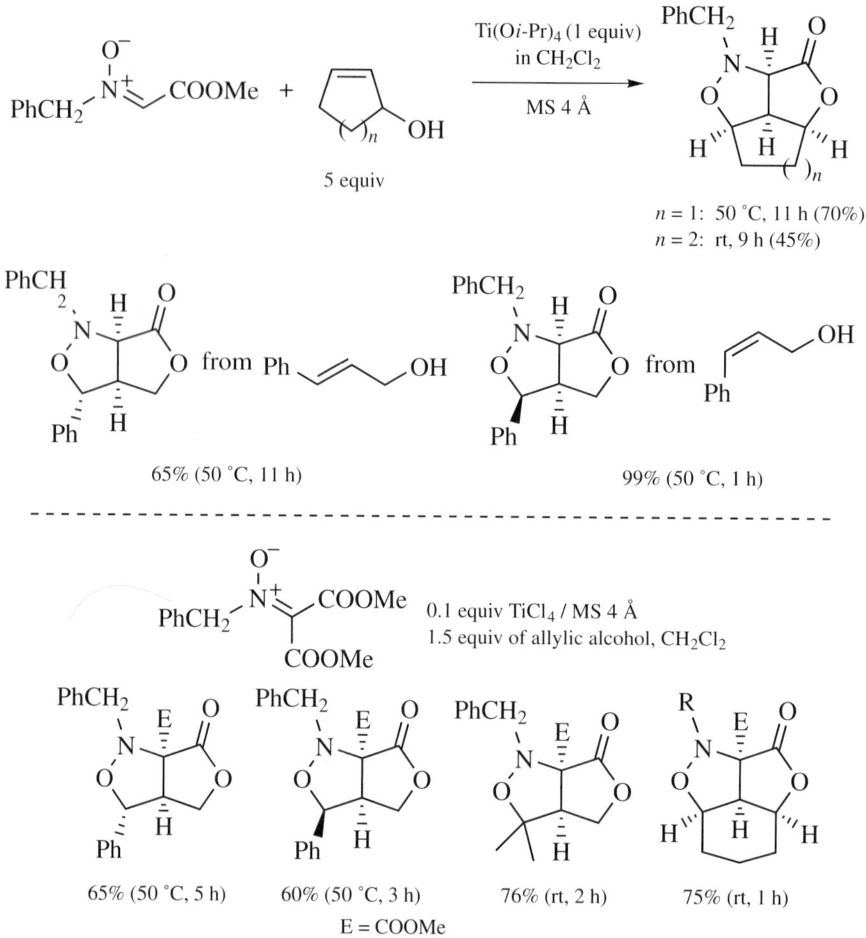

Scheme 11.52

formation of ring-closed derivatives of isoxazolidine-4-alcohol derivatives starting from a variety of allylic alcohols. In the presence of 4-Å molecular sieves, the amount of titanium tetrachloride catalyst can be reduced to 10 mol%. A large rate acceleration is achieved using a cleavable ester tether, which is an advantage of this tandem reaction.

The tandem transesterification/[3 + 2]-cycloaddition methodology is be a powerful synthetic tool, since it guarantees high diastereoselectivity even under thermal conditions. It has been successfully applied to synthetic work of the N-terminal amino acid component of Nikkomycin Bz (Scheme 11.53) (173). Thus, the sugar-based oxime is condensed with a glyoxylate hemiacetal to produce a chiral nitrone ester, which is then reacted with (*E*)-*p*-methoxycinnamyl alcohol in the presence of a catalytic amount of $TiCl_4$ at 100 °C. After the intramolecular cycloaddition, the

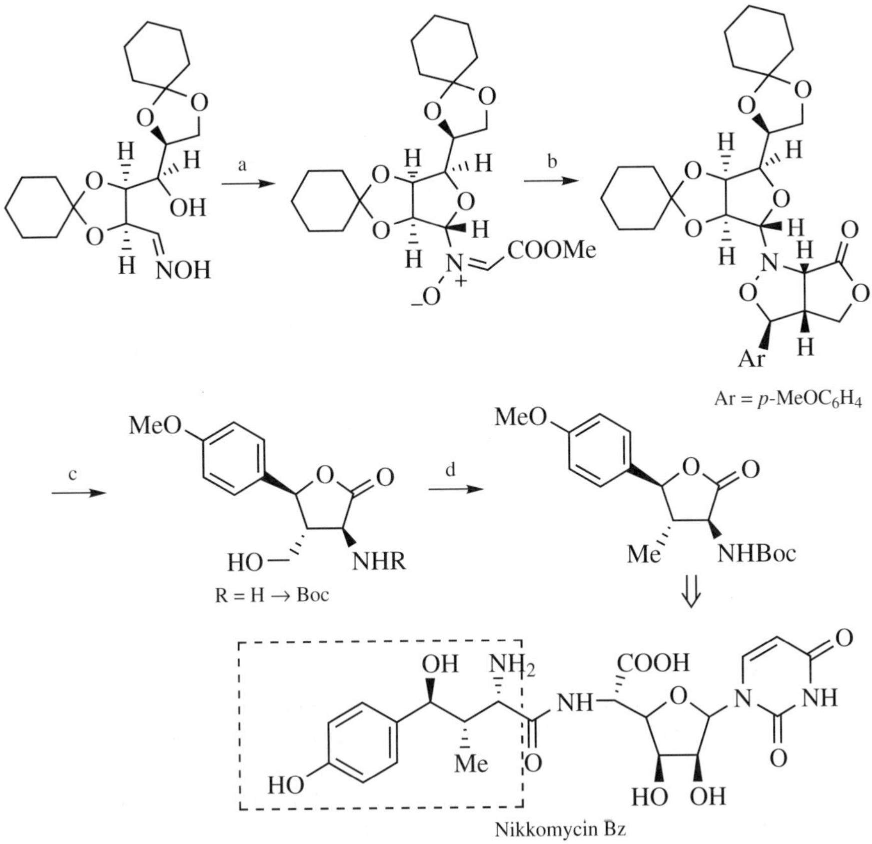

(a) Methyl glyoxylate hemiacetal, toluene reflux. (b) (*E*)-*p*-Methoxycinnamyl alcohol, cat $TiCl_4$, MS 4 Å, 100 °C, 75%. (c) $Mo(CO)_6$, $MeCN/H_2O$ (10:1), 1% HCl/MeCN; Boc_2O, MeCN, 53%. (d) MsCl , Et_3N in CH_2Cl_2, NaI, *n*-Bu_3SnH in 1,2-dimethoxyethane (DME), 80 °C, 65%.

Scheme 11.53

N—O bond of the isoxazolidine is reductively cleaved. A sequence of translactonization, mesylation, and homolytic reduction leads to the lactone form of the target amino acid.

Achiral ester-substituted nitrones as well as chiral nitrones can be employed in diastereoselective asymmetric versions of tandem transesterification/[3 + 2]-cycloaddition reactions, as shown in Scheme 11.54 (174). High diastereoselectivity and excellent chemical yields have been observed in the reaction with a (*Z*)-allylic alcohol having a chiral center at the α-position in the presence of a catalytic amount of $TiCl_4$. On the other hand, the reaction with an (*E*)-allylic alcohol having a chiral center at the α-position, under similar conditions, affords very low selectivities. Tamura et al. has solved this problem with a double chiral induction method. Thus, high diastereoselectivity has been attained by use of a chiral nitrone.

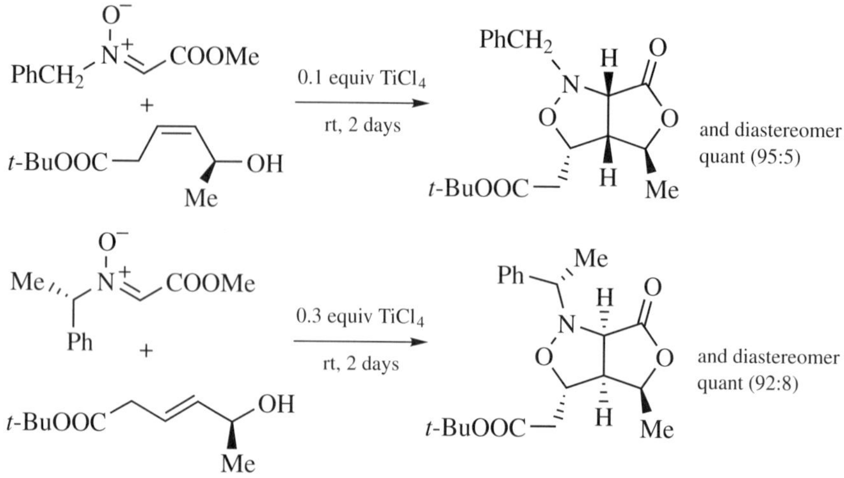

Scheme 11.54

11.3.5. Catalyzed Enantioselective Reactions

Among the various 1,3-dipolar cycloaddition reactions, nitrone cycloadditions have been the most widely studied in this half decade from the standpoint of enantioselective catalysis (6,13). Pioneers are both Jørgensen and co-workers (175) and Scheeren et al. (176). Although the reaction types are quite different from each other, both groups reported in 1994 the first successful catalyzed enantioselective nitrone cycloaddition reactions using both achiral nitrones and alkenes by the aid of homochiral Lewis acid catalysts. After pioneering contributions by the Jørgensen's group (177–184) as well as other groups including Furukawa and co-workers (185–190), Kobayashi and co-workers 191 (187), Kanemasa et al. 192 (188), Scheeren et al. 193, 194 (189,190), and others (195,196), these reactions have been elevated to a high level of diastereoselectivity, enantioselectivity, and catalytic efficiency. Details may be found Chapter 12 by Gothelf and Jørgensen.

11.4. REACTIONS INVOLVING OTHER 1,3-DIPOLES

11.4.1. Carbonyl Ylides

Suga et al. (197) reported the first stereocontrolled 1,3-dipolar cycloaddition reactions of carbonyl ylides with electron-deficient alkenes using a Lewis acid catalyst. Carbonyl ylides are highly reactive 1,3-dipoles and cannot be isolated. They are mainly generated through transition metal carbenoid intermediates derived *in situ* from diazo precursors by treatment with a transition metal catalyst. When methyl *o*-(diazoacetyl)benzoate is treated with *N*-methylmaleimide at reflux

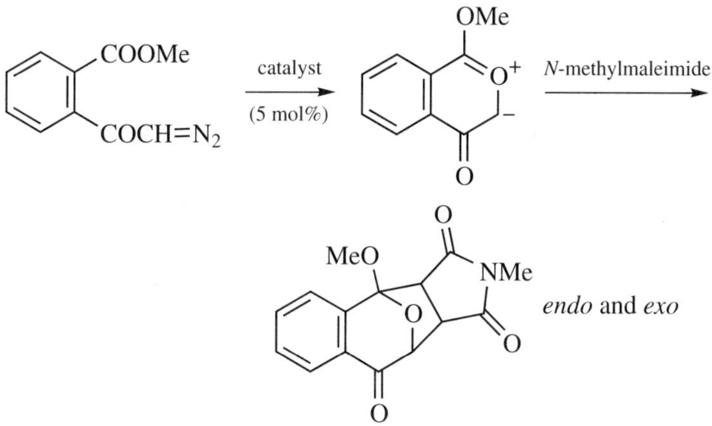

$Rh_2(OAc)_2$: reflux in benzene, 70% (*endo*/*exo* = 11:89)
CuCl: reflux in benzene, 81% (*endo*/*exo* = 26:74)
CuCl + $Yb(OTf)_3$: reflux in benzene, 52% (*endo*/*exo* = 94:6)

Scheme 11.55

in benzene in the presence of a catalytic amount (5 mol%) of $Rh_2(OAc)_2$ or CuCl, the catalyst for the generation of the carbonyl ylide via electrophilic attack of the carbenoid onto the ester carbonyl oxygen, a mixture of endo- and exo-cycloadducts are produced in good combined yields (Scheme 11.55). The exo-isomer is the major product of this reaction (endo/exo=11:89–26:74). However, with Lewis acidic transition metal catalysts such as $Cu(OTf)_2$ or CuOTf, the endo-isomer becomes the major product (endo/exo = 87:13–82:18). It is interesting that high endo-selectivity (endo/exo = 94:6) is observed when the same reaction is catalyzed by combined catalysts containing CuCl (5 mol%) and $Yb(OTf)_3$. This indicates that the cocatalyst $Yb(OTf)_3$ activates the cycloaddition reaction either by coordination to the maleimide carbonyl oxygen or the ylide carbonyl oxygen.

Reactions of the same carbonyl ylide intermediate with aldehydes are even more fruitful. The $Rh_2(OAc)_2$ catalyzed reaction proceeds at room temperature in the presence of 2 mol% of the catalyst, but the diastereoselectivity is disappointingly low (endo/exo = 49:51, Scheme 11.56). However, when 10 mol% of the cocatalyst $Yb(OTf)_3$ is added, the reaction becomes highly exo-selective (endo/exo = 3:97) (198). Suga has extended this Lewis acid catalyzed carbonyl ylide cycloaddition reaction to catalyzed asymmetric versions. The chiral cocatalyst employed is ytterbium(III) tris(*S*)-1,1′-binaphthyl-2,2′-diyl phosphonate, $Yb[(S)–BNP]_3$ (10 mol%). In the reaction of methyl *o*-(diazoacetyl)benzoate with benzyloxyacetaldehyde in the presence of $Rh_2(OAc)_2$ (2 mol%) at room temperature with the chiral Yb catalyst, the diastereoselectivity is low (endo/exo = 57:43) and the enantiopurity of the endo-cycloadduct is 52% ee.

Several examples are of enantioselective rearrangement reactions of oxonium, sulfonium, or selenonium ylides, generated by the catalytic decomposition of diazo compounds (199–203), are known. The first reports of enantioselective dipolar

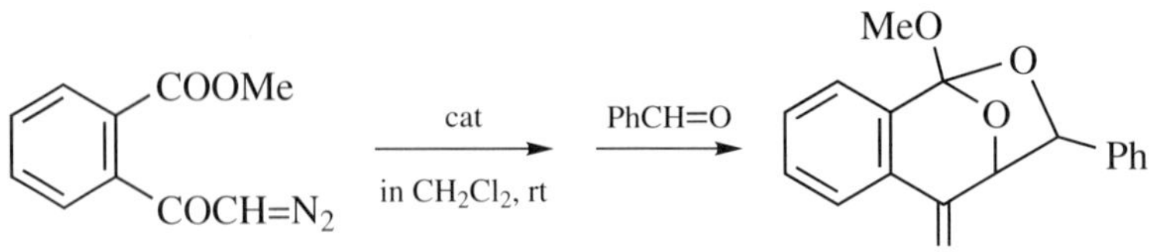

endo and *exo*

Catalyst:
$Rh_2(OAc)_2$: 2 mol%, 77% (*endo*/*exo* = 49:51)
$Rh_2(OAc)_2 + Yb(OTf)_3$: 2 mol% + 10 mol%, 89% (*endo*/*exo* = 3:97)

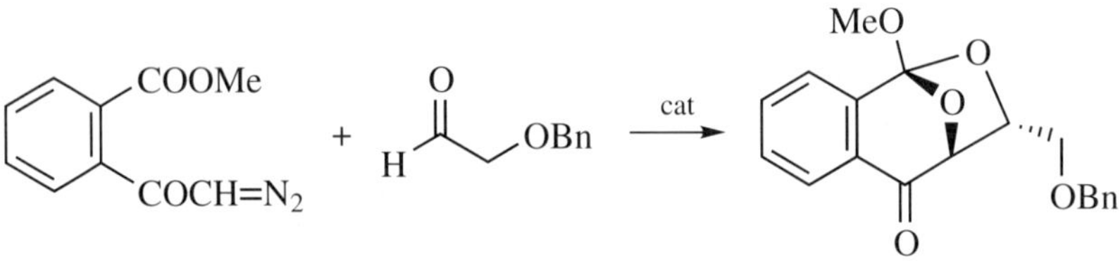

endo (52% ee)
and *exo*-isomer (14% ee)

Catalyst:
$Rh_2(OAc)_2 + Yb[(S)\text{-BNP}]_3$: 2 mol% + 10 mol%

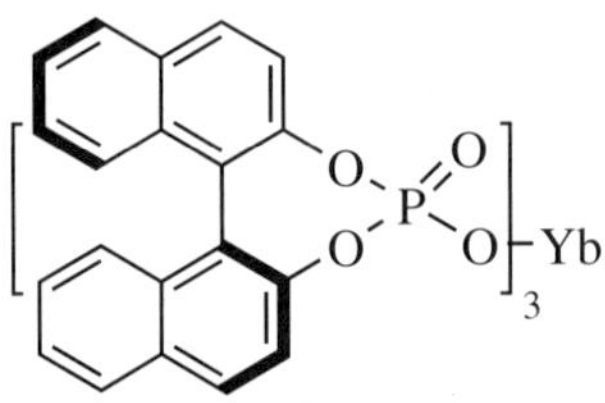

Scheme 11.56

cycloadditions of such species have been reported. Hodgson et al. (204) was the first to demonstrate enantioselective reactions of this type by application of Padwa's intramolecular carbonyl ylide trapping method (205). A diazodiketone ester having a tethered terminal alkene moiety was treated with a catalytic amount (1 mol%) of a proline-based chiral Rh(II) catalyst in hexane (Scheme 11.57). The resulting cycloadduct obtained in 76% shows a moderate enantiopurity of 53% ee.

Hashimoto and co-workers (206,207) recently published enantioselectivities of up to 92% ee in carbonyl ylide cycloadditions to acetylenic esters in the presence of a chiral rhodium catalyst (Scheme 11.58).

11.4.2. Nitrile Ylide Equivalents

Suga et al. (208,209) reported that 5-alkoxyoxazoles react with aldehydes giving 4,5-*cis*-2-oxazoline-4-carboxylates stereoselectively when activated with a

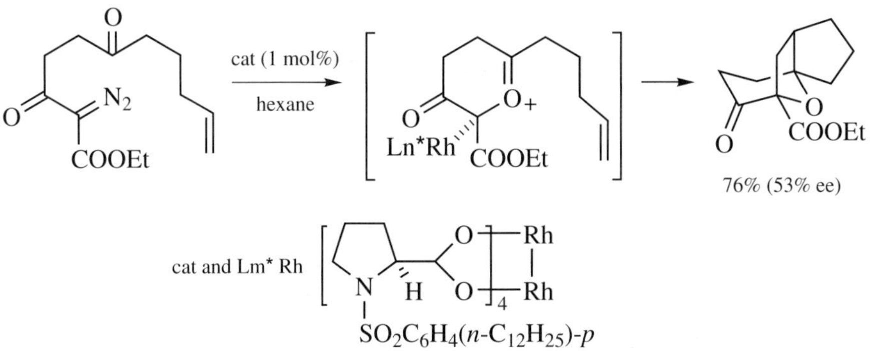

Scheme 11.57

stoichiometric amount of methylaluminum β-binaphthoxide as the Lewis acid. This reaction has recently been extended to a catalytic enantioselective version using an enantiopure methylaluminum β-binaphthoxide (Scheme 11.59) (210,211). Although the actual reacting species were not assigned, 5-alkoxyoxazoles behave as nitrile ylide 1,3-dipole equivalents in Lewis acid catalyzed reactions with aldehydes.

Ito et al. (212–216) developed new catalyzed asymmetric cycloaddition reactions of what are formally nitrile ylides (Scheme 11.60). Thus, methyl α-isocyanoalkanoates are allowed to react with benzaldehyde or acetaldehyde in the presence of 0.5–1.0 mol% of a chiral (aminoalkyl)ferrocenylphosphine–gold(I) complex to give optically active methyl 2-oxazoline-4-carboxylates with high enantioselectivity in quantitative yield. The oxazolines were converted into optically active β hydroxy-α-amino acid methyl esters. The selective formation of cis-isomers of 2-oxazoline-4-carboxylates in Suga's reactions is synthetically complimentary to the ferrocenylphosphine–gold(I) complex/catalyzed reactions by Ito. Grigg et al. (217) recently extended Ito's reaction to the Ag(I) ion-catalyzed

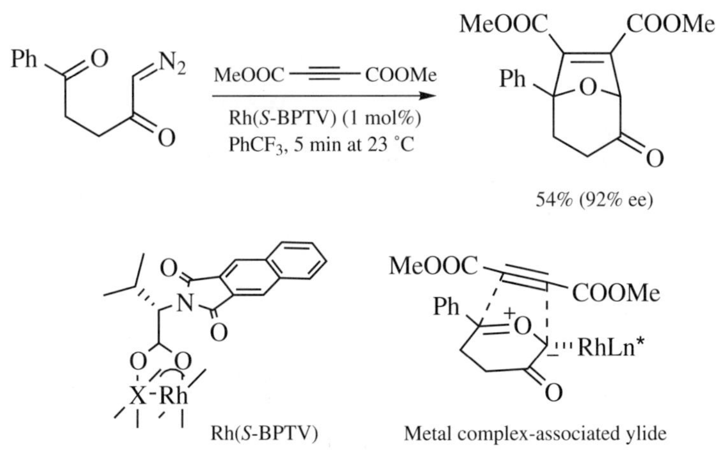

Scheme 11.58

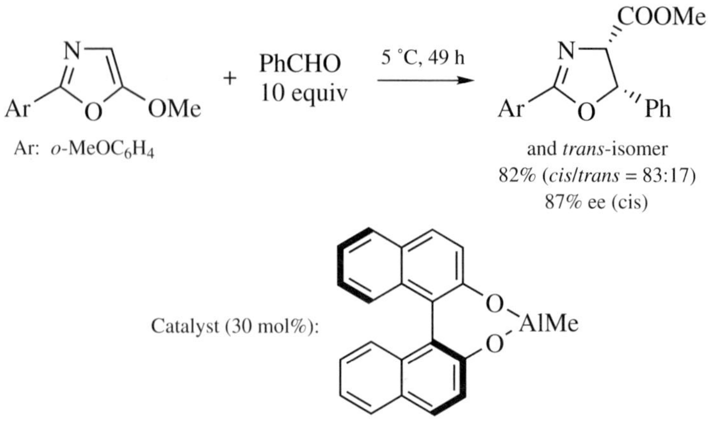

Scheme 11.59

1,3-dipolar cycloaddition of methyl isocyanoacetate with α,β-unsaturated carbonyl compounds (Scheme 11.60). The rate of this reaction can be enhanced effectively with a catalytic amount of silver(I) acetate (1 mol%). The Grigg et al. (217) proposed a reaction mechanism in which silver(I) acetate adds to the

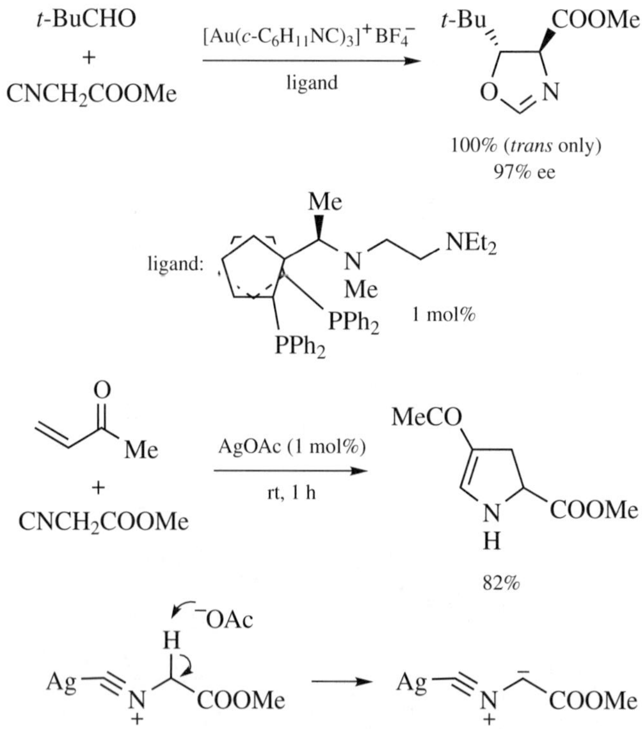

Scheme 11.60

isocyanoacetate carbon to form an organometallic intermediate that is subsequently deprotonated to generate a nitrile ylide 1,3-dipole. After the cycloaddition, double-bond migration via the imine–enamine tautomerism.

11.5. CONCLUSION

So far, a limited number of 1,3-dipoles have been used effectively in synthetic routes to stereochemically defined heterocycles, with the exception of nitrones (6,13) and azomethine ylides, although the use of external reagents to control stereochemistry and regiochemistry in the dipolar cycloadditions of such species is still relatively new, dating back 10–15 years. Nitrile oxides are beginning to show promise in such processes. Work on employing these and other dipoles in cycloadditions mediated by external reagents will likely continue at a rapid pace, since these processes are so powerful and there is much room for improvement using this strategy. Much work remains to be done on the external reagents themselves, especially Lewis acid catalysts in chiral form.

REFERENCES

1. A. Padwa, ed. *1,3-Dipolar Cycloaddition Chemistry*, Vols. 1 and 2, Wiley, New York, **1984**.
2. W. Carruthers, *Cycloaddition Reactions in Organic Sythesis*; Pergammon, Oxford, **1990**.
3. A. Padwa, in *Comprehensive Organic Synthesis*, Vol. 4, B. Trost, ed., Wiley, New York, **1991**, pp. 1069–1109.
4. P. A. Wade, in *Comprehensive Organic Synthesis*, B. Trost, ed., Wiley, New York, **1991**; Vol. 4, pp. 1111–1168.
5. R. Huisgen, "1,3-Dipolar Cycloadditions - Introduction, Survey, Mechanism", in *1,3-Dipolar Cycloaddition Chemistry*, Vol. 1, A. Padwa, ed., Wiley, New York, **1984**; Vol. 1, pp. 1–176.
6. K. V. Gothelf and K. A. Jørgensen, *Chem. Rev.*, **1998**, *98*, 863.
7. R. Huisgen, *Angew. Chem., Int. Ed. Engl.*, **1963**, *2*, 565.
8. P. Caramella and P. Grünanger, "Nitrile Oxides and Imines," in *1,3-Dipolar Cycloaddition Chemistry*, A. Padwa, ed., Wiley, New York, Vol. 1, **1984**, Chapter 3, pp. 291–392.
9. A. P. Kozikowski, *Acc. Chem. Res.*, **1984**, *17*, 410.
10. K. B. G. Torssell, *Nitrile Oxides, Nitrones and Nitronates in Organic Synthesis: Novel Strategies in Synthesis*, VCH Publishers, New York, **1988**.
11. S. Kanemasa and O. Tsuge, *Heterocycles*, **1990**, *30*, 719.
12. C. J. Easton, C. M. M. Hughes, G. P. Savage, and G. W. Simpson, *Adv. Heterocycl. Chem.*, **1994**, *60*, 261.
13. K. V. Gothelf and K. A. Jørgensen, *Chem. Commun.*, **2000**, 1449.
14. J. W. Lown, "Azomethine Ylides", in *1,3-Dipolar Cycloaddition Chemistry*, Vol. 1, A. Padwa, ed., J. Wiley, New York, **1984**, Chapter 6, pp. 653–732.
15. H. W. Heine and R. Peavy, *Tetrahedron Lett.*, **1965**, 3123.
16. R. Huisgen, W. Scheer, and H. Huber, *J. Am. Chem. Soc.*, **1967**, *89*, 1753.
17. R. Huisgen and H. Mäder, *J. Am. Chem. Soc.*, **1971**, *93*, 1777.
18. H. Hermann, R. Huisgen, and H. Mäder, *J. Am. Chem. Soc.*, **1971**, *93*, 1779.

19. O. Tsuge and S. Kanemasa, "Recent Advances in Azomethine Ylide Chemistry," in *Advances in Heterocyclic Chemistry*, Vol. 45, A. R. Katritzky, ed., Academic Press, San Diego, **1989**, pp. 232–349.
20. R. Grigg and J. Kemp, *J. Chem. Soc., Chem. Commun.*, **1977**, 125.
21. R. Grigg and J. Kemp, *J. Chem. Soc., Chem. Commun.*, **1978**, 109.
22. O. Tsuge, K. Ueno, and K. Oe, *Chem. Lett.*, **1979**, 1407.
23. O. Tsuge and K. Ueno, *Heterocycles*, **1982**, *19*, 1411.
24. R. Grigg, *Chem. Soc. Rev.*, **1987**, *16*, 89.
25. E. Vedejs and G. R. Martinez, *J. Am. Chem. Soc.*, **1979**, *101*, 6452.
26. K. Achiwa and M. Sekiya, *Chem. Lett.*, **1981**, 1213.
27. A. Padwa and Y.-Y. Chen, *Tetrahedron Lett.*, **1983**, *24*, 3447.
28. O. Tsuge, S. Kanemasa, A. Hatada, and K. Matsuda, *Chem. Lett.*, **1984**, 801.
29. A. Hosomi, Y. Sakata, and H. Sakurai, *Chem. Lett.*, **1984**, 1117.
30. E. Vedejs and F. G. West, *Chem. Rev.*, **1986**, *86*, 941.
31. E. Vedejs, "Nonstabilized Azomethine Ylides," in *Advances in Cycloaddition*, Vol. 2, D. P. Curran, ed., JAI Press Inc., Greenwich, **1988**, pp. 33–51.
32. N. Imai, Y. Terao, and K. Achiwa, *J. Org. Syn. Chem. Jpn.*, **1985**, *43*, 862.
33. P. N. Confalone and E. M. Huie, *J. Org. Chem.*, **1983**, *48*, 2994.
34. P. N. Confalone and E. M. Huie, *J. Am. Chem. Soc.*, **1984**, *106*, 7175.
35. M. Joucla, J. Mortier, and J. Hamelin, *Tetrahedron Lett.*, **1985**, *26*, 2775.
36. O. Tsuge, S. Kanemasa, M. Ohe, and S. Takenaka, *Chem. Lett.*, **1986**, 973.
37. O. Tsuge, S. Kanemasa, M. Ohe, K. Yorozu, S. Takenaka, and K. Ueno, *Chem. Lett.*, **1986**, 271; *Bull. Chem. Soc. Jpn.*, **1987**, *60*, 4067.
38. R. Grigg and S. Thianpatanagul, *J. Chem. Soc., Chem. Commun.*, **1984**, 180.
39. R. Grigg, M. F. Aly, V. Sridharan, and S. Thianpatanagul, *J. Chem. Soc., Chem. Commun.*, **1984**, 182.
40. M. Joucla and J. Mortier, *J. Chem. Soc., Chem. Commun.*, **1985**, 1566.
41. O. Tsuge, S. Kanemasa, M. Ohe, and S. Takenaka, *Bull. Chem. Soc. Jpn.*, **1987**, *60*, 4079.
42. R. Beugelmans, G. Negron, and G. Roussi, *J. Chem. Soc., Chem. Commun.*, **1983**, 31.
43. R. Beugelmans, L. Benadjila-Iguertsira, J. Chastanet, G. Negron, and G. Roussi, *Can. J. Chem.*, **1985**, *63*, 725.
44. S. Kanemasa and O. Tsuge "*N*-Metalated Azomethine Ylides," in *Advances in Cycloaddition*, Vol. 3, D. P. Curran, ed., JAI Press Inc., Greenwich, **1993**, pp. 99–159.
45. L. Casella, M. Gullotti, A. Pasini, and R. Psaro *Synthesis*, **1979**, 150.
46. L. Casella, M. Gullotti, and E. Melani, *J. Chem. Soc., Perkin Trans. 1*, **1982**, 1827.
47. R. Grigg, V. Sridharan, and S. Thianpatanagul, *J. Chem. Soc., Perkin Trans. 1*, **1986**, 1669.
48. T. Kauffmann, *Angew. Chem., Int. Ed. Engl.*, **1974**, *13*, 627.
49. T. Kauffmann, H. Berg, and E. Köppelmann, *Angew. Chem.*, **1970**, *82*, 396.
50. T. Kauffmann and R. Eidenschink, *Angew. Chem.*, **1971**, *83*, 794.
51. T. Kauffmann and E. Köppelmann, *Angew. Chem.*, **1972**, *84*, 261.
52. T. Kauffmann, H. Ahlers, A. Hamsen, H. Schulz, H.-J. Tilhard, and A. Vahrenhorst, *Angew. Chem.*, **1977**, *89*, 107.
53. T. Kauffmann, K. Habersaat, and E. Köppelmann, *Chem. Ber.*, **1977**, *110*, 638.
54. S. Sinbandhit and J. Hamelin, *J. Chem. Soc., Chem. Commun.*, **1977**, 768.
55. L. Vo-Quang and Y. Vo-Quang, *Tetrahedron Lett.*, **1978**, 4679.
56. K. Kamata, and M. Terashima, *Heterocycles*, **1980**, *14*, 205.
57. L. Vo-Quang, H. Gaessler, and Y. Vo-Quang, *Angew. Chem. Int. Ed. Engl.*, **1981**, *20*, 880.

58. L. Vo-Quang, Y. Vo-Quang, M. J. Pouet, and M. P. Simonnin, *Tetrahedron*, **1981**, *32*, 4343.

59. T. Kauffmann, H. Ahlers, K.-J. Echsler, H. Schulz, and H.-J. Tilhard, *Chem. Ber.*, **1985**, *118*, 4496.

60. W. H. Peason, M. A. Walters, and K. D. Oswell, *J. Am. Chem. Soc.*, **1986**, *108*, 2769.

61. T. Kauffmann, H. Berg, E. Ludorff, and A. Woltermann, *Angew. Chem.*, **1970**, *82*, 986.

62. T. Kauffmann and R. Eidenschink, *Angew. Chem.*, **1973**, *85*, 583.

63. W. H Pearson, D. P. Szura, and W. G. Harter, *Tetrahedron Lett.*, **1988**, *29*, 761.

64. W. H. Pearson, D. P. Szura, and M. J. Postich, *J. Am. Chem. Soc.*, **1992**, *114*, 1329.

65. W. H. Pearson and M. J. Postich, *J. Org. Chem.*, **1992**, *57*, 6354.

66. W. H. Pearson and E. P. Stevens, *Tetrahedron Lett.*, **1994**, *35*, 2641.

67. W. H. Pearson and M. J. Postich, *J. Org. Chem.*, **1994**, *59*, 5662.

68. W. H. Pearson and V. A. Jacobs, *Tetrahedron Lett.*, **1994**, *35*, 7001.

69. W. H. Pearson and F. E. Lovering, *Tetrahedron Lett.*, **1994**, *35*, 9173.

70. W. H. Pearson and F. E. Lovering, *J. Am. Chem. Soc.*, **1995**, *117*, 12336.

71. W. H. Pearson, N. S. Barta, and J. W. Kampf, *Tetrahedron Lett.*, **1997**, *38*, 3369.

72. W. H. Pearson and Y. Mi, *Tetrahedron Lett.*, **1997**, *38*, 5441.

73. W. H. Pearson and R. B. Clark, *Tetrahedron Lett.*, **1997**, *38*, 7669.

74. W. H. Pearson and F. E. Lovering, *J. Org. Chem.*, **1998**, *63*, 3607.

75. W. H. Pearson and E. P. Stevens, *J. Org. Chem.*, **1998**, *63*, 9812.

76. W. H. Pearson and Y. Len, *J. Org. Chem.*, **1999**, *64*, 688.

77. R. Grigg, J. Kemp, J. Malone, and A. Tangthongkum, *J. Chem. Soc., Chem. Commun.*, **1980**, 648.

78. O. Tsuge, S. Kanemasa, K. Yorozu, and K. Ueno, *Bull. Chem. Soc. Jpn.*, **1987**, *60*, 3359.

79. O. Tsuge, K. Ueno, S. Kanemasa, and K. Yorozu, *Bull. Chem. Soc. Jpn.*, **1986**, *59*, 1809.

80. O. Tsuge, K. Ueno, S. Kanemasa, and K. Yorozu, *Bull. Chem. Soc. Jpn.*, **1987**, *60*, 3347.

81. S. Kanemasa, T. Hayashi, J. Tanaka, H. Yamamoto, and T. Sakurai, *J. Org. Chem.*, **1991**, *56*, 4473.

82. S. Kanemasa, O. Uchida, and E. Wada, *J. Org. Chem.*, **1990**, *55*, 4411.

83. S. Kanemasa and H. Yamamoto, *Tetrahedron Lett.*, **1990**, *31*, 3633.

84. S. Kanemasa, H. Yamamoto, E. Wada, T. Sakurai, and K. Urushido, *Bull. Chem. Soc. Jpn.*, **1990**, *63*, 2857.

85. S. Kanemasa, M. Yoshioka, and O. Tsuge, *Bull. Chem. Soc. Jpn.*, **1989**, *62*, 2196.

86. O. Tsuge, S. Kanemasa, and M. Yoshioka, *J. Org. Chem.*, **1988**, *53*, 1384.

87. S. Kanemasa, M. Yoshioka, and O. Tsugc, *Bull. Chem. Soc. Jpn.*, **1989**, *62*, 869.

88. D. A. Barr, R. Grigg, H. Q. N. Gunaratne, J. Kemp, P. McMeekin, and V. Sridharan, *Tetrahedron*, **1988**, *44*, 557.

89. K. Amornraksa, D. Barr, G. Donegan, R. Grigg, P. Ratananukul, and V. Sridharan, *Tetrahedron*, **1989**, *45*, 4649.

90. D. A. Barr, R. Grigg, and V. Sridharan, *Tetrahedron Lett.*, **1989**, *30*, 4727.

91. S. Kanemasa, O. Uchida, E. Wada, and H. Yamamoto, H. *Chem. Lett.*, **1990**, 105.

92. A. Padwa, Y.-Y. Chen, U. Chiacchio, and W. Dent, *Tetrahedron*, **1985**, *41*, 3529.

93. J. Rouden, J. Boyer, and H.-P. Husson, *Tetrahedron Lett.*, **1989**, *30*, 5133.

94. A. G. H. Wee, *J. Chem. Soc., Perkin Trans. 1*, **1989**, 1363.

95. A. S. Anslow, L. M. Harwood, H. Phillips, and D. Watkin, *Tetrahedron Asymm.*, **1991**, *2*, 169.

96. S. Takano, Y. Iwabuchi, and K. Ogasawara, *J. Am. Chem. Soc.*, **1987**, *109*, 5523.

97. P. Garner, K. Sunitha, W.-B. Ho, W. J. Youngs, V. O. Kennedy, and A. Djebli, *J. Org. Chem.*, **1989**, *54*, 2041.

98. S. Takano, K. Samizu, and K. Ogasawara, *Chem. Lett.*, **1990**, 1239.

99. T. Takahashi, K. Kitano, T. Hagi, H. Nihonmatsu, and T. Koizumi, *Chem. Lett.*, **1989**, 597.

100. P. Garner and W.-B. Ho, *J. Org. Chem.*, **1990**, *55*, 3973.

101. M. Asami, H. Ohno, S. Kobayashi, and T. Mukaiyama, *Bull. Chem. Soc. Jpn.*, **1978**, *51*, 1869.

102. D. A. Bar, M. J. Dorrity, R. Grigg, J. F. Malone, J. Montgomery, S. Rajviroongit, and P. Stevenson, *Tetrahedron Lett.*, **1990**, *31*, 6569.

103. U. Schöllkopf, D. Pettig, and U. Busse, *Synthesis*, **1986**, 737.

104. D. Pettig and U. Schöllkopf, *Synthesis*, **1988**, 173.

105. U. Schöllkopf, D. Pettig, E. Schulze, M. Klinge, E. Egert, B. Benecke, and M. Noltemeyer, *Angew. Chem. Int. Ed. Engl.*, **1988**, *27*, 1194.

106. J. M. McIntosh and P. Mishra, *Can. J. Chem.*, **1986**, *64*, 726.

107. J. M. McIntosh and R. K. Leavitt, *Tetrahedron Lett.*, **1986**, *27*, 3839.

108. J. M. McIntosh, R.K. Leavitt, P. Mishra, K. C. Cassidy, J. E. Drake, and R. Chadha, *J. Org. Chem.*, **1988**, *53*, 1947.

109. J. M. McIntosh, K. C. Cassidy, and L. C. Matassa, *Tetrahedron*, **1989**, *45*, 5449.

110. A. E. Achqar, M. Boumzebra, M.-L. Roumestant, and P. Viallefont, *Tetrahedron*, **1988**, *44*, 5319.

111. S. Kanemasa, A. Tatsukawa, and E. Wada, *J. Org. Chem.*, **1991**, *56*, 2875.

112. H. Yamamoto, S. Kanemasa, and E. Wada, *Bull. Chem. Soc. Jpn.*, **1991**, *64*, 2739.

113. B. T. Golding and P. V. Ioannou, *Synthesis*, **1977**, 423.

114. S. Takano, A. Kurotaki, M. Takahashi, and K. Ogasawara, *Synthesis*, **1986**, 403.

115. N. Minami, S. S. Ko, and Y. Kishi, *J. Am. Chem. Soc.*, **1982**, *104*, 1109.

116. H. Matsunaga, T. Sakamaki, H. Nagaoka, and Y. Yamada, *Tetrahedron Lett.*, **1983**, *24*, 3009.

117. T. Suzuki, E. Sato, S. Kamada, H. Tada, and K. Unno, *J. Chem. Soc., Perkin Trans. 1*, **1986**, 387.

118. B. M. Trost, J. Lynch, P. Renaut, and D. H. Steinman, *J. Am. Chem. Soc.*, **1986**, *108*, 284.

119. B. M. Trost and S. M. Mignani, *Tetrahedron Lett.*, **1986**, *27*, 4137.

120. B. M. Trost, S. M. Mignani, and T. N. Nanninga, *J. Am. Chem. Soc.*, **1988**, *110*, 1602.

121. A. Tatsukawa, M. Dan, M. Obatake, K. Kawatake, T. Fukata, E. Wada, S. Kanemasa, and A. Kakehi, *J. Org. Chem.*, **1993**, *58*, 4221.

122. A. Tatsukawa, K. Kawatake, S. Kanemasa, and J. M. Rudzinski, *J. Chem. Soc., Perkin Trans. 2*, **1994**, 2525.

123. A. Dondoni, M. Pedro, and J. Oedna, *Tetrahedron Lett.*, **1991**, *32*, 3247 and references cited therein.

124. A. P. Kozikowski and A. K. Gohsh, *J. Am. Chem. Soc.*, **1982**, *104*, 5788.

125. A. P. Kozikowski and A. K. Gohsh, *J. Org. Chem.*, **1984**, *49*, 2762.

126. K. N. Houk, S. R. Moses, Y. Wu, N. G. Rondan, V. Jäger, R. Schoho, and F. R. Fronczek, *J. Am. Chem. Soc.*, **1984**, *106*, 3880.

127. K. N. Houk, H. Duh, Y. Wu, and S. R. Moses, *J. Am. Chem. Soc.*, **1986**, *108*, 2754.

128. P. Allway and R. Grigg, *Tetrahedron Lett.*, **1991**, *32*, 5817.

129. R. Grigg, *Tetraderon Asymm.*, **1995**, *6*, 2475.

130. S. Kanemasa, M. Nishiuchi, A. Kamimura, and K. Hori, *J. Am. Chem. Soc.*, **1994**, *116*, 2324.

131. S. Kanemasa, S. Kobayashi, M. Nishiuchi, H. Yamamoto, and E. Wada, *Tetrahedron Lett.*, **1991**, *32*, 6367.

132. S. Kanemasa, M. Nishiuchi, and E. Wada, *Tetrahedron Lett.*, **1992**, *33*, 1357.

133. S. Kanemasa and M. Nishiuchi, *Tetrahedron Lett.*, **1993**, *34*, 4011.

134. S. Kanemasa, K. Okuda, H. Yamamoto, and S. Kaga *Tetrahedron Lett.*, **1993**, *38*, 4095.

135. P. Micuch, L. Fisera, M. K. Cyranski, and T. M. Krygowski, *Tetrahedron Lett.*, **1997**, *40*, 167.

136. P. Micuch, L. Fisera, M. C. Cyranski, T. M. Krygowski, and J. Krajcik, *Tetrahedron*, **2000**, *56*, 5465.

137. P. C. B. Page, M. Purdie, and D. Lathbury, *Tetrahedron*, **1997**, *53*, 7365.

138. J. W. Bode and E. M. Carreir, *J. Am. Chem. Soc.*, **2001**, *123*, 3611.

139. J. W. Bode and E. M. Carreira, *J. Org. Chem.* **2001**, *66*, 6410.

140. H. Yamamoto, S. Watanabe, K. Kadotani, M. Hasegawa, M. Noguchi, and S. Kanemasa, *Tetrahedron Lett.*, **2000**, *41*, 3131.

141. H. Yamamoto, K. Zenitani, K. Fukushima, S. Watanabe, M. Noguchi, and S. Kanemasa, presented in the 79th Annual Meeting of Chemical Society of Japan (Kobe), **2001**, 2H6 12.

142. H. Yamamoto, M. Hasegawa, M. Harada, S. Watanabe, M. Noguchi, and S. Kanemasa, presented in the 79th Annual Meeting of Chemical Society of Japan (Kobe), **2001**, 2H6 13.

143. A. G. Meyer, C. J. Easton, S. F. Lincoln, and G. W. Simpson, *J. Chem. Soc., Chem. Commun.*, **1997**, 1517.

144. J. D. Toker, P. Wentworth, Jr., Y. Hu, K. N. Houk, and K. Janda, *J. Am. Chem. Soc.*, **2000**, *122*, 3244.

145. C. Raposo and C. S. Wilcox, *Tetrahedron Lett.*, **1999**, *40*, 1285.

146. R. H. Wallace and J. Liu, *Tetrahedron Lett.*, **1994**, *35*, 7493.

147. J. Liu, A. Eddings, and R. H. Wallace, *Tetrahedron Lett.*, **1997**, *38*, 6795.

148. K. Inomata and Y. Ukaji, *J. Syn. Org. Chem. Janan*, **1998**, *56*, 11.

149. Y. Ukaji, *J. Syn. Org. Chem. Janan*, **1999**, *57*, 581.

150. Y. Ukaji, K. Sada, and K. Inomata, *Chem. Lett.*, **1993**, 1847.

151. Y. Ukaji and K. Inomata, *Chem. Lett.*, **1996**, 455.

152. Y. Ukaji, K. Taniguchi, K. Sada, and K. Inomata, *Chem. Lett.*, **1997**, 547.

153. Y. Ukaji, Y. Shimizu, Y. Kenmotsu, A. Ahmed, and K. Inomata, *Chem. Lett.*, **1997**, 59.

154. Y. Yoshida, Y. Ukaji, S. Fujinami, and K. Inomata, *Chem. Lett.*, **1998**, 1023.

155. Y. Ukaji, M. Ima, T. Yamada, and K. Inomata, *Heterocycles*, **2000**, *52*, 563.

156. K. V. Gothelf and K. A. Jørgensen, *Acta Chem. Scand.*, **1996**, *50*, 652.

157. K. V. Gothelf and K. A. Jørgensen, *Chem. Commun.*, **2000**, 1449.

158. S. Kanemasa, T. Uemura, and E. Wada, *Tetrahedron Lett.*, **1992**, *33*, 7889.

159. M. Tiecco, L. Testaferri, M. Tingoli, L. Bagnoli, and F. Marini, *J. Chem. Soc., Perkin Trans. 1*, **1993**, 1989.

160. R. Shaw, D. Lathbury, M. Anderson, and T. Gallagher, *J. Chem. Soc., Perkin Trans. 1*, **1991**, 659.

161. M. Frederickson, R. Grigg, M. Thornton-Pett, and J. Redpath, *Tetrahedron Lett.*, **1997**, *38*, 7777.

162. I. S. Saba, M. Frederickson, R. Grigg, P. J. Dunn, and P. C. Levett, *Tetrahedron Lett.*, **1997**, *38*, 6099.

163. R. S. Kusurkar, M. S. Wadia, D. K. Bhosale. S. S. Tavale, and V. G. Puranik, *J. Chem. Res. (S)*, **1996**, 478.

164. M. Frederickson, R. Grigg, J. Redpath, and M. Thornton-Pett, *Tetrahedron*, **1994**, *50*, 5495.

165. M. Frederickson, R. Grigg, Z. Rankovic, M. Thornton-Pett, J. Redpath, and R. Crossley *Tetrahedron*, **1995**, *51*, 6835.

166. S. Kanemasa, T. Tsuruoka, and E. Wada, *Tetrahedron Lett.*, **1993**, *34*, 87.

167. S. Kanemasa, T. Tsuruoka, and H. Yamamoto, *Tetrahedron Lett.*, **1995**, *36*, 5019.

168. S. Kanemasa and T. Tsuruoka, *Chem. Lett.*, **1995**, 49.

169. O. Tamura, T. Kuroki, Y. Sakai, J. Takizawa, J. Yoshino, Y. Morita, N. Mita, K. Gotanda, and M. Sakamoto, *Tetrahedron Lett.*, **1999**, *40*, 895.

170. O. Tamura, T. Yamaguchi, K. Noe, and M. Sakamoto, *Tetrahedron Lett.*, **1993**, *34*, 4009.

171. O. Tamura, T. Okabe, T. Yamaguchi, K. Gotanda, K. Noe, and M. Sakamoto, *Tetrahedron*, **1995**, *51*, 107.

172. O. Tamura, T. Okabe, T. Yamaguchi, J. Kotani, K. Gotanda, K. Noe, and M. Sakamoto, *Tetrahedron*, **1995**, *51*, 119.

173. O. Tamura, N. Mita, N. Kusaka, H. Suzuki, and M. Sakamoto, *Tetrahedron Lett.*, **1997**, *38*, 429.

174. O. Tamura, N. Mita, T. Okabe, T. Yamaguchi, C. Fukushima, M. Yamashita, Y. Morita, N. Morita, H. Ishibashi, and M. Sakamoto, *J. Org. Chem.*, **2001**, *66*, 2602.
175. K. V. Gothelf and K. A. Jørgensen, *J. Org. Chem.*, **1994**, *59*, 5687.
176. J. P. G. Seerden, A. W. A. Scholte op Reimer, and H. W. Scheeren, *Tetrahedron Lett.*, **1994**, *35*, 4419.
177. K. V. Gothelf, R. G. Hazell, and K. A. Jørgensen, *J. Am. Chem. Soc.*, **1995**, *117*, 4435.
178. K. V. Gothelf and K. A. Jørgensen, *J. Org. Chem.*, **1995**, *60*, 6847.
179. K. V. Gothelf, I. Thomsen, and K. A. Jørgensen, *J. Am. Chem. Soc.*, **1996**, *118*, 59.
180. K. V. Gothelf, R. G. Hazell, and K. A. Jørgensen, *J. Org. Chem.*, **1996**, *61*, 346.
181. K. B. Jensen, K. V. Gothelf, and K. A. Jørgensen, *Helv. Chim. Acta*, **1997**, *80*, 2039.
182. K. V. Gothelf and K. A. Jørgensen, *J. Chem. Soc., Perkin Trans. 2*, **1997**, 111.
183. X. B. Jensen, K. V. Gothelf, R. G. Hazell, and K. A. Jørgensen, *J. Org. Chem.*, **1997**, *62*, 2471.
184. K. V. Gothelf, R. G. Hazell, and K. A. Jørgensen, *J. Org. Chem.*, **1998**, *63*, 5483.
185. K. Hori, H. Kodama, T. Ohta, and I. Furukawa, *Tetrahedron Lett.*, **1996**, *37*, 5947.
186. K. Hori, H. Kodama, T. Ohta, and I. Furukawa, *J. Org. Chem.*, **1999**, *64*, 5017.
187. K. B. Simonsen, P. Bayon, R. G. Hazell, K. V. Gothelf, and K. A. Jørgensen, *J. Am. Chem. Soc.*, **1999**, *121*, 3845.
188. K. B. Simonsen, K. V. Gothelf, and K. A. Jørgensen, *J. Org. Chem.*, **1998**, *63*, 7536.
189. K. B. Simonsen, K. A. Jørgensen, Q.-S. Hu, and L. Pu, *J. Chem. Soc., Chem. Commun.*, **1999**, 811.
190. K. B. Jensen, R. G. Hazell, and K. A. Jørgensen, *J. Org. Chem.*, **1999**, *64*, 2353.
191. M. Kawamura and S. Kobayashi, *Tetrahedron Lett.*, **1999**, *40*, 3213.
192. S. Kanemasa, Y. Oderaotoshi, J. Tanaka, and E. Wada, *J. Am. Chem. Soc.*, **1998**, *120*, 12355.
193. J.-P. G. Seerden, M. M. M. Kuypers, and H. W. Scheeren, *Tetrahedron: Asymm.*, **1995**, *6*, 1441.
194. J.-P. G. Seerden, M. M. M. Boeren, and H. W. Scheeren, *Tetrahedron*, **1997**, *53*, 11843.
195. Y. Ukaji, K. Taniguchi, K. Sada, and K. Inomata, *Chem. Lett.*, **1997**, 547.
196. M. Miura, M. Enna, K. Okuro, and M. Nomura, *J. Org. Chem.*, **1995**, *60*, 4999.
197. H. Suga, H. Ishida, and T. Ibata, *Tetrahedron Lett.*, **1998**, *39*, 3165.
198. H. Suga, A. Kakehi, S. Ito, K. Inoue, H. Ishida, and T. Ibata, *Org. Lett.*, **2000**, *2*, 3145.
199. N. McCarthy, M. A. McKervey, T. Ye, M. McCann, E. Murphy, and M. P. Doyle, *Tetrahedron Lett.*, **1992**, *33*, 5983.
200. Y. Nishibayashi, K. Ohe, and S. Uemura, *J. Chem. Soc., Chem. Commun.*, **1995**, 1245.
201. T. Fukuda and T. Katsuki *Tetrahedron Lett.*, **1997**, *38*, 3435.
202. N. Pierson, C. Fernandez-Garcia, and M. A. McKervey, *Tetrahedron Lett.*, **1997**, *38*, 4705.
203. J. S. Clark, M. Fretwell, G. A. Whitlock, C. J. Burns, and D. N. A. Fox, *Tetrahedron Lett.*, **1998**, *39*, 97.
204. D. M. Hodgson, P. A. Stupple, and C. Johnstone, *Tetrahedron Lett.*, **1997**, *38*, 6471.
205. (a) A. Padwa, M. A. Brodney, J. P. Marino, Jr, M. H. Osterhout, and A. T. Price, *J. Org. Chem.*, **1997**, *62*, 67. (b) A. Padwa, M. A. Brodney, J. P. Marino, J. P., Jr., and S. M. Sheehan, *J. Org. Chem.*, **1997**, *62*, 78.
206. S. Kitagaki, M. Anada, O. Kataoka, K. Matsuno, C. Umeda, N. Watanabe, and S.-I. Hashimoto, *J. Am. Chem. Soc.*, **1999**, *121*, 1417.
207. S. Kitagaki, M. Yasugahira, M. Anada, M. Nakajima, and S.-I. Hashimoto, *Tetrahedron Lett.*, **2000**, *41*, 5931.
208. H. Suga, X. Shi, and T. Ibata, *J. Org. Chem.*, **1993**, *58*, 7397.
209. H. Suga, H. Fujieda, Y. Hirotsu, and T. Ibata, *J. Org. Chem.*, **1994**, *59*, 3359.
210. H. Suga, K. Ikai, and T. Ibata, *Tetrahedron Lett.*, **1998**, *39*, 869.

211. H. Suga, K. Ikai, and T. Ibata, *J. Org. Chem.*, **1999**, *64*, 7040.

212. Y. Ito, M. Sawamura, and T. Hayashi, *J. Am. Chem. Soc.*, **1986**, *108*, 6405.

213. Y. Ito, M. Sawamura, and T. Hayashi, *Tetrahedron Lett.*, **1987**, *28*, 6215.

214. Y. Ito, M. Sawamura, and T. Hayashi, *Tetrahedron Lett.*, **1988**, *29*, 235.

215. Y. Ito, M. Sawamura, E. Shirakawa, K. Hayashizaki, and T. Hayashi, *Tetrahedron Lett.*, **1988**, *28*, 239.

216. Y. Ito, M. Sawamura, E. Shirakawa, K. Hayashizaki, and T. Hayashi, *Tetrahedron*, **1988**, *44*, 5253.

217. R. Grigg, M. I. Lansdell, and M. Thornton-Pett, *Tetrahedron*, **1999**, *55*, 2025.

CHAPTER 12

Asymmetric Reactions

Kurt Vesterager Gothelf and Karl Anker Jørgensen

Center for Metal Catalyzed Reactions, Department of Chemistry, Aarhus University, Denmark

12.1. Selectivity in 1,3-Dipolar Cycloadditions 818
12.2. Diastereoselective Reactions 822
  12.2.1. Chiral Dipoles 822
    12.2.1.1. Nitrones and Nitronates 822
    12.2.1.2. Nitrile Oxides 829
    12.2.1.3. Azomethine Ylides 829
    12.2.1.4. Other Dipoles 834
  12.2.2. Chiral Dipolarophiles 835
    12.2.2.1. Acyclic Allyl Alcohol Derivatives 835
    12.2.2.2. Cyclic Allyl Alcohol Derivatives 836
    12.2.2.3. Allyl Amine Derivatives 838
    12.2.2.4. Vinyl Acetals and Vinyl Aminals 839
    12.2.2.5. Vinyl Phosphine Oxides and Vinyl Sulfoxides 840
    12.2.2.6. Alkenes with More Distant Chiral Centers 842
    12.2.2.7. Heteroatom Dienophiles 844
  12.2.3. Intramolecular Reactions 845
    12.2.3.1. Nitrones and Nitronates 846
    12.2.3.2. Nitrile Oxides 848
    12.2.3.3. Other Dipoles 850
12.3. Diastereoselective Reactions of Substrates with Chiral Auxiliaries 851
  12.3.1. α,β-Unsaturated Esters 851
  12.3.2. α,β-Unsaturated Amides 853
  12.3.3. Other Auxiliaries Attached to Dipolarophiles 857
  12.3.4. Auxiliaries Attached to 1,3-Dipoles 860
12.4. Catalytic Asymmetric 1,3-Dipolar Cycloadditions 864
  12.4.1. Basic Aspects 864
  12.4.2. Nitrones 866
  12.4.3. Nitrile Oxides 882
  12.4.4. Azomethine Ylides 884
  12.4.5. Carbonyl Ylides 886
  12.4.6. Diazo Compounds 888
12.5. Conclusion 889

The Chemistry of Heterocyclic Compounds, Volume 59: Synthetic Applications of 1,3-Dipolar Cycloaddition Chemistry Toward Heterocycles and Natural Products, Edited by Albert Padwa and William H. Pearson.
ISBN 0-471-38726-6

Asymmetric synthesis has achieved a position as one of the most important areas of modern organic chemistry. During the past 20 years the number of publications in this area has been vast. On the pallet of organic reactions that are used in asymmetric synthesis, cycloadditions possess a prominent position, since they are some of the most efficient methods for creating new chiral centers with control of stereochemistry (1–4). The ability to introduce more than one new chiral center in a single step with control of both relative and absolute stereochemistry makes cycloaddition reactions highly attractive key reactions for stereoselective synthesis.

The carbo- and hetero-Diels–Alder reactions are excellent for the construction of six-membered ring systems and are probably the most commonly applied cycloaddition. The 1,3-dipolar cycloaddition complements the Diels–Alder reaction in a number of ways. 1,3-Dipolar cycloadditions are more efficient for the introduction of heteroatoms and are the preferred method for the stereocontrolled construction of five-membered heterocycles (1–4). The asymmetric reactions of 1,3-dipoles has been reviewed extensively by us in 1998 (5), and recently, Karlsson and Högberg reviewed the progress in the area from 1997 and until now (6). Asymmetric metal-catalyzed 1,3-dipolar cycloadditions have also been separately reviewed by us (7–9). Other recent reviews on special topics in asymmetric 1,3-dipolar cycloadditions have appeared. These include reactions of nitrones (10), reactions of cyclic nitrones (11), the progress in 1996–1997 (12), 1,3-dipolar cycloadditions with chiral allyl alcohol derivatives (13) and others (14,15).

The development within the area of asymmetric 1,3-dipolar cycloadditions since the first edition of this series is too extensive to be completely covered in this chapter and we have therefore chosen selected examples to illustrate the different aspects of the subject. The examples have been chosen on the basis of general importance and also to complement recent monographs in the area. Special attention will be given to recent developments within the area of metal-catalyzed reactions. Intramolecular 1,3-dipolar cycloadditions are only briefly described.

12.1. SELECTIVITY IN 1,3-DIPOLAR CYCLOADDITIONS

For all reactions of nonsymmetric 1,3-dipoles with nonsymmetric dipoles, a pair of regioisomers can be obtained (Scheme 12.1). The regioselectivity is highly substrate dependent and it is controlled by both electronic and steric factors (1,5). This subject is to complex and diverse to set out any general rules here. It is very often observed that 1,3-dipolar cycloadditions proceed with complete regioselec-

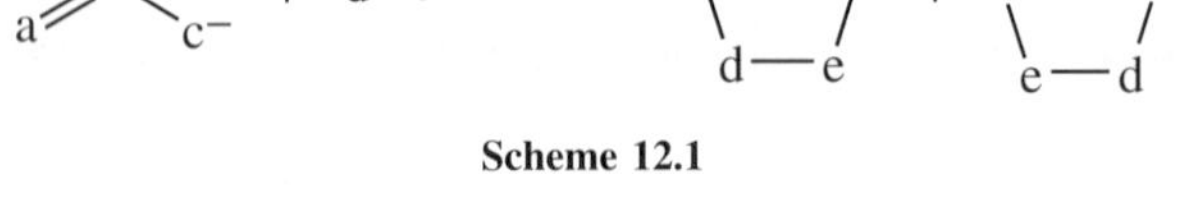

Scheme 12.1

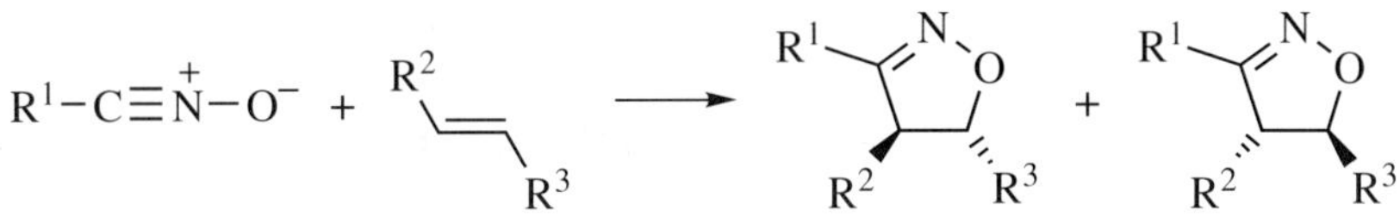

Scheme 12.2

tivity and throughout this chapter the described reactions proceed with complete regioselectivity unless otherwise mentioned.

In the following examination of the stereochemical aspects of 1,3-dipolar cycloadditions, it is assumed that the configuration of the alkene is conserved during the reaction, that is, reactions of trans alkenes leads to trans configuration of the former alkene substituents in the product. For 1,3-dipolar cycloadditions of propargyl–allenyl type 1,3-dipoles such as nitrile oxides, nitrile imines, and azides with alkenes, a maximum of two new stereocenters are formed and the number of possible diastereomers is thus limited to two (Scheme 12.2). If one or both of the starting materials are chiral, the selectivity of this reaction is referred to as diastereofacial selectivity. If the starting materials are both achiral, the products are enantiomers and the (catalyst induced) selectivity referred to as enantioselectivity. For the reactions of propargyl–allenyl type 1,3-dipoles such as nitrile ylides and diazo compounds with alkenes, four diastereomers may be obtained.

In reactions between nitrones and alkenes, up to three new contiguous stereocenters can be formed and a maximum of four diastereomers can be obtained (Scheme 12.3). There are two different kinds of diastereoselectivity to be

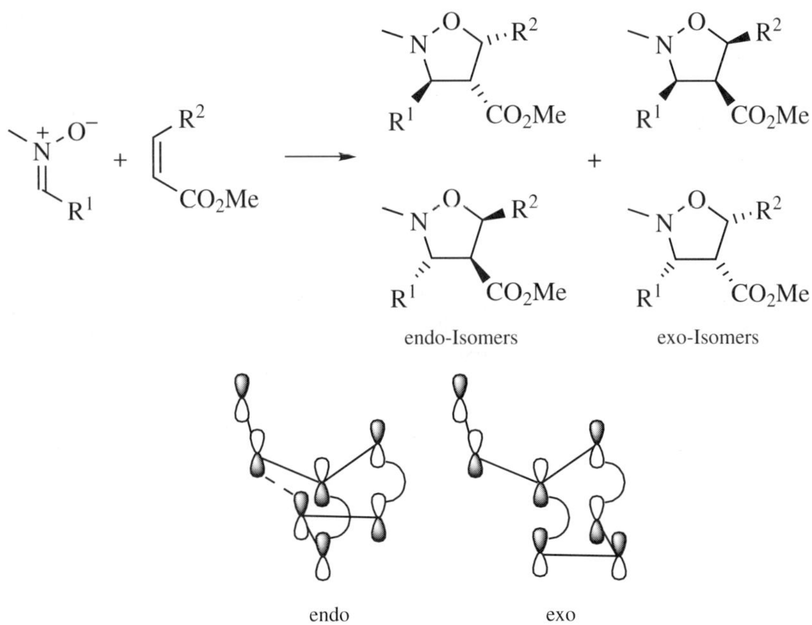

Scheme 12.3

considered in these reactions; the endo/exo selectivity and the diastereofacial selectivity (or enantioselectivity if the substrates are both achiral). The so-called endo-isomers arises from the reaction in which the nitrogen atom of the nitrone points toward a vicinal sp^2-hybridized substituent of the alkene. Whereas secondary π-orbital interactions are of importance for the diastereoselectivity of Diels–Alder reactions, such interactions in 1,3-dipolar cycloadditions is infinitesimal (16). The endo/exo-selectivity in 1,3-dipolar cycloadditions is primarily controlled by the structure of the substrates. Most acyclic aldo-nitrones exist in the (*Z*)-configuration, and most cyclic nitrones are locked in the more reactive (*E*)-configuration (17). However, some nitrones exist in equilibrium between the (*E*)- and (*Z*)-forms. Belonging to this class of nitrones are, for example, aldo-nitrones with an electron-withdrawing substituent at the carbon atom. If the nitrone exists as an equilibrating mixture of (*E*)- and (*Z*)-forms, the number of possible diastereoisomers resulting

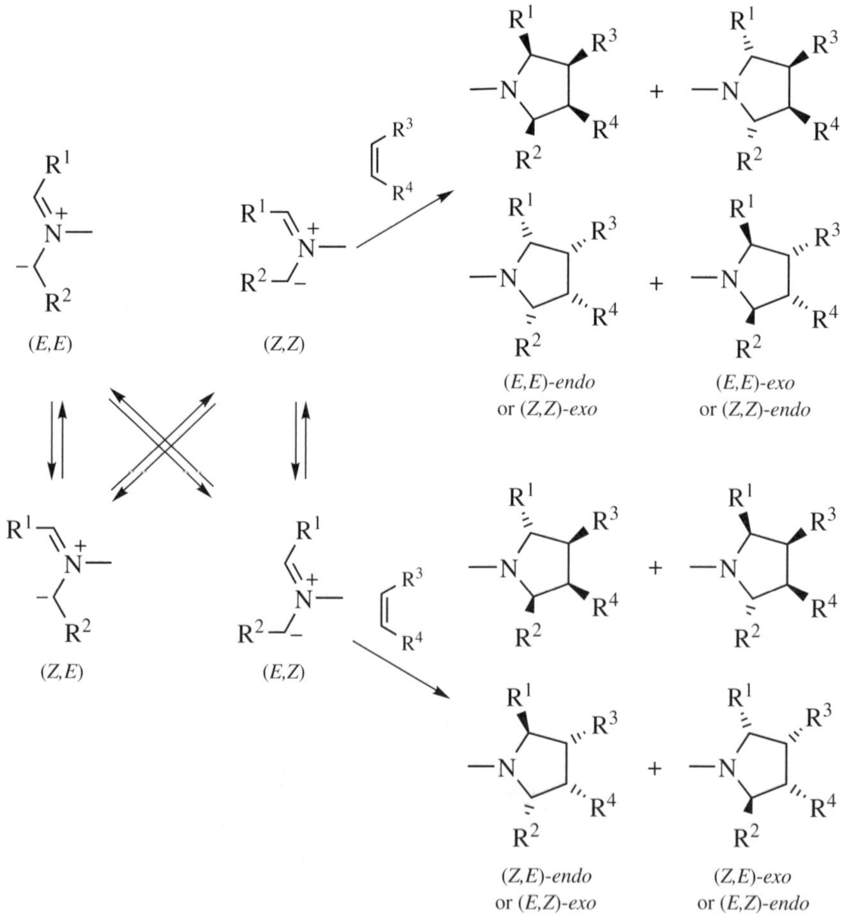

Scheme 12.4

from 1,3-dipolar cycloadditions with alkenes will still be limited to four, but in this case it makes no sense to use the endo/exo-nomenclature. Instead the terms cis/trans are used, defined relative to the configuration of the 3- and 4-substituents of the isoxazolidine ring.

The stereochemistry of 1,3-dipolar cycloadditions of azomethine ylides with alkenes is more complex. In this reaction, up to four new chiral centers can be formed and up to eight different diastereomers may be obtained (Scheme 12.4). There are three different types of diastereoselectivity to be considered, of which the two are connected. First, the relative geometry of the terminal substituents of the azomethine ylide determine whether the products have 2,5-cis or 2,5-trans conformation. Most frequently the azomethine ylide exists in one preferred configuration or it shifts between two different forms. The addition process can proceed in either an endo or an exo fashion, but the possible (*E*,*Z*) interconversion of the azomethine ylide confuses these terms to some extent. The endo-isomers obtained from the (*E*,*E*)-azomethine ylide are identical to the exo-isomers obtained from the (*Z*,*Z*)-isomer. Finally, the azomethine ylide can add to either face of the alkene, which is described as diastereofacial selectivity if one or both of the substrates are chiral or as enantioselectivity if the substrates are achiral.

The azomethine ylides applied in synthesis can be divided into two groups. The traditional azomethine ylide **2** is one that consists of a central nitrogen atom with three carbon substituents (Scheme 12.5). This type of azomethine ylide is most frequently formed by the condensation of an aldehyde with a secondary amine **1**. The second type are the so-called stabilized azomethine ylides **4**. They consist of a central nitrogen atom with two carbon substituents and the nitrogen atom also

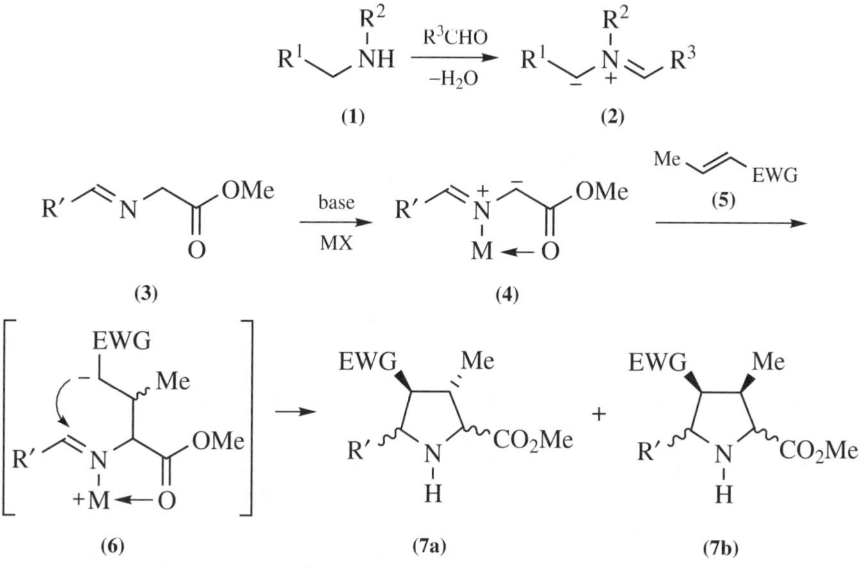

EWG = electron withdrawing group

Scheme 12.5

coordinates to a metal atom. Stabilized azomethine ylides **4** are commonly obtained by deprotonation of an amino ester derived precursor **3**. Deprotonation can be performed with a metal base or with a base such as triethylamine in the presence of a metal salt. The metalated azomethine ylide **4** is thus stabilized by bidentate chelation of the ester functionality and the nitrogen atom to the metal center. This type of azomethine ylide is often nucleophilic and reacts preferentially with electron-deficient alkenes such as **5**. In many cases, such reactions proceed via a two step Michael–Mannich type reaction instead of a 1,3-dipolar cycloaddition. Intermediate **6** may allow for rotation around the former alkene-bond leading to the possible formation of mixtures of 3,4-*trans* pyrrolidine **7a** and 3,4-*cis* pyrrolidine **7b**. This reaction allows, in principle, for formation of all possible diastereomeric products. However, this is never observed in practice. On the contrary, the reactions of stabilized azomethine ylides with alkenes often proceed with excellent diastereoselectivities.

12.2. DIASTEREOSELECTIVE REACTIONS

The most common method for inducing asymmetry in 1,3-dipolar cycloadditions is by the application of chiral 1,3-dipoles, chiral dipolarophiles, or both, the latter always being the case for intramolecular reactions (5). First the reaction of chiral 1,3-dipoles will be described, then the reactions of chiral dipolarophiles, and finally the intramolecular reactions. In this chapter we have chosen to treat the diastereoselective reactions employing chiral auxiliaries separately in Section 12.3.

12.2.1. Chiral Dipoles

Several different chiral 1,3-dipoles have been developed, especially for nitrones and azomethine ylides. In several cases, the chiral dipole has been developed specifically for the asymmetric synthesis of a target molecule and some examples of that will be given.

12.2.1.1. Nitrones and Nitronates

Among the most commonly applied chiral moiety for nitrones (2) is the *N*-α-methylbenzyl substituent (Scheme 12.6) (18–25). The nitrones **8** with this substituent are available from 1-phenethylamine, and the substituent has the advantage that it can be removed from the resulting isoxazolidine products **9** by hydrogenolysis. This type of 1,3-dipole has been applied in numerous 1,3-dipolar cycloadditions with alkenes such as styrenes (21,23), allyl alcohol (24), vinyl acetate (20), crotonates (22,25), and in a recent report with ketene acetals (26) for the synthesis of natural products. Reviewing these reactions shows that the α-methylbenzyl group

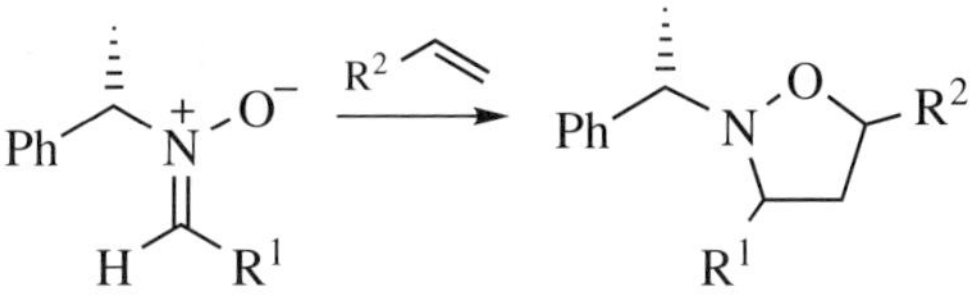

(8) (9)

Scheme 12.6

offers moderate chiral discrimination, since mixtures of diastereomers are generally obtained in reactions of this type of nitrone.

Sneider et al. (27,28) applied a familiar nitrone for the synthesis the immunosuppressant (−)-FR901483 (**14**) in a recent study (Scheme 12.7). The nitrone **12** is generated *in situ* from ketone **10** and the optically pure hydroxylamine **11** at 25 °C. The resultant nitrone **12** underwent a 1,3-dipolar cycloaddition reaction with ethyl acrylate in refluxing toluene to give the diastereomer **13** with 71% diastereomeric excess (de). In 22 synthetic steps including the 1,3-dipolar cycloaddition, the target molecule **14** was obtained.

Brandi and co-workers (29,30) studied the 1,3-dipolar cycloadditions of chiral α,β-dialkoxynitrones with vinylphosphine oxides (Scheme 12.8). The reaction of optically active **15** with vinylphosphine oxide **16a** gave a 65:14 mixture of the

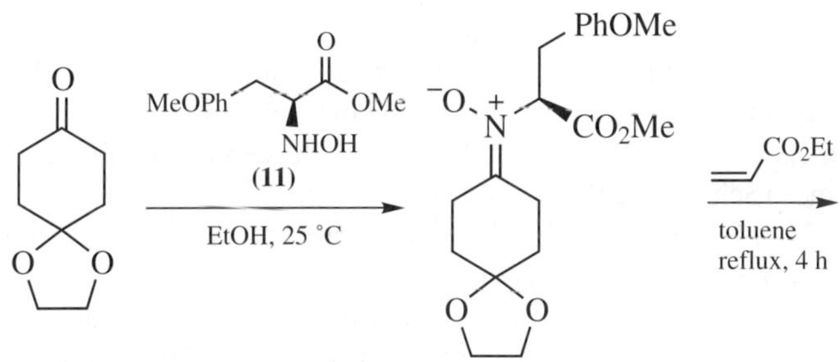

(10) (12)

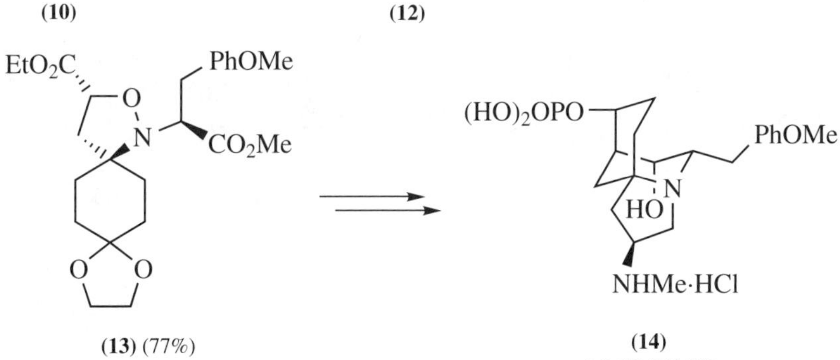

(13) (77%)
de = 71%

(14)
(−)-*FR 901483*

Scheme 12.7

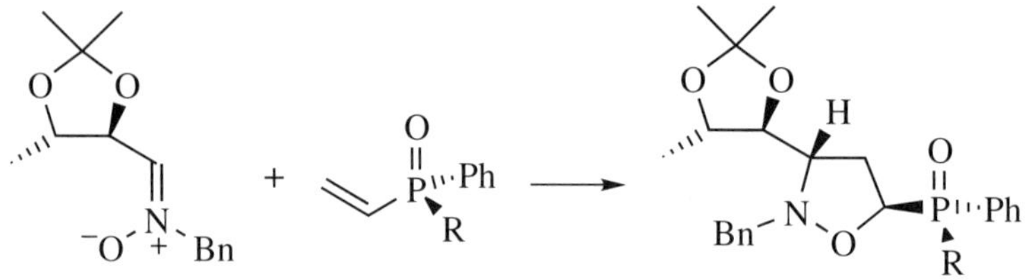

Scheme 12.8

endo/exo-isomers of **17a** along with 21% of the other regioisomer. The endo-isomer was formed with a high diastereofacial selectivity of 94% de. By the application of the optically active vinylphosphine oxide **16b**, the 1,3-dipolar cycloaddition with **15** proceeded to form **17b** with a high degree of endo-selectivity and without the appearance of the other regioisomer. The reaction also displayed excellent diastereofacial selectivity, since the product was obtained with 96% de. The reactions of other nitrones containing a vicinal dioxolane ring such as **17b** have been described in a recent experimental and theoretical study by Carda et al. (31).

Saito et al. (32) developed a tartaric acid derived chiral nitrone **18**. In the reaction of **18** with methyl crotonate **19**, the 1,3-dipolar cycloaddition product **20** was obtained in an endo/exo ratio of 10:1 and with high diastereofacial induction to give the endo-isomer (Scheme 12.9).

Other nitrones (**21–23**) having the chiral moiety located at the carbon atom have been applied in reactions with various alkenes (Scheme 12.10) (33–35). Nitrone **21** offered poor discrimination in 1,3-dipolar cycloadditions with benzyl crotonate, as all four diastereomers were obtained in both reactions (33). The fluorinated nitrone

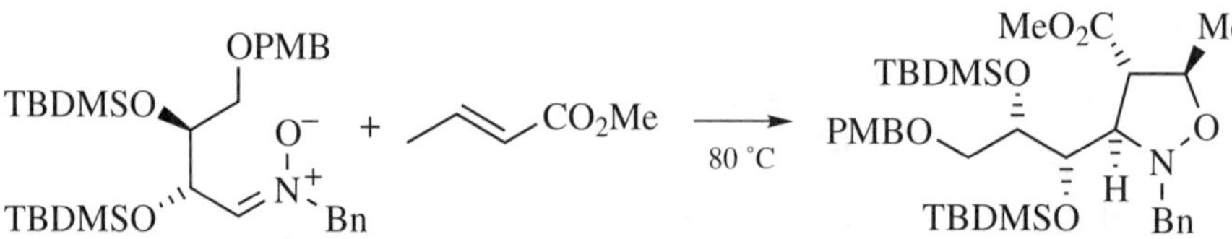

TBDMS = tert-butyl dimethylsilyl
Bn = benzyl
PMB = para-methoxy benzyl

Scheme 12.9

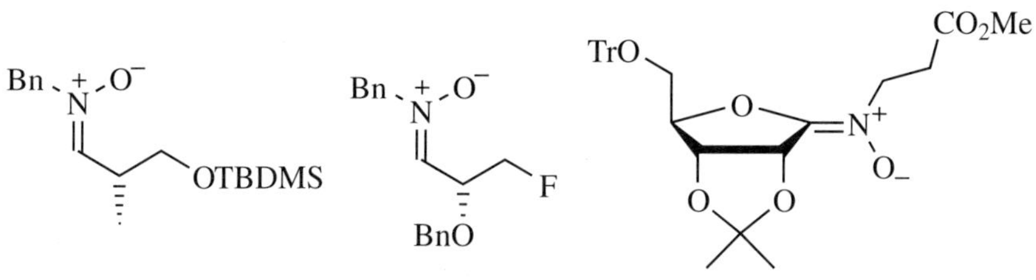

(21) (22) (23)

Scheme 12.10

22 led to a good endo/exo ratio and a high de in the reaction with diethyl fumerate (34). Nitrone **23** reacted with methyl acrylate to give a 44:21:0:0 ratio of the four possible diastereomers (35).

Mukai et al. (36,37) applied the chiral tricarbonyl(η^6-arene)chromium(0)-derived nitrone **24b** in 1,3-dipolar cycloadditions with various alkenes, such as styrene **25** (Scheme 12.11). The analogous nonmetallic nitrone **24a** was used in a reference reaction with **25**, giving the isoxazolidine **26a** with an endo/exo ratio of 82:18. By the application of nitrone **24b** in the 1,3-dipolar cycloaddition with **25**, the endo/exo-selectivity changed significantly to give *exo*-**26b** as the only observable product. The tricarbonylchromium moiety effectively shielded one face of the nitrone, leading to high diastereofacial selectivity. The product *exo*-**26b** was obtained with 96–98% de.

Cyclic chiral nitrones generally offer better stereoselectivity than their acyclic counterparts. A more efficient shielding of one of the nitrone faces is often obtained due to the more rigid conformation of the cyclic nitrones. Furthermore, in this approach, (*E/Z*)-interconversion is avoided and cyclic nitrones are often more reactive since they, depending on the substitution pattern, are usually locked in the

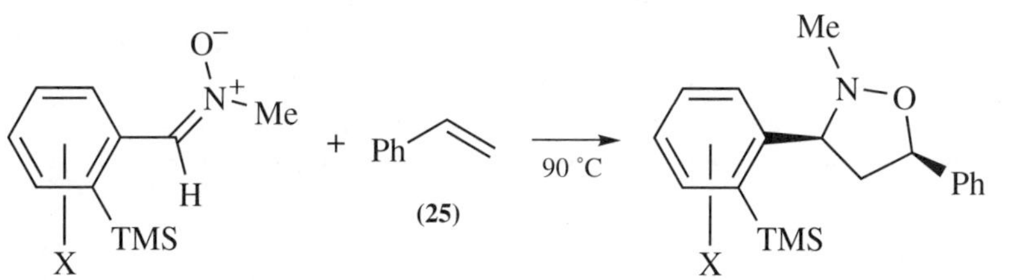

(**24a**): X absent
(**24b**): X = $Cr(CO)_3$

exo-(**26a**): X absent
(racemic)
endo/exo = 82:18

exo-(**26b**): X = $Cr(CO)_3$
endo/exo = <2:>98
de_{exo} = 96–98%

Scheme 12.11

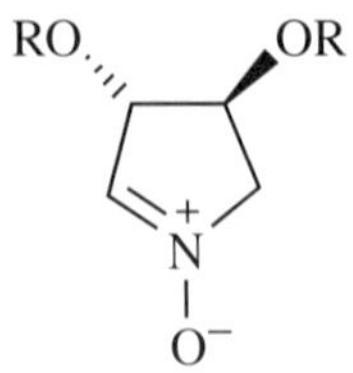
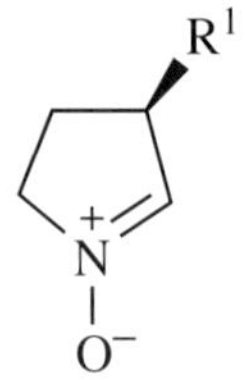
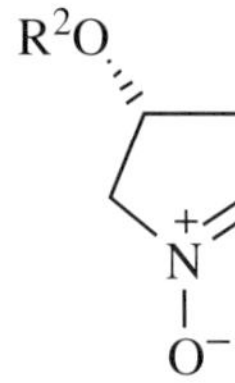

(27a): R = CH_2OMe
(27b): R = Me
(27c): R = *t*-Bu
(27d): R = CH_2Ph
(27e): R = $SiPh_2$*t*-Bu

(28a): R = O*t*-Bu
(28b): R = OBz
(28c): R = NBn_2

(29)

Bz = benzoyl

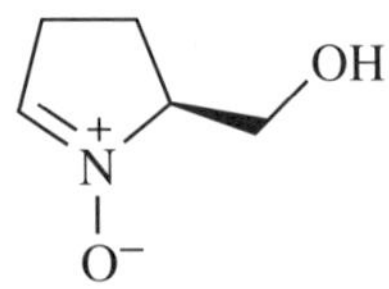

(30)

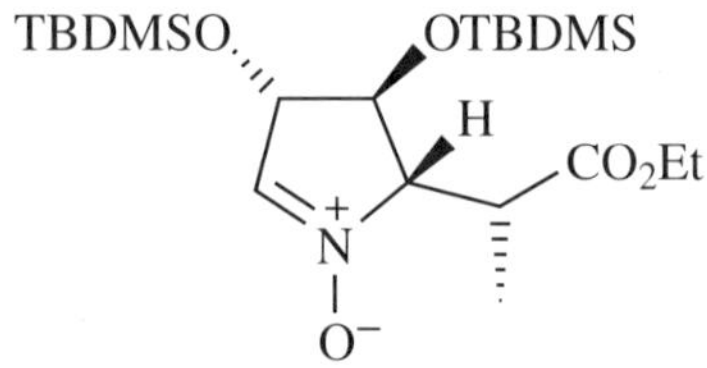

(31)

Scheme 12.12

more reactive (*E*)-configuration. Probably due to these findings there has been much activity in developing new cyclic nitrones.

Among the most commonly applied cyclic chiral nitrones are the pyrrolidine-derived nitrones. Brandi and co-workers (38–46) studied the application of the L-tartaric acid derived nitrones **27a–e** (Scheme 12.12). Nitrone **27e** shows a high degree of chiral discrimination in the 1,3-dipolar cycloaddition with methylenecyclopropane at room temperature, leading to one regioisomer in a yield of 75% with a de of 82% (38). This approach has successfully been applied in the synthesis of optically active hydroxylated indolizidines (38,39,41,47). Brandi and co-workers (43,44,46,48) furthermore described the synthesis and reactions of nitrones **28a–c**. The selectivities obtained using these nitrones are in most cases moderate to good and they have proven to be excellent building blocks for natural product synthesis. One of several examples of this is shown using nitrone **28b** in Scheme 12.13 (45). The 1,3-dipolar cycloaddition of **28b** with diethyl maleate gave a mixture of diastereomers of which the major isomer **32** was isolated in 56% yield. Another two synthetic steps furnished the alkaloid (−)-hastanecine (**33**).

Murahashi and co-workers (49) extensively studied the synthesis of nitrones such as **29** by a decarboxylative oxidation of proline derivatives (Scheme 12.12). However, these nitrones were primarily used in nucleophilic addition reactions rather than 1,3-dipolar cycloadditions. Others have synthesized cyclic nitrones **30** and **31** having a chiral center adjacent to the nitrogen atom (50,51). Saito and co-workers (51) applied nitrone **31** in reactions with fumaric and maleic acid

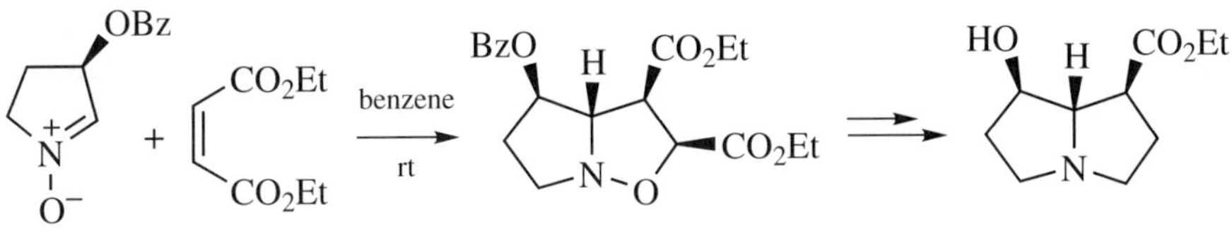

(28b) **(32)** (major) (56%) 4.5:1:1 **(33)** (−)-hastanecine

Scheme 12.13

derivatives at 70 °C in benzene and all reactions proceeded with complete regio- and diastereoselectivity.

The power of cyclic chiral nitrones in synthesis was demonstrated by Nagasawa et al. (52) by the synthesis of the enantiomerically pure pentacyclic guanidine derivative **38** (Scheme 12.14). The 1,3-dipolar cycloaddition of **27b** with **34** took

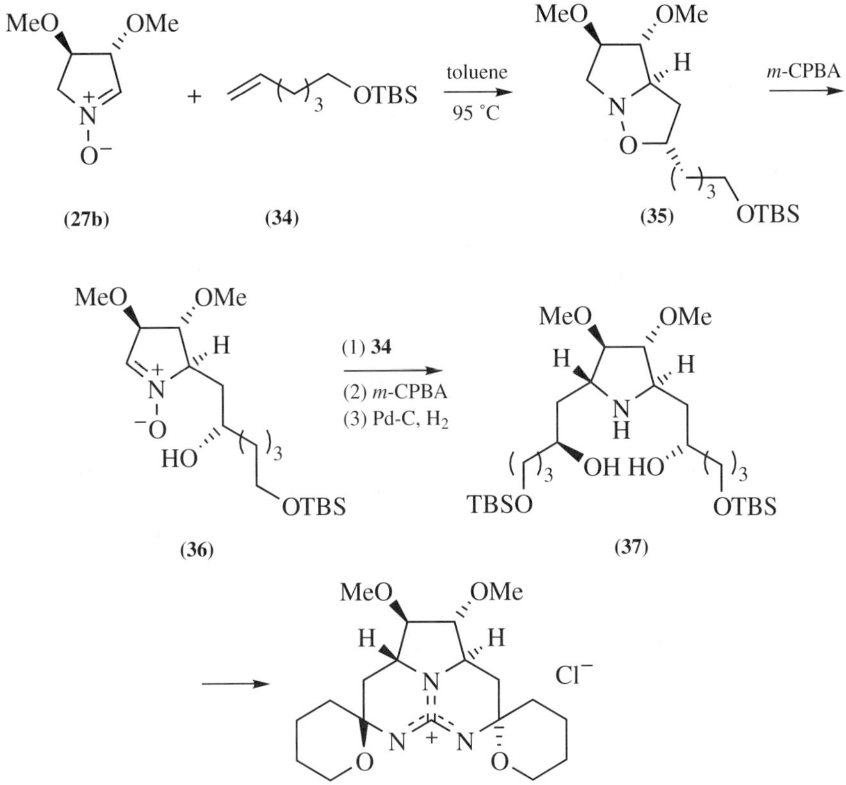

(27b) **(34)** **(35)** **(36)** **(37)** **(38)**

TBS = tributyl silyl
m-CPBA = meta-chloro peroxy benzoic acid

Scheme 12.14

place exclusively from the opposite side of the methoxy group in the α-position of the nitrone moiety to give **35**. The key reaction in this synthesis was an oxidative isoxazolidine cleavage oxidation with *m*-CPBA, which generates a new nitrone functionality. A subsequent 1,3-dipolar cycloaddition reaction, cleavage oxidation and reduction sequence furnishes the enantiomerically pure guanidine precursor **37**. In three additional steps, the guanidine derivative **38** was obtained in an overall yield of 24% from **27b**.

A chiral pipiridine-derived nitrone, bearing a silyloxy substituent at the chiral center in the 3-position, was applied in a reaction with allyl alcohol in the synthesis of the natural products (+)-febrifugine and (+)isofebrifugene (53). The selectivities obtained in this 1,3-dipolar cycloaddition were poor, but the isomeric product mixture could be separated. Two chiral cyclic nitrones **39** and **41** having an oxygen atom in the ring were described (Scheme 12.15). Langlois and co-workers (54–56) applied nitrone **39** in the synthesis of (−)-carbovir, a potential agent in treating acquired immune deficiency syndrome (AIDS). The 1,3-dipolar cycloaddition between cyclopentadiene and **39** proceeded at 40 °C with high regio- and diastereoselectivity, producing **40** as the only product isolated. In the succeeding synthetic steps toward (−)-carbovir, the camphor skeleton was recovered. In recent work by the same group, nitrone **39** was used in the synthesis of precursors of (+)-carpetimycin A and (+)-β-methylcarbapenem (57,58). Tamura et al. (59–61) prepared nitrone **41**, which was subjected to reaction with a series of alkenes. In the reaction between **41** and cyclopentene, isoxazolidine **42** was formed in 88% yield as a single isomer. The product **42** was converted into a bicyclic amino substituted lactone (61).

The reactions of chiral cyclic alkyl nitronates have been described (62–64). These nitronates are intermediates in a tandem[4+2]/[3+2] cycloadditions.

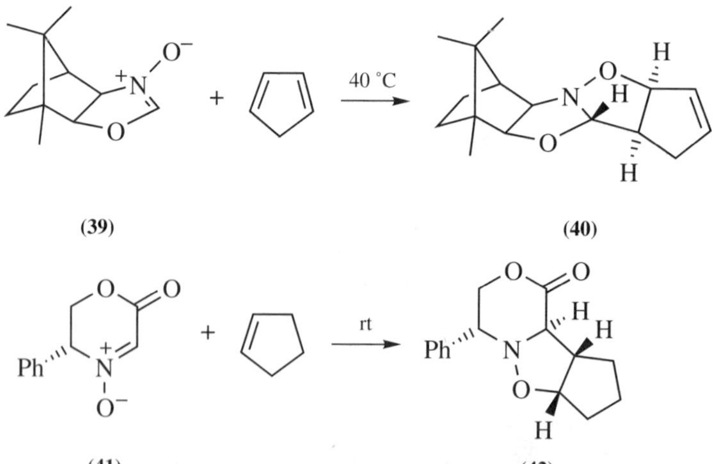

(39) **(40)**

(41) **(42)**

Scheme 12.15

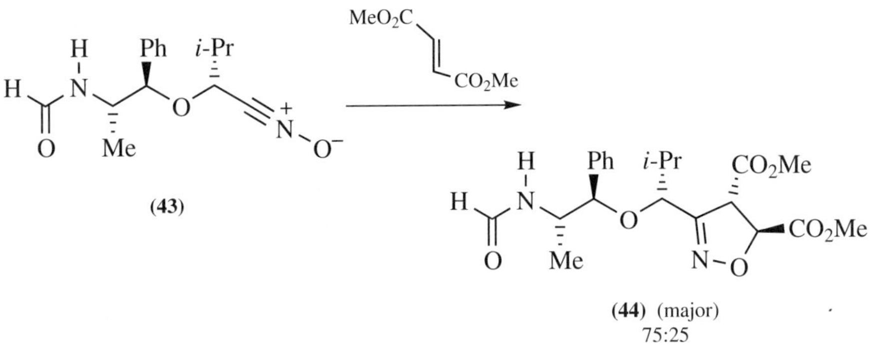

Scheme 12.16

12.2.1.2. *Nitrile Oxides*

Only a few reports have described the application of optically active nitrile oxides in 1,3-dipolar cycloadditions (65–70). A general trend for these reactions is that moderate-to-poor diastereoselectivities are obtained when it is attempted to control the stereoselectivity using a chiral nitrile oxide. In one of the few recent examples, the chiral nitrile oxide **43**, derived from *N*-formylnorephenedrine and 3-methylnitrobutene, was subjected to reaction with diethyl fumerate (Scheme 12.16) (69). Compound **44** was obtained as the major product of this reaction as a 75:25 mixture with its diastereomer.

12.2.1.3. *Azomethine Ylides*

Diastereoselective 1,3-dipolar cycloadditions of chiral azomethine ylides with alkenes leading to optically active products were first described in 1985 (71), and since then by several others. In a series of papers, Harwood and co-workers (72–81) described the development the (5*S*)-phenyl-morpholin-2-one azomethine ylide precursor **45** for reactions with alkenes (Scheme 12.17). For the reactions of the *in situ* generated azomethine ylide **46** (R^1=H or CO_2Me) with electron-deficient alkenes such as dimethyl fumerate, high selectivities could be obtained for product **47**, which was obtained in moderate to good yields (73,74). The products could be converted into proline derivatives. More recently, this approach was extended to heterodipolarophiles such as aldehydes (82) and imines (83). For the reactions with aldehydes, the same aldehyde R^1CHO was used for the formation of the azomethine ylide as well as the subsequent 1,3-dipolar cycloaddition. The reactions were conducted with various aldehydes (R^1=Ar, alkyl) and the azomethine ylide formation and cycloaddition proceeded to give single isomers of **48** in good-to-high yields. The products could be converted into the corresponding enantiomerically pure β-hydroxy-α-amino acids, and this approach was applied for the

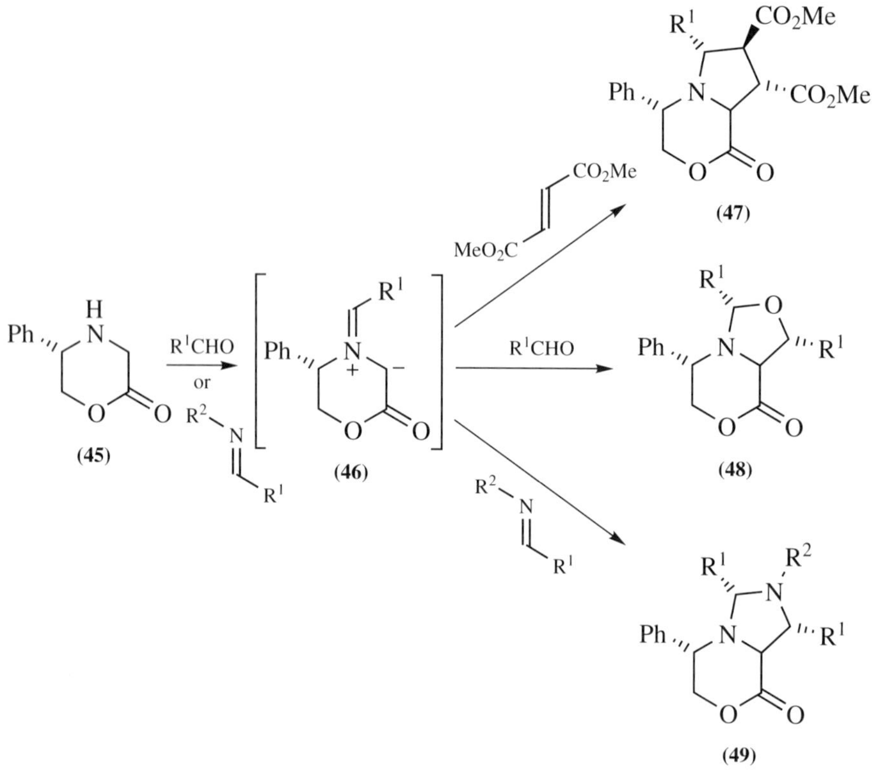

Scheme 12.17

synthesis of (+)-polyoxamic acid (82). It was demonstrated by the same authors that imines could be used as substrates for both the azomethine ylide formation and the 1,3-dipolar cycloaddition (83). These reactions also proceeded with excellent selectivities to give **49** (R^1=Ar, R^2=Bn, Me) as the only diastereomer. These products were converted into the corresponding enantiopure α,β-amino acids in high yields.

Husson and co-workers (84) investigated the 1,3-dipolar cycloaddition of acyclic chiral azomethine ylides derived from (−)-*N*-cyanomethyl-4-phenyl-1,3-oxazolidine with electron-deficient alkenes, and in some cases de >95% were obtained.

Another approach employing chiral acyclic azomethine ylides was published in two recent papers by Alcaide et al. (85,86). The azomethine ylide–silver complex (**51**) was formed *in situ* by reaction of the formyl-substituted chiral azetidinone (**50**) with glycine (or alanine) in the presence of AgOTf and a base (Scheme 12.18). Azomethine ylides formed in this manner were subjected to reaction with various electron-deficient alkenes. One example of this is the reaction with nitrostyrene, as illustrated in Scheme 12.18 (86). The reaction is proposed to proceed via a two step tandem Michael–Henry process in which the products **52a** and **52b** are isolated in a

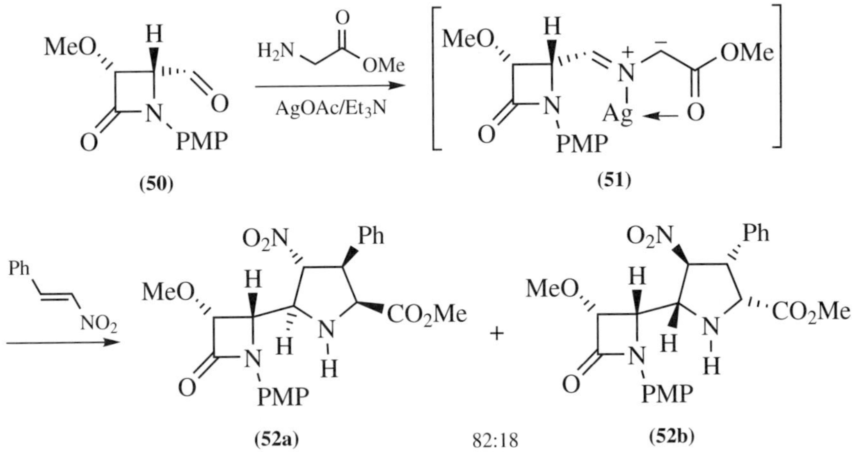

PMP = para-methoxy phenyl

Scheme 12.18

ratio of 82:18 (41 and 9% isolated yields), accompanied with 8% of another diastereomer. Compounds **52a** and **52b** are both endo-isomers and arise from an opposite facial approach of the alkene to **51**.

Azomethine ylides derived from (5*S*,6*R*)-2,3,5,6-tetrahydro-5,6-diphenyl-1,4-oxazin-2-one (**53**) and various aldehydes have been prepared by Williams and co-workers (87,88) (Scheme 12.19). In a recent communication they reported the application of the azomethine ylide **54** in the asymmetric total synthesis of spirotryprostatin B **56** (88). The azomethine ylide **54** is preferentially formed with (*E*)-geometry due to the bulkiness of the aldehyde substituent. The *in situ* formed azomethine ylide **54** reacted with ethyl oxindolylidene acetate to give the 1,3-dipolar cycloaddition adduct **55** in 82% yield as the sole isomer. This reaction, which sets four contiguous stereogenic centers, constructs the entire prenylated tryprophyl moiety of spirotryprostatin B (**56**), in a single step.

Other chiral azomethine ylide precursors such as 2-(*tert*-butyl)-3-imidazolidin-4-one have been tested as chiral controllers in 1,3-dipolar cycloadditions (89). 2-(*tert*-Butyl)-3-imidazolidin-4-one reacted with various aldehydes to produce azomethine ylides, which then were subjected to reaction with a series of different electron-deficient alkenes to give the 1,3-dipolar cycloaddition products in moderate diastereoselectivity of up to 60% de.

Chiral aziridines having the chiral moiety attached to the nitrogen atom have also been applied for diastereoselective formation of optically active pyrrolidine derivatives. In the first example, aziridines were used as precursors for azomethine ylides (90–95). Photolysis of the aziridine **57** produced the azomethine ylide **58**, which was found to add smoothly to methyl acrylate (Scheme 12.20) (91,93–95). The 1,3-dipolar cycloaddition proceeded with little or no de, but this was not surprising, as the chiral center in **58** is somewhat remote from the reacting centers

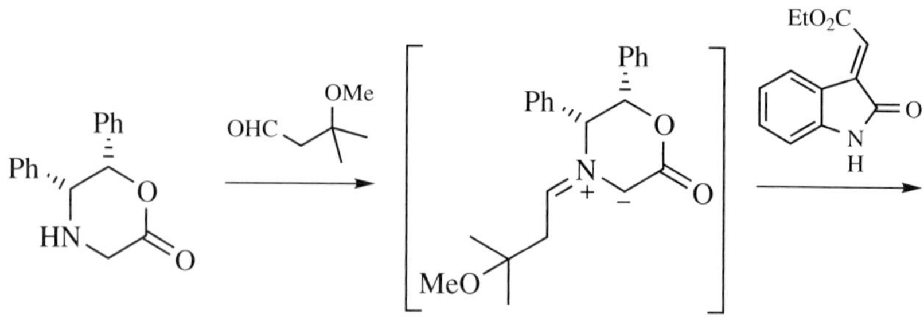

(53) (54)

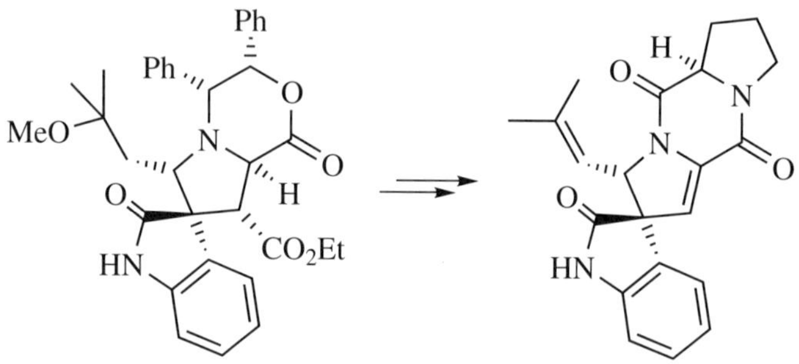

(55) sole isomer

(56) spirotryprostatin B

Scheme 12.19

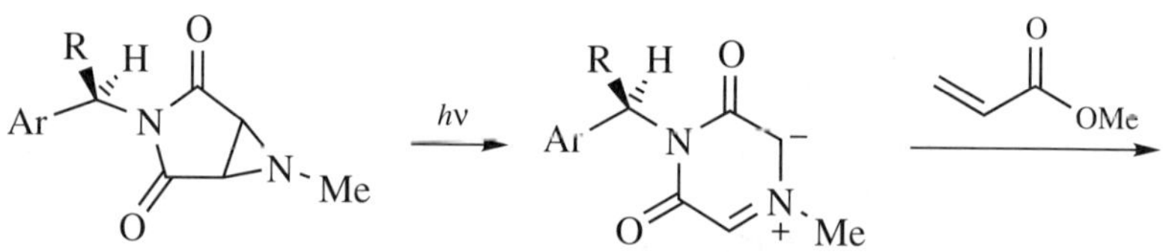

(57) (58)

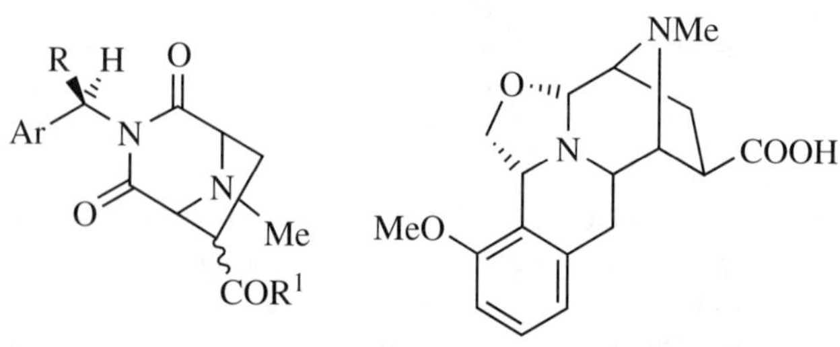

(59): R^1 = Me, endo/exo = 1:5, low de

(60) (−)-quinocarcin

Scheme 12.20

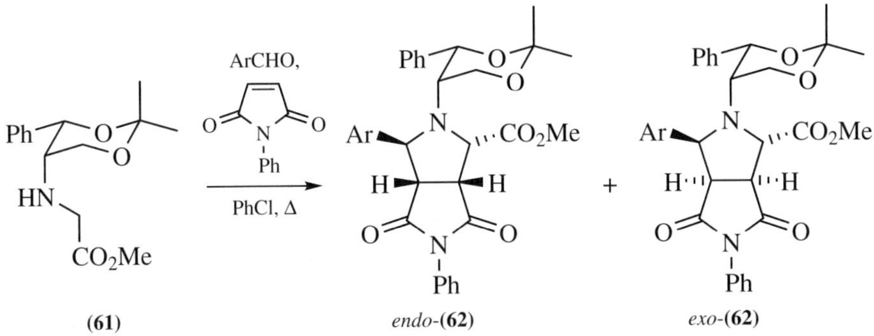

Scheme 12.21

of the azomethine ylide. The de can be improved significantly by application of the acrylate of Oppolzer's chiral sultam (see below). A 1,3-dipolar cycloaddition related to the one outlined in Scheme 12.20 has been used for the asymmetric synthesis and complete structure elucidation of (−)-quinocarcin **60** (91).

Enders et al. (96) recently described the application of the chiral azomethine precursor **61** (Scheme 12.21). The azomethine ylide was formed *in situ* by heating with different benzaldehydes. The reactions of four different azomethine ylides with *N*-phenyl maleimide led to the formation of *endo*-**62** and *exo*-**62** in ratios of 2:1 in very high yields. The diastereofacial selectivity was estimated to be >96% de for both products, since no other diastereomers were observed by proton nuclear magnetic resonance (^{1}H NMR) spectroscopy.

Azomethine ylides such as **64** can be generated from tertiary amine *N*-oxides (**63**) by reaction with lithium diisopropylamide (LDA) (Scheme 12.22) (97). Several different chiral *N*-substituted azomethine ylides were prepared in this manner. The best results were obtained when using **64** in 1,3-dipolar cycloaddition with alkenes, but the de values obtained of the product **65** were ≤60%.

Other 1,3-dipolar cycloadditions of chiral azomethine ylides with C_{60} (98) and reactions of chiral azomethine ylides derived from 1-benzyl-4-phenyl-2-imidazoline with different electron-deficient alkenes have been performed (99).

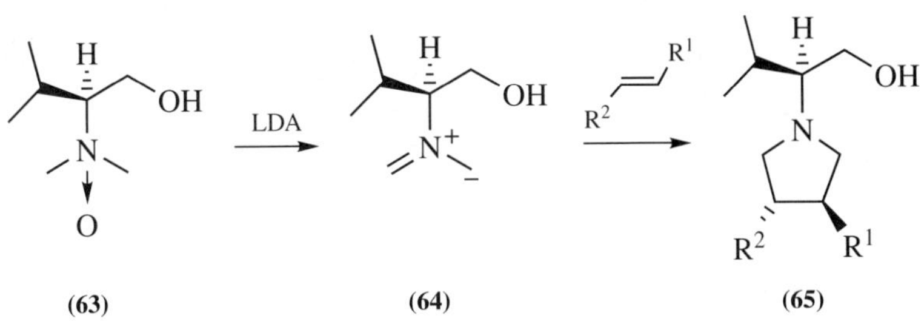

Scheme 12.22

12.2.1.4. Other Dipoles

The use of chiral azomethine imines in asymmetric 1,3-dipolar cycloadditions with alkenes is limited. In the first example of this reaction, chiral azomethine imines were applied for the stereoselective synthesis of C-nucleosides (100–102). Recent work by Husson and co-workers (103) showed the application of the chiral template **66** for the formation of a new enantiopure azomethine imine (Scheme 12.23). This template is very similar to the azomethine ylide precursor **52** described in Scheme 12.19. In the presence of benzaldehyde at elevated temperature, the azomethine imine **67** is formed. 1,3-Dipole **67** was subjected to reactions with a series of electron-deficient alkenes and alkynes and the reactions proceeded in several cases with very high selectivities. Most interestingly, it was also demonstrated that the azomethine imine underwent reaction with the electronically neutral 1-octene as shown in Scheme 12.23. Although a long reaction time was required, compound **68** was obtained as the only detectable regio- and diastereomer in 50% yield. This pioneering work demonstrates that there are several opportunities for the development of new highly selective reactions of azomethine imines (103).

Nitrile imines are related to azomethine imines, in the same manner as nitrile oxides are related to nitrones. In a single and recent report, the reactions of D-galactose derived chiral nitrile imines have been described (104). However, in reactions with nonchiral alkenes, no diastereoselection was obtained.

Padwa and Prein (105,106) applied chiral, but racemic, isomünchnone dipoles in diastereoselective 1,3-dipolar cycloadditions. The carbonyl ylide related isomünchnone derivative *rac*-**70** was obtained from the rhodium-catalyzed cyclization of diazo-derivative *rac*-**69** (Scheme 12.24) (105). The reactions of the *in situ* formed dipole with a series of alkenes was described and in particular the reaction with maleic acid derivatives **71a–c** gave rise to reaction with high selectivities. The tetracyclic products **72a–c** were all obtained in good yield with high endo/exo and diastereofacial selectivities. In another paper by the same authors, the reactions of racemic isomünchnones having an exo-cyclic chirality was described (106).

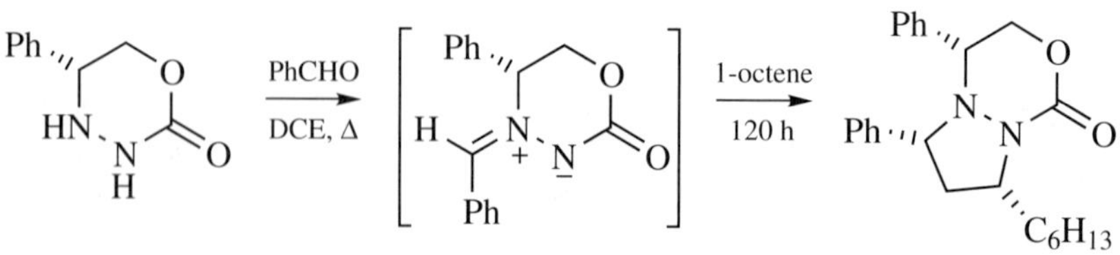

(66) **(67)** **(68)** 50% >97 de

DCE = dichloroethane

Scheme 12.23

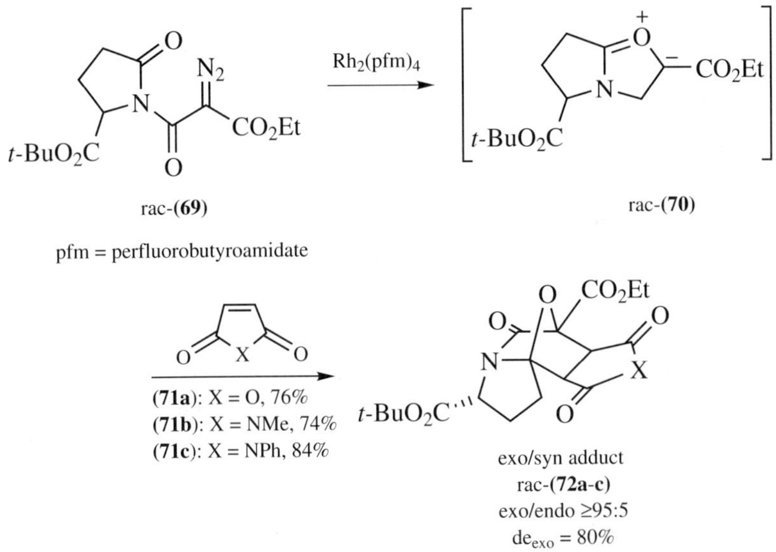

Scheme 12.24

12.2.2. Chiral Dipolarophiles

The 1,3-dipolar cycloadditions of 1,3-dipoles with chiral alkenes has been extensively reviewed and thus only selected examples will be highlighted here. We have chosen to divide this section on the basis of the different types of alkenes rather than on the basis of the type of 1,3-dipole. For 1,3-dipolar cycloadditions, as well as for other reactions, it is important that the chiral center intended to control the stereoselectivity of the reaction is located as close as possible to the functional group of the molecule at which the reaction takes place. Hence, alkenes bearing the chiral center vicinal to the double bond are most frequently applied in asymmetric 1,3-dipolar cycloadditions. Examples of the application of alkenes with the chiral center localized two or more bonds apart from the alkene will also be mentioned. Application of chiral auxiliaries for alkenes is very common and will be described separately in Section 12.3.

12.2.2.1. Acyclic Allyl Alcohol Derivatives

One of the classical ways to perform diastereoselective 1,3-dipolar cycloaddition is by the addition of a 1,3-dipole to an allyl alcohol derivative (65, 107–120). Very recently, a short review article was devoted to this area (13). Among the most commonly applied acyclic allyl alcohol derivatives are alkenes **73–75** (Scheme 12.25). These alkenes have been used in reactions with nitrones,

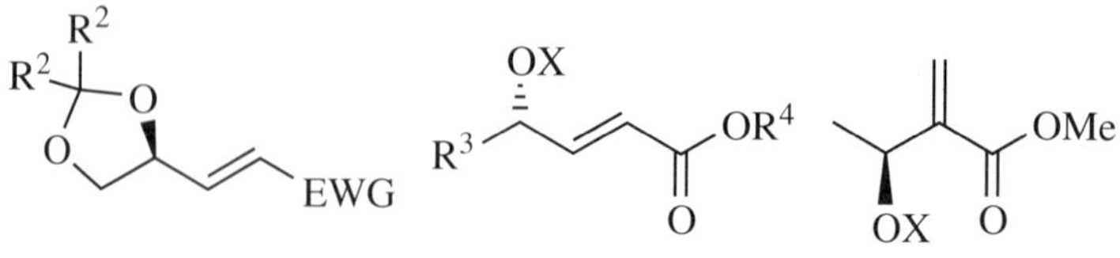

(**73**) (**74**): X = H, SiR_3, Bn, Alk (**75**)

R^2 = Me, $(CH_2)_5$

EWG = CO_2R, NO_2, SO_2Ph

Scheme 12.25

nitronates, nitrile oxides, azomethine ylides, and diazo compounds. The selectivities obtained in these reactions are highly substrate dependent. In some cases high diastereoselectivities are obtained, but only rarely is complete diastereoselection observed.

Saito et al. (32,121) developed a variety of tartaric acid derivatives, including C_2-symmetric chiral alkenes such as **76**. The 1,3-dipolar cycloaddition between **76** and **77** gave primarily *endo*-**78**. (Scheme 12.26) The diastereofacial selectivity of the reaction is excellent, as *endo*-**78** is obtained with >98% de. Other cyclic and acyclic nitrones have been employed in reactions with **76**, and in all cases, moderate to excellent endo/exo-selectivities and excellent diastereofacial selectivities were obtained (32,121). Three other research groups have applied various γ-hydroxylated α,β-unsaturated carbonyl compounds in related reactions with nitrones (122–124). However, the selectivities were somewhat lower than those obtained by Saito and et al. (32,121).

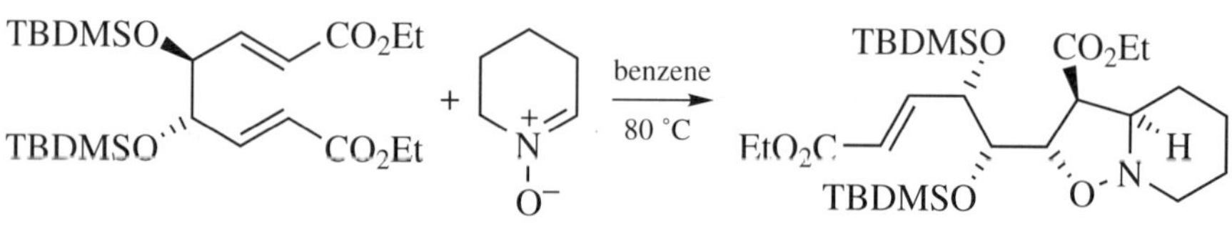

(**76**) (**77**) *endo*-(**78**)

endo/exo = 94:6

de_{endo} >98%

Scheme 12.26

The azomethine ylide derived from **79** has also been used in reactions with chiral (*E*)-γ-alkoxy-α,β-unsaturated esters **80** (Scheme 12.27). The corresponding tetrasubstituted pyrrolidines **81** were obtained with complete regiocontrol in fair to excellent de (125).

12.2.2.2. Cyclic Allyl Alcohol Derivatives

Numerous chiral cyclic allyl alcohol derivatives have been used as the chiral alkene part in 1,3-dipolar cycloadditions. In general, the more rigid conformational

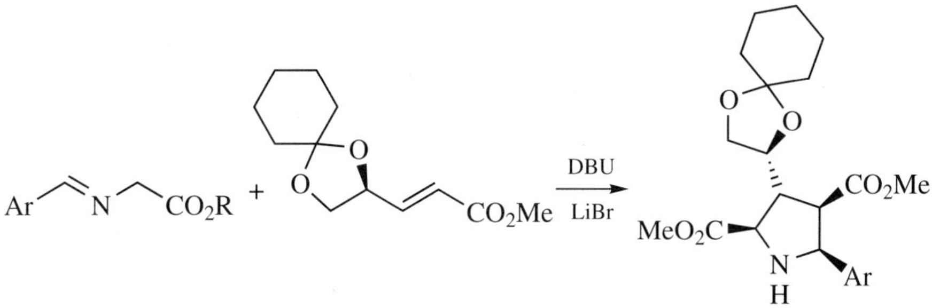

(79) (80) (81)
major stereoisomer
up to 92% de

Scheme 12.27

nature of cyclic alkenes bearing stereocenters offers better facial discrimination in the addition process. The sugar ene-lactone **82** was used by Chmielewski and co-workers (126–128) in reactions with *C*-methyl-nitrones (Scheme 12.28). The reaction proceeded in an endo-selective manner to the face of the alkene anti to the substituents of the lactone ring to give the isoxazolidine adduct as the sole product. The diene (1*R*)-acetoxy-(2*S*)-hydroxy-cyclohexa-3,5-diene (**83**) was applied in a 1,3-dipolar cycloaddition with diazomethane. In this reaction, a 3:1 mixture of two isomeric pyrazolines was obtained, arising from the two different double bonds in **83** (129). Paton and co-workers (130,131) applied the bicyclic lactone levoglucosenone **84** in reactions with various different 1,3-dipoles. In the reaction with *C,N*-diphenylnitrone, only one observable product was obtained. The structurally related chiral bicyclic lactone **85** has also been used in a reaction with a

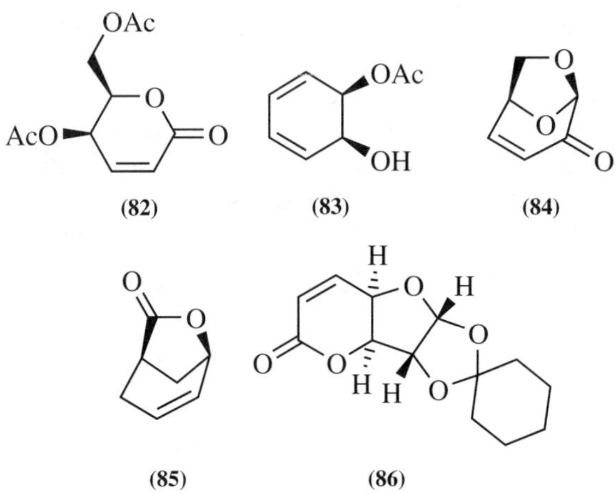

(82) (83) (84)
(85) (86)

Ac = acetyl

Scheme 12.28

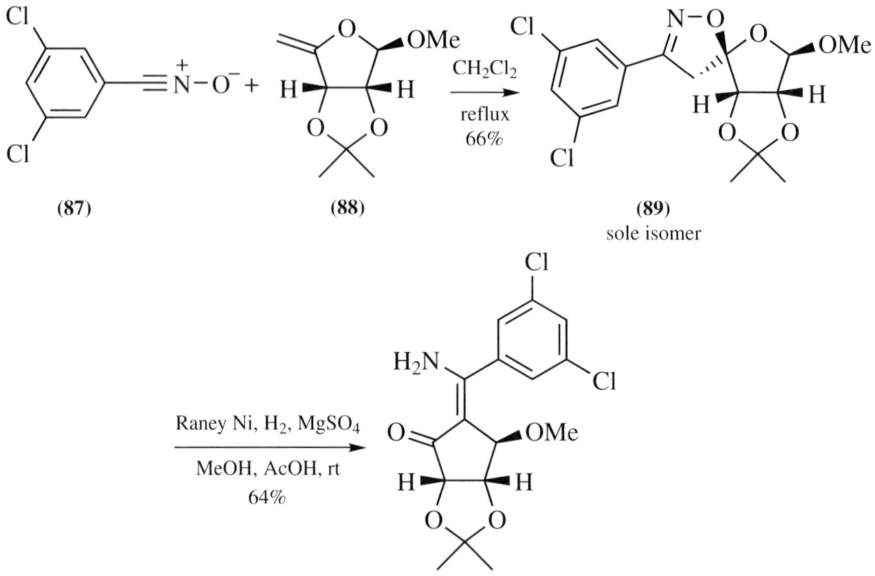

(87) **(88)** **(89)** sole isomer

(90)

Scheme 12.29

nitrile oxide (132). The 1,3-dipolar cycloaddition between **85** and the nitrile oxide derived from pyruvic aldehyde proceeds in refluxing toluene to give the main isoxazoline product in 45% yield along with 15 and 19% of other diastereo- and regioisomers, respectively. The tetracyclic lactone **86** was used by Trivedi and coworkers (120) in reactions with various nitrones. The reaction between **86** and *N*-methylphenyl nitrone gave a single product arising from the exo-approach of the nitrone to the bottom face of **86**.

The lactone **88** having an exo-cyclic double bond was applied in a 1,3-dipolar cycloaddition with nitrile oxides in recent work by Gallos et al. (Scheme 12.29) (133). The *spiro*-isoxazoline (**89**) was obtained as the sole diastereomer from the addition of the stable nitrile oxide **87**. The resulting adduct **89** was further subjected to N—O bond cleavage by hydrogenolysis, followed by a spontaneous cyclization to give the carbocyclic product **90** in 64% yield.

12.2.2.3. Allyl Amine Derivatives

Similarly, both acyclic and cyclic allyl amine derivatives have been applied in 1,3-dipolar cycloadditions (134–138). Langlois et al. (139) used α,β-unsaturated-γ-lactams derived from (*S*)-pyroglutaminol, such as **91** and **92**, in the 1,3-dipolar cycloaddition with the *N*-benzylnitrone derived from formaldehyde (Scheme 12.30). For compound **91**, one of the 1,3-dipolar cycloaddition product isomers obtained

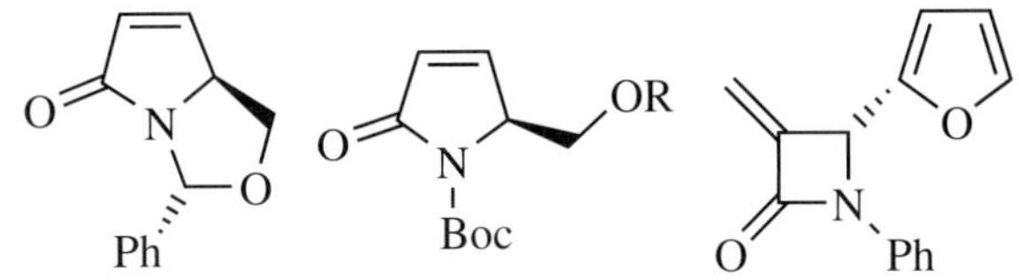

(91) **(92)** **(93)**

Boc = butoxycarbonyl

Scheme 12.30

was isolated in 75% yield, whereas only 5 and 3% of two other isomers were obtained. For lactam **92**, the 1,3-dipolar cycloaddition with a similar nitrone proceeded with high diastereoselection. The nitrone approaches anti to the bulky alkoxy substituent to give a single isomer of the cycloadduct (140). The β-lactam **93** was applied by Basak et al. (141) for the synthesis of 3-hydroxy-azetidin-2-ones. The reaction of **93** with an acyclic nitrone proceeded with high selectivity to give only one observable diastereomeric product.

12.2.2.4. Vinyl Acetals and Vinyl Aminals

Chiral alkenes derived from α,β-unsaturated aldehydes have also been applied in asymmetric 1,3-dipolar cycloadditions (142). Soucy et al. (142) used (−)-8-(benzylamino)menthol (**94**) and acrolein for the exclusive formation of **95** having an equatorial C(2) vinyl group (Scheme 12.31). The 1,3-dipolar cycloaddition of acetonitrile oxide with **95** gave **96** with a selectivity of >90% de.

Meyers and co-workers (143–146) studied the addition of azomethine ylides to the chiral bicyclic half-aminal **97** (Scheme 12.32). The diastereoselectivities of these reactions were highly dependent on the substituents of both **97** and

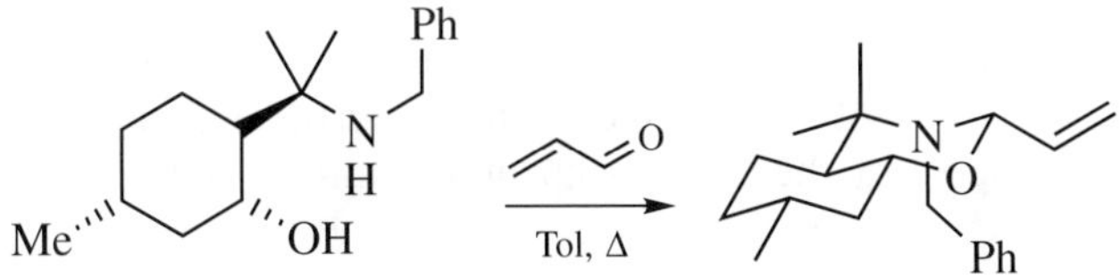

(94) **(95)**

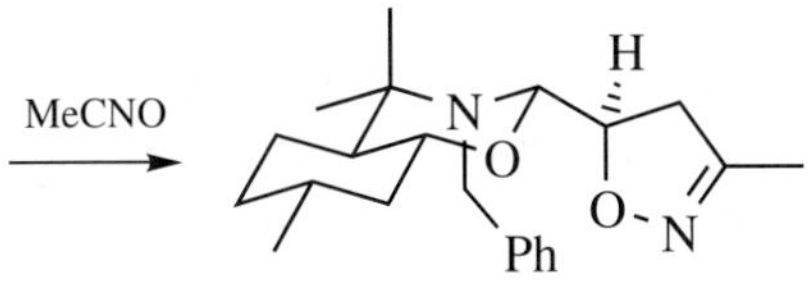

(96)
de >90%

Scheme 12.31

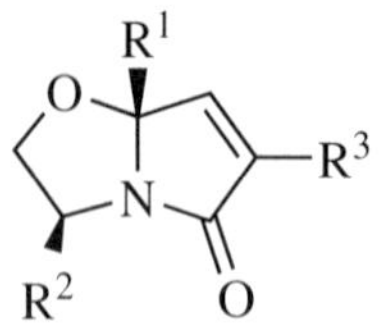

(97)

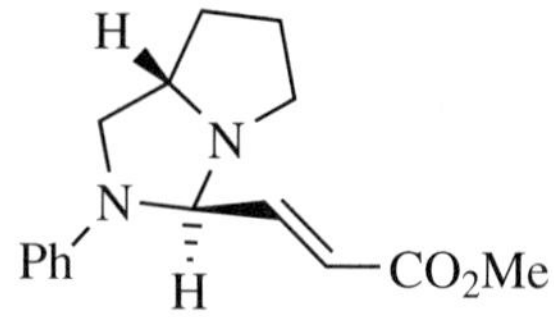

(98)

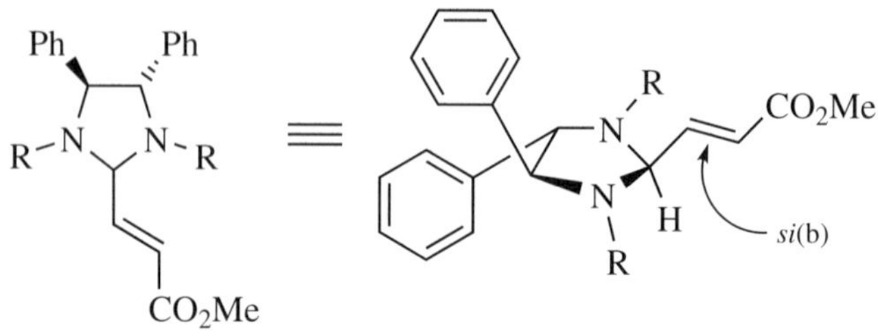

(99)

Scheme 12.32

the azomethine ylides applied. Kanamasa et al. (147–149) applied methyl (3*R*,7*aS*)-2-phenylperhydropyrrole[1,2-*c*]imidazole-3-(*E*)-propenoate (**98**) in 1,3-dipolar cycloadditions with azomethine ylides derived from glycine esters. Addition of the metal azomethine ylides to **98** proceeded in some cases with complete regio- and diastereocontrol by attack of the *syn*-azomethine ylide to the α-*si*-face of the alkene. This work has been extended to include α,β-unsaturated esters bearing a C_2-symmetric imidazolidine chiral controller at the β-position, such as **99** (147). The 1,3-dipolar cycloaddition of an azomethine ylide with **99** led to the formation of the cycloaddition product in high yield and with a satisfactory de of 92%. The reaction was found to be very dependent on the substituents in both the azomethine ylide and **99**. For reactions in which the (*R*)-substituent in **99** is methyl, high stereocontrol was occasionally achieved. The reaction proceeds under highest occupied molecular orbital $(HOMO)_{dipole}$–lowest unoccupied molecular orbital $(LUMO)_{alkene}$ control, with the dipole approaching the *si*-face of the alkene, as the *re*-face is hindered by the (*R*)-substituent as outlined for **99**.

12.2.2.5. Vinyl Phosphine Oxides and Vinyl Sulfoxides

Another type of chiral alkene applied in 1,3-dipolar cycloadditions are vinyl groups attached to chiral phosphine oxides or sulfoxides. Brandi et al. (150,151) used chiral vinyl phosphine oxide derivatives as alkenes in 1,3-dipolar cycloadditions with chiral nitrones. This group also studied reactions of achiral nitrones with chiral vinyl phosphine oxide derivatives. Using this type of substrate, fair endo/exo-selectivities were obtained. In reactions involving optically pure vinyl phosphine oxides, diastereofacial selectivities of up to 42% de were obtained. Chiral vinyl

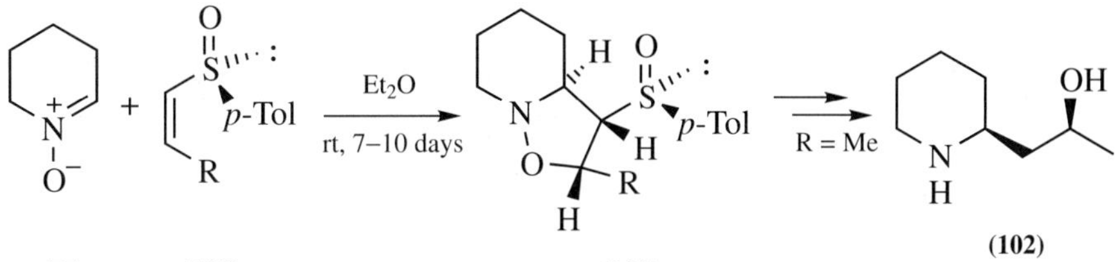

(77) (100) *exo*-(101) Yield = 85–95% de_{endo} = 82–96% (102) (+)-sedridine

Scheme 12.33

sulfoxides have been applied in 1,3-dipolar cycloadditions more frequently (152–164). Louis and Hootelé (152,156) have studied the 1,3-dipolar cycloadditions of vinyl sulfoxides **100** in 1,3-dipolar cycloadditions with the cyclic nitrone **77** (Scheme 12.33). The reactions proceeded with absolute exo-selectivity, and, especially when the substituent at the (*Z*)-alkene is phenyl, high de of 96%. In the case where R = Me (de = 82%) the product **101** has been applied in the synthesis of the natural product (+)-sedridine (**102**).

Fluoro-substituted chiral vinyl sulfoxides such as **103** have been used in 1,3-dipolar cycloadditions with various benzonitrile oxides (Scheme 12.34) (158). The reaction proceeded slowly at room temperature, however, after 5–10 days the isoxazoline (**104**) was obtained with excellent de in good yield. In some cases, the product tends to eliminate the 5-methoxy substituent of the isoxazoline, thus, after loss of two chiral centers, an isoxazole is obtained (158,159). Other chiral sulfinyl derivatives have also been used in 1,3-dipolar cycloadditions with nitrile oxides (160,161), and in one case a racemic vinyl phosphine was used in reactions with various nitrile oxides, but with moderate selectivities (151).

Diastereoselective reactions of azomethine ylides with chiral vinyl sulfoxides have also been conducted (Scheme 12.35) (162–164). The 1,3-dipolar cycloaddition of $(R)_S$-*p*-tolyl vinyl sulfoxide (**106**) with 1-methyl-3-oxidopyridinum (**105**) gave three of the four possible diastereomers, and one of these isomers **107** was used for the enantioselective synthesis of the (*1S*)-(−)-2α-tropanol **108** (162).

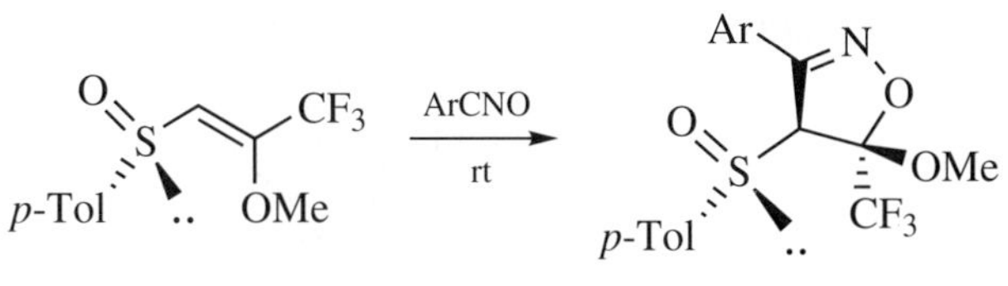

(103) (104) de = 100%

Scheme 12.34

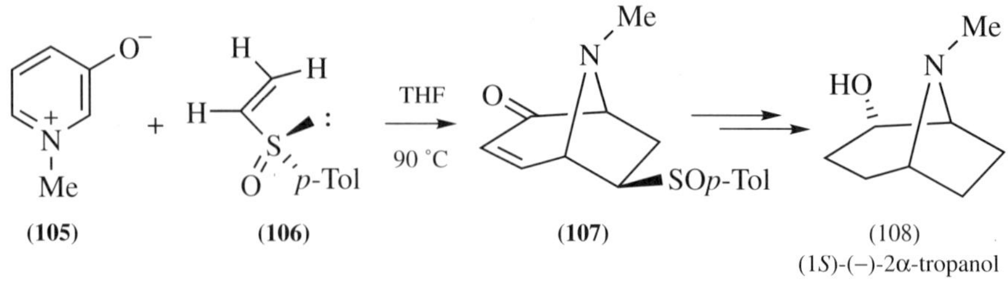

THF = tetrahydrofuron

Scheme 12.35

12.2.2.6. Alkenes with More Distant Chiral Centers

In all of the above reactions, a chiral center of the alkene was located in the allylic position. However, as shall be demonstrated next, more distant chiral centers may also lead to highly selective cycloadditions with 1,3-dipoles. In two recent papers, the use of exocyclic alkenes has been applied in reactions with *C,N*-diphenylnitrone (165,166). The optically active alkenes **109** obtained from (*S*)-methyl cysteine have been applied in reactions with nitrones, nitrile oxides, and azomethine ylides. The 1,3-dipolar cycloaddition of **109** (R=Ph) with *C,N*-diphenyl nitrone proceeded to give *endo*$_a$-**110** and *exo*$_a$-**110** in a ratio of 70:30 (Scheme 12.36). Both product isomers arose from attack of the nitrone **68** at the

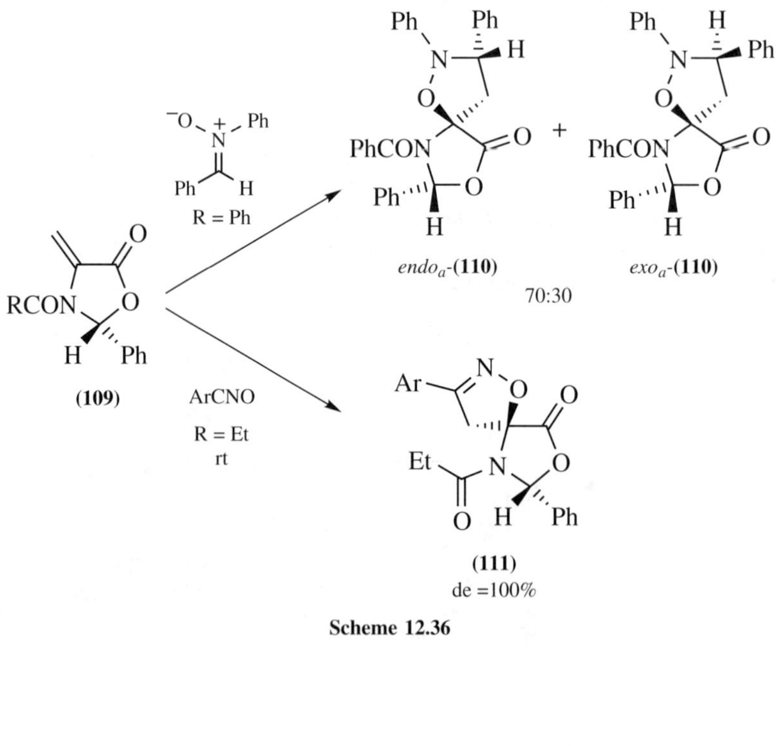

Scheme 12.36

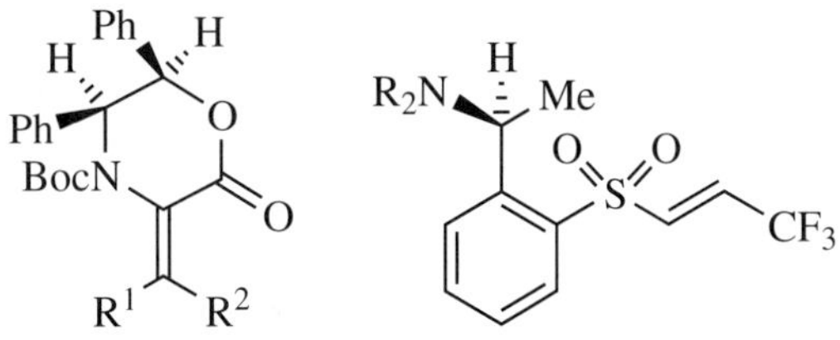

(112) **(113)**

Scheme 12.37

face of the alkene **109** anti to the phenyl substituent. The optically active alkene **109** (R=Me) also reacts with 3,5-dichlorobenzonitrile oxide at room temperature to furnish the *spiro*-isoxazoline (**111**) as the sole regio- and stereoisomer in 72% yield (Scheme 12.36) (167). Reactions of azomethine ylides with **109** have also been described, however, in these reactions, the Michael adducts were isolated and thus these reactions will not be described here (168).

Chiral exocyclic alkenes such as **112**, also having the chiral center two bonds away from the reacting alkene moiety, have been used in highly diastereoselective reactions with azomethine ylides, and have been used as the key reaction for the asymmetric synthesis of (*S*)-(−)-cucurbitine (Scheme 12.37) (169). The aryl sulfone **113** was used in a 1,3-dipolar cycloaddition reaction with acyclic nitrones. In **113**, the chiral center is located four bonds apart from alkene, and as a result, only moderate diastereoselectivities of 36–56% de were obtained in these reactions (170).

Nitrile oxides are relatively electron-deficient compounds that react smoothly with electron-rich vinyl ethers. Jenkins and co-workers (171) investigated the reactions of L-menthyl-, 8-phenylmenthyl- and (1*S*)-*endo*-bornyl vinyl ethers with nitrile oxides, but generally the de values were <33%. However, use of the chiral vinyl ether **114** in the reaction with a number of alkyl and benzonitrile oxides provided product **115** with de values of up to 66% (Scheme 12.38). It is proposed that the syn-staggered conformation is preferred for the vinyl ether **114** as indicated

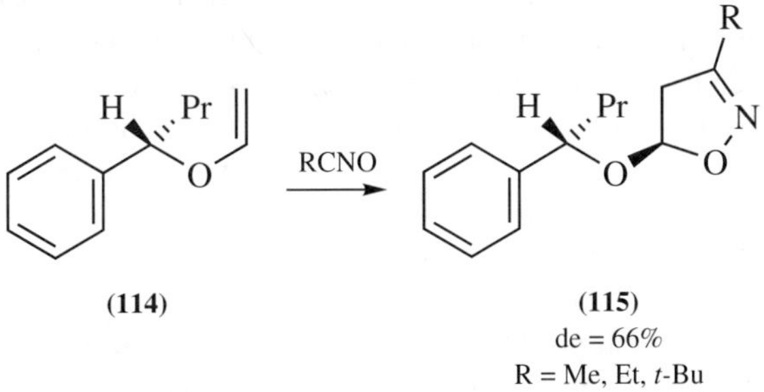

(114) **(115)**

de = 66%

R = Me, Et, *t*-Bu

Scheme 12.38

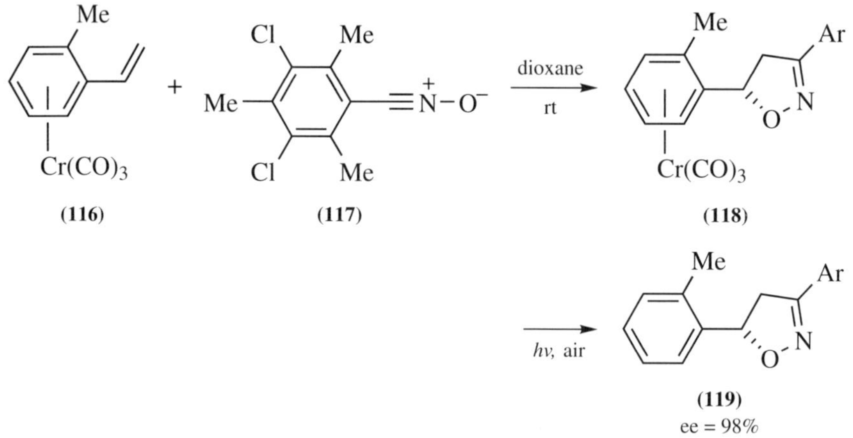

Scheme 12.39

in Scheme 12.39, and that the nitrile oxide attacks the double bond from the least hindered *re*-face, leading to diastereomer **115** as the major product.

The optically pure tricarbonyl chromium(0) complexes **116** have proven to offer an effective shielding of one of the faces of the alkene. Complex **116** was subjected to a 1,3-dipolar cycloaddition with the sterically crowded nitrile oxide **117** (Scheme 12.39) (172). The reaction proceeds at room temperature to give a 70% yield of **118**. After removal of the tricarbonylchromium moiety by a light induced oxidation with air, compound **119** was obtained with an optical purity of 98% enantiomeric excess (ee).

12.2.2.7. Heteroatom Dienophiles

A few reports of asymmetric 1,3-dipolar cycloadditions of azomethine ylides with chiral hetero-double bonds have been reported. Whereas Harwood and coworkers (82,83) used chiral azomethine ylides to conduct diastereoselective heterodienophile–azomethine ylide cycloadditions (see above). Viso et al. (173,174) used enantiopure chiral sulfimines such as **120** for reactions with a stabilized azomethine ylide **121** (Scheme 12.40). The dipole **121** was generated *in situ* from the corresponding amino ester-derived precursor and LDA. The reaction of **120** with **121** gave the corresponding products **122** (R=Bn, Me or *i*-Pr, Ar=Ph, *p*-NO_2Ph) in yields ranging from 53–80% and diastereoselectivities ranging from 90–96% de (173). Glycine-derived azomethine ylides failed to react under the above conditions. However, in the presence of $BF_3 \cdot OEt_2$, the reaction succeeded. In this reaction, the Mannich product was obtained as a stable intermediate (174). It was demonstrated that the product **122** (R=H, Bn, Ar=Ph) of both reactions could be converted into the corresponding 2,3-diaminoalcohols.

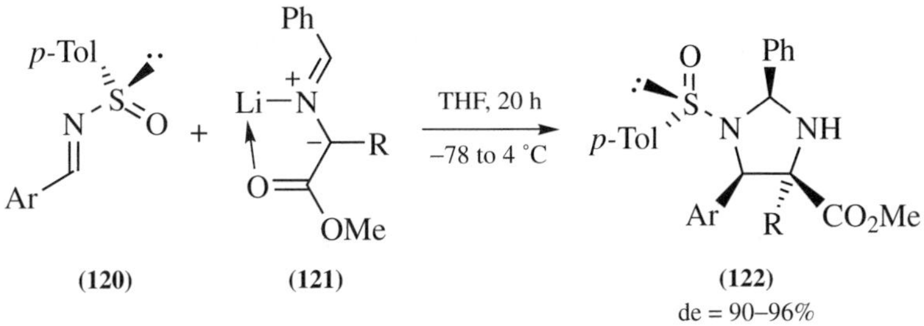

Scheme 12.40

12.2.3. Intramolecular Reactions

Intramolecular 1,3-dipolar cycloadditions of alkenyl nitrones have found broad application in organic synthesis (1). They have several advantages over the corresponding intermolecular reactions. Due to entropy factors, the activation barrier for the reaction is lower, allowing for lower reaction temperatures and for the use of dipoles and dipolarophiles of lower reactivity. The degree of freedom in the transition state of an intramolecular reaction is, of course, limited compared with the intermolecular reactions. Hence, a higher degree of regio-, endo/exo-, and diastereofacial selectivities are normally observed in intramolecular 1,3-dipolar cycloadditions. In intramolecular 1,3-dipolar cycloadditions the alkene part may be linked to a terminal atom, or if possible, to the central atom of the 1,3-dipole (Scheme 12.41). The latter case is only possible for allyl anion-type 1,3-dipoles with a central nitrogen atom, such as nitrones and azomethine ylides. The majority of the reported intramolecular 1,3-dipolar cycloadditions are those in which the alkene part is linked to the terminal atom of the dipole. This type of reactions gives rise to two regioisomers in which the five-membered ring product formed is either a bicyclo[X,3,0]- or a bicyclo[X,2,1]-compound (**A** and **B**). The most frequently observed product is the bicyclo[X,3,0]compound.

As mentioned in the beginning of this chapter, we have chosen to focus mainly on other areas than the intramolecular 1,3-dipolar cycloadditions, and thus only few examples of these processes will be highlighted here. For a more comprehensive coverage of this area, readers are directed to existing reviews (5,6).

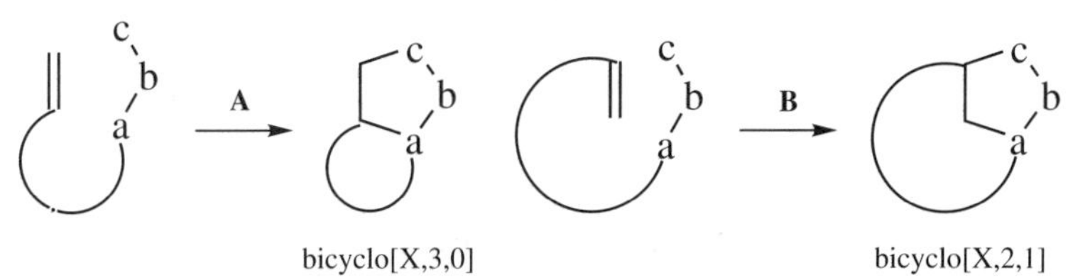

Scheme 12.41

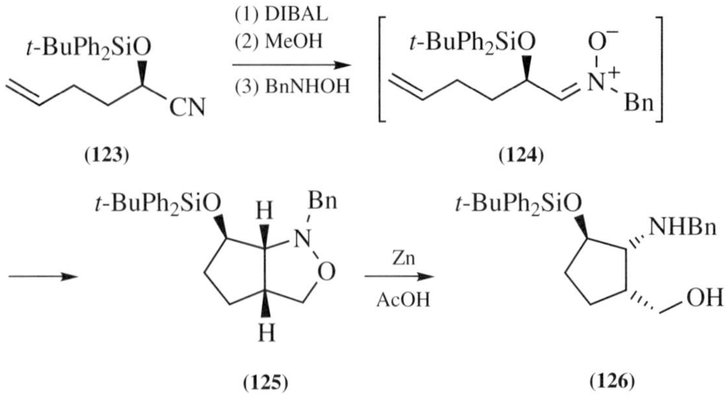

DIBAL = diisobutylaluminium hydride

Scheme 12.42

12.2.3.1. Nitrones and Nitronates

As for intermolecular 1,3-dipolar cycloadditions, the endo- and exo-isomers of each of the regioisomers can be formed in intramolecular reactions. In most cases, the exo-isomer is favored for steric reasons.

A classical method for controlling the stereoselectivity of intramolecular nitrone cycloadditions is to have a chiral center located on the chain between the nitrone and the alkene moiety (175–222). In a few other cases, the chirality is located outside the formed ring system (223–229). Marcus et al. (230) recently described an intramolecular 1,3-dipolar cycloadditions of the 5-alkenyl- and 6-alkenylnitrones **124** and **127** (Schemes 12.42 and 12.43). The chiral starting material **123** was obtained in 97% ee from an enzymatic cyanohydrin formation. Subsequent

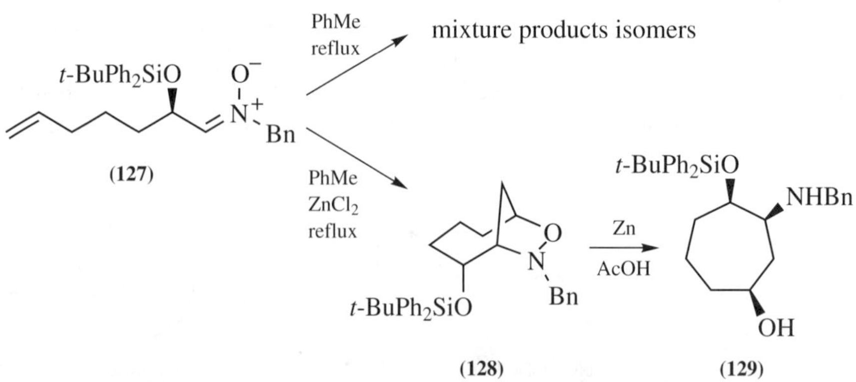

Scheme 12.43

reduction and condensation with benzyl hydroxylamine was performed in a one-pot procedure. The nitrone **124** was, however, never isolated, since it spontaneously undergoes intramolecular 1,3-dipolar cycloaddition to form **125**. The bicyclo[3.3.0] product **125** was obtained as the only observable regio- and stereoisomer, and an optical purity of 97% ee was found. Finally, the product **126** was obtained by a reduction.

The nitrone **127** (96% ee) containing an additional carbon atom was synthesized in a similar manner as **124**. However, for nitrone **127**, no spontaneous cyclization took place (Scheme 12.43). Upon heating of **127** in toluene, an inseparable mixture of product isomers was formed. It was also found that in the presence of a stoichiometric amount of $ZnCl_2$, the reaction proceeded to give the bicyclo[4.2.1] product **128** as the only observed product with close to complete retention of enantiopurity (ee = 94%). The authors propose a more polarized transition state of the Zn-mediated reaction with a well-developed positive charge on one of the carbon atoms of the alkene moiety to account for the inverted direction of the regioselectivity of the reaction (230). A subsequent reduction gave the functionalized cycloheptane **129**.

An alternative and elegant approach to bicyclo[3.3.0]isoxazolidines from alkenyl oximes was developed by Grigg (205) and applied in asymmetric reactions by Hassner et al. (206–209) and others (210). The optically active L-serine derived oxime **130** was proposed to be in a thermal tautomeric equilibrium with the nitrone tautomer **131**, which underwent an intramolecular 1,3-dipolar cycloaddition to form the product **132** in 80% yield as a single stereoisomer (Scheme 12.44) (209).

Alkenyl nitrones, having the alkene connected to the nitrone nitrogen atom, have been used in another approach to intramolecular reactions (231–235). Holmes and co-workers have this method for the synthesis of the alkaloid (−)-indolizidine 209B **137** (210,231). The alkenyl nitrone **134**, was obtained from the chiral hydroxylamine **133** and an aldehyde. In the intramolecular 1,3-dipolar cycloaddition, **135** was formed as the only isomer (Scheme 12.45). The diastereofacial selectivity was controlled by the favored conformation of the cyclohexane-like transition state in which the pentyl group was in a pseudoequatorial position, as indicated by **134**. Further transformation of **135** led to the desired product **137**.

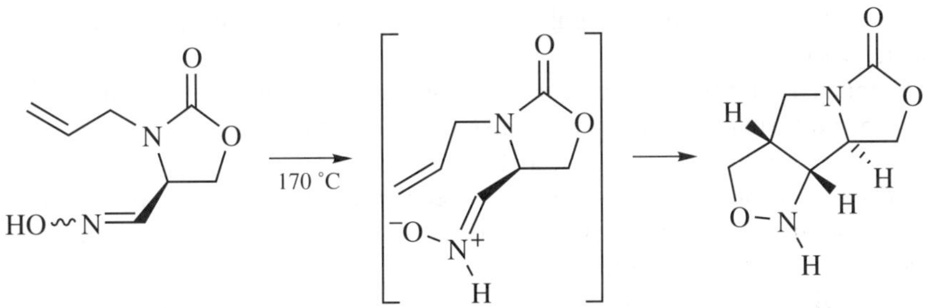

(130) **(131)** **(132)** sole product

Scheme 12.44

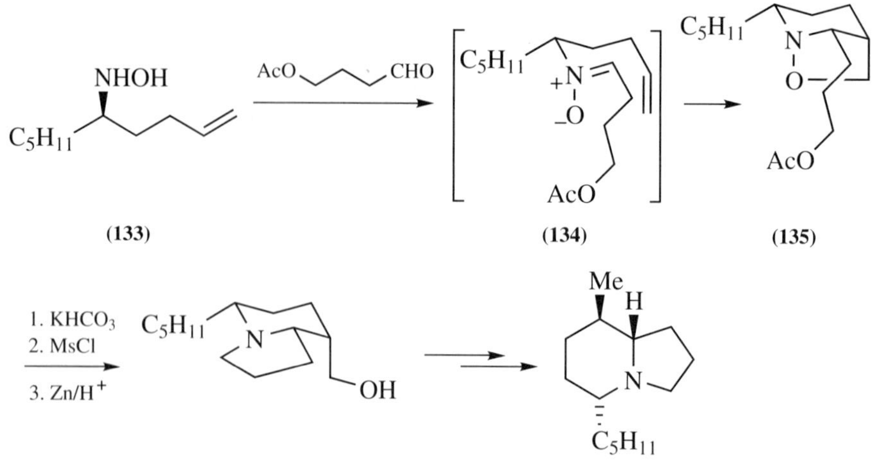

(133) (134) (135)

(136) (137)

(−)-indolizidine 209B

Scheme 12.45

Nitronates have also been applied in intramolecular 1,3-dipolar cycloaddition reactions. Denmark and Thorarensen (64) extensively studied the application of cyclic alkyl nitronates in tandem[4+2]/[3+2] cycloadditions of nitroalkanes. In most cases, the stereoselectivity of these reactions is directed by a chiral auxiliary and will thus be outlined in Section 12.3.4. The reader is also directed to the excellent chapter by Denmark in Chapter 2.

12.2.3.2. Nitrile Oxides

The use of alkenyl nitrile oxides is an effective method for the construction of bi- and polycyclic isoxazolines (2,4,200,236,237). Due to the rigid linear structure of the nitrile oxide, the reaction of alkenyl nitrile oxides almost always proceeds to give bicyclo[X,3,0] derivatives for X = 3–5. Most frequently, the diastereoselectivities are controlled by a chiral center on the link between the alkene and the dipole groups.

Compared to the intramolecular 1,3-dipolar cycloadditions of nitrones, the corresponding reactions of acyclic alkenyl nitrile oxides rarely gives complete diastereoselectivity, which can also be explained by the structure of the nitrile oxide. One exception of this is the intramolecular reactions of the D-glucose derived 5-alkenyloxime **138**, described by Gallos et al. (238) (Scheme 12.46). The nitrile oxide was obtained by the standard chlorination–elimination of oxime **138**. The benzyloxy groups of the carbon back-bone direct the facial selectivity of the 1,3-dipolar cycloaddition to give **139** as the only diastereomer. Subsequent reduction of the C=N double bond with $NaBH_3CN$ proceeded with complete diastereoselectivity and without reduction of the N–O bond to give **140**, having an

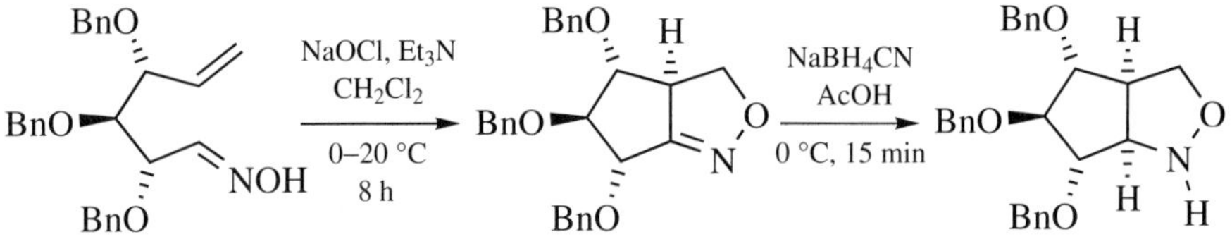

(138) (139) sole diastereomer (140)

Scheme 12.46

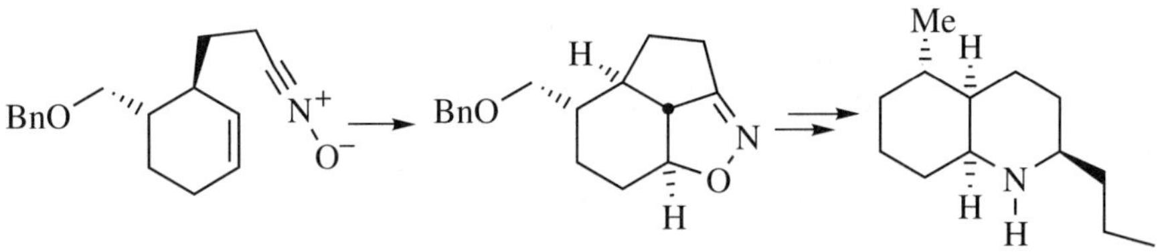

(141) (142) (143) (+)-pumiliotoxin C

Scheme 12.47

additional chiral center. It was also shown that the N—O bond of **140** could be cleaved by reduction with H_2 over Raney Ni.

For alkenyl nitrile oxides having the alkene in a cyclic structure, such as compound **141**, high diastereoselectivities can be obtained (Scheme 12.47). Compound **141** is formed *in situ*, and undergoes a spontaneous cyclization to furnish **142** as the sole diastereomer. Toyota et al. (239) used the tricyclic isoxazoline **143** in the synthesis of (+)-pumiliotoxin C.

The final example of the intramolecular 1,3-dipolar cycloadditions of nitrile oxides is the formation of the norbornadiene-derived tetracyclic adducts **146**, described by Tam and co-workers (240,241). The nitrile oxide **145**, formed from **144** by dehydration, can in principle give rise to four different cycloaddition products (three [2,3]-cycloaddition products). In practice, only diastereomer **146** was obtained. The reaction was used on substrates with a variety of different substituents (R=H, Me, hexyl, Cl, Br, CO_2Me, CH_2OMe), and in these cases, yields ranging between 66–89% were obtained (Scheme 12.48).

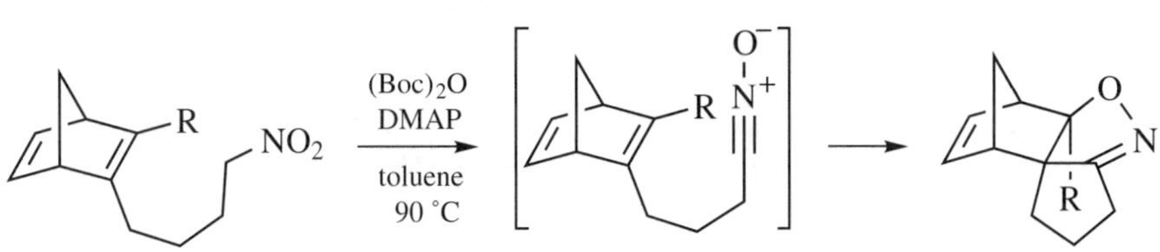

(144) (145) (146)

DMAP = dimethyl amino pyridine

Scheme 12.48

12.2.3.3. Other Dipoles

For intramolecular 1,3-dipolar cycloadditions, the application of nitrones and nitrile oxides is by far most common. However, in increasing frequency, cases intramolecular reactions of azomethine ylides (76,77,242–246) and azides (247–259) are being reported. The previously described intermolecular approach developed by Harwood and co-workers (76,77) has been extended to also include intramolecular reactions. The reaction of the chiral template **147** with the alkenyl aldehyde **148** led to the formation of the azomethine ylide **149**, which underwent an intramolecular 1,3-dipolar cycloaddition to furnish **150** (Scheme 12.49). The reaction was found to proceed with high diastereoselectivity, as only one diastereomer of **150** was formed. By a reduction of **150**, the proline derivative **151** was obtained.

Few asymmetric intermolecular 1,3-dipolar cycloaddition of azides are known, probably due to the relatively poor reactivity of this species. This lack of reactivity is compensated for by the favorable entropy of intramolecular reactions, and several examples of asymmetric intramolecular azide 1,3-dipolar cycloadditions have been described (247–259). Among the advantages of the azide dipole is its ease of preparation, its stability and the specificity of its reactions. One example of is the tandem Wittig-1,3-dipolar cycloaddition by Herdeis and Schiffer (252,260). A diastereomeric mixture of the half-acetal **152** was treated with a Wittig reagent to give **153** (Scheme 12.50). This intermediate could not be isolated because a intramolecular 1,3-dipolar cycloaddition followed immediately to give **154**. In

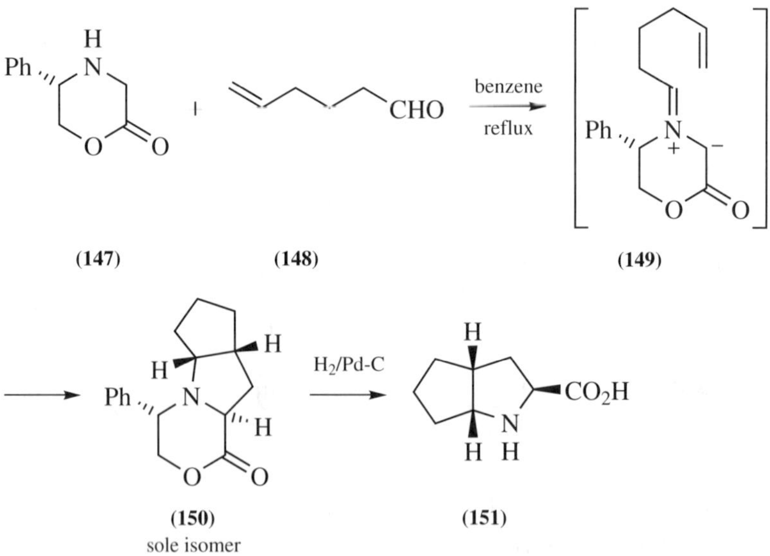

(147) **(148)** **(149)**

(150)
sole isomer

(151)

Scheme 12.49

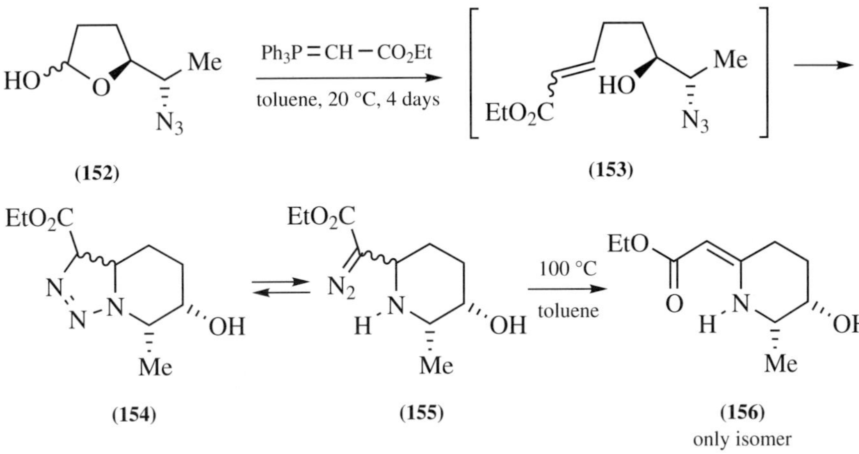

Scheme 12.50

the presence of traces of base such as triethylamine **154**, is in equilibrium with **155**. Heating of a mixture of **154** and **155** gave the stereochemically homogeneous product **156** after loss of nitrogen.

12.3. DIASTEREOSELECTIVE REACTIONS OF SUBSTRATES WITH CHIRAL AUXILIARIES

Chiral auxiliaries are very often applied for induction of asymmetry in 1,3-dipolar cycloadditions. For most of the reactions described in this section, recovery of the chiral auxiliary has been demonstrated, but for some reactions the chiral moiety has the potential to be recovered, although it was not performed. Most frequently, the chiral auxiliary is connected to the dipolarophile, very often as α,β-unsaturated esters or amides. In a few cases, auxiliaries have been attached to the 1,3-dipole.

12.3.1. α,β-Unsaturated Esters

The most commonly applied α,β-unsaturated ester auxiliary is the menthol group. It is inexpensive and easy to handle. Several different menthyl 2-alkenoates (**157**), in particular acrylates, have been applied in 1,3-dipolar cycloaddition reactions (Scheme 12.51). The major drawback of the menthyl ester auxiliary in 1,3-dipolar cycloadditions are the poor selectivities often associated with these reactions, except for reactions with azomethine ylides.

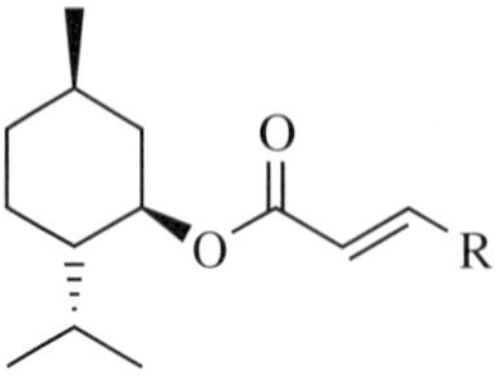

(157a): R = H
(157b): R = Me
(157c): R = $MenthO_2C$-

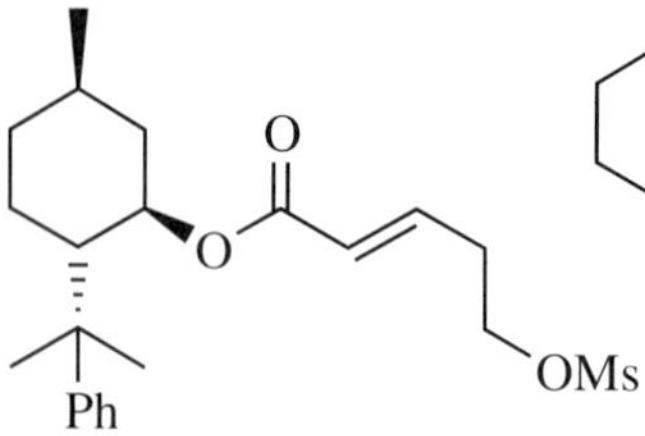

(158)

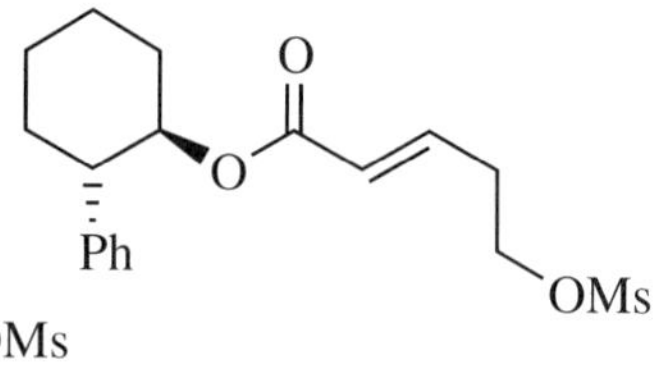

(159)

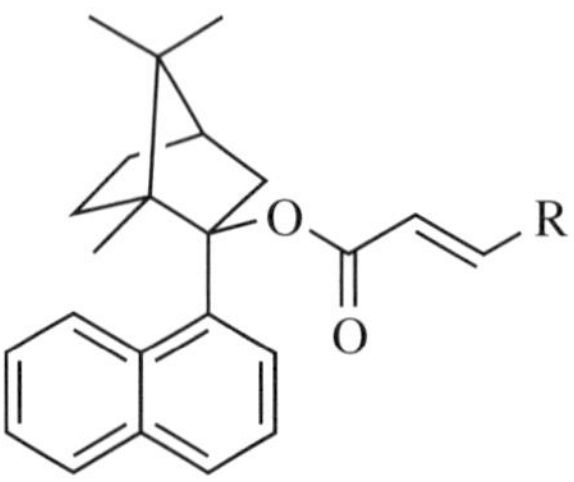

(160a): R = H
(160b): R = Me

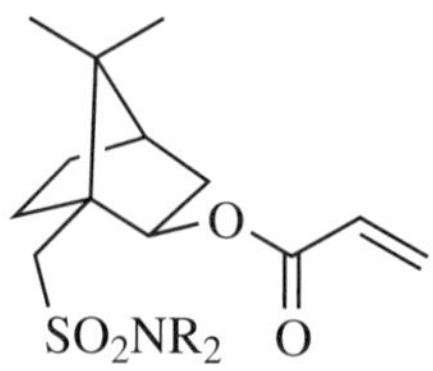

(161): R_2 = cyclohexyl

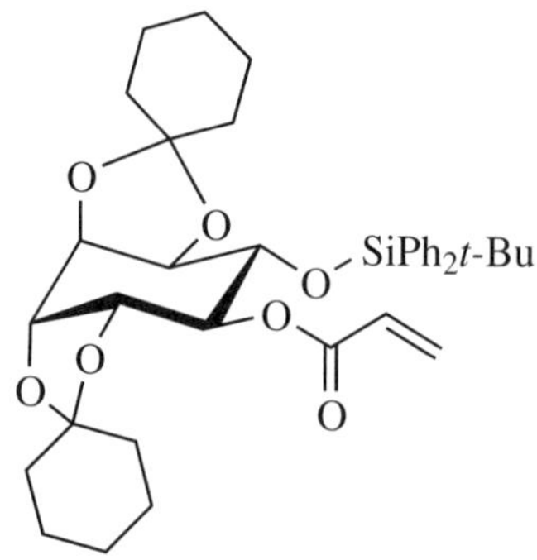

(162)

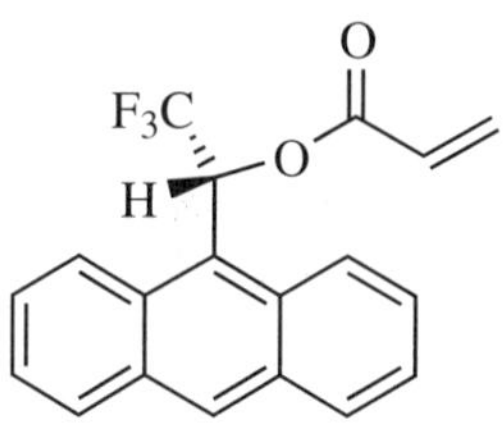

(163)

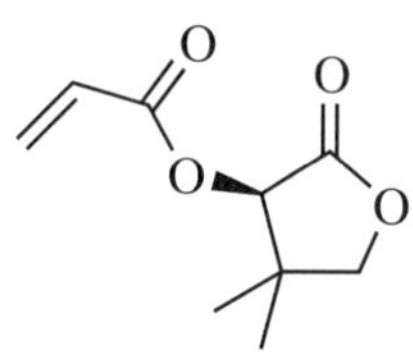

(164)

Scheme 12.51

Acrylates **157a** and crotonates **157b** have been applied in reactions with both acyclic and cyclic nitrones, but low selectivities were obtained (Scheme 12.51) (261,262). For the dimenthyl fumarate (**157c**) (R=$MenthO_2C$—), however, the cycloaddition with an acyclic nitrone gave complete stereoselection (120). Compounds **157a,b** have also been used in cycloadditions with nitrile oxides (263,264), although in these cases poor selectivities were observed. Similar poor selectivity was observed in the reactions of **157a** with diazo compounds (265). Much better selectivities were obtained in reactions of **157a** with azomethine ylides (Scheme 12.51). The reactions of metal-stabilized azomethine ylides (see above) with **157a** were described in a series of papers by Grieg and co-workers (266–270). In several reactions, only one diastereomeric product was obtained.

Brandi and co-workers (271) applied the familiar α,β-unsaturated esters **158** and **159** in reactions with cyclic nitrones. In these reactions, the isoxazolidine products were formed as intermediates, which immediately underwent N-alkylation to give tricyclic compounds. The reactions proceeded in both cases with moderate selectivities of 39% de for **158** and 57% de for **159**. Most remarkably, the reactions proceeded with opposite face selectivity.

The camphor-derived chiral acrylate **160a** was used in reactions with nitrones by Olsson (272) (Scheme 12.51). They observed low endo/exo-selectivity, but excellent diastereofacial discrimination in the reactions of cyclic nitrones with **160a**. They also studied the reactions of nitrile oxides with **160a,b**. Fair selectivity of up to 68–75% de was obtained. However, for the reaction of the crotonyl derivative **160b** with nitrile oxides, mixtures of regiomers were obtained.

The auxiliary acrylates **161** and **162** have been used in 1,3-dipolar cycloadditions with nitrile oxides. The camphor-derived acrylate **161** underwent a 1,3-dipolar cycloaddition with benzonitrile oxide with up to 56% de (Scheme 12.51) (263). The auxiliary in acrylate **162** is derived from naturally occurring L-quebrachitol, and provided an effective shielding of the *re*-face of the alkene in the reaction with benzonitrile oxide, as 90% de was obtained (273). Compound **163** was used in a reaction with the nitrone 1-pyrrole-1-oxide, and the reaction proceeded to give a complex mixture of products (274).

The chiral acrylate **164** was used in a 1,3-dipolar cycloaddition with a nitrile imine. Bis(trityl)nitrile imine was found to undergo a diastereoselective 1,3-dipolar cycloaddition with (*R*)-α-(acyloxy)-β,β-dimethyl-γ-butyrolactone **164** to give the 2-pyrazoline product with a de of 50% (Scheme 12.51) (275).

12.3.2. α,β-Unsaturated Amides

One of the most successful auxiliaries for α,β-unsaturated carbonyl compounds for not only 1,3-dipolar but also other cycloadditions is Oppolzer's chiral sultam (276). In particular, the acrylate **165** of Oppolzer's chiral sultam is one of the most frequently used substrates for asymmetric 1,3-dipolar cycloadditions, as shown in Scheme 12.52.

The reactions of nitrones with **165** have been described (277–279). In the approach described by Koskinen and co-workers (279), the bulky nitrone **166** was used in a reaction with **165** to give a 20:1 mixture of **167** and an unidentified diastereomer (Note: Opposite enantiomers are shown here). Reactions of less bulky nitrones gave lower selectivities (277,278). Kim et al. (280,281) described reactions of **165** with silyl nitronates (Scheme 12.52). The configuration of the direct isoxazolidine products was not determined. Instead, diastereoselectivities of 66–88% de of **169** were found after elimination of the silyloxy group. The reaction of various nitrile oxides proceeded to give the same isoxazoline products **169** as obtained for nitronates (Scheme 12.52). For the reactions of **165** with various alkyl and aryl nitrile oxides **170**, the products **169** were obtained with diastereoselectivities of 62–90% de (282–286). In a theoretical study, it was proposed that the

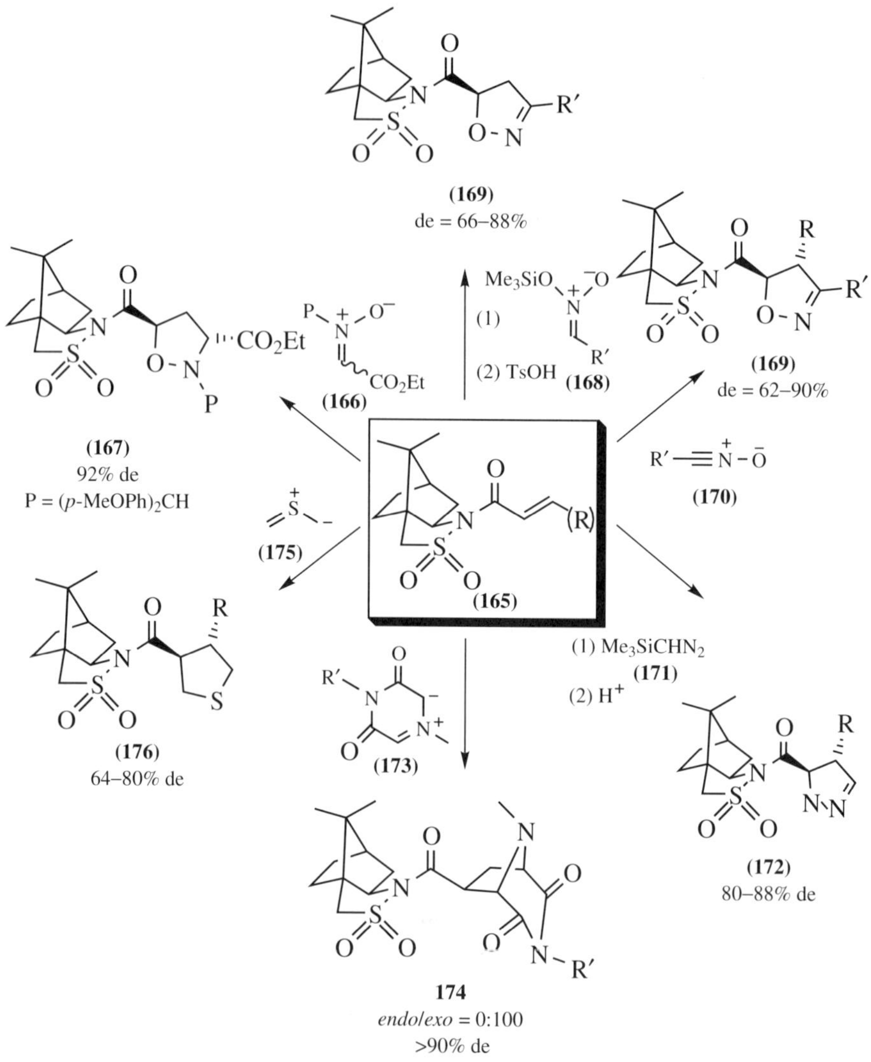

Scheme 12.52

preferred conformation of the acryloyl moiety in **165** is s-cis, where the carbonyl group points away from the sultam oxygen atom (282,283). The face selection of the nitrile oxide 1,3-dipolar cycloaddition with **165** cannot be explained in terms of the conventional face shielding by sterically bulky groups. Based on theoretical semiempirical and *ab initio* calculations, Kim et al. (283) suggested that the *si*-face is shielded by electrostatic repulsion between the sultam oxygen atoms and the oxygen atom of the incoming nitrile oxide. In calculations where the sultam oxygens were removed and the geometry of the acryloyl moiety was locked, no face selection was predicted. Curran and co-workers (284,285) applied Oppolzer's chiral sultam as the auxiliary for the preparation of (+)-hepialone and (−)-pestalotin.

Carriera and co-workers (287) described the 1,3-dipolar cycloaddition of trimethylsilyldiazomethane **171** with **165** (Note: Opposite enantiomers are shown here). The intermediate 1-pyrazoline obtained from this reaction rearranged after acidic work up to furnish the 2-pyrazolines **172** with 80–88% de (Scheme 12.52). By further conversion of these products, optically active azaprolines were synthesized.

Azomethine ylides have also been subjected to reactions with **165** (Scheme 12.52). Garner and Ho (288) developed the reaction of the photogenerated azomethine ylide **173** with **165** for the synthesis of quinocarcin. The reaction gave **174** with complete endo/exo selectivity and with more than 90% de. Other types of azomethine ylides have also been used in reactions with **165** and its derivatives (289,290).

Karlsson and Högberg (291,292) applied the thiocarbonyl ylide **175** in a diastereoselective 1,3-dipolar cycloaddition with **165**. The thiocarbonyl ylide was generated *in situ* by an elimination reaction. The reaction with **165** gave **176** (R=Bu, BnO, Ph) with selectivities of up to 64–80% de. Furthermore, the cycloaddition of a chiral galactose-derived nitrile imine with **165** has been reported (104).

For most of reactions of different 1,3-dipoles with Oppolzer's chiral sultam shown in Scheme 12.52, it has been demonstrated that the auxiliary can be recovered.

Wallace et al. (293) applied the boronic ester derivative **177** of Oppolzer's sultam acrylate for a special nitrile oxide cycloaddition (Scheme 12.53). This reaction has the advantage that it allows for the introduction of a hydroxy group in the 4-position of the isoxazoline product **178** after oxidative cleavage of the boronic

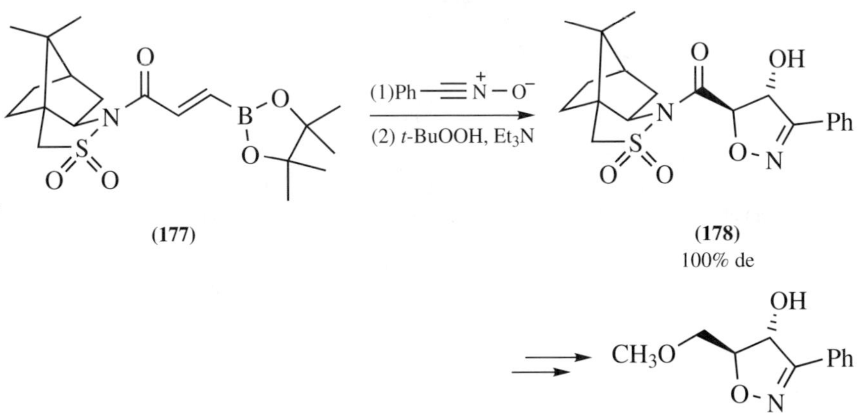

(177) **(178)** 100% de **(179)**

Scheme 12.53

ester. Furthermore, complete regio- and diastereoselectivity was obtained in this reaction. Protection of the hydroxy group in **178** followed by reduction with L-selectride, methylation, and deprotection furnished the isoxazoline product **179** under recovery of the chiral auxiliary.

A series of other α,β-unsaturated amide auxiliaries have been used for 1,3-dipolar cycloadditions, in particular for reactions of nitrile oxides (Scheme 12.54).

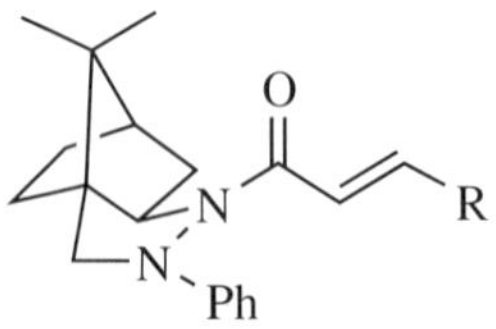

(180)
up to 98% de
with nitrile oxides (294)

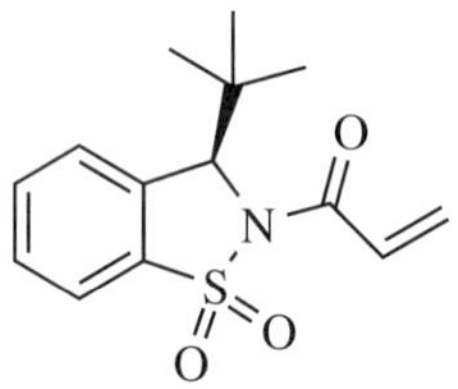

(181)
up to >94% de
with nitrile oxides (295)

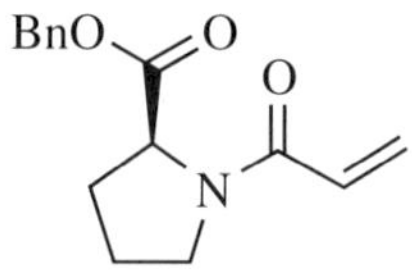

(182)
up to 50% de
with nitrile oxides (308)

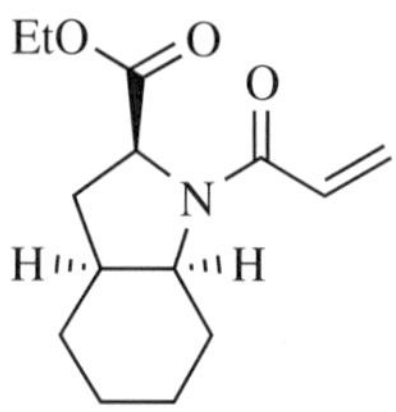

(183)
up to 90% de
with nitrile oxides (298)

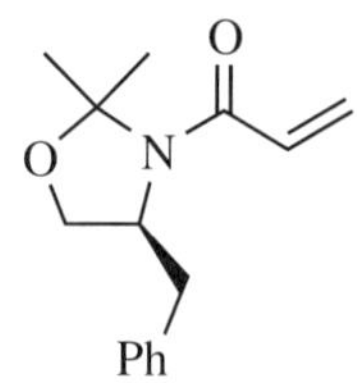

(184)
de = 86%
with nitrile oxides (299–301)

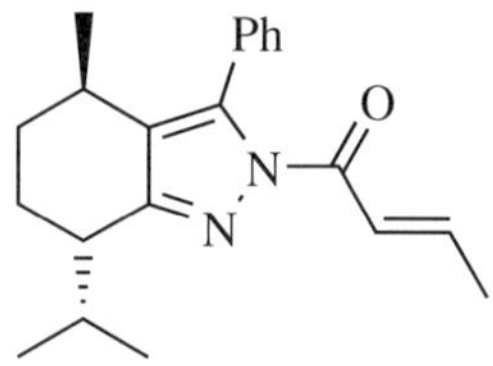

(185)
poor selectivities
with nitrile oxides (309)

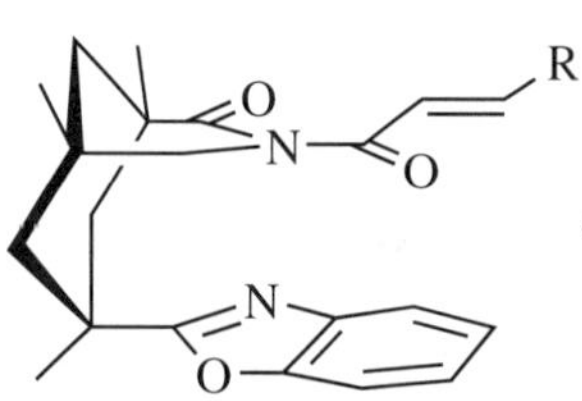

(186)
up to 100% de
with nitrile oxides (296)

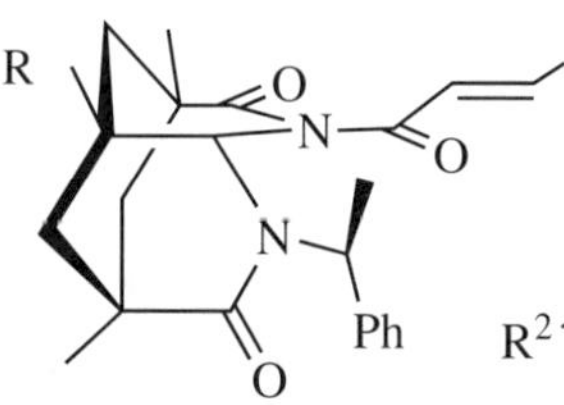

(187)
up to >98% de
with nitrile oxides (297)

up to 86% de
with silyl nitronates (302)

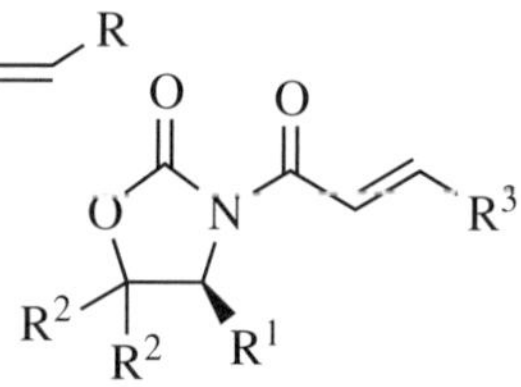

(188)
R^1 = Bn, R^2 = Me, R^3 = Me
up to 92% de
with nitrile oxide (303)

R^1 = *i*-Pr, R^2 = H, R^3 = Me
exo/endo = <1:>99
up to 100% de
with nitrones (16,277,304)

Various R^1 and R^3, R^2= H
up to 60% de
with azomethine ylides (305)

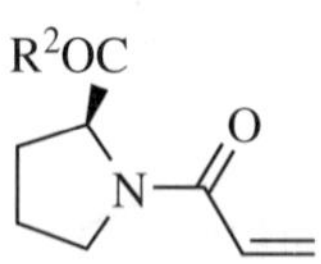

(189)
exo/endo = <1:>99
up to >98% de
with azomethine ylides (306,307)

Scheme 12.54

The α,β-unsaturated amides **180–188a** have all been used in 1,3-dipolar cycloadditions with nitrile oxides, and some of them represent the most diastereoselective reactions of nitrile oxides. The camphor derivative **180** of Chen and co-workers (294), the sultam **181** of Oppolzer et al. (295), and the two Kemp's acid derived compounds **186** (296) and **187** (297) described by Curran et al. (296) are excellent partners for diastereoselective reactions with nitrile oxides, as very high diastereoselectivities have been observed for all of them. In particular, compound **186** gave, with few exceptions, complete diastereoselection in reactions with a wide range of different nitrile oxides. Good selectivities were also observed when using compounds **183** (298) and **184** (299–301) in nitrile oxide cycloadditions, and they have the advantage that they are more readily available. Curran and co-workers also studied the 1,3-dipolar cycloaddition of **187** with silyl nitronates. However, compared to the reactions of nitrile oxides, lower selectivities of up to 86% de were obtained (302).

The amino acid derived chiral oxazolidinone **188** is a very commonly used auxiliary in Diels–Alder and aldol reactions. However, its use in diastereoselective 1,3-dipolar cycloadditions is less widespread. It has, however, been used with nitrile oxides, nitrones, and azomethine ylides. In reactions of **188** (R^1 = Bn, R^2 = Me, R^3 = Me) with nitrile oxides, up to 92% de have been obtained when the reaction was performed in the presence of 1 equiv of $MgBr_2$ (303). In the absence of a metal salt, much lower selectivities were obtained. The same observation was made for reactions of **188** (R^1 = Bn, R^2 = H, R^3 = Me) with cyclic nitrones in an early study by Murahashi et al. (277). In the presence of ZnI_2, endo/exo selectivity of 89:11 and up to 92% de was observed, whereas in the absence of additives, low selectivities resulted. In more recent studies, it has been shown for **188** (R^1 = *i*-Pr, R^2 = H, R^3 = Me) that, in the presence of catalytic amounts of MgI_2-phenanthroline (10%) (16) or $Yb(OTf)_3$(20%) (304), the reaction with acyclic nitrones proceeded with high yields and stereoselectivity. Once again, the presence of the metal salt was crucial for the reaction; no reaction was observed in their absence. Various derivatives of **188** were used in reactions with an unsubstituted azomethine ylide (305). This reaction proceeded in the absence of metal salts with up to 60% de. The presence of metal salts led to decomposition of the azomethine ylide.

The 1,3-dipolar cycloaddition of **189** with metalla-azomethine ylides was described by Waldmann et al. and others (306–309). In some cases, the reactions proceeded with almost complete endo/exo selectivity ($>99:<1$) and with de values of up to $>98\%$.

12.3.3. Other Auxiliaries Attached to Dipolarophiles

Chiral furanones (butanolides) such as **191** have been used as dipolarophiles in various 1,3-dipolar cycloadditions. The chiral 4-substituted butanolide **190** was prepared from **191** and the chiral auxiliary menthol (Scheme 12.55) (310,311). The single diastereomer **191** is obtained by crystallization and epimerization of the other diastereomer, as the amount of **191** in solution decreases. 1,3-Dipolar

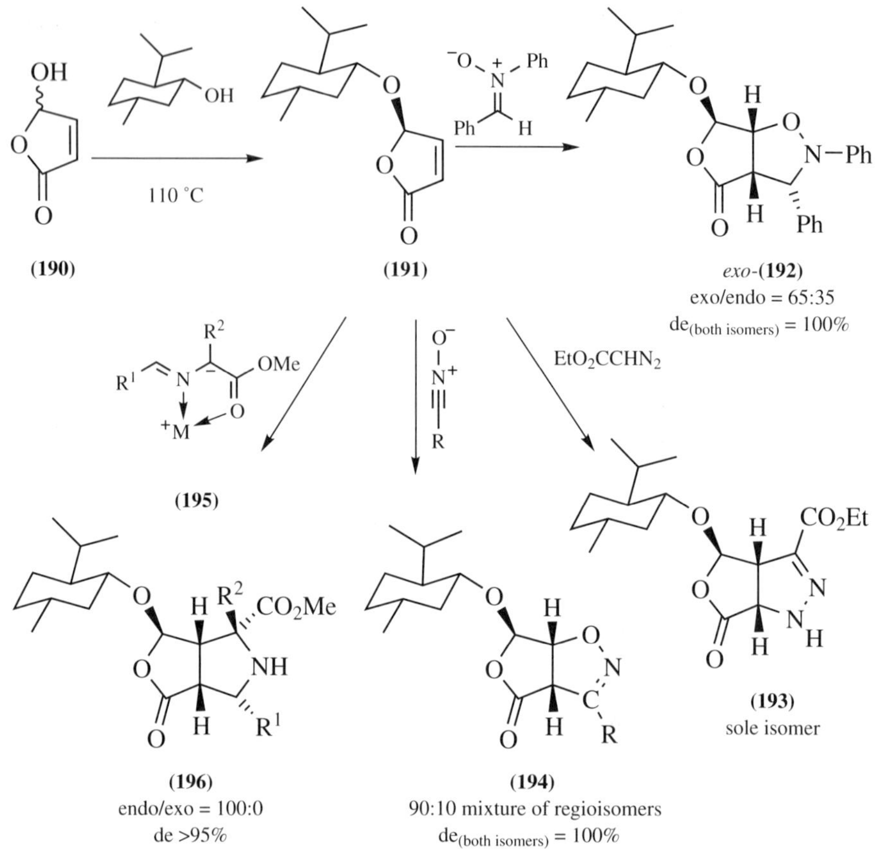

Scheme 12.55

cycloadditions of **191** have been performed with nitrones (310,311), diazoalkanes (311,312), nitrile oxides (310,311), and azomethine ylides (310,266), as shown in Scheme 12.55. For all reactions, the menthol moiety efficiently shields one face of the alkene, and nearly complete diastereoselectivity was observed. However, in the reaction between **191** and diphenyl nitrone, an exo/endo ratio of 65:35 of **192** was noted. In the reaction of nitrile oxides with **191**, mixtures of regioisomers were obtained.

The use of chiral vinyl ethers in 1,3-dipolar cycloadditions with nitrones allows for the subsequent removal and recovery of the chiral group. Using the chiral vinyl ether **197** and the cyclic nitrone **77**, the cycloaddition proceeded with high diastereoselectivity (Scheme 12.56). The endo/exo-selectivity was not given in this communication by Carruthers et al. (313), but this is of minor importance for the final outcome of this work, since one of the chiral centers was destroyed in the conversion of **198** into the final product **199**. The chiral auxiliary can by recovered in this reaction sequence, and **199** was obtained with an optical purity of $>95\%$ ee.

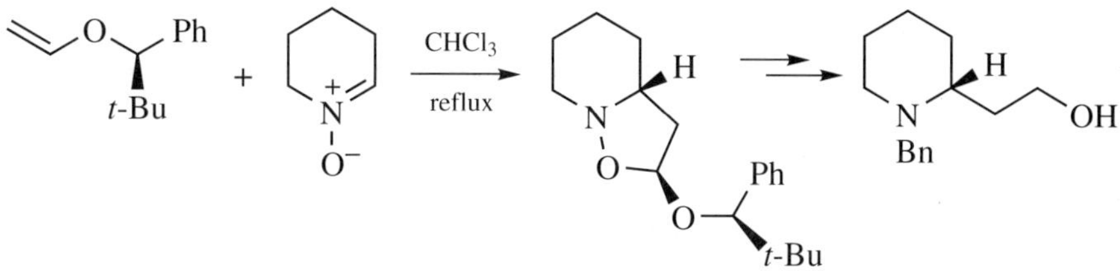

(197) (77) (198) (199)

Scheme 12.56

Barluenga et al. (314,315) used an α,β-unsaturated Fischer carbene complex **200** (Scheme 12.57) bearing a 9-phenylmenthol auxiliary in 1,3-dipolar cycloadditions with various 1,3-dipoles. Diastereoselective 1,3-dipolar cycloadditions of nitrile imines are relatively rare. However, compound **201** was used in a 1,3-cycloaddition with **200** (314). The direct adduct of the reaction was unstable. For this reason, an oxidation was promoted with pyridine *N*-oxide to give **202** in moderate to good yields. For a series of different aromatic substituents, the reactions proceeded with high regioselectivities in favor of **202** and with high de >90%. The 1,3-dipolar cycloaddition of diazo compounds with **200** has also been studied by this group, but the observed selectivities of these reactions were lower (315).

In three separate papers, the use of chiral boronic esters in 1,3-dipolar cycloadditions with nitrile oxides have been described (316–318). The reaction of **203** with nitrile oxides proceeded with low diastereoselectivities (Scheme 12.58).

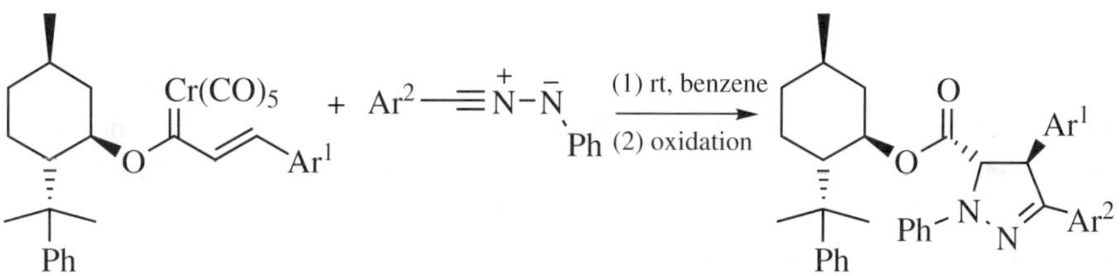

(200) (201) (202)

Scheme 12.57

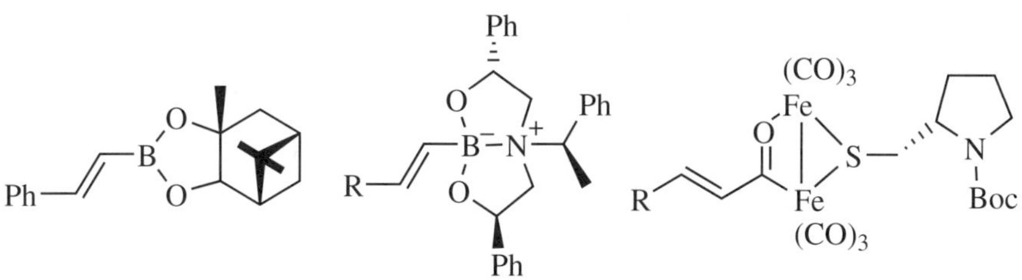

(203) (204) (205)

Scheme 12.58

The direct cycloaddition adduct was oxidized, resulting in the hydroxylated isoxazoline product (316). Better selectivities were obtained in 1,3-dipolar cycloadditions of **204** with nitrile oxides (317,318). The 1,3-dipolar cycloadditions proceeded with concomitant loss of the boron group to give the isoxazoline products in up to 74% ee (318). The alkene **204** was also tested in reactions with nitrones. The reactions proceeded with poor yields, but high selectivities were observed in two cases (318). Gilbertson et al. (319) investigated the use of chiral α,β-unsaturated hexacarbonyldiiron acyl complexes **205** as dipolarophiles in reactions with nitrones. Selectivities of up to >92% de were observed. The iron moiety was removed oxidatively after the cycloaddition and the thioester was hydrolyzed.

12.3.4. Auxiliaries Attached to 1,3-Dipoles

Garner et al. (90,320) used aziridines substituted with Oppolzer's sultam as azomethine ylide precursors. The azomethine ylide generated from **206** added to various electron-deficient alkenes, such as dimethyl maleate, *N*-phenylmaleimide, and methyl acrylate, giving the 1,3-dipolar cycloaddition product in good yields and up to 82% de (for *N*-phenylmaleimide). They also used familiar azomethine ylides formed by imine tautomerization (320). Aziridines such as **207** have also been used as precursors for the chiral azomethine ylides, but in reactions with vinylene carbonates, relatively low de values were obtained (Scheme 12.59) (92).

Oppolzer et al. (321) applied his own sultam as the auxiliary for a cyclic nitrone in the synthesis of (−)-allosedamine (Scheme 12.60). The enantiomerically pure nitrone **209** was synthesized from **208** by base treatment, attack of the enolate on 1-chloro-1-nitrosocyclohexane at the nitrogen atom, and subsequent elimination of chloride. Subsequent addition of aqueous HCl gave the cyclic nitrone **209**. The nitrone participated in a 1,3-dipolar cycloaddition with styrene, proceeding with complete exo-specificity. The product, **210**, was obtained with a de of 93%. Two further reaction steps yield the piperidine alkaloid (−)-allosedamine **211** in an overall yield of 21%.

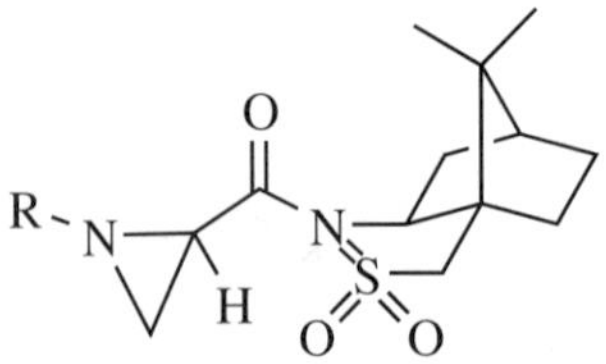

(206)

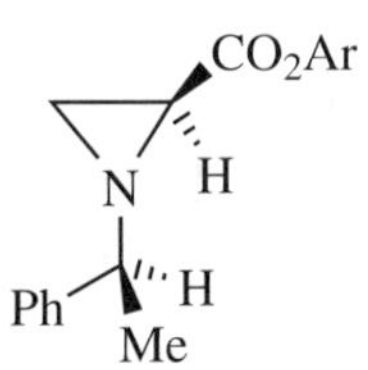

(207)

Scheme 12.59

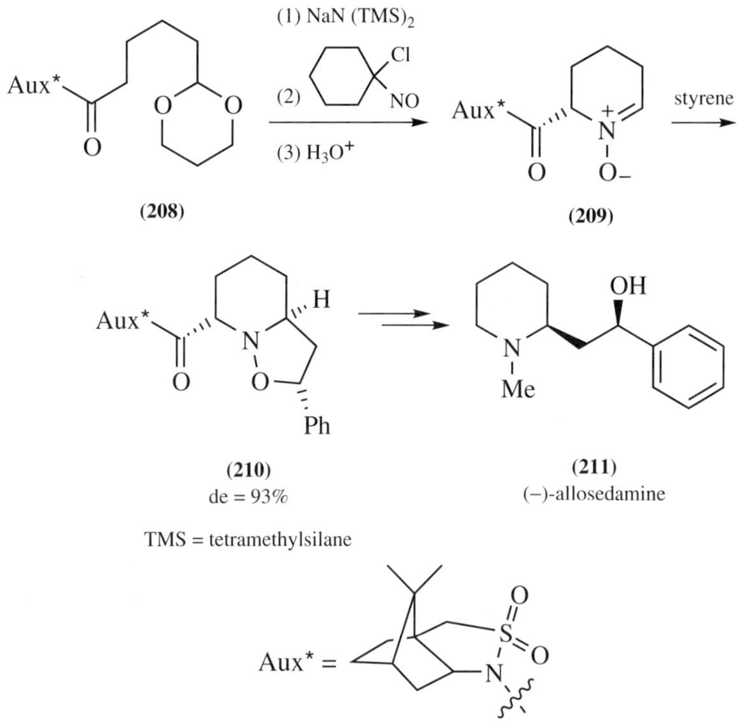

Scheme 12.60

Katagiri et al. (48,322,323) applied L-menthone and an *in situ* generated nitroso ketene for the synthesis of nitrone **212**, as shown in Scheme 12.61. The nitrone showed high selectivities in reactions with various allyl silanes. High pressure or the presence of $BF_3 \cdot Et_2O$ was required for the reaction to proceed. In the reaction of **212** with allyltrimethylsilane in the presence of $BF_3 \cdot Et_2O$, isoxazolidine (**213**) was obtained as the sole product. After hydrolysis and hydrogenolysis,

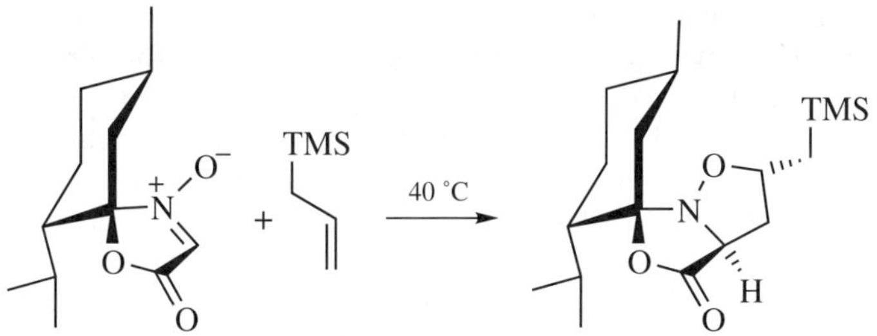

(212) **(213)**

Scheme 12.61

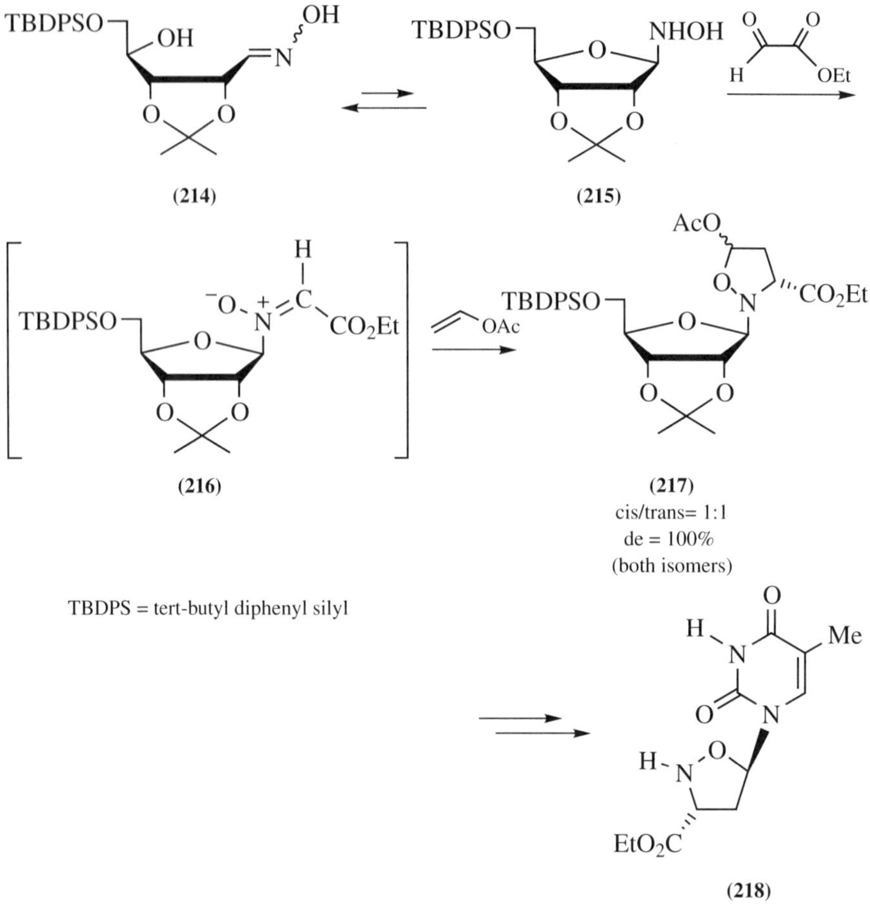

Scheme 12.62

enantiomerically pure α-amino acids were synthesized and the L-menthone was recovered. For reactions of chiral racemic allyl silanes, nitrone **212** in some cases reacted with only one of the isomers (i.e., kinetic resolution took place).

Vasella and co-workers (324–329) developed an elegant method for the introduction and recovery of sugar derivatives as auxiliaries for nitrones, and this methodology has later been used by others groups (330–332). Chiacchio et al. (332) applied this method for the synthesis of homochiral isoxazolidinylthymidines **218**. Oxime **214** is obtained from the corresponding protected carbohydrate; in this case ribofuranose (Scheme 12.62). The oxime is in equilibrium with hydroxylamine **215**, which is subjected to condensation with ethyl glyoxylate. The nitrone **215** underwent reaction with vinyl acetate *in situ* to form **217**. The cis/trans selectivity of the reaction was poor, as close to a 1:1 mixture of C(5)-epimers was obtained, however, the diastereofacial selectivity was high, since only one cis and one trans adduct were obtained. From either isomer of **217**, the nucleoside analogue **218** could be

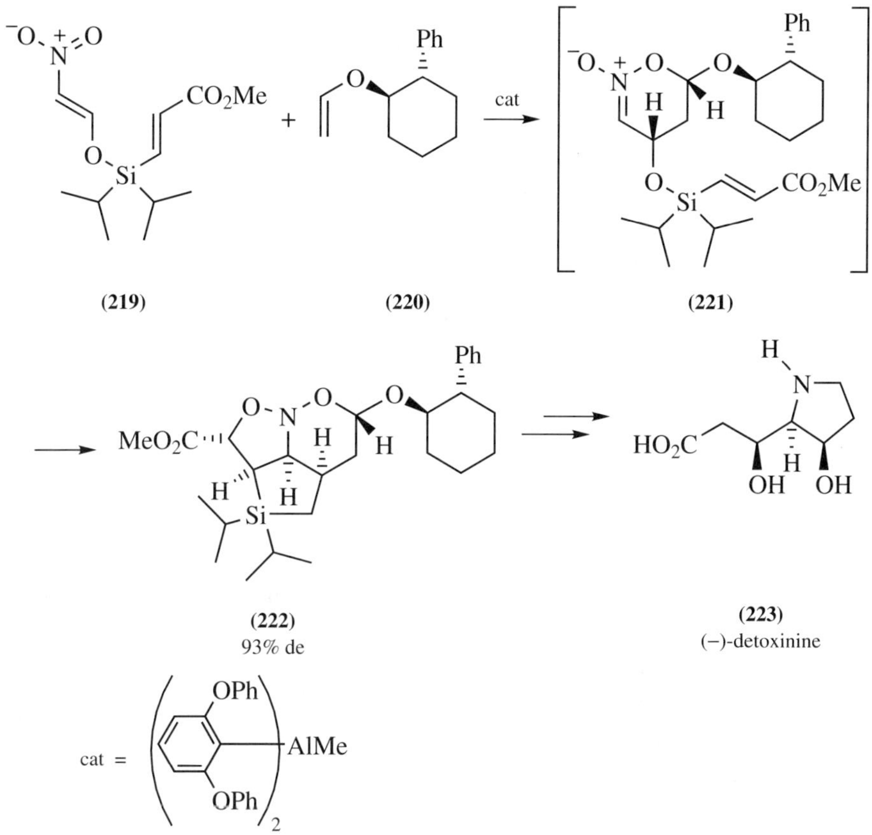

(219) (220) (221) (222) 93% de (223) (−)-detoxinine (224)

Scheme 12.63

obtained. First thymine was introduced, followed by a acid catalyzed hydrolysis of the sugar isoxazolidine to give homochiral **218** (332).

Cyclic alkyl nitronates may be used in tandem [4+2]/[3+2] cycloadditions of nitroalkanes, and this reaction has been extensively studied by Denmark et al. (64,333–335). In recent work, they developed the silicon-tethered heterodiene–alkene **219** (Scheme 12.63). Steric hindrance and the fact that both the nitroalkene and the α,β-unsaturated ester in **219** are electron deficient renders the possibility of self-condensation. Instead, **219** reacts with the electron-rich chiral vinyl ether **220** in the presence of the catalyst **224** to form the intermediate chiral nitronate **221**. The tandem reaction proceeds from **221** with an intramolecular 1,3-dipolar cycloaddition to form **222** with 93% de. Further synthetic steps led to the formation of (−)-detoxinine **223** (333). A similar type of tandem reaction has also been applied by Chattopadhyaya and co-workers (336), using 2′,3′-dideoxy-3′-nitro-2′,3′-didehydrothymidine as the starting material (336).

12.4. CATALYTIC ASYMMETRIC 1,3-DIPOLAR CYCLOADDITIONS

The development and application of catalytic enantioselective 1,3-dipolar cycloadditions is a relatively new area. Compared to the broad application of asymmetric catalysis in carbo- and hetero-Diels–Alder reactions (337,338), which has evolved since the mid-1980s, the use of enantioselective metal catalysts in asymmetric 1,3-dipolar cycloadditions remained almost unexplored until 1993 (5). In particular, the asymmetric metal-catalyzed reactions of nitrones with alkenes has received considerable attention during the past 5 years.

12.4.1. Basic Aspects

The relative frontier molecular orbital (FMO) energies of the reagents are very important for the catalytic control of 1,3-dipolar cycloadditions. In order to control the stereochemical outcome of a reaction with a substoichiometric amount of a ligand–metal catalyst, it is desirable that a large rate acceleration is obtained in order to assure that the reaction only takes place in the sphere of the metal and the chiral ligand. The FMO considerations will be outlined in the following using nitrones as an example.

Nitrones can be activated mainly in two different ways for the 1,3-dipolar cycloaddition with alkenes. In the reaction between a nitrone and an electron-deficient alkene, such as an α,β-unsaturated carbonyl compound (a normal electron-demand reaction), it is primarily controlled by the interaction between $HOMO_{nitrone}$-$LUMO_{alkene}$ (Scheme 12.64). By coordination of a Lewis acid (LA) catalyst to the α,β-unsaturated carbonyl compound, the $LUMO_{alkene}$ energy decreases and a better interaction with the nitrone can take place (16,17).

The other catalytic approach to the 1,3-dipolar cycloaddition is via inverse electron demand, in which the nitrone is activated for addition to an electron-rich alkene such as, for example, a vinyl ether (Scheme 12.64). In this scenario, the $FMOs_{alkene}$ have higher energies than the $FMOs_{nitrone}$, and the dominating interaction in such a reaction will be $LUMO_{nitrone}$–$HOMO_{alkene}$. In the presence of a LA catalyst, the nitrone can coordinate to the catalyst, leading to a decreased $LUMO_{nitrone}$ energy. The decreased energy gap between the two FMOs responsible for the dominating interaction may lead to an enhanced rate of the 1,3-dipolar cycloaddition.

One of the problems related to the LA induced activation of α,β-unsaturated carbonyl compounds for the reaction with a nitrone is the competitive coordination of the nitrone and the α,β-unsaturated carbonyl compound to the Lewis acid (Scheme 12.65). Calculations have shown that coordination of the nitrone to the LA is more feasible than a monodentate coordination of a carbonyl compound. However, this problem can be circumvented by the application of alkenes such as 3-alkenoyl-oxazolidinones, enabling a bidentate coordination to the LA, which is favored over the monodentate coordination to the nitrone.

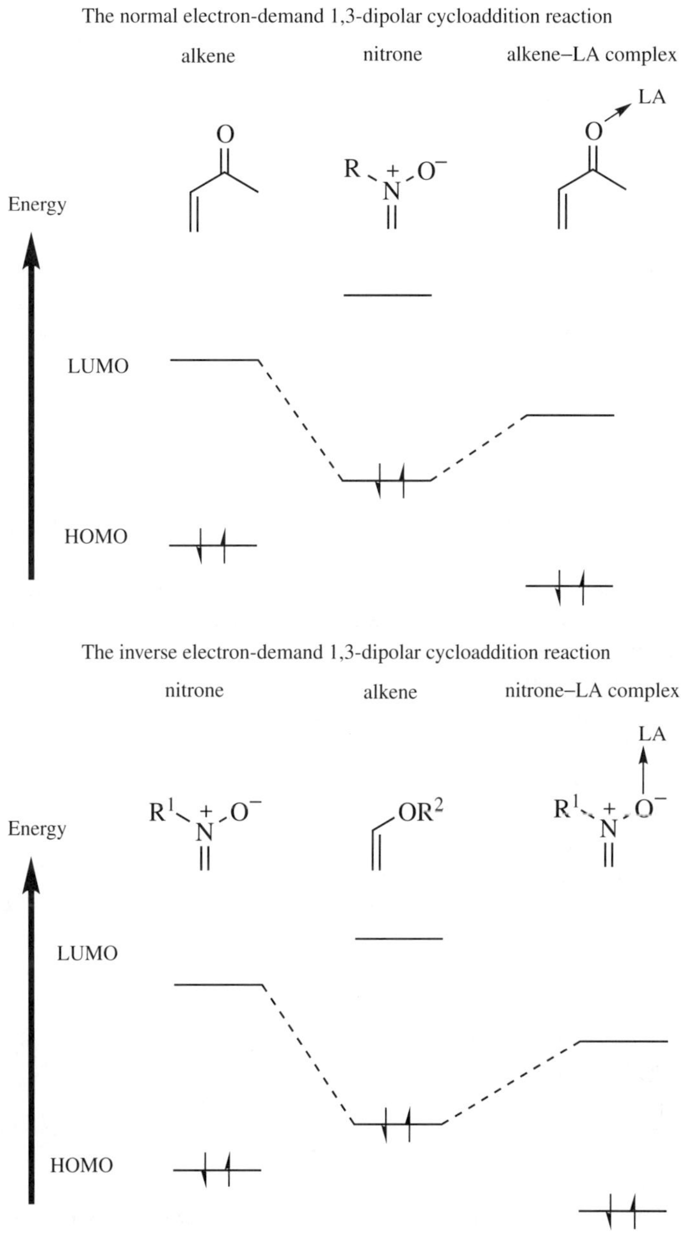

Scheme 12.64

These principles of activation and induction of asymmetry apply primarily to reactions of nitrones. For the reactions of other 1,3-dipoles, the catalyst-induced control of the enantioselectivity may in some cases be achieved by other principles. For the metal-catalyzed reactions of azomethine ylides, carbonyl ylides, and nitrile

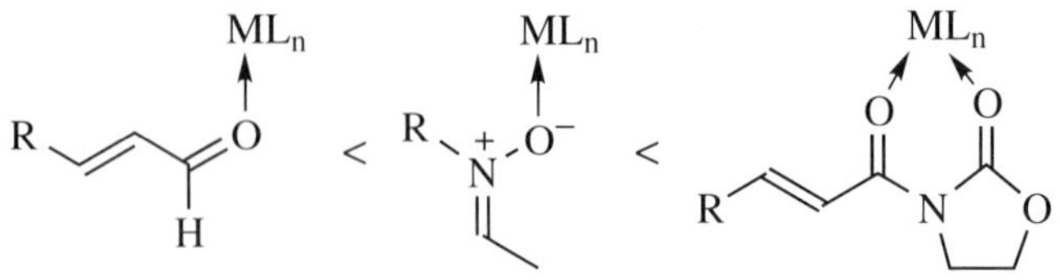

Scheme 12.65

oxides, the catalyst is crucial for the *in situ* formation of the 1,3-dipole from a precursor. After formation, the 1,3-dipole is coordinated to the catalyst due to a favored chelation and/or stabilization of the substrates that provide control of the enantioselectivity of the reaction.

12.4.2. Nitrones

The first catalytic asymmetric 1,3-dipolar cycloaddition of nitrones with alkenes was reported by Scheeren et al. in 1994 (339). *C,N*-Diphenylnitrone (**225a**) reacted with ketene acetals (**2**) in the presence of the amino acid derived oxazaborolidinones (**227**) as the catalyst (Scheme 12.66). In this reaction, an electron-rich alkene is involved, and the reaction proceeds by inverse electron demand. It was found that coordination of the nitrone to the boron of the LA strongly accelerated the 1,3-dipolar cycloaddition with ketene acetals. The reactions of **225a** with **226a**,**b**, catalyzed by 20 mol% of oxazaborolidinones such as **227a**,**b**, were carried out at −78 °C. In some reactions, fair enantioselectivities were induced by the catalysts, thus, **228a** was obtained with an optical purity of 74% ee, although in low yield. The reaction involving **226b** gave the C(3), C(4)-cis-isomer **228b** as the only diastereomer of the product with 62% ee.

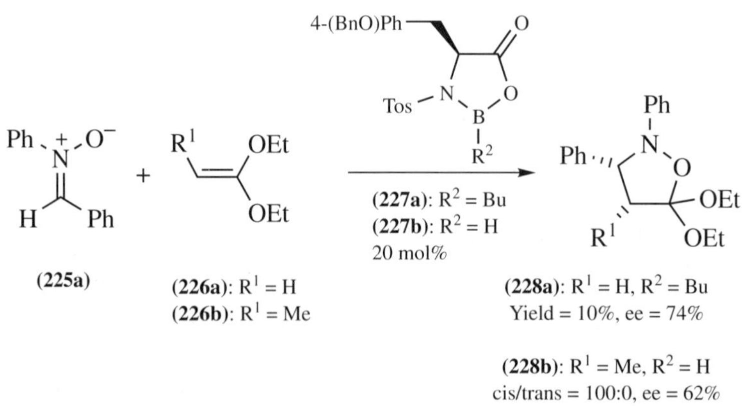

Tos = tosyl

Scheme 12.66

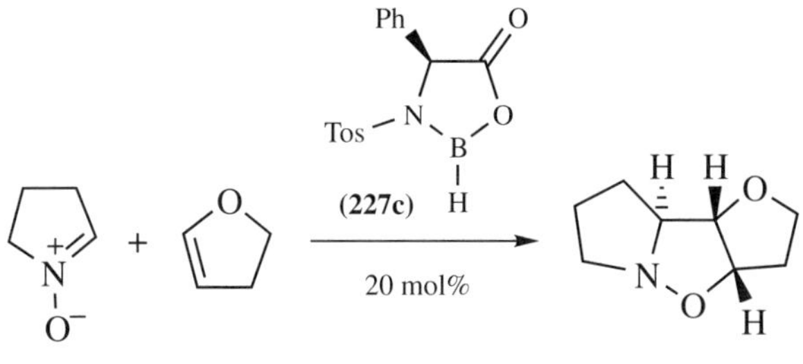

(229) **(230)** **(231)**
endo/exo = 0:100
ee = 38%

Scheme 12.67

This concept has been extended by Scheeren et al. (340), who investigated a series of derivatives of N-tosyl-oxazaborolidinones as catalysts for the 1,3-dipolar cycloaddition of **225a** with **226b**. Catalyst **227b** was synthesized from the corresponding amino acid and BH_3•THF. In this reaction, (−)-**228b** was obtained with 62% ee. If the catalyst was prepared from BH_3•SMe_2 and diphenyl ether was added, a remarkable reversal of the enantioselectivity of the reaction occurred, as (+)-**228b** was formed as the major isomer with 79% ee. In more recent work, the same research group has applied cyclic and acyclic vinyl ethers in the oxazaborolidinone catalyzed 1,3-dipolar cycloaddition with nitrones (341,342). The reaction of nitrone **229** with 2,3-dihydrofuran **230** with 20 mol% catalyst **227c** gave **231** in 56% yield as the sole diastereomer, although with an ee of 38% (Scheme 12.67).

The mechanism for a normal electron-demand boron catalyzed 1,3-dipolar cycloaddition has been investigated from a theoretical point of view (343). *Ab initio* calculations of the reaction between the nitrone $CH_2{=}N(O)H$ and acrolein in the presence of BH_3 showed that although the nitrone–BH_3 complex is predominantly formed at the early stage of the reaction, the acrolein–BH_3, which is a minor contributor, shows high rate acceleration, giving the electronically controlled *endo*-isoxazolidine–4-carbaldehyde complex as the major regio- and stereoisomeric cycloadduct. The latter reaction path has a 13.4 kcal/mol lower transition-state energy compared to the former one. The use of BF_3 as the LA for this reaction showed that the first step was the formation of the Michael-adduct complex followed by the cyclization step. These calculations indicated that the 1,3-dipolar cycloadditions of nitrones with electron-deficient alkenes may proceed via a stepwise mechanism when a strong Lewis is applied as the catalyst.

Chiral aluminium complexes have been used as catalysts for inverse electron-demand 1,3-dipolar cycloadditions of alkenes with nitrones, and the first contribution to this field was published in 1999 (344). The chiral AlMe–BINOL (BINOL=2,2′-bis(diphenylphosphino)-1,1′-binaphthyl) complexes **235** were excellent catalysts for the reaction between nitrone **225a** and vinyl ethers **232** (Scheme 12.68). The diastereo- and enantioselectivities are highly dependent on the chiral ligand. An exo/endo ratio of 73:27 was observed, and the exo-product was

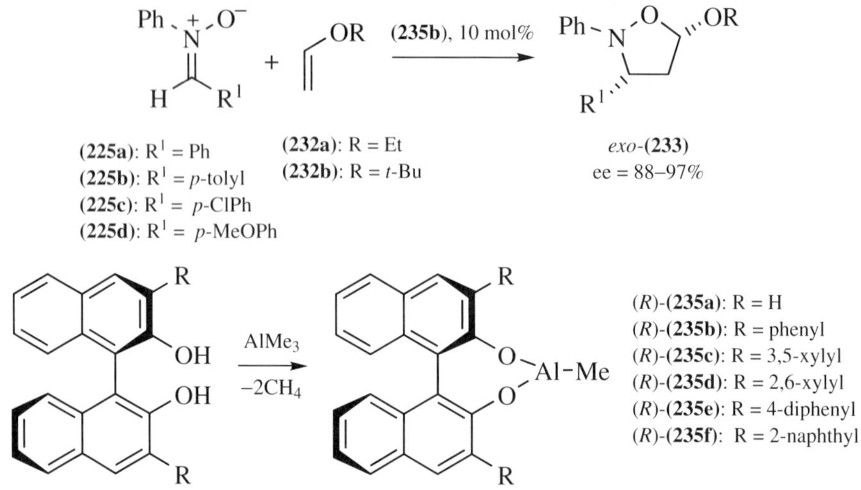

(234) **(235)**

Scheme 12.68

formed with a low ee of $<5\%$ when using **235a** as the catalyst, while application of catalysts with substituents in the 3,3′-positions of the ligand as in catalysts **235b–f** led to a remarkable improvement of the selectivities. In particular, complex **235b** has excellent properties, as the reaction performed in the presence 20 mol% of this catalyst gave *exo*-**233b** as the only observable diastereomer and the enantioselectivity of this product was 89% ee. The best results were obtained when using ethyl vinyl ether **232a**. The reactions between a series of nitrones **225a-d** with ethyl vinyl ether catalyzed by 10 mol% of **235b** all proceeded to give the corresponding products **233** with excellent exo-selectivities and with enantioselectivities of 88–97% ee (344).

The proposed mechanism for this highly enantioselective AlMe–BINOL catalyzed 1,3-dipolar cycloaddition was proposed as illustrated in Scheme 12.69 (344). The nitrone **225a** coordinates in the first step to the catalyst **235b** to form intermediate **236**. Intermediate **236**, which can account for the absolute stereoselectivity of this reaction, has the *si*-face shielded by the ligand, whereas the *re*-face remains available for reaction with ethyl vinyl ether **232a** as shown in the next step. The high exo-selectivity can also be accounted for by the model, as the ethoxy moiety of **232a** is pointing away from the nitrone N-phenyl group in the step in which **232a** approaches, as shown for complex **237**.

The 3,3′-cross-linked polymeric binaphthol ligand **238** in combination with $AlMe_3$ is also a highly selective catalyst for the 1,3-dipolar cycloaddition reaction in Scheme 12.69 (345). The only observable diastereomer resulting from the reactions was *exo*-**233**, which was obtained with an enantioselectivity of up to 99% ee using the aluminium catalyst of **238** (20 mol%). One of the advances of using a polymeric catalyst is the easy removal and recovery of the ligand from the

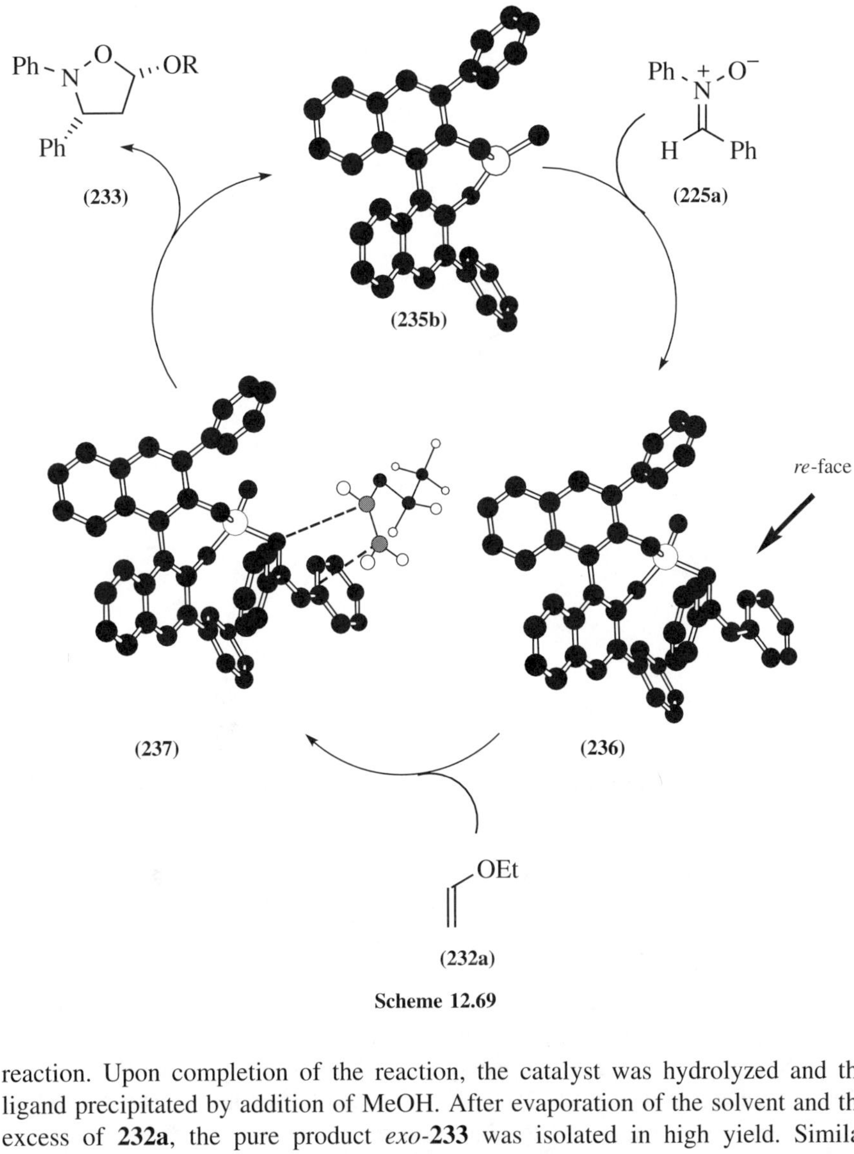

Scheme 12.69

reaction. Upon completion of the reaction, the catalyst was hydrolyzed and the ligand precipitated by addition of MeOH. After evaporation of the solvent and the excess of **232a**, the pure product *exo*-**233** was isolated in high yield. Similar excellent selectivities were obtained for reactions of other nitrones. Another important advantage of using the polymeric ligand **238** is, in addition to the easy purification of the product, that the ligand can be isolated and reused after the simple precipitation procedure (Scheme 12.70). In this manner a sample of the polymeric ligand was isolated and reused in four consecutive reactions of nitrone **225a** and ethyl vinyl ether. Both the yield and enantioselectivity of *exo*-**233a** showed only slight decreases after the ligand had been reused. The slight decrease was ascribed to the loss of small amounts of the ligand during the recycling procedure (345).

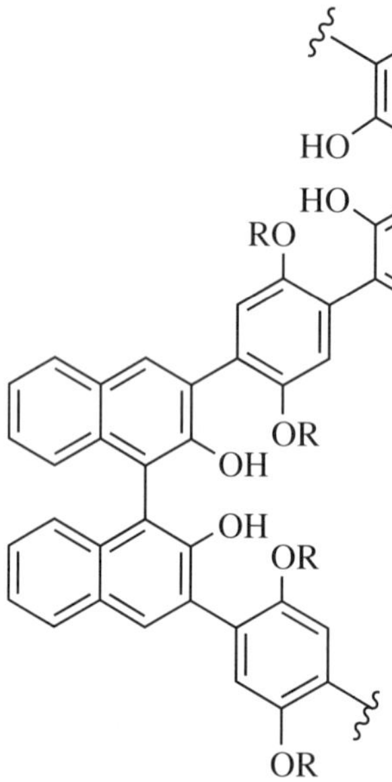

(238)
$R = C_6H_{13}$

Scheme 12.70

The chiral AlMe–3,3′-diaryl-BINOL complexes **235b–f** can also catalyze the 1,3-dipolar cycloaddition between the cyclic nitrones **239a** and ethyl vinyl ether **232a** (346). The use of the tetramethoxy-substituted derivative **235g** as the catalyst gave the cycloaddition adduct **240a** in 79% isolated yield. The diastereoselectivity was the same as in the acyclic case, giving an excellent ratio of *exo*-**240a** and *endo*-**240a** of >95:<5; *exo*-**240a** was obtained with up to 82% ee (Scheme 12.71). The high enantioselectivity of the exo-product opens up a new and readily accessible route to the enantioselective synthesis of interesting isoquinoline alkaloids.

Chiral bisoxazoline (BOX) magnesium complexes have also been used as catalysts for the 1,3-dipolar cycloaddition of nitrones with alkenes (16,304). The Ph–BOX–MgI_2 catalyst **242** can catalyze the 1,3-dipolar cycloaddition between **225a** and **242a,b** (Scheme 12.72). In the presence of 10 mol% of **242** (X=I), the reaction proceeded with good to high endo-selectivity and the endo-isomer was obtained in an enantioselectivity of up to 82%. In this case, the bidentate and electron-deficient alkenoyl-oxazolidinones are activated by the catalyst for reaction with the nitrone. Thus, contrary to the reactions catalyzed by the monodentate aluminium catalysts described above, the magnesium-catalyzed reaction proceeds according to the normal electron-demand concept.

When the 1,3-dipolar cycloaddition of **255a** with **241b** mediated by catalyst **242** (X=I) was performed in the absence of molecular sieves (MS) 4 Å, a remarkable reversal of enantioselectivity was observed, as the opposite enantiomer of *endo*-**243** was obtained (304). This reversal had not been observed for enantioselective

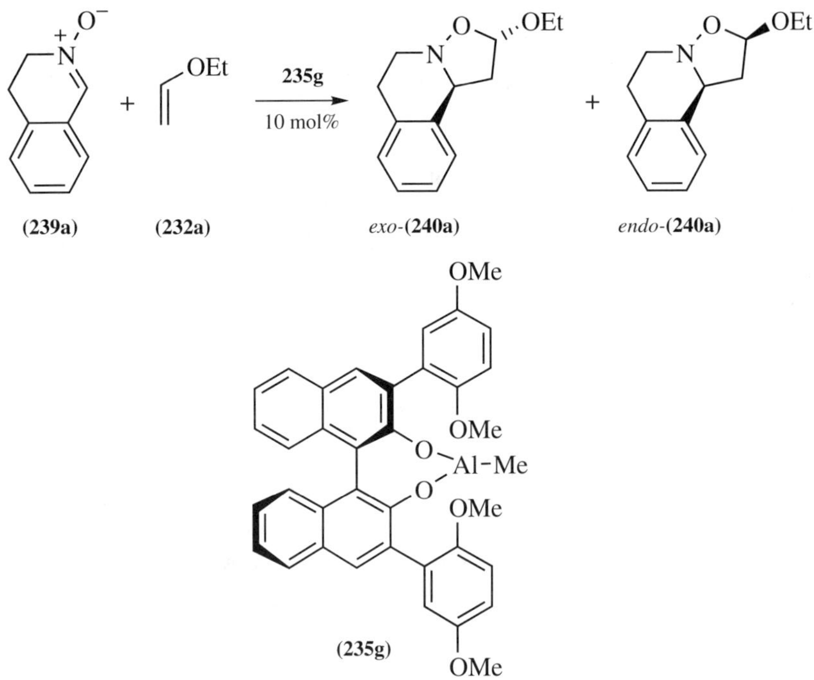

Scheme 12.71

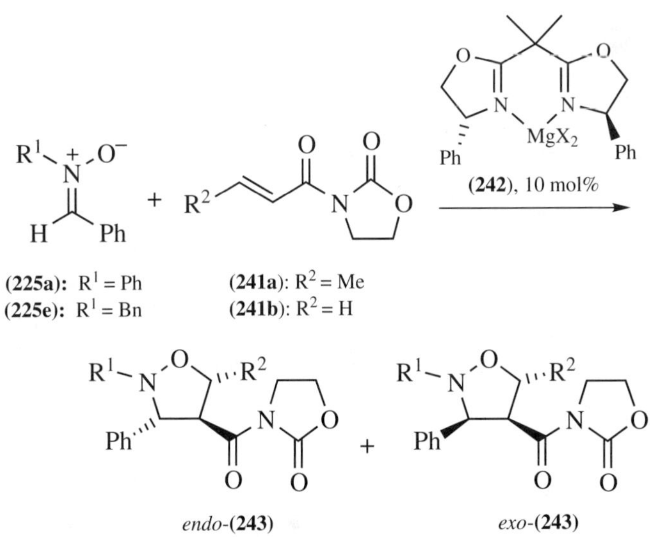

Scheme 12.72

catalytic reactions before, and the role of MS cannot simply be ascribed to the removal of water, since the application of 4 Å MS that were presaturated with water also induced the reversal of enantioselectivity. Desimoni and co-workers (347–349) also found that in addition to the presence of MS in the MgX_2–Ph–BOX catalyzed 1,3-dipolar addition shown in Scheme 12.72, the counterion for the magnesium catalyst also strongly affects the absolute stereoselectivity of the reaction. They applied the Ph–BOX–MgX_2 **242** ($X{=}ClO_4$) and (X=OTf) complexes and compared the results with the Ph–BOX–MgX_2 catalyst **242** (X=I). It was observed that both in the presence and absence of MS, the catalyst **242** ($X{=}ClO_4$) gave the opposite absolute configuration of the product as compared to the reaction of catalyst **242** (X=I). For catalyst **242** (X=OTf), the product was racemic in the presence of MS, whereas a high enantioselectivity of 86% ee was obtained in the absence of additives. The absolute configuration of the product of the reaction catalyzed by **242** (X=OTf) in the absence of MS was similar to that obtained with **242** ($X{=}ClO_4$) catalyst and opposite to that obtained with **242** (X=I).

Several chiral Ti(IV) complexes are efficient catalysts and have been applied to numerous reactions, especially in combination with the TADDOL **244** ligands (350). Chiral $TiCl_2$-TADDOLates were the first asymmetric catalysts to be applied in the normal electron-demand 1,3-dipolar cycloaddition of nitrones **225** with alkenoyl-oxazolidinones **241** (Scheme 12.73) (351). These substrates have turned

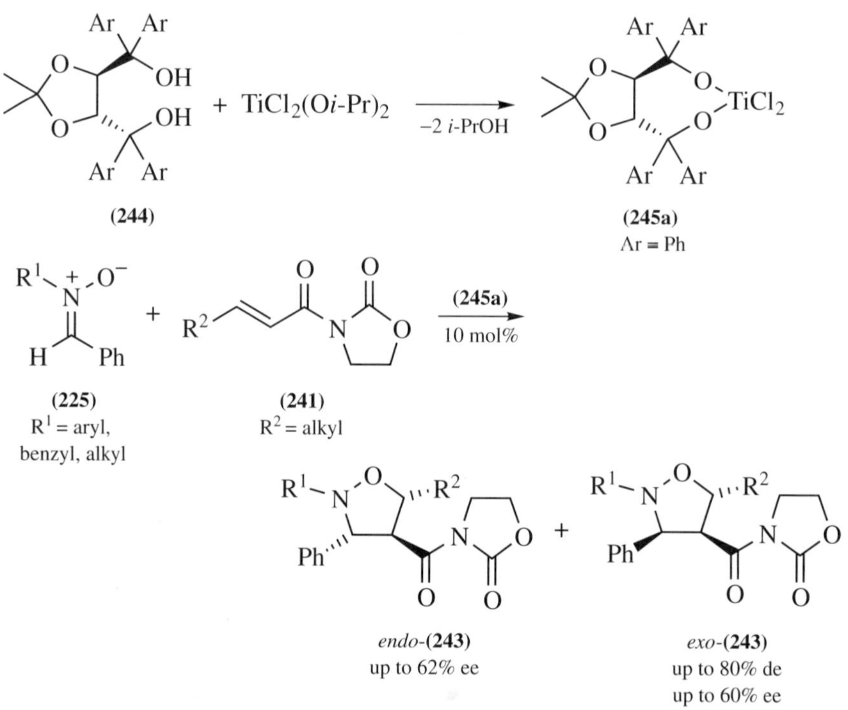

Scheme 12.73

out to be the model system of choice for most studies of metal-catalyzed normal electron-demand 1,3-dipolar cycloadditions of nitrones. The application of catalyst **245a** (10 mol%) in the reaction shown in Scheme 12.73 gave up to 94% yield of the cycloaddition products, and *exo*-**243** was formed as the major diastereomer (endo/exo 10:90). The ee of one of the isomers of *exo*-**243** was up to 60% ee.

The majority of $TiCl_2$-TADDOLate-catalyzed 1,3-dipolar cycloadditions are normally performed with oxazolidinone derivatives as auxiliaries for the alkenoyl moiety in order to obtain the favorable bidentate coordination of the substrate to the catalyst. However, the use of succinimide instead of the oxazolidinone auxiliary leads to an improvement of the stereoselection of the reaction (Scheme 12.74) (352). The succinimide derivatives **246a,b** are more reactive in the enantioselective 1,3-dipolar cycloaddition with nitrone **225a** catalyzed by $TiCl_2$-TADDOLate catalyst **245a** (5 mol%). The reaction of **225a** with **246a** proceeds well and after conversion of the unstable succinimide adduct into the amide derivative, the corresponding product **247** was obtained in an endo/exo ratio of <5:>95. The enantioselectivity of the reaction of 72% ee is also an improvement compared to the analogous reaction of the oxazolidinone derivative, and similar improvements could also be obtained in reactions of other nitrones with the succinimide-based auxiliary.

An important catalyst-substrate intermediate that applies to both the $TiCl_2$-TADDOLate catalyzed 1,3-dipolar cycloadditions and Diels–Alder reactions has been isolated and characterized (353). The crystalline compound **248** has been characterized by X-ray analysis, showing that the oxazolidinone is coordinated to the titanium center in a bidentate fashion (Scheme 12.75). The four oxygen atoms,

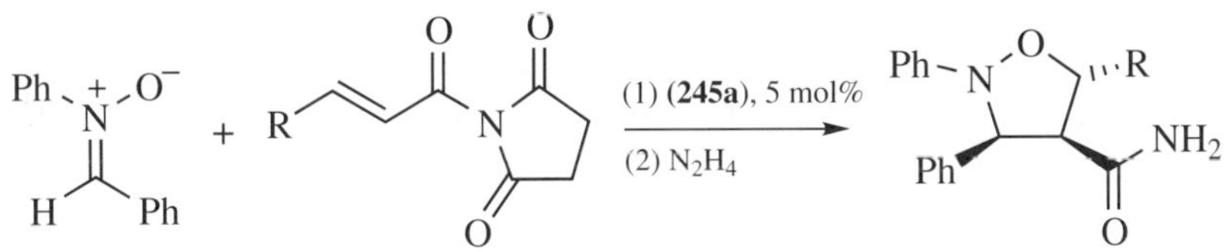

(225a) **(246a)**: R = Me **(246b)**: R = Pr *exo*-**(247)** up to >90% de up to 72% ee

Scheme 12.74

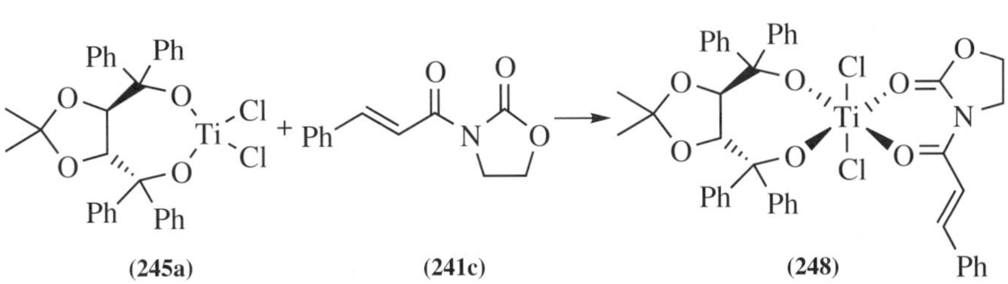

(245a) **(241c)** **(248)**

Scheme 12.75

two from the chiral ligand and two from **241c**, are located in a plane at the titanium center. The two chloride ligands are placed in the apical positions. This crystal structure provides valuable verification of the mechanism of the catalytic activation. To some extent, information can also be derived about the face shielding of the alkene by the ligand leading to the asymmetric addition of the nitrone. However, there are several other possible arrangements of the ligands around the titanium center, and whether structure **248** actually represents the reactive intermediate in the addition to the double bond has been the subject of some dispute (355–358).

Based on the investigation of the structure of intermediate **248**, the impact of the axial ligands on the endo/exo-selectivity in the 1,3-dipolar cycloaddition was studied by changing the axial chloride ligands in the $TiCl_2$-TADDOLate catalyst **245a** to bulkier ligands (359). The results of reactions between **225a** and **241a** in the presence of various TiX_2-TADDOLate catalysts are listed in Scheme 12.76. When **245b**, the bromide analogue of **245a** is used, the diastereoselectivity changes. This reaction proceeds with a low endo-selectivity, while the reaction performed in the presence of the bulky tosylate analogue **245d** takes place with complete endo selectivity and >90% ee of the 1,3-dipolar cycloaddition product. The successful application of the $Ti(Ts)_2$-TADDOLate catalyst **245d** was further extended for a series of reactions. For all eight reactions of nitrones **225** and alkenes **241** in which **245d** was applied as the catalyst, high endo-selectivity of >90% de was obtained. Most remarkably, >90% ee was obtained for all reactions involving nitrones with an aromatic R^1 substituent, whereas reactions with *N*-benzyl- and *N*-alkyl nitrones led to lower enantioselectivities (359).

The TiX_2-TADDOLate catalyzed 1,3-dipolar cycloadditions were extended to include an acrylate derivative (360). In the absence of a catalyst, the reaction between nitrones **225** and acryloyloxazolidinone **241b** proceeded to give a mixture all eight regio- and stereoisomers (Scheme 12.77). However, use of $Ti(OTs)_2$-

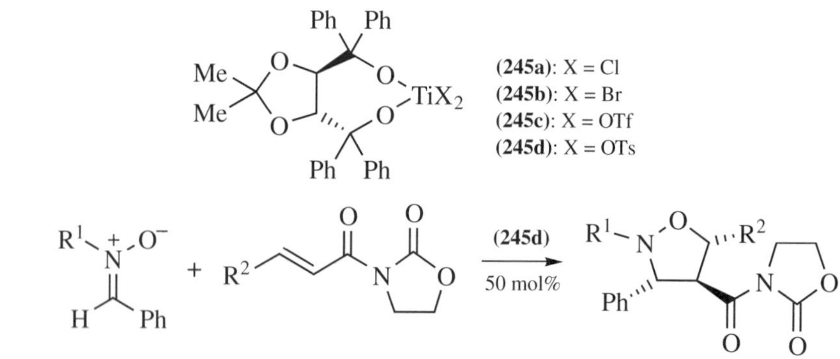

(225)
R^1 = aryl, benzyl, alkyl

(241)
R^2 = alkyl

endo-**(243)**
endo/exo = >95:<5
91–93% ee (R^1 = Ar)

Scheme 12.76

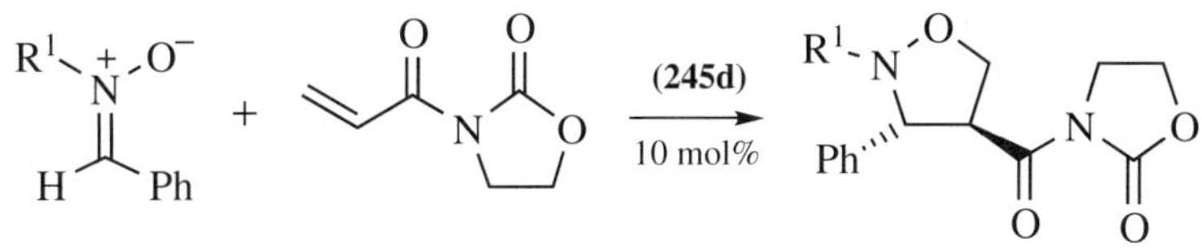

(225) (241b) *endo*-(243)
de = 92–100%
ee = 48–70%

Scheme 12.77

TADDOLate **245d** (10 mol%) as catalyst for the reaction of various nitrones **225** with alkene **241b** led to complete regioselectivity and high endo-selectivity in the reaction, and the endo-products **243** were obtained with 48–70% ee.

Seebach et al. (350), who first developed the TADDOL ligands, have also developed a number of polymer- and dendrimer-bound $TiCl_2$-TADDOLate catalysts derived from the monomeric TADDOLs (361). The use of catalysts derived from polymers and dendrimers of **249** and **250**, respectively, in the reactions between the nitrone **225a** and the alkene **241a** led to endo/exo-ratios of between 18:82 and 8:92, and enantioselectivities of up to 56% ee (Scheme 12.78). The enantioselectivities are thus slightly decreased compared to the similar reactions of

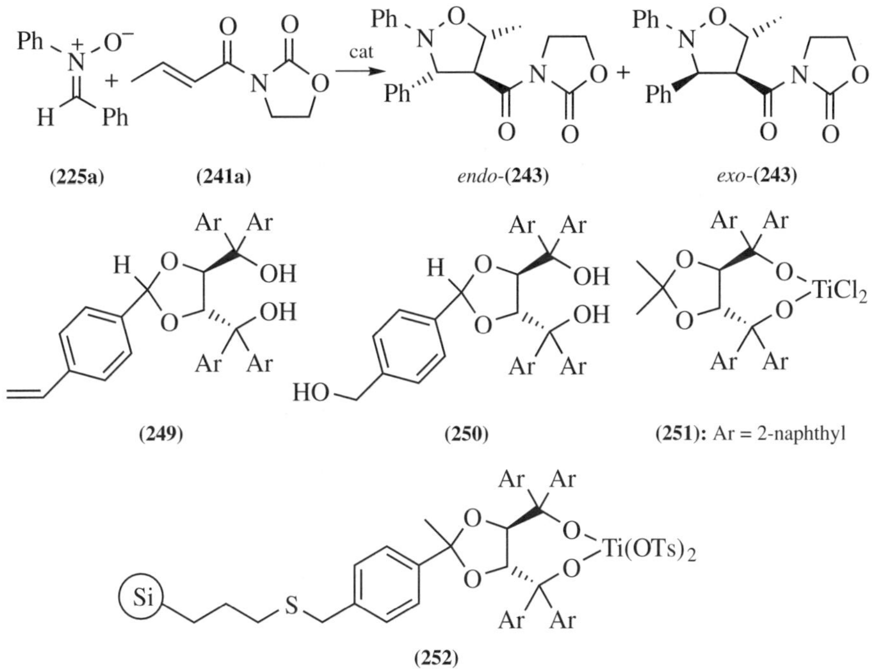

(225a) (241a) *endo*-(243) *exo*-(243)

(249) (250) (251): Ar = 2-naphthyl

(252)

Scheme 12.78

the homogeneous catalysts. They also performed an investigation of the relationship between the enantiomeric purity of the ligand of the homogeneous catalyst **251** with the products obtained in both the 1,3-dipolar cycloaddition between **225a** and **241a** and in the Diels–Alder reaction of **241a** with cyclopentadiene. Surprisingly, the 1,3-dipolar cycloaddition shows a linear relationship, whereas the Diels–Alder reaction shows a positive nonlinear relationship. In recent work, Seebach and co-workers (362) studied the use and reuse of Ti(OTs)$_2$-TADDOLate catalysts immobilized on porous silica gel. The selectivity obtained in the 1,3-dipolar cycloaddition between nitrone **225a** and **241a** catalyzed by **252** was only slightly lower compared to the corresponding homogeneous reaction (359). The same batch of the ligand in **252** could be used in four consecutive reactions with no significant loss of activity when the ligand was carefully washed between the reactions.

Bosnich and co-workers (363) used the chiral titanocene–(OTf)$_2$ complex **254** for the 1,3-dipolar cycloaddition between the cyclic nitrone **239a** and the ketene acetal **226c** for an inverse electron-demand reaction (Scheme 12.79). The reaction proceeded only in the presence of the catalyst, and a good cis/trans ratio of 8:92 was obtained using catalyst **254**, although only 14% ee was observed for the major isomer.

The inverse-electron demand 1,3-dipolar cycloaddition has also been pursued with other Ti(IV) complexes (364). The cycloaddition reaction of *C,N*-diphenyl nitrone to *tert*-butyl vinyl ether catalyzed by different bidentate C_2-symmetrical ligands gave moderate to good diastereoselectivity, and up to 41% ee was achieved.

Kanemasa et al. (365) showed that dibenzofuranyl-2,2′-bis(oxazoline) (DBFOX) is an excellent ligand for a variety of LAs (Scheme 12.80). This new type of catalyst

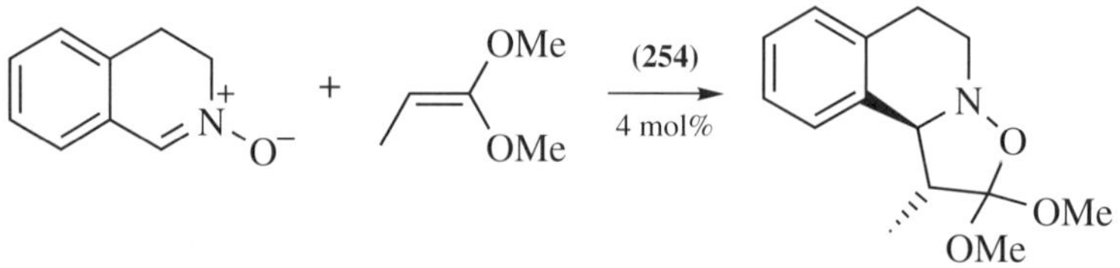

(239a) **(226c)** **(253)**

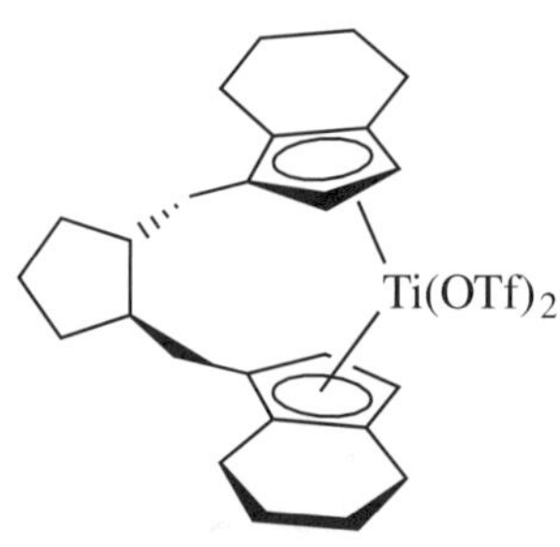

(254)

Scheme 12.79

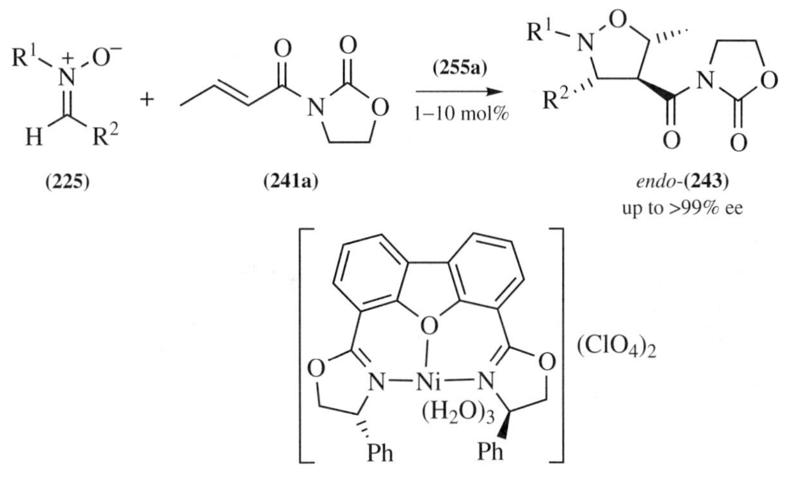

Scheme 12.80

has also been applied to 1,3-dipolar cycloadditions of nitrones, nitronates and diazo compounds. The reactions between different nitrones **225** and crotonoyl oxazolidinone **241a** in the presence of 10 mol% of the dicationic nickel complex **255a** was studied (366). Although long reaction times were required to obtain good yields, the reactions proceeded, in most cases, with very high endo-selectivities and, in several cases, >99% ee of the endo products **243**. So far this catalyst is undoubtedly the most selective catalyst for the normal electron-demand 1,3-dipolar cycloaddition between nitrones and alkenes, especially, with respect to the enantioselectivity of the reaction.

The enantioselective inverse electron-demand 1,3-dipolar cycloadditions of nitrones with alkenes described so far are catalyzed by metal complexes that favor a monodentate coordination of the nitrone, such as boron and aluminium complexes. However, the glyoxylate-derived nitrone **256** favors a bidentate coordination to the catalyst, and this nitrone is an interesting substrate, since the products that are obtained from the reaction with alkenes are masked α-amino acids (Scheme 12.81).

In order to control the stereochemistry of 1,3-dipolar cycloadditions involving this type of nitrone, the $Cu(OTf)_2$–BOX complex **238** was found to be the most suitable catalyst (Scheme 12.81) (367). The 1,3-dipolar cycloaddition of **256** with the electron-rich ethyl vinyl ether **232a** as the dipolarophile in the presence of 25 mol% of **258** proceeded at room temperature to give a high conversion, an exo/endo ratio of 84:16, and *exo*-**257** was obtained with up to 93% ee.

A model for the intermediate consisting of substrates **256** and **232a** coordinated to catalyst **258** is also outlined in Scheme 12.81. In the model **259**, the two triflate ligands are dissociated from copper. The ligands are arranged around copper as a trigonal bipyramid, and it should be noted that in this model, the oxygen atom of

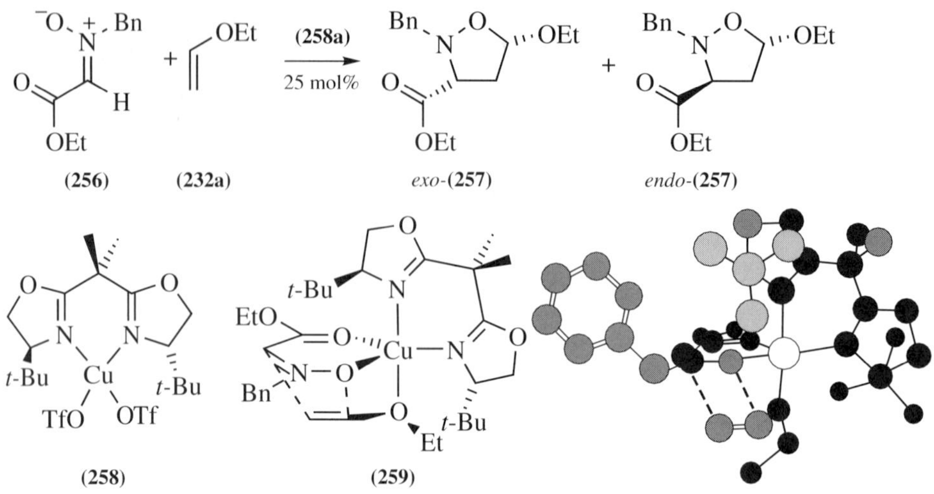

Scheme 12.81

the vinyl ether **232a** also coordinates to the metal center. However, another tetrahedral intermediate consisting of only the catalyst and the nitrone could also account for the absolute selectivity of the reaction.

Furukawa and co-workers (368,369) succeeded in applying the softer dicationic Pd-BINAP **260** as a catalyst for the 1,3-dipolar cycloaddition between **225** and **241a** (Scheme 12.82). In most cases, mixtures of *endo*-**243** and *exo*-**243** were obtained, however, enantioselectives of up to 93% ee were observed for reactions of some derivatives of **225**. A transition state structure has been proposed to account for the high selectivities obtained for some of the substrates (368). In the structure shown in Scheme 12.82, the two phosphorous atoms of the Tol-BINAP ligand and the two carbonyl oxygens of the crotonoyl oxazolidinone are arranged in a square-planar fashion around the palladium center. This leaves the *si*-face of the alkene available for the cycloaddition reaction, while the *re*-face is shielded by one of the Tol-BINAP tolyl groups.

Furukawa and co-workers (370) also used the above described palladium catalyst to the inverse electron-demand 1,3-dipolar cycloaddition of nitrones with vinyl ethers. However, all products obtained in this manner were racemic.

Ukaji et al. (371) developed a catalytic enantioselective 1,3-dipolar cycloaddition reaction of nitrones with allyl alcohol catalyzed by a zinc catalyst (Scheme 12.83). The zinc-catalyst complex, which was used in a stoichiometric amount, was generated from allyl alcohol, Et_2Zn, (*R,R*)-diisopropyl tartrate (DIPT) and EtZnCl. Addition of the nitrone **261a** led to primarily *trans*-**262a**, which was obtained in a moderate yield, although with an ee of up to 95%. Use of the nitrone **261b** led to higher yields of **262b** (47–68%), high trans-selectivities, and up to 93% ee. Compared to other asymmetric metal-catalyzed 1,3-dipolar cycloaddition reactions of nitrones, this reaction cannot be assigned as normal or inverse electron

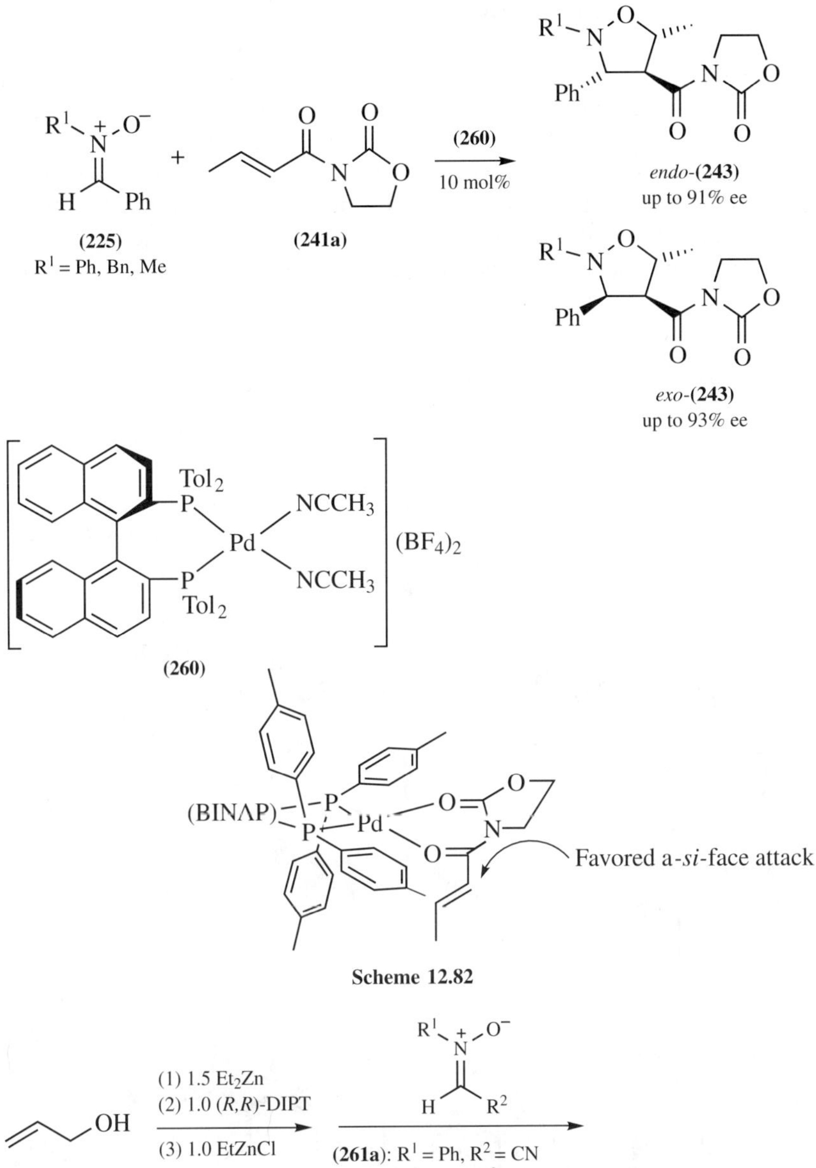

Scheme 12.82

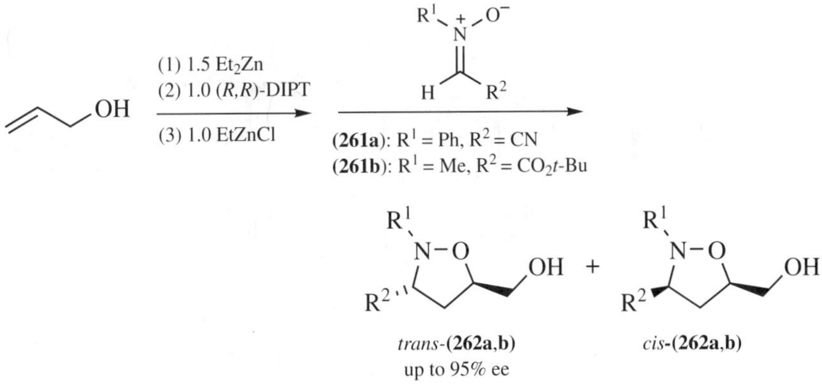

Scheme 12.83

demand. Rather, the reaction is controlled by the chelation of the substrates to the catalyst.

The application of two different chiral ytterbium catalysts **263** and **264** for the 1,3-dipolar cycloaddition was reported almost simultaneously by two independent research groups in 1997 (372,373). In both reports, it was observed that the achiral $Yb(OTf)_3$ and $Sc(OTf)_3$ salts catalyze the 1,3-dipolar cycloaddition between nitrones **225** and alkenoyl oxazolidinones **241** with endo-selectivity (Scheme 12.84). In the first study, 20 mol% of the $Yb(OTf)_2$–pyridine–bis(oxazoline) complex **263** was used as the catalyst for reactions of several derivatives of

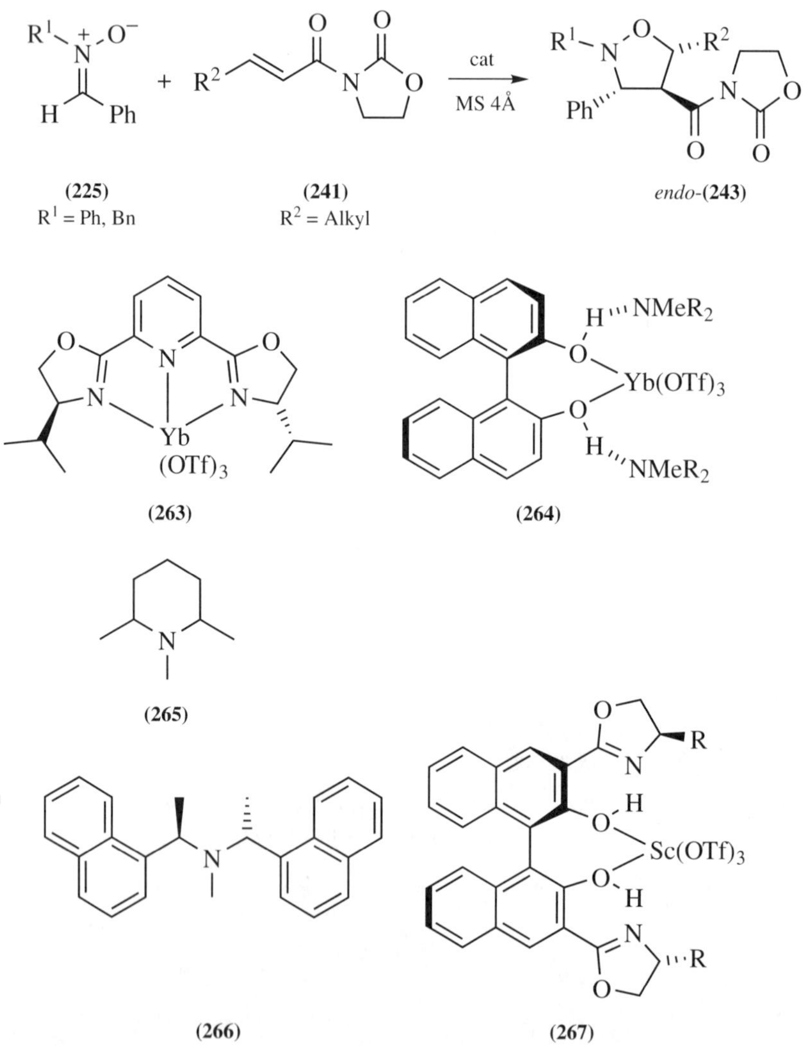

Scheme 12.84

225 and **241**. The reactions led to endo-selective 1,3-dipolar cycloadditions, giving products with up to 73% ee (372). In the other report, Kobayashi et al. (373) described a 1,3-dipolar cycloaddition catalyzed by 20 mol% of the $Yb(OTf)_3$–BINOL complex **264** in the presence of the achiral tertiary amine **265**. In this approach, the nitrone **225** was formed *in situ* from the respective aldehyde and hydroxylamine. High endo-selectivities were observed, and for one derivative, the product *endo*-**243** (R^1=Bn, R^2=Me) was obtained in 78% ee. In an extension of these investigations, the 1,3-dipolar cycloaddition was performed in the presence of 20 mol% of the catalyst **264** with 40 mol% of a the chiral amine **266** (374). By substituting the achiral amine **265** in **264** with the chiral amine **266**, the selectivity of the reaction was largely improved. For the reactions of derivatives of **225** and **241**, *endo*-**243** was obtained as a single diastereomer with enantioselectivities of up to 96% ee. Further investigation in this field by Kobayashi and co-workers (375) led to the finding that the absolute stereoselectivity of the reaction was reversed when the reaction was performed in the absence of 4-Å MS. This observation makes is analogous to the MgX_2-BOX catalyzed reactions described above (304). In the reaction catalyzed by **264** using **266** as the additive, *endo*-**243** (R^1=Bn, R^2=Me) was obtained in 96% ee in the presence of 4-Å MS. In the absence of MS, the opposite enantiomer was obtained in 50% ee. This inverse selectivity could be improved by using various N-oxides as a third additive (375).

A series of new BINOL–BOX ligands was developed by Kodema et al. (376) and used with $Sc(OTf)_3$ as the catalyst **267**. In a typical procedure, the 1,3-dipolar cycloaddition between **225** (R^1=Bn) and **241** (R^2=Me) was catalyzed by 20 mol% of **267**. The reaction proceeded in the presence of **267** (R=Ph) with high yields, endo-selectivity, and in up to 87% ee. In this study, it was also observed that the presence or absence of 4-Å MS could reverse the absolute induction of the reaction.

Kobayashi and Kawamura (374) used the catalytic enantioselective 1,3-dipolar cycloaddition for the synthesis of an optically active β-lactam (Scheme 12.85). The

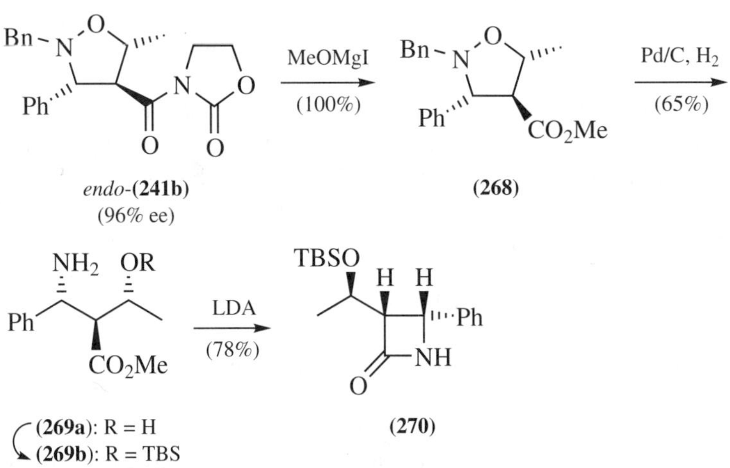

Scheme 12.85

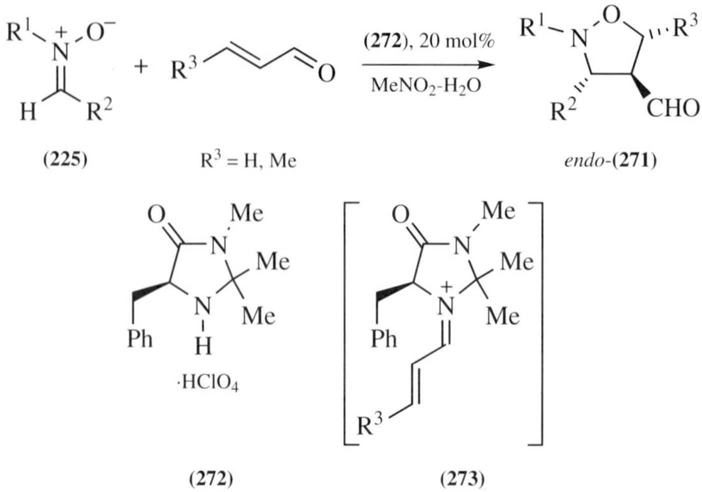

Scheme 12.86

isoxazolidine *endo*-**243b**, obtained in 96% ee from the Yb(OTf)$_3$-BINOL catalyzed 1,3-dipolar cycloaddition, was quantitatively converted into the ester derivative **268**. Hydrogenation over Pd/C opened the isoxazolidine ring and cleaved the N-benzyl moiety to give **269a**. Following silyl protection of the hydroxy group in **269a**, the final ring-closure was mediated by LDA to give the β-lactam **270** in high yield with conservation of optical purity (96% ee).

MacMillan and co-workers recently reported the first organocatalyzed 1,3-dipolar cycloaddition (377) as an extension of their work on Diels–Alder reactions (378). The reaction between the nitrones **225** and α,β-unsaturated aldehydes proceeded in the presence of 20 mol% of **272** to give *endo*-**271** in good yields, high endo-selectivities, and up to 99% ee (Scheme 12.86). One of the advantages of this organocatalyzed reaction over the normal electron-demand metal-catalyzed reactions is that it can be performed with α,β-unsaturated aldehydes. They are poor substrates in metal-catalyzed reactions due to preferential coordination of the LA to the nitrone in the presence of monodentate carbonyl compounds, as described in Section 12.4.1. Activation of the aldehyde substrate is achieved by formation of the iminium ion **273** with the chiral catalyst **272**, leading to decreased $LUMO_{alkene}$ energy. It was proposed that the nitrone approaches the *si*-face of the (*E*)-iminium ion **273** in an endo-fashion, leading to *endo*-**271**.

12.4.3. Nitrile Oxides

Ukaji and co-workers (379–381) described the first, and so far only, metal-catalyzed asymmetric 1,3-dipolar cycloaddition of nitrile oxides with alkenes. Upon treatment of allyl alcohol with diethylzinc and (*R,R*)-diisopropyl tartrate followed by the addition of diethylzinc and substituted hydroximoyl chlorides **274**, the isoxazolidines **275** are formed with enantioselectivities of up to 96% ee (Scheme 12.87).

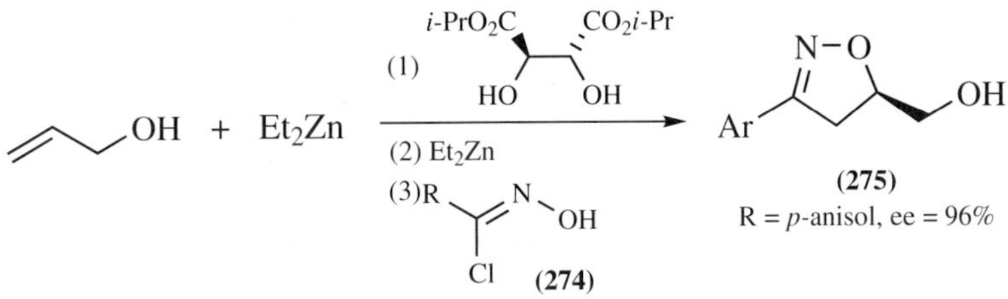

Scheme 12.87

They have also developed a catalytic version of the reaction in which the chiral ligand DIPT was used in 20 mol% (379–381). In spite of the reduction of the amount of the chiral ligand, enantioselectivities of up to 93% ee were obtained in this work. The addition of a small amount of 1,4-dioxane proved to be crucial for the enantioselectivity of the reaction. A proposal for the reaction mechanism is outlined in Scheme 12.88. Allyl alcohol, hydroximoyl chloride **274** and diethylzinc react to form **276**, which is mixed with the ligand and an additional amount of

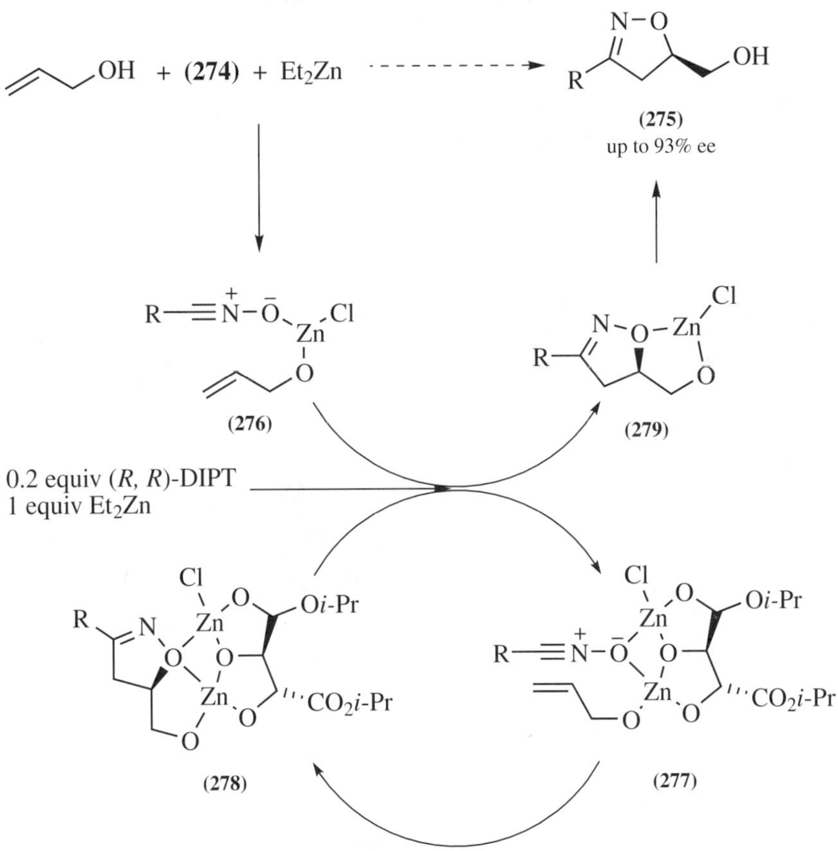

Scheme 12.88

diethylzinc to form **277**. The achiral complex **276** is apparently much less activated for a 1,3-dipolar cycloaddition compared to **277**, which controls the enantioselectivity of the reaction. The increased reactivity of **277** compared to **276** may be due to a ligand-accelerating effect of DIPT when coordinated to the metal. After formation of the isoxazoline **278**, the Zn-DIPT moiety of **278** participates in the catalytic cycle and **279** is formed. When the reaction is complete, **279** is hydrolyzed to give **275** in up to 93% ee (R = *t*-Bu).

The first antibody-catalyzed asymmetric 1,3-dipolar cycloaddition was reported recently by Janda and co-workers (382). The reaction of the relatively stable nitrile oxide **280** and dimethyl acrylamide **281** was catalyzed by antibody 29G12 having turnover numbers >50, and the product **282** was obtained in up to >98% ee (Scheme 12.89). The antibody 29G12 was formed for hapten **283** and coupled to a carrier protein by standard protocols. The hapten **283** contains no chiral center and therefore the immune system elicited a stereochemical environment capable of stabilizing the enantiomeric transition state leading to **282**.

12.4.4. Azomethine Ylides

The first report on metal-catalyzed asymmetric azomethine ylide cycloadditions appeared some years before this topic was described for other 1,3-dipolar cycloadditions (383). However, since then the activity in this area has been very limited

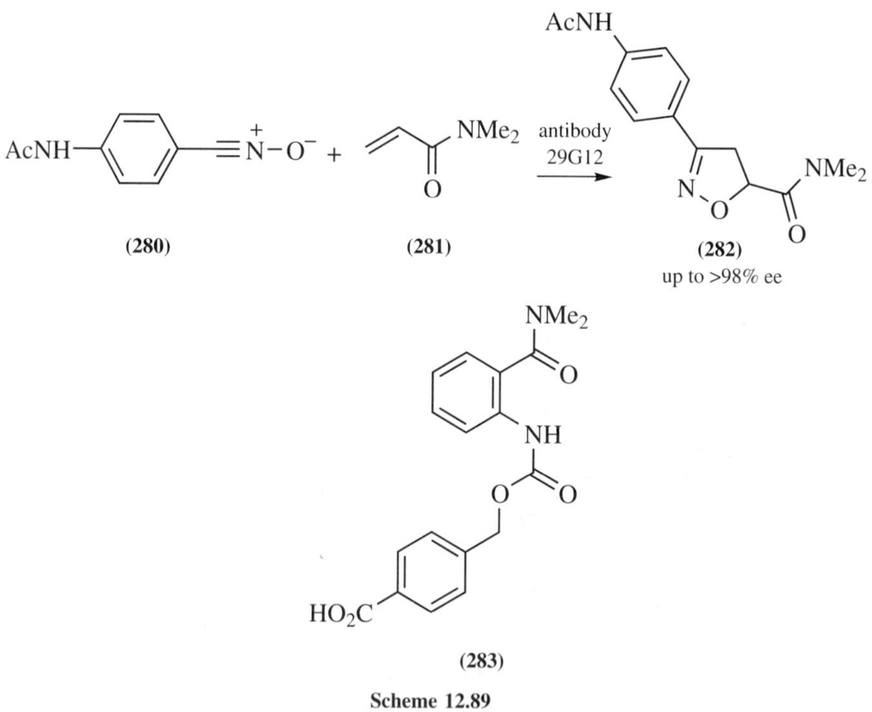

(283)

Scheme 12.89

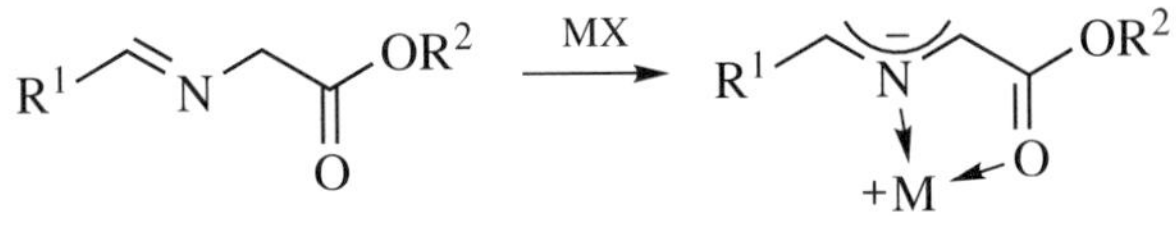

Scheme 12.90

in spite of the fact that azomethine ylides are often stabilized by metal salts, as shown in Scheme 12.90.

Grigg and co-workers (383) found that chiral cobalt and manganese complexes are capable of inducing enantioselectivity in 1,3-dipolar cycloadditions of azomethine ylides derived from arylidene imines of glycine (Scheme 12.91). This work was published in 1991 and is the first example of a metal-catalyzed asymmetric 1,3-dipolar cycloaddition. The reaction of the azomethine ylide **284a** with methyl acrylate **285** required a stoichiometric amount of cobalt and 2 equiv of the chiral ephedrine ligand. Up to 96% ee was obtained for the 1,3-dipolar cycloaddition product **286a**.

The yields of **286a** were in the range of 45–84%, and the highest ee values were obtained when Ar=2-napthyl, 4-BrC_6H_4, or 4-MeOC_6H_4 and when methyl acrylate was used as the solvent. Both the reaction time and ee are dependent on the counterion, as the use of CoF_2 as the catalyst leads to a very slow 1,3-dipolar cycloaddition with low chiral induction compared to the use of $CoCl_2$. A proposed working model for the asymmetric induction is shown in Scheme 12.92. The cis-arrangement of the methyl and phenyl group of the ligand in **287** results in a pseudo-equatorial conformation of the phenyl group and provides an effective shielding of one face of the dipole (383).

In a more recent publication by the same group, they mention that Ag(I) salts in combination with chiral phosphine ligands can catalyze the 1,3-dipolar cycloaddition involving the azomethine precursor **284b** and methyl vinyl ketone (Scheme 12.93) (384). The reaction, which presumably also requires a stoichiometric amount of the catalyst, proceeds to give **286b** in a good yield with 70% ee.

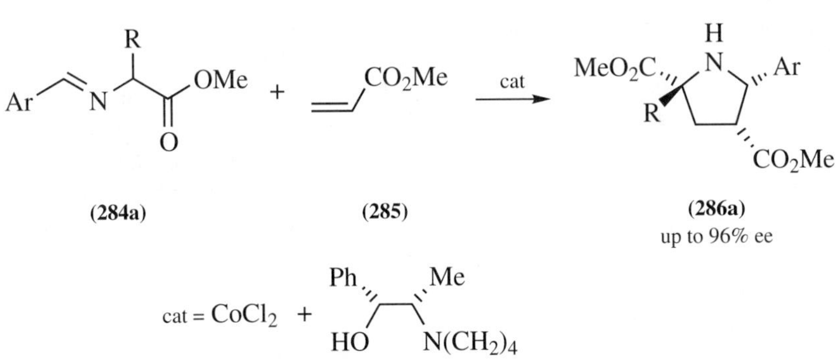

Scheme 12.91

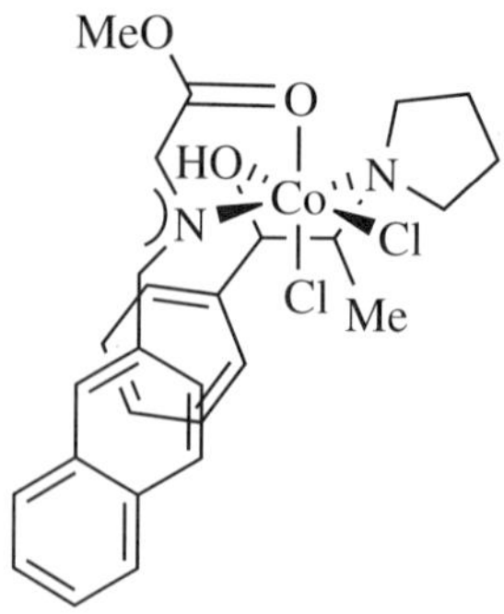

(287)

Scheme 12.92

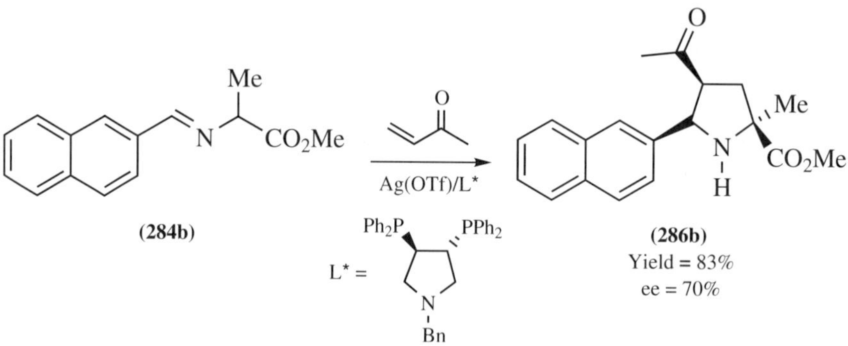

Scheme 12.93

12.4.5. Carbonyl Ylides

Carbonyl ylides are available from diazo compounds and rhodium carbenes (385,386). In the first metal-catalyzed asymmetric version of this reaction from 1997, Hodgson et al. (387) workers investigated the intramolecular 1,3-dipolar cycloaddition of the carbonyl ylide precursor **288** (Scheme 12.94). In the presence of 1 mol% of the catalyst **291** in a hydrocarbon solvent, the reaction proceeded to give **290** in 93% yield and with 52% ee. The exact nature of the interaction between the rhodium catalyst and the carbonyl ylide in the intermediate **289** is unknown. However, the catalyst–ligand induced control of enantioselectivity (387,388) and of the regioselectivity in earlier work (389,390) strongly indicates that the catalyst is in some manner associated with the 1,3-dipole in the cycloaddition step. In more recent work, the same authors tested a number of dirhodium tetrakis(1,1′-binaphthyl-2,2′-diylphosphonate) catalysts for the reaction (388). The best results were obtained with the catalyst **292** that had been developed for this reaction. The advantage of this ligand is that it has a high solubility in hexane, which is the

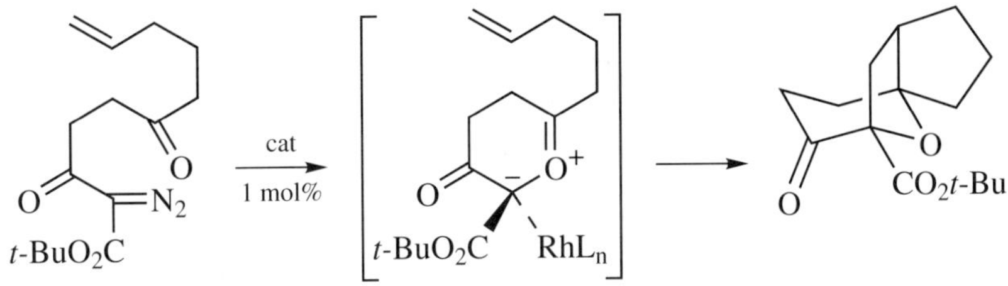

(288) **(289)** **(290)**

Cat. (**291**): ee = 52%

Cat. (**292**): ee = 90%

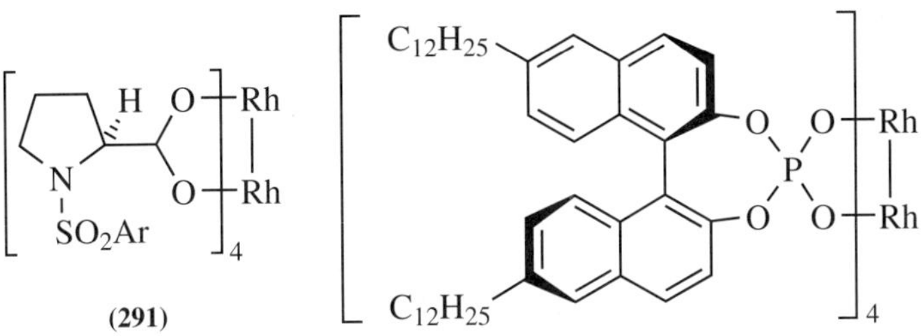

(291) **(292)**

Scheme 12.94

solvent of choice for this reaction. By the application of 1 mol% of catalyst **292** up to 90% ee of the tricyclic product **290** was obtained.

The intermolecular version of the above reaction has also been reported (391). In the first example, a rhodium-catalyzed carbonyl ylide cycloaddition with maleimide was studied. However, only enantioselectivities of up to 20% ee were obtained

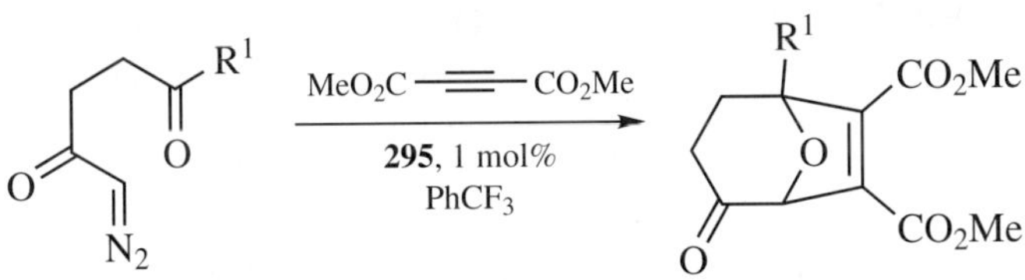

(293)
R^1 = Alkyl, aryl

(294)
up to 92% ee

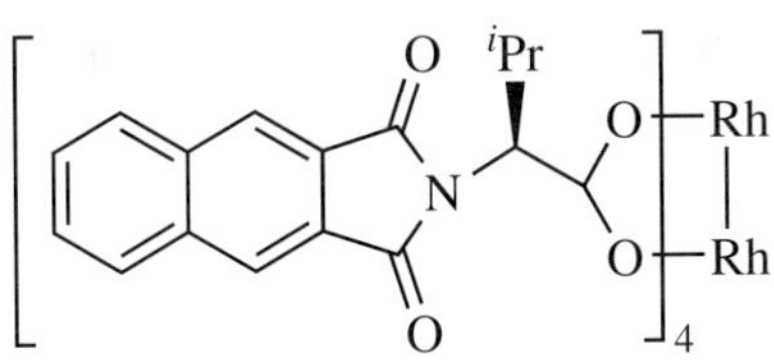

(295)

Scheme 12.95

(391,392). Hashimoto and co-workers were able to induce high enantioselectivities in the intermolecular carbonyl ylide reaction of the precursor **293** with dimethyl acetylenedicarboxylate (393,394). After optimizing the reaction conditions and testing a series of catalyst, they were able to obtain a series of products **294** in the reaction catalyzed by 1 mol% of **295** in moderate to good yields with enantioselectivities ranging from 80 to 92% ee (Scheme 12.95). It was important for both yield and enantioselectivity that the reaction was carried out in trifluoromethylbenzene as the solvent. They also showed that the reaction could be performed with substrates giving the corresponding bicyclo[2,2,1] and bicyclo[4,2,1] products with 68 and 80% ee, respectively. In a more recent report, they were able to obtain up to 93% ee using a similar catalyst (393).

12.4.6. Diazo Compounds

The 1,3-dipolar cycloaddition reaction of diazoalkanes with alkenes has also been reported (395). Kanemasa and Kanai (395) used the chiral DBFOX-Ph ligand with various metals such as Ni, Zn, and Mg for the preparation of **255a–c**. The reaction of TMS-diazomethane **171** with alkene **241** was catalyzed by 10 mol% of **255b** to afford the 1,3-dipolar cycloaddition product **296** in good yields and enantioselectivities of up to 99% ee (Scheme 12.96). Also, the Ni-catalyst **255a** and the Mg-catalyst **255c** were excellent catalysts for the reaction, resulting in >90% ee in both cases.

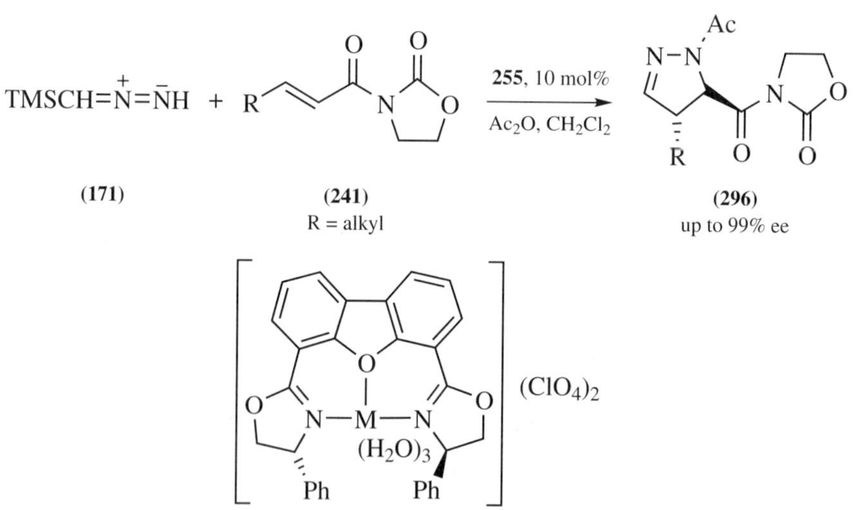

(255a): R=Ni
(255b): R=Zn
(255c): R= Mg

Scheme 12.96

12.5. CONCLUSION

Since the first edition of this book appeared in 1984, asymmetric reactions are probably the single field within the area of 1,3-dipolar cycloaddition area that has experienced the strongest growth. Concurrent with the increasing interest in asymmetric synthesis in general over the past 15 years, it has been verified that 1,3-dipolar cycloadditions are highly attractive reactions for introduction of new chiral centers with control of relative and absolute stereoselectivity. In Section 12.2, we selected a series of asymmetric diastereoselective 1,3-dipolar cycloadditions, demonstrating the wide range of this field and several applications in asymmetric synthesis. In Section 12.3, the application of chiral auxiliaries has been outlined. This approach is particularly well developed for dipolarophiles such as α,β-unsaturated esters and amides and in several cases excellent selectivities were obtained. The asymmetric metal-catalyzed 1,3-dipolar cycloadditions is a young and rapidly expanding area. As shown in Section 12.4, the majority of contributions deal with reactions of nitrones, where several different enantioselective approaches have been developed. For other types of 1,3-dipoles such as nitrile oxides, azomethine ylides, carbonyl ylides, and diazo compounds, a few examples have been reported and there is much room for new developments. Among the most recent trends is the successful application of organocatalysis and antibody catalysis for asymmetric 1,3-dipolar cycloadditions. In conclusion, asymmetric 1,3-dipolar cycloadditions are highly attractive reactions for the synthesis of enantiomerically enriched heterocycles and other heteroatom-containing molecules, and this area is rapidly evolving and provides many challenges for future work.

ACKNOWLEDGMENT

The authors are indebted to The Danish National Research Foundation for financial support.

REFERENCES

1. A. Padwa, ed., *1,3-Dipolar Cycloaddition Chemistry*, Vols. 1 and 2, Wiley, New York, **1984**.
2. K. B. G. Torssell, *Nitrile Oxides, Nitrones and Nitronates in Organic Synthesis*, VCH Weinheim, **1988**.
3. A. Padwa, in B. M. Trost, and I. Flemming, eds., *Comprehensive Organic Synthesis*, Vol. 4, Pergamon Press, Oxford, **1991**, p. 1069.
4. P. A. Wade, in *Comprehensive Organic Synthesis*, Vol. 4, B. M. Trost and I. Flemming, eds., Pergamon Press, Oxford, **1991**, p. 1111.
5. K. V. Gothelf and K. A. Jørgensen, *Chem. Rev.*, **1998**, *98*, 863.
6. S. Karlsson and H.-E. Högberg, *Org. Prep. Proc. Int.*, **2001**, *33*, 103.
7. K. V. Gothelf and K. A. Jørgensen, *Acta Chem. Scand.*, **1996**, *50*, 562.
8. K. V. Gothelf, K. B. Jensen, and K. A. Jørgensen, *Sci. Prog.*, **1999**, *82*, 327.

9. K. V. Gothelf and K. A. Jørgensen, *Chem. Commun.*, **2000**, 1449.
10. M. Frederickson, *Tetrahedron*, **1997**, *53*, 403.
11. P. De March, M. Figueredo, and J. Font, *Heterocycles*, **1999**, *50*, 1213.
12. C. P. Dell, *J. Chem. Soc. Perkin Trans. 1*, **1998**, 3973.
13. L. Raimondi and M Benaglia, *Eur. J. Org. Chem.*, **2001**, 1033.
14. E. Âbele and E. Lukevics, *Heterocycles*, **2000**, *53*, 2285.
15. G. Broggini and G. Zecchi, *Synthesis*, **1999**, 905.
16. K. V. Gothelf, R. G. Hazell, and K. A. Jørgensen, *J. Org. Chem.*, **1996**, *61*, 346.
17. . K. V. Gothelf, R. G. Hazell, and K. A. Jørgensen, *Acta Chem. Scand.*, **1997**, *51*, 1234.
18. T. Polonski and A. Chimiak, *Tetrahedron Lett.*, **1974**, 2453.
19. D. Keirs, D. Moffat, K. Overton, and R. Tomanek, *J. Chem. Soc. Perkin Trans. 1*, **1991**, 1041.
20. D. Keirs, D. Moffat, and K. Overton, *J. Chem. Soc. Chem. Commun.*, **1988**, 654.
21. C. Belzecki and I. Panfil, *J. Chem. Soc. Chem. Commun.*, **1977**, 303.
22. T. Kametani, T. Nagahara, and T. Honda, *J. Org. Chem.*, **1985**, *50*, 2327.
23. C. Belzecki and I. Panfil, *J. Org. Chem.*, **1979**, *44*, 1212.
24. C. M. Tice and B. Ganem, *J. Org. Chem.*, **1983**, *48*, 5048.
25. T. Kametani, S.-D. Chu, and T. Honda, *J. Chem. Soc. Perkin Trans. 1*, **1988**, 1593.
26. S. Jost, Y. Gimbert, A. E. Greene, and F. Fotiadu, *J. Org. Chem.*, **1997**, *62*, 6672.
27. B. B. Snider, H. Lin, and B. M. Foxman, *J. Org. Chem.*, **1998**, *63*, 6442.
28. B. B. Snider and H. Lin, *J. Am. Chem. Soc.*, **1999**, *33*, 7778.
29. A. Brandi, S. Cicchi, A. Goti, and K. M. Pietrusiewicz, *Tetrahedron: Asymmetry*, **1991**, *2*, 1063.
30. A. Goti, S. Cicchi, A. Brandi, and K. M. Pitrusiewicz, *Tetrahedron: Asymmetry*, **1991**, *2*, 1371.
31. M. Carda, R. Portolés, S. Uriel, J. A. Marco, L. R. Domingo, R. J. Zaragozà, and H. Röper, *J. Org. Chem.*, **2000**, *65*, 7000.
32. S. Saito, T. Ishikawa, N. Kishimoto, T. Kohara, and T. Moriwake, *Synlett*, 282 (**1994**).
33. T. Kametani, S. D. Chu, and T. Honda, *Heterocycles*, **1987**, *25*, 241.
34. L. Bruché, A. Arnone, P. Bravo, W. Panzeri, C. Pesenti, and F. Viani, *Eur. J. Org. Chem.*, **1999**, 1665.
35. M. Yokoyama, K. Sujino, M. Irie, N. Yamazaki, T. Hiyama, N. Yamada, and H. Togo, *J. Chem. Soc. Perkin Trans. 1*, **1991**, 2801.
36. C. Mukai, I. J. Kim, W. J. Cho, M. Kido, and M. Hanaoka, *J. Chem. Soc. Perkin . Trans. 1*, **1993**, 2495.
37. C. Mukai, W. J. Cho, I. J. Kim, and M. Hanaoka, *Tetrahedron Lett.*, **1990**, *31*, 6893.
38. F. M. Cordero, S. Cicchi, A. Goti, and A. Brandi, *Tetrahedron Lett.*, **1994**, *35*, 949.
39. A. Brandi, S. Cicchi, F. M. Cordero, R. Frignoli, A. Goti, S. Picasso, and P. Vogel, *J. Org. Chem.*, **1995**, *60*, 6806.
40. A. Brandi, S. Cicchi, A. Goti, M. Koprowski, and K. M. Pietrusiewicz, *J. Org. Chem.*, **1994**, *59*, 1315.
41. A. Goti, F. Cardona, A. Brandi, S. Picasso, and P. Vogel, *Tetrahedron: Asymmetry*, **1996**, *7*, 1659.
42. S. Cicchi, I. Höld, and A. Brandi, *J. Org. Chem.*, **1993**, *58*, 5274.
43. S. Cicchi, S. Crea, A. Goti, and A. Brandi, *Tetrahedron: Asymmetry*, **1997**, *8*, 293.
44. S. Valenza, F. M. Cordero, A. Brandi, A. Guidi, M. Altamura, A. Giolitti, F. Giuntini, F. Pasqui, A. R. Renzetti, and C. A. Maggi, *J. Org. Chem.*, **2000**, *65*, 4003.
45. A. Goti, S. Cichi, M. Caciarini, F. Cardona, V. Fedi, and A. Brandi, *Eur. J. Org. Chem.*, **2000**, 3633.
46. F. Cardona, S. Valenza, A. Goti, and A. Brandi, *Eur. J. Org. Chem.*, **1999**, 1219.
47. A. E. McCraig and R. H. Wightman, *Tetrahedron Lett.*, **1993**, *34*, 3939.
48. N. Katagiri, M. Okada, C. Kaneko, and T. Furuya, *Tetrahedron Lett.*, **1996**, *37*, 1801.

49. H. Ohtake, Y. Imada, and S.-I. Murahashi, *Bull. Chem. Soc. Jpn.*, **1999**, *72*, 2737.

50. M. Closa, P. De March, M. Figueredo, and J. Font, *Tetrahedron: Asymmetry*, **1997**, *8*, 1031.

51. T. Ishikawa, Y. Tajima, M. Fukui, and S. Saito, *Angew. Chem., Int. Ed. Engl.*, **1996**, *35*, 1863.

52. K. Nagasawa, A. Goergieva, and T. Nakata, *Tetrahedron*, **2000**, *56*, 187.

53. H. Ooi, A. Urushibara, T. Esumi, Y. Iwabuchi, and S. Hatakeyama, *Org. Lett.*, **2001**, *3*, 953.

54. T. Berranger, C. Andre-Barres, M. Kobayakawa, and Y. Langlois, *Tetrahedron Lett.*, **1993**, *34*, 5079.

55. T. Berranger and Y. Langlois, *Tetrahedron Lett.*, **1995**, *36*, 5523.

56. Y. Langlois, A. Pouilhes, C. Kouklovsky, J.-F. Morelli, A. Haudrechy, M. Kobayakawa, C. Andre-Berres, T. Berranger, and O. Dirat, *Bull. Soc. Chim. Belg.*, **1996**, *105*, 639.

57. M. Mauduit, C. Kouklovsky, and Y. Langlois, *Eur. J. Org. Chem.*, **2000**, 1595.

58. M. Mauduit, C. Kouklovsky, Y. Langlois, and C. Riche *Org. Lett.*, **2000**, *2*, 1053.

59. O. Tamura, K. Gotanda, R. Terashima, M. Kikuchi, T. Miyawaki, and M. Sakamoto *J. Chem. Soc. Chem. Commun.*, **1996**, 1861.

60. O. Tamura, S. Yoshida, H. Sugita, N. Mita, Y. Uyama, N. Morita, M. Ishiguro, T. Kawasaki, H. Ishibashi, and M. Sakamoto, *Synlett*, **2000**, 1553.

61. O. Tamura, K. Gotanda, J. Yoshino, Y. Morita, R. Terashima, M. Kikuchi, T. Miyawaki, N. Mita, M. Yamashita, H. Ishibashi, and M. Sakamoto *J. Org. Chem.*, **2000**, *65*, 8544.

62. L. R. Domingo, *Theor. Chem. Acc.*, **2000**, *104*, 240.

63. L. R. Domingo, M. T. Picher, and J. Andrés, *J. Phys. Chem.*, **1999**, *12*, 24.

64. S. E. Denmark and A. Thorarensen, *Chem. Rev.*, **1996**, *96*, 137.

65. K. N. Houk, S. R. Moses, Y. D. Wu, N. G. Randan, V. Jaeger, R. Schohe, and F. R. Fronczek, *J. Am. Chem. Soc.*, **1984**, *106*, 3880.

66. K. E. Larsen and K. B. G. Torssell, *Tetrahedron*, **1984**, *40*, 2985.

67. A. P. Kozikowski, Y. Kitagawa, and J. P. Springer, *J. Chem. Soc. Chem. Commun.*, **1983**, 1460.

68. J. M. J. Tronchet, A. Jotterand, N. Le Hong, M. F. Perret, S. Thorndahl-Jaccard, J. Tronchet, J. M. Chalet, M. L. Faivre, C. Hausser, and C. Sébastian, *Helv. Chim. Acta.*, **1970**, *53*, 1484.

69. D. Enders, A. Haertwig, and J. Runsink, *Eur. J. Org. Chem.*, **1998**, 1793.

70. G. J. McGarvey, J. A. Mathys, and K. J. Wilson, *Tetrahedron Lett.*, **2000**, *41*, 6011.

71. A. Padwa, Y. Y. Chen, U. Chiacchio, and W. Dent, *Tetrahedron*, **1985**, *41*, 3529.

72. A. S. Anslow, G. G. Cox, and L. M. Harwood, *Khim. Geterotsikl. Soedin.*, **1995**, 1393.

73. A. S. Anslow, L. M. Harwood, and I. A. Lilley, *Tetrahedron: Asymmetry*, **1995**, *6*, 2465.

74. L. M. Harwood and I. A. Lilley, *Tetrahedron: Asymmetry*, **1995**, *6*, 1557.

75. L. M. Harwood, A. C. Manage, S. Robin, S. F. G. Hopes, D. J. Watkin, and C. E. Williams, *Synlett*, **1993**, 777.

76. L. M. Harwood and L. C. Kitchen, *Tetrahedron Lett.*, **1993**, *34*, 6603.

77. L. M. Harwood and I. A. Lilley, *Tetrahedron Lett.*, **1993**, *34*, 537.

78. L. M. Harwood, J. Macro, D. Watkin, C. E. Williams, and L. F. Wong, *Tetrahedron: Asymmetry*, **1992**, *3*, 1127.

79. A. S. Anslow, L. M. Harwood, H. Phillips, D. Watkin, and L. F. Wong, *Tetrahedron: Asymmetry*, **1991**, *2*, 1343.

80. A. S. Anslow, L. M. Harwood, H. Phillips, and D. Watkin, *Tetrahedron: Asymmetry*, **1991**, *2*, 997.

81. A. S. Anslow, L. M. Harwood, H. Phillips, and D. Watkin, *Tetrahedron: Asymmetry*, **1991**, *2*, 169.

82. (a) D. Alker, G. Hamblett, L. M. Harwood, S. M. Robertson, D. J. Watkin, and C. E. Williams, *Tetrahedron*, **1998**, *54*, 6089. (b) L. M. Harwood and S. M. Robertson *Chem. Commun.*, **1998**, 2641.

83. D. Alker, L. M. Harwood, and C. E. Williams, *Tetrahedron Lett.*, **1998**, *39*, 475.

84. (a) P. Deprez, J. Rouden, A. Chiaroni, C. Riche, J. Royer, and H. P. Husson *Tetrahedron Lett.*, **1991**, *32*, 7531. (b) P. Deprez, J. Royer, and H. P. Husson *Tetrahedron: Asymmetry*, **1991**, *2*, 1189. (c) J. Rouden, J. Royer, and H. P. Husson *Tetrahedron Lett.*, **1989**, *30*, 5133.
85. B. Alcaide, P. Almendros, J. M. Alonso, and M. F. Aly, *Chem. Commun.*, **2000**, 485.
86. B. Alcaide, P. Almendros, J. M. Alonso, and M. F. Aly, *J. Org. Chem.*, **2001**, *66*, 1351.
87. R. M. Williams, W. Zhai, D. J. Aldous, and S. C. Aldous, *J. Org. Chem.*, **1992**, *57*, 6527.
88. P. R. Sebahar and R. M. Williams, *J. Am. Chem. Soc.*, **2000**, *122*, 5666.
89. J. F. Peyronel, S. Grisoni, B. Carboni, T. Courgeon, and R. Carrie, *Tetrahedron*, **1994**, *50*, 189.
90. P. Garner and O. Dogan, *J. Org. Chem.*, **1994**, *59*, 4.
91. P. Garner, W. B. Ho, and H. Shin, *J. Am. Chem. Soc.*, **1993**, *115*, 10742.
92. S. Takano, M. Moriya, and K. Ogasawara, *Tetrahedron: Asymmetry*, **1992**, *3*, 681.
93. P. Garner, W. B. Ho, S. K. Grandhee, W. J. Youngs, and V. O. Kennedy, *J. Org. Chem.*, **1991**, *56*, 5893.
94. P. Garner and W. B. Ho, *J. Org. Chem.*, **1990**, *55*, 3973.
95. P. Garner, K. Sunitha, and T. Shanthilal, *Tetrahedron Lett.*, **1988**, *29*, 3525.
96. D. Enders, I. Meyer, J. Runsink, and G. Raabe, *Tetrahedron*, **1998**, *54*, 10733.
97. G. Negron, G. Roussi, and J. Zhang, *Heterocycles*, **1992**, *34*, 293.
98. F. Novello, M. Prato, T. Da Ros, M. De Amici, A. Bianco, C. Toniolo, and M. Maggini, *J. Chem. Soc. Chem. Commun.*, **1996**, 903.
99. R. C. F. Jones, K. J. Howard, and J. S. Snaith, *Tetrahedron Lett.*, **1996**, *37*, 1707.
100. M. Zlicar, B. Stanovnik, and M. Tisler, *Tetrahedron*, **1992**, *48*, 7965.
101. M. Zlicar, B. Stanovnik, and M. Tisler, *J. Heterocycl. Chem.*, **1993**, *30*, 1209.
102. B. Stanovnik, B. Jelen, and M. Zlicar, *Farmaco*, **1993**, *48*, 231.
103. F. Roussi, A. Chauveau, M. Bonin, L. Micouin, and L.-P. Husson, *Synthesis*, **2000**, 1170.
104. A. Agócs, A. Bényei, L. Somogyi, and P. Herczegh, *Tetrahedron: Asymmetry*, **1998**, *9*, 3359.
105. A. Padwa and M. Prein, *J. Org. Chem.*, **1997**, *62*, 6842.
106. A. Padwa and M. Prein, *Tetrahedron*, **1998**, *54*, 6957.
107. M. Ito and C. Kibayashi, *Tetrahedron*, **1991**, *47*, 9329.
108. M. Ito, M. Maeda, and C. Kibayashi, *Tetrahedron Lett.*, **1992**, *33*, 3765.
109. H. Ina, M. Ito, and C. Kibayashi, *J. Chem. Soc. Chem. Commun.*, **1995**, 1015.
110. M. Ito and C. Kibayashi, *Tetrahedron Lett.*, **1990**, *31*, 5065.
111. H. Ina, M. Ito, and C. Kibayashi, *J. Org. Chem.*, **1996**, *61*, 1023.
112. M. Shimazaki, F. Okazaki, F. Nakajima, T. Ishikawa, and A. Ohta, *Heterocycles*, **1993**, *36* 1823.
113. R. Annunziata, M. Cinquini, F. Cozzi, P. Giaroni, and L. Raimondi, *Tetrahedron Lett.*, **1991**, *32*, 1659.
114. G. Galley, J. Liebscher, and M. Paetzel, *J. Org. Chem.*, **1995**, *60*, 5005.
115. M. Pätzel, G. Galley, P. G. Jones, and A. Chrapkowsky, *Tetrahedron Lett.*, **1993**, *34*, 5707.
116. K. N. Houk, H.-Y. Duh, Y.-D. Wu, and S. R. Moses, *J. Am. Chem. Soc.*, **1986**, *108*, 2754.
117. G. Galley, P. G. Jones, and P. Michael, *Tetrahedron: Asymmetry*, **1996**, *7*, 2073.
118. M. Diaz and R. M. Ortuno, *Tetrahedron: Asymmetry*, **1995**, *6*, 1845.
119. J. M. Jiménez, R. Casas, and R. M. Ortuno, *Tetrahedron Lett.*, **1994**, *35*, 5945.
120. S. Baskaran, C. Baskaran, P. J. Nadkarni, and G. K. Trivedi, *Tetrahedron*, **1997**, *53*, 7057.
121. S. Saito, T. Ishikawa, and T. Moriwake, *Synlett*, **1994**, 279.
122. F. Busque, P. De March, M. Figueredo, J. Font, M. Monsalvatje, A. Virgili, A. Alvarez-Larena, and J. F. Piniella, *J. Org. Chem.*, **1996**, *61*, 8578.
123. S. Baskaran and G. K. Trivedi, *J. Chem. Res., Synop.*, **1995**, 308.
124. R. Annunziata, M. Benaglia, M. Cinquini, and L. Raimondi, *Tetrahedron*, **1993**, *49*, 8629.

125. R. Annunziata, M. Cinquini, F. Cozzi, L. Raimondi, and T. Pilati, *Tetrahedron: Asymmetry*, **1991**, *2*, 1329.

126. I. Panfil, C. Belzecki, Z. Urbanczyk-Lipkowska, and M. Chmielewski, *Tetrahedron*, **1991**, *47*, 10087.

127. I. Panfil, C. Belzecki, M. Chmielewski, and K. Suwinska, *Tetrahedron*, **1989**, *45*, 233.

128. I. Panfil and M. Chmielewski, *Tetrahedron*, **1985**, *41*, 4713.

129. A. Patti, G. Nicolosi, M. Piattelli, and C. Sanfilippo, *Tetrahedron: Asymmetry*, **1995**, *6*, 2195.

130. A. J. Blake, T. A. Cook, A. C. Forsyth, R. O. Gould, and R. M. Paton, *Tetrahedron*, **1992**, *48*, 8053.

131. A. J. Blake, A. C. Forsyth, and R. M. Paton, *J. Chem. Soc. bChem. Commun.*, **1988**, 440.

132. A. P. Kozikowski and A. K. Ghosh, *J. Org. Chem.*, **1984**, *49*, 2762.

133. J. K. Gallos, T. V. Koftis, A. E. Koubis, and V. I. Moutsos, *Synlett*, **1999**, 1289.

134. S. Mzengeza and R. A. Whitney, *J. Chem. Soc. Chem. Commun.*, **1984**, 606.

135. S. Mzengeza, C. M. Yang, and R. A. Whitney, *J. Am. Chem. Soc.*, **1987**, *109*, 276.

136. S. Mzengeza and R. A. Whitney, *J. Org. Chem.*, **1988**, *53*, 4074.

137. W. J. Krol, S. Mao, D. L. Steele, and C. A. Townsend, *J. Org. Chem.*, **1991**, *56*, 728.

138. T. Naito, M. Ikai, M. Shirakawa, K. Kujimoto, I. Ninomiya, and T. Kiguchi, *J. Chem. Soc. Perkin Trans. 1*, **1994**, 773.

139. N. Langlois, N. V. Bac, N. Dahuron, J.-M. Delcroix, A. Deyine, D. Griffart-Brunet, A. Chiaroni, and C. Riche, *Tetrahedron*, **1995**, *51*, 3571.

140. N. Langlois, *Tetrahedron: Asymmetry*, **1998**, *9*, 1333.

141. A. Basak, H. M. M. Bdour, and G. Bhattacharya, *Tetrahedron Lett.*, **1997**, *38*, 2535.

142. C. Soucy, J.-E. Lacoste, and L. Breau, *Tetrahedron Lett.*, **1998**, *39*, 9117.

143. A. H. Fray and A. I. Meyers, *J. Org. Chem.*, **1996**, *61*, 3362.

144. A. H. Fray and A. I. Meyers, *Tetrahedron Lett.*, **1992**, *33*, 3575.

145. M. E. Kopach, A. H. Fray, and A. I. Meyers, *J. Am. Chem. Soc.*, **1996**, *118*, 9876.

146. A. I. Meyers and A. H. Fray, *Bull. Chim. Soc. Fr.*, **1997**, *134*, 283.

147. S. Kanemasa, T. Hayashi, J. Tanaka, H. Yamamoto, and T. Sakurai, *J. Org. Chem.*, **1991**, *56* 4473.

148. S. Kanemasa, H. Yamamoto, E. Wada, T. Sakurai, and K. Urushido, *Bull. Chem. Soc. Jpn.*, **1990**, *63*, 2857.

149. S. Kanemasa and H. Yamamoto, *Tetrahedron Lett.*, **1990**, *31*, 3633.

150. A. Brandi, S. Cicchi, A. Goti, K. M. Pietrusiewicz, M. Zablocka, and W. Wisniewski *J. Org. Chem.*, **1991**, *56*, 4383.

151. A. Brandi, P. Cannavo, K. M. Pietrusiewicz, M. Zablocka, and M. Wieczorek, *J. Org. Chem.*, **1989**, *54*, 3073.

152. C. Louis and C. Hootele, *Tetrahedron: Asymmetry*, **1997**, *8*, 109.

153. P. Bravo, L. Bruche, A. Farina, G. Fronza, S. V. Meille, and A. Merli, *Tetrahedron: Asymmetry*, **1993**, *4*, 2131.

154. T. Takahashi, A. Fujii, J. Sugita, T. Hagi, K. Kitano, Y. Arai, T. Koizumi, and M. Shiro, *Tetrahedron: Asymmetry*, **1991**, *2*, 1379.

155. T. Koizumi, H. Hirai, and E. Yoshii, *J. Org. Chem.*, **1982**, *47*, 4004.

156. H. Louis and C. Hootelé, *Tetrahedron: Asymmetry*, **1995**, *6*, 2149.

157. J. L. G. Ruano, A. Fraile, and M. R. Martin, *Tetrahedron: Asymmetry*, **1996**, *7*, 1943.

158. P. Bravo, L. Bruche, M. Crucianelli, A. Farina, S. V. Meille, A. Merli, and P. Seresini *J. Chem. Res., Synop.*, **1996**, 348.

159. P. Bravo, L. Bruche, D. Diliddo, and G. Resnati, *J. Chem. Res.*, **1993**, 346.

160. P. C. B. Page, M. Purdie, and D. Lathbury, *Tetrahedron*, **1997**, *53*, 1061.

161. A. Arnone, P. Bravo, L. Bruche, and P. Seresini, *J. Chem. Res., Synop.*, **1996**, 198.

162. T. Takahashi, K. Kitano, T. Hagi, H. Nihonmatsu, and T. Koizumi, *Chem. Lett.*, **1989**, 597.
163. G. L. Araldi, K. R. C. Prakash, C. George, and A. P. Kozikowski, *Chem. Commun.*, **1997**, 1875.
164. K. R. C. Prakash, M. Trzcinska, K. M. Johnson, and A. P. Kozikowski, *Bioorg. Med. Chem. Lett.*, **2000**, *10*, 1443.
165. A. Diaz-Ortiz, E. Diez-Barra, A. De la Hoz, P. Prieto, A. Moreno, F. Langa, T. Prange, and A. Neuman, *J. Org. Chem.*, **1995**, *60*, 4160.
166. S. G. Pyne, J. Safaei-G., B. W. Skelton, and A. H. White, *Aust. J. Chem.*, **1995**, *48*, 1511.
167. B. M. Kelly-Basetti, M. F. Mackay, S. M. Pereira, G. P. Savage, and G. W. Simpson *Heterocycles*, **1994**, *37*, 529.
168. A. Javidan, K. Schafer, and S. G. Pyne, *Synlett*, **1997**, 100.
169. R. M. Williams and G. J. Fegley, *Tetrahedron Lett.*, **1992**, *33*, 6755.
170. H. Tsuge, T. Okano, S. Eguchi, and H. Kimoto, *J. Chem. Soc. Perkin Trans. 1*, **1997**, 1581.
171. A. N. Boa, D. A. Dawkins, A. R. Hergueta, and P. R. Jenkins, *J. Chem. Soc. Perkin Trans. 1*, **1994**, 953.
172. C. Baldoli, P. Del Buttero, S. Maiorana, G. Zecchi, and M. Moret, *Tetrahedron Lett.*, **1993**, *34*, 2529.
173. A. Viso, R. F. Pradilla, C. G.-Strachan, M. Alonso, M. Martinez-Ripoll, and I. André *J. Org. Chem.*, **1997**, *62*, 2316.
174. A. Viso, R. F. Pradilla, A. García, M. Alonso, and C.-G. Strachan, *Synlett*, **1999**, 1543.
175. M. Ihara, M. Takahashi, K. Fukumoto, and T. Kametani, *J. Chem. Soc. Chem. Commun.*, **1988**, 9.
176. H. G. Aurich and F. Biesemeir, *Synthesis*, **1995**, 1171.
177. S. H. Kang and H. S. Lee, *Tetrahedron Lett.*, **1995**, *36*, 6713.
178. H. G. Aurich, F. Biesemeier, and M. Boutahar, *Chem. Ber.*, **1991**, *124*, 2329.
179. U. Chiacchio, F. Casuscelli, A. Corsaro, V. Librando, A. Rescifina, R. Romeo, and G. Romeo, *Tetrahedron*, **1995**, *51*, 5689.
180. H. G. Aurich and H. Köster, *Tetrahedron*, **1995**, *51*, 6285.
181. H. G. Aurich and J.-L. R. Quintero, *Tetrahedron*, **1994**, *50*, 3929.
182. R. Annunziata, M. Cinquini, F. Cozzi, P. Giaroni, and L. Raimondi, *Tetrahedron: Asymmetry*, **1990**, *1*, 251.
183. H. G. Aurich, G. Frenzen, and C. Gentes, *Chem. Ber.*, **1993**, *126*, 787.
184. A. Arnone, M. Cavicchioli, A. Donadelli, and G. Resnati, *Tetrahedron: Asymmetry*, **1994**, *5*, 1019.
185. S. Saito, T. Ishikawa, and T. Moriwake, *J. Org. Chem.*, **1994**, *59*, 4375.
186. M. Ihara, M. Takahashi, K. Fukumoto, and T. Kametani, *J. Chem. Soc. Perkin Trans. 1*, **1989**, 2215.
187. C. Baldoli, P. Del Buttero, E. Licandro, S. Majorana, and A. Papagni, *Tetrahedron: Asymmetry*, **1995**, *6*, 1711.
188. R. Patra, N. C. Bar, A. Roy, B. Achari, N. Ghoshal, and S. B. Mandal, *Tetrahedron*, **1996**, *52*, 11265.
189. T. K. M. Shing, C.-H. Wong, and T. Yip, *Tetrahedron: Asymmetry*, **1996**, *7*, 1323.
190. S. Datta, P. Chattopadhyay, R. Mukhopadhyay, and A. Bhattacharjya, *Tetrahedron Lett.*, **1993**, *34*, 3585.
191. A. Bhattacharjya, P. Chattopadhyay, A. T. McPhail, and D. R. McPhail, *J. Chem. Soc. Chem. Commun.*, **1990**, 1508.
192. P. M. Collins, M. S. Ashwood, H. Eder, S. H. B. Wright, and D. J. Kennedy *Tetrahedron Lett.*, **1990**, *31*, 2055.
193. S. W. Baldwin, R. B. McFadyen, J. Aube, and J. D. Wilson, *Tetrahedron Lett.*, **1991**, *32* 4431.
194. P. Bhattacharjee, A. Bhattacharjya, and A. Patra, *Tetrahedron Lett.*, **1995**, *36*, 4677.
195. T. K. M. Shing, D. A. Elsley, and J. G. Gillhouley, *J. Chem. Soc. Chem Commun.*, **1989**, 1280.
196. T. K. M. Shing, W.-C. Fung, and C.-H. Wong, *J. Chem. Soc. Chem. Commun.*, **1994**, 449.
197. R. J. Ferrier, R. H. Furneaux, P. Prasit, and P. C. Tyler, *J. Chem. Soc. Perkin Trans. 1*, **1983**, 1621.

198. R. J. Ferrier, and P. Prasit, *J. Chem. Soc. Chem. Commun.*, **1981**, 983.
199. S. Jiang, B. Mekki, G. Singh, and R. H. Wightman, *Tetrahedron Lett.*, **1994**, *35*, 5505.
200. N. O. Peet, E. W. Huber, and R. A. Farr, *Tetrahedron*, **1991**, *47*, 7537.
201. R. A. Farr, N. A. Peet, and M. S. Kaang, *Tetrahedron Lett.*, **1990**, *31*, 7109.
202. J. Rong, P. Roselt, J. Plavec, and J. Chattopadhyaya, *Tetrahedron*, **1994**, *50*, 4921.
203. A. Papchikhin, and J. Chattopadhyaya, *Tetrahedron*, **1994**, *50*, 5279.
204. R. Sustmann, *Heterocycles*, **1995**, *40*, 1.
205. R. Grigg, *Chem. Soc. Rev.*, **1987**, *16*, 89.
206. A. Hassner, S. Singh, R. Sharma, and R. Maurya, *Tetrahedron*, **1993**, *49*, 2317.
207. A. Hassner and R. Maurya, *Tetrahedron Lett.*, **1989**, *30*, 5803.
208. A. Hassner, R. Maurya, and E. Masko, *Tetrahedron Lett.*, **1988**, *29*, 5313.
209. A. Hassner, E. Falb, A. Nudelman, A. Albeck, and H. E. Gottlieb, *Tetrahedron Lett.*, **1994**, *35*, 2397.
210. U. Chiacchio, A. Corsaro, V. Pisthara, A. Rescifina, G. Romeo, and R. Romeo *Tetrahedron*, **1996**, *52*, 7875.
211. E. G. Baggiolini, H. L. Lee, G. Pizzolato, and M. R. Uskokovic, *J. Am. Chem. Soc.*, **1982**, *104* 6460.
212. H. G. Aurich, M. Geiger, M. Gentes, and H. Köster, *Tetrahedron Lett.*, **1996**, *37*, 841.
213. M. L. Mihailovic, L. Lorenc, and Z. Maksimovic, *Tetrahedron*, **1973**, *29*, 2683.
214. W. R. Roush and A. E. Walts, *J. Am. Chem. Soc.*, **1984**, *106*, 721.
215. D. Stanssens, D. Keukeleire, and M. Vandevalle, *Tetrahedron: Asymmetry*, **1990**, *1*, 547.
216. S. Majumdar, A. Bhattacharjya, and A. Patra, *Tetrahedron*, **1999**, *55*, 12157.
217. O. Tamura, N. Mita, T. Okabe, T. Yamaguchi, C. Fukushima, M. Yamashita, Y. Morita, N. Morita, H. Ishibashi, and M. Sakamoto, *J. Org. Chem.*, **2001**, *66*, 2602.
218. M. J. Uddin, M. Kikuchi, K. Takedatsu, K.-I. Arai, T. Fujimoto, J. Motoyoshiya, A. Kakehi, R. Iriye, H. Shirai, and I. Yamamoto, *Synthesis*, **2000**, 365.
219. J. Marco-Contelles and A. Opazo, *Tetrahedron Lett.*, **1999**, *40*, 4445.
220. N. C. Bar, R. Patra, B. Achari, and S. B. Mandal, *Tetrahedron*, **1997**, *53*, 4727.
221. A. Pal, A. Bhattacharjya, and R. Mukhopadhyay, *Tetrahedron Lett.*, **2000**, *41*, 10135.
222. H. G. Aurich, F. Biesemeier, and K. Harms *Liebigs Ann. Rec.*, **1997**, 469.
223. R. Annunziata, M. Cinquini, F. Cozzi, and L. Raimondi, *Tetrahedron*, **1987**, *43*, 4051.
224. R. Annunziata, M. Cinquini, F. Cozzi, and L. Raimondi, *J. Org. Chem.*, **1990**, *55*, 1901.
225. S. W. Baldwin, J. Aube, and A. T. McPhail, *J. Org. Chem.*, **1991**, *56*, 6546.
226. P. M. Wovkulich and M. R. Uskokovic, *Tetrahedron*, **1985**, *41*, 3455.
227. P. M. Wovkulich and M. R. Uskokovic, *J. Am. Chem. Soc.*, **1981**, *103*, 3956.
228. M. Frederikson, R. Grigg, Z. Rankovic, M. Thornton-Pett, J. Redpath, and R. Crossley, *Tetrahedron*, **1995**, *51*, 6835.
229. M. Frederickson, R. Grigg, J. Redpath, and M. Thornron-Pett, *Tetrahedron*, **1994**, *50*, 5495.
230. J. Marcus, J. Brussee, and A. Gen, *Eur. J. Org. Chem.*, **1998**, 2513.
231. A. L. Smith, S. F. Williams, A. B. Holmes, L. R. Hughes, C. Swithenbank, and Z. Lidert, *J. Am. Chem. Soc.*, **1988**, *110*, 8696.
232. A. B. Holmes, A. L. Smith, S. F. Williams, L. R. Huges, Z. Lidert, and C. Swithenbank, *J. Org. Chem.*, **1991**, *56*, 1393.
233. W. Oppolzer and M. Petrzilka, *Helv. Chim. Acta*, **1978**, *61*, 2755.
234. G. W. Gribble and T. C. Barden, *J. Org. Chem.*, **1985**, *50*, 5902.
235. W. P. Hems, C.-H. Tan, T. Stork, N. Feeder, and A. B. Holmes, *Tetrahedron Lett.*, **1999**, *40* 1393.
236. A. P. Kozikowski, *Acc. Chem. Res.*, 410, *1984*, *17*, 410.

237. R. Annunziata, M. Cinquini, F. Cozzi, A. Gilardi,, and A. Restelli, *J. Chem. Soc. Perkin Trans. 1*, **1985**, 2289.

238. J. K. Gallos, A. E. Koumbis, V. P. Xiraphaki, C. C. Dellios, and E. Coutouli-. Argyropoulou, *Tetrahedron*, **1999**, *55*, 15167.

239. M. Toyota, T. Asoh, and K. Fukumoto, *Heterocycles*, **1998**, *45*, 81.

240. C. Yip, S. Henderson, R. Jordan, and W. Tam, *Org. Lett.*, **1999**, *1*, 791.

241. C. Yip, S. Henderson, K. Tranner, and W. Tam, *J. Org. Chem.*, **2001**, *66*, 276.

242. S. Takano, Y. Iwabuchi, and K. Ogasawara, *J. Am. Chem. Soc.*, **1987**, *109*, 5523.

243. S. Takano, Y. Iwabuchi, and K. Ogasawara, *J. Chem. Soc. Chem. Commun.*, **1988**, 1204.

244. S. Takano, K. Samizu, and K. Ogasawara, *Chem. Lett.*, **1990**, 1239.

245. P. Garner, K. Sunitha, W. B. Ho, W. J. Youngs, V. O. Kennedy, and A. Djebli, *J. Org. Chem.*, **1989**, *54*, 2041.

246. P. Garner, P. B. Cox, J. T. Anderson, J. Protasiewicz, and R. Zaniewski, *J. Org. Chem.*, **1997**, *62*, 493.

247. N.-S. Kim, C. H. Kang, and J. K. Cha, *Tetrahedron Lett.*, **1994**, *35*, 3489.

248. J. R. Choi, S. Han, and J. K. Cha, *Tetrahedron Lett.*, **1991**, *32*, 6469.

249. R. B. Bennett III and J. K. Cha, *Tetrahedron Lett.*, **1990**, *31*, 5437.

250. R. B. Bennett III, J.-R. Choi, W. D. Montgomery, and J. K. Cha, *J. Am. Chem. Soc.*, **1989**, *111*, 2580.

251. J. G. Buchanan, A. R. Edgar, and B. D. Hewitt, *J. Chem. Soc. Perkin Trans. 1*, **1987**, 2371.

252. C. Herdeis and T. Schiffer, *Tetrahedron*, **1996**, *52*, 14745.

253. B. Bernet, A. R. C. B. Murty, and A. Vasella, *Helv. Chim. Acta*, **1990**, *73*, 940.

254. T. Hudlicky, G. Seoane, J. D. Price, and K. G. Gadamasetti, *Synlett*, **1990**, 433.

255. T. Hudlicky, J. O. Frazier, and J. D. Kwart, *Tetrahedron Lett.*, **1985**, *26*, 3523.

256. W. H. Pearson, *Tetrahedron Lett.*, **1985**, *26*, 3527.

257. W. H. Pearson, J. E. Celebuski, Y.-F. Poon, B. R. Dixon, and J. H. Glans, *Tetrahedron Lett.*, **1986**, *27*, 6301.

258. T. Hudlicky, H. Luna, J. D. Price, and F. Rulin, *J. Org. Chem.*, **1990**, *55*, 4683.

259. W. H. Pearson, S. C. Bergmeier, S. Degan, K.-C. Lin, Y.-F. Poon, J. M. Schkeryantz, and J. P. Williams, *J. Org. Chem.*, **1990**, *55*, 5719.

260. C. Herdies and T. Schiffer, *Tetrahedron Lett.*, **1999**, *55*, 1043.

261. T. Tejero, A. Dondoni, I. Rojo, A. Merchan, and P. Merino, *Tetrahedron*, **1997**, *53*, 3301.

262. P. J. Nadkarni, S. Baskaran, M. S. Savant, and G. K. Trivedi, *J. Chem. Res.*, **1998**, 370.

263. D. P. Curran, B. H. Kim, H. P. Piyasena, R. J. Loncharich, and K. N. Houk, *J. Org. Chem.*, **1987**, *52*, 2137.

264. T. Kametani, T. Nagahara, and M. Ihara, *J. Chem. Soc. Perkin Trans. 1*, **1981**, 3048.

265. F. J. Impastato, L. Barash, and H. M. Walborsky, *J. Am. Chem. Soc.*, **1959**, *81*, 1514.

266. D. M. Cooper, R. Grigg, S. Hargreaves, P. Kennewell, and J. Redpath, *Tetrahedron*, **1995**, *51*, 7791.

267. D. A. Barr, M. J. Dorrity, R. Grigg, S. Hargreaves, J. F. Malone, J. Montgomery, J. Redpath, P. Stevenson, and M. Thornton-Pett, *Tetrahedron*, **1995**, *51*, 273.

268. T. Coulter, R. Grigg, J. F. Malone, and V. Sridharan, *Tetrahedron Lett.*, **1991**, *32*, 5417.

269. D. A. Barr, M. J. Dorrity, R. Grigg, J. F. Malone, J. Montgomery, S. Rajviroongit, and P. Stevenson, *Tetrahedron Lett.*, **1990**, *31*, 6569.

270. R. Grigg, V. Sridharan, S. Suganthan, and A. W. Bridge, *Tetrahedron*, **1995**, *51*, 295.

271. F. M. Cordero, C. Fraggi, F. Sarlo, and A. Brandi, *Eur. J. Org. Chem.*, **2000**, 2595.

272. T. Olsson, K. Stern, G. Westman, and S. Sundell, *Tetrahedron*, **1990**, *46*, 2473.

273. T. Akiyama, K. Okada, and S. Ozaki, *Tetrahedron Lett.*, **1992**, *33*, 5763.

274. A. Carriere, A. Virgili, and M. Figueredo, *Tetrahedron: Asymmetry*, **1996**, *7*, 2793.

275. J.-L. Fauré, R. Réau, M. W. Wong, R. Koch, C. Wentrup, and G. Bertrand, *J. Am. Chem. Soc.*, **1997**, *119*, 2819.

276. W. Oppolzer, *Tetrahedron*, **1987**, *43*, 1969.

277. S. Murahashi, Y. Imada, M. Kohno, and T. Kawakami, *Synlett*, **1993**, 395.

278. T. Tejero, A. Dondoni, I. Rojo, A. Merchan, and P. Merino, *Tetrahedron*, **1997**, *53*, 3301.

279. T. Gefflaunt, U. Bauer, K. Airola, and M. P. Koskinen, *Tetrahedron: Asymmetry*, **1996**, *7*, 3099.

280. B. H. Kim, J. Y. Lee, K. Kim, and D. Whang, *Tetrahedron: Asymmetry*, **1991**, *2*, 27.

281. B. H. Kim and J. Y. Lee, *Tetrahedron: Asymmetry*, **1991**, *2*, 1359.

282. D. P. Curran, B. H. Kim, J. Daugherty, and T. Heffner, *Tetrahedron Lett.*, **1988**, *29*, 3555.

283. K. S. Kim, B. H. Kim, W. M. Park, S. J. Cho, and B. J. Mhin, *J. Am. Chem. Soc.*, **1993**, *115*, 7472.

284. D. P. Curran and T. A. Heffner, *J. Org. Chem.*, **1990**, *55*, 4585.

285. J. Zhang and D. P. Curran, *J. Chem. Soc. Perkin Trans. 1*, **1991**, 2627.

286. B. H. Kim, Y. J. Chung, G. Keum, J. Kim, and K. Kim, *Tetrahedron Lett.*, **1992**, *33*, 6811.

287. M. R. Mish, F. M. Guerra, and E. M. Carreira, *J. Am. Chem. Soc.*, **1997**, *119*, 8379.

288. P. Garner and W. B. Ho, *J. Org. Chem.*, **1990**, *55*, 3973.

289. S. Karlsson, F. Han, H.-E. Högberg, and P. Caldirola, *Tetrahedron: Asymmetry*, **1999**, *10* 2605.

290. G. Pandey, J. K. Laha, and A. K. Mohanakrishnan, *Tetrahedron Lett.*, **1999**, *40*, 6065.

291. S. Karlsson and H.-E. Högberg, *Org. Lett.*, **1999**, *1*, 1667.

292. S. Karlsson and H.-E. Högberg, *Synthesis*, **2000**, 1863.

293. R. H. Wallace, J. Liu, K. K. Zong, and A. Eddings, *Tetrahedron Lett.*, **1997**, *38*, 6791.

294. K. S. Yang, J.-C. Lain, C.-H. Lin, and K. Chen, *Tetrahedron Lett.*, **2000**, *41*, 1453.

295. W. Oppolzer, A. J. Kingma, and S. K. Pillai, *Tetrahedron Lett.*, **1991**, *32*, 4893.

296. D. P. Curran and M.-H. Yoon, *Tetrahedron Lett.*, **1997**, *53*, 1971.

297. D. P. Curran, K.-S. Jeong, T. A. Heffner, and J. Rebek, *J. Am. Chem. Soc.*, **1989**, *111*, 9238.

298. Y. H. Kim, S. H. Kim, and D. H. Park, *Tetrahedron Lett.*, **1993**, *34*, 6063.

299. S. Kanemasa and K. Onimura, *Tetrahedron*, **1992**, *48*, 8645.

300. S. Kanemasa, K. Onimura, E. Wada, and J. Tanaka, *Tetrahedron: Asymmetry*, **1991**, *2*, 1185.

301. S. Kanemasa and K. Onimura, *Tetrahedron*, **1992**, *48*, 8631.

302. J. A. Stack, T. A. Heffner, S. J. Geib, and D. P. Curran, *Tetrahedron*, **1993**, *49*, 995.

303. H. Yamamoto, S. Watanabe, K. Kadotani, M. Hasegawa, M. Noguchi, and S. Kanemasa, *Tetrahedron Lett.*, **2000**, *41*, 3131.

304. K. V. Gothelf, R. G. Hazell, and K. A. Jørgensen, *J. Org. Chem.*, **1998**, *63*, 5483.

305. Z. Ma, S. Wang, C. S. Cooper, A. K. L. Fung, J. K. Lynch, F. Plagge, and D. T. W. Chu, *Tetrahedron: Asymmetry*, **1997**, *8*, 883.

306. H. Waldmann, E. Blaeser, M. Jansen, and H.-P. Letschert, *Chem. Eur. J.*, **1995**, *1*, 150.

307. H. Waldmann, E. Blaeser, M. Jansen, and H. P. Letschert, *Angew. Chem., Int. Ed. Engl.*, **1994**, *33*, 683.

308. H. Waldmann, *Liebigs Ann. Chem.*, **1990**, 1013.

309. C. Kashima, K. Takahashi, I. Fukuchi, and K. Fukusaka, *Heterocycles*, **1997**, *44*, 289.

310. M. T. Rispens, E. Keller, B. De Lange, R. W. J. Zijlstra, and B. L. Feringa *Tetrahedron: Asymmetry*, **1994**, *5*, 607.

311. B. De Lange and B. L. Feringa, *Tetrahedron Lett.*, **1988**, *29*, 5317.

312. E. Keller, B. De Lange, M. T. Rispens, and B. L. Feringa, *Tetrahedron*, **1993**, *49*, 8899.

313. W. Carruthers, P. Coggins, and J. B. Weston, *J. Chem. Soc. Chem. Commun.*, **1991**, 117.

314. J. Barluenga, F. Fernández-Mari, R. González, E. Aguilar, G. A. Revelli, A. L. Viado F. J. Fananás, and B. Olano, *Eur. J. Org. Chem.*, **2000**, 1773.

315. J. Barluenga, F. Fernández-Mari, A. L. Viado, E. Aguilar, B. Olano, S. García-Granda, and C. Moya-Rubiera, *Chem. Eur. J.*, **2000**, *5*, 883.

316. J. Biao, Z. Ao and K. Ying, *Chin. J. Chem.*, **1999**, *17*, 293.

317. C. D. Davies, S. P. Marsden, and E. S. E. Stokes, *Tetrahedron Lett.*, **1998**, *39*, 8513.

318. C. D. Davies, S. P. Marsden, and E. S. E. Stokes, *Tetrahedron Lett.*, **2000**, *41*, 4229.

319. S. R. Gilbertson and O. D. Lopez, *J. Am. Chem. Soc.*, **1997**, *119*, 3399.

320. P. Garner, Ö. Dogan, W. J. Youngs, V. O. Kennedy, J. Protasiewicz, and R. Zaniewski *Tetrahedron*, **2001**, *57*, 71.

321. W. Oppolzer, J. Deerberg,, and O. Tamura, *Helv. Chim. Acta*, **1994**, *77*, 554.

322. N. Katagiri, M. Okada, Y. Morishita, and C. Kaneko, *J. Chem. Soc. Chem. Commun.*, **1996**, 2137.

323. N. Katagiri, M. Okada, Y. Morishita, and C. Kaneko, *Tetrahedron*, **1997**, *53*, 5725.

324. A. Vasella, *Helv. Chim. Acta*, **1977**, *60*, 1273.

325. A. Vasella, *Helv. Chim. Acta*, **1977**, *60*, 426.

326. A. Vasella and R. Voeffray, *J. Chem. Soc. Chem. Commun.*, **1981**, 97.

327. A. Vasella, R. Voeffray, J. Pless, and R. Huguenin, *Helv. Chim. Acta*, **1983**, *66*, 1241.

328. A. Vasella and R. Voeffray, *Helv. Chim. Acta*, **1982**, *65*, 1953.

329. R. Huber and A. Vasella, *Tetrahedron*, **1990**, *46*, 33.

330. K. Kasahara, H. Iida, and C. Kibayashi, *J. Org. Chem.*, **1989**, *54*, 2225.

331. H. Iida, K. Kasahara, and C. Kibayashi, *J. Am. Chem. Soc.*, **1986**, *108*, 4647.

332. U. Chiacchio, A. Corsaro, G. Gumina, A. Rescifina, D. Iannazzo, A. Piperno, G. Romeo and R. Romeo, *J. Org. Chem.*, **1999**, *64*, 9321.

333. S. E. Denmark, A. R. Hurd, and H. J. Sascha, *J. Org. Chem.*, **1997**, *62*, 1668.

334. S. E. Denmark and A. Thorarensen, *J. Am. Chem. Soc.*, **1997**, *119*, 125.

335. S. E. Denmark, D. L. Parker Jr., and J. A. Dixon *J. Org. Chem.*, **1997**, *62*, 435.

336. A. Papchikhin, P. Agback, J. Plavec, and J. Chattopadhyaya, *J. Org. Chem.*, **1993**, *58*, 2874.

337. H. B. Kagan and O. Riant, *Chem. Rev.*, **1992**, *92*, 1007.

338. K. A. Jørgensen, *Angew. Chem. Ind. Ed. Engl.*, **2000**, *39*, 3558.

339. J.-P. G. Seerden, A. W. A. Scholte op Reimer, and H. W. Scheeren, *Tetrahedron Lett.*, **1994**, *35*, 4419.

340. J.-P. G. Seerden, M. M. M. Kuypers, and H. W. Scheeren, *Tetrahedron: Asymmetry*, **1995**, *6*, 1441.

341. J.-P. G. Seerden, M. M. M. Boeren, and H. W. Scheeren, *Tetrahedron*, **1997**, *53*, 11843.

342. M. Meske, *J. Prakt. Chem.*, **1997**, *339*, 426.

343. J. Tanaka and S. Kanamasa, *Tetrahedron*, **2001**, *57*, 899.

344. K. B. Simonsen, P. Bayón, R. G. Hazell, K. V. Gothelf, and K. A. Jørgensen, *J. Am. Chem. Soc.*, **1999**, *121*, 3845.

345. K. B. Simonsen, K. A. Jørgensen, Q.-S. Hu, and L. Pu, *Chem. Commun.*, **1999**, 811.

346. K. B. Jensen, M. Roberson, and K. A. Jørgensen, *J. Org. Chem.*, **2000**, *65*, 9080.

347. G. Desimoni, G. Faita, A. Mortoni, and P. Righetti, *Tetrahedron Lett.*, **1999**, *40*, 2001.

348. S. Crosignani, G. Desimoni, G. Faita, S. Filippone, A. Mortoni, P. Righetti, and M. Zema, *Tetrahedron Lett.*, **1999**, *40*, 7007.

349. G. Desimoni, G. Faita, M. Mella, P. Righetti, and M. Zema, *Tetrahedron*, **1999**, *55*, 8509.

350. D. Seebach, A. K. Beck, and A. Heckel, *Angew. Chem. Int. Edit.*, **2001**, *40*, 92.

351. K. V. Gothelf and K. A. Jørgensen, *J. Org. Chem.*, **1994**, *59*, 5687.

352. K. B. Jensen, K. V. Gothelf, R. G. Hazell, and K. A. Jørgensen, *J. Org. Chem.*, **1997**, *62*, 2471.

353. K. V. Gothelf, R. G. Hazell, and K. A. Jørgensen, *J. Am. Chem. Soc.*, **1995**, *117*, 4435.

354. K. V. Gothelf and K. A. Jørgensen, *J. Org. Chem.*, **1995**, *60*, 6847.

355. K. V. Gothelf and K. A. Jørgensen, *J. Chem. Soc. Perkin Trans. 2*, **1997**, 111.

356. D. Seebach, R. Dahinden, R. E. Marti, A. K. Beck, D. A. Plattner, and F. N. Kühnle, *J. Org. Chem.*, **1995**, *60*, 1788.

357. C. Haase, C. R. Sarko, and M. DiMare, *J. Org. Chem.*, **1995**, *60*, 1777.

358. J. I. Garcia, V. Martinez-Merino, and J. A. Mayoral, *J. Org. Chem.*, **1998**, *63*, 2321.

359. K. V. Gothelf, I. Thomsen, and K. A. Jørgensen, *J. Am. Chem. Soc.*, **1996**, *118*, 59.

360. K. B. Jensen, K. V. Gothelf, and K. A. Jørgensen, *Helv. Chim. Acta*, **1997**, *80*, 2039.

361. D. Seebach, R. E. Marti, and H. W. Hintermann, *Helv. Chem. Acta*, **1996**, *79*, 1710.

362. A. Heckel and D. Seebach, *Angew. Chem. Int. Ed. Engl.*, **2000**, *39*, 163.

363. W. W. Ellis, A. Gavrilova, L. Liablesands, A. L. Rheingold, and B. Bosnich *Organometallics*, **1999**, *18*, 332.

364. P. Bayon, P. March, M. Espnosa, M. Figueredo, and J. Font, *Tetrathedron: Asymmetry*, **2000**, *11*, 1757.

365. S. Kanemasa, Y. Oderaotoshi, S. Sakaguchi, H.Yamamoto, J. Tanaka, E. Wada, and D. P. Curran, *J. Am. Chem. Soc.*, **1998**, *120*, 3074.

366. S. Kanemasa, Y. Oderaotoshi, J. Tanaka, and E. Wada, *J. Am. Chem. Soc.*, **1998**, *120*, 12355.

367. K. B. Jensen, R. G. Hazell, and K. A. Jørgensen, *J. Org. Chem.*, **1999**, *64*, 2353.

368. K. Hori, H. Kodama, T. Ohta, and I. Furukawa, *Tetrahedron Lett.*, **1996**, *37*, 5947.

369. K. Hori, H. Kodama, T. Ohta, and I. Furukawa, *J. Org. Chem.*, **1999**, *64*, 5017.

370. K. Hori, J. Ito, T. Ohta, and I. Furukawa, *Tetrahedron*, **1998**, *54*, 12737.

371. Y. Ukaji, K. Taniguchi, K. Sada, and K. Inomata, *Chem. Lett.*, **1997**, 547.

372. A. I. Sanchez-Blanco, K. V. Gothelf, and K. A. Jørgensen, *Tetrahedron Lett.*, **1997**, *38* 7923.

373. S. Kobayashi, R. Akiyama, M. Kawamura, and H. Ishitani, *Chem. Lett.*, **1997**, 1039.

374. S. Kobayashi and M. Kawamura, *J. Am. Chem. Soc.*, **1998**, *120*, 5840.

375. M. Kawamura and S. Kobayashi, *Tetrahedron Lett.*, **1999**, *40*, 3213.

376. H. Kodema, J. Ito, K. Hori, T. Otha, and I. Furukawa, *J. Organnomet. Chem.*, **2000**, *603*, 6.

377. S. J. Wendy, J. J. M. Wiener, and D. W. C. MacMillan, *J. Am. Chem. Soc.*, **2000**, *122*, 9874.

378. K. A. Ahrendt, C. J. Borths, and D. W. C. MacMillan, *J. Am. Chem. Soc.*, **2000**, *122*, 4243.

379. M. Shimizu, Y. Ukaji, and Inomata, *Chem. Lett.*, **1996**, 455.

380. Y. Ukaji, K. Sada, and K. Inomata, *Chem. Lett.*, **1993**, 1847.

381. Y. Yoshida, Y. Ukaji, S. Fujinami, and K. Inomata, *Chem. Lett.*, **1998**, 1023.

382. J. D. Toker, P. Wentworth, Jr., Y. Hu, K. N. Houk, and K. D. Janda, *J. Am. Chem. Soc.*, **2000**, *122*, 3244.

383. P. Allway and R. Grigg, *Tetrahedron Lett.*, **1991**, *32*, 5817.

384. R. Grigg, *Tetrahedron: Asymmetry*, **1995**, *6*, 2475.

385. M. P. Doyle and D. C. Forbes, *Chem. Rev.*, **1998**, *98*, 911.

386. A. Padwa and M. D. Weingarten, *Chem. Rev.*, **1996**, *96*, 223.

387. D. M. Hodgson, P. A. Stupple, and C. Johnstone, *Tetrahedron Lett.*, **1997**, *38*, 6471.

388. D. M. Hodgson, P. A. Stupple, and C. Johnstone, *Chem. Commun.*, **1999**, 2185.

389. A. Padwa, D. J. Austin, and S. F. Hornbuckle, *J. Org. Chem.*, **1996**, *61*, 63.

390. A. Padwa, D. J. Austin, S. F. Hornbuckle, and A. T. Price, *Tetrahedron Lett.*, **1992**, *33*, 6427.

391. H. Suga, H. Ishida, and T. Ibata, *Tetrahedorn Lett.*, **1998**, *39*, 3165.

392. H. Suga, A. Kakehi, S. Ito, K. Inoue, H. Ishida, and T. Ibata, *Org. Lett.*, **2000**, *2*, 3145.

393. S. Kitagaki, M. Anada, O. Kataoka, K. Matsuno, C. Umeda, N. Watabnabe, and S. Hashimoto, *J. Am. Chem. Soc.*, **1999**, *121*, 1417.

394. S. Kitagaki, M. Yasugahira, M. Anada, M. Nakajima, and S. Hashimoto, *Tetrahedron Lett.*, **2000**, *41*, 5931.

395. S. Kanemasa and T. Kanai, *J. Am. Chem. Soc.*, **2000**, *122*, 10710.

Index

ABC ring system, azomethine ylide intramolecular cycloadditions, 236–237
Ab initio calculations
nitrile imine structure, 476–478
nitrile oxide cycloadditions
chiral dipolarophiles, 387–393
dipolar/dipolarophile mechanism and relativity, 375–386
intramolecular cycloadditions, stereoselectivity, 414–416
olefin regioselectivity, 380–385
ABT-719 compound, azomethine ylide generation, asymmetric reaction, 197–200
Acetal derivatives
asymmetric reactions, 1,3-dipolar cycloadditions, 839–840
azomethine ylide generation, chiral dipolarophiles, asymmetric reaction, 193–200
nitrile oxide cycloadditions, olefin regioselectivity, 381–385
Acetone compounds, carbonyl ylides
generation, 262–263
structures, 257–259
Acetonitrile
(*N*-alkylideneamino)acetonitriles, irreversible metalation, 761–763
nitrile ylide structure, carbene reactions, 486–488
Acetylenes, thiocarbonyl ylides, thiophene derivatives, five-membered ring structure, 338–339
Achiral dipolarophiles, nitrile oxide cycloadditions
achiral olefins, 393–399
chiral nitrile oxides to olefins, 400, 403–406
intermolecular cycloadditions, "aldol" ring cleavage
achiral nitrile oxides-achiral olefins, 418–419
achiral nitrile oxides-chiral olefins, 419–424
relative reactivity, 376–380
Aci-nitro tautomers, nitronate formation, 84–86
Acromelic acid A, azomethine ylide intramolecular cycloadditions, 225–227
N-Acroyl-*(S)*-proline benzyl ester, azomethine ylide generation, asymmetric reaction, 200
Acrylamides, nitrile oxide cycloadditions
achiral olefins, 395–399
olefin regioselectivity, 382–385
Acrylates
alkyl nitronate intermolecular cycloadditions, 141–145
catalytic asymmetric nitrone reactions, 874–882
diastereoselective reactions, α,β-unsaturated esters, 852–853
Acyclic compounds
alkyl nitronates
intermolecular cycloadditions, substituent effects, 139–145
intramolecular cycloadditions, 148–153
preparation, 130–134
acylation, 133–134
alkylation, 130–132
trimethane derivatives, 132–133
spectroscopic data, 99–102
diastereoselective asymmetric reactions, allyl alcohol derivatives, 835–836
nitrile oxide intermolecular cycloadditions, "aldol" ring cleavage, 417–425
nitrone 1,3-dipolar cycloadditions, lactam formation, 10–11
Acyl-amino acids, mesoionic ring systems, münchnone cycloaddition reactions, 709–725
Acylation, acyclic nitronate preparation, 133–134
N-Acyliminium ions, isomünchnone cycloaddition, 734–735
S-Acylthioglycolic acid, mesoionic ring systems, thiomünchnone synthesis, 696–697
Acylsilanes, azide 1,3-dipolar cycloadditions, alkyne/nitrile reactions, 641–644
Adamantane compounds, diazoalkane cycloaddition reactions, carbon-heteroatom C=S double bonds, 571–575
Alcohols
magnesium ion-mediated reactions, nitrile oxide cycloadditions, 780
nitrile oxide intermolecular cycloadditions, aminoalcohol ring cleavage, 425–431
nitrile ylide nucleophile reactions, 526–527
thiocarbonyl ylide intermolecular reactions, 327–328

Aldehydes
azomethine ylide generation, 189–192
1,3-dipolar cycloadditions
asymmetric reactions, 831–833
external reagents, anti-selective Michael additions, 769–771
intramolecular cycloadditions, 221–227
metal-mediated reactions, 240–248
carbonyl ylides
1,3-dipolar cycloadditions, 805–806
directed syntheses, 281–284
generation, 273–274
chiral ylides
asymmetric reactions, 209–218
stereocontrol, 203–208
diazoalkane cycloaddition reactions, 545–549
carbon-heteroatom C=S double bonds, 571–575
nitrile oxides
cycloadditions, chiral nitrile oxides to olefins, 400–406
generation, 363–364
nitrile ylide 1,3-dipolar cycloadditions, 806–809
nitrone 1,3-dipolar cycloadditions, sugars, 25–29
"Aldol" ring cleavage
nitrile oxide intermolecular cycloadditions, isoxazoline intermediates, 417–425
achiral nitrile oxides/chiral olefins, 419–424
achiral nitrile oxides/olefins, 418–419
chiral nitrile oxides/achiral olefins, 424–425
chiral nitrile oxides/chiral olefins, 425
nitrile oxide intramolecular cycloadditions, isoazoline five-membered ring, 440–445
Aldoximes
nitrone 1,3-APT cycloaddition, 49–54
oxidation, nitrile oxide generation, 370–371
Alkaloids
alkyl nitronates, intramolecular cycloadditions, 158
azide 1,3-dipolar cycloadditions
alkyne/nitrile reactions, 639–644
natural synthesis, 652–664
diene double bonds, 666–671
azomethine ylide generation
intramolecular cycloadditions, 220–227
metal-mediated reactions, 240–248
silicon protocols, 171–181
carbonyl ylide cycloadditions, 284–292
chiral ylides, stereocontrol, 202–208
mesoionic ring systems
isomünchnones
cycloadditions, 732–735
synthesis, 693–695
thioisomünchnone cycloadditions, 741–742
nitrone 1,3-dipolar cycloadditions, 37–47
thiocarbonyl ylides, thiophene derivatives, five-membered ring structure, 336–339
Alkenes. *See also* Olefins.
acyclic nitronate preparation, trinitromethane derivatives, 132–133
alkyl nitronate intermolecular cycloadditions, 141–145
azide 1,3-dipolar cycloadditions
heterocyclic synthesis, 624–637
natural product synthesis, 649–664
azomethine ylide, intramolecular cycloadditions
aziridine precursors, 220–227
decarboxylative approach, 235–237
carbonyl ylides
1,3-dipolar cycloadditions, 804–806
oxadiazoline generation, 262–263
catalytic asymmetric nitrone reactions, 877–882
chiral ylides, asymmetric reactions, 212–218
allyl alcohol derivatives, 836–838
cyclic nitronate preparation, 135–136
diazoalkane cycloaddition reactions, 546–549
intramolecular cycloadditions, 598–601
structural properties, 549–552
1,3-dipolar cycloadditions, asymmetric reactions, 819–822
azomethine ylides, 830–833
catalytic nitrile oxide reactions, 882–884
catalytic reactions, 888
chiral dipolarophiles, 835–844
intramolecular cycloadditions, 847–848
intramolecular reactions, nitrile oxides, 849
intermolecular nitrone cycloaddition reactions, 60–68
intramolecular nitrone-alkene cycloadditions (INAC), 55–59
mesoionic ring systems
isomünchnone synthesis, 690–695
sydnone cycloadditions, 737–738
nitrile imine cycloadditions, six-atom or greater tethers, 515–516
nitrile oxide cycloadditions
chiral cycloalkene facial selectivity, 399–400
chiral dipolarophiles, 387–393
chiral nitrile oxides to olefins, 400–406
diastereoselectivity, 385–406
dipolarophile/dipole relative reactivity, 376–380
external reagents, 779–794
intramolecular cycloadditions, 410–416

magnesium ion-mediated reactions, 783–784, 786–789
olefin regioselectivity, 381–385
regioselectivity, intramolecular reactions, 407–410
nitrile ylides
carbene generation, 487–488
[2 + 3]-cycloaddition reactions, 496–502
electrocyclizations, 518–520
synthon generation, 481–485
nitroalkanes, [4 + 2] cycloaddition, 136–137
nitrone 1,3-dipolar cycloadditions
pyrrolidine nitrones, 14–18
reaction mechanisms, 795
sugars, 27–29
thiocarbonyl ylide structures, thiophene derivatives, five-membered ring structure, 331–339
Alkenyl nitrone, nitrone 1,3-dipolar cycloadditions, amino acids and peptides, 20–23
Alkenyloximes, nitrone 1,3-APT cycloaddition, 51–54
2-Alkylidene-1,3-oxathiole-4-carboxylate, thiocarbonyl ylides, 1,5-dipolar electrocyclization, five-membered heterocycles, 347–350
Alkylation
acyclic nitronate preparation, 130–132
cyclic nitronate preparation, 134–136
Alkyl compounds
azide 1,3-dipolar cycloadditions, heterocyclic synthesis, 631–637
nitrile ylide synthons, 481–485
α-(Alkylideneamino)esters, reversible metalation, 763–768
Alkylidenemalonates, azomethine ylide 1,3-dipolar cycloadditions, external reagents, enantiopure camphor imines, asymmetric Michael additions, 775–778
(*N*-Alkylideneamino)acetonitriles, irreversible metalation, azomethine ylide cycloadditions, 761–763
Alkylidene-phosphapyrazolines, diazoalkane cycloaddition reactions, carbon-heteroatom C=P bonds, 562–567
N-Alkylideneglycinates, azomethine ylide 1,3-dipolar cycloadditions, external reagents, 758–759
Alkyl nitronates
acyclic nitronate preparation, 130–134
acylation, 133–134
alkylation, 130–132
trimethane derivatives, 132–133
applications, 154–158
cyclic nitronate preparation, 134–137
alkylation, 134–136
nitroalkene [4 + 2] cycloaddition, 136–137
radical cyclization, 137
cycloadditions, 138–153
facial selectivity, 146–148
intermolecular, 138–145
intramolecular, 148–153
nitroso acetal functionalization, 153–154
structural properties
bonding, 96–97
nitroso acetals, spectroscopic data, 102–105
spectroscopic data, 97–102
stability, 95–96
Alkynes
alkyl nitronate intermolecular cycloadditions, 144–145
azide 1,3-dipolar cycloadditions
heterocyclic synthesis, 639–644
natural synthesis, 672–676
azomethine ylide intramolecular cycloadditions, 234–237
diazoalkane cycloaddition reactions, triple-bond systems, 581–586
mesoionic ring systems, münchnone cycloaddition reactions, 706–725
silyl nitronates, intramolecular cycloadditions, 124–125
Alkynyl aldehydes, nitrone 1,3-dipolar cycloadditions, lactam formation, 9–11
Alkynylhydroxylamines, nitrone 1,3-APT cycloaddition, 49–54
Allene-like skeleton, nitrile ylide structure, 475
Allenes
azide 1,3-dipolar cycloadditions, double-bonded diene conjugation, 637–639
nitrone 1,3-APT cycloaddition, 53–54
Allenic structure, nitrile imines, 478
D-Allosamine, nitrile oxide intermolecular cycloadditions, aminoalcohol ring cleavage, 430–431
Alloseamine compounds, 1,3-dipolar cycloadditions
chiral auxiliaries, 860–863
nitrones, 35–36
Allyl amine derivatives, asymmetric reactions, 1,3-dipolar cycloadditions, 838–839
Allyl glycines, nitrone 1,3-dipolar cycloadditions, 21–23
Allylic alcohols
catalytic asymmetric nitrone reactions, 878–882

Allylic alcohols (*Continued*)
diastereoselective asymmetric reactions
acyclic derivatives, 835–836
cyclic derivatives, 836–838
nitrile oxide cycloadditions
chiral dipolarophiles, 391–393
diastereoselectivity, 386
intramolecular cycloadditions, 411–416
magnesium ion-mediated reactions, 780–789
olefin regioselectivity, 384–385
nitrone 1,3-dipolar cycloadditions, magnesium ion-mediated reactions, 798–801
Allylic carbocations, azide 1,3-dipolar cycloadditions, heterocyclic synthesis, 644–645
Allyltrimethylsilane, nitrone 1,3-dipolar cycloadditions, 22
Aluminium complexes, 1,3-dipolar cycloadditions, catalytic asymmetric nitrone reactions, 867–882
AM1 calculations, nitronate structures, regioselectivity, 108–110
Amaryllidaceae alkaloids, nitrone 1,3-dipolar cycloadditions, 46–47
Amides
carbonyl ylide derivation, 276–278
diastereoselective reactions, α,β-amides, 853–857
nitrile ylides
electrocyclizations, diene-conjugated compounds, 521–524
synthon generation, 480–485
Amidines, chiral ylides, asymmetric reactions, 216–218
Amidoalcohol, nitrone 1,3-dipolar cycloadditions, 18–23
Amine compounds
azomethine ylide generation, 184–192
intramolecular cycloadditions, 230–237
nitrile imine nucleophile reactions, 528–530
nitrone 1,3-dipolar cycloadditions, 34–47
Amino acids
asymmetric reactions, 1,3-dipolar cycloadditions, vinyl aminals, 839–840
azomethine ylides, metal-mediated reactions, 240–248
chiral ylides, 209–218
mesoionic ring systems
münchnone cycloaddition reactions, 709–725
münchnone synthesis, 682–687
nitrile oxide intermolecular cycloadditions, 425–431
nitrile ylide synthons, 483–485
nitrone 1,3-dipolar cycloadditions, 18–23
nitrone transesterification-dipolar cycloadditions, 802–804
Aminoalcohol ring cleavage
nitrile oxide intermolecular cycloadditions, 425–431
nitrile oxide intramolecular cycloadditions
intermediate isoxazoline ring cleavage, 439–454
isoxazoline reduction, 454–458
β-Aminoketones, nitrone 1,3-dipolar cycloadditions, 35–36
Amino esters, azomethine ylides
1,3-dipolar cycloadditions
α-(alkylideneamino)esters, reversible metalation, 763–768
external reagents, anti-selective Michael additions, 770–771
generation, condensation protocols, 228–237
Aminoisomünchnones
cycloaddition reactions, 730–735
mesoionic ring systems, thioisomünchnone cycloaddition reactions, 739–742
Amino polyols, nitrile oxides
intermolecular cycloadditions, 425–431
intramolecular cycloadditions, aminoalcohol reduction, 456–458
Amyloglucosidase inhibitors, azide 1,3-dipolar cycloadditions, diene double bond reactions, 668–671
Andrachamine compounds, nitrone 1,3-dipolar cycloadditions, 46–47
Angiotensin converting enzyme (ACE)
azomethine ylide metal-mediated reactions, 240–241
carbonyl ylide alkaloid preparation, 289–292
Annulene compounds, azide 1,3-dipolar cycloadditions, heterocyclic synthesis, 630–636
Antibody catalysis, nitrile oxide cycloadditions, 791–792
Anti-selective Michael additions, azomethine ylides, 1,3-dipolar cycloadditions, external reagents, 768–771
1,3-APT cycloaddition, nitrone 1,3-dipolar cycloadditions
alkaloid compounds, 39–47
isoxazolidines, 48–54
Arene complexes, azomethine ylide metal-mediated reactions, 243–248
Aromaticity, nitrile oxide cycloadditions, dipolar/dipolarophile mehcanism and relativity, 376

Arsaalkenes, diazoalkane cycloadditions, carbon-heteroatom C=As double bonds, 568–570
Arteminolide, carbonyl ylides, oxidopyrylium ion cycloadditions, 304–305
Aryl compounds
aryl chlorocarbene, carbonyl ylide structures, 257–259
aryl nitronates, substituent effects, alkyl cycloaddition reactions, 139
aryl vinyl ether, carbonyl ylide nonmetallocarbenoid/metallocarbenoid syntheses, 306–309
azide 1,3-dipolar cycloadditions
alkyne/nitrile reactions, 641–644
double-bonded diene conjugation, 637–639
heterocyclic synthesis, 626–637
nitrile ylide synthons, 481–485
thiocarbonyl ylide heterocyclic synthesis, 1,5-dipolar electrocyclization, five-membered heterocycles, 350
Arylidenetetralones, carbonyl ylide directed syntheses, [1,3]-dipolar cycloadditions, 282–284
Asymmetric reactions
azomethine ylides, 193–218
chiral dipolarophiles, 193–200
chiral ylides
asymmetric template and auxiliary derivations, 208–218
inherent molecular stereocontrol, 200–208
1,3 dipolar cycloadditions, 772–774
catalytic 1,3-dipolar cycloadditions, 864–888
azomethine ylides, 884–886
carbonyl ylides, 886–888
diazo compounds, 888
mechanisms, 864–866
nitrile oxides, 882–884
nitrones, 866–882
diastereoselective reactions
chiral auxiliaries
1,3-dipoles, 860–863
dipolarophiles, 857–860
α,β-unsaturated amides, 853–857
α,β-unsaturated esters, 851–853
chiral dipolarophiles, 835–845
acyclic allyl alcohol derivatives, 835–836
alkenes, 842–844
allyl amine derivatives, 838–839
cyclic allyl alcohol derivatives, 836–838
heteroatomic dienophiles, 844–845
vinyl acetals and aminals, 839–840
vinyl phosphine oxides and sulfoxides, 840–842
chiral dipoles, 822–835
azomethine ylides, 829–833
enantiopure applications, 834–835
nitrile oxides, 829
nitrones/nitronates, 822–829
intramolecular reactions, 845–851
azide dipoles, 850–851
nitrile oxides, 848–850
nitrones/nitronates, 845–848
1,3-dipolar cycloaddition selectivity, 818–822
mechanisms, 818
nitrile oxide cycloadditions, diastereoselectivity, 386
nitrile ylide 1,3-dipolar cycloadditions, 807–809
nitrone 1,3-dipolar cycloadditions, magnesium ion-mediated reactions, 800–801
2-Azaallyl anions, azomethine ylide 1,3-dipolar cycloadditions, external reagents, 759
Azacyclobutadiene, diazoalkane cycloadditions, C=N bonds, 558–559
Azides
1,3-dipolar cycloadditions
asymmetric reactions, 819–822
intramolecular cycloadditions, 850–851
heterocyclic synthesis
alkene reactions, 624–637
alkynes and nitriles, 639–644
allylic carbocations, 644–645
dienes double-bond and conjugated systems, 637–639
fullerene reactions, 646–649
phosphorus ylides, 645–646
natural product synthesis
alkene reactions, 649–664
alkynes and nitriles, 672–676
diene double bonds, 664–671
structural properties, 623–624
Azidomethanes, azide 1,3-dipolar cycloadditions, heterocyclic synthesis, 630–636
Aziridines
azide 1,3-dipolar cycloadditions, natural product synthesis, 649–664
azomethine ylides
generation, asymmetric 1,3-dipolar cycloadditions, 831–833
intramolecular cycloadditions, precursors, 219–227
chiral ylides, stereocontrol, 207–208
diazoalkane intramolecular cycloaddition, multiple bonds, 604
1,3-dipolar cycloadditions, chiral auxiliaries, 860–863

Azirines, nitrile ylide structure, 475
carbene reactions, 486–488
[2 + 3]-cycloaddition reactions, 496–502
photolysis, 488–490
Azomethine ylides
asymmetric reactions, 193–218
chiral dipolarophiles, 193–200
chiral ylides
asymmetric template and auxiliary derivations, 208–218
inherent molecular stereocontrol, 200–208
diastereoselectivity, 829–833
1,3-dipolar cycloaddition reactions, 772–774
catalytic reactions, 884–886
1,3-dipolar cycloadditions, 821–822
enantiopure camphor imines, asymmetric Michael additions, 774–778
intramolecular cycloadditions, 850–851
development, 169–170
diastereoselective reactions, α,β-amides, 855–857
external reagents, 1,3-dipolar cycloaddition reactions, 757–778
anti-selective Michael additions, 768–771
asymmetric cycloadditions, 772–774
2-aazaallyl anions, 759
enantiopure camphor imines, asymmetric Michael additions, 774–778
enantioselective additions, 778
metal n-alkylideneglycinates, 758–759
N-metalated reactions, 759–778
α-(alkylideneamino)ester reversibility, 763–768
(N-alkylideneamine)acetonitrile irreversibility, 761–763
generation, 170–192
alternative procedures, 181–192
silicon-based protocols, 170–181
intramolecular cycloadditions, 219–239
alternative generation procedures, 237–239
aziridine precursors, 219–227
decarboxylative and amino ester condensation protocols, 228–237
metal-mediated reaction, 240–248
nitrile ylide synthon formation, 480–485
Azulenes, mesoionic ring systems, isomünchnone cycloadditions, 726–735

B3LYP/6-31G* calculations
alkyl nitronates, facial selectivity, 147–148
nitrile ylide structure, [2 + 3]-cycloaddition reactions, 498–502
nitronate structures, general reactivity, 106–114
Benzazocine derivatives, diazoalkane intramolecular cycloadditions, 596–601
Benzonitrile oxide, nitrile oxide cycloadditions, magnesium ion-mediated reactions, kinetic data, 784–786
Benzonitrilio methylide, nitrile ylide structure, 475
cycloaddition reactions, 500–502
Benzopyranyl nitrones, intramolecular nitrone-alkene cycloadditions (INAC), 56–59
Benzoquinones, mesoionic ring systems, münchnone cycloaddition reactions, 710–725
Benzothiazole derivatives, thiocarbonyl ylide structures, 316–317
Benzotriazole, azomethine ylide generation, silicon protocols, 177–181
Benzotropolone formation and reactivity, carbonyl ylides, 296–298
(S)-N-Benzylvalinol, azomethine ylide 1,3-dipolar cycloadditions, asymmetric reactions, 772–774
Bicyclic aminoketones, nitrone 1,3-dipolar cycloadditions, 17–18
Bicyclic imino acids and polyols, nitrile oxide intermolecular cycloadditions, 431–437
Biotin
azide 1,3-dipolar cycloadditions, natural synthesis, 660–664
nitrile oxide intramolecular cycloadditions
aminoalcohol-isoxazoline reduction, 454–458
polycyclic furoxans, 438–439
thiocarbonyl ylide structures, thiophene derivatives, five-membered ring structure, 332–339
1,2-Bis(trifluoromethyl)-fumaronitrile, thiocarbonyl ylide heterocyclic synthesis, seven-membered ring structure, 353–354
Bis(acetoxy)iodobenzene (BAIB), nitrone 1,3-dipolar cycloadditions, nucleoside antibiotics, 5–8
Bischler-Napieralski reaction, nitrile ylides, electrocyclizations, diene-conjugated compounds, 521–524
Bis-chloromethyl ether, carbonyl ylide generation, 262
Bis(trimethylsilyl)acetylene, thiocarbonyl ylides, thiophene derivatives, five-membered ring structure, 338–339
Bis[(trimethylsilyl)methyl]sulfoxides, thiocarbonyl ylide generation, 1,3-elimination reactions, 321

Boiling points
alkyl nitronates, 95–96
silyl nitronates, 86–87
Bonding
alkyl nitronates, 96–97
carbonyl ylide structures, 255–259
silyl nitronates, 87–88
Boron complexes, 1,3-dipolar cycloadditions, catalytic asymmetric nitrone reactions, 867–882
Boryl cation, silyl nitronates, 92–93
Bridged structures, alkyl nitronates, intramolecular cycloadditions, 149–153
Bromo compounds, azide 1,3-dipolar cycloadditions, heterocyclic synthesis, 624–637
Butoxycarbonyl (Boc), nitrile oxide generation, nitroalkene dehydration, 368–3870
2-(*tert*-Butyl)-3-imidazolidin-4-one, chiral ylides, stereocontrol, 207–208
Butyrolactones, nitrile oxide intermolecular cycloadditions, cycloaddition-reduction-hydrolysis sequence, 418–419

Caged compounds, mesoionic ring systems, münchnone cycloaddition reactions, 720–725
Camphor imines
diastereoselective reactions, α,β-unsaturated esters, 852–853
enantiopure compounds, asymmetric Michael additions, 774–778
Carbamoyl group, nitrile oxide generation, nitroalkene dehydration, 369–370
Carbenes
asymmetric reactions, chiral auxiliaries, 858–859
diazoalkane cycloaddition reactions, ketocarbene dipoles, 604–609
mesoionic ring systems, münchnone synthesis, 685–687
nitrile ylide structure, 474
generation reactions, 485–488
nonstabilized compounds, carbonyl ylide generation, 259–262
thiocarbonyl ylides, addition to C=S groups, 322–324
Carbenoids, thiocarbonyl ylides, addition to C=S groups, 322–324
Carbenoid strategy technique, thiocarbonyl ylide generation, thiophene derivatives, five-membered ring structure, 334–339
Carbocyclic structures
nitrile oxide intramolecular cycloadditions, stereoselectivity, 414–416
nitrone 1,3-dipolar cycloadditions, 6–8
Carbohydrate derivatives, nitrile oxide cycloadditions, intramolecular cycloadditions, 412–416
Carbon-13 nuclear magnetic resonance (^{13}C NMR)
alkyl nitronates
nitroso acetals, 103–105
spectroscopic data, 100–102
silyl nitronates
nitroso acetals, 94–95
spectroscopic data, 88–93
Carbon-heteroatom double bonds, diazoalkane cycloaddition reactions, 557–576
C=As bonds, 567–570
C=N bonds, 557–559
C=P bonds, 559–567
C=S bonds, 570–575
C=Se and C=Te bonds, 575–576
Carbonyl compounds
α-(alkylideneamino)esters, reversible metalation, 765–768
carbonyl ylides
1,3-dipolar cycloadditions, 804–806
catalytic asymmetric reactions, 886–888
diastereoselective reactions, α,β-amides, 853–857
diazoalkane cycloaddition reactions
carbon-heteroatom C=S double bonds, 572–574
intramolecular cycloadditions, 593–601
ketocarbene dipoles, 608–609
enantiopure camphor imines, asymmetric Michael additions, 774–778
mesoionic ring systems, isomünchnone synthesis, 688–695
nitrile oxide cycloadditions
intermolecular cycloadditions, "aldol" ring cleavage, 417–425
olefin regioselectivity, 382–385
nitrone 1,3-dipolar cycloadditions, magnesium ion-mediated reactions, 799–801
Carbonyl ylides
aldehyde derivation, 273–274
amide/urea derivatives, 276–278
cycloaddition reactions
alkaloid systems, 284–292
directed syntheses, 278–284
enantioselective syntheses, 298–302
illudins, pterosins, and ptaquilosides, 292–294
oxidopyrylium ion cycloadditions, 302–305

Carbonyl ylides (*Continued*)
tropolone and benzotropolone formation and reactivity, 296–298
zaragozic acids, 294–296
epoxide generation, 263–269
ester derivation, 274–276
ketone derivation, 269–273
metallocarbenoids
generation, 269
synthesis, 306–309
nonmetallocarbenoid precursors, synthesis using, 306–309
nonstable carbene generation, 259–262
oxadiazoline generation, 262–263
reactivity properties, 254–255
structural studies, 255–259
Carzinophilin, azide 1,3-dipolar cycloadditions, natural synthesis, 663–664
C=As bonds, diazoalkane cycloadditions, carbon-heteroatom double bonds, 567–570
Catalytic cycloadditions
asymmetric reactions, 1,3-dipolar cycloadditions, 864–888
azomethine ylides, 884–886
carbonyl ylides, 886–888
diazo compounds, 888
mechanisms, 864–866
nitrile oxides, 882–884
nitrones, 866–882
nitrile oxides
enantioselective reactions, 792–793
magnesium ion-mediated reactions, 787–788
nitrones, 34
enantioselective reactions, 804
Catalytic hydrogenolysis, alkyl nitronates, 154
C=C bonds, diazoalkane cycloaddition reactions, 541–557
alkenes, 549–552
diastereofacial selectivity, 552–557
dipolar properties, 543–549
intramolecular cycloadditions, 592–601
reactivity and regioselectivity, 541–543
C≡C bonds, intramolecular diazoalkane cycloaddition reactions, 601–604
Cephalotaxine analogues, azide 1,3-dipolar cycloadditions, natural synthesis, 649–664
Chain lengths, nitrile oxide intramolecular cycloadditions, regioselectivity, 407–410
Chelates
azomethine ylide 1,3-dipolar cycloadditions, external reagents
α-(alkylideneamino)esters, reversible metalation, 763–768
N-alkylideneglycinates, 758–759
nitrile oxide cycloadditions, magnesium ion-mediated reactions, 784
Chemoselectivity, nitrile oxide cycloadditions, magnesium ion-mediated reactions, 786
Chiral auxiliaries, diastereoselectivity, asymmetric reactions
1,3-dipoles, 860–863
dipolarophiles, 857–860
α,β-unsaturated amides, 853–857
α,β-unsaturated esters, 851–853
Chiral dipolarophiles
alkyl nitronates
facial selectivity, 146–148
intramolecular cycloadditions, 152–153
azomethine ylides
asymmetric reaction, 193–200
1,3-dipolar cycloadditions, asymmetric reactions, 772–774
enantiopure camphor imines, asymmetric Michael additions, 776–778
metal-mediated reactions, 243–248
diastereoselectivity, asymmetric reactions, 835–845
acyclic allyl alcohol derivatives, 835–836
alkenes, 842–844
allyl amine derivatives, 838–839
cyclic allyl alcohol derivatives, 836–838
heteroatomic dienophiles, 844–845
vinyl acetals and aminals, 839–840
vinyl phosphine oxides and sulfoxides, 840–842
diazoalkane cycloaddition reactions, diastereofacial selectivity, 552–557
mesoionic ring systems, isomünchnone cycloadditions, 729–735
nitrile oxide cycloadditions
achiral olefins, chiral auxiliaries, 393–399
cycloalkene facial selectivity, 399–400
diastereoselectivity, 386–393
intermolecular cycloadditions, "aldol" ring cleavage
chiral nitrile oxides-achiral olefins, 424–425
chiral nitrile oxides-chiral olefins, 425
intramolecular cycloadditions, stereoselectivity, 413–416
Lewis acid-mediated diastereoselective reactions, 789–790
olefins, chiral nitrile oxides, 400–406
relative reactivity, 377–380
nitrone 1,3-dipolar cycloadditions, magnesium ion-mediated reactions, 800–801

nitrone transesterification-dipolar cycloadditions, 803–804
silyl nitronates, facial selectivity, 121–123
Chiral dipoles, diastereoselectivity, asymmetric reactions, 822–835
azomethine ylides, 829–833
enantiopure applications, 834–835
nitrile oxides, 829
nitrones/nitronates, 822–829
Chiral enones, azomethine ylide generation, asymmetric reaction, 196–200
Chiral ylides, asymmetric reactions
asymmetric template and auxiliary derivations, 208–218
inherent molecular stereocontrol, 200–208
Chloramine-T, nitrile oxide generation, oxime one-pot operations, 368
Chloride ligands, catalytic asymmetric nitrone reactions, 874–882
Chlorosilanes, silyl nitronate preparation, 116–117
Chromium complexes
asymmetric diastereoselectivity, nitrone/nitronate chiral dipoles, 825–829
diazoalkane cycloaddition reactions, 554–557
mesoionic ring systems, münchnone synthesis, 687
nitrile ylide synthons, 484–485
Cis-fused heterobicyclic adducts, intramolecular nitrone-alkene cycloadditions (INAC), 57–59
Clavicipitic acid, azide 1,3-dipolar cycloadditions, natural product synthesis, 651–664
C=N bonds
diazoalkane cycloadditions, carbon-heteroatom double bonds, 557–559
thiocarbonyl ylides, heterocyclic synthesis, 343
^{13}C NMR. *See* Carbon-13 nuclear magnetic resonance (^{13}C NMR)
Coalescence temperature, silyl nitronates, 91–93
Cobalt complexes, azomethine ylide catalytic asymmetric reactions, 884–886
Compactin, nitrile oxide intramolecular cycloadditions, "aldol" ring cleavage, isoxazoline intermediates, 450–452
(–)-Coniine, nitrone 1,3-dipolar cycloadditions, 35–37
Conjugated thiocrabonyl ylide, 1,5-dipolar electrocyclization, five-membered heterocycles, 346–350
Cope elimination, nitrone 1,3-APT cycloaddition, 49–54
Copper-carbenoid species
carbonyl ylide 1,3-dipolar cycloadditions, 805–806
mesoionic ring systems, isomünchnone synthesis, 690–695
Coulomb repulsion, nitrile oxide cycloadditions, achiral olefins, 395–399
C=P bonds, diazoalkane cycloadditions, carbon-heteroatom double bonds, 559–567
Crambescidin cytotoxins, nitrone 1,3-dipolar cycloadditions, 44–47
Crinane, azide 1,3-dipolar cycloadditions, natural synthesis, 660–664
Crotonate compounds
asymmetric diastereoselectivity, nitrone/nitronate chiral dipoles, 824–829
azide 1,3-dipolar cycloadditions, natural synthesis, 657–664
diastereoselective reactions, αβ-unsaturated esters, 852–853
C=S bonds
diazoalkane cycloadditions, carbon-heteroatom double bonds, 570–575
thiocarbonyl ylides
carbene/carbenoid addition, 322–324
heterocyclic synthesis, 341–343
C=Se bonds, diazoalkane cycloaddition reactions, 575–576
C=Te bonds, diazoalkane cycloaddition reactions, 575–576
S-(–)-Cucurbitine, azomethine ylide generation, asymmetric reaction, 198–200
Cyano-substituted nitrile ylides
azide 1,3-dipolar cycloadditions, alkyne/nitrile reactions, 640–644
generation mechanism, 491–492
Cyathin compounds, carbonyl ylides, oxidopyrylium ion cycloadditions, 304–305
Cyclic derivatives, diastereoselective asymmetric reactions, allylic alcohols, 836–838
Cyclic nitronates
alkyl nitronates, 134–137
alkylation, 134–136
intermolecular cycloadditions
substituent effects, 139–145
unfunctionalized structure, 138–139
nitroalkene [4 + 2] cycloaddition, 136–137
radical cyclization, 137
spectroscopic data, 99–102
asymmetric diastereoselectivity, nitrone/nitronate chiral dipoles, 825–829
infrared absorbances, 98–102

Cycloaddition-modification-ring cleavage, nitrile oxides, 363
Cycloaddition reactions
alkyl nitronates, 138–153
applications, 154–158
facial selectivity, 146–148
intermolecular, 138–145
intramolecular, 148–153
azomethine ylides
generation protocols
alternative protocols, 181–192
chiral dipolarophiles, 193–200
silicon protocols, 170–181
intramolecular cycloadditions, 219–239
alternative generation procedures, 237–239
aziridine precursors, 219–227
decarboxylative and amino ester condensation protocols, 228–237
carbonyl ylides
alkaloid systems, 284–292
directed syntheses, 278–284
enantioselective syntheses, 298–302
epoxide generation, 265–269
illudins, pterosins, and ptaquilosides, 292–294
oxidopyrylium ion cycloadditions, 302–305
tropolone and benzotropolone formation and reactivity, 296–298
zaragozic acids, 294–296
cyclic nitronate preparation, [4 + 2] cycloaddition, 136–137
diazoalkanes
carbon-heteroatom double bonds, 557–576
C=As bonds, 567–570
C=N bonds, 557–559
C=P bonds, 559–567
C=S bonds, 570–575
C=Se and C=Te bonds, 575–576
C=C bonds, 541–557
alkenes, 549–552
diastereofacial selectivity, 552–557
dipolar properties, 543–549
reactivity and regioselectivity, 541–543
heteroatom-heteroatom double bonds, P=X bonds, 576–578
heterocumulenes, 578–581
intramolecular cycloadditions
C=C bonds, 592–601
C≡C bonds, 601–604
multiple bond reactions, 604
structural chemistry, 540–541
triple-bond systems, 581–592
alkynes, 581–586
heteroatom-heteroatom triple bonds, 586–587
nitriles, 586
phosphaalkynes, 587–592
mesoionic ring systems
1,3-dithiolium-4-olates, 742–744
isomünchnones, 725–735
miscellaneous mesoionic heterocycles, 745–746
münchnones, 700–725
1,3-oxathiazolium-5-olates, 745
1,3-oxathiolium-4-olates, 744
oxamünchnones, 742
sydnones, 735–738
1,3-thiazolium-4-thiolates, 742
thioisomünchnones, 739–742
thiomünchnones, 738
nitrile imines
[2 + 3]-cycloaddition reactions, 502–517
four-membered rings, 507–508
five-membered rings, 508–509
six-membered rings, 509–510
seven-membered rings, 510
exocyclic double bonds, 510–511
heterocyclic rings, 507
intramolecular cycloadditions, 511–516
three-atom tether, 512–513
our-atom tether, 513–514
five-atom tether, 514–515
six-or-greater atom tether, 515–516
selectivity, 505–507
stabilized nitrile imines, 502–504
tandem intermolecular-intramolecular cycloadditions, 516–517
transient formation, 494
nitrile oxides, 374–406
diastereoselectivity, 385–406
achiral olefins, with chiral auxiliaries, 393–399
chiral dipolarophiles, 386–393
facial selectivity, chiral alkenes, 399–400
olefin additions, chiral nitrile oxides, 400–406
dipole/dipolarophiles
mechanistic studies and calculations, 374–376
relative reactivity, 376–380
furoxan thermolysis, 371–374
intermolecular cycloadditions, 416–437
aminoalcohol ring cleavage, 425–431
isoxazoline compounds
"aldol" ring cleavage, intermediates, 417–425

achiral nitrile oxides/chiral olefins, 419–424
achiral nitrile oxides/olefins, 418–419
chiral nitrile oxides/achiral olefins, 424–425
mono- and bicyclic imino acids and polyols, 431–437
natural products, 416–417
intramolecular cycloadditions
regioselectivity, 407–410
stereoselectivity, 410–416
synthesis applications, 437–461
isoxazoline ring cleavage, 439–461
"aldol" reduction-hydrolysis, 440–454
γ-aminoalcohol, 454–458
isoxazolinium salts, transformation with, 458–461
polycyclic isoxazolines and furoxans, 437–439
olefin regioselectivity, 380–385
oxime one-pot operations, 367–368
nitrile ylides, synthon formation, 483–485
nitrile ylides, [2 + 3]-cycloaddition reactions, 496–502
nitronate structures
general reactivity, 105–114
mechanisms, 114
regioselectivity, 109–110
research background, 85–86
silyl nitronates, 117–125
facial selectivity, 120–123
intermolecular cycloadditions, 117–120
intramolecular cycloadditions, 123–125
stereoselectivity
intermolecular cycloadditions, 111–112
intramolecular cycloadditions, 112–114
nitrones
1,2-dipolar cycloaddition, 1–3
catalytic cycloadditions, 34
isoxazolidines, 47–68
1,3-APT process, 48–54
intermolecular reactions, 59–68
intramolecular cycloadditions
nitrone-alkene (INAC), 55–59
oxime-olefin, 54–55
lactams, 8–11
natural product formation, 3
nucleosides, 4–8
peptides and amino acids, 18–23
pyrrolidines, piperidines, and amines, 34–47
quinolizidines, indolizidines, and pyrrolizidines, 12–18
sugars, 24–29
sulfur- and phosphorus-containing compounds, 29–33
Cycloaddition-reduction-hydrolysis sequence
nitrile oxide intermolecular cycloadditions, "aldol" ring cleavage, isoxazoline intermediates, achiral nitrile oxides-achiral olefins, 418–419
nitrile oxide intramolecular cycloadditions, "aldol" ring cleavage, isoxazoline intermediates, 440–445
Cyclobutyl isoxazolidine adducts, nitrone 1,3-dipolar cycloadditions, 17–18
Cyclodextrins, nitrile oxide cycloadditions, olefin regioselectivity, 381–385
Cyclopentane products, nitrile oxide intramolecular cycloadditions, "aldol" ring cleavage, isoxazoline intermediates, 441–445
Cyclopentenoid compounds, azide 1,3-dipolar cycloadditions, heterocyclic synthesis, 625–637
Cyclopentyl isoxazolidine, intramolecular nitrone-alkene cycloadditions (INAC), 57–59
Cyclopropyl dipolarophiles, nitrone 1,3-dipolar cycloadditions, 15–18
Cycloreversion reactions
nitrile oxide generation, furoxan thermolysis, 371–374
thiocarbonyl ylide generation, 319–320
Cycloreversions, azomethine ylide generation, 189–192

Darlinine compounds, nitrone 1,3-dipolar cycloadditions, 35
Decahydrobenzocarbazoles, carbonyl ylide cycloadditions, 284–292
Decarboxylative techniques, azomethine ylide generation, 187–192
condensation protocols, 228–237
Dehydration mechanisms, nitroalkenes, nitrile oxide generation, 368–370
Dehydrochlorination, imidoyl chlorides, nitrile ylide generation, 490–491
Dehydrogenation techniques, azomethine ylide generation, 191–192
Dehydrohalogenation, hydroximoyl chloride, nitrile oxide generation, 364–368
one-pot oxime formation and cycloaddition, 367–368
oxime halogenation, 365–367
in situ generation, 367

Dendrobates alkaloids, nitrone 1,3-dipolar cycloadditions, 37–47
Desilylation
  azomethine ylide generation, 318
  nitrile ylide synthon formation, 480–485
  thiocarbonyl ylides, heterocyclic synthesis, 342–343
Diammoniumhexanitratocerate, aldoxime oxidation, nitrile oxide generation, 370–371
Diastereomeric ratio
  azomethine ylide generation, asymmetric reaction, 198–200
  chiral ylides, asymmetric reactions, 218
Diastereoselectivity
  asymmetric reactions
    chiral auxiliaries
      1,3-dipoles, 860–863
      dipolarophiles, 857–860
      α,β-unsaturated amides, 853–857
      α,β-unsaturated esters, 851–853
    chiral dipolarophiles, 835–845
      acyclic allyl alcohol derivatives, 835–836
      alkenes, 842–844
      allyl amine derivatives, 838–839
      cyclic allyl alcohol derivatives, 836–838
      heteroatomic dienophiles, 844–845
      vinyl acetals and aminals, 839–840
      vinyl phosphine oxides and sulfoxides, 840–842
    chiral dipoles, 822–835
      azomethine ylides, 829–833
      enantiopure applications, 834–835
      nitrile oxides, 829
      nitrones/nitronates, 822–829
    1,3-dipolar cycloadditions, 819–822
      azomethine ylides, 773–774, 829–833
        enantiopure camphor imines, asymmetric Michael additions, 778
    intramolecular reactions, 845–851
      azide dipoles, 850–851
      nitrile oxides, 848–850
      nitrones/nitronates, 845–848
  diazoalkane cycloaddition reactions, 552–557
  1,3-dipolar cycloadditions, external reagents, 756–757
  nitrile imine cycloadditions, 505–507
  nitrile oxide cycloadditions, 385–406
    achiral olefins, 394–399
      chiral auxiliaries, 393–399
    chiral dipolarophiles, 386–393
    chiral nitrile oxides to olefins, 400–406
    facial selectivity, chiral alkenes, 399–400
    intermolecular cycloadditions, imino acids and polyols, 431–437
    Lewis acid mediation, 789–790
    magnesium ion-mediated reactions, 780
    olefin additions, chiral nitrile oxides, 400–406
  nitrile ylides, cycloaddition reactions, 501–502
  nitrone transesterification-dipolar cycloadditions, 803–804
1,8-Diazabicyclo[5.4.0]undec-7-ene (DBU)
  (*N*-alkylideneamino)acetonitriles, irreversible metalation, 762–763
  azomethine ylide 1,3-dipolar cycloadditions, external reagents
    α-(alkylideneamino)esters, reversible metalation, 765–768
    anti-selective Michael additions, 769–771
    enantiopure camphor imines, asymmetric Michael additions, 775–778
  silyl nitronate preparation, 115–117
Diazadicyclo[5.3.0]decane, azomethine ylide intramolecular cycloadditions, 221–223
Diazepines
  diazoalkane intramolecular cycloadditions, 600–601
  nitrile imine cycloadditions, seven-membered rings, 510
Diazirine, carbonyl ylide structures, 257–259
  nonstabilized carbene generation, 259–262
Diazoacetacetamides, mesoionic ring systems, isomünchnone synthesis, 692–695
Diazoalkanes
  acyclic nitronate preparation, 130–132
  cycloaddition reactions
    carbon-heteroatom double bonds, 557–576
      C=As bonds, 567–570
      C=N bonds, 557–559
      C=P bonds, 559–567
      C=S bonds, 570–575
      C=Se and C=Te bonds, 575–576
    C=C bonds, 541–557
      alkenes, 549–552
      diastereofacial selectivity, 552–557
      dipolar properties, 543–549
      reactivity and regioselectivity, 541–543
    heteroatom-heteroatom double bonds, P=X bonds, 576–578
    heterocumulenes, 578–581
    intramolecular cycloadditions
      C=C bonds, 592–601
      C≡C bonds, 601–604
      multiple bond reactions, 604
    structural chemistry, 540–541
    triple-bond systems, 581–592

alkynes, 581–586
heteroatom-heteroatom triple bonds, 586–587
nitriles, 586
phosphaalkynes, 587–592
"ketocarbene dipoles" in hetereocyclic synthesis, 604–609
Diazo compounds. *See also* Diazoalkanes
catalytic asymmetric reactions, 888
mesoionic ring systems, isomünchnones
cycloadditions, 727–735
synthesis, 689–695
nitrile imine structure, 477–478
stabilized nitrile imine generation, 495–496
nitrile ylide generation
carbene reactions, 486–488
electrocyclizations, diene-conjugated compounds, 521–524
structural properties, 540–541
thiocarbonyl ylides
1,5-dipolar electrocyclization, five-membered heterocycles, 345–350
carbene/carbenoid addition, 322–324
thiirane three-membered rings, 330–331
Diazocumulenes, diazoalkane cycloaddition reactions
carbon-heteroatom C=P bonds, 562–567
dipole properties, 548–549
Diazomethane compounds
diazoalkane cycloaddition reactions, 543–549
thiocarbonyl ylide structures, 316–318
Diazophospholes, mesoionic ring systems, sydnone cycloadditions, 736–738
Diazo(trimethylsilyl)methane, diazoalkane cycloadditions, C=N bonds, 558–559
Dibenzofurzanyl-2,2′-bis(oxazoline)(DBFOX), catalytic asymmetric nitrone reactions, 876–882
1,4-Dicyanonaphthalene (DCN), azomethine ylide generation, silicon protocols, 175–181
Dicyclohexylcarbodiimide (DCC), mesoionic ring systems, münchnone synthesis, 682–687
Dideuteriodiazomethane, carbonyl ylide structures, 256–259
Diels-Alder reaction
carbonyl ylides
oxidopyrylium ion cycloadditions, 303–305
tropolone and benzotropolone formation and reactivity, 297–298
catalytic asymmetric nitrone reactions, 882
intermolecular nitrone cycloaddition reactions, 67–68
mesoionic ring systems, münchnone cycloaddition reactions, 723–725
Dienes
asymmetric reactions, heteroatom dienophiles, 844–845
azide 1,3-dipolar cycloadditions
double-bond reactions, 664–671
heterocyclic synthesis, double bonds, 637–639
nitrile ylide electrocyclizations, 520–524
Diethyl azodicarboxylate (DEAD)
acyclic nitronate preparation, 132
azomethine ylide generation, silicon protocols, 173–181
2,5-Dihydro-1,3,4-thiadiazoles, thiocarbonyl ylide structures, 317–318
cycloversion reactions, 319–320
seven-membered ring structure, 352–354
thiophene derivatives, five-membered ring structure, 31–339
Dihydroazepine, mesoionic ring systems, münchnone cycloaddition reactions, 719–725
Dihydrofuran derivatives, carbonyl ylide generation, 259–262
Dihydropyrimidines, carbonyl ylides, alkaloid preparation, 285–292
2,5-Dihydroxyoxazoles, azide 1,3-dipolar cycloadditions, heterocyclic synthesis, 627–637
Diisopropyl (R,R,)-(+)-tartrate (DIPT)
catalytic asymmetric reactions
nitrile oxides, 882–884
nitrones, 878–882
nitrile oxide cycloadditions, chiral dipolarophiles, 393
Dimerization reactions, thiocarbonyl ylide
heterocyclic synthesis, six-membered rings, 351
intermolecular cycloadditions, 328–329
4-Dimethylaminopyridine (DMAP), nitrile oxide generation, nitroalkene dehydration, 368–370
Dimethyl acetylenedicarboxylate (DMAD)
azide 1,3-dipolar cycloadditions, heterocyclic synthesis, 630–636
azomethine ylide generation
dipolar cascade, 185–192
oxazoline protocols, 181–192
silicon protocols, 170–181
carbonyl ylide generation
directed syntheses, [1,3]-dipolar cycloadditions, 281–284
enantioselective synthesis, 299–302

Dimethyl acetylenedicarboxylate (DMAD) (*Continued*)
ketone derivations, 270–273
mesoionic ring systems
cycloaddition reactions
1,3-dithiolium-4-olates, 742–744
isomünchnones, 725–735
miscellaneous mesoionic heterocycles, 745–746
münchnones, 700–725
1,3-oxathiazolium-5-olates, 745
1,3-oxathiolium-4-olates, 744
oxamünchnones, 742
sydnones, 735–738
1,3-thiazolium-4-thiolates, 742
thioisomünchnones, 739–742
thiomünchnones, 738
isomünchnone cycloaddition, 727–735
münchnone synthesis, 682–687
oxamünchnone cycloaddition reactions, 742
oxamünchnone synthesis, 699–700
sydnone cycloadditions, 735–738
1,3-thiazolium-4-thiolate synthesis, 699
nitrile imine structure, 477
nitrile ylides, cycloaddition reactions, 500–502
thiocarbonyl ylides, thiophene derivatives, five-membered ring structure, 338–339
N,N-Dimethylacrylamide, nitrile oxide cycloadditions, achiral olefins, 398–399
Dimethyl dicarboxylate (DMAD), carbonyl ylide generation, 259–262
Dimethylformamide (DMF), azomethine ylide generation, silicon protocols, 177–181
Dimethyl furmamate
azomethine ylide generation, silicon protocols, 175–181
chiral ylides, asymmetric reactions, 211–218
Dimethyl maleate, azomethine ylide generation, silicon protocols, 173–181
1,3-Dioxan-5-ylamines, chiral ylides, stereocontrol, 206–208
Dioxazaborocines, nitrile oxide cycloadditions, achiral olefins, 398–399
1,3-Dipolar cycloadditions
asymmetric reaction selectivity, 818–822
chiral auxiliaries, 860–863
azides
heterocyclic synthesis
alkene reactions, 624–637
alkynes and nitriles, 639–644
allylic carbocations, 644–645
dienes double-bond and conjugated systems, 637–639
fullerene reactions, 646–649
phosphorus ylides, 645–646
natural synthesis
alkene reactions, 649–664
alkynes and nitriles, 672–676
diene double bonds, 664–671
azomethine ylide generation, metal-mediated reactions, 240–248
carbonyl ylides
reaction mechanisms, 254–255
tandem directed syntheses, 278–284
diazoalkanes
carbon-heteroatom double bonds, 557–576
C=As bonds, 567–570
C=N bonds, 557–559
C=P bonds, 559–567
C=S bonds, 570–575
C=Se and C=Te bonds, 575–576
C=C bonds, 541–557
alkenes, 549–552
diastereofacial selectivity, 552–557
dipolar properties, 543–549
reactivity and regioselectivity, 541–543
heteroatom-heteroatom double bonds, P=X bonds, 576–578
heterocumulenes, 578–581
intramolecular cycloadditions
C=C bonds, 592–601
C≡C bonds, 601–604
multiple bond reactions, 604
structural chemistry, 540–541
triple-bond systems, 581–592
alkynes, 581–586
heteroatom-heteroatom triple bonds, 586–587
nitriles, 586
phosphaalkynes, 587–592
external reagents
azomethine ylides, 757–778
anti-selective Michael additions, 768–771
asymmetric cycloadditions, 772–774
2-aazaallyl anions, 759
enantiopure camphor imines, asymmetric Michael additions, 774–778
enantioselective additions, 778
metal n-alkylideneglycinates, 758–759
N-metalated reactions, 759–778
α-(alkylideneamino)ester reversibility, 763–768
(N-alkylideneamine)acetonitrile irreversibility, 761–763
carbonyl ylides, 804–806
nitrile oxides, 779–794

catalytic cycle, 787–788
chemoselectivity, 786
diastereocontrol, 780
enantioselective reactions, 793–794
kinetics, 784–786
Lewis acid-mediated diastereoselectivity, 789–790
magnesium ion-mediated reactions, 780–789
miscellaneous mediators, 790–793
reaction mechanisms, 779
regioselectivity, 780–782
related applications, 788–789
solvent effects, 782–784
nitrile ylide equivalents, 806–809
nitrones
magnesium ion-mediated reactions, 797–801
reaction mechanisms, 794–795
titanium ion-mediated reactions, 795–797
transesterification-diplar cycloadditions, 801–804
structural properties, 756–757
intermolecular nitrone cycloaddition reactions, 59–68
mesoionic ring systems
heterocyclic compounds, 745–747
isomünchnones
cycloadditions, 725–735
synthesis, 694–695
münchnone cycloaddition reactions, 703–725
1,3-oxathiolium-4-olates, 744
sydnone synthesis, 696
thioisomünchnones, 697–698, 740–742
nitrile oxides, 385–406
achiral olefins, with chiral auxiliaries, 393–399
chiral dipolarophiles, 386–393
facial selectivity, chiral alkenes, 399–400
intramolecular cycloadditions, stereoselectivity, 412–416
mechanistic studies/calculations, 374–376
olefin additions, chiral nitrile oxides, 400–406
structural chemistry, 362–363
nitrones, 1–3
thiocarbonyl ylide heterocyclic synthesis, seven-membered ring structure, 351–354
thiocarbonyl ylides, intermolecular reactions, 327–328
1,5-Dipolar electrocyclization, thiocarbonyl ylides, five-member heterocyclic synthesis, 345–350
[3 + 2]-Dipolar cycloadditions
acyclic nitronate preparation
acylation, 134
trinitromethane derivatives, 132–133
alkyl nitronate intermolecular cycloadditions, 143–145
azomethine ylides, external reagents, 2-azallyl anions, 759
carbonyl ylides, 255
chiral ylides, stereocontrol, 200–208
diazoalkane cycloaddition reactions, 544–549
alkyne triple-bond systems, 582–586
C=C intramolecular reactions, 596–601
heteroatom-heteroatom triple bonds, 586–587
nitronate structures, 85–86
alkyl nitroso acetals, spectroscopic data, 102–105
general reactivity, 105–114
nitrone transesterification-dipolar cycloadditions, 801–804
nitroso acetal functionalization, 125–127
silyl nitronates, 117
facial selectivity, 120–123
intermolecular cycloadditions, 117–120
intramolecular cycloadditions, 123–125
thiocarbonyl ylides
intermolecular reactions, 326–327
thiophene derivatives, five-membered ring structure, 331–339
Dipolarophiles
alkyl nitronates
facial selectivity, 146–148
intermolecular cycloadditions, 139–145
asymmetric cycloaddition reactions, chiral auxiliaries, 857–860
azide 1,3-dipolar cycloadditions, alkyne/nitrile reactions, 642–644
azomethine ylide generation
chiral dipolarophiles, asymmetric reactions, 193–200
intramolecular cycloadditions, 233–237
monosubstituted olefinic compounds, 182–192
silicon protocols, 170–181
carbonyl ylide structures, epoxide generation, 267–269
diazoalkane cycloaddition reactions, diastereofacial selectivity, 552–557
intermolecular nitrone cycloaddition reactions, 62–68
mesoionic ring systems
isomünchnones
cycloadditions, 727–735
synthesis, 691–695

Dipolarophiles (*Continued*)
münchnone cycloaddition reactions, 709–725
thioisomünchnone cycloaddition reactions, 739–742
nitrile oxide cycloadditions
mechanistic studies/calculations, 374–376
relative reactivity, 376–380
nitrile ylides, cycloaddition reactions, 499–502
nitronate structures
general reactivity, 106–114
intermolecular cycloadditions, stereoselectivity, 111–112
regioselectivity, 109–110
silyl nitronates
facial selectivity, 120–123
intermolecular cycloadditions, 118–120
intramolecular cycloadditions, 123–125
thiocarbonyl ylide structures, thiophene derivatives, five-membered ring structure, 333–339
Dipole cascade, azomethine ylide generation, 185–192
Dipole structures
diastereoselective asymmetric reactions
chiral dipoles, 822–835
azomethine ylides, 829–833
enantiopure applications, 834–835
nitrile oxides, 829
nitrones/nitronates, 822–829
nucleoside mechanisms, 834–835
diazoalkane cycloaddition reactions, 543–549
intramolecular cycloadditions, 600–601
ketocarbene dipoles, 604–609
nitrile oxide cycloaddition reactions
mechanistic studies and calculations, 374–376
relative reactivity, 376–380
Directed syntheses, carbonyl ylide cycloaddition, 278–284
Disilylamines, azomethine ylide generation, silicon protocols, 175–181
1,3-Dithiolanes, thiocarbonyl ylides
1,5-dipolar electrocyclization, five-membered heterocycles, 346–350
heterocyclic synthesis, 339–343
structural properties, 316
1,4-Dithianes, thiocarbonyl ylide heterocyclic synthesis, six-membered rings, 351
1,3-Dithiolium-4-olates, mesoionic ring systems, cycloaddition reactions, 742–744
Dittami protocol, carbonyl ylide, nonmetallocarbenoid/metallocarbenoid syntheses, 307–309
"Domino cycloaddition/*N*-acyliminium ion cyclization cascade," isomünchnones, 734–735
Double bonds
azide 1,3-dipolar cycloadditions, diene conjugation
heterocyclic synthesis, 637–639
natural product synthesis, 661–671
diazoalkane 1,3-dipolar cycloadditions
carbon-heteroatom double bonds, 557–576
C=As bonds, 567–570
C=N bonds, 557–559
C=P bonds, 559–567
C=S bonds, 570–575
C=Se and C=Te bonds, 575–576
heteroatom-heteroatom double bonds, P=X bonds, 576–578
nitrile imine [2 + 3]-cycloaddition reactions, exocyclic double bonds, 510–511

Electrocyclizations
carbonyl ylide structures, epoxide generation, 268–269
nitrile imines, 525–526
nitrile ylides, 517–525
alkene-conjugated ylides, 518–520
diene-conjugated ylides, 520–524
triene-conjugated ylides, 524–525
thiocarbonyl ylides
five-member heterocyclic synthesis, 345–350
intramolecular reactions, 324–325
Electrophiles, cyclic nitronate preparation, 135–136
Electrophilic capped carbonyl ylides, nonmetallocarbenoid/metallocarbenoid syntheses, 306–309
1,3-Elimination reactions, thiocarbonyl ylide generation, 320–321
Enamines
azide 1,3-dipolar cycloadditions, heterocyclic synthesis, 634–637
diazoalkane cycloaddition reactions, enaminoketones, 544–549
nitrile imine nucleophile reactions, 528–530
Enantiomer formation
asymmetric diastereoselectivity, nitrone/nitronate chiral dipoles, 827–829
azomethine ylide generation, chiral dipolarophiles, asymmetric reaction, 194–200
nitrile imine cycloadditions, 506–507
nitrile oxide cycloadditions, achiral olefins, 398–399

Enantiopure camphor imines, asymmetric Michael additions, 774–778
Enantioselectivity
  azomethine ylide 1,3-dipolar cycloadditions, asymmetric reactions, 778
  carbonyl ylide 1,3-dipolar cycloadditions, 805–806
    catalytic asymmetric reactions, 887–888
  carbonyl ylides, 298–302
  1,3-dipolar cycloadditions
    catalytic asymmetric nitrone reactions, 866–882, 868–882
    external reagents, 756–757
  nitrile oxide cycloadditions, 791–792
    catalyzed reactions, 792–793
  nitrone cycloadditions, catalyzed reactions, 804
Endocyclic reactions
  [4 + 2] cycloaddition, nitroalkenes, 136–137
  azomethine ylide generation
    asymmetric reaction, chiral dipolarophiles, 196–200
    metal-mediated reactions, 241–248
    silicon protocols, 175–181
  chiral ylides, asymmetric reactions, 200–218
  nitrile oxide intramolecular cycloadditions, regioselectivity, 407–410
  nitronate stereoselectivity, intramolecular cycloadditions, 112–114
  nitrone 1,3-dipolar cycloadditions, amine compounds, 39–47
  unsaturation, nitrone 1,3-dipolar cycloadditions, lactam formation, 10–11
Endo-isomers, asymmetric reactions, 1,3-dipolar cycloadditions, 820–822
Enolate compounds, α-(alkylideneamino)esters, reversible metalation, 763–768
Enone compounds, azide 1,3-dipolar cycloadditions, natural synthesis, 649–664
Epimerization reactions
  azomethine ylide 1,3-dipolar cycloadditions, external reagents, anti-selective Michael additions, 770–771
  enantiopure camphor imines, asymmetric Michael additions, 774–778
Epothilone compounds, nitrile oxide cycloadditions, magnesium ion-mediated reactions, 788–789
Epoxide compounds
  carbonyl ylide generation, 263–269
    nonmetallocarbenoid/metallocarbenoid syntheses, 307–309
  nitrile oxide intramolecular cycloadditions, "aldol" ring cleavage, isoxazoline intermediates, 448–452
Erythrose, nitrone 1,3-dipolar cycloadditions, 24–29
Eschenmoser-sulfide-contraction reaction, thiocarbonyl ylides, thiirane three-membered rings, 330–331
Esters
  amino esters
    α-(alkylideneamino)esters, reversible metalation, 763–768
    azomethine ylide condensation protocols, 228–237
  azomethine ylide 1,3-dipolar cycloadditions, asymmetric reactions, 773–774
  carbonyl ylide structures, 257–259
    derivative formations, 274–276
    directed syntheses, [1,3]-dipolar cycloadditions, 280–284
    enantioselective synthesis, 301–302
  diastereoselective reactions, α,β-unsaturated esters, 851–853
  nitrile oxide cycloadditions
    achiral olefins, 394–399
    magnesium ion-mediated reactions, 788–789
  nitrone 1,3-dipolar cycloadditions, transesterification-dipolar cycloadditions, 801–804
Ether compounds
  asymmetric reactions, chiral auxiliaries, 858–859
  carbonyl ylide generation
    directed syntheses, [1,3]-dipolar cycloadditions, 281–284
    nonmetallocarbenoid/metallocarbenoid syntheses, 306–309
  nitrile oxide generation, relative reactivity, 376–380
  nitrile oxide intramolecular cycloadditions, "aldol" ring cleavage, isoxazoline intermediates, 452
Ethyl (4*S*,2*E*)-3-(2,2-dimethyl-1,3-dioxolan-4-yl)acrylate, enantiopure camphor imines, asymmetric Michael additions, 776–778
N-Ethyl-*N'*-dimethylaminopropylcarbodiimide (EDC), mesoionic ring systems, münchnone synthesis, 682–687
Evan's auxiliary, azomethine ylide generation, asymmetric reaction, 198–200
Exchange reactions, silyl nitronates, 90–93

Exocyclic reactions
azomethine ylide generation
asymmetric reaction, chiral dipolarophiles, 196–200
metal-mediated reactions, 241–248
silicon protocols, 175–181
chiral ylides, asymmetric reactions, 200–218
1,3-dipolar cycloadditions, asymmetric reactions, alkene chiral centers, 842–844
nitrile imine cycloadditions, double bonds, 510–511
nitrile oxide intramolecular cycloadditions, regioselectivity, 407–410
nitronate stereoselectivity, intramolecular cycloadditions, 112–114
nitrone 1,3-dipolar cycloadditions, amine compounds, 38–47
Exo-dig processes, nitrone 1,3-APT cycloaddition, 49–54
External reagents, 1,3-dipolar cycloaddition reactions
azomethine ylides, 757–778
anti-selective Michael additions, 768–771
asymmetric cycloadditions, 772–774
2-aazaallyl anions, 759
enantiopure camphor imines, asymmetric Michael additions, 774–778
enantioselective additions, 778
metal n-alkylideneglycinates, 758–759
N-metalated reactions, 759–778
α-(alkylideneamino)ester reversibility, 763–768
(N-alkylideneamine)acetonitrile irreversibility, 761–763
carbonyl ylides, 804–806
nitrile oxides, 779–794
catalytic cycle, 787–788
chemoselectivity, 786
diastereocontrol, 780
enantioselective reactions, 793–794
kinetics, 784–786
Lewis acid-mediated diastereoselectivity, 789–790
magnesium ion-mediated reactions, 780–789
miscellaneous mediators, 790–793
reaction mechanisms, 779
regioselectivity, 780–782
related applications, 788–789
solvent effects, 782–784
nitrile ylide equivalents, 806–809
nitrones
magnesium ion-mediated reactions, 797–801
reaction mechanisms, 794–795
titanium ion-mediated reactions, 795–797
transesterification-dipolar cycloadditions, 801–804
structural properties, 756–757

Facial selectivity
alkyl nitronate cycloadditions, 146–148
azomethine ylide generation, chiral dipolarophiles, asymmetric reaction, 193–200
azomethine ylides
1,3-dipolar cycloadditions, asymmetric reactions, enantiopure camphor imines, asymmetric Michael additions, 778
metal-mediated reactions, 243–248
chiral ylides, stereocontrol, 202–208
diazoalkane cycloaddition reactions, 552–557
nitrile oxide cycloadditions
chiral cycloalkenes, 399–400
intramolecular cycloadditions, 410–416
silyl nitronates, 120–123
Five-atom tethers, nitrile imine intramolecular cycloadditions, 514–515
Five-membered rings
nitrile imine cycloadditions, 508–509
three-atom tethers, 512–513
nitrile oxide intramolecular cycloadditions, isoxazoline ring cleavage, 440–445
aminoalcohol reduction, 455–458
thiocarbonyl ylide heterocyclic synthesis
1,3-dithiolanes, 1,3-oxathioanes, and 1,3-thiazole derivatives, 339–343
1,5-dipolar electrocyclization, 345–350
heteroatoms, 344–345
thiophene derivatives, 331–339
Flash vacuum thermolysis
azomethine ylide
generation protocols, 183–192
intramolecular cycloadditions, aziridine precursors, 219–227
nitrile imine structure, 476–478
nitrile ylide structure, 474–475
Fluorine-mediated desilyation, azomethine ylide generation, 170–181
Formaldehyde, chiral ylides, 209–218
Four-atom tethers, nitrile imine intramolecular cycloadditions, 513–514
Four-membered rings, nitrile imine cycloadditions, 507–508
Frontier molecular orbital (FMO) theory
carbonyl ylide structures, 256–259
alkaloid preparation, 289–292
epoxide generation, 265–269

catalytic asymmetric 1,3-dipolar cycloadditions, 864–866
diazoalkane cycloaddition reactions
alkyne triple-bond systems, 582–586
reactivity and regioselectivity, 541–543
mesoionic ring systems
isomünchnones
cycloadditions, 727–735
synthesis, 692–695
münchnone cycloaddition reactions, 715–725
sydnone cycloadditions, 737–738
thioisomünchnone cycloaddition reactions, 739–742
nitrile oxide cycloadditions
dipolar/dipolarophile mehcanism and relativity, 375–376
olefin regioselectivity, 380–385
nitrile ylide structure, [2 + 3]-cycloaddition reactions, 497–502
nitronate structures
general reactivity, 106–114
regioselectivity, 108–110
nitrone 1,3-dipolar cycloadditions, 2–3
Fullerenes
azide 1,3-dipolar cycloadditions, heterocyclic synthesis, 646–649
diazoalkane cycloaddition reactions, 551–552
Fulleroproline compounds, chiral ylides, asymmetric reactions, 215–218
Fulvenes, mesoionic ring systems
isomünchnone cycloadditions, 726–735
münchnone cycloaddition reactions, 719–725
1,3,2-oxathiazolium-5-olate cycloadditions, 745
Furan-2-ylidene malonate, thiocarbonyl ylides, thiirane three-membered rings, 330–331
Furanones
asymmetric cycloaddition reactions, chiral auxiliaries, 857–860
diazoalkane cycloaddition reactions, 554–557
Furanoses, azide 1,3-dipolar cycloadditions, alkyne/nitrile synthesis, 675–676
Furans
mesoionic ring systems, isomünchnones
cycloadditions, 730–735
synthesis, 691–695
nitrile oxide intermolecular cycloadditions
aminoalcohol ring cleavage, 429–431
cycloaddition-reduction-hydrolysis sequence, 419
nitrile oxide intramolecular cycloadditions, 416
nitrile ylides, electrocyclizations, 518–520
Furoxans
alkyl nitronate intermolecular cycloadditions, 145
intramolecular nitrile oxide cycloadditions, 437–439
thermolysis, nitrile oxide generation, 371–374
Fused-mode cycloadditions, alkyl nitronates, intramolecular cycloadditions, 150–153

Galactose compounds, azide 1,3-dipolar cycloadditions, natural synthesis, 663–664
Ge=Ge bonds, diazoalkane cycloaddition reactions, 578
General reactivity. *See also* Relative reactivity
alkyl nitronates, intermolecular cycloadditions, 139–145
diazoalkane cycloaddition reactions, 541–543
nitronates, 105–114
regioselectivity, 108–110
stereoselectivity, 111–114
intermolecular cycloadditions, 111–112
intramolecular cycloadditions, 112–114
mechanisms, 114
silyl nitronates, intermolecular cycloadditions, 117–120
Glucose
azide 1,3-dipolar cycloadditions
alkyne/nitrile synthesis, 674–676
natural synthesis, 663–664
nitrile oxide intramolecular cycloadditions, "aldol" ring cleavage, isoxazoline intermediates, 441–445
Glycine imines, azomethine ylide 1,3-dipolar cycloadditions, external reagents, *N*-alkylideneglycinates, 758–759
Grignard reagents, nitrile oxide cycloadditions
chiral dipolarophiles, 392–393
olefin regioselectivity, 384–385
Guanidine compounds, nitrone 1,3-dipolar cycloadditions, 44–47

Hashimoto protocol
carbonyl ylide generation, enantioselective synthesis, 300–302
zaragozic acid carbonyl ylide derivatives, 294–296
Head-to-head dimerization, thiocarbonyl ylide heterocyclic synthesis, six-membered rings, 351
Head-to-tail dimerization, thiocarbonyl ylide heterocyclic synthesis, six-membered rings, 351

Hernaldulcin, nitrile oxide intramolecular cycloadditions, "aldol" ring cleavage, isoxazoline intermediates, 445–452
Heteroarsoles, diazoalkane cycloadditions, carbon-heteroatom C=As double bonds, 568–570
Heteroatomic structures
- asymmetric reactions, dienophiles, 844–845
- diazoalkane cycloaddition reactions
  - carbon-heteroatom double bonds, 557–576
    - C=As bonds, 567–570
    - C=N bonds, 557–559
    - C=P bonds, 559–567
    - C=S bonds, 570–575
    - C=Se and C=Te bonds, 575–576
  - heteroatom-heteroatom double bonds, 576–578
  - heteroatom-heteroatom triple bonds, 586–587
- nitrone 1,3-APT cycloaddition, 52–54
- thiocarbonyl ylides, five-membered rings, 344–345

Heterocumulenes
- diazoalkane cycloaddition reactions, 578–581
- thiocarbonyl ylides, heterocyclic synthesis, 341–343

Heterocyclic synthesis
- azide 1,3-dipolar cycloadditions
  - alkene reactions, 624–637
  - alkynes and nitriles, 639–644
  - allylic carbocations, 644–645
  - dienes double-bond and conjugated systems, 637–639
  - fullerene reactions, 646–649
  - phosphorus ylides, 645–646
- diazoalkane cycloaddition reactions, ketocarbene dipoles, 604–609
- mesoionic ring systems
  - isomünchnone cycloadditions, 728–735
  - miscellaneous compounds, 745–747
- nitrile imine cycloadditions, 507
- thiocarbonyl ylides
  - three-membered thiirane rings, 329–331
  - five-membered rings
    - 1,3-dithiolanes, 1,3-oxathioanes, and 1,3-thiazole derivatives, 339–343
    - 1,5-dipolar electrocyclization, 345–350
    - heteroatoms, 344–345
    - thiophene derivatives, 331–339
  - six-membered rings, 351
  - seven-membered rings, 351–355

Heterodipolarophiles
- carbonyl ylide generation, ketone derivation, 272–273
- silyl nitronates, dipolar cycloadditions, 119–120
- thiocarbonyl ylides, intermolecular reactions, 327

Heterophospholes, diazoalkane cycloaddition reactions
- carbon-heteroatom C=P bonds, 565–567
- triple bond systems, 588–592

HF/6-31G* calculations, nitronate structures, intermolecular cycloadditions, stereoselectivity, 111–112
Highest occupied molecular orbital (HOMO)
- azomethine ylide generation
  - intramolecular cycloadditions, 235–237
  - silicon protocols, 171–181
- carbonyl ylide structures, 256–259
  - alkaloid preparation, 290–292
  - epoxide generation, 265–269
- catalytic asymmetric 1,3-dipolar cycloadditions, 864–866
- chiral ylides, asymmetric reactions, 210–218
- diazoalkane cycloaddition reactions
  - alkyne triple-bond systems, 582–586
  - reactivity and regioselectivity, 541–543
- mesoionic ring systems
  - isomünchnone cycloaddition, 727–735
  - isomünchnone synthesis, 692–695
  - münchnone cycloaddition reactions, 709–725
- nitrile oxide cycloadditions, dipolar/dipolarophile mehcanism and relativity, 375–376
- nitrile ylide structure, [2 + 3]-cycloaddition reactions, 497–502
- nitronate structures
  - general reactivity, 106–114
  - regioselectivity, 108–110
- thiocarbonyl ylide heterocyclic synthesis, seven-membered ring structure, 351–354

(–)-Histrionicotoxin, nitrone 1,3-dipolar cycloadditions, 41–47
^{1}H NMR. *See* Proton nuclear magnetic resonance (^{1}H NMR)
Hodgson protocol, carbonyl ylide structures
- enantioselective synthesis, 299–302
- zaragozic acid carbonyl ylide derivatives, 295–296

Homochiral compounds, nitrile imine cycloadditions, five-atom tether, 515
Houk-Jäger model, nitrile oxide cycloadditions, chiral dipolarophiles, 390–393
Hydrazones, oxidation, transient nitrile imines, 494–495
Hydrazonyl chlorides, nitrile imines
- nucleophile reactions, 527–530

transient formation, 492–494
Hydrochloric acid, thiocarbonyl ylide generation, 1,3-elimination reactions, 321
1,4-Hydrogen shift, thiocarbonyl ylide intramolecular reactions, 325–326
Hydrogen bonding, nitrile oxide cycloadditions
chiral dipolarophiles, 390–393
olefin regioselectivity, 384–385
Hydroximoyl chloride dehydrohalogenation, nitrile oxide generation, 364–368
one-pot oxime formation and cycloaddition, 367–368
oxime halogenation, 365–367
in situ generation, 367
Hydroximoyl halide, nitrile oxide generation, 363–364
β-Hydroxy-α-amino acids, chiral ylides, asymmetric reactions, 213–218
β-Hydroxy carbonyl compounds, nitrile oxide intermolecular cycloadditions, "aldol" ring cleavage, 417–425
1-[2-Hydroxyethoxy-methyl]-6-(phenylthio)thymine (HEPT), azide 1,3-dipolar cycloadditions, alkyne/nitrile reactions, 642–644
β-Hydroxy ketone
isoxazolidine cyclization, 128–130
nitrile oxide intramolecular cycloadditions, intermediate isoxazoline ring cleavage, 439–454
nitroso acetal functionalization, 126–127
Hydroxylamines, intermolecular nitrone cycloaddition reactions, 64–68
3-Hydroxy-4-methylproline, nitrone 1,3-dipolar cycloadditions, 18–23
Hypobromite, nitrile oxide generation, oxime one-pot operations, 368

Illudin compounds, carbonyl ylide preparation, 292–294
Imidate compounds, nitrile ylide synthon formation, 480–485
Imidosulfoxides, mesoionic ring systems, isomünchnone synthesis, 692–695
Imidoyl chlorides, nitrile ylides
cycloaddition reactions, 500–502
dehydrochlorination, 490–491
electrocyclizations, diene-conjugated compounds, 521–524
Imine compounds
azide 1,3-dipolar cycloadditions, natural synthesis, 655–664
azomethine ylide generation
asymmetric reaction, 196–200
diastereoselective asymmetric reactions, 834–835
intramolecular cycloadditions, 228–237
enantiopure camphor imines, asymmetric Michael additions, 774–778
nitrile imines
[2 + 3]-cycloaddition reactions, 502–517
four-membered rings, 507–508
five-membered rings, 508–509
six-membered rings, 509–510
seven-membered rings, 510
exocyclic double bonds, 510–511
heterocyclic rings, 507
intramolecular cycloadditions, 511–516
three-atom tether, 512–513
our-atom tether, 513–514
five-atom tether, 514–515
six-or-greater atom tether, 515–516
selectivity, 505–507
stabilized nitrile imines, 502–504
tandem intermolecular-intramolecular cycloadditions, 516–517
electrocyclization reactions, 525–526
generation mechanisms
stabilized imines, 495–496
transient imines, 492–495
nucleophile reactions, 527–530
physical and spectroscopic properties, 475–478
nitrile ylide synthons, 479–485
Imino acids
azide 1,3-dipolar cycloadditions, natural synthesis, 653–664
mesoionic ring systems, münchnone cycloaddition reactions, 719–725
nitrile oxide intermolecular cycloadditions, isoxazoline route, 431–437
Iminopolyols, nitrile oxide intermolecular cycloadditions, isoxazoline route, 431–437
Immunosuppresive drugs, asymmetric diastereoselectivity, nitrone/nitronate chiral dipoles, 823–829
INDO calculations, nitronate structures
intermolecular cycloadditions, stereoselectivity, 111–112
regioselectivity, 108–110
Indolium salts, azomethine ylide generation, 239
Indolizidines
azide 1,3-dipolar cycloadditions
diene double bond reactions, 667–671
natural synthesis, 655–664

Indolizidines (*Continued*)
azomethine ylide generation, silicon protocols, 176–181
nitrone 1,3-dipolar cycloadditions, 12–18
Indoloquinones, azomethine ylide generation, 238–239
Infrared (IR) absorbances
alkyl nitronates, 97–102
nitroso acetals, 102–105
silyl nitronates, 89–90
Inside alkoxy effect
nitrile oxide cycloadditions
chiral dipolarophiles, 387–393
intramolecular cycloadditions, stereoselectivity, 413–416
nitrone 1,3-dipolar cycloadditions, 35–37
In situ techniques
azomethine ylide cycloadditions, asymmetric 1,3-dipolar cycloadditions, 833
mesoionic ring systems
isomünchnone cycloadditions, 731–735
münchnone cycloaddition reactions, 704–725
nitrile imines, transient formation, 493–494
nitrile oxide cycloadditions
dehydrohalogenation, hydroximoyl chloride, 367
Lewis acid-mediated diastereoselective reactions, 789–790
preparation, 362–363
Intermolecular cycloadditions
acyclic nitronate preparation, 131–132
alkyl nitronates, 138–145
carbonyl ylide generation
alkaloid systems, 284–292
amide derivations, 276–278
enantioselective synthesis, 300–302
ketone derivation, 269–273
mesoionic ring systems, isomünchnones, 727–735
nitrile oxides, 416–437
aminoalcohol ring cleavage, 425–431
isoxazoline compounds
"aldol" ring cleavage, intermediates, 417–425
achiral nitrile oxides/chiral olefins, 419–424
achiral nitrile oxides/olefins, 418–419
chiral nitrile oxides/achiral olefins, 424–425
mono- and bicyclic imino acids and polyols, 431–437
natural products, 416–417
nitronate stereoselectivity, 111–112
nitrone isoxazolidines, 59–68
silyl nitronates, 117–120
thiocarbonyl ylides, 326–329
1,3-addition reactions, 327–328
[3 + 2] cycloadditions, 326–327
dimerization reactions, 328–329
Intramolecular cycloadditions
acyclic nitronate preparation, 134
alkyl nitronates, 148–153
asymmetric reactions, diastereoselectivity, 845–851
azide dipoles, 850–851
nitrile oxides, 848–850
nitrones/nitronates, 845–848
azomethine ylides, 219–239
alternative generation procedures, 237–239
aziridine precursors, 219–227
decarboxylative and amino ester condensation protocols, 228–237
external reagents, anti-selective Michael additions, 770–771
carbonyl ylide generation
alkaloid systems, 284–292
amide/urea derivation, 276–278
enantioselective synthesis, 299–302
ester derivatives, 275–276
ketone derivation, 269–273
tropolone and benzotropolone formation and reactivity, 296–298
diastereoselectivity, asymmetric reactions, 845–851
azide dipoles, 850–851
nitrile oxides, 848–850
nitrones/nitronates, 845–848
diazoalkanes
C=C bonds, 592–601
C≡C bonds, 601–604
multiple bond reactions, 604
mesoionic ring systems
isomünchnones, 691–695, 732–735
münchnone cycloaddition reactions, 710–725
nitrile imines, 511–516
three-atom tether, 512–513
our-atom tether, 513–514
five-atom tether, 514–515
six-or-greater atom tether, 515–516
nitrile oxides
regioselectivity, 407–410
stereoselectivity, 410–416
synthesis applications, 437–461
isoxazoline ring cleavage, 439–461
"aldol" reduction-hydrolysis, 440–454
γ-aminoalcohol, 454–458

isoxazolinium salts, transformation with, 458–461
polycyclic isoxazolines and furoxans, 437–439
nitronate stereoselectivity, 112–114
nitrone 1,3-dipolar cycloadditions, pyrrolidines and piperidines, 37–47
nitrone-alkene cycloadditions (INAC), 55–59
oxime-olefin cycloadditions (IOOC), 54–55
silyl nitronates, 123–125
thiocarbonyl ylides, 324–326
electrocyclizations, 324–325
1,4-hydrogen shift, 325–326
thiophene derivatives, five-membered ring structure, 336–339
Intramolecular nitrile oxide cycloadditions (INOC), origins of, 363
Inverse electron demand
catalytic asymmetric 1,3-dipolar cycloadditions, 864–866
catalytic asymmetric nitrone reactions, 876–882
Iodohydrin, carbonyl ylide generation, 260–262
Ipalbidine, mesoionic ring systems, isomünchnone cycloadditions, 729–735
Irreversible metalation, (*N*-alkylideneamino) acetonitriles, 761–763
Isocyanates
mesoionic ring systems, isomünchnone cycloadditions, 731–735
nitrile ylide 1,3-dipolar cycloadditions, 807–809
Isoindolediones, mesoionic ring systems, münchnone cycloaddition reactions, 710–725
Isomer formation, azomethine ylide generation, chiral dipolarophiles, asymmetric reaction, 194–200
Isomerization barrier, silyl nitronates, 91–93
Isomünchnones
carbonyl ylide cycloaddition, alkaloid preparation, 285–292
diastereoselective asymmetric reactions, dipole structures, 834–835
mesoionic ring systems
cycloaddition reactions, 725–735
synthesis reactions, 687–695
Isonorborneol-tethered chiral dipolarophile, azomethine ylide generation, asymmetric reaction, 199–200
Isoquinolines
azide 1,3-dipolar cycloadditions, heterocyclic synthesis, 625–637
mesoionic ring systems, münchnone cycloaddition reactions, 710–725
nitrile imine cycloadditions, six-membered rings, 509–510
triene-conjugated nitrile ylides, electrocyclizations, 525
Isoquinolone compounds, azide 1,3-dipolar cycloadditions, heterocyclic synthesis, 632–637
Isoselenocyanates, diazoalkane cycloaddition reactions, heterocumulenes, 580–581
Isothiazole dioxides
azide 1,3-dipolar cycloadditions, double-bonded diene conjugation, 637–639
mesoionic ring systems
münchnone cycloaddition reactions, 714–725
1,3,2-oxathiazolium-5-olate cycloadditions, 745
Isothiocyanates, thiocarbonyl ylides, 1,5-dipolar electrocyclization, five-membered heterocycles, 347–350
Isoxazolidinylthymines, nitrone 1,3-dipolar cycloadditions, nucleoside antibiotics, 5–8
Isoxazolidines
applications, 127–130
nitrile oxide cycloadditions, achiral olefins, 396–399
nitrone 1,3-dipolar cycloadditions, magnesium ion-mediated reactions, 798–801
nitrone cycloaddition reactions, 47–68
1,3-dipolar cycloadditions, 2–3
lactam formation, 9–11
nucleosides, 4–8
sugars, 24–29
1,3-APT process, 48–54
intermolecular reactions, 59–68
intramolecular cycloadditions
nitrone-alkene (INAC), 55–59
oxime-olefin, 54–55
nitroso acetal functionalization, 125–127
silyl nitronates, nitroso acetals, 93–95
Isoxazoline compounds
alkyl nitronates, nitroso acetal functionalization, 153–154
nitrile oxide cycloadditions
chiral dipolarophiles, 391–393
enantiopure compounds, 792
external reagents, 779–794
intermolecular cycloadditions
"aldol" ring cleavage, intermediates, 417–425
achiral nitrile oxides/achiral olefins, 418–419
achiral nitrile oxides/chiral olefins, 419–424

Isoxazoline compounds (*Continued*)
chiral nitrile oxides/achiral olefins, 424–425
chiral nitrile oxides/chiral olefins, 425
aminoalcohol ring cleavage, 425–431
mono- and bicyclic imino acids and polyols, 431–437
natural products, 416–417
intramolecular cycloadditions
isoxazolinium salt transformation, 458–461
polycyclic synthesis, 437–439
reduction to γ-aminoalcohol, 454–458
ring cleavage, intermediate structures, 439–454
magnesium ion-mediated reactions, 781–782
olefin regioselectivity, 384–385
nitroso acetal functionalization, 125–127
silyl nitronates, applications, 128–130
Isoxazolinium salts, nitrile oxide intramolecular cycloadditions
intermediate isoxazoline ring cleavage, 439–454
isoxazoline transformation, 458–461

Kainoids, azomethine ylide generation, intramolecular cycloadditions, 225–227
Kanemasa protocol, nitrile oxide cycloadditions, chiral dipolarophiles, 391–393
Kemp's triacid, nitrile oxide cycloadditions, achiral olefins, 395–399
Ketene imines, thiocarbonyl ylide heterocyclic synthesis, seven-membered ring structure, 354
Ketenes
azide 1,3-dipolar cycloadditions, heterocyclic synthesis, 627–637
diazoalkane cycloaddition reactions, heterocumulenes, 581
Ketocarbene dipoles, diazoalkane cycloaddition reactions, 604–609
Ketones
carbonyl ylide generation, 269–273
directed syntheses, [1,3]-dipolar cycloadditions, 280–284
nitrone 1,3-dipolar cycloadditions, reaction mechanisms, 795
Kinetic data, nitrile oxide cycloadditions, magnesium ion-mediated reactions, 784–786

Lactams
asymmetric reactions, 1,3-dipolar cycloadditions, allyl amine derivatives, 838–839
azide 1,3-dipolar cycloadditions, heterocyclic synthesis, 628–637
azomethine ylide generation
intramolecular cycloadditions, 232–237
silicon protocols, 181
catalytic asymmetric nitrone reactions, 881–882
chiral ylides
asymmetric reactions, 217–218
stereocontrol, 203–208
mesoionic ring systems, isomünchnone synthesis, 689–695
nitrone 1,3-dipolar cycloadditions, 8–11
Lactones
chiral ylides, asymmetric reactions, allyl alcohol derivatives, 837–838
intermolecular nitrone cycloaddition reactions, 62–68
nitrile oxide intermolecular cycloadditions, achiral nitrile oxides to chiral olefin, 420–424
nitrile oxide intramolecular cycloadditions, "aldol" ring cleavage, 453–454
nitrone 1,3-dipolar cycloadditions, sugars, 26–29
Laser flash photolysis (LFP)
nitrile imine structure, 477
nitrile ylide structure
azirine photolysis, 489–490
carbene reactions, 486–488
[2 + 3]-cycloaddition reactions, 497–502
nucleophile reactions, 526–527
Lasubine II, nitrone 1,3-dipolar cycloadditions, 13–18
Lepadiformine, nitrone 1,3-dipolar cycloadditions, 43–47
Lewis acids
azomethine ylide 1,3-dipolar cycloadditions, external reagents, α-(alkylideneamino) esters, reversible metalation, 763–768
carbonyl ylide 1,3-dipolar cycloadditions, 804–806
catalytic asymmetric 1,3-dipolar cycloadditions, 864–866
nitrile oxide cycloadditions
diastereoselective reactions, 789–790
enantioselective reactions, 792–793
nitrone 1,3-dipolar cycloadditions, magnesium ion-mediated reactions, 799–801
nitrone cycloadditions
reaction mechanisms, 794–795
titanium ion-mediated reactions, 795–797
Ligand effects, carbonyl ylide generation, ketone derivation, 272–273

Lithium compounds
azomethine ylide 1,3-dipolar cycloadditions, external reagents
α-(alkylideneamino)esters, reversible metalation, 763–768
anti-selective Michael additions, 768–771
asymmetric cyloadditions, 772–774
2-azaallyl anions, 759
enantiopure camphor imines, asymmetric Michael additions, 775–778
diazoalkane cycloaddition reactions, heterocumulenes, 578–581
lithium diisopropylamide (LDA)
(*N*-alkylideneamino)acetonitriles, irreversible metalation, 761–763
azomethine ylide generation, 184–192
diazoalkane cycloaddition reactions, 544–549
lithium enolate reagent, intramolecular nitrone-alkene cycloadditions (INAC), 58–59
nitrile imines, hydrazone oxidation, 494–495
nitrile oxide intermolecular cycloadditions, aminoalcohol ring cleavage, 425–431
Lowest unoccupied molecular orbital (LUMO)
azomethine ylide generation
intramolecular cycloadditions, 235–237
silicon protocols, 171–181
carbonyl ylide structures, 257–259
alkaloid preparation, 290–292
epoxide generation, 265–269
catalytic asymmetric 1,3-dipolar cycloadditions, 864–866
chiral ylides, asymmetric reactions, 210–218
diazoalkane cycloaddition reactions
alkyne triple-bond systems, 582–586
reactivity and regioselectivity, 541–543
mesoionic ring systems
isomünchnone cycloaddition, 727–735
münchnone cycloaddition reactions, 703–725
nitrile oxide cycloadditions, dipolar/dipolarophile mechanism and reactivity, 375–386
nitrile ylide structure, [2 + 3]-cycloaddition reactions, 497–502
nitronate structures
general reactivity, 106–114
regioselectivity, 108–110
thiocarbonyl ylide heterocyclic synthesis, seven-membered ring structure, 351–354
Lupinine, mesoionic ring systems, isomünchnone synthesis, 693–695
Lycorane, azide 1,3-dipolar cycloadditions, natural product synthesis, 658–664
Lycorine, carbonyl ylide alkaloid preparation, 288–292
Macrocyclic compounds, nitrile oxide intramolecular cycloadditions, "aldol" ring cleavage, 453–454
Macrolides, nitrile oxide intramolecular cycloadditions, "aldol" ring cleavage, isoxazoline intermediates, 447–452
Magnesium alkoxides, nitrile oxide cycloadditions, magnesium ion-mediated reactions, kinetic data, 784–786
Magnesium bromide, nitrile oxide cycloadditions, achiral olefins, 398–399
Magnesium complexes, catalytic asymmetric nitrone reactions, 870–882
Magnesium ion-mediated reactions
nitrile oxide cycloadditions
applications, 788–789
catalytic cycle, 787–788
chemoselectivity, 786
diastereocontrol, 780
kinetics, 784–786
regiocontrol, 780–782
solvent effects, 782–784
nitrone 1,3-dipolar cycloadditions, 797–801
Mander's reagent, carbonyl ylide alkaloid preparation, 291–292
Manganese complexes, azomethine ylide catalytic asymmetric reactions, 885–886
Mannich cyclization, isomünchnone cycloaddition, 734–735
Mannoside compounds, nitrile oxide intramolecular cycloadditions, aminoalcohol reduction, 456–458
Manzamine A, azomethine ylide intramolecular cycloadditions, 236–237
Menthol groups, diastereoselective reactions, α,β-unsaturated esters, 851–853
Menthol-tethered chiral dipolarophile, azomethine ylide generation, asymmetric reaction, 199–200
Menthone compounds, 1,3-dipolar cycloadditions, chiral auxiliaries, 861–863
Merck protocol, zaragozic acid carbonyl ylide derivatives, 294–296
(–)-Mesembrine, azomethine ylide generation, intramolecular cycloadditions, 225–227
Mesoionic ring systems
classification, 681–682
cycloaddition reactions
1,3-dithiolium-4-olates, 742–744
isomünchnones, 725–735
miscellaneous mesoionic heterocycles, 745–746
münchnones, 700–725
1,3-oxathiazolium-5-olates, 745

Mesoionic ring systems (*Continued*)
1,3-oxathiolium-4-olates, 744
oxamünchnones, 742
sydnones, 735–738
1,3-thiazolium-4-thiolates, 742
thioisomünchnones, 739–742
thiomünchnones, 738
synthesis
isomünchnones, 687–695
münchnones, 682–687
oxamünchnones, 699–700
sydnones, 696
1,3-thiazolium-4-thiolates, 698–699
thioisomünchnones, 697–698
thiomünchnones, 696–697
Mesitylnitrile oxide, nitrile oxide cycloadditions, magnesium ion-mediated reactions, 788–789
Metallocarbenoids, carbonyl ylide generation, 269
novel synthetic processes, 306–309
Metal-mediated reactions
azomethine ylides, 240–248
1,3-dipolar cycloadditions, external reagents, 759–778
α-(alkylideneamino)ester reversibility, 763–768
(N-alkylideneamine)acetonitrile irreversibility, 761–763
N-alkylideneglycinates, 758–759
diazoalkane ketocarbene dipoles, 607–609
Methoxymethyl (MOM) groups, nitrone 1,3-dipolar cycloadditions, 12–13
Methyi *(E)*-4-oxo-2-propenoate, azomethine ylide 1,3-dipolar cycloadditions, asymmetric reactions, 772–774
Methyl acrylate
azomethine ylides
1,3-dipolar cycloadditions, external reagents, 759–778
α-(alkylideneamino)ester reversibility, 763–768
(N-alkylideneamine)acetonitrile irreversibility, 761–763
N-alkylideneglycinates, 758–759
generation, 186–192
diazoalkane cycloaddition reactions, 547–549
nitrile oxide cycloadditions, olefin regioselectivity, 380–385
nitrile ylide electrocyclizations, 519–520
silyl nitronates, intermolecular cycloadditions, 117–120
Methylene compounds
nitrile oxide cycloadditions, olefin regioselectivity, 384–385
nitrone 1,3-dipolar cycloadditions, lactam formation, 10–11
N-α-Methylbenzyl substituent, asymmetric diastereoselectivity, nitrone/nitronate chiral dipoles, 822–829
Methyl(naphthylideneamino)acetate, α-(alkylideneamino)esters, reversible metalation, 765–768
Methyl propiolate
azomethine ylide generation, 186–192
mesoionic ring systems
münchnone cycloaddition reactions, 706–725
sydnone cycloadditions, 735–738
Michael additions
asymmetric reactions, enantiopure camphor imines, 774–778
azomethine ylides, 1,3-dipolar cycloadditions
anti-selective external reagents, 768–771
asymmetric reactions, enantiopure camphor imines, 774–778
silyl nitronates, intramolecular cycloadditions, 124–125
Mintlactone, nitrile oxide intramolecular cycloadditions, "aldol" ring cleavage, isoxazoline intermediates, 446–452
Mitomycin A, azide 1,3-dipolar cycloadditions, natural synthesis, 652–664
Mitsunobu conditions
acyclic nitronate preparation, 132
cyclic nitronate preparation, 134–136
Molecular structure, chiral ylides, stereocontrol, 200–208
Mono-imino acids and polyols, nitrile oxide intermolecular cycloadditions, isoxazoline route, 431–437
Morita-Baylis-Hillman adducts, nitrile oxide cycloadditions, chiral dipolarophiles, 390–393
Morphinane compounds, azide 1,3-dipolar cycloadditions, heterocyclic synthesis, 630–636
Morpholinone
azomethine ylide generation, 187–192
chiral ylides, asymmetric reactions, 214–218
Münchnones, mesoionic ring systems
cycloaddition reactions, 700–725
synthesis mechanisms, 682–687
Muscone compounds, nitrile oxide intramolecular cycloadditions, "aldol" ring cleavage, 454

Naphthoquinone, azide 1,3-dipolar cycloadditions,
heterocyclic synthesis, 631–637
Naphthyridinomycin, azomethine ylide
generation, asymmetric reaction, 199–200
Naringinase, azide 1,3-dipolar cycloadditions,
alkyne/nitrile synthesis, 675–676
(+)-Negamycin, nitrone 1,3-dipolar
cycloadditions, 18–23
Neurokinin-A, nitrone 1,3-dipolar cycloadditions,
20–23
Nitrile compounds
azide 1,3-dipolar cycloadditions
heterocyclic synthesis, 639–644
natural synthesis, 672–676
azomethine ylide generation, silicon protocols,
171–181
carbonyl ylide generation, 277–278
diazoalkane cycloaddition reactions, alkyne
triple-bond systems, 586
thiocarbonyl ylides, heterocyclic synthesis, 343
Nitrile imines
asymmetric reactions, 1,3-dipolar
cycloadditions, 819–822
[2 + 3]-cycloaddition reactions, 502–517
four-membered rings, 507–508
five-membered rings, 508–509
six-membered rings, 509–510
seven-membered rings, 510
exocyclic double bonds, 510–511
heterocyclic rings, 507
intramolecular cycloadditions, 511–516
three-atom tether, 512–513
our-atom tether, 513–514
five-atom tether, 514–515
six-or-greater atom tether, 515–516
selectivity, 505–507
stabilized nitrile imines, 502–504
tandem intermolecular-intramolecular
cycloadditions, 516–517
electrocyclization reactions, 525–526
generation mechanisms
stabilized imines, 495–496
transient imines, 492–495
nucleophile reactions, 527–530
physical and spectroscopic properties, 475–478
Nitrile oxides
asymmetric reactions
alkene chiral centeres, 843–844
diastereoselectivity, 829
1,3-dipolar cycloadditions, 819–822
catalytic reactions, 882–884
intramolecular cycloadditions, 848–849
cycloaddition reactions, 374–406
diastereoselectivity, 385–406
achiral olefins, with chiral auxiliaries,
393–399
chiral dipolarophiles, 386–393
facial selectivity, chiral alkenes, 399–400
olefin additions, chiral nitrile oxides,
400–406
dipole/dipolarophiles
mechanistic studies and calculations,
374–376
relative reactivity, 376–380
olefin regioselectivity, 380–385
diastereoselective reactions, α,β-amides,
856–857
external reagents, 1,3-dipolar cycloaddition
reactions, 779–794
enantioselective reactions, 793–794
Lewis acid-mediated diastereoselectivity,
789–790
magnesium ion-mediated reactions,
780–789
catalytic cycle, 787–788
chemoselectivity, 786
diastereocontrol, 780
kinetics, 784–786
regioselectivity, 780–782
related applications, 788–789
solvent effects, 782–784
miscellaneous mediators, 790–793
reaction mechanisms, 779
generation methods, 363–374
aldoxime oxidation, 370–371
furoxan thermolysis, 371–374
hydroximoyl chloride dehydrohalogenation,
364–368
one-pot oxime formation and
cycloaddition, 367–368
oxime halogenation, 365–367
in situ generation, 367
nitroalkane dehydration, 368–370
intermolecular cycloadditions, 416–437
aminoalcohol ring cleavage, 425–431
isoxazoline compounds
"aldol" ring cleavage, intermediates,
417–425
achiral nitrile oxides/chiral olefins,
419–424
achiral nitrile oxides/olefins, 418–419
chiral nitrile oxides/achiral olefins,
424–425
mono- and bicyclic imino acids and
polyols, 431–437
natural products, 416–417

Nitrile oxides (*Continued*)
  intramolecular cycloadditions
    regioselectivity, 407–410
    stereoselectivity, 410–416
    synthesis applications, 437–461
      isoxazoline ring cleavage, 439–461
        "aldol" reduction-hydrolysis, 440–454
        γ-aminoalcohol, 454–458
        isoxazolinium salts, transformation with, 458–461
      polycyclic isoxazolines and furoxans, 437–439
  nitroso acetal functionalization, 126–127
  structural chemistry, 362–363
Nitrile ylides
  [2 + 3]-cycloaddition reactions, 496–502
  1,3-dipolar cycloadditions, 806–809
  electrocyclization reactions, 517–525
    alkene-conjugated ylides, 518–520
    diene-conjugated ylides, 520–524
    triene-conjugated ylides, 524–525
  generation mechanisms, 478–492
    azirine photolysis, 488–490
    carbene reactions, 485–488
    imidoyl chloride dehydrochlorination, 490–491
    synthon generation, 479–485
  imidoyl chloride dehydrochlorination, 490–491
  nucleophile reactions, 526–527
  physical and spectroscopic properties, 474–475
Nitroacetals
  mesoionic ring systems, münchnone cycloaddition reactions, 715–725
  nitrile oxide intermolecular cycloadditions, achiral nitrile oxides to chiral olefin, 421–424
Nitroalkanes
  acyclic nitronate preparation, 133–134
  nitrile oxide cycloadditions, chiral nitrile oxides to olefins, 400–406
  nitrile oxide generation, 363–364
    dehydration mechanisms, 368–370
  silyl nitronate preparation, 114–117
Nitroalkenes, [4 + 2] cycloaddition, 136–137
Nitroindole compounds, mesoionic ring systems, münchnone cycloaddition reactions, 716–725
Nitromethane
  alkyl nitronates, intermolecular cycloadditions, 138–145
  silyl nitronate preparation, 114–117
Nitronates
  alkyl nitronates
    acyclic nitronate preparation, 130–134
      acylation, 133–134
      alkylation, 130–132
    applications, 154–158
    cyclic nitronate preparation, 134–137
      alkylation, 134–136
      nitroalkene [4 + 2] cycloaddition, 136–137
      radical cyclization, 137
    cycloadditions, 138–153
      facial selectivity, 146–148
      intermolecular, 138–145
      intramolecular, 148–153
    nitroso acetal functionalization, 153–154
    structural properties
      bonding, 96–97
      nitroso acetals, spectroscopic data, 102–105
      spectroscopic data, 97–102
      stability, 95–96
  asymmetric reactions
    chiral auxiliaries, 863
    diastereoselectivity, chiral dipoles, 822–829
    intramolecular cycloadditions, 845–848
  classification, 84–86
  future research, 159
  general reactivity, 105–114
    regioselectivity, 108–110
    stereoselectivity, 111–114
      intermolecular cycloadditions, 111–112
      intramolecular cycloadditions, 112–114
      mechanisms, 114
  silyl nitronates
    applications, 127–130
    cycloadditions, 117–125
      facial selectivity, 120–123
      intermolecular cycloadditions, 117–120
      intramolecular cycloadditions, 123–125
    nitroso acetal functionalization, 125–127
    preparation, 114–117
    structural properties
      bonding, 87–88
      exchange mechanisms, 90–93
      nitroso acetals, spectroscopic data, 93–95
      spectroscopy, 88–90
      stability, 86–87
Nitrones
  asymmetric reactions
    catalytic asymmetric 1,3-dipolar cycloadditions, 866–882

chiral auxiliaries, 858–859
diastereoselectivity, chiral dipoles, 822–829
1,3-dipolar cycloadditions, 819–822
catalytic reactions, 864–866
chiral auxiliaries, 861–863
intramolecular cycloadditions, 845–848
cycloaddition reactions
1,2-dipolar cycloaddition, 1–3
catalytic cycloadditions, 34
isoxazolidines, 47–68
1,3-APT process, 48–54
intermolecular reactions, 59–68
intramolecular cycloadditions
nitrone-alkene (INAC), 55–59
oxime-olefin, 54–55
lactams, 8–11
natural product formation, 3
nucleosides, 4–8
peptides and amino acids, 18–23
pyrrolidines, piperidines, and amines, 34–47
quinolizidines, indolizidines, and pyrrolizidines, 12–18
sugars, 24–29
sulfur- and phosphorus-containing compounds, 29–33
diastereoselective reactions
α,β-amides, 853–857
α,β-unsaturated esters, 852–853
1,3-dipolar cycloaddition reactions, external reagents
magnesium ion-mediated reactions, 797–801
reaction mechanisms, 794–795
titanium ion-mediated reactions, 795–797
transesterification-dipolar cycloadditions, 801–804
Nitrosation compounds, mesoionic ring systems, sydnone synthesis, 696
Nitroso acetals
alkyl nitronates
applications, 154–158
functionalization, 153–154
spectroscopic data, 102–105
silyl nitronates
functionalization, 125–127
spectroscopic data, 93–95
thiocarbonyl ylides, five-membered rings, 344–345
Nonactin, silyl nitronates and formation of, 130
Nonmetallocarbenoid precursors, carbonyl ylide synthesis, 306–309
Nonstable carbenes, carbonyl ylide generation, 259–262
Norborene, nitrile oxide cycloadditions
magnesium ion-mediated reactions, 787–788
magnesium ion-mediated reactions, kinetic data, 784–786
Nuclear overhauser effect (NOE) experiments, mesoionic ring systems, münchnone cycloaddition reactions, 715–725
Nucleophile reactions
azomethine ylide generation, 181–192
nitrile imines, 527–530
nitrile ylides, 526–527
Nucleosides, nitrone 1,3-dipolar cycloadditions, 4–8

O-allyl carbohydrates, nitrone 1,3-dipolar cycloadditions, sugars, 28–29
Olefins. *See also* Alkenes.
acyclic nitronate preparation, trinitromethane derivatives, 132–133
alkyl nitronate intermolecular cycloadditions, 141–145
azide 1,3-dipolar cycloadditions
diene natural product synthesis, 664–671
heterocyclic synthesis, 627–637
double-bonded diene conjugation, 637–639
natural synthesis, 651–664
azomethine ylide generation
intramolecular cycloadditions, 231–237
metal-mediated reactions, 240–248
monosubstituted compounds, 182–192
silicon protocols, 177–181
carbonyl ylide structures
enantioselective synthesis, 298–302
epoxide generation, 267–269
oxidopyrylium ion cycloadditions, 303–305
diazoalkane cycloaddition reactions
intramolecular cycloadditions, 592–601
C=C bonds, 598–601
reactivity and regioselectivity, 541–543
mesoionic ring systems
münchnone cycloaddition reactions, 710–725
sydnone cycloadditions, 735–738
thioisomünchnone cycloaddition reactions, 739–742
nitrile imine intramolecular cycloadditions, three-atom tethers, 512–513
nitrile oxide cycloadditions
achiral olefins, chiral auxiliaries, 393–399
chiral dipolarophiles, 386–393
chiral nitrile oxides, 400–406
dipolarophile/dipole relative reactivity, 376–380

Olefins (*Continued*)
intermolecular cycloadditions, "aldol" ring cleavage
achiral nitrile oxides-achiral olefins, 418–419
achiral nitrile oxides-chiral olefins, 419–424
chiral nitrile oxides-achiral olefins, 424–425
chiral nitrile oxides-chiral olefins, 425
intramolecular cycloadditions, 410–416
regioselectivity, 380–385
silyl nitronates, intramolecular cycloadditions, 123–125
thiocarbonyl ylide structures, thiophene derivatives, five-membered ring structure, 331–339
Oligomeric isoxazolidines, intermolecular nitrone cycloaddition reactions, 67–68
Oligosaccharides, azide 1,3-dipolar cycloadditions, fullerene heterocyclic synthesis, 646–649
One-pot procedures
nitrile imines, transient formation, 493–494
nitrile oxide generation, 367–368
Organomercurial compounds, carbonyl ylide generation, 259–262
Organometallics, nitrile oxide generation, dehydrohalogenation, 367
Oxacyclic compounds, carbonyl ylides, cycloaddition, directed syntheses, 278–284
Oxadiazolines, carbonyl ylide generation, 262–263
Oxamünchnones, mesoionic ring systems
cycloaddition reactions, 742
synthesis, 699–700
1,3,2-Oxathiazolium-5-olates, mesoionic ring systems, cycloadditions, 745
1,3-Oxathiolan-5-ones
thiocarbonyl ylides, thiophene derivatives, five-membered ring structure, 337–339
thiocarbonyl ylide structures
1,5-dipolar electrocyclization, five-membered heterocycles, 346–350
cycloreversion reactions, 319–320
heterocyclic synthesis, 339–343
1,3-Oxathiolium-4-olates, mesoionic ring systems, cycloaddition reactions, 744
Oxazoles, nitrile ylide electrocyclizations, 519–520
Oxazolidinone
azomethine ylide intramolecular cycloadditions, 230–237
diastereoselective reactions, α,β-amides, 857
mesoionic ring systems, isomünchnone cycloadditions, 731–735
Oxazolidinone derivatives, catalytic asymmetric nitrone reactions, 873–882
Oxazolines
azomethine ylide generation, 181–192
flash vacuum thermolysis, 183–192
indoloquinone construction, 237–239
intermolecular nitrone cycloaddition reactions, 62–68
Oxazolones, nitrile imine cycloadditions, five-membered rings, 508–509
Oxazolonium salts, mesoionic ring systems, münchnone synthesis, 683–687
Oxidopyrylium ion, carbonyl ylide cycloadditions, 302–305
Oximes
acyclic nitronate preparation, 131–132
diazoalkane cycloadditions, C=N bonds, 558–559
halogenation, nitrile oxide generation, 365–368
intramolecular oxime-olefin cycloaddition (IOOC), 54–55
nitrile oxide intermolecular cycloadditions, chiral nitrile oxides to chiral olefins, 425
nitrile oxide intramolecular cycloadditions
"aldol" ring cleavage, isoxazoline intermediates, 446–452
polycyclic isoxazolines and furoxans, 437–439
nitrone 1,3-APT cycloaddition, isoxazolidines, 48–54
nitrone cycloadditions, reaction mechanisms, 795
nitrone transesterification-dipolar cycloadditions, 802–804
nitroso acetal functionalization, 126–127
silyl nitronates, spectroscopic data, 90
Oxirane geometry
carbonyl ylide, epoxide generation, 264–269
nonmetallocarbenoid/metallocarbenoid syntheses, 308–309
thiocarbonyl ylide structures, 317
Oxo-Brevicomin analogue, carbonyl ylide directed syntheses, [1,3]-dipolar cycloadditions, 279–284
Oxotetrahydroindoles, azomethine ylide intramolecular cycloadditions, 236–237
Oxygen compounds, mesoionic ring systems, münchnone cycloaddition reactions, 721–725

Padwa's intermediate complex, nitrile ylide structure, [2 + 3]-cycloaddition reactions, 497–502
Palladium complexes, catalytic asymmetric nitrone reactions, 877–882
Papuamine compounds, nitrile oxide intramolecular cycloadditions, aminoalcohol reduction, 455–458
Peptides
  mesoionic ring systems, münchnone synthesis, 683–687
  nitrone 1,3-dipolar cycloadditions, 18–23
Pestalotin compounds, nitrile oxide intermolecular cycloadditions, achiral nitrile oxides to chiral olefin, 422–424
Phenols, thiocarbonyl ylide intermolecular reactions, 327–328
3-Phenyl propionaldehyde, azomethine ylide generation, 187–192
Phenyl allyl ethers, azomethine ylide intramolecular cycloadditions, 231–237
Phenyl compounds, azide 1,3-dipolar cycloadditions, heterocyclic synthesis, 631–637
Phenylglyoxylic acid, azomethine ylide 1,3-dipolar cycloadditions, external reagents, *N*-alkylideneglycinates, 758–759
Phenylisocyanate, thiocarbonyl ylides, heterocyclic synthesis, 342–343
(–)-8-Phenylmenthyl, chiral ylides, stereocontrol, 203–208
N-Phenyl-1,2,4-triazoline-3,5-dione (PTAD), mesoionic ring systems, thioisomünchnone cycloaddition reactions, 739–742
N-Phenylmaleimide
  azomethine ylide generation, silicon protocols, 175–181
  carbonyl ylides
    akaloid preparation, 285–292
    nonstabilized carbene generation, 260–262
  chiral ylides, stereocontrol, 203–208
  mesoionic ring systems
    isomünchnones
      cycloadditions, 727–735
      synthesis, 688–695
    münchnone cycloaddition reactions, 709–725
    sydnone cycloadditions, 737–738
N-(Phenylsulfonyl)imines, mesoionic ring systems, münchnone cycloaddition reactions, 721–725
Phenyloxazine compounds, carbonyl ylides, alkaloid preparation, 286–292
Phorbol compounds
  carbonyl ylide generation
    directed syntheses, [1,3]-dipolar cycloadditions, 280–284
    oxidopyrylium ion cycloadditions, 302–305
  nitrile oxide intramolecular cycloadditions, isoxazoline intermediates, "aldol" ring cleavage, 452–453
Phosphaalkenes
  diazoalkane cycloadditions, carbon-heteroatom C=P bonds, 559–567
  mesoionic ring systems
    1,3-dithiolium-4-olate cycloadditions, 742–744
    isomünchnone cycloadditions, 726–735
    münchnone cycloaddition reactions, 722–725
    sydnone cycloadditions, 738
Phosphaalkynes, diazoalkane cycloaddition reactions, 587–592
Phosphate compounds
  azomethine ylide intramolecular cycloadditions, 231–237
  nitrile ylide structure, 475
N-Phosphino derivative, nitrile imine cycloadditions, 504
Phosphonium salts, mesoionic ring systems, münchnone cycloaddition reactions, 712–725
Phosphorus compounds
  diazoalkane cycloaddition reactions
    P=X bonds, 576–578
    triple bond systems, 587–592
  nitrone 1,3-dipolar cycloadditions, 29–33
  phosphorus ylides, azide 1,3-dipolar cycloadditions, heterocyclic synthesis, 645–646
Photoacoustic calorimetry (PAC), nitrile ylides, carbene generation, 487–488
Photoinduced electron transfer (PET), nitrile ylide azirine photolysis, 489–490
Phyllanthocin, nitrile oxide intermolecular cycloadditions, achiral nitrile oxides to chiral olefin, 420–424
Picryl compounds, azide 1,3-dipolar cycloadditions, heterocyclic synthesis, 634–637
Pinnaic acid, nitrone 1,3-dipolar cycloadditions, 39–47
π-bonds, mesoionic ring systems, isomünchnone cycloadditions, 732–735
Piperidines
  asymmetric diastereoselectivity, nitrone/nitronate chiral dipoles, 828–829
  nitrone 1,3-dipolar cycloadditions, 34–47

PM3 calculations, nitronate structures, regioselectivity, 108–110
Polycyclic systems
  azomethine ylide intramolecular cycloadditions, decarboxylative/amino ester condensation protocols, 228–237
  carbonyl ylides, tropolone and benzotropolone formation and reactivity, 297–298
  mesoionic ring systems, isomünchnone cycloadditions, 728–735
  nitrile oxide intramolecular cycloadditions, 437–439
  thiocarbonyl ylide generation, thiophene derivatives, five-membered ring structure, 336–339
Polyfluorinated esters, mesoionic ring systems, münchnone cycloaddition reactions, 703–725
Polyoxamic acid, chiral ylides, asymmetric reactions, 214–218
Polyoxins, nitrone 1,3-dipolar cycloadditions, nucleoside antibiotics, 4–8
(+)-Preussin, nitrone 1,3-dipolar cycloadditions, 35–36
Proline derivatives
  asymmetric diastereoselectivity
    azomethine ylides, 829–833
    nitrone/nitronate chiral dipoles, 826–829
  carbonyl ylide generation, 276–278
  mesoionic ring systems, münchnone cycloaddition reactions, 705–725
  nitrile oxide cycloadditions
    achiral olefins, 395–399
    intermolecular cycloadditions, imino acids and polyols, 434–437
Propargyl-like structure
  nitrile imines, 478
  nitrile ylides, 475
Prostaglandins, nitrile oxide intramolecular cycloadditions, "aldol" ring cleavage, isoxazoline intermediates, 440–445
Protoanemonin-based compounds, azomethine ylide generation, intramolecular cycloadditions, 229–237
Proton nuclear magnetic resonance (^{1}H NMR)
  alkyl nitronates, 99–102
    nitroso acetals, 103–105
  silyl nitronates
    nitroso acetals, 93–95
    spectroscopic data, 88–93
  thiocarbonyl ylide intramolecular reactions, 325–326
P=S bonds, diazoalkane cycloaddition reactions, 577–578
Pseudodistomins, nitrone 1,3-dipolar cycloadditions, 34–35
Pseudolaric acids, carbonyl ylide directed syntheses, [1,3]-dipolar cycloadditions, 282–284
Ptaquiloside compounds, carbonyl ylide preparation, 292–294
Pterosin compounds, carbonyl ylide preparation, 292–294
Ptilomycalin, nitrone 1,3-dipolar cycloadditions, 44–47
Pulsed-laser photolysis, nitrile ylides
  azirine interconversions, 488–490
  [2 + 3]-cycloaddition reactions, 496–502
Pummerer reaction, isomünchnone cycloaddition, 734–735
Push-pull ylides
  carbonyl ylides
    alkaloid preparation, 287–292
    amide/urea derivation, 276–278
    epoxide generation, 265–269
  thiocarbonyl ylide structures, 317–318
    1,5-dipolar electrocyclization, five-membered heterocycles, 350
P=X bonds, diazoalkane cycloaddition reactions, heteroatom-heteroatom bonding, 576–578
Pycnidione, carbonyl ylides, tropolone and benzotropolone formation and reactivity, 297–298
Pyranoses, azide 1,3-dipolar cycloadditions, alkyne/nitrile synthesis, 675–676
Pyrazines, nitrile imine cycloadditions, six-membered rings, 509–510
Pyrazoles, sydnone cycloadditions, 735–738
Pyrazolines
  diastereoselective reactions, α,β-amides, 855–857
  diazoalkane cycloaddition reactions, dipole structures, 543–549
  diazoalkane cycloadditions, carbon-heteroatom C=P bonds, 560–567
  nitrile imine cycloadditions, 506–507
Pyrenophorin, nitrile oxide intermolecular cycloadditions, achiral nitrile oxides to chiral olefin, 424–425
Pyridine compounds
  azide 1,3-dipolar cycloadditions, heterocyclic synthesis, 636–637
  mesoionic ring systems, münchnone cycloaddition reactions, 701–725

nitrile imine cycloadditions, six-membered rings, 509–510
Pyridones
azomethine ylide generation, asymmetric reaction, 197–200
mesoionic ring systems
isomünchnones
cycloadditions, 729–735
synthesis, 693–695
thioisomünchnone cycloaddition reactions, 739–742
Pyrimidines
azide 1,3-dipolar cycloadditions, alkyne/nitrile reactions, 642–644
mesoionic ring systems, isomünchnone cycloadditions, 730–735
nitrile imine cycloadditions, six-membered rings, 509–510
Pyrinodenims, intramolecular nitrone-alkene cycloadditions (INAC), 59
Pyroglutamates, azomethine ylide metal-mediated reactions, 241–248
Pyrone compounds, carbonyl ylides, oxidopyrylium ion cycloadditions, 305
Pyrrole compounds
(*N*-alkylideneamino)acetonitriles, irreversible metalation, 763
mesoionic ring systems
münchnone cycloaddition reactions, 701–725
münchnone synthesis, 685–687
Pyrrolidines
asymmetric diastereoselectivity, nitrone/nitronate chiral dipoles, 826–829
azide 1,3-dipolar cycloadditions, natural synthesis, 651–664
nitrone 1,3-dipolar cycloadditions, 34–47
Pyrrolizidines
azide 1,3-dipolar cycloadditions, diene double bond reactions, 667–671
azomethine ylide generation, silicon protocols, 176–181
nitrone 1,3-dipolar cycloadditions, 12–18
Pyrrolizines, mesoionic ring systems, münchnone cycloaddition reactions, 700–725
Pyrrolo[3,4-*c*]pyridine, azide 1,3-dipolar cycloadditions, heterocyclic synthesis, 624–637
Pyruvic acid, azomethine ylide 1,3-dipolar cycloadditions, external reagents, *N*-alkylideneglycinates, 758–759

Quinocarcin compounds, azide 1,3-dipolar cycloadditions, natural product synthesis, 658–664
Quinolines, nitrile imine cycloadditions, six-membered rings, 509–510
Quinolizidines, nitrone 1,3-dipolar cycloadditions, 12–18

Radical cyclization, cyclic alkyl nitronates, 137
Regioselectivity
azomethine ylide generation, 182–192
metal-mediated reactions, 241–248
carbonyl ylide structures, epoxide generation, 265–269
[4 + 2] cycloaddition, nitroalkenes, 136–137
diazoalkane cycloaddition reactions, 541–543
1,3-dipolar cycloadditions
asymmetric reactions, 818–822
external reagents, 756–757
α-(alkylideneamino)esters, reversible metalation, 767–768
mesoionic ring systems
isomünchnone synthesis, 692–695
münchnone cycloaddition reactions, 715–725
thioisomünchnone cycloaddition reactions, 739–742
nitrile imine cycloadditions, 505–507
nitrile oxide cycloadditions
antibody catalysis, 791–792
chiral dipolarophiles, 386–393
external reagents, 779–794
inclusion complex mediation, 790–793
intramolecular cycloadditions, 407–410
magnesium ion-mediated reactions, 780–782
olefin structures, 380–385
nitrile ylide
[2 + 3]-cycloaddition reactions, 498–502
synthon formation, 479–485
nitronate structures, 108–110
nitrone 1,3-dipolar cycloadditions
magnesium ion-mediated reactions, 797–801
titanium ion-mediated reactions, 796–797
thiocarbonyl ylide structures
heterocyclic synthesis, 341–343
thiophene derivatives, five-membered ring structure, 332–339
Reissert compounds, mesoionic ring systems, münchnone cycloaddition reactions, 723–725
Relative reactivity, nitrile oxide cycloadditions, dipolar/dipolarophile mehcanism and relativity, 376–380
Resin compounds, intermolecular nitrone cycloaddition reactions, 66–68

(+)-Retronecine, chiral ylides, stereocontrol, 202–208
Rhodium catalytic compounds
azomethine ylide generation, 184–192
carbonyl ylide generation
aldehyde derivation, 274
alkaloid preparation, 287–292
asymmetric reactions, 886–888
directed syntheses, [1,3]-dipolar cycloadditions, 278–284
enantioselective synthesis, 300–302
ketone derivation, 270–273
diazoalkane ketocarbene dipoles, 605–609
mesoionic ring systems
isomünchnones
cycloadditions, 727–735
synthesis, 688–695
thioisomünchnones
cycloaddition, 740–742
synthesis, 697–698
thiocarbonyl ylides, 1,5-dipolar electrocyclization, five-membered heterocycles, 347–350
Ribasine compounds, carbonyl ylide alkaloid preparation, 291–292
Ribavarin compounds, azide 1,3-dipolar cycloadditions
alkyne/nitrile reactions, 643–644
phosphorus ylide reactions, 645–646
Ring structure
azide 1,3-dipolar cycloadditions, heterocyclic synthesis, 625–637
nitrile oxide generation
dipolarophile/dipole relative reactivity, 376–380
intermolecular cycloadditions
"aldol" ring cleavage, 417–425
achiral nitrile oxides/achiral olefins, 418–419
achiral nitrile oxides/chiral olefins, 419–424
chiral nitrile oxides/achiral olefins, 424–425
chiral nitrile oxides/chiral olefins, 425
aminoalcohol ring cleavage, 425–431
intramolecular cycloadditions
intermediate isoxazoline ring cleavage, 439–454
regioselectivity, 407–410
nitrile ylides, electrocyclizations, diene-conjugated compounds, 522–524
thiocarbonyl ylide heterocyclic synthesis
three-membered thiirane rings, 329–331
five-membered rings
1,3-dithiolanes, 1,3-oxathioanes, and 1,3-thiazole derivatives, 339–343
1,5-dipolar electrocyclization, 345–350
heteroatoms, 344–345
thiophene derivatives, 331–339
six-membered rings, 351
seven-membered rings, 351–355
Ring structures, mesoionic ring systems
classification, 681–682
cycloaddition reactions
1,3-dithiolium-4-olates, 742–744
isomünchnones, 725–735
miscellaneous mesoionic heterocycles, 745–746
münchnones, 700–725
1,3-oxathiazolium-5-olates, 745
1,3-oxathiolium-4-olates, 744
oxamünchnones, 742
sydnones, 735–738
1,3-thiazolium-4-thiolates, 742
thioisomünchnones, 739–742
thiomünchnones, 738
synthesis
isomünchnones, 687–695
münchnones, 682–687
oxamünchnones, 699–700
sydnones, 696
1,3-thiazolium-4-thiolates, 698–699
thioisomünchnones, 697–698
thiomünchnones, 696–697

Samarium compounds
azomethine ylide metal-mediated reactions, 242–248
carbonyl ylides
nonmetallocarbenoid/metallocarbenoid syntheses, 307–309
nonstabilized carbene generation, 260–262
Sarain-A, azomethine ylide generation, intramolecular cycloadditions, 220–227
Sarkomycin derivatives, nitrile oxide intramolecular cycloadditions, "aldol" ring cleavage, isoxazoline intermediates, 440–445
Schönberg reaction, diazoalkane cycloadditions, carbon-heteroatom C=S double bonds, 570–575
Selenide cyclization, nitrone 1,3-APT cycloaddition, 52–54
Selenoketones, diazoalkane cycloaddition reactions, 575–576

Semiempirical calculations, nitronate structures, regioselectivity, 108–110
Sesquiterpenoid products
  carbonyl ylide preparation, 292–294
  diazoalkane cycloaddition reactions, intramolecular cycloadditions, 595–601
  nitrile oxide intramolecular cycloadditions, "aldol" ring cleavage, isoxazoline intermediates, 443–445, 448–452
Seven-membered rings
  diazoalkane intramolecular cycloadditions, 595–601
  nitrile imine cycloadditions, 510
    five-atom tether, 514–515
  nitrile oxide intramolecular cycloadditions, isoxazoline intermediates, 452–453
  thiocarbonyl ylide heterocyclic synthesis, 351–354
Seyferth reagents, thiocarbonyl ylides, thiirane three-membered rings, 330–331
(–)-Shikimic acid, intermolecular nitrone cycloaddition reactions, 65–68
Silicon nuclear magnetic resonance (^{29}Si NMR), silyl nitronates, 91–93
Silicon protocols, azomethine ylide generation, 170–181
  intramolecular cycloadditions, 223–227
Siloxyoazoles, mesoionic ring systems, münchnone synthesis, 684–687
Silver fluoride, azomethine ylide generation, silicon protocols, 170–181
Silylated thioethers, thiocarbonyl ylide generation, 1,3-elimination reactions, 320–321
Silyl groups
  carbonyl ylide generation, 261–262
    nonmetallocarbenoid/metallocarbenoid syntheses, 308–309
  nitrile imine structure, 476–478
  nitrone 1,3-dipolar cycloadditions, lactam formation, 9–11
α-Silylimidates, azomethine ylide generation, silicon protocols, 174–181
N-Silylmethyl thioureas, azomethine ylide generation, silicon protocols, 179–181
Silyl nitronates
  applications, 127–130
  cycloadditions, 117–125
    facial selectivity, 120–123
    intermolecular cycloadditions, 117–120
    intramolecular cycloadditions, 123–125
  nitroso acetal functionalization, 125–127
  preparation, 114–117
  structural properties
    bonding, 87–88
    exchange mechanisms, 90–93
    nitroso acetals, spectroscopic data, 93–95
    spectroscopy, 88–90
    stability, 86–87
N-Silyloxy-isoxazolidine (XX), nitroso acetal functionalization, 125–127
Silyloxy groups, zaragozic acid carbonyl ylide derivatives, 295–296
Si=Si bonds, diazoalkane cycloaddition reactions, 578
Six-membered ring structures
  nitrile imine cycloadditions, 509–510
    four-atom tether, 513–514
  nitrile oxide intramolecular cycloadditions, "aldol" ring cleavage, isoxazoline intermediates, 445–452
  thiocarbonyl ylide heterocyclic synthesis, 351
Skeleton rearrangements, thiocarbonyl ylide structures, 326
Slaframine compounds, azide 1,3-dipolar cycloadditions, natural synthesis, 657–664
N-Sodiated ylides, α-(alkylideneamino)esters, reversible metalation, 765–768
Sodium hypochlorite, nitrile oxide generation, oxime one-pot operations, 368
Solid-phase synthesis, intermolecular nitrone cycloaddition reactions, 63–68
Solvent effects, nitrile oxide cycloadditions, magnesium ion-mediated reactions, 782–784
Specionin, nitrile oxide intramolecular cycloadditions, "aldol" ring cleavage, isoxazoline intermediates, 442–443
Spectroscopic data
  alkyl nitronates, 97–102
  silyl nitronates, 88–90
    exchange reactions, 90–93
    nitroso acetals, 93–95
Spirocyclic nucleosides
  diazoalkane cycloaddition reactions, 546–549
  nitrone 1,3-dipolar cycloadditions, 7–8
Spirocyclic ring systems, nitrone 1,3-dipolar cycloadditions, 37–47
Spirocyclopropyl dipolarophiles
  isoxazolidines, intramolecular nitrone-alkene cycloadditions (INAC), 58–59
  nitrone 1,3-dipolar cycloadditions, 15–18
Spiro epoxide, carbonyl ylide generation, 264–269
Spiro mode intramolecular cycloadditions, alkyl nitronates, 152–153
Spirooxindole compounds, carbonyl ylide cycloadditions, 284–292

Spiropyrazolines, diazoalkane cycloaddition reactions, 554–557
Stability
alkyl nitronates, 95–96
nitrile imine generation mechanisms, 495–496
cycloaddition reactions, 502–504
silyl nitronates, 86–87
Stereoselectivity
asymmetric reactions, intramolecular cycloadditions, 845–848
azomethine ylide
intramolecular cycloadditions, aziridine precursors, 225–227
metal-mediated reactions, 241–248
chiral ylides, inherent molecular structure, 200–208
1,3-dipolar cycloadditions
asymmetric reactions, 819–822
external reagents, 756–757
anti-selective Michael additions, 768–771
mesoionic ring systems, isomünchnone cycloadditions, 729–735
nitrile imine intramolecular cycloadditions, three-atom tethers, 512–513
nitrile oxide cycloadditions
achiral olefins, 393–399
antibody catalysis, 791–792
chiral dipolarophiles, 389–393
diastereoselectivity, 386–406
external reagents, 779–794
intramolecular cycloadditions, 410–416, 411–416
nitrile ylide 1,3-dipolar cycloadditions, 806–809
nitronates, 111–114
asymmetric reactions, intramolecular cycloadditions, 846–848
intermolecular cycloadditions, 111–112
intramolecular cycloadditions, 112–114
mechanisms, 114
nitrone 1,3-dipolar cycloadditions, magnesium ion-mediated reactions, 797–801
nitroso acetal functionalization, 126–127
silyl nitronates, applications, 128–130
Steric hindrance
nitrile oxide cycloadditions, olefin regioselectivity, 381–385
silyl nitronates
facial selectivity, 121–123
intermolecular cycloadditions, 119–120
thiocarbonyl ylide heterocyclic synthesis, seven-membered ring structure, 351–354
Streptazolin, nitrile oxide intramolecular cycloadditions, "aldol" ring cleavage, isoxazoline intermediates, 443–445
Sugars
asymmetric reactions, 1,3-dipolar cycloadditions, chiral auxiliaries, 862–863
nitrile oxide intermolecular cycloadditions, 425–431
nitrone 1,3-dipolar cycloadditions, 24–29
N-Sulfinylamines, thiocarbonyl ylides, heteroatoms, five-membered rings, 344–345
Sulfur compounds
azomethine ylide generation
asymmetric reaction, 196–200
intramolecular cycloadditions, 233–237
nitrone 1,3-dipolar cycloadditions, 29–33
thiocarbonyl ylide generation
cycloreversion reactions, 319–320
heteroatoms, five-membered rings, 344–345
thiocarbonyl ylides, thiirane three-membered rings, 330–331
Sultam-derived dipolarophiles
azomethine ylides, asymmetric reaction, 199–200
chiral ylides, asymmetric templates and auxiliaries, 208–218
diastereoselective reactions, α,β-amides, 853–857
1,3-dipolar cycloadditions, chiral auxiliaries, 860–863
nitrile oxide cycloadditions, achiral olefins, 394–399
silyl nitronates
applications, 128–130
facial selectivity, 122–123
Superdipolarophiles, thiocarbonyl ylides
heterocyclic synthesis, 339–343
intermolecular reactions, 327
Swainsonine, azide 1,3-dipolar cycloadditions, natural synthesis, 652–664
Sydnones, mesoionic ring systems
cycloaddition reactions, 735–738
münchnone cycloaddition reactions, 703–725
synthesis, 696
Synthon formation, nitrile ylides, 479–485

TADDOL ligands, catalytic asymmetric nitrone reactions, 875–882
Talaromycin B, nitrile oxide intermolecular cycloadditions, "aldol" ring cleavage, isoxazoline intermediates, achiral nitrile oxides-achiral olefins, 418–419

Tandem coupling-cyclization
alkyl nitronates
applications, 154–158
facial selectivity, 146–148
intermolecular cycloadditions, 144–145
azomethine ylide generation, 233–237
carbonyl ylide generation
[1,3]-dipolar cycloaddition, 278–284
enantioselective synthesis, 298–302
ketone derivation, 270–273
cyclic nitronate preparation, 135–136
mesoionic ring systems, münchnone cycloaddition reactions, 720–725
nitrone transesterification-dipolar cycloadditions, 801–804
Tandem intermolecular-intramolecular synthesis
mesoionic ring systems, sydnone cycloadditions, 737–738
nitrile imines
cycloadditions, 516–517
five-atom tether, 514–515
transient formation, 493–494
Taxol precursors
carbonyl ylides, oxidopyrylium ion cycloadditions, 303–305
intramolecular nitrile oxide cycloadditions, 437–439
Telluroketones, diazoalkane cycloaddition reactions, 576
Tert-butylphosphaacetylene, diazoalkane cycloaddition reactions, phosphaalkynes, 589–592
Testosterone, nitrile oxide intramolecular cycloadditions, "aldol" ring cleavage, isoxazoline intermediates, 450–452
Tethered structures, nitrile imine intramolecular cycloadditions, 511–516
three-atom tether, 512–513
four-atom tether, 513–514
five-atom tether, 514–515
six-or-greater atom tether, 515–516
2,2,6,6-Tetramethyl-1-piperidinoxyl (TEMPO), nitrone 1,3-dipolar cycloadditions, nucleoside antibiotics, 5–8
3,4,5,6-Tetrahydro-1*H*-1,4-oxazin-2-one, chiral ylides, 209–218
Tetrafluoroborate salt, acyclic nitronate preparation, 130–132
Tetrahydrofuran (THF) derivatives
(*N*-alkylideneamino)acetonitriles, irreversible metalation, 761–763
carbonyl ylide generation, 274
carbonyl ylide structures, 256–259
nonstabilized carbene generation, 259–262
nitrile oxide cycloadditions, magnesium ion-mediated reactions, 782–784
Tetrahydropyridine *N*--oxide, nitrone 1,3-dipolar cycloadditions, 30–33
Tetrakis(binaphtholphosphate)dirhodium, diazoalkane ketocarbene dipoles, 606–609
Tetramethylsilane (TMS), nitrile ylide synthons, 482–485
Tetrazole compounds
azide 1,3-dipolar cycloadditions, alkyne/nitrile reactions, 644
diazoalkane cycloaddition reactions, heteroatom-heteroatom triple bonds, 586–587
1,-3-Thiazine derivatives, thiocarbonyl ylide heterocyclic synthesis, 351
1,3-Thiazole derivatives, thiocarbonyl ylide structures, heterocyclic synthesis, 339–343
1,3-Thiazolium-4-thiolates, mesoionic ring systems
cycloadditions, 742
synthesis, 698–699
Thiazolidine-4-carboxylic acid, azomethine ylide intramolecular cycloadditions, 228–237
Thiazolyl nitrones, 1,3-dipolar cycloadditions, 33
(+)-Thienamycin, nitrone 1,3-dipolar cycloadditions, lactam formation, 9–11
Thiiranes
diazoalkane cycloaddition reactions, carbon-heteroatom C=S double bonds, 574–575
thiocarbonyl ylide structures, 316
three-membered rings, 329–331
1,4-hydrogen shift, 325–326
intramolecular electrocyclizations, 324–325
Thiobenzophenone compounds, thiocarbonyl ylide structures, 318
Thiocamphor
thiocarbonyl ylide intermolecular reactions, 327–328
thiocarbonyl ylide intramolecular reactions, 1,4-hydrogen shift, 325–326
Thiocarbonyl ylides
chemical behavior
intermolecular reactions, 326–329
1,3-addition reactions, 327–328
[3 + 2] cycloadditions, 326–327
dimerization reactions, 328–329
intramolecular reactions, 324–326
electrocyclizations, 324–325
1,4-hydrogen shift, 325–326
skeleton rearrangements, 326
diastereoselective reactions, α,β-amides, 855–857

Thiocarbonyl ylides (*Continued*)
  generation, 319–324
    carbene/carbenoid additions, 322–324
    cycloreversion reactions, 319–320
    1,3-elimination reactions, 320–321
    thioether photochemical rearrangement, 324
    thioxonium salts, deprotonation, 321–322
  heterocyclic synthesis
    three-membered thiirane rings, 329–331
    five-membered rings
      1,3-dithiolanes, 1,3-oxathioanes, and 1,3-thiazole derivatives, 339–343
      1,5-dipolar electrocyclization, 345–350
      heteroatoms, 344–345
      thiophene derivatives, 331–339
    six-membered rings, 351
    seven-membered rings, 351–355
  nitrile imine cycloadditions, double bonds, 510–511
  structural properties, 316–318
Thioethers
  nitrile oxide cycloadditions, intramolecular cycloadditions, 411–416
  thiocarbonyl ylide generation
    three-membered thiirane rings, 329–331
    1,3-elimination reactions, 320–321
  thiocarbonyl ylides, photochemical rearrangement, 324
Thioisomünchnones
  mesoionic ring systems
    cycloaddition reactions, 739–742
    synthesis, 697–698
  thiocarbonyl ylide generation, 322
    thiophene derivatives, five-membered ring structure, 334–339
Thioketones, diazoalkane cycloadditions, carbon-heteroatom C=S double bonds, 570–575
Thiomünchnones, mesoionic ring systems
  cycloaddition reactions, 738
  synthesis, 696–697
Thionitrile ylides, synthon formation, 483–485
Thiophene derivatives
  mesoionic ring systems, 1,3-dithiolium-4-olate cycloadditions, 742–744
  nitrile imine intramolecular cycloadditions, three-atom tethers, 512–513
  thiocarbonyl ylide heterocyclic synthesis
    1,5-dipolar electrocyclization, five-membered heterocycles, 348–350
    five-membered ring structure, 331–339
Thiopivaldehyde, diazoalkane cycloadditions, carbon-heteroatom C=S double bonds, 571–575
Thiourea compounds, thiocarbonyl ylide structures, 317
Thioxonium salts, thiocarbonyl ylide generation
  deprotonation, 321–322
  skeletal rearrangements, 326
Three-atom tethers, nitrile imine intramolecular cycloadditions, 512–513
Three-membered ring structure, thiocarbonyl ylide heterocyclic synthesis, thiirane rings, 329–331
Tigliane diterpenes, carbonyl ylide directed syntheses, [1,3]-dipolar cycloadditions, 280–284
Titanium compounds
  azomethine ylide metal-mediated reactions, 245–248
  catalytic asymmetric nitrone reactions, 872–882
  nitrone 1,3-dipolar cycloadditions, titanium ion-mediated reactions, 795–797
  titanium ylides, α-(alkylideneamino)esters, reversible metalation, 767–768
N-Tosylation, azomethine ylide generation, chiral dipolarophiles, asymmetric reaction, 193–200
Tosylhydrazone sodium salts, diazoalkane intramolecular cycloadditions, 595–601
Tosylimines, mesoionic ring systems, münchnone cycloaddition reactions, 721–725
Tosyl oxazaborolidinones, 1,3-dipolar cycloadditions, catalytic asymmetric nitrone reactions, 867–882
Trans-2-methylene-1,3-dithiolane, nitrone 1,3-dipolar cycloadditions, 31–33
Transesterification-dipolar cycloadditions, nitrone 1,3-dipolar cycloadditions, 801–804
Trehazolin, nitrile oxide intramolecular cycloadditions, "aldol" ring cleavage, isoxazoline intermediates, 441–445
Triazoles
  azide structural properties, 623–624
  diazoalkane cycloaddition reactions, heterocumulenes, 580–581
  mesoionic ring systems, sydnone cycloadditions, 736–738
Triazolines, diazoalkane cycloadditions, C=N bonds, 558–559
Triazolines, mesoionic ring systems, thioisomünchnone cycloaddition reactions, 739–742
Tricyclic structures, nitrile oxide intramolecular cycloadditions, "aldol" ring cleavage, isoxazoline intermediates, 442–445
Triene-conjugated nitrile ylides, electrocyclizations, 524–525

Triethylamine, azomethine ylide 1,3-dipolar cycloadditions, external reagents
α-(alkylideneamino)esters, reversible metalation, 763–768
anti-selective Michael additions, 768–771
enantiopure camphor imines, asymmetric Michael additions, 775–778
Trifluoroacetylacetonitrile, nitrile imine nucleophile reactions, 529–530
Trifluoromethyl compounds
carbonyl ylide structures, 258–259
diazoalkane cycloaddition reactions, alkyne triple-bond systems, 584–586
nitrile oxide intermolecular cycloadditions, achiral nitrile oxides to chiral olefin, 424–425
Trihydroxyheliotridane-6,7-*O*-acetonide, azide 1,3-dipolar cycloadditions, diene double bond reactions, 667–671
Trimethylsilyl chloride (TMSCl), silyl nitronate preparation, 115–117
Trimethylsilyl methane compounds, diazoalkane cycloaddition reactions, alkyne triple-bond systems, 584–586
Trinitromethane, acyclic nitronate preparation, 132–133
Triple-bond systems, diazoalkane cycloaddition reactions, 581–592
alkynes, 581–586
heteroatom-heteroatom triple bonds, 586 587
intramolecular C≡C bonds, 601–604
nitriles, 586
phosphaalkynes, 587–592
Triquinane sesquiterpenes, nitrile oxide intramolecular cycloadditions, "aldol" ring cleavage, isoxazoline intermediates, 443–445
Tropolone formation and reactivity, carbonyl ylides, 296–298
Tropone compounds
isomünchnone cycloadditions, 726–735
thiomünchnone cycloaddition reactions, 738
Tungsten complexes
nitrile ylide synthons, 484–485
thiocarbonyl ylides, thiirane three-membered rings, 330–331
Twofold extrusion technique, thiocarbonyl ylide structures
cycloreversion reactions, 320
three-membered thiirane rings, 329–331
Two-phase methodology, thiocarbonyl ylides, carbene/carbenoid addition, 323–324
Tylophorine compounds, azide 1,3-dipolar cycloadditions, natural synthesis, 659–664

Ugi condensation, mesoionic ring systems, münchnone synthesis, 685–687
Unsaturated esters
α,β-unsaturated esters, diastereoselective reactions, 851–853
azomethine ylide generation, chiral dipolarophiles, asymmetric reaction, 193–200
Ureas
carbonyl ylide derivation, 276–278
mesoionic ring systems, isomünchnones, cycloadditions, 730–735
nitrile oxide generation, nitroalkene dehydration, 368–370

Vinyl acetals, asymmetric reactions, 1,3-dipolar cycloadditions, 839–840
Vinyl aminals, asymmetric reactions, 1,3-dipolar cycloadditions, 839–840
Vinylphosphine oxides, asymmetric diastereoselectivity, nitrone/nitronate chiral dipoles, 823–829
Vinyl phosphine oxides, asymmetric reactions, 1,3-dipolar cycloadditions, 840–842
Vinyl sulfoxides, asymmetric reactions, 1,3-dipolar cycloadditions, 840–842
Vitamin D, nitrile oxide intramolecular cycloadditions, "aldol" ring cleavage, isoxazoline intermediates, 446–452

Well-coordinating cations, nitrile oxide cycloadditions, diastereoselectivity, 386
WIN inhibitor, azide 1,3-dipolar cycloadditions, alkyne/nitrile synthesis, 672–676

X-ray crystallography
alkyl nitronates, 96–98
silyl nitronates, 87–88
facial selectivity, 122–123
Xylene, nitrile ylide structure, 474

Ylide compounds
azide 1,3-dipolar cycloadditions, phosphorus ylides, heterocyclic synthesis, 645–646
azomethine ylides
asymmetric reactions, 193–218
chiral dipolarophiles, 193–200
chiral ylides
asymmetric template and auxiliary derivations, 208–218

Ylide compounds (*Continued*)
inherent molecular stereocontrol, 200–208
development, 169–170
1,3-dipolar cycloadditions, external reagents, 757–778
anti-selective Michael additions, 768–771
asymmetric cycloadditions, 772–774
2-azaallyl anions, 759
enantiopure camphor imines, asymmetric Michael additions, 774–778
enantioselective additions, 778
metal *N*-alkylideneglycinates, 758–759
N-metalated reactions, 759–778
α-(alkylideneamino)ester reversibility, 763–768
(*N*-alkylideneamine)acetonitrile irreversibility, 761–763
generation, 170–192
alternative procedures, 181–192
silicon-based protocols, 170–181
intramolecular cycloadditions, 219–239
alternative generation procedures, 237–239
aziridine precursors, 219–227
decarboxylative and amino ester condensation protocols, 228–237
metal-mediated reaction, 240–248
carbonyl ylides
1,3-dipolar cycloadditions, 804–806
aldehyde derivation, 273–274
amide/urea derivatives, 276–278
cycloaddition reactions
alkaloid systems, 284–292
directed syntheses, 278–284
enantioselective syntheses, 298–302
illudins, ptwerosins, and ptaquilosides, 292–294
oxidopyrylium ion cycloadditions, 302–305
tropolone and benzotropolone formation and reactivity, 296–298
zaragozic acids, 294–296
epoxide generation, 263–269
ester derivation, 274–276
ketone derivation, 269–273
metallocarbenoid generation, 269
metallocarbenoid synthesis, 306–309
nonmetallocarbenoid precursors, synthesis using, 306–309
nonstable carbene generation, 259–262
oxadiazoline generation, 262–263
reactivity properties, 254–255
structural studies, 255–259
nitrile ylides
[2 + 3]-cycloaddition reactions, 496–502
1,3-dipolar cycloadditions, 806–809
electrocyclization reactions, 517–525
alkene-conjugated ylides, 518–520
diene-conjugated ylides, 520–524
triene-conjugated ylides, 524–525
generation mechanisms, 478–492
azirine photolysis, 488–490
carbene reactions, 485–488
imidoyl chloride dehydrochlorination, 490–491
synthon generation, 479–485
nucleophile reactions, 526–527
physical and spectroscopic properties, 474–475
thiocarbonyl ylides
chemical behavior
intermolecular reactions, 326–329
1,3-addition reactions, 327–328
[3 + 2] cycloadditions, 326–327
dimerization reactions, 328–329
intramolecular reactions, 324–326
electrocyclizations, 324–325
1,4-hydrogen shift, 325–326
skeleton rearrangements, 326
generation, 319–324
carbene/carbenoid additions, 322–324
cycloreversion reactions, 319–320
1,3-elimination reactions, 320–321
thioether photochemical rearrangement, 324
thioxonium salts, deprotonation, 321–322
heterocyclic synthesis
three-membered thiirane rings, 329–331
six-membered rings, 351
seven-membered rings, 351–355
five-membered rings
1,3-dithiolanes, 1,3-oxathioanes, and 1,3-thiazole derivatives, 339–343
1,5-dipolar electrocyclization, 345–350
heteroatoms, 344–345
thiophene derivatives, 331–339
structural properties, 316–318
Ytterbium complexes, catalytic asymmetric nitrone reactions, 880–882

Zaragozic acids, carbonyl ylide preparation, 294–296